George Simon Ohm (1787–1854)

The son of a locksmith, Ohm's greatest achievement was the discovery, published in his *Galvanic Circuit Investigated Mathematically*, that in any wire at uniform temperature the current that flows is directly proportional to the potential difference between its ends, the familiar Ohm's Law. Ohm's work was severely criticized, and he eked out a living as a private tutor for years before his finding began to gain acceptance.

Ohm – electric resistance

James Prescott Joule (1818–1889)

Compelled in youth by ill health to study at home, Joule devoted himself to scientific research. At the age of 20 he published a description of an electromagnetic engine that he had invented, and two years later described a method for ascertaining the mechanical equivalent of heat. The unit of its measurement became known as Joule's equivalent. His research in the correlation and conservation of energy brought him to the attention of Lord Kelvin, with whom he conducted considerable research.

Joule – energy

Wilhelm Eduard Weber (1804–1891)

Weber made measurements of the ratio of the electrostatic and electromagnetic systems of units. His work led to a definition of electric current in the electromagnetic system, based on the magnetic field produced by the current. When Weber protested the suppression of the constitution by the king of Hanovor in 1837, he was dismissed from the faculty of Gottingen.

Weber – magnetic flux

CIRCUITS:

Principles, Analysis and Simulation

Frank P. Yatsko
Pennsylvania State University

David M. Hata
Portland Community College

Saunders College Publishing
A Harcourt Brace Jovanovich College Publisher
Fort Worth Philadelphia San Diego New York Orlando
Austin San Antonio Toronto Montreal London Sydney Tokyo

Text Typeface: Times Roman
Compositor: General Graphic Services
Acquisitions Editor: Barbara Gingery
Developmental Editor: Alexa Barnes
Managing Editor: Carol Field
Project Editors: Margaret Mary Anderson and Becca Gruliow
Copy Editor: Suzanne Magida
Manager of Art and Design: Carol Bleistine
Art Director: Christine Schueler
Art and Design Coordinator: Caroline McGowan
Text Designer: Gene Harris
Cover Designer: Lawrence R. Didona
Text Artwork: Grafacon Inc.
Layout Artist: York Production Services
Director of EDP: Tim Frelick
Production Manager: Charlene Squibb
Product Manager: Monica Wilson

Cover Credit: Philip Habib/Tony Stone Worldwide

Printed in the United States of America

Circuits: Principles, Analysis and Simulation
0-03-000933-2

Library of Congress Catalog Card Number: 91-058034
234 061 98765432

To my wife Leonora,
daughters Francene and Paula,
and son Frank
FRANK P. YATSKO

To my wife Susan,
sons Jonathan and Aaron,
and daughter Abigail
DAVID M. HATA

PREFACE

What! Another electronic circuits textbook? There seem to be so many electronic circuits texts already on the market. Do we really need another one?

Yes, we think so, for the following reasons. First, the electronics industry has changed dramatically in the last forty years, and especially in the last decade. Some applications of these changes are microcomputers, space technology, robotics, and advanced forms of instrumentation. Yet the basic content of a course sequence in electronic circuits has not reflected this revolution. The topics that are emphasized are characteristic of an industry that is dominated by the power industry and the radio communications industry. Clearly, although these are still important players, they are not the dominant industry segments today. Computers, instrumentation, and data communications are large industry segments.

The concepts needed in the computer, instrumentation, and data communications industries center around information transfer and signal integrity, not single-frequency phasor analysis, resonance, and power factor.

Second, the computation tools that we have today have opened new opportunities to analyze, simulate, and create new circuits. These tools have also changed the way industry performs basic engineering, manufacturing, and troubleshooting functions. The introduction of these new tools into the study of electronic circuits has been cosmetic so far, e.g., the inclusion of an end-of-chapter section on simulation or a text appendix. Modern computational and computer-based tools should be incorporated throughout the curriculum, beginning with the first circuits course.

And finally, the electronic circuits course should lay a sound foundation for the courses that follow, presenting a point of view that is consistent with the way advanced circuits are studied. The traditional approach to teaching electronic circuits does not serve this function effectively. A common philosophical viewpoint of circuits should direct the entire curriculum.

Philosophy

The traditional way of viewing electronic circuits was to look at circuits as closed systems. That is, each circuit contained sources and circuit components. Voltages, currents, power, and resistances were computed within the boundaries of the closed system.

A more effective way of viewing circuits looks at a circuit as a system of building blocks that accepts an input signal, performs some operation on the signal, and produces an output signal that reflects the nature of the input signal and the signal processing operation performed by the circuit. This is precisely how we view advanced circuits such as amplifiers, filters, logic gates, and other system building blocks. Hence, in advanced courses, voltage dividers, *R-C* filters, and *R-L-C* circuits will form building blocks that will be used to build larger and more complex systems.

The personal computer will help the student to visualize circuit behavior. Using a software package called PSpice, circuits can be analyzed quickly and easily to see the effects of circuit modifications and parameter changes. The personal computer will also enable the student to perform advanced analysis techniques without the mathematical labor associated with on-computer applications of the techniques. As a result, we can expand the breadth and scope of the study of electronic circuits in the following ways:

- Students will be able to answer, ''What if..'' type questions by using the computer to simulate the behavior of electronic circuits.
- Students will be able to graph current and voltage waveforms easily and thus visualize circuit operation.
- Students will be able to visualize complex signals as a composite of sinusoidal signals of various amplitudes and frequencies.
- Students will emulate practices actually used in an industry that is rapidly moving away from hardware prototyping and toward computer-aided design, analysis, and testing techniques.

Audience and Prerequisites

Circuits: Principles, Analysis and Simulation is intended for students in electrical/electronic engineering technology programs at junior colleges, community colleges, technical institutes, and other institutions offering associate and baccalaureate degree programs. It can also be used for self-study and for training courses in industry.

The prime mathematics prerequisite for using this text is algebra. Especially important is the ability to manipulate terms in an equation in order to solve for any one of the terms. The solution of simultaneous equations is needed in Chapter 8. (Appendix A reviews this technique.) Right-angle functions of sine, cosine, and tangent are needed for ac circuit analysis in Chapter 15.

A scientific calculator is essential for performing computations quickly and accurately. We recommend that students use a scientific calculator that can perform standard trigonometric, logarithmic, and mathematical functions as well as specialized functions such as polar-to-rectangular conversion.

Access to a personal computer is also needed. The student version of PSpice is available at little or no cost. Students will also need a line-editing program or word-processing package in order to create source files for PSpice.

Organization and Content

Circuits: Principles, Analysis and Simulation is organized so that it can be used for a course sequence that covers topics traditionally covered in dc and ac circuits. However, the text can also be divided into three parts: resistive circuits (Chapters 1–10), transients in *R-C* and *R-L* circuits (Chapters 11–14) and reactive circuits (Chapters 15–22). A description of each of these parts is given below and suggested course outlines are shown in Figure 1.

Overview

Circuits: Principles, Analysis and Simulation begins with an overview of the development of technology and of the functions of the engineering technician and technologies. This material will be helpful to students trying to make career decisions.

Part I: Resistive Circuits

The resistive circuits section consists of Chapters 1–10. Chapter 2 introduces electrical parameters and includes discussions of the dangers of static electricity and the importance of safety practices. Chapter 3 presents the basic circuit and includes a discussion of the loss method of calculating efficiency.

Chapters 4 and 5 give an early introduction of time-varying signals and the transfer functions. From this point in the text, input signals are allowed to be time-varying waveforms, not just dc voltages and currents. Instruments (such as the signal generator and oscilloscope) that generate and display time-varying waveforms are introduced when appropriate so students visualize the operation of resistive circuits for a wider range of input signals. Chapter 4 also presents the most up-to-date method of rating lead acid batteries, reserve capacity and cold cranking current, along with the ampere-hour ratings.

For example, the two-resistor voltage divider circuit is viewed as an attenuator circuit. Whether the input signal is a dc voltage, sine wave, or pulse, the output signal will have the same characteristic shape, only with a decreased amplitude.

Circuit simulation techniques are introduced early and gradually. Chapter 6 of *Circuits: Principles, Analysis and Simulation* describes the capabilities of PSpice, how to write PSpice files, and how to perform dc and ac analysis using PSpice. At this point, the circuits are still simple and the simulation results are easily verified. This gentle approach helps students gain confidence in the simulation tool and prepares them to analyze larger and more complex circuits later in the course.

Part II: Transients in *R-L* and *R-C* Circuits

Chapters 11 through 14 cover the transient behavior of *R-C* and *R-L* circuits. As before, inputs to *R-C* and *R-L* circuits are allowed to be time-

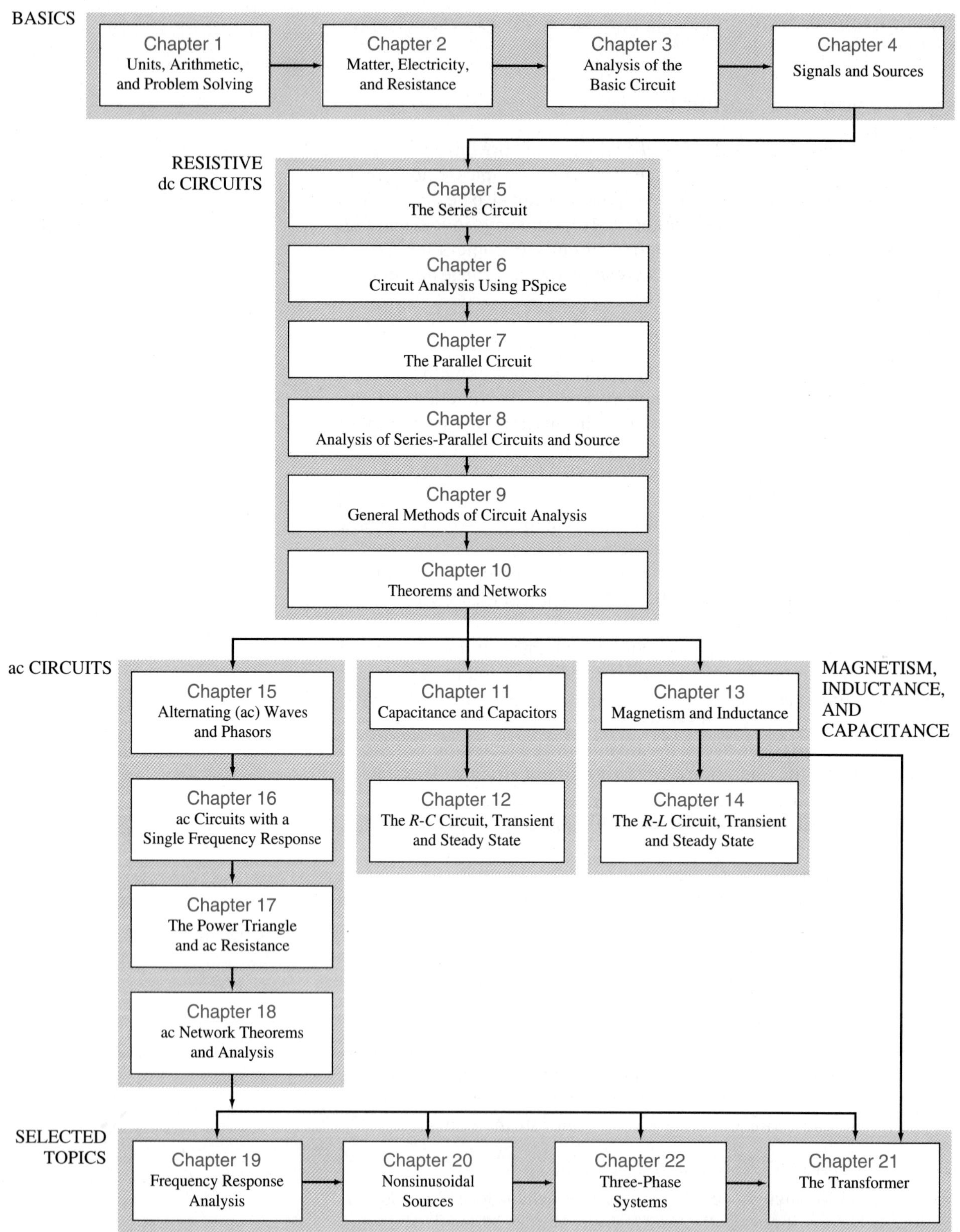

BASICS
Chapter 1
Units, Arithmetic, and Problem Solving
Chapter 2
Matter, Electricity, and Resistance
Chapter 3
Analysis of the Basic Circuit
Chapter 4
Signals and Sources
RESISTIVE dc CIRCUITS
Chapter 5
The Series Circuit
Chapter 6
Circuit Analysis Using PSpice
Chapter 7
The Parallel Circuit
Chapter 8
Analysis of Series-Parallel Circuits and Source
Chapter 9
General Methods of Circuit Analysis
Chapter 10
Theorems and Networks
ac CIRCUITS
Chapter 15
Alternating (ac) Waves and Phasors
Chapter 16
ac Circuits with a Single Frequency Response
Chapter 17
The Power Triangle and ac Resistance
Chapter 18
ac Network Theorems and Analysis
Chapter 11
Capacitance and Capacitors
Chapter 12
The R-C Circuit, Transient and Steady State
Chapter 13
Magnetism and Inductance
Chapter 14
The R-L Circuit, Transient and Steady State
MAGNETISM, INDUCTANCE, AND CAPACITANCE
SELECTED TOPICS
Chapter 19
Frequency Response Analysis
Chapter 20
Nonsinusoidal Sources
Chapter 22
Three-Phase Systems
Chapter 21
The Transformer

Figure 1 Flow diagram of chapter prerequisites

varying signals, namely pulses, instead of being created artificially by opening and closing switches. PSpice is used to perform transient analyses and the graphical output capability of PSpice (PROBE) is used to produce graphs of voltage and current waveforms.

Chapter 14 develops a method of graphical approximation to determine the area under curves for average value and effective value.

Part III: Reactive Circuits

Chapters 15–22 describe reactive circuits. The coverage of single-frequency analysis has decreased so that there is more room to discuss frequency response and spectral analysis of signals—topics that are essential to technicians and engineers today.

Chapter 18 is devoted to frequency response analysis techniques. Here PSpice and PROBE are used to obtain frequency response graphs. Chapter 19 discusses frequency response, Bode plots, and nonsinusoidal sources. PSpice is used to show the harmonic content of these signals. Chapter 20 includes a computer program for determining the terms in the Fourier series.

The text concludes with a look at transformers and three-phase systems. This chapter provides an introduction to ac power sources and completes our study of electronic circuits.

Figure 1 shows the suggested sequence of chapters. Chapters 1–10 should be studied in sequential fashion. Following Chapter 10, the course can take one of three directions. Chapters 19–22 can be studied in any sequence. This flexibility allows you to tailor the course to suit your needs and those of your students.

Features of the Text

The organization and pedagogical features of each chapter have been carefully crafted to motivate students in their study of electronic circuits. The extensive and consistent system of chapter pedagogy, numerous examples, problems and applications, the emphasis on problem-solving, the inclusion of new technologies, and the historical perspective on this field give students a broad and practical preparation in electronic circuits.

Chapter Pedagogy

Each chapter has an extensive pedagogical system to prepare students for what they will learn, reinforce key concepts and skills, and review the material thoroughly. The chapter begins with lists of key objectives and key terms and an introduction so students see the important topics and start building their technical vocabulary. Key terms are boldface in the text and listed in the margin as they are defined. Abundant diagrams and photographs motivate and explain the material. The examples, with their emphasis on problem-solving techniques, build students' confidence. Each chapter ends with a summary and a list of key equations for review, as well as numerous questions and problems.

Examples

More than 400 worked examples, an average of 19 per chapter, are included to show students how to apply the concepts, principles, and techniques presented in the chapter.

Emphasis on Problem-Solving

The most important skills students can learn at this stage are how to solve problems. Chapter 1 includes a section explaining the rudiments of problem-solving and demonstrating the strategy-solution format for the worked examples. Many of the examples throughout *Circuits: Principles, Analysis and Simulation* are worked using this unique format. The key steps of the solution are explained in the margin so students identify the problem-solving strategy and start developing strategies of their own.

Questions and Problems

The emphasis on problem-solving extends to the end-of-chapter question and problem sets. The questions are designed to test comprehension of the material. The problems give students practice applying their problem-solving skills to the concepts, techniques, and practices presented in the chapter. There are 500 questions in this text and almost 1000 problems. Answers to all of the questions and to odd-numbered problems appear at the back of the text.

Artwork and Photographs

Over 1000 diagrams and graphs have been included to clarify and motivate explanations. Drawing diagrams is also an important step in solving problems. Color is used to illustrate a key point or to emphasize changes.

The photographs in this text have been carefully selected to illustrate important equipment, technology, and applications and to show students practical applications of the theory being presented.

Scientific Calculators

Students are encouraged to have and use scientific calculators to make quick and accurate calculations. Calculator keys that may be new to the students are described where appropriate. Calculator keystrokes for both algebraic and RPN calculators are provided where appropriate to assist students in performing mathematical computations.

Integration of PSpice and PROBE

Because circuit simulation is such an important tool in both learning about electric circuits and using them in a practical setting, PSpice and PROBE are described and implemented through much of the text. Chapter 6 of *Circuits: Principles, Analysis and Simulation* introduces PSpice. Thereafter, separate sections with worked examples in PSpice and complete documentation are included at the end of all appropriate chapters. Graphing analysis using PROBE is included to help students visualize circuits better. All PSpice sections and problems are marked with an S for easy identification. For information about the PSpice disk, see the description under ANCILLARIES.

Instrumentation

At key points in the text, instruments are introduced and described briefly to support a laboratory component for the course. For information about the laboratory manual, see the description under ANCILLARIES.

Biographical and Historical Information

To help students gain a perspective about the field of electronics and electronic technology, we include historical material on the endpapers of the text about the development of the study of electricity and biographies of some of the most notable individuals involved.

Ancillaries

A variety of ancillaries written by the authors to accompany *Circuits: Principles, Analysis and Simulation* are available to the instructor and the student.

For the Instructor:

The **Instructor's Manual with Transparency Masters** contains complete solutions to all the end-of-chapter problems. (Answers to the odd-numbered problems and to all the end-of-chapter questions are given at the back of the text.) The Instructor's Manual also has 100 transparency masters that duplicate the important figures in the text, the ones most useful in the classroom.

The **Test Bank** provides sample tests for each chapter and answer keys. The test questions are similar to the examples and problems in the text and cover all the key concepts. The Test Bank is also available in computerized form for IBM® PC computers. The **ExaMaster™ Computerized Test Bank** allows you to preview, select, edit, and add items to tailor the tests to your needs and print up to 99 versions of the tests. **ExamRecord™** gradebook is also available.

A special **Spice Disk** contains files for all the Spice examples in the text. The disk helps students study these examples more easily and learn to use Spice more quickly.

For the Student:

The **Laboratory Manual,** by David Hata, contains 30 experiments keyed to the text and is available for sale to students. The experiments show students how to verify the theory, help build their skills using electronics equipment, provide special projects to apply the concepts, and provide work with Spice where appropriate.

Acknowledgments

We would like to express our appreciation to those individuals who served as reviewers for this project:

Robert L. Anderson	Purdue University—Calumet
Bill Barnes	County College of Morris
Randy Bedington	Catawba Valley Community College
Samuel Derman	City University of New York Queens College
Robert Ian Davis	Miami Dade Community College
Robert L. DeWitt	Augusta Technical Institute
Mike Ellis	Weber State College
Joseph T. Ennesser	DeVry Institute of Technology—Lombard
Earl T. Farley	Texas Tech University
Thomas Gendrachi	University of Hartford
Tony Gundrum	College of Lake County
Bernard Guss	Pennsylvania State University—New Kensington
Roger D. Hack	Purdue University—Fort Wayne
Gordon Haggety	North Central Technical College—Wausau
Gerald Jensen	Western Iowa Tech Community College
Bradley E. Jenkins	St. Petersburg Junior College
Edward P. Kearney	Waterbury State Technical College
Stephen J. Kuyath	Central Piedmont Community College
Vincent J. Loizzo	DeVry Institute of Technology—Chicago
William Maxwell	Nashville State Technical Institute
Robert Mobley	Pellissippi State Technical Community College
Leonard Perry	University of Cincinnati
Edward R. Peterson	Arizona State University
Roy J. Powell	Chattanooga State Technical Community College
Mason Rittman	Texas A&M University
Parker M. Tabor	Greenville Technical College
Neal Voke	Triton College
Klaus Wuersig	SUNY College of Technology—Alfred
Steve Yelton	Cincinnati Technical College

The preparation and production of a textbook requires the talents and efforts of many people. We would like to acknowledge the contributions of the following individuals: Barbara Gingery, Senior Acquisitions Editor; Alexa Barnes, Developmental Editor; Margaret Mary Anderson, Senior Project Editor; Becca Gruliow, Project Editor; Christine Schueler, Art Director and Charlene Squibb, Production Manager.

FRANK P. YATSKO

DAVID M. HATA

September 1991

TO THE STUDENT

Whether you have decided to study electronic circuits as the beginning of your preparation for a career in the field of electronics or as an interesting avocation, we are glad that you have selected this textbook. Some new features have been included in this text that heretofore have not been part of an electronic circuits textbook. It is our hope that these features will not only lay a solid foundation for your further study in the field, but will also make your study of electronic circuits more interesting and enjoyable.

This textbook will help you visualize the operation of electronic circuits as system building blocks. In this context, each circuit will have an input signal, which is then processed or acted upon in some fashion by the circuit, producing an output signal. In an actual system, this output signal becomes the input signal for the next stage in the system and the processing of the signals in the system continues until the desired result is obtained.

Therefore, our goal is to help you visualize current and voltage signals in an electronic circuit. These signals are represented by waveforms. In the laboratory, we will use a personal computer to simulate or predict circuit behavior and to plot current and voltage waveforms. Then you will prototype or build the actual circuit and, using function generators to generate input signals and oscilloscopes to view signals within the circuit, you can determine if the circuit is performing as it should or not.

This textbook will present the theory behind the operation of circuit elements and circuits themselves. In order to gain maximum benefit from the textbook, we offer the following suggestions:

- Determine what you should learn from the chapter by reading the chapter objectives.
- Become familiar with the terminology in the chapter by reviewing the list of key words.
- Read through the chapter several times, once for an overview and then for understanding.
- Review the list of equations, paying particular attention to the context in which each can be used.
- Follow the steps in the examples, paying close attention to the logical progression to a final solution.
- Answer the questions at the end of the chapter to test your comprehension of the material presented.
- Work the end-of-the-chapter problems to hone your problem-solving skills.

The laboratory exercises are equally important. It will be through the laboratory experience that you will develop the key occupational skills needed by electronics technicians. You will be asked to construct circuits and to use the analysis techniques, including computer simulation, and the measurement techniques available to you to determine whether the circuit is performing correctly or not.

Some suggestions to help you use your laboratory time efficiently and effectively are:

- Prepare for the laboratory exercise to be performed by reading the lab assignment before coming to lab.
- Visualize the steps to be performed.
- Begin documenting the experiment by drawing schematics and constructing data tables.
- Predict the results that you expect to obtain in each part of the experiment.
- Construct the circuit before coming to lab so that lab time can be spent testing, measuring, and troubleshooting the circuit.
- Document every step performed and record all measurements in a laboratory notebook.
- Strive to solve problems yourself, but seek the instructor's help when you get stuck.

The study of electricity and electronics is a fascinating path that can lead to a rewarding and challenging career. The field is ever-changing and new devices, processes, and practices are constantly emerging.

We wish you well in your study and sincerely hope that this textbook is a good beginning.

FRANK P. YATSKO

DAVID M. HATA

September 1991

CONTENTS OVERVIEW

CONTENTS

CIRCUITS:

Principles, Analysis and Simulation

THE FIELD OF ELECTRICITY AND ELECTRONICS

CHAPTER OBJECTIVES

After completing this chapter, you should be able to:

1. List some milestones in the history of electronics and briefly explain the significance of each.
2. Define the terms engineering technologist and engineering technician.
3. List the types of skills required of engineering technologists and technicians.
4. List some of the tools of the trade used by engineering technologists and technicians.

KEY TERMS

1. Brattain, Bardeen, and Shockley
2. engineering technologist
3. engineering technician
4. function generator
5. hardware
6. microprocessor
7. operating systems software
8. oscilloscope
9. power supply
10. programming languages
11. programmable logic devices
12. prototyping
13. software
14. SPICE
15. voltmeter

1. A Historical Perspective

The field of electronics as we know it has a relatively recent history. In fact, the last 40 years have held a mind-boggling array of advances, which has changed the way we live, work, and play.

Our awareness of electricity and all of the forces of nature developed out of man's innate curiosity about how materials in our world behaved. The unusual properties of the naturally magnetic material, magnetite, were known to the early Chinese and Greeks. The Chinese are credited with the invention of the magnetic compass. The Greeks gave us our word for magnetism and electricity, named after the Greek word for amber.

The static electric phenomena were easy to observe. However, as long as early investigations were restricted to these static phenomena, electricity and magnetism remained only curiosities. Dynamic electric phenomena are essential for any significant use of electricity. We are familiar with Benjamin Franklin's famous kite experiment, which proved that lightning was nothing more than a big electric spark.

One of the early milestones was the development of the first battery by Alessandro Volta in 1800. This early design consisted of zinc and copper discs separated by cardboard soaked in water or salt water. Improvements to this battery were quickly made. Our modern-day dry cell is a direct derivative of an early cell called a Leclanche cell. In this cell, the liquid electrolyte of earlier cell designs was replaced by a paste or jelly, the porous pot by a muslin bag, and the glass jar by the zinc rod, which had been shaped into a can into which everything else was placed. The battery was encased in steel and sealed to prevent leaks. The battery produced 1.5 volts.

Twenty years later in 1820, Hans Christian Oersted, a professor at the University of Copenhagen, discovered the intimate link between magnetism and electricity. The 1800's gave us the work of well-known scientists such as Ampere, Ohm, and Henry. Joseph Henry, a man of modest financial means living in the then small town of Albany, New York, demonstrated the principle of the electromagnetic telegraph. About a mile of wire was strung around his classroom. The wire connected a battery at one end to an electromagnet at the other. When the magnet was energized, it repelled a small permanent magnet, which then struck a bell, giving an audible sound. What Henry demonstrated was that mechanical motion could be produced at great distance by means of electricity. What developed out of his work was the telegraph, which enabled communication across vast distances.

The other major discovery of the 1800's was electromagnetic induction. It is the scientific base for the development of power generation and its use, transmission lines, radio, and early electronics. Faraday and Maxwell were two of the scientists who contributed to this discovery. In fact, Faraday's work contributed so heavily to the field of electrical engineering that he has been called the "patron saint of electrical engineers."

Another milestone in the use of electricity was the development of the incandescent lamp. The work of Thomas Edison, known as the "Wizard of Menlo Park," is familiar to most of us. His laboratory was a prototype of the modern industrial laboratory, where he invented the best of the early lamp. It was Edison's foresight to design a complete lighting system that propelled the widespread use of electricity. His system included dynamos, a distribution system, lamp fittings, switches, cables, fuses, meters, and everything else that was needed to provide electric lighting for an area of a city. And so was born the General Electric Company in 1892.

The turn of the century also saw the work of Hertz and Marconi produce the beginning of radiotelegraphy. It is to Marconi that we owe a great debt for the early application of electromagnetic waves to communications. He was born into a well-to-do family in Bologna, Italy, and despite failing to gain his matriculation exams to the University of Bologna, gained access to the laboratory of Augusto Righi. From Righi, Marconi learned how to generate, radiate, and detect electromagnetic waves in an effort to communicate over greater and greater distances.

However, communication remained in its infancy until the development of the vacuum tube diode just after the turn of the century. Working independently, Fleming and De Forest pioneered this development. The early diodes and triodes were not a great commercial success, but World War I encouraged the development and application of these vacuum tube devices. Over 1 million were used in that war.

The superheterodyne radio circuit was patented in 1920. Its undisputed inventor was Edwin Armstrong. This important patent was acquired by Westinghouse and later Radio Corporation of America (RCA).

Broadcasting mushroomed, and in 1922, there were 30 licensed radio stations. Two years later, there were over 500 stations. At this time, well-known broadcasting companies were born—the National Broadcasting Company (NBC), the Columbia Broadcasting Company (CBS), and in Great Britain, the British Broadcasting Company (BBC).

World War II again spurred research and development in electronics. For example, the cathode ray tube, or CRT, was improved for radar use. After the war, the television was invented, and now we consider a television receiver an essential item of household equipment, and many homes have one in several rooms.

It was also at this time that Tektronix, Inc., a company in Portland, Oregon, developed the first **oscilloscope** with a triggered and calibrated sweep, an improvement over earlier designs by RCA and Dumont. The ability to portray time-varying voltages as graphs on the face of a CRT provided a most useful tool to the electrical engineer. No longer were electrical engineers totally reliant on analog meters.

oscilloscope

What history of electronics would be complete without highlighting the discovery of the transistor by John **Bardeen** (an experimental and theoretical physicist), Walter **Brattain** (a physical chemist), and William **Shockley** (a circuit expert) at Bell Labs in 1947?

Bardeen
Brattain
Shockley

Shockley predicted that it ought to be possible to modulate the conductivity of a thin layer of semiconductor by the application of an external field to produce amplification. Bardeen found he could explain the effects by assuming the presence of energy states on the semiconductor surface. And Brattain oversaw the construction of that first "transfer resistor" or "transistor."

The 1950's saw a phenomenal change in electronics as the transistor gradually ousted the electron tube from many applications. Imaginations ran wild in the early 1950's as scientists and engineers began to envision complete circuits on a tiny piece of silicon. By 1953, the first patent for an integrated circuit was filed by Harwick Johnson of RCA. Johnson's patent was for a "semiconductor phase shift oscillator and device." However, it was J. S. Kilby of Texas Instruments who made the first integrated circuit. Integrated circuit manufacturing blossomed in the 1960's, and by the end of the decade we had complete logic families, operational amplifiers, and an array of devices that filled data books.

Finally, computers benefited from all of these developments. From the initial work of Blaise Pascal who built the first practical calculator in 1642, to the work of Herman Hollerith with his punched card machines in the 1890's, computers have migrated from the mechanical world to the electronic world. And as each advance was made, computers became smaller, faster, and more powerful.

microprocessor

From such early computers as ENIAC and MANIAC, we now have computers many times more powerful sitting on our desk tops. The advent of the **microprocessor,** developed by Intel Corporation in the early 1970's, has not only put a computer on our desk, but a computer in our car, our kitchen, and our toys. From the work of three men at Bell Labs, we now have microprocessor chips with a million transistors on one little piece of silicon. It is truly amazing when you think about it. And it has not been that long.

2. A Look Forward

Looking to the future is a bit risky, but there are some trends that are worth noting. First, the role of the technician and the skill-set required by the technician continue to evolve. The early 1980's saw the demise of the bench technician in many companies. Automated test stations have taken over the tasks that technicians performed in calibrating and testing circuit boards. But with advances in technology, new roles have emerged, which will be discussed in the next section.

hardware
software

Instruments and systems will gain more functionality—and much of this will come not through **hardware**—but through **software**. Look at the trend in oscilloscopes, from the early models such as the Tektronix 561 and 465 dual-trace oscilloscopes, to the new oscilloscopes, which provide sampling, processing, and signal-analysis capabilities. Front panel switches no longer set the function. Instead, functionality is provided through software and selected through a system of menus.

The personal computer (PC) will become, if it has not already, an integral part of the technician's workbench. The PC will serve as a multipurpose tool that will run design, test, analysis, and simulation software as well as word-processing, spreadsheet, and other general communications packages. It will most likely be connected to a network allowing the use of electronic mail and access to data bases and other forms of knowledge.

We are in the "information age," and the technician will become increasingly involved in the management of information. That information may be contained in design files that describe a new product, the test data from the manufacturing floor, and servicing data from the field.

Systems will continue to be designed to do more in less time. We have seen clock speeds increase in our PC from 4 MHz to 16 MHz and beyond. The same increase in operating speed will occur in other products as well. But there is a limit to the clock frequency with conventional circuit configurations. New circuit operating modes that will operate asynchronously are currently being developed at research labs.

In light of these observations, let us take a look at the role of the technician and the tools of the trade in the next two sections.

3. What is an Electronics Technician/Technologist?

engineering technologist
engineering technician

An **engineering technologist** is a graduate of a four-year bachelor's degree program in engineering technology. An **engineering technician** is a graduate of a two-year associate of applied science degree program, or a person who has received comparable education through on-the-job experience and training programs. Many associate of applied science programs have linkages with upper-division programs, which lead to the bachelor's degree and are often referred to as "2+2" programs. These 2+2 programs offer the technician the education needed for career advancement.

Engineering technologists work in professional-level positions alongside engineers and technicians, translating concepts into working systems. They supervise the activities of technicians and tradespersons to implement the technical aspects of projects and systems. Assignments may include the preparation of plans and specifications, the design of standardized subsystems, and the supervision of manufacturing activities.

On the other hand, engineering technicians perform tasks following procedures set down by engineering standards. They are not expected to be responsible for design/manufacturing changes or judgments requiring marked deviations from accepted procedures, without consultation with their supervisors. Tasks may include data collection, testing, troubleshooting, system installation, and documentation of test results.

Your educational preparation for future jobs as engineering technologists and technicians will include the study of mathematics, science, and engineering technology courses, as well as general education courses with a strong emphasis in writing and speaking skills. Electronic/electrical engineering technology programs will include technical courses in circuits, devices, digital systems, linear systems, microprocessor technology, and other specialties. Many programs will increasingly emphasize software—both programming languages and operating systems software. Bachelor's degree programs will allow you to specialize in telecommunications, manufacturing, design automation, technical writing, computer systems, and many other areas.

But there are other areas to consider as possible careers. Mask design is an area where one can apply electronics knowledge to the design and layout of integrated circuits. Or, if you have a bent for technical writing, as a technical writer you can work with engineers to produce the manuals, technical articles, and other materials that will support a product. An applications specialist answers questions from customers on the use of company products. These are just a few of the new jobs that have emerged as technology has advanced. Each requires an understanding of fundamental electronics, software, and scientific concepts—plus the ability to effectively communicate, both verbally and in written form.

The types of jobs performed by technicians are expanding and becoming more varied. The possibilities are limitless. But one must begin somewhere, and the usual starting place is a first circuits course. Before we discuss the study of circuits, let's look at the tools-of-the-trade. What tools will you have at your disposal as you perform your job, and how have those tools changed over the years?

4. Tools of the Trade

voltmeter
power supply
function generator

In the 1960's and 1970's, the technician's tools-of-the-trade included the **voltmeter** or multimeter, **power supply, function generator,** oscilloscope, and possibly a counter/timer. Of course, each specialty had its own set of unique instruments, but these instruments formed the heart of the instrumentation package each technician was expected to use.

The instruments were stand-alone instruments and for the most part were non-intelligent—meaning the decision-making functions were performed by hardware. You could walk up to these instruments and with a basic knowledge of their functions and the front panel markings you could figure out how to operate them. Behind each knob was a shaft that went back to a switch that turned on the appropriate function.

Today, many instruments are intelligent instruments. They include self-test features, detection of operator errors, and decision making capabilities. Furthermore, instruments now include the ability to communicate with computers so data can be stored for analysis. Computers can also configure instruments for specific test functions. The front panel is also different. Keys have replaced the knobs. Now when a certain function is desired, the proper key is pressed. Instead of a shaft controlling a switch, the key closure is decoded in software and the proper function is activated.

Instead of front panel markings, many instruments now are menu-driven. The function is chosen from a list shown on the instrument's CRT. Hence, instruments have greater functionality without overloading the front panel with switches. Each key handles more than one function. Take a look at your hand-held calculator as an example of multifunction keys.

The biggest change occurring is the incorporation of the computer into the instrumentation at the workbench. The computer controls instruments, acquires data, analyzes the data, and formats the results in the most useful form, all at the touch of the proper keys. A myriad of software tools is at your disposal to perform the desired tasks. The technician must find the most efficient path through operating systems, applications, and test software.

prototyping

Let's look at another commonplace activity, **prototyping**. Prototyping is the process of building a model of a new design. In the past, prototyping meant making a few units that would be used to see if the design worked as specified, so wire wrapping and other techniques were used to build the circuit.

With increased operating speeds, wire wrapping no longer works, and making printed circuit boards is expensive and the result is hard to change. Furthermore, the whole process is very time-consuming. With the tremendous competition to get products on the market and the decreasing life cycle of products, wire wrapping is no longer acceptable.

Today, prototyping is done with engineering workstations using sophisticated design automation software. Concepts are input into workstations in schematic form and translated into data files, which can be used by other software packages. One of those packages performs circuit simulation, so the technician can use the computer to simulate the operation of the circuit. The results can be analyzed, design changes entered, and additional simulation performed. Hence, many iterations in the design cycle can be performed in a fraction of the time it would take to make a hardware prototype. This is a big advantage to companies.

If you get the opportunity, go to one of the regional electronics shows and tour the exhibit area. You will be amazed at the many new products being displayed. The tools-of-the-trade will continue to expand, making the technician's job easier and more productive.

5. The Study of Circuits: The First Step

The study of circuits is usually the first course in an associate degree program in electronic/electrical engineering technology. Circuits are building blocks for systems and hence are needed to understand the functioning of complex systems. Circuits are composed of smaller units called components. These components can be resistive, capacitive, inductive, or one of many different types of active devices such as transistors and integrated circuits.

In our study of circuits, we will view circuits from a signal-processing viewpoint. A block diagram of our circuit model is shown in Figure 1. The model consists of a circuit, which performs a defined signal-processing function, an input signal, and an output signal. The input signal enters the circuit at the input terminals, is acted upon or processed by the circuit, and exits the circuit at the output terminals. The input signal can be a constant voltage, a sine or square wave, or another complex signal. The processing activity may include amplification, attenuation, filtering, integration, or one of many other functions.

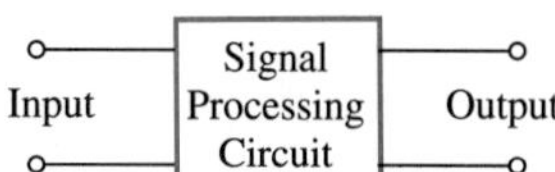

Figure 1. Block diagram of a system component.

A system, then, is the combination of these building blocks to form a configuration that will perform the desired overall process. Understanding the functioning of the individual circuits will aid you in tracing the signal through the system, from one circuit block to the next. Furthermore, you will know the form of the signal at the inputs and outputs of the circuits so that you can use one of the instruments at your workbench to measure the signal to see if the circuit is performing the expected function. This ability to accurately predict signals at intermediate points is a requisite to successful troubleshooting.

In this textbook, we will only consider the passive building blocks of circuits, leaving the study of active devices to be covered subsequently. We will learn how to identify components, read their component values, and predict their operating characteristics.

Basic circuit configurations will be studied. These configurations will include series and parallel connections. Analysis techniques such as Thevenin's and Norton's Theorems will reduce complex circuits into these simpler forms for ease in analysis.

Basic measurements will include the measurement of voltage, current, resistance, period, and frequency. We will not limit the measurement of parameters to the voltmeter but will include the oscilloscope early in the text.

SPICE

We will add simulation of circuits, not because the circuits are too complex to analyze by hand, but to gain familiarity with simulation tools like **SPICE** so that it can be used in this and subsequent courses. SPICE will also facilitate a discussion of the spectral content of signals, a topic that until now has been too mathematically laborious to be included in a first circuits course. We'll let the computer worry about the complexity of the mathematics while we will study the general concept of spectral analysis.

There will be some relationships and formulas that you will have to memorize. There is no way around that. But for the most part, knowing a few basic relationships will enable the student to derive the other needed relationships. Hence, we won't have to memorize all of the equations used in this text.

What we will strive to do is to develop mental pictures of circuit operation. SPICE will aid us in doing this as we view the simulation of circuits.

Once we have mastered passive circuits, we can add active devices to our circuits and proceed with our study of electronics. But just like a runner beginning a race, we must get out of the starting-block. So let's discuss how to use this book to aid you in getting started in electronics.

7. How to Use this Book

This textbook consists of 22 chapters that include the major topics normally found in a first circuit course sequence. Each chapter consists of a number of subsections related to the major concept being discussed in the chapter.

Each chapter begins with a list of objectives. These objectives will tell you what we hope you will learn from the chapter. After completing the chapter, return to these objectives to see if you have been successful. Your hard work and input from your classroom instructor are vital to the learning process.

The chapter also includes a list of key words. These are new terms that are introduced in the chapter. Each field has its own vocabulary, and to communicate ideas efficiently requires the use of accepted terminology. Return to the key words list after completing the chapter and check whether you have added each term to your technical vocabulary.

Each section within a chapter contains an explanation of a technique or concept. Examples are included to illustrate the use of a technique, equation, or concept.

At the end of the chapter is a brief summary of the major concepts presented in the chapter. A list of equations reviews the key relationships presented in the chapter.

To help you practice the techniques presented, or to check for understanding, questions and problems are included. The questions will enable you to check your understanding of the concepts. The problems provide practice in performing certain types of analyses. The problem section is divided according to the chapter section divisions. Hence, each section contains a group of problems designed to help in your application of the concepts in that section. Some answers are provided at the end of the text so that you can check some of your answers.

Electronics is best learned by applying the knowledge that you have studied. Like any activity, the more you do it and exercise the skills that you have acquired, the better you will become in performing your work responsibilities, so practice, practice, practice.

We hope you have a successful course, which will lay the foundation for a long and rewarding career.

SUMMARY

1. The short history of electronics has included many milestones that have affected the skills required of engineering technologists and technicians, the tools that they use, and the types of systems on which they work.
2. In the future, the role and skill-set of engineering technologists and technicians will continue to evolve.
3. The engineering technologist is a graduate of a four-year bachelor's degree program in engineering technology, while the engineering technician is a graduate of an associate of applied science program.
4. The tools-of-the-trade will continue to include stand-alone instruments such as multimeters, power supplies, function generators, and oscilloscopes, but newer instruments will most likely interface with computers.
5. Software will become an important tool for the engineering technologist/technician.
6. The study of electric circuits is the normal starting point for educational programs for engineering technologists and technicians.

QUESTIONS

1. List three major milestones in the history of electronics and briefly explain the significance of each.
2. What impact did the invention of the microprocessor have on the field of electronics?
3. Define the terms engineering technologist and engineering technician.
4. Describe the skills needed by engineering technologists/technicians.
5. Name some tools-of-the-trade used by engineering technologists and technicians.
6. Write a news article on the development of the transistor.
7. Trace the development of vacuum tubes prior to World War II.
8. What is an engineering workstation and what impact does it have on the way new products are being designed?
9. How have recent developments in electronics technology affected your life? What changes would you predict for the future?

CHAPTER 1

UNITS, ARITHMETIC, AND PROBLEM SOLVING

CHAPTER OBJECTIVES

After completing this chapter, you should be able to:

1. Use the International System of Units correctly.
2. List the prefix names and use the prefixes in problem solving.
3. Effectively use conversion factors and round off numbers.
4. Perform calculations and present the result to the correct number of significant figures.
5. Write decimal numbers in scientific notation.

$$(5 \times 10^{-2} \frac{m}{s})(3.28 \frac{ft}{m}) = ?$$

6. Change numbers represented in scientific notation to their decimal equivalent.
7. Use scientific notation in problem solving.
8. Apply the problem-solving procedure to word-statement problems.

KEY TERMS

1. ampere
2. approximate number
3. base unit
4. candela
5. conversion factor
6. derived unit
7. exact number
8. International System of Units
9. kelvin
10. kilogram
11. meter
12. mole
13. prefix
14. radian
15. rounding off
16. scientific notation
17. second
18. significant figure
19. steradian
20. supplementary unit
21. system of units
22. unit of measure

INTRODUCTION

As a student and a technician, you will need to make measurements and perform calculations. The results will be meaningless unless they are numerically and dimensionally correct and have the proper accuracy. For instance, what can one conclude if told that an electric current is 5? One must ask, 5 what? Or, how can the length of an item be expressed to the nearest centimeter if one is not sure of the number of meters?

Dimensional and numerical correctness comes from an understanding of units, prefixes, conversion factors, and scientific notation. Proper accuracy comes from an understanding of significant figures, rounding off, and arithmetic operations with approximate numbers. This chapter provides the material needed to help you handle numbers and measurements properly.

The last part of the chapter deals with procedures for problem solving. Two procedures for developing a path from the knowns to the wanted quantities are explained. The common denominator to all word-statement problems is identified, and the reader is shown how to apply it. You will see that problem solving is something that can be learned, not something that only gifted students can do.

1.1 The International System of Units (SI)

A unit of measurement is a definite quantity that is recognized and accepted as a standard of measurement. Some units that might be familiar to you are the second, meter, and the liter. Units must be included when working with electrical concepts and analyzing circuits. The numerical values have no meaning without the units.

For measurements to be meaningful, a unit of a quantity must represent the same amount, regardless of where the measurement is made. This is ensured by the use of standards. These standards are used to check or calibrate other measuring devices. One standard, the kilogram, is the mass of a cylinder of a specific material with accurately measured dimensions at a given temperature. International physical standards are maintained by the International Bureau of Weights and Measures in France. Secondary physical standards are kept at the National Bureau of Standards in Washington, DC.

units of measurement
systems of units
International System of Units (SI)

Units of measurement are grouped into **systems of units**. These are groups of units that are related in some way. Most countries use the **International System of Units (SI)**, a metric system that was adopted by 36 countries at the 11th General Conference on Weights and Measures held in France in 1960. The United States is slowly changing from the British System (foot, pounds, etc.) to SI (meter, kilogram, second, etc.). This change began when Congress passed the Metric Conversion Act in 1975. Many electrical quantities are the same in the British and SI systems, so the conversion to SI is not difficult. This text uses SI units except where a non-SI unit is still commonly used in the electrical field. One of these units is the circular-mil, which is still used for the area of a round conductor.

base units

SI has seven base units, two supplementary units, and many derived units. **Base units** are well-defined units that by convention are dimensionally independent. These base units are combined to form the

derived units. **Supplementary units** are two units that are used to measure plane and solid angles. The relation of these units to each other is shown in Figure 1.1.

derived units
supplementary units

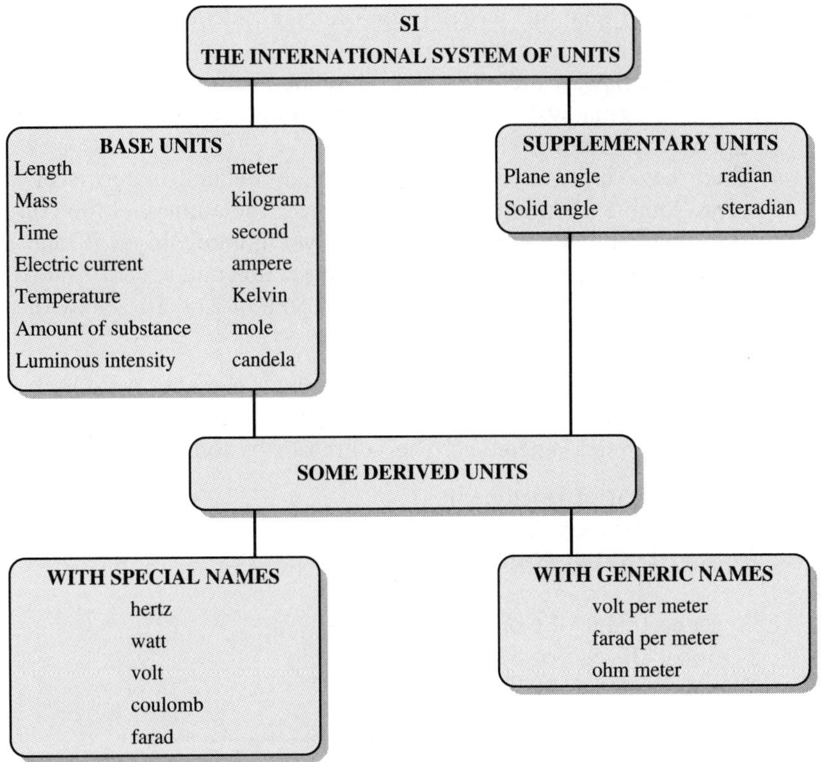

Figure 1.1 The relationships among the various units of SI.

The base units are as follows:

The **kilogram.** This is the unit of mass. A kilogram is approximately the mass of a 2.2 pound weight on earth. Its symbol is "kg." **kilogram**

The **second.** This is the unit of time. Its symbol is "s." **second**

The **ampere.** This is the unit of electric current. It is 1.25 times the current in a 100-watt incandescent lamp. Its symbol is "A." **ampere**

The **meter.** This is the unit of length. The length of one meter is about 1.1 yards. Its symbol is "m." **meter**

The **kelvin.** This is the unit of temperature. Its symbol is "K." However, the degree Celsius (°C) is more commonly used. **kelvin**

The **candela.** This is the unit of luminous intensity. Its symbol is "cd." **candela**

The **mole.** This is the amount of substance in terms of the number of atoms. Its symbol is "mol." **mole**

The two supplementary units are:

The **radian.** This is the unit of measure of a plane angle. A radian is approximately 57.3 degrees. Its symbol is "rad." **radian**

The **steradian.** This is the unit of measure of a solid angle. Its symbol is "sr." **steradian**

These base units are combined to give the derived units. Some of these units and their combinations can be seen in Table 1.1.

Some advantages of SI are:

1. Since each quantity has only one recognized unit, there is less chance of misunderstanding.
2. The units have a unique and well-defined set of symbols.
3. It is a decimal system. Multiples of SI units are related to each other by a factor of 10.
4. Each base unit, except the kilogram, is defined in terms of a reproducible phenomenon. For instance, the ampere is the constant current which, if maintained in two infinitely long, straight, parallel conductors of negligible cross section and placed 1 meter apart in a vacuum, would produce a force of 2×10^{-7} newtons per meter of length between them.

Other definitions and explanations are available in metric handbooks. These are published by the United States Government Printing Office and various technical societies. The addresses of some sources are:

Superintendent of Documents
U.S. Government Printing Office
Washington, DC 20402

U.S. Department of Commerce
National Bureau of Standards
Washington, DC 20234

TABLE 1.1 Some SI Derived Units

Quantity	Unit	Symbol	Formula
Force	newton	N	$kg \cdot m/s^2$
Energy	joule	J	N·m
Power	watt	W	J/s
Frequency	hertz	Hz	1/s
Electric charge	coulomb	C	A·s
Electric potential	volt	V	J/C or W/A
Electric resistance	ohm	Ω	V/A
Capacitance	farad	F	C/V
Magnetic flux	weber	Wb	V·s
Magnetic flux density	tesla	T	Wb/m^2
Inductance	henry	H	Wb/A
Electric conductance	siemens	S	A/V or 1/Ω
Electric field strength	volt per meter	$\mathscr{E}$	V/m
Electric permittivity	farad per meter	ϵ	F/m
Electric resistivity	ohm meter	ρ	Ω·m

The Institute of Electrical and Electronics Engineers
345 East 47th Street
New York, NY 10017

American Society of Mechanical Engineers
345 East 47th Street
New York, NY 10017

American Society for Engineering Education
National Center for Education
One Dupont Circle
Washington, DC 20046

It is easier to understand SI units if they are compared to the units in other systems. Table 1.2 compares the SI unit to the unit in the MKS

TABLE 1.2 A Comparison of Units of Various Systems

Quantity	SI	CGS	FPS	MKS	American Engineering
Force	newton (N)	dyne (dyn)	pound$_f$ (lb$_f$)	kilogram$_f$ (kg$_f$)	pound$_f$ (lb$_f$)
Length	meter (m)	centimeter (cm)	foot (ft)	meter (m)	foot (ft)
Time	second (s)	second (s)	second (s)	second (s)	second (s)
Mass	kilogram (kg$_m$)	gram (g)	slug ———	——— (kg$_f$·s^2/m)	pound$_m$ (lb$_m$)
Energy	joule (J)	erg (erg)	foot pound$_f$ (ft·lb$_f$)	——— (m·kg$_f$)	foot pound$_f$ (ft·lb$_f$)
Power	joules per second (J/s)	ergs per second (ergs/s)	foot pound$_f$ per second (ft·lb$_f$/s)	meter kilogram$_f$ per second (m·kg$_f$/s)	foot pound$_f$ per second (ft·lb$_f$/s)
Area	meter^2 (m^2)	centimeter^2 (cm^2)	foot^2 (ft^2)	meter^2 (m^2)	foot^2 (ft^2)
Volume	meter^3 (m^3)	centimeter^3 (cm^3)	foot^3 (ft^3)	meter^3 (m^3)	foot^3 (ft^3)
Pressure	pascal (Pa)	dyne per centimeter^2 (dyn/cm^2)	pound$_f$ per foot^2 (lb$_f$/ft^2)	kilogram$_f$ per meter^2 (kg$_f$/m^2)	pound$_f$ per foot^2 (lb$_f$/ft^2)

lb$_f$ is the gravitational force of 1 pound.
lb$_m$ is the mass that will experience a gravitational force of 1 pound at sea level and 45 deg latitude.
kg$_f$ is the mass that will experience a force of 1 newton at sea level and 45 deg latitude.
1 slug is the mass that will experience a gravitational force of 32.174 pounds at sea level and 45 deg latitude.

(meter, kilogram, second), CGS (centimeter, gram, second), FPS (foot, pound, second), and the American Engineering Systems.

Some systems use the kilogram and pound for a force while others use it for a mass. As a result, the units are often used incorrectly. Note that the weight content of many food items is given in grams or kilograms. Although there is not much difference in the two values for common applications, the difference can be critical in some applications. The two values will be the same at sea level and at a 45 degree latitude.

1.2 Prefixes and Using SI

The quantities in the first column of Table 1.3 are called prefixes. A

prefix

prefix represents a power of 10. For instance, ''kilo'' in kilogram indicates that the unit is a 1000 or 10^3 multiple of the gram. ''Mega'' represents a 1,000,000 or 10^6 multiple. A few more familiar terms that use prefixes are kilogram, centimeter, and microwave. Less common examples are nanofarad, microfarad, and gigahertz.

Why use a prefix? Several good reasons include: They require less space to write the quantity, there is less chance of making a writing error, and there is less chance of the term being misunderstood. Table 1.3 gives some useful information about prefixes. Since you will be meeting them throughout the text and in your work as a technician, it is to your benefit to study the table.

TABLE 1.3 Prefix Symbols and Multipliers

Name	Symbol	Pronunciation	Multiplier	Exponent
femto	f	fem' toh	0.000000000000001	10^{-15}
pico	p	peek' oh	0.000000000001	10^{-12}
nano	n	nan' oh (as in ant)	0.000000001	10^{-9}
micro	μ	micr' oh (as in microphone)	0.000001	10^{-6}
milli	m	as in military	0.001	10^{-3}
centi	c	as in sentiment	0.01	10^{-2}
deci	d	as in decimal	0.1	10^{-1}
deka	da	deck' a (as in about)	10	10^{1}
hecto	h	heck' toe	100	10^{2}
kilo	k	kill' oh	1000	10^{3}
mega	M	as in megaphone	1,000,000	10^{6}
giga	G	jiga	1,000,000,000	10^{9}
tera	T	as in terrace	1,000,000,000,000	10^{12}
peta	P	pet' ah (as in pet)	1,000,000,000,000,000	10^{15}
exa	E	x' ah	1,000,000,000,000,000,000	10^{18}

EXAMPLE 1.1

Express the following quantities using an appropriate prefix.

a. 6 million watts
b. 5 thousand watts
c. 3 hundredths of a meter
d. 8 thousandths of a second
e. 5 millionths of a farad

Solution

a. The prefix for million is "mega." So, six million watts is 6 megawatts or 6 MW.
b. The prefix for thousand is "kilo." So, five thousand watts is 5 kilowatts or 5 kW.
c. The prefix for hundredths is "centi." So, three hundredths of a meter is 3 centimeters or 3 cm.
d. The prefix for thousandths is "milli." So, eight thousandths of a second is 8 milliseconds or 8 ms.
e. The prefix for millionths is "micro." So, five millionths of a farad is 5 microfarads or 5 μF.

EXAMPLE 1.2

How much of each unit is represented by the terms below?

a. 3 milliamperes
b. 5 picofarads
c. 2 centimeters
d. 8 megawatts

Solution

a. 3×0.001 amperes $= 0.003$ amperes
b. 5×0.000000000001 farads $= 0.000000000005$ farads
c. 2×0.01 meters $= 0.02$ meters
d. $8 \times 1{,}000{,}000$ watts $= 8{,}000{,}000$ watts

Sometimes it is necessary to take a quantity that is given in one prefix and express it in another. For instance, how many centimeters are there in 500 millimeters? To express one prefix in terms of another, the ratio of the multipliers listed in Table 1.3 will be used. The procedure for doing so is shown in the next examples.

EXAMPLE 1.3

The current in a lamp is 500 microamperes. How many milliamperes is that?

Solution

A comparison of the multipliers in Table 1.3 shows that the multiplier for "micro" is 0.001 times that for "milli." That is,

$$\frac{0.000001}{0.001} = 0.001$$

Therefore,

$$\mu A = 0.001 \times mA$$

So

$$500\ \mu A = (500)(0.001) = 0.5\ mA$$

EXAMPLE 1.4

Express 5.2 megawatts in kilowatts.

Solution

The multiplier for megawatt in Table 1.3 is 1000 times as great as that for kilowatt. That is,

$$\frac{1{,}000{,}000}{1000} = 1000$$

Therefore,

$$kW = 1000 \times MW$$

So

$$5.2\ MW = (1000)(5.2) = 5200\ kW$$

To use SI correctly, specific rules must be followed. A few of the more important ones are:

1. The degree unit Celsius is the only unit name that is capitalized.
2. Metric unit symbols named after persons are the only ones capitalized.
3. Prefix symbols are lowercase except for the symbols for mega (M), giga (G), tera (T), peta (P), and exa (E).
4. The symbol is the same for one or several units (e.g., m not ms, kg not kgs). But unit names are pluralized (e.g., meters, kilograms).
5. A prefix alone should not be used to indicate a quantity (e.g., kilohms, not kilos).
6. A period should not be used after a symbol except at the end of a sentence.
7. A space should always be left between numerals and symbols. (Example: 10 V, not 10V.)
8. A zero is used before a decimal quantity that is less than a whole unit.

9. A division of symbols is expressed by using a slash or a solid line. The word "per" is used only with the names of the units. (Example: m/s or $\frac{\text{m}}{\text{s}}$, not m per s.)
10. A raised dot between the symbols is used to show a product of two unit symbols. (Example: N · m for newton-meter.)

Other rules will be introduced as necessary. A complete set of the rules can be found in many metric handbooks printed by technical societies and the United States Government. Refer to Section 1.1 for the names of some sources.

1.3 Conversion Factors

Quantities are often expressed in units different from the quantities needed for the mathematical operation. Sometimes these units will be used in the same system. That is, the length might be given in centimeters when the answer is needed in meters. At other times, the units might be in different systems. Length might be given in feet when it is needed in meters.

conversion factor

Changing from one unit to another for the same quantity is done with the use of a **conversion factor**. A conversion factor is a number that specifies the amount of one unit that is the same as another unit. Examples are: 12 items per dozen, 100 cents per dollar, and 24 hours per day. A conversion factor is simply a multiplier. A partial list of conversion factors is given in Table 1.4. Others will be introduced as needed.

Usually, the conversion procedure can be done in two steps:

a. Find the factor that lists the number of the wanted units that are in the given unit.
b. Multiply the given unit by the conversion factor.

The rules will be applied in the next two examples.

TABLE 1.4 Conversion Factors

Length	
0.0394 in/mm	2.540 cm/in (1)
0.394 in/cm	12 in/ft (1)
3.281 ft/m	0.305 m/ft
0.621 mile/km	3.000 ft/yd (1)
1.094 yd/m	5280 ft/mile (1)
Mass	
0.00221 lb_m/g	28.350 g/oz
2.205 lb_m/kg	0.454 kg/lb_m
35.274 oz/kg	14.594 kg/slug
0.0685 slug/kg	

TABLE 1.4 Conversion Factors (cont.)

Area	
10.764 ft^2/m^2	0.7854 square mil/circular mil
1.196 yd^2/m^2	6.542 cm^2/in^2
1,550.016 in^2/m^2	0.093 m^2/ft^2
Force	
0.000036 oz/dyne	1,000,000 dyne/N (1)
0.225 lb_f/N (2)	4.448 N/lb_f
Volume	
0.001 m^3/liter (1)	0.946 liter/qt
35.315 ft^3/m^3	0.0283 m^3/ft^3
1.308 yd^3/m^3	0.134 ft^3/gal
16.387 cm^3/in^3	4 qt/gal
Energy	
0.0000001 J/erg (1)	3,600,000 J/kWh
0.738 (ft·lb)/J	1.356 J/(ft·lb)
0.278 kWh/MJ	
Power	
0.738 (ft·lb)/s/W	746 W/hp
1.341 hp/kW	550 (ft·lb)/s/hp
Pressure	
144 (lb/ft^2)/(lb/in^2)	0.0209 (lb/ft^2)/pascal
14.696 (lb/in^2)/(1 atm)	1 (N/m^2)/pascal
Time	
60 minutes/h (1)	60 s/minute (1)
3600 s/h (1)	
Angle	
57.296 degrees/rad	
60 minutes/degree (1)	
Temperature	
1.8 Fahrenheit degrees/Celsius degree	
Temperature in °C $= \frac{5}{9}$ (°F − 32)	
Temperature in °F $= \frac{9(°C)}{5} + 32$	

(1) these are exact values
(2) a gravitational force of 1 lb.
1 lb_m is the mass that will experience a gravitational force (weight) of 1 lb at sea level and 45 deg latitude where the acceleration due to gravity is 32.174 ft/s^2.
1 slug is the mass that will experience a gravitational force (weight) of 32.174 lb at sea level and 45 deg latitude.

EXAMPLE 1.5

A bar is 2 meters in length. What is the length in feet?

Solution

Table 1.4 gives 3.281 feet per meter (3.281 ft/m). Then

$$(2 \cancel{\text{m}})\left(3.281\ \frac{\text{ft}}{\cancel{\text{m}}}\right) = 6.562\ \text{ft}$$

Note that the meter unit in ft/m cancels the meter unit in 2 m. The unwanted units should always cancel if the conversion is done correctly.

EXAMPLE 1.6(a)

How many horsepower (hp) are there in 2238 watts?

Solution

Table 1.4 gives 746 watts in 1 hp but we need the number of hp in 1 watt. Since 746 watts equals 1 hp, 1 watt must equal 1/746 hp.

Then

$$(2238\ \cancel{\text{watts}})\left(\frac{1\ \text{hp}}{746\ \cancel{\text{watts}}}\right) = 3\ \text{hp}$$

When a single conversion factor is not available, several of them might have to be used. Such is the case in Example 1.6(b).

EXAMPLE 1.6(b)

How many centimeters are there in 2 feet?

Solution

No cm/ft conversion factor is given in Table 1.4, so a combination of conversion factors must be used. A combination must be used that will result in the canceling of unwanted units. One such combination is in/ft and cm/in.

$$(2\ \cancel{\text{ft}})\left(12\ \frac{\cancel{\text{in}}}{\cancel{\text{ft}}}\right)\left(2.54\ \frac{\text{cm}}{\cancel{\text{in}}}\right) = 60.96\ \text{cm}$$

Since the equations in the examples have the same units on both sides of the equation sign, they are dimensionally consistent. If an equation is not dimensionally consistent, it cannot be correct. Recognizing this fact can help you check a solution. However, be careful! A dimensionally consistent equation does not always mean that the equation is correct.

1.4 Significant Figures and Rounding Off

Numbers used in engineering calculations are either exact or approximate. An **exact number** has no uncertainty in it, and is usually obtained

exact number

by a count or definition. The number of squares in Figure 1.2(a) is an exact number. There will be 12 squares every time a count is made.

approximate number

An **approximate number** has uncertainty in it. It is usually obtained by a measuring process, or by combining approximate numbers. The reading on the meter scale in Figure 1.2(b) has uncertainty in it. Each reading will probably give a different value because some of the digits must be estimated.

When dealing with approximate numbers, we should include only the significant figures. Otherwise, the number will appear to be more accurate than it is. A **significant figure** is one that has a reasonable amount of certainty. The last digit that can be estimated with reasonable certainty in Figure 1.2(b) is the 2 in 12. Since the digits after that are questionable, they are not significant.

The digits that are significant in a written approximate number are:

a. All non-zero digits. For example, 1, 3, and 4 in the number 1340.
b. Zeros that are not used to locate the decimal point. For example, the 0 in 104.
c. Zeros that are after the decimal point if there are some non-zero digits before the decimal point. For example, the 0 in 1.0.
d. Zeros used to locate the decimal point can be significant in some cases. If so, they should be overscored. For example, the 0's in $5\overline{00}$ and $5\overline{000}$ are significant.

A few examples will help to clarify these rules.

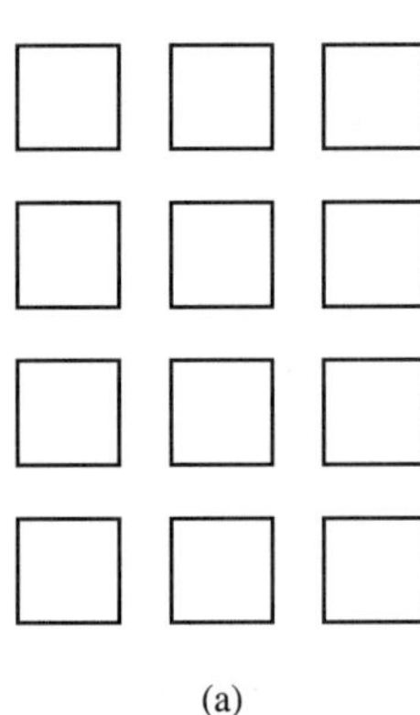

(a)

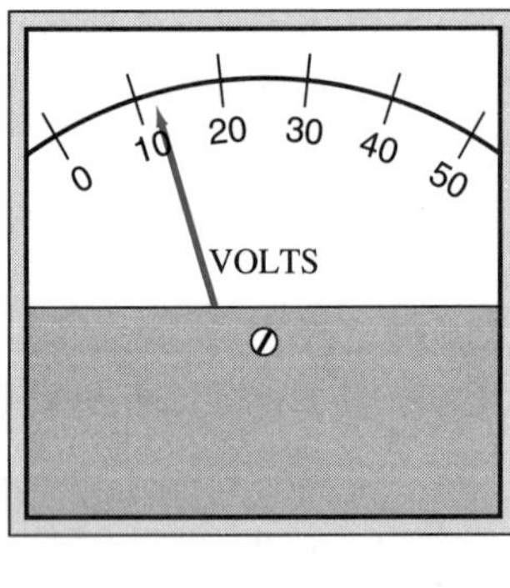

(b)

Figure 1.2 An example of an exact number and an approximate number. (a) The count of the squares will always be the same (12) so it is an exact number. (b) The reading on the meter, which has an estimated digit, is an approximate number.

EXAMPLE 1.7

How many significant digits are there in each of the following approximate numbers?

a. 745
b. 0.006
c. 0.0605
d. $3\overline{000}$

Solution

a. 745 All the digits are non-zero, so there are three significant digits.
b. 0.006 The zeros are used to locate the decimal point, which leaves only one significant digit.
c. 0.0605 There are three significant digits—the 6, the 5, and the zero between them.
d. $3\overline{000}$ The overscore indicates that the zeros are significant so there are four significant digits.

rounding off

When there are too many digits in a number, the unwanted ones are removed by rounding off. **Rounding off** means dropping the unwanted digits so that the remaining number is as near to the original as possible.

For example, 3.162 can be rounded off to 3.16, 3.2, or 3. The choice depends on how many digits are wanted. The procedure for rounding off is as follows:

1. All digits to the right of the last wanted digit are dropped. If any dropped digits are to the left of the decimal point, they are replaced by zeros.
2. If the digit after the last remaining digit is 5 or greater, add 1 to the last digit. Otherwise, leave the last remaining digit unchanged.

The procedure will be used in the following example.

EXAMPLE 1.8

Round off each number to the specified digits.

a. 53,567 to three digits
b. 0.13748 to three digits
c. 35,655 to three digits
d. 5.275 to one digit
e. 4.650 to two digits

Solution

a. The fourth digit is greater than 5 so 1 is added to the third digit. Since the 6 and 7 are to the left of the decimal point, zeros must be added:
   Answer: 53,600
b. The fourth digit is less than 5 so the third digit is not changed:
   Answer: 0.137
c. The fourth digit is a 5 so the third digit is increased by 1.
   Answer: 35,700
d. The second digit is less than 5 so the first digit is left unchanged.
   Answer: 5
e. The third digit is 5 so the second digit is increased by 1.
   Answer: 4.7

Rounding off can be done automaticallty on electronic calculators such as those shown in Figure 1.3. The calculator in Figure 1.3(a) uses

(a)

(b)

Figure 1.3 Electronic calculators such as these can make calculating easier and faster. (a) The solar-powered TI-30SLR+ uses algebraic logic. (b) The Hewlett-Packard 42S uses Reverse Polish Logic and has more features. (a, Photo courtesy of Texas Instruments; b, photo courtesy of Hewlett-Packard Company)

algebraic logic while the one shown in Figure 1.3(b) uses Reverse Polish Notation (RPN). For algebraic logic, the entries are made in the same order as when writing them. For RPN, the numbers are entered first, then the operation is specified. For example, the addition of 3 and 4 is

In general, rounding off is done by using the FIX key together with a number to specify the number of digits after the decimal point. A typical rounding-off operation is performed in Example 1.9.

EXAMPLE 1.9

Round off the number 53.1487 to four digits.

Solution:

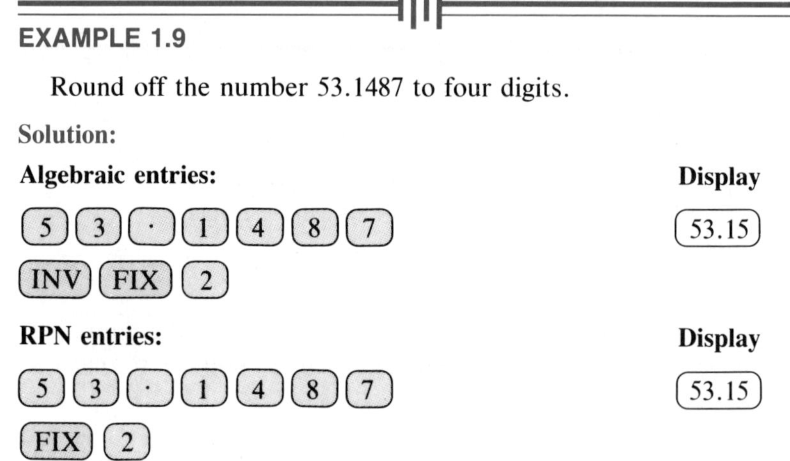

Algebraic entries: 5 3 · 1 4 8 7, INV FIX 2 — **Display** 53.15

RPN entries: 5 3 · 1 4 8 7, FIX 2 — **Display** 53.15

The entries for rounding off can differ for other calculators. Refer to your calculator manual for specific instructions.

1.5 Operating with Approximate Numbers

Results of calculations with approximate numbers should not have more significant figures than justified by the numbers that are used in the calculations. For example, adding the approximate numbers 2.62 and 3.7 gives 6.32. This answer is numerically correct, but its accuracy is not. The answer should not be expressed to two places after the decimal point when one of the parts is expressed to only one place after the decimal point.

What is the correct number of digits for the answer with approximate numbers? When adding or subtracting, the answer should be carried to the last complete column. When multiplying or dividing, the result is given the same number of significant digits as the number with the least number of significant digits in it. In all operations, extra digits are dropped and the remaining number is rounded off. The next examples will help to explain the procedure.

EXAMPLE 1.10

Perform the indicated operations for each of the following. Express the answer to the correct number of significant digits.

a. Add 2.23, 7.08, 5.4, and 8.132.
b. Add 15.826 and 3.51.
c. Subtract 3.82 from 9.065.

Solution

a.

```
   2.23
   7.08
   5.4   ←——Digit is missing in this column.
   8.132
  ------
  22.842
```

The last complete column is the tenths column. Rounding off the answer to the tenths gives 22.8.

b.

```
  15.826
   3.51  ←——Digit is missing in this column.
  ------
  19.336
```

The last complete column is the hundredths column. Rounding off to that column gives 19.34.

c.

```
   9.065
  -3.82  ←——Digit is missing in this column.
  ------
   5.245
```

The last complete column is the hundredths column. Rounding off to that column gives 5.25.

EXAMPLE 1.11

Express the results of the indicated operations to the correct number of significant digits.

a. 8.542×2.3
b. 8.02×10.535
c. 2500/2.5
d. 2050/25

Solution

a. $8.542 \times 2.3 = 19.6466$

8.542 has four significant digits while 2.3 has only two significant digits. Hence, the answer should have two significant digits. Rounded off, the answer becomes 20.

b. $8.02 \times 10.535 = 84.4907$

10.535 has five significant digits while 8.02 has only three significant digits. Hence, the answer should have three significant digits. Rounded off, the answer becomes 84.5.

c. $2500/2.5 = 1000$

2500 has two significant digits. 2.5 also has two significant digits. Hence, the answer should have two significant digits and is $1\bar{0}00$.

d. $2050/25 = 82$

2050 has three significant digits. 25 has only two significant digits. Hence, the answer should have two significant digits. It is 82.

Since exact numbers do not have any uncertainty, their result can include all digits. Rounding off of these numbers is done using a commonsense approach. A number such as 1.25683202 might be rounded off to 1.26, and 0.23528 to 0.24.

1.6 Scientific Notation

Writing numbers in decimal form when they are very large or very small is unwieldy, makes arithmetic operations difficult, and increases the chances of making errors. For these types of numbers, scientific notation is a more suitable form. This is a form where a number is written as a number between 1 and 10, multiplied by 10 raised to some power. Examples of some numbers written in scientific notation are 5×10^3, 1.2×10^5, and 2.3×10^{-2}.

The number 154 will be used to show the relation between the decimal form and scientific notation. The number 154 can be written as 1.54×100. Since 100 is equivalent to 1×10^2, the number 154 can also be written as 1.54×10^2, which is the scientific notation form. For the number 154, the equivalence of 100 and 10^2 was used, but other numbers use other equivalences. A listing of the more common equivalences is given in Table 1.5.

TABLE 1.5 Scientific Notation Equivalents of Decimal Numbers

Decimal Number	Scientific Notation Form
1	1×10^0
10	1×10^1
100	1×10^2
1000	1×10^3
10,000	1×10^4
100,000	1×10^5
1,000,000	1×10^6
0.1	1×10^{-1}
0.01	1×10^{-2}
0.001	1×10^{-3}
0.0001	1×10^{-4}
0.00001	1×10^{-5}
0.000001	1×10^{-6}

The general form for scientific notation is as follows:

$$M \times 10^n = N \quad \textbf{(1.1)}$$

Where:

N is the decimal form of the number
M is a digit between 1 and 10, called the coefficient, and
n is an integer power of 10.

EXAMPLE 1.12

Write the following in scientific notation:

a. 56,500
b. 0.0045

Solution

a. 56,500 can be represented as $5.65 \times 10{,}000$. Since 10,000 is 10^4,

$$56{,}500 = 5.65 \times 10^4$$

b. 0.0045 can be represented as 4.5×0.001. Since 0.001 is 10^{-3},

$$0.0045 = 4.5 \times 10^{-3}$$

Electronic calculators such as those in Figure 1.3 can be used to convert from decimal to scientific notation and back again. But remember, an incorrect entry or key punch will given an incorrect answer. Understanding the conversion procedure can help you to check your results. The keypunch sequence for numbers in Example 1.12 is as follows:

a. 56,500

Algebraic	**Display**
[INV] [SCI] [3] [5] [6] [5] [0] [0] =	5.65 04

This display means 5.65×10^4. The 3 after SCI fixes the number of digits in the display. If 4 were entered, the display would have been 5.650   04.

RPN

[5] [6] [5] [0] [0] [INV] [SCI] [2]	5.65 04

The 2 after SCI fixes the number of digits after the decimal point. If 3 were entered, the display would have been 5.650   04.

b. 0.0045

Algebraic	**Display**
[INV] [SCI] [3] [·] [0] [0] [4] [5] =	4.50 −03

This display means 4.50×10^{-3}. As in (a), the number 3 fixes the number of digits.

RPN **Display**

4.50 −03

The 2 after SCI fixes the number of digits after the decimal point.

The calculator entries shown here are for two specific calculators. You should keep in mind that the entries might vary slightly for other models. For instance, some calculators use a second function key instead of the Inverse Key. Others, such as the Sharp, use an F ⟨−⟩ E key to convert from one form to the other. You should refer to your calculator's instruction manual for the exact procedure.

Now that the relation between the decimal form and scientific notation is understood, another method for making the conversion will be examined. This method is more mechanical and does not require converting the decimal number to its power of 10 equivalent each time. The procedure is as follows:

1. Move the decimal point left or right in the decimal form to get a number between 1 and 10.
2. Let n be equal to the number of places that the decimal point is moved.
3. Make the sign of n positive (+) if M is smaller in value than the original number. It is negative (−) if M is larger in value than the original decimal form number. This is because $M \times 10^n$ must equal the decimal form. So, if M is smaller, 10^n must be greater than 1. This requires that n be positive. If M is greater, 10^n must be smaller than 1. This requires a negative exponent.

Now let's move on to some examples.

EXAMPLE 1.13

Write the following in scientific notation.

a. 56,500
b. 0.0045

Solution

a. The decimal must be moved four places to the left to give 5.65. Therefore, $n = 4$.
5.65 is smaller in value than 56,500 so n must be positive.

$$56{,}500 = 5.65 \times 10^4$$

b. Now the decimal point must be moved three places to the right to give 4.5.
Therefore, $n = 3$.
4.5 is larger in value than 0.0045 so n must be negative.

$$0.0045 = 4.5 \times 10^{-3}$$

Numbers can also be changed from scientific notation to decimal notation. This is done by reversing the procedure.

EXAMPLE 1.14

Write the following numbers in decimal form.

a. 2.48×10^5
b. 5.12×10^{-3}

Solution

a. $10^5 = 100{,}000$.
So,

$$2.48 \times 10^5 = 2.48 \times 100{,}000 = 248{,}000$$

b. $10^{-3} = 0.001$.
So,

$$5.12 \times 10^{-3} = 5.12 \times 0.001 = 0.00512$$

The keystroke sequence for part (a) of Example 1.14 is as follows:

Algebraic	**Display**
[2] [·] [4] [8] [EXP] [5] [INV] [FIX] [2]	[248000.00]

The 2 after FIX fixes the number of digits after the decimal point.

RPN	**Display**
[2] [·] [4] [8] [EEX] [5] [FIX] [2]	[248000.00]

The number 2 after FIX fixes the number of digits after the decimal point.

Scientific notation can also be used to show the number of significant zeros in a number. This is done by retaining only the significant zeros. For example, if 53,000 is significant to three digits, it is written as 5.30×10^4. If 53,000 is significant to four digits, it is written as 5.300×10^4. And, if 53,000 is significant to five digits, it becomes 5.3000×10^4.

1.7 Arithmetic Operations with Scientific Notation

Writing numbers in scientific notation is of little value if one cannot perform arithmetic operations with them. The procedures for multiplying, dividing, adding, and subtracting will now be presented.

Multiplication and Division

The coefficients are multiplied to obtain the coefficient of the product. The exponents are algebraically added to obtain the exponent of 10 in the answer.

$$(A \times 10^m)(B \times 10^n) = (A \times B) \times 10^{m+n} \quad \textbf{(1.2)}$$

Where:

A and B are the coefficients
m and n are the exponents of 10

EXAMPLE 1.15

Perform the multiplication for each of the following:

a. $(2 \times 10^3)(3 \times 10^2)$
b. $(4 \times 10^2)(1.5 \times 10^{-4})$
c. $(3 \times 10^{-2})(2.5 \times 10^{-4})$

Solution

a. $(2 \times 10^3)(3 \times 10^2) = (2 \times 3) \times 10^{(3+2)} = 6 \times 10^5$
b. $(4 \times 10^2)(1.5 \times 10^{-4}) = (4 \times 1.5) \times 10^{(2+(-4))} = 6 \times 10^{-2}$
c. $(3 \times 10^{-2})(2.5 \times 10^{-4}) = (3 \times 2.5) \times 10^{(-2+(-4))} = 7.5 \times 10^{-6}$

EXAMPLE 1.16

Perform the following multiplication:

$$(4 \times 10^3)(8 \times 10^2)$$

Solution

$$(4 \times 10^3)(8 \times 10^2) = (4 \times 8) \times 10^{(3+2)} = 32 \times 10^5$$

Since $32 = 3.2 \times 10^1$,

$$32 \times 10^5 = 3.2 \times 10^6$$

Next, we have the more general case, which is the multiplication of several numbers.

EXAMPLE 1.17

Perform the following multiplication:

$$(5 \times 10^2)(4 \times 10^3)(8 \times 10^2)$$

Solution

$$\begin{aligned}(5 \times 10^2)(4 \times 10^3)(8 \times 10^2) &= (5 \times 4 \times 8) \times 10^{(2+3+2)} \\ &= 160 \times 10^7 \\ &= 1.6 \times 10^2 \times 10^7 \\ &= 1.6 \times 10^{(2+7)} \\ &= 1.6 \times 10^9\end{aligned}$$

The multiplication of numbers in scientific notation can also be done using a calculator. The keypunch sequence for Problem 1.15(b) is as follows:

Algebraic Entry

Set the display for scientific notation using,

[INV] [SCI] [3]

The "3" will give three digits in the display. The entries for multiplication are:

[4] [EXP] [2] [×] [1] [·] [5] [EXP] [4] [+/−] [=]

The display will be 6.00 −02 which is 6.00×10^{-2}.

RPN Entries

Set the display for scientific notation using

[INV] [SCI] [3]

This will give three digits after the decimal point. The entries for multiplication are:

[4] [EEX] [2] [ENTER] [1] [·] [5] [EEX] [4]
[CHS] [×]

The display will be 6.000 −02 which is 6.000×10^{-2}.

As before, the procedure for your calculator might be a little different, so refer to the instruction manual.

Division—The coefficients are divided to obtain the coefficient of the quotient. The exponent in the denominator is algebraically subtracted from the exponent in the numerator to obtain the exponent of 10 in the answer.

$$(A \times 10^m)/(B \times 10^n) = (A/B) \times 10^{(m-n)} \quad \textbf{(1.3)}$$

EXAMPLE 1.18

Perform the division for each of the following:

a. $(6 \times 10^4)/(3 \times 10^2)$
b. $(9 \times 10^3)/(3 \times 10^{-5})$
c. $(8 \times 10^{-2})/(4 \times 10^{-4})$

Solution

a. $(6 \times 10^4)/(3 \times 10^2) = (6/3) \times 10^{(4-2)} = 2 \times 10^2$
b. $(9 \times 10^3)/(3 \times 10^{-5}) = (9/3) \times 10^{(3-(-5))} = 3 \times 10^8$
c. $(8 \times 10^{-2})/(4 \times 10^{-4}) = (8/4) \times 10^{(-2-(-4))} = 2 \times 10^2$

EXAMPLE 1.19

Perform the following division:

$$(4 \times 10^2)/(8 \times 10^4)$$

Solution

$(4 \times 10^2)/(8 \times 10^4) = (4/8) \times 10^{(2-4)}$
$= 0.5 \times 10^{-2}$
$= 5 \times 10^{-1} \times 10^{-2}$
$= 5 \times 10^{(-1+(-2))}$
5×10^{-3}

Next we have an example with multiple operations.

EXAMPLE 1.20

Perform the following divisions:

a. $$\frac{(2.4 \times 10^6)}{(6 \times 10^5)(5 \times 10^3)}$$

b. $$\frac{(2.4 \times 10^6)}{(6 \times 10^{-5})(5 \times 10^3)}$$

Solution

a. $$\frac{(2.4 \times 10^6)}{(6 \times 10^5)(5 \times 10^3)} = \frac{2.4 \times 10^6}{30 \times 10^8}$$
$= 0.08 \times 10^{-2}$
$= 8 \times 10^{-2} \times 10^{-2}$
$= 8 \times 10^{-4}$

b. $$\frac{(2.4 \times 10^6)}{(6 \times 10^{-5})(5 \times 10^3)} = \frac{2.4 \times 10^6}{30 \times 10^{-2}}$$
$= 0.08 \times 10^8$
$= 8 \times 10^{-2} \times 10^8$
8×10^6

Addition and Subtraction

The exponents must be the same for all numbers.

$$(A \times 10^n) \pm (B \times 10^n) = (A \pm B) \times 10^n \quad \textbf{(1.4)}$$

EXAMPLE 1.21

Perform the following addition:

$$(5 \times 10^3) + (2 \times 10^3) + (4 \times 10^3)$$

Solution

The exponents are the same, $n = 3$. So, no changes are needed.

$$(5 \times 10^3) + (2 \times 10^3) + (4 \times 10^3)$$
$$= (5 + 2 + 4) \times 10^3$$
$$= 11 \times 10^3$$
$$= 1.1 \times 10^4$$

When the numbers being added or subtracted have different exponents, the exponents must first be made the same.

EXAMPLE 1.22

Perform the following addition:

$$(5 \times 10^2) + (6 \times 10^3)$$

Solution

The larger number is 6×10^3, so we will express it in terms of 10^2. According to Section 1.6, 6×10^3 is equivalent to 60×10^2. Then

$$(5 \times 10^2) + (60 \times 10^2) = (5 + 60) \times 10^2$$
$$= 65 \times 10^2 = 6.5 \times 10^3$$

EXAMPLE 1.23

Perform the following addition:

$$(3 \times 10^{-2}) + (2 \times 10^2)$$

Solution

The number 2×10^2 is larger than the number 3×10^{-2}, so its exponent will be changed to 10^{-2}. Changing the power from $+2$ to -2 is a change of four, so the decimal point must be moved four places. Since -2 is smaller in value than $+2$, the decimal point must be moved to the right to make 2 larger. So

$$2 \times 10^2 = 20{,}000 \times 10^{-2}$$

Then

$$(3 \times 10^{-2}) + (2 \times 10^2) = (3 \times 10^{-2}) + (20{,}000 \times 10^{-2})$$
$$= 20{,}003 \times 10^{-2}$$
$$= 2.0003 \times 10^2$$

1.8 A Procedure for Problem Solving

While working with electrical circuits, both in the classroom and in industry, you will have to analyze circuits and engage in problem solving. In the classroom, the problems are provided by texts such as this one. In practice, they are the result of the projects to which you are assigned. Unfortunately, many students have never been shown how to develop an orderly or organized procedure to use in problem solving. Some depend on memorization, while others use a hit-or-miss approach. Still others simply never even get beyond the single-equation, single-unknown type of problem. If problem solving is to be successful, an organized approach must be developed. The remainder of this section deals with developing such an approach.

There are four basic steps in solving a problem. They are:

1. Read the problem statement carefully, noting the quantities that are given and those that must be found. Some will be given directly while others might be obtained from the problem conditions.
2. Draw a diagram where appropriate and label all the parts and quantities.
3. Use individual relationships or groups of them to form a path linking the wanted quantities to the known quantities. In many problems, the unknowns in the equations can be used to form the path to the answer.
4. Once the number of independent equations in Step 3 is equal to the number of unknowns, solve the equations.

These steps will now be applied to the solution to some problems.

EXAMPLE 1.24

If the density of copper is 8.93 g/cm^3, what is the length of a piece of copper wire that has a mass of 50 g and a diameter of 0.254 cm?

Solution

The knowns are all given directly.

A diagram is not needed for this problem.

A study of the statement leads to the following sequence as a possible path.

First, the volume of the wire can be found using the mass and density. Next, the area is found using the diameter and the formula for the area of a circle. Then, the length is found using the relation between the area, length, and volume.

The solution is:

1. Volume = mass/density = (50 g)/(8.93 g/cm^3)
 = 5.6 cm^3
2. Area = $(\pi/4)D^2$ = $(\pi/4)(0.254\text{ cm})^2$
 = 0.051 cm^2
3. Since
 volume = length × area
 length = volume/area

Substituting in the known values,

$$\text{length} = 5.6\text{ cm}^3/0.051\text{ cm}^2 = 109.8\text{ cm}$$

In this solution, the path came from a study of the problem statement. Another source is the equations themselves. The unknowns in one equation can be used to indicate which equation must be used next. This is a more mechanical method than the first, but it does work. Example 1.22 will be worked using this method.

Since the length is wanted, the first equation must be one that has length in it.

$$(1)\ \text{volume} = \text{length} \times \text{area}$$

This equation has two unknowns, volume and area, in addition to the length. This indicates that equations having volume and area in them must follow. An equation with volume in it is

$$(2)\ \text{volume} = \text{mass/density}$$

An equation having area in it is still needed. One that can be used is for the area of a circular cross section. It is

$$(3)\ \text{area} = (\pi/4)(\text{diameter})^2$$

We now have three equations and only three unknowns, so the equations can be solved. The solution is

From (2),

$$\begin{aligned}\text{volume} &= (50\ \text{g})/(8.93\ \text{g/cm}^3)\\ &= 5.6\ \text{cm}^3\end{aligned}$$

From (3),

$$\begin{aligned}\text{area} &= (\pi/4)(0.254\ \text{cm})^2\\ &= 0.051\ \text{cm}^2\end{aligned}$$

Then, from (1),

$$\begin{aligned}\text{length} &= \text{volume/area}\\ &= (5.6\ \text{cm}^3)/(0.051\ \text{cm}^2)\\ &= 109.8\ \text{cm}\end{aligned}$$

Since equations (2) and (3) each had only one unknown in them, they could have been solved when writttten.

A comparison of the two procedures shows that the first path begins at one end and works back to the wanted quantity. In the second procedure, the path starts with the wanted quantity and works through the other relations.

Which procedure should be used? Well, probably a combination of the two is the best. Try to mentally form the path, but use the equations if you reach a point where you cannot determine what should come next. A second example will be worked using both methods.

EXAMPLE 1.25

Some of the relations that are used in electric circuit analysis are $V = R_{eq}I$, $R_{eq} = R_1 + R_2$, and $P = VI$, where V is the voltage, I is the current,

R_1 and R_2 are the resistances in the circuit, P is the power, and R_{eq} is the equivalent resistance of the circuit. Using these relations, find the power in the circuit if V is 120 volts, R_1 is 20 ohms, and R_2 is 40 ohms.

Solution

First Procedure:

First we can find the equivalent resistance, then find the current using R_{eq} and V. Once the current is known, the power can be calculated using V and I. The solution is

$$R_{\text{eq}} = R_1 + R_2 = 20\ \Omega + 40\ \Omega = 60\ \Omega$$
$$I = E/R_{\text{eq}} = 120\ \text{V}/60\ \Omega = 2\ \text{A}$$
$$P = VI = (120\ \text{V})(2\ \text{A}) = 240\ \text{W}$$

Second Procedure:

Since the power is wanted, an equation including power must be written. It is

$$(1)\ P = VI$$

The voltage is known as 120 volts, but the current (I) is unknown. So, the next equation should include the current (I). From the given relations,

$$(2)\ V = IR_{\text{eq}}$$

This equation introduces the unknown R_{eq}. So, an equation including R_{eq} must be next. Again, from the given relations,

$$(3)\ R_{\text{eq}} = R_1 + R_2$$

At this point, there are three equations and three unknowns, so the equations can be solved.

From (3),

$$R_{\text{eq}} = R_1 + R_2 = 20\ \Omega + 40\ \Omega = 60\ \Omega$$

From (2),

$$I = V/R_{\text{eq}} = 120\ \text{V}/60\ \Omega = 2\ \text{A}$$

From (1),

$$P = VI = (120\ \text{V})(2\ \text{A}) = 240\ \text{W}$$

EXAMPLE 1.26

The length of a rectangular field is double the width, and the numerical value of the area is six times the numerical value of the perimeter. What is the area and the perimeter if the length and width are measured in meters?

Solution

No diagram is needed.

There are no knowns given. However, there are two conditions given, the length-width relation and the area-perimeter relation.

A study of the problem leads to the following sequence as a possible path.

a. Write the length in terms of the width.
b. Write the perimeter and area in terms of the width.
c. Write the area in terms of the perimeter.
d. Solve the width from the two equations for area.

Using the foregoing approach, the equations are:

$$(1)\quad P = 6\,W$$
$$(2)\quad A = 2\,W^2$$
$$(3)\quad A = 6\,P$$

Substituting the expression for P from (1) into (3) gives

$$A = 6(6\,W) = 36\,W$$

From (2),

$$A = 2\,W^2 = 36\,W$$

Dividing both sides of the equation by W gives

$$2\,W = 36 \qquad \text{or} \qquad W = 18$$

Since the units are meters,

$$W = 18 \text{ meters}$$

From (1), $P = 6\,W$, so

$$P = 6(18 \text{ m}) = 108 \text{ m}$$

From (2), $A = 2\,W^2$, so

$$A = 2(18 \text{ m})^2 = 648 \text{ m}^2$$

A check of the answers shows that $A = 6\,P$.

Could the sequence have been developed using the unknowns in the equations? Yes, and we will now see how that is done.

The first equation will be one for the perimeter, P, of a rectangle:

$$(1)\ P = 2\,L + 2\,W$$

This equation has L and W in it so equations containing them are needed. From the conditions of the problem,

$$(2)\ L = 2\,W$$

Another equation with L in it can be found using the area relation. For a rectangle, it is

$$(3)\ A = L\,W$$

This equation introduced the unknown, A, so an equation that includes area is needed. Again, from the problem statement,

$$(4)\ A = 6\,P$$

There are now four equations and four unknowns, so the set of equations can be solved.

Equations (1) and (2) can be combined to give Equation (1) of the preceding solution.

Equations (3) and (2) can be combined to give Equation (2) of the preceding solution.

Equation (4) is left as it is, giving us the three equations that were used in the preceding solution. The solving of the set then follows as before. You might note that the length could also have been calculated with the equations.

In any problem solution, documentation of your work is important. As a student, it permits you to check your work and makes it easier to understand. As a technician, it can provide proof that a certain process or analysis was made at a certain time. This can be used as legal evidence to support the claim of your employer, or even yourself.

Acceptable documentation includes neat diagrams, labeled quantities, letter equations, number substitutions, units, explanations where needed, and an organized presentation. The student should strive to develop habits that will result in such documentation.

SUMMARY

1. Groups of units of measurements that are related by derivation or some other way make up a system of units. The International System of Units (SI) is used in most of the world.
2. SI consists of base units, derived units, and supplementary units. Derived units are formed by the combination of base units.
3. Multiples and submultiples of a unit can be expressed by using either a letter or symbol, called a prefix, before the unit name or symbol.
4. A conversion factor is used to change a quantity expressed in one unit of measure to an equivalent representation in another unit of measure.
5. Engineering calculations can have exact numbers and approximate numbers in them. Exact numbers have no uncertainty. In approximate numbers, the last digit is uncertain because it is an estimated digit.
6. Rounding off is a procedure that is used to reduce the digits in a number. The last digit is either increased by one, or kept the same, so that the number is nearest in value to the original number.
7. Significant figures are used to represent an accepted degree of certainty. Results of operations with approximate numbers must be rounded off to obtain a result that does not imply a greater accuracy than the accuracy of any of the individual factors.
8. Scientific notation is a form that expresses a number as a digit

between 1 and 10, multiplied by 10 raised to some power. Using scientific notation takes up less space and reduces the chance of errors in representing numbers.

9. Problem analysis and problem solving requires an orderly and organized approach. In such an approach, the unknowns in the equations can be used to form a path leading to the solution.

EQUATIONS

1.1	$M \times 10^n = N$	The form for scientific notation	29
1.2	$(A \times 10^m)(B \times 10^n) = (A \times B) \times 10^{m+n}$	Multiplication of two numbers in scientific notation	32
1.3	$(A \times 10^m)/(B \times 10^n) = (A/B) \times 10^{m-n}$	Division of two numbers in scientific notation	33
1.4	$(A \times 10^n) + (B \times 10^n) = (A + B) \times 10^n$	Addition/subtraction of two numbers in scientific notation	34

QUESTIONS

1. What is meant by a "unit of measurement"?
2. What is meant by a system of units?
3. What is the name of the system of metric units to which the United States is changing?
4. What are the names of the three types of units in SI?
5. What are some examples of each type of unit in SI?
6. What is the name of the term used to express a multiple or submultiple of 10?
7. What is the multiple or submultiple represented by each of the following prefixes: mega, nano, micro, milli, and centi?
8. What are the three incorrect uses of SI in the following sentence? "The winner averaged .15 meter per second on a day when the temperature was 30 degrees celsius."
9. What is the name of a multiplier that is used to change quantities of one unit to an equivalent quantity of another unit?
10. Is the following dimensionally correct? Why or why not?

$$(15 \text{ mi/hr})(1 \text{ hr}/3600 \text{ s})(5280 \text{ ft/mi}) = 22 \text{ ft}$$

11. What type of number is obtained from a measurement, approximate or exact?
12. What is meant by a significant digit?
13. When can a zero be considered to be a significant digit?
14. What is the difference between rounding off and dropping digits?
15. When rounding off 3.546 to two digits, should the 5 be left unchanged, increased by one, or decreased by one?
16. What determines the number of digits in the answer when adding or subtracting two approximate numbers?
17. What determines the number of digits in the answer when adding or subtracting two approximate numbers?
18. Which one of the following is the correct form of scientific notation: 5500, 55×10^2, or 5.5×10^3?

19. What is the sign of the exponent for the scientific notation form of a number greater than 10?
20. What is the sign of the exponent of the scientific notation form of a number smaller than 1?
21. How is the magnitude of the exponent in scientific notation obtained?
22. How is the exponent obtained when two scientific notation numbers are multiplied?
23. How is the exponent obtained when two scientific notation numbers are divided?
24. How do you change numbers from decimal form to scientific notation form on your calculator?
25. Which logic system does your calculator use, algebraic or RPN?
26. What is the reason for using word-statement problems in a technical text?
27. What are the four steps in the procedure for problem solving?

PROBLEMS

SECTION 1.1 The International System of Units (SI)

1. Which of the following units are base units? Derived units?

(a) kilogram (d) meter (g) ampere
(b) joule (e) volt (h) candela
(c) second (f) coulomb (i) ohm

2. What is the symbol for each of the following units?

(a) volt (e) meter
(b) henry (f) ohm
(c) watt (g) coulomb
(d) joule

3. What unit is represented by the following symbols?

(a) m (d) N
(b) s (e) J
(c) Hz

SECTION 1.2 Prefixes and Using SI

4. Write each of the following using (a) the prefix name with the unit name and (b) the prefix symbol with the unit symbol.

(a) 0.001 grams (c) 0.000000001 farad
(b) 1,000,000 volts (d) 1,000 watts

5. Repeat Problem 4 for the following:

(a) 1,000,000,000 hertz (c) 0.000000000001 farad
(b) 0.000001 ampere (d) 0.01 meter

6. Write the numerical equivalent of each of the following, without using the prefix.

   (a) 20 centimeters (c) 560 megawatts
   (b) 3.5 kilohertz (d) 25 milliamperes

7. Write the prefix name and symbol for each of the following:

   (a) 1,000,000 (d) 0.001
   (b) 1000 (e) 0.000001
   (c) 1,000,000,000 (f) 0.000000001

8. Write each of the following using symbols:

   (a) meters per second (c) grams per kilogram
   (b) volt-amperes (d) newton-meters

9. Write each of the following using the unit name:

   (a) J/C (c) V·s
   (b) V/A (d) A·s

10. Determine the equivalent quantity for each of the items listed:

    (a) 500 millimeters to centimeters
    (b) 350 milliwatts to microwatts
    (c) 20 nanofarads to picofarads

11. Repeat Problem 10 for the following:

    (a) 0.001 gigahertz to kilohertz
    (b) 0.005 microfarads to picofarads
    (c) 6400 kilowatts to megawatts

SECTION 1.3 Conversion Factors

12. A car travels at 50 miles per hour. How fast is this in meters per second?
13. The length of a power transmission line is 3300 feet. How many meters is this?
14. A spring exerts a force of 25 pounds on a door. What is the force in newtons?
15. What is the length of a 100-yard football field in (a) miles and (b) meters?
16. How many pounds of force are equivalent to a force of 3×10^7 dynes?
17. A field has an area of 100 sq. ft. Express the area in (a) sq. meters and (b) sq. kilometers.
18. Convert the power output of a 200-hp motor to (a) foot pounds per minute and (b) watts.
19. The temperature on a warm sunny day is 85 degrees Fahrenheit. What is the temperature in degrees Celsius?
20. An electric motor is rated at ¾ horsepower. What is its rating in watts?

SECTION 1.4 Significant Figures and Rounding Off

21. How many significant digits are there in each of the following approximate numbers?

(a) 153 (c) 0.0021
(b) 53,000 (d) 0.5002

22. Repeat Problem 21 for the following:

(a) 29,000 (c) 3.002
(b) 5.86 (d) 38.602

23. Round off each of the following approximate numbers to two significant digits:

(a) 0.02683 (d) 37.023
(b) 5.246 (e) 2.57
(c) 3.141 (f) 3.653

24. Repeat Problem 23 for the following:

(a) 39.374 (c) 0.2248
(b) 14.693 (d) 0.738

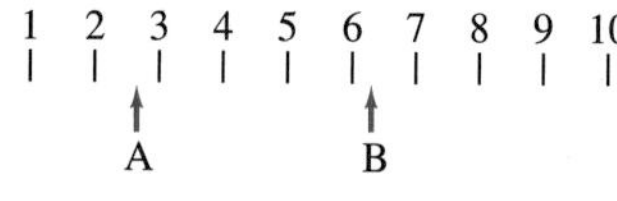

Figure 1.4

25. If only one estimated digit is acceptable as a significant figure, what is the correct reading for the pointer in Position *A* of Figure 1.4? In Position *B*?

SECTION 1.5 Operating with Approximate Numbers

26. Perform the indicated operations with the approximate numbers. Write the answer to the correct number of significant digits.

(a) $8.3 + 0.514 + 25.08$ (b) $9.25 - 3.1$

27. Perform the indicated operations with the approximate numbers. Write the answer to the correct number of significant digits.

(a) $25 + 6.904 + 253.29$ (b) $1.046 - 0.71$

28. Repeat Problem 27 for the following:

(a) $53.1 + 37.92 + 28$ (b) $2.0573 - 0.53$

29. Repeat Problem 27 for the following:

(a) $3.764 + 2.81 + 0.1$ (b) $272.16 - 26.815$

30. Perform the indicated operations with the approximate numbers. Express the answer to the correct number of significant digits.

(a) $(6.25)(5.21)$ (c) $(0.784)/(5.0)$
(b) $(11.56)(4.25)$ (d) $(805)/(2.6)$

31. Repeat Problem 30 for the following:

(a) $(0.526)(55)$ (c) $(8.26)/(1532)$
(b) $(11.56)(0.013)$ (d) $(55.38)/(0.51)$

32. Repeat Problem 30 for the following:

    (a) (0.003)(5.21) (c) (0.025)/(5.0)
    (b) (0.128)(4.25) (d) (7770)/(2.22)

SECTION 1.6 Scientific Notation

33. Write the following numbers in scientific notation:

    (a) 1600 (b) 25,000 (c) 0.25

34. Repeat Problem 33 for the following:

    (a) 0.0004 (b) 185 (c) 0.053

35. Write the following numbers in scientific notation:

    (a) 0.0035 (b) 155,000 (c) 25.8

36. Repeat Problem 35 for the following:

    (a) 0.006 (b) 16,750 (c) 0.00008

37. Write the following in decimal form:

    (a) 5×10^3 (c) 4×10^{-2}
    (b) 2.5×10^6 (d) 8.1×10^{-4}

38. Write the following in decimal form:

    (a) 1.2×10^{-5} (c) 3.6×10^{-2}
    (b) 9.5×10^3 (d) 8.1×10^6

39. Write the following in scientific notation so as to show the number of significant digits.

    (a) 76,000 (c) $25\overline{0{,}000}$
    (b) $1\overline{000}$ (d) 120

SECTION 1.7 Arithmetic Operations with Scientific Notation

40. Perform the multiplication in each of the following. Express the answer in scientific notation.

    (a) $(5 \times 10^5)(4 \times 10^4)$ (c) $(3.5 \times 10^{-4})(2.5 \times 10^{-8})$
    (b) $(2.5 \times 10^{-2})(5 \times 10^{-3})$ (d) $(1.2 \times 10^6)(2.4 \times 10^{-2})$

41. Repeat Problem 40 for each of the following:

    (a) $(1.5 \times 10^2)(1.8 \times 10^4)$ (c) $(3.5 \times 10^{-4})(3.5 \times 10^{-8})$
    (b) $(2.2 \times 10^3)(1.8 \times 10^4)$ (d) $(1.2 \times 10^6)(4.2 \times 10^{-2})$

42. Repeat Problem 40 for the following:

    (a) $(2.5 \times 10^2)(6 \times 10^3)$ (b) $(3 \times 10^{-4})(3 \times 10^{-2})$

43. Repeat Problem 40 for the following:

    (a) $(1.6 \times 10^{-3})(4.8 \times 10^{5})$ (b) $(5.6 \times 10^{-4})(3.6 \times 10^{-12})$

44. Perform the division in each of the following. Express the answer in scientific notation.

    (a) $(3.8 \times 10^{5})/(1.9 \times 10^{3})$ (c) $(3.6 \times 10^{4})/(9 \times 10^{2})$
    (b) $(3.0 \times 10^{-4})/(6 \times 10^{6})$ (d) $(1.6 \times 10^{-8})/(4 \times 10^{4})$

45. Repeat Problem 44 for the following:

    (a) $(2.4 \times 10^{3})/(4 \times 10^{-5})$ (c) $(7.5 \times 10^{-5})/(5 \times 10^{-3})$
    (b) $(3.6 \times 10^{3})/(6.0 \times 10^{5})$ (d) $(42 \times 10^{-6})/(7 \times 10^{4})$

46. Repeat Problem 44 for the following:

    (a) $(8.4 \times 10^{2}) / (2.1 \times 10^{-4})$ (b) $(5.6 \times 10^{-3}) / (6 \times 10^{-4})$

47. Repeat Problem 44 for the following:

    (a) $(48 \times 10^{3}) / (1.6 \times 10^{-4})$ (b) $(2 \times 10^{-4}) / (50 \times 10^{2})$

48. Perform the addition or subtraction as indicated. Express the result in scientific notation.

    (a) $(5 \times 10^{3}) + (6 \times 10^{2})$ (b) $(5 \times 10^{5}) - (3 \times 10^{4})$

49. Repeat Problem 48 for the following:

    (a) $(2.5 \times 10^{5}) + (3.1 \times 10^{3})$ (b) $(1.6 \times 10^{-2}) - (4 \times 10^{-4})$

50. Repeat Problem 48 for the following:

    (a) $(6 \times 10^{2}) - (3 \times 10^{-2})$ (b) $(2 \times 10^{-3}) + (2 \times 10^{-2})$

51. Repeat Problem 48 for the following:

    (a) $(4 \times 10^{5}) + (3 \times 10^{6})$ (b) $(6 \times 10^{-4}) - (3 \times 10^{-3})$

SECTION 1.8 A Procedure for Problem Solving

52. Which of the following sets have enough equations to solve for all the unknowns?

    (a) $4I_1 + 2I_2 = 3I_3$
    $2I_1 + 6I_2 = 0$

    (b) $30V_1 + 60V_2 = 10$
    $12V_1 - 18V_2 = 6$

    (c) $-6V_1 + 3V_2 = 6$
    $3V_1 - 18V_2 + 9V_3 = 12$
    $9V_2 - 21V_3 = 6$

    (d) $4I_1 - 3I_2 = 12$
    $2I_2 + 4I_3 = 15$

53. Repeat Problem 52 for the following:

(a) $2X + 3Y - Z = 6$
$4X + Y = 5$

(b) $6X + 4Y - Z = 4$
$2X - Y + Z = 3$
$X + Y - 2Z = -3$

54. Repeat Problem 52 for the following:

(a) $2X - Y + Z = 3$
$5X + 3Y = 12$
$2Y - 3Z = -5$

(b) $X - 2Y = 0$
$4X - 2Y + Z = 0$
$2X - 22Y - 3Z = 0$

55. The number of 10-ohm resistors in a box was equal to the number of 25-ohm resistors. There were five times as many 5-ohm resistors as 25-ohm resistors. If the total number of resistors was 1400, how many resistors of each kind were there?

56. The resistance of resistor #1 is twice that of resistor #2 and 10 ohms more than that of resistor #3. If the total resistance is 265 ohms, how many ohms is each resistor?

57. The total weight of three boxes is 15 newtons. Box A weighs twice as much as box B, and box C weighs as much as box A plus box B. What is the weight of each box?

58. One assembly line in a television manufacturing plant uses 60 percent of 20-inch tubes and 40 percent of 27-inch tubes. A second line uses 30 percent of 20-inch tubes and 70 percent of the 27-inch tubes. How many sets can each line assemble if there is a total of 2250 tubes and the number of 27-inch tubes is twice the number of 20-inch tubes?

59. A machine can sort a batch of resistors in 7.5 minutes. A second machine can do the sorting in 4 minutes. How long will it take to sort the resistors if both machines are used together?

60. A manufacturer purchased 50,000 resistors of three resistance groups. The first group cost $5.00 per hundred. The second group cost $4.50 per hundred. And the third group cost $5.80 per hundred. There are twice as many resistors in the second group as in the first group. How many are there in each group if the total cost was $2390?

CHAPTER 2

MATTER, ELECTRICITY, AND RESISTANCE

CHAPTER OBJECTIVES

After completing this chapter, you should be able to:

1. Define matter and describe its structure.
2. Describe the atom and give the characteristics of its parts.

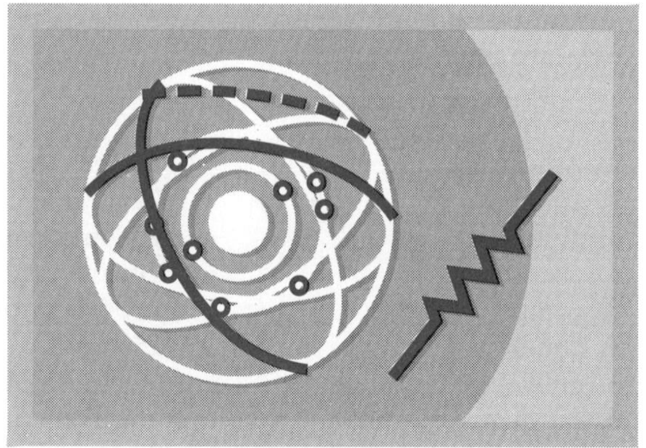

3. Apply Coulomb's Law to determine the force between charged bodies.
4. Prevent damage to parts from electrostatic discharge.
5. Define current and explain it in terms of the atom's electrons.
6. Define the unit of current and compare electron and conventional current.
7. Define the potential difference, emf, and the unit of each.
8. Calculate the resistance of conductors whose shape and material are known.
9. Apply the resistance-temperature relations to problems involving the effect of temperature on resistance.
10. Describe the different types of resistors.
11. Explain the difference between linear and nonlinear resistors.
12. Use the American Wire Gage Table to determine resistance, length, or area of a conductor.
13. Use the resistor color code to determine resistance and tolerance.
14. Define conductors, insulators, and semiconductors.
15. Give an application of the conductor, insulator, and semiconductor materials.
16. Understand the hazards of working around electricity and know the safety procedures to observe.

KEY TERMS

1. absolute zero
2. American Wire Gage
3. ampere
4. atom
5. circular mil
6. compound
7. conductance
8. conductivity
9. conductor material
10. conventional current
11. coulomb
12. Coulomb's Law
13. doping
14. electric charge
15. electric current
16. electric shock
17. electromotive force (emf)
18. electron
19. electron current
20. electrostatic discharge
21. element
22. free electron
23. inert elements
24. inferred absolute zero
25. insulator
26. ion
27. ionic current
28. linear resistance
29. matter
30. mil
31. molecule
32. multiplier band
33. neutron
34. nonlinear resistance
35. nucleus
36. ohm
37. potential difference
38. potentiometer
39. power rating
40. proton
41. relative conductivity
42. resistance
43. resistivity
44. resistor
45. Resistor Color Code
46. rheostat
47. semiconductor material
48. shell
49. shell diagram
50. static electricity
51. subshells
52. superconductivity
53. temperature coefficient of resistance
54. tolerance band
55. valence electron
56. volt

INTRODUCTION

Flip a switch and a room is illuminated. Turn on another switch and a picture appears on a television screen, and food is cooked in an oven heated by electricity. What is electricity, how does it act, and how is it conducted? These are just a few of the questions answered in this chapter.

The structure of the atom explains the nature of electricity. We see how the potential difference provides the energy to move the electrons. Conductors, either wire or printed circuits, form the paths for the electrons. Insulators keep the electrons from flowing into other conductors. They act as guard rails. The moving electrons release their energy at the point of use. That energy can be in the form of heat, light, magnetism, force, or chemical action.

Energy is lost in the conductors because they have resistance. The oven becomes warm because the heating element has resistance. What is resistance? What affects it? What are some forms of the devices designed to have resistance? You will know the answers after completing this chapter. This chapter also explains why wire is made with a round cross section, as well as the meaning of the colored bands on some resistors.

But electricity is a Jekyll and Hyde! It also can cause damage. Static electricity can break down sensitive electronic parts. An electrical current, which passes through one's body, can cause serious injury—or even death. Only by learning about electricity can we put it to use without suffering any serious consequences.

2.1 Matter and the Molecule

Circuit analysis does not require knowledge of matter or the structure of the atom. However, familiarity with these items will help one to understand the nature of electric current. (Also, it will help one understand the reason for using certain materials in parts of an electric circuit.) For instance, why is copper used instead of rubber to connect a light bulb to a battery? Or, why is plastic instead of aluminum wrapped around the wires of a lamp cord?

Matter is any substance that has weight and occupies space. As you look at objects, you see that matter can exist as a liquid, a solid, or a gas. The water that you drink, the copper from which wire is made, and the hydrogen used to inflate balloons are examples of matter.

matter

All matter is composed of one or more basic substances called **elements.** There are more than 100 elements. They are substances that cannot be chemically decomposed or formed by the chemical union of other substances. They are the building blocks of all matter. Water contains the elements hydrogen and oxygen.

elements

Most of the matter which exists today is in a form called a **compound.** A compound is matter formed by the chemical union of several elements. Ordinary table salt, water, sulfuric acid, and carbon dioxide are compounds. A characteristic of a compound is that it can be decomposed into the elements that form it. For example, table salt can be decomposed into sodium, and chlorine and water will yield hydrogen and oxygen.

compound

Compounds can be divided into small particles called molecules. A **molecule** is the smallest particle of a compound that can exist and still retain the physical and chemical properties of the compound such as density, weight, odor, taste, and hardness. Chemical properties are those that involve a change in composition. These are properties such as the reaction caused by heating a material in air, or treating it with an acid.

molecule

2.2 The Atom, Charge, and Coulomb's Law

Molecules are made up of atoms. An **atom** is the smallest particle of an element that can enter into chemical reactions with other particles. It was named after the Greek word "atomos" meaning "something that cannot be cut," or "indivisible."

atom

Atoms are very small and have a diameter of about 10^{-8} centimeters. How small is this? Well, about 200 million atoms will fit across the diameter of a 1-cent coin! Although atoms cannot be seen with an optical microscope, their behavior can be studied with the use of an electron microscope.

The theories on atomic fusion, fission, and electron current are relatively new. The concept of the atom dates back as far as 500 B.C., when the Greek philosopher Democritus developed an atomic theory. However, it was not until the late 19th and early 20th centuries that theorems were proposed and experiments conducted that resulted in the atomic structure as we know it now.

electrons
protons
neutrons
nucleus

The atom is made up of three main types of particles: **electrons, protons,** and **neutrons.** The protons and neutrons are at the center core called the **nucleus.** The nucleus also contains some smaller particles called mesons, neutrinos, and hyperons. The electrons whirl about the nucleus in specific orbits or shells. A pictorial diagram of this structure is shown in Figure 2.1. The diameter of the atom is 10^4 times as large as the diameter of the nucleus. The study of the atom showed that protons and electrons exert forces on each other. These forces are accounted for by assigning a quantity called electricity or **electric charge** to the particles. A negative (−) charge is assigned to the electrons and a positive (+) charge to the protons.

electric charge

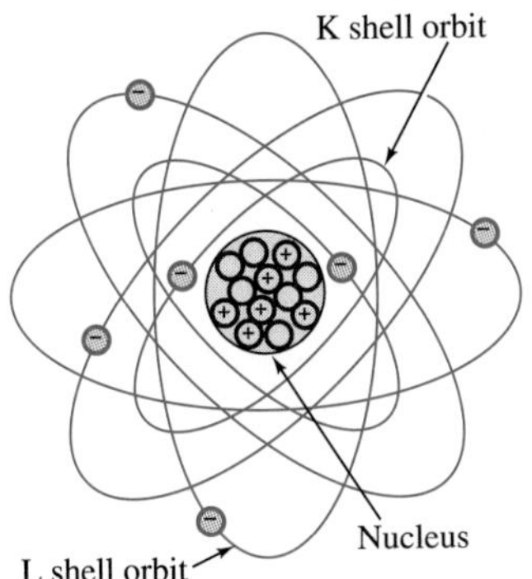

Figure 2.1 The pictorial diagram of the atom shows the electrons orbiting around the nucleus.

"Like charged" particles repel each other and "unlike charged" particles attract each other. This action is shown in Figure 2.2. Thus, electrons are repelled by other electrons, but are attracted to protons. You can observe this force by rubbing two glass rods on silk and placing them near each other. The rods will be repelled because of like charges. On the other hand, if a rubber rod is rubbed on silk and brought near the rubbed glass rod, the two rods will be attracted. That is because the charge is positive (+) on the glass rod and negative (−) on the rubber rod.

In 1900, Sir J. J. Thompson (1865–1940), an English physicist, and Sir John Townsend (1865–1957) experimentally determined that the charge on 1 electron is 1.602×10^{-19} **coulombs** (C). This unit is named after Charles A. Coulomb (1736–1806), a French physicist.

coulombs

Later, Sir Thompson and E. Rutherford (1871–1937), a British physicist, conducted experiments that determined the charge and mass of the proton. Its charge is 1.602×10^{-19} coulombs. The mass is 1.673×10^{-27} kg. Thus the proton's charge is opposite and equal to the charge on the electron. But its mass is about 1836 times that of the electron, whose mass is 9.109×10^{-31} kg. The final particle, the neutron, was found to have no charge, and a mass of 1.675×10^{-27} kg. This discovery was made by Sir James Chadwick in 1938. He later received a Nobel Prize for this discovery.

Charles A. Coulomb conducted some experiments to measure the force between charged particles. He found that the force was stronger for (a) smaller separations and (b) greater amounts of charge. He developed a relation, which is known as **Coulomb's Law.** In equation form, it is

$$F = \frac{k\, Q_1 Q_2}{r^2} \tag{2.1}$$

where:

k is the constant equal to $9.0 \times 10^9\ \text{N}\cdot\text{m}^2/\text{C}^2$,
Q_1, Q_2 are the charges of each particle, in coulombs
r is the distance between the charged particles, in meters
and F is the force between the particles, in newtons.

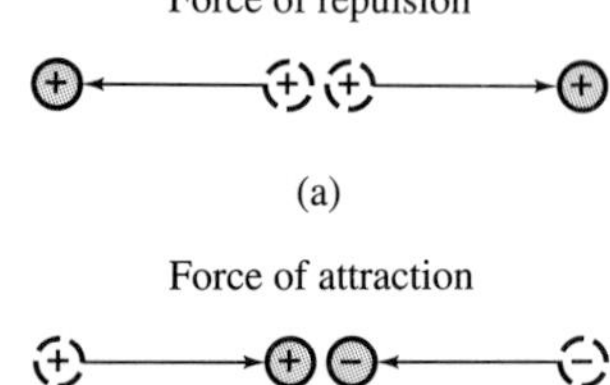

Figure 2.2 Charged bodies exert a force on each other. (a) If the charges are of similar polarity, the force repels the bodies. (b) If the charges are of opposite polarity, the objects will be attracted to each other.

Using this relationship, we can calculate the force between two charged objects separated by a known distance. Consider the following example.

EXAMPLE 2.1

a. What is the force between a negative charge of 10×10^{-6} coulombs and a positive charge of 4×10^{-6} coulombs that are 2 meters apart?
b. Are the charges attracted or repelled?

Solution

a. Using Equation 2.1:

$$F = \frac{9 \times 10^9 \text{ N·m}^2/\text{C}^2 \,(10 \times 10^{-6}\text{ C})(4 \times 10^{-6}\text{ C})}{(2\text{ m})^2}$$
$$= \frac{360 \times 10^{-3}\text{ N·m}^2}{4\text{ m}^2}$$
$$F = 0.09 \text{ newtons}$$

b. The charges are opposite charges, so the force is one of attraction.

Although the atom is three-dimensional, it can be studied more easily with a **shell diagram** as that in Figure 2.3. This diagram shows the atom's electrons orbiting in one plane around the nucleus. Some observations made from the diagram are:

shell diagram

1. The electrons whirl about the nucleus in set orbits.
2. The orbits are divided into groups called shells. These are marked as K, L, M, N, O, P, and Q.
3. The maximum number of orbits in a group is four. These are marked as *s*, *p*, *d*, and *f*.
4. The separation is smaller between the orbits in a group than between the shells.

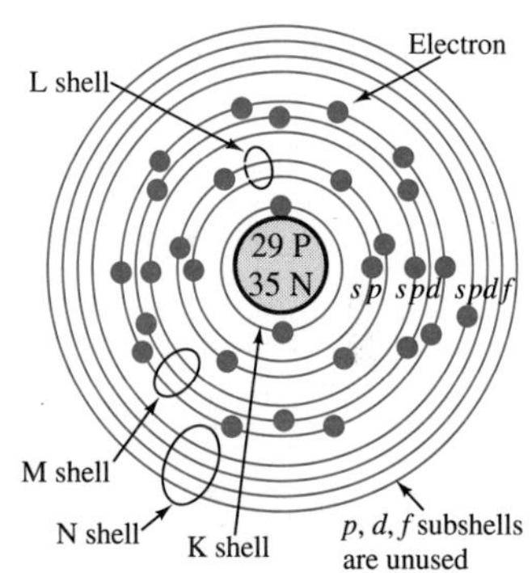

Figure 2.3 The shell diagram shjows the structure of the atom in a single plane, which is simpler to study.

As the electrons whirl, they remain in their orbit. This is because the attraction by the nucleus balances the outward pull.

Each group of orbits is called a **shell.** The shell nearest the nucleus is the K shell. Successive outward shells are L, M, N, O, P, and Q. Each shell can hold a maximum number of electrons, given by $2n^2$, where n is the shell number. The K, or number 1 shell, can hold two electrons. The L, or number 2 shell, can hold eight. The limits for other shells are listed in Table 2.1.

shell

TABLE 2.1 Electron Capacity of Shells

Shell	Capacity	Shell	Capacity
K	2	O	50
L	8	P	72
M	18	Q	98
N	32		

The first three shells in Table 2.1 must be filled before any electrons go into the fourth shell. After that, the combinations become complex. One example is the uranium atom. Its first four shells are filled. However, the fifth, sixth, and seventh shells have only 21, 9, and 2 electrons.

subshells

The orbits in the shells are called **subshells** and are marked as *s*, *p*, *d*, and *f*. Note that the copper atom does not have any *f* subshell electrons. The number of electrons in each subshell is also limited. There can be 2 in the *s* subshell, 8 in the *p* subshell, 10 in the *d* subshell, and 14 in the *f* subshell. The outer shell cannot have more than eight electrons. Having eight electrons makes an element chemically stable. It will not react with other elements. Such elements are called **inert elements.** Two inert elements are argon and neon.

inert elements

2.3 Static Electricity

static electricity

Static electricity, or electrostatic charging, is the accumulation of charge on an object. This charge can result in a high voltage being stored on your body. Charge voltages as high as 10 kV are not unusual. Static electricity can be produced by rubbing a hard rubber rod with fur. Also, your body becomes charged with static electricity when you walk across a synthetic-material rug with leather-soled shoes.

electrostatic discharge

Electrostatic discharge (ESD) occurs when the charged body comes in contact with another object. The current from the discharge lasts only a short time, but can be very large.

At one time ESD was only a novelty, or minor annoyance. Now, with the use of smaller semiconductor devices and low-power logic chips, ESD presents a major problem. The high voltage or current can damage these parts. Damage can happen at any time the part is handled. This can be during assembly, testing, inspection, and normal movement of the part. Or even worse, a weakened part can fail in the field.

Industry spends millions of dollars to prevent damage caused by ESD. The methods fall into the following categories.

1. Preventing the buildup of charge. This can be done by wearing cotton clothing instead of synthetic materials, maintaining a high relative humidity, spraying areas with antistatic spray, using antistatic paint, and using antistatic materials for floor coverings.
2. Draining off the charge. Conductive mats, touch plates, wrist straps, and conductive chairs provide a path for the charge to drain off to ground.
3. Neutralize the charge. Ionizers that provide positive and negative ions neutralize the charge.
4. Reduce the effect of the discharge. The current during ESD can be limited by coating work surfaces with a material whose conductance is small enough to limit the current.

Static-free work stations such as the one in Figure 2.4 incorporate some of the measures in the preceding list. As a minimum, one should wear wrist straps and discharge any probe or tool before touching a

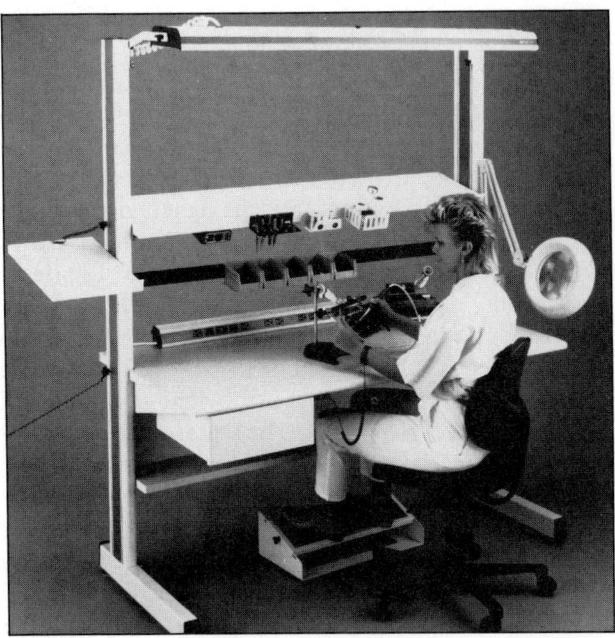

Figure 2.4 This and similar workstations are designed to prevent the buildup of static electricity and damage to electronic parts. (Photo courtesy of EPSCO, Inc.)

sensitive device. Touching the probe or tool to a grounded point will discharge it.

Shielding and grounding are two techniques used to prevent ESD from damaging electronic equipment. Another method is the use of protective devices such as capacitors, varistors, and silicon avalanche suppressors to limit the voltage to a safe level. They are connected across the circuit or device that is being protected.

2.4 Electric Current

Electric current is the movement of charged particles, either positive or negative, in some general direction in a material. The most familiar form of current is that in metals, such as copper. Since the charged particles are the negative electrons, this current is called **electron current.**

electric current

electron current

Now we will consider the relation of current to the structure of the atom. The electrons in the outer shells, being farther away from the nucleus, are not as strongly bound to it. Also, outer subshells that do not contain the eight electrons have a greater tendency to attract and lose electrons.

These electrons in the outer shell are called **valence electrons.** Valence electrons that have been freed from their shell are called **free electrons.** The electrical properties of the element are related to both the number of valence electrons and their distance from the nucleus. For example, copper, widely used in electrical devices and connections, has only one

valence electrons
free electrons

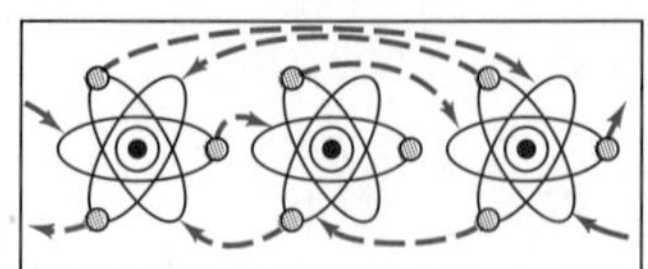

No potential applied

(a)

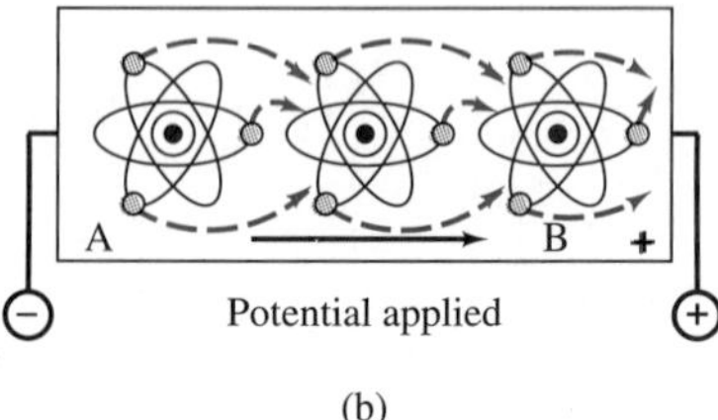

(b)

Figure 2.5 The movement of electrons in metal. (a) With no potential applied, the movement is random, and there is no current. (b) An applied potential causes the movement to have some general direction, so there is current.

free electron; however, it is in the N shell. Thus, only a relatively small amount of energy is needed to move the electron.

With no external energy applied, these free electrons are acted upon only by adjoining atoms. Their movement is from atom to atom in a random manner. There is no general direction, as seen in Figure 2.5(a). Suppose energy is applied to the section so that the left end of the section has a more negative (−) charge on it and the right end has a more positive (+) charge on it. The electrons follow Coulomb's Law and they move in a general direction from left to right, as shown in Figure 2.5(b). An electric current now exists in the section.

An electron does not have to move from point A to B instantaneously. Instead, there can be a bumping of electrons. The displacement of the electrons on the left results in the ones at the right moving. This is like the water flow when a faucet is opened. The water comes out of the faucet instantaneously, but the water from the source does not reach that faucet until a later time.

Current was named in honor of Andre Marie Ampere, a French physicist (1775–1836). The unit of current is the **ampere** (A). The ampere is a base unit in SI.

ampere

In terms of charge, 1 ampere is the movement of charge at a rate of 1 coulomb, or 6.242×10^{18} electrons per second (see Figure 2.6). The direction of the current is shown with an arrow. The symbol for current is "*I*." Current is measured with a measuring instrument called an ammeter.

In equation form, the current-charge relation is:

$$I = Q/t \tag{2.2}$$

where:

I is the current in amperes,
Q is the charge in coulombs,
and t is the time in seconds.

EXAMPLE 2.2

Fifty coulombs of charge move past a point in 10 seconds. What is the current?

Solution

$I = Q/t$
$I = (50 \text{ coulombs}/10 \text{ seconds})$
$I = 5 \text{ amperes}$

Time = 0 seconds

(a)

0 1 2 3 4

A

6.242×10^{18} electrons

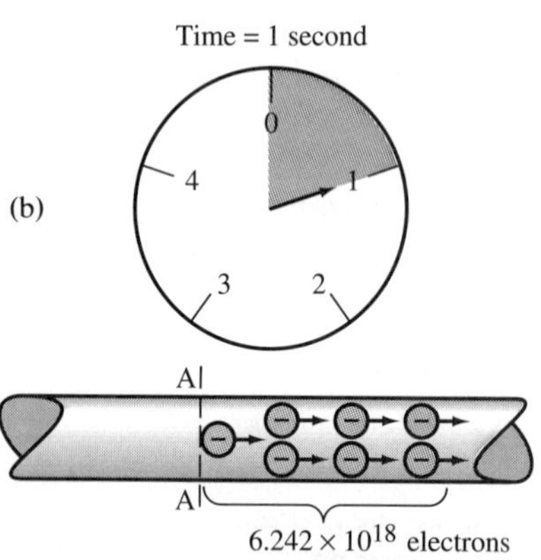

Figure 2.6 One ampere of current is the movement of 6.242×10^{18} electrons in 1 second in some general direction.

The charged particles in metal elements are generally electrons. However, in gases, some liquids, and semiconductor materials, charge carriers can be both positive and negative particles. In the gas-filled electron tube of Figure 2.7, electrons are emitted from a heated cathode.

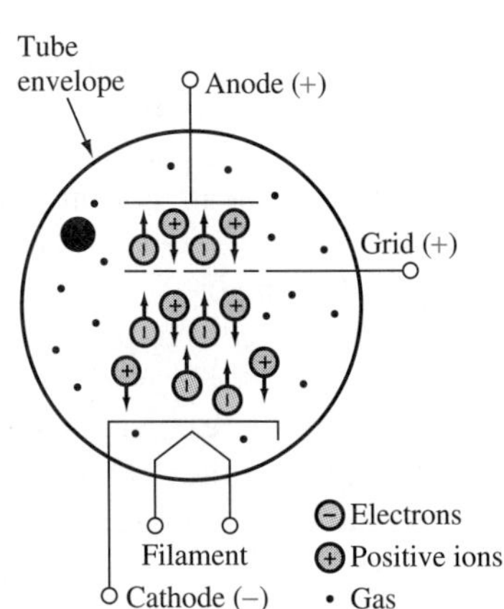

Figure 2.7 The current in this electron tube consists of electron movement, as well as the movement of positive charges, called ions.

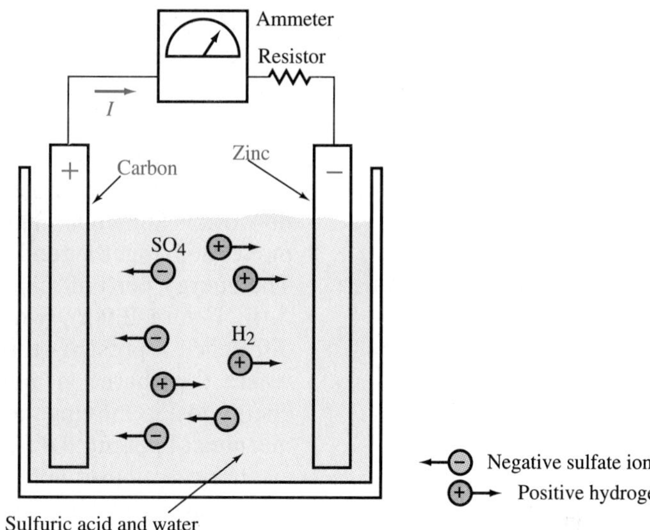

Figure 2.8 Ionic current also exists in a liquid such as the electrolyte of the lead-acid automobile battery.

The electrons are attracted to the positive anode. As they move, these electrons collide with the molecules of argon or helium gas in the tube. The collisions transfer energy to the gas molecules, freeing some of the electrons from the atoms. These electron-deficient atoms now have a positive charge. They are called positive **ions.** The positive ions are attracted by the negative cathode and move in that direction.

ions

The current made up of the positive ions is called **ionic current.** Another common example of ionic current is the neon tube that is used in display lighting and outdoor signs. This tube utilizes the ionic current in neon. Ionic current can also be found in some liquid solutions. For example, a solution of sulfuric acid and water produces positive hydrogen ions and negative sulfate ions. If a zinc cathode with a negative charge on it and a positive-charged carbon anode are placed in this solution, the hydrogen ions will be attracted to the zinc cathode. Meanwhile, the sulfate ions are attracted to the carbon anode. This is shown in Figure 2.8.

ionic current

Benjamin Franklin and other early experimenters of electricity considered it to be a flow of some sort of fluid. Franklin originated the theory that the flow of this fluid was from the more positive-charged point to the more negative charged point. The eventual development of the electron theory proved that theory to be incorrect. Instead, current is the movement of electrons moving from the more negative point to the more positive point. So now there are two theories of current that are in use. To distinguish between the two, Franklin's current is called **conventional current** and the electron flow is called **electron current.**

conventional current
electron current

Either current can be used when analyzing circuits. This text will use conventional current mainly because that form is used in most aspects of practical applications. Therefore, unless otherwise stated, current will mean conventional current.

2.5 Potential Difference and Electromotive Force (emf)

Consider the spheres A and B in Figure 2.9(a). They will be attracted to each other in accordance with Coulomb's Law. Because of this attraction, there is a potential to do work. By definition, this potential to do work is known as energy. If the spheres are freed, the attraction will move them together and the potential energy will be expended as work. The energy per unit charge is called **potential difference.**

potential difference

In Figure 2.9(a), where the spheres cannot move, the potential difference represents the energy stored to do work. In Figure 2.9(b), where the charge is moving, the potential difference represents the energy lost in moving the charge from one point to the other. This dual meaning of potential difference must be kept in mind when dealing with electricity. Its use will be seen in later chapters.

volt

The unit in which potential difference is measured is the **volt.** This unit is named in honor of Allesandro Volta (1745–1827), an Italian scientist who discovered the voltaic cell. Potential difference is commonly called voltage and can be measured with a voltmeter. One volt of potential difference exists between two points when 1 joule of work must be expended to move 1 coulomb of charge from one point to the other. One joule is approximately the work expended to move 1 pound through a distance of three-fourths of a foot.

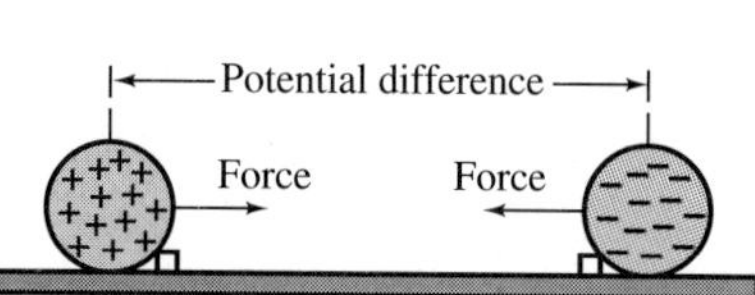

(a)

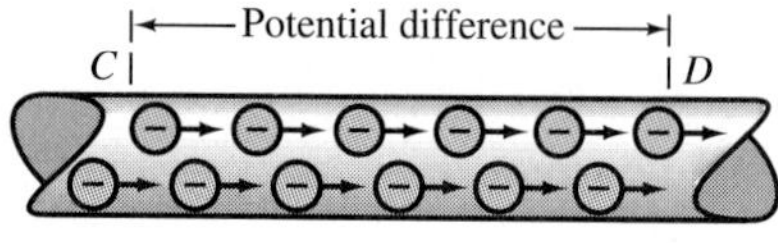

(b)

Figure 2.9 Charge at rest and in motion represents a potential difference. (a) The accumulation of charge on two bodies represents a potential to do work, with a resulting potential difference. (b) The work done in moving an amount of charge also represents a potential difference between two points.

In equation form, we have:

$$V = \mathcal{W}/Q \tag{2.3}$$

where:

V is the potential difference,
$\mathcal{W}$ is the work in joules,
and Q is the charge in coulombs.

EXAMPLE 2.3

Ten joules of work is needed to move 2 coulombs of charge from one point to another. What is the potential difference between the two points?

Solution

$$V = \mathcal{W}/Q$$
$$= 10 \text{ joules}/2 \text{ coulombs}$$
$$V = 5 \text{ volts}$$

Another belief held by early experimenters was that the current in a conductor was maintained by a force. They likened it to the way water flow is maintained in a pipe. The name given to this force is **electromotive force** (emf). The devices that were connected to maintain the current were quite naturally called sources of electromotive force. Although the

electromotive force (emf)

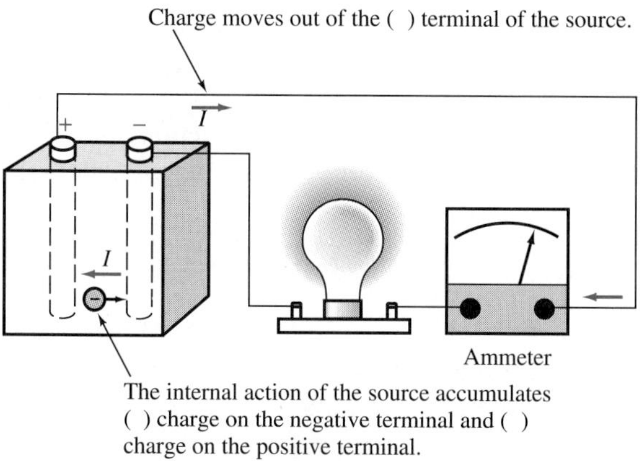

Figure 2.10 Connecting a conductive path between the terminals of a source of emf results in current.

discovery of the electron theory of current showed that emf is the electrical energy due to the accumulation of charge, the use of the terms electromotive force and source of electromotive force continued.

Sources of emf maintain a positive charge at one terminal and a negative charge at the other. This is done by converting some other form of energy to electrical energy. A common source of emf is the flashlight battery.

The drawing in Figure 2.10 shows a source of emf with a lamp connected across its terminals. The charge on the terminals will result in electron movement. The direction of the current will be out of the positive (+) terminal, through the wire, and into the negative (−) terminal. The current will continue as long as the source is able to keep the (+) and (−) charges on its terminals—that is—maintain a potential difference across the terminals. If the wire is disconnected from the source, charges collect on each terminal. This continues until the repelling action of the charge prevents the movement of additional charge in the source. This accumulation of charge now represents a potential to do work.

The unit for emf is the volt, the same as for potential difference. When used for a source of emf, the unit indicates the amount of energy the source can supply. The emf of a typical flashlight battery is about 1.5 volts. The emf of a house outlet is 120 volts.

2.6 Resistance and Resistivity

The charge moving in a conductor collides with other charges and is sometimes moved in different directions. Also, its movement can be a "stop and go" one. These actions tend to oppose the movement of the charge. This opposition to the movement is known as **resistance.** The unit for resistance is the **ohm** (Ω), named after George Simon Ohm (1787–1854), a German physicist. One ohm is the resistance that will have a potential difference of 1 volt across it when the current is 1 ampere. A 100-watt light bulb has a resistance of about 140 ohms when

resistance
ohm

operating. Without resistance, there would be no need for energy to move the charge, no conversion of electrical energy into heat energy, and no potential difference between two points in a conductor.

conductance

The reciprocal of resistance is called **conductance** (G). Conductance is a measure of how well a conductor permits current to flow through it. Its unit is the siemens (S).

$$G = 1/R \tag{2.4}$$

where:

G is the conductance, in siemens,
and R is the resistance, in ohms.

EXAMPLE 2.4

The resistance of a length of wire is 2 ohms. Find its conductance.

Solution

$$\begin{aligned} G &= 1/R \\ &= 1/2 \text{ ohm} \\ G &= 0.5 \text{ S} \end{aligned}$$

resistivity

The **resistivity** (ρ) of a material is the resistance of a section one unit in length with a cross section of one unit of area. The SI unit for resistivity is the ohm-meter (Ω-m). It is the resistance of a cube whose sides are each 1 meter. The more common unit is the circular mil-ohms per foot (CM-Ω/ft). It is the resistance of a section that is 1 foot long and has a

circular mil

cross-sectional area of 1 **circular mil**. One circular mil represents the area in a circle having a diameter of 1 mil (1/1000th of an inch).

Table 2.2 list values of resistivity for some materials. The values are

TABLE 2.2 Resistivity of Materials at 20°C*

	Resistivity (ρ)	
Material	***Ohm-meter***	***CM-ohms per ft***
Silver	1.59×10^{-8}	9.565
Copper	1.724×10^{-8}	10.371
Gold	2.44×10^{-8}	14.678
Aluminum	2.824×10^{-8}	16.988
Tungsten	5.6×10^{-8}	33.688
Nickel	7.8×10^{-8}	46.922
Iron	10×10^{-8}	60.157
Constantin	49×10^{-8}	294.77
Nichrome	100×10^{-8}	601.57

*Values may vary depending on the purity of the material.

typical, but can vary, depending on the purity of the material and temperature. For most materials, it increases as the temperature increases. The values of resistivity can be used to compare how well materials will conduct.

The relation of resistance to material, length, and cross-sectional area of a conductor was discovered by George Simon Ohm. His experiments showed that resistance is (a) directly proportional to resistivity and length, and (b) inversely proportional to the cross-sectional area. For example, if length or resistivity doubles, the resistance of the material will double. On the other hand, if the cross-sectional area doubles, the resistance will be halved. These relations are illustrated by the conductors in Figure 2.11.

In equation form, Ohm's results are:

$$R = \rho\, L/A \tag{2.5}$$

where:

- R is the resistance, in ohms,
- ρ is the resistivity of the material in ohm-meters, or circular mil-ohms per foot,
- A is the cross-sectional area. It is in meter^2 if ρ is in ohm-meters, and circular mils if ρ is in circular mil-ohms per foot,
- and L is the length. It is in meters if ρ is in ohm-meters, and feet if ρ is in circular mil-ohms per foot.

To use Equation 2.5, one must be able to calculate the number of circular mils in an area. The relation for a round conductor is:

$$\text{CM} = (\text{DM})^2 \tag{2.6}$$

where:

- CM is the number of circular mils
- and DM is the diameter in mils. One mil is 1/1000th of an inch.

EXAMPLE 2.5

How many circular mils are in a circle having a diameter of 0.5 inch?

Solution

$$\text{DM} = (0.5\text{ in})(1000\text{ mils/in}) = 500\text{ mils}$$
$$\text{CM} = (500)^2$$
$$\text{CM} = 2.5 \times 10^5$$

For a section that is not round, the relation is:

$$\text{CM} = (4/\pi)(SM) \tag{2.7}$$

where:

- CM is the number of circular mils
- and SM is the area in square mils.

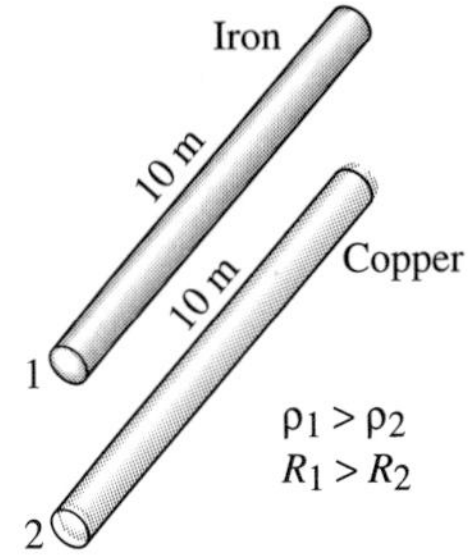

(a) Effect of material on resistance

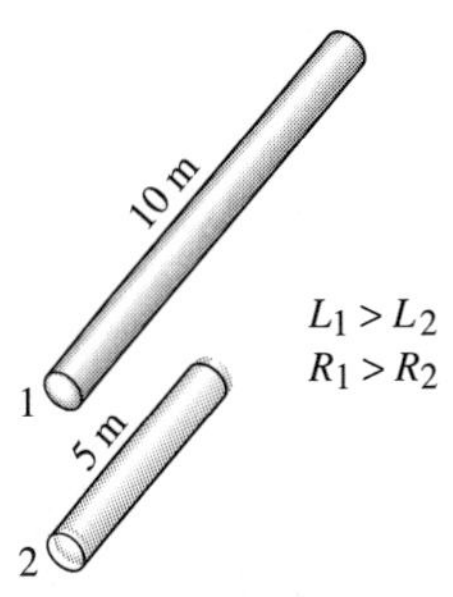

(b) Effect of length on resistance

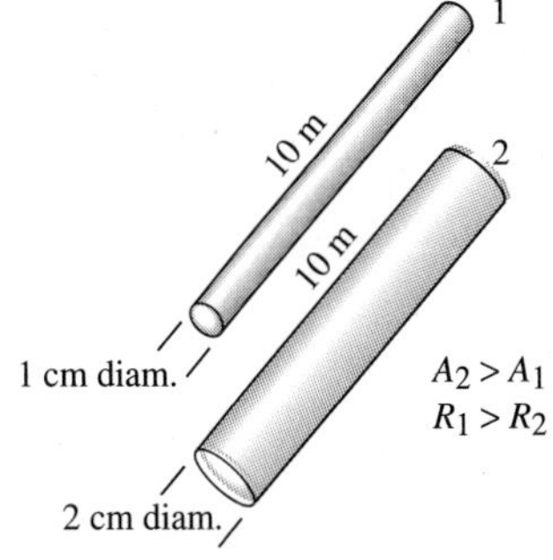

(c) Effect of area on resistance

Figure 2.11 Resistance depends on the material, length, and cross-sectional area. (a) Different materials have different values of resistance. (b) Resistance is directly proportional to the length of the material. (c) Resistance is inversely proportional to the cross-sectional area of the material.

EXAMPLE 2.6

Find the number of circular mils in:
a. A square with sides of 0.5 inch.
b. A triangle having a base of 0.5 inch and a height of 1 inch.

Solution

a. SM = length $\times$ width
= (0.5 in)(1000 mils/in)(0.5 in)(1000 mils/in)
= 2.5×10^5 square mils
CM = $(4/\pi)$(SM)
= $(4/\pi)(2.5 \times 10^5$ sq. mils)
CM = 3.18×10^5 circular mils
b. SM = (1/2) base $\times$ height
base = (0.5 in)(1000 mils/in) = 500 mils
height = (1 in)(1000 mils/in) = 1000 mils
SM = (1/2)(500 mils)(1000 mils) = 2.5×10^5 sq. mils
CM = $(4/\pi)(2.5 \times 10^5$ sq. mils)
CM = 3.18×10^5 cir. mils

EXAMPLE 2.7

A round copper conductor has a 0.0508-inch diameter. What is the resistance of 100 feet at 20°C?

Solution

The diameter is in inches and it is a round conductor, so the resistivity in circular mil-ohms per foot will be used.

$$A = (\text{DM})^2$$
$$= ((0.0508 \text{ in})(1000 \text{ mils/in}))^2$$
$$= 2580.64 \text{ CM}$$

Using Equation 2.5, we can now calculate the resistance,

$$R = \rho\, L/A$$
$$R = ((10.371 \text{ CM-ohm/ft})(1000 \text{ ft}))/(2580.64 \text{ CM})$$
$$R = 4.019 \text{ ohms}$$

EXAMPLE 2.8

A round copper conductor has a diameter of 1.04 centimeters. What is the resistance of 500 meters at 20°C?

Solution

Since the dimensions are in meters and centimeters, the resistivity in ohm-meters will be used.

$$R = \rho\, L/A$$
$$A = \pi\, D^2/4$$
$$= (\pi)(1.04 \times 10^{-2}\text{m})^2/4$$
$$= 8.49 \times 10^{-5}\ \text{m}^2$$
$$R = (1.724 \times 10^{-8}\ \Omega\text{-m})(500\ \text{m})/(8.49 \times 10^{-5}\ \text{m}^2)$$
$$R = 0.102\ \text{ohm}$$

2.7 The Effect of Temperature on Resistance

When the temperature changes, the resistivity of a material and its resistance change. The resistance of most metals increases as their temperature increases. The resistance of carbon and semiconductor materials such as silicon decreases as the temperature increases. This change in resistance is useful when it is used to determine the temperature of an oven, and to limit current. It can be disastrous if it causes a malfunction of the electronic control circuits in a space vehicle.

The resistance of most metals changes with temperature, as shown in the curve in Figure 2.12. This type of curve is obtained experimentally. An interesting part of the curve is the part near 273°C (absolute zero). **Absolute zero** is the temperature at which a substance has no heat and the movement of molecules stops. In this region, the resistance of a conductor approaches 0 ohms. This condition of 0 ohms is called **superconductivity,** which was discovered in 1911 by Heike Kamerlingh Onnes, a Dutch physicist. Onnes was able to induce a current in a ring of mercury at near absolute-zero temperature and maintain it for several hours without any source of emf. This principle of superconductivity is being explored for use in power transmission and large electromagnets. Recently, superconductivity has been reached in semiconductors at temperatures above absolute zero.

absolute zero

superconductivity

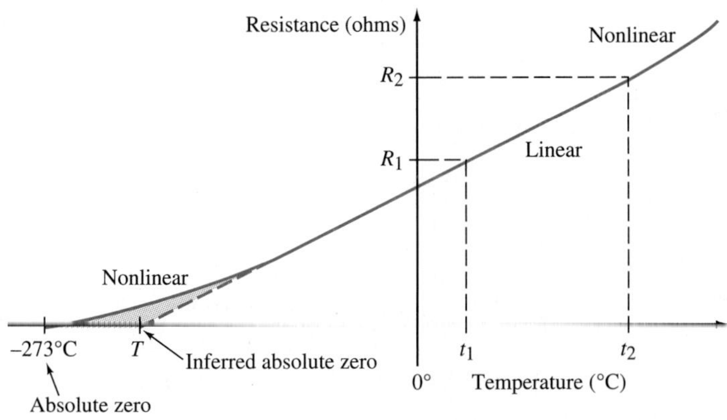

Figure 2.12 The resistance of most metals increases with temperature, and is 0 ohms at absolute zero.

To design circuits that will operate properly over a range of temperature, the designer must know how much change will occur. Equations 2.8 and 2.9 calculate the change in resistance. They are obtained by applying the right triangle relations to the curve of Figure 2.12. This gives

$$\frac{R_2}{|T_i| + t_2} = \frac{R_1}{|T_i| + t_1} \tag{2.8}$$

$$R_2 = R_1[1 + \alpha_1(t_2 - t_1)] \tag{2.9}$$

where:

t_1 is the lower temperature in °C
t_2 is the higher temperature in °C
R_2 and R_1 are the resistances at t_2 and t_1 in ohms
α_1 is the temperature coefficient of resistance at temperature t_1 in $°C^{-1}$

and T_i is the inferred absolute-zero temperature in °C. The parallel lines signify the absolute value.

The **temperature coefficient of resistance** (α) is the amount of resistance change per ohm of resistance for each degree change in temperature. Its unit is 1/°C, $°C^{-1}$, or per °C. From Table 2.3 we see that the change for

TABLE 2.3 Temperature Coefficient of Resistance at 20°C and Inferred Absolute Zero

Material	Temperature Coefficient (α) per °C	Inferred Absolute Zero Temperature °C
Silver	0.0038	−243.2
Copper	0.00393	−234.5
Gold	0.0034	−274.1
Aluminum	0.0039	−236.4
Tungsten	0.0045	−202.2
Nickel	0.006	−146.7
Iron	0.006	−146.7
Constantin	0.000008	−124.980
Nichrome	0.00017	−5862.4
Carbon*	−0.0005	—
Lead	0.004	−230
Manganin	0.0002	−4980
Mercury	0.00089	−1103.6
Platinum	0.003	−313.3

*The carbon curve does not follow the metal curve.

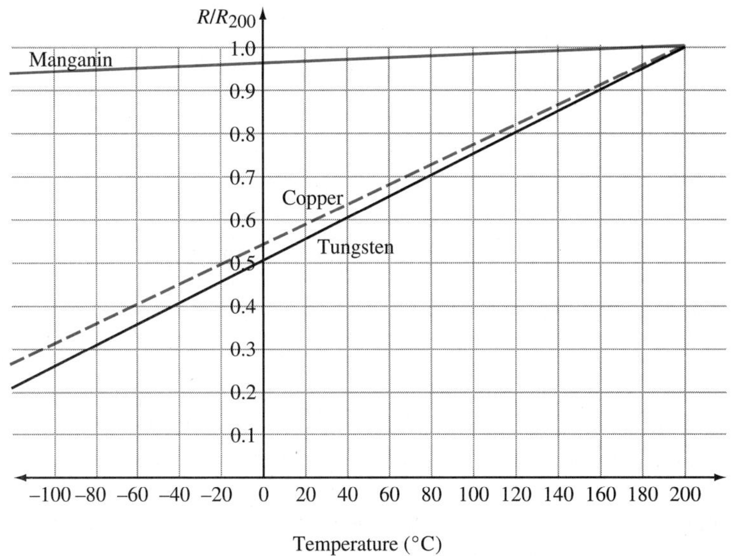

Figure 2.13 The rate of change of resistance with temperature is not the same for all conducting materials.

nickel is double that for platinum. The change for nichrome is smaller than that for copper. The relative change for some materials can be seen in Figure 2.13.

inferred absolute zero

The **inferred absolute zero** temperature is the temperature at which the extended curve crosses the temperature axis. It is useful in deriving some resistance-temperature relations.

Table 2.3 lists values of α at 20°C and the inferred absolute-zero temperature for some materials. As for relative resistivity, the actual values can vary.

Equations 2.8 and 2.9 will now be used to find the effect of temperature on resistance.

EXAMPLE 2.9

A piece of tungsten has a resistance of 300 ohms at 20°C. What is the resistance at 40°C?

Solution

Table 2.3 gives α at 20°C as $0.0045°\text{C}^{-1}$

$$\begin{aligned} R_2 &= R_1[1 + \alpha_1(t_2 - t_1)] \\ &= 300 \text{ ohms } [1 + 0.0045°\text{C}^{-1}(40°\text{C} - 20°\text{C})] \\ &= 300 \text{ ohms } [1 + 0.09] \\ R_2 &= 327 \text{ ohms} \end{aligned}$$

EXAMPLE 2.10

The resistance of a copper conductor is 3 ohms at 30°C. What is the resistance at 60°C?

Solution

Since neither temperature is the 20°C for which α is given in Table 2.3, Equation 2.9 will be used:

$$\frac{R_1}{t_1 + |T_i|} = \frac{R_2}{t_2 + |T_i|}$$

$$\frac{3}{30°\text{C} + 234.5°\text{C}} = \frac{R_2}{60°\text{C} + 234.5°\text{C}}$$

$$R_2 = 3.34 \text{ ohm}$$

Most tables list α at either 20°C or 0°C. Unfortunately, t_1 is not always 20°C or 0°C. For instance, we might know the resistance at 60°C and want to determine its value at 100°C. Using α at 20°C or 0°C will result in some error because α varies with temperature. If the error is not acceptable, or if α is not known, we can use Equation 2.10 to calculate α. It is also derived from the curve in Figure 2.12.

$$\alpha_1 = \frac{1}{|T_i| + t_1} \quad \textbf{(2.10)}$$

where:

α_1 is the temperature coefficient of resistance at temperature t_1

$|T_i|$ is the absolute value of the inferred zero temperature in °C

EXAMPLE 2.11

Calculate the value of α for copper at 20°C and 100°C.

Solution

$$\alpha_1 = \frac{1}{t_1 + |T_i|}$$

At 20°C,

$$\alpha_1 = \frac{1}{20°\text{C} + 234.5°\text{C}}$$

$$\alpha_1 = 0.00393°\text{C}^{-1}$$

At 100°C,

$$\alpha_1 = \frac{1}{100°\text{C} + 234.5°\text{C}}$$

$$\alpha_1 = 00299°\text{C}^{-1}$$

This example shows that α is not a constant for a material but varies with temperature.

2.8 The Resistor and Types of Resistors

A **resistor** is a device designed to have a specific amount of resistance. Some functions of resistors are to limit current, divide the voltage, control the charging and discharging of a capacitor, and to obtain heat from electricity.

resistor

In addition to having the desired resistance, a resistor must also be able to get rid of, or dissipate, the heat that is produced in it. If not, its resistance will change, or worse yet, it will be damaged. The heat-dissipating capacity of a resistor is called its **power rating.** Manufacturers give this rating in watts (W). This unit will be discussed in more detail in Chapter 3. The greater the power rating, the more heat the resistor can dissipate.

power rating

Resistors are made in many forms and sizes. They can be grouped according to (1) construction, (2) resistance values available, and (3) voltage-current characteristics. Each of these will now be explained.

Resistors According to Construction

Carbon Composition. This is one of the most common types of resistors. It is made of a mixture of carbon or graphite and ceramic formed into a slug. Its general appearance and size are shown in Figure 2.14. The desired resistance is obtained by adjusting the composition of the mixture. The slug is put into an insulating case and leads are attached to it, as shown in Figure 2.15. These resistors are available in values up to

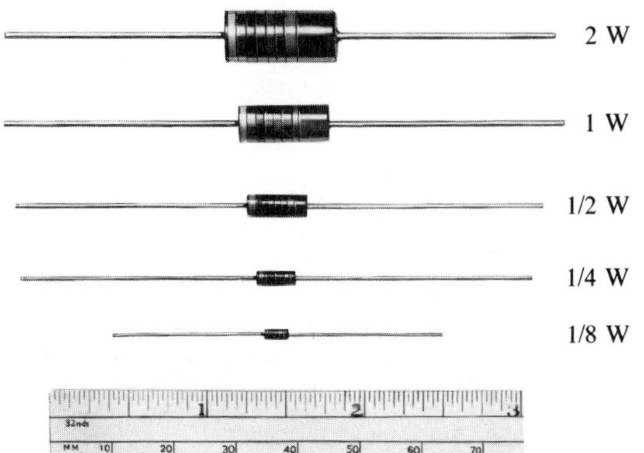

Figure 2.14 The carbon-composition resistor is small, inexpensive, and comes in many resistance values and power ratings. (Photo courtesy of Allen-Bradley Company)

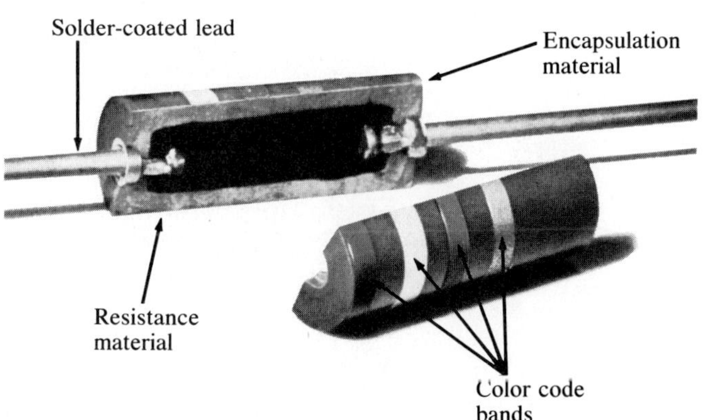

Figure 2.15 A cross section of a carbon-composition resistor. The carbon-composition slug in the center of the assembly provides the desired resistance. (Photo courtesy of Allen-Bradley Company)

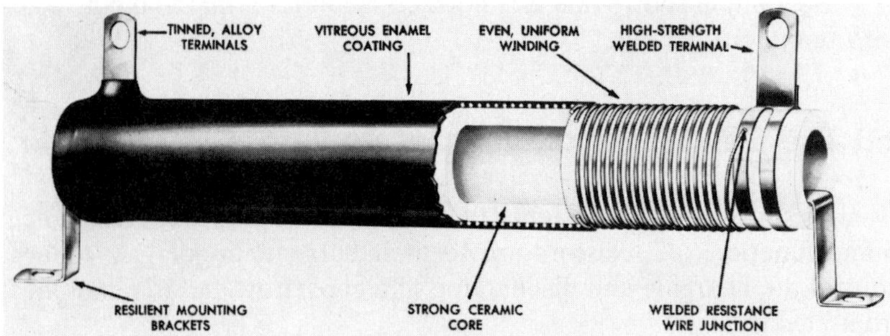

Figure 2.16 Tubular wirewound resistors such as this one are used in many high-power applications. (Photo courtesy of Ohmite Manufacturing Company)

22 megohms with commercial tolerances of ±5%, ±10%, and ±20%. The resistors are relatively inexpensive and can be produced at a fast rate in large quantities. These are generally low-power resistors. They are made in ratings of 1/8 watt to several watts.

Wirewound. These resistors are formed by winding resistance wire around a form of insulating material, usually ceramic insulation. The form can be tubular, as shown in Figure 2.16, or a flat strip. Manganin or some other material that has a low temperature coefficient of resistance is used. This produces better temperature stability than for carbon-composition resistors. On the negative side, the inductance of the turns can cause some problems at very high frequencies.

Thin Film. These resistors are made by depositing a thin layer of resistance material on an insulating core, glass, or ceramic. A diagram of a typical film resistor's construction is shown in Figure 2.17. This yields low values of resistance. Etching the film into a spiral, as on the metal-oxide film resistor shown in Figure 2.18, gives higher resistance. These resistors are small and take up very little space. Using a metal-oxide film instead of carbon permits higher operating temperatures. The power rating ranges from 1/8 watt to 1 or 2 watts. They are low-power resistors.

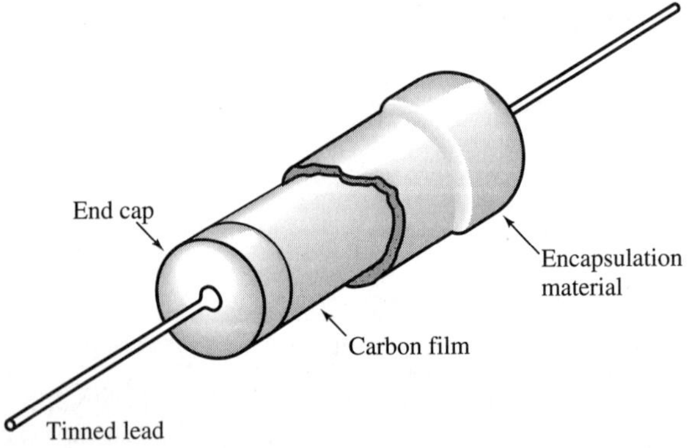

Figure 2.17 The carbon-film resistor is made by depositing a carbon film on a ceramic core.

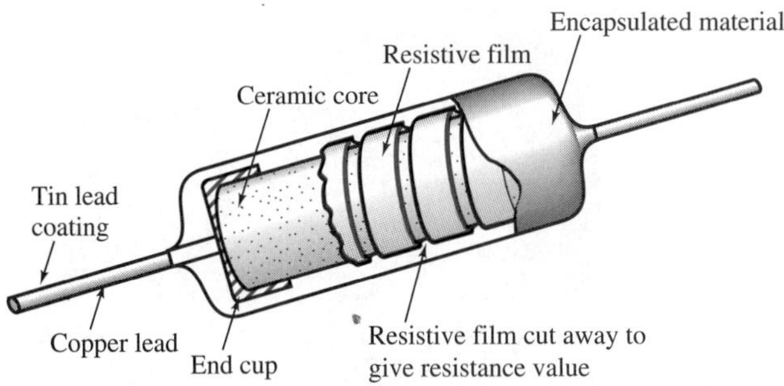

Figure 2.18 The metal-film resistor consists of a metal resistive film on a ceramic core. The film is helixed to give a spiral band of resistive material.

Metal film resistors can be made to smaller tolerances than carbon-composition resistors. Because of this, and other improved characteristics, they are gradually replacing the carbon-composition resistor in many applications.

Resistance Values Available

Fixed Resistor. These devices are resistors that have only one resistance value. They are available in carbon composition, wirewound, and thin-film construction. The resistors shown in Figure 2.14–2.18 are fixed resistors.

Tapped Resistors. These devices are resistors with terminals permanently connected at different points along the resistive material. These taps give several values of resistance. The resistor in Figure 2.19 is a tapped resistor. The resistance available is between any two taps, or between a tap and one of the end terminals. These resistors are available only in the wirewound type.

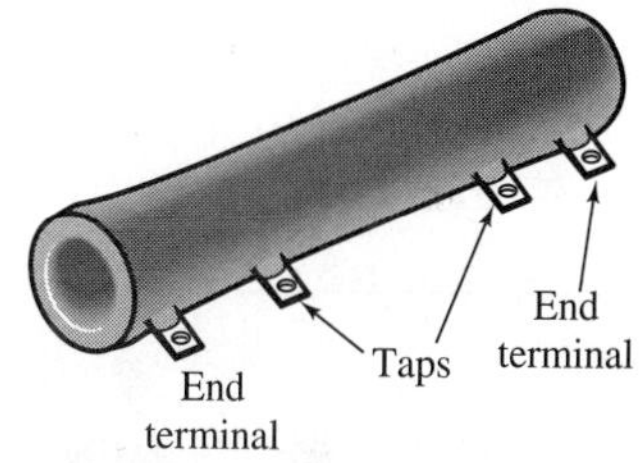

Figure 2.19 A tapped resistor provides several values of resistance from one unit.

Variable Resistors. These devices are resistors whose resistance can be easily changed by moving a sliding contact. They include rheostats and potentiometers.

rheostat

The **rheostat** is a variable resistor that is used to control current. Some rheostats are made in a long, cylindrical form as in Figure 2.20(a). Others are made in a circular form as in Figure 2.20(b). One has a sliding contact while the other has a rotating contact. As the sliding contact is moved, the resistance is changed. This is because the length of wire between the end terminal and the contact is changed. Any value between 0 ohms and the total value of the resistance can be obtained.

The rheostat has three terminals but only two need to be connected. They are the center one and one of the end ones. The other end terminal is sometimes connected to the center terminal.

potentiometer

The **potentiometer** is a three-terminal variable resistor that is used to obtain different voltages from a fixed voltage. A potentiometer looks

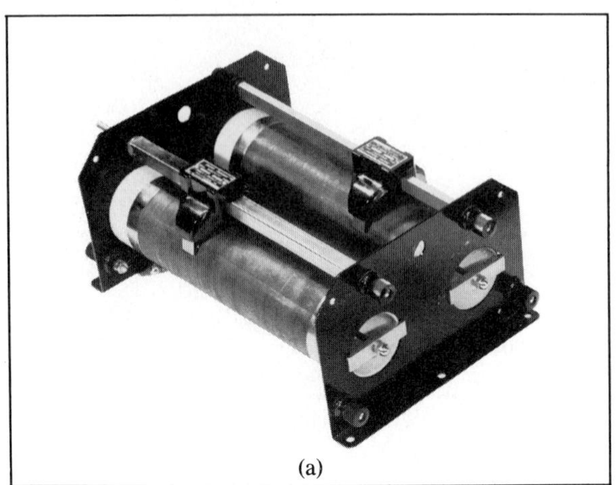

(a)

(b)

Figure 2.20 Rheostats are made in two forms, circular and linear. They are used as variable resistors in higher current applications. (a) A slidewire rheostat. (b) A circular rheostat. (a, photo courtesy of Biddle Instruments, b, photo courtesy of Ohmite Manufacturing Company)

similar to a circular rheostat but the wire is not as heavy. Figure 2.21 shows several types of potentiometers. Some uses for potentiometers are as volume controls, tone controls, light dimmers, and stereo speaker balancing. In use, the two end terminals of the potentiometer are connected to the fixed source. The load is connected to the center terminal and the end terminal goes to the negative terminal of the source. This connection will give output voltages from 0 volts to the full voltage of the source.

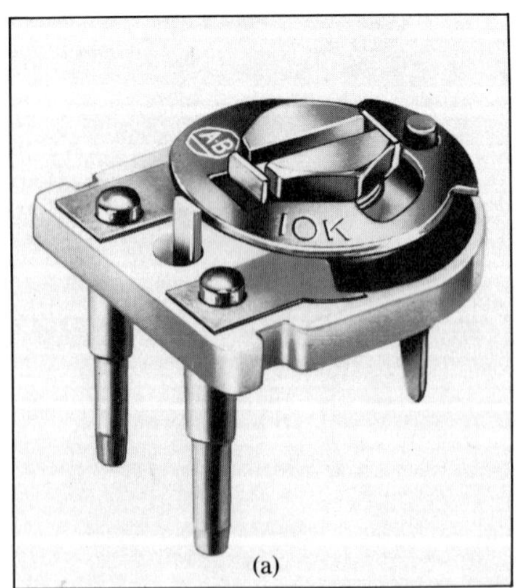

(a)

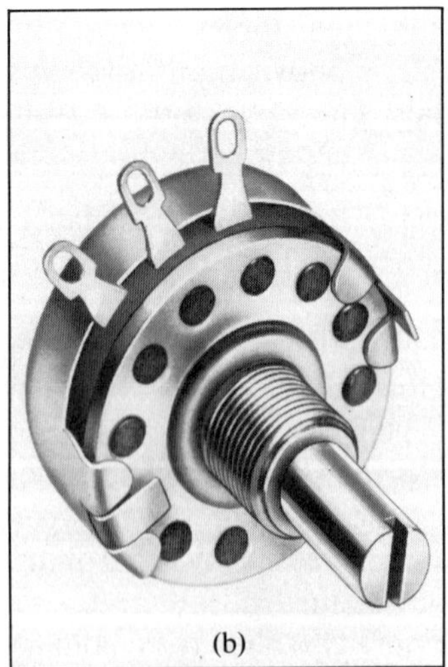

(b)

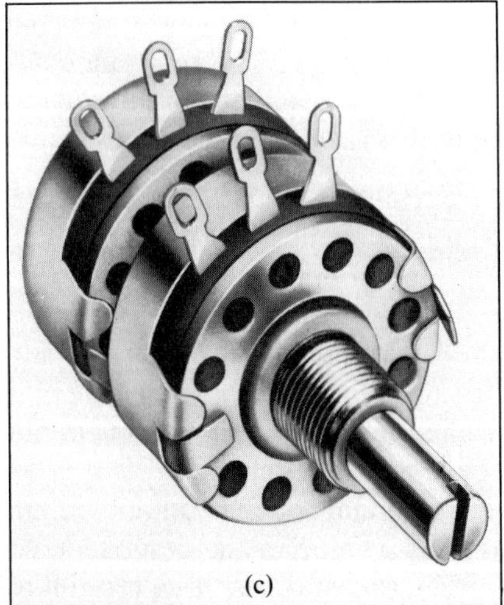

(c)

Figure 2.21 A potentiometer is used to control voltage as in a volume or tone control. (a) A trimmer potentiometer. (b) A single unit potentiometer. (c) a dual unit potentiometer. (Photos courtesy of Allen-Bradley Company)

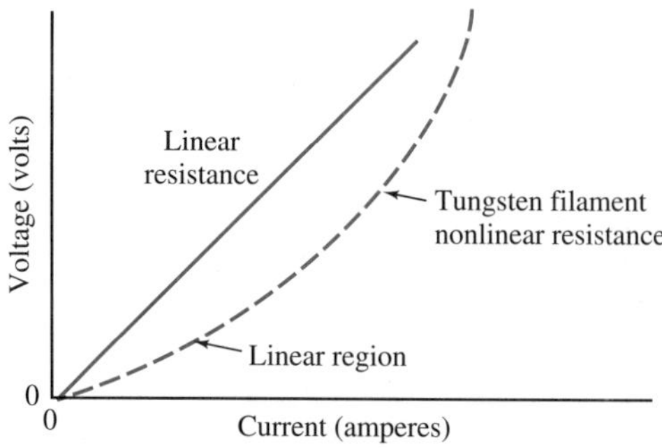

Figure 2.22 A comparison of linear and nonlinear resistance voltage-current characteristics.

Voltage-Current Characteristics

linear resistance

Linear. A **linear resistance** is one that has the same resistance value within its current range. The voltage-current curve will be a straight line. The curve for a tungsten filament in Figure 2.22 includes both linear and nonlinear regions. Metals such as manganin and constantin have very little change in resistance.

nonlinear resistance

Nonlinear. A **nonlinear resistance** is one that changes in resistance value with current. The $V - I$ curve will not be a straight line. Nonlinear resistors have many applications. The thermistor, whose resistance increases with temperature, can be used to limit surge currents and to protect meters. The varistor, whose resistance changes with voltage, is used to protect electronic equipment from high voltages.

The resistors described in this and the preceding section are discrete devices. That is, they are separate units. Another form of resistor is used in integrated circuits. The integrated circuit resistor is formed by chemically altering the structure of the substrate material to make it have resistance. The resistor is part of the material (chip) and cannot be separated. This type of resistor permits miniaturization or subminiaturization of the total package. They are low-power resistors and are used mostly in electronic circuits.

Some of the resistors are made in the surface mount package. Two resistance network packages are shown in Figure 2.23. The tabs serve as contacts and mounting pads. Some packages have J leads as in Figure 2.23(a). Others use the gull wing leads. These extend out from the package at 90 deg. Surface mount packages do not need through-holes on the printed circuit board because they mount on the surface of the board. This elimination of holes and small size results in lower cost, improved quality, and miniaturization. Typical size reduction of 30–70 percent can be realized.

(a)

(b)

Figure 2.23 Surface mount components use several lead shapes. (a) *J*-lead. (b) Gull Wing lead. (Photos courtesy of Bourns Networks, Inc.)

Surface mount parts are placed on the board by machines such as that shown in Figure 2.24. The machine can pick and place several thousand parts in an hour. The assembly of a single-sided board follows the steps in the flowchart of Figure 2.25. Using surface mount devices has resulted in smaller boards, improved electrical performance, lower manufacturing costs, automated assembly, higher resistance to shock and vibration, and physical standardization.

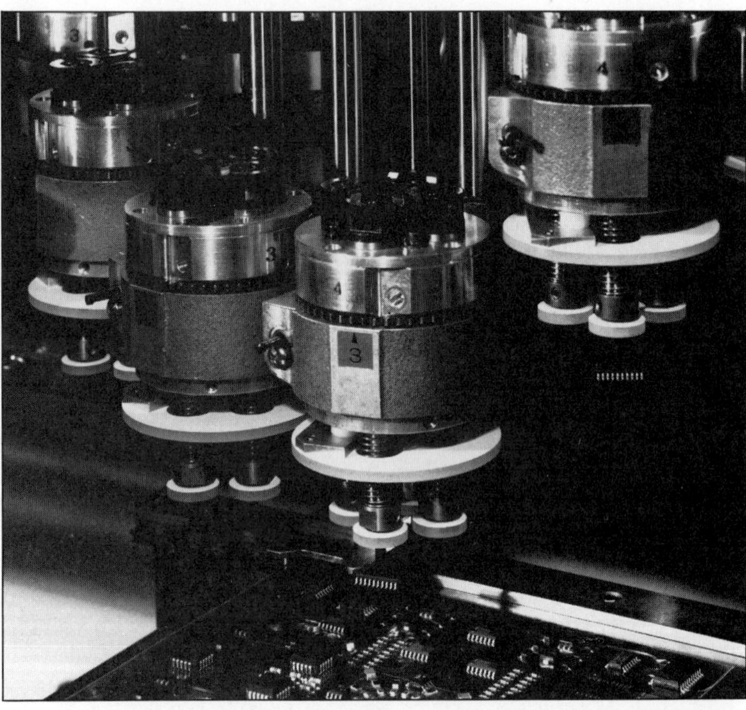

(a)

(b)

Figure 2.24 Automation reduces the cost of mounting and connecting surface mount components. (a) This Pick and Place machine can mount hundreds of components per minute. (b) A surface mount component is placed on a printed circuit board by one of the heads. (Photos courtesy of Universal Instruments Corporation)

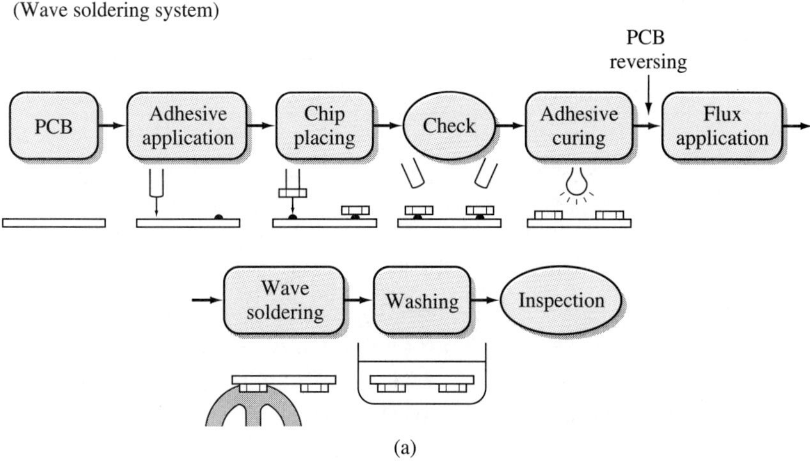

(a)

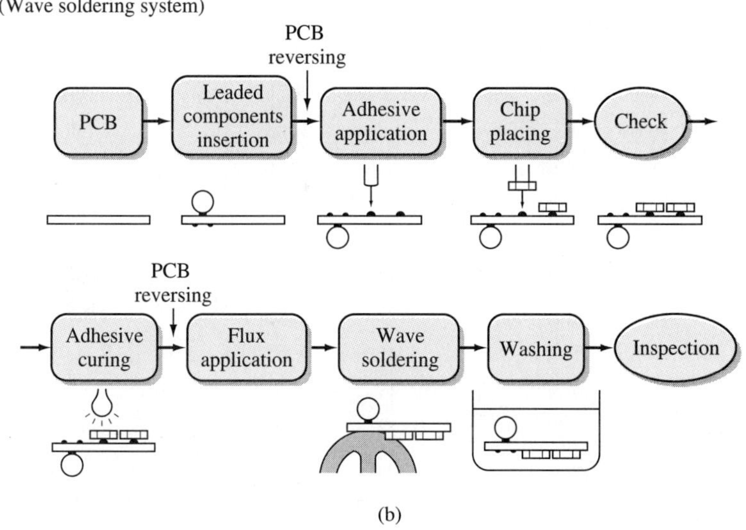

(b)

Figure 2.25 Typical production line for the assembly of surface mount circuits. (a) Chip placing on a single side. (b) Chip placing on both sides. (Diagram courtesy of Murata Erie North America)

2.9 American Wire Gage and Resistor Color Code

As you come in contact with conductors and resistors, you might notice that:

1. Conductors come in different sizes, and wire specifications usually include a number such as 12, 22, 28, and so forth.
2. Some resistors have colored bands on them.

American Wire Gage

The numbers for the wire are the **American Wire Gage** (AWG) sizes. Table 2.4 lists the numbers and characteristics up to AWG 38. The table actually continues to 56.

TABLE 2.4 American Wire Gage Table*

(Annealed Copper at 20°C)						
Gage	*Diameter (mils)*	*Area (Cir. Mils)*	*Ohms per 1000 ft*	*Diameter (mm)*	*Area (mm^2)*	*Ohms per kilometer*
0000	460.0	211,600	0.04901	11.68	107.2	0.160
000	409.6	167,800	0.06182	10.40	85.01	0.202
00	364.8	133,100	0.07793	9.266	67.43	0.255
0	324.9	105,600	0.09825	8.252	53.49	0.322
1	298.3	83,690	0.1239	7.348	42.41	0.406
2	257.6	66,360	0.1563	6.543	33.62	0.512
3	229.4	52,620	0.1971	5.827	22.67	0.646
4	204.3	41,740	0.2485	5.189	21.15	0.815
5	181.9	33,090	0.3134	4.620	16.77	1.028
6	162.0	26,240	0.3952	4.115	13.30	1.297
7	144.3	20,820	0.4981	3.665	10.55	1.634
8	128.5	16,510	0.6281	3.264	8.367	2.061
9	114.4	13,090	0.7925	2.906	6.631	2.600
10	101.9	10,380	0.9988	2.588	5.261	3.277
11	90.7	8,230	1.26	2.30	4.17	4.14
12	80.8	6,530	1.59	2.05	3.31	5.21
13	72.0	5,180	2.00	1.83	2.63	6.56
14	64.1	4,110	2.52	1.63	2.08	8.28
15	57.1	3,260	3.18	1.45	1.65	10.4
16	50.8	2,580	4.02	1.29	1.31	13.2
17	45.3	2,050	5.05	1.15	1.04	16.6
18	40.3	1,620	6.39	1.02	0.823	21.0

The sizes were set up for the convenience of the user and the manufacturer. Each smaller diameter is the next smaller size in drawing the wire. Diameters for #0000 and #36 are arbitrarily selected as 0.4600 inches and 0.005 inches. Other diameters were calculated based on the ratio of $39/\sqrt{(0.4600/0.005)}$ or 1.1229322 between one gage number and the next smaller number.

The wire used for conductors in a house is about #12 or #14; radio and electronic equipment use about #22 or larger gage size.

When using Table 2.4, your attention is called to the following three points of interest.

1. The diameter approximately doubles for each decrease of six gage numbers. It approximately halves for each increase of six gage numbers.

TABLE 2.4 American Wire Gage Table* (*continued*)

(Annealed Copper at 20°C)

Gage	*Diameter (mils)*	*Area (Cir. Mils)*	*Ohms per 1000 ft*	*Diameter (mm)*	*Area (mm^2)*	*Ohms per kilometer*
19	35.9	1,200	8.05	0.912	0.653	26.4
20	32.0	1,020	10.1	0.813	0.519	33.2
21	28.5	812	12.8	0.724	0.412	41.9
22	25.3	640	16.2	0.643	0.324	53.2
23	22.6	511	20.3	0.574	0.259	66.6
24	20.1	404	25.7	0.511	0.205	84.2
25	17.9	320	32.4	0.455	0.162	106
26	15.9	253	41.0	0.404	0.128	135
27	14.2	202	51.4	0.361	0.102	169
28	12.6	159	65.3	0.320	0.0804	214
29	11.3	128	81.2	0.287	0.0647	266
30	10.0	100	104	0.254	0.0507	340
31	8.9	79.2	131	0.226	0.0401	430
32	8.0	64	162	0.203	0.0324	532
33	7.1	50.4	206	0.180	0.0255	675
34	6.3	39.7	261	0.160	0.0201	857
35	5.6	31.4	331	0.142	0.0159	1,090
36	5.0	25.0	415	0.127	0.0127	1,360
37	4.5	20.2	512	0.114	0.0103	1,680
38	4.0	16.0	648	0.102	0.00811	2,130

*Values are for α_{20} of 0.00393°C^{-1} and a resistivity of 1.7241×10^{-8} ohm-meters.

2. The resistance approximately doubles for each increase of three gage numbers. It approximately halves for each decrease of three gage numbers.
3. The resistance of 1000 feet of #10 wire is approximately 1 ohm.

Some practical applications of the AWG Table are made in Examples 2.12 and 2.13.

EXAMPLE 2.12

A coil is made from 3500 feet of #22 copper wire. What is its resistance at 20°C?

Solution

The table gives R = 16.2 ohms per 1000 feet of #22 wire. For 3500 feet,

$$R = (3500 \text{ ft})(16.2 \text{ ohms}/1000 \text{ ft})$$
$$R = 56.7 \text{ ohms}$$

EXAMPLE 2.13

Sixteen hundred feet of copper wire at 20°C is used to connect a transformer to a house. The resistance cannot be greater than 1.6 ohms. What gage number must it be?

Solution

(1.6 ohms)/(1600 ft) = 0.001 ohm/ft.
(0.001 ohms/ft)(1000 ft) = 1 ohm/1000 ft.
#10 wire has 0.9988 ohm/1000 ft.
#11 wire has 1.26 ohm/1000 ft, which is greater than 1 ohm. Hence, use AWG #10 wire.

A single copper conductor in free air at temperature of 30°C and having an insulation rating of 60°C should be able to safely carry the currents listed in Table 2.5 without overheating.

TABLE 2.5 Copper Conductor Current Capacity

Gage #	Amperes
0000	300
000	260
00	225
0	196
1	165
2	140
3	120
4	105
6	80
8	55
10	40
12	25
14	20

Some conductors are made of several smaller-diameter conductors twisted together. These are called stranded conductors. Electric lamp cords and extension cords are two common examples of stranded conductors. Using several strands results in greater flexibility than using a single, large-diameter conductor.

Stranded conductors are specified by an AWG number and the number of strands. For instance, the conductor AWG 24 7/32 consists of seven strands of AWG #32 conductors. The total area of the stranded conductor will be equivalent or slightly greater than the area of the AWG # of the conductor. The seven strands have a total area of 432.47 *CM* compared to 404 *CM* for #24 conductor.

Resistor Color Code

Some resistors use colored bands, called a **Resistor Color Code**, as a coding mechanism to give the nominal value of the resistor. Some band arrangements are shown in Figure 2.26. They are the commercial 4 and 5 band system and the military system. The figure also gives the significance of each band. The number represented by the multiplier band is the power of 10, by which the first two digits are multiplied. The actual resistance of the resistor can differ from the nominal value by a ± percent given by the **tolerance band.** Thus, a ± 10 percent, 1000-ohm resistor can have a value between 900 and 1100 ohms.

tolerance band

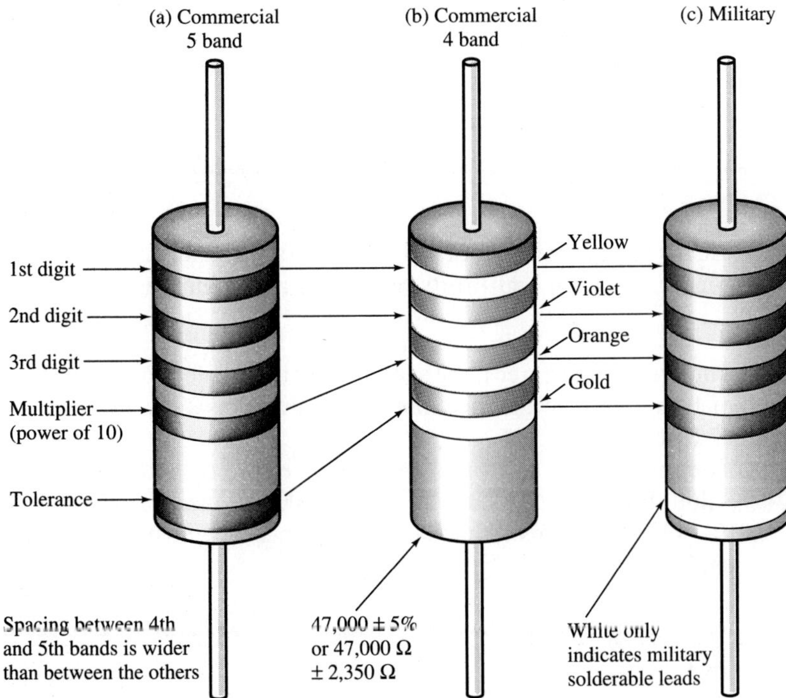

Figure 2.26 Three resistor color code systems. (a) Commercial 5 band. (b) Commercial 4 band. (c) Military.

TABLE 2.6 Resistor Color Code

Color	Number	Multiplier	% Tolerance	Failure Rate % per 1000 Hours
Black	0	1	—	—
Brown	1	10^1	±1%	1%
Red	2	10^2	±2%	0.1%
Orange	3	10^3	—	0.01%
Yellow	4	10^4	—	0.001%
Green	5	10^5	—	—
Blue	6	10^6	—	—
Violet	7	10^7	—	—
Grey	8	10^8	—	—
White	9	10^9	—	—
Silver	—	0.1	±10%	—
Gold	—	0.01	±5%	—
None	—	—	±20%	—

The number value for each color is given in Table 2.6. These values will now be used to find the resistance for some color combinations.

EXAMPLE 2.14

A resistor has color bands of red, violet, orange, and silver. Find: (a) its nominal resistance and (b) its maximum and minimum values.

Solution

a. The first digit is 2 (red), the second digit is 7 (violet), and the power of 10 is 3 (orange). Hence, the nominal value is 27×10^3 ohms, or 27,000 ohms.
b. The tolerance is ±10% (silver). Hence, the resistor can have a value between 27,000 − (0.1)(27,000) ohms and 27,000 + (0.1)(27,000) ohms. The resistance can be between 24,300 and 29,700 ohms.

EXAMPLE 2.15

Suppose you were building a circuit and the schematic diagram called for a 150-ohm, ±10% resistor. What color band would you look for as you sort through your collection of resistors?

Solution

The first digit in 150 ohms is a "1," which is represented with a "brown" band. The second digit is a "5," which is represented by a "green" band. The multiplier is 10^1, which is represeted by a "brown" band. And finally, the tolerance band must show $\pm 10\%$, a "silver" band.

Hence, you would look for a band sequence of "brown, green, brown, silver."

In the military specification, the fifth band gives the percent of failures in 1000 hours of operation. Each color represents the percent listed in Table 2.5. Knowing these percentages lets the designer predict whether the resistor will hold up for the application.

A 1500-ohm resistor with a -20 percent tolerance could have the same value as a 1000-ohm resistor with a $+20$ percent tolerance. This overlapping is reduced by not making every value in all tolerances. The made values are called preferred values and are listed in Table 2.7 for the four-band commercial resistors.

Even with the preferred values, some overlapping still results. The graphic in Figure 2.27 shows the overlapping for the preferred values of 3300–5100 ohms. Other values will have a similar pattern.

TABLE 2.7 Preferred Values of Resistors for Percent Tolerance

Values of First Two Digits	% Tolerance in Which Made
10	5, 10, 20
11	5
12	5, 10
13	5
15	5, 10, 20
16	5
18	5, 10
20	5
22	5, 10, 20
24	5
27	5, 10
30	5
33	5, 10, 20
36	5
39	5, 10
43	5
47	5, 10, 20
51	5
56	5, 10
62	5
68	5, 10, 20
75	5
82	5, 10
91	5

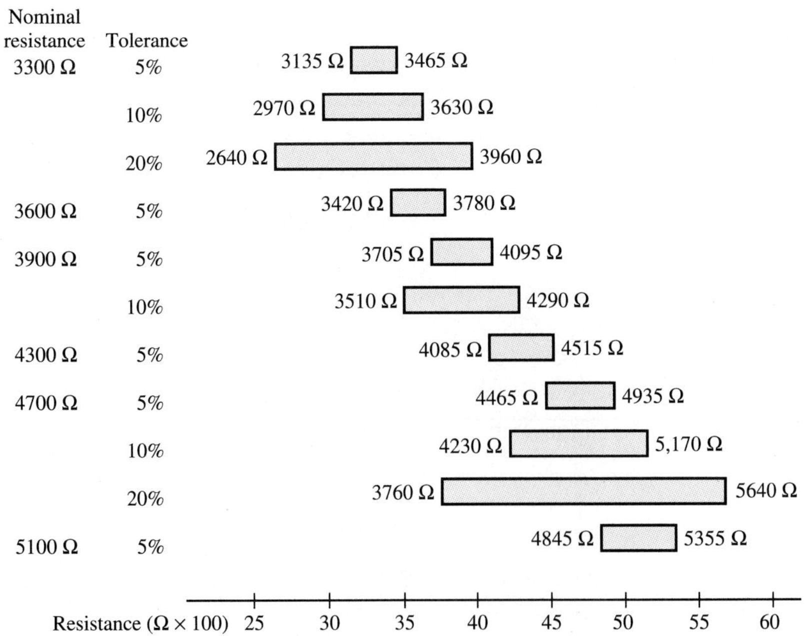

Figure 2.27 The combination of nominal value and tolerance results in overlapping of resistance values.

2.10 Conductors, Insulators, and Semiconductors

Materials that are used in electrical circuits and devices can be grouped into three main categories: (a) conductors, (b) insulators, and (c) semiconductors.

Conductors

These are materials in which the free electrons can be moved with little energy. Some conductor materials are silver, copper, and aluminum.

conductor materials

Conductor materials are generally used to connect one part of the circuit to another. For instance, conductor materials connect the electric power-generating plant to the customer's load. They also connect the batteries in a radio to the transistors and other components.

The ideal conductor offers no opposition to the movement of the electrons. That is, no energy is needed to make the electrons drift, and all of the energy from the source will be delivered to the load. Unfortunately, the ideal conductor does not yet exist; however, scientists are currently developing the ideal conductor.

conductivity

Conductivity is a measure of how easily the electrons can be made to move in a conductor. Silver, copper, and aluminum have a high conductivity. A more common and useful characteristic than conductivity is

relative conductivity

relative conductivity. Relative conductivity is the ratio of a material's conductivity to that of a standard reference. The reference material is given the relative conductivity of one. Such a comparison is shown in Table 2.8. This table uses copper as the reference. Some tables use silver. According to the relative conductivity values, silver has a higher conductivity than copper, and aluminum's conductivity is lower. A larger relative conductivity means high conductivity.

TABLE 2.8 Relative Conductivity

Material	Relative Conductivity
Silver	1.08
Copper	1.00
Gold	0.707
Aluminum	0.611
Tungsten	0.308
Nickel	0.221
Iron	0.172
Lead	0.082
Mercury	0.0177
Nichrome	0.0172
Carbon	0.0038–0.0013

The conductivity of a material should not be the only factor to consider when selecting a material for a conductor. Other factors include cost, weight, strength, resistance to the surrounding, and the change in conductivity that will occur in use. Copper is generally the most widely used material. This is because of its high conductivity, relatively low cost, malleability, and ductility. Silver is expensive, but it is used to plate surfaces where a good contact is needed. Tungsten is a poorer conductor than copper, but its hardness and resistance to erosion from arcing make it suitable for use in switch contacts where arcing can exist. Carbon is also a poorer conductor than copper, but it is used in motor brushes since it is soft. This prevents damage to the commutator material. Also, its conductivity increases as its temperature increases, which is a desirable characteristic in this application.

Insulators

insulator

An **insulator** is a material that, for all practical purposes, has no free electrons. Insulating materials have a very low conductivity and are generally nonmetals. Some examples of insulating materials are glass, mica, paraffin, hard rubber, oil, porcelain, polyvinyl chloride, and Teflon. Pure water is also an insulator. However, it becomes a conductor if it has some minerals or salts in it.

Insulator materials keep charge from flowing into unwanted areas. For instance, the polyvinyl chloride insulation on an electrical lamp cord prevents the bare conductors from touching. If they touch, the electric current would go from one conductor to the other instead of to the lamp. Insulation is also used to prevent a person from touching a bare conductor. As you will learn in the next section, this can result in electric shock.

Semiconductors

Semiconductors normally are neither good conductors nor good insulators. However, certain conditions such as placing them in heat or light introduce enough energy to make some free electrons move. Thus, while they initially behaved like insulators, they now act like conductors. Semiconductor materials have many applications in devices such as transistors, diodes, silicon-controlled rectifiers, and the light-emitting diodes (LED). LEDs are used in some calculators and watches to form the display for numbers. Gallium phosphide, gallium arsenide, and gallium arsenide phosphide are used for light-emitting diodes. The junction glows when a voltage is applied across the diode. This conversion of electrical energy to light energy is known as electroluminescence.

doping

semiconductor material

Semiconductor materials, which are used in transistors, are treated to provide either positive or negative charges for current. This treatment is known as **doping.** In one case, impurities are added to give a deficiency of electrons in the atom. These impurities are called acceptors. This **semiconductor material** is known as P-type. In the other case, impurities are added to give more free electrons. These impurities are called donors. This type of semiconductor material is known as N-type. In the

P-type semiconductor material, the carriers are positive charges. They are called holes. In the N-type semiconductor material, the carriers are negatively charged electrons.

A diode is formed by making a junction between a P-type and an N-type material. Current will flow only if the P-type material is at a more positive potential than the N-type. The transistor has either a P-type section between two N-type sections (NPN) or an N-type section between two P-type sections (PNP). Because of the junctions, a small current in the center section can control large currents in the outer section. This characteristic is used in amplification and rectification.

2.11 Electrical Safety

electric shock

Working with electricity is not a matter to be taken lightly. A voltage applied across parts of your body results in a current. The effect of the current is called **electric shock** and can range from a mild sensation to death. Although shock can cause death, it is also used to restore regularity to a heartbeat when the heart is fibrillating.

The exact effect of shock depends on the amount of current and the path. Current through the heart area will have a more serious effect than through a hand or foot. The chart in Figure 2.28 gives the range of the effect. However, keep in mind that the shown effects will vary.

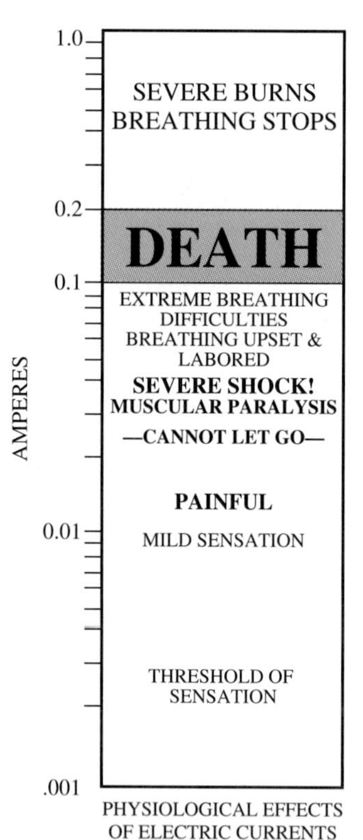

Figure 2.28 The effect of electrical shock depends on the amount of current and the path through the body. (Chart courtesy of Graymark International, Inc.)

Touching the terminals of a source of emf, touching a part of an energized circuit and a grounded part, letting your fingers come in contact with the test probe when measuring voltage, and even touching the metal chassis of test equipment are some actions that can result in shock. However, the chances of electric shock can be reduced by using some safe practices. Some of these are:

1. Disconnect the circuit from the source if possible.
2. Remove rings, watches, and any metal objects that might touch a part of an energized circuit.
3. Know the location of safety devices such as circuit breakers and fire extinguishers.
4. Work with one hand in your pocket so you cannot touch two parts.
5. When using a test instrument, avoid touching the test probes.
6. Do not stand on wet surfaces or work with wet hands when working with high voltages.
7. Remove the charge from capacitors by connecting a wire or other connector across the two terminals.
8. Do not assume that someone has opened a switch. Check it yourself.
9. Replace power cords that are frayed or worn.
10. Use three-prong plugs on all power cords.
11. Do not touch the metal parts of two test instruments at the same time.
12. Do not engage in practical jokes or horseplay.

Some electrical devices such as lamps and heaters become hot during operation. Also, a high current in a resistor can increase its temperature. Thus, even though there is no electrical shock, one can experience a burn from touching a part of the circuit. The effect of the burn can be just as bad as that of electrical shock.

SUMMARY

1. Matter has weight and occupies space. It can be broken down into one or more of the following: atom, element, molecule, and compound.
2. A charged body exerts a force on other charged bodies. Like charges repel and opposite charges atttract. Coulomb's Law relates the force to the charge and the separation.
3. An atom is made up of electrons, protons, and neutrons. The electrons whirl in orbits around the protons and neutrons at the center. Electrons have a negative charge and protons have a positive charge. Neutrons do not have any charge.
4. Charge accumulating on an object forms static electricity. Static electricity can damage electronic circuits when discharged through them.
5. Electric current is the movement of charge in some general direction. It is measured in amperes. The direction of electron current is out of the more negative terminal of the source. The direction of conventional current is out of the more positive terminal.
6. Potential difference represents the energy available to move charge. It is measured in volts.
7. Resistance is the opposition to current and is measured in ohms. The resistance of a unit section of material is called resistivity.
8. The resistance of a material is directly proportional to resistivity and length. It is directly proportional to the cross-sectional area.
9. A change in temperature causes a change in resistance. The temperature coefficient of resistance is the amount of change per ohm, per degree of change in temperature.
10. Some types of resistors are carbon-composition, metal-film, and wirewound. These can be grouped into fixed, tapped, adjustable, and variable. The potentiometer and rheostat are two types of variable resistors.
11. The resistance of a linear resistor does not change with current, whereas a nonlinear one does change with current. Its *V-I* curve will not be a straight line.
12. The diameters of wire conductors are set according to the American Wire Gage. This standard also gives the area and the resistance per 1000 feet for the wire.
13. The Resistor Color Code is a method of using colored bands to indicate the resistance value. The bands represent the digits, multiplier, and tolerance.

14. Conductors are materials in which free electrons can be moved with little energy. Insulators are materials that require large amounts of energy to move the electrons.
15. Semiconductor materials are neither good conductors nor good insulators.

EQUATIONS

	Equation	Description	Page
2.1	$F = k\dfrac{Q_1Q_2}{r^2}$	Coulomb's law	52
2.2	$I = \dfrac{Q}{t}$	Current and charge relation	56
2.3	$V = \dfrac{\mathcal{W}}{Q}$	Potential difference and energy relation	58
2.4	$G = \dfrac{1}{R}$	Relation of conductance to resistance	60
2.5	$R = \rho\dfrac{L}{A}$	Relation of resistance to material and dimension	61
2.6	$\text{CM} = (\text{DM})^2$	Area in circular mils	61
2.7	$\text{No. of CM} = (\text{SM})\left(\dfrac{4}{\pi}\right)$	Conversion of square mils to circular mils	61
2.8	$\dfrac{R_1}{t_1 + \lvert T_i\rvert} = \dfrac{R_2}{t_2 + \lvert T_i\rvert}$	Effect of temperature on resistance using the inferred absolute zero temperature	64
2.9	$R_2 = R_1[1 + \alpha_1(t_2 - t_1)]$	Effect of temperature on resistance using the temperature coefficient of resistance	64
2.10	$\alpha = \dfrac{1}{t_1 + \lvert T_i\rvert}$	The temperature coefficient of resistance	66

QUESTIONS

1. What is the definition of matter?
2. Explain the relation of the atom, element, and molecule to matter.
3. What are the three main particles of the atom?
4. Describe the shell structure of the atom.
5. Why aren't the negative electrons in the positive nucleus?
6. Will the force on two charged bodies increase or decrease if they are brought closer together?
7. What is the name of the law that is used to calculate the force between charged bodies?
8. What is a valence electron?
9. What is a free electron?

10. How can a person generate some static electricity?
11. What are the four general methods of preventing or controlling static electricity?
12. What is meant by electrostatic discharge?
13. Why has electrostatic discharge become a matter of concern in electronics?
14. Explain electric current in terms of charge.
15. How did the term "conventional current" originate?
16. Does a 6-volt battery have more or less potential difference than a 1.5-volt battery?
17. How did the concept of "source of emf" originate?
18. What is the name of the unit of area that is used for round conductors?
19. How is the area in Question 18 related to the diameter of the conductor?
20. What is a mil?
21. How many circular mils are there in an area of 1 square mil?
22. What is meant by resistivity?
23. What are the two units of resistivity?
24. If the length of a conductor is doubled, how much change will there be in the resistance?
25. If the area of a conductor is doubled, how much change will there be in the resistance?
26. Does the resistance of a light bulb increase or decrease when it is operating?
27. What is the meaning of the temperature coefficient of resistance?
28. What is superconductivity and at what temperature does it occur?
29. What is the difference between a linear resistance and a nonlinear one?
30. What are two semiconductor devices that have nonlinear resistance characteristics?
31. What are three types of resistors according to construction?
32. What are some differences between the construction of the carbon-composition resistor and a metal-film resistor?
33. What are three types of resistors according to the resistance available?
34. What is the difference between a rheostat and a potentiometer?
35. Would one use a rheostat or a potentiometer to limit the current?
36. Is the resistance of a #22 AWG conductor greater or smaller than that of a #10 AWG conductor?
37. What do the first three bands stand for in the four-band resistor color code?
38. Which band gives the percent tolerance?
39. What is the number value for each color of the resistor color code?
40. What would be the color of the multiplier band for a 4700-ohm resistor?
41. Will there be any current in a piece of string when connected to a source of emf? Why?
42. What if the string in Question 41 is replaced with a piece of copper wire?
43. Why are wires of a lamp cord covered with plastic and not aluminum foil?
44. Is nichrome a better conductor than copper?
45. Why doesn't a bird on a high voltage wire experience electric shock?
46. What are ten safety practices that will help prevent electric shock?

PROBLEMS

SECTION 2.2 The Atom, Charge, and Coulomb's Law

1. How many electrons are needed to have a charge of 1 coulomb?

2. Silver has 47 electrons in its atom. How many coulombs of charge are on the 47 electrons?

3. Copper has 35 neutrons, 29 protons, and 29 electrons in its atom. What is the mass of the atom?

4. What is the mass of the aluminum atom nucleus if it has 14 neutrons and 13 protons?

5. Two charged particles are separated by 0.2 meters. The charge is $+5 \times 10^{-6}$ coulombs on one, and $+4 \times 10^{-6}$ coulombs on the other.

   (a) Find the force acting on the particles.

   (b) Are the particles attracted or repelled?

6. How far apart must two objects with an 8×10^{-6} coulomb charge on each be placed so that the force on them is 9 newtons?

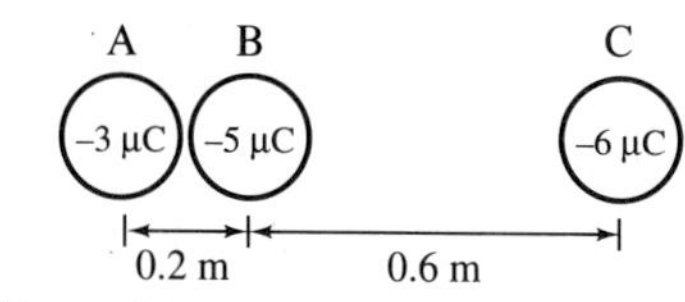

Figure 2.29

7. Three charged particles are placed as shown in Figure 2.29.

   (a) What is the force on particle B?

   (b) What is the direction of the force?

8. There is a force of 18 mN between two charged particles. What will the force be if the distance between them is tripled?

9. If the electron in the hydrogen atom is 0.53×10^{-10} meters away from the nucleus, how much electrical force does the proton exert on the electron?

SECTION 2.4 Electric Current

10. How many amperes of current are there when 5 coulombs of charge are transferred in 0.5 seconds?

11. The current in a wire is 2 amperes. How many coulombs of charge are being transferred per second?

12. Which condition will result in a larger current: 10 coulombs transferred in 5 seconds or 5 coulombs transferred in 10 seconds?

13. How many seconds are needed to transfer 20 coulombs if the current is 5 amperes?

14. How many electrons are transferred in Problem 11?

15. A calculator battery is charged for 8 hours with a current of 200 mA. How many coulombs of charge are transferred into the battery?

SECTION 2.5 Potential Difference and Electromotive Force (emf)

16. What is the potential difference between two points if 3 joules of work are needed to move 1 coulomb of charge between the points?

17. How many joules of work are needed to move 5 coulombs between two points when the potential difference is 20 volts?

18. What is the potential difference between two points if 50 joules of work are needed to maintain a current of 2 amperes for 2 seconds?

19. How many joules of work are done by a 12-volt source of emf to maintain a 2-ampere current for 6 seconds?

20. Thirty joules of work are done to maintain 2 amperes of current for 5 seconds between two points. What is the potential difference between the points?

SECTION 2.6 Resistance and Resistivity

21. What is the circular-mil area of a round conductor with the following diameter?

    (a) 0.15 inch (c) 2.3 mm
    (b) 0.2 inch (d) 0.32 mm

22. What is the circular-mil area of a round conductor with the following diameter?

    (a) 0.5 inch (c) 4.62 mm
    (b) 0.005 inch (d) 0.127 mm

23. What is the diameter of the circles that have the following circular-mil area?

    (a) 2500 CM (c) 1600 CM
    (b) 900 CM (d) 1225 CM

24. A rectangular cross-sectional conductor has a width of 1.5 inches and is 0.25 inch thick. What is its area in circular-mils?

25. What is the circular-mil area of a rectangular cross-sectional conductor that has a width of 0.5 inch and a thickness of 0.1 inch?

26. What must the diameter of a round conductor be to have the same circular-mil area as a 1-inch × 1-inch square conductor?

27. What is the area in circular-mils of a conductor with a cross section as shown in Figure 2.30?

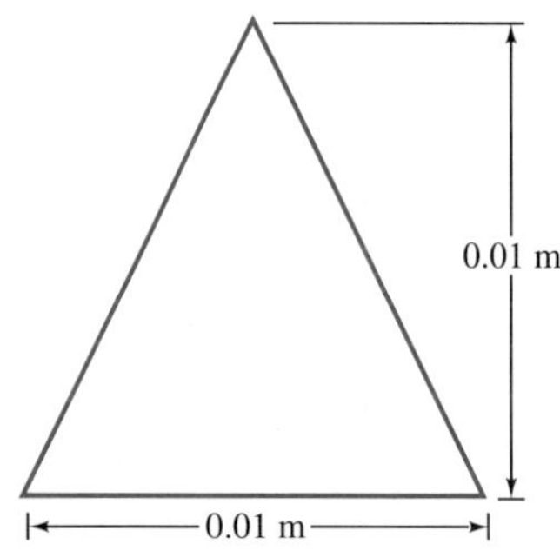

Figure 2.30

28. What is the resistance of 800 feet of round copper conductor at 20°C if its diameter is 0.2576 inch?

29. What is the resistance of 300 meters of round aluminum conductor at 20°C if its diameter is 10.4 mm?

30. A coil of wire has a resistance of 3 ohms at 20°C. If the diameter of the wire is 0.102 inch, how many feet of wire are on the coil?

31. A round copper conductor with a 4.62-mm diameter has a resistance of 5 ohms at 20°C. What is the length in meters?

32. A copper conductor with a 0.4-inch diameter and a resistance of 1.5 ohms at 20°C is replaced with a copper conductor of the same length that has a resistance of 0.8 ohm. What is the diameter of the second conductor?

33. A round conductor 2000 meters long, with a diameter of 10.7 mm, has a resistance of 0.628 ohm at 20°C. Of what material is the conductor made?

34. What is the square mil area of a 300-meter square copper conductor that has a resistance of 0.2 ohm at 20°C?

SECTION 2.7 The Effect of Temperature on Resistance

35. A tungsten filament lamp has a resistance of 20 ohms at 20°C. What is its resistance at 800°C?

36. A nichrome heater element in an electric iron has a resistance of 10 ohms at 20°C, and 13 ohms when operating. What is the operating temperature of the heater element?

37. A copper conductor has a resistance of 1.5 ohms at 100°C. What will be its resistance at 20°C?

38. What is the temperature of a soldering iron if its nichrome heating element has a resistance of 6 ohms at 20°C and 7 ohms when hot?

39. What is the resistance of an aluminum conductor at 80°C if its resistance at 30°C is 0.5 ohm?

40. A copper conductor has a resistance of 2 ohms at 40°C and 3.5 ohms at some higher temperature. What is the temperature at 3.5 ohms?

41. A conductor has a resistance of 0.04 ohm at 20°C and 0.046 ohm at 70°C. Of what material is the conductor made?

SECTION 2.8 The Resistor and Types of Resistors

42. Given the following voltage and current readings, is the resistor linear or nonlinear?

V (volts)	0	12.0	16.0	20.0	24.0	28.0	32.0
I (amperes)	0	0.3	0.4	0.5	0.6	0.7	0.8

43. Repeat Problem 42 for the following voltage and current readings.

V (volts)	0	4.5	8.0	12.5	18.0	24.5	32.0
I (amperes)	0	0.3	0.4	0.5	0.6	0.7	0.8

44. Repeat Problem 42 for the following voltage and current readings.

V (volts)	0	30	40	50	60	70	80
I (amperes)	0	0.45	0.8	1.25	1.8	2.45	3.2

SECTION 2.9 American Wire Gage and Resistor Color Code

45. The following lengths of wire are used in the wiring of a house. What is the resistance of each at 20°C?

    (a) 300 meters of #16 copper wire
    (b) 1200 meters of #10 copper wire
    (c) 200 feet of #12 copper wire

46. Which one of the following will have the highest resistance at 20°C?

    (a) 2500 feet of #20 copper wire
    (b) 150 meters of #32 copper wire
    (c) 75 meters of #4 copper wire

47. The conductor from a service panel to a light has two 50-foot lengths of #12 copper wire (100 feet total). What is its resistance at 20°C? What would be the resistance if #10 wire were used?

48. A 500-meter copper conductor cannot have a resistance of more than 4 ohms at 20°C. What is the largest gage number wire that can be used?

49. How many meters of #24 copper conductor will have a resistance of 5 ohms at 20°C?

50. What is the diameter in millimeters of a 2000-meter conductor that has a resistance of 5.2 ohms at 20°C?

51. Use the values for #36 conductor in the Wire Gage Table to calculate the resistivity of copper.

52. Resistors with the following color bands are in a circuit. What is the nominal resistance and tolerance, in ohms, of each?

 (a) gray, red, black, gold

 (b) orange, blue, red, silver

 (c) red, red, green

53. A customer bought some resistors having the following color bands. What is the nominal resistance and tolerance, in ohms, of each?

 (a) brown, gray, blue, silver

 (b) orange, white, orange, gold

 (c) red, violet, red, gold

54. It is desired to purchase the following resistors. What are the color bands that one should look for?

 (a) 3.3 ohms, ±20% (b) 12,000 ohms, ±5% (c) 820 ohms, ±10%

55. What are color bands in the commercial four-band color code for each of the following resistors?

 (a) 220 ohms, ±20% (b) 11 kilohms, ±5% (c) 6.8 megohms, ±10%

56. What color is the tolerance band for resistors that have the following values?

 (a) Value according to color bands 1.8 ohms, measured 1.9 ohms

 (b) Value according to color bands 47,000 ohms, measured 54,000 ohms

57. Repeat Problem 56 for the following values.

 (a) Value according to color bands 1500 ohms, measured 1560 ohms

 (b) Value according to color bands 270 ohms, measured 251 ohms

58. The color bands on a carbon-composition resistor represent 4700 ohms and the tolerance band is gold. The measured resistance is 4982 ohms. Is the resistance within the tolerance limits?

59. What are the minimum and maximum resistance values that a carbon-composition resistor can have if the color of the bands are orange, white, red, and silver?

60. Repeat Problem 59 for a color band sequence of blue, gray, yellow, and no tolerance band.

CHAPTER 3

ANALYSIS OF THE BASIC CIRCUIT

CHAPTER OBJECTIVES

After completing this chapter, you should be able to:

1. Understand the connections in a circuit diagram.
2. Interpret a block diagram and follow the signal through it.
3. Determine the polarity, voltage rises, and voltage drops.
4. Make measurements with a voltmeter and an ammeter.

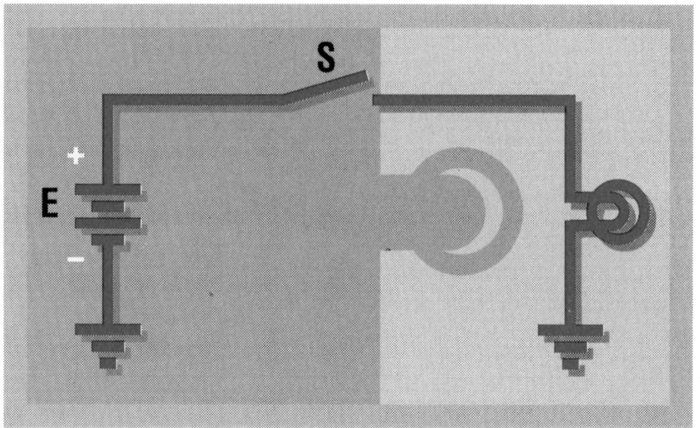

5. Use Ohm's Law to determine resistance, current, or voltage.
6. Calculate power in a device, or supplied by a source in a dc circuit.
7. Measure power with a wattmeter.
8. Calculate energy and the cost of energy use.
9. Determine the efficiency of a device or a system.
10. Explain the meaning of an open and short circuit.

KEY TERMS

1. ammeter
2. block diagram
3. circuit breaker
4. circuit diagram
5. efficiency
6. electric circuit
7. energy
8. fuse
9. graphic symbol
10. horsepower
11. joule
12. kilowatt-hour
13. letter symbol
14. loading effect
15. long shunt
16. Ohm's law
17. open circuit
18. polarity
19. power
20. short circuit
21. short shunt
22. switch
23. voltage drop
24. voltage rise
25. voltmeter
26. watt
27. wattmeter

INTRODUCTION

This chapter deals with diagrams, quantities, and basic relations. These diagrams and relations will be encountered throughout the text. The chapter explains the construction and use of the circuit diagram. The block diagram is also presented.

When studying or troubleshooting a circuit, terminal polarity, voltage rise and voltage drop, measuring voltage, current, and power can all be involved. This includes connecting the meters and using additional information.

Our life is affected in many ways by power, energy, and efficiency. The nameplate of most appliances lists the power rating for the appliance. We pay the electric utility company for the energy that we use. Devices that have a high efficiency will make better use of the power put into the device.

Ohm's Law provides a simple but important relation among voltage, current, and resistance. It will be used to obtain many other relations.

Last, but not least, the meaning and effect of open and short circuits are presented. We will consider the damage they can cause as well as how they can be used for our benefit.

Figure 3.1 Graphic and letter symbols similar to this for the resistor are used to represent parts in a circuit diagram.

3.1 Circuit Diagrams and Electronic Symbols

electric circuit

An **electric circuit** is a group of electrical parts connected to form a path through which electric charge can move. The connections can be made of wire, as in a house; in the material itself, as in an integrated circuit; or with flat strips, as in the printed circuit board.

Since a circuit must be wired, and sometimes repaired, some means of showing the connections and the parts is needed. One such means is the circuit diagram. Such a diagram is also called a schematic.

circuit diagram
graphic symbol
letter symbol

A **circuit diagram** shows the parts and their interconnection using graphic and letter symbols. A **graphic symbol** is a simple line diagram that represents a part. The graphic symbol for a resistor is shown in Figure 3.1. A **letter symbol** is a letter of the alphabet that represents a quantity. R is the letter symbol for resistance. A partial list of symbols and their meaning is given in Tables 3.1 and 3.2 and others will be introduced as needed. A more complete listing can be found in various handbooks. These symbols have been standardized through the efforts of various organizations such as the Institute of Electrical and Electronics Engineers. Thus, each symbol always has the same meaning. Subscripts are used to distinguish between symbols of the same kind. A subscript is a small letter or number placed alongside the letter symbol. The "1" in R_1 is a subscript.

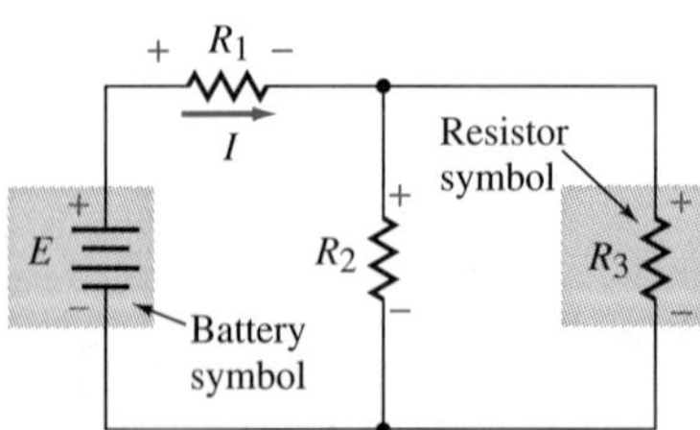

Figure 3.2 This basic circuit demonstrates the use of graphic and letter symbols.

A basic circuit diagram is drawn in Figure 3.2. In that circuit, the symbol with the parallel lines represents a battery. The letter E represents the emf of the battery. Interconnections are shown as lines. A more

TABLE 3.1 Graphic Symbols

RESISTOR	
Fixed	
Variable	
Tapped	
CAPACITOR	
Fixed	
Variable	
WIRES	
Connected	
Unconnected	
Unconnected	
INDUCTOR	
Fixed air core	
Fixed iron core	
Tapped	
Variable	
GROUND	
Transformer (Iron core)	
Autotransformer	
Circuit breaker	
dc Generator	G
ac Generator	G
dc Motor	M
ac Motor	M

Voltmeter	V
Ammeter	A
Wattmeter	W
Alternating current source	
Cell	
Battery	
Switch, single pole, single throw	
Switch, single pole, double throw	
Switch, double pole, double throw	
Switch, double pole, single throw	
Fuse	
Lamp, incandescent	
Lamp, pilot light	
Current source	
Semiconductor diode	A, K
Silicon controlled rectifier	A, K, G
Bipolar Junction Transistor (PNP)	E, C, B
Bipolar Junction Transistor (NPN)	E, C, B
Field Effect Transistor (FET) N channel	D, S, G
Field Effect Transistor (FET) P channel	D, S, G

TABLE 3.2 Letter Symbols

ampere	A	current	*I*
volt	V	resistance	*R*
henry	H	reactance	*X*
ohm	Ω	impedance	*Z*
farad	F	power	*P*
watt	W	energy	$\mathcal{W}$
siemen	S	charge	*Q*
kilowatt	kW	source emf	*E*
kilowatt-hour	kWh	direct current	dc
hertz	Hz	alternating	ac
coulomb	C	current	
newton	N	weber	Wb
joule	J	conductance	*G*
electromotive force	emf	inductance	*L*
frequency	f	capacitance	*C*

Figure 3.3 A transistor amplifier circuit, another example of the use of graphic and letter symbols.

complex diagram, that of a transistor amplifier circuit, is shown in Figure 3.3.

As a technician, you will be exposed to other types of diagrams. Some of these are the block diagram, pictorial diagram, and wiring diagram. The block diagram will be described in Section 3.2.

3.2 The Block Diagram

block diagram

A **block diagram** divides sections of the circuit into blocks according to function, and uses arrows to show the flow of power or signal through the circuit. This diagram is useful when the circuit is being designed or its operation is being studied. One advantage of this diagram is that the designer can first set up the general circuit according to the function that the circuit must perform. Then, the actual circuits for each block can be developed.

Let us now interpret the block diagram for a simple dc power supply shown in Figure 3.4. The path of the power through the diagram will now be traced. The alternating current enters the variable transformer at the left. The transformer is used to vary the voltage. Next, the signal passes through the rectifier where it is changed from ac to dc. After the rectifier, the signal passes through the filter circuit, which smoothes out the

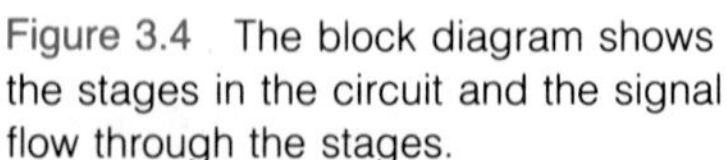

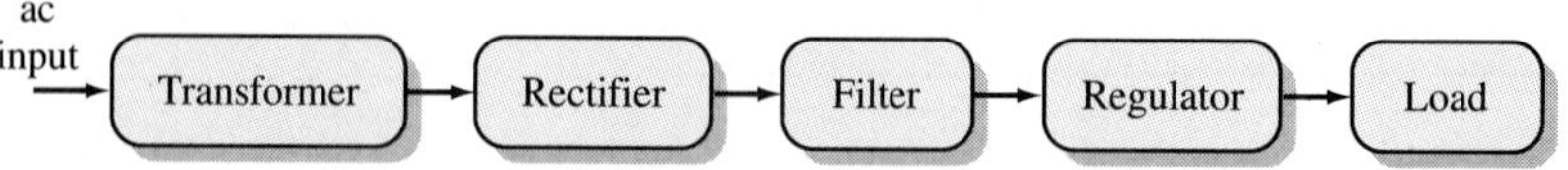

Figure 3.4 The block diagram shows the stages in the circuit and the signal flow through the stages.

fluctuations in the voltage to give a more uniform dc voltage. The last section is the regulator. This unit reduces the change in voltage due to a change in the current output. Finally, the signal reaches the load.

Notice that the block diagram does not give any information about the types of circuits in the blocks, the parts, or the connections. It shows only what each block does, and the path of the power flow through the system. The user can see the big picture without getting lost in the details of the circuit.

3.3 Polarity, Voltage Rise, and Voltage Drop

The theory that electrical current is a fluid led to the conclusion that the pressure must decrease as the current passes through a path. The point of higher pressure was thus marked with a plus (+) sign; the point of lower pressure with a minus (−) sign. This designation of a point being positive or negative is known as **polarity.**

polarity

By convention, the current leaves the positive terminal of a source. Note the (+) and (−) markings on a battery the next time that you use one. However, for a resistor or other part of a circuit, the positive terminal is the one at which the current enters. This agrees with the concept of a pressure drop in a circuit.

The polarity markings for a simple circuit are shown in Figure 3.5. Terminal B is more negative than terminal A, but more positive than terminal D.

The polarity of points in an energized circuit can be found using a voltmeter. Either an analog or digital voltmeter is connected across the two points as shown in Figure 3.6. An analog meter is one with a pointer and a scale. A digital meter is one with a number display. Each has some advantages and disadvantages. The test leads should be connected to obtain an up-scale deflection or a positive digital reading. When this is done, the positive point is the point that is connected to the lead from the positive terminal of the voltmeter. Some meters have this terminal

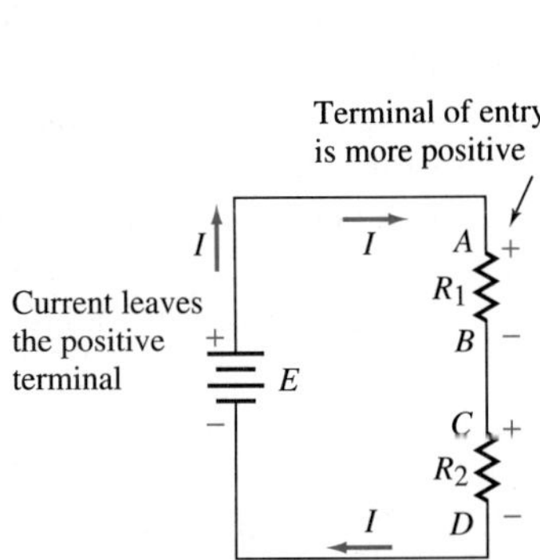

Figure 3.5 The current exits the (+) terminal of the source; the entry terminal of the resistors is more positive than the exit terminal.

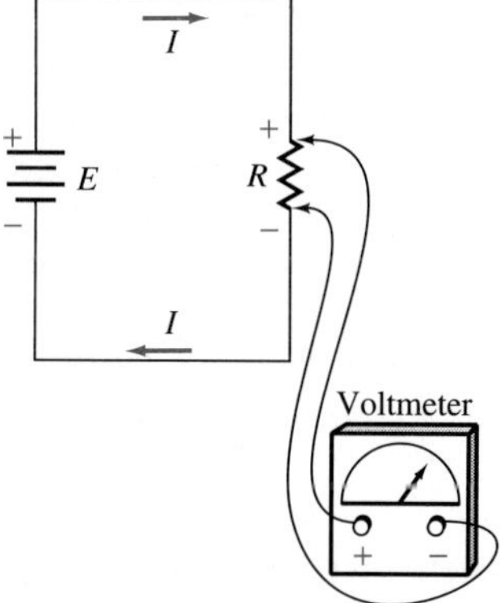

Figure 3.6 A voltmeter can be used to determine terminal polarity by connecting it to get up-scale deflection and observing which terminal is connected to the (+) and (−) terminals of the meter.

marked with a plus (+) sign. Others have a red-colored terminal or lead. Still others have the positive terminal marked as "dc volts."

voltage drop

voltage rise

The terms voltage drop and voltage rise are often used when working with electrical circuits. A **voltage drop** means that one point is less positive than another point. There is a voltage drop across the bulb in a flashlight equal to the battery voltage. A **voltage rise** means that one point is more positive than another. There is a voltage rise from the (−) terminal of the battery to the (+) terminal.

Voltage rises and drops can be determined by noting the polarity of the two points. It is always a voltage rise from the negative (−) point to the positive (+) point. It is always a voltage drop from the positive (+) point to the negative (−) point. An example will help to explain this.

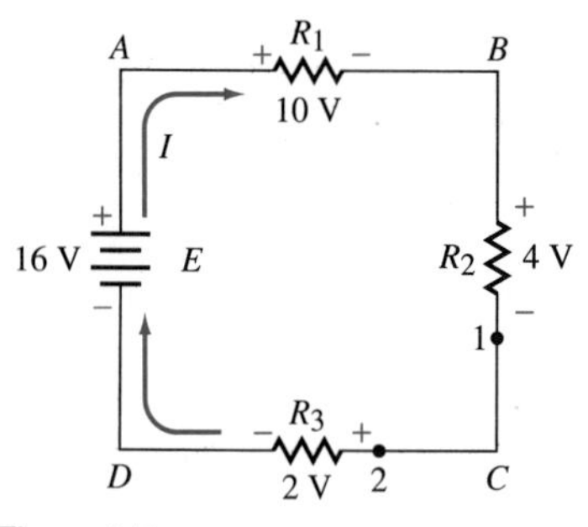

Figure 3.7

EXAMPLE 3.1

In Figure 3.7, determine whether there is a voltage rise or a voltage drop from:

a. C to B
b. C to D
c. A to B

Solution

The direction of the current gives the resistor terminal polarities as drawn in the circuit.

a. From C to B is from (−) to (+). Therefore, there is a voltage rise.
b. From C to D is from (+) to (−). Therefore, there is a voltage drop.
c. From A to B is from (+) to (−). Therefore, there is a voltage drop.

Before leaving this figure, consider Point 1 and Point 2. The markings seem to indicate that Point 2 is more positive than Point 1. Actually, this is not so because the polarities marked are not across those points. They are for the voltages across R_2 and R_3. It is common practice to consider a source voltage as a rise and the voltage across other parts of a circuit as drops. However, whether there is a rise or a drop depends on which point is taken as a reference. In Figure 3.7, there is a 10-volt rise from B to A. But, there is a 10-volt drop from A to B. This will become important later in the analysis of complex circuits.

3.4 Measuring Voltage, Current, and Resistance

voltmeter

ammeter

Voltage, or potential difference, is measured with an instrument called a **voltmeter.** Current, in amperes, is measured with an instrument called an **ammeter.** Smaller units of current are measured with

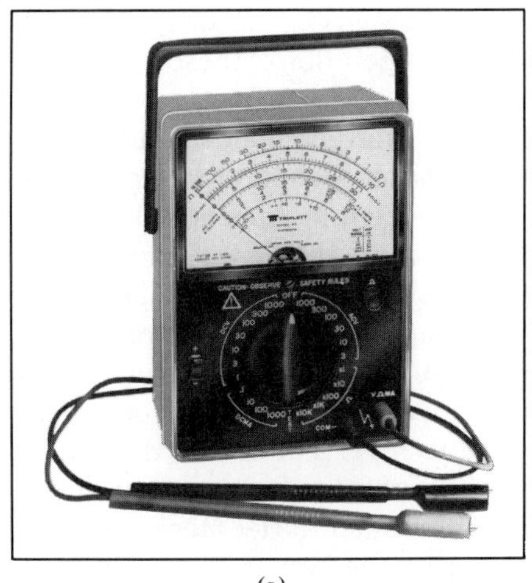

(a) (b)

Figure 3.8 Multimeters can be used to measure voltage, current, and resistance. (a) This VOM has an analog display with linear and nonlinear scales. (b) This DMM has a digital display and separate function and range selectors. (a, Photo courtesy of Triplett Corp.; b, furnished courtesy of the Simpson Electric Company)

milliammeters. Voltmeters and ammeters can be part of a multimeter as in Figure 3.8, or separate units such as those in Figures 3.9.

The multimeter in Figure 3.8(a) is a volt-ohm-milliammeter (VOM), while the one in Figure 3.8(b) is a digital multimeter (DMM). The VOM has an analog display consisting of a scale and a pointer. The DMM gets its name from the digital display, which is similar to that on an electronic calculator. The meter shown has a 3½ digit display. This means that four digits can be displayed if the extreme left digit is a "1." Otherwise, only three digits will show on the display. For instance, four digits will show for 1500, but only three digits for 500.

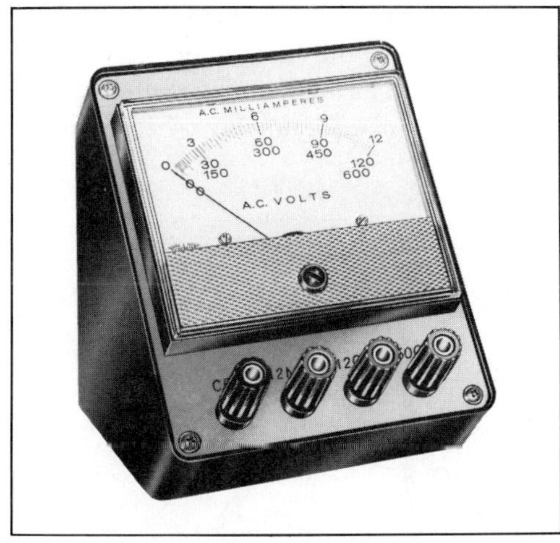

(a) (b)

Figure 3.9 Meters such as these are often used to measure current and voltage. (a) A three-range meter that measures voltage and current. (b) A three-range meter that measures current. (Photos courtesy of Triplett Corp.)

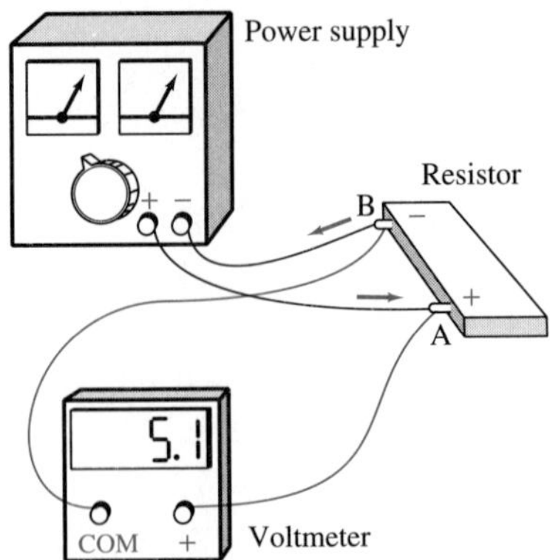

Figure 3.10 Measuring voltage across two points, *A* and *B*. The meter is connected across the points with the (+) lead connected to the more positive point (*A*).

Both meters make use of voltage-divider and current-divider circuits to reduce large voltages and currents. The operation of these circuits are explained in Chapter 8.

Making a measurement requires the correct connection, proper range, and correct reading of the value. The first condition is explained in this section. The last two are treated in detail in Appendix D.

The voltmeter should be connected across the circuit as shown in Figure 3.10, where it is parallel to the resistor. The (+) terminal of the voltmeter must be connected to the most positive end of the resistor. If not, the pointer of an analog meter will deflect down-scale, while a negative digital readout will result on a digital meter. The proper ammeter connection is shown in Figure 3.11. The ammeter must be connected so that the current in it is the current to be measured. It is in series with the circuit and the current must enter the (+) terminal of the ammeter. Connecting the ammeter across the circuit will damage the meter. Switching the leads to the terminals of either meter will put a minus sign (−) to the left of the digital display. On an analog display, the pointer will deflect to the left (as for the voltmeter) and will be damaged. When the approximate value of voltage or current is not known, starting at the highest range is advisable. The range should then be changed to obtain the most digits, or greatest power deflection.

loading effect

Both meters change the circuit conditions when connected. This change makes the measured value smaller than the actual value. This effect of the meters on the circuuit is known as the **loading effect.** It can be reduced by using high-resistance voltmeters and low-resistance ammeters.

Two important characteristics of any meter are accuracy and resolution. Accuracy specifies the possible difference between the reading on the meter and the actual value. The actual voltage for a reading of 50 volts on the 100-V range of a ±1 percent meter can be between 49 and 51 volts. For a ±5 percent meter, it can be between 45 and 55 volts. The number of volts or amperes that the percent represents will be smaller if a smaller range is used. However, that will increase the loading effect.

Resolution specifies the smallest unit that the meter can measure. A display of 2.111 A has a resolution of 1 mA. A display of 2.1 A has a resolution of 0.1 A.

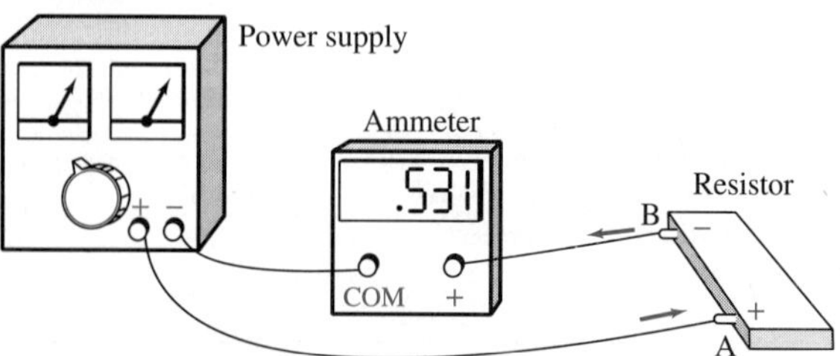

Figure 3.11 Measuring current. The meter is connected in series with the circuit so that the current enters the (+) terminal of the meter.

The ideal voltmeter choice is the meter with the highest resistance, smallest percent accuracy, and resolution to the smallest unit. The ammeter of choice is a meter with the lower resistance, smallest percent accuracy, and resolution to the smallest unit. Unfortunately, one meter might not have all three characteristics and a compromise must often be made.

Normally, the circuit must be broken to insert the ammeter. However, when measuring an ac current, a clamp-on ammeter such as the one shown in Figure 3.12 avoids breaking the circuit. The meter shown also measures voltage when connected using leads as for a normal voltmeter. Unfortunately, the meter will not measure dc current.

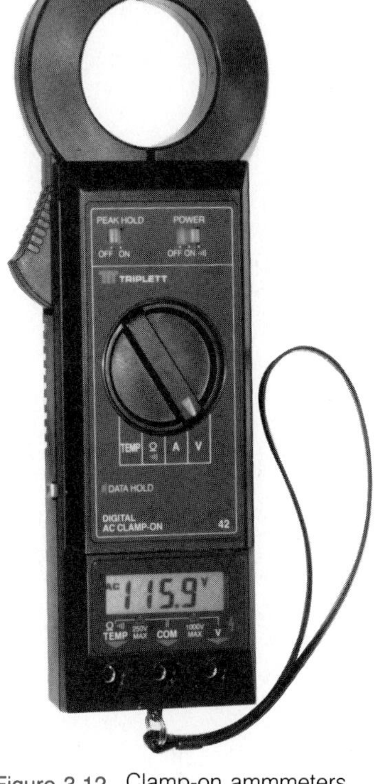

Figure 3.12 Clamp-on ammmeters such as the one shown here eleminate the need to break a line in the circuit. (Photo courtesy of Triplett Corp.)

EXAMPLE 3.2

A $\pm 5\%$ voltmeter indicates 30 volts. What is the minimum and maximum that the actual voltage can be if (a) the 40-volt range was used and (b) if the 160-volt range was used?

Solution

a. 5 percent equals 0.05.
$(0.05) \times 30\text{ V} = 1.5\text{ V}$
max. $= 30\text{ V} + 1.5\text{ V}$
max. $= 31.5\text{ V}$
min. $= 30\text{ V} - 1.5\text{ V}$
min. $= 28.5\text{ V}$

b. $(0.05) \times 160\text{ V} = 8\text{ V}$
max. $= 30\text{ V} + 8\text{ V}$
max. $= 38\text{ V}$
min. $= 30\text{ V} - 8\text{ V}$
min. $= 22\text{ V}$

The example shows that the lower range reduces the uncertainty in the measurement.

3.5 Ohm's Law

During his study of electricity, George Simon Ohm, for whom the unit of resistance was named, formed the exact relation among voltage, current, and resistance. Namely, the voltage across the resistance is equal to the resistance multiplied by the current. This relation is now known as **Ohm's Law**. In equation form, it is:

Ohm's Law

$$V = I \times R \tag{3.1}$$

where:

V is the voltage across the resistor, in volts
R is the resistance of the resistor, in ohms
and I is the current in the resistor, in amperes.

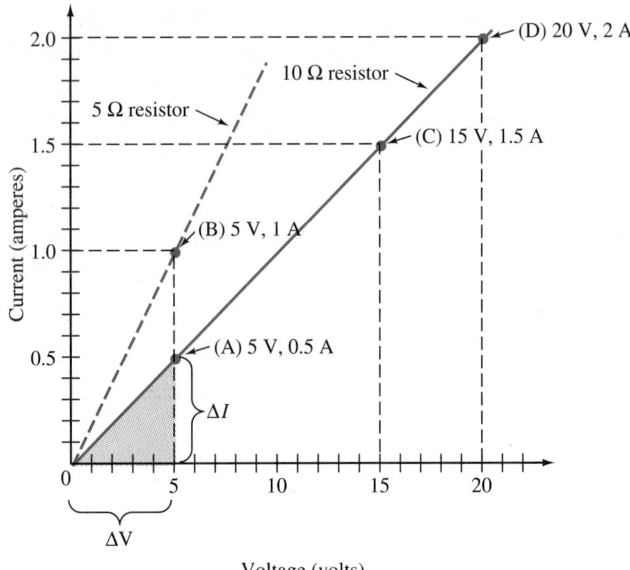

Figure 3.13 The relation of Ohm's Law ($V = RI$) is presented by the V-I curves for two values of resistance.

Equation 3.1 shows that the current is (a) directly proportional to the voltage and (b) inversely proportional to the resistance. These proportionalities can be seen in the curves of Figure 3.13. Thus, tripling the voltage, from 5 volts to 15 volts, triples the current in the 10-ohm resistor. Doubling the resistance, from 5 ohms to 10 ohms, halves the current from an initial value of 1 A to 0.5 A.

Resistance can be measured with an ohmmeter or a VOM and DMM such as those shown in Figure 3.8. Since the operation and controls vary with models and manufacturers, one should consult the meter manual for specific operating instructions. However, there are some general guidelines to follow when measuring resistance. They are:

1. Do not measure resistance of a resistor that has a voltage applied across its terminals.
2. If the resistor is in a circuit, at least one end should be disconnected from the circuit.
3. Do not hold the bare probes with your fingers. This will put your body's resistance in parallel with the resistance being measured.
4. Select a range that will give the maximum divisions per ohm for the analog display and the maximum digits for the digital display.
5. Use the zero adjust knob to set the pointer at zero with the leads short circuited for the VOM.

A study of the curves in Figure 3.13 shows that the resistance is the reciprocal of the slope of the curves. In Figure 3.13,

$$\text{Slope} = \frac{\Delta I}{\Delta V} = \frac{I_A}{V_A}$$

but, by Ohm's Law,

$$\frac{I_A}{V_A} = \frac{1}{R} \quad \text{or} \quad R = \frac{V_A}{I_A}$$

Thus, the resistance is the reciprocal of the slope.

A large resistance will have a steep curve while a smaller resistance will have a more horizontal curve.

Ohm's Law also holds true for a nonlinear resistance. However, the resistance at a given voltage will now be equal to the reciprocal of the slope of a line drawn from (0,0) through the point on the curve.

Ohm's Law can be used to find any one of the three quantities, V, I, or R, when the other two are known. Many more complex circuit relations are derived from Ohm's Law. Therefore, understanding its application is important. Some ways in which Ohm's Law can be used are shown in the following examples.

EXAMPLE 3.3

Damage to the insulation on the wire of a resistor caused some of the turns to touch each other. What will be the effect on the current in the resistor?

Solution

The touching turns will reduce the resistance. Since

$$I = V/R$$

the current will become larger.

EXAMPLE 3.4

The current in a resistor is 2 amperes when connected to a 24-volt source. How many ohms is the resistance?

Solution

Noting that the voltage across the resistor is E, we have

$$E = I \times R$$

Substituting, we get

$$24 \text{ V} = 2 \text{ A} \times R$$

Solving for resistance,

$$R = 24 \text{ V}/2 \text{ A}$$
$$R = 12 \ \Omega$$

EXAMPLE 3.5

The current in each resistor of the circuit in Figure 3.14 is 3 amperes. Use Ohm's Law to find the voltage between points 1 and 2.

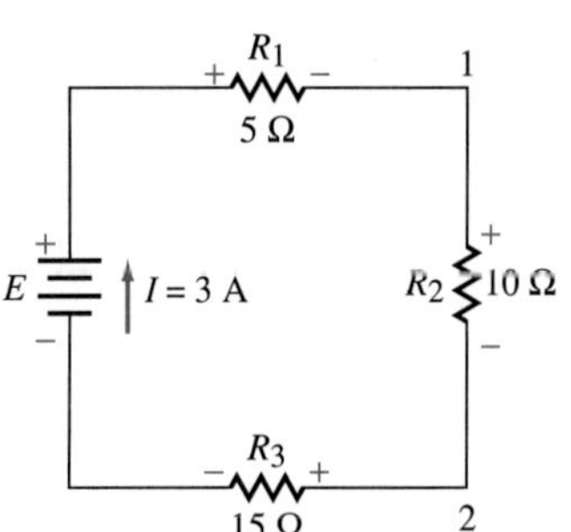

Figure 3.14

Solution

The circuit diagram shows that the resistance between points 1 and 2 is 10 ohms. Hence,

$$V = I \times R$$
$$= 3\ \text{A} \times 10\ \Omega$$
$$V = 30\ \text{V}$$

3.6 Power in the Electric Circuit

power

Power is the rate at which work is done or energy is converted. Work means the movement of a force through a distance. An example of work is the lifting of a weight. The force is equal to the weight that is moved a distance when lifted.

In electric circuits, a force is needed to overcome the attraction of the electron in its atom. Since current is the net movement of electrons, work is done to maintain the current.

watt

The unit of power in SI units is the **watt** (W) named after James Watt (1736–1819), a Scottish engineer who invented the steam engine. One watt of power is the equivalent of doing work or converting energy at the rate of 1 joule per second. Recall that 1 joule represents moving a force of 1 newton through a distance of 1 meter. In more familiar terms, 1 watt of power is the equivalent of lifting approximately 44 pounds of weight a distance of 1 foot in a time of 1 minute.

horsepower

A more common unit of power is the **horsepower** (hp). This is the unit in the British Gravitational System of Units. It is more commonly used to express the mechanical power of a motor. One horsepower is the equivalent of 746 watts of power.

The movement of the electrons through a resistance causes a temperature rise as the electrical energy is converted to heat. This heat is "given off" or "dissipated." Therefore, the term "dissipated" is often used when speaking of power in a resistor. Recall the term "power rating" was introduced in Chapter 2, Section 2.8. A resistor can be damaged if it cannot give off the heat. Therefore, its power rating should be greater than the power that must be dissipated.

The power for some common tools and appliances is given in Table 3.3. Why not examine the nameplates on some of your appliances to see their power rating?

Since power is the rate of doing work, it seems logical that higher currents or potential differences represent more power. Combining Equation 2.2, $Q = I t$, and Equation 2.3, $V = \mathcal{W}/Q$, proves this to be true. The combination gives $\mathcal{W}/t = V I$. But $\mathcal{W}/t$ is the rate of doing work, so

$$P = V \times I \qquad \textbf{(3.2)}$$

where:

P is the power, in watts
V is the voltage, in volts
and I is the current, in amperes.

TABLE 3.3

Item	Power (W)
Four-slice toaster	1500
Ceiling light bulb	100
Hand hair dryer	1250–1500
Digital clock radio	10
Calculator	0.15
Night light	7.5
Quartz heater	1500

From Equation 3.2, you can see that the power depends on both the current and the voltage in the circuit.

EXAMPLE 3.6

The current in the circuit in Figure 3.15 is 5 amperes. Find (a) the power dissipated by resistor R_1 and (b) the power supplied by the source.

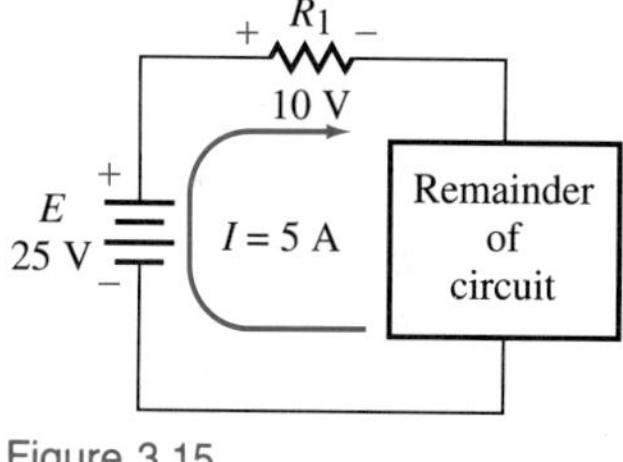

Figure 3.15

Solution

a. The voltage across resistor R_1 and the current in R_1 are known. Hence,

$$\begin{aligned} P &= V \times I \\ &= 10 \text{ V} \times 5 \text{ A} \\ P &= 50 \text{ W} \end{aligned}$$

b. Now, the source voltage and the source current are known. Hence,

$$\begin{aligned} P &= E \times I \\ &= 25 \text{ V} \times 5 \text{ A} \\ P &= 125 \text{ W} \end{aligned}$$

The source supplies more power than is dissipated in the resistor. The difference, 75 watts, is dissipated in the network to the right of the resistor.

EXAMPLE 3.7

A resistor dissipates 10 watts when the current is 0.5 amperes. What is the voltage across the resistor?

Solution

$$P = V \times I$$

Solving this equation for V yields

$$\begin{aligned} V &= P/I \\ &= 10\ \text{W}/0.5\ \text{A} \\ V &= 20\ \text{V} \end{aligned}$$

One limitation in the use of Equation 3.2 is the need to break the circuit to insert an ammeter. How does one do that with a printed circuit, or a part that is enclosed? Fortunately, Ohm's Law can be used to derive another relation. Substituting $I = V/R$ into Equation 3.2 gives

$$\begin{aligned} P &= V \times I = (V)(V/R) \\ P &= V^2/R \end{aligned} \quad \textbf{(3.3)}$$

where:

P is the power, in watts
V is the voltage across the resistor, in volts
and R is the resistance, in ohms.

Finally, a third power relation can be obtained by substituting $V = I \times R$ into Equation 3.2. This gives,

$$\begin{aligned} P &= V \times I \\ &= (I \times R)(I) \\ P &= I^2R \end{aligned} \quad \textbf{(3.4)}$$

where:

P is the power, in watts
I is the current, in amperes
and R is the resistance, in ohms.

EXAMPLE 3.8

A 12-volt lighter in an automobile has a resistance of 4 ohms when operating.

a. How much power is dissipated in the lighter?
b. What is the current in the lighter?

Solution

a. Since the voltage and the resistance are given, Equation 3.3 will be used.

$$\begin{aligned} P &= V^2/R \\ &= (12\ \text{V})^2/4\ \Omega \\ P &= 36\ \text{W} \end{aligned}$$

b. Here, V, I, and R are known, so either Equation 3.2 or 3.4 can be used. Using Equation 3.2 gives

$$P = V \times I$$

Solving this equation for current yields

$$I = P/V$$

Substituting known values into this equation gives

$$I = 36\ \text{W}/12\ \text{V}$$
$$I = 3\ \text{A}$$

Checking this answer with Ohm's Law gives

$$V = R \times I \qquad \text{or} \qquad I = V/R$$

Substituting known values into this equation yields

$$I = 12\ \text{V}/4\ \Omega$$
$$I = 3\ \text{A}$$

A check using Equation 3.4 gives

$$P = I^2 R$$
$$= (3\ \text{A})^2\ (4\ \Omega)$$
$$P = 12\ \text{W}$$

EXAMPLE 3.9

A 40-ohm resistor dissipates 10 watts. Find:

a. The voltage across the resistor.
b. The current in the resistor.

Solution

a. Since the power and resistance are given, Equation 3.3 will be used.

$$P = V^2/R \qquad \text{or} \qquad V = \sqrt{P \times R}$$

Hence,

$$V = \sqrt{(10\ \text{W})(40\ \Omega)}$$
$$V = 20\ \text{V}$$

b. V and I are now known, so Equation 3.2 will be used.

$$P = V \times I \qquad \text{or} \qquad I = P/V$$

Substituting known values yields

$$I = 10\ \text{W}/20\ \text{V}$$
$$I = 0.5\ \text{A}$$

3.7 The Wattmeter

Power in a dc or low-frequency ac circuit is measured with a **wattmeter.** Power at radio frequencies is measured by devices that become hot from the radio frequency power. One of these devices is the thermistor, discussed in Chapter 2.

wattmeter

Figure 3.16 A 150-watt dynamometer type wattmeter. Wattmeters have voltage terminals and current terminals. (Photo courtesy of Triplett Corp.)

An example of a wattmeter is shown in Figure 3.16. This one has an analog display, but digital meters are also available. Note that the meter has both current and voltage terminals. Also, one terminal is a different color than the other. Some meters use a ± marking instead of colors.

The proper connection of the wattmeter is shown in Figure 3.17. For an up-scale deflection, the current must enter either both ± terminals or both unmarked terminals. The **short shunt** connection in Figure 3.17(a)

short shunt

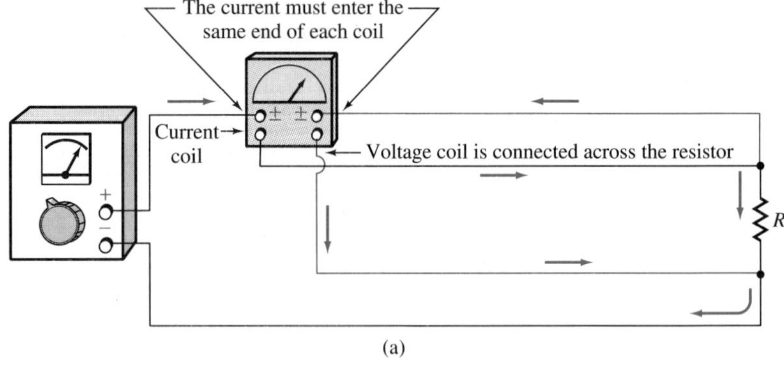

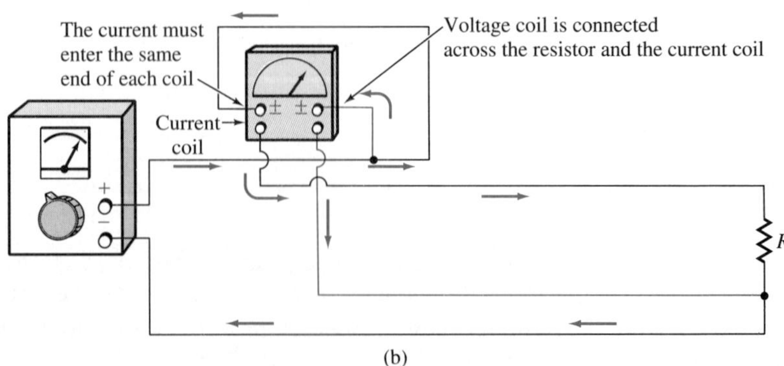

Figure 3.17 To measure power, the wattmeter must be connected so the current enters or leaves the ± terminals of both coils. (a) In the short-shunt connection, the voltage terminals are connected across the resistor. (b) In the long-shunt connection, the voltage terminals are connected across the resistor plus the meter current coil.

has the voltage coil connected directly across the load. The indicated power will include the power in the voltage coil. The **long shunt** connection in Figure 3.17(b) has the voltage coil connected across the load plus the current coil. For that connection, the indicated power will include the power in the current coil.

long shunt

Each connection in Figure 3.17 gives some error in the measurement. It can be shown that the ratio of the actual power to the measured power in the long-shunt connection is given by

$$\frac{P_{\text{act}}}{P_{\text{meas}}} = \frac{R_L}{R_A + R_L} \tag{3.5}$$

For the short-shunt connection, it is

$$\frac{P_{\text{act}}}{P_{\text{meas}}} = \frac{R_V}{R_V + R_L} \tag{3.6}$$

where:

P_{act} is the power in R_L, in watts
P_{meas} is the indication on the meter, in watts
R_L is the resistance in which the power is being measured, in ohms
R_A is the resistance of the current coil, in ohms
and R_V is the resistance of the voltage coil, in ohms.

The characteristics for the connections are summarized in Table 3.4.

As a general rule, the long-shunt connection gives acceptable results for most loads. This is because of the small resistance of the current coil compared to the load resistance. For instance, a 100-W, 125-V bulb has a resistance of about 156 ohms. The current coil resistance of 0.01 ohms is quite small compared to that.

TABLE 3.4 Wattmeter Connection Characteristics

Connection	Voltage Coil	Current Coil	Power Error	Terminal Connection	R_L for Smallest Error
Long shunt	Measures current coil plus load voltage	Measures load current	Power in current coil	± to ± or unmarked to unmarked	When $R_L >>> R_A$
Short shunt	Measures load voltage	Measures voltage coil plus load current	Power in voltage coil	± to unmarked	When $R_V >>> R_L$

EXAMPLE 3.10

The resistance of a wattmeter voltage coil is 11,000 ohms and that of the current coil is 0.01 ohm. What is the ratio of P_{act}/P_{meas} when measuring power in a 50,000-ohm resistor (a) for a long shunt and (b) for a short shunt?

Solution

a. Long shunt:

$$\frac{P_{act}}{P_{meas}} = \frac{R_L}{R_A + R_L} = \frac{50{,}000\ \Omega}{0.01\ \Omega + 50{,}000\ \Omega}$$

$$\frac{P_{act}}{P_{meas}} = 1$$

b. Short shunt:

$$\frac{P_{act}}{P_{meas}} = \frac{R_V}{R_V + R_L} = \frac{11{,}000\ \Omega}{11{,}000\ \Omega + 50{,}000\ \Omega}$$

$$\frac{P_{act}}{P_{meas}} = 0.18$$

For this load, the long-shunt connection should be used.

EXAMPLE 3.11

Repeat Example 3.10 for a 10-ohm load.

Solution

a. Long shunt:

$$\frac{P_{act}}{P_{meas}} = \frac{R_L}{R_A + R_L} = \frac{10\ \Omega}{0.01\ \Omega + 10\ \Omega}$$

$$\frac{P_{act}}{P_{meas}} = 0.999$$

b. Short shunt:

$$\frac{P_{act}}{P_{meas}} = \frac{R_V}{R_V + R_L} = \frac{11{,}000\ \Omega}{11{,}000\ \Omega + 10\ \Omega}$$

$$\frac{P_{act}}{P_{meas}} = 0.999$$

Here, the results are the same to three digits. A greater difference would result if either R_L were smaller or R_A were larger.

Some wattmeters have the coil connections made internally. They have two sets of terminals, one set marked "Line" and the other marked "Load."

The wattmeter of Figure 3.16 can also measure ac power because it has the dynamometer movement of Figure 3.18. The fields of the coils interact to turn the shaft. When ac power is measured, the current changes in both coils at the same time. As a result, the direction of the turning does not change, so the pointer still deflects up-scale.

A wattmeter can be damaged if either current, voltage, or power is above its rating. Unfortunately, the first two conditions can happen while the power is still below the rated value, and the reading is below the maximum. Thus, one is not aware of possible damage until it is too late. A good practice is to be aware of the current and voltage values when measuring power.

Wattmeters are also made in multicurrent, multivoltage, and multipower ranges. The basic connection remains the same, but one must become familiar with the range and scale selection procedure for the meter.

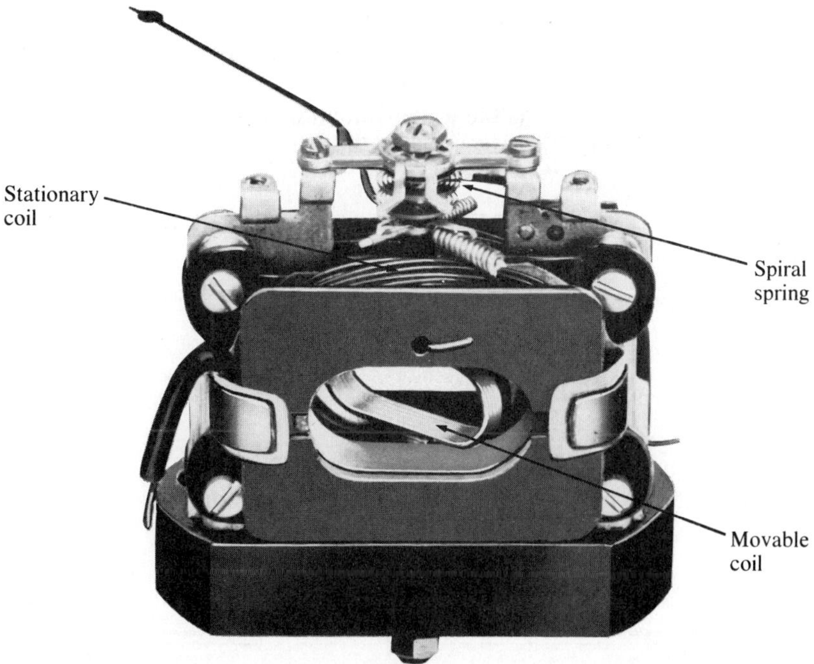

Figure 3.18 The dynamometer movement will measure both ac and dc power. (Photo courtesy of the Simpson Electric Co.)

3.8 Energy in the Electric Circuit

energy

Energy is the capacity to do work. The spring in Figure 3.19 has this capacity because it is stretched. In this condition, it has potential energy. This is the potential to do work. When the string is cut, the spring will come together and pull the block to the left. The potential energy is now expended as work in moving the block.

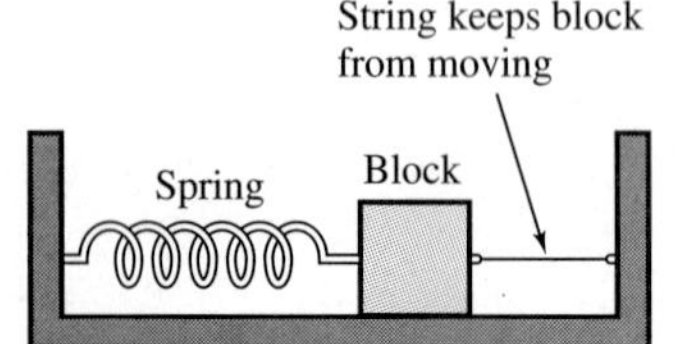

Figure 3.19 The energy in the stretched spring is an example of stored energy. Cutting the string will release the spring, causing the block to move.

The source of emf in an electric circuit provides the energy that is needed to move the electrons. In a resistive circuit, all of the energy provided by the source is converted to heat energy. This conversion is analogous to the conversion of heat energy by friction when two objects are rubbed together.

The energy in an electric circuit is easily calculated when the power is known. This is because power and energy are related in the same way that power and work are related. That is, energy is the product of power and time. In equation form,

$$\mathcal{W} = P \times t \tag{3.7}$$

where:

$\mathcal{W}$ is the energy, in joules
P is the power, in watts
and t is the time, in seconds.

joule

kilowatt-hour

Although the **joule** is the unit for energy in the International System of Units, electric utilities use a more practical and larger unit, the kilowatt-hour (kWh). One **kilowatt-hour** of energy is the equivalent of 1000 watts of power maintained for a period of 1 hour. This can be any equivalent combination of power and time whose product is 1 kWh. This unit is 3,600,000 times as large as the joule. Equation 3.7 will be in kWh if the time is in hours and the power is in kilowatts.

EXAMPLE 3.12

How many joules of energy are needed to operate a 60-watt lamp for 10 minutes?

Solution

$$\mathcal{W} = P \times t$$

where the time must be in seconds.

Substituting known quantities into the energy equation yields

$$\mathcal{W} = (60 \text{ watts})(10 \text{ minutes})(60 \text{ sec/min})$$
$$\mathcal{W} = 36{,}000 \text{ J}$$

EXAMPLE 3.13

How many kilowatt-hours of energy are used by a 100-watt lamp operating for one day?

Solution

$$t = (1 \text{ day}) \times (24 \text{ hours/day})$$
$$t = 24 \text{ hours}$$

and

$$P = (100 \text{ W})(1 \text{ kW}/1000 \text{ W})$$
$$= 0.1 \text{ kW}$$

Hence,

$$\mathcal{W} = P \times t$$
$$= (0.1 \text{ kW}) \times (24 \text{ h})$$
$$\mathcal{W} = 2.4 \text{ kWh}$$

In the preceding two examples, the power was constant for the time that the energy was calculated. This is not usually the case in actual practice. For instance, we might operate a 100-watt lamp for a 10-minute interval, and a 500-watt curling iron for another 10 minutes. For this type of operation, Equation 3.7 must be changed to the form in Equation 3.8.

$$\mathcal{W}_t = P_1t_1 + P_2t_2 + \cdots + P_nt_n \quad \textbf{(3.8)}$$

where:

$\mathcal{W}_t$ is the energy used in the total time.
$P_1t_1, P_2t_2, \ldots, P_nt_n$ is the energy used in each interval in which the power is constant.

Equation 3.8 will give the energy in kilowatt-hours if the power is in kilowatt-hours and the time is in hours.

EXAMPLE 3.14

A customer operates an electric load as follows:
1000 watts for 4 hours,
3000 watts for 2 hours,
500 watts for 1 hour,
0 watts for 1 hour.
How much energy is used in the 8-hour period?

Solution

The time intervals are 4 hours, 2 hours, 1 hour, and 1 hour.

$$\mathcal{W}_t = P_1t_1 + P_2t_2 + P_3t_3 + P_4t_4$$
$$\mathcal{W}_1 = P_1t_1 = (1000 \text{ watts})(4 \text{ hours}) = 4000 \text{ Wh}$$
$$\mathcal{W}_2 = P_2t_2 = (3000 \text{ watts})(2 \text{ hours}) = 6000 \text{ Wh}$$
$$\mathcal{W}_3 = P_3t_3 = (500 \text{ watts})(1 \text{ hour}) = 500 \text{ Wh}$$
$$\mathcal{W}_4 = P_4t_4 = (0 \text{ watts})(1 \text{ hour}) = 0 \text{ Wh}$$

Summing the individual energy equations yields

$$\mathcal{W}_t = 4000 \text{ Wh} + 6000 \text{ Wh} + 500 \text{ Wh} + 0 \text{ Wh}$$
$$\mathcal{W}_t = 10{,}500 \text{ Wh}$$

Converting this quantity to kilowatt-hours can be done as follows,

$$\mathcal{W}_t = (10{,}500 \text{ Wh})(1 \text{ kWh}/1000 \text{ Wh})$$
$$\mathcal{W}_t = 10.5 \text{ kWh}$$

Obviously, this method becomes very lengthy for many time intervals. Fortunately, electric utility companies do not have to keep records of the energy used in each time interval; instead they measure it with a watt-hour meter. Such a meter is shown in Figure 3.20. This meter has a rotating aluminum disc whose speed of rotation depends on the power. The greater the power, the faster the disc rotates and more turns are made in the time interval. The meter has calibrated dials that convert the number of turns into the amount of energy that is used in the time interval. Electric power companies charge their customers for energy by the kilowatt-hour. For example, a company's rate might be \$0.05 per kilowatt-hour. The total cost for energy use can then be found by multiplying the rate by the number of kilowatt-hours that were used.

EXAMPLE 3.15

What is the cost of operating the load in Example 3.12 when the rate is \$0.06 per kilowatt-hour?

Solution

$$\text{cost} = (\text{rate})(\text{kWh})$$
$$= (\$0.06/\text{kWh})(10.5 \text{ kWh})$$
$$\text{cost} = \$0.63$$

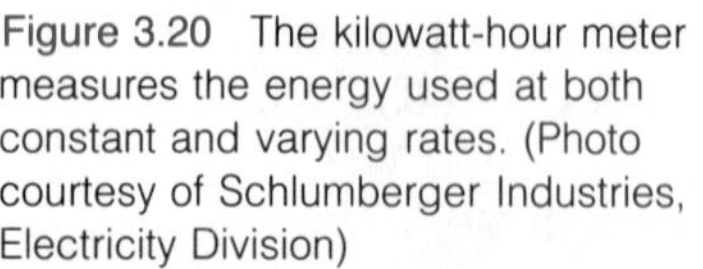

Figure 3.20 The kilowatt-hour meter measures the energy used at both constant and varying rates. (Photo courtesy of Schlumberger Industries, Electricity Division)

The rate is usually set so that it becomes smaller as the energy use increases. Example 3.16 illustrates that type of situation.

EXAMPLE 3.16

A customer uses 150 kWh of energy. What is the cost, if the rate is

$1.00 for the first 15 kWh
$0.06 for the next 25 kWh
$0.04 for the next 60 kWh
and $0.03 for each kWh above 100 kWh?

Solution

The 150 kWh gives

15 kWh @ $1.00 each	= $15.00
25 kWh @ $0.06 each	= $ 1.50
60 kWh @ $0.04 each	= $ 2.40
50 kWh @ $0.03 each	= $ 1.50
COST	= $20.40

3.9 Efficiency

efficiency

Efficiency is a measure of how much useful energy is obtained from the energy put into a device.

Efficiency is more commonly given in terms of the power rather than the energy and is usually expressed as percent. The symbol for efficiency is η.

In equation form, efficiency as a percent is

$$\eta = \frac{(P_{out})(100)}{(P_{in})} \tag{3.9}$$

where:

η is the percent efficiency
P_{out} is the useful power output, in any power unit
P_{in} is the power input, in the same unit as P_{out}

If you do not multiply by 100, the efficiency is expressed in decimal form. This decimal form is useful in calculating the total efficiency of several devices in cascade. This will be discussed later in this section.

The efficiency relation will now be applied to an electric motor in Example 3.17.

EXAMPLE 3.17

A dc motor has a power output of 3/4 horsepower.
The current is 6 amperes when connected to a 120-volt source. Find:
a. the power input, in watts

b. the power output, in watts
c. the percent efficiency

Solution

a. $P_{in} = VI = (120\text{ V})(6\text{ A})$
$P_{in} = 720\text{ W}$
b. $P_{out} = (3/4\text{ hp})(746\text{ W/hp})$
$P_{out} = 559.5\text{ W}$
c. $\eta = (P_{out}/P_{in})(100\%)$
$= (559.5\text{ W}/720\text{ W})(100\%)$
$\eta = 77.7\%$

Only 77.7 percent of the input power is available as useful power out. What happened to the difference? Some was converted to heat in the resistance of the windings. Some was used to overcome the friction in the parts of the motor. The remainder was used to overcome other losses.

In any device, the power out equals the power in minus the losses. Using this relation in Equation 3.9 gives two more forms of the efficiency equations. They are:

$$\eta = \left(\frac{P_{in} - P_{lost}}{P_{in}}\right)100\% \quad \textbf{(3.10)}$$

$$\eta = \left(\frac{P_{out}}{P_{out} + P_{lost}}\right)100\% \quad \textbf{(3.11)}$$

where:

η, P_{in}, and P_{out} have the same meaning as in Equation 6.8,

and P_{lost} is the power lost in the unit, in the same units as P_{in} and P_{out}.

When using the loss method, a device does not have to be operated at rated conditions. This is useful because sometimes that is not possible or test equipment is not available. However, the losses can usually be determined from tests at below-rated conditions.

EXAMPLE 3.18

A 50-horsepower motor has 5000 watts of losses. Find the percent efficiency of the motor.

Solution

Since the losses and the output power are given, Equation 6.10 will be used.

$$P_{out} = (50\text{ hp})\left(746\,\frac{\text{W}}{\text{hp}}\right) = 37{,}800\text{ W}$$

$$\eta = \frac{(37{,}800\text{ W})(100\%)}{(37{,}800\text{ W} + 5000\text{ W})}$$

$$\eta = 88.3\%$$

EXAMPLE 3.19

A transformer has 3000 watts of losses in it when the input is 36,000 watts. Find the percent efficiency of the transformer.

Solution

Since the losses and the input are given, Equation 3.10 will be used.

$$\eta = \frac{(P_{\text{in}} - P_{\text{lost}})(100\%)}{P_{\text{in}}}$$

$$\eta = \frac{(36{,}000 \text{ W} - 3000 \text{ W})(100\%)}{36{,}000 \text{ W}}$$

$$\eta = 91.6\%$$

Many systems are made up of several devices or sections connected in cascade as in Figure 3.21. The dc power supply provides the dc voltage for the motor. Some power is lost in each of the devices and the system has an overall percent efficiency. The overall percent efficiency can be calculated using the percent efficiency for each device to obtain the power input to the next device until the final output is reached. A simpler method is to use Equation 3.12. It eliminates the need to calculate the input and output power for each section.

$$\eta_t = (\eta_1)(\eta_2 \cdots (\eta_n)(100\%) \quad \textbf{(3.12)}$$

where:

η_t is the overall percent efficiency
η_1, η_2, and so forth are the efficiencies (in decimals) of each section

$$\%\eta_t = (\eta_1 \times \eta_2 \times \cdots \times \eta_n) \times 100$$

Figure 3.21 The overall efficency of a system can be found by taking the product of the efficiencies of the parts.

EXAMPLE 3.20

A transformer, power supply, and a 1/2-horsepower motor are connected together. The percent efficiencies are as follows: transformer 80%; power supply 90%; motor 80%.

a. Find the percent efficiency of the system.
b. Find the input power to the transformer.

Solution

a. The efficiencies in decimal form are 0.8, 0.9, and 0.8

$$\eta_t = (0.8)(0.9)(0.8)(100\%)$$
$$\eta_t = 57.6\%$$

b. Since η and the power output are known, Equation 6.8 can be used to find the power input.

$$\eta_t = \frac{(P_{out})(100\%)}{P_{in}}, \quad \text{so} \quad P_{in} = \frac{(P_{out})(100\%)}{\eta_t}$$

$$P_{out} = (0.5 \text{ hp})(746 \text{ W/hp}) = 378 \text{ W}$$

$$P_{in} = \frac{(378 \text{ W})(100\%)}{57.6\%} = 656.25 \text{ W}$$

$$P_{in} = 656.25 \text{ W}$$

Efficiency lets us select the devices that either give the most power output for a given input, or the least power input for a given output. Another way in which efficiency is used is in estimating the operating cost of motors, appliances, and so forth. Society has become more efficiency conscious in the past decade. For instance, appliances now have efficiency numbers (EER) and automobile shapes are designed to give the best fuel economy.

3.10 Open and Short Circuits

open circuit

An **open circuit** is a condition where the circuit path is broken. The circuit resistance is then infinite ohms. Current in the circuit must be 0 amperes.

Some open circuits are caused by the breakdown of a part of a circuit, as when the filament of a bulb burns out. Also, too much current in a resistor can cause it to burn open from the heat. Or, a cold solder joint can result in an open circuit. Open circuits caused by faults generally do not damage the circuit. This is because they disconnect the parts from the circuit. However, if the open-circuit voltage is high, it can present a dangerous condition. Also, the open circuit can prevent the circuit from serving the normal function.

Some open circuits are caused by devices designed to open the circuit. Three such devices are the switch, the fuse, and the circuit breaker. A **switch** is a device designed to connect and disconnect parts of a circuit. A common application is the switch used to turn the lights in a room off and on. A **fuse** is a nonreusable device designed to open a circuit when the current exceeds some value. A high current melts a metal element in the fuse and opens the circuit. A **circuit breaker** is a reusable device that also opens the circuit at high currents. Some circuit breakers are activated by temperature rise. Others are activated by an electromagnet. Fuses and circuit breakers must be selected to open the circuit in the desired time.

switch

fuse

circuit breaker

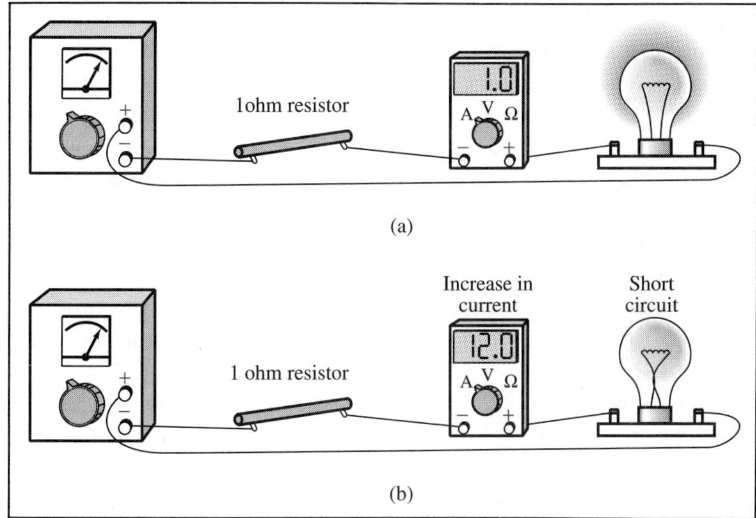

Figure 3.22 A short circuit in part of a circuit can result in a high current. (a) With no short circuit, the current is 1 ampere. (b) With a short circuit in the lamp, the current increases to 12 amperes.

For instance, meter protection requires a fast-acting fuse. On the other hand, a slow-acting fuse should be used in a motor circuit. If not, the starting surge current will cause the fuse to open. The choice of a fuse or a circuit breaker depends on factors such as the amount of current, space, weight, accessibility for changing, response time, and cost.

short circuit

An ideal **short circuit** is a condition where zero resistance is connected across two points of a circuit. The circuit resistance is then 0 ohms. The current in the circuit will be some high value. Some ways by which short circuits can be caused are: solder contacting two parts of a circuit, positioning a part so that it touches another, and damage to the wire insulation from heat. House wiring often becomes short-circuited when the insulation is damaged. Although an ideal short circuit has zero resistance, a practical short circuit can have some small resistance.

The high current in short circuits caused by faults can damage a circuit. Consider the circuit that is shown in Figure 3.22. When the lamp is short-circuited, the source current will be large. This would damage the source if some device such as the fuse is not used to open the circuit.

Short circuits can also serve some useful purpose, for example, when an ammeter is short-circuited during the starting of a motor. This protects the motor from the high surge current.

SUMMARY

1. The connections of a circuit are shown with a circuit diagram. Graphic and letter symbols are used to represent parts and electrical quantities.
2. The block diagram is used to show the functions of sections of a circuit and the signal flow through the circuit.

3. The terminal of a circuit where the current enters is said to be more positive than that where the current exits.
4. There is a voltage drop when the second point is more negative than the first. There is a voltage rise when the second point is more positive than the first.
5. Voltage is measured with a voltmeter. It is connected across the circuit and should have a high resistance.
6. Current is measured with an ammeter. It is connected in series with the circuit and should have a low resistance.
7. Resistance can be measured with an ohmmeter or a multimeter.
8. Resolution and percent accuracy are two characteristics of the meters.
9. Current, voltage, and resistance are related by Ohm's Law. The voltage is equal to the product of the current and resistance.
10. Power is the rate of doing work and is equal to the product of voltage and current. Two other relations are $P = I^2R$ and $P = V^2/R$. The unit of power is the watt.
11. The wattmeter is an instrument that measures power. It can be connected in long shunt or short shunt.
12. Energy is the product of power and time. The joule is the unit in SI. The more common unit is the kilowatt-hour.
13. All of the power put in a device does not end up as useful power. Efficiency is the ratio of useful power output to total power input.
14. Zero resistance between two points represents a short circuit. The high currents that result from a short circuit might cause damage.
15. A break in a circuit represents an open circuit. It will have an infinite resistance. Open circuits generally do not cause damage to circuit elements but can prevent a circuit from serving its function.

EQUATIONS

3.1	$V = IR$	Ohm's Law	99
3.2	$P = IV$	Relation of power to voltage and current	102
3.3	$P = \dfrac{V^2}{R}$	Relation of power to voltage and resistance	104
3.4	$P = I^2R$	Relation of power to current and resistance	104
3.5	$\dfrac{P_{\text{act}}}{P_{\text{meas}}} = \dfrac{R_L}{R_A + R_L}$	The ratio of power for the long-shunt connection of the wattmeter	107
3.6	$\dfrac{P_{\text{act}}}{P_{\text{meas}}} = \dfrac{R_V}{R_V + R_L}$	The ratio of power for the short-shunt connection of the wattmeter	107
3.7	$\mathcal{W} = Pt$	Energy equation	110
3.8	$\mathcal{W}_t = P_1t_1 + P_2t_2 + \cdots + P_nt_n$	Energy equation for a varying energy use	111

3.9	$\eta = \left(\dfrac{P_{out}}{P_{in}}\right) \times 100\%$	Efficiency equation	113
3.10	$\eta = \dfrac{(P_{in} - P_{lost})}{P_{lost}} \times 100\%$	Efficiency in terms of input power and losses	114
3.11	$\eta = \left(\dfrac{P_{out}}{P_{out} + P_{lost}}\right) \times 100\%$	Efficiency in terms of output power and losses	114
3.12	$\eta_T = (\eta_1)(\eta_2) \cdots (\eta_n) \times 100\%$	Efficiency equation for a system	115

QUESTIONS

1. What is the difference between the function of a circuit diagram and a block diagram?
2. What does a graphic symbol represent?
3. What type of symbol is used to represent a quantity such as current?
4. When are subscripts used with a letter symbol?
5. Which terminal of a resistor is more negative if the current is from terminal A to terminal B?
6. In Question 5, is it a voltage rise, or a voltage drop, from A to B?
7. How can a voltmeter be used to determine terminal polarity?
8. Which meter should be connected across the circuit: the voltmeter or the ammeter?
9. Which meter should be connected in series with the circuit, the voltmeter or the ammeter?
10. All other things being equal, which voltmeter will give a measured voltage nearest to the actual voltage without the meter in the circuit: one that has a 2-megohm resistance or one that has a 6-megohm resistance?
11. Which meter gives the smaller uncertainty, a $\pm 1\%$ meter on the 100-V range, or a $\pm 5\%$ meter on the 10-V range?
12. Which meter has the best resolution, one that reads 5.15 V or one that reads 5.2 V?
13. All other things being equal, which ammeter will give the more correct measurement: one that has a 1-ohm internal resistance, or one that has a 0.1-ohm internal resistance?
14. What will happen to an ammeter if its is connected across a resistor in a circuit?
15. The digital read-out on a voltmeter has a minus sign. What does this mean?
16. Which one of the following correctly states Ohm's Law, $V = R/I$, $V = RI$, or $V = I/R$?
17. A 3-V flashlight bulb is placed in a 6-V flashlight. According to Ohm's Law, will the bulb be brighter or dimmer?
18. What is the meaning of power?
19. Which of the following is the unit for power: the joule, the foot-pound, or the watt?
20. Which one of the following correctly states the relation for power in a dc circuit, $P = VI$, $P = V/I$, or $P = I/V$?
21. What is the difference between the long-shunt connection and the short-shunt connection?
22. The resistance of a wattmeter current coil is 0.05 ohms, and that of the voltage coil is 10,000 ohms. Which connection will give the most correct power measurement for a 5000-ohm load, long shunt or short shunt?

23. What can cause the pointer of a wattmeter to deflect down-scale?
24. How is energy related to power?
25. What are two units for energy?
26. A device with an efficiency of 90 percent has a power output of 50 W. What will happen to the power output if the efficiency changes to 80 percent?
27. Explain the effect of losses on efficiency.
28. Can the efficiency of a system be greater than the efficiency of any part of the system? Explain why or why not.
29. An electrician accidentally placed a screwdriver across two terminals of a circuit. Is this a short circuit or an open circuit?
30. A light bulb is removed from its socket. Does this cause a short circuit or an open circuit?
31. What are two devices that are used to create an open circuit?
32. What is the resistance of an ideal short circuit?
33. What is the resistance of an ideal open circuit?

PROBLEMS

SECTION 3.1 Circuit Diagrams and Electronic Symbols

1. Draw the graphic symbol for each of the following:

   (a) air core inductor
   (b) fuse
   (c) ammeter
   (d) single-pole, double-throw switch

2. What is the letter symbol for each of the following?

   (a) current
   (b) reactance
   (c) power
   (d) frequency
   (e) hertz

3. What is the subscript in each of the following?

   (a) R_2
   (b) E_A
   (c) L_2
   (d) C_3

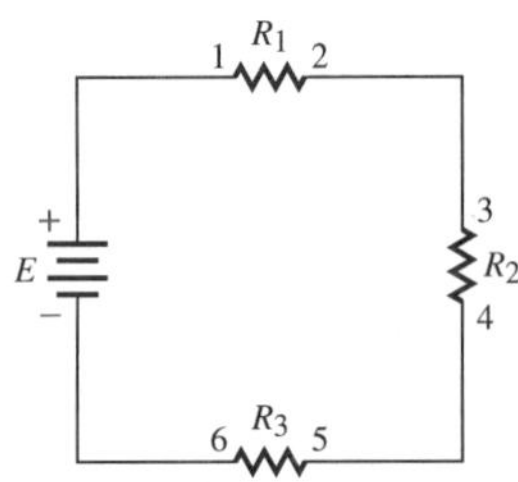

Figure 3.23

SECTION 3.3 Polarity, Voltage Rise, and Voltage Drop

4. Mark the proper polarity, + or −, on the terminal of each resistor and the source of emf in the circuit of Figure 3.23. Is it a voltage rise, or voltage drop?

   (a) from 1 to 2
   (b) from 6 to 5
   (c) from 5 to 1

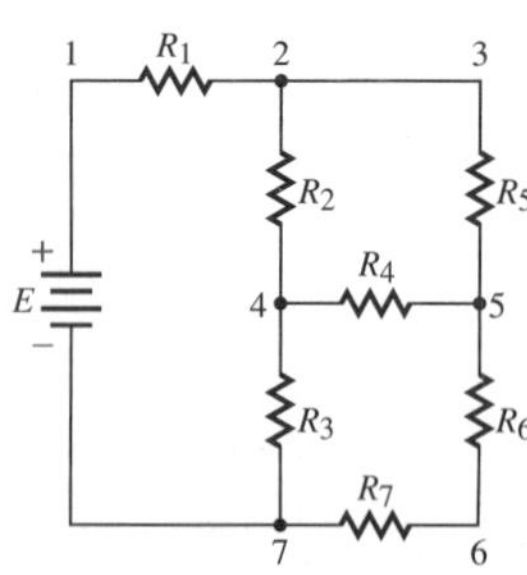

Figure 3.24

5. What is the polarity of each of the resistor terminals in the circuit of Figure 3.24 for each of the following conditions?

   (a) a 3-volt rise from 5 to 3
   (b) a 4-volt rise from 2 to 4
   (c) a 1-volt rise from 4 to 5

6. Place the proper polarity marking in the circuit of Figure 3.25 for the following conditions:

   (a) a 12-volt rise from 3 to 2

   (b) an 8-volt drop from 1 to 2

   (c) a 120-volt rise from 4 to 1

7. Work Problem 6 for the following conditions:

   (a) a 10-volt drop from 3 to 4

   (b) a 6-volt drop from 1 to 2

   (c) a 5-volt drop from 1 to 4

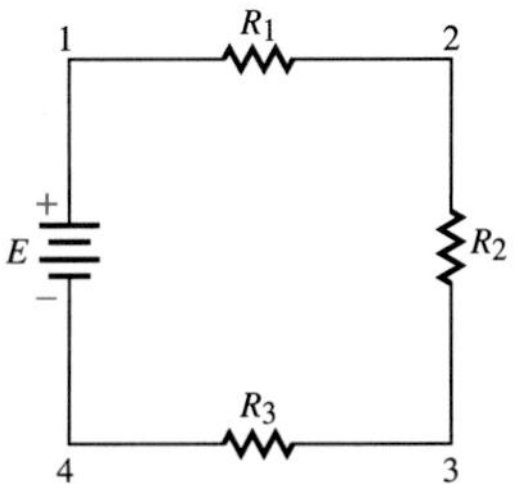

Figure 3.25

SECTION 3.4 Measuring Voltage, Current, and Resistance

8. How many volts does a ±5% accuracy represent on a 300-volt scale?

9. A 2-A full-scale ammeter has an accuracy of ±2% and indicates 1.2 A. What are the minimum and maximum that the actual current can be?

10. A voltage must be known to within 1 volt. What should be the accuracy if the 50-volt range is used?

11. A ±2% voltmeter indicates 120 volts on the 160-volt range. What are the minimum and maximum values that the actual voltage can be?

12. Calibration of an ammeter shows it to read 2.6 amperes on the 5-ampere range when the actual current is 2.5 amperes. What is the minimum ±% tolerance that the meter can be?

SECTION 3.5 Ohm's Law

13. What is the current in a 600-ohm resistor when it is connected to a 120-volt dc source?

14. An electric lamp is connected to a 120-volt dc source. What is the resistance if the current is 0.5 ampere?

15. The current in a 24-ohm heating coil is 5 amperes. What is the voltage across the coil?

16. The current in a resistor is 2 amperes when it is connected to a 24-volt dc source. What will the current be if the resistor is connected to a 120-volt dc source?

17. The current in a resistor is 1.2 amperes when connected to a 120-volt dc source. When a second resistor is connected to the same source, the current is one-half that in the first resistor. What is the resistance of the second resistor?

18. The current in a heater is 5 amperes when connected to a 120-volt source. Some of the turns short circuit and increase the current to 6 amperes. What is its resistance now?

19. A tapped resistor connected across a 24-volt source gives the following voltages:

(−) terminal to Tap 1	18 V
Tap 1 to Tap 2	12 V
Tap 2 to (+) terminal	4 V

What is the resistance between each set of taps if the current is 0.5 ampere?

20. If 24 volts across a resistor results in 2 amperes of current, what voltage is needed for 6 amperes?

SECTION 3.6 Power in the Electric Circuit

21. An electric heater draws 10 amperes of current when connected to a 115-volt dc source. How many watts of power are dissipated in the heater?

22. A 50-ohm resistor is connected to a 20-volt dc source. How many watts of power are dissipated in the resistor?

23. The power dissipated in a 500-ohm resistor is 20 watts when connected to a dc source. What is the voltage across the terminals of the resistor?

24. How many watts of power are dissipated in a 50-ohm resistor when the current in the resistor is 2 amperes?

25. What is the resistance of a 50-watt light bulb when connected to a 120-volt dc source?

26. A 50-ohm resistor is rated at 100 watts. Will it be operating above its power rating when it is connected to a 100-volt dc source?

27. How many watts should a 500-ohm resistor be able to dissipate so that it will not be damaged when there is 0.5 ampere of current in it?

28. What is the current in a 100-watt light bulb when it is operating at 125 volts?

29. A 1000-ohm resistor rated at 2 watts is connected across 120 volts. Will it be damaged?

30. In a 1500-watt heater operating at 125 volts, what is the minimum range that an ammeter should have to measure the current?

SECTION 3.7 The Wattmeter

31. A wattmeter indicates 100 watts when connected in long shunt across a 200-ohm resistance. What is the actual power in the resistor if R_A of the wattmeter is 0.1 ohms and R_V is 22,000 ohms?

32. How many watts would the meter in Problem 31 indicate if it is connected in short shunt?

33. Given that R_V is 5000 ohms and R_A is 0.01 ohms, find the ratio of P_{act}/P_{meas} for a wattmeter in (a) long shunt and (b) short shunt, across a 10,000-ohm resistor.

34. Derive Equations 3.5 and 3.6.

SECTION 3.8 Energy in the Electric Circuit

35. A motor draws 6.2 amperes of current when connected to 120 volts.

    (a) What is the power input in watts?

    (b) How much energy does it use if it is operated for 1/2 hour?

36. A 600-watt heating is operated for 5 hours. How many joules of energy are used in that time?

37. How long will it take a 100-watt lamp to use the same amount of energy as a 25-watt lamp operating for 5 hours?

38. How many kilowatt-hours of energy are used by an electric heater that operates at 1500 watts for 6 hours, 0 watts for 8 hours, and 500 watts for 10 hours?

    (a) How much energy is used by the unit in the 24-hour period?

    (b) If the cost of energy is 0.02 dollars per kWh, how much will it cost to operate the unit for the 24 hours?

39. The energy rate is as follows:

    $1.00 for the first 10 kWh

    5 cents for each of the next 25 kWh

    3 cents for each of the next 60 kWh

    2 cents for each kWh above 95 kWh

    What is the operating cost for 150 kWh of energy use?

40. A lamp uses 0.9 kilowatt-hours of energy in an 8-hour period. What was the average power in the lamp during that period?

SECTION 3.9 Efficiency

41. A 1/2-horsepower motor has a percent efficiency of 90%.

    (a) What is the power input to the motor?

    (b) What is the current in the motor if it is connected to a 120-volt source?

42. A motor drives a dc generator whose percent efficiency is 80%. The percent efficiency of the motor is 87%. What is the power output of the generator when the motor power input is 1000 watts?

43. A 1/3-horsepower motor has a current of 3 amperes when connected to a 120-volt source. What is the percent efficiency of the motor?

44. The losses in a transformer were measured as being 100 watts. What will the percent efficiency of the transformer be when supplying a 10-ampere load at 120 volts?

45. How many watts of power are lost in a motor if the input current is 4 amperes when connected to a 120-volts source and the percent efficiency of the motor is 85%?

46. The losses in a transformer are 6000 watts when supplying 40 kW of power to a load. Determine the percent efficiency of the transformer.

47. A motor draws 60 A when connected to a 220-volt source. If the losses are 2000 W, find

    (a) the percent efficiency

    (b) the horsepower output of the motor.

CHAPTER 4

SIGNALS AND SOURCES

CHAPTER OBJECTIVES

After completing this chapter, you should be able to:

1. Describe two methods of supplying dc power to a circuit.
2. Determine the amplitude, period, duty cycle, pulse width, and frequency of pulse waveforms.
3. Determine the amplitude, period, and frequency of a sinusoidal waveform.
4. Describe the function of the controls on a function generator.

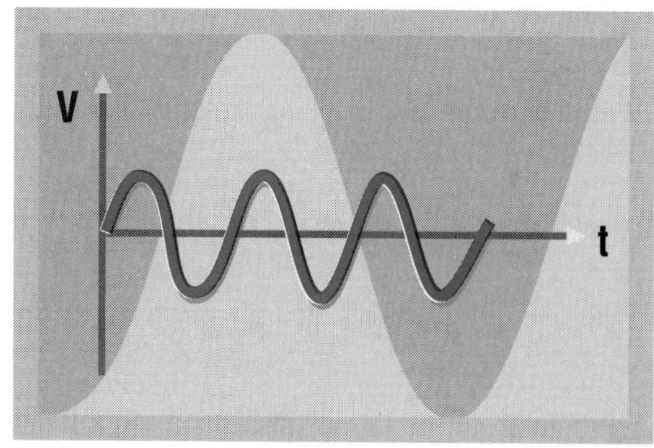

5. Describe the function of the main controls on an oscilloscope.
6. Apply Ohm's Law to time-varying signals.

KEY TERMS

1. ac source
2. ampere hour capacity (Ah)
3. amplitude
4. battery
5. cell
6. charging
7. charging rate
8. cold cranking performance
9. coupling
10. current drain
11. current output
12. current source
13. cutoff voltage
14. dc offset
15. dc power supply
16. dc voltage
17. duty cycle
18. electrodes
19. electrolyte
20. falling edge
21. floating output
22. focus
23. frequency
24. function generator
25. intensity
26. lead-acid cell
27. line regulation
28. nickel-cadmium cell (nicad)
29. operating temperature
30. oscilloscope
31. period
32. primary cells
33. probe
34. probe compensation
35. pulses
36. pulse width
37. reserve capacity
38. ripple factor
39. rise time
40. rising edge
41. rms value
42. secondary cells
43. shelf time
44. sine wave
45. sinusoidal signal
46. square wave
47. trace rotation
48. triangle wave
49. trigger
50. voltage output
51. voltage regulation

INTRODUCTION

This chapter describes a number of different power sources and signal sources. To function properly, circuits must be powered. In most systems, circuits are powered by dc or constant voltage sources. These voltages are provided by power supplies that convert the ac voltage present at the standard wall outlet into the dc voltages required by the circuit. If systems are operated in remote areas, or if they must be portable, sources such as solar cells and batteries are used to power the circuits.

Circuits and systems process a variety of signals. The information contained in these signals is used to record events, make decisions, and control processes. Information may be coded in the shape of the signal, its amplitude, its frequency, or any number of other parameters.

As an example, let us take a look at an electrocardiogram (EKG). The human body is an excellent conductor of electricity. As the heart beats, a characteristic pattern of electrical activity is produced during the rhythmic contractions of the heart, which force the blood through the intricate pathway of vessels that make up the circulatory system. Electrodes on the surface of the skin pick up the electrical echoes of the electrical currents in the heart. The signals picked up by the electrodes are very small—millivolts in amplitude. A patient monitor is used to sense, amplify, and display a picture representing the electrical activity of the heart.

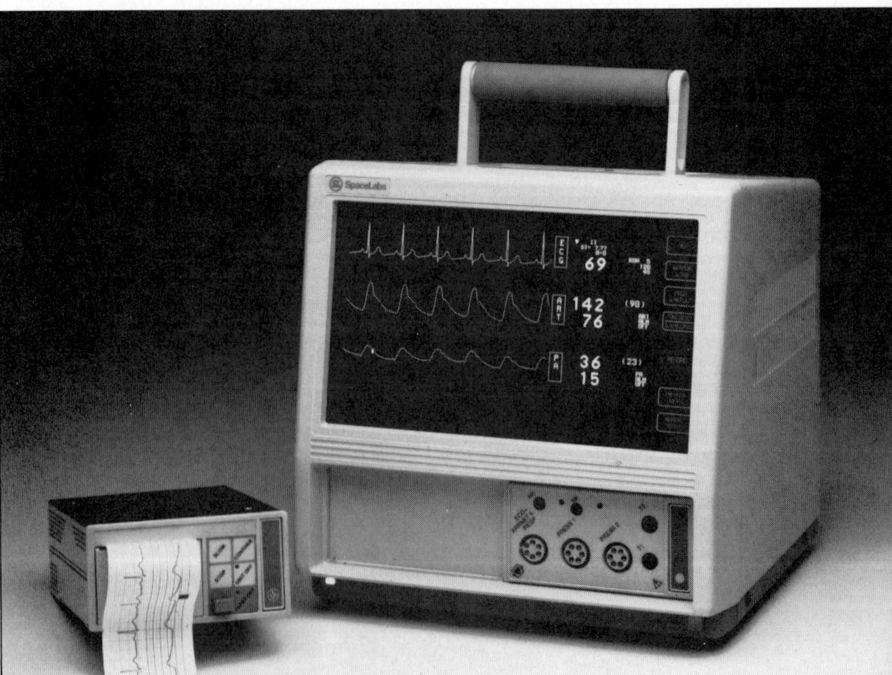

Figure 4.1 A patient monitor. (Courtesy of SpaceLabs, Redmond, Washington)

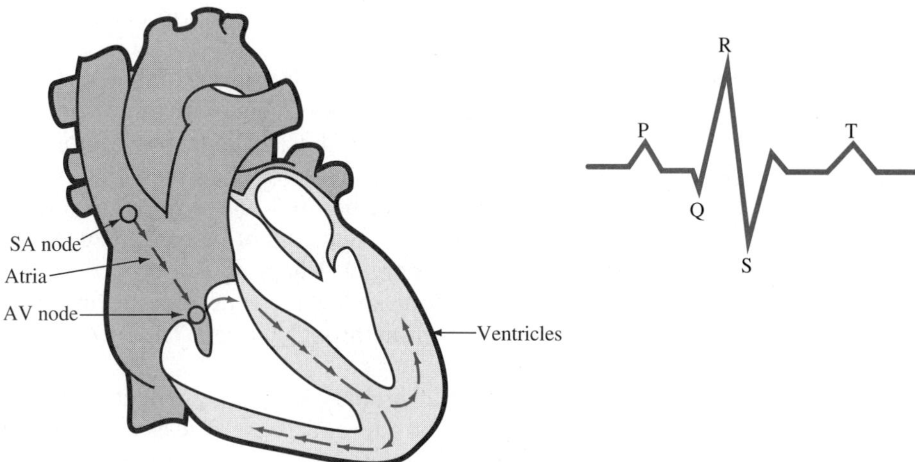

Figure 4.2 The EKG signal. The arrows show the spread of electrical activity from the SA node to the AV node, and then through the ventricles.

The shape and frequency of the EKG signal carry the information needed by the doctor or nurse to make an accurate diagnosis of heart function. A normal EKG signal, shown in Figure 4.2, consists of three parts: the P wave, the QRS complex, and the T wave. The P wave charts the electrical activity through the atria. The QRS complex then tracks the spread of the activity through the ventricles. And finally, the T wave is produced as the electrical activity in the ventricles ends. All this activity occurs in the span of 10–20 milliseconds.

Changes in the shape and frequency of the EKG signal picked up by the patient monitor are telltale signs of abnormal functioning of the heart. Doctors and nurses are trained to interpret these changes and respond with the appropriate medical treatment necessary to help the patient.

In Section 4.7, the oscilloscope will be described. The oscilloscope serves the same type of function for the technician or engineer that the patient monitor serves the doctor or nurse. The oscilloscope displays a picture of signals within a circuit. By using the information contained in the displayed picture, the engineer and technician can determine the proper or improper functioning of the circuit and take appropriate action.

We will begin our study of sources and signals with devices that provide a constant dc voltage. The flashlight battery is familiar to all of us. The dc power supply will become familiar as you gain laboratory experience.

4.1 Primary Cells

Cells produce an emf through a chemical reaction. In this reaction, chemical energy is converted to electrical energy. A device that uses the chemical reaction to produce an emf is called a **cell**. Sometimes a cell is

cell

battery incorrectly called a **battery**. A battery is actually a group of cells connected together.

dc source A cell is a **dc source** . A dc source is one that provides a voltage whose magnitude and direction do not change with time. These cells are used to operate such devices as radios, calculators, motor-driven toys, and portable tools. They also can provide emergency power when other sources of power cannot be used or are not available.

primary cells **secondary cells** **charging** **Primary cells** are cells that cannot be recharged. **Secondary cells** are cells that can be recharged. **Charging** is a procedure that restores the cell's original condition so that it can provide an emf again. It is done by passing a reverse current through the cell. The current reverses the cell's chemical reaction.

electrolyte **electrodes** The main parts of primary cells are an electrolyte, a negative electrode, a positive electrode, and a case. The **electrolyte** is a paste or liquid that permits ionic conduction between the electrodes. The **electrodes** are conducting material that react chemically with the electrolyte. The chemical reaction forms positive and negative ions. The parts of the cell are placed in a sealed case. This case permits the cell to be used in any position.

Cells can be made from a wide variety of materials. Take, for example, the TWO POTATO* clock shown in Figure 4.3. Each potato forms a cell; the two cells are connected in series to produce an emf or voltage large enough to power the digital clock. Two types of electrodes

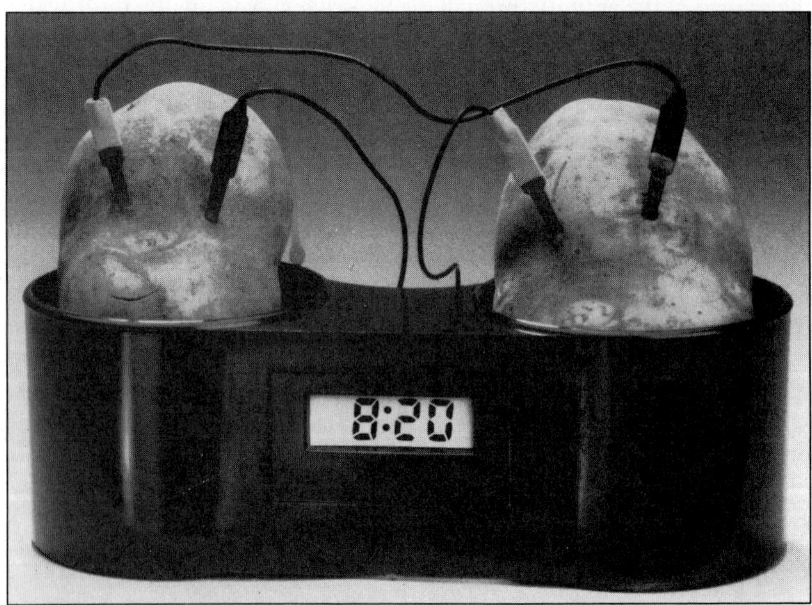

Figure 4.3 A novelty cell that uses a potato for the electrolyte and two metals for electrodes. (Photo and print courtesy of Skilcraft, a Division of Monogram Models, Inc.)

*Trademark of Monogram Models.

are used. One electrode is made of copper and the other is made of zinc. Potatoes are not the only electrolyte source that can be used with the TWO POTATO clock. Apples, cucumbers, lemons, potted plants, and soda pop will also work.

The principal types of primary cells are (1) carbon-zinc, (2) zinc-chloride, (3) alkaline-manganese dioxide, (4) mercuric oxide, (5) silver oxide, (6) lithium, and (7) zinc-air. The electrochemical system for each will now be described.

1. **Carbon-Zinc.** The electrolyte is a paste of ammonium chloride and zinc chloride (see Figure 4.4) It has a manganese- dioxide paste for the positive electrode. The carbon rod in the center merely serves as a terminal. The zinc case is the negative electrode.

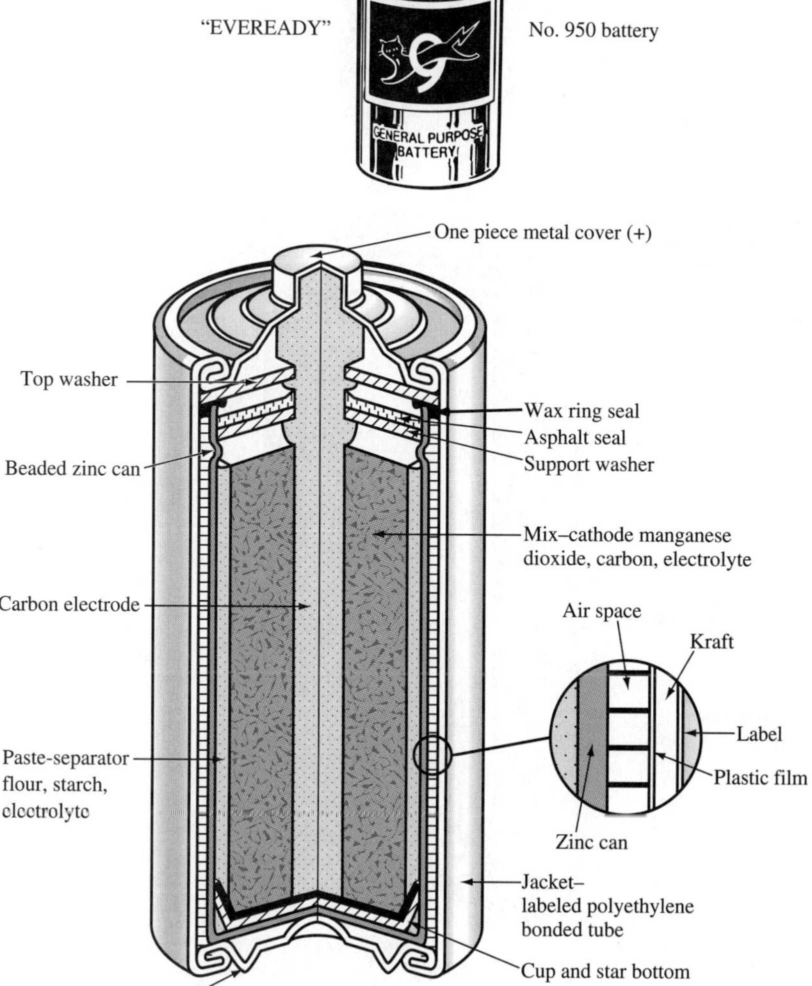

Figure 4.4 A cutaway view of a general-purpose zinc-carbon cell. (Photo courtesy of Eveready Battery Co.)

2. **Zinc-Chloride.** The electrolyte has more manganese dioxide than that in the carbon-zinc cell. Otherwise, its construction is similar to the carbon-zinc cell.
3. **Alkaline-Manganese Dioxide.** The electrolyte is a potassium hydroxide paste. The electrodes are of the same materials as in the carbon-zinc cell. A typical alkaline-manganese dioxide cell has the construction shown in Figure 4.5.
4. **Mercuric Oxide.** This cell also has a potassium hydroxide electrolyte. The zinc case is the negative electrode. The positive electrode is mercuric oxide.

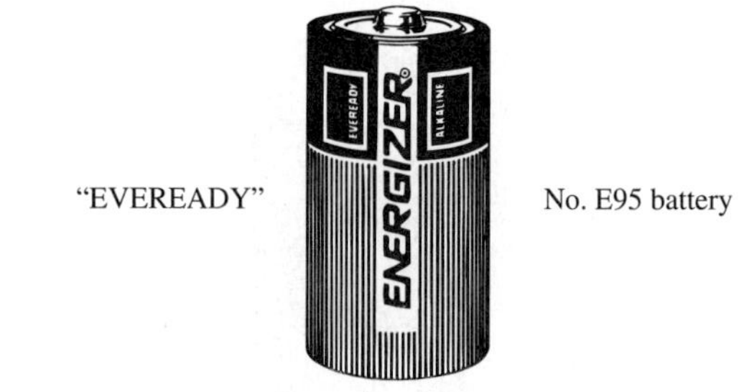

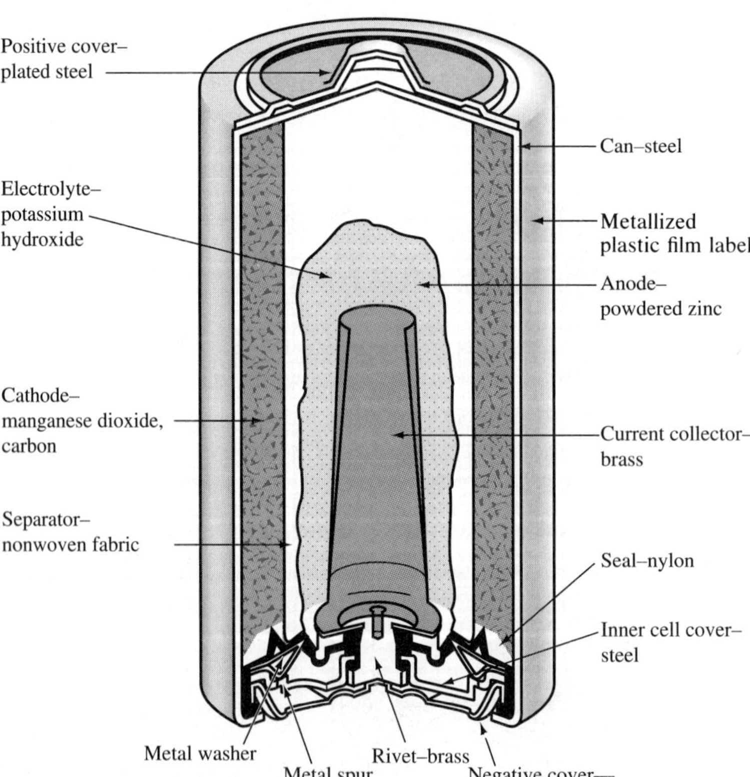

Figure 4.5 A cutaway view of a cylindrical energizer alkaline cell. (Photo courtesy of Eveready Battery Co.)

5. **Silver Oxide.** The electrolyte and negative electrode are of the same materials as in the mercuric oxide cell. The positive electrode is the silver oxide.
6. **Lithium.** The electrolytes are lithium salts in an organic solvent. The negative electrode is lithium. The positive electrode is iron sulfide, copper oxide, or some other metal halide, sulfide, or oxide.
7. **Zinc-Air.** This cell has a potassium hydroxide and zinc oxide electrolyte. The anode is zinc. The cathode is a mixture of carbon, teflon, and manganese dioxide. The construction of the cell and some typical zinc-air cells are shown in Figure 4.6 (a) and (b). Zinc-air cells are unique in that the chemical action does not start until air contacts the cathode. This is done by breaking a seal that covers a hole in the cell. Thus, the cell will not deteriorate in storage. The emf of the cell is about 1.4 volts. This cell is well suited for devices that are used frequently, like paging devices and hearing aids.

Each primary cell has characteristics that make it best suited for certain applications. For example, the carbon-zinc cell has a low purchase price. It is good for low-current, intermittent-use applications such as radios, toys, flashlights, and paging systems.

On the other hand, alkaline cells have a long shelf life. They are good for applications that call for higher currents. Lithium cells have found wide use in watches. They can supply a higher power per unit size than other cells. Thus, they can be made in the small sizes required for watches. Lithium cells are made with voltage outputs of 1.6–3.7 volts. The voltage rating depends on the electrode-electrolyte system that is used.

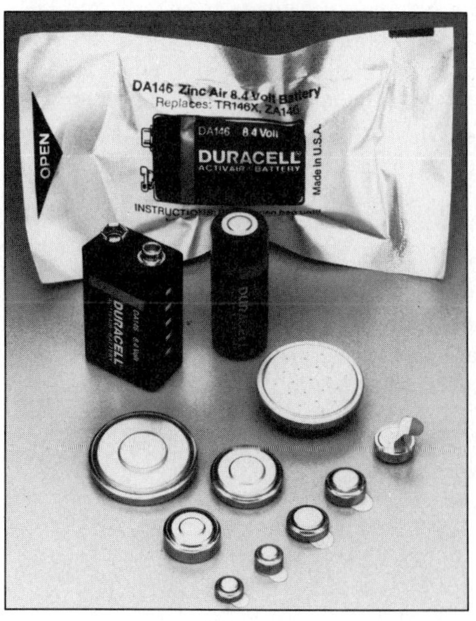

(a)

(b)

Figure 4.6 Zinc-air cells are used in hearing aids, paging equipment, and medical equipment. (a) Some sizes and shapes of zinc-air cells. (b) A cutaway view of a zinc-air cell. (Photos courtesy of Duracell Inc.)

TABLE 4.1* Cell Application Comparison Table

System	Type	Features	Recommended Applications
Carbon-zinc	Primary	Low cost, gradually sloping discharge curve, variety of shapes and sizes, decrease in efficiency at high current drains, poor low-temperature performance	Radios, barricade flashers, marine depth finders, toys, lighting systems, signaling circuits, novelties, flashlights, paging, laboratory instruments
Zinc chloride	Primary	Gradually sloping discharge curve, low-temperature performance and service capacity at moderate-to-high current drains, better than carbon-zinc, decrease in efficiency at high current drains, but not to the extent of carbon-zinc, lesser variety of shapes and sizes than carbon-zinc	Cassette players and recorders, calculators, motor-driven toys, radios, clocks, video games
Alkaline-manganese dioxide	Cylindrical primary	Gradually sloping discharge curve, better low-temperature performance, service maintenance, and lower impedence than carbon-zinc. Competitive to carbon-zinc in terms of cost per hour of use on moderate-to-high current drains	Radios (particularly high current drain), shavers, electronic flash, lighting systems, movie cameras, tape recorders, television sets, walkie-talkies, cassette players and recorders, calculators, motor-driven toys, clocks, heavy duty lighting, camera motor drive, any high current drain, heavy discharge schedule use
	Miniature MnO_2 primary	Lower cost and energy density than silver or mercuric oxide, gradually sloping discharge curve, service maintenance equivalent to silver oxide	Calculators, novelties, toys, clocks, watches, cameras
Mercuric oxide	Primary	Higher ampere hours per cu. in. than silver oxide, higher energy density than alkaline or carbon-zinc, better service maintenance than carbon-zinc	Secondary voltage standard, walkie-talkies, paging, radiation detection, test equipment, hearing aids, watches, calculators, microphones, cameras
Silver oxide	Primary	Relatively flat discharge curve, higher operating voltage and service maintenance than mercury, equivalent energy density to mercury	Hearing aids, reference voltage source, cameras, instruments, watches, calculators
Nickel cadmium	Secondary	Sealed maintenance-free construction, relatively flat discharge curve, good high-temperature performance, more competitive in terms of cost per hour of use than carbon-zinc, high resistance to shock and vibration, variety of shapes and sizes, long cycle life, higher effective capacitance than other systems, constant current charging, high initial cost, only fair charge retention	Portable hand tools and appliances, shavers, toothbrushes, photoflash equipment, dictating machines, movie cameras, instruments, portable communication equipment, tape recorders, radios, television sets, cassette players and recorders, calculators, R/C models

*Courtesy of Eveready Battery Co.

Characteristics and application comparisons of some types of cells are shown in Table 4.1.

Cells are made in various sizes, a few of which can be seen in Figure 4.7. The operating time of the cell is related to its size, as shown in Table 4.2. The larger size cells can supply current for a longer time. This is because they have a larger amount of electrolyte and electrode material. The comparison is for a load of 80 mA. Other loads would change the time, but the order will remain the same. Primary cells generally have a higher operating cost than secondary cells. In spite of this, they are still widely used. They are easy to replace and have a lower first cost.

4.2 Secondary Cells

Secondary cells can be recharged and reused over and over again. This makes their long-term cost usually lower than that of primary cells. In some cases, the recharging can be repeated several hundred times.

lead-acid cell

The most common secondary cell is the **lead-acid cell**. It uses a water and sulfuric acid solution for the electrolyte. The positive electrode is lead peroxide and the negative electrode is metallic lead. This cell is used to make batteries that are used in automobiles, boats, and electric carts. These lead-acid batteries can supply several hundred amperes while

Figure 4.7 Cells are made in many sizes and shapes. (Photo courtesy of Eveready Battery Co.)

TABLE 4.2* Average Service Time for Different Sizes of Alkaline Cells

Current Drain of 80 mA		
Size Cell	*Weight (oz.)*	*Time (Hours)*
AAA	0.41	6.2
AA	0.78	15
C	2.22	45
D	4.41	105

* Courtesy Eveready Battery Co.

starting a car. They can be recharged and provide a very reliable, low-cost, and trouble-free source of power.

The liquid electrolyte in the lead-acid battery prevents it from being used in all positions. This problem is eliminated in some batteries by

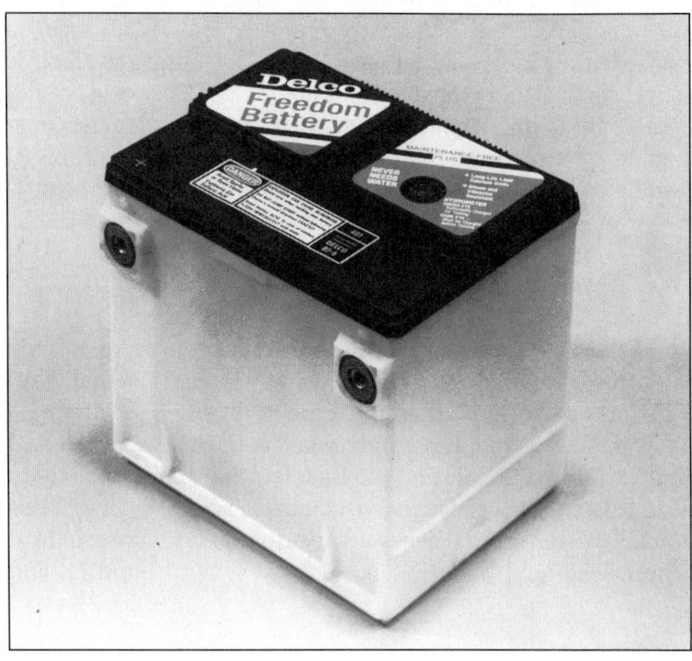

(a)

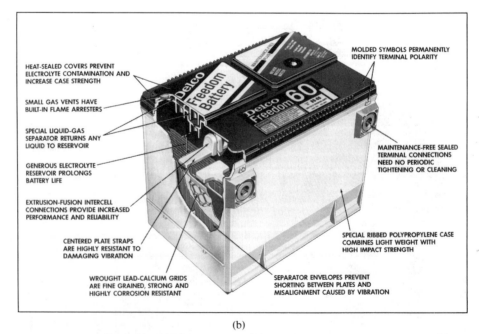

(b)

Figure 4.8 The lead-acid battery can provide large currents and is rechargeable. (a) A typical lead-acid battery. (b) A cutaway view of the lead-acid battery. (Photos courtesy of Delco Remy Div., GMC)

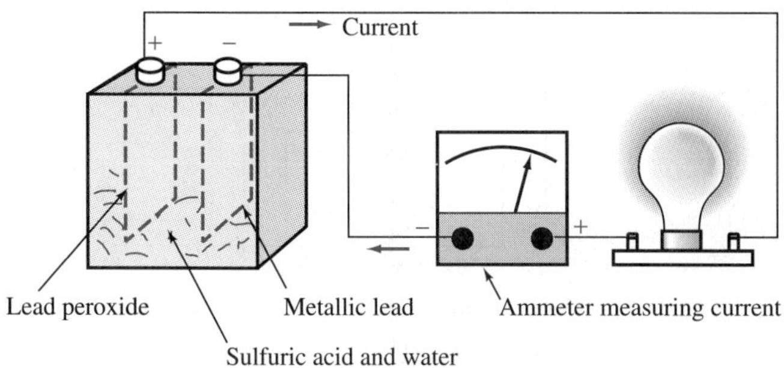

Figure 4.9 A lead-acid battery discharges when connected to an electrical load such as a lamp.

using a gel-type electrolyte. The emf of the cell is about 2.1 volts; a standard 12-volt battery uses six cells to provide a no-load emf of about 12.6 volts. A typical lead-acid battery and its cutaway view can be seen in Figure 4.8 (a) and (b).

The function of each part of the battery is as follows: The case provides the support for the other parts in the battery. The electrolyte is a solution of sulfuric acid and water. A fully charged battery has a specific gravity of 1.265 at 26.7°C (80°F). The specific gravity can be measured with a hydrometer. The positive plate has some lead peroxide pressed into it, while the negative plate has metallic lead pressed into it. Fine amounts of other metals such as antimony are added to the lead to improve its casting. The plates are prevented from touching by separators.

The electrolyte action on the electrodes leaves a positive charge on the lead peroxide grid and a negative charge on the metallic one. When fully charged, the positive grid is a reddish brown color. The negative grid is a gray color. Drawing current from the battery causes some changes (see Figure 4.9). The sponge lead in the negative plate and lead peroxide of the positive plate are changed to lead sulfate, taking on a whitish color. The chemical reaction dilutes the sulfuric acid, changing it to water. The changes in the parts of the cell are

lead + lead peroxide + sulfuric acid
yields
lead sulfate + water + electrical energy

These changes are expressed by the chemical equation:

$$Pb + PbO_2 + 2\ H_2SO_4 \longrightarrow 2\ PbSO_4 + 2\ H_2 + O_2$$

Eventually, the dilution of the electrolyte and the lead sulfate prevent further chemical reaction. The battery is then unable to deliver a useful voltage. It is said to be discharged. The state of the battery can be checked by measuring its voltage when drawing current, or by checking its specific gravity. A fully discharged battery has a specific gravity of 1.120 at 26.7°C (80°F).

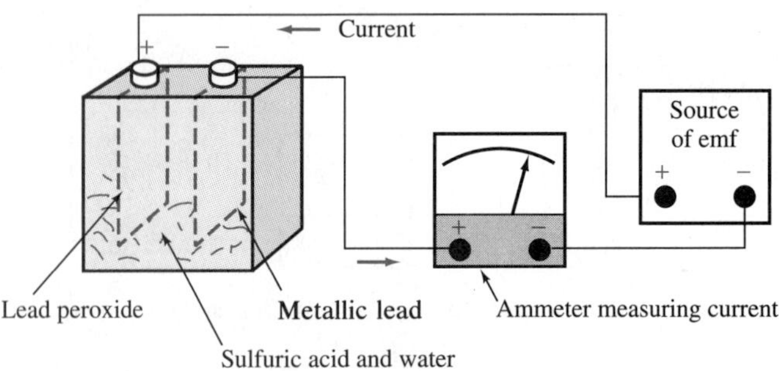

lead sulfate + water + electrical energy yields lead + lead peroxide + sulfuric acid

Figure 4.10 A lead-acid battery can be charged by connecting it to a source of emf.

The cells can be restored to a useful condition with an external source of emf, as shown in Figure 4.10. The direction of current through the cell is now reversed. This reverse current restores the plates and electrolyte to their original condition. The chemical reactions are now reversed. The complete process is

lead sulfate + water + electrical energy
yields
lead + lead peroxide + sulfuric acid

A secondary cell must be charged at a controlled rate; otherwise, the plates can overheat and buckle, or the electrolyte can start to boil. Either of these conditions will damage the cell.

During the charging process, hydrogen and oxygen are given off as gases. This causes a loss of some of the electrolyte and a buildup of pressure in the cell. This increase in pressure and the possibility of the gas being ignited by a spark create a hazardous condition. Some cells have removable vent caps to relieve the pressure and to add water.

As battery technology advanced, the buildup of gas and the loss of liquid was decreased by replacing the antimony in the grids with other metals such as calcium or strontium. Batteries of this type are called "maintenance-free." They do not have any vent caps. However, they still have a small vent to prevent pressure buildup from changes in atmospheric conditions.

nickel-cadmium cell

The **nickel-cadmium cell**, commonly known as the "nicad" battery, is another secondary cell, which has become popular in the past few years. It has a cadmium positive electrode and a nikelic hydroxide negative electrode, as shown in Figure 4.11. The electrolyte is an aqueous solution of potassium hydroxide. Nicad batteries can be recharged several hundred times. A charging unit, shown in Figure 4.12, can recharge the nickel-cadmium cells around it.

A nicad battery has good high- and low-temperature characteristics and a relatively flat discharge curve. It is competitive in operating cost with the carbon-zinc cell. On the negative side, its initial cost is high, and it has only fair charge retention. The nicad battery is used in portable

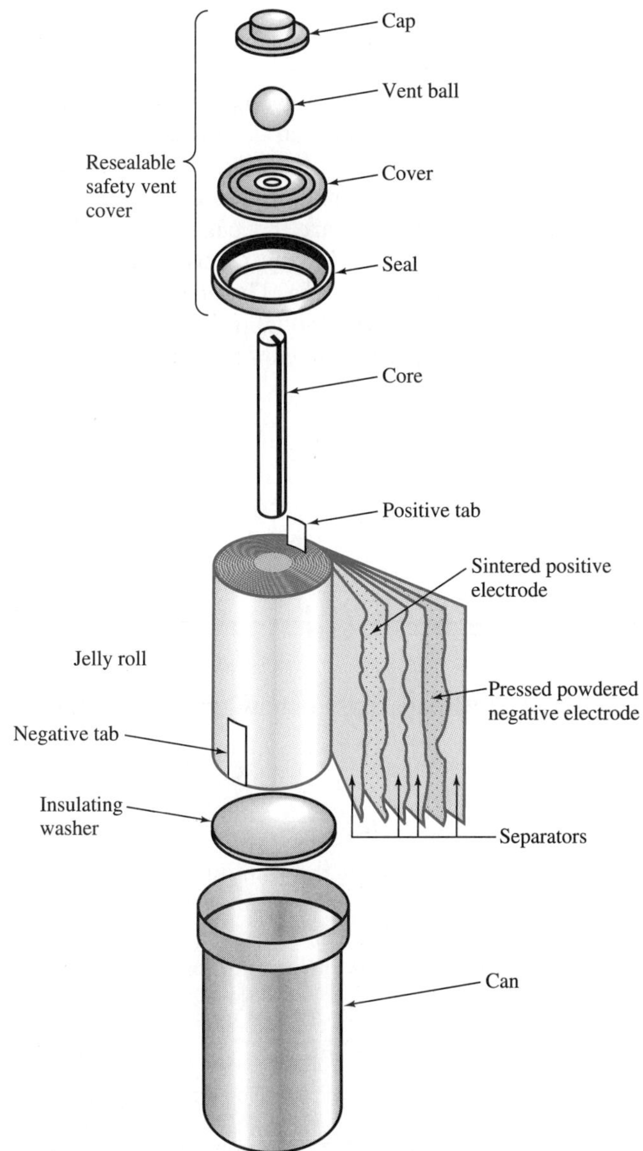

Figure 4.11 An expanded view of the nickel-cadmium rechargeable cell. (Photo courtesy of Eveready Battery Co.)

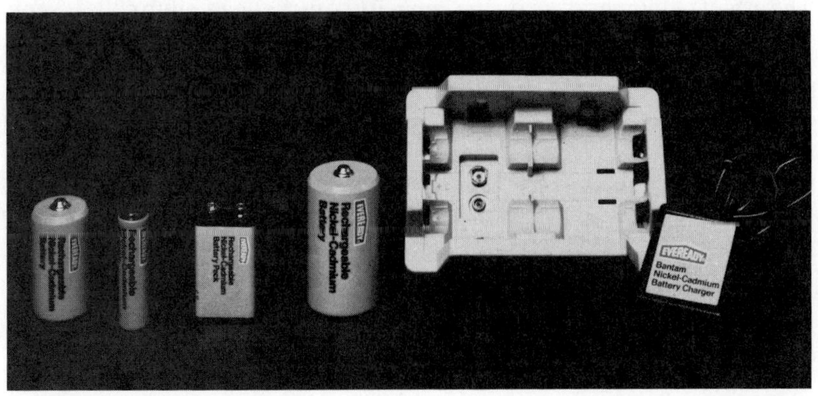

Figure 4.12 These nickel-cadmium cells can be recharged in the charging unit show here. (Photo courtesy of Eveready Battery Co.)

tools and appliances, tape recorders, calculators, toys, electronic photo flashes, and standby power systems.

ampere hour capacity

The emf of a cell is only one factor to consider when selecting a cell or battery. Other factors to be considered are the amount of current the battery can supply and the length of time that current can be supplied. For example, a car battery must maintain more than 100 amperes for a long enough period of time to start the car's motor. The battery's capacity to supply current for a period of time while having a usable emf is called its **ampere hour capacity** (Ah). The ampere hour is one of the units whose use is continued even though it is not in the International System of Units. This ampere hour capacity is equal to the current multiplied by time.

$$\text{Ah} = I \times t \tag{4.1}$$

where Ah is ampere hours
I is the current, in amperes
and t is the time, in hours

EXAMPLE 4.1

A battery has an 80-Ah capacity. For how long can it supply 2 amperes of current?

Solution

$$\text{Ah} = I \times t$$

By rearranging the equation, we obtain an expression for time t,

$$t = (\text{Ah}/I)$$

Substituting the known values into the equations yields

$$t = (80\ \text{Ah}/2\ \text{A})$$
$$t = 40\ \text{h}$$

The ampere-hour capacity changes with temperature and current. The effect of temperature on the capacity of nickel-cadmium cells is seen in Figure 4.13. At −10°C, the cell has only 70 percent of its capacity at

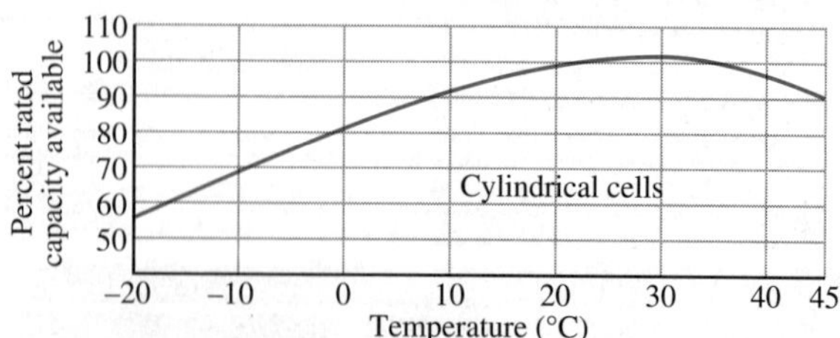

Figure 4.13 A change in temperature affects the capacity of the nickel-cadmium cell. (Graph courtesy of Eveready Co.)

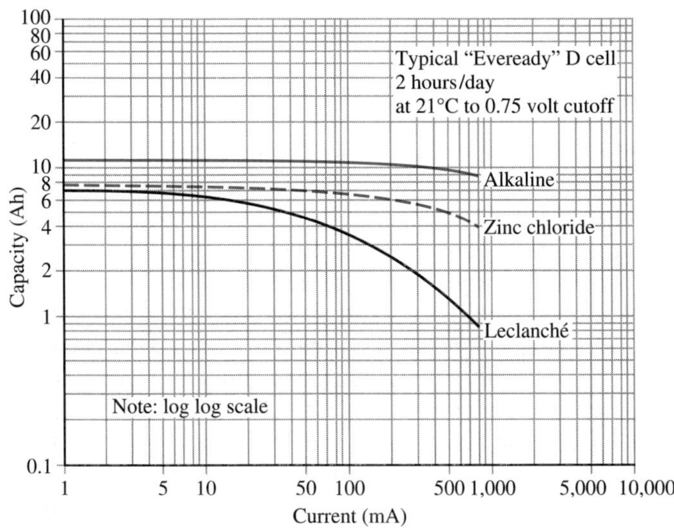

Figure 4.14 The capacity of a chemical cell is also affected by the current drain. (Graph courtesy of Eveready Battery Co.)

20°C. This effect can be noticed when starting a car on a cold day. The motor just does not turn over as easily as it does on a warm morning.

The cell current drain also affects the ampere hour capacity, as seen in Figure 4.14. The alkaline cell and the zinc-chloride cell have a relatively constant capacity. The carbon-zinc (Le Clanche) cell capacity drops off rather sharply. Some of this change is due to the increase in temperature of the cell materials, which is caused by the current. Larger currents result in a higher temperature. The ampere hour capacity can also be used to indicate how much electricity a battery has supplied. For instance, a battery that has been connected to a 5-ampere load for 3 hours has supplied 15 ampere hours of electricity.

EXAMPLE 4.2

A battery that is rated at 10 ampere hours is connected to a circuit that draws 500 mA of current for 8 hours.

a. How many ampere hours are supplied?
b. What percent of the capacity has been used?

Solution:

a. Since

$$\text{Ah} = I \times t$$

the number of ampere hours supplied can be found by substituting the known quantities for current and time,

$$\begin{aligned} \text{Ah} &= (500 \text{ mA}) \times (8 \text{ hours}) \\ &= (0.5 \text{ A}) \times (8 \text{ h}) \\ \text{Ah} &= 4 \text{ Ah} \end{aligned}$$

b. The percent of the capacity that has been consumed is defined as

$$\% \text{ used} = (\text{Ah used/Ah capacity}) \times 100\%$$
$$= (4 \text{ Ah}/10 \text{ Ah}) \times 100\%$$
$$\% \text{ used} = 40\%$$

As the automobile's complexity and electrical needs increased, the Ah rating became inadequate. It has been replaced by two more meaningful ratings. They are cold cranking performance and reserve capacity.

cold cranking performance

The **cold cranking performance** is specified as cold cranking amperes (CCA). It is the current that a battery can supply for 30 seconds at 0°F (−17.78°C) while maintaining a voltage of 1.2 volts per cell or higher. This test simulates the demands placed on the battery when starting the vehicle in cold weather. Some manufacturers also provide a CCA rating at 0°C. Typical ratings for passenger car applications range from about 300–600 amperes.

reserve capacity

The **reserve capacity** measures the ability of the battery to supply a vehicle's minimum electrical needs if the alternator fails. The minimum load is the current for the motor, low-beam lights, windshield wipers, and defroster. More specifically, it is the number of minutes that a new fully charged battery takes to discharge to 1.75 volts per cell at a load of 25 amperes. In general, batteries that have a higher CCA and reserve capacity rating will also have a higher ampere hour capacity.

The amount of electrical energy that chemical cells can provide is limited. This amount is affected by any of the following conditions.

shelf time

1. **Shelf Time.** The time that a battery is in storage before it is used. Storage at lower temperatures slows down the effect for most batteries. A typical lead-acid battery loses its charge, as shown in Figure 4.15. One way to avoid this loss is to add the electrolyte when the battery is put into use.

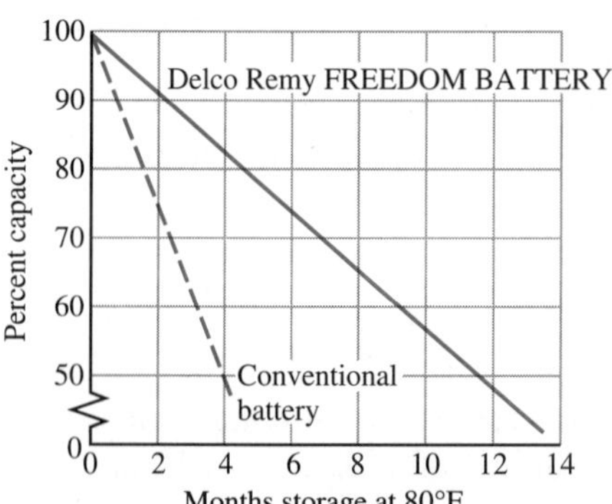

Figure 4.15 The effect of shelf time for a lead-acid battery. (Graph courtesy of Delco Remy Div., GMC)

2. **Cutoff Voltage.** The voltage below which the battery is no longer useful. A value near the emf of the new battery results in a shorter useful life.

cutoff voltage

3. **Current Drain.** The current that the battery is expected to supply. Exceeding the cell's rated value will cause damage to the electrolyte and electrodes.

current drain

4. **Duty Cycle.** The sequence of ON and OFF periods during operation. Shorter ON time with longer OFF time gives the battery a chance to rest and increases the useful life.

duty cycle

5. **Operating Temperature.** The temperature at which the battery operates. Very high or very low temperatures will decrease the useful life. The power of a lead acid battery drops from 100 percent at about 27°C to 40 percent at about −18°C.

operating temperature

6. **Charging Rate.** The rate at which energy is restored to the secondary battery. Rapid charging can damage the cells.

charging rate

4.3 The dc Power Supply

Cells and batteries are used for powering portable equipment or equipment located in places where ac sources are not available. When ac power is available, power can be supplied by a unit called a **dc power supply**. These supplies convert the ac voltage present at the outlet into a constant dc voltage.

dc power supply

A typical laboratory power supply is shown in Figure 4.16, which is a Hewlett Packard 6227 DUAL DC POWER SUPPLY. This power supply is capable of providing both a negative and a positive voltage. Each of these voltages can be adjusted by turning the appropriate control on the front panel.

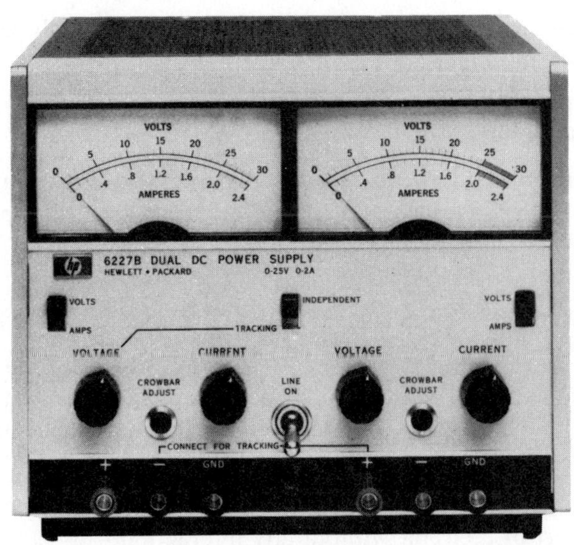

Figure 4.16 A dc power supply. This supply provides two outputs of 0–25 V each, at currents up to 2 A. (Photo courtesy of Hewlett-Packard Co.)

Figure 4.17 A constant current supply. These supplies are used for semiconductor development and component testing. (Photo courtesy of Hewlett-Packard Co.)

An ideal power supply should provide an output voltage that does not change with current. In real life, the output voltage of a power supply will exhibit some fluctuation. This will cause the amount of current drawn from the supply to change. Some supplies are designed to maintain a fixed current even though the output voltage changes. These supplies are called **current sources**, as shown in Figure 4.17.

current sources

The dc sources vary from very basic to very complex. As a result, output voltage is not always the only thing to be considered. The following list contains other important characteristics.

current output

1. **Current Output.** The amount of current that the supply can provide under continuous use.

voltage output

2. **Voltage Output.** The maximum voltage that can be obtained from the supply at zero current.

voltage regulation

3. **Voltage Regulation.** A measure of the output voltage change caused by a change in the current. It is usually expressed as a percent. A smaller percentage indicates a smaller change in voltage for a given change in current.

line regulation

4. **Line Regulation.** A measure of the output voltage change caused by a change in the input line voltage.

ripple factor

5. **Ripple Factor.** The ratio of the ac voltage in the output to the dc voltage in the output. A pure dc voltage has a ripple factor of zero.

floating output

6. **Floating Output.** A type of output that has ungrounded positive (+) and negative (−) terminals. This permits connecting the source to any part of a circuit without any interaction.

Additional features, which can enter into the selection of a dc power supply are remote sensing, remote programming, output impedance, current limiting, and the mechanical features of the power supply.

Using a laboratory power supply involves adjusting the voltage and current settings for the application at hand. The following example will help illustrate the setup procedure.

Suppose that the circuit to be powered requires a voltage of +5 volts and a current of 500 mA. Before the power is supplied to the circuit, the unit should be switched ON and the voltage control adjusted to 5 volts. Some power supplies have meters that can be used, but to obtain a more accurate voltage measurement, a digital voltmeter can be used to monitor the power supply's output voltage.

The current limit can then be set. On the HP 6227B DUAL DC POWER SUPPLY, the current output range is 0–2 amperes. Since the output current needed is approximately 600 mA, the current control should be placed at its midrange position. This will set a current limit just above the required current level. In the event that a short is present in the circuit, the current limit circuit will shut down the power supply. Once the power supply settings are made, the circuit can be connected and the voltage setting should be checked. The voltmeter connections are shown in Figure 4.18.

Many products use modular power supplies to supply the required voltages needed. For example, an IBM personal computer requires a power supply that produces four different output voltages. If the cover of an IBM computer is removed, the power supply sits in one corner of the case in an enclosed box. On its top or side will be the ratings of the various output terminals.

Modular supplies are manufactured by companies specializing in this construction. These supplies come in various sizes and output configurations and ratings. For the general consumer, when a power supply fails, the entire module is replaced. It might also be necessary to replace a power supply with another supply, which has higher ratings when additional circuits are added to the system (e.g., interface cards to a personal computer or the addition of a hard disk).

Modular supplies are rated in terms of their output voltages and the power that they can deliver. For example, the power supply in the IBM

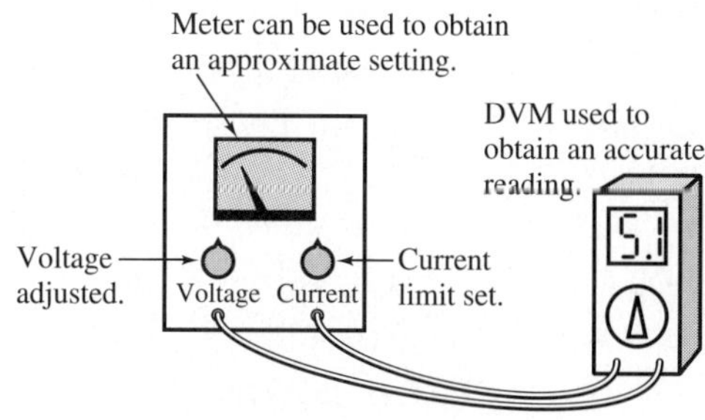

Figure 4.18 Setting up a power supply.

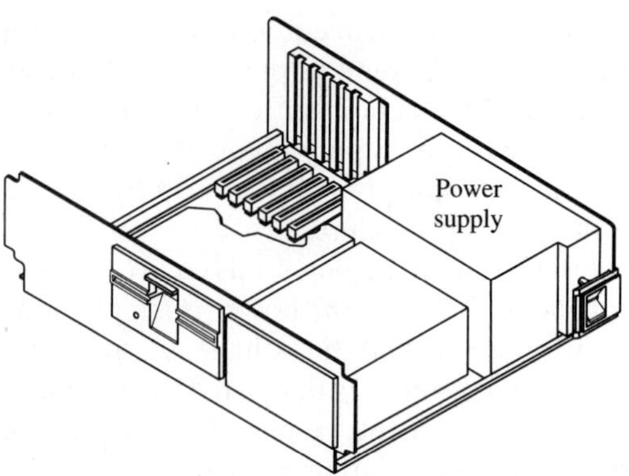

Figure 4.19 A modular power supply used in a personal computer.

personal computer (Figure 4.19) takes the voltage present at the wall outlet, nominally in the range of 100–125 VAC @ 50/60 Hz, and produces output voltages of +5 volts, −5 volts, +12 volts, and −12 volts. These output voltages are present on specific connector terminals within the IBM PC. Table 4.3 lists output voltages, their maximum current, and power rating of an IBM personal computer. The maximum power that the power supply is able to deliver is 63.45 watts; it is the sum of the power capable of being delivered by each output terminal.

The enormous variety of power supply tasks has given rise to an incredible proliferation of circuits. Each type of power supply has certain advantages and some drawbacks. When troubleshooting electronic systems, the first thing to check are the power supply output voltages. The system will not work if the power supply is malfunctioning.

TABLE 4.3 IBM PC Power Supply Connections

Output Voltage	Maximum Current	Power Rating
+12 V	0.20 A	2.4 W
−12 V	0.25 A	3.0 W
−5 V	0.30 A	1.5 W
+5 V	1.93 A	9.65 W
+5 V	1.93 A	9.65 W
+5 V	1.93 A	9.65 W
+12 V	0.90 A	10.80 W
+5 V	0.60 A	3.0 W
+12 V	0.90 A	10.8 W
+5 V	0.60 A	3.0 W
	TOTAL POWER	63.45 W

A simple voltmeter check will suffice at this point. If the correct voltages are present, the supply is working and the troubleshooting process can focus on the rest of the circuit.

More difficult troubleshooting problems arise when power supplies produce noise on the power buses, or they fail to adequately filter out transients that occur on the power distribution network. Capturing and displaying these transients may require special equipment such as storage oscilloscopes and transient recorders.

Up to this point, our discussion of dc power sources has focused on dc voltage sources. Another class of dc power sources is the dc current source. Just as the dc voltage source produces a constant voltage for varying levels of load current, the dc current source produces a constant level of output current for varying load conditions.

An ideal current source has an infinite internal resistance. Hence, the circuit resistance is always small compared to the internal resistance, and the resulting source current is limited mainly by the internal resistance of the source. However, practical current sources have a large, but not infinite internal resistance. Hence, practical current sources cannot maintain a constant source current when the load resistance approaches the value of the internal source resistance.

Before leaving this section, it is appropriate to mention some other sources of dc power. These are the thermocouple, piezoelectric crystal, and solar cell.

A thermocouple is made of two different metals joined by welding. When this junction is exposed to a high temperature, an emf is generated. Thermocouples are low-voltage, low-power devices. One application is to generate a voltage to operate the temperature control system of a heating system. Since the thermocouples are excited by the furnace pilot light, the system is independent of power failure.

The piezoelectric crystal generates an emf when pressure is applied to it. Common applications are in microphones and the pickup cartridges of a record player.

The solar cell is another common source of dc power. The solar cell generates an emf when exposed to light. These cells are commonly found in small consumer products such as calculators, toys, and small appliances. In these low-power applications, single solar cells made out of silicon generate voltages of 0.5 volts, and deliver a current of 100 milliamperes. However, when combined in series to produce higher voltages and in parallel to deliver greater amounts of current, solar cells produce electricity in places where other forms of generation are impractical. For example, satellites are powered by solar cells over the long periods that they must remain in space. Remote sensing equipment can also use solar cells to charge secondary cells, and to operate monitoring equipment again over long periods of time.

4.4 The ac Sources

An ac source is one that provides a time-varying voltage which repeats itself at regular intervals. The most common ac source is the common wall outlet. Available at this outlet is an ac voltage of approximately 120–125 volts (measured in rms units) at a frequency of 60 Hertz, or 60 cycles per second.

Many appliances use the ac voltage directly from the wall outlet. Household lighting is a case in point. The ac voltage causes an ac current to flow through the light bulb, which heats the filament to produce light. Other appliances like a toaster or iron also have filaments that convert the ac current flowing through them to heat, which is then used to brown the toast or press the clothes.

Other appliances, instruments, and systems must first convert the ac voltage to dc before powering the electronic circuits that will perform the desired function. Take a personal computer, for example. If you remove the cover from a personal computer, you will see an enclosed box, which is the dc power supply. The power supply is essentially the ac-to-dc converter.

The ac voltage at the wall outlet is a time-varying voltage. Its magnitude changes between a maximum positive voltage and maximum negative voltage 60 times per second. Its amplitude can be expressed in rms, peak, and peak-to-peak values. Let's examine this ac voltage in more detail.

Sinusoidal Signals

sinusoidal signals

sine wave

Figure 4.20 illustrates the process of **sinusoidal signals**. The emf produced has a characteristic shape that could be described mathematically using the **sine wave** function. Hence, we get the terminology, "sinusoidal waveform." Figure 4.20 shows the correspondence be-

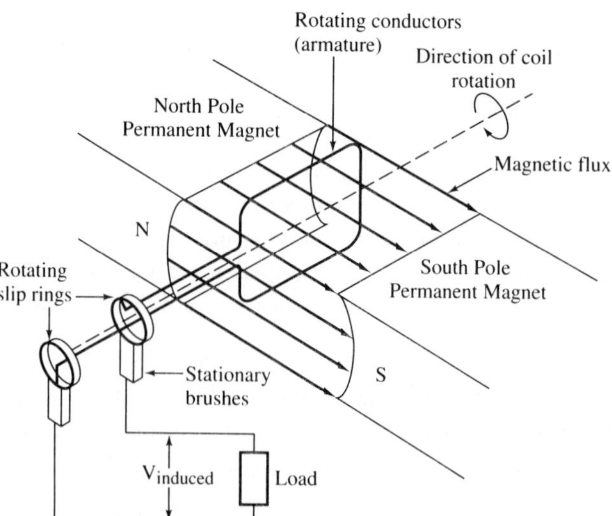

Figure 4.20 Producing a sinusoidal signal by rotating a coil through a magnetic field.

tween the emf waveform produced and the position of the conductor as it was rotated through the magnetic field of the generator.

At 0 degrees, the direction of the conductor is parallel to the magnetic field lines; hence, the emf produced is zero. As the conductor rotates, the conductor begins to cross some of the magnetic field lines. As it does so, a small emf is produced. As the conductor rotation approaches 90 degrees, the maximum rate of crossing magnetic field lines occurs and the largest emf results. Rotation from 90 degrees to 180 degrees results in a decrease in the emf until the conductor is again traveling parallel to the magnetic lines of force at 180 degrees. Rotation from 180 degrees to 360 degrees results in an induced emf of opposite polarity. If the first half of the rotation is considered to have a positive polarity, then the rotation through the second half of the cycle results in a negative polarity for the induced emf.

The power that the electric company delivers to your home or business is generated in this manner. The turbines, which are driven by water rushing through a hydroelectric dam, or steam in a nuclear or coal-fired power plant, turn generators that produce a sinusoidally varying voltage and current. The power contained in this voltage and current is then transmitted to residential users or large industrial customers.

A waveform that follows the shape of the sine function is said to be ''sinusoidal.'' Table 4.4 lists the values of the sine function from 0–360 degrees. If these values are plotted, the graph shown in Figure 4.21 is obtained. All sinusoidal waveforms have this characteristic shape even though they may differ in amplitude and the time it takes to complete a cycle.

Let us examine sinusoidal waveforms in more detail. Plotting several cycles of a sinusoidal voltage or current as a function of time gives us a

TABLE 4.4 Sine Function Values*

Degrees	Radius	Sin (θ)
0	0	0.0000
30	$\pi/6$	0.5000
60	$\pi/3$	0.8660
90	$\pi/2$	1.0000
120	$2\pi/3$	0.8660
150	$5\pi/6$	0.5000
180	π	0.0000
210	$7\pi/6$	−0.5000
240	$4\pi/3$	−0.8660
270	$3\pi/2$	−1.000
300	$5\pi/2$	−0.8660
330	$11\pi/6$	−0.5000
360	2π	0.0000

*A more complete table of values for the sine function can be found in a handbook of mathematical tables and formulas.

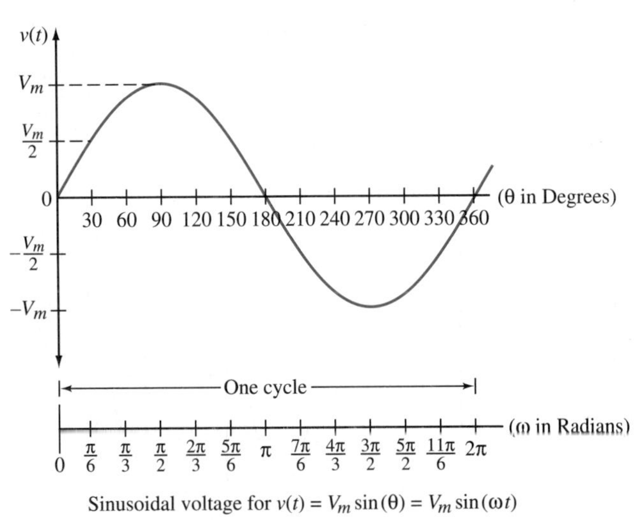

Figure 4.21 A sinusoidal voltage, $v(t) = V_m \sin(\theta) = V_m \sin(\omega t)$.

graph like the one shown in Figure 4.22. One parameter, which is associated with sine waves, or any repetitive signal for that matter, is **period**. The period of a signal is the time it takes to complete one cycle. A cycle is defined as the time it takes to go from one point on the waveform to the next identical point on the waveform—that is— the next point on the waveform that has the same amplitude and slope. From the period of the waveform, we can then calculate the number of cycles per second or frequency by taking the reciprocal of the period. That is,

period

$$f = 1/T \tag{4.2}$$

where

f is the frequency, in Hertz
and T is the period, in seconds

To measure the period of the signal, two consecutive zero-crossing points A and A′ can be selected. The time between points A and A′ is labeled T_1 and measures 2 milliseconds. Points B and B′ could also be used to determine the period of the waveform. Both points have a magnitude of 0.75 V_p and have a positive slope. The time difference between these two points, T_2, measures 2 milliseconds. And finally, the negative peaks could be used. Again, the time difference between points C and C′ measures 2 milliseconds. All three pairs of points on the waveform yield the same period measurement.

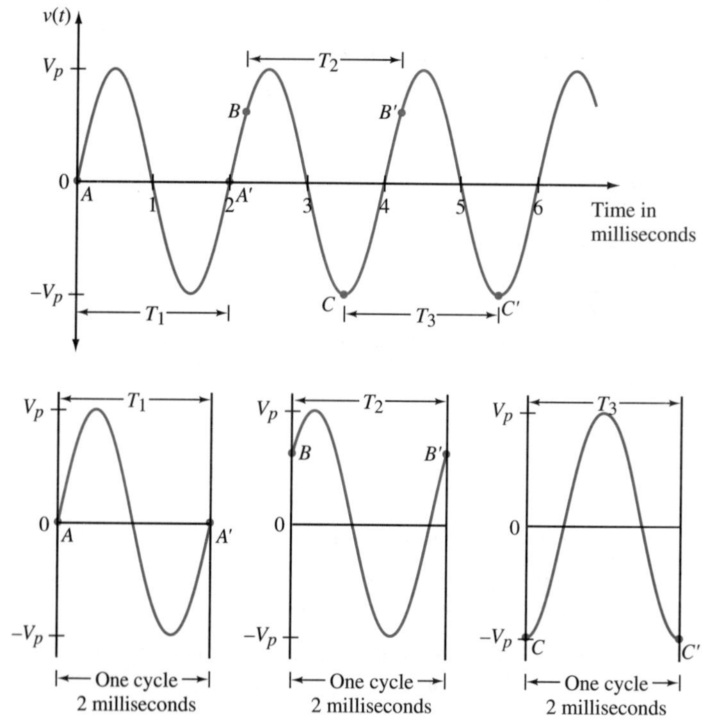

Figure 4.22 Selecting identical points on a sinusoidal waveform to measure the period.

EXAMPLE 4.3

Suppose that we are given the voltage waveforms shown in Figure 4.23. Determine the period and frequency of each signal.

Solution

Waveform A:

$$\text{Period (by measurement on the graph)} = 1 \text{ ms}$$
$$f = 1/T = 1/(1 \text{ ms})$$
$$f = 1 \text{ kHz}$$

Waveform B:

$$\text{Period (by measurement on the graph)} = 0.5 \text{ ms}$$
$$f = 1/T = 1/(0.5 \text{ ms})$$
$$f = 2 \text{ kHz}$$

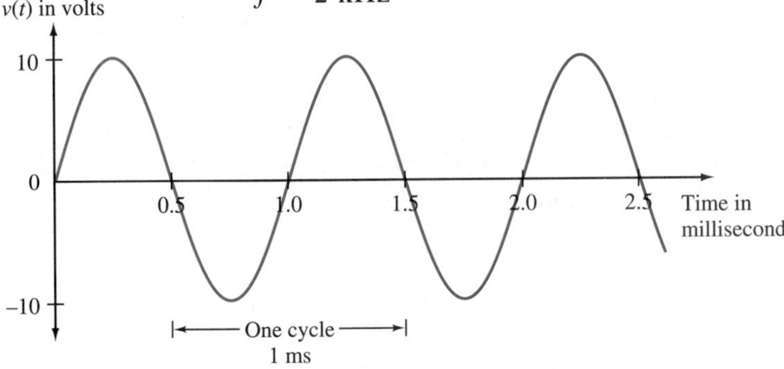

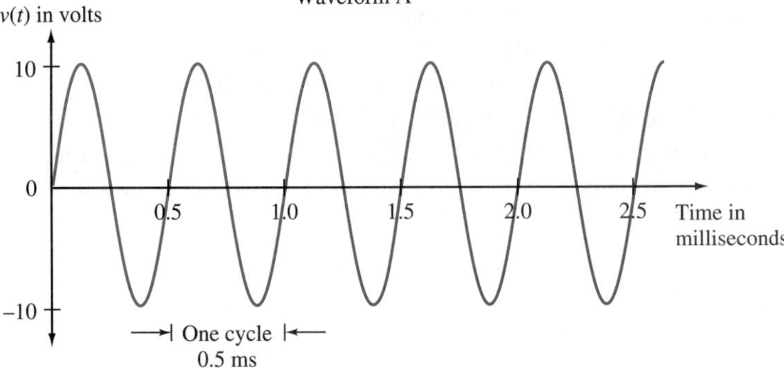

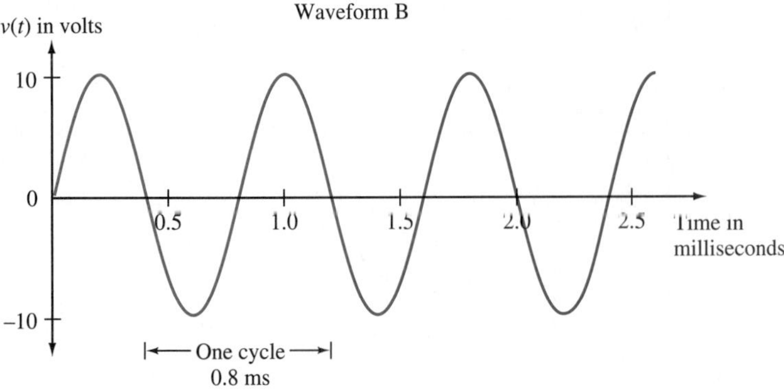

Figure 4.23 Sinusoidal signals of varying frequencies.

Waveform C:

$$\text{Period (by measurement on the graph)} = 0.8 \text{ ms}$$
$$f = 1/T = 1/(0.8 \text{ ms})$$
$$f = 1.25 \text{ kHz}$$

Amplitude

amplitude

There are two ways of specifying the **amplitude** of a sinusoidally varying signal. One way is to give the peak-to-peak or peak value of the signal. The peak-to-peak value is easily obtained from a amplitude versus time graph of the signal, as can be seen on the oscilloscope.

Consider the signal shown in Figure 4.24. The positive peak voltage is +10 volts. The negative peak voltage is −10 volts. The difference between these two values is 20 volts. Hence, the signal's amplitude is said to be "20 volts peak-to-peak."

The peak value of the signal shown in Figure 4.24 is one-half the peak-to-peak value, or 10 volts peak.

rms value

The second method of specifying the amplitude of a sinusoidal signal uses the **root-mean-square (rms) value**. This value gives the amplitude of a dc voltage that produces an equivalent amount of power to be delivered to a load.

For example, the 100-watt light bulb is rated at 120 volts. The 120-volt rating tells you that the bulb can be connected to a standard wall outlet that delivers 120 volts. The 120-volt rating is an rms value for the sinusoidally varying, 60-Hz voltage that is present. If we were to take the same 100-watt, 120-volt light bulb and connect it to a dc power supply or 110 volt battery, the same amount of power would be dissipated by the bulb and its brightness should be the same in both cases.

The instantaneous value of the 120-volt, 60-Hz voltage actually varies between +170 volts and −170 volts. The peak-to-peak amplitude of the

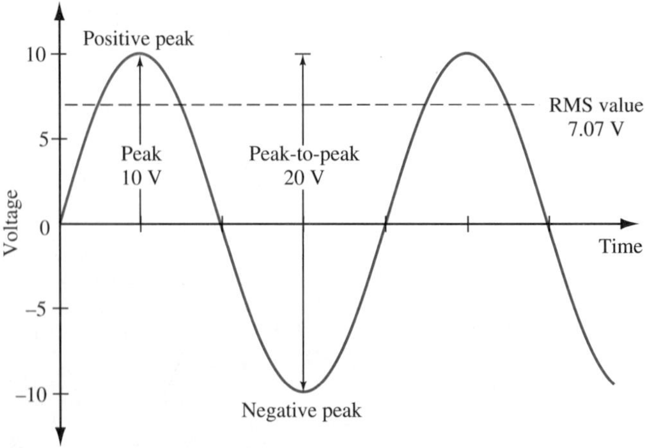

Figure 4.24 A 20 V_{p-p} sinusoidal signal.

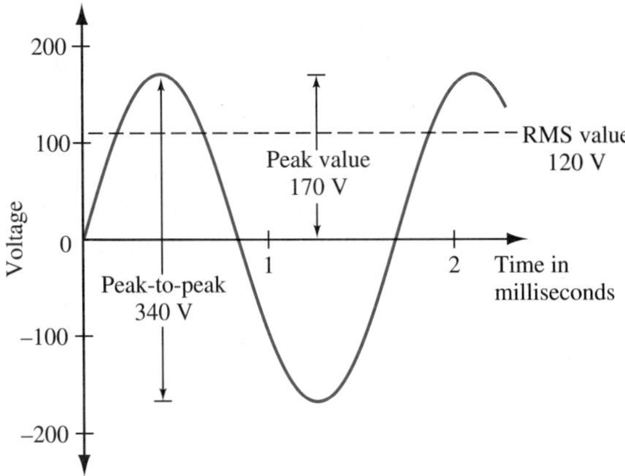

Figure 4.25 A 120 V_{rms}, 60-Hz signal.

wall outlet voltage is 340 volts. Figure 4.25 shows the relationship among the peak-to-peak, peak, and rms values for the 120-volt wall outlet potential. The relationship between the peak value and the rms value can be stated as follows:

$$\text{rms value} = 0.707 \times \text{peak value} \tag{4.3}$$

EXAMPLE 4.4

Determine the rms value of the signal shown in Figure 4.26.

Solution

From the graph, the peak value of the signal is measured to be 15 volts. Using this value, we can calculate the rms value by multiplying the peak value by 0.707.

$$V_{rms} = 0.707 \times 15 \text{ volts}$$
$$V_{rms} = 10.5 \text{ V}$$

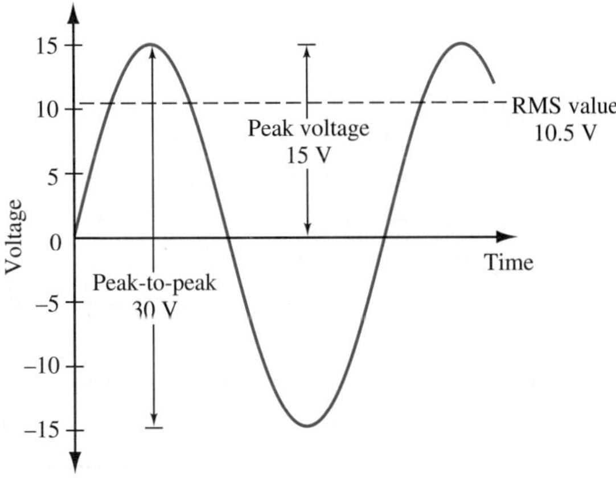

Figure 4.26 A 30 $V_{p\text{-}p}$ sinusoidal signal.

Digital multimeters and voltmeters, in their ac volt setting, measure the rms value of the signal being monitored. A reading will be produced for nonsinusoidal waveforms if they are used as inputs to the digital multimeter or voltmeter, but the reading will be inaccurate on most meters. The one exception is for those meters that are true rms instruments. These instruments will accurately measure the rms value of nonsinusoidal signals.

Looking ahead, in Section 4.6, we will learn how to generate sinusoidal signals using an instrument called a "function generator." Then in Section 4.7, the operation and capabilities of an oscilloscope will be described. The oscilloscope, or scope, is still the main instrument used to analyze and monitor signals within an electronic system.

4.5 Pulse Waveforms

pulses

Digital systems use a different type of signal to convey information and to control events. These signals take the form of **pulses.** Figure 4.27 shows the Intel 8088 microprocessor "Read Bus Cycle" timing diagram. The timing diagram shows the necessary relationship of many signals if the 8088 microprocessor is to operate correctly. Understanding these diagrams is crucial to one's ability to analyze and troubleshoot microprocessor systems.

In this section, we will examine three types of waveforms: step functions, single pulses, and repetitive pulse waveforms. These waveforms, as contrasted with the sinusoidal signal that we studied in the previous sections, consist of abrupt transitions from one level to another.

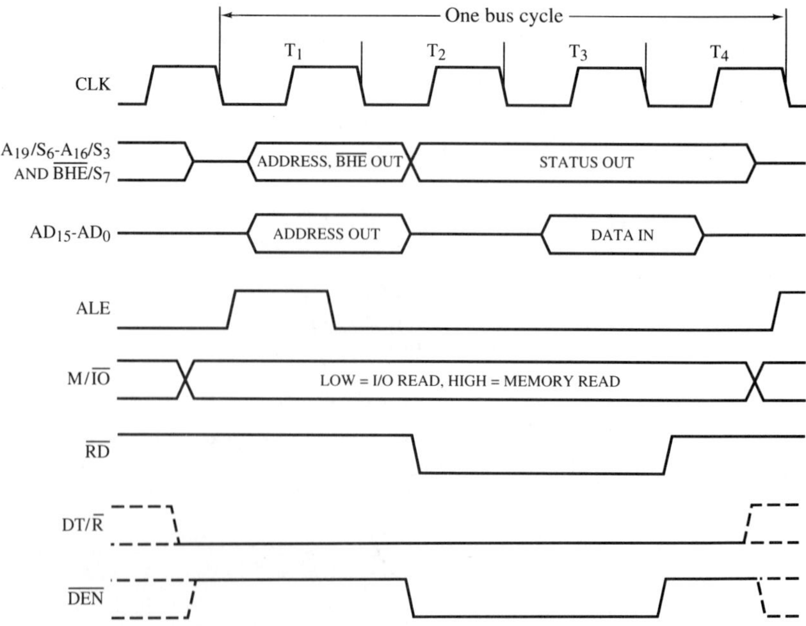

Figure 4.27 The timing sequence for a "Read Bus Cycle" for an Intel 8086 microprocessor.

Step Functions

An ideal step function is an abrupt change from one voltage or current level to another level, at a specified point in time. Consider the signal in Figure 4.28(a). The signal has been at level V_1 for some time. At a certain point in time, the signal makes an abrupt change to a new level, V_2. If V_2 is more positive than V_1, the function is said to be a positive-going step. The magnitude of the step is the difference between level V_2 and level V_1.

A negative-going step occurs when level V_1 is more positive than level V_2. Again, it is assumed that the signal has been at level V_1 for some time and then abruptly changes to level V_2. This is illustrated in Figure 4.28(b).

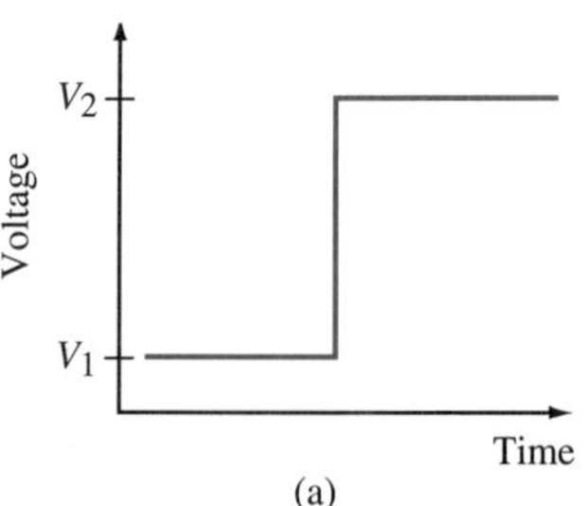

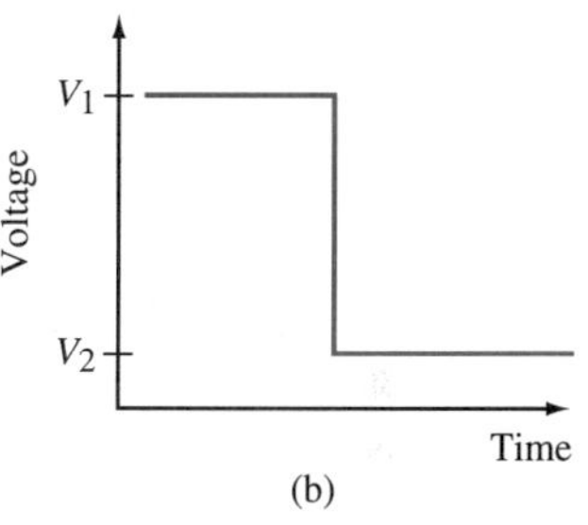

Figure 4.28 (a) A positive-going step function. (b) A negative-going step function.

A flashlight will serve as an illustration of a step function. Assume that the flashlight is initially turned OFF. With the switch open, the battery is not connected to the bulb. Hence, with no applied voltage, the current through the bulb is zero and the bulb is not lit. At some point in time, the switch on the flashlight is moved to its ON position. The circuit is now complete and the voltage produced by the battery is dropped across the bulb. The resulting current produces heat in the filament of the bulb, and the bulb glows, which produces light.

Turning on the flashlight can be described graphically as a step function. Figure 4.29 shows an initial level V_1 of 0 volts when the flashlight is OFF. Once the flashlight is turned ON, the voltage jumps to 3 volts, assuming we are using two "C" or "D" cells in our flashlight. The voltage across the bulb can be described as a 3-volt, positive-going step function from 0 volts to +3 volts.

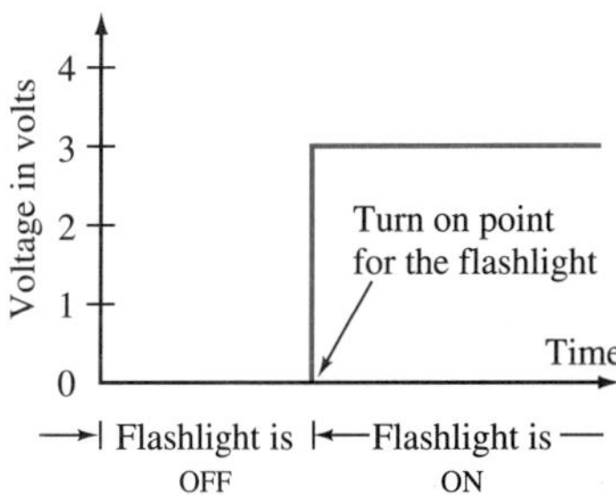

Figure 4.29 A graphical representation of the voltage across a flashlight bulb, before and after the switch is turned ON.

Let us consider another example of generating a step function. The circuit shown in Figure 4.30(a) shows the connection of a 1-kilohm resistor and a single-pole, single-throw (SPST) switch between a +10-volt source and ground. Let us assume that the switch is initially in the "open" position. What type of input signal is generated when the switch is closed? With the switch in the "open" position, no current can flow through the 1-kilohm resistor and switch. With zero current in the resistor, the voltage drop across the 1-kilohm resistor is zero. Subtracting the resistor voltage from the supply voltage produces a voltage at Point A of +10 volts. When the switch is closed, current flows through the 1-kilohm resistor and closed switch to ground. The entire 10 volts is

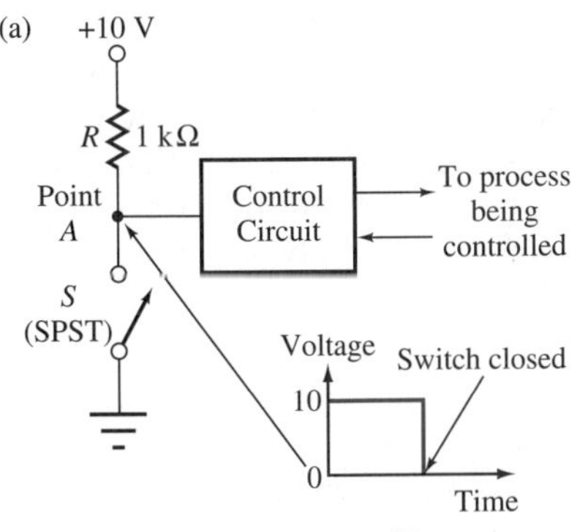

Figure 4.30 Switch input circuit that generates a step function input when the switch is closed.

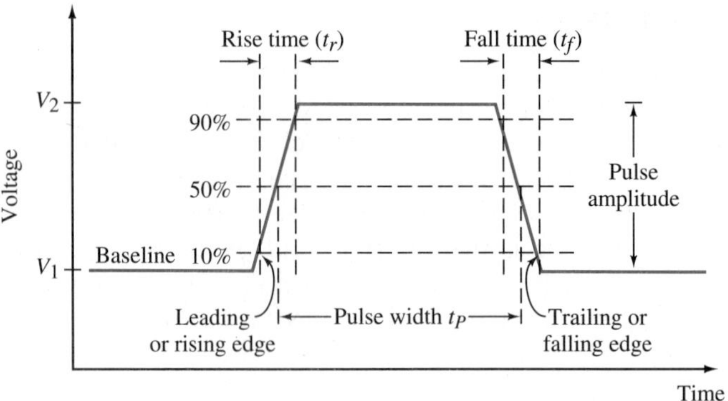

Figure 4.31 Pulse waveform parameters.

dropped across the resistor and the potential at Point A is zero volts. Essentially, the closed switch shorts Point A to ground. Representing the voltage at Point A in graphical form produces the signal shown in Figure 4.30(b). The signal at Point A is a negative-going step function, which goes from +10 volts to 0 volts.

Pulse Waveforms

A pulse waveform consists of two transitions or steps as shown in Figure 4.31. The signal is initially at a baseline voltage, V_1. At time T_1, the signal changes quickly from V_1 to a new level V_2. The voltage then remains at level V_2 for a length of time called the **pulse width**. At the end of the pulse width, the signal then returns to its initial level, V_1. Hence, the pulse width is the time that the pulse is present.

pulse width

Ideally, the transitions from V_1 to V_2 and back again would happen instantaneously. In actuality, some time is required to make these transitions. The transition on the leading edge of the pulse is called the "rise time." Rise time is measured between the 10–90 percent points on the leading or rising edge. Likewise, the "fall time" is measured between the 90–10 percent points on the **falling edge** of the pulse.

falling edge

Practical pulses exhibit additional characteristics such as preshoot, overshoot, and ripple or ringing. Preshoot occurs just prior to the leading or **rising edge** of the waveform. Overshoot occurs when the signal "overshoots" the pulse amplitude before it settles down to a stable value. The **ripple factor** or ringing results in oscillation of the signal around the stable pulse amplitude. These attributes of practical pulse waveforms will not be discussed here, but rather will be reserved for your advanced circuits and digital systems courses of study.

rising edge

ripple factor

Repetitive Pulse Waveforms

Repetitive pulse waveforms are used as clock and timing signals. Returning to our earlier computer example, we noted in Figure 4.27 that the Intel 8088 requires a clock signal. This clock signal defines time states that make up each bus cycle.

Like sinusoidal signals, repetitive pulse waveforms are defined by the parameters: period, **frequency**, and peak-to-peak amplitude. In addition, two new parameters will be added to describe the pulse waveform. They are baseline and duty cycle.

frequency

Consider the repetitive pulse waveform shown in Figure 4.32. The waveform shown has a period of 2 microseconds. Taking the reciprocal of the period, we can calculate the frequency of the signal:

$$f = 1/T = 1/(2\ \mu s)$$
$$f = 500\ \text{kHz}$$

The peak-to-peak amplitude of the waveform is 2.5 volts (3.0 volts − 0.5 volts). The pulses are riding on a baseline voltage of 0.5 volts. The rise time and fall time are assumed to be very short compared to the pulse width, and are shown as vertical lines in the graph.

Duty cycle is defined as the percentage of time the signal spends producing the pulse compared to the period of the waveform. Duty cycle is defined as

$$\text{duty cycle} = (T_p/T) \times 100\% \qquad \textbf{(4.4)}$$

where T_p is the pulse width
and T is the period of the waveform

For the waveform shown in Figure 4.32, the pulse width is 1 microsecond while the period is 2 microseconds. Therefore, the duty cycle is

$$\text{duty cycle} = (1\ \mu s/2\ \mu s) \times 100\%$$
$$\text{duty cycle} = 50\%$$

A repetitive pulse waveform with a 50 percent duty cycle is called a **square wave**.

square wave

Viewing pulses is usually accomplished with an oscilloscope. The adage that "a picture is worth a thousand words" is true here. Being able to view the shape of the pulse waveforms yields much information to the engineer and technician.

For some applications, viewing pulse waveforms with a digital counter and logic probe can also provide useful diagnostic information. The

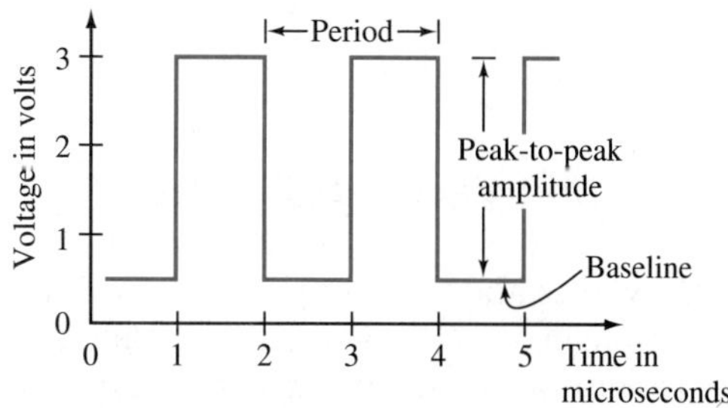

Figure 4.32 A repetitive pulse waveform.

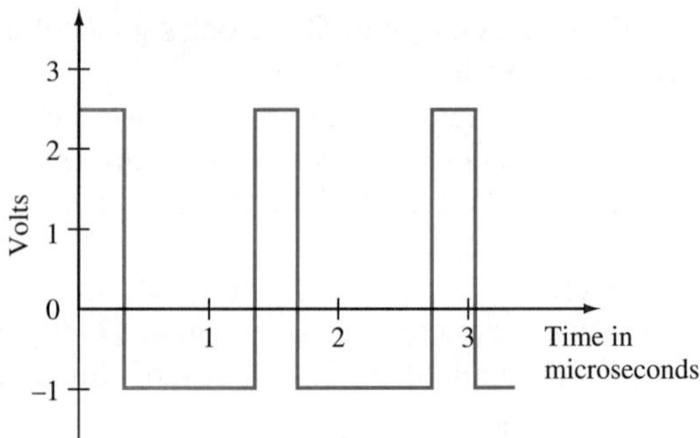

Figure 4.33 A repetitive pulse waveform.

digital counter and logic probe will be covered in greater detail in the digital systems course later in the curriculum.

EXAMPLE 4.5

Describe in words the waveform shown in Figure 4.33.

Solution

The amplitude can be determined by measuring the voltage difference between the baseline value (−1 volt) and the peak of the pulses (+2.5 volts). Hence, the amplitude of the pulses is 3.5 volts and the pulses are positive-going.

The frequency of the signal can be determined by first measuring the period and then taking the reciprocal of the period. By picking two identical points on the waveform (usually either the leading or trailing edges), the period is determined to be 1.4 microseconds. Hence, the frequency is

$$f = (1/T) = 1/(1.4\ \mu s)$$
$$f = 714\ \text{kHz}$$

The baseline value for the waveform is −1 volt.

And finally, the duty cycle is:

$$\begin{aligned}\text{duty cycle} &= (T_p/T) \times 100\% \\ &= (0.4\ \mu s/1.4\ \mu s) \times 100\% \\ \text{duty cycle} &= 28.6\%\end{aligned}$$

4.6 Generating Time-Varying Signals

function generator

Generating sine waves and repetitive pulse waveforms in the laboratory is easily accomplished using an instrument called a **function generator**. A function generator is capable of generating a number of wave shapes, such as sine, square, and triangular.

There are a number of standard, front-panel controls on a function generator: AMPLITUDE, FREQUENCY and MULTIPLIER, FUNCTION, and OFFSET. Understanding the parameter controlled by each of these controls enables you to successfully set up the function generator to produce a given signal.

The Tektronix FG 503 Function Generator, shown in Figure 4.34, is designed to produce low distortion sine, square, and triangle waveforms in the frequency range of 1 Hz–3 mHz. A variable dc offset of ±5 V is also provided. For more advanced applications, a +2.5-V square-wave trigger output and a voltage-controlled frequency (VCF) input control are available at the front panel. For now, we will not concern ourselves with the trigger output and VCF control inputs, but will focus our attention on the amplitude, frequency and multiplier, and function controls to produce the desired signal at the output terminal. Figure 4.35 shows the position of the front panel controls on the FG 503.

Basic Operation

The basic operation of the function generator as a signal source is very straightforward. The first thing that you need to know is the characteristics of the signal to be produced:

1. Function or waveshape
2. Frequency
3. Amplitude
4. dc offset

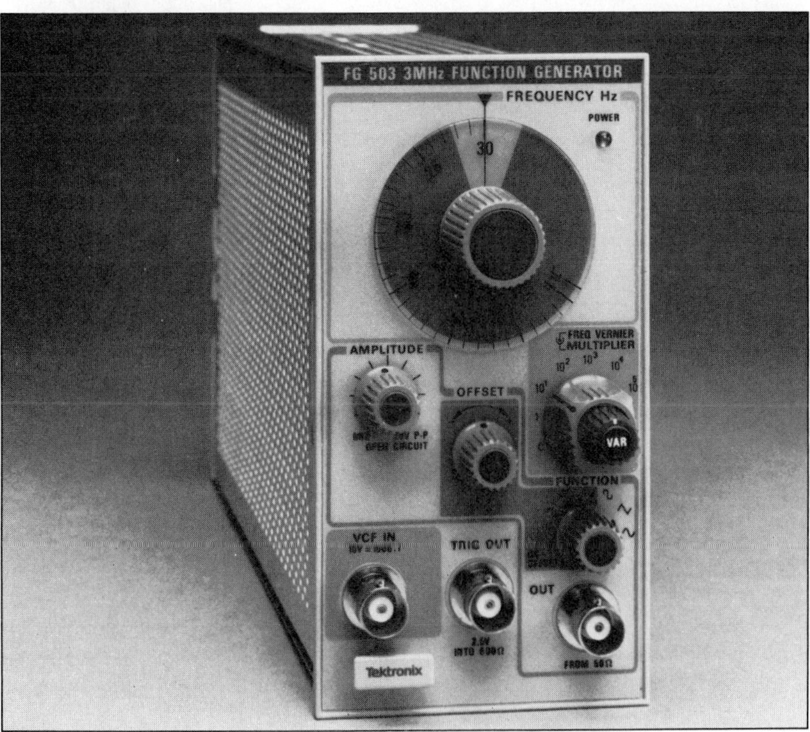

Figure 4.34 The Tektronix FG 503 Function Generator. (Used with permission of Tektronix, Inc.)

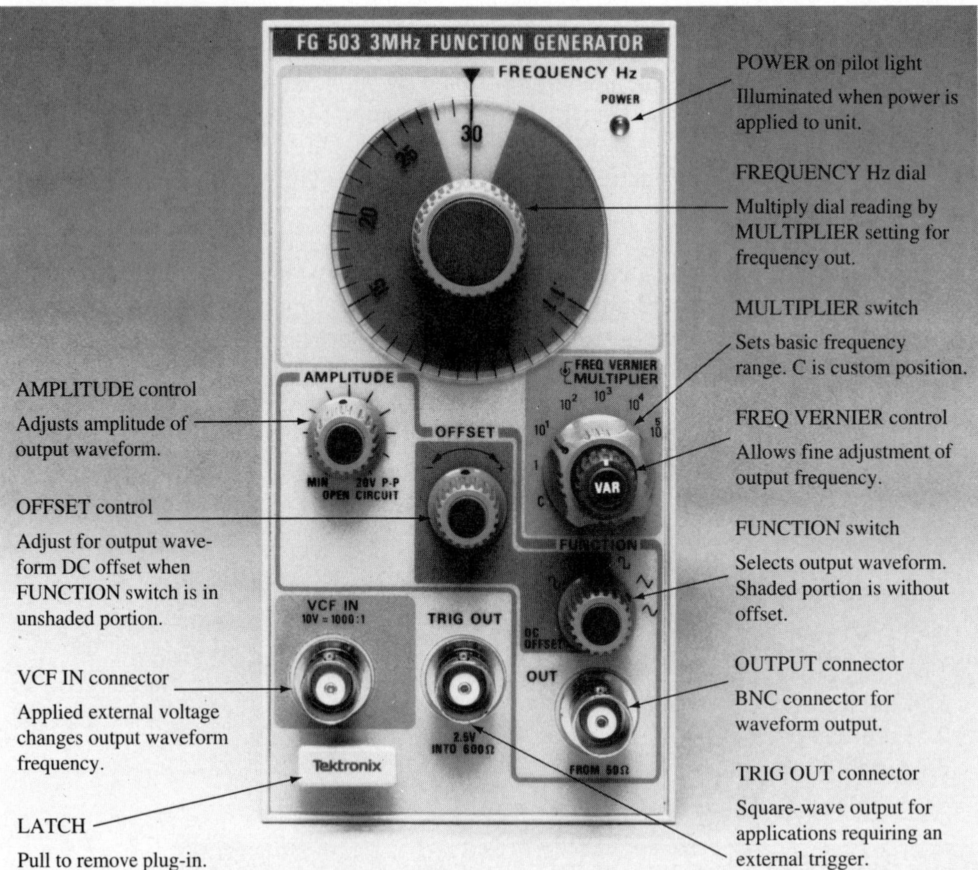

Figure 4.35 The front panel controls of a Tektronix FG 503 Function Generator. (Used with permission of Tektronix, Inc.)

Waveshape

triangle wave

There are only three basic waveshapes that the FG 503 (Figure 4.35) can produce: sine, square, and **triangle**. Other makes and models of function generators can generate additional waveshapes.

Notice that there are two identical sets of waveforms near the FUNCTION selector switch on the front panel. The three waveforms in the lightly shaded area allow use of the OFFSET control; the other three do not. The OFFSET control will be described after the AMPLITUDE control.

Frequency

The frequency of the desired signal is selected by using two controls: FREQUENCY (large dial) and MULTIPLIER (switch). The large dial is calibrated from 1–30, providing a continuous range of settings between the minimum and maximum settings. The MULTIPLIER switch, on the other hand, has specific settings, each being an integer power of 10 (e.g., 10^2, 10^3).

For example, if a 5000-Hz signal is needed, the large dial would be set to 5 by matching 5 on the dial to the arrow at the top of the front panel. The pointer on the MULTIPLIER control is then set to 10^3. The combination of these two settings yields a frequency of 5×10^3 or 5000 Hz. These settings for the FG 503 are shown in Figure 4.36.

Note that there may be two equivalent settings for some frequencies. Take, for example, a frequency of 10 kHz. One setting would be "1" on the FREQUENCY dial with a MULTIPLIER of 10^4. An equivalent setting would be "10" on the FREQUENCY dial with a MULTIPLIER of 10^3.

Amplitude

The output amplitude can be set to a value up to 20 $V_{p\text{-}p}$ into an open circuit, or 10 $V_{p\text{-}p}$ into a 50-ohm load. From the minimum setting (full counterclockwise position), as the control is rotated clockwise, the amplitude of the output signal will gradually increase until the maximum setting is reached.

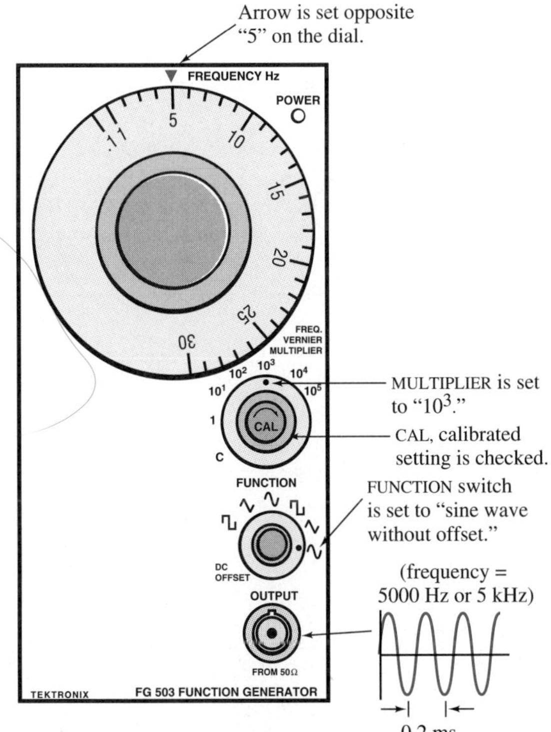

Figure 4.36 Frequency and multiplier settings for the FG 503 Function Generator to produce a 5-kHz sine wave.

dc Offset

dc offset

The FG 503 will produce at least a ±7.5-V offset into an open circuit or ±3.75-V offset into a 50-ohm load. Essentially, the **dc offset** adds a dc voltage to the time-varying signal being produced. Figure 4.37 illustrates the insertion of a dc offset to a sinusoidally varying signal.

Basic Operation

Given the parameters of the signal to be generated, the following procedure can be used to set up the function generator:

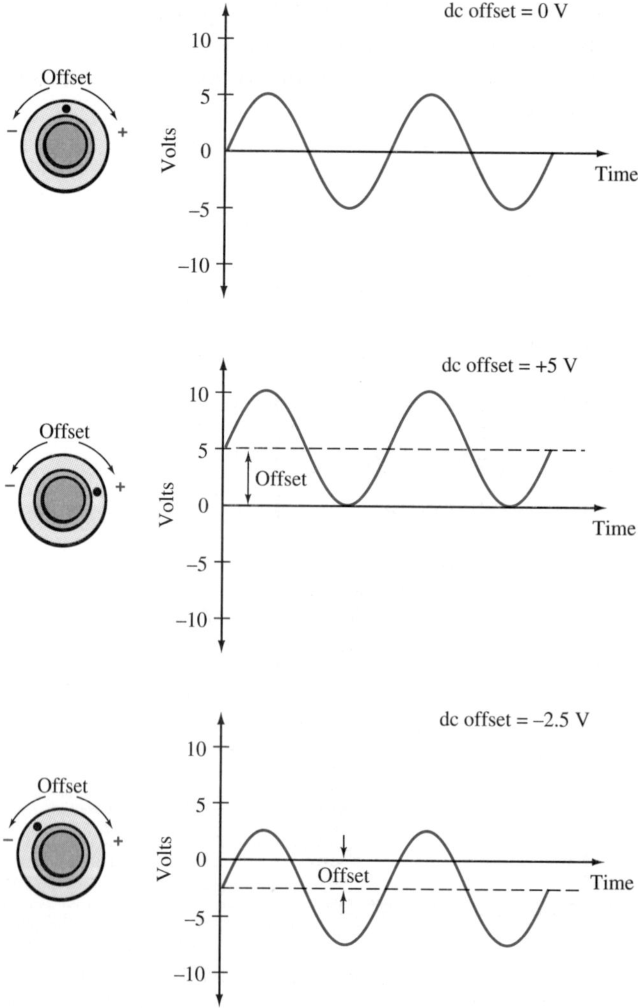

Figure 4.37 Examples of dc offset settings.

1. Set the amplitude control to its minimum setting, which is the full counterclockwise setting.
2. Select the desired waveform, with or without dc offset.
3. Select the desired frequency with the MULTIPLIER switch and the FREQUENCY Hz dial. The output frequency is calibrated when the FREQUENCY VERNIER control is in the full clockwise position.
4. Connect the load to the OUTPUT connector and adjust the AMPLITUDE control for the desired peak-to-peak output amplitude. A voltmeter or oscilloscope is needed to measure the amplitude of the output signal while the AMPLITUDE control is adjusted.
5. If a dc offset is required, adjust the OFFSET control to position the dc level (baseline) of the output signal above or below 0 volts, as desired.

EXAMPLE 4.6

Set up the FG 503 to produce a 2-kHz triangle wave with a peak-to-peak amplitude of 8 volts with zero offset.

Solution

The graph of the signal is shown in Figure 4.38. To set up the FG 503,

1. Set the amplitude control to its minimum setting, which is the full counterclockwise rotation.
2. Set the FUNCTION switch to the triangle wave symbol without offset.
3. Rotate the FREQUENCY Hz dial to "2" (the first small division to the right of the "1"), placing it opposite the arrow, and set the MULTIPLIER to 10^3. Check to make sure the FREQUENCY VERNIER is in the CAL position.
4. Connect the load to the OUTPUT. Connect a scope to monitor the output and gradually increase the AMPLITUDE control until the output signal measures 8 $V_{p\text{-}p}$.

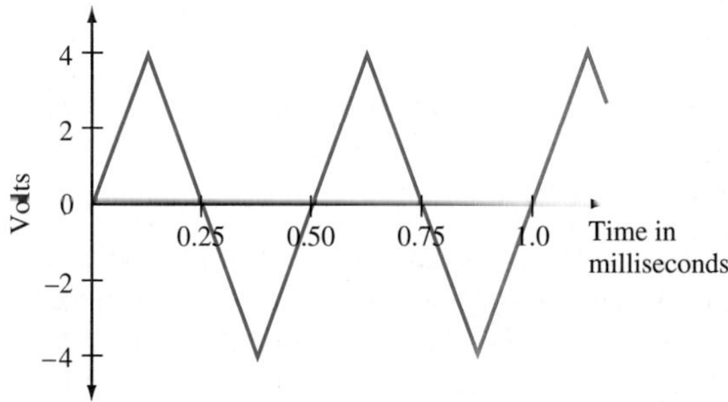

Figure 4.38 A 2-kHz triangular waveform.

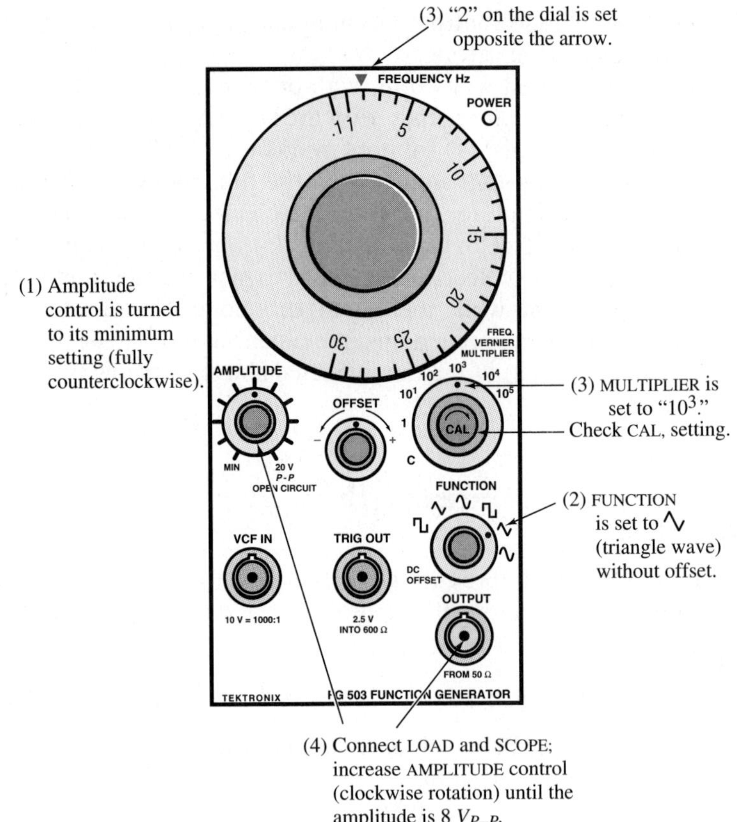

Figure 4.39 Front panel controls for the FG 503 Function Generator.

Figure 4.39 shows the front-panel control settings to produce the desired signal.

EXAMPLE 4.7

Set up the FG 503 Function Generator to produce a 10-Hz square wave, which goes from 0 volts to +4 volts.

Solution

Again, let us begin by graphing the desired signal. The graph is shown in Figure 4.40

To set up the FG 503,

1. Set the AMPLITUDE control to its minimum setting.
2. Set the FUNCTION switch to the square wave symbol with dc offset.
3. Set the FREQUENCY Hz dial to "10" and the MULTIPLIER switch to "1." Check to make sure the FREQUENCY VERNIER control is in the CAL position.

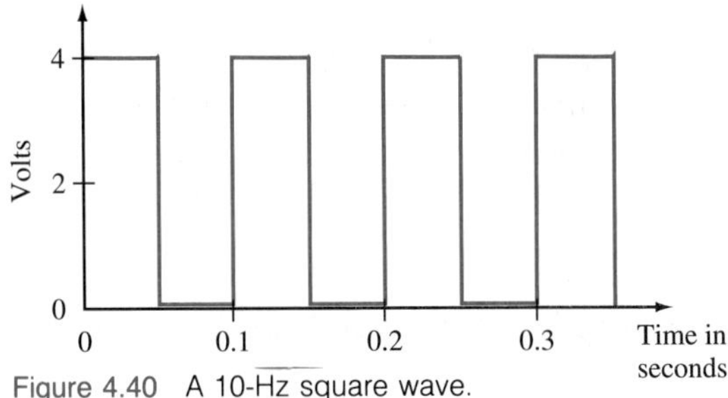

Volts
4
2
0
0
0.1
0.2
0.3
Time in seconds

Figure 4.40 A 10-Hz square wave.

4. Connect the load and a scope to the OUTPUT terminal, and gradually increase the AMPLITUDE control until the peak-to-peak amplitude of the square wave is 4 volts.
5. Finally, adjust the DC OFFSET control until the square wave fits between 0 volts and +4 volts. The final settings are shown in Figure 4.41.

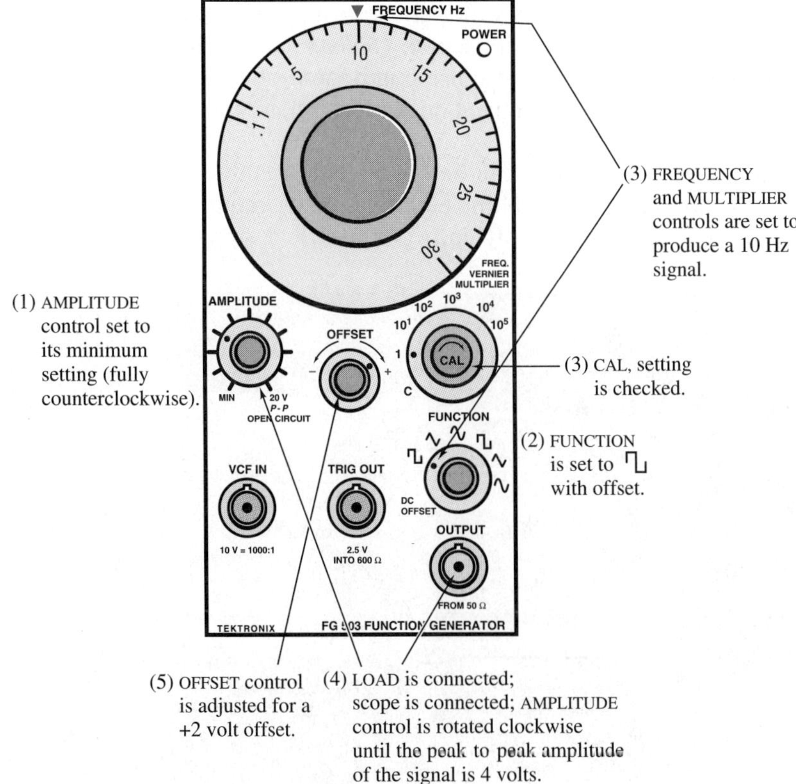

FREQUENCY Hz
POWER
10
15
5
20
.1
25
30
FREQ. VERNIER MULTIPLIER
AMPLITUDE
OFFSET
CAL
MIN
20 V P-P OPEN CIRCUIT
FUNCTION
VCF IN
TRIG OUT
DC OFFSET
OUTPUT
10 V = 1000:1
2.5 V INTO 600 Ω
FROM 50 Ω
TEKTRONIX
FG 503 FUNCTION GENERATOR
(1) AMPLITUDE control set to its minimum setting (fully counterclockwise).
(3) FREQUENCY and MULTIPLIER controls are set to produce a 10 Hz signal.
(3) CAL, setting is checked.
(2) FUNCTION is set to with offset.
(5) OFFSET control is adjusted for a +2 volt offset.
(4) LOAD is connected; scope is connected; AMPLITUDE control is rotated clockwise until the peak to peak amplitude of the signal is 4 volts.

Figure 4.41 Front panel controls for the FG 503.

The function generator you use in the laboratory may not be the Tektronix FG 503 Function Generator. However, the one you use will have controls similar in function to the AMPLITUDE, FREQUENCY AND MULTIPLIER, FUNCTION, and DC OFFSET controls on the FG 503. Consult your operator's manual for the position and operation of these controls.

4.7 Measuring and Displaying Time-Varying Signals

The Oscilloscope

oscilloscope

The **oscilloscope** is the most important general-purpose test instrument, which an engineer or technician will use. Almost any physical phenomena can be measured with the two-dimensional graph drawn by an oscilloscope. In this section, we will begin the process of correctly learning to use the oscilloscope as a versatile measurement tool.

An example of an oscilloscope, which you might use in the laboratory, is shown in Figure 4.42. In most applications, the scope shows you a graph of voltage (on the vertical axis) and time (on the horizontal axis). This graphical representation of a signal presents far more information than is available from any other single test instrument. For example, from one display on an oscilloscope, you can find the dc component, ac component, frequency, and noise component of a signal. The oscilloscope lets you see everything at once rather than requiring you to make many separate tests.

Most electrical signals can be connected directly to the scope with either probes or cables. Nonelectrical signals require transducers to convert one kind of energy into an electrical signal. Speakers and

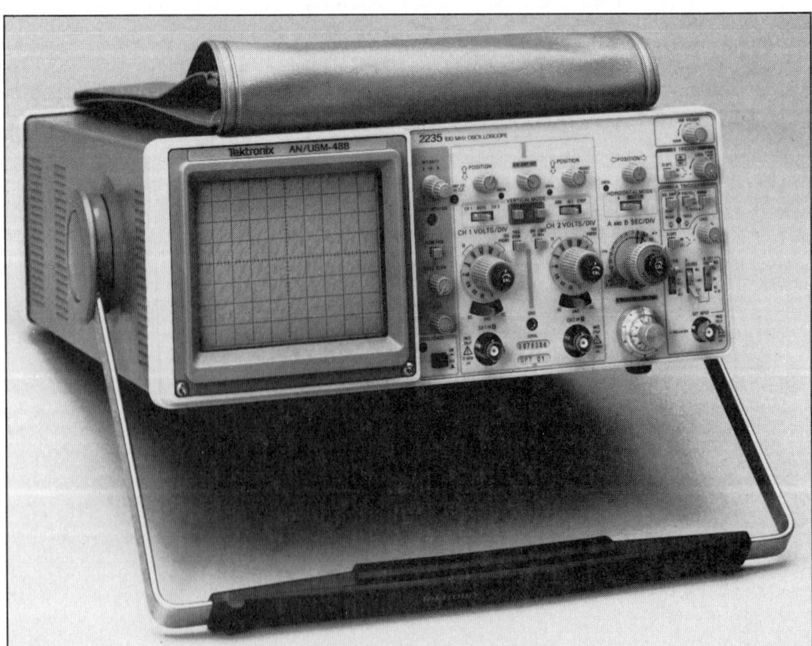

Figure 4.42 The Tektronix 2235 oscilloscope. (Used with permission of Tektronix, Inc.)

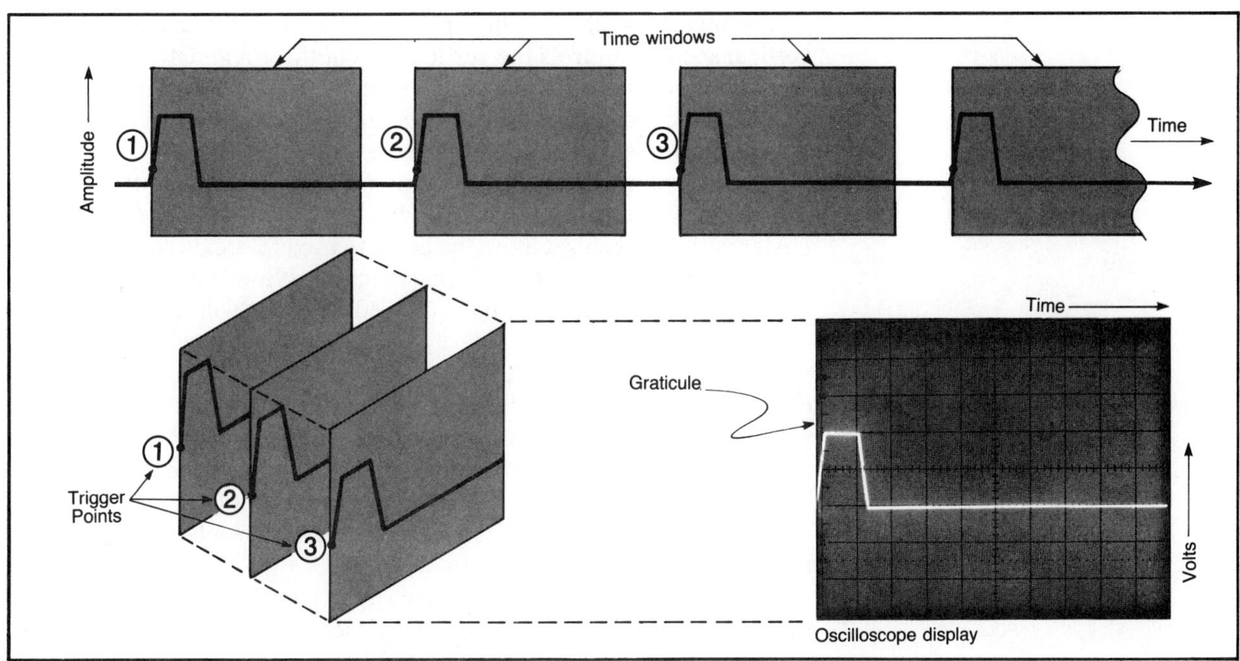

Figure 4.43 Construction of a crt display by capturing and overlaying successive windows of time. (Copyright 1978, "Basic Oscilloscope Operation," Figure 3, Page 3, Tektronix, Inc. Used with permission.)

microphones are two examples of transducers; speakers change electrical energy to sound waves, and microphones change sound waves to time-varying voltages. Other transducers convert mechanical stress, heat, light, and pressure into electrical signals.

Basically, the oscilloscope creates a CRT display by capturing and successively overlaying successive time windows of the signal. Each window starts at the same point on the waveform, called the trigger point, as shown in Figure 4.43.

Making measurements with an oscilloscope is easier if you understand how a scope works. Figure 4.44 shows a simplified block diagram of a oscilloscope. The block diagram consists of four functional blocks: the vertical section, the trigger section, the horizontal section, and the display section. Each is named for what it does.

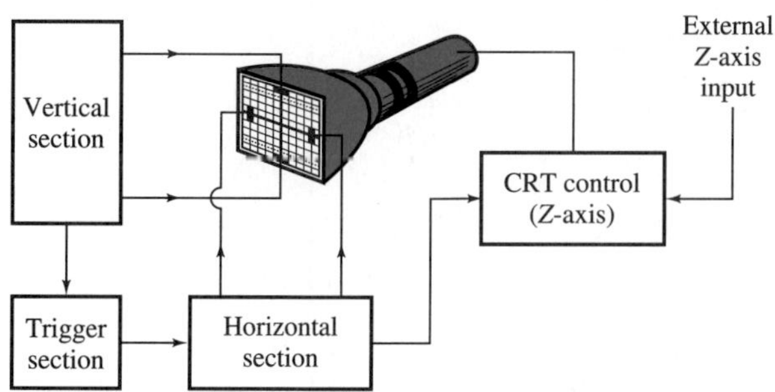

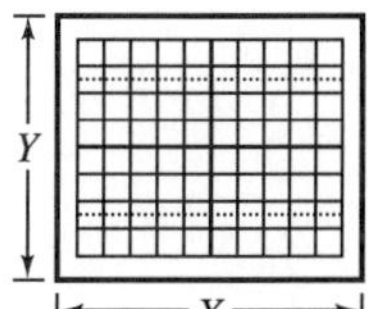

Figure 4.44 The display system of an oscilloscope. (Copyright 1983, "The XYZs of Using a Scope," Figure 3, Page 5, Tektronix, Inc. Used with permission.)

The incoming signal is first processed by the vertical section. The vertical section controls the vertical axis of the graph. Often the incoming signal must be amplified or attenuated to cause the right amount of vertical deflection of the electron beam that draws the waveform on the CRT screen.

The horizontal and trigger sections work together to control the left-to-right movement of the electron beam across the face of the CRT. Faster varying signals require the beam to travel faster across the face of the CRT than slower varying signals require. If too slow a sweep rate is used, too many cycles will be displayed and they will appear squished together; it will be impossible to distinguish individual cycles of the signal. On the other hand, if the writing rate is too fast, only a small portion of one cycle will be displayed.

intensity
focus
trace rotation

The trigger section determines when a time window begins. Front panel controls let you select the amplitude and slope of the input signal at the trigger point. The display section consists of the CRT and its associated drive circuitry. This section controls the **intensity**, **focus**, and **trace rotation** of the displayed waveform.

Now let us examine these features in greater detail. The objective of this section is not to make you expert users of the oscilloscope, but to give you some basic operating instructions so that you can begin displaying signals and interpreting oscilloscope displays.

The Vertical Section

coupling

The vertical section, as its name implies, supplies the vertical deflection voltages that will move the beam up and down as it moves across the face of the CRT. Important controls for the vertical section include **coupling**, position, vertical sensitivity, and operating modes, which are shown in Figure 4.45.

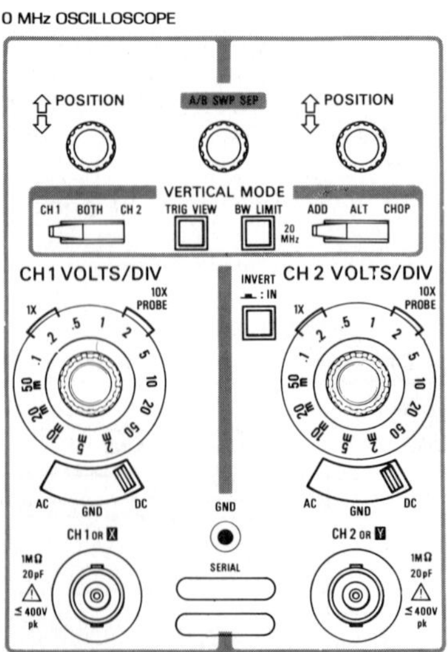

Figure 4.45 The vertical controls of the Tektronix 2235 oscilloscope. (Copyright 1983, "2235 Oscilloscope Operators Instruction Manual," Figure 2-4, Tektronix, Inc. Used with permission.)

The scope's POSITION controls lets you move the entire waveform up and down on the screen for effective viewing of the waveform. There is a vertical POSITION control for each channel.

Input coupling is controlled by a switch. It usually has three positions: DC, AC, and GND (ground). DC input coupling lets you see all of the input signal. AC coupling blocks out the constant signal components and permits only the alternating components to be displayed. The third setting, GND, disconnects the input signal from the vertical system and makes a triggered display show the scope's chassis ground. This yields a ground or 0-volt reference level on the CRT.

The sensitivity of each channel is controlled by the VOLTS/DIV switch. Having different sensitivities allows a wider range of signals, millivolts to volts, to be displayed. By rotating the VOLTS/DIV switch, the scale factor changes. The scale factor for each switch setting is written next to the switch. One caution needs to be mentioned here. The probe that you use, a 1× or 10×, may affect the scale factor.

Vertical operating modes include: Channel 1 alone, Channel 2 alone, both channels either in the alternate or chopped mode, and both channels algebraically summed. To display either channel by itself so that only one waveform appears on the CRT screen, use one of the first two settings. To display two signals, set the vertical operating mode switch to BOTH. You must also select either ALT (alternate) or CHOP (chopped). In the ALT mode, the two signals are drawn alternately; that is, the scope completes a sweep on Channel 1, then sweeps Channel 2, and then repeats the process. In the CHOP mode, the scope draws a small part of Channel 1 and then a small part of Channel 2, switching back and forth as it travels across the face of the CRT. The scope does this so fast that you cannot see the gaps in the waveform, so it appears to be drawn continuously.

The horizontal section controls the rate at which the electron beam moves across the face of the CRT. Front panel controls for the horizontal section include: horizontal position, horizontal mode, and SEC/DIV.

The Horizontal Section

The horizontal position controls, shown in Figure 4.46, work like the vertical position controls. They will allow you to position the waveform

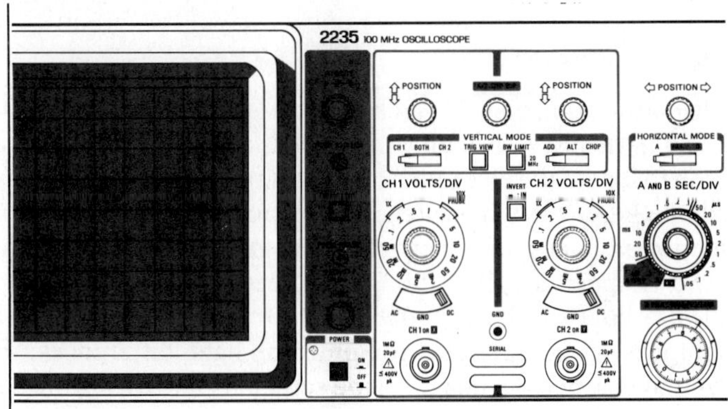

Figure 4.46 The horizontal controls for the Tektronix 2235 oscilloscope. (Copyright 1983, "2235 Oscilloscope Operators Instruction Manual," Figure 2-5, Tektronix, Inc. Used with permission.)

by moving it left or right on the screen. They will also allow you to position a waveform so that a particular point on the waveform lines up with a major division on the graticule. Here we will use the normal setting on the horizontal operating mode switch. Advanced features will include intensified and delayed-sweep operations.

The sweep speed is controlled by the SEC/DIV switch. Changing the SEC/DIV switch allows you to look at longer or shorter time intervals of the input signal. Settings will range from microseconds per division to seconds per division.

The Trigger Section

trigger

The **trigger** section, shown in Figure 4.47, tells the scope when to start drawing the signal. When this happens is important for a number of reasons. First, the image that you see is a composite of many traces overlaid on top of each other. Therefore, to get a single image, each waveform must start on the same point on the input signal. Second, if the waveforms being displayed are to be meaningful, they must have the same time relationship to each other.

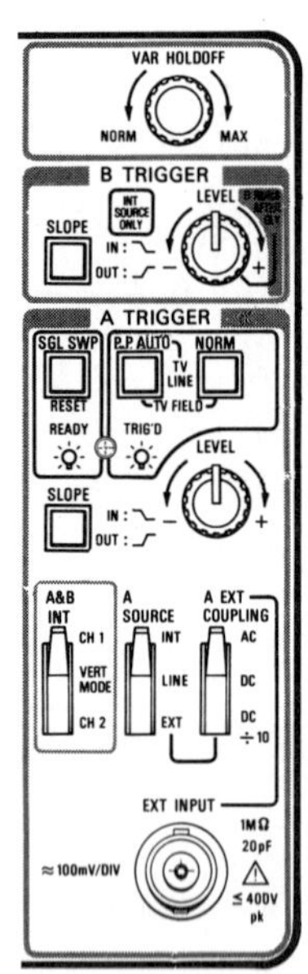

Figure 4.47 The trigger controls for the Tektronix 2235 oscilloscope. (Copyright 1983, "2235 Oscilloscope Operators Instruction Manual," Figure 2-6, Tektronix, Inc. Used with permission.)

To properly set the trigger for the scope, you must first select the trigger signal using the source switches. The trigger signal may be the signal on one of the input channels, or it could be a third or external signal that is connected to the external input connector. Next you must select the trigger voltage level and slope.

The SLOPE control determines whether the trigger point is on the rising or falling edge of the waveform. The LEVEL control determines where on that edge the trigger point occurs.

Trigger operating modes include AUTO and NORM. In the AUTO (automatic) mode, a trigger starts the sweep. When the sweep ends, a timer begins to run. If another trigger is not found before the timer runs out, a trigger is generated anyway, causing the bright baseline to appear when there is no waveform on the channel.

In the NORM mode of operation, there will only be a sweep if there is a signal to trigger the sweep. If such a trigger signal is absent, then the screen will appear blank with no signal displayed. To begin, leave the mode switch set to AUTO, so that there will always be a signal displayed on the baseline trace.

The Display Section

The display section controls include BEAM FINDER, TRACE ROTATION, FOCUS, and INTENSITY. The BEAM FINDER is a convenience control that allows you to locate the electron beam anytime it is off the screen. When you push the BEAM FINDER switch, you reduce the vertical and horizontal deflection voltages and override the INTENSITY control so that the beam always appears within the CRT screen.

INTENSITY controls the brightness of the trace. It allows you to adjust the brightness to compensate for ambient light differences. FOCUS allows you to sharpen the image on the screen by narrowing the beam of electrons. TRACE ROTATION allows you to align the horizontal deflection of the trace with the fixed graticule. This adjustment should not have to be done often, but the earth's magnetic field does affect the trace alignment and this control may come in handy.

Probes

probes

Probes are usually taken for granted, but in reality they must be selected carefully, handled with respect, and adjusted for proper operation.

Most measurements you will make with an oscilloscope will require an attenuator probe. An attenuator probe attenuates or reduces the magnitude of the incoming signal before it is sent to the vertical amplifier section of the oscilloscope. The most common attenuation factor is 10× (read ''ten times) passive probes. A 10× probe reduces the amplitude of the incoming signal and the circuit loading by 10:1.

probe compensation

Before you use an attenuator probe to make a measurement, you must make sure that the probe is compensated. **Probe compensation** failure is a most common mistake in making oscilloscope measurements. An uncompensated probe will distort or change the shape of the incoming signal. To insure that this does not happen, a small capacitor is placed either in the probe tip or in the probe connector, which attaches the probe cable to the vertical input connector on the oscilloscope. Figure 4.48 shows how a small screwdriver can be used to adjust the capacitor.

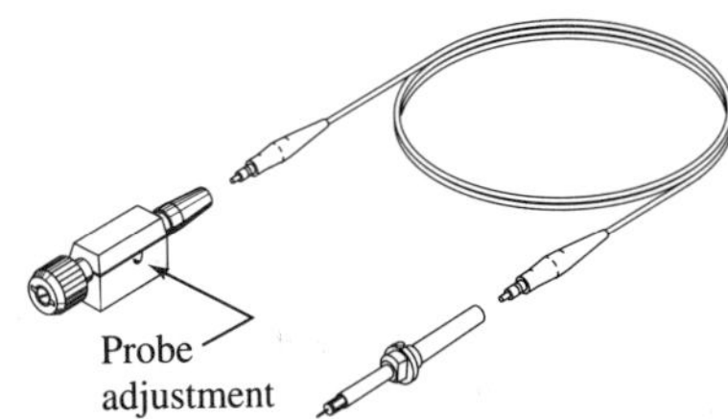

Figure 4.48 Probe compensation adjustment.

To determine if the probe is compensated, a square wave is applied to the probe tip. Figure 4.49 shows the waveforms displayed for a undercompensated, overcompensated, and compensated probe. Note that the compensated waveform has nice vertical sides and square corners. An undercompensated probe exhibits rounding on the leading and trailing edges of the waveform. An overcompensated probe shows peaking or overshoot on both edges of the pulse waveform.

Getting Started

A frequent source of inaccuracies in making oscilloscope measurements is forgetting to check the controls to make sure they are where you think they are. You should run through a checklist to make sure the settings are proper. After you have used an oscilloscope for a while, this procedure becomes automatic; checking the controls should always be routine.

Here are some things to check.

- Check all the vertical system controls.

    Variable controls (VOLTS/DIV) should be in their detente positions.

    Make sure CH 2 is not inverted (unless you want it to be inverted.)

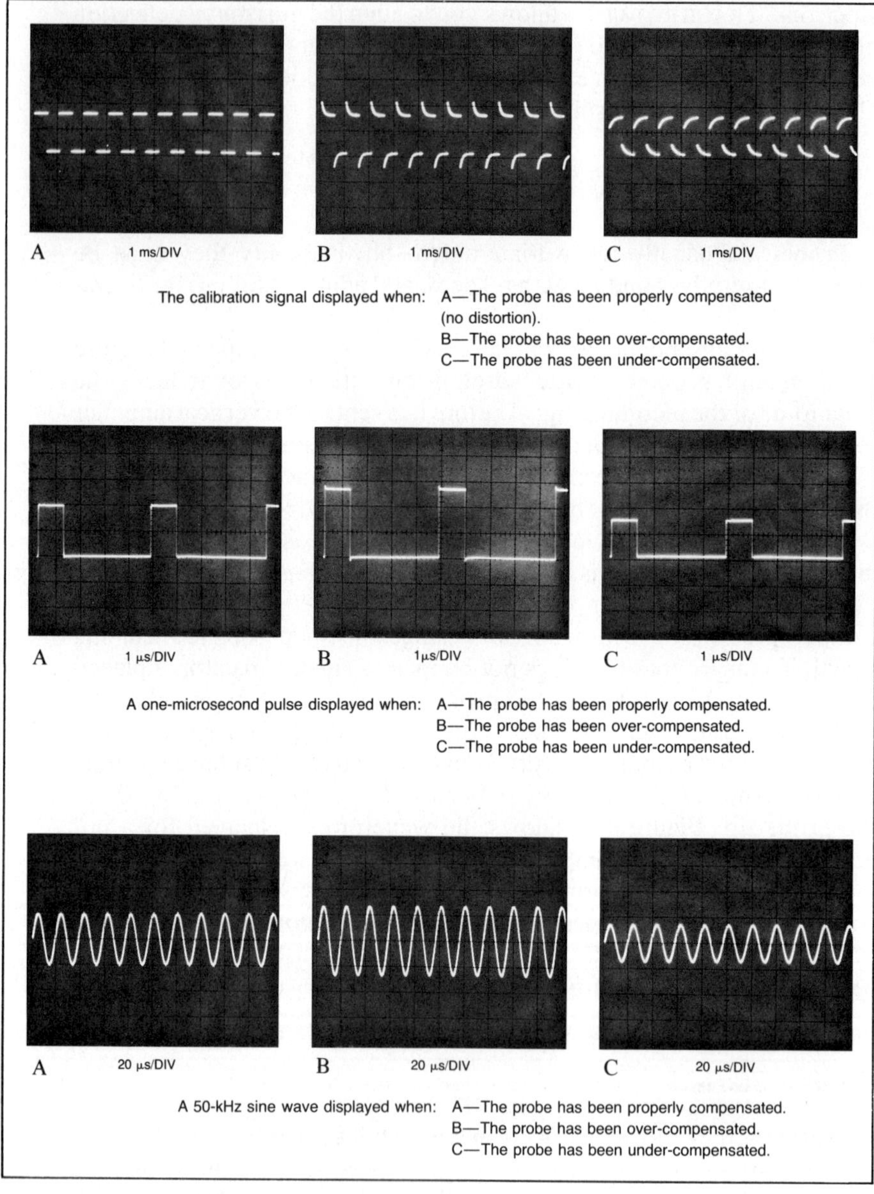

Figure 4.49 Effect of probe compensation on various signals. (Copyright 1978, "Basic Oscilloscope Operation," Figure 24, Page 18, Tektronix, Inc. Used with permission.)

Check the vertical mode switches to make sure the signal from the proper channel(s) will be displayed.

Set the VOLTS/DIV switch at its maximum setting if the magnitude of the signal is unknown.

Check the input coupling lever or switch.

- Check the horizontal system.

  Make sure magnification is OFF.

  The variable SEC/DIV switch should be in the detente or calibrated position.

- Check the trigger system controls.

  Set the mode switch to AUTO.

  Select the proper coupling.

This is not an exhaustive set of checks. Your oscilloscope may have additional features, which will have to be checked before a measurement can be made. Read your operator's manual for proper instructions.

4.8 Ohm's Law Applied to Time-Varying Signals

Now that we can generate time-varying signals and measure them, let us take another look at Ohm's Law. We will consider two types of inputs, a sine wave and the square wave. Let's take the sine wave first.

Figure 4.50(a) shows a function generator connected to a 1000-ohm resistor. The controls on the function generator are set to produce a sinusoidal signal whose frequency is 50 Hz and whose amplitude is 5 volts rms. An ac voltmeter is connected across the 1000-ohm resistor. The voltmeter will read 5 volts.

We can find the current flowing through 1000-ohm resistor by applying Ohm's Law:

$$I_{\text{rms}} = (5\ \text{V}_{\text{rms}})/1000\ \Omega$$
$$I_{\text{rms}} = 5\ \text{mA}$$

Note that the units must be in the same form for current and voltage when applying Ohm's Law. In this case, we were given the voltage in its rms units. Therefore, the computed value for current will also be in rms units.

Let us replace the voltmeter with an oscilloscope and take a look at the type of information that we can obtain. Figure 4.50(b) shows how to connect the scope. Whereas the voltmeter gave a single number representing the rms voltage of the applied signal, the oscilloscope will draw a graph of the voltage waveform as a function of time.

The voltage waveform is shown in Figure 4.50. By counting the number of divisions from the 0 volts axis to the positive peak and multiplying by the VOLTS/DIV setting, we obtain a peak amplitude of 7.0 volts. The negative peak has the same magnitude except that it is negative.

To obtain the waveform for the current using Ohm's Law, we will need to divide the voltage waveform by R, where R is 1000 ohms. Since R is a number, the current waveform will be a scaled-down version of the voltage waveform. In this case, the current waveform will be a sinusoidal

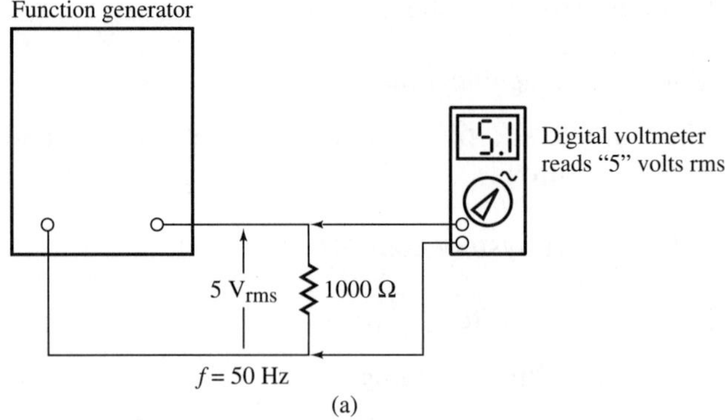

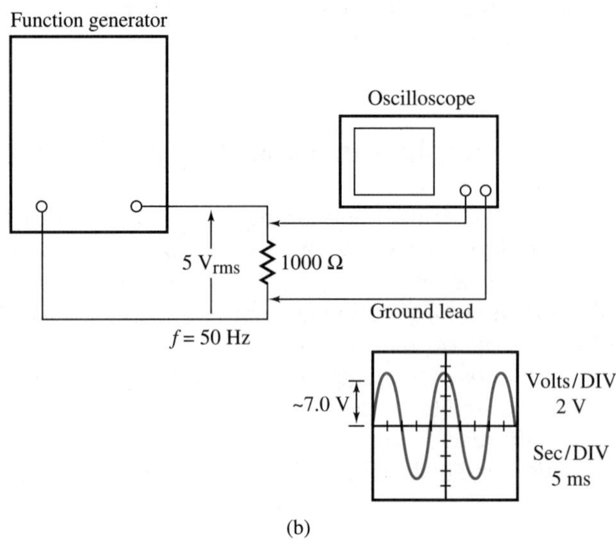

Figure 4.50 Comparison of oscilloscope and voltmeter measurements.

signal of amplitude 7.0 mA. The voltage and current waveforms are shown in Figure 4.51.

When working with time-varying waveforms, it is useful to state Ohm's Law in terms of instantaneous values—that is—values for voltage and current at a specific point in time.

$$I(t) = V(t)/R \tag{4.5}$$

where $I(t)$ is the instantaneous value of current at time $= t$ seconds

$V(t)$ is the instantaneous value of voltage at time $= t$ seconds

and R is the resistance, in ohms

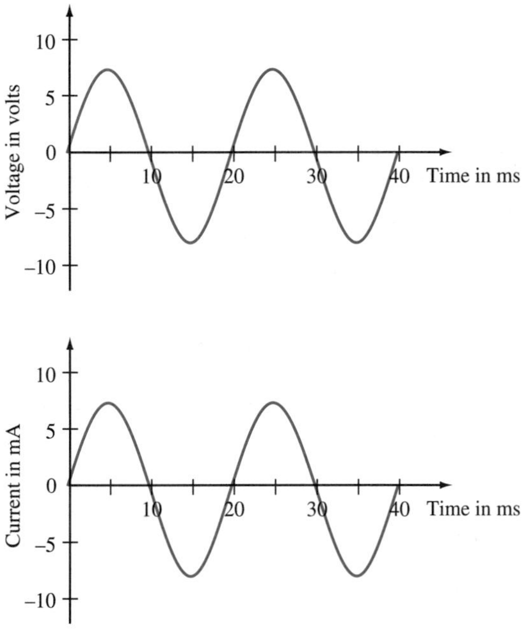

Figure 4.51 Voltage and current waveforms for a 5-V_{rms}, 50-Hz sinusoidal voltage applied across a 1-k ohm resistor.

EXAMPLE 4.8

A 10-volt peak-to-peak, 250-Hz triangular wave is applied across a 500-ohm resistor. Assume that at $t = 0$ seconds, the amplitude of the triangular wave is 0 volts and the slope is positive. What is the magnitude of the current flowing through the 500-ohm resistor at $t = 0, 1, 2, 3,$ and 4 milliseconds?

Solution

To begin our solution, let us graph the triangular wave. Since the frequency of the signal is 250 Hz, the period can be found by taking the reciprocal of the frequency as follows:

$$T = 1/f = 1/250 \text{ Hz}$$
$$T = 4 \text{ ms}$$

If the waveform starts out at 0 volts at $t = 0$ seconds, then at $t = 1$ ms, the triangular wave is at its positive peak, +5 volts. The waveform then decreases in amplitude, and at $t = 2$ ms, the waveform has returned to 0 volts. The waveform continues to move in the negative direction until $t = 3$ ms. At $t = 3$ ms, the waveform has reached its most negative point and the amplitude is −5 volts. The signal then increases, and at $t = 4$ ms,

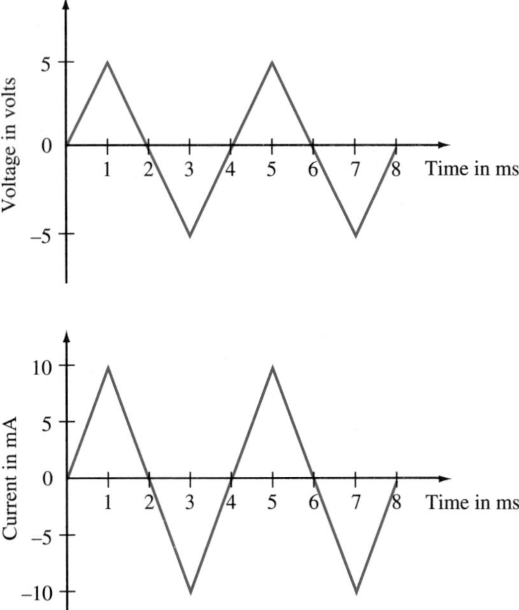

Figure 4.52 Current and voltage waveforms for Example 4.8.

the magnitude is again back to zero. All this explanation can be expressed in one drawing as shown in Figure 4.52.

By applying the definition of Ohm's Law that uses instantaneous values, we obtain the following results:

$$
\begin{aligned}
&\text{At } t = 0 \text{ ms,} && I(t = 0 \text{ ms}) = 0 \text{ V}/500\ \Omega = 0 \text{ mA} \\
&\text{At } t = 1 \text{ ms,} && I(t = 1 \text{ ms}) = +5 \text{ V}/500\ \Omega = 10 \text{ mA.} \\
&\text{At } t = 2 \text{ ms,} && I(t = 2 \text{ ms}) = 0 \text{ V}/500\ \Omega = 0 \text{ mA.} \\
&\text{At } t = 3 \text{ ms,} && I(t = 3 \text{ ms}) = -5 \text{ V}/500\ \Omega = -10 \text{ mA.} \\
&\text{At } t = 4 \text{ ms,} && I(t = 4 \text{ ms}) = 0 \text{ V}/500\ \Omega = 0 \text{ mA.}
\end{aligned}
$$

Now let us consider a pulse input. In this case, we can divide the time axis into discrete time segments. During each time segment, the input signal will have a constant amplitude. Hence, each time segment is similar to the dc input case we studied earlier. Let us see how this works out.

EXAMPLE 4.9

Suppose a 0–10 volt, positive-going, 5-kHz square wave is applied to a 2-kilohm resistor. What will be the resulting current waveform?

Solution

Again, let us begin by graphing the applied voltage. Assume that the square wave makes its positive transition at time $t = 0$ seconds. Hence,

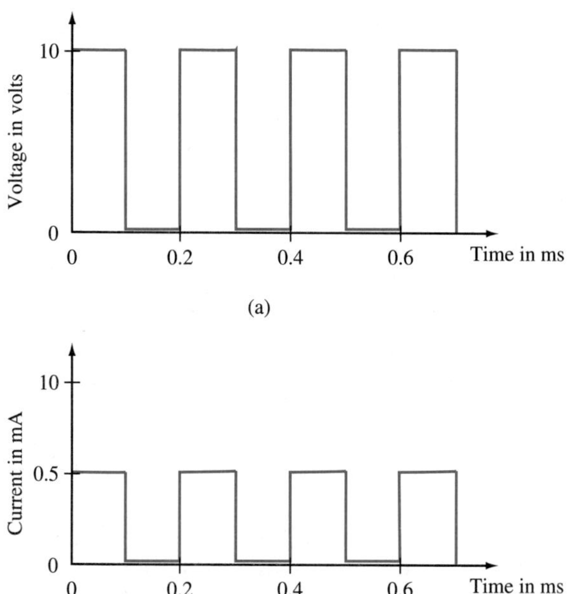

Figure 4.53 Current and voltage waveforms for Example 4.9.

the applied voltage is at +10 volts from time $t = 0$ seconds until time $t = 0.1$ ms. At $t = 0.1$ ms, the voltage falls abruptly to 0 volts and remains at that level until time $t = 0.2$ ms. At $t = 0.2$ ms, the voltage changes abruptly from 0 volts to +10 volts and the cycle repeats. The graph of the waveform is shown in Figure 4.53.

Let us divide the time axis into two time segments, $t = 0.1$ ms and $t = 0.2$ ms. For the first interval between 0 and 0.1 ms, the applied voltage is a constant +10 volts. Hence, the current is also at a constant level and can be calculated as follows:

$$I(t = 0\text{–}0.1\text{ ms}) = +10\text{ V}/2000\ \Omega$$
$$I(t = 0\text{–}0.1\text{ ms}) = +5\text{ mA}$$

For the second interval, from $t = 0.1$ ms–$t = 0.2$ ms, the applied voltage is 0 volts. Hence, the current during this interval is:

$$I(t = 0.1\text{ ms–}0.2\text{ ms}) = 0\text{ V}/2000\ \Omega$$
$$I(t = 0.1\text{ ms–}0.2\text{ ms}) = 0\text{ mA}$$

Using this information, the current waveform can be constructed.

The process can be reversed. That is, if the current waveform and resistance value are known, Ohm's Law can be used to determine the voltage waveform. From Ohm's Law, we know that $V(t) = I(t) \times R$.

Hence, the voltage waveform has the same shape as the current waveform, and is scaled by a factor R.

SUMMARY

1. dc power can be supplied by two means: primary and secondary cells, and dc power supplies.
2. ac power can be supplied by transforming the line voltage supplied through the building electrical system.
3. Repetitive signals are characterized by their shape, frequency, and amplitude.
4. Signals can be generated by a function generator.
5. The frequency of a signal is equal to the number of complete cycles the signal makes in 1 second.
6. A voltmeter measures the rms value of a sinusoidal signal, while an oscilloscope graphs voltage as a function of time, allowing us to view the exact shape of the signal.
7. Ohm's law applies equally well to dc and time-varying signals.

EQUATIONS

4.1	$\text{Ah} = I \times t$	Ampere hour capacity of a cell	138
4.2	$f = 1/T$	Frequency is inversely related to the period of a repetitive signal	148
4.3	$\text{rms value} = 0.707 \times \text{peak value}$	The rms value of a sinusoidal signal is related to the peak amplitude	151
4.4	$\text{Duty cycle} = \dfrac{T_p}{T} \times 100\%$	Duty cycle of a repetitive pulse waveform	155
4.5	$I(t) = V(t)/R$	Ohm's Law relation for instantaneous values	172

QUESTIONS

1. What type of cells can be recharged, the primary cell or the secondary cell.
2. Name four types of primary cells.
3. What are the four basic parts of a chemical cell?
4. Which of the following cells can be recharged: carbon-zinc, zinc chloride, or nickel-cadmium?
5. A toy car motor needs an emf of 3 volts to operate. Will one zinc-carbon battery be enough?

6. Explain the meaning of the term "ampere hour."
7. List five factors that affect the useful life of a cell.
8. You have a choice of charging a battery in 30 minutes at 20 amperes, or 600 minutes at 1 ampere. Which method is better for the battery?
9. Explain the relationship between frequency and period for a repetitive signal.
10. If you double the period of a signal, how will the frequency of the signal change?
11. What is the relationship between the rms value of a sinusoidal signal and its peak amplitude?
12. Between what two points do you measure the rise time of a pulse waveform?
13. List the steps required to set the controls of a function generator to produce a given waveform of specified amplitude and frequency.
14. What type of information about a signal does an oscilloscope provide?
15. What is the difference between ac and dc coupling of the input signal on an oscilloscope?
16. Why is triggering so important when using an oscilloscope?

PROBLEMS

SECTION 4.1 and 4.2 Primary and Secondary Cells

1. The battery in a golf cart has a 400 ampere hour rating. For how many hours will the cart operate if it has a 5-ampere motor?

2. A flashlight battery can supply 120 mA for 20 hours. What is the ampere hour rating of the battery?

3. A 100 ampere hour battery is operated at 2 amperes for 20 hours. If the charging current is 5 amperes, how long will it take to restore it to the fully charged condition?

4. A 12.2-volt, 3-Ah nickel-cadmium battery must be charged in 20 minutes. If a 40-V charger with an 0.4-ohm internal resistance is to be used, what additional series resistance is needed, and what must be its power rating? (Assume that the voltages remain constant during the charging period.)

SECTION 4.3 The dc Power Supply

5. A dc power supply can supply a maximum current of 5 amperes at a fixed output voltage of 15 volts. What is the power rating of the dc power supply?

6. Show how two 10-volt dc power supplies can be connected to produce a +10-volt potential and a −10-volt potential with respect to a common ground terminal.

7. Draw a schematic diagram showing how to connect two 12-volt dc power supplies to produce a −24-volt dc potential with respect to a ground reference point.

SECTION 4.4 ac Sources

8. An ac voltage alternates 300 times in 1 minute. What is the frequency? What is the period?

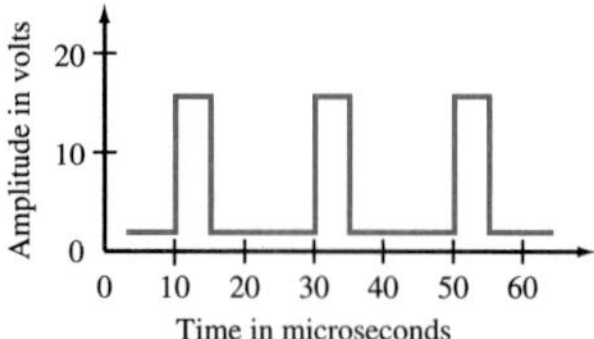

Figure 4.54

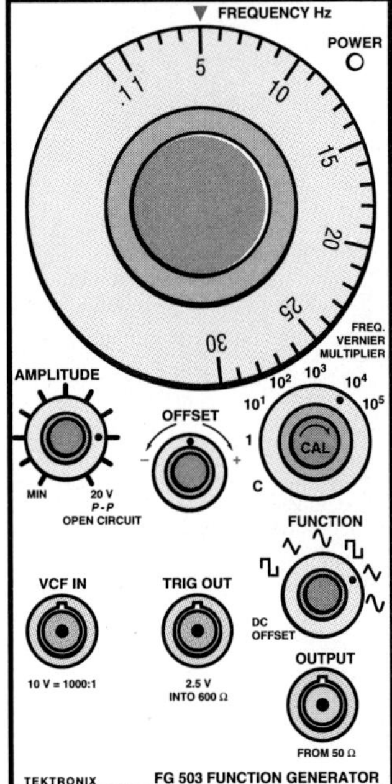

Figure 4.55

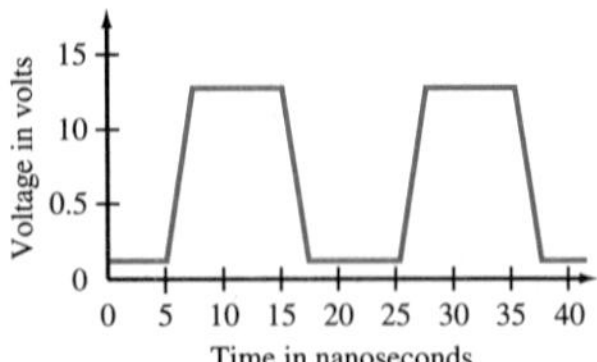

Figure 4.56

9. A sine wave has an rms value of 57 volts. What is the peak-to-peak amplitude of the same signal?

10. Determine the frequency, duty cycle, and average value of the waveform shown in Figure 4.54.

SECTION 4.5 Pulse Waveforms

11. Graph a repetitive pulse waveform of amplitude 8 volts riding on a 0-volt dc level. The waveform has a frequency of 4.0 kHz and a duty cycle of 20 percent.

12. Graph a repetitive pulse waveform of amplitude −6 volts riding on a −2 volt dc level. The duration of the pulse is 25 microseconds and the duty cycle of the waveform is 20 percent.

13. Indicate the effect (increase, decrease, or remain the same) of the following changes on the duty cycle of a repetitive pulse waveform:

    (a) Halve the frequency of the waveform while keeping the pulse duration constant

    (b) Decrease the amplitude of the pulses

    (c) Decrease the pulse duration while keeping the frequency the same

SECTION 4.6 Generating Time-Varying Signals

14. Given the settings on the FG 503 Function Generator shown in Figure 4.55, sketch the waveform that the function generator is producing. Assume an amplitude setting of 15 V and zero volts offset.

15. Describe how you would configure the FG 503 to produce a 2.5-kHz sinusoidal waveform with an amplitude of 5 volts peak and a 0-volt dc component.

16. Describe the setup procedure to produce a 4-volt peak square wave riding on a 0-volt dc level. Assume a frequency setting of Hz.

SECTION 4.7 Measuring and Displaying Time-Varying Signals

17. Given the waveform shown in Figure 4.56, determine the period, frequency, rise time, fall time, and duty cycle.

18. Describe how you would configure a dual trace oscilloscope to display a 120-mVp-p sinusoidal waveform having a frequency of 30 kHz and a 0-volt dc level. Sketch the picture of the signal that you would see on the oscilloscope screen.

19. Describe the setup procedure to display the waveform generated in Problem 15.

20. Describe the setup procedure to display the waveform generated in Problem 16.

SECTION 4.8 Ohm's Law Applied to Time-Varying Signals

21. A 12-V_p sinusoidal waveform having a frequency of 750 Hz is applied across a 1.5-kilohm resistor. Draw the corresponding current waveform. Assume that the amplitude is 0 and the slope is positive at time = 0 seconds.

22. A 20-volt peak-to-peak triangular waveform having a 5-volt dc component and a frequency of 4 kHz is applied across a 4.7-kilohm resistor. Draw the voltage waveform and the corresponding current waveform. Use the same time axis and draw the current waveform directly under the voltage waveform, being careful to show the exact time relationship between the two waveforms.

23. An 8-V_p sinusoidal waveform is applied to a 500-ohm resistor. If the frequency of the waveform is 60 Hz and the dc component is 0 volts, draw a graph of power dissipated versus time for two complete cycles.

CHAPTER 5

CHAPTER OBJECTIVES

After completing this chapter, you should be able to:

1. Determine which parts of a circuit are connected in series.
2. Apply Kirchhoff's Voltage Law to a series circuit.
3. Calculate the equivalent resistance of a series circuit.
4. Use the voltage divider rule to calculate voltages in a series circuit.
5. Calculate the power in a series circuit.
6. Express voltages using the double subscript notation.
7. Perform a complete analysis of a series circuit.

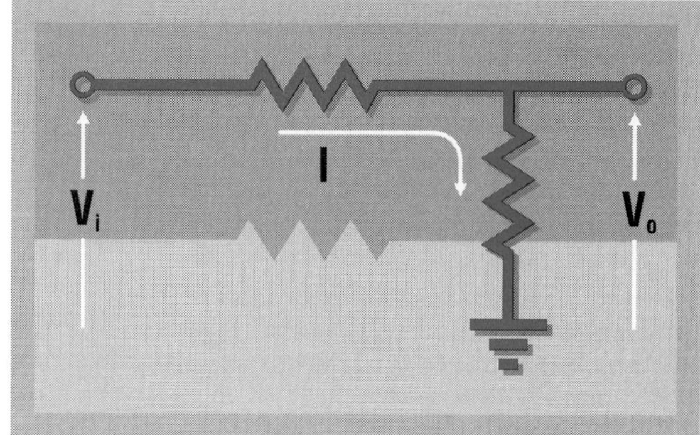

KEY TERMS

1. algebraic sum
2. branch point
3. double subscript notation
4. equivalent resistance
5. Kirchhoff's Voltage Law
6. series circuit
7. transfer function
8. voltage divider rule

INTRODUCTION

Having learned about the basic quantities and relations in a simple circuit, the series connection will now be studied. This is a circuit in which the parts are connected so that the current is the same in each part.

Series circuits are used to limit or interrupt current, protect devices from damage, and reduce voltage. A switch, connected in series with a lamp, provides a means of turning the lamp on and off. A resistor, connected in series with a motor, reduces the voltage to the desired level. And a fuse, connected in series with a power circuit, protects the circuits from damage that would be caused by excessive current flow.

This chapter presents the basic relations for the series circuit, a technique for analyzing the circuit, and some specific applications of the series circuit.

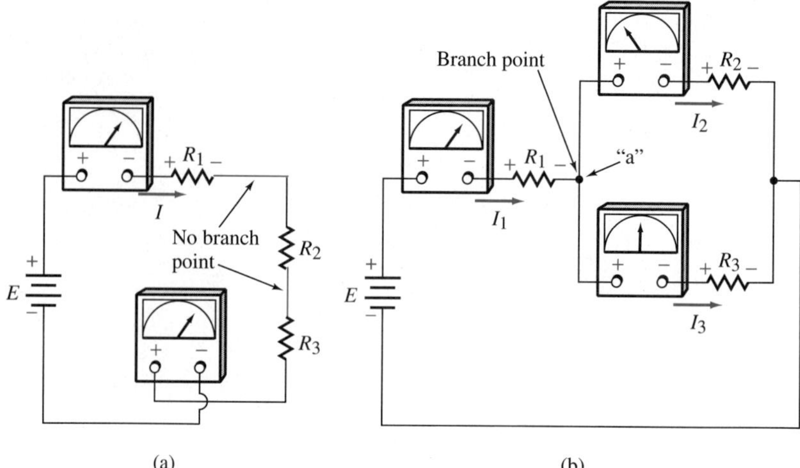

Figure 5.1 Two examples of resistor circuits. (a) These three resistors are in series. The meters show the same current in each one. (b) These three resistors are not in series. There is a branch point at "a."

5.1 Circuit Examples and Current in the Series Circuit

Practical circuits can have several parts in them, each of which can be connected in many different combinations. One of these combinations is the series circuit.

series circuit
branch point

A **series circuit** does not have any branch points between the parts in the circuit. A **branch point**, or node, is a junction of three or more parts. Since there are no branch points, the current in each part of the circuit will be the same. The resistors R_1, R_2, and R_3 in Figure 5.1(a) are connected in series. The current in R_1, R_2, and R_3 must be the same because there is no other place between the resistors where charge can be added or removed. However, R_1 and R_2 in Figure 5.1(b) are not connected in series. In this case, a part of the current is in R_2 and the remainder goes to R_3 at the branch point located at Point "a."

EXAMPLE 5.1

Which two resistors are connected in series in the circuit of Figure 5.2?

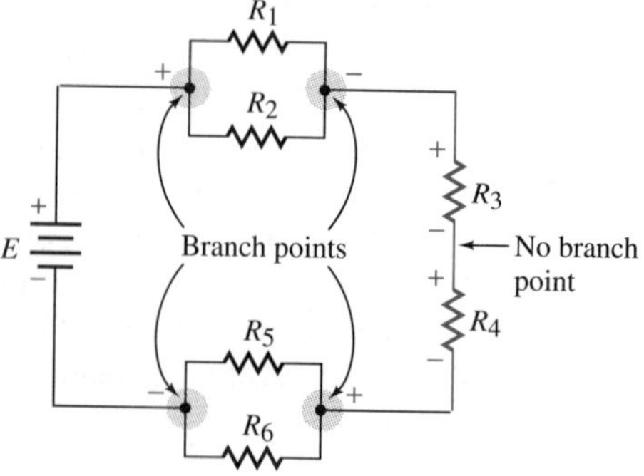

Figure 5.2

Solution

Resistors R_3 and R_4 are the only resistors without a branch point between them, so they are connected in series.

A change in the current or voltage in any part of a series circuit affects the current and voltage in all parts of the circuit. This characteristic can be useful, but can also cause problems. In Figure 5.3, the fuse will open when the current is too high. Since the fuse is in series with the lamp, the lamp current must also drop to 0 amperes when the fuse opens. Thus, the lamp and the circuit are protected from damage from high currents.

On the other hand, this characteristic is a problem in the lamp circuit of Figure 5.4. When the filament of one lamp opens, the current in all parts of the circuit becomes 0 amperes, just as when the fuse opened. Since every lamp becomes dark, each lamp must now be checked to find the defective one. This can be a lengthy process when there are many lamps.

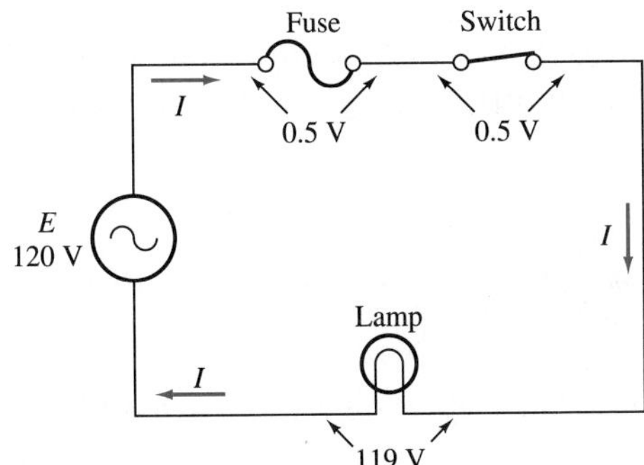

Figure 5.3 The lighting circuit is a series circuit.

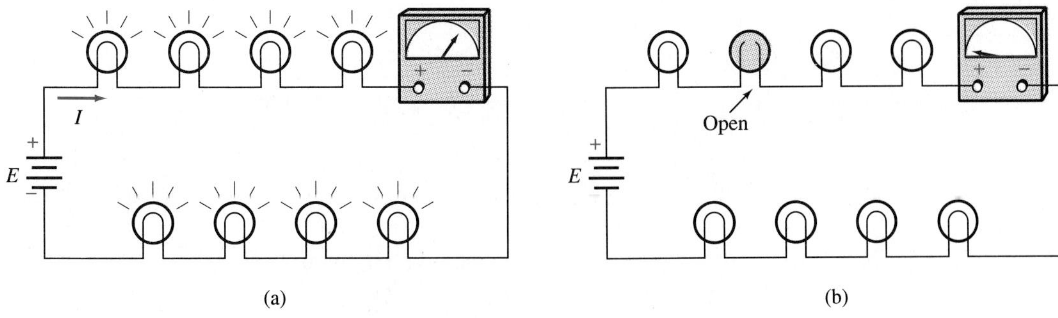

Figure 5.4 A change in any part of a series circuit affects the other parts of the circuit. (a) All lamps are lit. (b) One filament opens; all lamps go out.

5.2 Kirchhoff's Voltage Law

Kirchhoff's Voltage Law

algebraic sum

In a closed loop of a circuit, the algebraic sum of the voltage rises and drops equals zero. This relation is known as **Kirchhoff's Voltage Law** (KVL). It was formulated circa 1842 by Gustav Kirchhoff, a German scientist (1824–1887). The **algebraic sum** means that the rises and the drops must be given opposite signs when they are added. In Figure 5.3, the closed loop is made up of the source, the fuse, the switch, and the lamp. The voltage rise is E and the voltage drops are the voltages across the fuse, switch, and lamp. Hence, the voltage rise E, 120 V, must equal the sum of the three voltage drops: -0.5 V, -0.5 V, and -119 V. In equation form, the law written for the series circuit is:

$$E - V_1 - V_2 - V_3 - \cdots - V_n = 0 \quad \textbf{(5.1)}$$

where:

E is the source voltage, in volts

and $V_1, V_2, V_3, \ldots, V_n$ are the voltages across the resistors, in volts

Kirchhoff's Voltage Law is also expressed as "The sum of the rises in a closed loop equals the sum of the drops." Equation 5.1 then becomes

$$E = V_1 + V_2 + V_3 + \cdots + V_n \quad \textbf{(5.2)}$$

Kirchhoff's Voltage Law gives us a tool to analyze circuits with several parts in them by relating what is happening in parts of the circuit to the total circuit.

EXAMPLE 5.2

A resistor is connected in series with a lamp (Figure 5.5) to maintain 6 volts across the lamp. Find V_x and R_x.

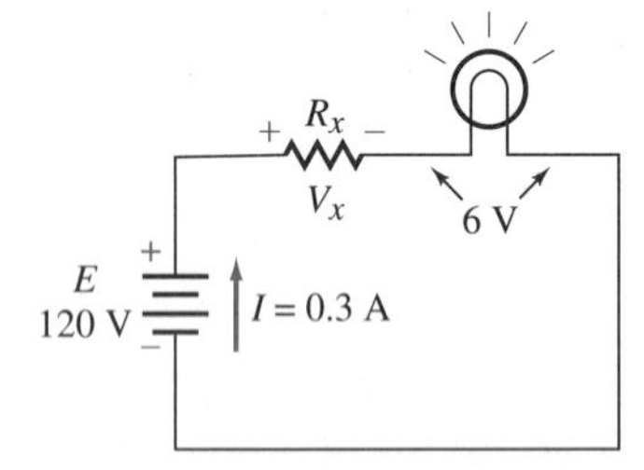

Figure 5.5

Solution

Since two of the three voltages are known, Equation 5.1 can be used. The total voltage across the circuit is E, so by KVL,

$$E - V_x - V_{\text{lamp}} = 0$$

or

$$V_x = E - V_{\text{lamp}}, \qquad V_x = 120\text{ V} - 6\text{ V}$$
$$V_x = 114\text{ V}$$

Using Ohm's Law for R_x,

$$R_x = V_x/I$$
$$= 114\text{ V}/0.3\text{ A}$$
$$R_x = 380\ \Omega$$

EXAMPLE 5.3

Ten 12-volt lamps are connected in series to a voltage source. What is the source voltage required for each bulb to be at its rated voltage?

Solution

Since each lamp has the same voltage, Equation 5.1 becomes

$$E = 10 \text{ lamps} \times V_1$$

or

$$E = 10 \text{ lamps} \times 12 \text{ V/lamp}$$
$$E = 120 \text{ V}$$

The voltage across a series circuit is divided among the parts of the circuit. This voltage-division characteristic is sometimes used to get different voltages from a fixed voltage source. The circuits used for that are called *voltage dividers*. An examination of the circuit in Figure 5.6 will bring out some of the voltage relations in the series circuit.

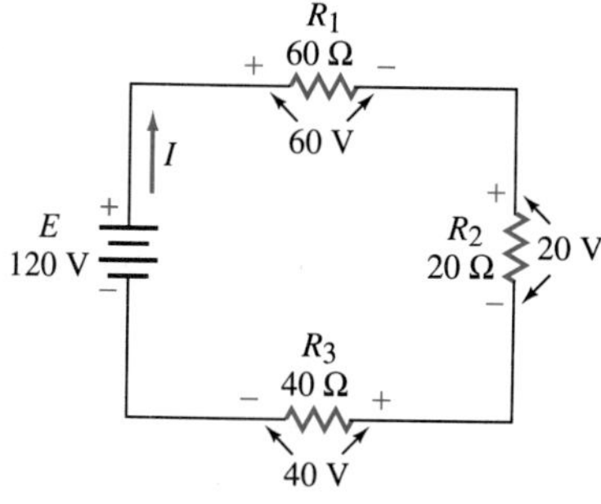

Figure 5.6 The voltage is greater across the larger resistance, and the algebraic sum of the rises and drops around the loop equals zero.

From Figure 5.6 it can be seen that the larger resistances have larger voltages across them. The 60-ohm resistor has 60 volts across it while the 20-ohm resistor has 20 volts across it. This voltage represents the work per unit charge that is needed to make the charge move through the resistances. Since greater resistance requires more work, the larger resistance will have a greater voltage across it.

Second, the voltage across a part of a circuit divided by the resistance of that part of the circuit is the same for all parts of the circuit. This is because V/R equals I, and I is the same for each resistor. In equation form,

$$V_1/R_1 = V_2/R_2 = \cdots = V_n/R_n \tag{5.3}$$

where:

$R_1, R_2, \ldots, R_n$ are the resistances of each resistor, in ohms

and $V_1, V_2, \ldots, V_n$ are the voltages across each resistor, in volts

EXAMPLE 5.4

The voltage across R_1 in the circuit of Figure 5.7 is 10 volts. Find:
(a) the voltage across R_2
(b) the current in the circuit.

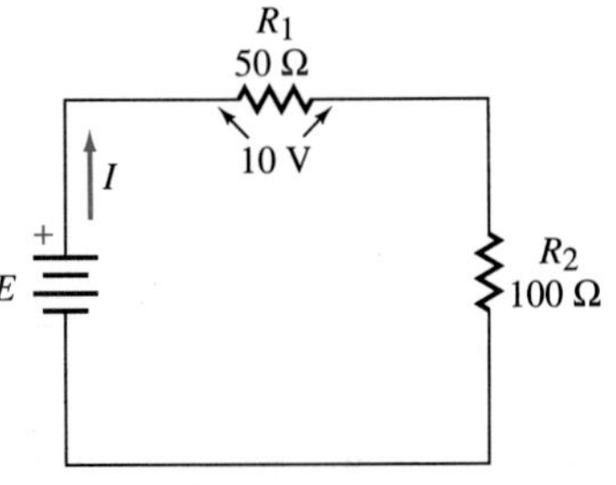

Figure 5.7

Solution

$$V_2/R_2 = V_1/R_1$$

so

$$V_2 = V_1(R_2/R_1)$$

then

$$V_2 = (10\ \text{V})(100\ \Omega/50\ \Omega)$$
$$V_2 = 20\ \text{V}$$

Ohm's Law can be used for part (b) of Example 5.4:

$$V_1 = IR_1$$

or

$$I = V_1/R_1$$
$$= 10\ \text{V}/50\ \Omega$$
$$I = 0.2\ \text{A}$$

5.3 Equivalent Resistance

equivalent resistance

The circuit of Figure 5.6 can be replaced by one resistance so that the current will be the same as in the original circuit. This resistance is the equivalent resistance of the circuit. **Equivalent resistance** is the resistance that draws the same amount of current and has the same voltage across it when connected in place of the original circuit. An experimental procedure for finding the equivalent resistance is shown in Figure 5.8.

When resistance R_x in Figure 5.8(b) is adjusted so that the current is the same as in the circuit of Figure 5.8(a), its resistance is equal to the equivalent resistance of the circuit. The value of R_x can now be determined by either:

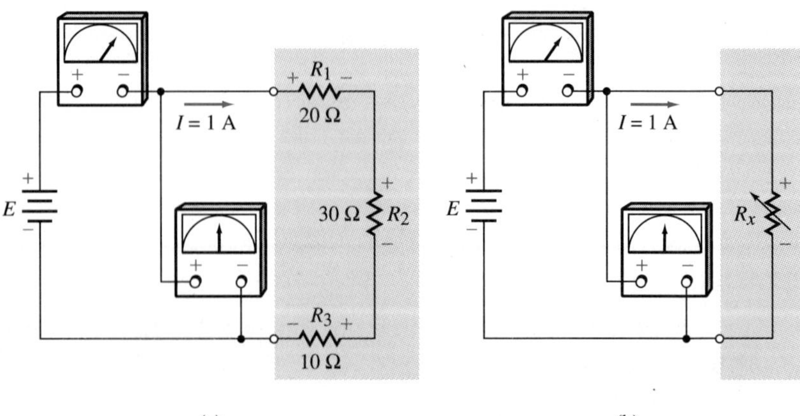

Figure 5.8 The circuits are equivalent when the current is the same in both. (a) A series circuit. (b) R_x is the equivalent resistance.

(a) Removing R_x from the circuit and measuring its resistance, if it is a linear resistance, or
(b) Using the voltmeter and ammeter readings and Ohm's Law relation for the entire circuit. This method gives the equation:

$$R_{eq} = V/I \tag{5.4}$$

where:

R_{eq} is the equivalent resistance of the circuit, in ohms
V is the voltage across the series circuit, in volts
and I is the current in the circuit, in amperes

EXAMPLE 5.5

The current in a series circuit is 2 amperes when connected to a 120-volt source. What is the equivalent resistance of the circuit?

Solution

$$R_{eq} = V/I$$

In this case, V is equal to E so

$$R_{eq} = 120 \text{ V}/2 \text{ A}$$
$$R_{eq} = 60\ \Omega$$

EXAMPLE 5.6

Find the equivalent resistance of the part between Points A and B in the circuit of Figure 5.9.

Solution

The voltage across the two points is 10 volts, and the current is 1 ampere. Putting these values into Equation 5.4 gives

$$R_{eq} = V/I$$
$$= 10 \text{ V}/1 \text{ A}$$
$$R_{eq} = 10\ \Omega$$

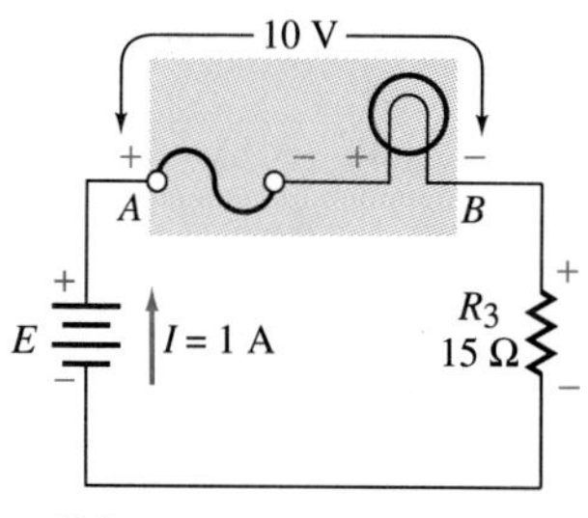

Figure 5.9

These methods of finding R_{eq} are suitable only for circuits that are already connected. Since connecting every circuit is not practical, a method that uses only the values of the resistances in the circuit would be better. Fortunately, such a method is available. To find it, consider applying Kirchhoff's Voltage Law to a series circuit of three resistors—R_1, R_2, and R_3. It gives:

$$E - V_1 - V_2 - V_3 = 0$$

but
$$V_1 = IR_1$$
$$V_2 = IR_2$$

and

$$V_3 = IR_3$$

so

$$E - IR_1 - IR_2 - IR_3 = 0$$

Dividing both sides by I gives

$$E/I - (IR_1)/I - (IR_2)/I - (IR_3)/I = 0$$

but E/I is the equivalent resistance.
So

$$R_{eq} = R_1 + R_2 + R_3$$

Thus, we see that the equivalent resistance of the series circuit is equal to the sum of the resistances in the circuit. In general, for "n" resistors in a series circuit,

$$R_{eq} = R_1 + R_2 + \cdots + R_n \qquad \textbf{(5.5)}$$

where:

R_{eq} is the equivalent resistance, in ohms
and $R_1, R_2, \ldots, R_n$ are the resistance values of each resistor, in ohms

Connecting the resistors in series is analogous to increasing the length of a conductor. Recall that $R = (\rho L)/A$.

With Equation 5.5, we can find the equivalent resistance of circuits that are not connected. It also eliminates any errors introduced in the measurements. All that is needed are the values of resistance in the circuit. The equation can also be used for circuits that are already built, if the resistance values are known.

EXAMPLE 5.7

Use Equation 5.5 to find the equivalent resistance of the circuit shown in Figure 5.8(a).

Solution

$$\begin{aligned} R_{eq} &= R_1 + R_2 + R_3 \\ &= 20\ \Omega + 30\ \Omega + 10\ \Omega \\ R_{eq} &= 60\ \Omega \end{aligned}$$

Note that the equivalent resistance of a series circuit is *greater than* any of the resistances in the circuit. This relation can be used to check the accuracy of any calculation.

EXAMPLE 5.8

Find the equivalent resistance and the current in a series circuit of a 50-Ω, a 100-Ω, and a 90-Ω resistor connected across a 120-V source.

Solution

Since R_1, R_2, and R_3 are given, Equation 5.4 will be used.

$$\begin{aligned} R_{eq} &= R_1 + R_2 + R_3 \\ &= 50\ \Omega + 100\ \Omega + 90\ \Omega \\ R_{eq} &= 240\ \Omega \end{aligned}$$

With V and R_{eq} known, Equation 5.4 can be used to find I.

$$R_{eq} = V/I$$

so

$$\begin{aligned} I &= V/R_{eq} \\ &= 120\ \text{V}/240\ \Omega \\ I &= 0.5\ \text{A} \end{aligned}$$

It is interesting to note that circuits can have the same equivalent resistance even though they have different values of resistance in them. Figure 5.10(a)–(d) shows some examples of that. The equivalent resistance of each circuit is 30 ohms.

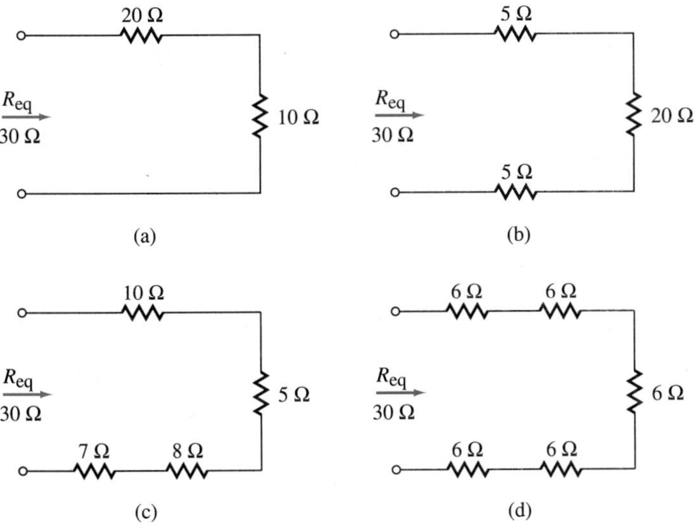

Figure 5.10 Various combinations of resistance values can have the same equivalent resistance. R_{eq} = 30 ohms.

Before leaving this section, we will examine the conditions where R_{eq} is ∞ ohms (an open circuit) and where R_{eq} is 0 ohms (short circuit).

When R_{eq} is infinite, the circuit current will be 0 amperes, making the potential difference across parts of the circuit 0 volts. All of the voltage in the circuit will be across the open ends. If this voltage is high, a hazardous condition can exist. For example, removing the bulb from a socket in a circuit such as that in Figure 5.4 leaves a hazardous open circuit in the socket. Generally, open circuits do not damage the circuit because there is no current.

A short circuit in part of a series circuit decreases the equivalent resistance, causing an increase in current. This increase can affect the operation of the circuit and even damage it, which might result in an open circuit. For instance, short circuiting a lamp in a house circuit will increase the current, causing the fuse or circuit breaker to open. If there were no fuse or breaker, the high current would increase the temperature of the circuit and might cause a fire.

The next example will show the effect of an open and a short circuit on the operation of a series circuit.

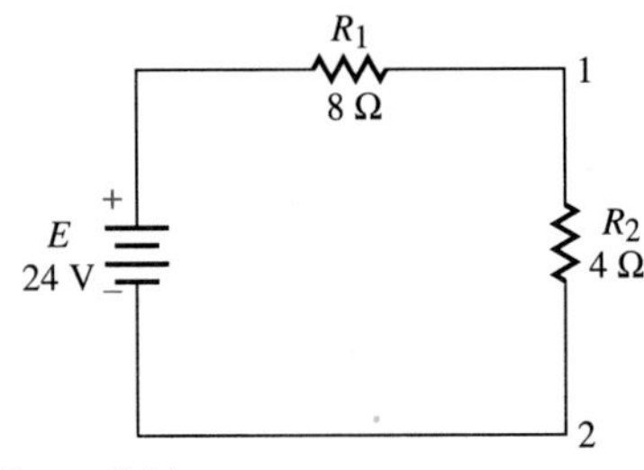

Figure 5.11

EXAMPLE 5.9

For the circuit of Figure 5.11, find:

a. The voltage across Points 1–2.
b. The current in the circuit.
c. Items (a) and (b) if the 4-ohm resistor becomes an open circuit.
d. Items (a) and (b) if the 4-ohm resistor is short circuited.

Solution

a.

$$V_{R2} = \frac{ER_2}{R_{eq}}$$

$$R_{eq} = R_1 + R_2 = 8\ \Omega + 4\ \Omega$$

$$R_{eq} = 12\ \Omega$$

so

$$V_{R2} = \frac{(24\ \text{V})(4\ \Omega)}{12\ \Omega}$$

$$V_{R2} = 8\ \text{V}$$

b.

$$I = \frac{E}{R_{eq}} = \frac{24\ \text{V}}{12\ \Omega}$$

$$I = 2\ \text{A}$$

c. With the open circuit,

$$I = 0\ \text{amperes}$$

$$V = \text{the source voltage}$$

so

$$V = 24\ \text{V}$$

d. With the short circuit,

$$R_{eq} = R_1 + R_2 = 8\ \Omega + 0\ \Omega = 8\ \Omega$$

$$I = \frac{E}{R_{eq}} = \frac{24\ \text{V}}{8\ \Omega}$$

$$I = 3\ \text{A}$$

$$V = IR_2 = (3\ \text{A})(0\ \Omega)$$

$$V = 0\ \text{V}$$

5.4 The Voltage Divider Rule

In the previous section, the voltage across any part of the series circuit was found by using Ohm's Law. This required that the current be known. The **voltage divider rule** lets us find voltages without knowing the current. The rule states that the voltage across a part of a series circuit is equal to the resistance of that part multiplied by the voltage across the series circuit, and divided by the equivalent resistance of the series circuit. The equation form of the voltage divider rule is:

voltage divider rule

$$V_x = (R_x\ V)/R_{eq} \qquad \textbf{(5.6)}$$

where:

V_x is the voltage across a part of the series circuit, in ohms
R_x is the resistance of that part, in ohms
R_{eq} is the equivalent resistance of the series circuit, in ohms
and V is the voltage across the series circuit, in volts

Equation 5.6 is obtained by applying Ohm's Law to the resistor R_x and to the series circuit. Consider a series circuit with a voltage of V volts across it. Then, for the resistor R_x in the circuit,

$$V_x = IR_x \qquad \text{but} \qquad I = V/R_{eq}$$

so

$$V_x = (VR_x)/R_{eq}, \text{ which is Equation 5.6.}$$

EXAMPLE 5.10

A series circuit consisting of a 40-Ω, a 60-Ω, and a 20-Ω resistor is connected to a 120-V source.

a. Use the voltage divider rule to find the voltage across the 60-ohm resistor.
b. Check the answer using series circuit theory.

Solution

a. The voltage across R_{eq} is E. So from Equation 5.6,

$$V_2 = (R_2E)/R_{eq}$$
$$R_{eq} = R_1 + R_2 + R_3$$
$$= 40\ \Omega + 60\ \Omega + 20\ \Omega$$
$$R_{eq} = 120\ \Omega$$

so

$$V_2 = (60\ \Omega)\ (120\ \text{V})/120\ \Omega$$
$$V_2 = 60\ \text{V}$$

b. Using Ohm's Law,

$$V_2 = R_2I$$
$$V_2 = (60\ \Omega)(I)$$

But from Equation 5.1,

$$I = E/R_{eq}$$
$$= 120\ \text{V}/120\ \Omega$$
$$I = 1\ \text{A}$$

so

$$V_2 = (60\ \Omega)(1\ \text{A})$$
$$V_2 = 60\ \text{V}$$

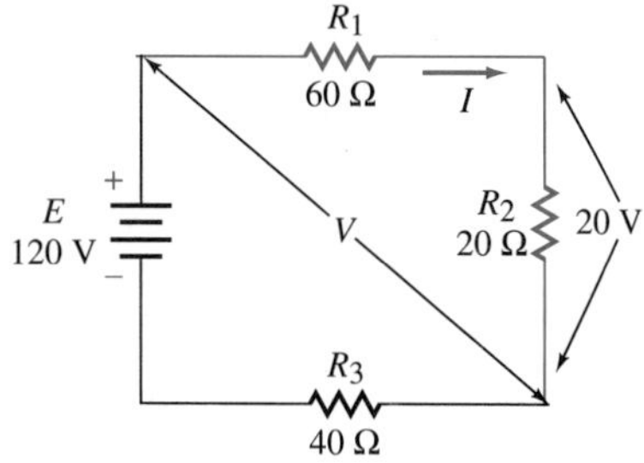

Figure 5.12

EXAMPLE 5.11

Use Equation 5.6 to find the voltage V in the circuit of Figure 5.12.

Solution

The series group is $R_1 + R_2$. Since V_2 and R_2 are known, these can be used as V_x and R_x in Equation 5.6:

$$V_x = (R_x V)/R_{eq} \quad \text{or} \quad V_2 = (R_2 V)/(R_1 + R_2)$$

Solving for V yields

$$V = (V_2(R_1 + R_2))/R_2$$
$$= (20\ \text{V}\ (60\ \Omega + 20\ \Omega))/20\ \Omega$$
$$V = 80\ \text{V}$$

A voltage divider is an example of a practical application of the voltage divider rule. Figure 5.13(a)–(c) shows the circuit of a voltage divider used to supply two loads and several other voltage dividers.

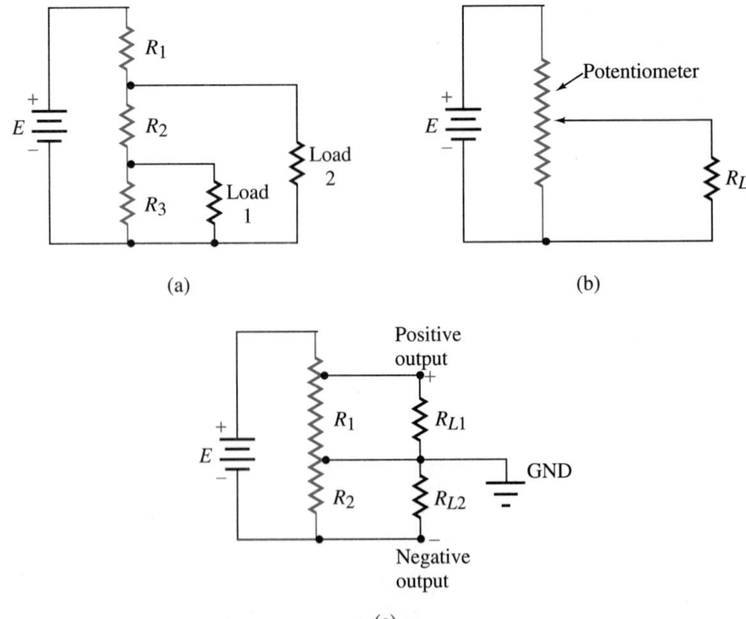

Figure 5.13 A voltage divider provides one or more voltages from a fixed voltage. (a) A three-resistor divider that provides two voltages. (b) A potentiometer divider that provides an output from 0 volts to that of the source. (c) A divider that provides a negative and positive voltage.

Voltage dividers are generally used where small load currents (mA) are needed. They eliminate the need for multiple sources in a circuit. They are also less costly and smaller in size than a voltage source.

A potentiometer can be used as variable output divider while a tapped resistor can give several fixed outputs. Volume controls and tone controls are two examples of voltage dividers.

The loaded divider is discussed in Chapter 8; now we examine the unloaded divider. The results using unloaded conditions are often satisfactory because of the small load currents.

EXAMPLE 5.12

Design a voltage divider of 1000 ohms total resistance to give the output voltages of Figure 5.14, when connected across a 24-V source. Also, calculate the power in each section of the divider.

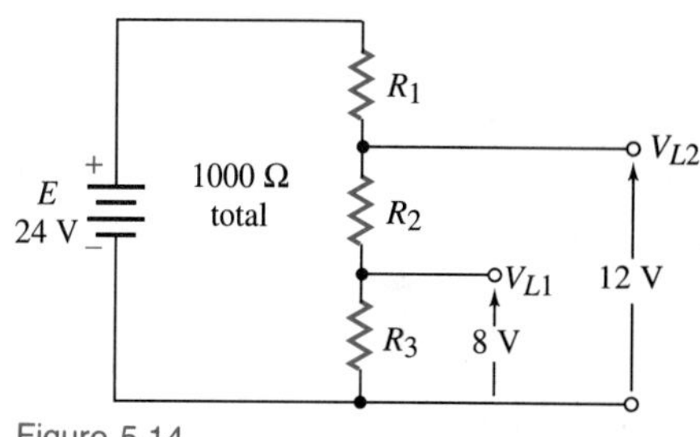

Figure 5.14

Solution

The design requires determining the resistance of sections R_1, R_2, and R_3. The resistance between V_{L1} and ground is R_3 so

a. $V_{L1} = \dfrac{ER_3}{R_{eq}}$

which gives

$$R_3 = \frac{(V_{L1})R_{eq}}{E}$$

or

$$R_3 = \frac{(8\text{ V})(1000\ \Omega)}{24\text{ V}}$$
$$R_3 = 333.3\ \Omega$$

The potential difference across R_2 is

$$V_{R2} = V_{L2} - V_{L1}$$
$$= 12\text{ V} - 8\text{ V} \quad \text{or} \quad 4\text{ V}$$

Then

$$4\text{ V} = E\,(R_2/R_{eq})$$

so

$$R_2 = [(4\text{ V})\,(R_{eq})]/E$$

or

$$R_2 = [(4\text{ V})(1000\ \Omega)]/24\text{ V}$$
$$R_2 = 166.67\ \Omega$$
$$R_{eq} = R_1 + R_2 + R_3$$

so

$$R_1 = R_{eq} - R_2 - R_3$$
$$= 1000\ \Omega - 166.67\ \Omega - 333.3\ \Omega$$
$$R_1 = 500\ \Omega$$

b. $P_1 = V_{R1}{}^2/R_1$

but

$$V_{R1} = E - V_{L2} = 24\text{ V} - 12\text{ V} = 12\text{ V}$$

so

$$P_1 = (12\text{ V})^2/500\ \Omega$$
$$P_1 = 0.288\text{ W}$$
$$P_2 = V_{R2}{}^2/R_2$$
$$= (4\text{ V})^2/166.67\ \Omega$$
$$P_2 = 0.096\text{ W}$$
$$P_3 = V_{R3}{}^2/R_3$$
$$= (8\text{ V})^2/333.3\ \Omega$$
$$P_3 = 0.192\text{ W}$$

The two preceding examples illustrate the use of voltage dividers to divide a known voltage into smaller voltages. This is commonly done with dc voltages. What happens when a time-varying signal is applied to a series network?

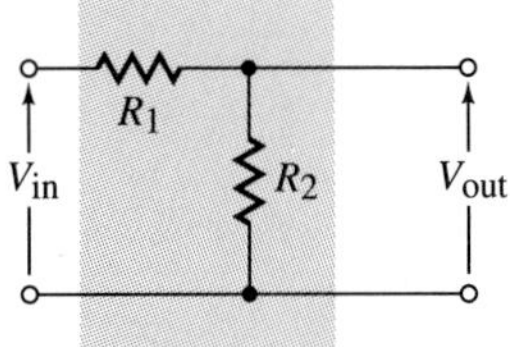

Signal processing circuit

Figure 5.15

Let us return to the simplest case, the two-resistor series circuit. We will use this circuit to develop a way of looking at circuits. In Figure 5.15, an input signal is applied to the resistive circuit. The circuit will interact with the input signal, processing the signal in some manner. The output signal will be a resultant waveform produced by the interaction of the input signal and the processing circuit.

In the case of the two-resistor voltage divider, the output voltage can be described as

$$V_{out} = I \times R_2.$$

We also know that

$$I = V_{in}/(R_1 + R_2)$$

Substituting the equation for I into the equation for V_{out} yields

$$V_{out} = (V_{in}/(R_1 + R_2)) \times R_2$$

Using simple algebraic rules for manipulating the equation, we obtain

$$V_{out} = (R_2/(R_1 + R_2)) \times V_{in}$$

or

$$V_{out}/V_{in} = R_2/(R_1 + R_2).$$

transfer function

The ratio of V_{out} to V_{in} is called a **transfer function**. Note that the transfer function, in this case, depends only on the values of the circuit elements. Note also that the transfer function is a constant.

Knowing the transfer function for a circuit is a great aid in analyzing the behavior of a circuit given a specific input signal. Consider the following examples.

EXAMPLE 5.13

Analyze the circuit of Figure 5.16. Assume that the input is a 10-volt peak-to-peak 1-kHz sine wave.

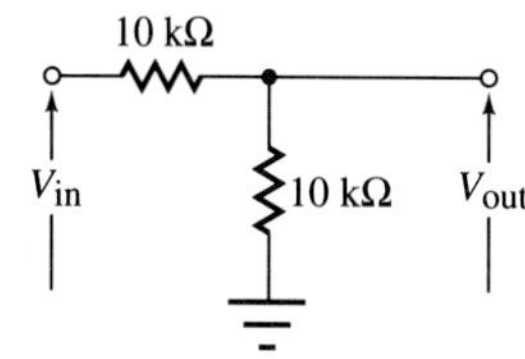

Figure 5.16

Solution

The transfer function can be written for the voltage divider circuit,

$$V_{out}/V_{in} = (10\ \text{k}\Omega/(10\ \text{k}\Omega + 10\ \text{k}\Omega))$$
$$V_{out}/V_{in} = 0.5$$

What this tells us is that the circuit will attenuate (make the signal smaller) by a factor of 0.5, or one-half.

Hence, for our 10-volt (peak-to-peak) sine wave input, the output signal will be exactly one-half the magnitude. The output signal will have the same time relationship to the input signal—that is—the peaks and valleys will occur at the same time. The output will be a 5-volt (peak-to-peak) sine wave.

EXAMPLE 5.14

A 10-volt input signal is to be attenuated to 100 mV. Could a resistive voltage divider be used to accomplish this reduction?

Solution

Yes, a resistive voltage circuit could be used. The transfer function for a two-resistor divider circuit would be

$$V_{out}/V_{in} = 100\ \text{mV}/10\ \text{V}$$
$$V_{out}/V_{in} = 0.01$$

Hence, the ratio of resistances is

$$R_2/(R_1 + R_2) = 0.01$$

or

$$R_2 = 0.01\ (R_1 + R_2)$$

or

$$0.99\ R_2 = 0.01\ R_1$$
$$R_2 \approx 0.01\ R_1$$

Therefore, R_2 must be approximately 0.01 R_1. At this point, without any other information on which to make the selection of resistance values, we can pick a value for R_1 and calculate the needed value of R_2. For example, if we pick a value of 100 kilohms for R_1, then R_2 must be 100 kΩ × 0.01, or 1 kilohm. There are many resistance pairs that will work, as long as the ratio of resistors is 0.01.

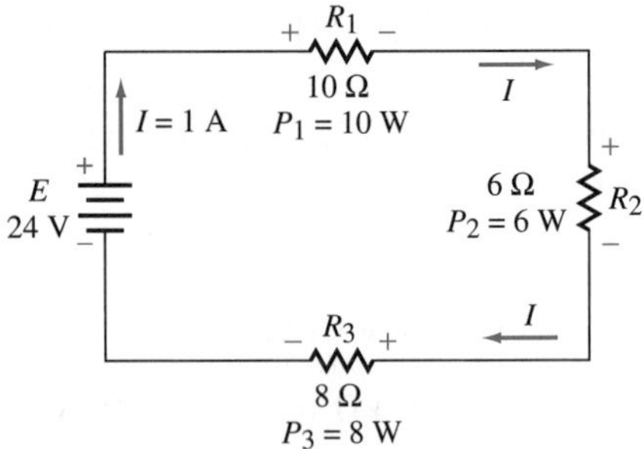

Figure 5.17 Power dissipated in a series circuit can be found in several ways. (a) $P = EI$. (b) $P = I^2R_{eq}$. (c) $P = E^2/R_{eq}$. (d) $P = P_1 + P_2 + P_3$.

5.5 Power and Energy in a Series Circuit

Although power and energy are not as commonly measured as voltage and current, they can be important quantities. For instance, a source must have a power rating large enough for the circuit to which it is connected. Also, a part in a circuit must be able to get rid of the energy that is converted to heat. If it does not, damage can result.

In the circuit of Figure 5.17, power is dissipated in each resistor. Since this power is supplied by the source, the power supplied is equal to the power dissipated.

Since a series circuit can be replaced by an equivalent resistance, the power in that resistance must be equal to that in the original circuit.

Applying the relations of the power equations from Chapter 3 to the equivalent resistance gives

$$P = VI \qquad \textbf{(5.7)}$$

$$P = I^2R_{eq} \qquad \textbf{(5.8)}$$

$$P = V^2/R_{eq} \qquad \textbf{(5.9)}$$

where:

P is the power dissipated or supplied, in watts
I is the current in the circuit, in amperes
V is the voltage across the circuit, in volts
and R_{eq} is the equivalent resistance of the circuit, in ohms

EXAMPLE 5.15

What is the power in the series circuit of Figure 5.17?

Solution

$$P = VI$$

Since V is E,

$$P = (24 \text{ V})(1 \text{ A})$$
$$P = 24 \text{ W}$$

Using Equation 5.7,

$$P = V^2/R_{eq}$$

But V is E, and

$$R_{eq} = 10\ \Omega + 6\ \Omega + 8\ \Omega$$
$$R_{eq} = 24\ \Omega$$

so

$$P = (24\ \text{V})^2/24\ \Omega$$
$$P = 24\ \text{W}$$

Using Equation 5.8,

$$P = I^2R_{eq}$$
$$P = (1\ \text{A})^2(24\ \Omega)$$
$$P = 24\ \text{W}$$

Next consider the power in part of a series circuit.

EXAMPLE 5.16

What is the power in the resistance combination of R_2 and R_3 in Figure 5.17?

Solution

$$R_{eq} = R_2 + R_3$$

Hence

$$R_{eq} = 6\ \Omega + 8\ \Omega$$
$$R_{eq} = 14\ \Omega$$

The voltage across the two resistances is

$$V = IR_{eq}$$
$$= (1\ \text{A})(14\ \Omega)$$
$$V = 14\ \text{V}$$

From Equation 5.7,

$$P = VI$$
$$= (14\ \text{V})(1\ \text{A})$$
$$P = 14\ \text{W}$$

Using Equation 5.9,

$$P = V^2/R_{eq}$$
$$= (14\ \text{V})^2/14\ \Omega$$
$$P = 14\ \text{W}$$

Using Equation 5.8,

$$P = I^2R_{eq}$$
$$= (1\ \text{A})^2(14\ \Omega)$$
$$P = 14\ \text{W}$$

EXAMPLE 5.17

The lines connecting a 120-V dc source to a heater have a resistance of 0.5 ohm. If the heater current is 5 A, find:

a. The source power.
b. The power lost in the lines.
c. The efficiency of the system.

Solution

a. $P_s = EI = (120\text{ V})(5\text{ A})$

$P_s = 600\text{ W}$

b. $P_{\text{lost}} = I^2R$

$= (5\text{ A})^2(0.5\ \Omega)$

$P_{\text{lost}} = 12.5\text{ W}$

c. $$\eta = \frac{(P_s - P_{\text{lost}})}{P_s} \times 100\%$$

$$= \frac{600\text{ W} - 12.5\text{ W}}{600\text{ W}} \times 100\%$$

$\eta = 97.9\%$

Another useful power relation is between the total power and the power in each part of the circuit. This relation is derived for the circuit of Figure 5.17:

$$P = I^2R_{\text{eq}} \quad \text{but} \quad R_{\text{eq}} = R_1 + R_2 + R_3$$

so

$$P = I^2(R_1 + R_2 + R_3)$$
$$= I^2R_1 + I^2R_2 + I^2R_3$$

Since I^2R_1, I^2R_2, and I^2R_3 are the power in each resistor, the power in the circuit is equal to the sum of the power dissipated in each resistor. In equation form,

$$P = P_1 + P_2 + P_3 + \cdots P_n \tag{5.10}$$

where:

P is the power dissipated in the circuit, in watts
$P_1, P_2, P_3, \ldots, P_n$ is the power dissipated in each part of the series circuit, in watts

EXAMPLE 5.18

Use Equation 5.10 to find the power in the circuit of Figure 5.17 and in the resistance combination of R_2 and R_3.

Solution

$$P = P_1 + P_2 + P_3$$

Since

$$P_1 = I_1^2R = (1\ \text{A})^2(10\ \Omega) = 10\ \text{W}$$
$$P_2 = I_2^2R = (1\ \text{A})^2(6\ \Omega) = 6\ \text{W}$$
$$P_3 = I_3^2R = (1\ \text{A})^2(8\ \Omega) = 8\ \text{W}$$

so

$$P = 10\ \text{W} + 6\ \text{W} + 8\ \text{W}$$
$$P = 24\ \text{W}$$

The power in the combination of R_2 and R_3 is equal to P_2 plus P_3. This gives $P_{2+3} = 6\ \text{W} + 8\ \text{W} = 14\ \text{W}$. These values are the same as those obtained in the preceding example.

EXAMPLE 5.19

How much power is supplied by the source in the circuit of Figure 5.18?

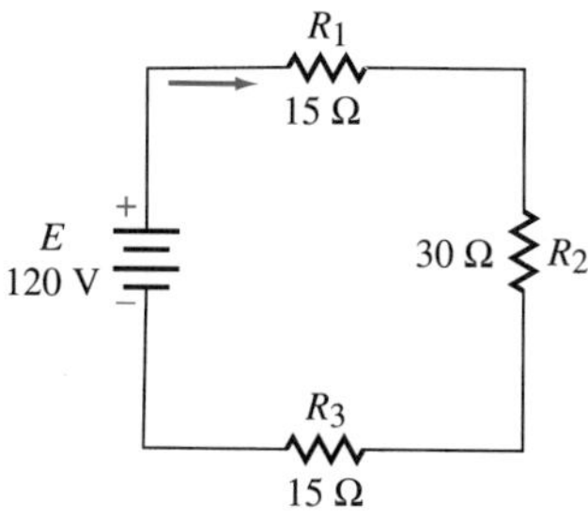

Figure 5.18

Solution

The resistance values and the source voltage are given, so Equation 5.9 will be used.

$$P = E^2/R_{eq}$$

Since

$$R_{eq} = R_1 + R_2 + R_3$$
$$= 15\ \Omega + 30\ \Omega + 15\ \Omega$$
$$R_{eq} = 60\ \Omega$$

we can compute the power,

$$P = (120\ \text{V})^2/60\ \Omega$$
$$P = 240\ \text{W}$$

The solution could also have been obtained using Equation 5.7 or 5.8. However, the current would have to be calculated first.

The electrical energy supplied by a source is converted to other forms—for example—heat in a resistance or light in a bulb. Just as for power, the energy supplied must equal the energy used. Energy, as learned in the basic circuit, is the product of power and time. Therefore, multiplying Equations 5.7–5.10 by time will give the energy supplied or used. That is,

$$\mathcal{W} = VIt \quad \textbf{(5.11)}$$
$$\mathcal{W} = I^2R_{eq}\,t \quad \textbf{(5.12)}$$
$$\mathcal{W} = V^2t/R_{eq} \quad \textbf{(5.13)}$$
$$\mathcal{W} = \mathcal{W}_1 + \mathcal{W}_2 + \cdots + \mathcal{W}_n \quad \textbf{(5.14)}$$

where:

$\mathcal{W}$ is the energy supplied or used, in joules
$\mathcal{W}_1$ etc. are the energy in each part of the circuit, in joules
t is the time, in seconds
and V, I, and R_{eq} have the same meaning as in Equations 5.7, 5.8, and 5.9

Expressing the time in hours and the power in kilowatts gives the energy in kilowatt-hours, which is the more common unit used by electric utility companies.

EXAMPLE 5.20

The current in a lamp is 0.5 ampere when connected to a 25-volt source. How much energy is used in 30 minutes in (a) joules and (b) kilowatt-hours?

Solution

a. Using Equation 5.11,

$$\mathcal{W} = VIt$$

Since V is equal to E and I is equal to 0.5 ampere,

$$\mathcal{W} = (25\ \text{V})\ (0.5\ \text{A})(30\ \text{min})(60\ \text{s/min})$$
$$\mathcal{W} = 22{,}500\ \text{J}$$

b. Again, using Equation 5.11,

$$\mathcal{W} = (25\ \text{V})(0.5\ \text{A})(1\ \text{kW}/1000\ \text{W})(30\ \text{min})(1\ \text{hr}/60\ \text{min})$$
$$\mathcal{W} = 0.00625\ \text{kWh}$$

EXAMPLE 5.21

Work example 5.20 using the relation in Equation 5.14.

Solution

Using Equation 5.14,

$$\mathcal{W} = \mathcal{W}_1 + \mathcal{W}_2 \quad \text{and} \quad \mathcal{W}_1 = P_1\, t$$

and

$$P_1 = I_1^2\, R_1$$
$$= (0.5\ \text{A})^2(10\ \Omega)$$
$$P_1 = 2.5\ \text{W}$$

Likewise, $\mathcal{W}_2 = P_2 t$

and

$$P_2 = I^2{}_2\, R_2$$
$$= (0.5\ \text{A})^2(40\ \Omega)$$
$$P_2 = 10\ \text{W}$$

Since the time is in minutes, it must be converted to seconds. So

$$\mathcal{W}_1 = (2.5\ \text{W})(30\ \text{min})(60\ \text{s/min}) = 4500\ \text{J}$$
$$\mathcal{W}_2 = (10\ \text{W})(30\ \text{min})(60\ \text{s/min}) = 18{,}000\ \text{J}$$

and finally,

$$\mathcal{W} = 4500\ \text{J} + 18{,}000\ \text{J}$$
$$\mathcal{W} = 22{,}500\ \text{J}$$

This answer agrees with the one obtained in Example 5.20.

5.6 Double Subscript Notation

Consider the potential difference across R_1 in the circuit of Figure 5.19. Up to this point, it was expressed as V_{R1} or V_R and was always given a positive sign. However, a voltage means a potential difference between two points. Therefore, a voltage is a potential at one point as measured from a second reference point. The voltage between points A and B can also be expressed as V_{AB} or V_{BA}, using a double subscript. This is the **double subscript notation** and means the voltage at the first subscript point as measured from the second subscript point. Since Point A is more positive than Point B and the potential difference is 5 volts, V_{AB} is $+5$ volts while V_{BA} is -5 volts.

double subscript notation

The second subscript is often left out on schematics because all voltage are specified as measured from ground. This is a point, usually the metal chassis, taken as the reference, or 0-volt point. The ground symbol is shown at Point G in Figure 5.19.

In Figure 5.19, the voltage at Point A is 12 volts more positive than ground. Its double subscript expression is $V_A = 12$ V. Schematics of electronic circuits usually have the voltages marked on them. These provide information that a technician can use when troubleshooting the circuit.

Double subscript notation is also useful when determining the voltage across points that span more than one element. The procedure for doing that will be shown in Example 5.22.

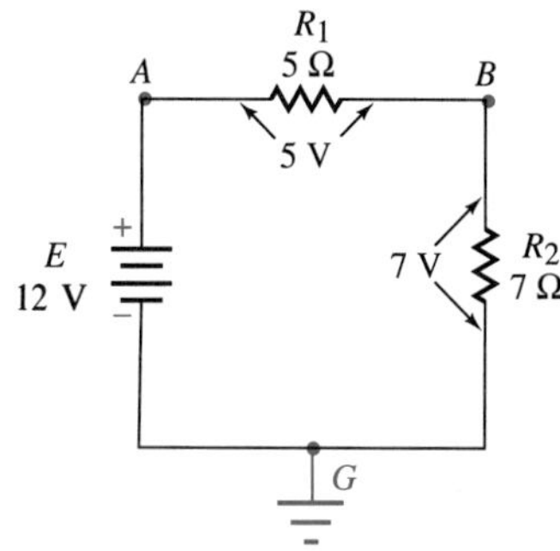

Figure 5.19 V_{AG} and V_{BG} use double subscripts to express the voltage of A and B as measured with respect to G.

EXAMPLE 5.22

Determine V_{AB}, V_{CB}, V_A, V_B, and V_{AC} in the circuit of Figure 5.20.

Solution

V_{AB} is the voltage across R_1 and Point A is more positive than Point B. So

$$V_{AB} = +6\ \text{volts}$$

V_{CB} is the voltage across R_2 and Point C is more negative than Point B.

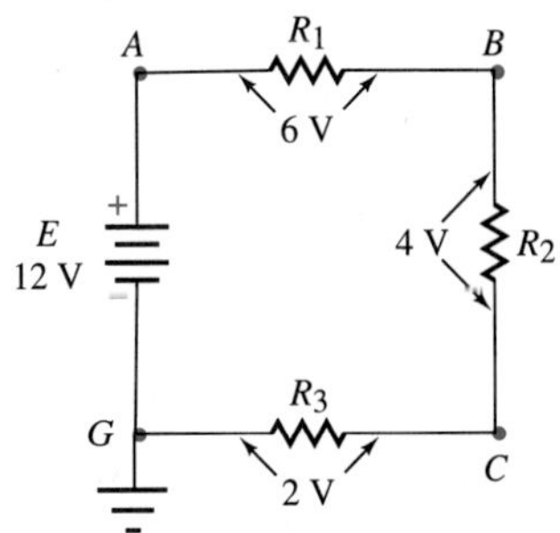

Figure 5.20

So

$$V_{CB} = -4 \text{ volts}$$

V_A is measured from ground and is the voltage across the source. Point A is more positive than ground so

$$V_A = +12 \text{ volts}$$

V_B is the voltage measured from ground. To get that voltage, we must start at ground and add the voltages in a path to Point B. That is,

$$\begin{aligned} V_B &= V_{CG} + V_{BC} \\ &= +2 \text{ V} + 4 \text{ V} \\ V_B &= +6 \text{ V} \end{aligned}$$

Also,

$$\begin{aligned} V_B &= V_{AG} + V_{BA} \\ &= +12 \text{ V} - 6 \text{ V} \\ V_B &= +6 \text{ V} \end{aligned}$$

To get V_{AC}, we must start at Point C and add the voltages in a path to Point A. That is,

$$\begin{aligned} V_{AC} &= V_{AB} + V_{BC} \\ V_{AC} &= +6 \text{ V} + 4 \text{ V} \\ V_{AC} &= +10 \text{ V} \end{aligned}$$

Also,

$$\begin{aligned} V_{AC} &= V_{AG} + V_{GC} \\ &= +12 \text{ V} - 2 \text{ V} \\ V_{AC} &= +10 \text{ V} \end{aligned}$$

The circuit for this example is a series circuit, but the same procedure can be used for any circuit. That is, start at the second subscript and add the voltages in a path to the first subscript.

5.7 Analysis of Series Circuits

The analysis of series circuits requires that a path between the known and unknown quantities be found. The series circuit gives us a chance to use the problem-solving procedure that was explained in Chapter 1. The steps in that procedure are:

1. Read the statement and note the quantities that are given and those that must be found. Some quantities will be given directly while others might be obtained from the problem conditions.
2. Draw a diagram where appropriate and label all parts, quantities, and relevant points in the circuit.
3. Use individual relationships or groups of them to form a path linking the wanted quantities to the given quantities. The unknowns in the equations can be used as a guide.

4. When the number of independent equations is equal to the number of unknowns in the equations, solve the set.

EXAMPLE 5.23

Find the power dissipated in R_2 of Figure 5.21.

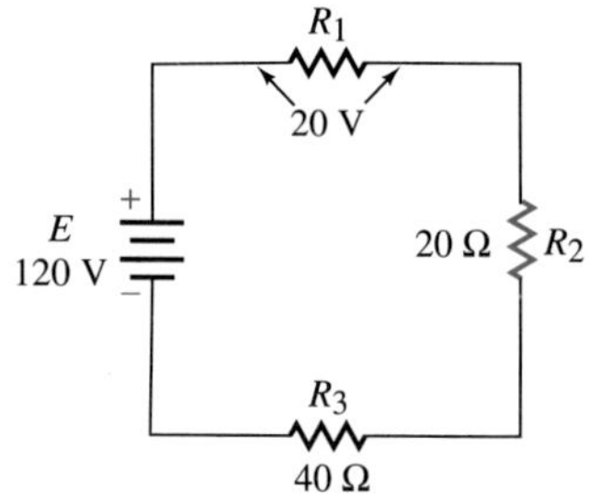

Figure 5.21

Solution

One possible path for the solution is:

1. Find R_1 using the voltage divider rule.
2. Find the equivalent resistance.
3. Use R_{eq} to calculate the circuit current.
4. Calculate the power in R_2 using I^2R_2.

Another path is:

1. Use KVL to find the voltage across the R_2, R_3 combination.
2. Use the voltage from Step 1 and $R_2 + R_3$ to calculate the current in the circuit.
3. Calculate the power using I^2R_2.

Suppose that the path is not obvious. What happens then? Well, why not use the unknowns in the equations as a guide? That is done as follows.

Since P_2 is wanted, our first equation must include P_2. Using the power relation for P_2 gives

$$(1)\quad P_2 = I^2R_2$$

This equation has the unknown I in it. So our next equation should include I. Using Ohm's Law,

$$(2)\quad E = IR_{eq}$$

Equation (2) has R_{eq} in it. So the next equation should be for R_{eq}. For a series circuit,

$$(3)\quad R_{eq} = R_1 + R_2 + R_3$$

Since Equation (3) introduced the unknown R_1, we must write an equation with R_1 in it. Applying Ohm's Law to R_1 gives

$$(4)\quad V_1 = IR_1$$

There are no new unknowns and the number of independent equations now equals the number of unknowns. The set can be solved. Using the method of substitution that is explained in the Appendix gives

$$\text{From (4)}\quad R_1 = 20\text{ V}/I$$

Substituting this into (3) gives three equations. They are:

1. $P_2 = I^2\,(20\ \Omega)$
2. $120\text{ V} = IR_{eq}$
3. $R_{eq} = (20\text{ V}/I) + 60\ \Omega$

From (2), $R_{eq} = 120\ \text{V}/I$. Substituting this into (3) gives

$$120\ \text{V}/I = (20\ \text{V}/I) + 60\ \Omega$$

or

$$I = 100\ \text{V}/60\ \Omega$$
$$I = 1.67\ \text{A}$$

Putting the value of I in (1) gives

$$P_2 = (1.67\ \text{A})^2(20\ \Omega)$$
$$P_2 = 55.78\ \text{W}$$

This choice of equations is only one of several possibilities. An alternate combination is:

1. $P_2 = V_2 I$
2. $E - V_2 - V_1 - V_3 = 0$ or $100 - V_2 - V_3 = 0$
3. $V_2 = IR_2$, or $V_2 = I(20\ \Omega)$
4. $V_3 = IR_3$, or $V_3 = I(40\ \Omega)$

Equation 2 included the unknown V_2 but not I. It also indicated that an equation with V_3 in it must be written. The unknown I was included in Equation 3. Then, Equation 4 was used to take care of V_3.

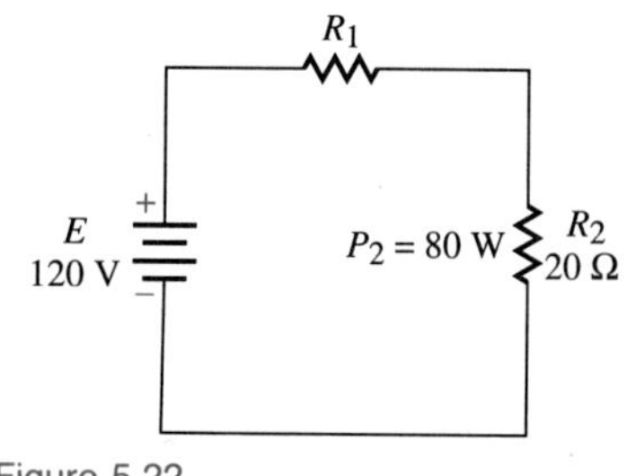

Figure 5.22

EXAMPLE 5.24

How much energy is used in the circuit of Figure 5.22 in 3 hours (a) in joules and (b) in kilowatt-hours?

Solution

A possible path for the solution is:

1. Use the power and resistance of R_2 to find the current.
2. Use the current, source voltage, and time to find the energy.

Using the equations to complete that path gives the following steps:

a. Since energy is wanted, it must be included in the first equation.

$$(1)\quad \mathscr{W} = EIt$$

The next equation should include the unknown I. Using the power and resistance for R_2 gives

$$(2)\quad I = \sqrt{P_2/R_2}$$

There are now two independent equations and two unknowns so the set can be solved. Equation (2) can be solved because it has only one unknown.

$$I = \sqrt{80\ \text{W}/20\ \Omega}$$
$$I = 2\ \text{A}$$

Putting this value of I in Equation (1) gives

$$\mathcal{W} = (120\text{ W})(2\text{ A})(3\text{ hr})(3600\text{ s/hr})$$
$$\mathcal{W} = 2.592 \times 10^6\text{ J}$$

b. $\mathcal{W} = (120\text{ W})(2\text{ A})(3\text{ hr})(1\text{ kW}/1000\text{ W})$
$\mathcal{W} = 0.72\text{ kWh}$

EXAMPLE 5.25

Find the circuit current, circuit power, and voltage across R_2 in Figure 5.23.

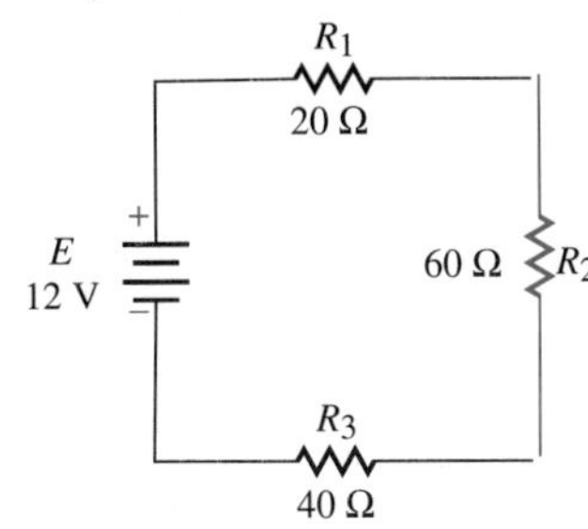

Figure 5.23

Solution

The first equation should have the current in it. Applying Ohm's Law to the circuit gives

$$(1)\quad E = IR_{eq} \qquad (I, R_{eq})$$

According to the unknowns, the next equation should have R_{eq} in it. Using Equation 5.5, we obtain

$$(2)\quad R_{eq} = R_1 + R_2 + R_3$$

There are two equations and two unknowns. They can be solved for I. From (2) $R_{eq} = 20\ \Omega + 60\ \Omega + 40\ \Omega = 120\ \Omega$. Putting this into (1) gives,

$$I = E/R_{eq}$$
$$= 12\text{ V}/120\ \Omega$$
$$I = 0.1\text{ A}$$

Since equations 1 and 2 did not include power and V_2, more equations must be used. Using the power relation for the circuit gives

$$(3)\quad P = EI$$
$$= (12\text{ V})(0.1\text{ A})$$
$$P = 1.2\text{ W}$$

Finally, applying Ohm's Law to R_2 gives

$$(4)\quad V_2 = R_2I$$
$$= (60\ \Omega)(0.1\text{ A})$$
$$V_2 = 6\text{ V}$$

Equations (3) and (4) had only one unknown so they are solved without writing other equations.

Another possible path is:

1. Use $P = E^2/R_{eq}$ to find the circuit power.
2. Use the circuit power and current to find the source voltage.
3. Use Ohm's Law to find the voltage across R_2.

5.8 Series Circuit Applications

The Common Flashlight

The common flashlight serves as our first series circuit application. A flashlight consists of three components: batteries, which serve as the

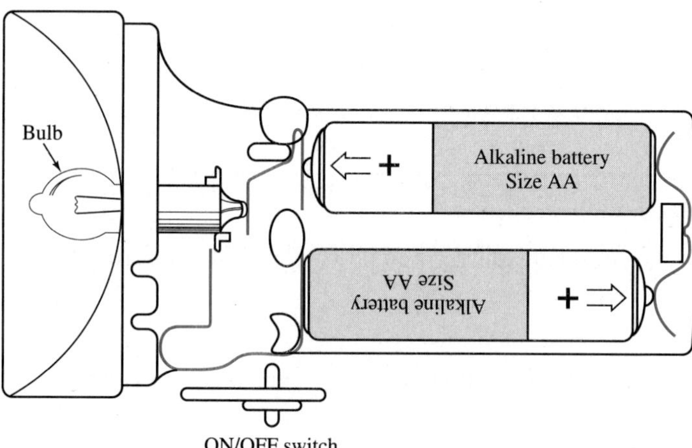

Figure 5.24 The diagram of a common flashlight.

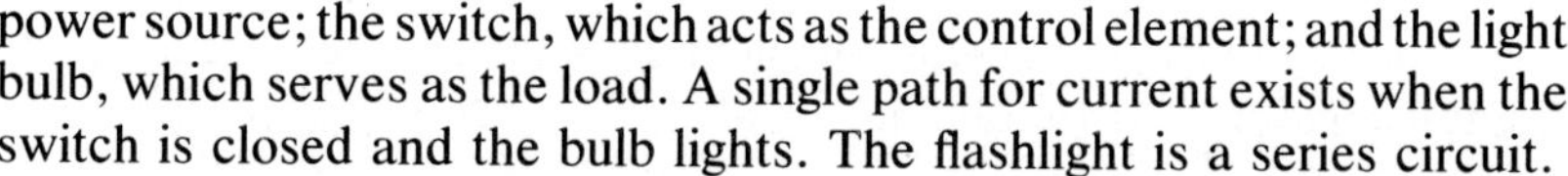
power source; the switch, which acts as the control element; and the light bulb, which serves as the load. A single path for current exists when the switch is closed and the bulb lights. The flashlight is a series circuit.

Let us examine the two operating modes of the flashlight. When the switch is open, the flashlight is said to be OFF. The batteries are ready to supply power to the bulb. The switch, however, represents a very high resistance when the contacts of the switch are open. Hence, the total resistance of the circuit is very large and for all practical purposes, the current is zero. With zero current flowing, no heat is produced in the filament of the bulb, and no light is produced.

When the switch is closed, the resistance between the two contacts of the switch is practically zero. Hence, the total resistance of the circuit is essentially the resistance of the light bulb. The voltage of the batteries is applied across the terminals of the light bulb. According to Ohm's Law, the current flowing equals the total battery voltage, divided by the resistance of the light bulb.

The ratings of the light bulb must match the ratings of the power source—the batteries in this case. If a light bulb with different ratings were substituted for the original light bulb in the flashlight, the bulb would either glow dimmer or, even worse, burn out because the current flow exceeds the maximum current rating of the bulb.

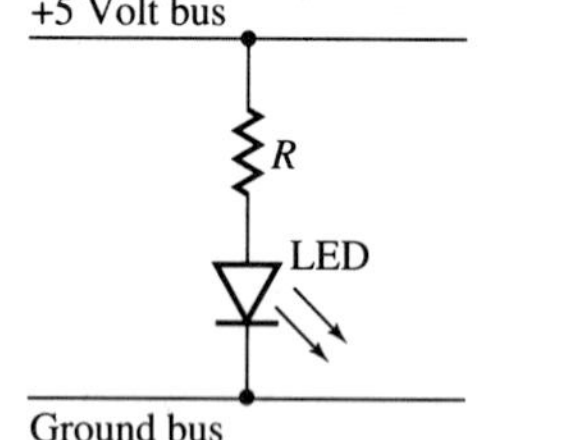

Figure 5.25 An LED indicator circuit.

Current Limiting

Suppose that in an electronic system of some type, we want to construct an indicator circuit, which will tell us when the power is ON. A light-emitting diode or LED will serve this purpose. An LED, however, has certain ratings that must not be exceeded. For example, the LED has a maximum current rating of about 20 mA for common red LEDs. The normal operating range for diode current is in the range of 10–20 milliamperes. The forward voltage drop across the red LED is about 1.6 volts and is determined by the gallium arsenide semiconductor material, which is used to make the diode. The forward resistance of the diode is very small, a few tens of ohms. The reverse resistance of the diode is very large, tens of thousands of ohms.

With this background information, let us return to our project. To limit the current to a level in the safe operating range of the diode, a limiting resistance must be placed in series with the diode. The current flowing in the diode-resistor series combination is equal to the potential difference between the power and ground buses, +5 volts in this case, divided by the total resistance of the diode-resistor combination.

EXAMPLE 5.26

What value of limiting resistance should be used if the diode is to operate at a current of 15 mA?

Solution

The value of resistance can be found if the voltage across the resistor and the current flowing through the resistor are known. The current is given as 15 mA. Hence, we only need to find the resistor voltage.

By examining the circuit, we note that the total voltage across the diode-resistor series combination is +5 volts. The 5-volt potential must be divided between the two elements. The diode voltage and the resistor voltage must add up to +5 volts. The forward voltage drop of the red LED is approximately 1.6 volts. Therefore,

$$V_{\text{resistor}} = 5 \text{ volts} - 1.6 \text{ volts}$$
$$V_{\text{resistor}} = +3.4 \text{ volts}$$

We can now apply Ohm's Law and calculate the needed value of resistance.

$$R_{\text{limiting}} = V_{\text{resistor}}/I_{\text{resistor}}$$
$$= 3.4 \text{ volts}/15 \text{ mA}$$
$$R_{\text{limiting}} = 227 \text{ ohms}$$

The closest color-coded value is 220 ohms.

What if the circuit is constructed and the diode is inadvertently put in backward? If this is the case, the reverse resistance of the diode plus the limiting resistance will be a large value. According to Ohm's Law, a fixed voltage divided by a very large resistance yields a very small current. In this case, the current will be insufficient to cause the LED to light. In fact, a common red LED will not produce visible light until a level of around 8 mA is reached.

What would happen in the LED circuit if the limiting resistance were left out and the diode were connected directly across the two buses? We stated earlier that the forward resistance of the diode is very small. According to Ohm's Law, a fixed voltage divided by a small resistance yields a large current. That is just what would happen. The current would increase and exceed the maximum rating of the diode. The brightness of the bulb would increase and then go out. The diode, because of excessive current flow, would open up just like a fuse under overload conditions.

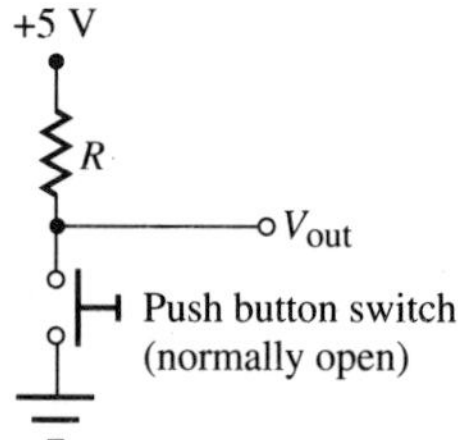

Figure 5.26 A push button switch circuit.

Pull-Up Resistor

Consider another situation, which is rather common in the real world. A common mechanical switch needs to be connected to a control circuit so that when the push-button switch is open, the voltage produced is +5 volts, or the supply voltage. When the switch is closed, the voltage produced should be ground potential, or 0 volts.

The interface circuit for the mechanical switch is shown in Figure 5.26. The resistor serves two functions. When the switch is open, the resistor acts like a "pull-up" resistor. By this we mean that the output voltage will be "pulled-up" to the supply voltage. This occurs because the open switch is a very large resistance. Hence, the current is zero for all practical purposes. With zero current flowing through the resistor, the voltage drop across the resistor is also zero. Consequently, the voltage on both sides of the resistor must be the same if no potential difference exists across the resistor and V_{out} is equal to +5 volts.

When the switch is closed, the output terminal is connected to ground. The resistor now serves in a current limiting capacity. The total voltage is dropped across the resistor. The current can then be found by applying Ohm's Law. Consider the following example.

EXAMPLE 5.27

In Figure 5.26, assume that $R = 1\ \text{k}\Omega$ and the bus voltage is +5 volts. How much current will flow when the switch is closed?

Solution

When the switch is closed, the total resistance of the circuit is slightly higher than 1kΩ. Using Ohm's Law,

$$I = +5\ \text{volts}/1\ \text{k}\Omega$$
$$I = 5\ \text{mA}$$

An inspection of electrical and electronic circuits will show many other applications of the series connection. Some of these applications are:

1. A fuse is found in almost every electronic instrument. The fuse protects the transformers, diodes, and other parts of the instrument from being damaged by high currents. A high current will cause the fuse to open. Since the fuse is in series with the other circuits, when the fuse blows, the current is reduced to 0 amperes.
2. A diode can be connected in series with an ammeter to protect the meter from damage that might result if it is connected in reverse polarity. The diode prevents current from passing in the reverse direction through the meter.
3. Switches are used to connect and disconnect circuits from sources of emf and from each other.

4. Resistors can be connected in series to provide one or more voltages from a fixed voltage. These circuits are known as voltage divider circuits.
5. A resistor connected in series with a capacitor determines the time needed for the capacitor voltage to change a certain amount. An example of this is the time delay circuit that keeps the garage light on after the door is closed.
6. A thermal fuse is connected in series with the heating element in some electric heaters. Excess heat will cause the fuse to open and turn off the heater. Thermal fuses are also used to protect motors from overheating.
7. The current in a part of a circuit is viewed on an oscilloscope by connecting the oscilloscope terminals across a small resistor that is connected in series with the circuit.
8. To measure the current, an ammeter is connected in series with a circuit.

SUMMARY

1. A series resistor circuit is a group of resistors connected so that there are no branch points between any two of them.
2. The current is the same in all parts of a series circuit, but the voltage divides, with larger resistances having larger voltages.
3. A change in any part of a series circuit causes a change in the current and voltage in all parts.
4. Kirchoff's Voltage Law states that the algebraic sum of the voltages in a closed loop is equal to zero.
5. A series circuit can be replaced by an equivalent resistance that is equal to the sum of the resistors in the circuit without changing the current in the circuit.
6. The voltage divider rule states that the voltage across a part of a series circuit is equal to the resistance of the part, multiplied by the total voltage, and divided by the equivalent resistance.
7. The transfer function of a circuit is the ratio of the output voltage divided by the input voltage.
8. The power in a series circuit can be found if any two of the quantities I, V, or R_{eq} are known.
9. The power in a series circuit is also equal to the sum of the power in the parts of the circuit.
10. Since the energy is power multiplied by time, the relations in points 8 and 9 also hold true for energy if the time is known.
11. The double subscript notation, which expresses the voltage at one point as measured from another point, gives the magnitude and polarity.
12. Some applications of series circuits are current limiting, voltage division, time delays, and current interruption.

EQUATIONS

No.	Equation	Description	Page
5.1	$E - V_1 - V_2 - \cdots - V_n = 0$	Kirchhoff's Voltage Law	184
5.2	$E = V_1 + V_2 + \cdots + V_n$	Another form of Kirchhoff's Voltage Law	184
5.3	$\frac{V_1}{R_1} = \frac{V_2}{R_2} = \cdots = \frac{V_n}{R_n}$	The relation of the voltage-resistance ratio for parts of a series circuit	185
5.4	$R_{eq} = \frac{V}{I}$	The relation of the equivalent resistance and current in a circuit	187
5.5	$R_{eq} = R_1 + R_2 + \cdots + R_n$	The equivalent resistance of a series circuit	188
5.6	$V_x = \frac{(R_x V)}{R_{eq}}$	The voltage divider equation	191
5.7	$P = VI$	Power in a series circuit in terms of the voltage and current	196
5.8	$P = I^2 R_{eq}$	Power in a series circuit in terms of the current and equivalent resistance	196
5.9	$P = \frac{V^2}{R_{eq}}$	Power in a series circuit in terms of the voltage and equivalent resistance	196
5.10	$P = P_1 + P_2 + \cdots + P_n$	Power in a series circuit in terms of the power in parts of the circuit	198
5.11	$\mathcal{W} = VIt$	Energy in a series circuit in terms of voltage, current, and time	199
5.12	$\mathcal{W} = I^2 R_{eq} t$	Energy in a series circuit in terms of current, equivalent resistance, and time	199
5.13	$\mathcal{W} = \frac{V^2 t}{R_{eq}}$	Energy in a series circuit in terms of voltage, equivalent resistance, and time	199
5.14	$\mathcal{W} = \mathcal{W}_1 + \mathcal{W}_2 + \cdots + \mathcal{W}_n$	Energy in a series circuit in terms of the energy in parts of the circuit	199

QUESTIONS

1. How would you determine which parts of a circuit are connected in series?
2. Does the fact that two resistors each have 5 amperes of current in them mean that they are connected in series?
3. What is the name of the circuit law that relates the voltage across parts of the circuit to the total voltage?
4. Name three procedures that determine the equivalent resistance of a series circuit.
5. How does short circuiting one of the resistors in a series circuit affect the equivalent resistance?
6. What will happen to the current in a series circuit if one of the resistors is short circuited? Open circuited?
7. What advantages does the voltage divider rule have over using Ohm's Law?
8. What four quantities are related in the voltage divider rule?
9. How can the power in the series circuit be determined when the power in each part of it is known?
10. What will happen to the power in a series circuit if one of the resistors is short circuited? Open circuited?
11. Which draws more current? a 50-W, 120-V bulb, or a 75-W, 125-V bulb?
12. How are energy and power related?
13. If V_{AB} is 8 volts, what is V_{BA}?
14. What is the voltage of the ground point in a schematic?
15. What is the procedure for determining the voltage at one point in a circuit as measured from a second point?
16. What are three general uses of the series connection?
17. Which characteristic of the series circuit enables a fuse to protect a circuit from damage?
18. Why is it a poor practice to replace a fuse with a piece of conductor?
19. How can a voltmeter be used to determine which bulb has an open filament in a series string of bulbs?

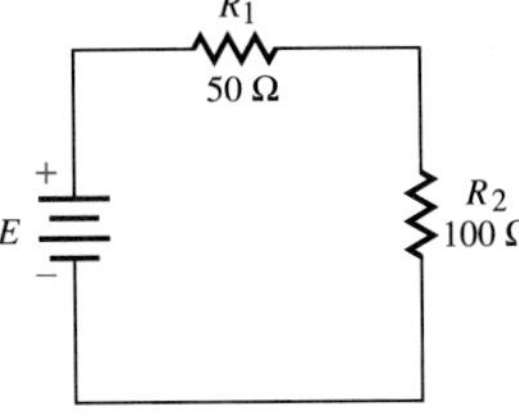

Figure 5.27

PROBLEMS

SECTION 5.2 Kirchhoff's Voltage Law

1. How many volts are across R_1 when the voltage across R_2 is 80 volts in the circuit of Figure 5.27?

2. What is the voltage across the 600-ohm and 400-ohm resistors in the circuit in Figure 5.28?

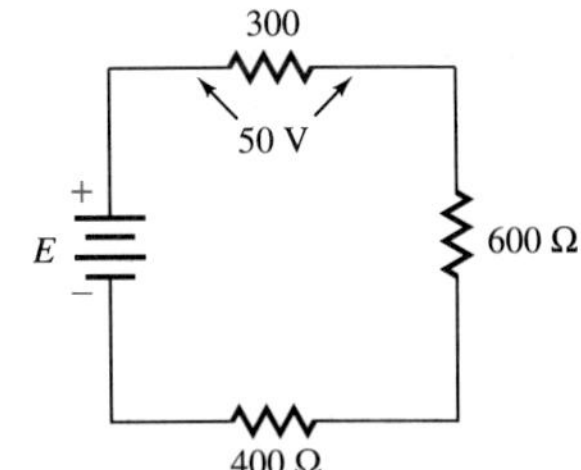

Figure 5.28

3. Using Figure 5.29, find:

   (a) The voltage across R_1 and R_3.

   (b) The resistance of R_3.

4. Three lamps are connected in series. The voltage is 50 volts across L_1, 30 volts across L_2, and 60 volts across L_3.

   (a) What is the source voltage?

   (b) If the resistance of L_1 is 80 ohms, what is the resistance of L_2 and L_3?

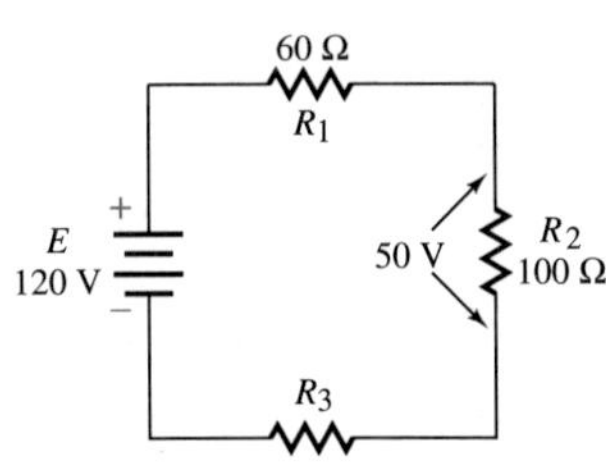

Figure 5.29

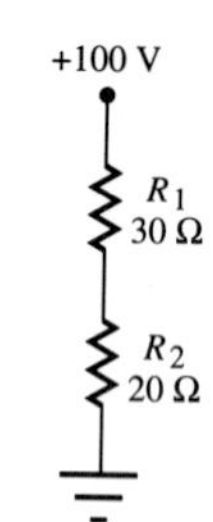

Figure 5.30

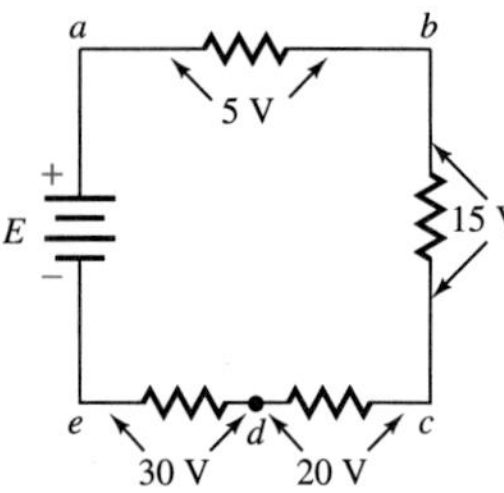

Figure 5.31

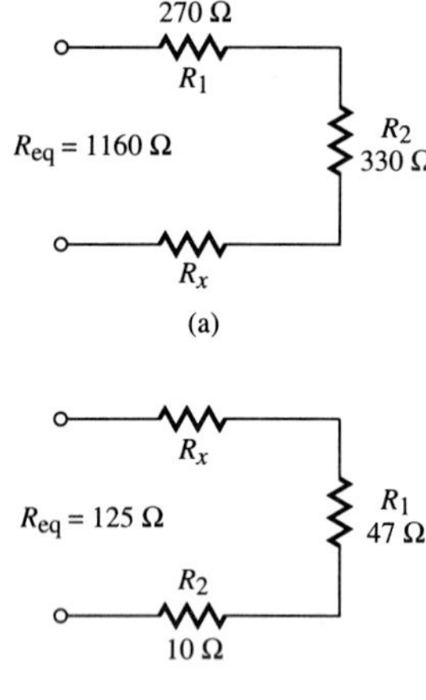

Figure 5.32

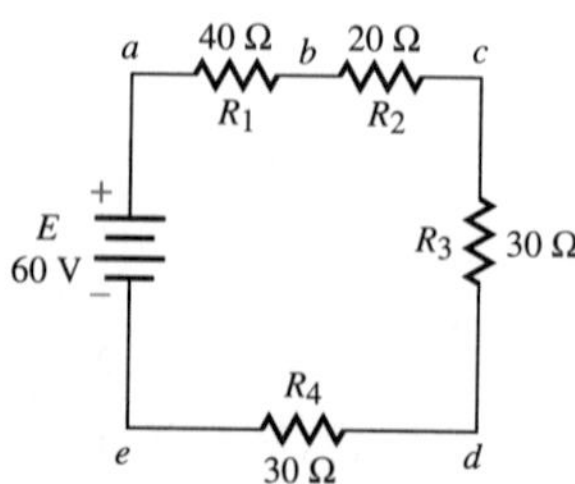

Figure 5.33

5. A series circuit of five 500-ohm resistors is connected across a 120-volt source.

   (a) How many volts are across each resistor?

   (b) What would be the voltage across each resistor if each one were 1000 ohms? 10,000 ohms?

6. How many volts are across R_2 in the circuit in Figure 5.30?

7. In Figure 5.31, how many volts will a voltmeter indicate when connected across:

   (a) a and c?

   (b) c and e?

   (c) a and d?

   (d) a and e?

SECTION 5.3 Equivalent Resistance

8. Determine the equivalent resistance of the following networks:

   (a) 120-, 150-, and 180-ohm resistors connected in series

   (b) 22- and 68-ohm resistors connected in series.

9. Determine the value of R_x in the networks of Figure 5.32(a) and (b).

10. Three resistors are connected in series. The first resistor is 820 ohms, the second resistor is 330 ohms, and the third resistor is 1000 ohms. What is the equivalent resistance of the network?

11. How many 270-ohm resistors must be connected in series to obtain an equivalent resistance of 1350 ohms?

12. You have one each of the following resistors: 5-ohm, 10-ohm, 15-ohm, 20-ohm, 25-ohm, and 30-ohm. What are the possible series combinations that can be used to get an equivalent resistor of 30-ohms? You can use a resistor in more than one combination but cannot use it more than once in a combination.

SECTION 5.4 The Voltage Divider Law

13. A 30-ohm, 20-ohm, and 10-ohm resistor are connected in series. The series circuit is connected to a 60-volt source. Use the voltage divider rule to find the voltage across the 30-ohm resistor.

14. Use the voltage divider rule to find the voltage across points a, c; c, d; and c, e in the circuit of Figure 5.33.

15. A 50-ohm, 100-ohm, and 150-ohm resistor are connected in series. Use the voltage divider rule to find the voltage across the series group when the voltage across the 100-ohm resistor is 70 volts.

16. Use the voltage divider rule to find the resistance of R_x in Figure 5.34.

SECTION 5.5 Power and Energy in a Series Circuit

17. Three resistors R_1, R_2, and R_3 are connected in series to a source.

(a) How much power does the source supply if the power is 60 W in R_1, 80 W in R_2, and 100 W in R_3?

(b) How much energy in kilowatt-hours is used in each resistor and in the total circuit if it is connected to the source for 3 hours?

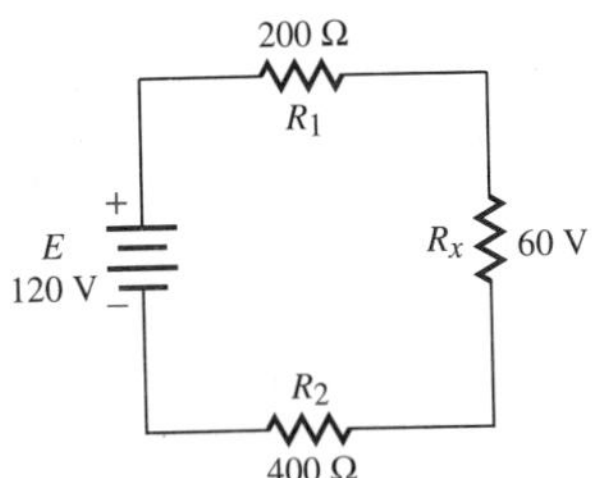

Figure 5.34

18. A source supplies 200 watts of power to a series circuit of three resistors. Sixty watts are dissipated in the first resistor. Ninety watts are dissipated in the second one. How many watts are dissipated in the third one?

19. In the circuit of Figure 5.35, find:

(a) The power dissipated in the circuit.

(b) The energy used by the circuit if it is connected to the source for 8 hours.

20. Repeat Problem 19 for the circuit of Figure 5.36.

21. Three resistors R_1, R_2, and R_3 are connected in series to a source. The energy used is 1.2 kWh in R_1, 1.6 kWh in R_2, and 2.4 kWh in R_3.

(a) How many watts of power are dissipated in each resistor in 8 hours?

(b) How many watts of power are supplied by the source?

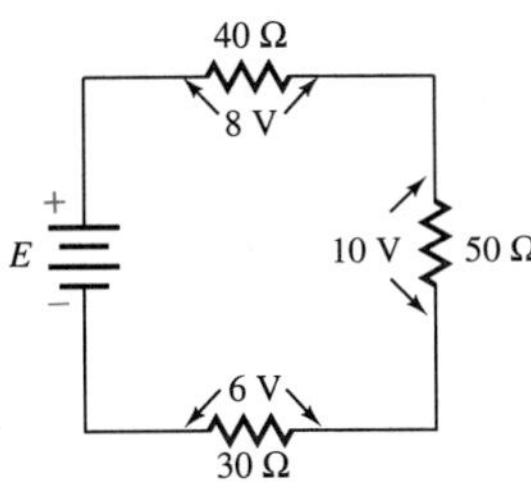

Figure 5.35

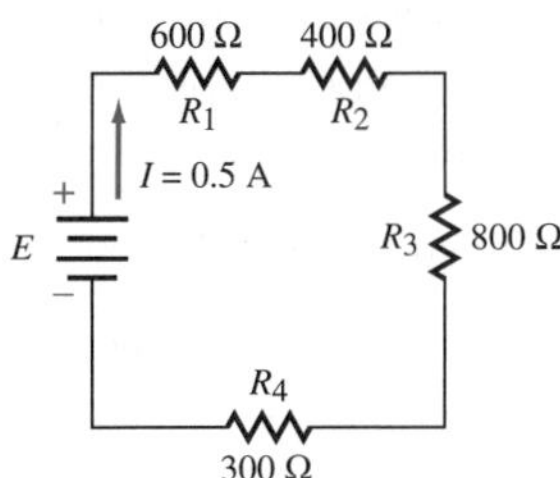

Figure 5.36

SECTION 5.6 Double Subscript Notation

22. For each of the following, which terminal is the more positive one?

(a) $V_{AC} = -3$ V

(b) $V_{BD} = 8$ V

(c) $V_{AB} = 2$ V

(d) $V_{BA} = -6$ V

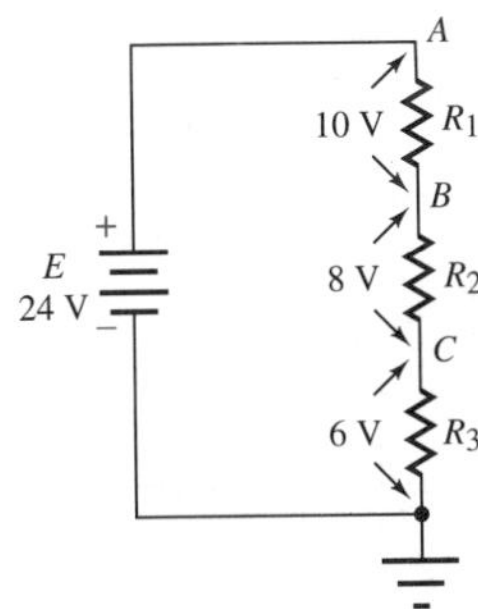

Figure 5.37

23. Determine V_{AB}, V_{CA}, and V_B for the voltage divider circuit shown in Figure 5.37.

24. How many volts is the source E in the circuit of Figure 5.38 if V_{BA} is -4 V?

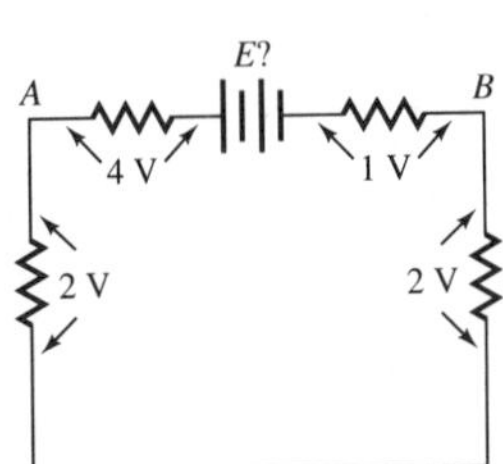

Figure 5.38

SECTION 5.7 Analysis of Series Circuits

25. A series circuit consisting of a 200-ohm and a 300-ohm resistor is connected to a 100-V source. Determine the following:

(a) The current in the network.

(b) The voltage across each resistor.

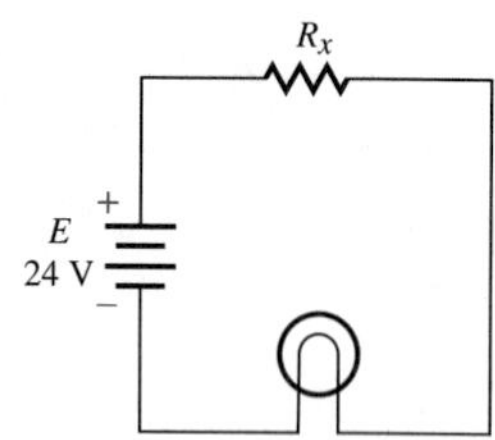

Figure 5.39

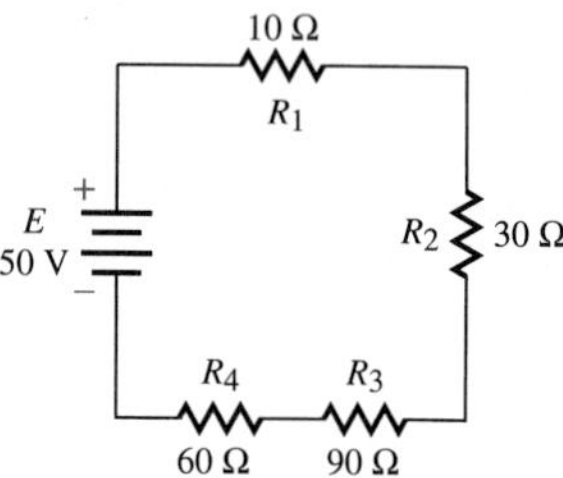

Figure 5.40

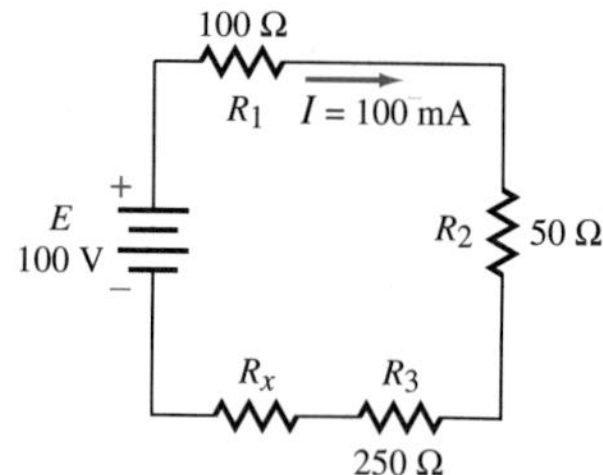

Figure 5.41

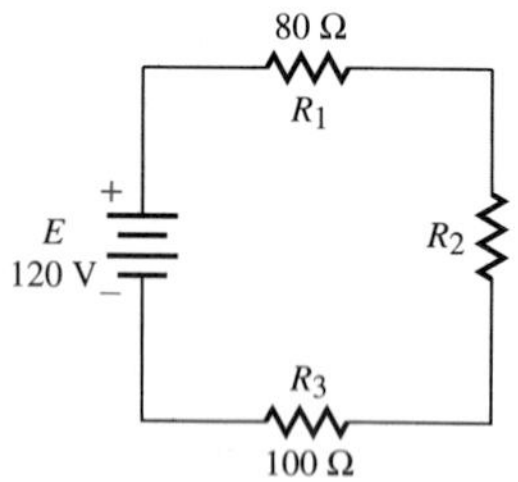

Figure 5.42

(c) The power that is dissipated in the network.

(d) Repeat part (c) using Equation 5.9

26. A lamp rated at 12 volts and 0.1 ampere is connected in a circuit as shown in Figure 5.39.

 (a) What should be the value of R_x so that the lamp operates at rated voltage and current?

 (b) How many joules of energy does the lamp use in 8 hours?

27. An electric heater draws 10 amperes of current. It is connected to a 120-volt source. The fuse has a resistance of 0.1 ohm and the switch has a resistance of 0.2 ohm.

 (a) What is the voltage at the heater?

 (b) How much power does the heater use?

 (c) What is the heater resistance when operating?

28. Given the circuit shown in Figure 5.40, determine:

 (a) The equivalent resistance of the network.

 (b) The voltage across the 10-ohm resistor.

 (c) The power in the 30-ohm resistor.

 (d) The voltage across the 60-ohm resistor using the voltage divider equation.

29. A 50-ohm resistor is connected to a circuit as shown in Figure 5.41.

 (a) Determine the value of R_x.

 (b) Determine the voltage across the 50-ohm resistor.

 (c) Determine the power supplied by the source.

 (d) Determine (b) and (c) if R_1 is short circuited.

30. The power dissipated in the circuit in Figure 5.42 is 30 watts. Find:

 (a) The current in the circuit.

 (b) The resistance of R_2.

 (c) The voltage across each resistor.

 (d) The kilowatt-hour of energy used in the circuit in 3 hours.

31. Determine the resistance that must be connected in series with a 19-ohm resistance so that the current is 20 mA when the circuit is connected to a 10-volt source.

32. The current in a series circuit is 5 mA when connected to a 40-volt source. If one resistance is 2800 ohms, how many ohms is in the rest of the circuit?

CHAPTER 6

CIRCUIT ANALYSIS USING PSPICE

CHAPTER OBJECTIVES

After completing this chapter, you should be able to:

1. Create a source file for input to PSpice.
2. Translate a schematic diagram into a PSpice description.
3. Perform a dc analysis of a resistive circuit using PSpice.
4. Perform an ac analysis of a resistive circuit using PSpice.
5. Interpret the results of a PSpice analysis.

KEY TERMS

1. ac analysis
2. component line
3. dc analysis
4. device name
5. node
6. operating point
7. simulation
8. source file

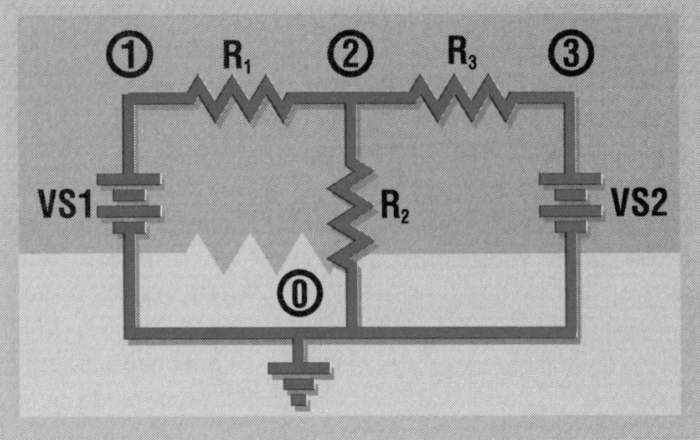

6

INTRODUCTION

Breadboarding in the 1960's and 1970's consisted of physically building a circuit with actual components, and then methodically testing the prototype to see if it did indeed meet design specifications. The process was time-consuming and delays were common as the circuit board was manufactured, or as the circuit was connected using wire-wrapping methods.

If an error occurred, the process started all over again with a new prototype and further delays. These delays are unacceptable in today's competitive international market. A delay of a month may mean the loss of market share and millions of dollars in research and development costs.

Furthermore, circuits are being built with application specific integrated circuits (ASICs); these circuits do not currently exist and may not exist until the product nears production. Hence, the ability to simulate the behavior of circuits is crucial to developing products in a timely manner.

In this chapter, we introduce you to a circuit simulation software package called PSpice. PSpice allows you to simulate your circuit designs before committing them to hardware. Using PSpice, you can determine the circuit's response to different inputs, different signal frequencies, and the effects of noise in a circuit. In effect, you will be

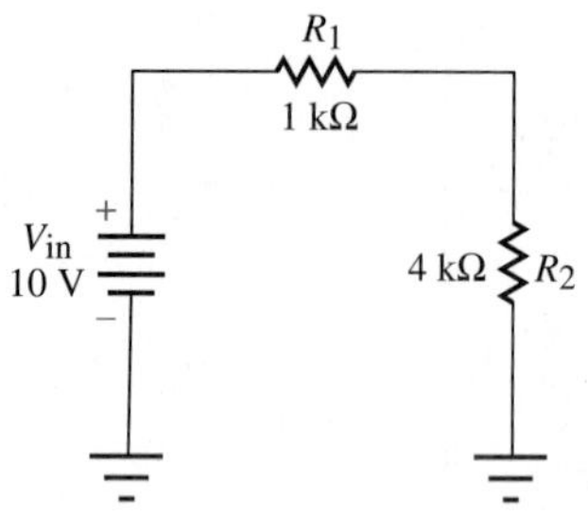

Figure 6.1 Schematic of a voltage divider circuit.

breadboarding and testing your circuit inside the computer.

PSpice is a member of the SPICE family of circuit simulators. The programs come from the SPICE2 circuit simulation program, which was developed at the University of California at Berkeley during the early 1970's. The generality and speed of SPICE2 led to its becoming the de facto standard for analog circuit simulation.

At this point in your study of electronics, it may seem like overkill to use PSpice to study simple dc circuits. These circuits can be easily analyzed by hand with the help of a scientific calculator. However, our objective is to illustrate the use of a circuit simulator in an environment where we can easily test its validity. This will build your confidence in PSpice, and in its ability to act like the real circuit.

S 6.1 PSpice Circuit Simulator

Let us begin with a familiar circuit, one that we have analyzed before. The circuit shown in Figure 6.1 is a voltage divider circuit powered by a single voltage source. Suppose that we are asked to find the voltage across R_2.

To use the PSpice circuit simulator, the process is:

source file

1. Create a **source file** containing the circuit description and PSpice commands.
2. Run the source file through the PSpice circuit simulator.
3. If the run contains errors, correct the errors and run the source file through the PSpice simulator again.
4. Interpret the output file produced by the circuit simulator.

node

To describe the circuit, the first step is to number all of the circuit nodes. For PSpice, a **node** is a connection point where two or more circuit components are connected together. The node numbers do not need to be integers, as in some versions of SPICE. Node names may be any alphanumeric string, up to 131 characters long. The reference or ground node *must be labeled Node 0*. Furthermore, the node numbers do not need to be sequential. Figure 6.2 shows one choice of node labels for the circuit shown in Figure 6.1.

Before we create a source file to describe this circuit, we need to know the rules used by PSpice.

Rule 1. PSpice Statements can be either upper case or lower case letters.

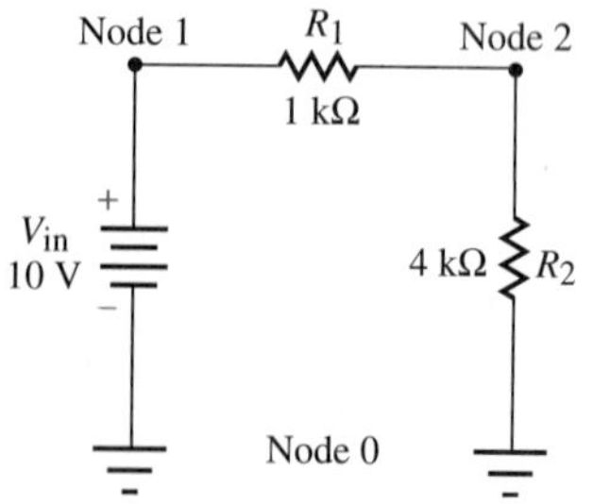

Figure 6.2 Labeling the nodes in the voltage divider circuit.

EXAMPLE 6.1

The circuit shown in Figure 6.2 is driven by a dc voltage source. On the schematic, the source is labeled V_{in}. When naming this source, we can use either "`VIN`" or "`vin`." Since the PSpice simulator disregards upper/lowercase letters, PSpice considers both names to be the same.

Rule 2. Names for circuit components must *start* with a letter; after that, they can contain either letters or numbers. Names can be up to 131 characters long, but for practical considerations, limit them to a length of eight characters or less.

EXAMPLE 6.2

There are two resistors shown in Figure 6.1. Since the schematics contains the names R_1 and R_2 for these two resistors, we can use the names "`R1`" and "`R2`." Or we can rename them in our PSpice source file as "`RESISTOR1`" and "`RESISTOR2`" or any other pair of names we might want to use.

Rule 3. Values are written in standard floating-point notation, with optional scale and units suffixes. The scale suffixes follow standard scientific convention, and they multiply the number that they follow. In addition, unit suffixes are allowed. Any letter that is not a scale suffix may be used as a unit suffix. Unit suffixes are ignored by PSpice.

PSpice scale suffixes:

$F = 10^{-15}$	$M = 10^{-3}$
$P = 10^{-12}$	$K = 10^{3}$
$N = 10^{-9}$	$MEG = 10^{6}$
$U = 10^{-6}$	$G = 10^{9}$
$MIL = 25.4 \times 10^{-6}$	$T = 10^{12}$

EXAMPLE 6.3

A resistor has a value of 1500 ohms. How would 1500 ohms be represented in PSpice?

Solution

There are several ways of representing 1500 ohms in PSpice. They are:

```
1500     1.5E3     1.5K     .0015MEG
```

component line

Rule 4. Each device in the circuit is represented in the source file by a component line, which does not begin with period. Each **component line** has a similar format:

The device name, followed by
two or more node names, followed by
a model name (not all devices have this), followed by
0 or more values.

device name

The first letter of the **device name** tells you what kind of device it is. For example, resistors must start with "R" and capacitors with a "C." The type of device then determines what parameters must be included in the rest of the line—for example—how many nodes it is connected to, whether it needs a model name, and the component's value.

For each component, PSpice uses the convention shown in Figure 6.3. The element is connected between two nodes labeled N_1 or N^+ and N_2 or N^-. The positive current flows in the direction of the voltage drop, that is, from the more positive node N^+ toward the more negative node N^-.

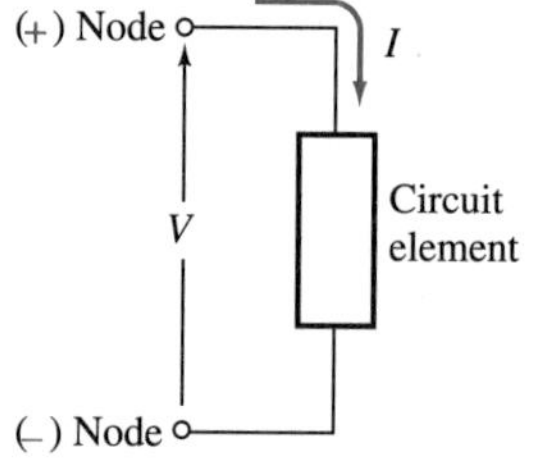

Figure 6.3 Device conventions for voltage and current in PSpice.

Rule 5. A device description for a resistor has the following form:

```
R<name> <(+) node> <(−) node> <value>.
```

Note: Quantities enclosed by "< >" symbols are required items.

The format for describing a resistor may look a little bewildering at this point. Let's take a closer look at each part of the description. The resistor name must begin with "R." Using the schematic in Figure 6.1, the resistor R_1 can be given the name "R1" and resistor R_2 can be given the name "R2."

The (+) and (−) nodes define the polarity meant when the resistor has a positive voltage across it. Positive current flows from the (+) node through the resistor to the (−) node. When entering the circuit description, you do not have to know the current direction. For example, for the two resistors in Figure 6.2, resistor R_1 is connected between Node 1 and Node 2 and resistor R_2 is connected between Node 2 and Node 0. Hence, for resistor R_1, Node 1 can be the (+) node, and Node 2 can be the (−) node. Likewise, for resistor R_2, Node 2 can be the (+) node, and Node 0 can be the (−) node.

For our current study, we will not use the optional [(model)name]. In a more detailed study of resistors in a circuit, you can use this parameter to describe the behavior of resistors over a certain temperature range.

The last entry on the device line is the component value. For the circuit shown in Figure 6.2, resistor R_1 has a value of 1 kilohm and resistor R_2 has a value of 4 kilohm.

Now we are ready to write the two device lines for resistors R_1 and R_2.

EXAMPLE 6.4

Write the two device lines describing the two resistors in Figure 6.2.

Solution

```
R<name>  <(+) node>  <(−) node>  <value>
  R1         1           2           1K
  R2         2           0           4K
```

> **Rule 6.** The general form for a dc voltage source is
>
> `V<name>  <(+) node>  <(−) node>  [[DC] <value>].`
>
> Note: the "( )" symbol is a comment while the "[ ]" symbol indicates an optional quantity.

Describing a dc voltage source is very similar to describing a resistor. Both are two terminal devices and have a specific value. Again, let us return to the circuit shown in Figure 6.2. The dc voltage source is connected between Node 1 and Node 0. Node 1 is the positive terminal of the voltage source. The magnitude of the source is 10 volts. Hence, the device line can be written as follows:

```
V<name>  <(+) node>  <(−) node>  [[DC] <value>]
  Vin        1           0           10volts
```

> **Rule 7.** Comment lines are marked by an asterisk in the first column, and may contain any text or message.

This rule allows you to write messages that will help in labeling the source file or identifying what is done in different parts of the analysis.

The source file is constructed in the following manner:

- The first line is the title line and may contain any type of text. This title line will be used at the beginning of the output file to identify the results.
- Next comes the description of the circuit. Each circuit component gets a component line.

- Next come the command lines. If no command lines are included, only a dc bias analysis is performed.
- The last line must be ".END".

It should be noted at this point that the number of blanks between items in each line is not significant, except in the title line. Tabs and commas are equivalent to blanks.

Now we are ready to create a source file. Again, we will use the circuit shown in Figure 6.1 as our example. We need to create a title line to describe our circuit. Let us use the following title line:

```
Analysis of Circuit 6.1.
```

There are three circuit elements, so three separate lines must be used to describe the wiring of the circuit.

```
Vin  1  0  10volts
R1   1  2  1K
R2   2  0  4K
```

This last line in our source file will be

```
.END
```

Putting this together into one listing, we get the source file shown in Figure 6.4.

```
Analysis of Circuit 6.1
VIN  1  0  10Volts
R1   1  2  1K
R2   2  0  4K
.END
```

Figure 6.4 Source file for a PSpice simulation of a simple voltage divider circuit.

EXAMPLE 6.5

Write a PSpice source file for the circuit shown in Figure 6.5.

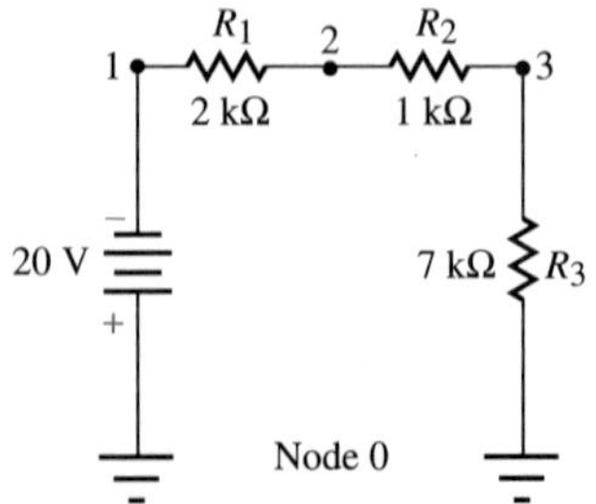

Figure 6.5 Circuit for Example 6.5.

Solution

```
DC Analysis--Circuit in Figure 6.5
V    1  0  -20
R1   1  2  2K
R2   2  3  1K
R3   3  0  7K
.END
```

Figure 6.6 PSpice source file for the circuit shown in Figure 6.5.

S 6.2 A dc Analysis Using PSpice

The PSpice simulator provides eight types of analyses which include:

- dc sweep of an input voltage/current source, a model parameter, or temperature.
- Bias point or operating point calculation.
- Frequency response calculation.
- Total and individual noise calculation.
- dc sensitivity calculation.
- Small-signal transfer function (Thevenin equivalent) calculation.
- Transient response, or behavior over time.
- Fourier components of the transient response.

operating point
dc analysis

We will begin our use of PSpice with the bias point or **operating point** calculation. This calculation performs a **dc analysis** when no other commands are specified; it is the basis for the frequency response, noise, dc sensitivity, and small-signal transfer function calculations.

To illustrate the operating point calculation, let us use the source file shown in Figure 6.4 to perform a PSpice simulation. The command used will vary, depending on the availability of a hard disk. The command you use may be similar to the following:

```
A>PSPICE CIRCUIT4 <RETURN>
```

where the file `CIRCUIT4.CIR` exists on disk.

simulation

The results of the **simulation** will be found in an output file having the same name as the source file except that the extension will be "`.OUT`." The `CIRCUIT4.OUT` file is shown in Figure 6.7.

The first entry that you will note is the header. It is the same as the title line in the source file, "`DC Analysis--Circuit in Figure 6.1`." Following this entry, the output file tells you the type of calculation performed, "`Small Signal Bias Solution`."

The results are presented in the following order. First, each node voltage is presented. Since the circuit includes only two nodes, a voltage is computed for Node 1 and for Node 2, 10 volts and 8 volts, respectively.

The next result is the voltage source current. Note that the value has a negative sign. Remembering the conventions presented in Figure 6.3, the negative sign tells us that the current flows in the opposite direction

```
DC Analysis--Circuit in Figure 6.1
**** SMALL SIGNAL BIAS SOLUTION      TEMPERATURE = 27.000 DEG C
*******************************************
NODE VOLTAGE NODE VOLTAGE NODE VOLTAGE NODE VOLTAGE
(   1) 10.0000 (   2) 8.0000
VOLTAGE SOURCE CURRENTS
NAME      CURRENT
VIN      -2.000E-03
TOTAL POWER DISSIPATION 2.00E-02 WATTS
```

Figure 6.7 The PSpice output file.

```
DC Analysis--Circuit in Figure 6.5
V    1  0  -20
R1   1  2  2K
R2   2  3  1K
R3   3  0  7K
.END
```

```
DC Analysis--Circuit in Figure 6.5
SMALL SIGNAL BIAS SOLUTION    TEMPERATURE = 27.000 DEG C
*************************************
NODE VOLTAGE NODE VOLTAGE NODE VOLTAGE NODE VOLTAGE
(   1) -20.0000 (   2) -16.0000 (   3) -14.0000
VOLTAGE SOURCE CURRENTS
NAME      CURRENT
V         2.000E-03
TOTAL POWER DISSIPATION    4.00E-02 WATTS
```

Figure 6.8 PSpice output file for a simulation of the circuit shown in Figure 6.5.

as the direction shown in Figure 6.3; that is, the current flows from the N^+ terminal through the external circuit and back to the N^- terminal.

The last computation is the total power dissipated in the circuit. In this example, the total power dissipated is 2.00E-02 watts or 20 milliwatts.

What if we want to change a component value? All we need to do is return to our word processor or editor and make the necessary change in our source file, and then run the source file through the PSpice simulator again to get our new results. This interactivity makes PSpice a very useful tool.

Let us try another example. This time we will run the source file for the circuit shown in Figure 6.5 through the PSpice simulator. The results are shown in Figure 6.8.

In this case, the input voltage at Node 1 is negative with respect to the ground node. Hence, the three node voltages are negative and in the right proportion for each resistor.

The source current in this is positive, indicating that the current flows into the N^+ terminal, which is the counterclockwise direction in the circuit. The total power consumption is 4.00E-02 watts or 40 milliwatts. This checks with a value obtained by multiplying the source voltage, 20 volts, by the current supplied by the source, 2 milliamperes, or 40 milliwatts.

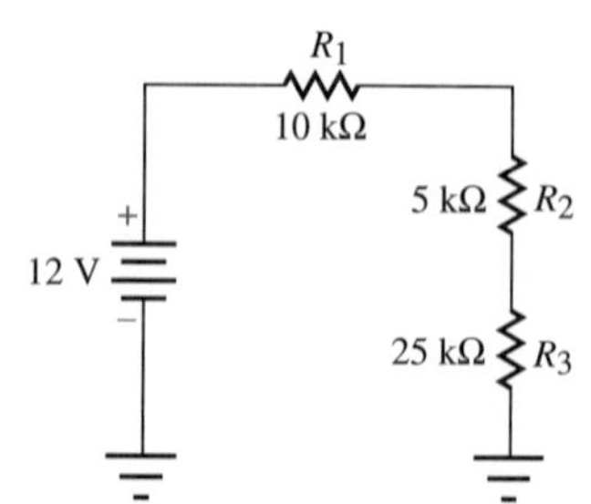

Figure 6.9 Circuit diagram of a three-resistor series circuit.

A word of caution should be noted here. PSpice is a great tool for analyzing circuits. However, it is imperative that you have a solid foundation in circuit analysis techniques so that you can sense when the simulator is producing erroneous results. Do not automatically assume that the circuit simulator has given you correct results.

Consider the following situation. The circuit shown in Figure 6.9 is to be analyzed. A source file is created and the results obtained are shown in Figure 6.10. Just looking at the node voltages raises a red flag. The voltages at Nodes 2 and 3 are identical, yet the schematic shows 5

```
DC Analysis--Circuit in Figure 6.9
V    1  0  12
R1   1  2  10K
R2   2  3  5M
R3   3  0  25K
.END

DC Analysis--Circuit in Figure 6-9
**** SMALL SIGNAL BIAS SOLUTION TEMPERATURE = 27.000 DEG C
*****************************************
NODE VOLTAGE NODE VOLTAGE NODE VOLTAGE NODE VOLTAGE
(   1) 12.0000 (   2) 8.5714 (   3) 8.5714
VOLTAGE SOURCE CURRENTS
NAME       CURRENT
V          -3.429E-04
TOTAL POWER DISSIPATION    4.11E-03 WATTS
```

Figure 6.10 PSpice output file for the circuit shown in Figure 6.9.

kilohms between them. The results just cannot be correct. What would cause the voltage at Nodes 2 and 3 to be the same? To have no voltage drop between these two nodes, either the current would have to be zero or the resistance between the two nodes would have to be zero. The results indicate that the current is non-zero and a voltage drop exists across the other two resistors.

The second possibility of zero or near-zero ohms between Nodes 2 and 3 takes us back to the source file or the circuit description at the beginning of the output file. What value was entered for R_2? The value entered on the `R2` line is `5M`. The suffix M stands for 10^{-3}, making the value of this component 5×10^{-3} instead of 5×10^3. Returning to our word processor or editor, we can easily correct the source file, enter the correct value of resistance for R_2, and rerun the source file through the PSpice circuit simulator.

S 6.3 An ac Analysis Using PSpice

PSpice can be used to analyze resistive circuits driven by an ac voltage or current source. As an example, consider the circuit shown in Figure 6.11.

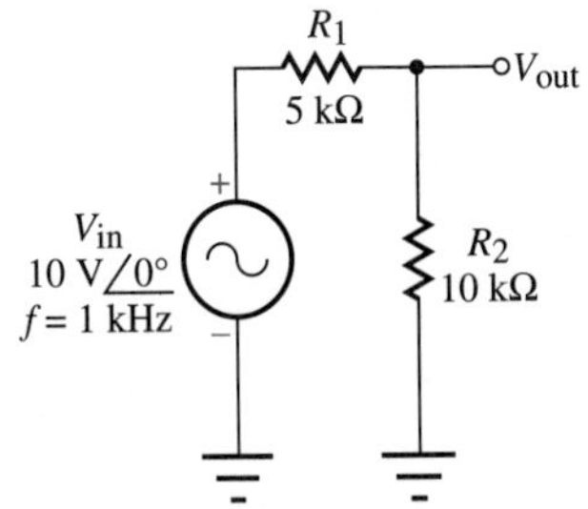

Figure 6.11 A resistive voltage divider driven by an ac source.

The circuit is driven by a sinusoidal signal whose peak amplitude is 10 volts. The input signal has a 0-volt dc component. The frequency of the input signal is 1 kHz.

The circuit is a voltage divider consisting of two resistors, R_1 and R_2. The output signal is measured across R_2.

To create a source file to describe this circuit and the analysis to be performed, we will need to add two new features of PSpice: (1) the definition of ac sources and (2) the control statements that perform the ac analysis.

ac analysis

To perform an **ac analysis**, the circuit must contain at least one ac source, which seems only reasonable. An ac source is defined as follows:

```
V<name> <(+) node> <(-)node> [AC <(magnitude)value>
                    <(Phase)value>]
```

where:

V<name> is the name of the voltage source,
<(+)node> is the name of the (+) node,
<(−)node> is the name of the (−) node,
<(magnitude)value> is the peak amplitude of the voltage source,
and <(phase)value> is the phase angle of the input signal in degrees.

EXAMPLE 6.6

Write a component line for a PSpice file to describe the ac source in Figure 6.11.

Solution

The ac source is connected between Node 1 and Node 0 (ground). Node 1 will be assigned the (+) node. The ac signal has a peak amplitude of 10 volts. Since no phase angle is given, it can be assumed to be zero.

Hence, the component line can be written as

```
VIN 1 0 AC 10 0
```

In Example 6.6, the phase angle could have been omitted. If it is omitted, the default value is 0 degrees.

An ac current source has the same format,

```
I<name> <(+)node> <(-)node> [AC <magnitude)value>
                    <(phase)value>]
```

EXAMPLE 6.7

Identify the current source described by the following component line:

```
IAC  2 3 AC 0.001  90
```

Solution

The name of the ac current source is "IAC." The first letter "I" indicates that the source is a current source.

The current source is connected between Node 2 and Node 3. Node 2 is the (+) node since it is listed first.

The "AC" indicates that this is an ac source whose peak amplitude is 0.001 ampere or 1 mA, with a phase angle of 90 degrees.

The .AC control statement is used to calculate the frequency response of a circuit over a range of frequencies. The general form of this control statement is

```
.AC [TYPE] <(points)value> <(start frequency)value>
    <(end frequency)value>
```

The TYPE of sweep must be specified. It can be one of three types, identified by the keywords: LIN, OCT, or DEC. These types of sweeps are defined as follows:

LIN Linear sweep. The frequency is swept linearly from the starting frequency to the ending frequency. The parameter, <(points)value>, is the total number of calculation points in the sweep.

OCT Sweep by octaves. The frequency is swept logarithmically by octaves. An octave is a doubling of the frequency. The parameter, <(points)value>, is the number of calculation points per octave.

DEC Sweep by decades. The frequency is swept logarithmically by decades. A decade is a factor of 10 increase in the frequency. The parameter, <(points)value> is the number of calculation points per decade.

The start frequency defines the first frequency used to analyze the circuit. The end frequency must be greater than or equal to the start frequency, and both frequencies must be greater than zero. The whole sweep may be one point, if you wish, and is accomplished by setting the start frequency equal to the ending frequency. Also, for a single frequency analysis, it does not matter which sweep type is used.

To get an output from an ac sweep analysis, .PRINT, or .PLOT, statements must be used. The .PRINT statement allows results from a dc, ac, noise, or transient analysis to be output in the form of tables, referred to as print tables. The general form of the .PRINT statement is

```
.PRINT [TYPE] [(Output Variable)]
```

where

```
[TYPE] is one of the following: DC, AC, NOISE, or TRAN.
```

Following the analysis type is a list of output variables. There is no limit to the number of output variables. The values of the output variables are printed as a table, with each column corresponding to one output variable. In addition, each output variable can be modified to obtain special values, according to the following classifications:

VR	Real Part
VI	Imaginary Part
VM	Magnitude, V
VP	Phase, in degrees
VDB	$20 \log_{10} V$

EXAMPLE 6.8

Write a `.PRINT` statement for the voltage across R_2 in Figure 6.11 for an ac analysis.

Solution

The `.PRINT` statement would be written as follows:

```
.PRINT AC VM(2)   VP(2)
```

where:

`VM(2)` is the magnitude of the signal at Node 2

and `VP(2)` is the phase angle of the signal at Node 2.

Now we are ready to put together a source file for an ac analysis of the circuit shown in Figure 6.11. The source file is shown in Figure 6.12.

Running the source file through PSpice produces an output file. This output file is shown in Figure 6.13.

The voltage at Node 2 is listed under the `VM(2)` column and is calculated to be 6.667 volts. By inspection of the circuit, we can verify

```
AC ANALYSIS FOR CIRCUIT 6.11
VIN     1    0    AC  10  0
R1      1    2    5K
R2      2    0    10K
.AC     LIN  1    1K  1K
.PRINT  AC   V(2)
.END
```

Figure 6.12 PSpice source file for the circuit shown in Figure 6.11.

```
AC ANALYSIS FOR CIRCUIT 6.11
**** AC ANALYSIS       TEMPERATURE = 27.000 DEG C
**************************************************
FREQ      V(2)
1.000E+03       6.667E+00
```

Figure 6.13 PSpice output file for the circuit shown in Figure 6.11.

```
AC ANALYSIS FOR CIRCUIT 6.11
VIN      1    0    AC    10   0
R1       1    2    5K
R2       2    0    10K
.AC      DEC  4    100   100K
.PRINT   AC   V(2)
.END

AC ANALYSIS FOR CIRCUIT 6.11
**** AC ANALYSIS      TEMPERATURE = 27.000 DEG C
****************************************
FREQ      V(2)
1.000E+02    6.667E+00
1.778E+02    6.667E+00
3.162E+02    6.667E+00
5.623E+02    6.667E+00
1.000E+03    6.667E+00
1.778E+03    6.667E+00
3.162E+03    6.667E+00
5.623E+03    6.667E+00
1.000E+04    6.667E+00
1.778E+04    6.667E+00
3.162E+04    6.667E+00
5.623E+04    6.667E+00
1.000E+05    6.667E+00
```

Figure 6.14 PSpice source and output files for a simulation over a range of frequencies.

that this is the correct value.

Suppose we want to analyze the behavior of the circuit shown in Figure 6.11 over the range of frequencies from 100 Hz–100 kHz, three decades of frequency. The only statement that we would have to change is the `.AC` command statement line. Let's make four calculations per decade. The new `.AC` command statement would be written as follows:

```
.AC  DEC  4  100  100K
```

The results of the PSpice simulation are shown in Figure 6.14.

Under the `FREQ` column are listed the frequency points at which `VM(2)` and `VP(2)` were calculated. Note that there are four points per decade.

All of the values under the `VM(2)` volumn are the same, namely 6.667 volts. For a purely resistive circuit, there are no frequency dependent components and hence, the voltage across R_2 does not change with frequency.

S 6.4 Graphing Waveforms Using PSpice

PSpice can also be used to generate graphs of voltage and current waveforms in a circuit. This feature is similar to using an oscilloscope to

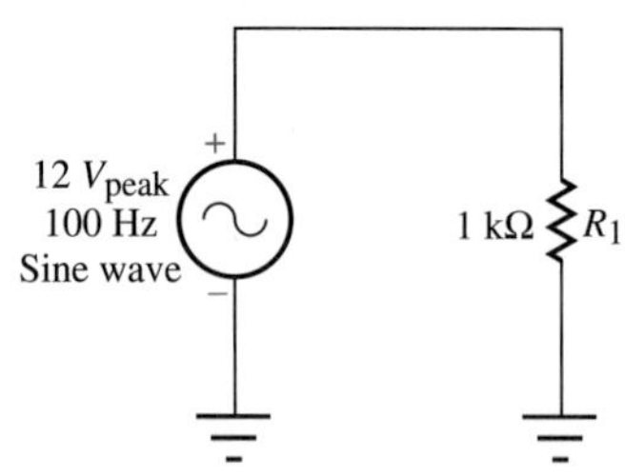

Figure 6.15 Series circuit with an ac source and single resistor.

measure the actual voltage or current waveform in a circuit.

To use PSpice to generate graphs of voltage or current waveforms, we have to do two things. First, we need to learn how to specify independent voltage and current sources; second, we need to learn how to use the transient or time-domain analysis feature in PSpice.

Let us begin with specifying voltage sources. The circuit shown in Figure 6.15 consists of an independent voltage source. This voltage source generates a sine wave whose amplitude is 12 volts peak and frequency is 100 Hz.

An independent voltage source when used with transient analysis has the following form:

```
V<name> <(+) node> <(−) node> + [(transient
        specification)]
```

The transient specification must be one of the following:

`EXP(<parameters>)`	for an exponential waveform
`PULSE(<parameters)>`	for a pulse waveform
`PWL(<parameters)>`	for a piecewise-linear wavefore
`SFFM(<parameters)>`	for a frequency-modulated waveform
`SIN(<parameters)>`	for a sinusoidal waveform

For now, we will only use the specification for the sinusoidal waveform. The general form of the specification is:

```
SIN (<voff> <vampl> <freq> <td> <df> <phase>)
```

where `<voff>` is the offset voltage, `<vampl>` is the peak amplitude of the voltage, `<freq>` is the frequency of the sine wave, `<td>` is the delay, `<df>` is the damping factor, and `<phase>` is the phase angle.

This specification looks complicated; do not let it scare you. We are going to simplify the specification by using the default values for delay, damping factor, and phase. The default values are zero for all three parameters. Furthermore, we will also set the offset voltage to zero. This leaves only the peak amplitude of the source and the frequency that will be non-zero. The result will be a nondamped sine wave, which is what we want in this case.

Hence, we can specify the 12-volt, 100-Hz sinusoidal source as,

```
VIN 1 0 SIN(0 12V 100Hz).
```

The voltage source is named `VIN` and is connected between Node 1 and Node 0, with Node 1 being the + node. `SIN` specifies a sinusoidal voltage source whose parameters tell us that the offset voltage is zero, the peak amplitude is 12 volts, and the signal frequency is 100 Hz.

Now we will discuss the `.TRAN` statement. The `.TRAN` statement causes a transient analysis to be performed on the circuit. The transient analysis calculates the circuit's behavior over time. We start at time = 0 seconds, and end with the final value that you will specify.

The general form of the .TRAN statement is:

```
.TRAN  <(print step) value>  <(final time) value>
```

There are several other parameters that can be used in a .TRAN statement. But for now, let's keep things simple.

For the circuit shown in Figure 6.15, let us analyze the behavior of the circuit over two cycles of the input sine wave. Remember that the length of one cycle is the period of the waveform. If we know the frequency, we can take the reciprocal of the frequency to find the period of the waveform. The reciprocal of 100 Hz is 10 milliseconds. Hence, we want to analyze the behavior of the circuit over two cycles or two periods, a total time of 20 milliseconds.

Also, we need to specify the print step value. How many data points do we want the simulator to calculate and print the values for, in our two-period time interval? For this example, let us print a value every 0.1 milliseconds. This should give us enough data points to have a good idea of how the circuit is behaving, without generating reams of output or making the simulation take too long.

We can tell the PSpice simulator to do this with the following statement:

```
.TRAN  0.1MS  10MS
```

And finally, we will need .PROBE. Probe is the graphics post-processor for PSpice. Probe lets you look at various results of a simulation using graphics, both on the display and on your printer. In effect, Probe is your "software oscilloscope." It will draw waveforms of signals in your circuit, much like an oscilloscope draws a graph of the signal connected to the oscilloscope probe.

Now we are ready to construct our input file. It is shown in Figure 6.16.

Again, we begin with our title line. It is followed by two lines that specify the type and value of the two components in the circuit. Then comes the .TRAN statement, .PROBE statement, and .END statement.

The graphs produced by .PROBE are shown in Figure 6.17. These graphs are not generated automatically. You have to tell .PROBE what signals you want graphed. You can use the commands within .PROBE to modify the graphs to suit your needs.

The top graph is the graph of the voltage at Node 1. This is the waveform generated by our sinusoidal source. In our .TRAN statement, we specified a time period of two cycles. Our graph shows exactly two cycles.

The current waveform for resistor R1 is shown in the bottom graph. The current waveform has a peak amplitude of 12 milliamperes. We can easily verify that this is the correct value by applying Ohm's Law.

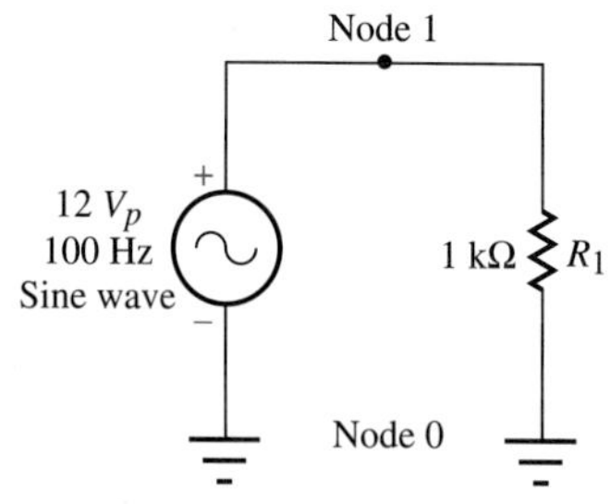


```
Analysis of Figure 6.15
VIN  1  0  SIN(0 12V 100HZ)
R1   1  0  1K
.TRAN 0.1MS 20MS
.PROBE
.END
```

Figure 6.16 Node labeling and source file for a PSpice simulation.

12V
0V
-12V
v(1)
12mA
0mA
-12mA
0ms 5ms 10ms 15ms 20ms
i(r1)
Time

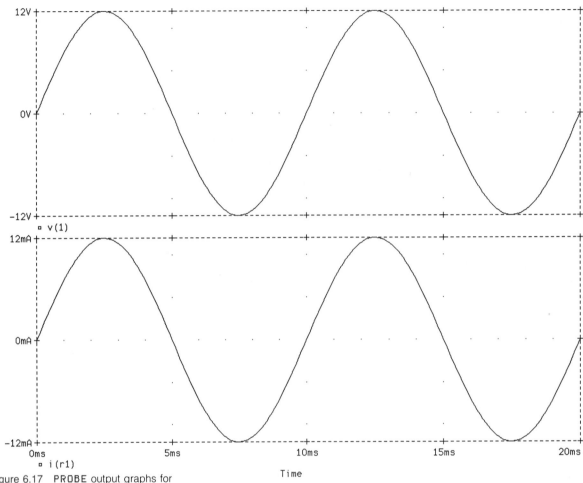

Figure 6.17 PROBE output graphs for the voltage and current associated with resistor R1 in Figure 6.15.

Dividing the input voltage, 12 volts peak, by the resistance, 1 kilohm, gives a peak amplitude for the current of 12 milliamperes. Our results are confirmed.

Now, let us use the simulator to analyze a series circuit.

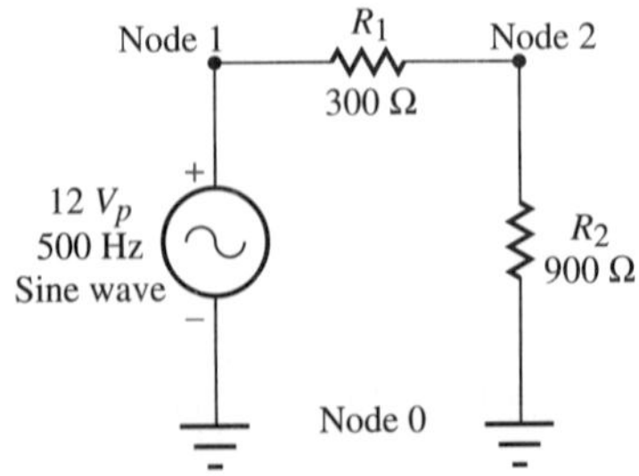

Figure 6.18 A sinusoidally driven voltage divider circuit.

EXAMPLE 6.9

Use PSpice to obtain a graph of the voltage and current for resistor R2 in the circuit shown in Figure 6.18. Graph two complete cycles.

Solution

The circuit consists of three elements. The node between the input source and resistor R1 will be named Node 1. The connection point between R1 and R2 will be named Node 2. The ground point is always named Node 0.

To construct an input file, the first thing you need is a title line. For example,

```
Analysis of the Circuit in Figure 6.18
```

Next come the component lines. The voltage source is a sinusoidal signal whose amplitude is 12 volts peak and frequency is 500 Hz. The period of the 500-Hz signal is 2 milliseconds. Two periods equals 4 milliseconds.

```
VIN 1 0 SIN(0 12V 500HZ)
```

The resistor lines are straightforward:

```
R1 1 2 300
R2 2 0 900
```

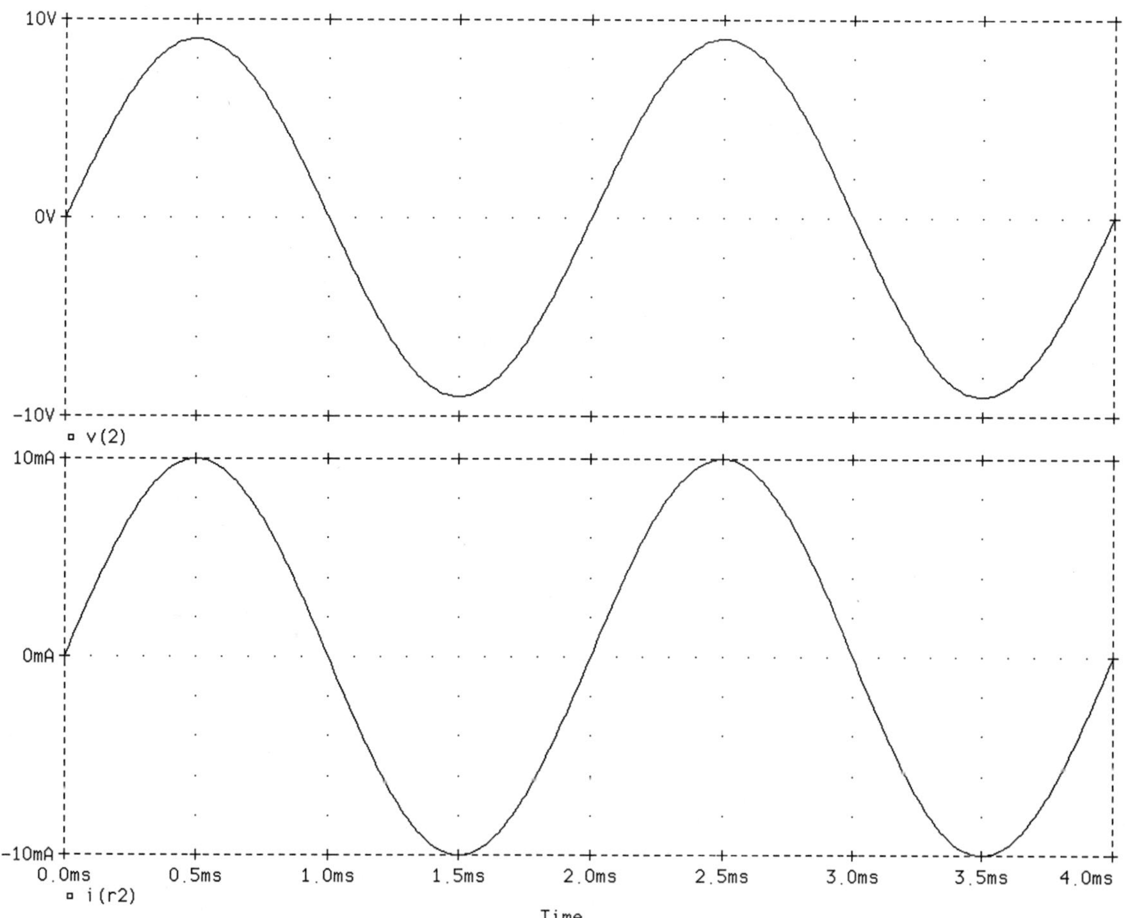

Figure 6.19 PROBE output graphs for the voltage and current associated with resistor R2 in Figure 6.18.

For the .TRAN statement, the final time value will be 4 milliseconds. You choose the print step value. For this example, we will pick 0.01 milliseconds.

The final two statements are:

```
.PROBE
.END
```

Putting all of these statements together and running the file through the simulator produces the results shown in Figure 6.19. The top graph shows the voltage waveform for resistor R2 and the bottom graph shows the current waveform.

Again, let us run a check on the validity of our results. From our knowledge of series circuits, we know that the current flowing through the circuit is equal to the applied voltage divided by the total resistance. The total resistance in this circuit is the sum of the two resistance values, 300 + 900 ohms, or 1200 ohms. Dividing the applied voltage, 12 volts peak, by 1200 ohms, gives 10 milliamperes peak. This is precisely the result that is produced by the simulator.

The voltage across R2 is given by the voltage divider formula.

$$V_{R2} = (12\ V_p)\left(\frac{900\ \Omega}{300\ \Omega + 900\ \Omega}\right)$$

$$V_{R2} = 9 \text{ volts}$$

The graph of V(2), which is the voltage across R2, shows a peak amplitude of 9 volts.

We have shown you how to use the simulator to graph voltage and current waveforms that are produced by time-varying input signals. Using the other transient specifications, we can generate other types of input signals, which will be addressed in later chapters.

SUMMARY

1. PSpice, in one of a number of different versions, can be used to analyze electronic circuits.
2. A source file must be created. This file contains the circuit description and PSpice commands that control the types of analyses to be performed.
3. PSpice contains specific rules for naming components, identifying component values, and invoking commands.
4. A device description for a resistor requires the two node numbers and component value.
5. A device description for a dc voltage source requires the two node numbers and the magnitude of the voltage.
6. The source file must begin with a title line and end with an .END line.
7. Source files are created using an editor or word processor package.
8. A dc analysis of a circuit produces the following results: node

voltages, voltage source current, and the power dissipated in the circuit.
9. An ac analysis will compute the magnitude and phase of voltages and/or currents in a circuit.
10. The results from an ac analysis can be printed using the .PRINT command or plotted using the .PLOT command.
11. Transient analysis can be used as a "software oscilloscope" to obtain graphs of voltage and current waveforms in a circuit.

EQUATIONS

1.	R<name> <(+)node> <(−)node> <value>	Resistor Component Description	220
2.	V<name> <(+)node> <(−)node> [DC <value>]	dc Voltage Source Description	221
3.	V<name> ((+)node> <(−)node> [AC <(magnitude)value> <(phase)value>]	ac Voltage Source Description	226
4.	I<name> ((+)node> <(−)node> [AC <(magnitude)value> <(phase)value>]	ac Current Source Description	226
5.	.AC [TYPE] <(points)value> <(start fre-quency)value> <(end frequency)value>]	ac Analysis Command	227
6.	.PRINT [TYPE] [(output variable)]	Print Command	227
7.	V<name> <(+) node> <(−) node> + [(transient specification)]	Source for Transient Analysis	230
8.	V<name> <(+) node> <(−) node> SIN(<voff> <vampl> <freq>)	Sinusoidal Voltage Source (Short Form)	230
9.	.TRAN <(print step) value> <(final time) value>	Transient Statement	231
10.	.PROBE	Graphics Processor	231

QUESTIONS

1. When describing a circuit, connection points are given node labels. What label must be given the ground node?
2. Give the rules for naming circuit components.
3. What do the following suffixes mean in PSpice: F, P, N, U, MIL, M, K, MEG, G, and T?
4. Name the parameters that must be given in a resistor component line.
5. Name the parameters that must be given in a dc voltage source component line.
6. Describe the major parts of a PSpice source file.
7. What information results when a circuit is analyzed using the dc analysis capability of PSpice?
8. Name the parameters that must be given to describe an ac voltage or current source.
9. What parameters must be in an ac analysis command statement?
10. What information is produced by a PSpice ac analysis?
11. What parameters must be included in a .PRINT command?
12. What parameters must be included in a .TRAN statement?

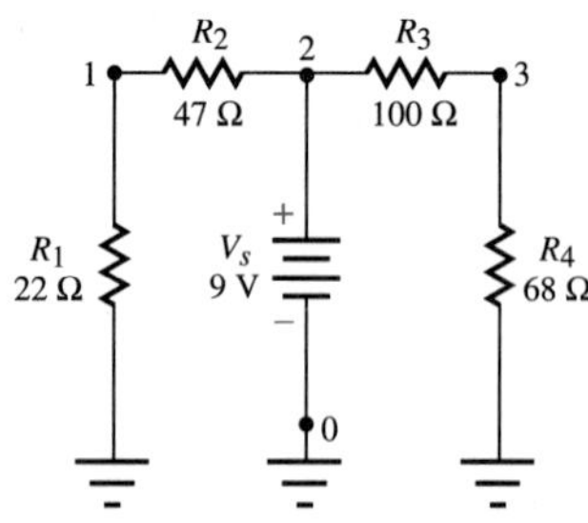

Figure 6.20

```
FILE FOR PROBLEM 5
VIN   1  0  30
R1    1  2  300
R2    2  0  150
R3    1  2  200
.END
```

Figure 6.21

```
FILE FOR PROBLEM 6
V1    4  0  15
V2    3  0  5
V3    2  0  -10
R1    2  1  1K
R2    3  1  2K
R3    4  1  3K
.END
```

Figure 6.22

PROBLEMS

SECTION 6.1 PSpice Circuit Simulator

1. Write the following component values in proper form for use in a PSpice source file:

   (a) 15×10^{-3} V

   (b) 75 microamperes

   (c) 4.7 kilohms

   (d) 0.001 microfarad

   (e) 25,000 Hz

2. Write a PSpice component line for resistors R_1 and R_2 in Figure 6.20. Assign Node 1 as the (+) node.

3. Write a PSpice component line for resistors R_3 and R_4 in Figure 6.20. Assign Node 3 as the (+) node.

4. Write a PSpice component line for the voltage source in Figure 6.20. Assign Node 2 as the (+) node.

5. Draw a schematic of the circuit described by the PSpice source file listed in Figure 6.21.

6. Draw the schematic of the circuit described by the PSpice source file listed in Figure 6.22.

SECTION 6.2 A dc Analysis Using PSpice

Ⓢ 7. Use PSpice to calculate the dc node voltages and current in the circuit shown in Figure 6.23.

Ⓢ 8. Use PSpice to calculate the dc node voltages and current in the circuit shown in Figure 6.24.

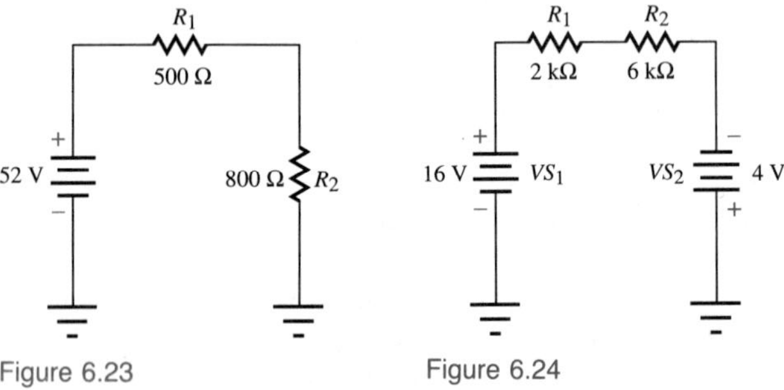

Figure 6.23

Figure 6.24

9. The circuit shown in Figure 6.25 was analyzed using PSpice. The `OUTPUT` file is shown in Figure 6.26. Are the results correct?

10. The circuit shown in Figure 6.27 was analyzed using PSpice. The `OUTPUT` file is shown in Figure 6.28. Are the results correct?

```
CIRCUIT DESCRIPTION
***************************************
VIN    1  0  50V
R1     1  2  10K
R2     2  3  5
R3     3  0  25K
.END
```

Figure 6.25

```
SMALL SIGNAL BIAS SOLUTION       TEMPERATURE = 27.000 DEG C
****************************************
NODE VOLTAGE NODE VOLTAGE NODE VOLTAGE NODE VOLTAGE
(   1) 50.0000 (   2) 35.7160 (   3) 35.7090
VOLTAGE SOURCE CURRENTS
NAME       CURRENT
VIN       -1.428E-03
TOTAL POWER DISSIPATION 7.14E-02 WATTS
```

Figure 6.26

```
CIRCUIT DESCRIPTION
***************************************
V1     1  2  10V
V2     3  0  25V
R1     1  0  5K
R2     2  3  5K
.END
```

Figure 6.27

```
SMALL SIGNAL BIAS SOLUTION       TEMPERATURE = 27.000 DEG C
****************************************
NODE VOLTAGE NODE VOLTAGE NODE VOLTAGE NODE VOLTAGE
(   1) 17.5000 (   2) 7.5000 (   3) 25.0000
VOLTAGE SOURCE CURRENTS
NAME       CURRENT
V1       -3.500E 03
V2       -3.500E-03
TOTAL POWER DISSIPATION 1.23E-01 WATTS
```

Figure 6.28

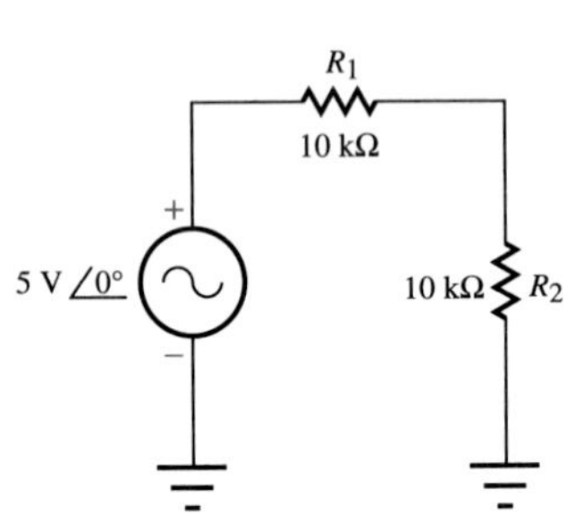

Figure 6.29

SECTION 6.3 An ac Analysis Using PSpice

11. Draw the waveform described by each of the following PSpice statements:

    (a) `VS 2 3 AC 5`

    (b) `VIN 1 0 15`

    (c) `V4 4 2 AC 0.005 0`

    (d) `VAB 6 3 AC 0.25 45`

12. Write a PSpice statement to describe the following ac voltage source:

    (a) Voltage source V_3 connected between Node 8 and Node 5 whose peak amplitude is 200 mV and phase is 30 deg.

    (b) Voltage source, V_{source}, connected between Node 2 and ground whose amplitude is 10 V_{rms} and phase is 0 deg.

    (c) Voltage source V_{in} is a 20 volt peak-to-peak sine wave riding on a 5-volt dc level. The source is connected between Node 1 and Node 3.

13. Determine the frequencies at which calculations will be performed if the following `.AC` command statements are used:

    (a) `.AC LIN 11 100HZ 200HZ`

    (b) `.AC OCT 1 5K 5K`

    (c) `.AC DEC 5 1K 10K`

Ⓢ 14. Use PSpice to find the voltage across each resistor R_3 in Figure 6.29.

Ⓢ 15. Use PSpice to calculate the node voltages in the circuit shown in Figure 6.30.

Ⓢ 16. Use PSpice to show that the circuit shown in Figure 6.31 has a constant output voltage over the frequency range, 1 kHz–100 kHz.

Ⓢ 17. Use the `.PRINT` statement to create a table that lists frequency, output magnitude, and output phase for the circuit shown in Figure 6.31.

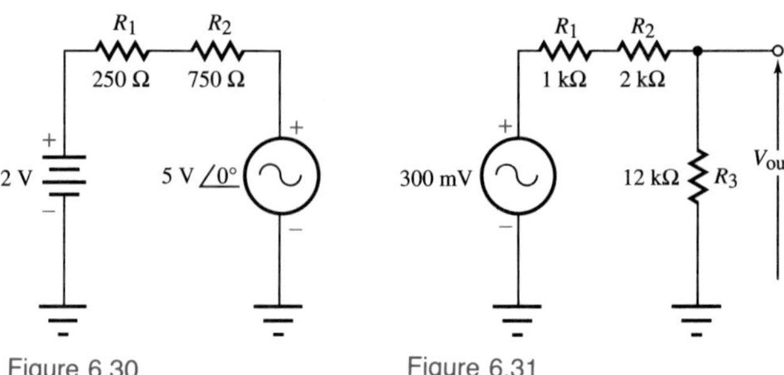

Figure 6.30

Figure 6.31

SECTION 6.4 Graphing Waveforms Using PSpice

Ⓢ 18. Use PSpice to graph the voltage across the two resistors in the circuit shown in Figure 6.32. Graph one complete input cycle.

Ⓢ 19. Use PSpice to graph the voltage across each of the three resistors in the circuit shown in Figure 6.33. Graph two complete input cycles.

Ⓢ 20. Use PSpice to graph the output waveform in Figure 6.34. Graph at least two input cycles.

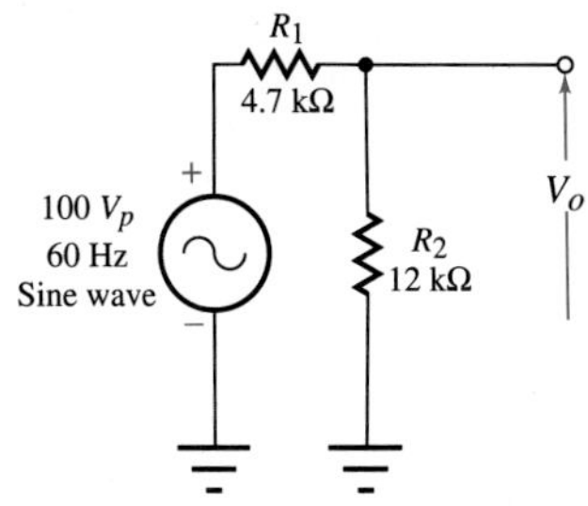

Figure 6.32

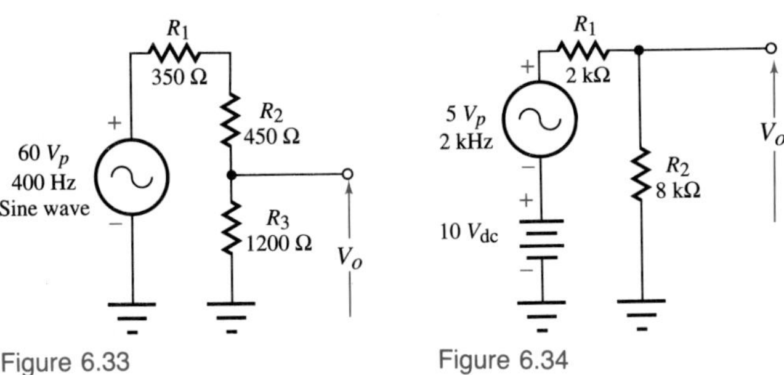

Figure 6.33

Figure 6.34

CHAPTER 7

THE PARALLEL CIRCUIT

CHAPTER OBJECTIVES

After completing this chapter, you should be able to:

1. Recognize the parts of a circuit that are connected in parallel.
2. Apply Kirchhoff's Current Law to a parallel circuit.
3. Calculate the equivalent resistance of a parallel circuit.
4. Use the current divider rule to calculate the current in parts of a parallel circuit.
5. Calculate the power in a parallel circuit.
6. Calculate the energy in a parallel circuit.

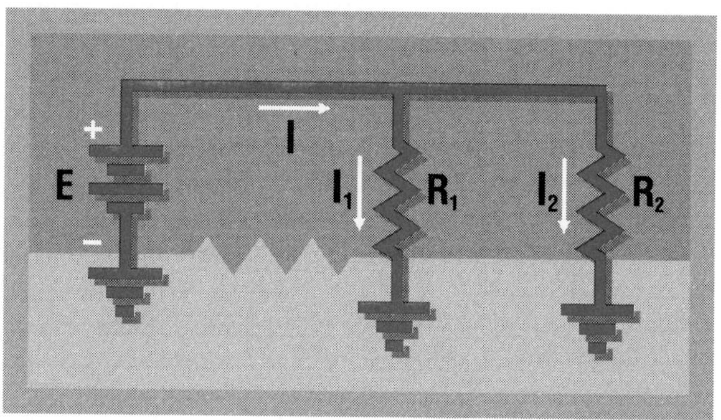

7. Use SPICE to perform the analysis of a parallel circuit.
8. Perform a complete analysis of a parallel circuit using your calculator to solve the necessary equations.

KEY TERMS

1. branch
2. current divider rule
3. equivalent resistance
4. Kirchhoff's Current Law
5. node
6. parallel circuit
7. shunt resistor

INTRODUCTION

This chapter presents the second basic circuit connection, the parallel connection. Resistors are connected in parallel when they are connected across the same two points in a circuit.

The parallel circuit can be used to divide current, protect meters from damage, obtain a needed resistance value, and obtain a higher current capacity than that available with standard resistors.

The parallel connection is suitable for many applications because a change in one branch has little or no effect on the other branches.

7.1 Circuit Examples and Voltage in a Parallel Circuit

Before we begin a detailed study of parallel circuits, let us again draw on our personal experiences. You know some of the properties of parallel circuits already.

The wiring in your house or apartment is wired in a parallel arrangement. The main power comes into your dwelling at a single point. From the distribution panel, a number of circuits run to the various parts of the house. One circuit may run to the living room, another to the kitchen, and several to the bedrooms. Separate circuits may also be dedicated to the hot water heater, clothes washer, range, and clothes dryer. From a practical viewpoint, these circuits are wired in parallel because the line resistance is small. A **parallel circuit** is one that has two or more elements connected across the same two points.

parallel circuit

For the moment, let us consider only the 120-volt circuits in your dwelling. Groups of these 120-volt circuits are connected in parallel—that is—they are connected between the same two points: a hot wire in the distribution system and the neutral connection. Modern wiring rules call for a third wire, the ground connection, which is used for safety reasons.

Figure 7.1 shows the power distribution scheme used in residential wiring.

To guide your thinking, answer the following questions about how your house is wired.

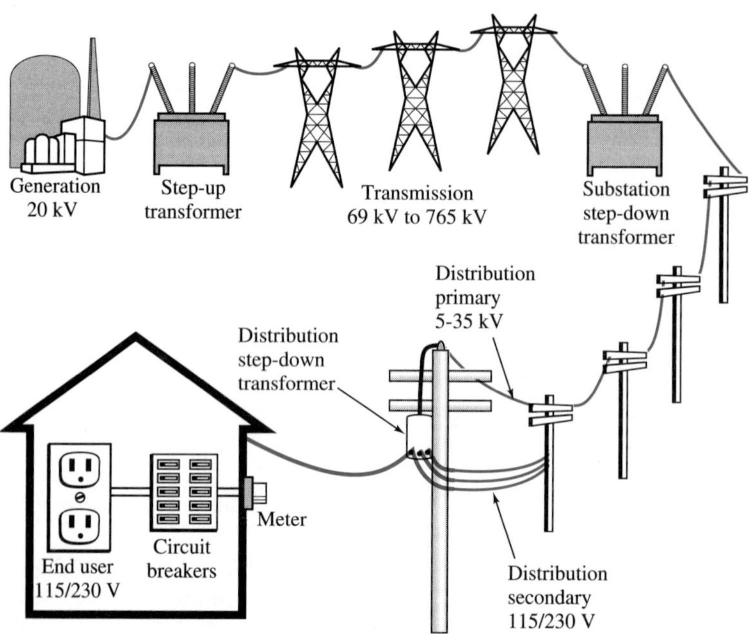

Figure 7.1 Power distribution in residential electrical systems.

When you plug an appliance cord into an outlet, what voltage do you expect to be connected to?

Do you expect to find the same voltage at each wall outlet, or will the voltage vary from outlet to outlet?

Suppose two lamps are connected to the same circuit. Assume that one lamp is ON and the other lamp is OFF. When the second lamp is turned ON, what will happen to the brightness of the lamp that was initially ON? Will it get brighter, dimmer, stay the same, or turn OFF?

How many electrical devices can be plugged into one household circuit at one time?

What happens when the fuse or circuit breaker opens?

Now that you have had a chance to think about these questions, let us compare our common experiences. In this comparison, we will assume that our two houses are wired the same.

When an appliance cord is plugged into an outlet, it connects the appliance to a 120-volt, 60-Hz source. This same voltage is present at all standard wall outlets. Therefore, we can make the observation that all appliances connected in parallel are connected to the same voltage. Note: Ranges and clothes dryers are connected to a different potential, which use a different type of outlet.

If two lamps are connected in parallel, with one lamp ON and the other one OFF, the brightness of the first lamp should not change when the second lamp is turned ON. By turning ON the second lamp, the voltage supplied to the first lamp will not change. Hence, the current in the first lamp remains constant and the brightness of the bulb remains the same. Can you imagine the conflicts that would arise if every time an appliance was turned on, the voltage delivered to all other appliances connected to that circuit changed?

This characteristic of "no change in voltage" makes the parallel connection well suited for electrical distribution systems, house wiring, and automotive lighting systems. In these applications, normal load changes do not cause serious changes or possible disruptions in the service of the other branches. This can be observed by noting that a light or appliance can be turned ON or OFF without affecting other lights or appliances. However, for larger currents, the voltage drop in the wires reduces the voltage to the branches. For instance, operating an air conditioner can sometimes reduce the voltage enough to dim the lights or shrink the picture on a television set.

The number of electrical devices connected to a circuit depends on the current requirements of each device. If devices require large amounts of current, then very few devices can be connected to a household circuit, usually protected at 15 amperes. As you add devices to a circuit, the total current flowing in the circuit increases. When the total current exceeds the current limit, the fuse or circuit breaker opens the circuit.

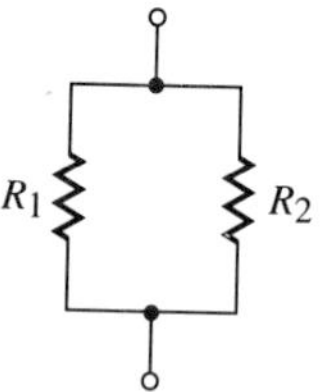

Figure 7.2 Two resistive elements connected in parallel.

When a circuit breaker or fuse opens, it breaks the circuit and stops all current flow in the circuit. The fuse or circuit breaker is placed in series with the parallel circuit; all of the current must flow through the series protective element.

The power distribution scheme used in an instrument uses the same approach, except that the voltages distributed to the various parts of the instrument are dc voltages. Let us use a personal computer as an example. The power cord brings the ac voltage from the wall outlet to the computer. A power supply in the computer converts the ac voltage to multiple dc voltages. From the outputs of the power supply, these voltages are distributed to different parts of the computer, in parallel. On each circuit board, the voltages are then distributed to each integrated circuit, again in parallel manner.

branch

To begin our study of parallel circuits, let us study the most elemental parallel circuit, which consists of two resistive elements, as shown in Figure 7.2. Each element creates a path for current to flow in a parallel circuit. Each path in a parallel group is called a **branch**. The circuit in Figure 7.3(a) has two branches. Each branch is connected across Points A and B. Figure 7.3(b) shows how the parallel connection is usually drawn. The connecting lines have zero resistance, so electrically the resistors are connected across the same two points. Another parallel connection is shown in Figure 7.3(c). R_2 and R_3 are connected across Points B and C, so they form a parallel group. Since R_1 is not connected across Points B and C, it is not part of the parallel group.

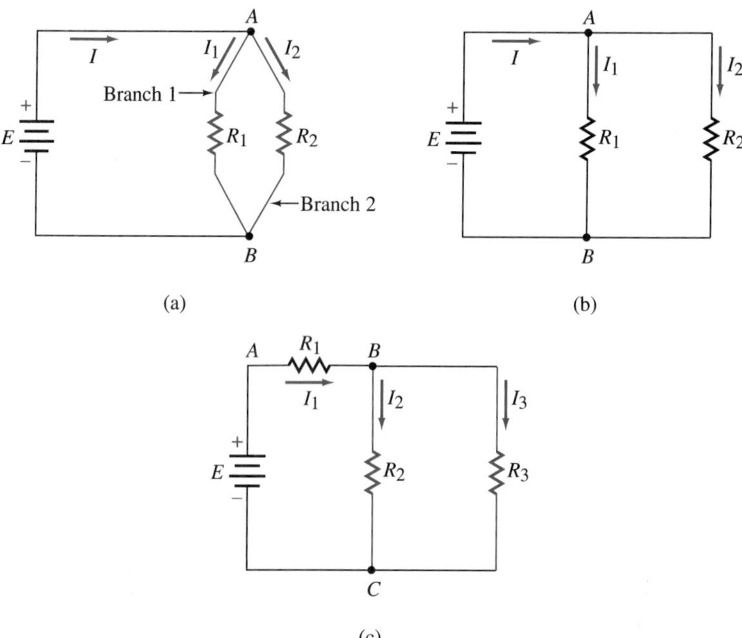

Figure 7.3 Parallel branches are connected across the same two points. (a) R_1 and R_2 are connected in parallel. (b) The more common method of showing the parallel connection. (c) R_2 and R_3 are connected in parallel, but R_1 is not in parallel with them.

EXAMPLE 7.1

Which two resistors are connected in parallel in the circuit of Figure 7.4?

Solution

R_2 and R_3 are connected across Points B and C, R_1 is connected across Points A and B, and R_4 is connected across Points C and D. R_2 and R_3 are the only resistors connected across the same two points. Therefore, only R_2 and R_3 are in parallel.

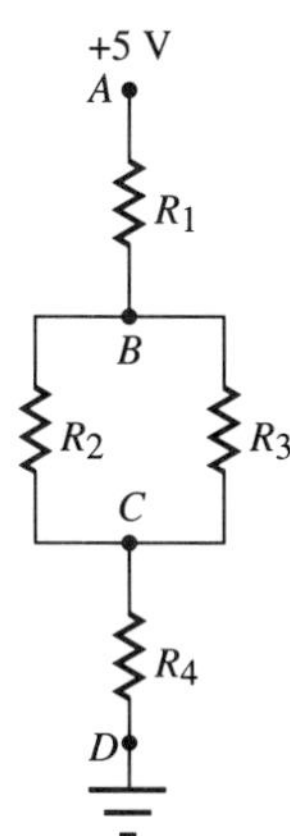

Figure 7.4

If circuit elements are connected across the same points, the voltage across each parallel element must be the same. Therefore, when the voltage across one branch changes, the voltage across the other parallel branches must also change.

7.2 Kirchhoff's Current Law

Although the voltage across each branch of a parallel circuit is the same, the currents are not the same. The total current divides among the branches, as in Figure 7.5(a). Larger currents will be found in branches

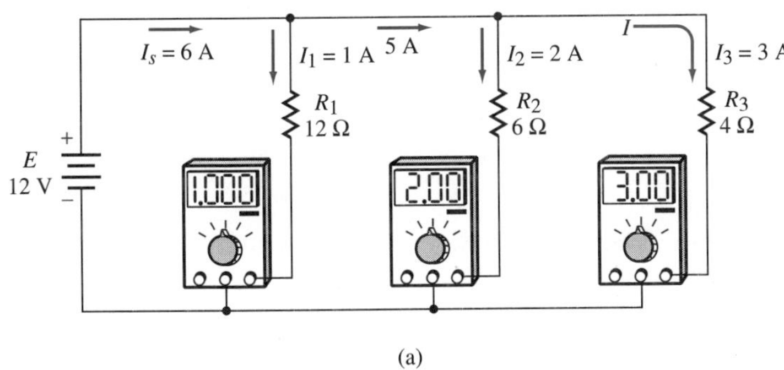

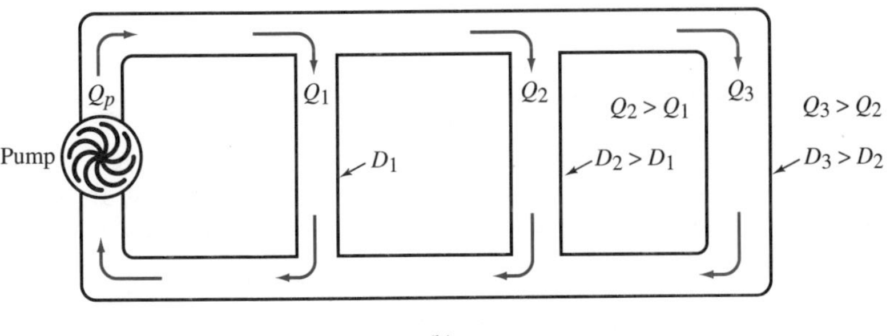

Figure 7.5 The division of current in a parallel circuit is similar to the division of water in the pipes of a water system. (a) The current divides with more current in the smaller resistance. (b) The water divides with greater flow in the larger pipes.

having smaller resistance. For instance, if the resistance in one branch is three times that in another, the current in the smaller resistance will be three times that in the larger one. The actual values of the currents can be obtained by applying Ohm's Law to each branch.

This division of current is similar to the division of water in your house. Just as the total amount of water entering the water meter must equal the amount of water leaving the open faucets, the total current entering the parallel circuit must equal the sum of the currents in the individual branches. This relationship is called **Kirchhoff's Current Law**. Kirchhoff's Current Law states that the algebraic sum of the currents at a node must equal zero. For Kirchhoff's Current Law, a **node** is a junction of three or more paths. Kirchhoff's Current Law for the parallel circuit is analogous to Kirchhoff's Voltage Law for series circuits.

Kirchhoff's Current Law

node

When applying Kirchhoff's Current Law, the first step is to establish a convention for current with respect to the node. Is current into the node assumed to be positive? If so, then current out of the node will be assigned a negative sign. The other convention for current is equally valid. That is, current out of the node could be assigned a positive value while current into the node is assigned a negative value. Either convention works equally well. The only restriction is to be consistent once a convention for current is chosen.

Let us assume that current into the node is assigned a positive value. Then, Kirchhoff's Current Law can be stated:

$$I_s - I_1 - I_2 - \cdots - I_n = 0 \qquad \textbf{(7.1)}$$

where I_s is the current entering the group, in any unit of current, and I_1, I_2, ... , I_n are the currents in each branch, in the same units as I_s.

Note that had we chosen the other convention, and assigned a positive value to current leaving the node, Kirchhoff's Current Law would have been written:

$$-I_s + I_1 + I_2 + \cdots + I_n = 0$$

Note that this equation is Equation 7.1 multiplied by (-1); it is another form of Equation 7.1.

Kirchhoff's Current Law is also sometimes stated as "The sum of the currents entering a node is equal to the sum of the currents leaving the node." Equation 7.1 then becomes

$$I_s = I_1 + I_2 + \cdots + I_n \qquad \textbf{(7.2)}$$

EXAMPLE 7.2

A 30-W, 60-W, and 75-W lamp are connected in parallel, as seen in Figure 7.6. Use Kirchhoff's Current Law to find the current I_s.

Solution

The current I_s is the current entering the parallel group, so

$$I_s - I_1 - I_2 - I_3 = 0$$

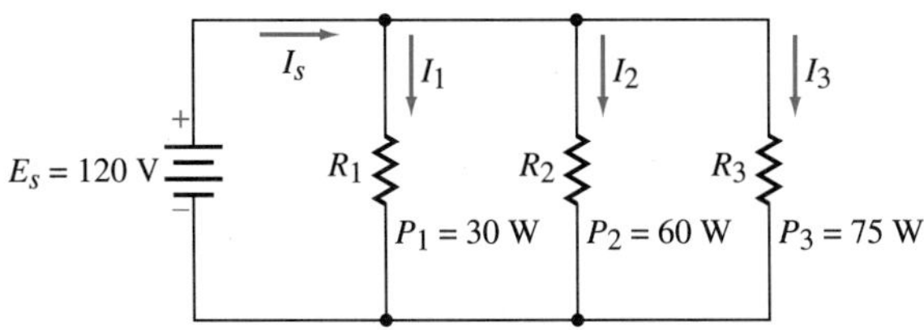

Figure 7.6

By the power relation for a single resistor,

$$I_1 = P_1/E = 30\text{ W}/120\text{ V} = 0.25\text{ A}$$
$$I_2 = P_2/E = 60\text{ W}/120\text{ V} = 0.50\text{ A}$$
$$I_3 = P_3/E = 75\text{ W}/120\text{ V} = 0.625\text{ A}$$

then

$$I_s = 0.25\text{ A} + 0.50\text{ A} + 0.625\text{ A}$$
$$I_s = 1.375\text{ A}$$

EXAMPLE 7.3

Use Kirchhoff's Current Law to find the value of I_2 in Figure 7.7.

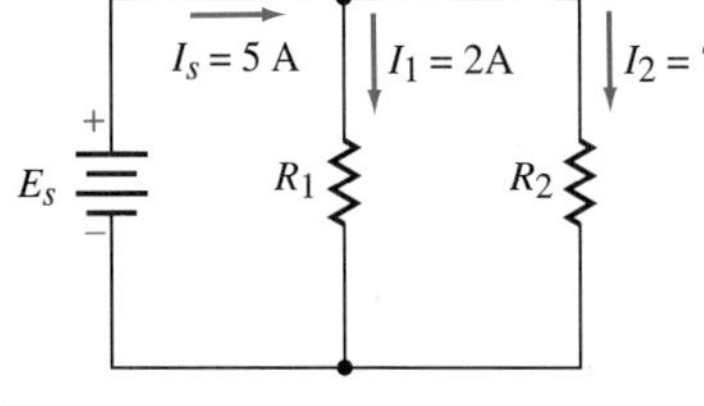

Figure 7.7

Solution

Applying Kirchhoff's Current Law,

$$I_s - I_1 - I_2 = 0$$

Rearranging this equation to solve for I_2, we obtain

$$I_2 = I_s - I_1$$

so

$$I_2 = 5\text{ A} - 2\text{ A}$$
$$I_2 = 3\text{ A}$$

Kirchhoff's Current Law also holds for circuits with a time-varying source. The values obtained are pulse values, instantaneous values, and so forth, depending on the units used to describe the input signal.

EXAMPLE 7.4

The pulse input of Figure 7.8 is applied to a parallel circuit consisting of an 8-ohm resistor and a 12-ohm resistor.

a. What is the current in each resistor?
b. What is the source current?
c. Draw the waveforms of the three currents.

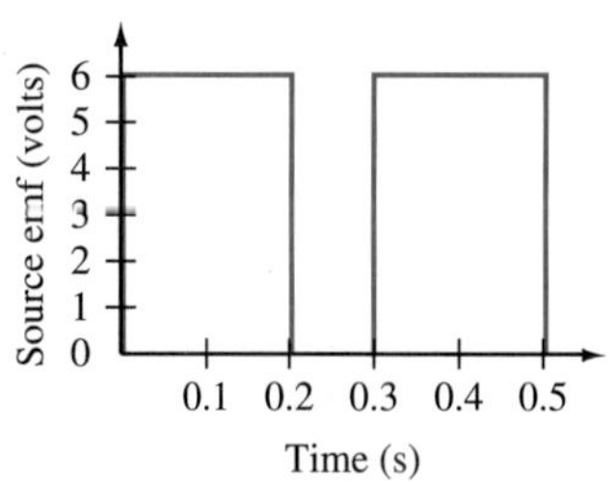

Figure 7.8

Solution

a.

$$I_1 = E/R_1 = 6\text{ V}/8\ \Omega$$
$$I_1 = 0.75\text{ A}$$
$$I_2 = E/R_2 = 6\text{ V}/12\ \Omega$$
$$I_2 = 0.5\text{ A}$$

b.

$$I_s = I_1 + I_2 = 0.75\text{ A} + 0.5\text{ A}$$
$$I_s = 1.25\text{ A}$$

c. The source current and the current in each resistor are shown in Figures 7.9(a)–(c).

7.3 Equivalent Resistance

equivalent resistance

Recall from Chapter 5 that **equivalent resistance** is the resistance that draws the same amount of current as the original circuit draws, when connected to the same voltage. This equivalent resistance can be measured, calculated from voltage and current values, or calculated using resistance values. Using voltage and current measurements for a parallel circuit yields an equation similar to that for the series circuit. It is

$$R_{eq} = V/I \tag{7.3}$$

where R_{eq} is the equivalent resistance, in ohms; V is the voltage across the circuit, in volts; and I is the current in the circuit, in amperes.

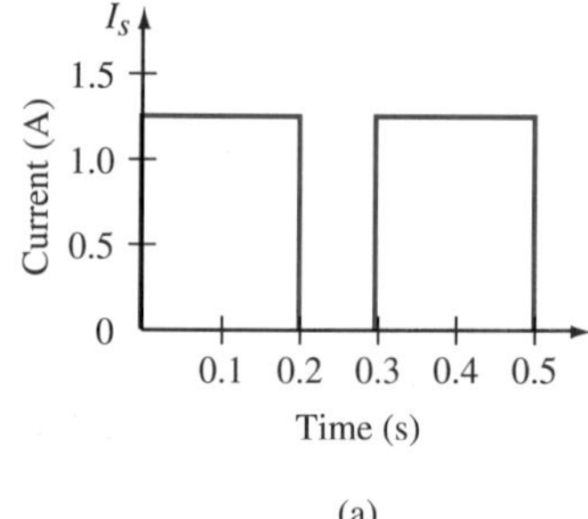

(a)

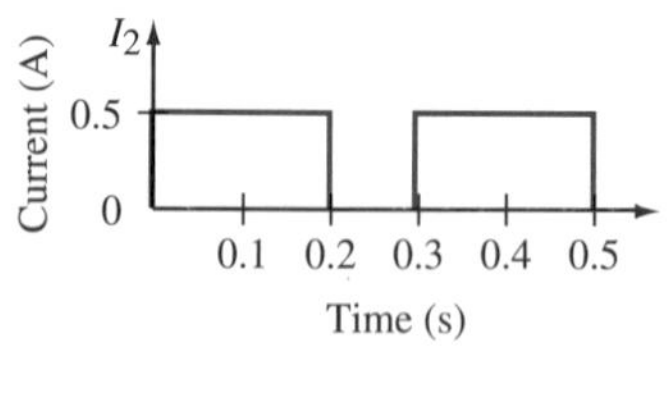

(b)

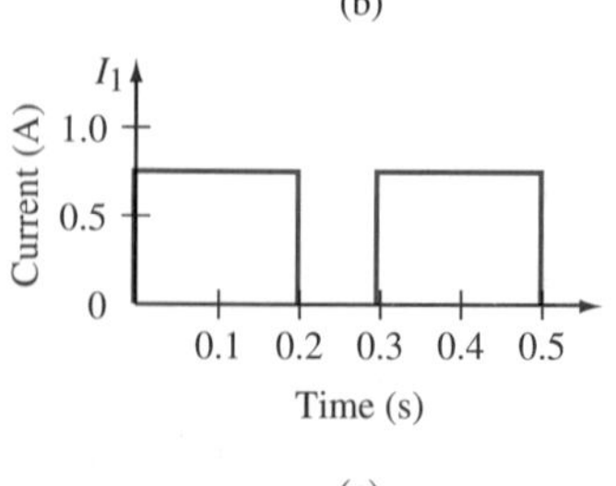

(c)

Figure 7.9

EXAMPLE 7.5

The current in an electric heater is 3 amperes when connected to a 120-volt source. What is its equivalent resistance?

Solution

From Equation 7.3,

$$R_{eq} = V/I$$
$$= 120\text{ V}/3\text{ A}$$
$$R_{eq} = 40\ \Omega$$

Example 7.5 asks a very important question. From what standpoint is equivalence viewed? The answer to that question is "from the viewpoint of the power source." From that viewpoint, the circuit looks like an equivalent resistance of 40 ohms, because the power source supplies 3 amperes of current at an applied voltage of 120 volts. In the actual circuit, many components might be used to make up the circuit, but collectively, the circuit looks like one resistance of 40 ohms. It is often helpful to imagine yourself to be a circuit component and view the rest of the circuit from that component's viewpoint. Imagine that you are the power supply, or resistor R_1.

As with the series circuit, calculating R_{eq} from measured voltage and current values is suitable only for circuits that already exist. Also, errors can be introduced by the measurement and the resistance of the meters. These undesirable factors can be avoided by calculating the equivalent resistance with Equation 7.4. It uses the values of the resistance in the circuit.

$$\frac{1}{R_{eq}} = \frac{1}{R_1} + \frac{1}{R_2} + \frac{1}{R_3} + \cdots + \frac{1}{R_n} \qquad \textbf{(7.4)}$$

where R_{eq} is the equivalent resistance of the circuit in ohms and R_1, R_2, R_3, ..., R_n are the resistances of each branch, in ohms.

The way in which Kirchhoff's Current Law and Ohm's Law were used to obtain Equation 7.4 will now be shown. In the circuit of Figure 7.10,

$$R_{eq} = E/I$$

Then, using Kirchhoff's Current Law,

$$I - I_1 - I_2 - I_3 = 0$$

Also, from Ohm's Law,

$$I_1 = E/R_1$$
$$I_2 = E/R_2$$
$$I_3 = E/R_3$$

so

$$I - E/R_1 - E/R_2 - E/R_3 = 0$$

or

$$I/E = (1/R_1 + 1/R_2 + 1/R_3)$$

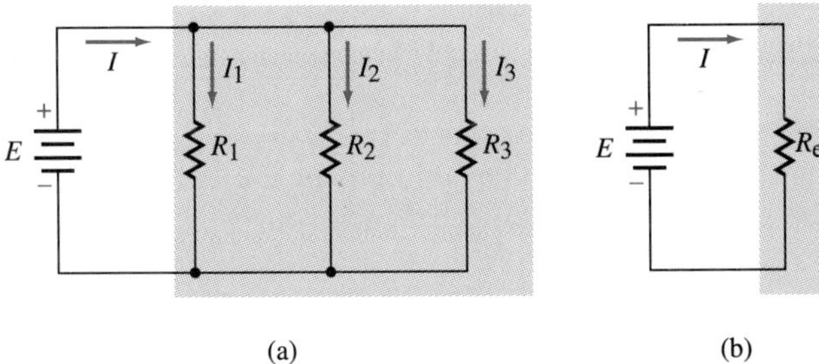

Figure 7.10 A parallel circuit can be replaced by an equivalent resistance so that the current is the same in each circuit. (a) The parallel circuit. (b) The equivalent circuit where $1/R_{eq} = 1/R_1 + 1/R_2 + 1/R_3$

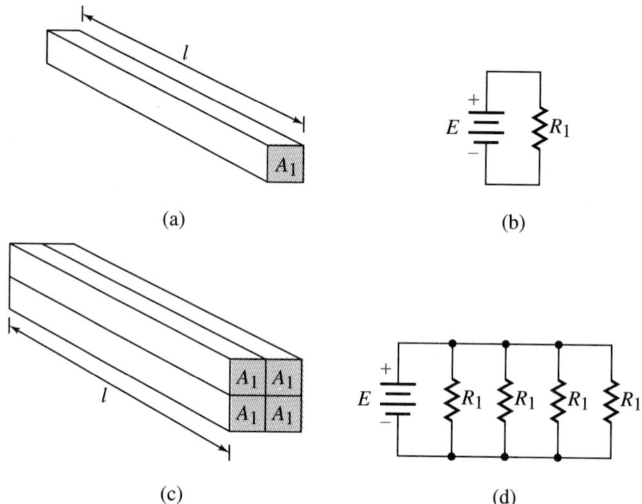

Figure 7.11 Adding parallel paths has the same effect on resistance as increasing the area of a conductor. (a) $R = \rho l/A_1$. (b) $R = R_1$. (c) $R = \rho l/4A_1$. (d) $R = R_1/4$.

From Equation 7.3,

$$I/E = 1/R_{eq}$$

so

$$1/R_{eq} = 1/R_1 + 1/R_2 + 1/R_3,$$

which is the same as Equation 7.4 for three branches.

Connecting resistors in parallel offers more paths for the current to flow between two points. This is similar to increasing the cross-sectional area of a conductor, as shown in Figure 7.11. In Chapter 2 you learned that increasing the area decreased the resistance. Therefore, adding parallel paths decreases the equivalent resistance. Also, the equivalent resistance of a parallel circuit must be smaller than the resistance of any of the individual parallel paths. That can be used as a check when calculating the equivalent resistance.

EXAMPLE 7.6

A 10-ohm, 20-ohm, and 100-ohm resistor are connected in parallel. What is the equivalent resistance of the group?

Solution

Using Equation 7.4,

$$1/R_{eq} = 1/R_1 + 1/R_2 + 1/R_3$$

and substituting the resistances values yields

$$1/R_{eq} = 1/10\ \Omega + 1/20\ \Omega + 1/100\ \Omega$$
$$1/R_{eq} = 0.16\ \text{S}$$
$$R_{eq} = 6.25\ \text{ohms}$$

Note that R_{eq} is smaller than the smallest resistance in the group, namely 10 ohms. This should always be so, and can be used to check your answer.

The algebraic and RPN calculator keystrokes for the solution are:

Quantity	Algebraic	RPN	DISPLAY
R_{eq}	1 0 1/x + 2 0 1/x + 1 0 0 1/x = 1/x	1 0 1/x ENTER 2 0 1/x + 1 0 0 1/x + 1/x	6.25 00

Equation 7.4 can be put into a more convenient form when there are only two branches. It is

$$R_{eq} = \frac{R_1R_2}{R_1 + R_2} \tag{7.5}$$

where R_{eq} is the equivalent resistance of the two branches, in ohms, and R_1 and R_2 are the resistances of each branch, in ohms.

When the resistance in each branch is the same, Equation 7.4 becomes

$$R_{eq} = R/n \tag{7.6}$$

where R is the resistance in a branch and n is the number of branches.

Equations 7.5 and 7.6 are obtained from Equation 7.4 as follows. For two resistors,

$$\frac{1}{R_{eq}} = \frac{1}{R_1} + \frac{1}{R_2} \qquad \text{or} \qquad \frac{1}{R_{eq}} = \frac{R_2 + R_1}{R_1R_2}$$

Inverting both sides of the equation gives

$$R_{eq} = \frac{R_1R_2}{R_1 + R_2}$$

which is Equation 7.5.

For resistors of equal resistance of R ohms, Equation 7.4 becomes

$$\frac{1}{R_{eq}} = \frac{1}{R_1} + \frac{1}{R_2} + \cdots + \frac{1}{R_n} = n\left(\frac{1}{R}\right)$$

Inverting both sides of the equation gives

$$R_{eq} = R/n$$

which is Equation 7.6.

EXAMPLE 7.7

A 60-ohm and 120-ohm resistor are connected in parallel. What is the equivalent resistance?

Solution

Since there are only two resistances, Equation 7.5 will be used.

$$R_{eq} = \frac{R_1 R_2}{R_1 + R_2}$$

$$R_{eq} = \frac{(60\ \Omega)(120\ \Omega)}{60\ \Omega + 120\ \Omega}$$

$$R_{eq} = 40\ \Omega$$

Checking this with Equation 7.4 gives,

$$1/R_{eq} = 1/R_1 + 1/R_2$$

$$1/R_{eq} = 1/60\ \Omega + 1/120\ \Omega$$

$$1/R_{eq} = 0.025\ \text{S}$$

$$R_{eq} = 40\ \Omega$$

The algebraic and RPN calculator keystrokes for the solution are:

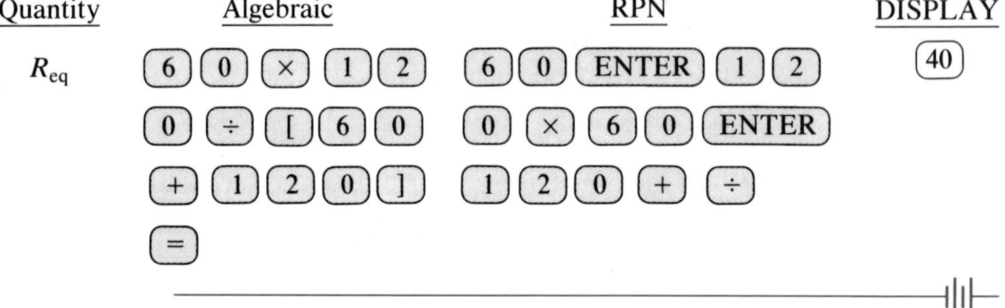

Quantity	Algebraic	RPN	DISPLAY
R_{eq}	6 0 × 1 2 0 ÷ [6 0 + 1 2 0] =	6 0 ENTER 1 2 0 × 6 0 ENTER 1 2 0 + ÷	40

Conductance is the reciprocal of resistance. Therefore, Equation 7.4 can be written as

$$G_{eq} = G_1 + G_2 + G_3 + \cdots + G_n \tag{7.7}$$

where G_{eq} is the equivalent conductance, in Siemens, and $G_1, G_2, G_3, \ldots, G_n$ are the conductances of each branch, in Siemens.

The equivalent resistance can then be found by using

$$R_{eq} = 1/G_{eq}$$

EXAMPLE 7.8

Work Example 7.5 using Equation 7.7.

Solution

Since

$$G_{eq} = G_1 + G_2 + G_3$$

and

$$G_1 = 1/R_1 = 1/10\ \Omega = 0.1\ \text{S}$$
$$G_2 = 1/R_2 = 1/20\ \Omega = 0.05\ \text{S}$$
$$G_3 = 1/R_3 = 1/100\ \Omega = 0.01\ \text{S}$$

we obtain by substitution,

$$G_{eq} = 0.1\ \text{S} + 0.05\ \text{S} + 0.01\ \text{S}$$
$$G_{eq} = 0.16\ \text{S}$$

or

$$R_{eq} = 1/G_{eq}$$
$$R_{eq} = 6.25\ \Omega$$

Equations 7.3–7.7 can also be used for circuits with time-varying waveforms as sources. If pulse values are used, they will also yield pulse values.

Parallel circuits can also have the same value of equivalent resistance, even though the circuits are different. Some examples of such circuits are shown in Figure 7.12. In each example, from the viewpoint of the

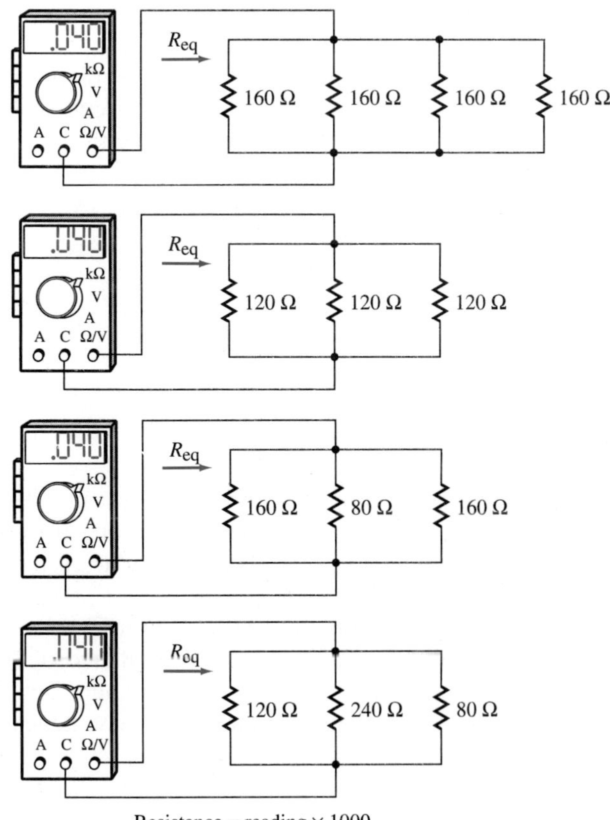

Figure 7.12 Circuits that have the same equivalent resistance do not have to be identical.

terminals on the left side of the circuit, the equivalent resistance is 40 ohms for each circuit. Indeed, if we constructed each circuit and then measured the resistance at the terminals to the circuit, in each instance the ohmmeter would read the same resistance value, in this case, 40 ohms.

The effect of a short circuit or an open circuit in one or more branches will now be examined. For a short circuit, R equals 0 ohms, which results in the equivalent resistance being 0 ohms. If the circuit is connected directly to a source, the high current either opens a fuse or damages the source. If there is another circuit between the parallel circuit and the source, the other circuit can be damaged. For example, a diode in a power supply can be damaged by a short circuit in the load. The higher currents also cause larger voltage drops in parts of the circuit. These larger drops reduce the voltage in other parts. Also, the voltage across branches that are in parallel with the short circuit will be 0 volts.

For an open circuit in a branch, R equals infinite ohms, causing an increase in the equivalent resistance. Normally, this does not damage the circuit, but it might result in a change in voltage in other parts of the circuit. This can affect the normal operation of the circuit. It is interesting to note that the short circuit can result in damage in both the series circuit and the parallel circuit.

7.4 The Current Divider Rule

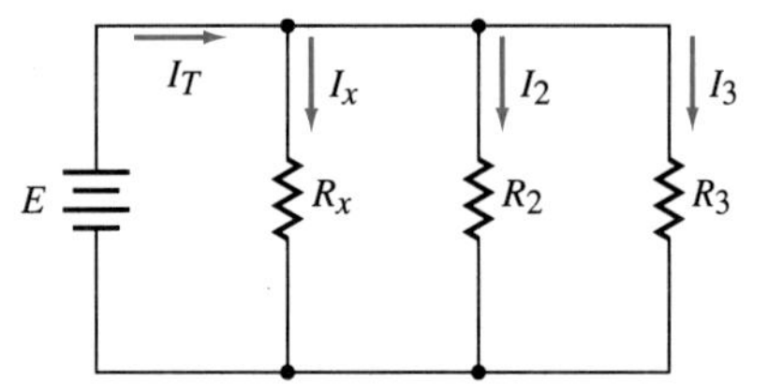

Figure 7.13 The current in any branch can be calculated using the current divider equation, $I_x = (I_T R_{eq})/(R_x)$.

To calculate the current using Ohm's Law requires knowing the voltage across the resistor. An analysis of the circuit in Figure 7.13 provides a method that does not require knowing the voltage.

First, by Ohm's Law for R_x,

$$E = I_x R_x \qquad \text{and} \qquad E = I_T R_{eq}$$

so

$$I_x R_x = I_T R_{eq}$$

Rearranging this equation gives

$$I_x = \frac{I_T R_{eq}}{R_x} \tag{7.8}$$

where I_x is the current in one branch of an n-branch parallel circuit, in amperes; I_T is the current entering the parallel circuit, in amperes; R_x is the resistance of the branch, where I_x is in ohms; and R_{eq} is the equivalent resistance of the parallel branches, in ohms.

current divider rule

Thus, the current in one branch of parallel branches is equal to the total current, multiplied by the equivalent resistance of the parallel branches, and divided by the resistance in the branch. This relation is called the **current divider rule**. It corresponds to the voltage divider rule for the series circuit.

EXAMPLE 7.9

Use the current divider rule to find (a) the current in R_2 and (b) the current in R_1 of Figure 7.13. Let $I_T = 4$ amperes, $R_x = 30$ ohms, $R_2 = 60$ ohms, and $R_3 = 60$ ohms.

Solution

a. For this part, I_2 is I_x, R_2 is R_x, and I_T is 4 amperes. Then,

$$I_2 = (R_{eq}/R_2)I_T$$

$$\frac{1}{R_{eq}} = \frac{1}{R_1} + \frac{1}{R_2} + \frac{1}{R_3}$$

$$= \frac{1}{30\ \Omega} + \frac{1}{60\ \Omega} + \frac{1}{60\ \Omega} = 0.067\ \text{S}$$

$$R_{eq} = 15\ \Omega$$

so

$$I_2 = \frac{(4\ \text{A})(15\ \Omega)}{(60\ \Omega)}$$

$$I_2 = 1\ \text{A}$$

b. For this part, I_1 is I_x, R_1 is R_x, and I_T is 4 A. Then,

$$I_1 = \frac{(4\ \text{A})(15\ \Omega)}{30\ \Omega}$$

$$I_1 = 2\text{A}$$

The algebraic and RPN calculator keystrokes for the solution for I_2 are:

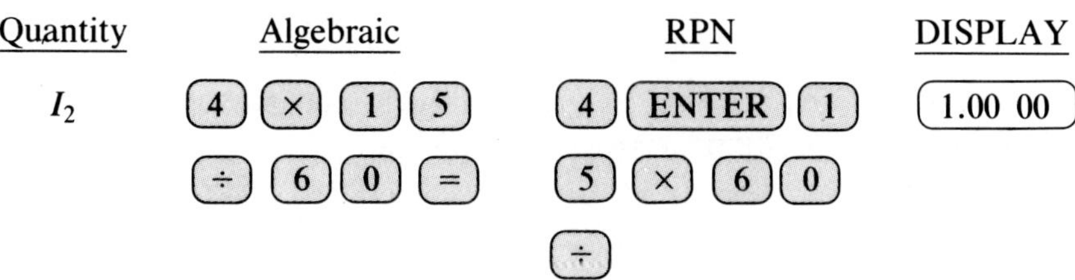

Quantity
Algebraic
RPN
DISPLAY
I_2
4 × 1 5 ÷ 6 0 =
4 ENTER 1 5 × 6 0 ÷
1.00 00

The keystrokes for I_1 follow the same sequence, with 60 ohms replaced by 30 ohms.

EXAMPLE 7.10

Find the current I_1 in the circuit in Figure 7.14.

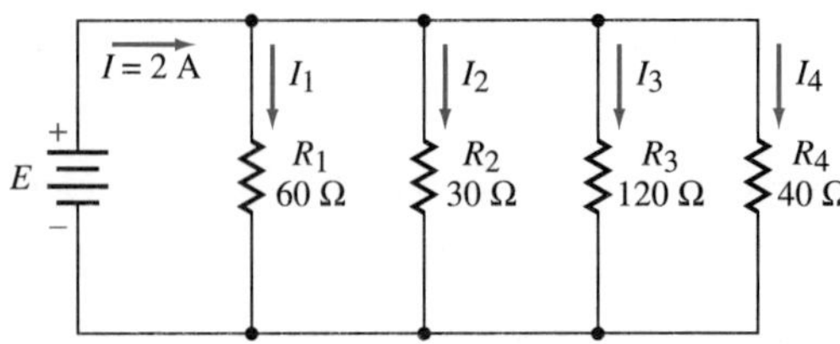

Figure 7.14

Solution

$$I_x = (I_T R_{eq})/R_x$$
$$I_1 = I_x$$
$$\frac{1}{R_{eq}} = \frac{1}{60\ \Omega} + \frac{1}{30\ \Omega} + \frac{1}{120\ \Omega} + \frac{1}{40\ \Omega} = \frac{10}{120}\ \text{S}$$
$$R_{eq} = 12\ \Omega$$

so

$$I_1 = \frac{(2\ \text{A})(12\ \Omega)}{60\ \Omega}$$
$$I_1 = 0.4\ \text{A}$$

When there are only two parallel branches, Equation 7.8 can be written as:

$$I_x = \frac{R_p I_T}{R_p + R_x} \quad \textbf{(7.9)}$$

where:

I_x, I_T, and R_x have the same meaning as in Equation 7.8
and R_p is the resistance of the second branch, in ohms

Equation 7.9 is a form that lets one visually estimate the value of the current more easily than does Equation 7.8.

EXAMPLE 7.11

The current entering a parallel circuit of a 300-ohm and a 100-ohm resistor is 0.8 ampere. What is the current in the 300-ohm resistor? What will be the current in the 300-ohm resistor if the 100-ohm resistor is short-circuited? If the 100-ohm resistor is open-circuited?

Solution

$$I_x = \frac{R_p I_T}{R_p + R_x}$$

$$I_{300} = \frac{(100\ \Omega)(0.8\ \text{A})}{100\ \Omega + 300\ \Omega}$$

$$I_{300} = 0.20\ \text{A}$$

For R_{100} short-circuited, $R_p = 0$ ohms. So

$$I_{300} = \frac{(0\ \Omega)(0.8\ \text{A})}{0\ \Omega + 300\ \Omega}$$

$$I_{300} = 0\ \text{A}$$

For R_{100} open-circuited, $R_p = \infty$ ohms. So

$$I_{300} = \frac{(\infty\ \Omega)(0.8\ \text{A})}{\infty\ \Omega + 300\ \Omega}$$

This expression gives the value of

$$I_{300} = 0.8\ \text{A}$$

7.5 Power and Energy in the Parallel Circuit

Our concerns with power in a parallel circuit are the same as for the series circuit. These are: (1) the amount of power dissipated in each circuit component and (2) the total amount of power supplied by the source. The first item sets the power rating of resistors. The second item sets the capacity of the voltage source.

In Chapter 5, we saw that the power in a circuit and in its equivalent circuit must be the same. Applying this to the parallel circuit gives the same equations as for power in a series circuit. They are:

$$P = I^2 R_{eq} \tag{7.10}$$

$$P = V^2 / R_{eq} \tag{7.11}$$

$$P = VI \tag{7.12}$$

where P is the total power dissipated in the circuit, in watts; I is the total current in the circuit, in amperes; V is the voltage across the parallel circuit, in volts; and R_{eq} is the equivalent resistance of the circuit, in ohms.

A note of caution here. To use these relations for time-varying sources, the effective value of current or voltage must be used. This is because the waveform does not have the same value for the entire time. The effective value is discussed in Chapter 15, Section 15.5.

EXAMPLE 7.12

Use Equations 7.10–7.12 to find the power dissipated in the circuit of Figure 7.15.

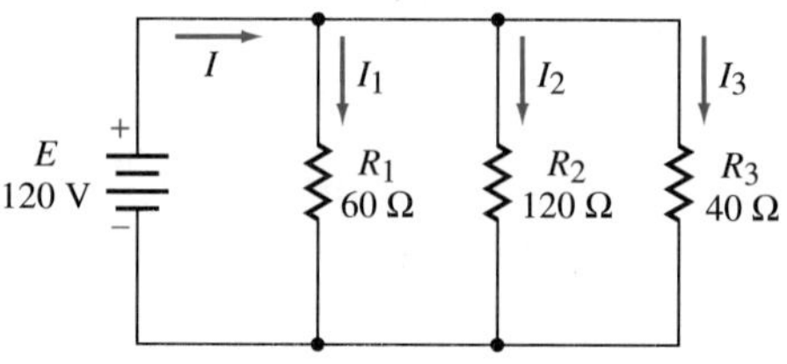

Figure 7.15

Solution

Using

$$P = I^2R_{\text{eq}}$$

$$\frac{1}{R_{\text{eq}}} = \frac{1}{R_1} + \frac{1}{R_2} + \frac{1}{R_3}$$

$$\frac{1}{R_{\text{eq}}} = \frac{1}{60\ \Omega} + \frac{1}{120\ \Omega} + \frac{1}{40\ \Omega} = 0.05\ \text{S}$$

$$R_{\text{eq}} = 1/0.05\ \text{S}$$

$$R_{\text{eq}} = 20\ \Omega$$

By Ohm's Law,

$$I = E/R_{\text{eq}}$$
$$= 120\ \text{V}/20\ \Omega$$
$$I = 6\ \text{A}$$

so

$$P = (6\ \text{A})^2(20\ \Omega)$$
$$P = 720\ \text{W}$$

Using $P = V^2/R_{\text{eq}}$

The voltage across the circuit is 120 volts. So

$$P = (120\ \text{V})^2/20\ \Omega$$
$$P = 720\ \text{W}$$

Using $P = VI$

$$\frac{1}{R_{\text{eq}}} = \frac{1}{R_1} + \frac{1}{R_2} + \frac{1}{R_3}$$

$$\frac{1}{R_{eq}} = \frac{1}{60\ \Omega} + \frac{1}{120\ \Omega} + \frac{1}{40\ \Omega} = 0.05\ \text{S}$$
$$R_{eq} = 1/0.05\ \text{S}$$
$$R_{eq} = 20\ \Omega$$

then

$$I = (120\ \text{V})/(20\ \Omega)$$
$$I = 6\ \text{A}$$

and

$$P = (120\ \text{V})(6\ \text{A})$$
$$P = 720\ \text{W}$$

Each equation gives the same result. The difference is in the amount of calculations that are required.

The total power dissipated in Example 7.12 was 720 watts. Now consider the power dissipated in each resistor.

$$P_1 = E^2/R_1 = (120\ \text{V})^2/60\ \Omega = 240\ \text{W}$$
$$P_2 = E^2/R_2 = (120\ \text{V})^2/120\ \Omega = 120\ \text{W}$$
$$P_3 = E^2/R_3 = (120\ \text{V})^2/40\ \Omega = 360\ \text{W}$$

Comparing the total power dissipated to that dissipated in each resistor shows that the total power dissipated is equal to the sum of the power dissipated in each resistor. They both are 720 W. This is the same relationship as for the series circuit. For n branches,

$$P = P_1 + P_2 + \cdots + P_n \qquad \textbf{(7.13)}$$

where P is the power dissipated in the circuit, in watts, and $P_1, P_2, \ldots, P_n$ is the power dissipated in each resistance, in watts.

The relation in Equation 7.13 is useful for calculating the power in a distribution system or circuit when the power in parts of the circuit is known.

The proof of Equation 7.13 can be obtained by analyzing the circuit shown in Figure 7.16. In that circuit

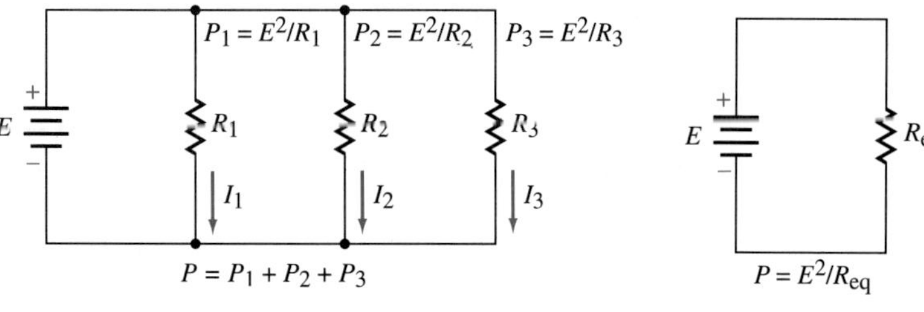

Figure 7.16 The total power in a parallel circuit can be obtained by taking the sum of the power in each part.

$$P_1 = E^2/R_1$$
$$P_2 = E^2/R_2$$
$$P_3 = E^2/R_3$$

then

$$P_1 + P_2 + P_3 = E^2(1/R_1 + 1/R_2 + 1/R_3)$$

but

$$1/R_1 + 1/R_2 + 1/R_3 = 1/R_{eq}$$

so

$$P_1 + P_2 + P_3 = E^2/R_{eq}$$

From Equation 7.11, E^2/R_{eq} is the circuit power, so $P_1 + P_2 + P_3 = P$, which is the same as Equation 7.13 for three branches. Thus, the total power in a parallel circuit equals the sum of the power in each part of the circuit. Equation 7.13 can also be used for circuits with time-varying sources.

EXAMPLE 7.13

The loads in a house at a given time are shown in Figure 7.17. How much power is being used in the house?

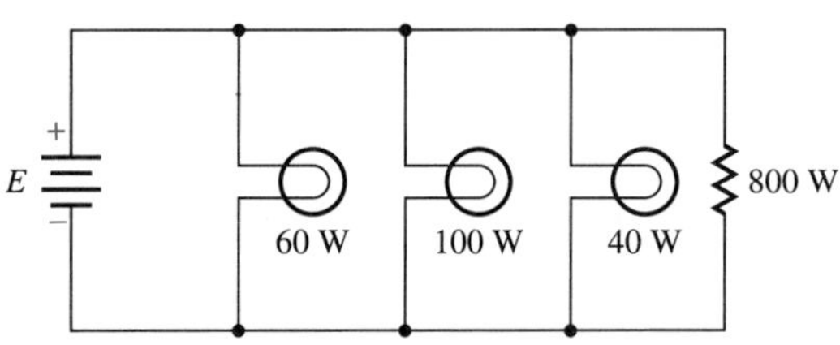

Figure 7.17

Solution

$$P = P_1 + P_2 + P_3 + P_4$$
$$P = 60\text{ W} + 100\text{ W} + 40\text{ W} + 800\text{ W}$$
$$P = 1000\text{ W}$$

EXAMPLE 7.14

Resistor R_3 in the circuit of Figure 7.18 has a power rating of 2 watts. The total power dissipated in the circuit is 3.25 watts. Is R_3 operating above its rating?

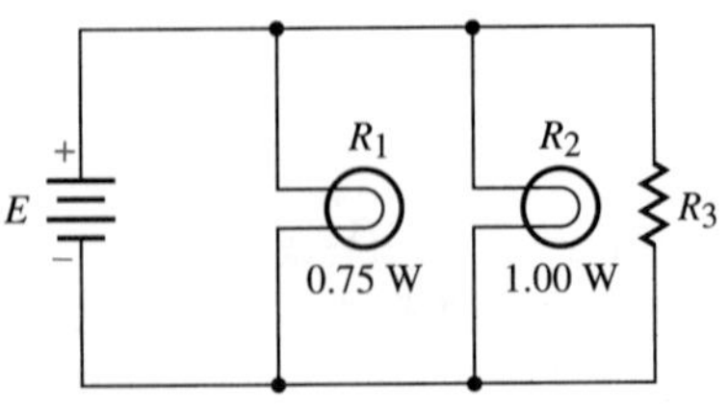

Figure 7.18

Solution

$$P = P_1 + P_2 + P_3$$

so

$$P_3 = P - P_1 - P_2$$
$$= 3.25\text{ W} - 0.75\text{ W} - 1.00\text{ W}$$
$$P_3 = 1.50\text{ W}$$

It is not above its power rating.

Our second concern with power is the amount supplied by the source. When a parallel circuit has only one source in it, all of the power must come from that source. Therefore, the source power is equal to the circuit power. Any of the equations used to find the circuit power can also be used to find the source power.

EXAMPLE 7.15

How much power does the source supply in the circuit of Figure 7.19?

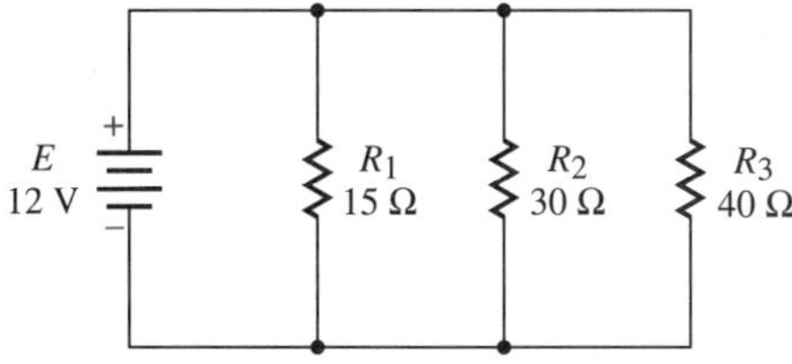

Figure 7.19

Solution

Since the source voltage and circuit resistance values are given, Equation 7.11 will be used.

$$P_s = E^2/R_{eq}$$

where

$$1/R_{eq} = 1/R_1 + 1/R_2 + 1/R_3$$

Substituting known values for R_1, R_2, and R_3 yields

$$1/R_{eq} = 1/15\ \Omega + 1/30\ \Omega + 1/40\ \Omega = 0.125\text{ S}$$
$$R_{eq} = 1/0.125\text{ S} = 8\ \Omega$$

then

$$P_s = E^2/R_{eq} = (12\text{ V})^2/8\ \Omega$$
$$P_s = 18\text{ W}$$

Equations 7.10 and 7.12 can also be used if the current I_s is first calculated.

Recall that energy is power multiplied by time, and the time is the same for each part of the circuit, and for the total circuit. That is, if the charge is moving in one resistor for 10 seconds, it is also moving in the total circuit for 10 seconds. Thus, energy relations for the parallel circuit are the same as for the series circuit. They are

$$\mathcal{W} = Pt \tag{7.14}$$

$$\mathcal{W} = \mathcal{W}_1 + \mathcal{W}_2 + \cdots + \mathcal{W}_n \tag{7.15}$$

where $\mathcal{W}$ is the energy in the circuit, or supplied by the source, in joules; V is the voltage across the parallel circuit, in volts; I is the current in the parallel circuit, in amperes; R_{eq} is the equivalent resistance of the parallel circuit, in ohms; and $\mathcal{W}_1, \mathcal{W}_2, \ldots, \mathcal{W}_n$ is the energy in each resistor, in joules.

If the energy is wanted in kilowatt-hours, one should use power in kilowatts and time in hours.

The power in Equation 7.14 can be found using Equation 7.10, 7.11, or 7.12.

EXAMPLE 7.16

How much energy is used in the circuit of Figure 7.20 in 10 minutes?

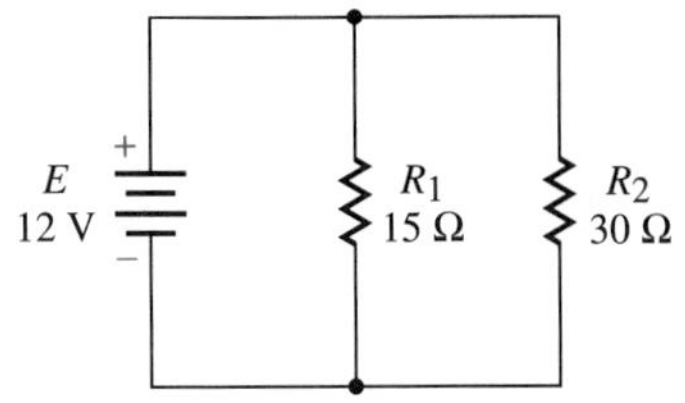

Figure 7.20

Solution

Using Equation 7.14,

$$\mathcal{W} = Pt$$
$$P = V^2/R_{eq}$$

but

$$R_{eq} = (R_1R_2)/(R_1 + R_2)$$
$$= ((15\ \Omega)(30\ \Omega))/(15\ \Omega + 30\ \Omega)$$
$$R_{eq} = 10\ \Omega$$

so

$$P = (12\ \text{V})^2/10\ \Omega = 14.4\ \text{W}$$
$$t = (10\ \text{min})(60\ \text{s/min}) = 600\ \text{s}$$

then

$$\mathcal{W} = (14.4\ \text{W})(600\ \text{s})$$
$$\mathcal{W} = 8640\ \text{J}$$

Using Equation 7.15,

$$\mathcal{W} = \mathcal{W}_1 + \mathcal{W}_2$$

Since

$$\mathcal{W}_1 = P_1 t \qquad \text{and} \qquad P_1 = V_1^2/R_1$$

so

$$\mathcal{W}_1 = [(12\ \text{V})^2/15\ \Omega]\ [600\ \text{s}]$$
$$= 5760\ \text{J}$$
$$\mathcal{W}_2 = P_2\, t$$

but

$$P_2 = V_2^2/R_2$$

so

$$\mathcal{W}_2 = [(12\ \text{V})^2/30\ \Omega]\ [600\ \text{s}]$$
$$\mathcal{W}_2 = 2880\ \text{J}$$

then

$$\mathcal{W} = 5760\ \text{J} + 2880\ \text{J}$$
$$\mathcal{W} = 8640\ \text{J}$$

EXAMPLE 7.17

How much energy is supplied to the house in Example 7.13, if the loads are operated for 3 hours (a) in joules and (b) in kilowatt hours?

Solution

a. $\mathcal{W} = Pt$

$$= (1000\ \text{W})(3\ \text{hr})\left(3600\ \frac{\text{s}}{\text{hr}}\right)$$

$$\mathcal{W} = 10{,}800{,}000\ \text{J}$$

b. $\mathcal{W} = Pt$

$$= (1000\ \text{W})(3\ \text{hr})\left(\frac{1\ \text{kW}}{1000\ \text{W}}\right)$$

$$\mathcal{W} = 3\ \text{kWh}$$

7.6 Analysis of the Parallel Circuit

The complete analysis of the parallel circuit can require the use of several of the concepts presented in this chapter. The decisions that must be made are the same as for the series circuit. Some of these decisions are:

Which concept to use?

How many of them are needed?

How do you put them together to obtain enough equations?

How to solve the equations?

This section presents several examples that will show the general procedures used in the analysis. These examples not only show the use of the concepts, but also give some explanation of the analysis.

EXAMPLE 7.18

The power in the circuit of Figure 7.21 is 1020 watts. Find the resistance of R_2.

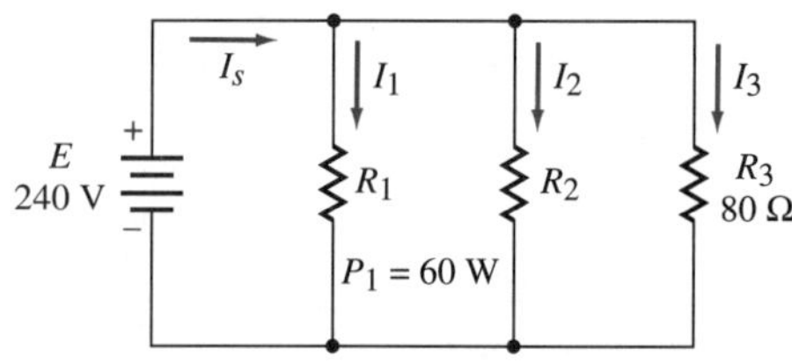

Figure 7.21

Solution

A review of the given quantities might suggest the following as one approach to solving for R_2.

1. Calculate the power in R_3.
2. Calculate the power in R_2 using the total power and the power in R_1 and R_3.
3. Calculate R_2 using the power in R_2 and the voltage across it.

Strategy

Find the power in R_3.

$$P_3 = E^2/R_3$$
$$= (240\text{ V})^2/80\ \Omega$$
$$P_3 = 720\text{ W}$$

Use Equation 7.12 to find the power in R_2.

$$P_2 = P_T - P_1 - P_3$$
$$= 1020\text{ W} - 60\text{ W} - 720\text{ W}$$
$$P_2 = 240\text{ W}$$

Find R_2 using power and voltage.

$$R_2 = E^2/P_2$$
$$= (240\text{ V})^2/240\text{ W}$$
$$R_2 = 240\ \Omega$$

Let us take a look at how the equations can be used as a guide in this example.

Since R_2 is wanted, the first equation will include it. Using the power relation,

$$R_2 = E^2/P_2$$

This equation introduces P_2, so the next equation should have P_2 in it.

$$P_T = P_1 + P_2 + P_3$$

No new unknowns are introduced, so the equations can be solved.

$$P_2 = P_T - P_1 - P_3$$
$$P_2 = 1020\ \text{W} - 60\ \text{W} - 720\ \text{W}$$
$$P_2 = 240\ \text{W}$$

so

$$R_2 = (240\ \text{V})^2/240\ \text{W}$$
$$R_2 = 240\ \Omega$$

EXAMPLE 7.19

Find the source current, the current in each branch, and the resistance of R_1 in the circuit of Figure 7.22.

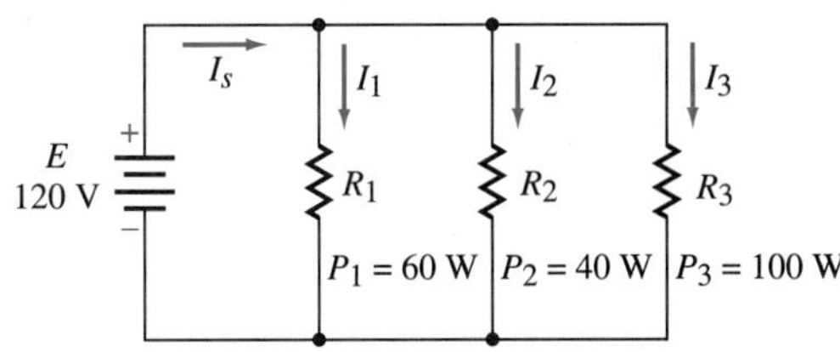

Figure 7.22

Solution

A review of the knowns and unknowns shows the following to be a possible sequence:

1. Calculate the total power using the power in each branch.
2. Calculate the source current using the total power and source voltage.
3. Calculate the current in each branch using the branch power and voltage.
4. Calculate R_1 using the branch power and voltage.

Strategy

Find the total power.

$$P_T = P_1 + P_2 + P_3$$
$$P_T = 60\ \text{W} + 40\ \text{W} + 100\ \text{W} = 200\ \text{W}$$

Find the current.

$$I_s = P_T/E = 200\ \text{W}/120\ \text{V}$$
$$I_s = 1.67\ \text{A}$$
$$I_1 = P_1/E = 60\ \text{W}/120\ \text{V}$$
$$I_1 = 0.5\ \text{A}$$
$$I_2 = P_2/E = 40\ \text{W}/120\ \text{V}$$
$$I_2 = 0.33\ \text{A}$$
$$I_3 = P_3/E = 100\ \text{W}/120\ \text{V}$$
$$I_3 = 0.833\ \text{A}$$

Use the power and voltage to find R_1.

$$P_1 = E^2/R_1$$

so

$$R_1 = E^2/P_1$$

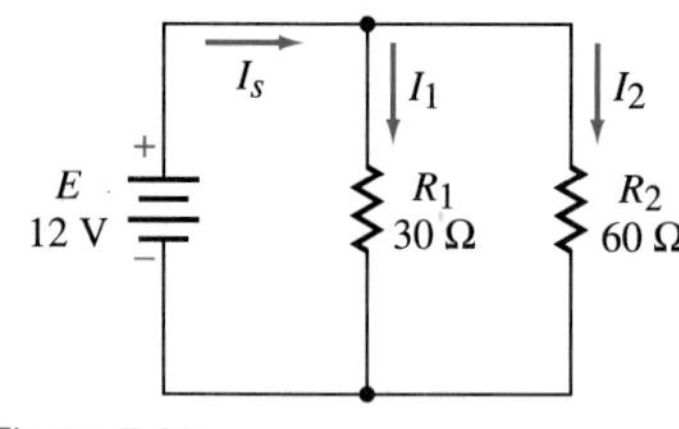

Figure 7.23

$$R_1 = (120\text{ V})^2/60\text{ W}$$
$$R_1 = 240\ \Omega$$

EXAMPLE 7.20

How many joules of energy are used in the circuit of Figure 7.23 in 4 hours?

Solution

Strategy

A review of the problem shows the following as a sequence.

1. Calculate the energy in each resistor.
2. Calculate the total energy from the sum of the energy in each part.

Find the energy in R_1.

$$\mathcal{W}_1 = P_1 t = (E^2/R_1)(t)$$
$$= \frac{(12\text{ V})^2(4\text{ hr})(3600\text{ sec/hr})}{30\ \Omega}$$
$$\mathcal{W}_1 = 69{,}120\text{ J}$$

Find the energy in R_2.

$$\mathcal{W}_2 = P_2 t = (E^2/R_2)(t)$$
$$= \frac{(12\text{ V})^2(4\text{ hr})(3600\text{ sec/hr})}{60\ \Omega}$$
$$\mathcal{W}_2 = 34{,}560\text{ J}$$

Add the two energies to get the total energy.

$$\mathcal{W} = \mathcal{W}_1 + \mathcal{W}_2$$
$$= 69{,}120\text{ J} + 34{,}560\text{ J}$$
$$\mathcal{W} = 103{,}680\text{ J}$$

The solution sequences presented here are only a few of several that can be used. However, for all sequences, the general procedure is the same. That is, relate the unknowns to the knowns, and form a path to the answer. In many cases, the unknowns in the equation will guide you to the next equation.

S 7.7 SPICE and Parallel Circuits

In this section, we will use PSpice to simulate the behavior of parallel circuits driven by a dc source and an ac source. The output obtained from a dc analysis will be a printout of voltage and current values. For the analysis of a parallel circuit driven by an ac source, we will use the graphical output capability of PSpice.

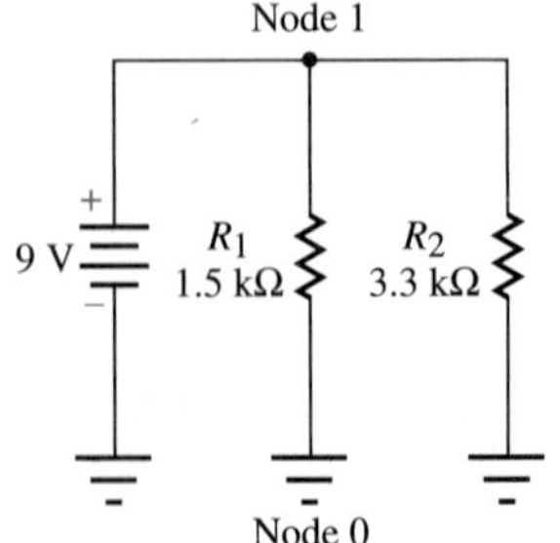

Figure 7.24

EXAMPLE 7.21

Use PSpice to calculate the branch currents for the parallel circuit shown in Figure 7.24.

Solution

First, we must create a source file. We will give this file the name EX7-21.CIR.

The first step is to number the nodes in the circuit, remembering that the reference or ground node is always numbered "0."

Since there are only two nodes in the circuit, the second node is given the number "1."

The complete source file is:

```
PARALLEL CIRCUIT - EXAMPLE 7.21
VIN  1   0   9
R1   1   0   1.5K
R2   1   0   3.3K
.OPTIONS NOPAGE
.DC VIN 9 9 9
.PRINT DC I(R1) I(R2) I(VIN)
.END
```

Let us examine this source file, line by line. The first line is always the title line. Here we have titled this file "PARALLEL CIRCUIT - EXAMPLE 7.21." The second line describes the voltage source in our circuit. Since this circuit element is a voltage source, the device name must begin with the letter V. The name we have given the voltage source is "VIN" for voltage input. VIN is connected between Nodes 1 and 0, the reference node, and has a value of 9 volts (the unit, volts, is assumed).

The third line in the file describes circuit element R1. The first letter of the device name is the letter R, so we know that this element is a resistor. The name R1 will help us distinguish this element from other resistors in the circuit. Element R1 is connected between Node 1 and Node 0 and has a value of 1.5 kΩ.

The fourth line describes element R2. This element name also indicates that the element is a resistor, is connected between Node 1 and Node 0, and has a value of 3.3 kΩ.

Note that each circuit element requires one line in the file. To this point, we have described the interconnections between circuit elements and have given each one a value. Now we are ready to specify the type of analysis that we would like to have performed. The .OPTIONS NOPAGE command is used to conserve paper. It will print the output without paginations. The .DC VIN 9 9 9 line calls for a dc analysis to be performed. It must begin with .DC. The next item in the command is the name of the dc source, VIN in this case. Following the name of the voltage source are three parameters: the start value, the end value, and the increment value. Since we are only interested in one voltage, 9 volts, we can make the start and end values the same, namely 9 volts. This will calculate circuit voltages and currents at only one value of VIN, VIN = 9 volts. In this case, the increment size is immaterial.

The .PRINT DC line will cause the program to output values for the parameters listed in the command. In this case, we want dc values for the following parameters:

The first thing in the OUTPUT file is a listing of the circuit description. It is a copy of your source file. The results of the simulation are then printed out. In our .PRINT command, we asked for the current through R1, `I(R1)`, the current through R2, `I(R2)`, and the current supplied by the source VIN, `I(VIN)`.

`I(R1)` is `6.000E-03` amperes or 6 milliamperes.
`I(R2)` is `2.727E-03` amperes or 2.727 milliamperes.
`I(VIN)` is $-$`8.727E-03` amperes or -8.727 milliamperes.

Note that the current `I(VIN)` is negative since the current flows out of the positive terminal of the voltage source, through the parallel resistive circuit, and then back to the negative terminal of the voltage source. Note also that the Kirchhoff's Current Law holds—that is—the current supplied by the voltage source equals the sum of the branch currents in the parallel circuit. That is, the sum of the currents at Node 1 is zero.

```
PARALLEL CIRCUIT EXAMPLE 7.21
**** CIRCUIT DESCRIPTION
****************************************************

VIN    1    0    9
R1     1    0    1.5K
R2     1    0    3.3K
.OPTIONS NOPAGE
.DC VIN 9 9 9
.PRINT DC I(R1)  I(R2)  I(VIN)
.END

**** DC TRANSFER CURVES       TEMPERATURE = 27.000 DEG C

VIN     I(R1)      I(R2)      I(VIN)
  9.000E+00  6.000E-03  2.727E-03  -8.727E-03
```

Figure 7.25 PSpice output file.

`I(R1)` The current through resistor R1.

`I(R2)` The current through resistor R2.

`I(VIN)` The current supplied by the voltage source.

The `.END` statement closes the file and should always be the last statement in your file.

Now, we are ready to simulate our circuit using PSpice. Returning to DOS, we can type

```
PSpice EX7-21
```

The resulting OUTPUT file, `EX7-21.OUT`, contains the information shown in Figure 7.25.

Now let us use PSpice to examine a parallel circuit driven by an ac source.

EXAMPLE 7.22

Use PSpice to graph the current waveforms for the three branch currents in Figure 7.26. Compare the phase relationships between the branch currents and the input signal.

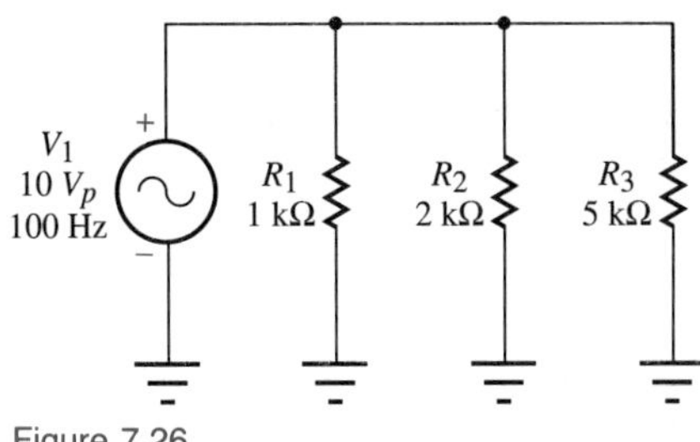

Figure 7.26

Solution

Since we have two nodes, we will label the reference or ground node as Node 0 and the other node as Node 1. Now, to create our source file, we can open a file that we will arbitrarily call "EX7-22.CIR."

The first line is always our title line:

```
PARALLEL CIRCUIT, EXAMPLE 7.22
```

The following lines will describe our circuit.

Voltage source, V1, is connected between Nodes 1 and 0 and is a sine wave with offset 0 volts, peak amplitude of 10 volts, and a frequency of 100 Hz. Voltage source, V1, is described by the line,

```
V1  1  0  SIN  (0  10V  100HZ).
```

The three resistors are described by three element lines:

```
R1  1  0  1K
R2  1  0  2K
R3  1  0  5K
```

The OPTIONS line will specify NOPAGE to conserve trees. The .TRAN line will allow us to graph the currents and voltages as functions of time. The .TRAN line specifies the time interval to be graphed and the increment size. For this example, let us assume that we want to graph one input cycle or 10 milliseconds. The increment size we will select is 0.1 millisecond. Hence, we will have 100 calculation points during one cycle. The larger the number of calculation points, the more detail in the graph, but it takes longer to compute all of those points. So, it's a trade-off that you will have to consider.

The .TRAN line looks like this:

```
.TRAN  0.1MS  10MS
```

The .PROBE command calls up the graphics program within PSpice. It allows you to select the variables that you want plotted and control the construction of the plots.

```
.PROBE
```

And the last line in our file is the .END command.

```
.END
```

The complete file is:

```
PARALLEL CIRCUIT EXAMPLE 7.22
V1  1  0  SIN(0  10V  100HZ)
R1  1  0  1K
R2  1  0  2K
R3  1  0  5K
.OPTIONS NOPAGE
.TRAN  0.1MS  10MS
.PROBE
.END
```

Key in this file and save it to disk. Run the file through PSpice. PROBE will be called automatically. Once you are in PROBE, you can add and delete plots, change the scale, obtain a hard copy, and so forth.

Figure 7.27 shown three graphs. Figure 7.27(a) shows a graph of the input voltage. It was obtained by adding a graph and selecting V(1) as the

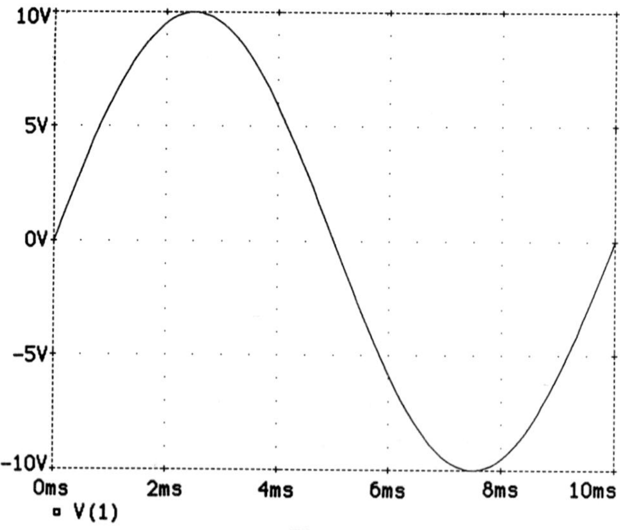

(Figure 7.27a)

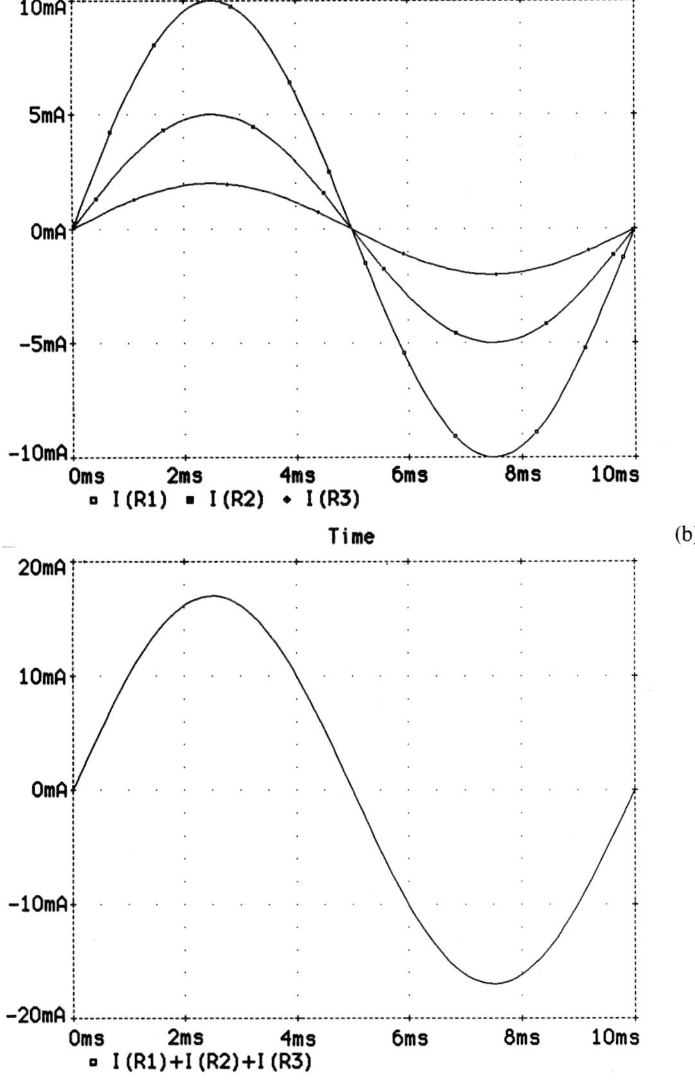

Figure 7.27 (a) Input voltage waveform. (b) Branch current waveforms. (c) Total current waveform.

parameter to be plotted. Figure 7.27(b) shows all three branch currents on one plot. The graph was obtained by removing the V(1) trace, adding a trace, and selecting I(R1) as the variable to be plotted. The process was repeated for I(R2) and then I(R3). And finally, the total of the branch currents was obtained. Again, all previous traces must be deleted. Once this is done, we can add a trace and specify I(R1) + I(R2) + I(R3) as the parameter to be plotted. Note that PSpice will evaluate the expression at each calculation point and then plot the sum.

By looking at the graphs obtained, it is apparent that the three branch currents are in phase with the input signal. Both the peaks and valleys occur at the same point in time. We can also note that the current waveform for R1, the 1 kΩ resistor, is 10 milliamperes. The current waveform for R2, the 2 kΩ resistor, has a peak of 5 milliamperes, one half the value obtained for resistor R1. And resistor R3, the largest resistance value, has the smallest peak current, only 2 milliamperes.

This gives you two examples that illustrate the capability of PSpice to analyze parallel circuits. Being able to obtain graphical outputs from PSpice gives us the same type of tool as the oscilloscope sitting on our lab bench. The advantage of using PSpice is that we can analyze the circuit before we actually build it. We will be using PSpice again in Chapter 8 on series-parallel circuits, so remember what we have just done.

7.8 Applications of the Parallel Circuit

shunt resistor

A practical application of a parallel circuit is the ammeter. Since the movement is a low current device, a resistor is connected in parallel with the movement to bypass the excess current. This resistor is called a **shunt resistor**. The resistance value is selected so that the meter movement deflects the pointer full-scale when the current entering the parallel combination is the full-scale value. The value of this shunt resistance can be found using parallel circuit theory. Applying Ohm's Law to the shunt resistor in Figure 7.28 gives

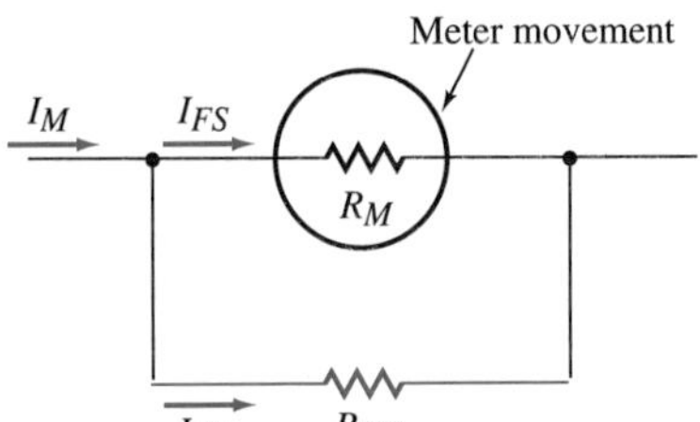

Figure 7.28 A meter movement can be used to measure large currents by placing a resistor in parallel with the movement.

$$R_{SH} = \frac{\text{Voltage across } R_{SH}}{\text{Current in } R_{SH}}$$

or

$$R_{SH} = \frac{R_M I_{FS}}{I_M - I_{FS}} \tag{7.16}$$

where I_M is the desired range; I_{FS} is the current needed for full-scale deflection on the movement, in the same units as I_M; R_M is the resistance of the movement, in ohms; and R_{SH} is the shunt resistance, in ohms.

Equation 7.16 indicates that large-range meters use a small shunt resistance. Since the total meter resistance is R_{SH} in parallel with R_M, large-range meters also have a small resistance.

EXAMPLE 7.23

A meter movement has a resistance of 50 ohms and a full-scale deflection of 50 microamperes. Find the shunt resistance needed to use the movement in a 5-mA meter.

Solution

$$R_{SH} = \frac{R_M I_{FS}}{I_M - I_{FS}}$$

$$= \frac{(50)(50 \times 10^{-6}\ \text{A})}{(5 \times 10^{-3}\ \text{A}) - (50 \times 10^{-6}\ \text{A})}$$

$$R_{SH} = 0.505\ \Omega$$

The shunt resistance is always smaller than the movement's resistance for multiplier values greater than two.

The electrical loads in a house or other building are often considered to be connected in parallel. The small resistance of the conductors is usually neglected when calculating the effect on the system. Although the voltage is ac, the dc relations apply if the loads are resistive.

EXAMPLE 7.24

A 25-W lamp, a 75-W lamp, a 1500-W toaster, and a 2000-W heater are operating at the same time from a 120-V source in a house.

(a) What is the power supplied by the source?

(b) What is the current, in amperes?

(c) What will be the power if the element in the 1500-W toaster burns open?

Solution

a. $P_s = P_1 + P_2 + P_3 + P_4$

$= 25\ \text{W} + 75\ \text{W} + 1500\ \text{W} + 2000\ \text{W}$

$P_s = 3600\ \text{W}$

b. $P = VI \quad \text{so} \quad I = P/V$

$I = 3600\ \text{W}/120\ \text{V}$

$I = 30\ \text{A}$

c. $P_s = 25\ \text{W} + 75\ \text{W} + 0\ \text{W} + 2000\ \text{W}$

$P_s = 2100\ \text{W}$

The circuitry in some lights used on Christmas trees is another application of the parallel circuit. The lamps are connected to form parallel circuits so that if one lamp burns out, the others will remain lit. This feature eliminates the need to check each bulb to find the one with the open filament.

Sources can also be connected in parallel to increase the current capacity. Two 5-ampere sources can be connected in parallel to provide 10 amperes of current. The sources can be batteries, power supplies, or transformers.

Diodes and other devices can be connected in parallel to increase the current and power capacity. For instance, two 0.25-watt resistors will be able to handle 0.50 watts of power. Another example of parallel circuits is the use of two similar circuits for redundancy. For instance, the space shuttle uses several computers connected in parallel in many of its critical systems. Thus, should one computer fail, the parallel one will still provide the data needed.

A computer and its peripherals are connected in parallel when they are plugged into a multiple outlet strip. Another parallel connection is the loudspeaker of a stereo unit. Usually, several speakers are connected to the output of the amplifier. These are only a few of the applications of the parallel circuit. Others exist in dc circuits as well as in ac circuits.

SUMMARY

1. A parallel circuit consists of two or more branches connected across the same two points.
2. Kirchhoff's Current Law states that the algebraic sum of the current at a junction or node in a circuit is equal to zero.
3. The reciprocal of the equivalent resistance of a parallel circuit is equal to the sum of the reciprocals of the resistance in each branch. Equivalent resistance is always smaller than the smallest branch resistance.
4. The current divider rule states that the current in a branch of a parallel circuit is equal to the current entering the circuit, multiplied by the equivalent resistance of the branches, divided by the resistance in the branch.
5. The power in a parallel circuit is equal to the sum of the power in each branch.
6. The energy in a parallel circuit is equal to the sum of the energy in each branch.
7. The parallel circuit can be analyzed by determining the wanted quantity, and using the equations to guide you through the analysis.
8. The ammeter shunt resistance, electrical distribution systems, and Christmas tree lights are only three examples of the applications of parallel circuits.
9. All of the relations except power and energy can be used with resistive circuits and time-varying sources. They will give the pulse or instantaneous value, depending on which one is used to describe the source.

EQUATIONS

7.1	$I_s - I_1 - I_2 - \cdots - I_n = 0$	Kirchhoff's Current Law for current at a node.	246
7.2	$I_s = I_1 + I_2 + \cdots + I_n$	An alternate form of Kirchhoff's Current Law.	246

7.3	$R_{eq} = V/I$	Equivalent resistance of a circuit from voltage and current.	248
7.4	$\frac{1}{R_{eq}} = \frac{1}{R_1} + \frac{1}{R_2} + \cdots + \frac{1}{R_n}$	Equivalent resistance of resistors connected in parallel.	249
7.5	$R_{eq} = \frac{R_1R_2}{R_1 + R_2}$	Equivalent resistance of two resistors connected in parallel.	251
7.6	$R_{eq} = R/n$	Equivalent resistance of n equal resistors connected in parallel.	251
7.7	$G_{eq} = G_1 + G_2 + \cdots + G_n$	Equivalent conductance of a parallel circuit.	252
7.8	$I_x = I_T \frac{I_T R_{eq}}{R_x}$	Current divider equation for any number of resistors connected in parallel.	254
7.9	$I_x = \frac{R_p I_T}{R_p + R_x}$	Current divider equation for two resistors connected in parallel.	256
7.10	$P = I^2 R_{eq}$	Power in terms of current and equivalent resistance.	257
7.11	$P = V^2/R_{eq}$	Power in terms of voltage and equivalent resistance.	257
7.12	$P = VI$	Power in terms of voltage and current.	257
7.13	$P = P_1 + P_2 + \cdots + P_n$	Relation of circuit power to the power in the parts of the circuit.	259
7.14	$\mathcal{W} = P\,t$	Energy in terms of voltage, current, and time.	262
7.15	$\mathcal{W} = \mathcal{W}_1 + \mathcal{W}_2 + \cdots + \mathcal{W}_n$	Relation of circuit energy to the energy in the parts of the circuit.	262
7.16	$R_{SH} = \frac{R_M I_{FS}}{(I_M - I_{FS})}$	Shunt resistance of an ammeter.	272

QUESTIONS

1. How can one tell if resistors are connected in parallel?
2. A parallel circuit of two branches is connected to a constant voltage source of 24 V. What effect will opening one path have on the voltage across the other path?
3. What effect does opening one path of a parallel circuit have on the current in the other path?
4. What effect does short circuiting one path of a parallel circuit have on the current in the other path?
5. What characteristics of the parallel circuit makes it suitable for use in electrical distribution systems?

6. A 5-ohm, 10-ohm, and 15-ohm resistor are connected in parallel. In which resistor will the current be the smallest? In which resistor will the current be the largest?
7. What does Kirchhoff's Current Law state?
8. What effect does adding a parallel path have on the equivalent resistance of a parallel circuit? What effect does removing a path have?
9. What is the current divider equation for circuits having more than two parallel paths?
10. Three 10-watt loads are connected in parallel. How much power is in the parallel circuit?
11. What is the relation between the energy in a parallel circuit and the energy in each branch?
12. A 2-A ammeter is made from a 1-mA movement. How much current must be bypassed?
13. The equivalent resistance of three equal resistances connected in parallel is 15 ohms. What will be the equivalent resistance if one resistor is disconnected?
14. The current in one branch of a parallel circuit increased while the current in another branch did not change. What could have happened to the circuit?
15. How can turning ON an electric iron cause the picture on the television to get smaller?
16. One of the lamps in a house burns out. What effect will this have on the brightness of the other lamps that are ON?
17. How will the brightness of four 100-W lamps connected in parallel to 120 volts compare to the brightness of the same four lamps connected in series across 120 volts?

PROBLEMS

SECTION 7.2 Kirchhoff's Current Law

1. The currents in a three-branch parallel circuit are $I_1 = 1$ A, $I_2 = 2$ A, and $I_3 = 3$ A. How many amperes is the source current?

2. The source current in a parallel circuit of two lamps is 3 A and the current in one branch is 1 A. What is the current in the second branch?

3. A 240-ohm soldering iron and a 60-ohm electric heater are connected in parallel across a 120-volt source.

   (a) What is the current in each item?

   (b) What is the source current?

   (c) What will the source current be if the 60-ohm heater becomes open-circuited?

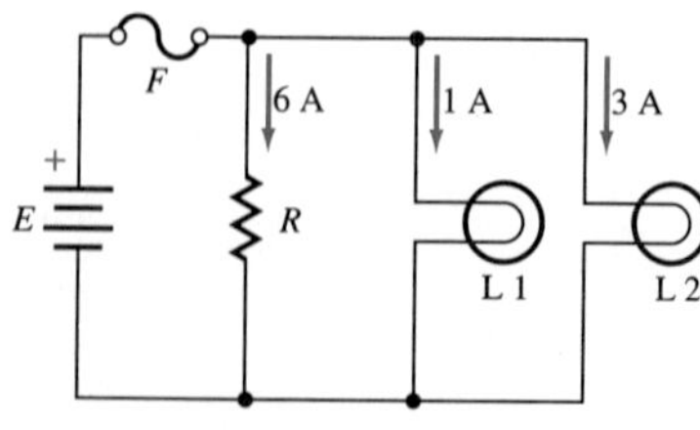

Figure 7.29

4. The source current in a three-branch circuit is 0.6 A, and the branch currents are 0.2 A, 0.1 A, and 0.3 A. If the smallest resistance is 20 ohms

   (a) What is the resistance in the other branches?

   (b) How many volts is the source?

5. What should the current rating be of the fuse to keep it from opening in the circuit of Figure 7.29?

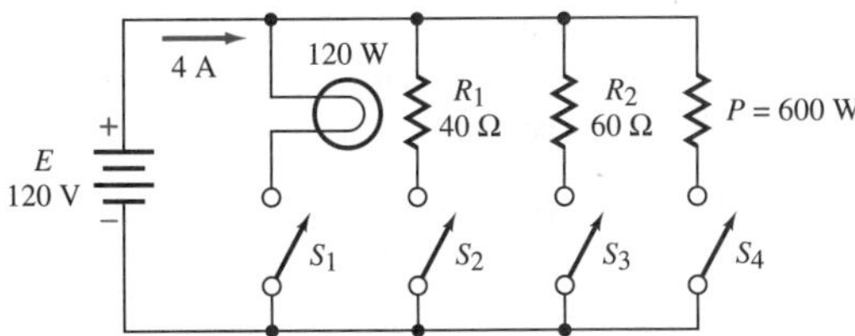

Figure 7.30

6. Which switches must be closed in Figure 7.30 to give the source current shown?

SECTION 7.3 Equivalent Resistance

7. What is the equivalent resistance of a parallel circuit that draws 0.5 amperes when connected to a 120-volt source?

8. A 600-ohm, 400-ohm, and 480-ohm resistor are connected in parallel. What is the equivalent resistance of the circuit?

9. A 30-ohm and 60-ohm resistor are connected in parallel to a 24-volt source. How much resistance must be connected in parallel with them if the current entering the three-branch circuit is to be 2.4 amperes?

10. What will be the equivalent resistance if a 320-ohm resistor is connected in parallel with the circuit of Problem 8?

11. Use Equation 7.4 to find the equivalent resistance of a 48-ohm resistor in parallel with a 64-ohm resistor.

12. Use Equation 7.4 to find the resistance of a 600-ohm resistor in parallel with a 300-ohm resistor.

13. The equivalent resistance of two resistors in parallel is 400 ohms. If one resistor is 600 ohms, what is the resistance of the other resistor?

14. How many 600-ohm resistors must be put in parallel for an equivalent resistance of 100 ohms?

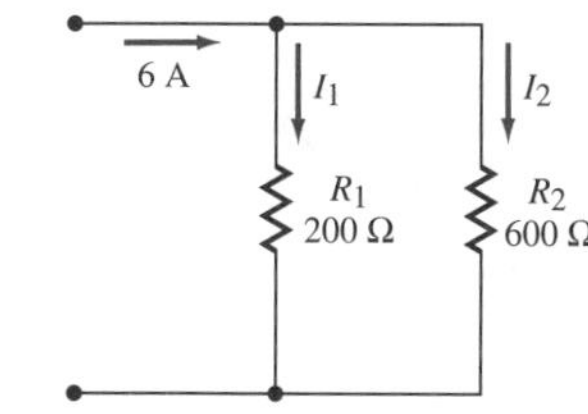

Figure 7.31

SECTION 7.4 The Current Divider Rule

15. Use the current divider equation to find:

    (a) the current in R_1 of Figure 7.31.

    (b) the current in R_2.

16. Work Problem 15 for $R_1 = 180\,\Omega$, $R_2 = 240\,\Omega$, and a source current of 500 mA.

17. Find the current I_s in Figure 7.32, using the current divider equation.

18. Use the current divider rule in Figure 7.33 to find:

    (a) the value of R_2.

    (b) the current in R_2.

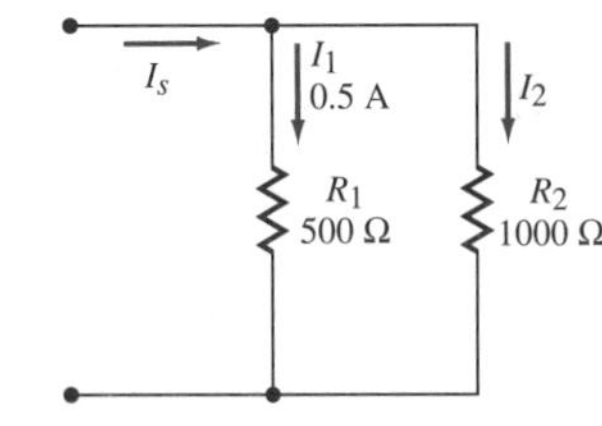

Figure 7.32

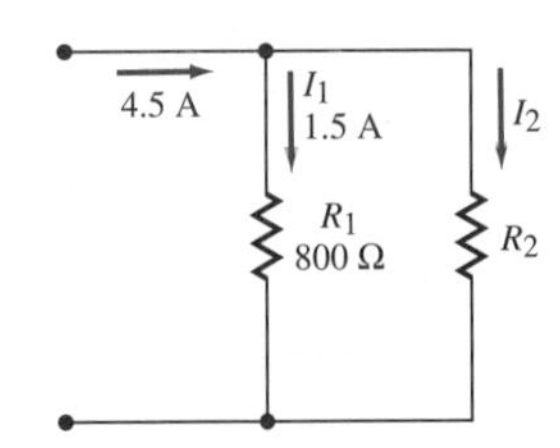

Figure 7.33

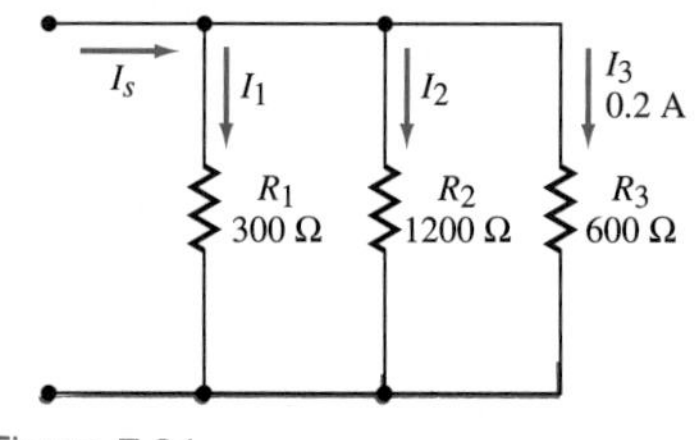

Figure 7.34

19. A 200-ohm, 250-ohm, and 1000-ohm resistor are connected in parallel across a source. The source current is 6 A. Use Equation 7.8 to find the current in the 200-ohm resistor.

20. A 120-ohm, 240-ohm, and 480-ohm resistor are connected in parallel across a source. The source current is 500 mA. Use Equation 7.8 to find the current in the 240-ohm resistor.

21. Use Equation 7.8 to find:

    (a) the current I_s.

    (b) the current in the 300-ohm resistor.

    (c) the current in the 1200-ohm resistor in Figure 7.34.

22. A 240-ohm, 120-ohm, and a 480-ohm resistor are connected in parallel across a source. The source current is 600 mA. Use Equation 7.9 to find the current in the 120-ohm resistor.

23. The equivalent resistance of two parallel resistors is 200 ohms. If one of the resistors is 600 ohms, what is the resistance of the second resistor?

24. Derive Equation 7.8 for a three-branch circuit.

SECTION 7.5 Power and Energy in the Parallel Circuit

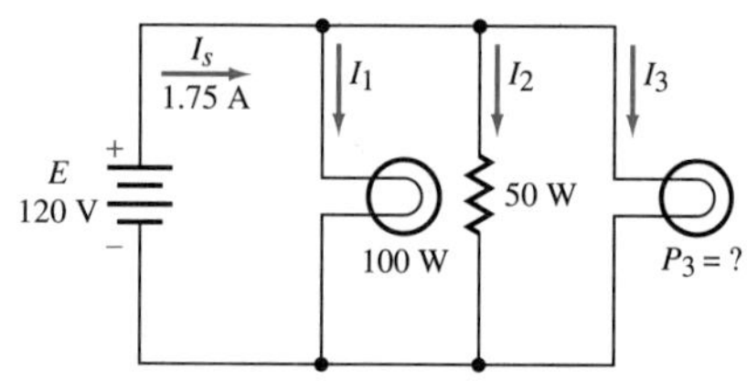

Figure 7.35

25. What is the power input to a motor if the current is 2 A when the circuit is connected to a 120-V source?

26. What is the voltage across a parallel circuit if the current is 500 mA and the power is 60 watts?

27. A 200-watt lamp, a 600-watt heater, and a 500-watt curling iron are connected in parallel across a 120-volt source.

    (a) What is the power in the circuit?

    (b) What is the source current?

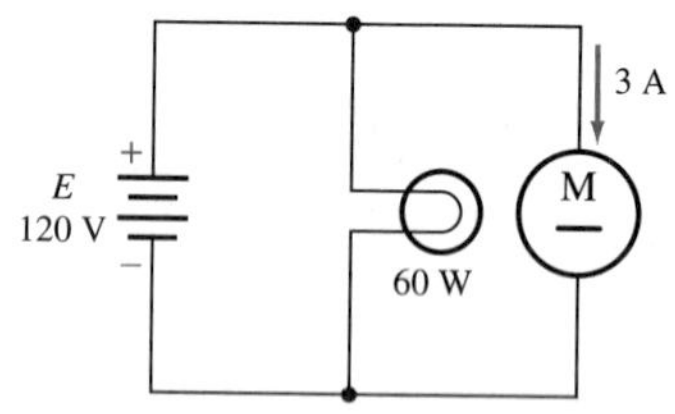

Figure 7.36

28. How many watts of power are dissipated in the lamp in branch 3 of Figure 7.35? What is the current in each branch?

29. A 200-ohm and 100-ohm resistor are connected in parallel across a source. The source current is 1.5 A. How many watts of power are dissipated in the circuit?

30. How much power is supplied by a 120-volt source connected to an electric heater that draws 10 amperes?

31. The circuit of Figure 7.36 operates for 3 hours. How much energy is used

    (a) in joules?

    (b) in kilowatt hours?

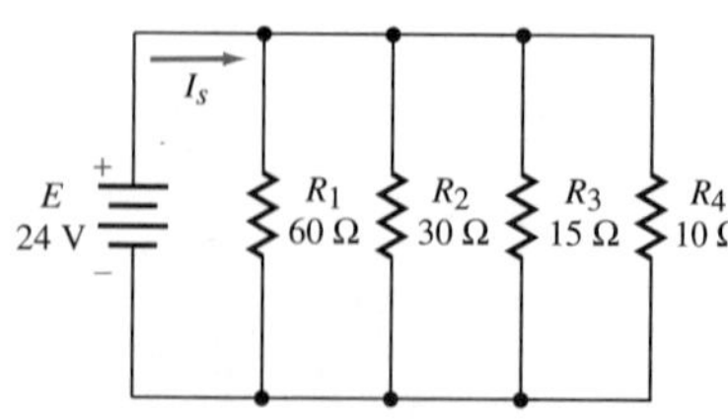

Figure 7.37

32. Four 100-watt bulbs are connected in parallel across a 120-volt source. How much energy does the source supply in 5 hours

    (a) in joules?

    (b) in kilowatt hours?

SECTION 7.6 Analysis of the Parallel Circuit

33. For the circuit of Figure 7.37, find:
    (a) The source current.
    (b) The power supplied by the source.
    (c) The current in each branch.
    (d) The power in each branch.

34. For the circuit of Figure 7.38, find:
    (a) The current I_s.
    (b) The resistance R_1 and R_2.
    (c) The power dissipated in the circuit.

35. For the circuit of Figure 7.39, find:
    (a) The current in each branch.
    (b) The source voltage.
    (c) The energy in joules for 3 hours.

36. For the circuit of Figure 7.40, find:
    (a) The current I_s.
    (b) The power in the circuit.
    (c) Current in each branch.

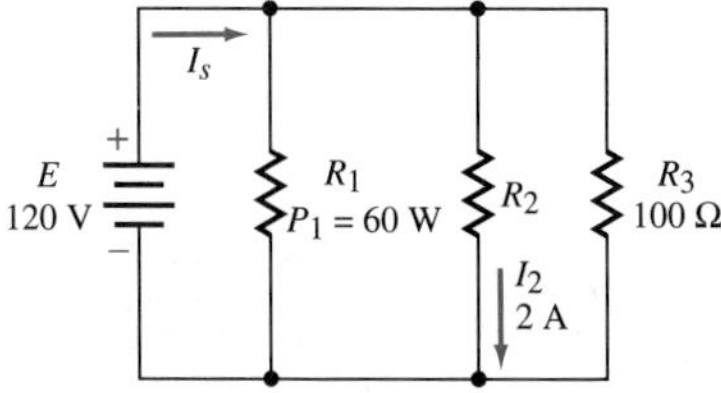

Figure 7.38

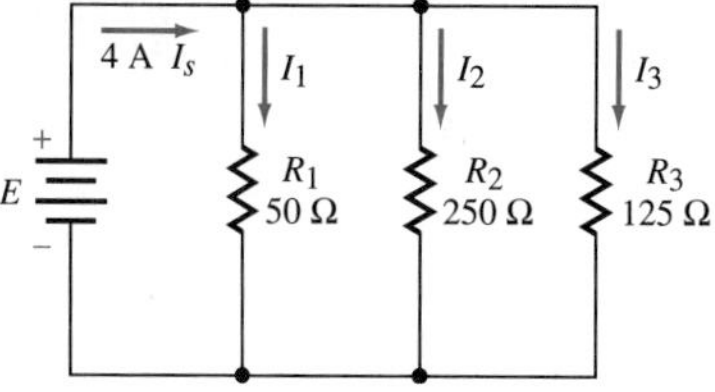

Figure 7.39

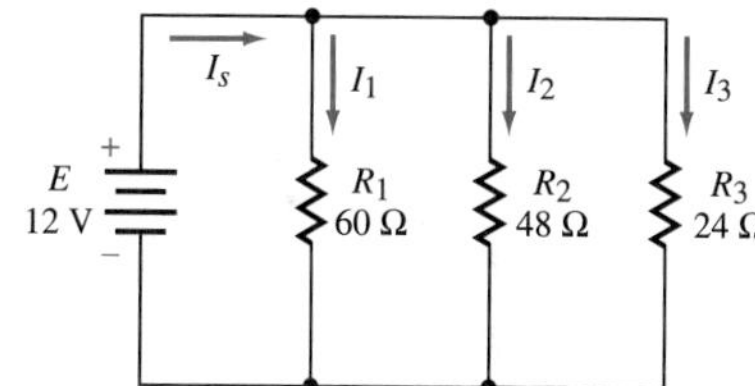

Figure 7.40

SECTION 7.7 SPICE and Parallel Circuits

Ⓢ 37. Use PSpice to find the branch currents, the total current, and the total power dissipated in the circuit shown in Figure 7.41. Check your results to make sure the simulation is correct.

Ⓢ 38. Use PSpice to find the branch currents, the total current, and the total power dissipated in the circuit shown in Figure 7.42. Check your results to make sure the simulation is correct.

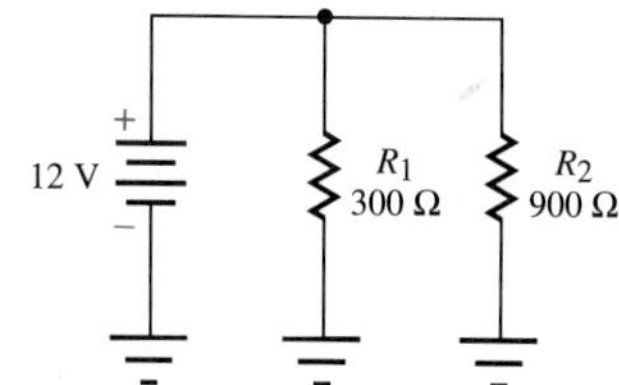

Figure 7.41

SECTION 7.8 Applications of the Parallel Circuit

39. Determine the shunt resistance needed to make a 0–10-mA milliammeter from a 1-mA movement that has a 50-ohm resistance.

40. Work Problem 39 for a 5-A ammeter from a 50-mA movement that has a resistance of 10 ohms.

41. The current in a 0.5-ohm shunt resistance is 1.99 amperes when a 10-mA movement deflects full-scale. What is the resistance of the movement?

42. What must the resistance of a 0.5-mA movement be to make it into a 10-mA milliammeter with a 0.8-ohm shunt resistance?

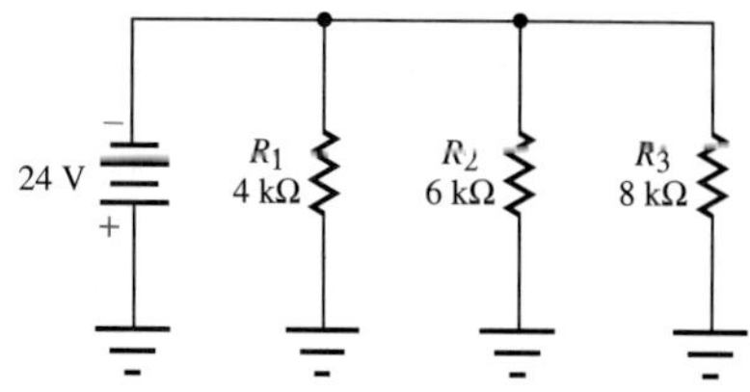

Figure 7.42

CHAPTER 8

ANALYSIS OF SERIES-PARALLEL CIRCUITS AND SOURCES

CHAPTER OBJECTIVES

After completing this chapter, you should be able to:

1. Recognize a series-parallel circuit.
2. Analyze a series-parallel circuit for currents, voltages, and other quantities.
3. Determine the source voltage and/or component values needed to obtain the specified operating conditions.
4. Analyze a voltage-divider circuit.
5. Design a voltage divider to provide specified operating conditions.
6. Calculate the error caused by the loading effect of a meter.

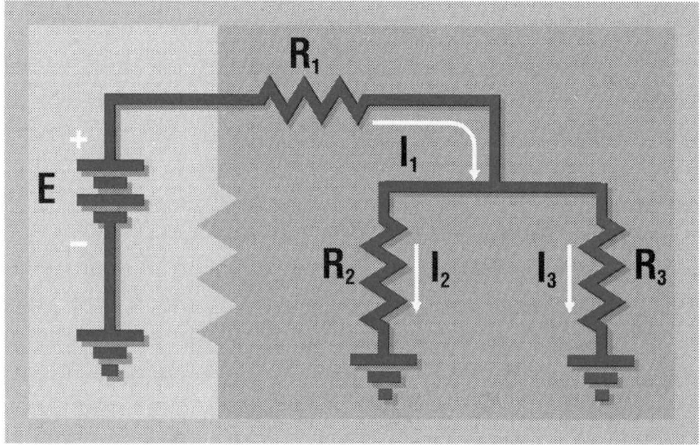

7. Calculate the internal resistance of a source, and determine its effect on the source voltage.
8. Calculate the voltage regulation of a source.
9. Determine the parameters for a voltage and current source, and convert from one to the other.
10. Connect sources in series, parallel, and series-parallel and determine their characteristics.
11. Use SPICE to calculate the operating point of a series-parallel circuit.

KEY TERMS

1. analysis
2. bleeder current
3. bleeder resistor
4. current source
5. internal resistance
6. loading effect
7. Millman's Theorem
8. series-parallel circuit
9. synthesis
10. voltage divider
11. voltage regulation

INTRODUCTION

Your study of schematics and electronic circuits will introduce you to many connections that do not fall into either the basic series or the basic

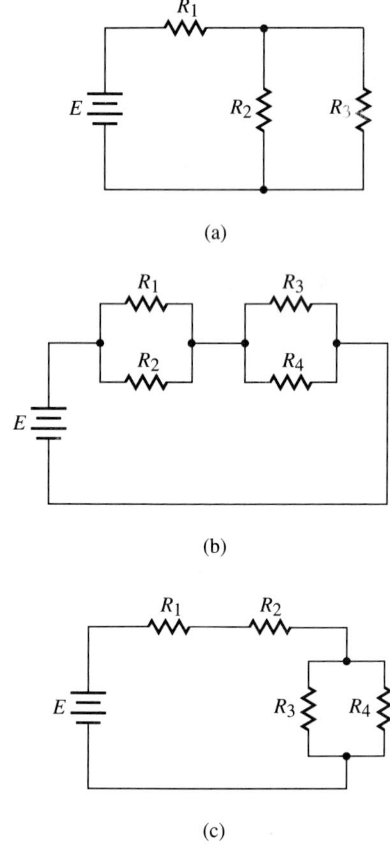

Figure 8.1 Three series-parallel circuits. (a) The parallel group of R_2 and R_3 is in series with R_1. (b) The parallel group of R_1 and R_2 is in series with the parallel group of R_3 and R_4. (c) The series group of R_1 and R_2 is in series with the parallel group of R_3 and R_4.

parallel categories. Some of these circuits will be of the series-parallel type. This is a circuit that is made up of groups of series or parallel circuits arranged in series and parallel arrangements.

The power distribution system in a house is a series-parallel circuit. In it, the series connection of the fuse and switch are connected in series with the parallel circuit of the lamps, appliances, and other household devices.

Circuits of electronic equipment also include many series-parallel circuits. These circuits provide the proper operating characteristics by dividing both the voltage and currents.

What provides the voltages for the many transistors in a piece of electronic equipment? Most of the time, voltage dividers are used. These are low-cost, small-size, resistor configurations that divide a fixed voltage. Volume controls, tone controls, and light dimmers are just a few uses for voltage dividers. Their circuits are series-parallel circuits. The analysis and design of a voltage divider will help you to understand this useful device.

Part of making a measurement is knowing how the instrument affects the measurement. Ammeters and voltmeters can change the circuit conditions, causing errors in the measurement.

Selecting a power supply requires more than satisfying only the voltage requirement. How much will the voltage change when current is supplied? What is the regulation of the source? How can a voltage source be represented by an equivalent current source? These quantities will be studied in the analysis of sources.

Sources of emf can be interconnected to provide more voltage, more current, or both. How to make the connections and what output to expect are presented in this chapter.

Finally, you will be introduced to the use of PSpice for analyzing the series-parallel circuit.

8.1 The Circuit and Its Equivalent Resistance

series-parallel circuit

The series-parallel circuit is another form of circuit configuration that is often found in electronic equipment. A **series-parallel circuit** is one in which electrical components are connected in series and parallel groups. Three examples of such a circuit and their descriptions are shown in Figure 8.1.

Many series-parallel circuits are treated as parallel circuits because of the small resistance of some parts. One example is the circuit shown in Figure 8.2. The combined resistance of the wire, fuse, and switch is normally small compared to the resistance in the branches. Ignoring them usually does not result in too much of an error.

A series-parallel circuit also has an equivalent resistance. Unfortunately, we do not have any single equation for the equivalent resistance. This is because the circuits can have many different configurations. The equivalent resistance can be obtained by replacing series and parallel groups by their equivalent resistance until only one resistor remains.

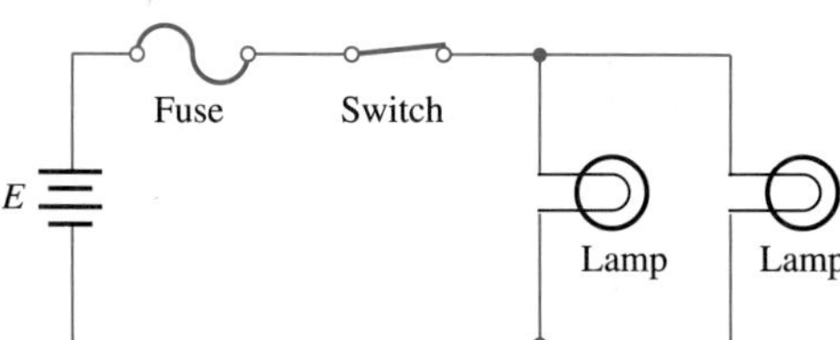

Figure 8.2 A distribution circuit such as this is often treated as a parallel circuit because of the small resistance in the switch, fuse, and wires.

This can be done by an inspection of the circuit, or as explained in the following Steps 1 through 4. Inspection is the shorter of the two methods, but requires an understanding of the circuit. On the other hand, the second method is more mechanical. As a starter, why not use the second method until the ability to use the first is developed? The steps in the second method are:

1. Draw the circuit, label all quantities, and use letters to mark points between resistors.
2. Replace each series group by its equivalent resistance.
3. Replace each parallel group by its equivalent resistance, then redraw the circuit.
4. Repeat Steps 2 and 3 until there is only one resistance. This is the equivalent resistance of the circuit.

EXAMPLE 8.1

Find the equivalent resistance of the circuit in Figure 8.3 by
a. inspection.
b. steps.

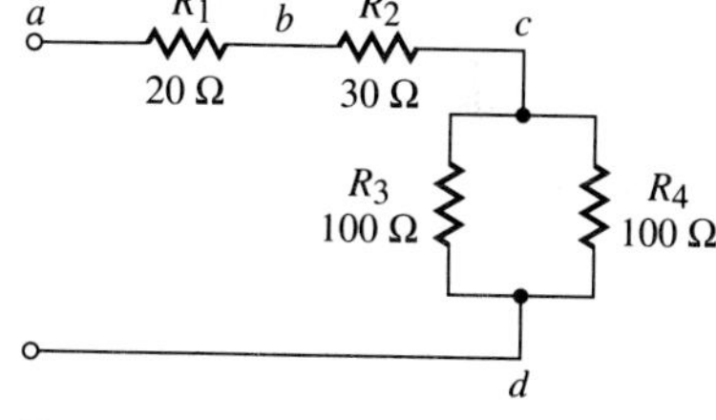

Figure 8.3

Solution

a. The points have already been marked. An inspection of the circuit shows that R_1, R_2, and the parallel group of R_3 and R_4 are in series. So

$$R_{eq} = R_1 + R_2 + (R_3 \| R_4)$$

The two vertical lines stand for "in parallel with," which means that R_3 is connected in parallel with R_4.

$$R_3 \| R_4 = \frac{R_3 R_4}{R_3 + R_4} = \frac{(100\ \Omega)(100\ \Omega)}{100\ \Omega + 100\ \Omega} = 50\ \Omega$$

so

$$R_{eq} = 20\ \Omega + 30\ \Omega + 50\ \Omega$$

$$R_{eq} = 100\ \Omega$$

b.
1. The circuit is labeled.

Strategy
Find the resistance of the series resistors.

2. R_1 and R_2 are in series. Therefore, they can be replaced by an equivalent resistance,

$$R_{ac} = R_1 + R_2 = 20\ \Omega + 30\ \Omega = 50\ \Omega$$

Find the resistance of parallel resistors.

3. R_3 and R_4 are in parallel. Their equivalent resistance is:

$$R_{cd} = \frac{R_3 R_4}{R_3 + R_4} = \frac{(100\ \Omega)(100\ \Omega)}{100\ \Omega + 100\ \Omega} = 50\ \Omega$$

The circuit is shown in Figure 8.4.

Find the resistance of series resistors. There are no parallel resistors.

4. In the redrawn circuit, R_{ac} and R_{cd} are in series. Their equivalent resistance is

$$R_{ad} = R_{ac} + R_{cd}$$
$$= 50\ \Omega + 50\ \Omega = 100\ \Omega$$

The redrawn circuit in Figure 8.5 has only one resistance so

$$R_{eq} = R_{ad}$$
$$R_{eq} = 100\ \Omega$$

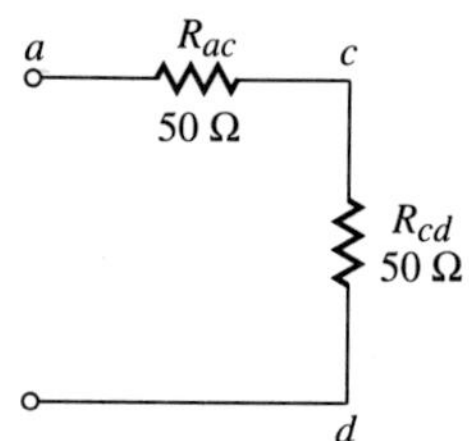

Figure 8.4

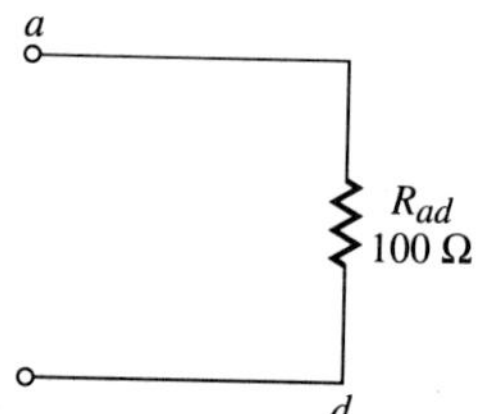

Figure 8.5

Now, let us try another example.

EXAMPLE 8.2

Find the equivalent resistance of the circuit shown in Figure 8.6 by
a. inspection.
b. steps.

Solution

Again, the circuit has been marked.

a. The series combination of R_{ab} and R_{bd} is in parallel with R_1. So

$$R_{eq} = (R_{ab} + R_{bd})\|R_1$$

or

$$R_{eq} = (R_{ab} + R_{bd})R_1/(R_{ab} + R_{bd} + R_1)$$

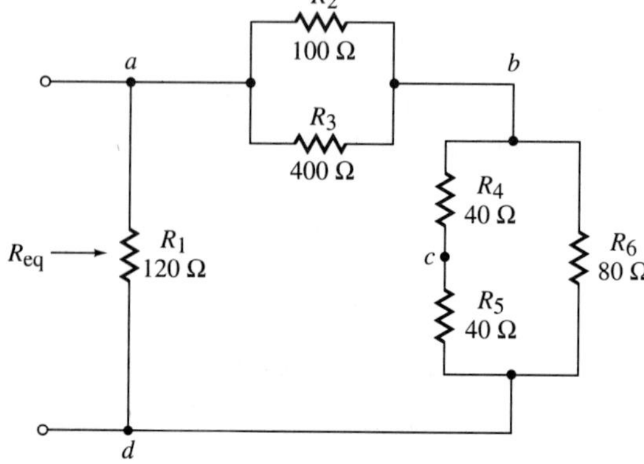

Figure 8.6

To find R_{eq}, we need the values for R_{ab} and R_{bd}. The resistance values of R_{ab} and R_{bd} must be computed using the resistance values given. To compute R_{ab},

$$\begin{aligned} R_{ab} &= R_2 \| R_3 \\ &= (R_2 R_3)/(R_2 + R_3) \\ &= (100\ \Omega)(400\ \Omega)/(100\ \Omega + 400\ \Omega) \\ R_{ab} &= 80\ \Omega \end{aligned}$$

To compute R_{bd},

$$\begin{aligned} R_{bd} &= (R_4 + R_5) \| R_6 \\ &= (R_4 + R_5)(R_6)/(R_4 + R_5 + R_6) \\ &= ((40\ \Omega + 40\ \Omega)(80\ \Omega))/(40\ \Omega + 40\ \Omega + 80\ \Omega) \\ R_{bd} &= 40\ \Omega \end{aligned}$$

Now we have all of the resistance values needed to compute R_{eq}. Putting these in the expression for R_{eq} gives

$$\begin{aligned} R_{eq} &= (80\ \Omega + 40\ \Omega)(120\ \Omega)/(80\ \Omega + 40\ \Omega + 120\ \Omega) \\ R_{eq} &= 60\ \Omega \end{aligned}$$

b. The first step is to combine those resistors that are in simple series and parallel combinations. R_4 is in series with R_5. Hence,

$$R_{bcd} = R_4 + R_5 = 40\ \Omega + 40\ \Omega = 80\ \Omega$$

R_2 and R_3 are in parallel. Hence,

$$\begin{aligned} R_{ab} &= (R_2 R_3)/(R_2 + R_3) \\ &= (100\ \Omega)(400\ \Omega)/(100\ \Omega + 400\ \Omega) \\ R_{ab} &= 80\ \Omega \end{aligned}$$

The circuit is shown in Figure 8.7.

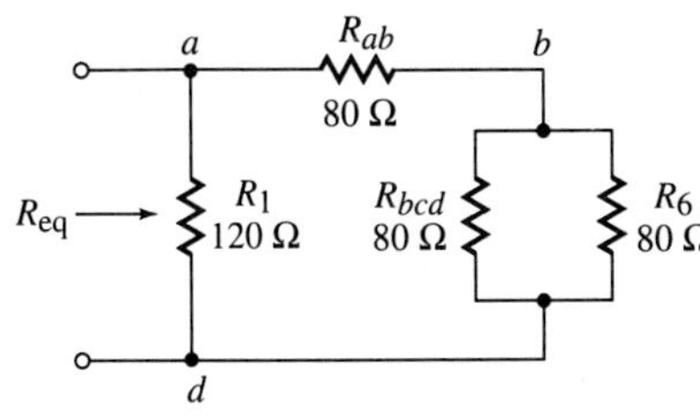

Figure 8.7

R_{bcd} is in parallel with R_6 so we can replace this parallel combination with an equivalent resistance called R_{bd}.

$$\begin{aligned} R_{bd} &= (R_{bcd})(R_6)/(R_{bcd} + R_6) \\ &= (80\ \Omega)(80\ \Omega)/(80\ \Omega + 80\ \Omega) \\ R_{bd} &= 40\ \Omega \end{aligned}$$

The redrawn circuit is shown in Figure 8.8.

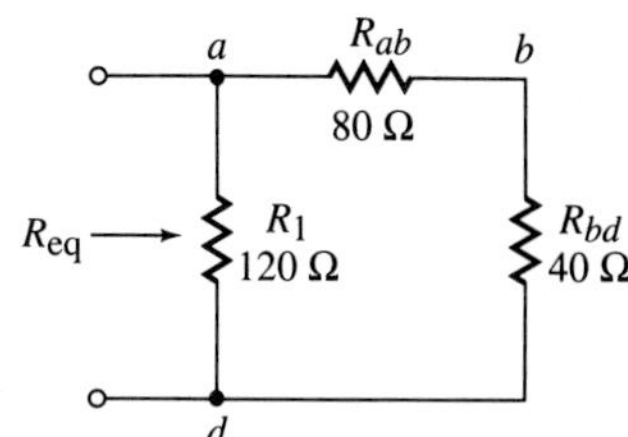

Figure 8.8

R_{ab} is in series with R_{bd}. So

$$\begin{aligned} R_{abd} &= R_{ab} + R_{bd} \\ &= 80\ \Omega + 40\ \Omega \\ R_{abd} &= 120\ \Omega \end{aligned}$$

The redrawn circuit is shown in Figure 8.9.

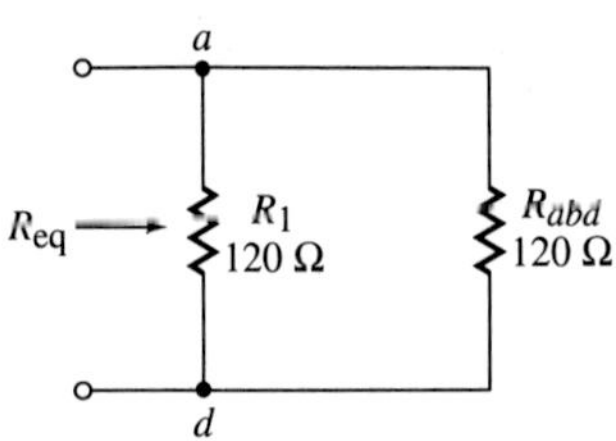

Figure 8.9

And finally, R_{abd} is in parallel with R_1. The equivalent of these two resistors finally leaves only one resistor, which is R_{eq}.

$$R_{eq} = (R_1)(R_{abd})/(R_1 + R_{abd})$$

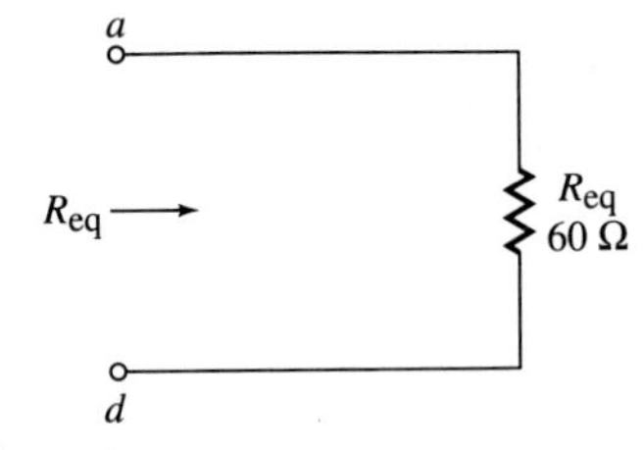

Figure 8.10

$$= (120\ \Omega)(120\ \Omega)/(120\ \Omega + 120\ \Omega)$$

$$R_{eq} = 60\ \Omega$$

The circuit is shown in Figure 8.10.

8.2 Illustrative Problems

analysis

synthesis

This section deals with the analysis and synthesis of some typical series-parallel circuits. **Analysis** means finding the conditions in the circuit when the parts, source voltage, and component values are known. **Synthesis** is finding the circuit component values needed to obtain specified conditions.

The key to analysis and synthesis is to develop a sequence that relates the wanted quantities to the given quantities. Since the conditions can change, each sequence can be different. If the path cannot be visualized, the relations can be used as a guide. The unknowns in each relation steer you to the next concept to use. Now, some examples are worked.

EXAMPLE 8.3

What is the source current, the current I_4, and the voltage across R_5 in the circuit shown in Figure 8.11?

Solution

The circuit, with currents and points marked, is shown in Figure 8.12. One sequence can be:

a. Calculate the equivalent resistance.

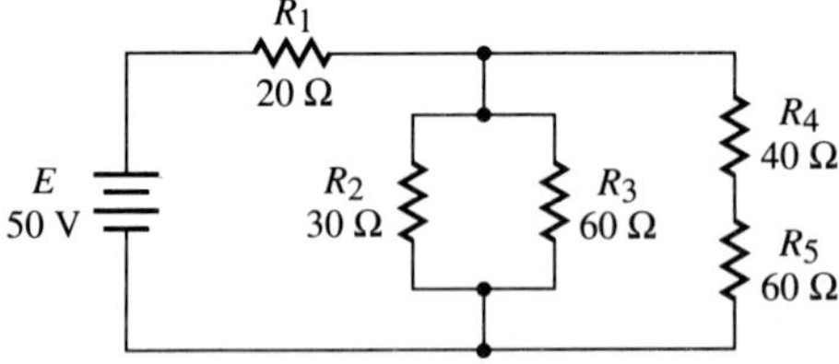

Figure 8.11

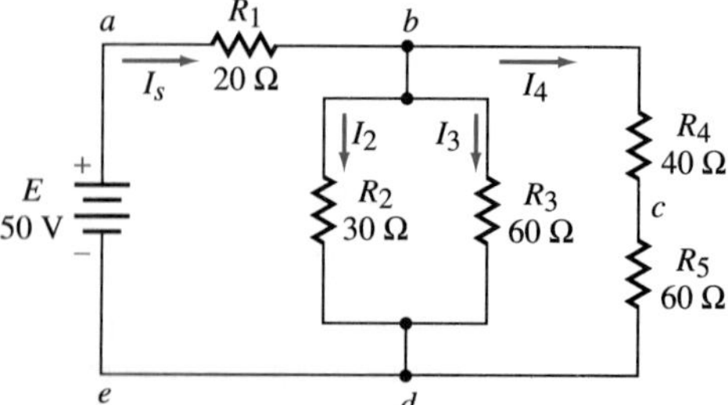

Figure 8.12

b. Use R_{eq} and E to calculate I_s.
c. Use the current divider equation to calculate I_4.
d. Use Ohm's Law to calculate V_5.

Strategy

Write the equation for R_{eq}.

$$R_{eq} = R_1 + R_{bd} \| R_{bcd} \qquad (R_{bd}, R_{bcd})$$

$$R_{bd} \text{ is } R_2 \| R_3$$

or

$$R_{bd} = (R_2 R_3)/(R_2 + R_3)$$

$$R_{bd} = \frac{(30\ \Omega)(60\ \Omega)}{30\ \Omega + 60\ \Omega}$$

$$R_{bd} = 20\ \Omega$$

R_{bcd} is R_4 in series with R_5. So

$$R_{bcd} = 40\ \Omega + 60\ \Omega$$

$$R_{bcd} = 100\ \Omega$$

Then

$$R_{eq} = 20\ \Omega + \frac{(20\ \Omega)(100\ \Omega)}{20\ \Omega + 100\ \Omega}$$

$$R_{eq} = 36.67\ \Omega$$

Use Ohm's Law to calculate I_s.

$$I_s = E/R_{eq}$$

so

$$I_s = 50\ \text{V}/36.67\ \Omega$$

$$I_s = 1.36\ \text{A}$$

Use the current divider equation to find I_4.

$$I_4 = (I_s R_{bd})/(R_{bd} + R_{bcd})$$

$$I_4 = \frac{(1.36\ \text{A})(20\ \Omega)}{20\ \Omega + 100\ \Omega}$$

$$I_4 = 0.23\ \text{A}$$

then

Write Ohm's Law for V_5.

$$V_5 = (0.23\ \text{A})(60\ \Omega)$$

$$V_5 = 13.8\ \text{V}$$

Next, the same problem will be worked using the equations to form the path. First,

$$I_s = V_1/R_1$$

Since the unknown in this equation is V_1, our next equation will include it. Using the voltage-divider equation gives

$$V_1 = \frac{(E)(R_1)}{R_1 + R_{bd}}$$

This equation has the unknown R_{bd} in it. From the circuit connection,

$$R_{bd} = R_2 \| R_3 \| (R_4 + R_5)$$

There are three equations and three unknowns, so I_s and V_1 can be found.

To get V_5, an equation for it must be used. Again, using the voltage-divider relation gives

$$V_5 = \frac{(V_{bd})(R_5)}{R_4 + R_5}$$

The unknown in this equation is V_{bd}, so the next equation should include it. We can use Kirchhoff's Voltage Law

$$E - V_1 - V_{bd} = 0$$

These two equations have two unknowns, so V_5 and V_{bd} can be calculated. The equations have formed the path to the solution.

EXAMPLE 8.4

Given the circuit in Figure 8.13, find the current in each resistor, the voltage across each resistor, and the power supplied by the source.

Solution

The circuit, with currents and points marked, is shown in Figure 8.14. A study of the problem might show the following as one approach.

a. Calculate the equivalent resistance.
b. Using Ohm's Law, R_{eq}, and E, calculate the current, I_s.
c. Use the current divider relation to calculate the current in R_3 and R_4.
d. Use Kirchhoff's Current Law to calculate the current in R_2.
e. Use Ohm's Law to get the voltage across each resistor.
f. Use the power relation to calculate the source power.

a. Calculate R_{eq}:

$$R_{eq} = R_1 + R_{bd}$$
$$R_{bd} = R_2 \| R_{bcd}$$
$$R_{bcd} = R_3 + R_4$$

This set of equations can now be solved because they equal the number of unknowns.

From the third equation in the preceding set,

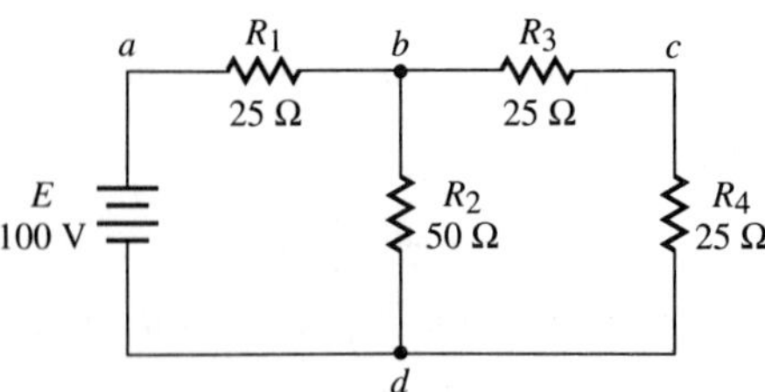

Figure 8.13

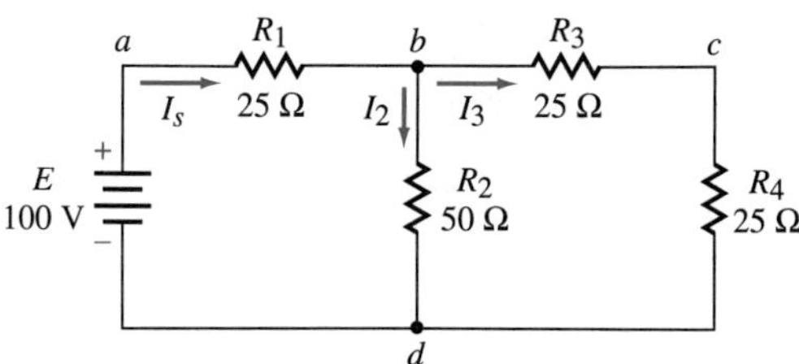

Figure 8.14

$$R_{bcd} = 25\ \Omega + 25\ \Omega = 50\ \Omega$$

From the second equation in the preceding set,

$$R_{bd} = \frac{(50\ \Omega)(50\ \Omega)}{50\ \Omega + 50\ \Omega} = 25\ \Omega$$

then

$$R_{eq} = 25\ \Omega + 25\ \Omega$$
$$R_{eq} = 50\ \Omega$$

b. Calculate the source current:

$$I_s = E/R_{eq}$$

so

$$I_s = 100\ \text{V}/50\ \Omega$$
$$I_s = 2\ \text{A}$$

I_1 is also I_s, so $I_1 = 2$ A.

c. Use the current divider rule for branches *bd* and *bcd* to get I_2.

$$I_2 = \frac{I_s R_{bcd}}{R_{bd} + R_{bcd}} = \frac{(2\ \text{A})(50\ \Omega)}{50\ \Omega + 50\ \Omega}$$
$$I_2 = 1\ \text{A}$$

d. Apply Kirchhoff's Current Law to branches *bd* and *bcd* to get I_3.

$$I_s - I_2 - I_3 = 0$$

or

$$I_3 = I_s - I_2 = 2\ \text{A} - 1\ \text{A}$$
$$I_3 = 1\ \text{A}$$

e. Apply Ohm's Law for R_1, R_2, R_3, and R_4 to get V_1, V_2, V_3, and V_4.

$$V_1 = I_s R_1 = (2\ \text{A})(25\ \Omega)$$
$$V_1 = 50\ \text{V}$$
$$V_2 = I_2 R_2 = (1\ \text{A})(50\ \Omega)$$
$$V_2 = 50\ \text{V}$$
$$V_3 = I_3 R_3 = (1\ \text{A})(25\ \Omega)$$
$$V_3 = 25\ \text{V}$$

$$V_4 = I_3R_4 = (1\text{ A})(25\ \Omega)$$
$$V_4 = 25\text{ V}$$

f. Finally, use the power relation to get the power.

$$P = EI = (100\text{ V})(2\text{ A})$$
$$P = 200\text{ W}$$

Again, the equations are used to develop the sequence. We arbitrarily start with the voltage-divider equation for V_2.

$$V_2 = \frac{ER_{bd}}{R_1 + R_{bd}}$$
$$R_{bd} = R_2 \| (R_3 + R_4)$$

This set of equations can be solved for V_2.
Writing Kirchhoff's Voltage Law gives

$$V_1 = E - V_2$$

Since V_1 is the only unknown, it can be calculated.
Next, going to I_s and using Ohm's Law gives

$$I_s = V_1/R_1$$

Again, there is only one unknown, so I_s can be calculated.
Writing the voltage-divider relation for V_3 gives

$$V_3 = \frac{R_3V_{bd}}{R_3 + R_4}$$

V_{bd} equals V_2, so V_3 can be calculated.
Going to V_4 and using Kirchhoff's Voltage Law gives

$$V_4 = V_{bd} - V_3$$

V_4 can be calculated since it is the only unknown. Finally, for power

$$P = EI_s$$

Again, the equations developed the path to the solution. Other sequences can be developed by a different choice of equations.

When a circuit is being designed, the operating conditions are known but the values of the elements must be determined. This is called circuit synthesis. Mathematically, it amounts to having a different unknown. Otherwise, the procedure is the same as for the other examples. Some synthesis problems are now worked.

EXAMPLE 8.5

The relay coil in the circuit of Figure 8.15 must have 25 volts across it for the relay to operate. Find the value of R_x.

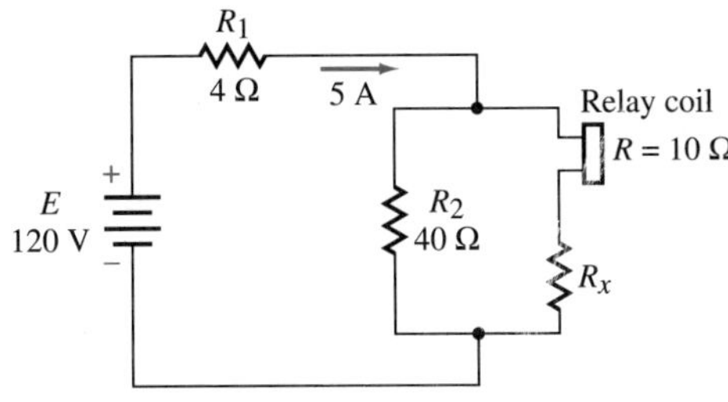

Figure 8.15

Solution

The redrawn and labeled circuit is shown in Figure 8.16.

Since R_x is wanted, the starting point is a concept that has R_x in it. Some of the possibilities are Ohm's Law, equivalent resistance, current divider rule, and voltage-divider rule. Starting with the current divider rule,

1. $I_x = \dfrac{R_2 I_s}{R_2 + (R_x + R_c)} \qquad (I_x, R_x)$

I_x is in this equation, so a relation with I_x in it should be next. Applying Ohm's Law to the coil gives

2. $V_c = I_x R_c$

The equations can now be solved because they equal the number of unknowns.

From Equation 2,

$$I_x = 25 \text{ V}/10 \ \Omega = 2.5 \text{ A}$$

Putting 2.5 A in Equation 1 gives

$$2.5 \text{ A} = \frac{(40 \ \Omega)(5 \text{ A})}{40 \ \Omega + (R_x + 10 \ \Omega)}$$

so

$$R_x = \frac{(40 \ \Omega)(5 \text{ A}) - 125 \text{ V}}{2.5 \text{ A}}$$

$$R_x = 30 \ \Omega$$

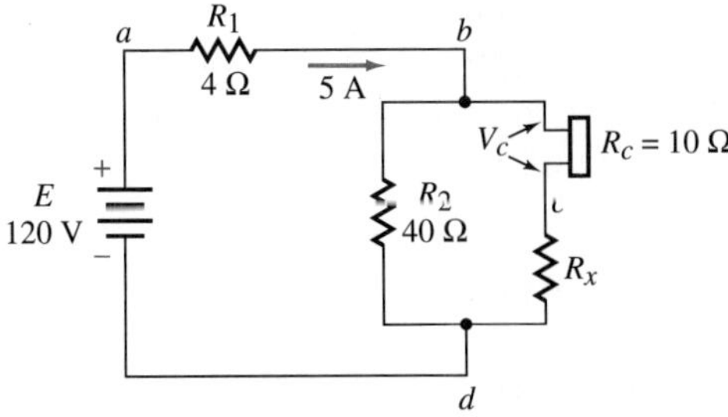

Figure 8.16

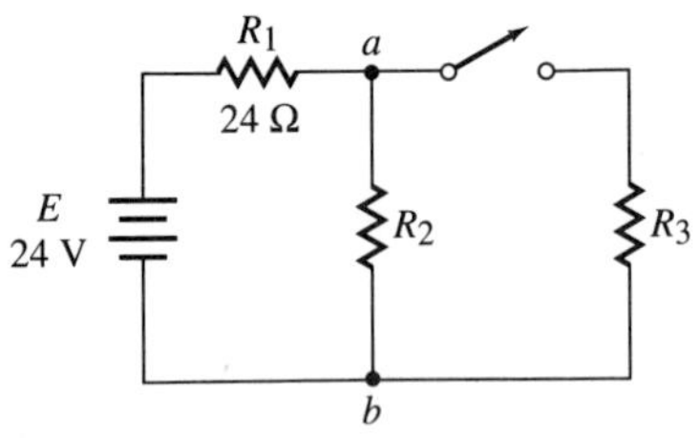

Figure 8.17

EXAMPLE 8.6

The voltage across a–b in Figure 8.17 is 8 volts with R_3 out of the circuit, and 6 volts with R_3 in the circuit. Determine the value of R_2 and R_3.

Solution

With R_3 out of the circuit,

$$V_2 = \frac{R_2 E}{R_1 + R_2}$$

or

$$R_2 = \frac{V_2 R_1}{E - V_2} = \frac{(8 \text{ V})(24\ \Omega)}{24 \text{ V} - 8 \text{ V}}$$

$$R_2 = 12\ \Omega$$

With R_2 in the circuit,

$$V_2 = \frac{(R_2\|R_3)(E)}{R_1 + R_2\|R_3}$$

Rearranging this equation gives

$$R_2\|R_3 = \frac{V_2 R_1}{E - V_2} = \frac{(6 \text{ V})(24\ \Omega)}{24 \text{ V} - 6 \text{ V}}$$

$$R_2\|R_3 = 8\ \Omega$$

$$R_2\|R_3 = \frac{R_2 R_3}{R_2 + R_3}$$

so

$$\frac{(12\ \Omega)R_3}{12\ \Omega + R_3} = 8\ \Omega$$

or

$$R_3 = \frac{(8\ \Omega)(12\ \Omega)}{12\ \Omega - 8\ \Omega}$$

$$R_3 = 24\ \Omega$$

The solution of these examples shows that the choice of the sequence affects the length of the solution. As your problem-solving ability improves, you will be able to make a better selection. This will lead to simpler and shorter solutions.

8.3 The Loaded Voltage Divider

voltage divider

Having mastered the analysis of the series-parallel circuit, we now examine the loaded voltage-divider circuit. A **voltage divider** is a series circuit used to divide a fixed or time-varying voltage into one or more fixed or variable voltages. The fixed voltage can be from a power supply or a low-power signal input. Some applications of voltage dividers are volume controls; tone controls; light dimmers, which use semiconductor devices such as silicon-controlled rectifiers (SCRs); and power-supply output controls.

Three basic voltage-divider circuits are shown in Figure 8.18. In Figure 8.18(a), the two fixed resistors divide the input voltage into a fixed output voltage. This division is done by a tapped resistor in Figure 8.18(b). In Figure 8.18(c), a potentiometer divides the input voltage into a variable output. This divider can provide a voltage from 0 volts to the input value.

Figure 8.19 shows a schematic with some voltage dividers in it. Using the dividers eliminates the need for three sources.

The unloaded divider is a series circuit, which was analyzed in Chapter 5. When a load resistor is added, this circuit can be analyzed just as the previous series-parallel circuits were analyzed.

First, let us explain the operation of the divider. A fixed voltage is applied across the total resistance as in Figure 8.18(a). The output voltage is taken across part of the resistance. The source current enters the upper resistor R_A and divides between R_B and R_L at Point b. The resistor R_B that is in parallel with the load resistor is the **bleeder resistor**. The current in it is the **bleeder current**. The bleeder current helps to keep the load voltage change small as the load current changes. Consider the circuit of Figure 8.18(a) with a bleeder resistor. In Figure 8.18(a), V_L goes from 40 volts at no load to about 28.57 volts at load. This is a 1.4:1 ratio. Without the bleeder, R_L will vary from 100 volts at no load to 50 volts at load. This is a 2:1 ratio. Such large changes might damage the circuit or affect its operation.

Because the bleeder power does not contribute to the load power, it

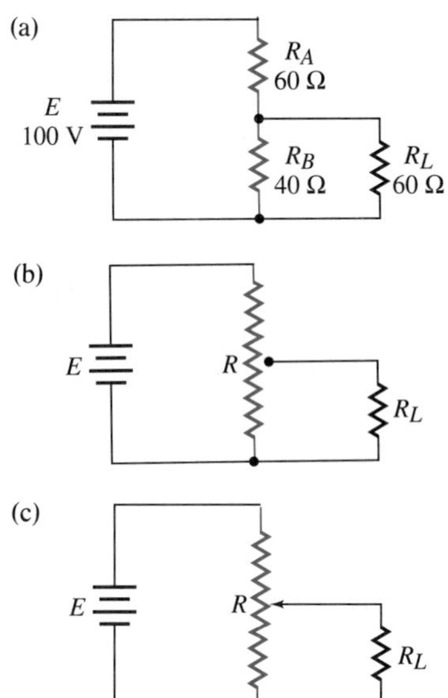

Figure 8.18 Three voltage-divider configurations. (a) A single-output divider using two fixed resistors. (b) A single-output divider using a tapped resistor. (c) A variable-output divider using a potentiometer.

bleeder resistor
bleeder current

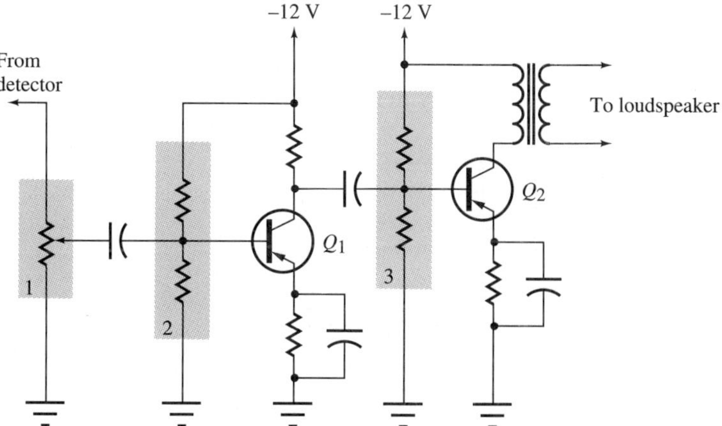

Figure 8.19 Voltage dividers are used in this transistor amplifier circuit. Divider 1 is used to control the input to the transistor amplifier. Divider 2 is used to obtain the bias voltage for transistor Q_1. Divider 3 is used to obtain the bias voltage for transistor Q_2.

represents lost power. As a rule of thumb, the bleeder current should be about 10 percent of the load current. However, the actual value will depend on the desired characteristic. If good stability is needed, larger values should be used. On the other hand, if efficiency is more important, a lower ratio should be used. Therefore, the choice depends on which has the highest priority, other factors being equal.

The transfer function for the single-output voltage divider is obtained by using the voltage-divider rule. For the unloaded voltage divider, it is:

$$V_{out}/V_{in} = R_B/(R_A + R_B) \quad \textbf{(8.1)}$$

where V_{out} is the output voltage in volts, V_{in} is the divider input voltage in volts, R_B is the bleeder resistance in ohms, and R_A is the resistance of the upper part of the divider in ohms.

For the loaded divider,

$$\frac{V_{out}}{V_{in}} = \frac{(R_B \| R_L)}{R_A + (R_B \| R_L)}$$

so

$$\frac{V_{out}}{V_{in}} = \frac{(R_B R_L)}{(R_A R_B + R_A R_L + R_B R_L)} \quad \textbf{(8.2)}$$

where V_{out}, V_{in}, R_B, and R_A have the same meaning as in Equation 8.1 and R_L is the resistance of the load connected to the divider in ohms.

A voltage divider is now analyzed to obtain its characteristics.

EXAMPLE 8.7

The divider in Figure 8.20 is used to provide the dc voltage to a transistor. Find:

a. load voltage, V_L.
b. bleeder current, I_B.
c. % efficiency.
d. power in R_A and R_B.
e. power rating of the divider.

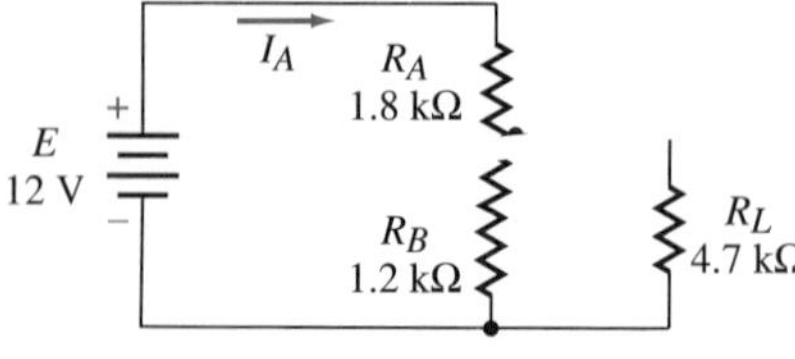

Figure 8.20

Solution

The sequence for the solution is:

a. Use the transfer function to calculate V_L.

b. Use Ohm's Law and V_L to calculate I_B.
c. Calculate P_{out} using V_L and R_L.
d. Calculate P_{in} using E and R_{eq}.
e. Calculate efficiency using P_{out} and P_{in}.
f. Calculate the power in R_B using V_B and R_B.
g. Calculate the power in R_A using $P_T = P_A + P_B + P_L$.

The divider circuit for the solution is shown in Figure 8.21.

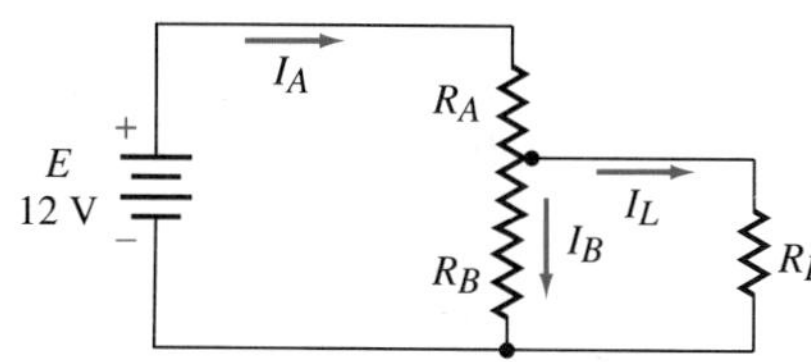

Figure 8.21

a. Using the transfer function,

Strategy

Select a relation that includes one of the wanted quantities.

$$\frac{V_L}{V_{in}} = \frac{R_B R_L}{R_A R_B + R_A R_L + R_B R_L}$$

$$\frac{V_L}{V_{in}} = \frac{(1200\ \Omega)(4700\ \Omega)}{(1800\ \Omega)(1200\ \Omega) + (1800\ \Omega)(4700\ \Omega) + (1200\ \Omega)(4700\ \Omega)}$$

$$V_L = 4.16\ \text{V}$$

b. Since V_L is also across R_B,

Solve for I_B.

$$I_B = V_L/R_B$$
$$= 4.16\ \text{V}/1200\ \Omega$$
$$I_B = 3.47 \times 10^{-3}\ \text{A}$$

c. % Efficiency = $(P_L/P_{in}) \times 100\%$. Some relations with P_L and P_{in} are needed.

Use the equation for efficiency.

$$P_L = V_L^2/R_L$$
$$= (4.16\ \text{V})^2/(4700\ \Omega)$$
$$P_L = 3.68 \times 10^{-3}\ \text{W}$$

and

$$P_{in} = E^2/R_{eq}$$

The relation for P_{in} indicates that R_{eq} is needed.

Let the equation steer you to the next step.

$$R_{eq} = R_A + R_B \| R_L$$
$$= 1800\ \Omega + \frac{(1200\ \Omega)(4700\ \Omega)}{1200\ \Omega + 4700\ \Omega}$$
$$= 2756\ \Omega$$

so

$$P_{in} = (12\ \text{V})^2/2756\ \Omega$$
$$P_{in} = 5.22 \times 10^{-2}\ \text{W}$$

$$\% \text{ efficiency} = \frac{(3.68 \times 10^{-3} \text{ W})}{(5.22 \times 10^{-2} \text{ W})} \times 100\%$$

$$\% \text{ efficiency} = 7.05\%$$

Find the power in R_B.

d. $P_B = V_L^2/R_B$

$$= (4.16 \text{ V})^2/1200\ \Omega$$

$$P_B = 1.44 \times 10^{-2} \text{ W}$$

e. Since

$$P_{\text{in}} = P_A + P_B + P_L,$$

$$P_A = P_{\text{in}} - P_B - P_L$$

$$= 5.22 \times 10^{-2} \text{ W} - 1.44 \times 10^{-2} \text{ W} - 3.68 - 10^{-3} \text{ W}$$

$$P_A = 3.41 \times 10^{-2} \text{ W}$$

Since the divider is made of one resistor, the resistors must be selected for the current in the R_A part. Based on that, the power rating must be at least

$$P_T = I_A^2(R_A + R_B)$$

$$I_A = V_A/R_A = (E - V_B)/R_A$$

or

$$I_A = \frac{12 \text{ V} - 4.16 \text{ V}}{1800\ \Omega} = 4.36 \times 10^{-3} \text{ A}$$

so

$$P_T = (4.36 \times 10^{-3} \text{ A})^2(1800\ \Omega + 1200\ \Omega)$$

$$= 57 \times 10^{-3} \text{ W}$$

The algebraic and RPN calculator keystrokes for the solution are:

Quantity	Algebraic	RPN	Display
V_L $\frac{V_{\text{in}}R_BR_L}{R_AR_B + R_AR_L + R_BR_L}$	1 2 0 0 × 4 7 0 0 × 1 2 × (1 8 0 0 × 1 2 0 0 + 1 8 0 0 × 4 7 0 0 + 1 2 0 0 × 4 7 0 0) =	1 2 0 0 ENTER 4 7 0 0 × 1 2 × 1 8 0 0 ENTER 1 2 0 0 × 1 8 0 0 ENTER 4 7 0 0 × + 1 2 0 0 ENTER 4 7 0 0 × + ÷	4.16 00
I_B V_B/R_B	4 · 1 6 ÷ 1 2 0 0 =	4 · 1 6 ENTER 1 2 0 0 ÷	3.47 −03
P_L $\frac{V_L^2}{R_L}$	4 · 1 6 x² 4 7 0 0 =	4 · 1 6 x² 4 7 0 0	3.68 −03

Quantity	Algebraic	RPN	Display
R_{eq}	1 8 0 0 + [	1 8 0 0 ENTER	
	1 2 0 0 × 4 7	1 2 0 0 ENTER	
$R_A + R_B \| R_L$	0 0 ÷ [1 2 0	4 7 0 0 × 1	
	0 + 4 7 0 0]	2 0 0 ENTER	
	=	4 7 0 0 + ÷ +	2.756 03
P_{in}	1 2 x^2 ÷	1 2 ENTER x^2	
E^2/R_{eq}	2 7 5 6 =	2 7 5 6 ÷	5.23 −02
	3 · 6 8 EXP	3 · 6 8 EEX	
%Efficiency	3 ± ÷ 5 · 2	3 CHS ENTER 5	
$= (P_L)(100\%)/P_{in}$	3 EXP 2 ± ×	· 2 3 EEX	
	1 0 0 =	2 CHS ÷ 1 0 0 ×	7.04 00
P_B	4 · 1 6 x^2 ÷	4 · 1 6 x^2	
V_B^2/R_B	1 2 0 0 =	1 2 0 0 ÷	1.44 −02
P_A	5 · 2 3 EXP	5 · 2 3 EEX	
	2 ± − 1 · 4	2 CHS ENTER 1 ·	
$P_{in} - P_B - P_L$	4 EXP 2 ± −	4 4 EEX 2 CHS	
	3 · 6 8 EXP 3	− 3 · 6 8 EEX	
	± =	3 CHS −	3.42 −02

The total divider is not dissipating 57 mW because the current in the R_B part is less than 4.35 mA. The actual power dissipated in the divider is $P_A + P_B$, which is 48.5 mW.

Dividers such as the one in Figure 8.22(a) can provide voltages for two loads. This type of divider is suitable for use where the load currents do not change much. It is less costly than using two dividers or two sources. On the negative side, a change in the current in one load causes a change in the voltage of the other load. Also, the loads are not electrically isolated from each other. If resistance R_{L2} is made smaller, the current I_2 will increase. This causes a larger voltage drop in R_A, which makes V_{L2} smaller.

Another interesting divider circuit is shown in Figure 8.22(b). This one provides a positive and a negative voltage with respect to the ground point at Point B. This type of output is frequently used for electronic circuits. The divider is connected to a source that has a floating output,

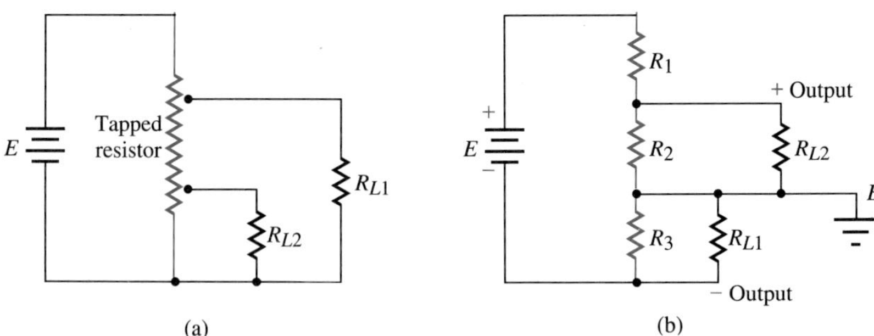

Figure 8.22 Voltage dividers such as these can provide several voltages. (a) This tapped-resistor divider provides the voltage for two loads. (b) This fixed-resistor divider provides a positive and a negative voltage.

which means that neither of the output terminals is directly connected to a ground point. Although these dividers are more complex than the basic one in Figure 8.18, each one is a series-parallel circuit and is analyzed as such.

Some characteristics that must be considered when selecting or designing a voltage divider are:

a. Some power is lost in the divider.
b. The entire source current is not available for the load because of the bleeder resistor.
c. The current capacity is limited by the power dissipation rating of the divider resistors.
d. Designing for smaller voltage changes with load changes results in a lower efficiency.
e. Changes in the current of one load of a multitap divider affect the voltage at the other loads.

Despite these undesirable characteristics, voltage dividers are ideal for use where the load current is relatively small, and low weight, small size, and low cost are important.

The steps in the design of a voltage divider can be summarized as follows:

1. Determine the conditions that must be satisfied. These can be load voltage, efficiency, resistance values, load current, and so forth.
2. Draw the circuit showing the resistances, currents, voltages, and so forth.
3. Apply the concepts of series-parallel circuits to determine the unknown quantities.

In the next example, we design a voltage divider.

EXAMPLE 8.8

Design a single-output voltage divider to supply 400 mA at 12 volts, from a 24-volt supply with a bleeder current of 0.08 A.

Solution

The divider configuration is similar to that in Figure 8.18(a).

To complete the design, R_A, R_B, and the power rating of each section must be determined.

Starting with Ohm's Law for R_B,

$$R_B = V_B/I_B$$

Since V_B is also V_L,

$$R_B = 12\ \text{V}/0.08\ \text{A}$$
$$R_B = 150\ \Omega$$

Now, applying Ohm's Law to R_A,

$$V_A = I_A R_A$$
$$R_A = V_A/I_A$$

From Kirchhoff's Current Law,

$$I_A = I_B + I_L$$
$$= 0.08\ \text{A} + 0.4\ \text{A}$$
$$I_A = 0.48\ \text{A}$$

And from Kirchhoff's Voltage Law,

$$V_{\text{in}} = V_A + V_B$$

so

$$V_A = 24\ \text{V} - 12\ \text{V}$$
$$V_A = 12\ \text{V}$$

then

$$R_A = 12\ \text{V}/0.48\ \text{A}$$
$$R_A = 25\ \Omega$$

To find the power, some concept with power in it must be used.

For R_A,
$$P_A = I_A^2 R_A$$
$$P_A = (0.48\ \text{A})^2(25\ \Omega)$$
$$P_A = 5.76\ \text{W}$$

For R_B,
$$P_B = I_B^2 R_B$$
$$= (0.08\ \text{A})^2(150\ \Omega)$$
$$P_B = 0.96\ \text{W}$$

Looking back, one can see that each equation indicated what should be in the next expression. Since the priorities were the load voltage and bleeder current, the efficiency and voltage regulation that result from the design will have to be accepted.

8.4 The Loading Effect of Meters

loading effect

As a technician, there will be times when you will measure voltage and current in a circuit. Therefore, it is important for you to understand how the insertion of meters can change the conditions in the circuit. This effect is known as the **loading effect**. Because of this loading effect, a 100 percent accurate meter can give incorrect results.

Digital meters have a resistance of 10 megohms or more, so their loading effect is small. However, the analog multimeter (VOM) has a resistance of about 3.2 megohms on the 160-volt range and has a larger loading effect. This effect worsens as the range is decreased.

Although the loading effect might be negligible in power circuits, it can be significant in electronic circuits where large resistance values are used. Let us now examine loading and see how it affects the measurement.

Consider the circuit in Figure 8.23. By analysis, 48 volts should be across R_2, but the voltmeter in Figure 8.23 indicates less than that voltage. Why? It is because the resistance of the meter is connected in parallel with R_2 and reduces the resistance across Points A and B. Since $V_{AB} = (R_{AB}E)/(R_1 + R_{AB})$, the voltage across A–B must be smaller. The indicated voltage is always smaller than the actual voltage. How much smaller? The amount depends on the ratio of the meter resistance to the resistance of R_2. Large ratios give small differences.

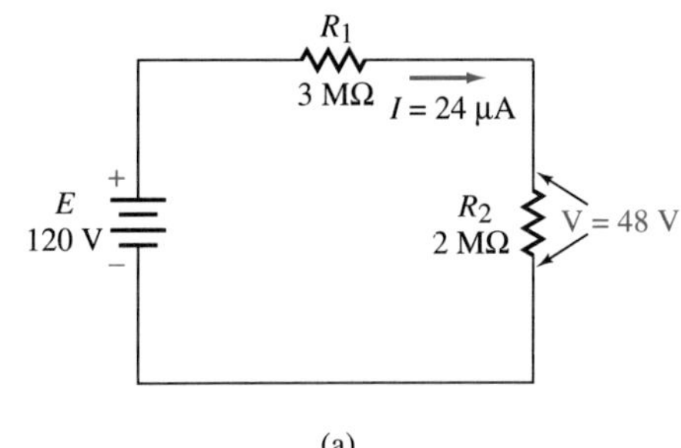

(a)

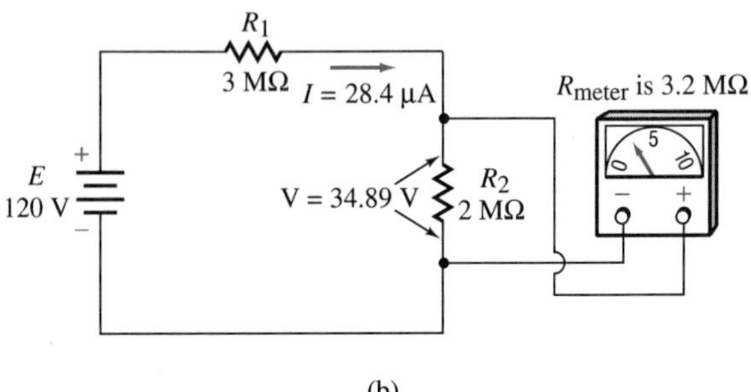

(b)

Figure 8.23 Connecting a voltmeter in a circuit changes the conditions of the circuit. (a) The voltage is 48 volts without the voltmeter. (b) Connecting the voltmeter reduces the voltage to 34.89 volts.

EXAMPLE 8.9

A series circuit of a 2-MΩ and a 3-MΩ resistor is connected across a 120-volt source. The voltage across the 2-MΩ resistor is measured with a 3.2-MΩ voltmeter. Find:

a. the voltage indicated on the meter.
b. the actual voltage.

Solution

a. With the voltmeter connected, the resistance of the parallel combination is

$$R = \frac{(2 \times 10^6\ \Omega)(3.2 \times 10^6\ \Omega)}{2 \times 10^6\ \Omega + 3.2 \times 10^6\ \Omega}$$

$$R = 1.23 \times 10^6\ \Omega$$

then

$$V = \frac{(1.23 \times 10^6\ \Omega)(120\ \text{V})}{3 \times 10^6\ \Omega + 1.23 \times 10^6\ \Omega}$$

$$V = 34.89\ \text{V}$$

b. Without the voltmeter connected,

$$V = \frac{(2 \times 10^6\ \Omega)(120\ \text{V})}{2 \times 10^6\ \Omega + 3 \times 10^6\ \Omega}$$

$$V = 48\ \text{V}$$

The indicated voltage is 13.11 volts less than the actual voltage. This is a difference of about 27 percent. Using a 10-MΩ meter would reduce the difference to about 5.1 volts. Using even a higher resistance voltmeter would further reduce the loading effect.

Another effect of loading is the increase in current caused by it. If large enough, this increase can damage parts of a circuit. An inspection of Example 8.9 shows that the voltmeter increases the current in the 3-MΩ resistor by 4.36 microamperes.

Next, consider the effect of the ammeter on the circuit of Figure 8.24(a). The current should be 0.5 A, but the ammeter in Figure 8.24(b) indicates less than that. Again, why? This time it is because the ammeter increases the resistance of the circuit. With a larger resistance, the current is smaller. The indicated current is always smaller than the actual current. The actual difference depends on the ratio of the meter resistance to the circuit resistance. A small ratio gives a small difference. This is just the opposite of the effect of the voltmeter ratio.

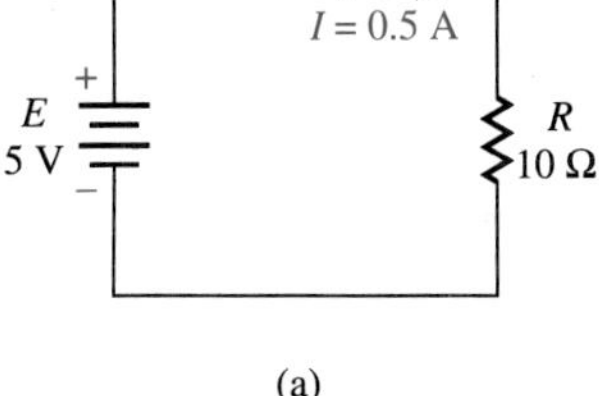

(a)

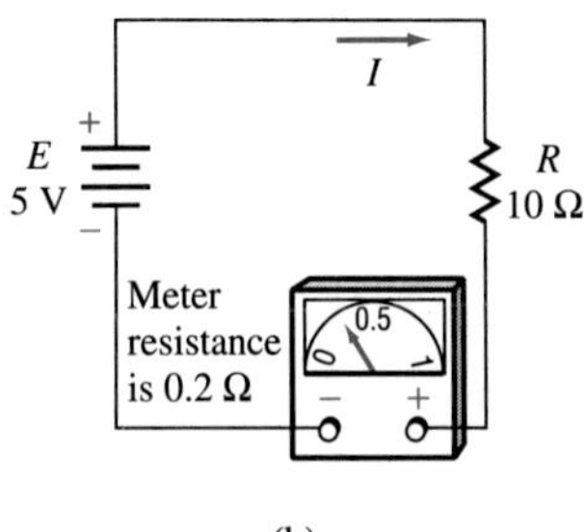

(b)

Figure 8.24 Connecting an ammeter in a circuit changes the conditions of the circuit. (a) The current is 0.5 ampere without the ammeter. (b) Connecting the ammeter reduces the current to 0.49 ampere.

EXAMPLE 8.10

A 10-ohm resistor is connected to a 1.6-volt source. The current is measured with a meter whose resistance is 2 ohms. Find:

a. the current indicated on the meter.
b. the actual current without the meter.

Solution

a.

$$I = E/R_{eq}$$
$$R_{eq} = R + R_{meter} = 10\ \Omega + 2\ \Omega = 12\ \Omega$$

so

$$I = 1.6\ \text{V}/12\ \Omega$$
$$I = 0.133\ \text{A}$$

b.

$$I = E/R$$
$$= 1.6\ \text{V}/10\ \Omega$$
$$I = 0.16\ \text{A}$$

The loading effect results in a decrease of 27 mA. This is a difference of about 17 percent.

Generally, the loading effect of the ammeter will not damage the circuit. However, it can affect the circuit operation. Using a lower-resistance ammeter reduces the error in measurement.

EXAMPLE 8.11

An electronic circuit has a series combination of a 3-megohm resistor and a 2-megohm resistor connected across 24 volts. Find:

a. the actual voltage across the 2-megohm resistor.
b. the voltage indicated on a 10-megohm digital multimeter.
c. the voltage indicated on a 0.8-megohm VOM.

Solution

a. Using the voltage-divider equation,

$$V_{2M} = \frac{(24\ \text{V})(2 \times 10^6\ \Omega)}{3 \times 10^6\ \Omega + 2 \times 10^6\ \Omega}$$
$$V_{2M} = 9.6\ \text{V}$$

b. The 10-megohm meter connected across the 2-megohm resistor gives a resistance of

$$R = \frac{(2 \times 10^6\ \Omega)(10 \times 10^6\ \Omega)}{2 \times 10^6\ \Omega + 10 \times 10^6\ \Omega}$$

$$R = 1.67 \times 10^6\ \Omega$$

so

$$V_{2M} = \frac{(24\ \text{V})(1.67 \times 10^6\ \Omega)}{3 \times 10^6\ \Omega + 1.67 \times 10^6\ \Omega}$$

$$V_{2M} = 8.58\ \text{V}$$

c. The 0.8-megohm VOM connected across the 2-megohm resistor gives a resistance of

$$R = \frac{(2 \times 10^6\ \Omega)(0.8 \times 10^6\ \Omega)}{2 \times 10^6\ \Omega + 0.8 \times 10^6\ \Omega}$$

$$R = 0.57 \times 10^6\ \Omega$$

so

$$V_{2M} = \frac{(24\ \text{V})(0.57 \times 10^6\ \Omega)}{3 \times 10^6\ \Omega + 0.57 \times 10^6\ \Omega}$$

$$V_{2M} = 3.83\ \text{V}$$

Although the DMM gave slightly better results, neither meter is suitable for this measurement.

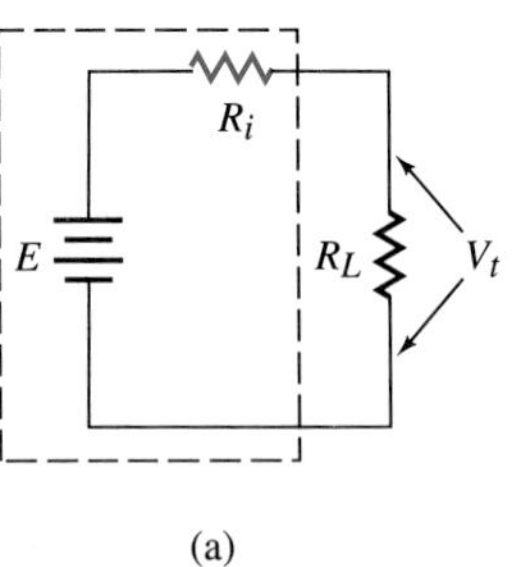

(a)

(b)

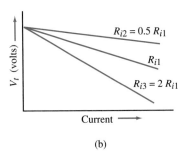

Figure 8.25 The internal resistance of a dc source makes the output voltage drop with current. (a) The graphic symbol for a practical voltage source. (b) The effect of several values of internal resistance on the output voltage.

8.5 Internal Resistance and Voltage Regulation

The circuits studied thus far included ideal sources of emf. Their output was the same for all currents. Most practical sources have a drop in voltage when supplying current. The graphic symbol for a practical source is seen in Figure 8.25(a). The series resistor represents the internal resistance of the source.

internal resistance

Internal resistance is the resistance within the source that accounts for the drop in voltage when supplying current. The internal resistance of a source can be constant or can vary, depending on the source. The internal resistance of a generator is the resistance of the armature windings, brushes, and contact resistance. A chemical cell has internal resistance in its electrolyte and electrodes. A solar cell has internal resistance in the semiconductor materials.

How does the internal resistance affect the terminal voltage of a source? To find out, consider the source in Figure 8.25(a). Applying Kirchhoff's Voltage Law to the circuit gives

$$E - IR_i - V_t = 0$$

so

$$V_t = E - IR_i \tag{8.3}$$

where V_t is the terminal voltage of the source in volts, at current I; I is the source current, in amperes; E is the terminal voltage of the source at no current, in volts; and R_i is the internal resistance, in ohms.

This equation, plotted for several values of R_i, is shown in Figure 8.25(b). You can see that the larger R_i source has a larger drop in voltage for the same current. Sources with small values of R_i have the least change in voltage as the current changes.

EXAMPLE 8.12

A cell has an open-circuit voltage of 6 volts and an internal resistance of 0.5 ohm. What will be the terminal voltage at 2 amperes of current?

Solution

$$V_t = E - IR_i$$
$$= 6\text{ V} - (2\text{ A})(0.5\ \Omega)$$
$$V_t = 5\text{ V}$$

The internal resistance of some sources can be obtained from manufacturers' specifications. For other sources, such as a generator, it can be measured with an ohmmeter. But, for many sources, it either cannot be measured or the measured value will be incorrect. For instance, the resistance of a dc power supply when measured on a bridge can be different than the amount indicated by the change in voltage when supplying current. For sources such as a battery, using an ohmmeter to measure the resistance can damage the meter. Fortunately, R_i can be found with data from a simple test. This test consists of measuring the no-load voltage and the voltage and current at the same load condition. Then, these values can be used in Equation 8.4 to get R_i. This equation is simply Equation 8.3 rearranged.

$$R_i = (E - V_t)/I \qquad \textbf{(8.4)}$$

where R_i, E, V_t, and I have the same meaning as in Equation 8.3.

Equation 8.4 shows that the magnitude of R_i is the slope of a line drawn from the no-load voltage to the load voltage. If the curve is linear, R_i is constant. If the curve is nonlinear, R_i is not constant. Then, the value of R_i holds only for the current at that particular load voltage.

EXAMPLE 8.13

A cell has an open-circuit voltage of 12 volts. The current is 2 amperes when a 4-ohm load is connected to the cell. It is 1.0 ampere when connected to an 8-ohm load. Is R_i constant?

Solution

$$R_i = (E - V_L)/I$$

At 2 amperes,

$$V_L = IR_L$$
$$V_L = (2\ \text{A})(4\ \Omega)$$
$$= 8\ \text{V}$$

and

$$R_i = (12\ \text{V} - 8\ \text{V})/2\ \text{A}$$
$$R_i = 2\ \Omega$$

At 1.0 amperes,

$$V_L = I\,R_L = (1.0\ \text{A})(8\ \Omega)$$
$$V_L = 8\ \text{V}$$

and

$$R_i = (12\ \text{V} - 8\ \text{V})/1\ \text{A}$$
$$R_i = 4\ \Omega$$

Since the two values of R_i are different, R_i is not constant.

As noted before, the internal resistance can cause the voltage to drop below a needed value. Therefore, when specifying or selecting a source for an application, we must know what voltage the source will provide under actual operating conditions. One way to determine the voltage is by connecting a load to the source and measuring the voltage. This is not always very practical, since many sources might have to be tested; testing the sources requires that all of the sources have to be available.

Another way to obtain the voltage is from the voltage-regulation specification. **Voltage regulation** is a measure of how well a source maintains a terminal voltage as the current changes. It is usually expressed as a percent. In equation form,

voltage regulation

$$\%\text{VR} = \frac{(V_{NL} - V_L)}{V_L} \times 100\% \qquad \textbf{(8.5)}$$

where V_{NL} is the terminal voltage at no current, in volts, and V_L is the terminal voltage at a current, in volts.

Ideally, we would want a source with 0 percent voltage regulation. Such a source would have no change in voltage when a load is connected to it. Also, such a source would have no internal resistance. Some practical sources approach the ideal by the use of electronic voltage-regulating circuits.

EXAMPLE 8.14

The output of a source is 20 volts at no load and 10 volts when supplying 5 amperes of current. What is the percent VR of the source?

Solution

$$\%\text{VR} = \frac{(V_{NL} - V_L)}{V_L} \times 100\%$$
$$= \frac{(20\text{ V} - 10\text{ V})}{10\text{ V}} \times 100\%$$
$$\%\text{VR} = 100\%$$

The algebraic and RPN keystrokes for the solution are:

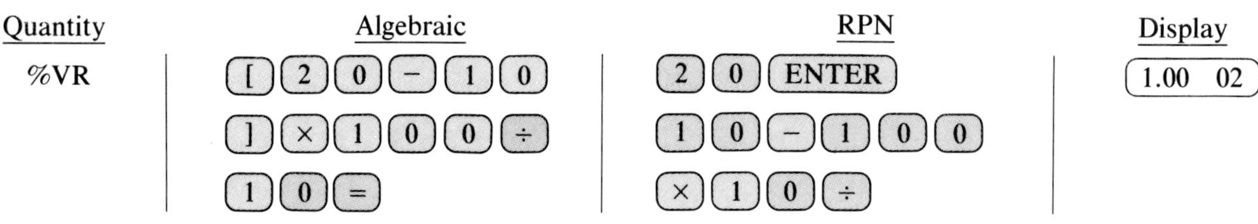

Quantity	Algebraic	RPN	Display
%VR	[2 0 − 1 0] × 1 0 0 ÷ 1 0 =	2 0 ENTER 1 0 − 1 0 0 × 1 0 ÷	1.00 02

EXAMPLE 8.15

The current in a circuit will vary from 0–3 amperes during operation. The voltage limits are 22–24 volts. A source is available that has a no-load voltage of 24 volts and a %VR of 20%. Will it be satisfactory?

Solution

The required voltage regulation will be calculated and compared to that of the available source.

$$\%\text{VR} = \frac{(V_{NL} - V_L)}{V_L} \times 100\%$$

For the required source,

$$\%\text{VR} = \left(\frac{24\text{ V} - 22\text{ V}}{22\text{ V}}\right) \times 100\%$$
$$\%\text{VR} = 9.09\%$$

This is less than that of the available source, so the source is not satisfactory.

EXAMPLE 8.16

A power supply has a no-load voltage of 12 volts and a %VR of 90% at 2 amperes of current. What will be the terminal voltage at 2 amperes?

Solution

$$\%\text{VR} = \frac{(V_{NL} - V_L)}{V_L} \times 100\%$$

or

$$V_L = \frac{V_{NL}}{(\%\text{VR}/100\%) + 1}$$

$$V_L = \frac{12 \text{ V}}{(90\%/100\%) + 1}$$

$$V_L = 6.32 \text{ V}$$

Since the voltage drop is caused by the internal resistance, voltage regulation is also related to the internal resistance. Larger internal resistances give larger changes and larger percent voltage regulations.

When comparing the percent voltage regulation of several sources, it is important that the comparison be made for the same no-load voltage and load current. Otherwise, misleading conclusions might be obtained.

8.6 Current Sources and Source Conversion

Some electronic circuits require a source that maintains a constant current as the load changes. These are often called **current sources**. One such source was shown in Chapter 4, Figure 4.17. Actually, a practical current source can have some change. The graphic symbol for the ideal current source is shown in Figure 8.26(a). Its internal resistance is infinite ohms. The practical source shown in Figure 8.26(b) has a shunt resistor, R_p, added. Its internal resistance is R_p.

current sources

Sources with large values of R_p maintain a more constant current. In fact, a simple current source can be made by placing a large resistor in series with a voltage source. This is the opposite of the constant voltage source, where a small internal resistance gave less change in voltage. The large resistance keeps the effect of a change in load resistance small.

Short-circuit and open-circuit data will provide the values of I and R_p. For a short circuit, all of the current "I" is in the short circuit. So

$$I = I_{sc} \tag{8.6}$$

When the terminals are open-circuited, all of the current "I" is in R_p. So $V_{oc} = I R_p$. But $I = I_{sc}$, so

$$R_p = V_{oc}/I_{sc} \tag{8.7}$$

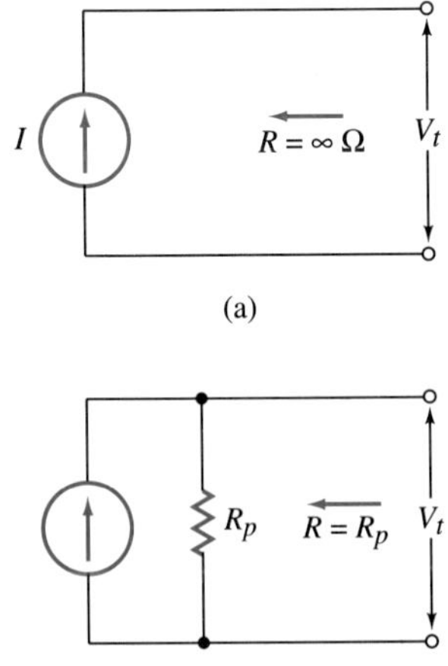

Figure 8.26 The graphic representation of a current source. (a) The ideal source provides a constant current. (b) The practical source includes a shunt resistor and the current varies with load resistance.

The voltage source in Figure 8.27(a) and current source in Figure 8.27(b) will be equivalent if their terminal voltages are the same for all currents. That is, they have the same V_{oc} and I_{sc}. Applying the short-circuit and open-circuit conditions to both sources yields

$$E = IR_p \tag{8.8}$$

$$R_i = R_p \tag{8.9}$$

$$I = E/R_i \tag{8.10}$$

where E is the constant voltage in the voltage source in volts, I is the constant current in the current source in amperes, R_i is the internal resistance of the voltage source in ohms, and R_p is the shunt resistance of the current source in ohms.

The first two equations can be used to convert a current source to a voltage source. The last two equations can be used to convert a voltage source to a current source. For both conversions, the polarity and current directions must yield the same polarity at the terminals.

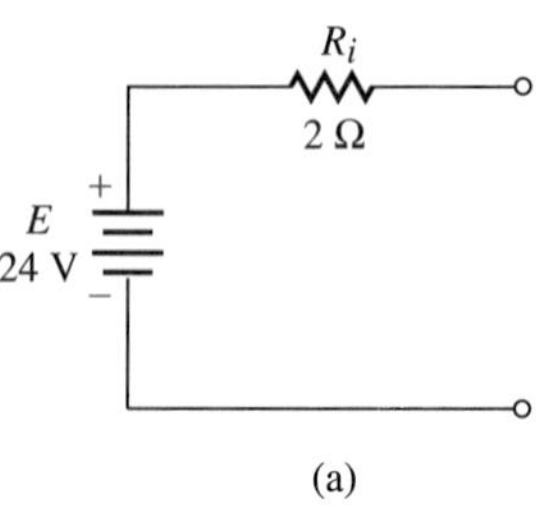

(a)

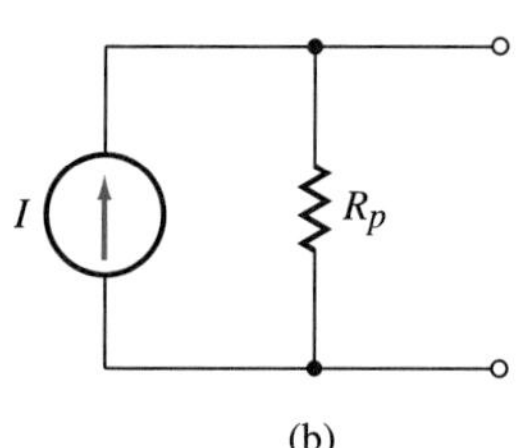

(b)

Figure 8.27

EXAMPLE 8.17

Convert the voltage source of Figure 8.27(a) to a current source.

Solution

$$I = E/R$$
$$= 24\text{ V}/2\ \Omega$$
$$I = 12\text{ A}$$
$$R_p = R_i$$
$$R_p = 2\ \Omega$$

The source with the correct current direction is drawn in Figure 8.27(b). I is 12 A and R_p is 2 Ω.

EXAMPLE 8.18

Convert the current source of Figure 8.28(a) to a voltage source.

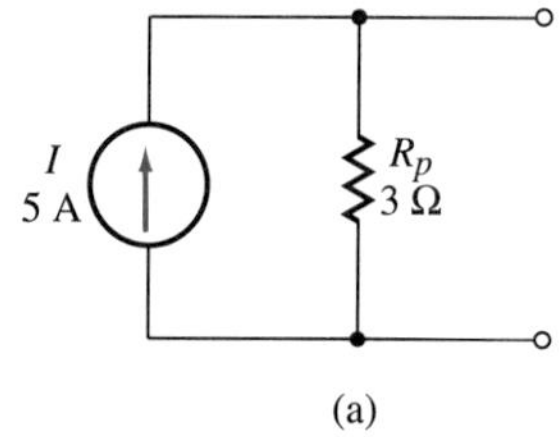

(a)

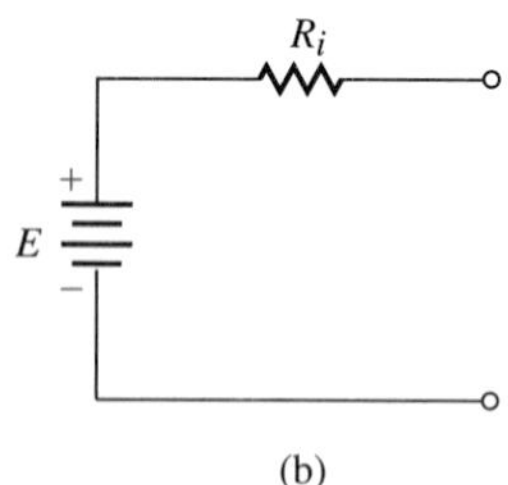

(b)

Figure 8.28

Solution

$$E = IR_p$$
$$= (5\text{ A})(3\ \Omega)$$
$$E = 15\text{ V}$$
$$R_i = R_p$$
$$R_i = 3\ \Omega$$

The source with the correct terminal polarity is drawn in Figure 8.28(b). E is 15 V and R_i is 3 Ω.

EXAMPLE 8.19

What is the current in R_L when connected to the current source shown in Figure 8.29?

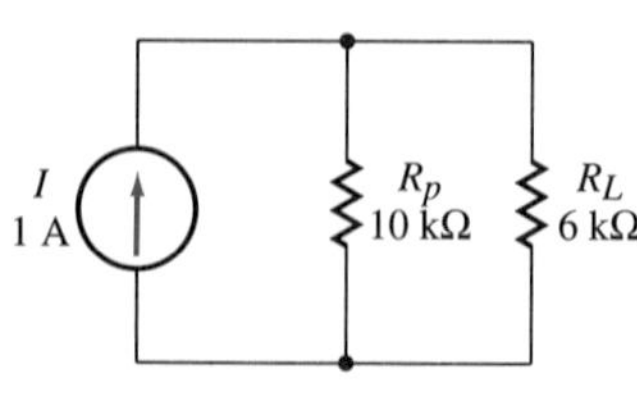

Figure 8.29

Solution

R_p and R_L are in parallel, so by the current divider relation,

$$I_L = \frac{R_p I_s}{R_p + R_L} = \frac{(10 \times 10^3\ \Omega)(1\ \text{A})}{10 \times 10^3\ \Omega + 6 \times 10^3\ \Omega}$$

$$I_L = 0.625\ \text{A}$$

8.7 Interconnection of Sources

Sources of emf can be interconnected to supply more voltage, more current, or both. Some devices in which this is done are the flashlight and the lead-acid automobile battery. When more voltage is needed, the sources must be connected in series as in Figure 8.30. Note that the (+) terminal of one source is connected to the (−) terminal of another. This leaves one (+) and (−) terminal to which the load is connected. Although the series connection increases the voltage, it does not in-

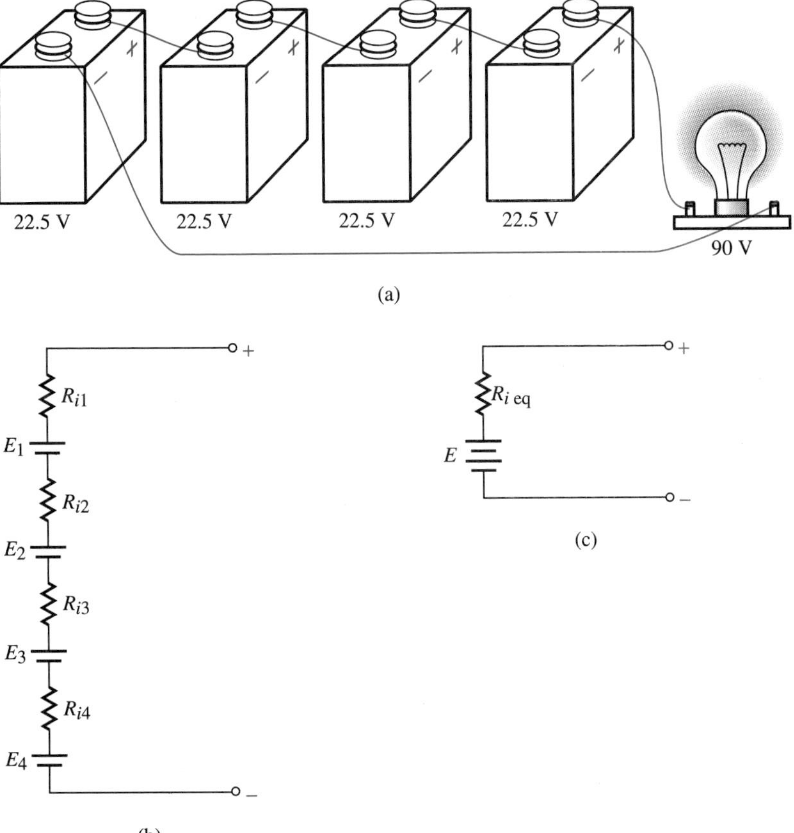

Figure 8.30 Cells are connected in series to obtain a larger voltage. (a) The (+) terminal of one cell is connected to the (−) terminal of the next cell. (b) The graphic representation of the series connection. (c) The graphic representation of the equivalent source.

crease the current capacity. The current capacity is that of the lowest capacity source.

If each source has the same internal resistance and open-circuit voltage, the characteristics of the series combination are:

$$E = nE_1 \tag{8.11}$$

$$R_{ieq} = nR_{i1} \tag{8.12}$$

$$V_t = E - I\,R_{ieq} \tag{8.13}$$

where E is the open-circuit voltage of the combination in volts, E_1 is the open-circuit voltage of one source in volts, n is the number of sources, and V_t is the terminal voltage in volts for the current I in amperes.

EXAMPLE 8.20

Three 1.5-volt cells, each with an R_i of 0.5 ohm, are connected in series. Find:

a. the open-circuit voltage and interval resistance of the combination.
b. the terminal voltage when connected to a 3-ohm resistor.

Solution

Since the cells have the same characteristics,

$$E = nE_1$$
$$= (3)(1.5 \text{ V})$$
$$E = 4.5 \text{ V}$$
$$R_{ieq} = R_i n = (3)(0.5\ \Omega)$$
$$R_{ieq} = 1.5\ \Omega$$
$$V_t = E - I\,R_{ieq}$$

but

$$I = \frac{E}{R_{ieq} + R_L}$$
$$= \frac{4.5 \text{ V}}{1.5\ \Omega + 3\ \Omega}$$
$$I = 1 \text{ A}$$

so

$$V_t = 4.5 \text{ V} - (1 \text{ A})(1.5\ \Omega)$$
$$V_t = 3 \text{ V}$$

If the sources are not similar, Equations 8.11–8.13 become

$$E = E_1 + E_2 + \cdots + E_n \tag{8.14}$$

$$R_{ieq} = R_{i1} + R_{i2} + \cdots + R_{in} \tag{8.15}$$

EXAMPLE 8.21

What is the load voltage for the series combination shown in Figure 8.31?

Solution

$$\begin{aligned} E &= E_1 + E_2 + E_3 \\ &= 12\text{ V} + 8\text{ V} + 4\text{ V} \\ E &= 24\text{ V} \end{aligned}$$

but

$$\begin{aligned} R_{ieq} &= R_1 + R_2 + R_3 \\ &= 2\ \Omega + 1\ \Omega + 1\ \Omega \\ R_{ieq} &= 4\ \Omega \end{aligned}$$

so

$$\begin{aligned} V_t &= 24\text{ V} - (0.5\text{ A})(4\ \Omega) \\ V_t &= 22\text{ V} \end{aligned}$$

Figure 8.31

When more current capacity is needed, the sources must be connected in parallel as in Figure 8.32. Here, the (+) terminals are connected together and the (−) terminals are connected together. The load is connected across a (+) and a (−) terminal.

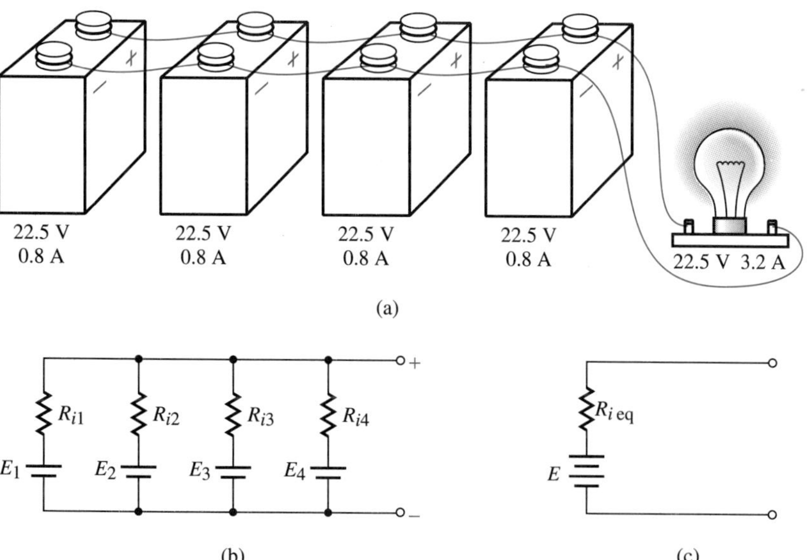

Figure 8.32 Cells are connected in parallel to obtain a higher current capacity. (a) The (+) terminals of the sources are connected together and the (−) terminals are connected together. (b) The graphic representation of the parallel combination. (c) The graphic representation of the equivalent voltage source.

Connecting sources in parallel increases the current capacity. For "n" similar sources, it is n times the capacity of one source. The voltage is that of one source.

If each source has the same open-circuit voltage and internal resistance, the characteristics of the parallel combination are

$$E = E_1 \tag{8.16}$$

$$R_{ieq} = R_{i1}/n \tag{8.17}$$

$$V_t = E - I_L R_{ieq} \tag{8.18}$$

where E is the open-circuit voltage of the combination in volts, R_i is the internal resistance of one source in ohms, n is the number of sources, R_{ieq} is the internal resistance of the combination in ohms, V_t is the terminal voltage when supplying current, and I_L is the current supplied by the combination in amperes.

EXAMPLE 8.22

Three 1.5-V cells, each with an R_i of 0.3 ohm, are connected in parallel. Find:

a. the open-circuit voltage
b. the voltage when supplying 0.5 ampere of current.

Solution

$$E = E_1$$

$$E = 1.5 \text{ V}$$

$$V_t = E - I_L R_{ieq}$$

$$R_{ieq} = R_{i1}/n = 0.3\ \Omega/3 = 0.1\ \Omega$$

so

$$V_t = 1.5 \text{ V} - (0.5 \text{ A})(0.1\ \Omega)$$

$$V_t = 1.45 \text{ V}$$

Generally, connecting sources with different characteristics in parallel is not advised. This results in circulating currents that can damage a source. Should different sources be connected in parallel, equations 8.16–8.18 become

$$E = \left(\frac{E_1}{R_{i1}} + \frac{E_2}{R_{i2}} + \cdots + \frac{E_n}{R_{in}}\right)(R_{ieq}) \tag{8.19}$$

$$\frac{1}{R_{ieq}} = \frac{1}{R_{i1}} + \frac{1}{R_{i2}} + \cdots + \frac{1}{R_{in}} \tag{8.20}$$

$$V_t = \left(\frac{E_1}{R_{i1}} + \frac{E_2}{R_{i2}} + \cdots + \frac{E_n}{R_{in}}\right)(R_{ieq} \| R_L) \qquad \textbf{(8.21)}$$

where R_{ieq}, E, and V_t have the same meaning as in Equations 8.16–8.18 and $E_1, E_2, \ldots, E_n$ are the open-circuit voltages of each source, in volts.

The relations in Equations 8.19–8.21 are obtained by converting the voltage sources to current sources, using Millman's Theorem to combine the current sources, then converting the current source to a voltage source.

Millman's Theorem

Millman's Theorem states that a number of parallel current sources can be replaced by a single current source. The current is equal to the algebraic sum of the source currents. The internal resistance is equal to the parallel equivalent of the sources' internal resistance.

EXAMPLE 8.23

What are the terminal voltage and the open-circuit voltage of the parallel sources in Figure 8.33?

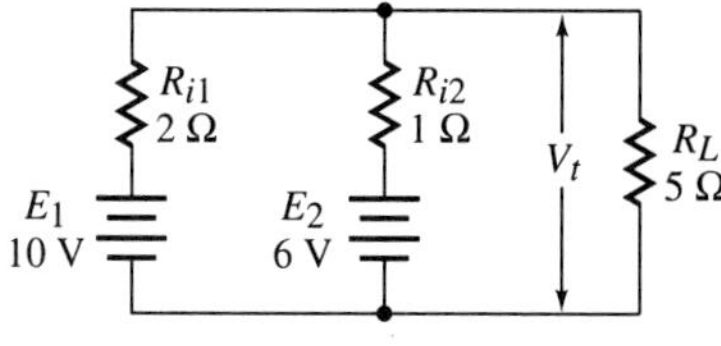

Figure 8.33

Solution

For the terminal voltage,

$$V_t = \left(\frac{E_1}{R_{i1}} + \frac{E_2}{R_{i2}}\right)(R_{ieq} \| R_L)$$

$$R_{ieq} = \frac{R_{i1}R_{i2}}{R_{i1} + R_{i2}} = \frac{(2\ \Omega)(1\ \Omega)}{2\ \Omega + 1\ \Omega}$$

$$R_{ieq} = 0.67\ \Omega$$

$$R_{ieq} \| R_L = \frac{(R_{ieq}(R_L)}{R_{ieq} + R_L} = \frac{(0.67\ \Omega)(5\ \Omega)}{0.67\ \Omega + 5\ \Omega} = 0.59\ \Omega$$

so

$$V_t = \left(\frac{10\ \text{V}}{2\ \Omega} + \frac{6\ \text{V}}{1\ \Omega}\right)(0.59\ \Omega)$$

$$V_t = 6.49\ \text{V}$$

For the open-circuit voltage,

$$E = \left(\frac{E_1}{R_{i1}} + \frac{E_2}{R_{i2}}\right)(R_{ieq}) = \left(\frac{10\ \text{V}}{2\ \Omega} + \frac{6\ \text{V}}{1\ \Omega}\right)(0.67\ \Omega) = 7.37\ \text{V}$$

$$E = 7.37\ \text{V}$$

When both voltage and current must be increased, a series-parallel connection must be used. For example, the Eveready 706, 6-V battery has four parallel groups, each one having four sources in series.

The terminal voltage and other characteristics of the series-parallel source can be found by applying the series and parallel connection concepts.

EXAMPLE 8.24

Find the no-load voltage and the voltage at 2 amperes for the combination shown in Figure 8.34.

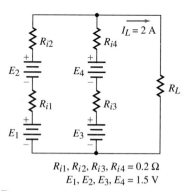

$R_{i1}, R_{i2}, R_{i3}, R_{i4} = 0.2\ \Omega$
$E_1, E_2, E_3, E_4 = 1.5\ \text{V}$

Figure 8.34

Solution

Applying the series relation to the series branches gives:
For Branch A:

$$E_A = E_1 + E_2 = 1.5\ \text{V} + 1.5\ \text{V}$$
$$E_A = 3\ \text{V}$$
$$R_{ieq} = R_{i1} + R_{i2} = 0.2 + 0.2$$
$$R_{ieq} = 0.4\ \Omega$$

Branch B will be the same as Branch A.
For the two parallel branches, at no load,

$$E = E_A = E_B$$
$$E = 3\ \text{V}$$
$$R_{ieq} = R_{iA} \| R_{iB}$$
$$= \frac{(0.4\ \Omega)(0.4\ \Omega)}{0.4\ \Omega + 0.4\ \Omega}$$

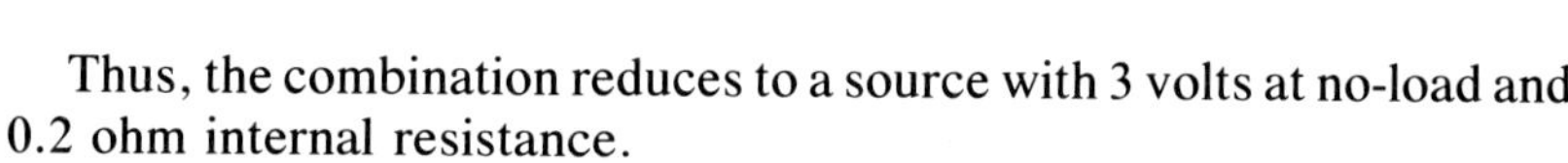

Thus, the combination reduces to a source with 3 volts at no-load and 0.2 ohm internal resistance.

At 2 amperes,

$$V_t = E - I_L R_{ieq}$$
$$V_t = 3\ \text{V} - (2\ \text{A})(0.2\ \Omega)$$
$$V_t = 2.6\ \text{V}$$

S 8.8 SPICE and Series-Parallel Circuit

Performing a dc analysis of a series-parallel circuit using PSpice is similar to our PSpice analyses of series and parallel circuits. The steps used in creating a source file are the same. Here are two examples.

EXAMPLE 8.25

Using PSpice, find the node voltages and currents flowing through each circuit component for the circuit shown in Figure 8.35.

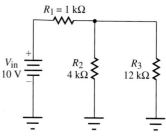

$R_1 = 1\ \text{k}\Omega$, V_{in} 10 V, R_2 4 kΩ, R_3 12 kΩ

Figure 8.35

Solution

The node assignments are shown in Figure 8.36. There are three nodes in this circuit. Node 0 is assigned to the reference or ground node. Node 1 can be assigned to the positive terminal of the voltage source, and Node 2 to the junction of the three resistors.

The desired outputs from our analysis are the voltage at Node 2 and the currents through three resistors—I(R1), I(R2), and I(R3).

The SPICE source file can be written as follows:

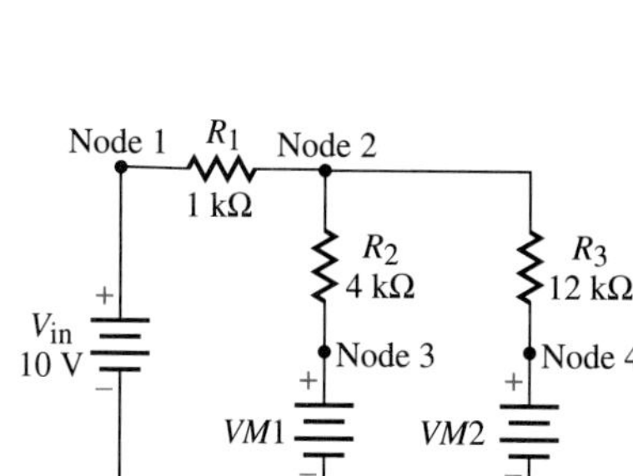

Figure 8.36 Node assignments.

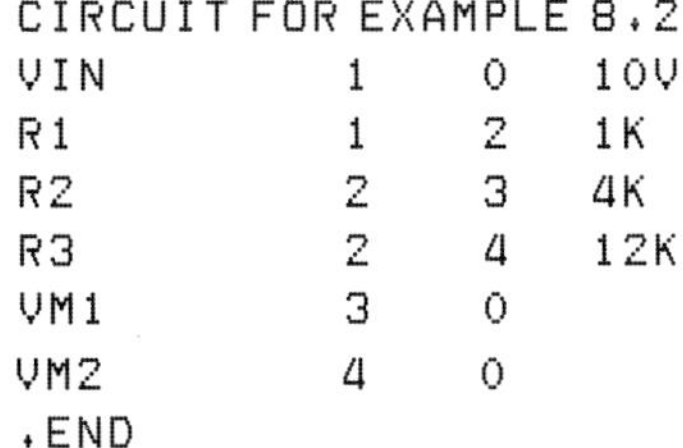

```
CIRCUIT FOR EXAMPLE 8.25
VIN     1   0   10V
R1      1   2   1K
R2      2   3   4K
R3      2   4   12K
VM1     3   0
VM2     4   0
.END
```

Let us examine each line in our file. The first line is our title line. We can give the file any name that we desire. The second line specifies our input source. In this case, the input is a 10-volt dc voltage. This input source is connected between Node 1 and Node 0. The next three lines specify the connections and values of the three resistors making up the circuit. For example, resistor R1 is connected between Node 1 and Node 2 and has a value of 1 kilohm.

The two voltage sources, VM1 and VM2, each with a value of zero volts, are placed in series with R2 and R3, respectively. When PSpice specifies the dc analysis, the current flowing through each voltage source will automatically be calculated.

And, of course, the .END is the last command and completes our file.

The results of a PSpice simulation are shown in Figure 8.37. Three currents are given in the OUTPUT file. The current associated with VIN is the total current flowing in the circuit and is the current that flows through resistor R_1. The negative sign indicates that the current flows from Node 1 to Node 2. Note that Kirchhoff's Current Law is verified. That is,

$$I_{VIN} + I_{VM1} + I_{VM2} = 0$$

And finally, the total power dissipated in the circuit is 25 mW.

```
PSPICE CIRCUIT EXAMPLE 8.25
****      SMALL SIGNAL BIAS SOLUTION      TEMPERATURE = 27.000 DEG C
****************************************

 NODE     VOLTAGE     NODE     VOLTAGE     NODE     VOLTAGE     NODE     VOLTAGE

(   1)    10.0000    (   2)     7.5000    (   3)     0.0000    (   4)     0.0000

VOLTAGE SOURCE CURRENTS
NAME          CURRENT

VIN          -2.500E-03
VM1           1.875E-03
VM2           6.250E-04

TOTAL POWER DISSIPATION     2.50E-02   WATTS
```

Figure 8.37 Output file.

EXAMPLE 8.26

Perform a dc analysis for the circuit shown in Figure 8.38.

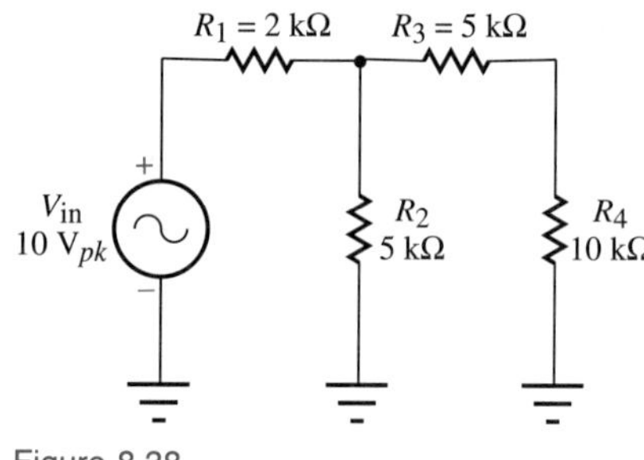

Figure 8.38

Solution

Again, we start by assigning an identifier to each node. In this circuit, there are four nodes. If we also want to find the currents in R_2 and the R_3–R_4 series combination, we need to add two additional nodes and two 0-volt sources to act as ammeters. The node assignments are shown in Figure 8.39.

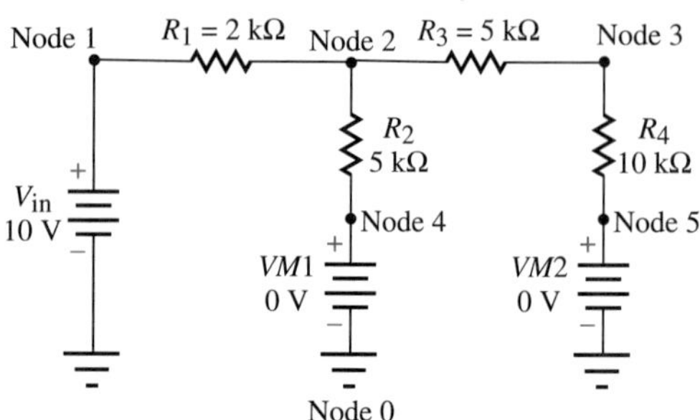

Figure 8.39 Node assignments.

The source file can now be written.

```
CIRCUIT FOR EXAMPLE 8.26
VIN        1     0     10
R1         1     2     2K
R2         2     4     5K
R3         2     3     5K
R4         3     5     10K
VM1        4     0
VM2        5     0
.END
```

The results of the simulation are shown in Figure 8.40.

There are three major nodes in this circuit. Each node voltage is given. Then the three currents are listed. Again, it is always a good idea to ask if the results obtained make sense.

```
PSPICE CIRCUIT EXAMPLE 8.26
****      SMALL SIGNAL BIAS SOLUTION     TEMPERATURE = 27.000 DEG C
****************************************

 NODE      VOLTAGE      NODE      VOLTAGE      NODE      VOLTAGE      NODE      VOLTAGE

(    1)    10.0000     (    2)    6.5217     (    3)    4.3478     (    4)    0.0000
(    5)     0.0000

VOLTAGE SOURCE CURRENTS
NAME          CURRENT

VIN          -1.739E-03
VM1           1.304E-03
VM2           4.348E-04

TOTAL POWER DISSIPATION     1.74E-02    WATTS
```

Figure 8.40 Output file.

SUMMARY

1. Many practical circuits have groups of series and/or parallel resistors connected in series and/or parallel. Such circuits are series-parallel circuits.
2. The equivalent resistance of a series-parallel circuit is found by replacing each series or parallel combination until only one resistance remains.
3. The key to the analysis of a series-parallel circuit is to develop a sequence that relates the wanted quantities to the given ones. The unknowns in the equations can be used as a guide to the solution.
4. The synthesis of a series-parallel circuit requires the selection of circuit relations to provide enough equations to solve for the wanted

quantities. Here too, the unknowns can be used as a guide.

5. Voltages in electronic and other low-current circuits are obtained from voltage dividers. These can provide single outputs, multiple outputs, and a positive and negative voltage.
6. The part of the voltage divider that does not have any load current in it is the bleeder resistor. Smaller values of bleeder resistance provide a more stable output but result in a lower efficiency.
7. Connecting an ammeter or voltmeter in a circuit changes the conditions in the circuit. This loading effect can result in incorrect measurements and even damage to parts of the circuit.
8. Practical sources of emf have internal resistance. This internal resistance results in a voltage drop when the source supplies current.
9. The graphic symbol for a practical voltage source is an ideal source with a series resistor to represent the internal resistance.
10. The voltage regulation of a source of emf is a measure of how much the voltage changes when the source supplies current.
11. The graphic symbol for an ideal current source is a constant current generator. It has a resistance of infinite ohms.
12. The practical current source graphic symbol is a constant current generator with a shunt resistance.
13. Conversion from one source to another is possible by selecting the values that give the same terminal voltages for all currents.
14. Sources of emf and power supplies can be connected in series for more voltage, in parallel for more current capacity, and in series-parallel for both.
15. PSpice can be used for the analysis of series-parallel circuits, voltage dividers, and interconnected sources.

EQUATIONS

8.1	$\frac{V_{\text{out}}}{V_{\text{in}}} = \frac{R_B}{R_A + R_B}$	Transfer function for the unloaded voltage divider.	294
8.2	$\frac{V_{\text{out}}}{V_{\text{in}}} = \frac{R_B R_L}{R_A R_B + R_A R_L + R_B R_L}$	Transfer function for the loaded voltage divider.	294
8.3	$V_t = E - IR_i$	Terminal voltage of a source.	303
8.4	$R_i = \frac{E - V_t}{I}$	Internal resistance of a source.	304
8.5	$\%\text{VR} = \frac{(V_{NL} - V_L)}{V_L} \times 100\%$	Percent voltage regulation of a source.	305
8.6	$I = I_{sc}$	Relation of current source current to short-circuit current.	307
8.7	$R_p = \frac{V_{oc}}{I_{sc}}$	Relation of the shunt resistance of a current source to the open-circuit voltage and short-circuit current.	307

8.8	$E = IR_p$	Relation of the emf of a voltage source to the current source quantities.	308
8.9	$R_i = R_p$	Relation of the resistance of a voltage source to the shunt resistance of a current source.	308
8.10	$I = \dfrac{E}{R_i}$	Relation of the current of a current source to the quantities of a voltage source.	308
8.11	$E = nE_1$	The open-circuit voltage of series-connected sources having the same voltage.	310
8.12	$R_{ieq} = nR_{i1}$	The internal resistance of series-connected sources having the same internal resistance.	310
8.13	$V_t = E - IR_{ieq}$	The terminal voltage of series-connected sources.	310
8.14	$E = E_1 + E_2 + \cdots + E_n$	The open-circuit voltages of series-connected sources with different open-circuit voltages.	310
8.15	$R_{ieq} = R_{i1} + R_{i2} + \cdots R_{in}$	The internal resistance of series-connected sources with different internal resistances.	310
8.16	$E = E_1$	The open-circuit voltage of parallel-connected sources having the same voltage.	312
8.17	$R_{ieq} = \dfrac{R_{i1}}{n}$	The shunt resistance of parallel-connected sources having the same internal resistance.	312
8.18	$V_t = E - I_L R_{ieq}$	The terminal voltage of parallel-connected sources.	312
8.19	$E = \left(\dfrac{E_i}{R_{i1}} + \dfrac{E_2}{R_{i2}} + \cdots + \dfrac{E_n}{R_{in}}\right)(R_{ieq})$	The open-circuit voltage of parallel-connected sources having different open-circuit voltages and internal resistances.	312
8.20	$\dfrac{1}{R_{ieq}} = \dfrac{1}{R_{i1}} + \dfrac{1}{R_{i2}} + \cdots + \dfrac{1}{R_{in}}$	The shunt resistance of parallel-connected sources having different internal resistances.	312
8.21	$V_t = \left(\dfrac{E_1}{R_{i1}} + \dfrac{E_2}{R_{i2}} + \cdots + \dfrac{E_n}{R_{in}}\right)(R_{ieq} \| R_L)$	The terminal voltage of parallel-connected sources with different open-circuit voltages and internal resistance.	313

QUESTIONS

1. How does one go about calculating the equivalent resistance of a series-parallel circuit?
2. Why is it not possible to develop one equation for the equivalent resistance of a series-parallel circuit?
3. What is the difference between synthesis and analysis?
4. The no-load voltage of Source A is 24 volts and drops to 20 volts at 2 amperes. Source B drops to 22 volts at 2 amperes. Which source has the larger internal resistance? Which source has the larger percent voltage regulation?
5. Which part of a voltage divider has the largest current in it: the bleeder resistance or the top part?
6. A voltage divider with a bleeder resistance of 500 ohms and a no-load voltage of 8 volts provides 6 volts to a load. What would happen to the load voltage if the bleeder resistance were 100 ohms and the no-load voltage was kept at 8 volts?
7. How will a decrease in the load resistance affect the output voltage of a voltage divider?
8. What are some applications that use voltage dividers?
9. What would be one reason for not using a voltage divider to supply an output of 10 amperes?
10. A voltage divider supplies 6 volts to a load. What would happen to the voltage if the bleeder resistor opened? If it were short-circuited?
11. A voltage divider is connected to a 12-volt source. What is the range of voltages that can be obtained from the divider?
12. What is meant by the loading effect of a meter?
13. Does the loading effect of an ammeter result in a reading lower than the current in the circuit without the meter connected?
14. Is it better to use a voltmeter that has a very high resistance or one that has a low resistance?
15. Is it better to use an ammeter that has a very high resistance or one that has a low resistance?
16. How can parts of a circuit be damaged by the loading effect of a voltmeter?
17. Does the loading effect of a voltmeter result in a reading lower than, or higher than, the voltage that is across the points without the meter connected?
18. Voltmeter A indicates 5 volts across two points. Voltmeter B indicates 5.1 volts. Assuming each meter is calibrated and reads correctly, which reading is nearest to the voltage across the points?
19. How can the internal resistance be obtained from voltage and current measurements?
20. Will the voltage regulation of a source increase or decrease if the internal resistance is made smaller?
21. Why is it not recommended to measure the internal resistance of a battery with an ohmmeter?
22. How will a decrease in the load resistance affect the terminal voltage of a power supply?
23. Does the internal resistance of a battery increase or decrease as the battery is used?

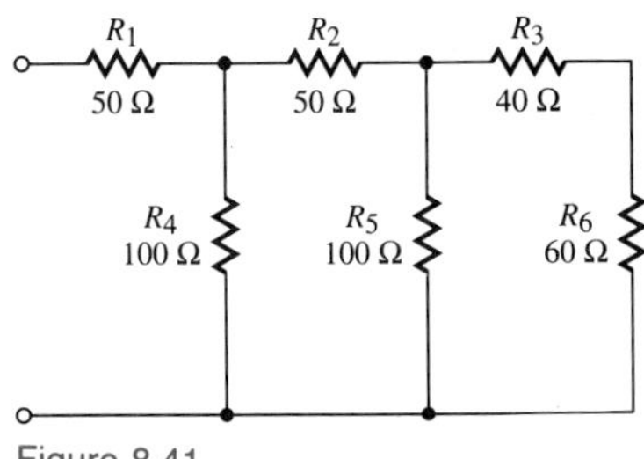

Figure 8.41

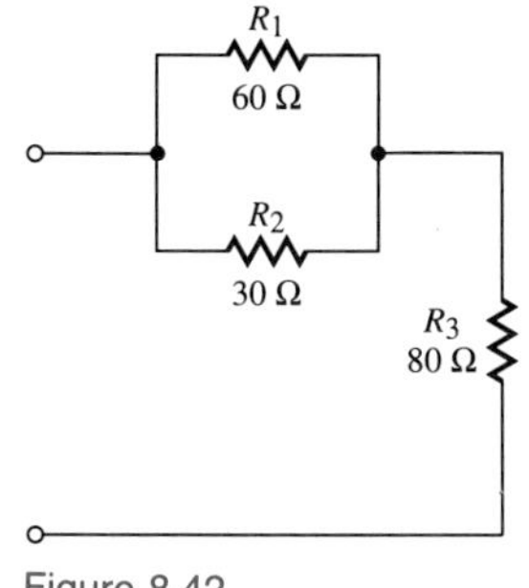

Figure 8.42

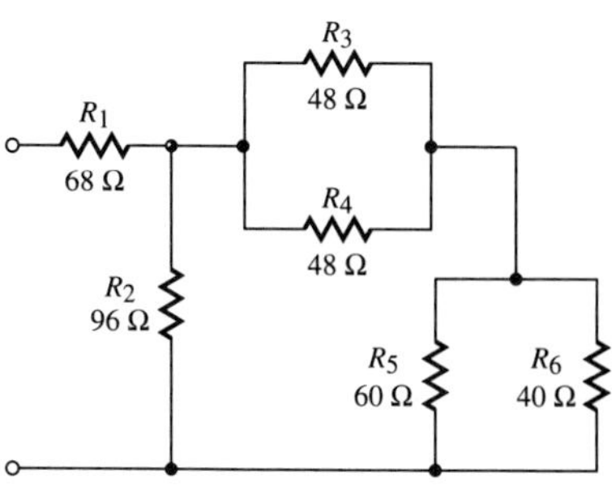

Figure 8.43

24. What effect does the internal resistance of a source have on the terminal voltage?
25. What makes up the internal resistance of a generator?
26. What makes up the internal resistance of a battery?
27. What can one say about the internal resistance of a voltage source if the slope of the V–I curve is constant?
28. What relation between the terminal voltage of a current source and a voltage source must exist if they are to be equivalent?
29. Why can it be said that the current in R_p of the current source is 0 amperes when the terminals are short-circuited?
30. What kind of source connection should be used to get 3 volts from 1.5-volt cells?
31. What kind of source connection should be used to get a higher current capacity?
32. Draw the connection diagram to obtain 4.5 volts and three times current capacity of a 1.5-volt cell from a group of 1.5-volt cells.

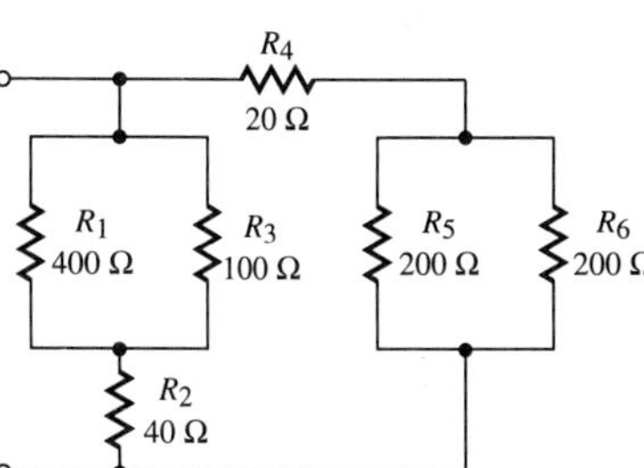

Figure 8.44

PROBLEMS

SECTION 8.1 The Circuit and Its Equivalent Resistance

1. Find the equivalent resistance of the circuit in Figure 8.41.
2. Find the equivalent resistance of the circuit in Figure 8.42.
3. What is the equivalent resistance of the circuit in Figure 8.43?
4. What is the equivalent resistance of the circuit in Figure 8.44?
5. How many ohms is R_3 if the equivalent resistance of the circuit in Figure 8.45 is 12 ohms?
6. Determine the resistance of R_L in Figure 8.46 if R_{eq} is 1300 ohms.
7. The equivalent resistance of the circuit in Figure 8.47 was 75 ohms before the lamp was damaged. What was the resistance of the lamp?

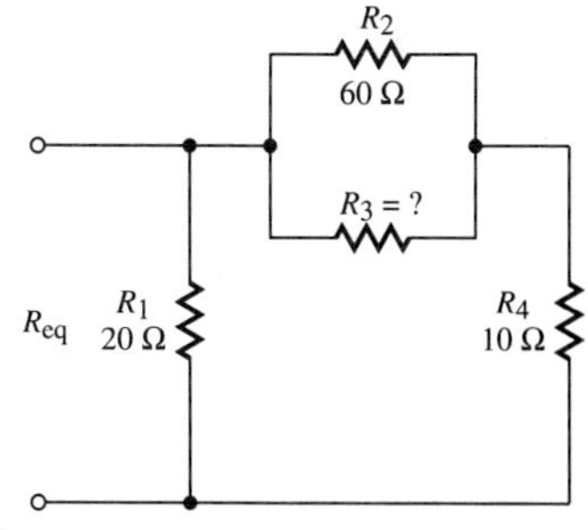

Figure 8.45

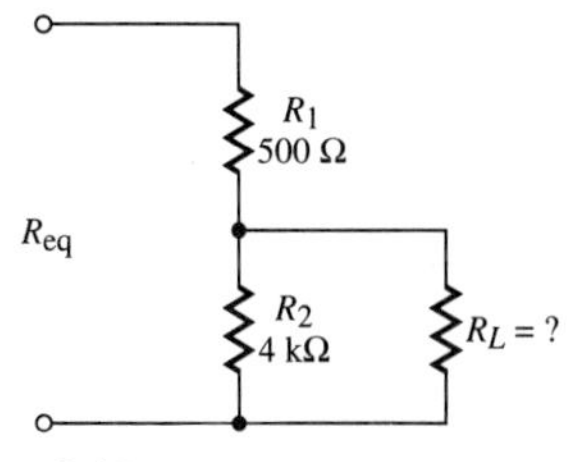

Figure 8.46

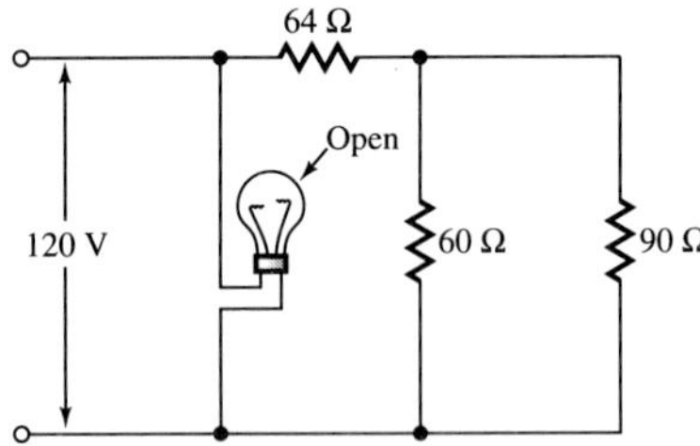

Figure 8.47

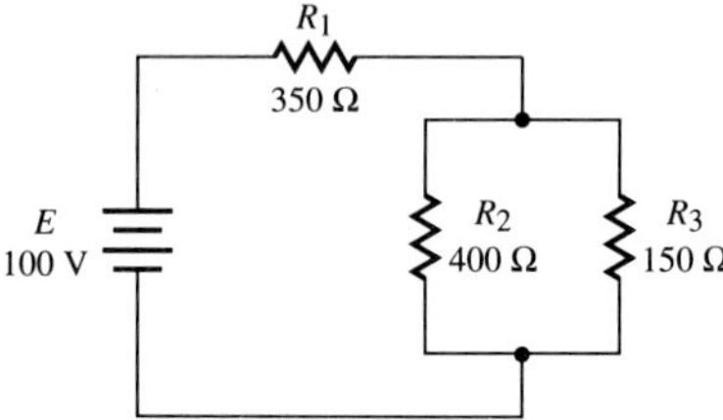

Figure 8.48

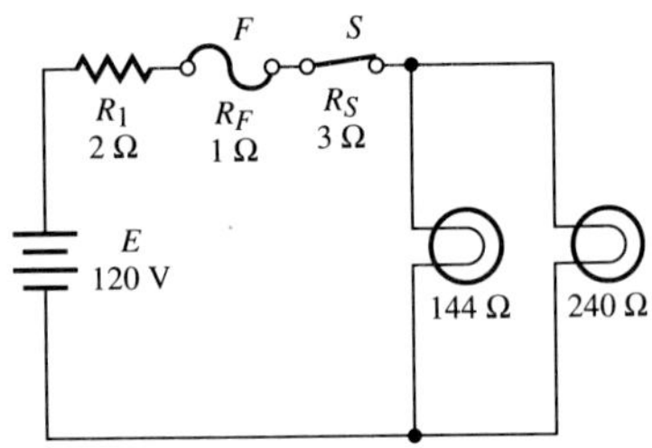

Figure 8.49

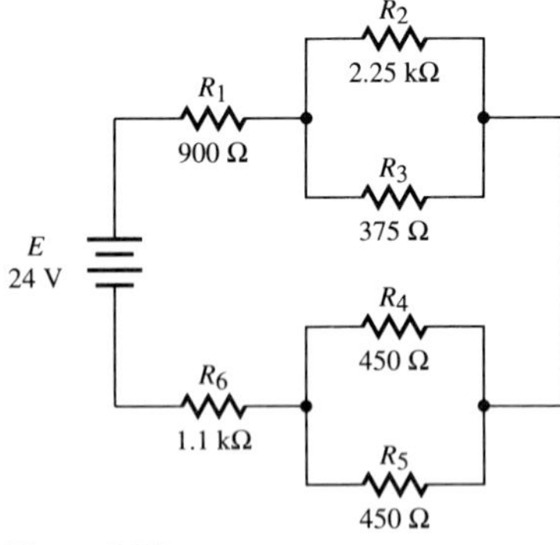

Figure 8.50

SECTION 8.2 Illustrative Problems

8. In the circuit in Figure 8.48, what is:

   (a) The source current.
   (b) The voltage across R_1.
   (c) The power in R_2.

9. Two lamps are connected in a circuit, as shown in Figure 8.49. The switch resistance is 3 ohms and the fuse resistance is 1 ohm. Find:

   (a) The minimum current rating that the fuse should have for the load.
   (b) The voltage across the lamps.
   (c) The power supplied by the source.
   (d) The power dissipated in each lamp.

10. Given the circuit in Figure 8.50, find:

    (a) The voltage across R_2.
    (b) The current in R_4.
    (c) The power dissipated in the circuit.

11. In the circuit in Figure 8.51, find:

    (a) The source current.
    (b) The voltage across each resistor.
    (c) The current in each resistor.
    (d) The power supplied by the source.

12. A 6-volt bulb and a 6-volt motor are to be operated from a 12-volt battery. If the connection is as shown in Figure 8.52, what must be the value of R_x?

13. The voltage across the lamp in Figure 8.53 must be at least 25 volts for it to light.

    (a) Is the lamp lit?
    (b) How much power is supplied by the source?
    (c) What is the voltage across R_5?

14. In the circuit of Figure 8.54, find:

    (a) The equivalent resistance.
    (b) The voltage across R_5.

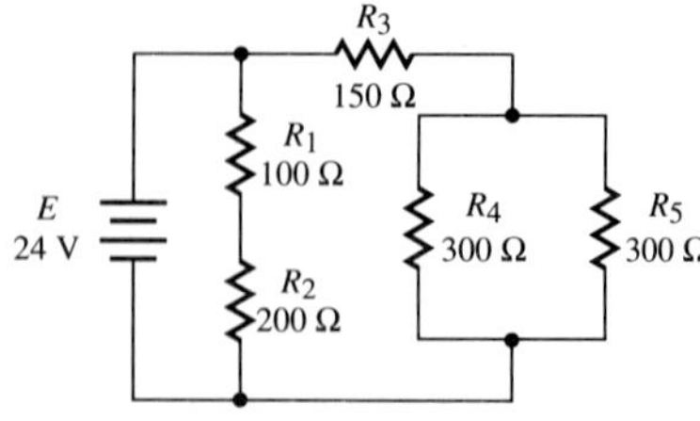

Figure 8.51

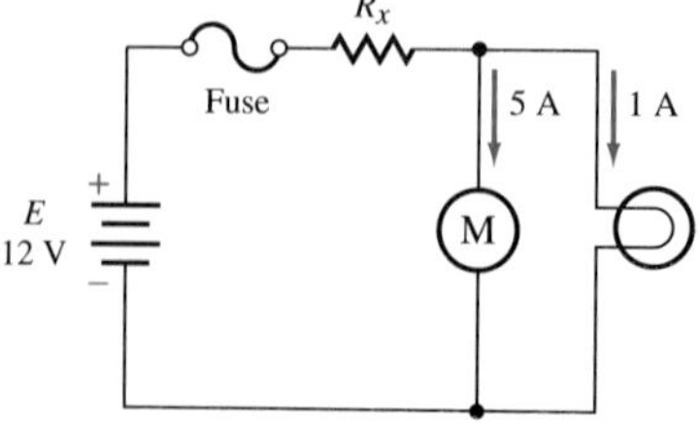

Figure 8.52

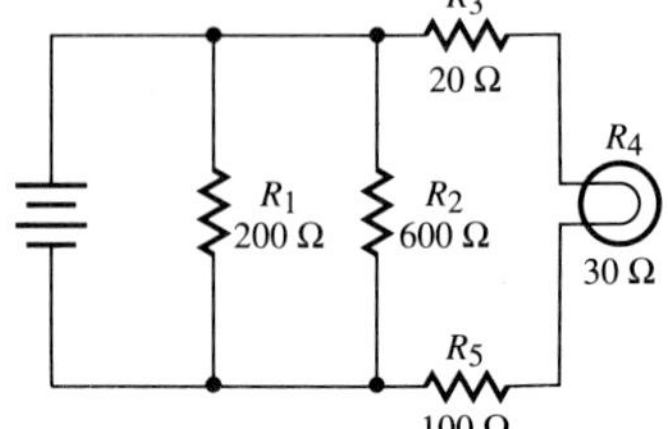

Figure 8.53

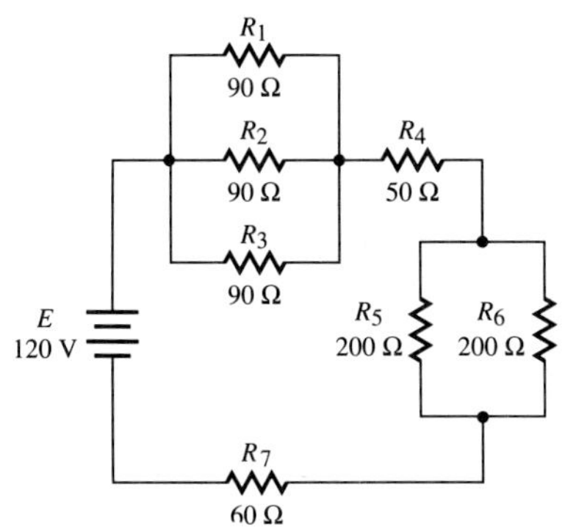

Figure 8.54

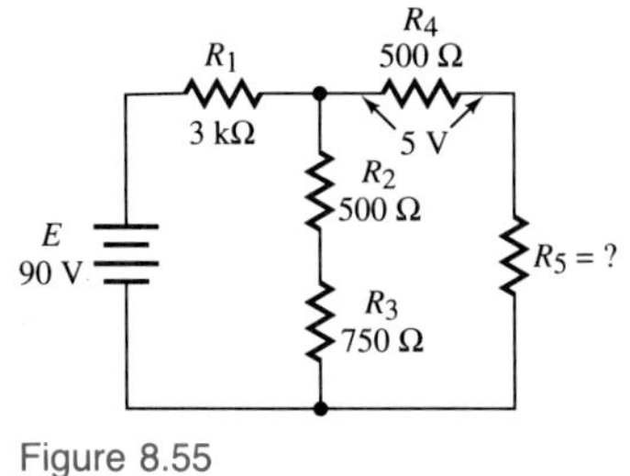

Figure 8.55

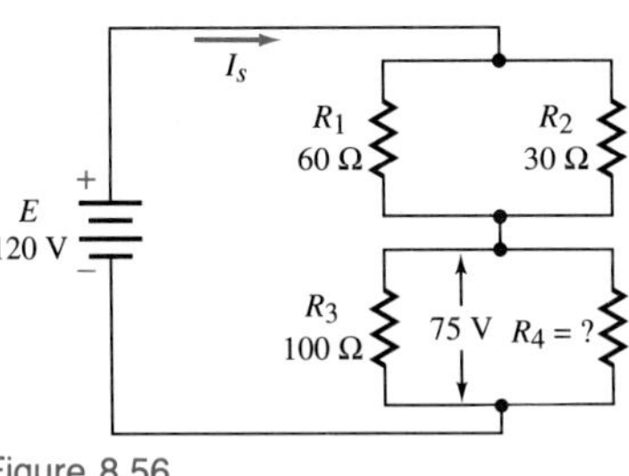

Figure 8.56

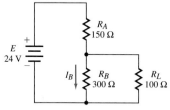

(c) The current in R_3.

(d) The power dissipated in the R_1, R_2, R_3 group.

(e) The power dissipated in the R_5, R_6 group.

15. What must be the resistance of R_5 in the circuit in Figure 8.55?

16. In the circuit in Figure 8.56, find:

(a) The value of R_4.

(b) The current I_s.

SECTION 8.3 The Loaded Voltage Divider

17. In the voltage divider in Figure 8.57, find:

(a) The voltage across R_L.

(b) The current supplied by the source.

(c) The bleeder current.

(d) The power rating of R_A and R_B.

18. A potentiometer is connect as a voltage divider as in the circuit shown in Figure 8.58. Find:

(a) The source current.

(b) The resistance of R_A.

(c) The power rating of the potentiometer.

(d) The % efficiency of the divider.

19. A voltage divider is connected as shown in Figure 8.59. Find:

(a) The value of R_L.

(b) The voltage across R_L.

(c) The power lost in the divider.

20. A 500-ohm potentiometer in Figure 8.60 is connected to give 24 volts across a resistor at 0.167 ampere of current. What will the voltage be if some of R_L shorts out, making the current 0.2 A?

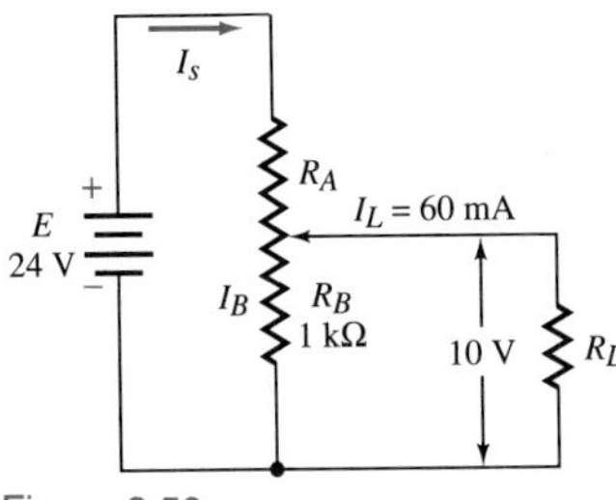

Figure 8.57

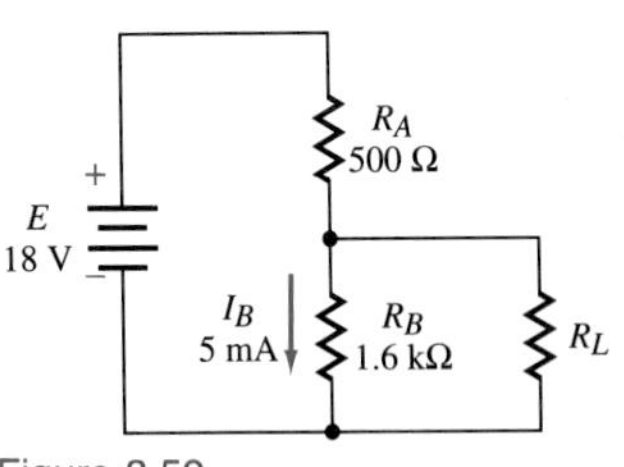

Figure 8.58

Figure 8.59

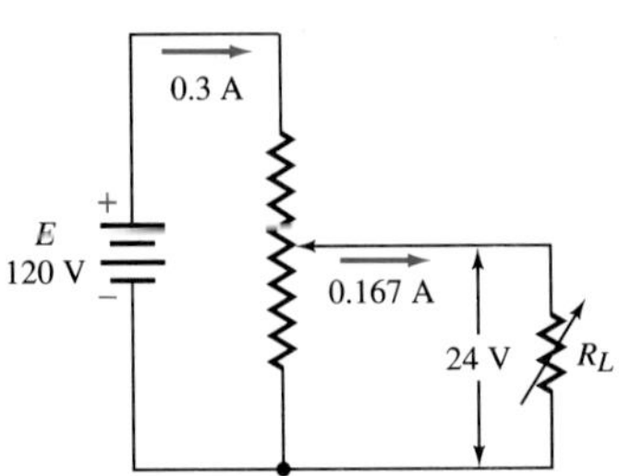

Figure 8.60

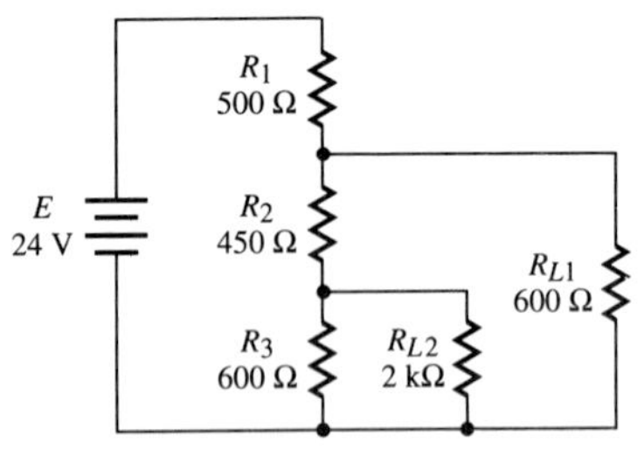

Figure 8.61

21. A tapped resistor is used to provide voltage for two loads, as shown in Figure 8.61. Find:

    (a) The voltage across each load.
    (b) The current in each load.
    (c) The power in each section of the divider.
    (d) The voltage acros R_{L2} if R_{L1} opens.

22. Design a single-output voltage divider for the following conditions:

Total resistance	180 ohms
Input voltage	12 volts
Load voltage	6 volts at 50 mA

    What must be the power rating of the R_A part of the divider?

23. Design a voltage divider for the following conditions:

Load voltage	8 volts at 100 mA
Bleeder current	16 mA
Input voltage	16 volts

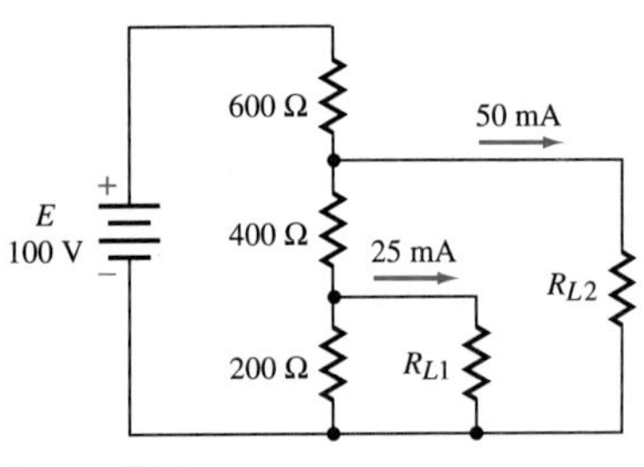

Figure 8.62

24. For the voltage divider in Figure 8.62, find:
    (a) The voltage across each load.
    (b) Power supplied by the source.
    (c) The power rating of the resistor if it is a tapped resistor.

SECTION 8.4 The Loading Effect of Meters

25. A series circuit of a 5000-ohm resistor and a 10,000-ohm resistor is connected across a 120-volt source. The voltage across the 10,000-ohm resistor is measured with a 0.8-MΩ voltmeter.

    (a) What is the actual voltage across the resistor without the meters in the circuit?
    (b) How many volts does the meter indicate?

26. Work Problem 25 for a 3.2-MΩ voltmeter.

27. A 30-ohm resistor and a 5-ohm milliammeter are connected in series across a 6-volt source. Find:

    (a) The current indicated on the meter.
    (b) The current if the ammeter were not in the circuit.

28. Work Problem 27 for a 10-ohm milliammeter.

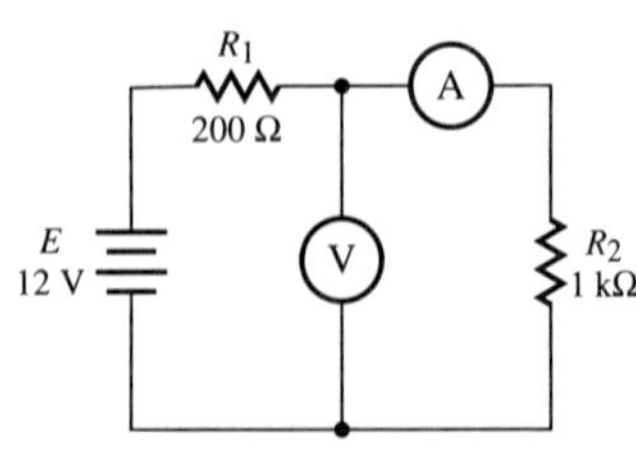

Figure 8.63

29. A 10-ohm milliammeter and a 16-kilohm voltmeter are used to measure current and voltage in the circuit of Figure 8.63.

    (a) What will be the readings on the meters?
    (b) What are the actual current and voltage in the circuit (with the meters out of the circuit)?

SECTION 8.5 Internal Resistance and Voltage Regulation

30. A power supply has an open-circuit voltage of 120 volts and an internal resistance of 10 ohms. What will the terminal voltage be when supplying 2 amperes of current?

31. A dc supply with an internal resistance of 5 ohms has terminal voltage of 20 volts at 0.8 ampere of current. What will the voltage be when the load is disconnected?

32. A source supplies 0.5 ampere of current at 40 volts. Its open-circuit voltage is 45 volts. What is the internal resistance of the source?

33. A source has a constant internal resistance. At 5 amperes, its voltage is 100 volts. At 10 amperes, its voltage is 80 volts. What is its internal resistance?

34. What is the percent voltage regulation of a generator if the voltage drops from 24 to 20 volts when the load is connected?

35. What must the voltage regulation of a 60-volt source be if its voltage is to stay about 50 volts at 3 amperes of current?

36. What will be the no-load voltage of a source if it has a 5% voltage regulation and a load voltage of 100 volts?

37. A 120-volt generator has an internal resistance of 6 ohms. What is its voltage regulation at 2 amperes of current?

38. A dc power supply has a no-load voltage of 60 volts and a voltage regulation of 10% at 0.5 ampere. Find:

   (a) Its voltage at 0.5 ampere.

   (b) Its internal resistance.

SECTION 8.6 Current Sources and Source Conversion

39. Convert the current source of Figure 8.64 to a voltage source.

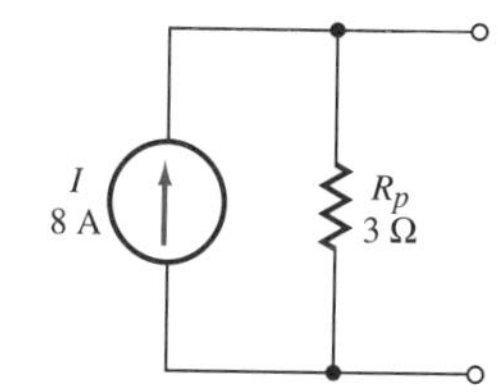

Figure 8.64

40. Repeat Problem 39 for R_p = 4 ohms, I = 6 amperes, and the direction of I reversed from that in Problem 39.

41. Convert the voltage source of Figure 8.65 to a current source.

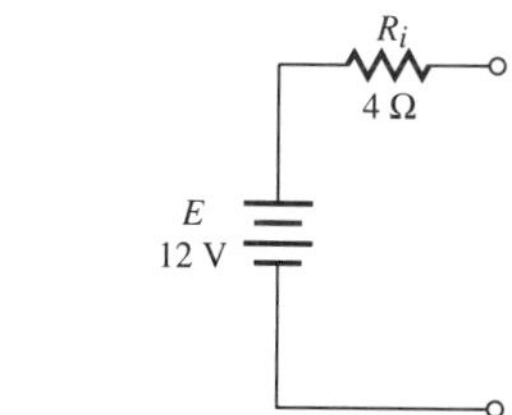

Figure 8.65

42. Repeat Problem 41 for R_i = 2 ohms, E = 24 volts, and the polarity of E reversed from that in Problem 41.

43. Measurements on a power supply gave the following data:

Short-circuit current	10 A
Open-circuit voltage	60 V

   (a) Determine the values of I and R_p for the equivalent current source.

   (b) Determine the values of E and R_i for the equivalent voltage source.

44. The short-circuit current for a power supply is 3 amperes and it has an internal resistance of 2 ohms.

   (a) Determine the values of E and R_i for the equivalent voltage source.

   (b) Determine the values of I and R_p for the equivalent current source.

45. The no-load voltage of a source is 24 volts. Its voltage at 4 amperes is 20 volts.

   (a) Determine the values of E and R_i for the equivalent voltage source.

   (b) Determine the values of I and R_p for the equivalent current source.

46. Convert the sources in Figure 8.66 so that the circuit consists of only voltage sources.

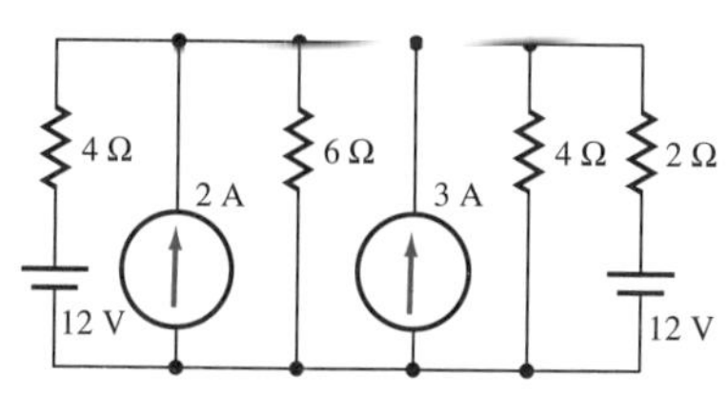

Figure 8.66

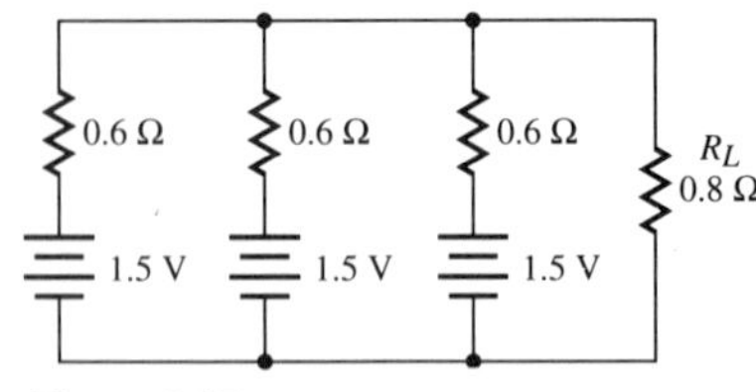

Figure 8.67

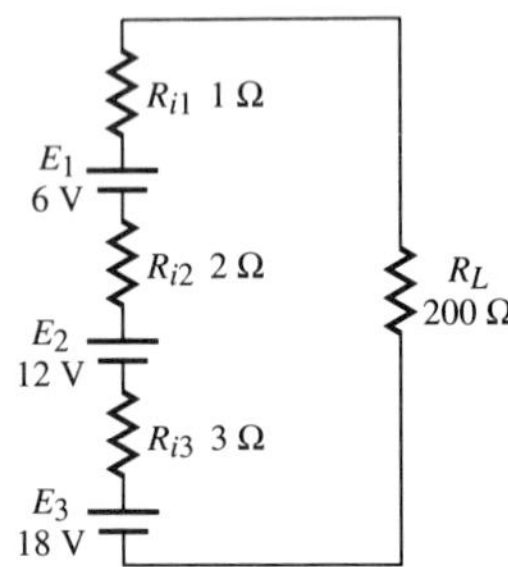

Figure 8.68

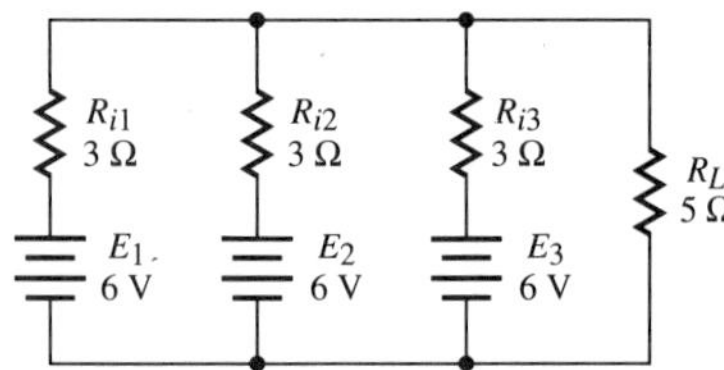

Figure 8.69

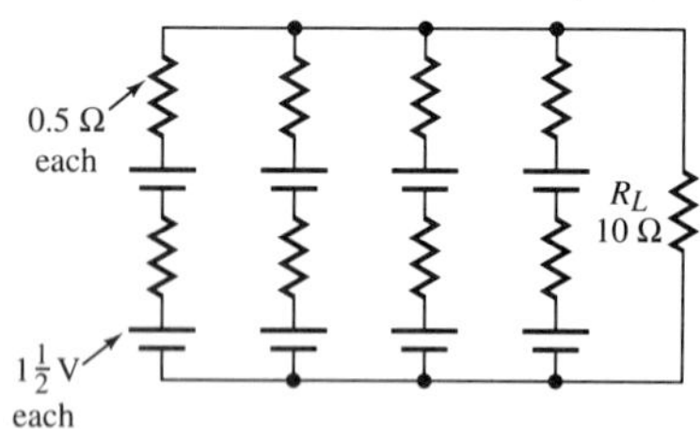

Figure 8.70

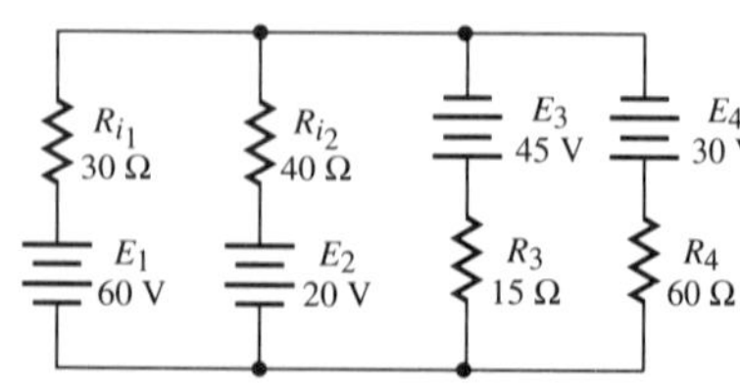

Figure 8.71

47. Convert the sources in Figure 8.67 so that the circuit consists of only current sources and R_L.

SECTION 8.7 Interconnection of Sources

48. What is the open-circuit emf of three 1.5-volt cells, each with an R_i of 0.8 ohm:

    (a) Connected in series?
    (b) Connected in parallel?

49. In the circuit of Figure 8.68, find:

    (a) The voltage across R_L.
    (b) The source current.
    (c) The maximum current that can be supplied if each cell is rated at 0.8 ampere.

50. In the circuit of Figure 8.69,

    (a) What is the voltage across R_L?
    (b) What is the maximum current that can be supplied if each cell is rated at 1 ampere?

51. Two 12-volt, 20-ampere, 40 ampere-hour batteries with an R_i of 8 ohms each are connected in parallel to a 36-ohm load. Find:

    (a) The voltage across the load.
    (b) The current in the load.
    (c) The hours for which the current can be supplied.

52. Repeat Problem 51 for the batteries connected in series.

53. For the circuit of Figure 8.70, find:

    (a) The equivalent source for the battery connection shown.
    (b) The voltage across R_L.
    (c) The maximum current that can be supplied by the group if each cell is rated at 0.5 ampere.

54. Reduce the parallel branch circuit in Figure 8.71 to a single voltage source.

55. Reduce the parallel branch circuit in Figure 8.72 to a single voltage source.

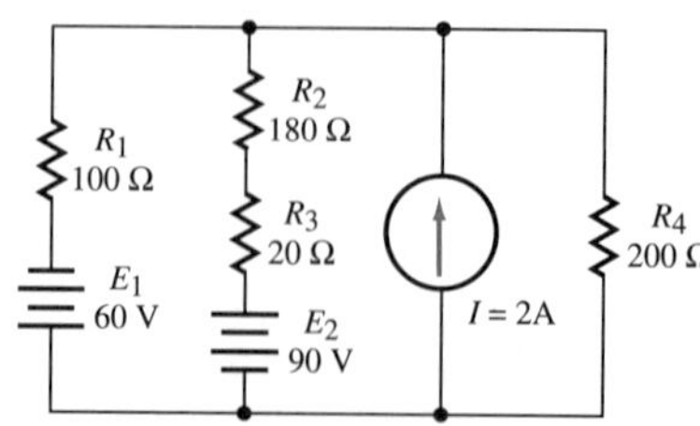

Figure 8.72

56. A lead-acid battery consists of six cells connected in parallel as shown in Figure 8.73. Find:

    (a) The voltage across the lamp when the current is 2 amperes.

    (b) The power dissipated in the lamp.

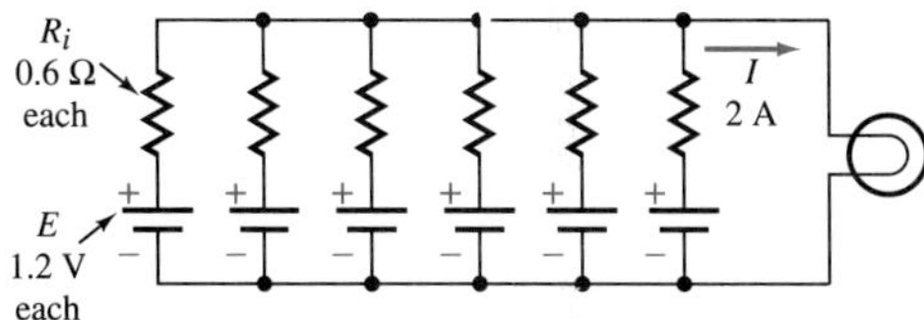

Figure 8.73

57. Find the voltage across R_4 in the circuit of Figure 8.74.

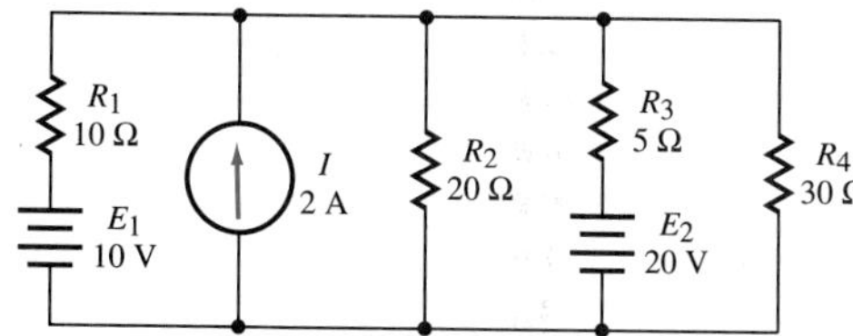

Figure 8.74

58. Find:

    (a) The voltage across R_6 in the circuit of Figure 8.75.

    (b) The current in R_6.

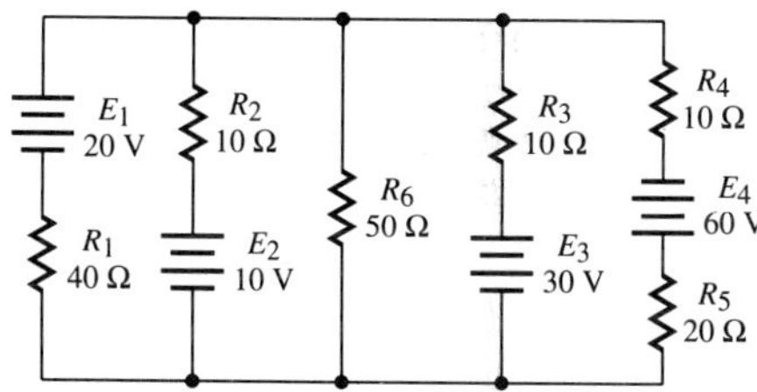

Figure 8.75

SECTION 8.8 SPICE and Series-Parallel Circuits

Ⓢ 59. Using PSpice, find the voltage drop across each resistor and the current flowing through each resistor for the circuit shown in Figure 8.41 when connected to a 10-V source.

Ⓢ 60. Using PSpice, find the voltage drop across each resistor and the current flowing through each resistor for the circuit shown in Figure 8.43 when connected to a 10-V source.

Ⓢ 61. Using PSpice, find the voltage drop across each resistor and the current flowing through each resistor for the circuit shown in Figure 8.44 when connected to a 10-V source.

Ⓢ 62. Using PSpice, find the voltage drop across each resistor and the current flowing through each resistor for the circuit shown in Figure 8.54.

CHAPTER 9

GENERAL METHODS OF CIRCUIT ANALYSIS

CHAPTER OBJECTIVES

After completing this chapter, you should be able to:

1. Write the equations for circuit analysis by the loop current, branch current, and nodal analysis methods.
2. Calculate the loop currents, branch currents, and node voltages.
3. Use the results of the analysis methods to obtain other quantities in the circuit.

KEY TERMS

1. branch current method
2. loop
3. loop current method
4. node
5. nodal method
6. node voltage
7. reference node

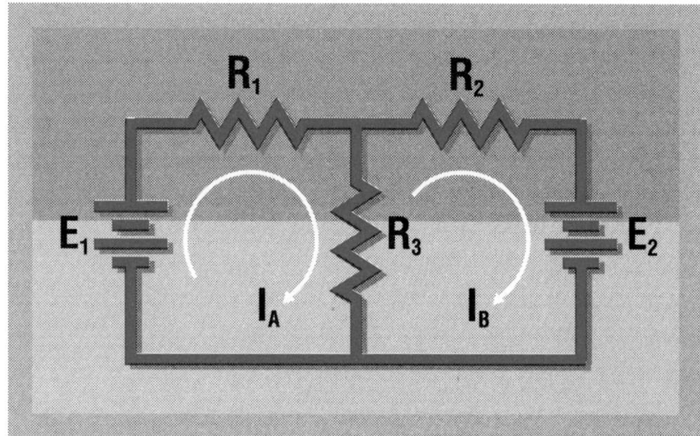

INTRODUCTION

How does one analyze a circuit that is neither a series, parallel, nor series-parallel circuit? How about circuits with several sources of emf? Fortunately, several methods are available for use with any of these types of circuits. Some, like the loop current method, are based on assumed loop currents. Others, like the nodal analysis method, provide the node voltages first. And others, like the branch current method, give us the actual currents in the branches.

Which methods to use, and how to use them, are explained in this chapter. You will be introduced to loop currents, node voltages, and reference nodes. After completing this chapter, you should be prepared to analyze any type of dc circuit.

9.1 Loop Current Analysis

The methods of Chapter 8 might not be adequate when a circuit has several sources, or when resistors do not form series or parallel combinations. For these circuits, several other methods are available. These methods can also be used for series-parallel and single-source circuits. The first of these, the mesh or loop current method, is presented in this section.

The **loop current method** is a method that uses an assumed current in each loop. A **loop** is a path formed by tracing a path from one point,

loop current method
loop

around the circuit, and back to the initial point. First, Kirchhoff's Voltage Law is applied to the loops to develop a series of equations that are used to determine the loop currents. The branch currents are then found by combining the loop currents in the branch. Finally, Ohm's Law and other relations are used to find other quantities. The steps in applying the mesh method of analysis are:

1. Change any current source to a voltage source.
2. Identify the loops in the circuit and let their number be N.
3. Draw loop currents clockwise in N–1 loops. If one views the circuit as a window, N–1 loops are the number of panes.
4. Mark the terminal polarity of each element in the loop as follows: (a) *sources*: small line (−), large line (+); (b) *resistors*: current enters (+), current leaves (−). If a resistor has two loop currents in it, mark the polarity for each current.
5. Apply Kirchhoff's Voltage Law to (N–1) loops, writing the resistor voltage as the product of the loop currents and the resistance. The rise or drop from any mutual loop current must also be included.
6. Solve the set of loop equations for the loop currents. If the loop currents come out negative, use the negative value in all calculations.
7. After every loop current is calculated, reverse the direction of any negative current and change the polarities accordingly.
8. For a branch that has only one loop current, the branch current is the loop current. For a branch having two loop currents, the branch current is the algebraic sum of the loop currents.
9. Use the basic circuit concepts to find other quantities.

The following example shows the application of this procedure.

EXAMPLE 9.1

Write the loop equations for the circuit in Figure 9.1.

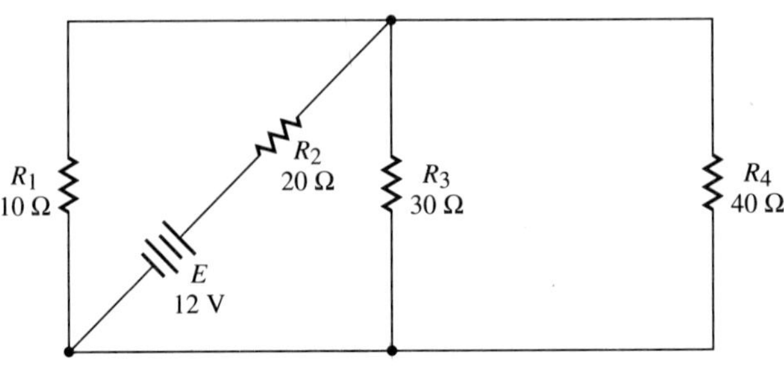

Figure 9.1

Strategy **Solution**

Steps 1–3 The circuit has four loops, so 4 − 1 = 3. The clockwise loop currents I_A, I_B, and I_C, and the terminal polarity are shown in Figure 9.2.

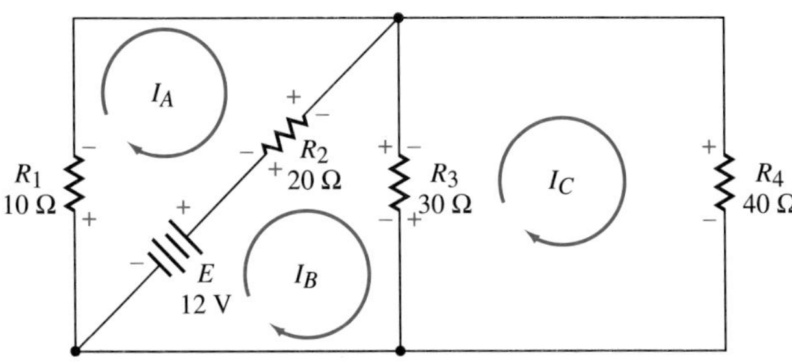

Figure 9.2

Applying Kirchhoff's Voltage Law clockwise around Loop A gives Step 4

$$-I_AR_1 - I_AR_2 + I_BR_2 - E = 0\text{ V}$$
$$-I_A(10\ \Omega) - I_A(20\ \Omega) + I_B(20\ \Omega) - 12\text{ V} = 0\text{ V}$$

or

$$-(30\ \Omega)I_A + (20\ \Omega)I_B - 12\text{ V} = 0\text{ V}$$

For Loop B:

$$+E - I_BR_2 + I_AR_2 - I_BR_3 + I_CR_3 = 0\text{ V}$$
$$+12\text{ V} - I_B(20\ \Omega) + I_A(20\ \Omega) - I_B(30\ \Omega)$$
$$+ I_C(30\ \Omega) = 0\text{ V}$$

or

$$+(20\ \Omega)I_A - (50\ \Omega)I_B + (30\ \Omega)I_C + 12\text{ V} = 0\text{ V}$$

For Loop C:

$$-I_CR_3 + I_BR_3 - I_CR_4 = 0\text{ V}$$
$$-I_C(30\ \Omega) + I_B(30\ \Omega) - I_C(40\ \Omega) = 0\text{ V}$$

or

$$+(30\ \Omega)I_B - (70\ \Omega)I_C = 0\text{ V}$$

EXAMPLE 9.2

Find the currents, voltages, and power in the circuit of Figure 9.3.

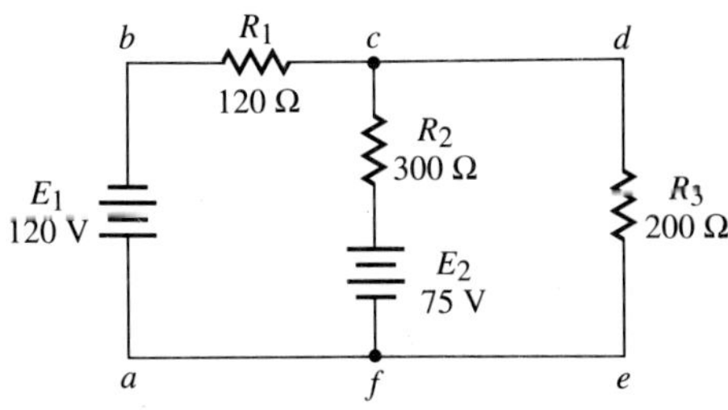

Figure 9.3

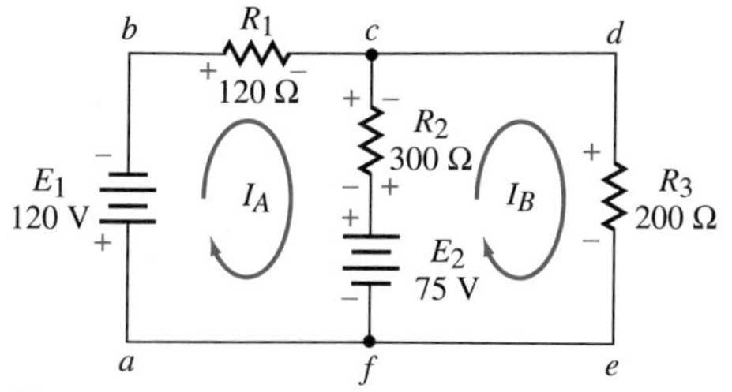

Figure 9.4

Strategy

The loop currents are drawn clockwise and polarities are marked.

A drop is when one moves from a + to a −.

Solution

Step 1. The loops are around paths *abcfa*, *fcdef*, and *abcdefa*.

Step 2. *N*–1 is two so the outer loop will not be used. The loop currents are I_A and I_B.

Step 3. The polarity of each terminal is shown in Figure 9.4.

Step 4. For Loop *A*, moving clockwise, the terms are $-E_1$, $-V_1$, $-V_2$ due to I_A, $+V_2$ due to I_B, and $-E_2$. So

a. $-E_1 - I_AR_1 - I_AR_2 + I_BR_2 - E_2 = 0\text{ V}$

$$-120\text{ V} - I_A(120\ \Omega) - I_A(300\ \Omega) + I_B(300\ \Omega) - 75\text{ V} = 0\text{ V}$$

or

$$(-420\ \Omega)I_A + (300\ \Omega)I_B - 195\text{ V} = 0\text{ V}$$

b. For Loop *B*, moving clockwise, the terms are, $+E_2$, $-V_2$ due to I_B, $+V_2$ due to I_A, and $-V_3$. So

$$+E_2 - I_BR_2 + I_AR_2 - I_BR_3 = 0\text{ V}$$
$$+75\text{ V} - I_B(300\ \Omega) + I_A(300\ \Omega) - I_B(200\ \Omega) = 0\text{ V}$$

or

$$+75\ V + (300\ \Omega)I_A - (500\ \Omega)I_B = 0\text{ V}$$

The number of equations equals the number of unknowns so the set can be solved.

Step 5. There are two equations and two unknowns, so I_A and I_B can be found. Solving the equations gives

$$I_A = -0.625\text{ A}$$

Putting this value in the equation for Loop *B* gives

$$I_B = \frac{(300\ \Omega)(-0.625\text{ A}) + 75\text{ V}}{500\ \Omega}$$

$$I_B = -0.225\text{ A}$$

The negative signs mean that the direction of I_A and I_B is the opposite of that assigned.

Step 6. The circuit with the correct current directions and polarity is drawn in Figure 9.5.

The loop currents are combined to get the branch current.

Step 7. In R_2, I_A is up and I_B is down, so

$$I_2 = I_A - I_B$$
$$= 0.625\text{ A} - 0.225\text{ A}$$

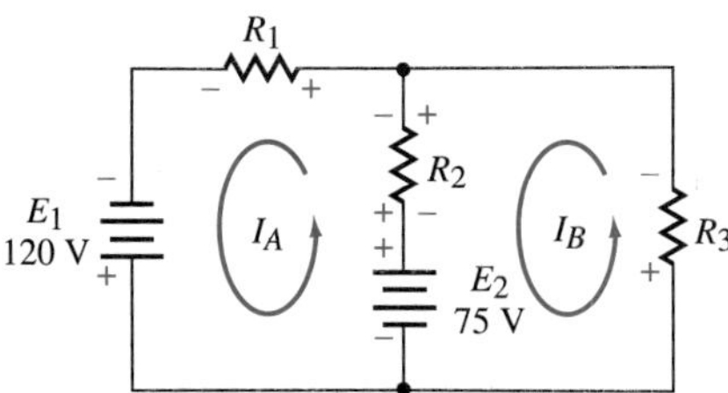

Figure 9.5

$$I_2 = 0.4 \text{ A up}$$

The current in R_1 is I_A so $I_1 = 0.625$ A to the left. The current in R_3 is I_B so $I_3 = 0.225$ A upward.

Step 8.

Ohm's Law will give us the voltage.

$$V_1 = I_1R_1$$
$$= 0.625 \text{ A} \times 120 \ \Omega$$
$$V_1 = 75 \text{ V}$$
$$V_2 = I_2R_2$$
$$= 0.4 \text{ A} \times 300 \ \Omega$$
$$V_2 = 120 \text{ V}$$
$$V_3 = I_3R_3$$
$$= 0.225 \text{ A} \times 200 \ \Omega$$
$$V_3 = 45 \text{ V}$$
$$P_T = P_1 + P_2 + P_3$$

The total power relation is used for circuit power.

$$P_1 = I_1^2R_1$$
$$= (0.625 \text{ A})^2 \times 120 \ \Omega$$
$$P_1 = 46.88 \text{ W}$$
$$P_2 = I_2^2R_2$$
$$= (0.4 \text{ A})^2 \times 300 \ \Omega$$
$$P_2 = 48 \text{ W}$$
$$P_3 = I_3^2R_3$$
$$= (0.225 \text{ A})^2 \times 200 \ \Omega$$
$$P_3 = 10.13 \text{ W}$$
$$P_T = 46.88 \text{ W} + 48 \text{ W} + 10.12 \text{ W}$$
$$P_T = 105 \text{ W}$$

In the single-source circuit, the source supplies the power dissipated in the circuit. In the multisource circuit, a source supplies power when the current leaves the (+) terminal (discharging). When the current leaves the (−) terminal (charging), the source uses power. Therefore, the circuit power in a multisource circuit does not always equal the sum of the source powers. Instead, it is the power supplied minus the power used by each source. In this circuit,

$$P_{s1} = E_1 I_1 = 120 \text{ V} \times 0.625 \text{ A} = 75 \text{ W}$$
$$P_{s2} = E_2 I_2 = 75 \text{ V} \times 0.4 \text{ A} = 30 \text{ W}$$

According to the direction of I_1 and I_2, each source supplies power. So

$$P_T = P_{s1} + P_{s2}$$
$$= 75 \text{ W} + 30 \text{ W}$$
$$P_T = 105 \text{ W}$$

This is the same as the power dissipated in the circuit.

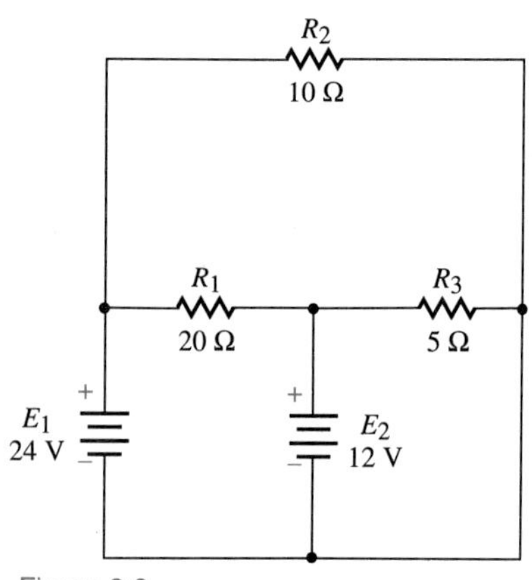

Figure 9.6

EXAMPLE 9.3

For the circuit of Figure 9.6, find:
a. the current in each resistor.
b. the voltage across R_2.
c. the power dissipated in the resistance of the circuit.

Solution

1. There are four loops, the three panes, and the outer loop, so $N = 4$.
2. The loop currents, I_A, I_B, and I_C for N–1 loops and terminal polarity are shown in Figure 9.7.
3. In Loop A, moving clockwise, the terms are $+E_1$, $-V_1$ due to I_A, $+V_1$ due to I_B, and $-E_2$. So

$$+E_1 - I_A R_1 + I_B R_1 - E_2 = 0 \text{ V}$$
$$+24 \text{ V} - I_A(20\ \Omega) + I_B(20\ \Omega) - 12 \text{ V} = 0 \text{ V}$$

or

$$-(20\ \Omega)I_A + (20\ \Omega)I_B + 12 \text{ V} = 0 \text{ V}$$

Figure 9.7

4. In Loop B, moving clockwise, the terms are $-V_2$, $-V_3$ due to I_B, $+V_3$ due to I_C, $-V_1$ due to I_B, and $+V_1$ due to I_A. So

$$-I_B R_2 - I_B R_3 + I_C R_3 - I_B R_1 + I_A R_1 = 0 \text{ V}$$

or

$$-I_B(10\ \Omega) - I_B(5\ \Omega) + I_C(5\ \Omega) - I_B(20\ \Omega) + I_A(20\ \Omega) = 0 \text{ V}$$

or

$$+(20\ \Omega)I_A - (35\ \Omega)I_B + (5\ \Omega)I_C = 0 \text{ V}$$

5. For Loop C, moving clockwise, the terms are $+E_2$, $-V_3$ due to I_C, and $+V_3$ due to I_B. So

$$+E_2 - I_C R_3 + I_B R_3 = 0 \text{ V}$$

or

$$+12 \text{ V} + (5 \ \Omega)I_B - (5 \ \Omega)I_C = 0 \text{ V}$$

6. There are now three equations and three unknowns, so the set can be solved for I_A, I_B, and I_C. The solution is done using determinants, so the equations are changed to:

$$\begin{aligned} -20I_A + 20I_B + 0I_C &= -12 \\ +20I_A - 35I_B + 5I_C &= 0 \\ 0I_A + 5I_B - 5I_C &= -12 \end{aligned}$$

By determinants,

$$I_A = \frac{D_A}{D}, \qquad I_B = \frac{D_B}{D}, \qquad I_C = \frac{D_C}{D}$$

so

$$I_A = \frac{\begin{bmatrix} -12 & +20 & 0 \\ 0 & -35 & +5 \\ -12 & +5 & -5 \end{bmatrix}}{\begin{bmatrix} -20 & +20 & 0 \\ +20 & -35 & +5 \\ 0 & +5 & -5 \end{bmatrix}} = \frac{(-3300) - (-300)}{(-3500) - (-2500)} = 3 \text{ A}$$

$$I_B = \frac{\begin{bmatrix} -20 & -12 & 0 \\ +20 & 0 & +5 \\ 0 & -12 & -5 \end{bmatrix}}{\begin{bmatrix} 20 & +20 & 0 \\ +20 & -35 & +5 \\ 0 & +5 & -5 \end{bmatrix}} = \frac{(0) - (2400)}{(-3500) - (-2500)} = 2.4 \text{ A}$$

$$I_C = \frac{\begin{bmatrix} -20 & +20 & -12 \\ +20 & -35 & 0 \\ 0 & +5 & -12 \end{bmatrix}}{\begin{bmatrix} -20 & +20 & 0 \\ +20 & -35 & +5 \\ 0 & +5 & -5 \end{bmatrix}} = \frac{(-9600) - (-4800)}{(-3500) - (-2500)} = 4.8 \text{ A}$$

Since the loop currents are positive, the directions are as assigned.

7. For the branch currents:
   The current in R_2 is I_B.

$$I_2 = 2.4 \text{ A}.$$

   The current in R_1 is $I_A - I_B$. It is

$$I_1 = 3 \text{ A} - 2.4 \text{ A}$$
$$I_1 = 0.6 \text{ A} \quad \text{in the direction of } I_A.$$

   The current in R_3 is $I_C - I_B$. It is

$$I_3 = 4.8 \text{ A} - 2.4 \text{ A}$$
$$I_3 = 2.4 \text{ A} \quad \text{in the direction of } I_C.$$

8. Applying Ohm's Law to R_2 gives

$$V_2 = I_2R_2 = (2.4 \text{ A})\ (10\ \Omega)$$
$$V_2 = 24 \text{ V}$$
$$P_T = P_1 + P_2 + P_3$$
$$P_1 = I_1^2R_1 = (0.6 \text{ A})^2\ (20\ \Omega)$$
$$P_1 = 7.2 \text{ W}$$
$$P_2 = I_2^2R_2 = (2.4 \text{ A})^2\ (10\ \Omega)$$
$$P_2 = 57.6 \text{ W}$$
$$P_3 = I_3^2R_3 = (2.4 \text{ A})^2\ (5\ \Omega)$$
$$P_3 = 28.8 \text{ W}$$
$$P_T = 7.2 \text{ W} + 57.6 \text{ W} + 28.8 \text{ W}$$
$$P_T = 93.6 \text{ W}$$

Now that the basic principle behind the loop current method has been explained, a more mechanical procedure for developing the equations can be examined. This procedure does not require marking of terminal polarity nor determining rises and drops. Once the loop currents are drawn in a clockwise direction, the terms are as follows:

a. Loop Current Terms

   + (loop current) × (sum of the resistance in the loop)

b. Mutual Loop Current Terms

– (mutual loop current) × (sum of the resistance in the mutual branch)

c. Loop Source Terms

Value of source emf with the sign of the terminal at which the loop current enters the source

The equation is formed by setting the sum of the terms in a–c equal to 0.

This procedure gives the correct relation between rises and drops, but it is important to recognize that it assigns a (–) to drops and a (+) to rises for a counterclockwise summation.

EXAMPLE 9.4

Write the loop equations for the circuit in Figure 9.8.

Solution

The circuit with the clockwise loop currents I_A, I_B, and I_C is shown in Figure 9.9.

The terms in the equations are

Loop	Loop Current Terms	Mutual Loop Current Terms	Source Terms
A	$+I_A(R_1 + R_2)$	$-I_BR_2$	$+E_1$
B	$+I_B(R_2 + R_3 + R_4)$	$-I_AR_2$, $-I_CR_4$	$-E_1$, $+E_2$
C	$+I_C(R_4 + R_5)$	$-I_BR_4$	$-E_2$

Setting the sum of the terms equal to zero gives

1. $I_A(R_1 + R_2) - I_BR_2 + E_1 = 0\text{ V}$

$$I_A(10\ \Omega + 20\ \Omega) - I_B(20\ \Omega) + 10\text{ V} = 0\text{ V}$$

or

$$(30\ \Omega)I_A - (20\ \Omega)I_B + 10\text{ V} = 0\text{ V}$$

2. $I_B(R_2 + R_3 + R_4) - I_AR_2 - I_CR_4 - E_1 + E_2 = 0\text{ V}$

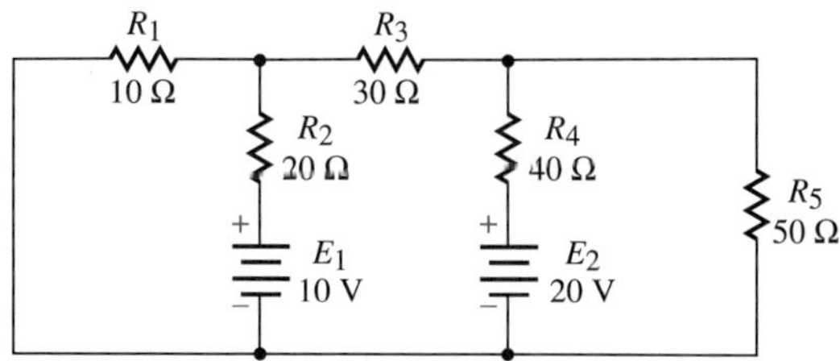

Figure 9.8

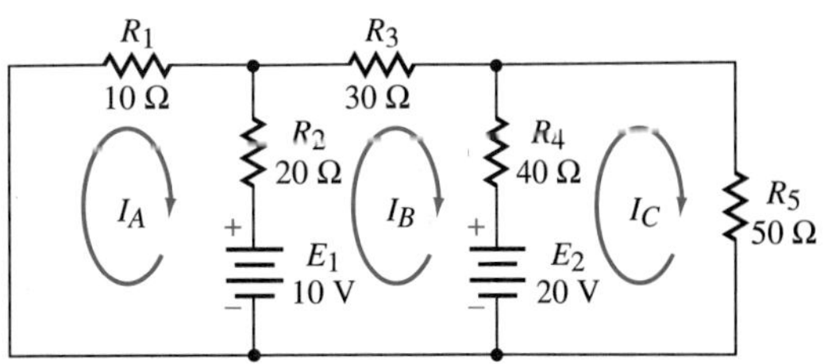

Figure 9.9

$$I_B(20\ \Omega + 30\ \Omega + 40\ \Omega) - I_A(20\ \Omega) - I_C(40\ \Omega) - 10\ \text{V} + 20\ \text{V} = 0\ \text{V}$$

or

$$-(20\ \Omega)I_A + (90\ \Omega)I_B - (40\ \Omega)I_C + 10\ \text{V} = 0\ \text{V}$$

3. $I_C(R_4 + R_5) - I_BR_4 - E_2 = 0\ \text{V}$

$$I_C(40\ \Omega + 50\ \Omega) - I_B(40\ \Omega) - 20\ \text{V} = 0\ \text{V}$$

or

$$-(40\ \Omega)I_B + (90\ \Omega)I_C - 20\ \text{V} = 0\ \text{V}$$

EXAMPLE 9.5

Determine the loop currents I_A, I_B, and I_C in Figure 9.10.

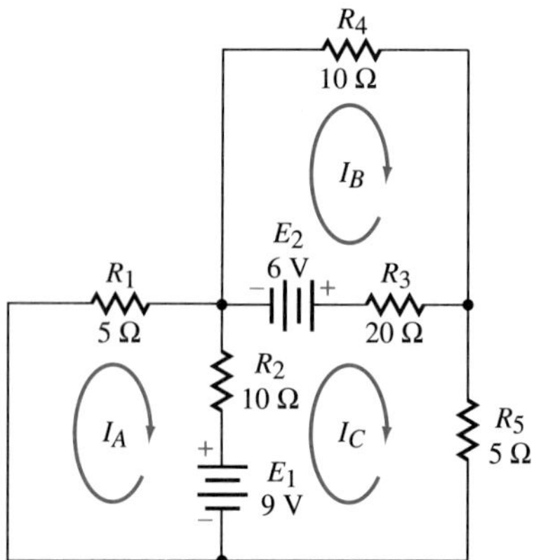

Figure 9.10

Solution

The loop currents have already been drawn, so we must find the terms. The terms in the equations are

Loop	Loop Current Terms	Mutual Loop Current Terms	Source Terms
A	$+I_A(R_1 + R_2)$	$-I_CR_2$	$+E_1$
B	$+I_B(R_3 + R_4)$	$-I_CR_3$	$+E_2$
C	$+I_C(R_2 + R_3 + R_5)$	$-I_AR_2,\ -I_BR_3$	$-E_1,\ -E_2$

Setting the sum of the terms equal to 0 gives

1. $I_A(R_1 + R_2) - I_CR_2 + E_1 = 0\ \text{V}$

$$I_A(5\ \Omega + 10\ \Omega) - I_C(10\ \Omega) + 9\ \text{V} = 0\ \text{V}$$

or

$$(15\ \Omega)I_A - (10\ \Omega)I_C + 9\ \text{V} = 0\ \text{V}$$

2. $I_B(R_3 + R_4) - I_CR_3 + E_2 = 0\ \text{V}$

$$I_B(20\ \Omega + 10\ \Omega) - I_C(20\ \Omega) + 6\ \text{V} = 0\ \text{V}$$

or

$$(30\ \Omega)I_B - (20\ \Omega)I_C + 6\ \text{V} = 0\ \text{V}$$

3. $I_C(R_2 + R_3 + R_5) - I_AR_2 - I_BR_3 - E_1 - E_2 = 0\ \text{V}$

$$I_C(10\ \Omega + 20\ \Omega + 5\ \Omega) - I_A(10\ \Omega) - I_B(20\ \Omega) - 9\ \text{V} - 6\ \text{V} = 0\ \text{V}$$

or

$$-(10\ \Omega)I_A - (20\ \Omega)I_B + (35\ \Omega)I_C - 15\ \text{V} = 0\ \text{V}$$

The next step is to arrange the equation terms for solving by determinants. This gives

$$15I_A + 0I_B - 10I_C = -9$$
$$0I_A + 30I_B - 20I_C = -6$$
$$-10I_A - 20I_B + 35I_C = 15$$

$$I_A = \frac{D_A}{D}, \quad I_B = \frac{D_B}{D}, \quad I_C = \frac{D_C}{D}$$

$$I_A = \frac{\begin{bmatrix} -9 & 0 & -10 \\ -6 & +30 & -20 \\ +15 & -20 & +35 \end{bmatrix}}{\begin{bmatrix} +15 & 0 & -10 \\ 0 & +30 & -20 \\ -10 & -20 & +35 \end{bmatrix}} = \frac{(-10{,}650) - (-8100)}{(15{,}750) - (9000)} = -0.378\ \text{A}$$

$$I_B = \frac{\begin{bmatrix} +15 & -9 & -10 \\ 0 & -6 & -20 \\ -10 & +15 & +35 \end{bmatrix}}{\begin{bmatrix} +15 & 0 & -10 \\ 0 & +30 & -20 \\ -10 & -20 & +35 \end{bmatrix}} = \frac{(-4950) - (-5100)}{(15{,}750) - (9000)} = +0.022\ \text{A}$$

$$I_C = \frac{\begin{bmatrix} +15 & 0 & -9 \\ 0 & +30 & -6 \\ -10 & -20 & +15 \end{bmatrix}}{\begin{bmatrix} +15 & 0 & -10 \\ 0 & +30 & -20 \\ -10 & -20 & +35 \end{bmatrix}} = \frac{(6750) - (4500)}{(15{,}750) - (9000)} = +0.333\ \text{A}$$

Since I_A is negative, the correct direction is opposite to the assigned direction.

EXAMPLE 9.6

The bridge circuit in Figure 9.11 is balanced by varying R_4. Use the loop method to determine R_4 for balance. That is, zero current in R_5.

Solution

Since there are four loops, the equations must be written for three. The circuit with clockwise currents I_A, I_B, and I_C is shown in Figure 9.12.

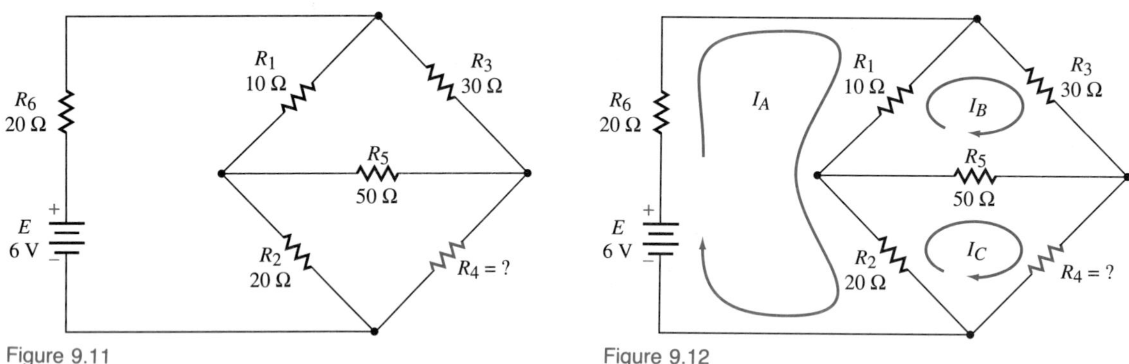

Figure 9.11

Figure 9.12

For I_{R5} to be 0 amperes, I_B must equal I_C. Why? The terms in the loop equations are:

Loop	Loop Current Terms	Adjacent Loop Current Terms	Source Terms
A	$I_A(R_6 + R_1 + R_2)$	$-I_BR_1,\ -I_CR_2$	$-E$
B	$I_B(R_1 + R_3 + R_5)$	$-I_AR_1,\ -I_CR_5$	–
C	$I_C(R_2 + R_4 + R_5)$	$-I_AR_2,\ -I_BR_5$	–

The complete equations are:

$$I_A(20\ \Omega + 10\ \Omega + 20\ \Omega) - I_B(10\ \Omega) - I_C(20\ \Omega) - 6\text{ V} = 0\text{ V}$$

or

$$(50\ \Omega)I_A - (10\ \Omega)I_B - (20\ \Omega)I_C - 6\text{ V} = 0\text{ V}$$

$$I_B(10\ \Omega + 30\ \Omega + 50\ \Omega) - I_A(10\ \Omega) - I_C(50\ \Omega) = 0\text{ V}$$

or

$$-(10\ \Omega)I_A + (90\ \Omega)I_B - (50\ \Omega)I_C = 0\text{ V}$$

$$I_C(20\ \Omega + 50\ \Omega + R_4) - I_A(20\ \Omega) - I_B(50\ \Omega) = 0\text{ V}$$

or

$$-(20\ \Omega)I_A - (50\ \Omega)I_B + (70\ \Omega + R_4)I_C = 0\text{ V}$$

Writing $70\ \Omega + R_4$ as R_x, and arranging the equations for determinants gives

$$\begin{aligned} 50I_A - 10I_B - 20I_C &= 6\text{ V} \\ -10I_A + 90I_B - 50I_C &= 0\text{ V} \\ -20I_A - 50I_B + R_xI_C &= 0\text{ V} \end{aligned}$$

$$I_B = \frac{D_B}{D}, \qquad I_C = \frac{D_C}{D}$$

Since D is the same for I_B and I_C, I_B will be equal to I_C if $D_B = D_C$.

$$D_B = \begin{bmatrix} +50 & +6 & -20 \\ -10 & 0 & -50 \\ -20 & 0 & R_x \end{bmatrix} = 6000 + 60\ R_x$$

$$D_C = \begin{bmatrix} +50 & -10 & +6 \\ -10 & +90 & 0 \\ -20 & -50 & 0 \end{bmatrix} = 13{,}800$$

so $6000 + 60R_x = 13{,}800$

$$R_x = \frac{13{,}800 - 6000}{60} = 130\ \Omega$$

$$R_4 + 70\ \Omega = 130\ \Omega$$

$$R_4 = 60\ \Omega$$

This agrees with the condition for a balanced bridge. That is,

$$\frac{R_1}{R_2} = \frac{R_3}{R_4}$$

9.2 Branch Current Analysis

A method of circuit analysis that use branch currents instead of loop currents is the **branch current method**. One advantage of this method is that it gives the circuit currents directly. However, because the loop equations are less than the number of unknowns, it is necessary to apply Kirchhoff's Current Law at the nodes. For this method, a **node** must be a junction of three or more branches.

branch current method

node

Using the actual branch currents increases the number of equations that are solved simultaneously. A solution of Example 9.1 would require the simultaneous solving of three equations as compared to two for the mesh analysis. The number of equations for any circuit will be equal to the number of branch currents. The steps in the branch current method of analysis are:

1. Identify the loops, the nodes, the branches, and designate the number of each. This number will be referred to as N.
2. Draw a current in each branch, arbitrarily assigning a direction. If the direction is incorrect, the calculated value will be negative.
3. Mark the polarity of the resistors and sources as follows: (a) *sources*: small line (−), large line (+); (b) *resistors*: current enters (+), current leaves (−).
4. Apply Kirchhoff's Voltage Law to N–1 loops, writing the potential across each resistance as the product of the branch current and the resistance. Give each term the sign on the element's terminal that is first encountered as the path is traced around the loop. This will

automatically give the correct relation between rises and drops. However, a (+) will be assigned to drops and a (−) to rises.

5. Apply Kirchhoff's Current Law to N–1 nodes. Currents entering a node are positive and currents out of a node are negative.
6. Solve the equations from Steps 4 and 5 for the branch currents.
7. If any branch current comes out negative, use the negative value in all calculations. When every branch current has been calculated, reverse the direction of the negative currents and change the polarities according to the correct direction. This will give the correct direction and polarities.
8. Use the basic circuit concepts to find other quantities.

The next example shows the application of the branch current procedure.

EXAMPLE 9.7

Find the currents, voltages, and power in the circuit of Figure 9.13.

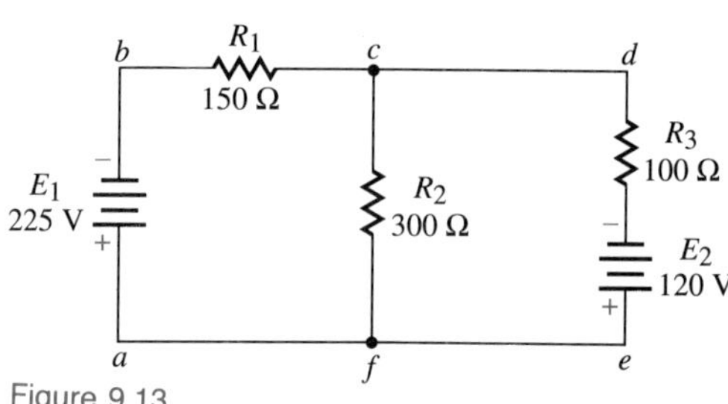

Figure 9.13

Strategy

Identify the loops, branches, and nodes.

The direction of the current is arbitrarily selected.

Algebraically sum the voltages in each loop.

Solution

Step 1. There are three loops. They are around paths *abcfa*, *fcdef*, and *abcdefa*. There are three branches. They are paths *fabc*, *fc*, and *fedc*. There are two nodes. They are at *c* and *f*.

Step 2,3. The branch currents I_1, I_2, and I_3 are drawn, and the polarity of each terminal is shown in Figure 9.14.

Step 4. For Loop A, the terms are $+E_1$, $+V_1$, and $+V_2$. So

a. $E_1 + I_1R_1 + I_2R_2 = 0$

or

$$+225\text{ V} + (150\ \Omega)I_1 + (300\ \Omega)I_2 = 0$$

For Loop B, the terms are $-V_2$, $-V_3$, and $-E_2$. So,

b. $-I_2R_2 - I_3R_3 - E_2 = 0$

or

$$-(300\ \Omega)I_2 - (100\ \Omega)I_3 - 120\text{ V} = 0$$

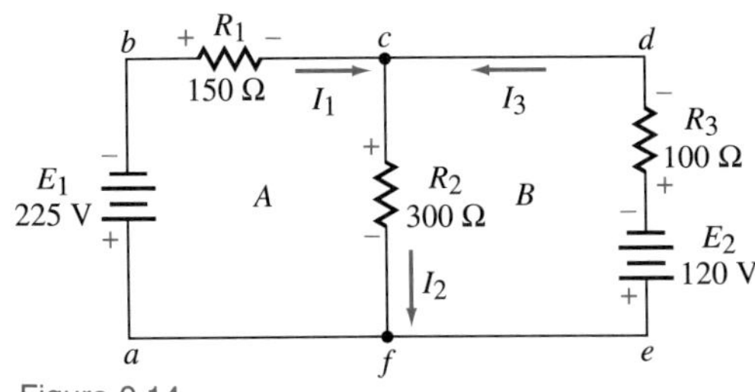

Figure 9.14

The equations for $N-1$ loops have been written. Since there are three unknowns, we must next write the node equations.

Step 5. At Node C, I_1 and I_3 enter the node, and I_2 is out of the node. So

Write Kirchhoff's Current Law for the node.

c. $I_1 + I_3 - I_2 = 0$

There are now three equations with three unknowns.

The number of equations must be equal to the number of unknowns.

a. $+225 + (150\ \Omega)I_1 + (300\ \Omega)I_2 = 0$
b. $-120\text{ V} - (300\ \Omega)I_2 - (100\ \Omega)I_3 = 0$
c. $I_1 - I_2 + I_3 = 0$

Step 6. The equations, arranged for solution by determinants are

Solve the equations.

$$\begin{aligned} +150\,I_1 + 300\,I_2 + \ \ 0\,I_3 &= -225 \\ 0\,I_1 - 300\,I_2 - 100\,I_3 &= +120 \\ I_1 - \quad I_2 + \quad I_3 &= \quad 0 \end{aligned}$$

$$I_1 = \frac{D_1}{D}, \qquad I_2 = \frac{D_2}{D}, \qquad I_3 = \frac{D_3}{D}$$

$$I_1 = \frac{\begin{bmatrix} -225 & +300 & 0 \\ +120 & -300 & -100 \\ 0 & -1 & +1 \end{bmatrix}}{\begin{bmatrix} +150 & +300 & 0 \\ 0 & -300 & -100 \\ +1 & -1 & +1 \end{bmatrix}} = \frac{(67{,}500) - (13{,}500)}{(-75{,}000) - (15{,}000)}$$

$$I_1 = \frac{54{,}000}{-90{,}000} = -0.6\text{ A}$$

$$I_2 = \frac{\begin{bmatrix} +150 & -225 & 0 \\ 0 & +120 & -100 \\ +1 & 0 & +1 \end{bmatrix}}{\begin{bmatrix} +150 & +300 & 0 \\ 0 & -300 & -100 \\ +1 & -1 & +1 \end{bmatrix}} = \frac{+40{,}500 - 0}{-90{,}000} = -0.45\text{ A}$$

$$I_3 = \frac{\begin{bmatrix} +150 & +300 & -225 \\ 0 & -300 & +120 \\ +1 & -1 & 0 \end{bmatrix}}{\begin{bmatrix} +150 & +300 & 0 \\ 0 & -300 & -100 \\ +1 & -1 & +1 \end{bmatrix}} = \frac{+36{,}000 - (+49{,}500)}{-90{,}000}$$

$$I_3 = +0.15 \text{ A}$$

Step 7. The circuit is drawn in Figure 9.15 showing the correct current directions and polarity.

Use circuit concepts to find other quantities.

Step 8. Applying Ohm's Law to each resistor gives

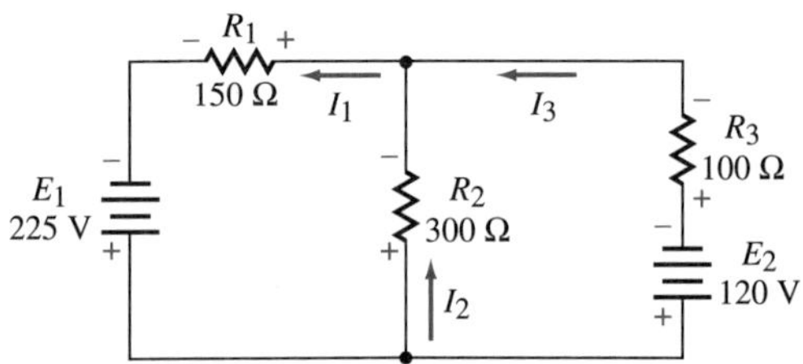

Figure 9.15

$$\begin{aligned} V_1 &= I_1R_1 \\ &= 0.6 \text{ A} \times 150\ \Omega \\ V_1 &= 90 \text{ V} \\ V_2 &= I_2R_2 \\ &= 0.45 \text{ A} \times 300\ \Omega \\ V_2 &= 135 \text{ V} \\ V_3 &= I_3R_3 \\ &= 0.15 \text{ A} \times 100\ \Omega \\ V_3 &= 15 \text{ V} \end{aligned}$$

The power in the three resistors can be found by using the power relation.

$$\begin{aligned} P_T &= P_1 + P_2 + P_3 \\ P_1 &= I_1^2R_1 \\ &= (0.6 \text{ A})^2 \times 150\ \Omega \\ &= 54 \text{ W} \\ P_2 &= I_2^2R_2 \\ &= (0.45 \text{ A})^2 \times 300\ \Omega \\ &= 60.75 \text{ W} \\ P_3 &= I_3^2R_3 \end{aligned}$$

$$= (0.15\ \text{A})^2 \times 100\ \Omega$$
$$= 2.25\ \text{W}$$
$$P_T = 54\ \text{W} + 60.75\ \text{W} + 2.25\ \text{W}$$
$$P_T = 117.00\ \text{W}$$

EXAMPLE 9.8

Use the branch current method to find the following quantities in Figure 9.16:

a. the current in R_4 and R_5.
b. the voltage across R_4 and R_5.
c. the power supplied by the source.
d. the efficiency of the divider.

Solution

1. There are three inner loops and one outer loop, five branches, and two nodes.
2, 3. The loops, nodes, branch currents, and terminal polarity are shown in Figure 9.17. Since there is only one source, the actual current direction will be as shown.
4. Applying Kirchhoff's Voltage Law to the loops gives
For A: The terms are

$$-E + I_1R_1 + I_2R_2 + I_3R_3 = 0\ \text{V}$$

or

$$-24\ \text{V} + (30\ \Omega)I_1 + (200\ \Omega)I_2 + (100\ \Omega)I_3 = 0\ \text{V}$$

For B: The terms are

$$-I_3R_3 + I_5R_5 = 0\ \text{V}$$

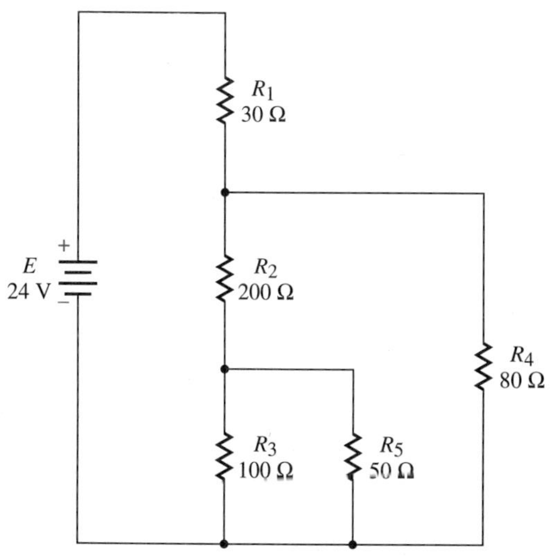

Figure 9.16

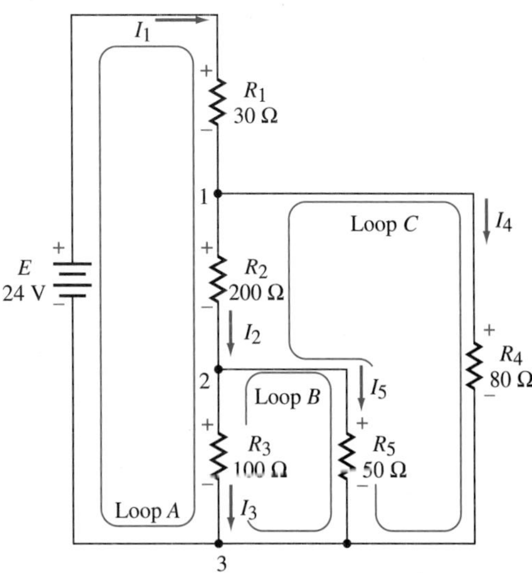

Figure 9.17

or

$$-(100\ \Omega)I_3 + (50\ \Omega)I_5 = 0\ \text{V}$$

For C: The terms are

$$-I_5R_5 - I_2R_2 + I_4R_4 = 0\ \text{V}$$

or

$$-(50\ \Omega)I_5 - (200\ \Omega)I_2 + (80\ \Omega)I_4 = 0\ \text{V}$$

5. Applying Kirchhoff's Current Law to (3 − 1) or 2 nodes gives

$$\text{At 1:} \quad I_1 - I_2 - I_4 = 0\ \text{A}$$
$$\text{At 2:} \quad I_2 - I_3 - I_5 = 0\ \text{A}$$

We now have five equations and five unknowns, so the set can be solved.

6. The five equations are reduced to three for solving by determinants. The general procedure is to substitute the node equations into the loop equations.

$$\text{From Node 1,} \quad I_4 = I_1 - I_2$$
$$\text{From Node 2,} \quad I_5 = I_2 - I_3$$

Putting these into the Loop B and Loop C equations gives

For A: $-24\ \text{V} + (30\ \Omega)I_1 + (200\ \Omega)I_2 + (100\ \Omega)I_3 = 0\ \text{V}$

For B: $+ (50\ \Omega)I_2 - (150\ \Omega)I_3 = 0\ \text{V}$

For C: $+ (80\ \Omega)I_1 - (330\ \Omega)I_2 + (50\ \Omega)I_3 = 0\ \text{V}$

Solving these equations with any of the methods in the Appendix yields

$$I_1 = 2.68 \times 10^{-1}\ \text{A}$$
$$I_2 = 6.84 \times 10^{-2}\ \text{A}$$
$$I_3 = 2.28 \times 10^{-2}\ \text{A}$$

Then, from the node equations,

$$\begin{aligned} I_4 &= I_1 - I_2 \\ &= 2.68 \times 10^{-1}\ \text{A} - 6.84 \times 10^{-2}\ \text{A} \\ I_4 &= 2.0 \times 10^{-1}\ \text{A} \\ I_5 &= I_2 - I_3 \\ &= 6.84 \times 10^{-2}\ \text{A} - 2.28 \times 10^{-2}\ \text{A} \\ I_5 &= 4.56 \times 10^{-2}\ \text{A} \end{aligned}$$

7. Applying Ohm's Law to each load resistor gives

$$\begin{aligned} V_5 &= I_5R_5 \\ &= (4.56 \times 10^{-2}\ \text{A})(50\ \Omega) \\ V_5 &= 2.28\ \text{V} \end{aligned}$$

$$V_4 = I_4R_4$$
$$= (2.0 \times 10^{-1}\ \text{A})(80\ \Omega)$$
$$V_4 = 16.0\ \text{V}$$

The source power is

$$P = E\,I_1$$
$$= (24\ \text{V})(2.68 \times 10^{-1}\ \text{A})$$
$$P = 6.43\ \text{W}$$

The efficiency is

$$\eta = (P_{out}/P_{in}) \times 100\%$$
$$P_{out} = P_4 + P_5$$
$$P_4 = V_4I_4 = (16.0\ \text{V})(0.2\ \text{A}) = 3.2\ \text{W}$$
$$P_5 = V_5I_5 = (2.28\ \text{V})(0.0456\ \text{A}) = 0.10\ \text{W}$$

so

$$P_{out} = 3.2\ \text{W} + 0.10\ \text{W} = 3.3\ \text{W}$$

and

$$\eta = (3.3\ \text{W}/6.43\ \text{W}) \times 100\%$$
$$\eta = 51.32\%$$

9.3 Nodal Analysis

A third method of circuit analysis is the **nodal method**. This method is used in many circuit analysis programs such as PSpice. In this method, the branch currents are written in terms of the difference between node voltages and the parts between the nodes. **Node voltage** means the voltage of the node as measured from a **reference node**. Kirchhoff's Current Law is then used to find the node voltages, and these voltages are used to find the branch currents. Ohm's Law and other circuit concepts can then be used to find other quantities. Since voltages are found first, this method lends itself to analyses where voltage is needed. One such example is the Thevenin equivalent, explained in Section 10.5.

nodal method

node voltage

reference node

The method presented here combines the branch current equations and Kirchhoff's Voltage Law in one step. It also eliminates the need to convert voltage sources to current sources. Finally, it gives the branch currents directly without having to combine a source current and a resistor current. The following list explains the procedure:

1. Identify the nodes and branches.
2. Select any node as a reference node and mark it with the ground symbol. Since this node is a reference, its voltage is 0 volts. The voltages at the other nodes are taken with respect to this node. Mark the other nodes using a letter such as A, B, and so forth, and voltages using subscript such as, V_A, V_B, and so forth.

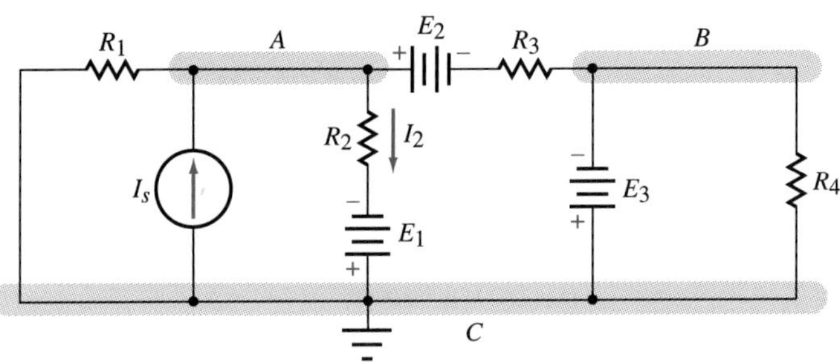

Figure 9.18

3. Write an expression for each node, except the reference node, to include the terms in a nodal equation as described for Node A of Figure 9.18.
   a. Node terms

   + (Node voltage) × (sum of the reciprocals of the resistance in each branch connected to the node)

   $+ \ V_A(1/R_1 + 1/R_2 + 1/R_3)$

   b. Adjacent node terms

   – (Adjacent node voltage) × (reciprocal of the resistance in the branch between the nodes)

   $- \ V_B \ (1/R_3)$

   c. Source terms

   (Source voltage) × (reciprocal of the sum of the resistance in the branch)
   This term is given the sign of the source terminal nearest the node.

   $$E_2(1/R_3), \qquad -E_1(1/R_2)$$

   d. Current source value
   Current into the node is given a (+) sign. Current away from the node is given a (−) sign.

   $$+ \ I_s$$

4. Set the sum of the terms in 3(a) and (b) equal to the sum of the terms in 3(c). For Node A of Figure 9.18, the expression is

   $$V_A(1/R_1 + 1/R_2 + 1/R_3) - V_B(1/R_3)$$
   $$= E_3(1/R_3) - E_2(1/R_2) + I_s$$

5. Solve the equations of Step 4 for the node voltages. If any node voltage is negative, use that negative value in the calculations.
6. Arbitrarily assign a direction for the current in a branch. Then, write an equation for each branch current in terms of the node voltages at the two ends, and the rises and drops in the branch. Start at any end and move toward the other node. For the R_2 branch

of Figure 9.18 and I_2 downward, V_2 is a drop and E_1 is a rise, so

$$V_A - I_2R_2 + E_1 = 0 \qquad \text{or} \qquad I_2 = (V_A + E_1)/R_1$$

If the calculated current is negative, its direction is opposite of that assigned.

7. Use the basic circuit concepts to find other quantities using basic circuit concepts.

The following example shows how these steps are applied.

EXAMPLE 9.9

Write the nodal equations and the equations for the current in R_1 of the circuit in Figure 9.19.

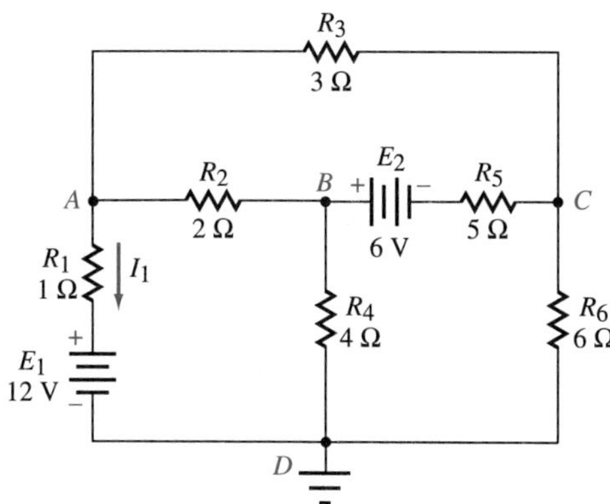

Figure 9.19

Solution

The circuit has four nodes, marked A, B, C, and D. Node D will be the reference node, and equations are written for A, B, and C. The terms in the equations are:

Node	Node Terms	Adjacent Node Terms	Source Terms
A	$V_A\left(\frac{1}{R_1} + \frac{1}{R_2} + \frac{1}{R_3}\right)$	$-V_B\left(\frac{1}{R_2}\right)$, $-V_C\left(\frac{1}{R_3}\right)$	$E_1\left(\frac{1}{R_1}\right)$
B	$V_B\left(\frac{1}{R_2} + \frac{1}{R_4} + \frac{1}{R_5}\right)$	$-V_A\left(\frac{1}{R_2}\right)$, $-V_C\left(\frac{1}{R_5}\right)$	$E_2\left(\frac{1}{R_5}\right)$
C	$V_C\left(\frac{1}{R_3} + \frac{1}{R_5} + \frac{1}{R_6}\right)$	$-V_B\left(\frac{1}{R_5}\right)$, $-V_A\left(\frac{1}{R_3}\right)$	$-E_2\left(\frac{1}{R_5}\right)$

The complete equations are

(1).
$$V_A\left(\frac{1}{R_1} + \frac{1}{R_2} + \frac{1}{R_3}\right) - V_B\left(\frac{1}{R_2}\right) - V_C\left(\frac{1}{R_3}\right) = E_1\left(\frac{1}{R_1}\right)$$

$$V_A\left(\frac{1}{1\ \Omega} + \frac{1}{2\ \Omega} + \frac{1}{3\ \Omega}\right) - V_B\left(\frac{1}{2\ \Omega}\right) - V_C\left(\frac{1}{3\ \Omega}\right) = 12\ \text{V}\left(\frac{1}{1\ \Omega}\right)$$

or

$$11V_A - 3V_B - 2V_C = 72\ \text{V}$$

(2).
$$V_B\left(\frac{1}{R_2} + \frac{1}{R_4} + \frac{1}{R_5}\right) - V_A\left(\frac{1}{R_2}\right) - V_C\left(\frac{1}{R_5}\right) = E_2\left(\frac{1}{R_5}\right)$$

$$V_B\left(\frac{1}{2\ \Omega} + \frac{1}{4\ \Omega} + \frac{1}{5\ \Omega}\right) - V_A\left(\frac{1}{2\ \Omega}\right) - V_C\left(\frac{1}{5\ \Omega}\right) = 6\ \text{V}\left(\frac{1}{5\ \Omega}\right)$$

or

$$-10V_A + 19V_B - 4V_C = 24\ \text{V}$$

(3).
$$V_C\left(\frac{1}{R_3} + \frac{1}{R_5} + \frac{1}{R_6}\right) - V_B\left(\frac{1}{R_3}\right) - V_B\left(\frac{1}{R_5}\right) = -E_2\left(\frac{1}{R_5}\right)$$

$$V_C\left(\frac{1}{3\ \Omega} + \frac{1}{5\ \Omega} + \frac{1}{6\ \Omega}\right) - V_A\left(\frac{1}{3\ \Omega}\right) - V_B\left(\frac{1}{5\ \Omega}\right) = -6\ \text{V}\left(\frac{1}{5\ \Omega}\right)$$

or

$$-10V_A - 6V_B + 21V_C = 36\ \text{V}$$

To write the equation for the current in R_1, I_1 is assigned a downward direction. Then starting at A and moving toward D, R_1 is a drop and E_1 is also a drop.

So

$$V_A - I_1R_1 - E_1 = 0\ \text{V}$$

or

$$I_1 = \frac{V_A - E_1}{R_1}$$

$$I_1 = \frac{V_A - 12\ \text{V}}{1\ \Omega}$$

or

$$I_1 = (V_A - 12)\ \text{A}$$

Once V_A is determined from the solution of the nodal equations, the value of I_1 can be calculated.

EXAMPLE 9.10

Find the currents, voltages, and power in the circuit of Figure 9.20.

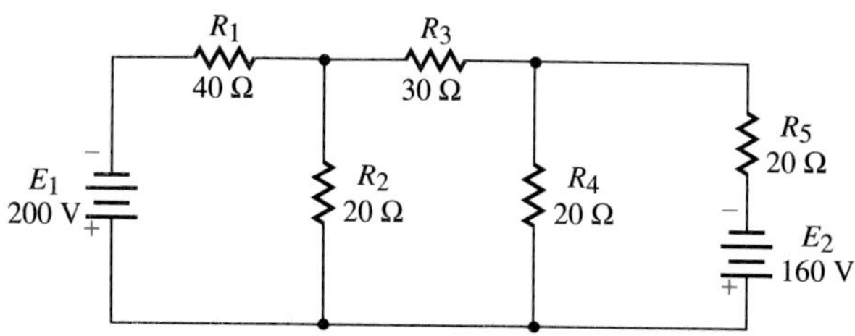

Figure 9.20

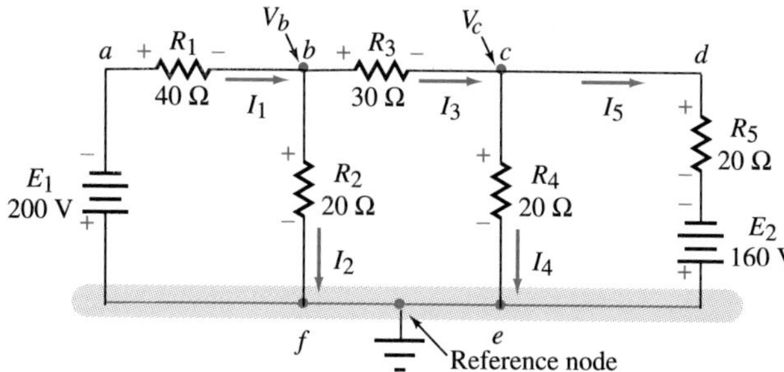

Figure 9.21

Solution

Strategy

Step 1. There are three nodes. They are at *b*, *c*, and the bottom line. There are five branches. They are *fab*, *fb*, *bc*, *ec*, and *cde*. These are shown in Figure 9.21.

Step 2. The node at the bottom line will be taken as the reference node. The other nodes are marked *b* and *c*.

Any node can be taken as the reference node.

Step 3. The terms in the equation are

Node	Node Terms	Adjacent Terms	Source Terms
B	$V_b\left(\frac{1}{R_1} + \frac{1}{R_2} + \frac{1}{R_3}\right)$	$-V_c\left(\frac{1}{R_3}\right)$	$-E_1\left(\frac{1}{R_1}\right)$
C	$V_c\left(\frac{1}{R_3} + \frac{1}{R_4} + \frac{1}{R_5}\right)$	$-V_b\left(\frac{1}{R_3}\right)$	$-E_2\left(\frac{1}{R_5}\right)$

Step 4. The complete expressions are

1. $$V_b\left(\frac{1}{R_1} + \frac{1}{R_2} + \frac{1}{R_3}\right) - V_c\left(\frac{1}{R_3}\right) = -E_1\left(\frac{1}{R_1}\right)$$

$$V_b\left(\frac{1}{40\ \Omega} + \frac{1}{20\ \Omega} + \frac{1}{30\ \Omega}\right) - V_c\left(\frac{1}{30\ \Omega}\right) = \left(-200\text{ V}\right)\left(\frac{1}{40\ \Omega}\right)$$

or

$$13\ V_b - 4\ V_c = -600\text{ V}$$

2. $V_c\left(\frac{1}{R_3} + \frac{1}{R_4} + \frac{1}{R_5}\right) - V_b\left(\frac{1}{R_3}\right) = -E_2\left(\frac{1}{R_5}\right)$

$$V_c\left(\frac{1}{30\ \Omega} + \frac{1}{20\ \Omega} + \frac{1}{20\ \Omega}\right) - V_b\left(\frac{1}{30\ \Omega}\right) = -160\ \text{V}\left(\frac{1}{20\ \Omega}\right)$$

or

$$-V_b + 4\ V_c = -240\ \text{V}$$

Step 5. Solving these by determinants gives

$$V_b = D_b/D, \qquad V_c = D_c/D$$

so

$$V_B = \frac{\begin{bmatrix} -600 & -4 \\ -240 & 4 \end{bmatrix}}{\begin{bmatrix} 13 & -4 \\ -1 & 4 \end{bmatrix}} = \frac{-2400 - (960)}{52 - (4)} = -70\ \text{V}$$

$$V_C = \frac{\begin{bmatrix} 13 & -600 \\ -1 & -240 \end{bmatrix}}{\begin{bmatrix} 13 & -4 \\ -1 & 4 \end{bmatrix}} = \frac{-3120 - (600)}{52 - (4)} = -77.5\ \text{V}$$

Step 6. For the I_1 branch, starting at Point f, and the current as shown, $0 - E_1 - I_1 R_1 = V_b$ so

$$I_1 = \frac{-E_1 - V_b}{R_1}$$

$$I_1 = \frac{-200\ \text{V} - (-70\ \text{V})}{40\ \Omega}$$

$$I_1 = -3.25\ \text{A}$$

The $(-)$ sign means that the direction is opposite of that assigned. For I_2 branch, starting at Point f, and the current as shown,

$$0 + I_2 R_2 = V_b$$

so

$$I_2 = \frac{V_b}{R_2}$$

$$I_2 = \frac{-70\ \text{V}}{20\ \Omega}$$

$$I_2 = -3.5\ \text{A}$$

The direction of I_2 is also opposite of that assigned. For the I_3 branch, starting at b, and the current as shown,

$$V_b - I_3R_3 = V_c$$

so

$$I_3 = \frac{V_b - V_c}{R_3}$$

$$= \frac{-70\text{ V} - (-77.5\text{ V})}{30\ \Omega}$$

$$I_3 = 0.25\text{ A}$$

The direction is as assigned. For the I_4 branch, starting at f, and the current as shown,

$$0 + I_4R_4 = V_c$$

so

$$I_4 = \frac{V_c}{R_4}$$

$$= \frac{-77.5\text{ V}}{20\ \Omega}$$

$$I_4 = -3.875\text{ A}$$

The direction of I_4 is also opposite of that assigned. For the I_5 branch, starting at f, and the current as shown,

$$0 - E_2 + I_5R_5 = V_c$$

so

$$I_5 = \frac{V_C + E_2}{R_5}$$

$$= \frac{-77.5\text{ V} + 160\text{ V}}{20\ \Omega}$$

$$I_5 = 4.125\text{ A}$$

The direction is as assigned.

Step 7. Applying Ohm's Law to each resistor,

$$V_1 = I_1R_1 = 3.25\text{ A} \times 40\ \Omega$$

$$V_1 = 130\text{ V}$$

$$V_2 = I_2R_2 = 3.5\text{ A} \times 20\ \Omega$$

$$V_2 = 70\text{ V}$$

$$V_3 = I_3R_3 = 0.25\text{ A} \times 30\ \Omega$$

$$V_3 = 7.5\text{ V}$$

$$V_4 = I_4R_4 = 3.875\text{ A} \times 20\ \Omega$$

$$V_4 = 77.5\text{ V}$$

$$V_5 = I_5R_5 = 4.125\text{ A} \times 20\ \Omega$$

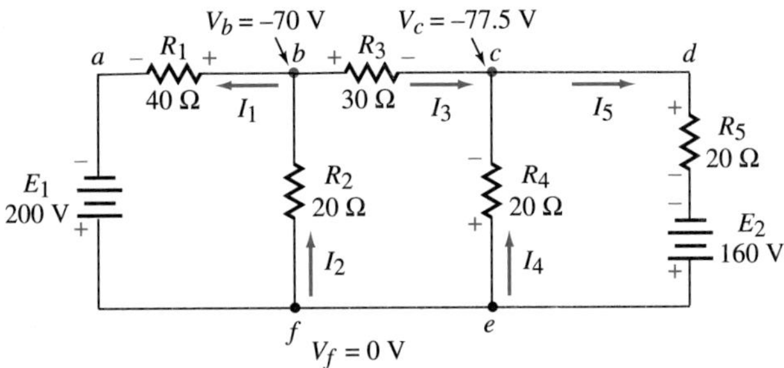

Figure 9.22

$$V_5 = 82.5 \text{ V}$$

Finally, the total power relation will be used to calculate the circuit power.

$$P_T = P_1 + P_2 + P_3 + P_4 + P_5$$
$$P_1 = V_1 I_1 = (130 \text{ V})(3.25 \text{ A}) = 422.5 \text{ W}$$
$$P_2 = V_2 I_2 = (70 \text{ V})(3.5 \text{ A}) = 245 \text{ W}$$
$$P_3 = V_3 I_3 = (7.5 \text{ V})(0.25 \text{ A}) = 1.88 \text{ W}$$
$$P_4 = V_4 I_4 = (77.5 \text{ V})(3.875 \text{ A}) = 300.31 \text{ W}$$
$$P_5 = V_5 I_5 = (82.5 \text{ V})(4.125 \text{ A}) = 340.31 \text{ W}$$
$$P_T = 422.5 \text{ W} + 245 \text{ W} + 1.88 \text{ W} + 300.31 \text{ W} + 340.31 \text{ W}$$
$$P_T = 1310 \text{ W}$$

The circuit is drawn in Figure 9.22 showing the correct current directions and terminal polarity.

EXAMPLE 9.11

Use the Nodal method to find the following quantities in the circuit of Figure 9.23:

a. the voltage V_{BC}.

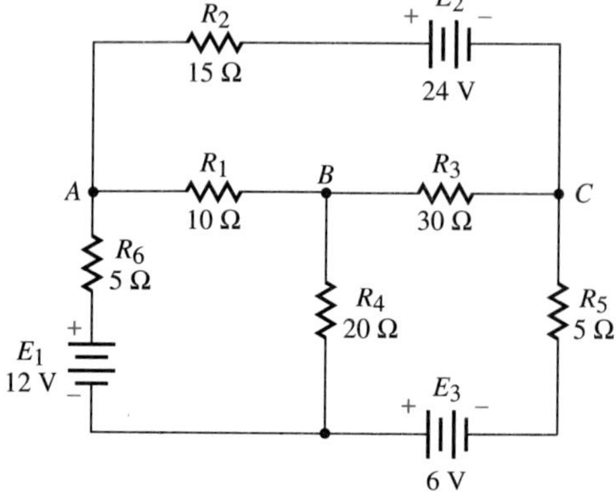

Figure 9.23

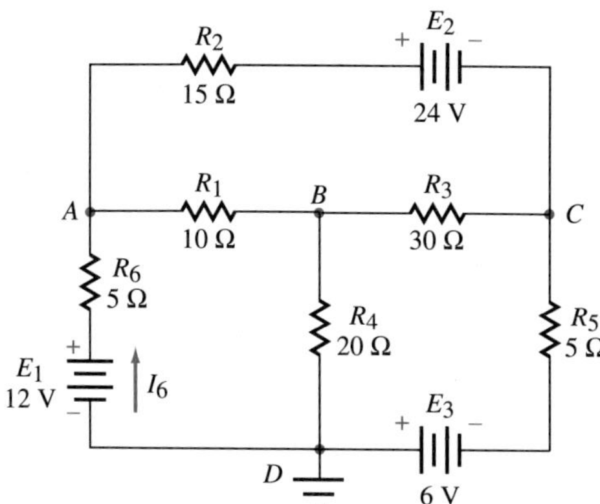

Figure 9.24

b. the current in R_6.

Solution

Step 1. There are four nodes and six branches. The nodes are marked A, B, C, and D in Figure 9.24.

Step 2. Node D will be the reference node.

Step 3. The terms in the node equations are

Node	Node Terms	Adjacent Terms	Source Terms
A	$V_A\left(\frac{1}{R_1}+\frac{1}{R_2}+\frac{1}{R_6}\right)$	$-V_B\left(\frac{1}{R_1}\right),\ -V_C\left(\frac{1}{R_2}\right)$	$E_1\left(\frac{1}{R_1}\right),\ E_2\left(\frac{1}{R_2}\right)$
B	$V_B\left(\frac{1}{R_1}+\frac{1}{R_3}+\frac{1}{R_4}\right)$	$-V_B\left(\frac{1}{R_1}\right),\ -V_C\left(\frac{1}{R_3}\right)$	—
C	$V_C\left(\frac{1}{R_2}+\frac{1}{R_3}+\frac{1}{R_5}\right)$	$-V_A\left(\frac{1}{R_2}\right),\ -V_B\left(\frac{1}{R_3}\right)$	$-E_2\left(\frac{1}{R_2}\right),\ -E_3\left(\frac{1}{R_5}\right)$

Step 4. The complete expressions are

$$(1)\quad V_A\left(\frac{1}{R_1}+\frac{1}{R_2}+\frac{1}{R_6}\right)-V_B\left(\frac{1}{R_1}\right)-V_C\left(\frac{1}{R_2}\right)$$
$$=E_1\left(\frac{1}{R_6}\right)+E_2\left(\frac{1}{R_2}\right)$$
$$V_A\left(\frac{1}{10\ \Omega}+\frac{1}{15\ \Omega}+\frac{1}{5\ \Omega}\right)-V_B\left(\frac{1}{10\ \Omega}\right)-V_C\left(\frac{1}{15\ \Omega}\right)$$
$$=12\ \text{V}\left(\frac{1}{5\ \Omega}\right)+24\ \text{V}\left(\frac{1}{15\ \Omega}\right)$$

or

$$11V_A\ \text{A}-3V_B\ \text{A}-2V_C\ \text{A}=120\ \text{A}$$

$$(2)\quad V_B\left(\frac{1}{R_1}+\frac{1}{R_3}+\frac{1}{R_4}\right)-V_A\left(\frac{1}{R_1}\right)-V_C\left(\frac{1}{R_3}\right)=0\ \text{A}$$

$$V_B\left(\frac{1}{10\ \Omega} + \frac{1}{30\ \Omega} + \frac{1}{20\ \Omega}\right) - V_A\left(\frac{1}{10\ \Omega}\right) - V_C\left(\frac{1}{30\ \Omega}\right) = 0\ \text{A}$$

or

$$-6V_A\ \text{A} + 11V_B\ \text{A} - 2V_C\ \text{A} = 0$$

(3) $V_C\left(\frac{1}{R_2} + \frac{1}{R_3} + \frac{1}{R_5}\right) - V_A\left(\frac{1}{R_2}\right) - V_B\left(\frac{1}{R_3}\right)$

$= -E_2\left(\frac{1}{R_2}\right) - E_3\left(\frac{1}{R_5}\right)$

$V_C\left(\frac{1}{15\ \Omega} + \frac{1}{30\ \Omega} + \frac{1}{5\ \Omega}\right) - V_A\left(\frac{1}{15\ \Omega}\right) - V_B\left(\frac{1}{30\ \Omega}\right)$

$= -24\ \text{V}\left(\frac{1}{15\ \Omega}\right) - 6\ \text{V}\left(\frac{1}{5\ \Omega}\right)$

or

$$-2V_A\ \text{A} - V_B\ \text{A} + 9V_C\ \text{A} = -84\ \text{A}$$

Step 5. The solution by determinants is

$$V_A = D_A/D, \qquad V_B = D_B/D, \qquad V_C = D_C/D$$

$$V_A = \frac{\begin{bmatrix} +120 & -3 & -2 \\ 0 & +11 & -2 \\ -84 & -1 & +9 \end{bmatrix}}{\begin{bmatrix} +11 & -3 & -2 \\ -6 & +11 & -2 \\ -2 & -1 & +9 \end{bmatrix}} = \frac{11{,}376 - (2088)}{1065 - (228)} = 11.10\ \text{V}$$

$$V_B = \frac{\begin{bmatrix} +11 & +120 & -2 \\ -6 & 0 & -2 \\ -2 & -84 & +9 \end{bmatrix}}{\begin{bmatrix} +11 & -3 & -2 \\ -6 & +11 & -2 \\ -2 & -1 & +9 \end{bmatrix}} = \frac{-528 - (-4632)}{837} = 4.90\ \text{V}$$

$$V_C = \frac{\begin{bmatrix} +11 & -3 & +120 \\ -6 & +11 & 0 \\ -2 & -1 & -84 \end{bmatrix}}{\begin{bmatrix} +11 & -3 & -2 \\ -6 & +11 & -2 \\ -2 & -1 & +9 \end{bmatrix}} = \frac{-9444 - (-4152)}{837} = -6.32\ \text{V}$$

Since V_{BC} is $V_B - V_C$,

$$V_{BC} = 4.90 \text{ V} - (-6.32 \text{ V})$$
$$V_{BC} = 11.22 \text{ V}$$

Node B is at a more positive potential than Node C.

Step 6. If the current in R_6 is taken as upward, and starting at Node D,

$$0 + E_1 - I_6R_6 = V_A$$

so

$$I_6 = \frac{E_1 - V_A}{R_6} = \frac{12 \text{ V} - 11.10 \text{ V}}{5 \ \Omega}$$
$$I_6 = 0.18 \text{ A}$$

The current is in the assigned direction because it is positive.

EXAMPLE 9.12

Use nodal analysis to find the voltage of Terminal A as measured from B in Figure 9.25.

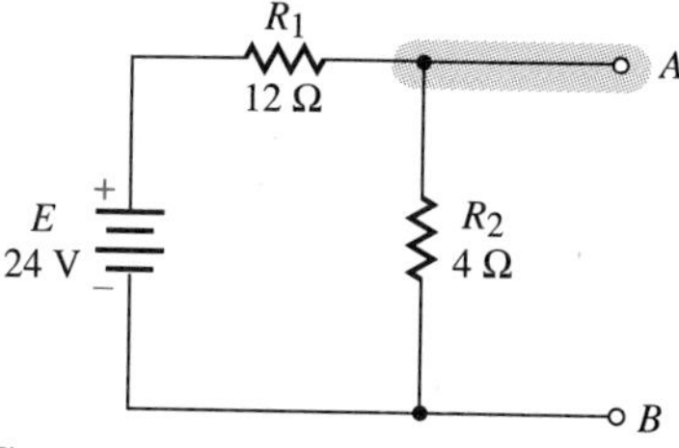

Figure 9.25

Solution

Terminal A is at the same voltage as the junction of R_1 and R_2. Although the junction has only two branches, the nodal equation can be written for it. The expression is

$$V_A\left(\frac{1}{R_1} + \frac{1}{R_2}\right) = E\left(\frac{1}{R_1}\right)$$
$$V_A\left(\frac{1}{12 \ \Omega} + \frac{1}{4 \ \Omega}\right) = (24 \text{ V})\left(\frac{1}{12 \ \Omega}\right)$$

or

$$4 \ V_A = 24 \text{ V}$$

so

$$V_A = 6 \text{ V}$$

SUMMARY

1. The loop current, branch current, and nodal methods of analysis are used with circuits that have several sources or are not series-parallel.
2. The loop current method uses an assumed current in each loop. Kirchhoff's Voltage Law is applied to N–1 loops, and branch currents are found from the loop currents.
3. The branch current method uses a branch current in each branch.

Kirchhoff's Voltage Law is applied to N–1 meshes. Kirchhoff's Current Law is applied to N–1 nodes.

4. In the nodal method, the node voltages are found first. These are used to find the branch currents.
5. The format approach for the loop and nodal analysis methods provides a mechanical procedure for developing the equations.

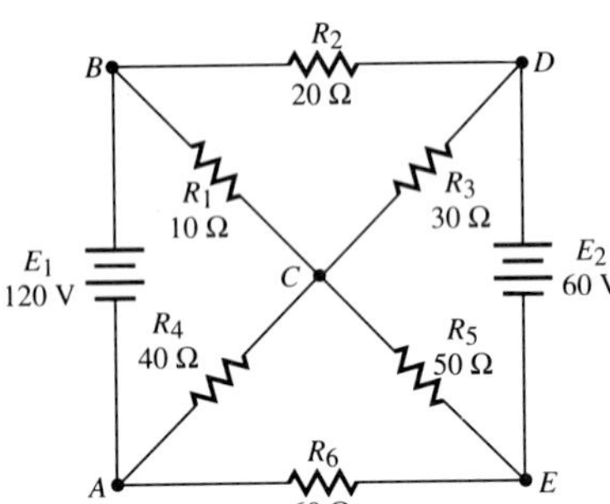

Figure 9.26

QUESTIONS

1. What are three methods of analysis that can be used for circuits containing several sources, or circuits that are neither series, parallel, nor series-parallel?
2. To how many loops must Kirchhoff's Voltage Law be applied in a circuit of four loops in the loop current method and the branch current method?
3. Can a branch current be greater than a loop current?
4. How is a branch current found once the loop currents are known?
5. What is the significance of a negative sign for a loop or branch current?
6. How can one find the equivalent resistance of a circuit that is not a series, parallel, or series-parallel connection?
7. Which method of analysis expresses the branch currents in terms of the node voltages?
8. What is the voltage of the reference node in the nodal analysis method?
9. What will be the direction of a current between Node A and Node B if the voltage of Node A is -8 V and that of Node B is -10 V?
10. Will changing the reference node change the calculated magnitudes of the current in a branch?
11. How many node equations must be written for a circuit with five nodes?
12. In a branch with only resistance, the current direction is out of the more positive node. Does this have to be so if there is a source in the branch?
13. The currents in two branches of a node are out of the node. What will be the direction of the current in the third branch?

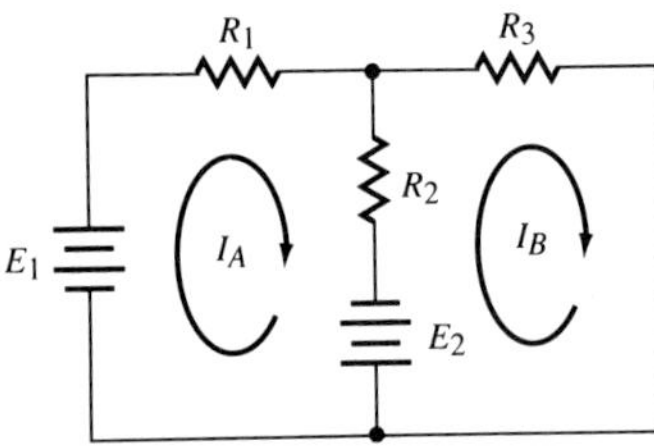

Figure 9.27

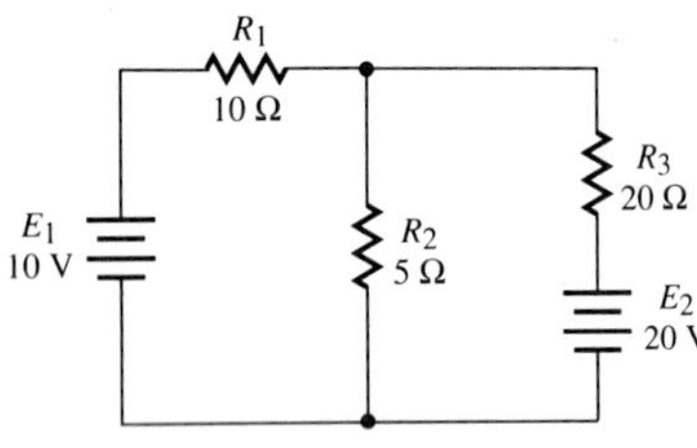

Figure 9.28

PROBLEMS

SECTION 9.1 Loop Current Analysis

Solve the problems in this section by using the loop method.

1. Write the loop equations for the circuit of Figure 9.26.
2. Determine the current and its direction for each branch of Figure 9.27 if I_A was calculated to be 2 amperes and I_B to be -5 amperes.
3. For the circuit of Figure 9.28, find:

   (a) The current in each resistor.

   (b) The voltage across each resistor.

   (c) The total power dissipated in the resistance of the circuit.

4. In the two-load voltage divider in Figure 9.29, find:

   (a) The voltage across R_{L1} and R_{L2}.

   (b) The % efficiency of the divider.

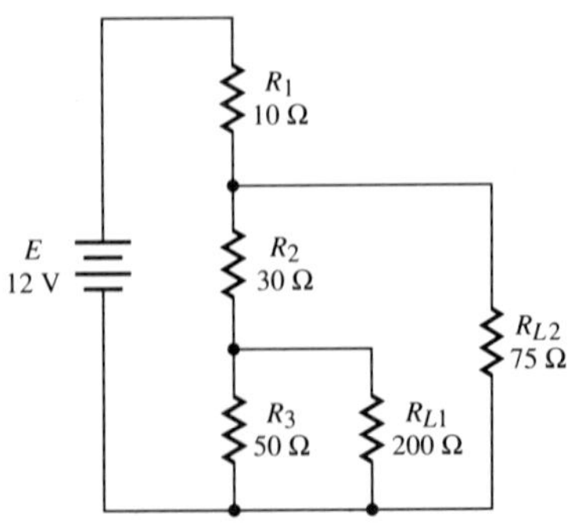

Figure 9.29

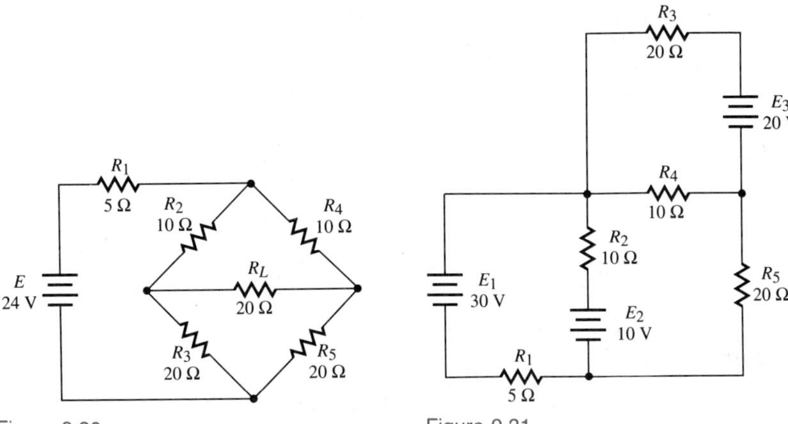

Figure 9.30

Figure 9.31

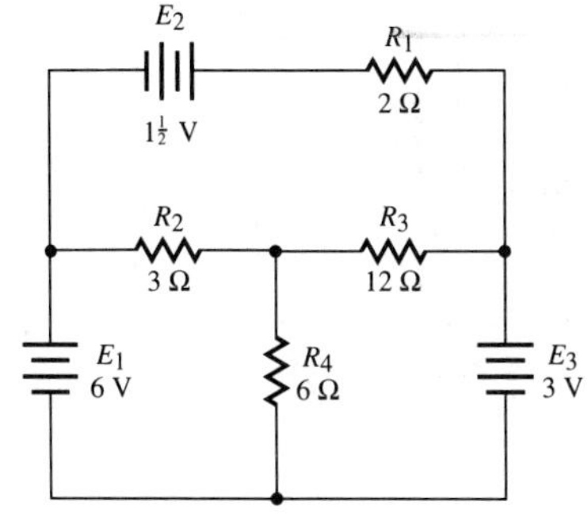

Figure 9.32

(c) The voltage across R_L if R_{L2} is disconnected.

5. In the bridge circuit for Figure 9.30, find the voltage across R_L.

6. For the circuit of Figure 9.31, find:

   (a) The current in each resistor.

   (b) The voltage across each resistor.

   (c) The power rating that R_4 must have to prevent damage to it.

7. For the circuit of Figure 9.32, find:

   (a) The current in each resistor.

   (b) The voltage across each resistor.

   (c) The current in each source.

8. What is the resistance of R_2 in Figure 9.33 if I_2 is 4.5 amperes? Does E_2 supply or use power?

9. For the circuit of Figure 9.34, find:

   (a) The current in R_1, R_2, and R_3.

   (b) The voltage across R_4.

SECTION 9.2 Branch Current Analysis

Solve the following problems using the branch current method.

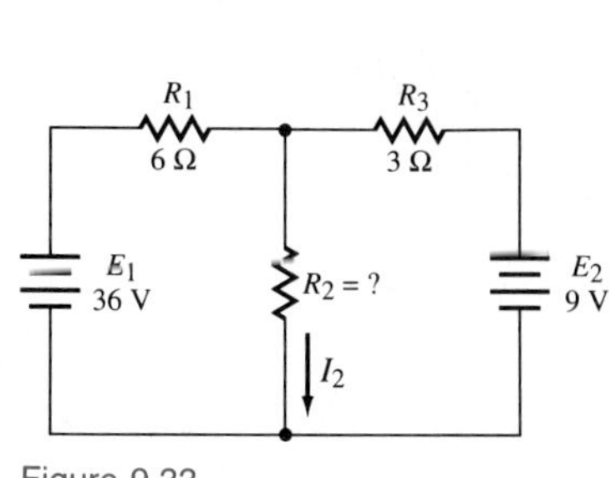

Figure 9.33

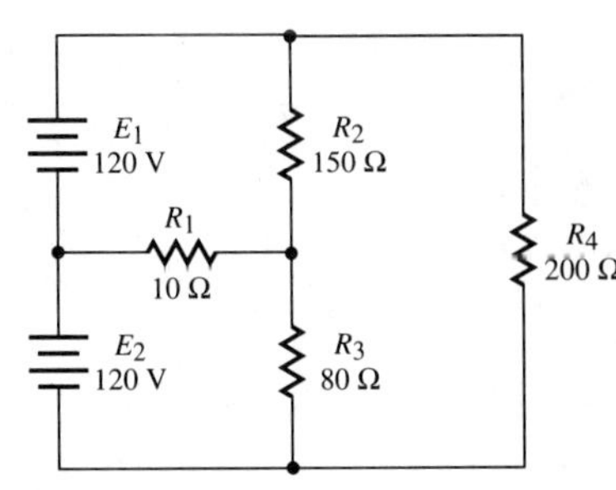

Figure 9.34

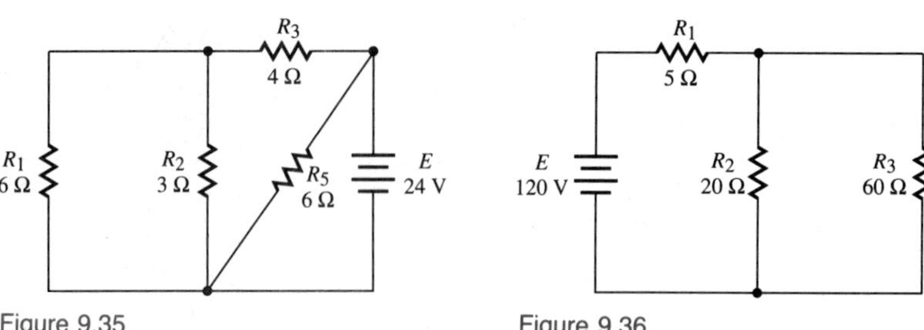

Figure 9.35 Figure 9.36

10. Write the loop equations for the circuit of Figure 9.35 using the branch current method.
11. Determine the power in R_2 of Figure 9.36 using the branch current method.
12. Work Problem 3 using the branch current method.
13. For the circuit of Figure 9.37, find:
    (a) The current in each resistor.
    (b) The voltage across each resistor.
    (c) The power dissipated in the circuit.

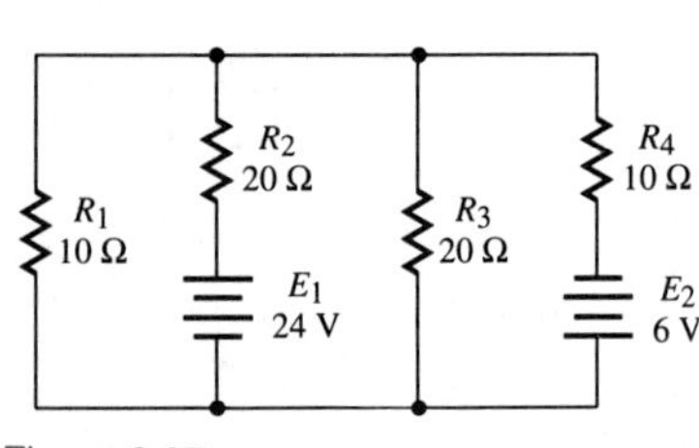

Figure 9.37

14. Work Problem 4 using the branch current method.
15. Work Problem 5 using the branch current method.
16. Work Problem 6 using the branch current method.
17. Work Problem 7 using the branch current method.
18. Work Problem 8 using the branch current method.

SECTION 9.3 Nodal Analysis

Solve the following problems using the nodal method.

19. Write the nodal equations for the circuit of Problem 1 using A as the reference node. Assume 20 ohms of internal resistance for E_1 and E_2.
20. Determine the current in each resistor of Figure 9.38 for the node voltage given.

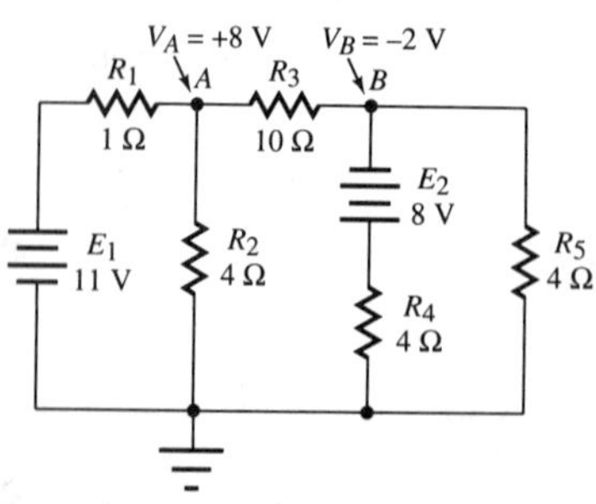

Figure 9.38

21. Work Problem 3 using nodal analysis.

22. In Figure 9.39, find:

 (a) The current in each resistor.

 (b) The voltage across R_3.

 (c) The power in the circuit.

23. Find V_{L1}, V_{L2}, and the source current for the circuit of Problem 4.

24. Work Problem 6 using nodal analysis.

25. In Figure 9.40, find:

 (a) The current in E_1 and E_3.

 (b) The voltage across R_4.

26. Work Problem 7 using nodal analysis.

27. Work Problem 8 using nodal analysis.

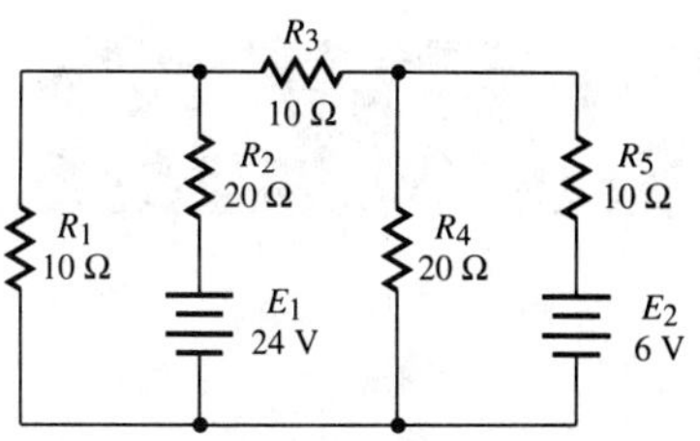

Figure 9.39

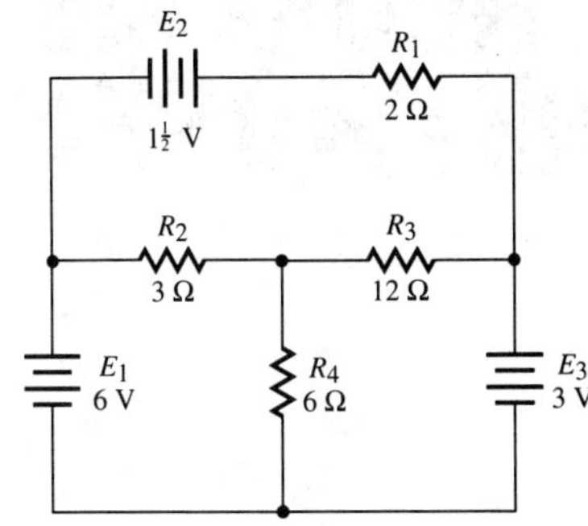

Figure 9.40

CHAPTER 10

THEOREMS AND NETWORKS

CHAPTER OBJECTIVES

After completing this chapter, you should be able to:

1. Explain the difference between a linear and a nonlinear resistance, and a unilateral and a bilateral resistance.
2. Effectively apply the Superposition Theorem to perform the analysis of a multisource circuit.
3. Determine the Thevenin and Norton equivalents of a circuit.
4. Make the conversion between the Thevenin and Norton equivalent circuit.
5. Effectively apply the maximum power transfer theorem to determine source and load characteristics.
6. Describe the relation of power and efficiency to the ratio of load and source resistance.
7. Describe the Delta, Wye, Pi, and Tee network configurations.

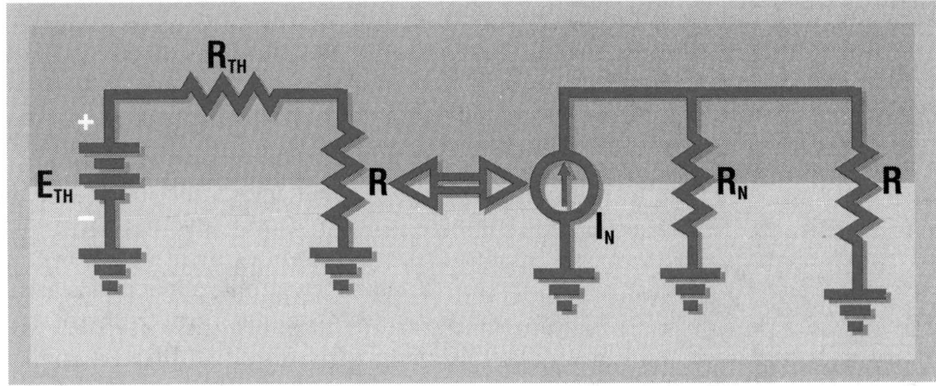

8. Make the conversion between Delta and Wye networks.

KEY TERMS

1. bilateral resistance
2. Delta network
3. linear resistance
4. maximum power transfer theorem
5. Norton's Theorem
6. superposition theorem
7. Thevenin's Theorem
8. Wye network

INTRODUCTION

Many theorems for electrical circuits were developed during the study of electricity. Some of these theorems such as superposition, Thevenin's, Norton's, and maximum power transfer are presented in this chapter.

These theorems can be used to simplify a circuit, provide a relation between electrical quantities, or replace a device by a circuit. For example, a transistor is replaced by a combined Thevenin and Norton circuit.

With Thevenin's and Norton's Theorems, a circuit can be replaced by a single source and resistor. This is useful where repetitive calculations are needed.

Superposition shows us how currents and voltages of the sources in a circuit can be combined to give the actual currents and voltages. This theorem also helps us to understand how a waveform can be made up of several individual waveforms.

The maximum power transfer theorem deals with power transfer from a source to a load. This transfer can be critical in electronic applications where only a small amount of power is available.

Finally, the Delta and Wye networks and their conversion are explained. These networks are used in power distribution systems. They also are found in electronic applications as Pi and Tee networks. Being able to convert from one form to the other often helps in calculating power loads, or in analyzing the operation of an electronic circuit.

10.1 The Superposition Theorem

superposition theorem

linear resistance

bilateral resistance

The **superposition theorem** states that the current and voltage at any point in a multisource circuit of linear bilateral resistances are the algebraic sum of the current and voltage from each source placed in the circuit one at a time, while the others are replaced by their internal resistance. A **linear resistance** is one whose resistance does not change with current—that is—V/I is constant. A **bilateral resistance** is one in which the resistance is the same for both directions of current through the resistor.

An added feature of the superposition theorem is that the calculations for one source are relatively easy. The analysis can be done by using any of the methods presented in past sections. On the negative side, the calculations are lengthy since they must be done for each source; keeping track of the directions and polarities might become difficult.

The sources in the circuit need not all be dc or all ac ones; they can be both ac and dc. For example, electronic amplifiers have a dc bias voltage in them along with the ac signal, the output of a dc power supply has ac and dc in it, and the signals used in television and electronic communications are combinations of several voltages.

The superposition theorem is applied as follows:

1. Replace all sources except one by their internal resistance.
2. Calculate the currents and voltages for that one source. Note the current directions and terminal polarities.
3. Repeat Steps 1 and 2 for each source.
4. Determine the currents by algebraically adding the currents due to each source.
5. Determine the voltages by either algebraically adding the voltages for each source, or using the total current and Ohm's Law.
6. Use basic circuit concepts to find other quantities.

The following examples illustrate the application of the superposition theorem.

EXAMPLE 10.1

Use superposition to find the currents, voltages, and power in the circuit of Figure 10.1.

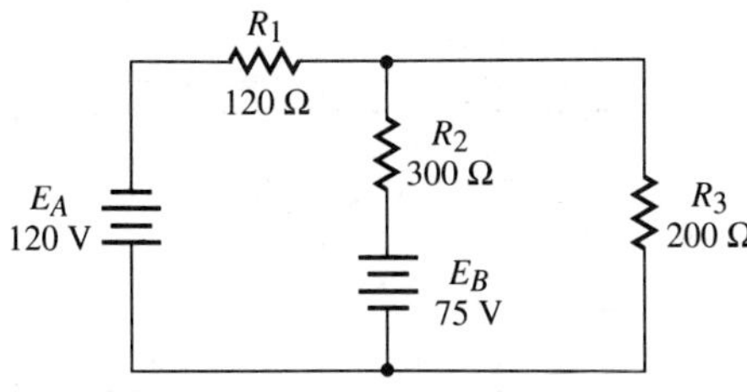

Figure 10.1

Solution

Strategy

The effect of Source A is considered first.

Step 1. E_B is replaced by a short circuit because no internal resistance is shown. The circuit is shown in Figure 10.2. Since there is only one source, the current direction is as shown.

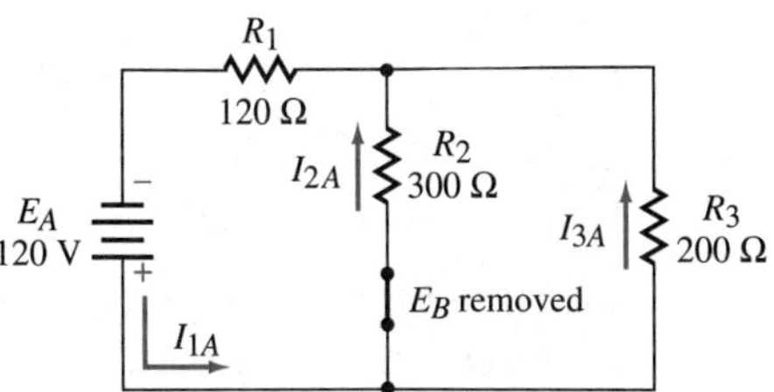

Figure 10.2

Step 2. This is a series-parallel circuit, so for Source A,

$$I_{1A} = \frac{E_A}{R_{eqA}}$$

$$\begin{aligned} R_{eqA} &= R_1 + R_2 \| R_3 \\ &= R_1 + \frac{R_2 R_3}{R_2 + R_3} \\ &= 120\ \Omega + \frac{(300\ \Omega)(200\ \Omega)}{300\ \Omega + 200\ \Omega} \\ &= 240\ \Omega \end{aligned}$$

so,

The direction of the current must also be noted.

$$I_{1A} = \frac{120\text{ V}}{240\ \Omega}$$

$$I_{1A} = 0.5\text{ A down}$$

Using the current divider rule gives

$$\begin{aligned} I_{2A} &= \left(\frac{R_3}{R_2 + R_3}\right) I_{1A} \\ &= \left(\frac{200\ \Omega}{300\ \Omega + 200\ \Omega}\right)(0.5\text{ A}) \end{aligned}$$

$$I_{2A} = 0.2\text{ A up}$$

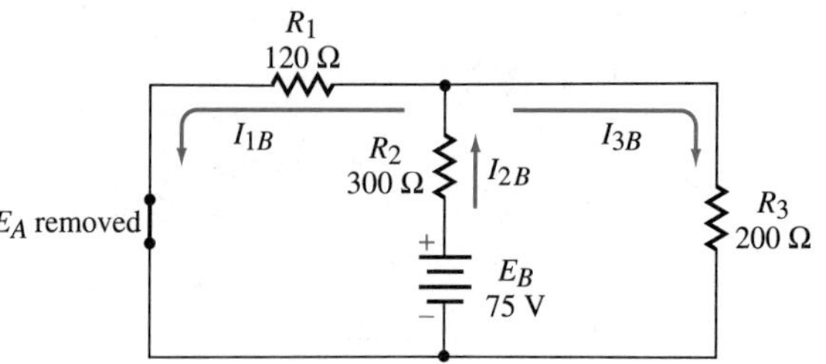

Figure 10.3

$$I_{3A} = \left(\frac{R_2}{R_2 + R_3}\right)I_{1A}$$

$$= \left(\frac{300\ \Omega}{300\ \Omega + 200\ \Omega}\right)(0.5\ \text{A})$$

$$I_{3A} = 0.3\ \text{A up}$$

$$V_{1A} = I_{1A}\,R_1 = (0.5\ \text{A})(120\ \Omega) = 60\ \text{V}$$

$$V_{2A} = I_{2A}\,R_2 = (0.2\ \text{A})(300\ \Omega) = 60\ \text{V}$$

$$V_{3A} = I_{3A}\,R_3 = (0.3\ \text{A})(200\ \Omega) = 60\ \text{V}$$

Next, the effect of Source B is considered.

Step 3. Next, E_A is replaced by a short circuit, as in Figure 10.3. This is also a series-parallel circuit, so for E_B,

$$I_{2B} = \frac{E_B}{R_{\text{eq}}}$$

$$R_{\text{eq}B} = R_2 + R_1\|R_3$$

$$= R_2 + \frac{R_1R_3}{R_1 + R_3}$$

$$R_{\text{eq}B} = 300\ \Omega + \frac{(120\ \Omega)(200\ \Omega)}{120\ \Omega + 200\ \Omega}$$

$$= 375\ \Omega$$

so

$$I_{2B} = 75\ \text{V}/375\ \Omega$$

$$I_{2B} = 0.2\ \text{A up}$$

Using the current divider rule gives

$$I_{1B} = \left(\frac{R_3}{R_1 + R_3}\right)(I_{2B})$$

$$= \left(\frac{200\ \Omega}{120\ \Omega + 200\ \Omega}\right)(0.2\ \text{A})$$

$$I_{1B} = 0.125\ \text{A down}$$

$$I_{3B} = \left(\frac{R_1}{R_1 + R_3}\right)(I_{2B})$$

$$= \left(\frac{120\ \Omega}{120\ \Omega + 200\ \Omega}\right)(0.2\ \text{A})$$

$$I_{3B} = 0.075 \text{ A down}$$
$$V_{1B} = I_{1B} R_1 = (0.125 \text{ A})(120\ \Omega) = 15 \text{ V}$$
$$V_{2B} = I_{2B} R_2 = (0.2 \text{ A})(300\ \Omega) = 60 \text{ V}$$
$$V_{3B} = I_{3B} R_3 = (0.075 \text{ A})(200\ \Omega) = 15 \text{ V}$$

Step 4. By superposition, I_{1A} and I_{1B} are in the same direction,

The currents must be combined to get the current with all sources in the circuit.

$$I_1 = I_{1A} + I_{1B}$$
$$= 0.5 \text{ A} + 0.125 \text{ A}$$
$$I_1 = 0.625 \text{ A in the direction of } I_{1A}$$

I_{2A} and I_{2B} are in the same direction, so

$$I_2 = I_{2A} + I_{2B}$$
$$= +0.2 \text{ A} + 0.2 \text{ A}$$
$$I_2 = +0.4 \text{ A in the direction of } I_{2A}$$

I_{3A} and I_{3B} are in opposite directions, so

$$I_3 = I_{3A} - I_{3B}$$
$$= 0.3 \text{ A} - 0.075 \text{ A}$$
$$I_3 = 0.225 \text{ A in the direction of } I_{3A}$$

Step 5. By superposition, the voltages are:
For V_1, the polarity is the same for both sources, so

Use superposition to calculate voltage.

$$V_1 = V_{1A} + V_{1B}$$
$$= 60 \text{ V} + 15 \text{ V}$$
$$V_1 = 75 \text{ V}$$

The left terminal is negative.
For V_2, the polarity is the same for both sources, so

$$V_2 = V_{2A} + V_{2B}$$
$$= 60 \text{ V} + 60 \text{ V}$$
$$V_2 = 120 \text{ V}$$

The bottom terminal is positive.
For V_3, the polarities are different and V_{3A} is larger, so

$$V_3 = V_{3A} - V_{3B}$$
$$= 60 \text{ V} - 15 \text{ V}$$
$$V_3 = 45 \text{ V}$$

The bottom terminal is positive. The correct directions and polarities are shown in Figure 10.4. Note that these results are the same as from the loop method.
Sometimes it is less confusing to find the voltage by using Ohm's Law and the total currents. For example,

$$V_2 = I_2 R_2$$
$$= (0.4 \text{ A})(300\ \Omega)$$

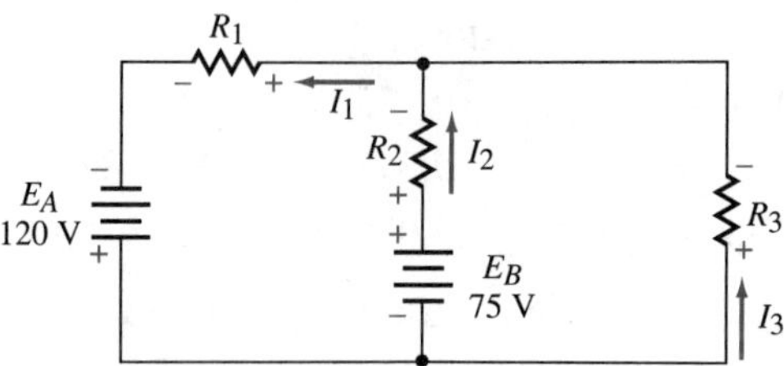

Figure 10.4

$$V_2 = 120 \text{ V}$$

Use the total current to calculate power.

Step 6. $(I_{1A} + I_{2A})^2$ is not the same as $I_{1A}^2 + I_{2A}^2$, so superposition cannot be used to calculate power. Power is calculated using either the total current or the total voltage. For this example,

$$P_T = P_1 + P_2 + P_3$$
$$P_1 = I_1^2 R_1$$
$$= (0.625 \text{ A})^2 (120 \ \Omega)$$
$$P_1 = 46.88 \text{ W}$$
$$P_2 = I_2^2 R_2$$
$$= (0.4 \text{ A})^2 (300 \ \Omega)$$
$$P_2 = 48 \text{ W}$$
$$P_3 = I_3^2 R_3$$
$$= (0.225 \text{ A})^2 (200 \ \Omega)$$
$$P_3 = 10.13 \text{ W}$$
$$P_T = 46.88 \text{ W} + 48 \text{ W} + 10.13 \text{ W}$$
$$P_T = 105 \text{ W}$$

Also, since both sources supply power for this circuit,

$$P_T = E_A I_1 + E_B I_2$$
$$= (120 \text{ V})(0.625 \text{ A}) + (75 \text{ V})(0.4 \text{ A})$$
$$= 75 \text{ W} + 30 \text{ W}$$
$$P_T = 105 \text{ W}$$

EXAMPLE 10.2

A voltmeter is used in the circuit of Figure 10.5 to show the effect of light on a photocell. The resistance of the cell is 5 ohms under light. Use superposition to determine the reading on the meter.

Solution

Source E_2 is replaced by a short circuit. The circuit is not a series-parallel circuit so the loop method is used. The circuit showing the loops, nodes, and terminal polarity is drawn in Figure 10.6.

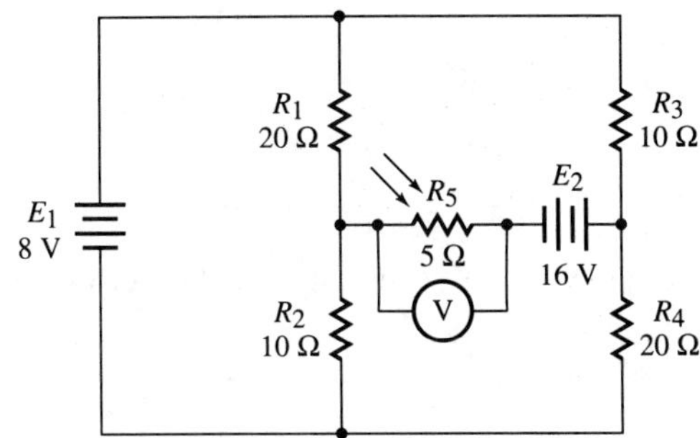

Figure 10.5

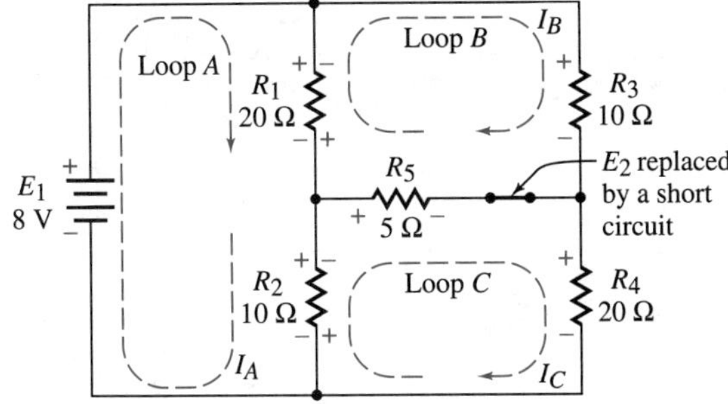

Figure 10.6

For Loop A:

$$I_A(R_1 + R_2) - I_BR_1 - I_CR_2 - E_1 = 0 \text{ V}$$

or

$$-8 \text{ V} + (30 \ \Omega)I_A - (20 \ \Omega)I_B - (10 \ \Omega)I_C = 0 \text{ V}$$

For Loop B:

$$I_B(R_1 + R_3 + R_5) - I_AR_1 - I_CR_5 = 0 \text{ V}$$

or

$$+ (35 \ \Omega)I_B - (20 \ \Omega)I_A - (5 \ \Omega)I_C = 0 \text{ V}$$

For Loop C:

$$I_C(R_2 + R_5 + R_4) - I_AR_2 - I_BR_5 = 0 \text{ V}$$

or

$$+ (35 \ \Omega)I_C - (10 \ \Omega)I_A - (5 \ \Omega)I_B = 0 \text{ V}$$

Solving the three loop equations gives

$$I_A = 5.8 \times 10^{-1} \text{ A}$$
$$I_B = 3.63 \times 10^{-1} \text{ A}$$
$$I_C = 2.18 \times 10^{-1} \text{ A}$$

The signs are all positive so the assumed directions are correct.

The current in R_5 is found by algebraically combining the loop currents.

For source E_1,

$I_5 = I_C - I_B$

$= 2.18 \times 10^{-1}$ A $- 3.63 \times 10^{-1}$ A

$I_5 = -1.45 \times 10^{-1}$ A. The $(-)$ means that its direction is that of I_B.

Next, E_1 is replaced by a short circuit. The circuit is as in Figure 10.7(a).

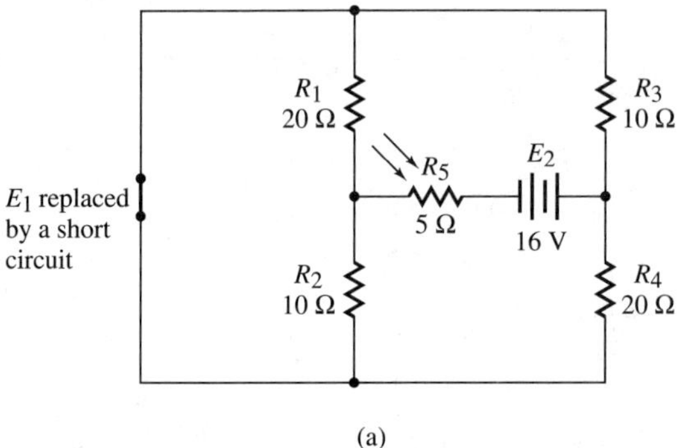

(a)

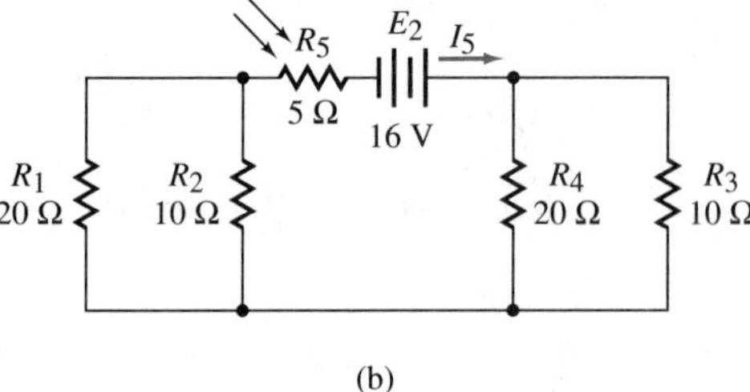

(b)

Figure 10.7

An inspection of the circuit, redrawn in Figure 10.7(b), shows that the short circuit has put R_1 in parallel with R_2, and R_3 in parallel with R_4. The circuit is now a series-parallel circuit. For it,

$$I_s = \frac{E_2}{R_{eq}}$$

but

$$R_{eq} = R_5 + R_1 \| R_2 + R_3 \| R_4$$

$$R_1 \| R_2 = \frac{R_1 R_2}{R_1 + R_2}$$

$$= \frac{(20\ \Omega)(10\ \Omega)}{20\ \Omega + 10\ \Omega}$$

$$= 6.67\ \Omega$$

$$R_3 \| R_4 = \frac{R_3 R_4}{R_3 + R_4}$$

$$= \frac{(10\ \Omega)(20\ \Omega)}{10\ \Omega + 20\ \Omega}$$
$$= 6.67\ \Omega$$

so

$$R_{eq} = 5\ \Omega + 6.67\ \Omega + 6.67\ \Omega$$
$$= 18.34\ \Omega$$

For source E_2,

$$I_5 = \frac{16\ \text{V}}{18.34\ \Omega}$$
$$I_5 = 8.72 \times 10^{-1}\ \text{A to the right}$$

By superposition, with both sources in the circuit,

$$I_5 = 8.72 \times 10^{-1}\ \text{A} - 1.45 \times 10^{-1}\ \text{A}$$
$$I_5 = 7.27 \times 10^{-1}\ \text{A to the right}$$

then

$$V_5 = I_5R_5$$
$$= (7.27 \times 10^{-1}\ \text{A})(5\ \Omega)$$
$$V_5 = 3.63\ \text{V}$$

The voltmeter indicates 3.63 V.

EXAMPLE 10.3

A source generates a pulse as shown in Figure 10.8(a). It is desired to change this to Figure 10.8(b). What type of voltage must be combined with it?

Solution

By superposition, the pulse voltage combined with the added voltage must give the voltage in Figure 10.8(b). That is,

$$V_1 + V_x = V_2$$

From t = 0 ms to t = 5 ms,

$$V_1 + V_x = 10\ \text{volts}$$

so

$$V_x = 10\ \text{V} - 5\ \text{V} = 5\ \text{V}$$

From t = 5 ms to t = 10 ms,

$$V_1 + V_x = 0\ \text{V}$$
$$-5\ \text{V} + V_x = 0\ \text{V}$$

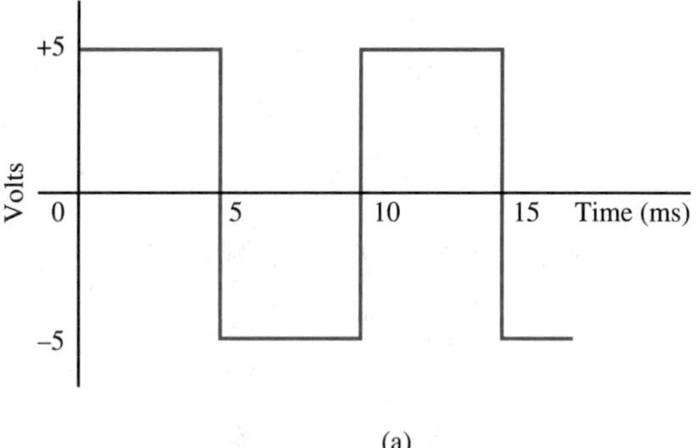

(a)

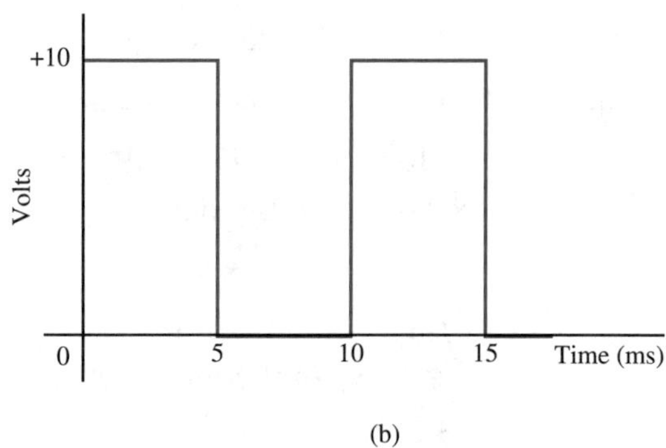

(b)

Figure 10.8

so

$$V_x = 0\text{ V} + 5\text{ V} = 5\text{ V}$$

The voltage must be a dc voltage of 5 volts to give the pulse in Figure 10.8(b).

The next example presents a different application of superposition. In it, the emf of a source is determined when one of the branch currents is known.

EXAMPLE 10.4

What are the emf and terminal polarity of E_B in Figure 10.9 if the current in R_2 is 1 ampere as shown?

Solution

Since I_2 is the combination of I_{2A} and I_{2B}, finding I_{2B} will let us determine the polarity and magnitude of E_B. E_B is short-circuited to consider the effect of E_A. The circuit is shown in Figure 10.10(a).

$$I_{1A} = \frac{E_A}{R_{\text{eq}A}}$$

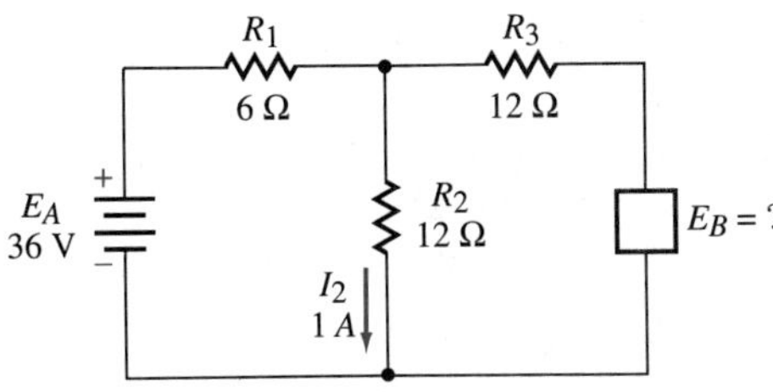

Figure 10.9

$$
\begin{aligned}
R_{eqA} &= R_1 + R_2 \| R_3 \\
&= 6\ \Omega + 12\ \Omega \| 12\ \Omega \\
&= 12\ \Omega
\end{aligned}
$$

so

$$
\begin{aligned}
I_{1A} &= \frac{36\ \text{V}}{12} \\
I_{1A} &= 3\ \text{A} \\
I_{2A} &= \frac{(I_{1A})\,(R_3)}{R_2 + R_3} \\
&= \frac{(3\ \text{A})\,(12\ \Omega)}{12\ \Omega + 12\ \Omega} \\
I_{2A} &= 1.5\ \text{A}
\end{aligned}
$$

Since I_2 is smaller than I_{2A}, I_{2B} must be opposing it. The lower terminal of E_B is the positive terminal.

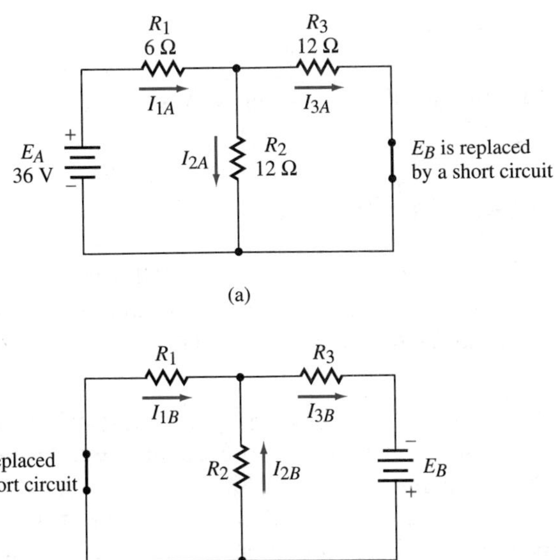

Figure 10.10

$$I_2 = I_{2A} - I_{2B}$$

so

$$I_{2B} = I_{2A} - I_2$$
$$= 1.5\text{ A} - 1\text{ A}$$
$$I_{2B} = 0.5\text{ A}$$

The circuit with only E_B active is shown in Figure 10.10(b). I_{2B} must be related to E_B in some manner to determine E_B. Applying Kirchhoff's Voltage Law to the circuit gives

$$V_{R2} = I_{2B}R_2 = (0.5\text{ A})(12\ \Omega) = 6\text{ V}$$

But by the voltage-divider rule,

$$V_{R2} = \frac{(R_1\|R_2)(E_B)}{R_3 + (R_1\|R_2)}$$

so

$$R_1\|R_2 = \frac{(6\ \Omega)(12\ \Omega)}{6\ \Omega + 12\ \Omega}$$
$$= 4\ \Omega$$

then

$$6\text{ V} = \frac{(4\ \Omega)(E_B)}{12\ \Omega + 4\ \Omega}$$

or

$$E_B = \frac{(6\text{ V})(16\ \Omega)}{4\ \Omega}$$
$$E_B = 24\text{ V}$$

E_B is 24 V with the lower terminal positive.

10.2 Thevenin's Theorem

Thevenin's Theorem

Thevenin's Theorem, named after M. L. Thevenin, a French engineer, states that any two terminal, linear circuits can be replaced by a constant voltage source and a series resistance. This theorem is especially useful for:

1. Reducing the calculations where repetitive calculations are made. An example is calculating the output of a circuit for many values of load resistance.
2. Simplifying the solution of R-C and R-L transient circuits.
3. Replacing transistors and other devices with an electrical circuit.

The emf of the Thevenin source is equal to the open-circuit voltage across the two terminals, as in Figure 10.11(a). Open-circuit voltage refers to the voltage that exists across the terminals when the load is disconnected from the terminals. The resistance of the Thevenin equivalent is equal to the resistance looking into the terminals with all sources replaced by their internal resistance, as in Figure 10.11(b). The open-circuit voltage and resistance can be calculated or measured.

The procedure for finding the Thevenin equivalent of a circuit is:

1. Identify the circuit that is to be Theveninized and the load that is connected to it.
2. Disconnect the load from the circuit that is to be Theveninized.
3. Use circuit concepts to find the voltage across the open-circuited two terminals. This is E_{TH}.
4. Find the resistance looking into the two terminals with the sources replaced by their internal resistance. This is R_{TH}.
5. Reconnect the load to the Thevenin equivalent and make any required analysis of the load conditions.

These steps are now used to solve the following example.

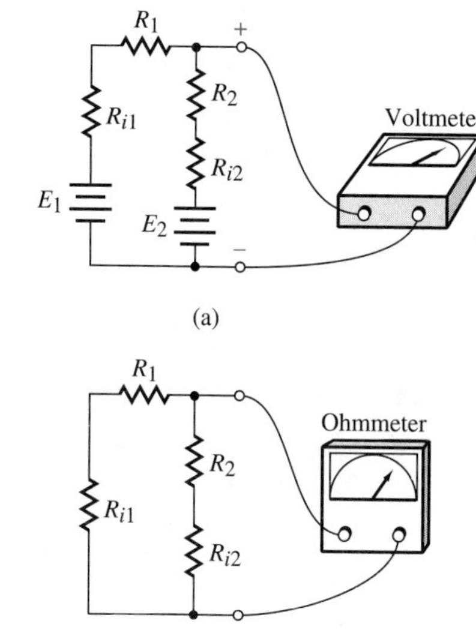

Figure 10.11 The relation of E_{TH} and R_{TH} to the circuit being Theveninized. (a) E_{TH} is equal to the voltage across the open-circuited terminals. (b) R_{TH} is the resistance looking into the output terminals with the sources replaced by their internal resistance.

EXAMPLE 10.5

Find the Thevenin equivalent of the circuit in the colored section of Figure 10.12.

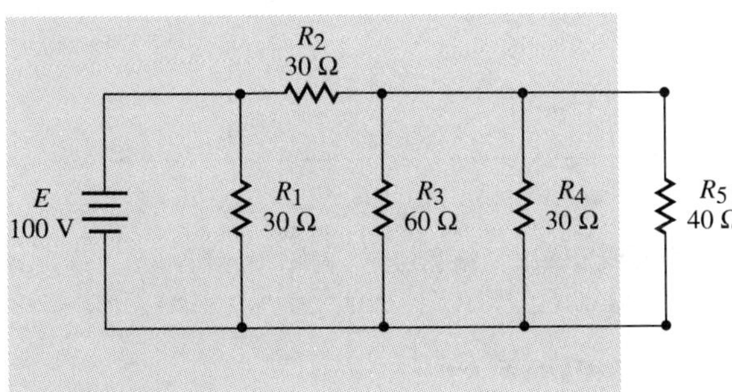

Figure 10.12

Solution

Step 1. The circuit is in the colored section and the load is R_5.

Steps 2, 3. The circuit for finding E_{TH} is shown in Figure 10.13.

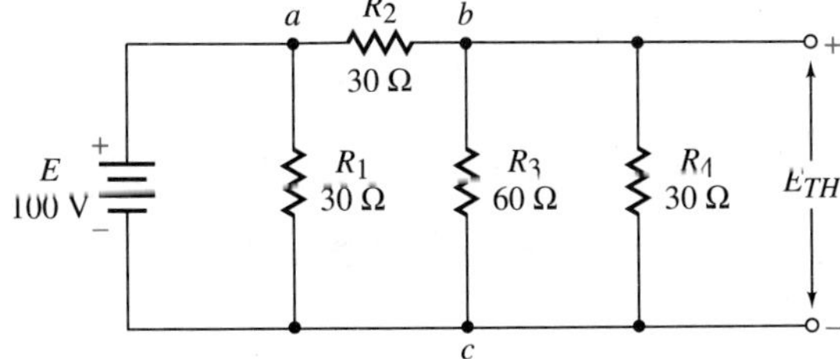

Figure 10.13

The open-circuit voltage is across b–c.

Since R_{bc} and R_{ab} make a series circuit connected across E, the voltage-divider rule can be used to give

$$E_{\text{TH}} = \frac{(R_{bc})(E)}{R_{ab} + R_{bc}}$$

but

$$R_{bc} = R_3 \| R_4$$

$$= \frac{R_3 R_4}{R_3 + R_4}$$

$$= \frac{(60\ \Omega)(30\ \Omega)}{60\ \Omega + 30\ \Omega}$$

$$R_{bc} = 20\ \Omega$$

so

$$E_{\text{TH}} = \frac{(20\ \Omega)(100\ \text{V})}{30\ \Omega + 20\ \Omega}$$

$$E_{\text{TH}} = 40\ \text{V}$$

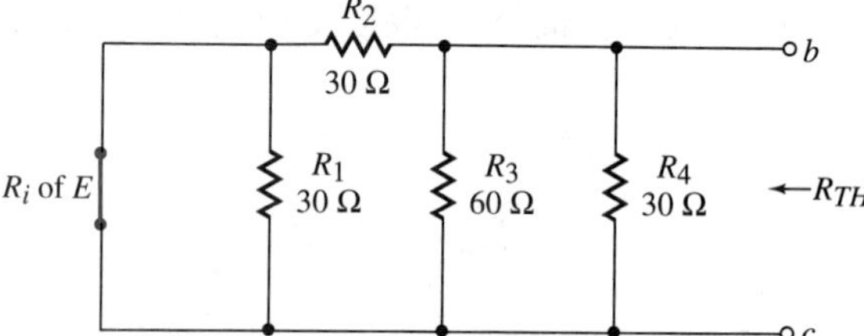

Figure 10.14

Step 4. The circuit for finding R_{TH} is shown in Figure 10.14. Replacing E by a short circuit also shorts out R_1, so

$$R_{\text{TH}} = R_2 \| R_3 \| R_4$$

$$\frac{1}{R_{\text{TH}}} = \frac{1}{R_2} + \frac{1}{R_3} + \frac{1}{R_4}$$

$$= \frac{1}{30\ \Omega} + \frac{1}{60\ \Omega} + \frac{1}{30\ \Omega}$$

$$= \frac{5}{60\ \Omega}$$

$$R_{\text{TH}} = \frac{60\ \Omega}{5}$$

$$R_{\text{TH}} = 12\ \Omega$$

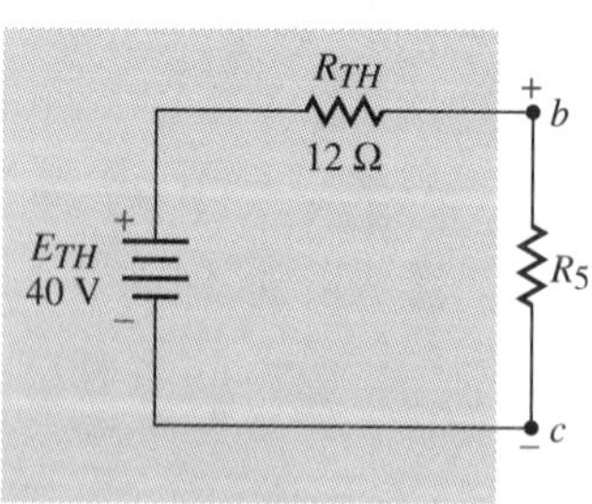

Figure 10.15

Step 5. The Thevenin equivalent with R_5 reconnected is shown in Figure 10.15.

EXAMPLE 10.6

Use the Thevenin equivalent to find the voltage across A–B for an R_L of 10 ohms and 20 ohms in the circuit of Figure 10.16.

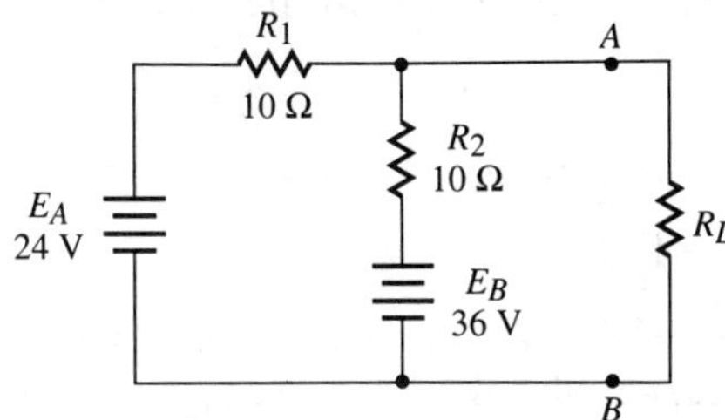

Figure 10.16

Solution

Step 1. The circuit being Theveninized is that to the left of A–B, and the load is R_L.

Steps 2,3. The circuit for finding E_{TH} is drawn in Figure 10.17.

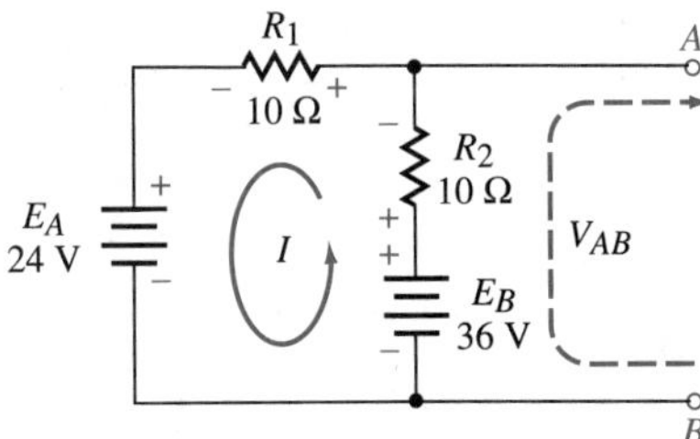

Figure 10.17

In this example, the terminal polarity is not as obvious as in Example 10.5. Is $A\,(+)$ or is $B\,(+)$? In these types of situations, it is helpful to use the double subscript notation for the terminal voltage. Starting at Point B, the rises and drops along any continuous path to A are algebraically added to give V_{AB}. Thus, for the current direction shown:

Through Path E_B, R_2,

$$V_{AB} = E_B - IR_2$$

Through Path E_A, R_1,

$$V_{AB} = E_A + IR_1$$

Since "I" is the unknown, an equation that includes it is needed. Kirchhoff's Voltage Law will be used to give

$$E_A + IR_1 + IR_2 - E_B = 0$$

or

$$I = \frac{E_B - E_A}{R_1 + R_2}$$
$$= \frac{36\text{ V} - 24\text{ V}}{10\ \Omega + 10\ \Omega}$$
$I = 0.6$ A. The current is in the assigned direction.

and

$$V_{AB} = 36\text{ V} - (0.6\text{ A})(10\ \Omega)$$
$$V_{AB} = 30\text{ V}$$

Terminal A is at $+30$ V with respect to terminal B. Since V_{AB} is the open-circuit voltage, it is also the value of E_{TH}. E_{TH} could also have been calculated using superposition or other methods of circuit analysis.

Step 4. The circuit for finding R_{TH} is as in Figure 10.18.

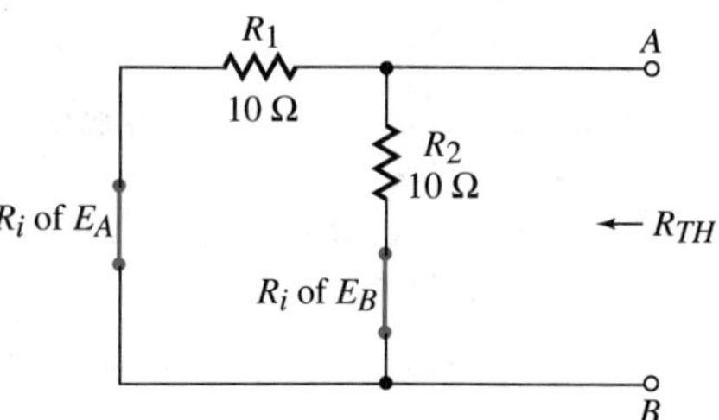

Figure 10.18

$$R_{\text{TH}} \text{ is } R_2 \| R_1$$

so

$$R_{\text{TH}} = \frac{(10\ \Omega)(10\ \Omega)}{10\ \Omega + 10\ \Omega}$$
$$R_{\text{TH}} = 5\ \Omega$$

Step 5. The Thevenin equivalent with R_L reconnected is drawn in Figure 10.19.

For this circuit, the voltage-divider equation gives

$$V_L = \frac{R_L E_{\text{TH}}}{R_L + R_{\text{TH}}} \quad \text{or} \quad V_L = \frac{30 R_L}{R_L + 5\ \Omega}$$

For $R_L = 10\ \Omega$,

$$V_L = \frac{(30\text{ V})\ (10\ \Omega)}{10\ \Omega + 5\ \Omega}$$
$$V_L = 20\text{ V}$$

For $R_L = 20\ \Omega$,

$$V_L = \frac{(30\text{ V})\ (20\ \Omega)}{20\ \Omega + 5\ \Omega}$$
$$V_L = 24\text{ V}$$

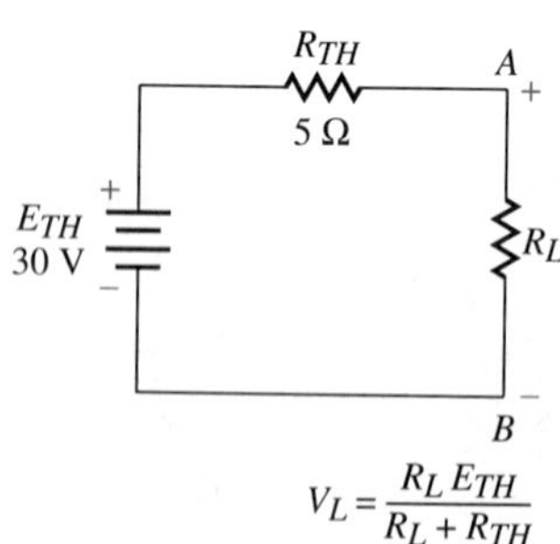

Figure 10.19

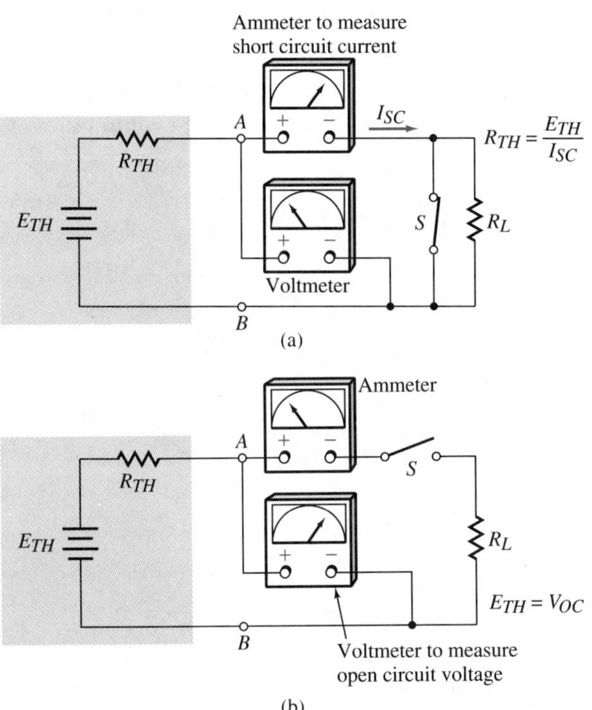

Figure 10.20 Open-circuit voltage and short-circuit current can also be used to Theveninize a circuit. (a) E_{TH} is the voltage, either measured or calculated, with the terminals open-circuited. (b) Dividing the short-circuit current into the open-circuit voltage gives R_{TH}.

Had the circuit not been Theveninized, a separate analysis would have to be made for each value of R_L. You can imagine how lengthy and error-prone this can be for a three-loop circuit.

An examination of the Thevenin equivalent in Figure 10.20 results in another method of determining the Thevenin equivalent.

When terminals A and B are short-circuited, as in Figure 10.20(a), $V = 0$ and $E_{\text{TH}} = I_{sc}R_{\text{TH}}$. When the terminals are open-circuited, as in Figure 10.20(b), $E_{\text{TH}} = V_{oc}$. Therefore,

$$R_{\text{TH}} = \frac{V_{oc}}{I_{sc}} \quad \textbf{(10.1)}$$

$$E_{\text{TH}} = V_{oc} \quad \textbf{(10.2)}$$

where V_{oc} is the open-circuit voltage of the original circuit in volts, I_{sc} is the short-circuit current in the original circuit in amperes, R_{TH} is the Thevenin resistance in ohms, and E_{TH} is the Thevenin source emf in volts.

If the circuit is built, the relations in Equations 10.1 and 10.2 can also be used to determine E_{TH} and R_{TH} from laboratory tests when only the load terminals are accessible. Also, R_{TH} can be measured directly by an ohmmeter if the sources can be replaced by their internal resistance.

Equation 10.1 can also be used to get E_{TH} when I_{sc} is easier to calculate than E_{TH}.

EXAMPLE 10.7

Measurements on the circuit of Figure 10.21 are $V_L = 20$ V with S open, $I_L = 0.5$ A with S closed. Find the Thevenin equivalent for the circuit.

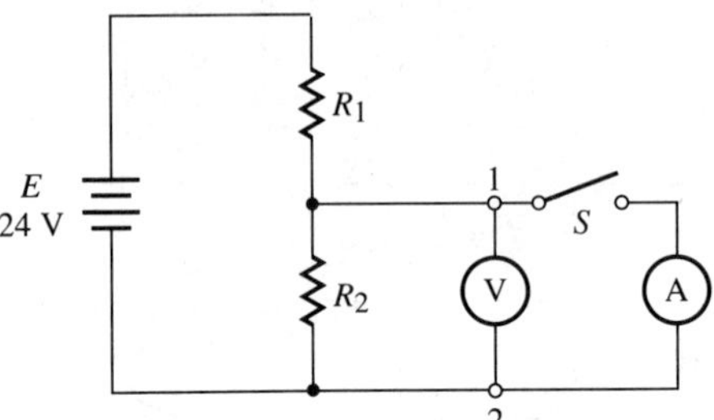

Figure 10.21

Solution

$$E_{\text{TH}} = V_{oc}$$
$$E_{\text{TH}} = 20 \text{ V}$$

Using Equation 10.1 gives

$$R_{\text{TH}} = \frac{E_{\text{TH}}}{I_{sc}} = \frac{20 \text{ V}}{0.5 \text{ A}}$$
$$R_{\text{TH}} = 40 \ \Omega$$

The Thevenin Equivalent circuit is shown in Figure 10.22.

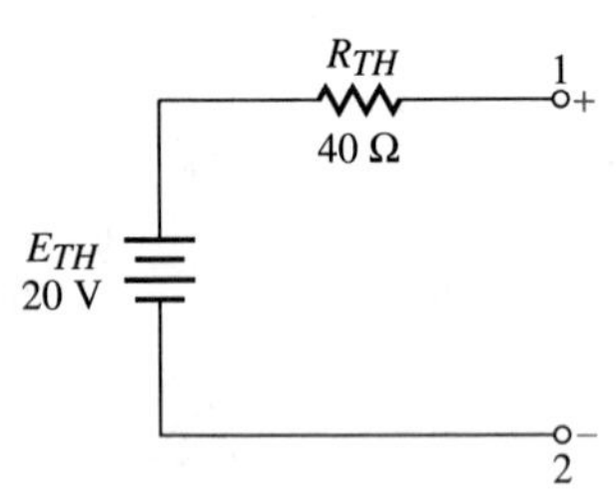

Figure 10.22

EXAMPLE 10.8

Determine the Thevenin equivalent of the circuit shown in Figure 10.23.

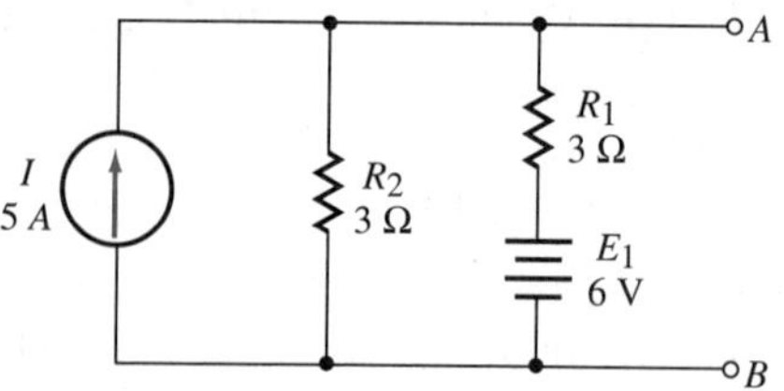

Figure 10.23

Solution

For this solution, Millman's Theorem from Chapter 9 is used. Converting E_1 to a current source gives

$$I_1 = \frac{E_1}{R_1} = \frac{6 \text{ V}}{3} = 2 \text{ A up}$$

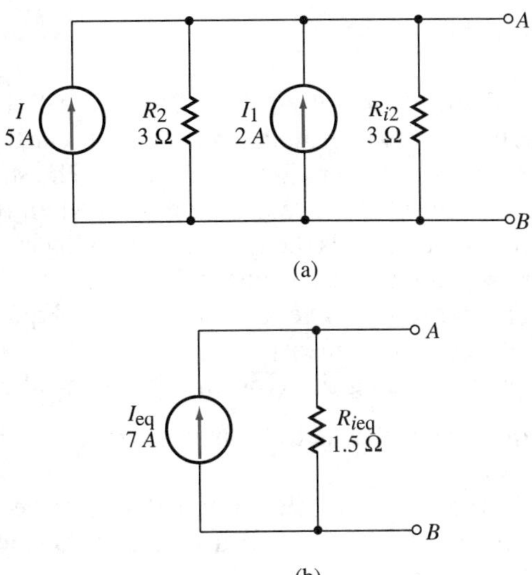

Figure 10.24

$$R_{i1} = R_1 = 3\ \Omega$$

The circuit is shown in Figure 10.24(a).
The equivalent current source for Figure 10.24(a) is

$$\begin{aligned} I_{\text{eq}} &= I + I_1 \\ &= 5\text{ A} + 2\text{ A} \\ &= 7\text{ A up} \\ R_{i\text{eq}} &= R_2 \| R_{i1} \\ &= \frac{(3\ \Omega)(3\ \Omega)}{3\ \Omega + 3\ \Omega} \\ &= 1.5\ \Omega \end{aligned}$$

The equivalent current source is shown in Figure 10.24(b). From that circuit,

$$\begin{aligned} V_{AB} &= (I_{\text{eq}})(R_{i\text{eq}}) \\ &= (7\text{ A})(1.5\ \Omega) \\ &= 10.5\text{ V} \end{aligned}$$

hence

$$\begin{aligned} E_{\text{TH}} &= 10.5\text{ V} \\ R_{\text{TH}} &= R_{i\text{eq}} \end{aligned}$$

so

$$R_{\text{TH}} = 1.5\ \Omega$$

The Thevenin equivalent circuit is shown in Figure 10.25.

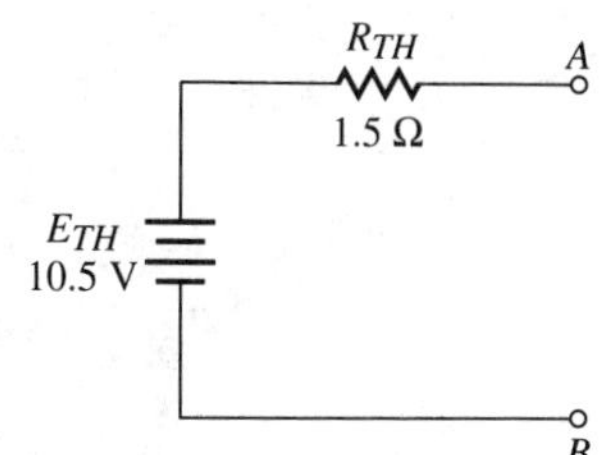

Figure 10.25

10.3 Norton's Theorem

Norton's Theorem

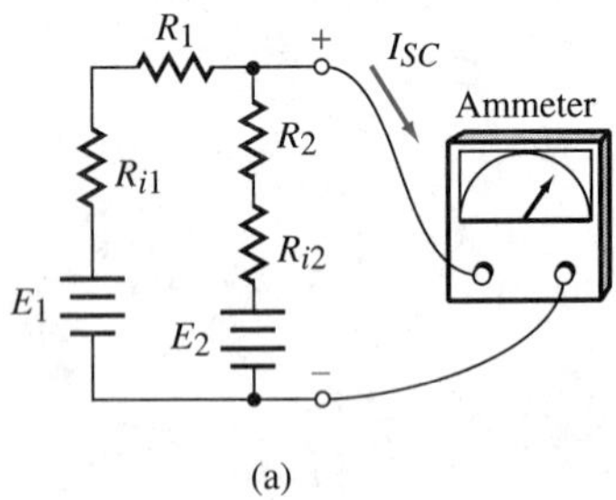

(a)

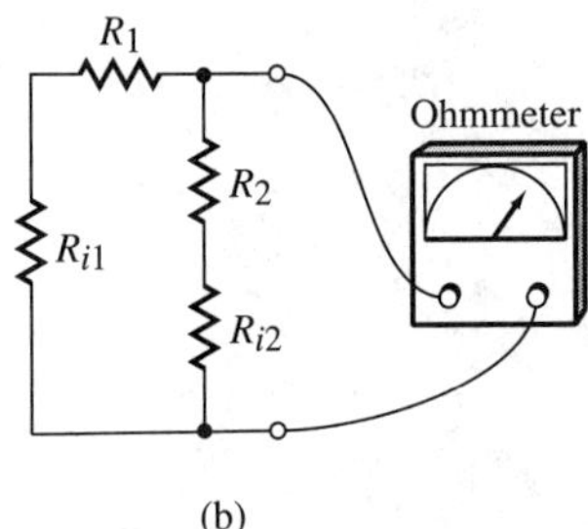

(b)

Figure 10.26 The relation of I_N and R_N to the circuit being Nortonized. (a) I_N is equal to the current with the terminal short-circuited. (b) R_N is the resistance looking into the output terminals with the sources replaced by their internal resistance.

Norton's Theorem, named after E. L. Norton, a Bell Laboratories Scientist, states that any two terminal linear circuits can be replaced by a constant current source and a parallel resistance. The source's current must equal the current at the circuit's terminals when they are short-circuited. The resistance equals the resistance looking into the terminals with all sources replaced by their internal resistance. The theorem is useful for the same reasons as Thevenin's Theorem. Figure 10.26 shows how these quantities can be measured.

The procedure for finding the Norton equivalent of a circuit is:

1. Identify the circuit that is to be Nortonized, and the load that is connected to it.
2. Disconnect the load from the circuit that is to be Nortonized.
3. Short circuit the terminals and use circuit concepts to find the short-circuit current. This is I_N.
4. Open the terminals, replace the sources by their internal resistance, and find the resistance looking into the terminals. This is R_N.
5. Reconnect the items of Step 2 and make any required analysis.

These steps are applied in the following example.

EXAMPLE 10.9

Find the Norton equivalent for the circuit in the colored section of Figure 10.27.

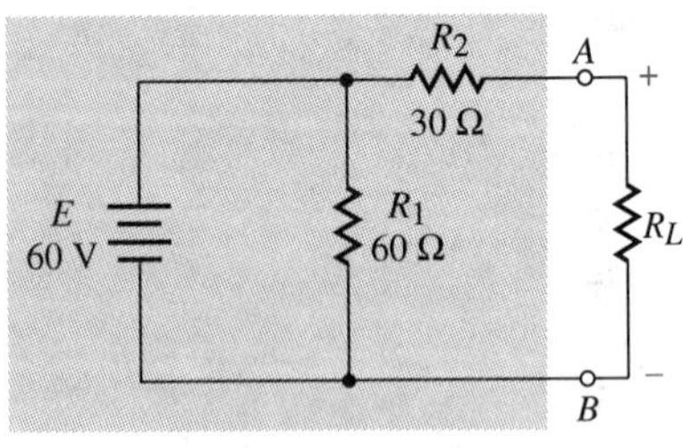

Figure 10.27

Solution

Step 1. The circuit is in the colored section and the load is R_L.

Steps 2, 3. The circuit for finding I_N is as in Figure 10.28. With A–B short-circuited, R_2 is connected across E so

$$I_{sc} = E/R_2 = 60\text{ V}/30\ \Omega = 2\text{ A}$$

then

$$I_N = 2\text{ A}$$

Step 4. The circuit for finding R_N is shown in Figure 10.29. Replacing E by its R_i, a short circuit, also shorts out R_1.

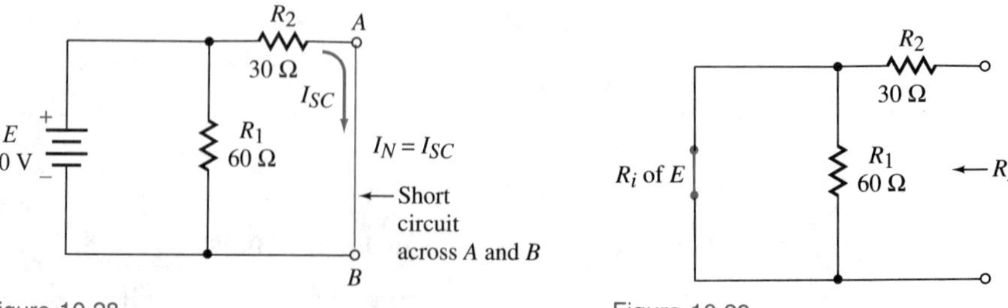

Figure 10.28

Figure 10.29

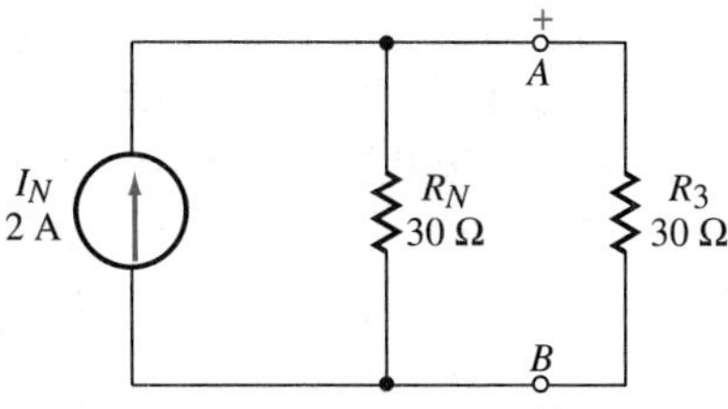

Figure 10.30

So,

$$R_N = R_2$$
$$R_N = 30\ \Omega$$

Step 5. Figure 10.30 shows the Norton equivalent circuit, with R_3 reconnected.

Just as E_{TH}, R_{TH}, and I_{sc} were related to the Thevenin equivalent, V_{oc}, R_N, and I_{sc} are also related in the Norton equivalent.

If the terminals are open-circuited, as in Figure 10.31, R_L will be

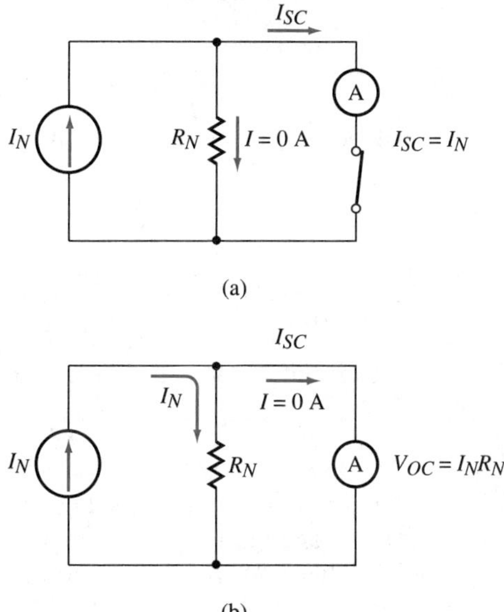

Figure 10.31 Short-circuit current and open-circuit voltage can be used to Nortonize a circuit. (a) I_N is the current, either measured or calculated, with the terminals open-circuited. (b) Dividing the open-circuit voltage by the short-circuit current gives R_{TH}.

infinity and V_L will be V_{oc}, and $V_{oc} = I_N R_N$. When the terminals are short-circuited, all of the current, I_N, is in the short circuit. So

$$R_N = V_{oc}/I_{sc} \tag{10.3}$$
$$I_N = I_{sc} \tag{10.4}$$

where V_{oc} is the open-circuit voltage in the original circuit in volts, I_{sc} is the short-circuit current in the original circuit, also I_N, and R_N is the Norton resistance, in ohms.

This relation lets us find I_N or R_N when V_{oc} is easier to find than one of the two quantities. It also provides a way to find the Norton equivalent from open-circuit and short-circuit measurements on a circuit.

EXAMPLE 10.10

Find the Norton equivalent of the circuit in Figure 10.32 using Equations 10.3 and 10.4.

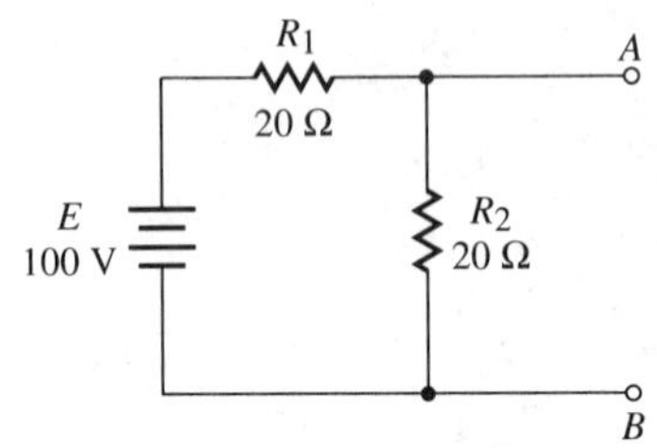

Figure 10.32

Solution

V_{oc} is the voltage across R_2.

R_2 and R_1 are connected in series across E. So using the voltage-divider relation gives

$$V_{oc} = \frac{R_2 E}{R_1 + R_2} = \frac{20\ \Omega}{20\ \Omega + 20\ \Omega} \times 100\ \text{V} = 50\ \text{V}$$

With A–B shorted,

$$I_{sc} = E/R_1 = 100\ \text{V}/20\ \Omega = 5\ \text{A}$$
$$I_N = 5\ \text{A}$$

so

$$R_N = V_{oc}/I_{sc} = 50\ \text{V}/5\ \text{A}$$
$$R_N = 10\ \Omega$$

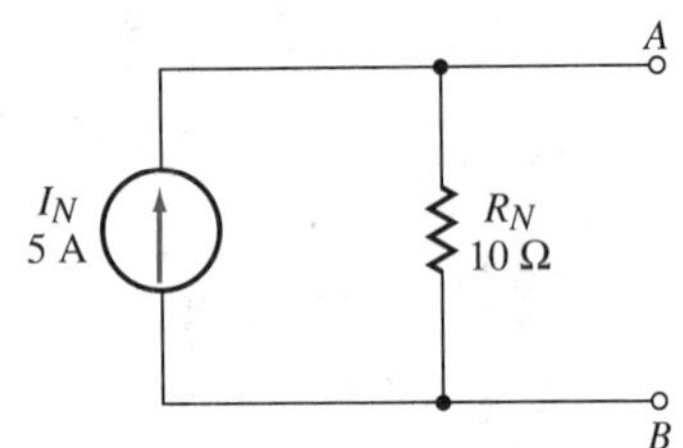

Figure 10.33

The Norton equivalent circuit is as in Figure 10.33.

If necessary, Thevenin equivalents can be changed to Norton's and vice versa. The rules for the change are the same as those for source conversions in Chapter 8. This then results in the following relations between the two circuits.

$$E_{TH} = I_N R_N \tag{10.5}$$
$$R_N = R_{TH} \tag{10.6}$$
$$I_N = E_{TH}/R_{TH} \tag{10.7}$$

The first two equations are used to convert from Norton's to Thevenin's. The last two equations are used to convert from Thevenin's to Norton's.

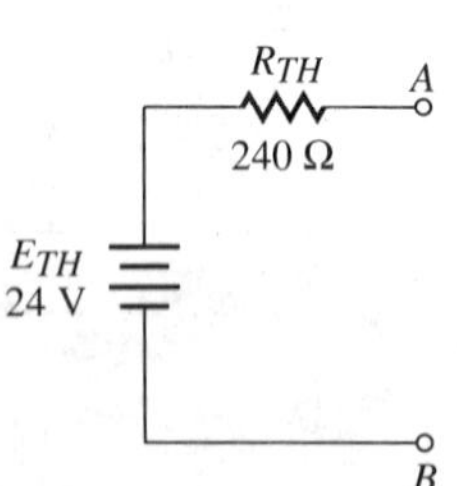

Figure 10.34

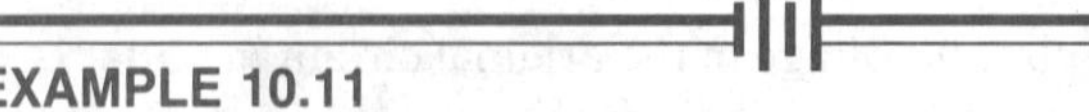

EXAMPLE 10.11

Find the Norton equivalent for the circuit shown in Figure 10.34.

Solution

$$I_N = E_{TH}/R_{TH} = 24\ \text{V}/240\ \Omega$$
$$I_N = 0.1\ \text{A}$$
$$R_N = R_{TH} = 240\ \Omega$$
$$R_N = 240\ \Omega$$

The Norton equivalent with the correct current direction can be seen in Figure 10.35.

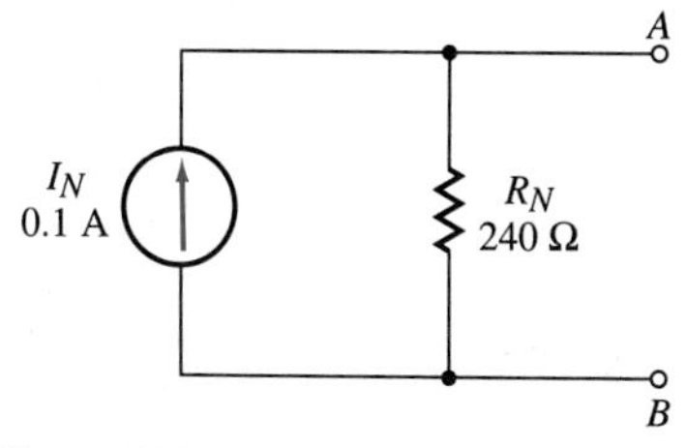

Figure 10.35

S 10.4 Maximum Power Transfer

The electric utility providing energy to its customers, an antenna connected to a receiver, and an amplifier connected to a speaker have one thing in common. Each one transfers energy and power to a load. But there is one difference: the utility is concerned with maximum efficiency while the others are concerned with maximum power transfer.

When is maximum power transfer more important than efficiency? One example is the transfer of power to a receiver from an antenna that receives a low-power signal from a transmitter. Since the power is low, we want to get as much of it as possible to the receiver. This section deals with interaction between load, source, and power transfer.

The curve in Figure 10.36 shows that the load power is a maximum when R_L equals R_i. For students familiar with calculus, the value of R_L at maximum power can also be found by taking the derivative of the expression for P_L with respect to R_L and setting it equal to zero. P_L then drops off for values of R_L that are smaller or larger than R_i.

maximum power transfer theorem

This condition is defined by the **maximum power transfer theorem.** The theorem states, "maximum power is transferred from a source to a

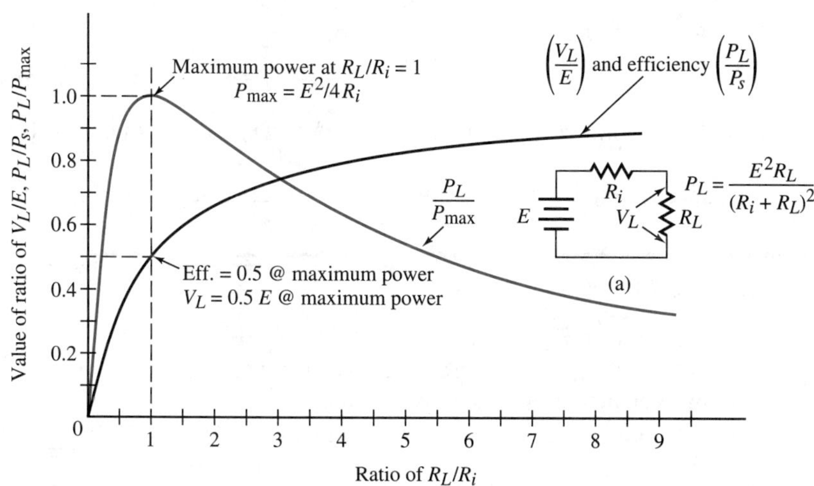

Figure 10.36 These curves show how the ratios V_L/E, P_L/P_s, and P_L/P_{max} vary with the ratio R_L/R_i.

load when the load resistance equals the source's internal resistance.'' The source can be a voltage source, an amplifier, or another circuit. The relation for a single source connected to a single resistor is now examined.

The power in R_L in the circuit of Figure 10.36(b) is

$$P = I^2R_L$$

but

$$I = E/(R_L + R_i)$$

so

$$P = \frac{E^2R_L}{(R_L + R_i)^2}$$

It can be shown by calculus or inserting values of R_L that maximum power transfer results when

$$R_L = R_i \qquad \textbf{(10.8)}$$

Putting this relation in the expression for power gives

$$P_{max} = E^2/(4R_L) \qquad \textbf{(10.9)}$$

where E is the open-circuit voltage in volts, R_i is the internal resistance of the source in ohms, and P_{max} is the maximum power that can be transferred in watts.

Next, let us take a look at the load voltage. In general,

$$V_L = E - IR_i \qquad \text{or} \qquad V_L = E - \frac{ER_i}{R_L + R_i}$$

At maximum power transfer $R_L = R_i$ so

$$V_L = E - \frac{ER_i}{2R_i}$$

Since R_i equals R_L,

$$V_L = E/2 \qquad \textbf{(10.10)}$$

where V_L is the load voltage, in volts, and E is the source's open-circuit voltage, in volts.

The curve in Figure 10.36 shows V_L for other ratios of R_L/R_i.

EXAMPLE 10.12

For the circuit of Figure 10.37, find:

a. the value of R_L for maximum power transfer.
b. the power at maximum power transfer.
c. the voltage across R_L at maximum power.

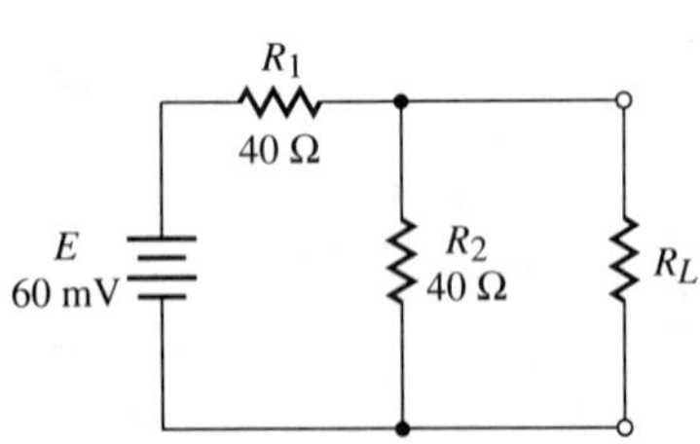

Figure 10.37

Solution

a. At maximum power transfer, $R_L = R_i$. To find R_i, the circuit will be put into the same form as Figure 10.36(a) by Theveninizing it. Then, $R_i = R_{TH}$.

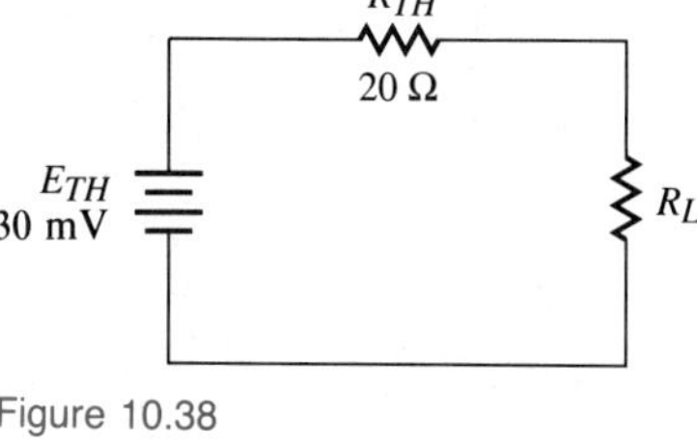

Figure 10.38

$$E_{TH} = \left(\frac{R_2}{R_1 + R_2}\right)E$$

$$= \left(\frac{40\ \Omega}{40\ \Omega + 40\ \Omega}\right)(60\ \text{mV})$$

$$E_{TH} = 30\ \text{mV}$$

$$R_{TH} = R_2 \| R_1$$

$$R_{TH} = \frac{(40\ \Omega)(40\ \Omega)}{40\ \Omega + 40\ \Omega}$$

$$R_{TH} = 20\ \Omega$$

so

$$R_L = 20\ \Omega$$

The circuit is shown in Figure 10.38.

b. For maximum power transfer, R_L must equal R_{TH}, which is 20 Ω. The open-circuit voltage is the same as E_{TH} so

$$P_L = (E_{TH})^2/(4\ R_i)$$

$$P_L = (0.030\ \text{V})^2/(4 \times 20\ \Omega)$$

$$P_L = 1.125 \times 10^{-5}\ \text{W}$$

and

c.

$$V_L = E_{TH}/2 = 30\ \text{mV}/2$$

$$V_L = 15\ \text{mV}$$

EXAMPLE 10.13

At maximum power transfer, 8 watts of power is delivered to an 8-ohm loudspeaker. How many watts will be delivered to the two speakers if another 8-ohm speaker is connected across the present speaker?

Solution

$$P_{max} = \frac{E^2}{4R_L}$$

so

$$E = \sqrt{(P_{max})(4)(R_L)}$$

$$= \sqrt{(8\ \text{W})(4)(8\ \Omega)}$$

$$= 16\ \text{V}$$

At maximum power transfer, $R_i = R_L$ so

$$R_i = 8\ \Omega$$

With two speakers connected,

$$R_L = 8\ \Omega \| 8\ \Omega = 4\ \Omega \qquad \text{and} \qquad P_L = I^2 R_L$$

but

$$I = E/R_{eq} = E/(R_i + R_L)$$
$$P_L = I^2 R_L$$

so

$$P_L = \left(\frac{E}{R_{eq}}\right)^2 R_L$$
$$P_L = \frac{(16\ \text{V})^2}{(8\ \Omega + 4\ \Omega)^2}(4\ \Omega)$$
$$P_L = 7.11\ \text{W}$$

Adding the second loudspeaker reduced the power output of the circuit.

EXAMPLE 10.14

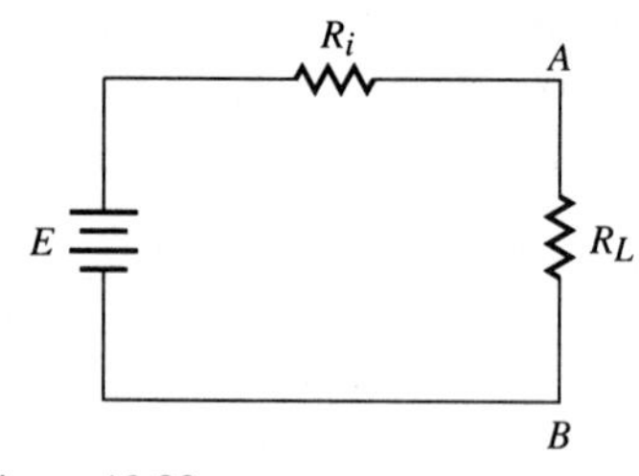

Figure 10.39

The voltage across A–B in Figure 10.39 with R_L disconnected is 12 volts. The short-circuit current is 3 amperes. Determine:

a. the resistance of R_i.
b. the power in R_L at maximum power transfer.

Solution

$$R_i = R_{TH} = V_{oc}/I_{sc}$$
$$R_i = 12\ \text{V}/3\ \text{A}$$
$$R_i = 4\ \Omega$$
$$E_{TH} = V_{oc}$$
$$E_{TH} = 12\ \text{V}$$

At maximum power transfer,

$$P_{max} = \frac{(E_{TH})^2}{4\ R_L} = \frac{(12\ \text{V})^2}{4\ (4\ \Omega)}$$
$$P_{max} = 9\ \text{W}$$

Matching the load to the source, that is, making $R_L = R_i$, is very seldom done in power applications but is more commonly used in electronic applications. Some examples are the matching of loudspeakers to an amplifier and antennas to receivers. Matching not only results in maximum power, but can also reduce the distortion of the output signal.

In general, the load power peaks, then drops off, and the percent efficiency increases as R_L/R_i increases. The variation of these characteristics is shown in the curves in Figure 10.36. From these curves, it is apparent that the relation of internal resistance and load resistance must be considered in power transfer applications.

Electric utilities do not design for maximum transfer because of the low percent efficiency that exists at maximum power transfer. Although this can be accepted when it is offset by the need for maximum power transfer, it is not justified in other applications. How low is the efficiency? We now examine that characteristic.

The power developed by the source in Figure 10.36 is

$$I^2(R_i + R_L) \quad \text{or} \quad 2\,I^2R_L$$

at maximum transfer. The load power is I^2R_L. Therefore,

$$\begin{aligned} \%\text{ Efficiency} &= \left(\frac{P_L}{P_{\text{source}}}\right) \times 100\% \\ &= \left(\frac{I^2R_L}{2\,I^2R_L}\right) \times 100\% \\ \eta &= 50\% \end{aligned}$$

This level of efficiency is impractical from an economical and equipment standpoint to justify operating at maximum power transfer. The generating equipment would have to be rated at double the power needed by the customers. On the other hand, the utilities would be paid for only one-half of the power because of the 50 percent efficiency.

Before leaving the topic of maximum power transfer, an example of how to use PSpice to determine the value of load resistance to achieve maximum power transfer is illustrated.

To determine the value of R_L for maximum power transfer, PSpice can be used to graph power dissipated in the load as a function of R_L. Hence, in our analysis, a range of R_L values must be investigated.

To do this in PSpice, the .MODEL statement will be used in conjunction with the .DC statement. The .MODEL statement identifies a set of parametric values used to describe the operational characteristics of the device. The syntax for the statement is

```
.MODEL    <name> <type name> ([<parameter name> =
                 <value>])
```

where

<name> is the "model name" for the device
<type name> is the device type description
<parameter name> is the name of the part
<value> is the value of the part

For example,

```
.MODEL    RMOD    RES(R=1K)
```

describes a resistor, R, whose value is 1 kilohm, using a "RE" model named "RMOD."

The .DC statement can then be used to sweep component values by sweeping the model parameters. For example,

```
.DC    RES    RMOD(R)    1K    5K    100
```

will sweep the "R" parameter of the "RES" model named "RMOD," starting with 1 kilohm, in increments of 100 ohms, until R reaches 5 kilohms.

Now, let us put these statements to use in the following example.

EXAMPLE 10.15

Find the value of R_2 for maximum power transfer in the circuit shown in Figure 10.40. R_1 equals 3 kΩ.

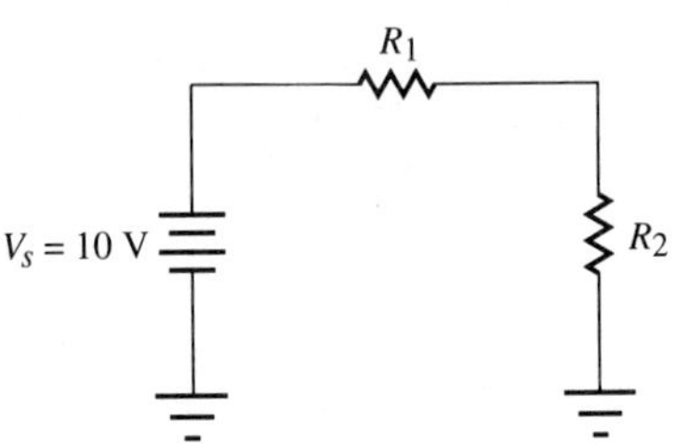

Figure 10.40 Circuit for demonstrating maximum power transfer.

Solution

First the nodes must be assigned a name. The reference node is node 0. The node between the (+) terminal of the source and $R1$ is node 1. The node between $R1$ and $R2$ is node 2. The negative terminal of the source and one end of $R2$ are connected to the reference node.

The PSpice file can then be constructed. First comes the title line:

```
MAXIMUM POWER TRANSFER EXAMPLE
```

Next comes the device statements for the voltage source, VS, and the two resistors, R1 and R2:

```
VS    1    0    10V
R1    1    2    3K
R2    2    0    RMOD 1
```

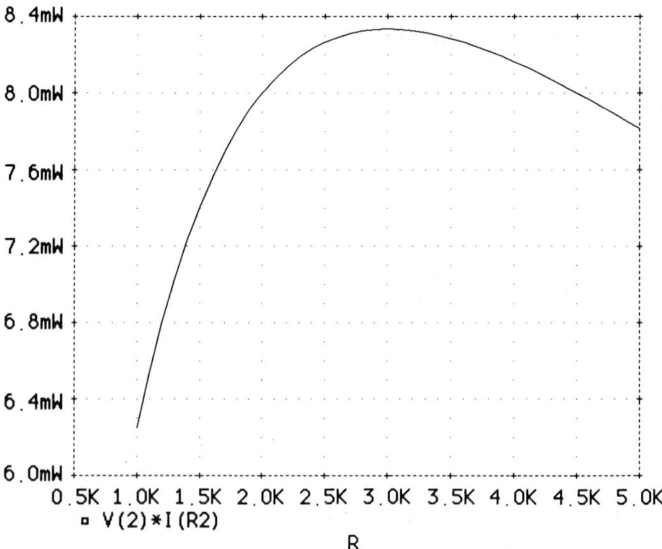

Figure 10.41 PROBE output graph of power as a function of resistance R_2.

Note, the resistance value of R2 is described by a model named RMOD.

The .MODEL statement follows

```
.MODEL    RMOD    RES(R=1K)
```

and then the .DC statement

```
.DC    RES    RMOD(R)    1K    5K    100
```

and finally closing the file with

```
.END
```

The complete file is

```
MAXIMUM POWER TRANSFER EXAMPLE
VS     1    0    10V
R1     1    2    3K
R2     2    0    RMOD  1
.MODEL      RMOD   RES(R=1K)
.DC         RES    RMOD(R)  1K  5K  100
.END
```

Running the file through PSpice and using the .PROBE graphics processor yields the graph shown in Figure 10.41. The graph plots the power dissipated in R_L, namely V(2)*I(R2), as a function of R_L.

By visually inspecting the power curve, the maximum power transfer point can be determined. It is the point where the power graph peaks at R = 3 kilohms. The power dissipation for this resistance value is 8.33 mW. This is the maximum power transfer point.

10.5 Delta and Wye Networks

Two network shapes that are used in electrical circuits are the **Delta (Δ) and Wye (Y) networks** shown in Figure 10.42.

In electronic circuits, these networks are often seen as Pi (Π) and Tee (T). Inverting the Delta and spreading the sides gives the Π shape in Figure 10.43. Spreading the branches of the Wye gives the T shape.

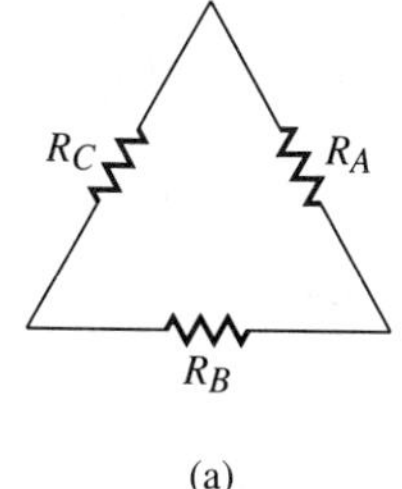

(a)

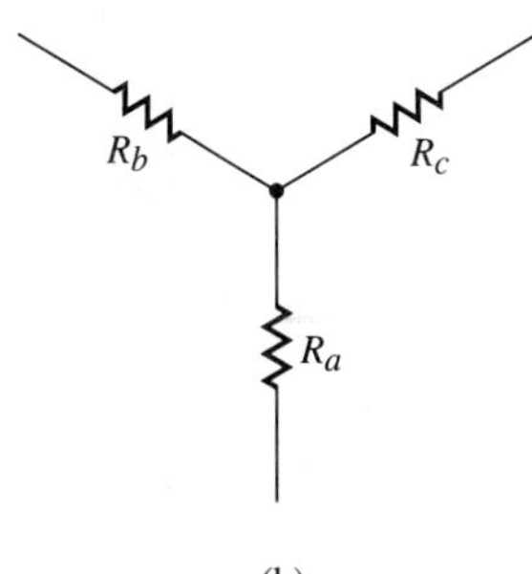

(b)

Figure 10.42 Two common network configurations are the Delta and the Wye. (a) A Delta network. (b) A Wye network.

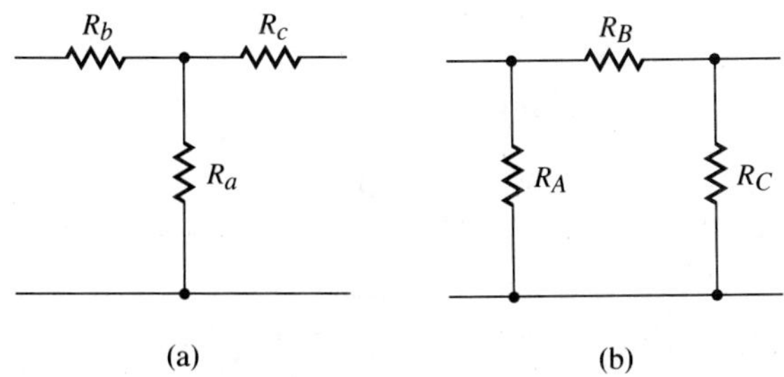

(a) (b)

Figure 10.43 The Delta and Wye networks are also known as the Pi and the Tee networks. (a) The Pi network. (b) The Tee network.

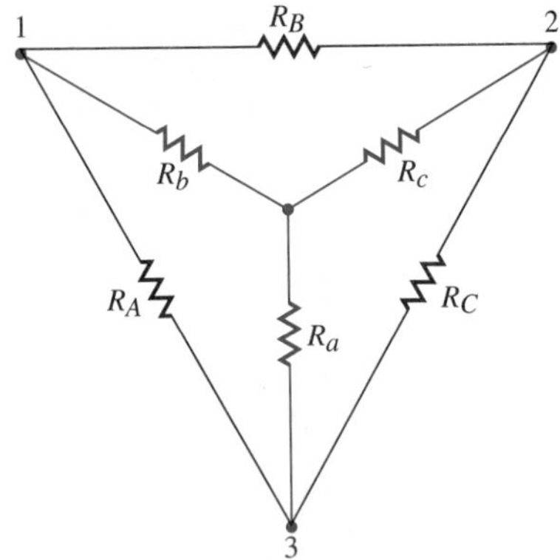

Figure 10.44 A Delta network and a Wye network can be equivalent if the resistances have the proper values.

(a) $R_a = \dfrac{R_A R_C}{R_A + R_B + R_C}$,

$R_b = \dfrac{R_A R_B}{R_A + R_B + R_C}$,

$R_c = \dfrac{R_B R_C}{R_A + R_B + R_C}$.

(b) $R_A = \dfrac{R_a R_b + R_a R_c + R_b R_c}{R_c}$,

$R_B = \dfrac{R_a R_b + R_a R_c + R_b R_c}{R_a}$,

$R_C = \dfrac{R_a R_b + R_a R_c + R_b R_c}{R_b}$.

This section considers the conversion from one network to its equivalent. One reason for doing this is to simplify load calculations. Later, in your study of three-phase power distribution systems, you will see that three-phase sources and loads are often connected as Wye's and Delta's.

The graphic relation between the Delta and its equivalent Wye network is shown in Figure 10.44. The two are equivalent if the resistance across any two terminals of one is equal to the resistance across the same two terminals of the other. The conversions are now considered.

Delta to Wye

The general relation is

$$R_Y \text{ branch} = \frac{\text{product of the Delta adjacent sides}}{\text{sum of the Delta sides}}$$

For the network of Figure 10.44,

$$R_a = \frac{R_A R_C}{R_A + R_B + R_C} \tag{10.11}$$

$$R_b = \frac{R_A R_B}{R_A + R_B + R_C} \tag{10.12}$$

$$R_c = \frac{R_B R_C}{R_A + R_B + R_C} \tag{10.13}$$

If $R_A = R_B = R_C$, Equations 10.11–10.13 reduce to

$$R_Y = \frac{R_\Delta}{3} \tag{10.14}$$

EXAMPLE 10.16

A Delta network has a 60-ohm resistance in each side. What is the branch resistance in the equivalent Wye?

Solution

Since $R_A = R_B = R_C$,

$$R_Y = \frac{R_A}{3} = \frac{60\ \Omega}{3}$$

$$R_Y = 20\ \Omega$$

EXAMPLE 10.17

Convert the Delta network of Figure 10.45 to an equivalent Wye network.

Solution

The Wye network is shown imposed on the Delta network in Figure 10.46.

Using Equations 10.11–10.13 gives

$$R_a = \frac{R_A R_C}{R_A + R_B + R_C} = \frac{(18\ \Omega)(24\ \Omega)}{18\ \Omega + 8\ \Omega + 24\ \Omega}$$

$$R_a = 8.64\ \Omega$$

$$R_b = \frac{R_A R_B}{R_A + R_B + R_C} = \frac{(18\ \Omega)(8\ \Omega)}{18\ \Omega + 8\ \Omega + 24\ \Omega}$$

$$R_b = 2.88\ \Omega$$

$$R_c = \frac{R_B R_C}{R_A + R_B + R_C} = \frac{(8\ \Omega)(24\ \Omega)}{18\ \Omega + 8\ \Omega + 24\ \Omega}$$

$$R_c = 3.84\ \Omega$$

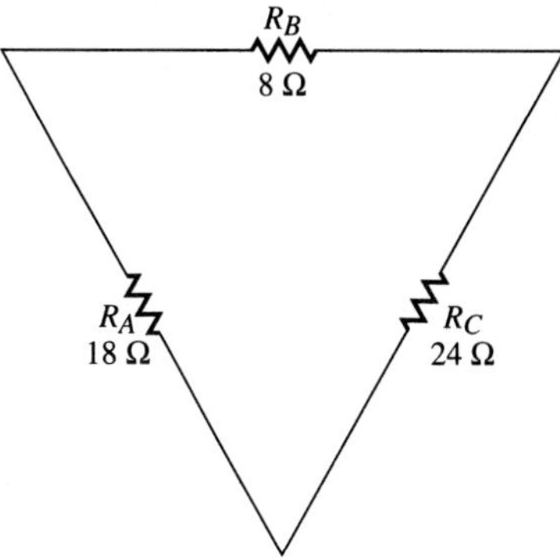

Figure 10.45

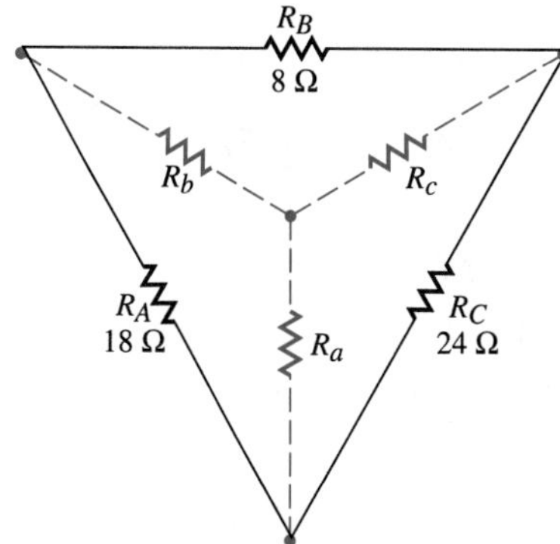

Figure 10.46

Wye to Delta

The general relation is

$$R_{\text{SIDE}} = \frac{\text{sum of the products of pairs of Wye legs}}{\text{resistance of the opposite Wye legs}}$$

For the network of Figure 10.44,

$$R_A = \frac{R_a R_b + R_a R_c + R_c R_b}{R_c} \qquad \textbf{(10.15)}$$

$$R_B = \frac{R_a R_b + R_a R_c + R_c R_b}{R_a} \qquad \textbf{(10.16)}$$

$$R_C = \frac{R_a R_b + R_a R_c + R_c R_b}{R_b} \qquad \textbf{(10.17)}$$

If $R_a = R_b = R_c$,

$$R_\Delta = 3\ R_Y \qquad \textbf{(10.18)}$$

EXAMPLE 10.18

The resistance in each leg of a Wye network is 30 ohms. What is the resistance in each side of the equivalent Delta?

Solution

Since $R_a = R_b = R_c$,

$$R_\Delta = 3\,R_Y = (3)\,(30\ \Omega)$$
$$R_\Delta = 90\ \Omega$$

EXAMPLE 10.19

Convert the Wye network of Figure 10.47 to an equivalent Delta network.

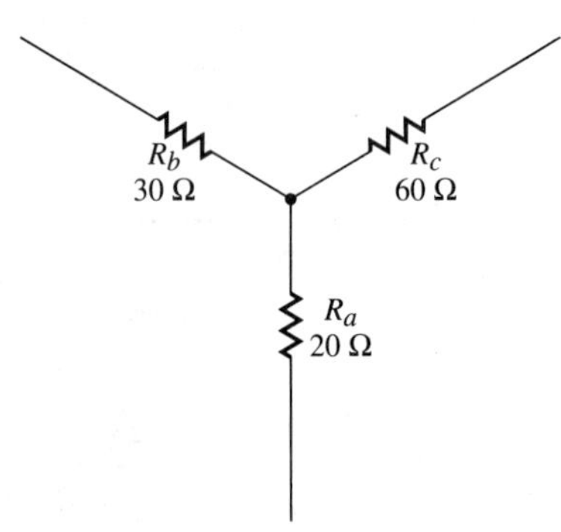

Figure 10.47

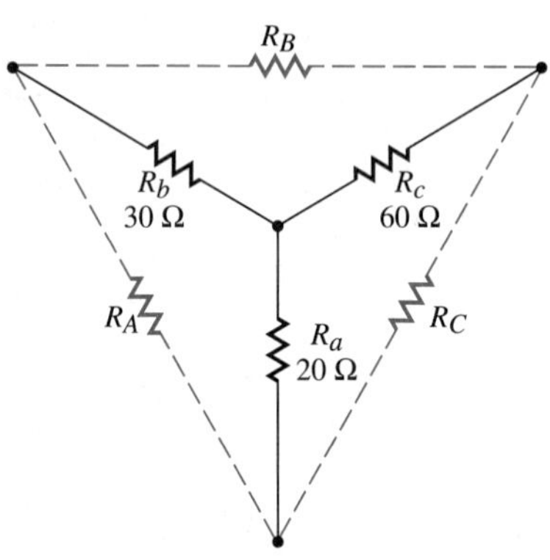

Figure 10.48

Solution

The Wye network is shown imposed on the Delta network in Figure 10.48.

Using Equations 10.15–10.17 gives

$$R_A = \frac{R_aR_b + R_aR_c + R_bR_c}{R_c}$$
$$= \frac{(20\ \Omega)(30\ \Omega) + (20\ \Omega)(60\ \Omega) + (30\ \Omega)(60\ \Omega)}{60\ \Omega}$$
$$= \frac{3600\ \Omega^2}{60\ \Omega}$$
$$R_A = 60\ \Omega$$
$$R_B = \frac{R_aR_b + R_aR_c + R_bR_c}{R_a}$$
$$= \frac{3600\ \Omega^2}{20\ \Omega}$$
$$R_B = 180\ \Omega$$
$$R_C = \frac{R_aR_b + R_aR_c + R_bR_c}{R_b}$$
$$= \frac{3600\ \Omega^2}{30\ \Omega}$$
$$R_C = 120\ \Omega$$

These examples showed the procedure for making the conversion. The procedure for applying the conversion when the network is part of a circuit is as follows:

1. Identify the Wye or Delta network that is to be converted.
2. Mark the points where the points of the Delta or legs of the Wye network are connected.
3. Make the conversion.

4. Connect the equivalent to the points in Step 2.

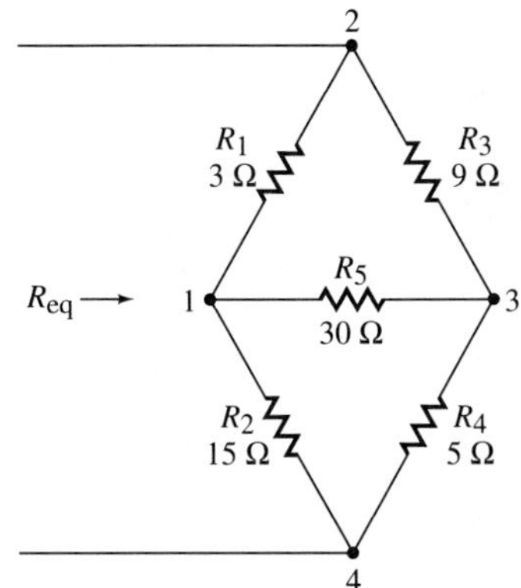

Figure 10.49

EXAMPLE 10.20

Use the conversion to change the circuit of Figure 10.49 into a series-parallel circuit and determine the equivalent resistance of the circuit.

Solution

Converting the bottom Delta network, connected to 1, 3, and 4, results in the series-parallel circuit of Figure 10.50.

Figure 10.51 shows the equivalent Wye network imposed on the Delta network.

Applying the conversion equations gives

$$R_a = \frac{R_2 R_4}{R_2 + R_4 + R_5}$$

$$R_a = \frac{(15\ \Omega)(5\ \Omega)}{15\ \Omega + 5\ \Omega + 30\ \Omega}$$

$$= 1.5\ \Omega$$

$$R_b = \frac{R_2 R_5}{R_2 + R_4 + R_5}$$

$$= \frac{(15\ \Omega)(30\ \Omega)}{50\ \Omega}$$

$$= 9\ \Omega$$

$$R_c = \frac{R_5 R_4}{R_2 + R_4 + R_5}$$

$$= \frac{(30\ \Omega)(5\ \Omega)}{50\ \Omega}$$

$$= 3\ \Omega$$

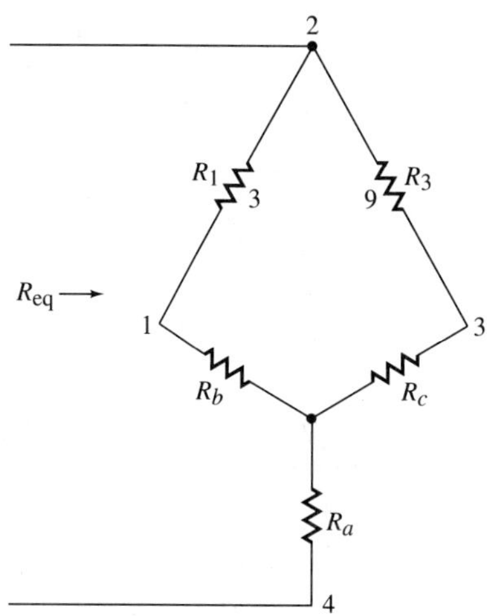

Figure 10.50

The equivalent resistance of the circuit in Figure 10.50 is

$$R_{eq} = R_a + (R_1 + R_b)\|(R_3 + R_c)$$

$$R_1 + R_b = 3\ \Omega + 9\ \Omega = 12\ \Omega$$

$$R_3 + R_c = 9\ \Omega + 3\ \Omega = 12\ \Omega$$

so

$$R_{eq} = 1.5\ \Omega + \frac{(12\ \Omega)(12\ \Omega)}{12\ \Omega + 12\ \Omega}$$

$$R_{eq} = 7.5\ \Omega$$

Thus, we see that by using the conversion, we are able to make a series-parallel circuit.

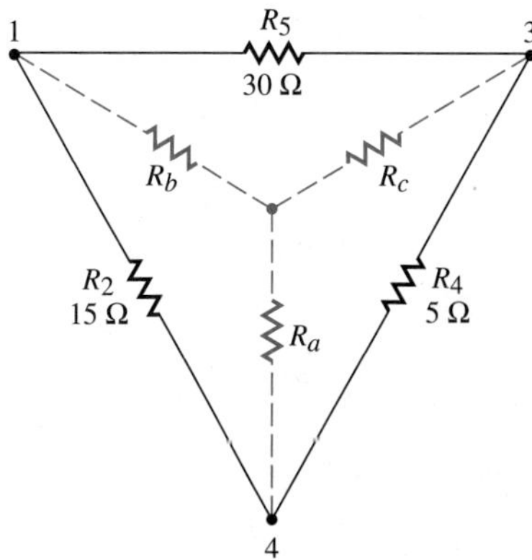

Figure 10.51

SUMMARY

1. In the superposition theorem, the current and voltage of interest are found for each source while the others are replaced by their internal resistance. The quantities are then combined algebraically to obtain the currents and voltages with all sources in the circuit.
2. The power in a circuit cannot be found by using superposition to combine the power from each source.
3. Thevenin's Theorem provides a means of reducing a linear, bilateral element circuit to one voltage source and one resistance.
4. Norton's Theorem provides a means of reducing a linear, bilateral element circuit to one current source and one resistance.
5. Each Thevenin circuit has a Norton equivalent, and each Norton circuit has a Thevenin equivalent.
6. A source transfers maximum power to a load when the load resistance has the same value as the source internal resistance.
7. The power transferred to a load at maximum power transfer is $E^2/(4\ R_L)$, and the load voltage is $E/2$.
8. At maximum power transfer, 50 percent of the source power is transferred to the load.
9. Each Delta network has an equivalent Wye network, and each Wye network has an equivalent Delta network.

EQUATIONS

10.1	$R_{\mathrm{TH}} = \dfrac{V_{oc}}{I_{sc}}$	Thevenin resistance from open- and short-circuit values.	379
10.2	$E_{\mathrm{TH}} = V_{oc}$	Thevenin emf from open-circuit voltage.	379
10.3	$R_{\mathrm{N}} = \dfrac{V_{oc}}{I_{sc}}$	Norton resistance from open- and short-circuit values.	383
10.4	$I_{\mathrm{N}} = I_{sc}$	Norton current from short-circuit current.	383
10.5	$E_{\mathrm{TH}} = I_{\mathrm{N}}R_{\mathrm{N}}$	Relation of Thevenin emf to Norton values.	384
10.6	$R_{\mathrm{N}} = R_{\mathrm{TH}}$	Relation of Norton resistance to Thevenin resistance.	384
10.7	$I_{\mathrm{N}} = \dfrac{E_{\mathrm{TH}}}{R_{\mathrm{TH}}}$	Relation of Norton current to Thevenin values.	384

10.8	$R_L = R_i$	Relation of R_L to R_i at maximum power transfer.	386
10.9	$P_{max} = \dfrac{E^2}{4R_L}$	Load power at maximum power transfer.	386
10.10	$V_L = \dfrac{E}{2}$	Load voltage at maximum power transfer.	386
10.11	$R_a = \dfrac{R_A R_C}{R_A + R_B + R_C}$	The equation for converting the Delta resistance to one branch of the Wye.	392
10.12	$R_b = \dfrac{R_A R_B}{R_A + R_B + R_C}$	The equation for converting the Delta resistance to the second branch of the Wye.	392
10.13	$R_c = \dfrac{R_B R_C}{R_A + R_B + R_C}$	The equation for converting the Delta resistance of the third resistance of the Wye.	392
10.14	$R_Y = \dfrac{R_\Delta}{3}$	The relation of the Wye resistance to the Delta resistance when the Delta sides are of equal resistance.	392
10.15	$R_A = \dfrac{R_a R_b + R_a R_c + R_c R_b}{R_c}$	The equation for converting the Wye resistance to one side of the Delta.	393
10.16	$R_B = \dfrac{R_a R_b + R_a R_c + R_c R_b}{R_a}$	The equation for converting the Wye resistance to the second side of the Delta.	393
10.17	$R_C = \dfrac{R_a R_b + R_a R_c + R_c R_b}{R_b}$	The equation for converting the Wye resistance to the third side of the Delta.	393
10.18	$R_\Delta = 3\ R_Y$	The relation of the Delta resistance to the Wye resistance when the Wye legs are of equal resistance.	393

QUESTIONS

1. Why does the Superposition Theorem give incorrect results if any part in the circuit is nonlinear or unilateral?
2. Why is the power in a part of a circuit incorrect if it is obtained by superposition of the power from each source?
3. When determining the current from one source by superposition, what is done with the other sources?
4. What is done with the sources in a circuit when finding R_N or R_{TH}?
5. What are two applications in which the Thevenin equivalent of a circuit is helpful?

6. Can circuits having different resistances and sources have the same Thevenin equivalent circuit?
7. Can circuits with different resistances and sources have the same Norton equivalent circuit?
8. How are the open-circuit and short-circuit values used to find the parts in the Thevenin and Norton circuits?
9. What is the relation between the load resistance and the source resistance at maximum power transfer?
10. What percent of the generated power is transferred to the load at maximum power transfer?
11. What happens to the power that is not transferred to the load at maximum power transfer?
12. A 10-ohm load resistor results in maximum power transfer. In which direction will efficiency and power change if the resisance is reduced to 5 ohms?
13. Why is maximum power transfer not desirable in the transmission of electric power for residential and industrial use?
14. What is the percent efficiency at maximum power transfer?
15. What percent of the source open-circuit voltage is across the load at maximum power transfer?
16. What conditions must be met for a Delta and Wye network to be equivalent?
17. What is the name given to the Delta network when used in electric circuits?
18. What is the name given to the Wye network when used in electronic circuits?

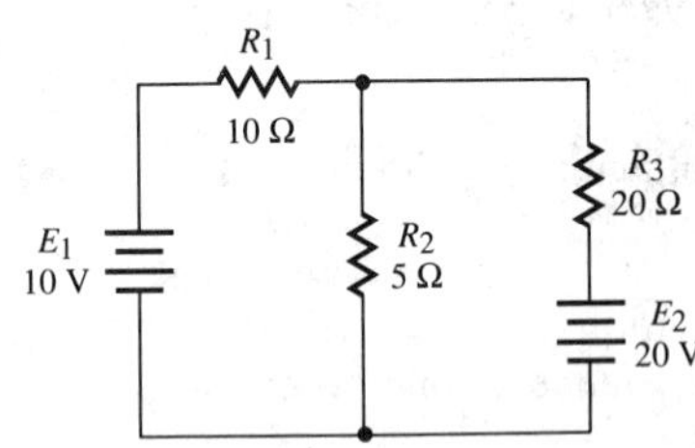

Figure 10.52

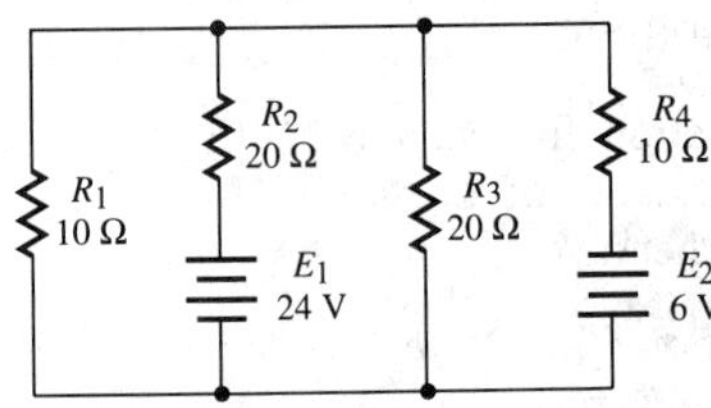

Figure 10.53

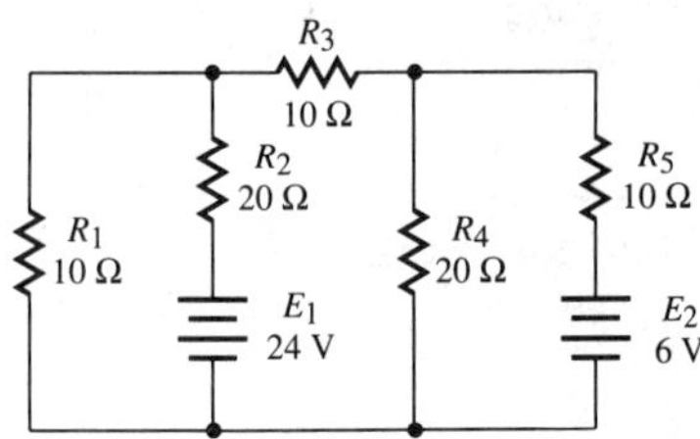

Figure 10.54

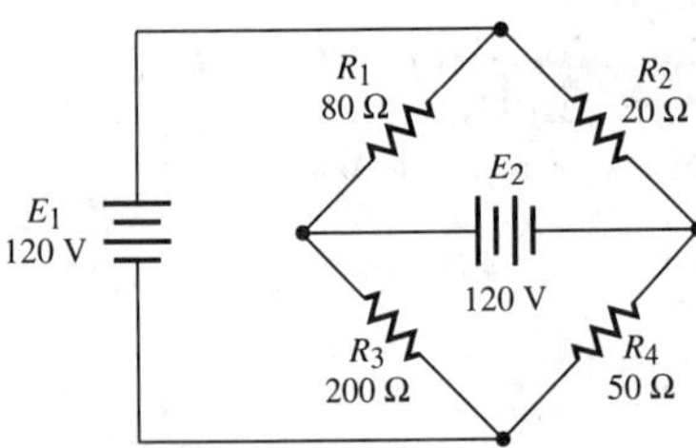

Figure 10.55

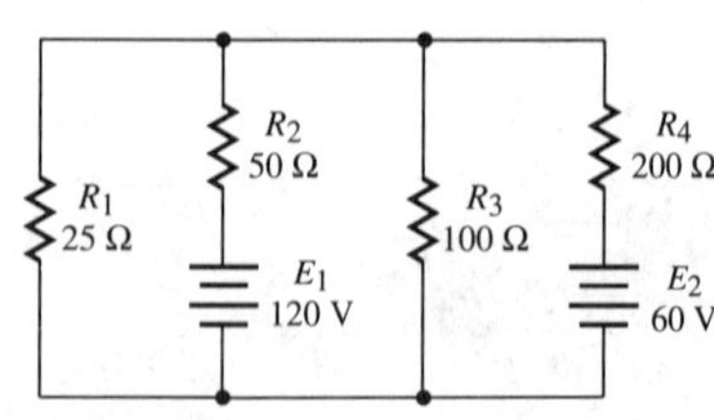

Figure 10.56

PROBLEMS

SECTION 10.1 The superposition Theorem

Solve the problems in this section by using superposition.

1. For the circuit of Figure 10.52, find:
   (a) The current in each resistor.
   (b) The voltage across each resistor.
   (c) The total power in the resistance of the circuit.
2. For the circuit of Figure 10.53, find:
   (a) The current in each resistor.
   (b) The voltage across each resistor.
   (c) The power dissipated in the circuit.
3. In Figure 10.54, find:
   (a) The current in each resistor.
   (b) The voltage across R_3.
   (c) The power in the circuit.
4. In Figure 10.55, find:
   (a) The voltage across R_3.
   (b) The current in R_1.
5. In the circuit of Figure 10.56, find:

(a) The voltage across R_4.
(b) The current in R_4.
(c) The power dissipated in the circuit.

6. What is the resistance of R_2 in Figure 10.57 if I_2 is 4.5 amperes? Does E_2 supply or use power?

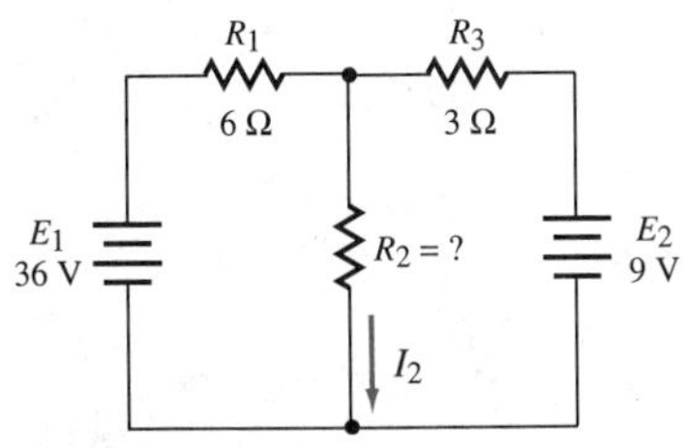

Figure 10.57

SECTION 10.2 Thevenin's Theorem

Solve the problems in this section using Thevenin's Theorem.

7. Replace the circuit connected to R_3 of Problem 1 by its Thevenin equivalent.
8. Find the Thevenin equivalent of the circuit of Figure 10.58 looking into Terminals A–B.
9. Find the voltage across R_5 and the power in R_5 for the circuit of Figure 10.59 using Thevenin's Theorem.
10. Find the voltage across R_3 in Figure 10.60 for $R_3 = 10$, 20, and 30 ohms.
11. Work Problem 6 using the Thevenin equivalent circuit.
12. A circuit has an open-circuit voltage of 12 V and a short-circuit current of 6 A. Determine the values of E_{TH} and R_{TH} for the Thevenin equivalent circuit.
13. The open-circuit voltage of a circuit is 8 volts. The current is 2 A when a 2-ohm resistor is connected to it. Determine the values of E_{TH} and R_{TH} for the Thevenin equivalent circuit.

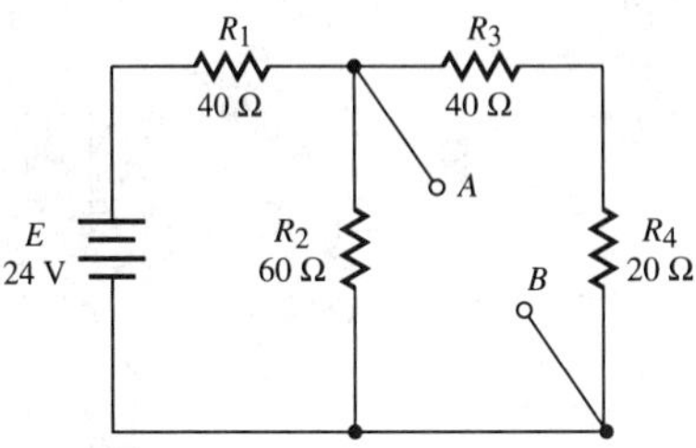

Figure 10.58

SECTION 10.3 Norton's Theorem

Solve the following problems using Norton's Theorem.

14. Find the Norton equivalent for the circuit connected to R_3 in Problem 1.
15. Convert the Thevenin equivalent of Problem 7 to the Norton equivalent.
16. Find the Norton equivalent of the circuit of Figure 10.61 looking into Terminals A–B.
17. Find the voltage across R_5 and the power in R_5 for the circuit of Problem 9 using Norton's Theorem.
18. Find the voltage across R_6 in Figure 10.62 for $R_6 = 20$, 50, and 100 ohms.
19. Find R_N and I_N if the open-circuit voltage is 120 V and the short-circuit current is 40 A.

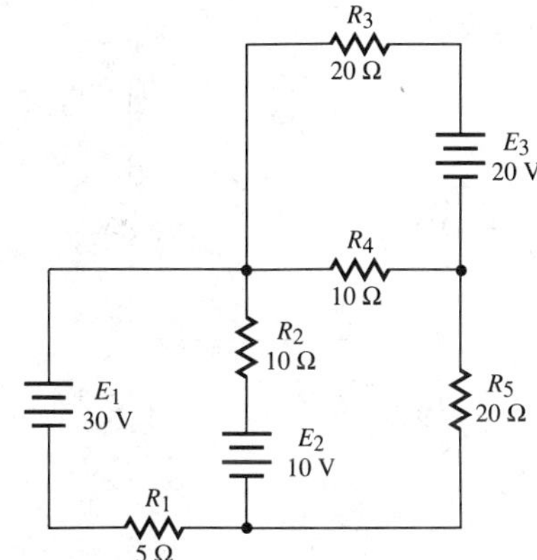

Figure 10.59

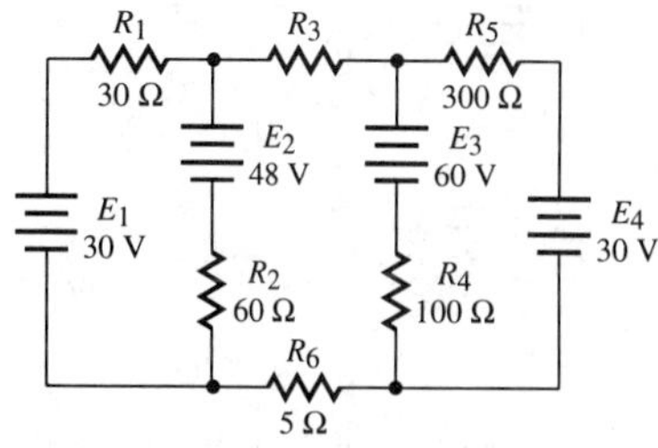

Figure 10.60

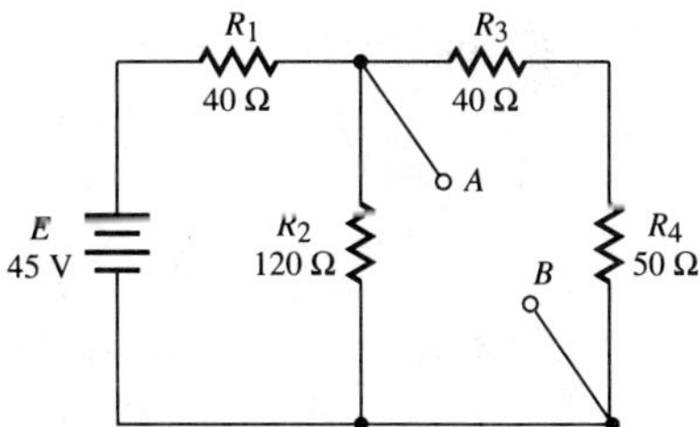

Figure 10.61

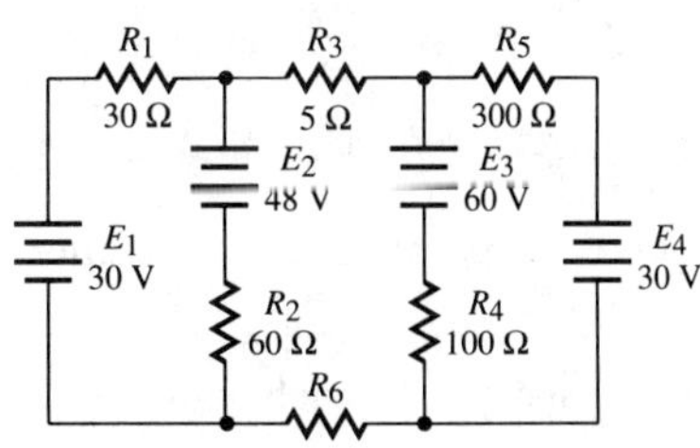

Figure 10.62

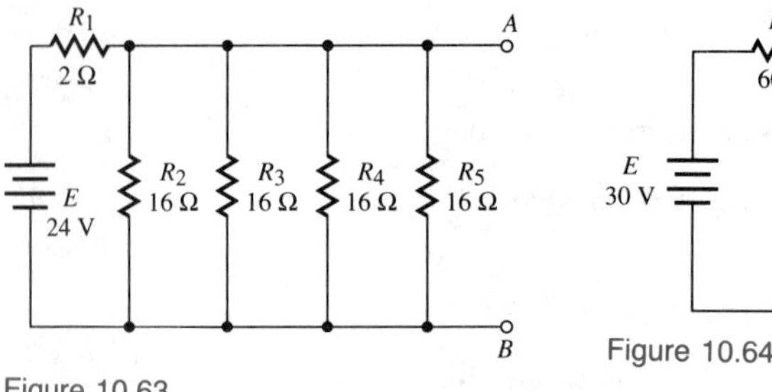

Figure 10.63

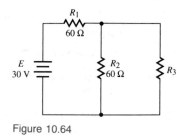

Figure 10.64

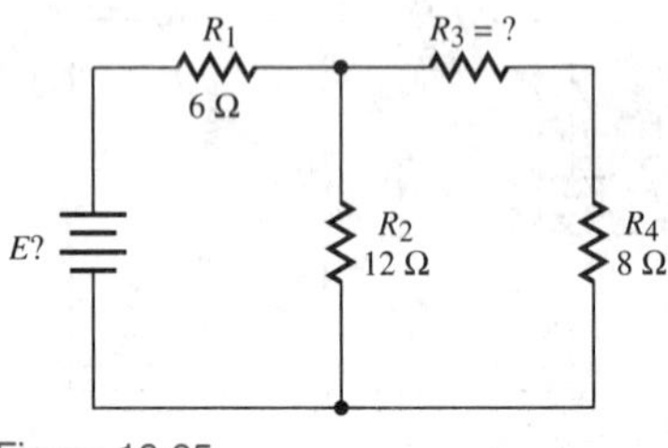

Figure 10.65

20. Determine the open-circuit voltage and short-circuit current for the circuit of Figure 10.63 and use them to find R_N and I_N.

SECTION 10.4 Maximum Power Transfer

Solve the problems in this section by using maximum power transfer.

21. What must be the value of R_3 to get maximum power transfer to it in Figure 10.64?

22. The power in R_4 of Figure 10.65 at maximum power transfer is 18 watts. Find:

 (a) The resistance of R_3.

 (b) The emf of E.

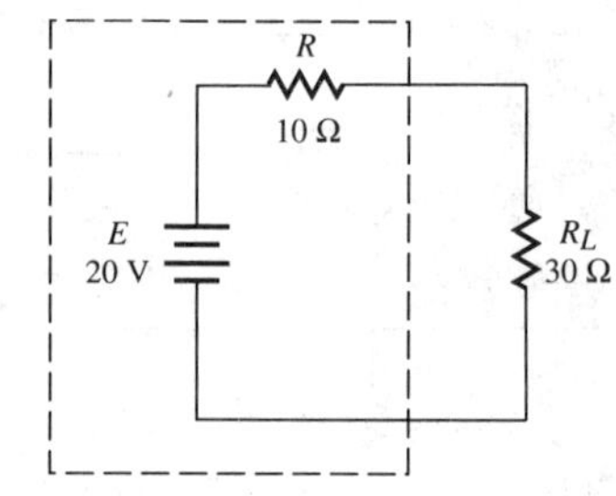

Figure 10.66

23. A microphone has an open-circuit voltage output of 0.1 V and a resistance of 600 ohms. Find:

 (a) The maximum power that it can deliver to an amplifier at matched conditions.

 (b) The power that it can deliver to a 300-ohm amplifier.

24. The circuit in the dashed block of Figure 10.66 is connected to the 30-ohm resistor. Using the curves of Figure 10.36, find:

 (a) The power delivered to the 30-ohm resistor.

 (b) The voltage across the resistor.

 (c) The % efficiency.

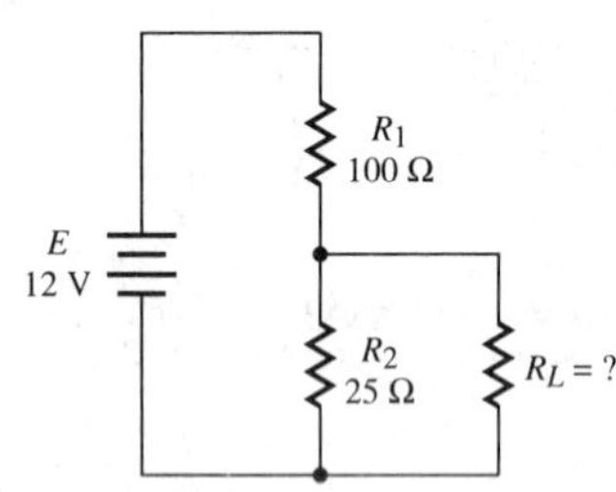

Figure 10.67

25. An amplifier delivers maximum power when connected to a 5-ohm loudspeaker. The power is 5 W. How much power will be delivered if the loudspeaker is replaced by an 8-ohm loudspeaker?

26. What is the voltage across R_L in Figure 10.67 when it is set for maximum power transfer?

(S) 27. Use PSpice to determine the value of R_L in Problem 26.

SECTION 10.5 Delta and Wye Networks

Solve the problems in this section by using the Delta and Wye networks.

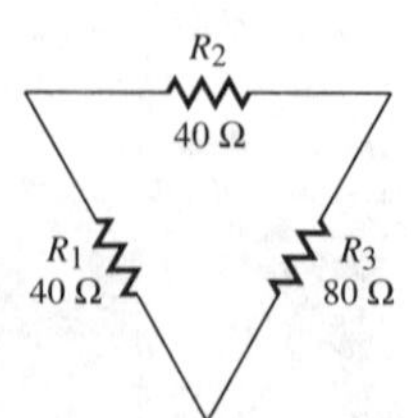

Figure 10.68

28. A Delta network has 60 ohms of resistance in each side. Determine the resistance in each leg of the equivalent Wye.

29. Convert the Delta network of Figure 10.68 to its equivalent Wye.

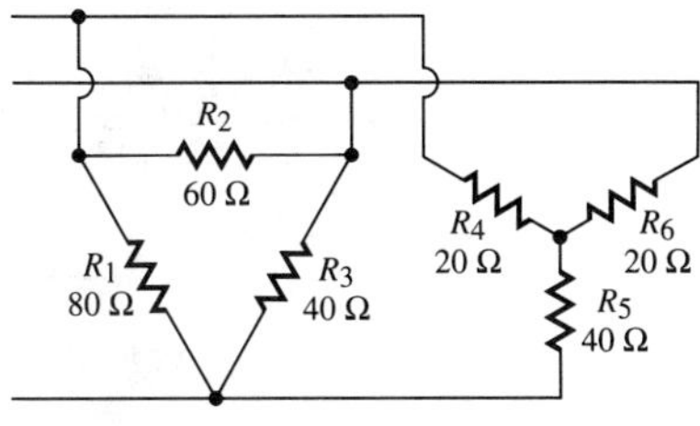

Figure 10.69

30. Convert the circuit of Figure 10.69 to an equivalent Delta circuit.
31. A Wye network has 80 ohms of resistance in each leg. Determine the resistance in each side of the equivalent Delta network.
32. Convert the Wye network of Figure 10.70 to its equivalent Delta network.
33. Convert the circuit of Figure 10.69 to an equivalent Wye network.
34. Reduce the circuit of Figure 10.71 to a series-parallel circuit and determine its equivalent resistance.

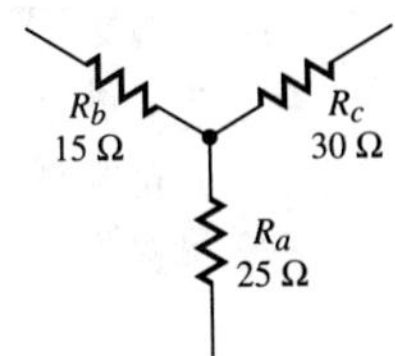

Figure 10.70

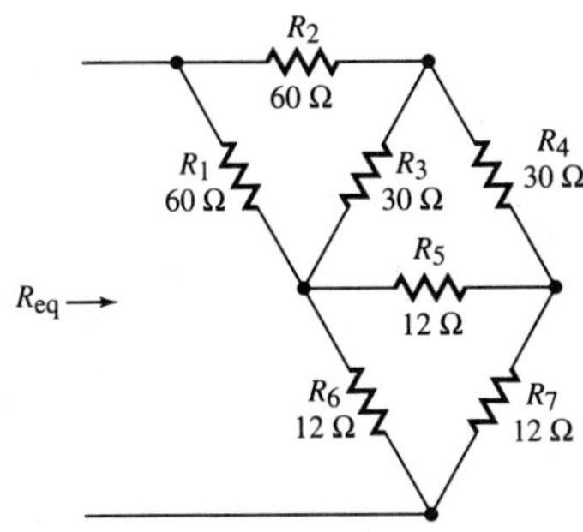

Figure 10.71

CHAPTER 11

CAPACITANCE AND CAPACITORS

CHAPTER OBJECTIVES

After completing this chapter, you should be able to:

1. Define capacitance and the unit of capacitance.
2. Explain the electrostatic field, its characteristics, and field intensity.
3. Explain what a capacitor is, name its parts, and describe the charging process.
4. Define permittivity and relative permittivity.

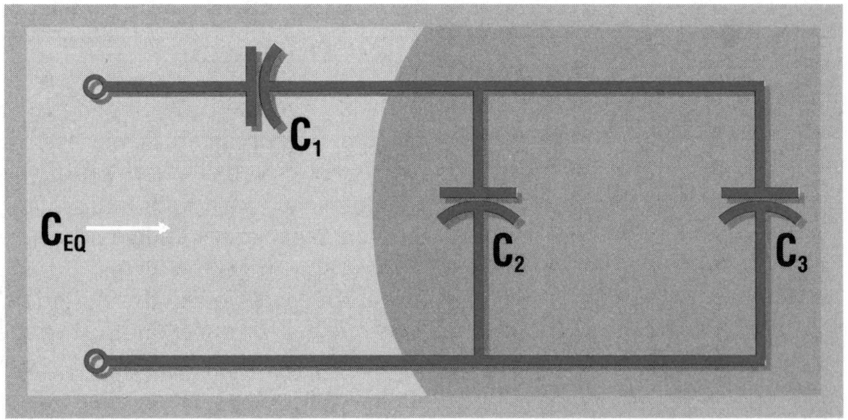

5. List the three factors that affect capacitance, and calculate the capacitance of a parallel plate capacitor.
6. Determine the maximum voltage that a capacitor can safely withstand.
7. Name the types of capacitors, describe their basic construction, and give some characteristics of each.
8. Perform the analysis of series and parallel capacitor circuits for charge, equivalent capacitance, and breakdown voltage.

KEY TERMS

1. breakdown voltage
2. capacitance
3. capacitor
4. ceramic capacitor
5. charged capacitor
6. dielectric
7. dielectric constant
8. dielectric strength
9. discharged capacitor
10. electrolytic capacitor
11. electrostatic field
12. equipotential lines
13. equivalent capacitance
14. farad
15. field intensity
16. fixed capacitor
17. flux density
18. leakage current
19. leakage resistance
20. mica capacitor
21. oil capacitor
22. paper capacitor
23. parallel capacitor connection
24. permittivity
25. plastic film capacitor
26. relative permittivity
27. series capacitor connection
28. surge breakdown voltage
29. temperature coefficient
30. trimmer capacitor
31. varactor
32. variable capacitor
33. working voltage

INTRODUCTION

Lights remain bright for minutes after the switch is opened. Pictures are taken with a camera whose flash does not use a bulb. High-frequency signals are directed to one set of speakers and low-frequency ones to another set. These and many other occurrences are made possible through capacitance and capacitors.

The capacitor is the second circuit component which will be studied. It is simply two conducting materials separated by an insulating layer. Some capacitors are cylindrical and resemble resistors. Others range in size from tiny ceramic disks to larger box-size capacitors used in power applications.

What criteria are used when selecting a capacitor? What type of capacitor is best suited for an application? What are the circuit connections for capacitors? What affects capacitance? And what are the negative effects of capacitance? These are some of the questions that are answered in this chapter.

11.1 Capacitance

capacitance

farad

Another property that is used to obtain certain characteristics from an electrical circuit is capacitance. **Capacitance** is the property by which two conducting materials separated by an insulator can store charge. The unit for capacitance is the **farad** (F), named in honor of Michael Faraday.

The property of capacitance is useful in electric circuits because it opposes a change in voltage, stores energy, and blocks dc current. Each of these properties will be studied in the next chapter. Some specific applications of capacitance are time delay circuits, waveshaping circuits, phase shifting, electronic flashes, filters, dc isolation, and ac coupling between sections of an electronic circuit.

One farad of capacitance will store 1 coulomb of charge when 1 volt of emf is applied across the conducting surfaces. Most practical applications use capacitances in the order of microfarads (μF) or less. The symbol used for capacitance is C.

The relation of capacitance to charge and voltage is given by

$$C = Q/V \tag{11.1}$$

where C is the capacitance, in farads; Q is the charge, in coulombs; and V is the potential difference between the conducting materials, in volts.

EXAMPLE 11.1

Two plates, 1 millimeter apart, have a capacitance of 6 microfarads. How much charge is stored when 12 volts are applied across the plates?

Solution

$$C = Q/V \qquad \text{or} \qquad Q = CV$$

Substituting the known values for C and V gives

$$Q = (6 \times 10^{-6}\ \text{F})(12\ \text{V})$$
$$= 72 \times 10^{-6}\ \text{C}$$
$$Q = 72\ \mu\text{C}$$

There is a small amount of capacitance between all conductive materials separated by an insulator. This capacitance is called "stray capacitance." Stray capacitance can cause interaction between parts of a circuit at high frequencies. This interaction is called "crosstalk." It can affect the characteristics of a circuit and cause problems. In extreme cases, the capacitance can result in a short circuit, even though there is no physical connection of the equipment.

Fortunately, stray capacitance can be reduced by following certain practices. They are:

1. Use connecting leads that are as short as possible.
2. Run leads at an angle to each other instead of parallel.
3. Use shielded leads.
4. Separate the leads by as much distance as possible.

11.2 The Electrostatic Field

Capacitance is related to the **electrostatic field**. This is the region around a charge in which a force is exerted on other charges. Since the electrostatic field is around a nonmoving charge, the field has the name electro (electric) and static (not moving).

electrostatic field

The shape of the electrostatic field between two spheres of opposite charge is shown in Figure 11.1(a). The lines between the spheres represent the path that a hypothetical charge of no mass would take while moving between the spheres. Figure 11.1(b) shows the shape of the field between two parallel plates. Here, the lines are straight between the plates with some outward bending at the edges.

The voltage along the lines can be measured using appropriate instruments. It varies from 0 volts at the bottom plate, to the voltage of the top plate. The lines that join points in the field, which have the same voltage, result in the dashed lines, which are called **equipotential lines**. Note that these are evenly spaced for the parallel plate capacitor but not for the two spheres in Figure 11.1(a).

equipotential lines

Some characteristics of the electrostatic field are:

1. The direction of the lines is that in which a positive charge will move in the field.

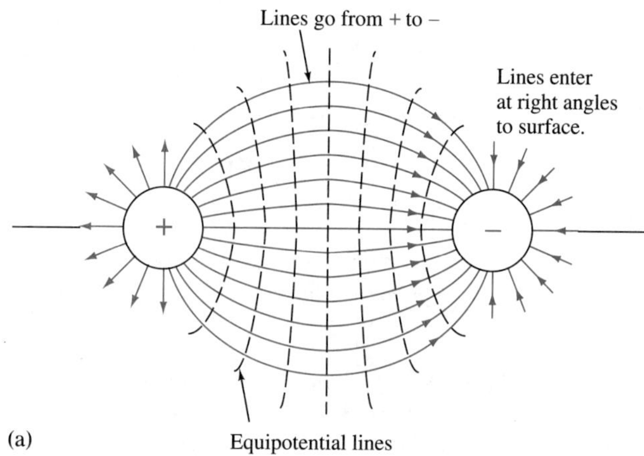

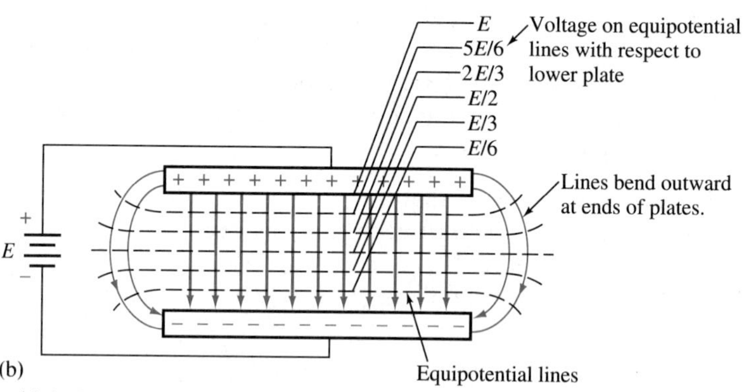

Figure 11.1 Examples of electrostatic fields. (a) The field between oppositely charged particles. (b) The field between two parallel plates.

2. The lines start at the positive charge and end at the negative charge.
3. The lines enter the charged surface at right angles to it.

field intensity

In accordance with Coulomb's Law, a force acts on a charge placed in the field. The force per unit positive charge is called the **field intensity** ($\mathscr{E}$). The more practical unit for $\mathscr{E}$ is volts per meter. A field intensity of 2 newtons per coulomb means that a charge of 1 coulomb would be acted on by a force of 2 newtons.

Since the equipotential lines for the parallel plates are evenly spaced, the field intensity between them is uniform. It is given by

$$\mathscr{E} = \frac{V}{d} \qquad \textbf{(11.2)}$$

where $\mathscr{E}$ is the field intensity, in volts per meter; V is the potential difference between the plates, in volts; and d is the distance between the plates, in meters.

The lines must start on a positive charge and end on a negative charge; the total number of lines or flux is numerically equal to the charge in coulombs. If the flux is represented by γ,

$$\gamma = Q \tag{11.3}$$

where γ is the number of lines, in coulombs; and Q is the charge in coulombs.

The **flux density** represents the number of lines per unit area. This results in

flux density

$$D = \frac{\gamma}{A} \tag{11.4}$$

where D is the flux density, in coulombs per square meter; γ is the flux, in coulombs, and A is the area enclosing the lines, in square meters.

11.3 The Capacitor

A **capacitor** is a device that is built to have a specific amount of capacitance. It is shown on schematics by the symbols in Figure 11.2. The basic parts of a capacitor are two conducting surfaces separated by an insulating material. The insulating material is called the **dielectric**. Some typical plate and dielectric arrangements are shown in Figure 11.3.

capacitor

dielectric

Having seen the structure of the capacitor, we now study its behavior, using Figure 11.4.

With the switch open, the plates are assumed to be electrically neutral. On the closing of the switch, as in Figure 11.4(a), the emf of the source causes an excess of electrons to be on the lower plate, and an excess of positive charge on the upper plate. This leaves a positive charge on the upper plate and a negative charge on the lower plate, as in Figure 11.4(b). The movement lasts for only a short time because there is no movement of electrons through the dielectric. The electron accumulation on the plates results in an electrostatic field between the plates that

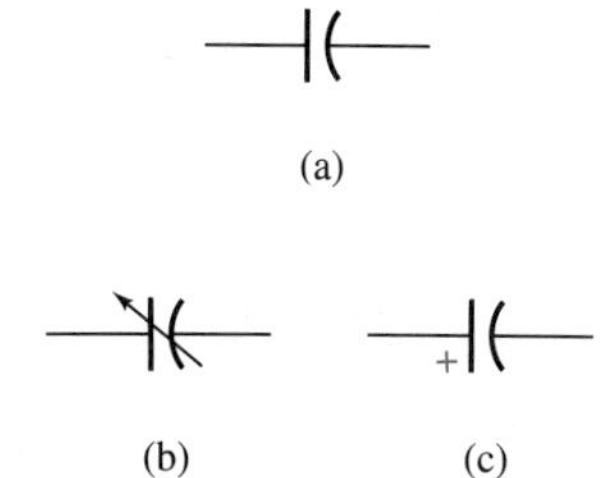

Figure 11.2 The graphic symbols for a capacitor. (a) Fixed capacitor. (b) Variable capacitor. (c) Polarized capacitor.

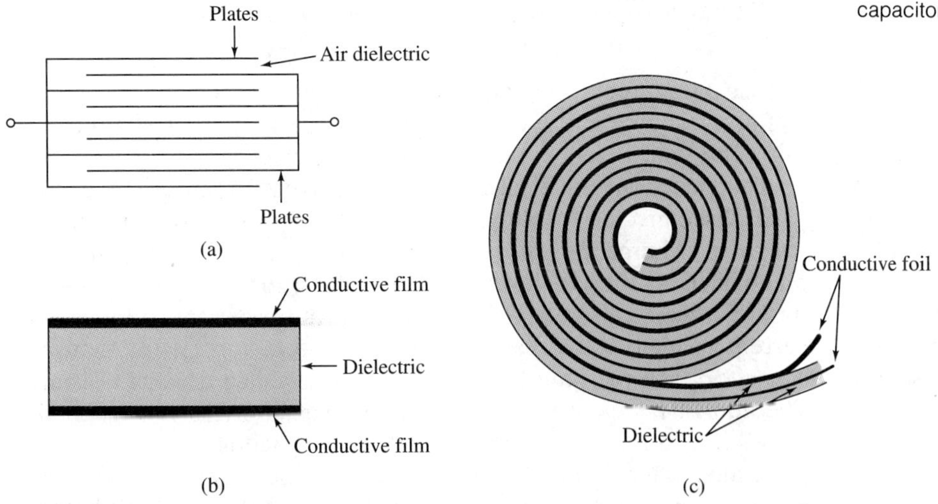

Figure 11.3 Some examples of the dielectric and plate arrangement. (a) A multiple plate arrangement. (b) Metallic film deposited on the dielectric. (c) The dielectric and foil are rolled together.

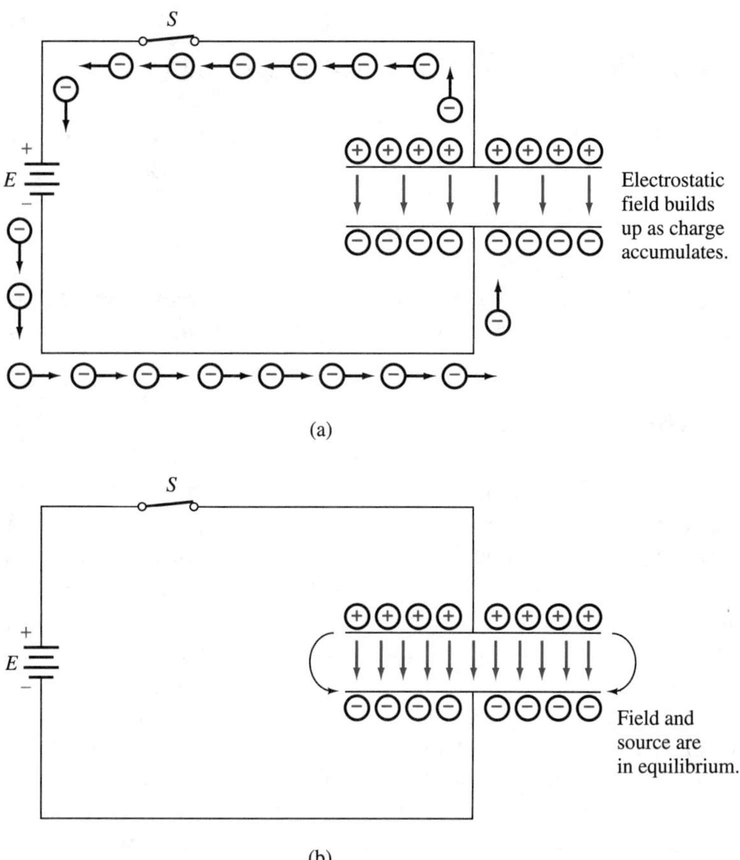

Figure 11.4 The movement of the charge to the plates of the capacitor and the resulting electrostatic field. (a) The source causes the positive charge to accumulate on the top plate and the negative charge to accumulate on the bottom plate. (b) The electrostatic field and the source emf are in equilibrium. The movement of the charge has stopped; the voltage on the capacitor equals E.

oppose the source. Eventually the force of the electrostatic field and the emf of the source reach a state of equilibrium, as in Figure 11.4(b). For this condition, the capacitor is a **charged capacitor**.

charged capacitor

A capacitor with an ideal dielectric would remain charged indefinitely when disconnected from the source of emf, since there is no path by which the charge can move from one plate to the other. However, practical dielectrics are not perfect insulators, so there is a small current through them. This current is known as **leakage current**. If the leakage current is high, the capacitor will act more as a resistor than as a capacitor. The resistance of the dielectric is called **leakage resistance**. The equivalent circuit for the practical capacitor, as in Figure 11.5, shows it as a resistor in parallel with the capacitor.

leakage current

leakage resistance

In most capacitors, this resistance is in the order of megohms. A discharge path can also be formed by the accumulation of dirt between the capacitor's terminals. These leakage conditions provide a path for the charge to transfer from one plate to the other, and the plates are

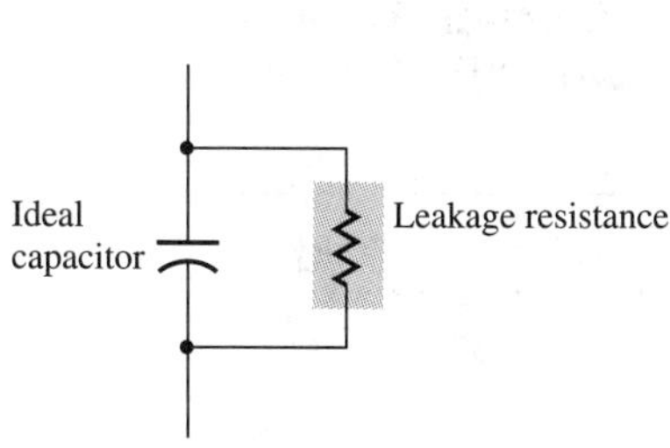

Figure 11.5 The graphic symbol for a practical capacitor includes a parallel leakage resistance.

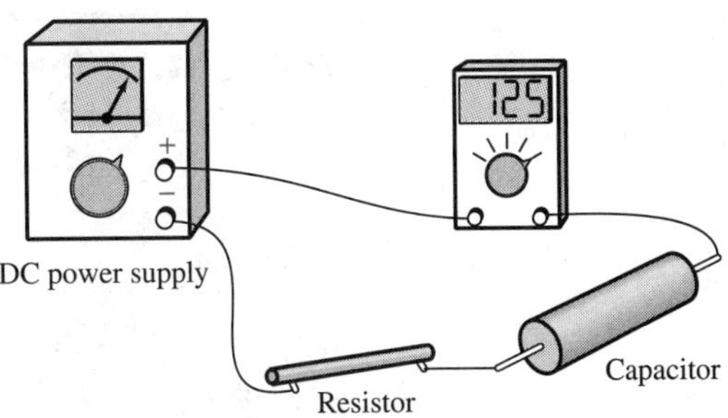

Figure 11.6 The voltage of dc power supply results in a leakage current, which can be measured on the microammeter.

returned to their neutral condition. For this condition, the capacitor is a **discharged capacitor**.

discharged capacitor

Resistance and leakage current can be measured using special instrumentation such as an impedance bridge and meters. However, a simple check can be made using an ohmmeter or DMM to measure resistance. When the meter is connected to the capacitor, the instantaneous reading should be 0 ohms. Then, as the capacitor charges, the reading should indicate some high resistance. Unfortunately, this test does not test the capacitor at a voltage, so it is not foolproof.

It is a wise practice to discharge a capacitor by placing a small resistance across its terminals before using it. In fact, higher capacitance units are often stored with a shorting strip across the terminals. This prevents any accumulation of charge from the atmosphere.

11.4 Permittivity and Relative Permittivity

Permittivity is a measure of how easily the lines of an electrostatic field can be set up in a material. The symbol for permittivity is ϵ, and it is expressed in farads per meter. Quantitatively, permittivity is the ratio of the flux density to the field intensity. Or

permittivity

$$\epsilon = \frac{D}{\mathscr{E}} \qquad \textbf{(11.5)}$$

where ϵ is the permittivity, in farads per meter; D is the flux density; in coulombs per square meter, and $\mathscr{E}$ is the field intensity, in volts per meter.

Air is not commonly used as the dielectric for capacitors. Some other dielectrics are materials such as mica, ceramic, oil, paper, and glass. These materials have higher permittivities than air, thus providing more capacitance. They also provide part of the capacitor's supporting structure.

The ratio of a material's permittivity to that of a vacuum is defined as its **relative permittivity** (ϵ_r). Another name for it is the **dielectric constant**. The dielectric constants for various materials are listed in Table 11.1. These are typical, and the actual value can vary because of temperature

relative permittivity
dielectric constant

TABLE 11.1 The Relative Permittivity (Dielectric Constant) of Materials

Material	Relative Permittivity
Air	1.0006
Bakelite	7
Ceramic titanium dioxide	100
Strontium titanate	7500
Glass	7.5
Mica	5
Oil	4
Paper	2.5
Plastic film	2.5
Porcelain	6
Rubber	3
Teflon	2

impurities and other factors. The permittivity of a material is given by

$$\epsilon = \epsilon_o \epsilon_r \qquad \textbf{(11.6)}$$

where ϵ is the permittivity of the dielectric, in farads per meter; ϵ_o is the permittivity of a vacuum, 8.85×10^{-12} farads per meter; and ϵ_r is the relative permittivity, without any units.

EXAMPLE 11.2

What is the permittivity of mica?

Solution

$$\epsilon = \epsilon_o \epsilon_r$$

From Table 11.1, $\epsilon_r = 5$.
Therefore

$$\epsilon = (8.85 \times 10^{-12}\ \text{F/m})(5)$$
$$\epsilon = 4.425 \times 10^{-11}\ \text{F/m}$$

11.5 Factors that Affect Capacitance

Section 11.4 explained that the dielectric affects the capacitance. Capacitance is also affected by the distance between the plates, and the area of the plates. The exact effect of these can be seen from the following:

By the definition of capacitance,

$$C = \frac{Q}{V}$$

But, from Equation 11.3, Q is equal to the total lines of flux, so $C = \gamma/V$.

Also, from Equations 11.4 and 11.5,

$$\gamma = DA \quad \text{and} \quad \epsilon = D/\mathscr{E}$$

Combining these gives

$$C = \frac{\epsilon \mathscr{E} A}{V}$$

But from Equation 11.2,

$$\mathscr{E} \text{ is } V/D, \quad \text{so} \quad C = \frac{\epsilon V A}{dV} = \frac{\epsilon A}{d}$$

Thus, changing the plate area or the spacing changes the capacitance.

Writing ϵ in terms of ϵ_o and ϵ_r gives the equation for the capacitance of a parallel plate capacitor as being:

$$C = \frac{\epsilon_o \epsilon_r A}{d} \tag{11.7}$$

where C is the capacitance, in F; ϵ_o is the permittivity of a vacuum, 8.85×10^{-12} F/m; ϵ_r is the relative permittivity of the dielectric; A is the area that is common to both plates, in m^2; and d is the distance between the plates, in m.

EXAMPLE 11.3

What is the capacitance of the parallel plate capacitor in Figure 11.7?

Solution

$$C = \frac{\epsilon_o \epsilon_r A}{d} = \frac{(8.85 \times 10^{-12}\ \text{F/m})(5)(0.05\ \text{m})(0.08\ \text{m})}{0.001\ \text{m}}$$

$$C = 177 \times 10^{-12}\ \text{F}$$

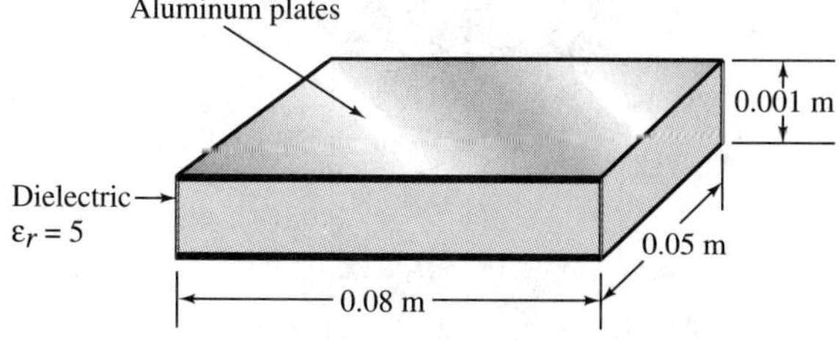

Figure 11.7

EXAMPLE 11.4

A parallel plate capacitor has a spacing of 0.2 mm and a dielectric of paper. What is the plate area if the capacitance is 0.6 picofarad?

Solution

$$C = \frac{\epsilon_o \epsilon_r A}{d}$$

so

$$\begin{aligned} A &= \frac{Cd}{\epsilon_o \epsilon_r} \\ &= \frac{(0.6 \times 10^{-12}\ \text{F})\ (0.2 \times 10^{-3}\ \text{m})}{(8.85 \times 10^{-12}\ \text{F/m})\ (2.5)} \\ A &= 5.4 \times 10^{-6}\ \text{m}^2 \end{aligned}$$

Although the form of the equation varies with the capacitor configuration, the general relation given in Equation 11.7 still holds. Namely, the capacitance is dependent on the dielectric, the space between the plates, and the area. This general relation is shown in Equation 11.8 for a coaxial cable.

A coaxial cable is a cable used to carry high-frequency signals from an antenna to a receiver, and is also used for other high-frequency applications. It is made of two concentric conductors, as shown in Figure 11.8. The capacitance for this configuration is given by

$$C = \frac{2.41 \times 10^{-5}\ \epsilon_r}{\log_{10}(r_2/r_1)}\ \mu\text{F/m} \qquad \textbf{(11.8)}$$

where C is the capacitance for each meter of conductor, in μF/m; ϵ_r is the relative permittivity of the dielectric between the conductors; r_2 is the radius of the outer conductor, in meters; and r_1 is the radius of the inner conductor, in meters.

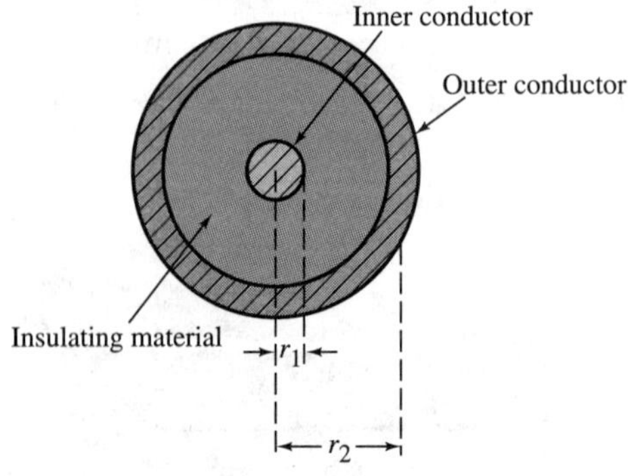

Figure 11.8 A cross section of a coaxial conductor.

EXAMPLE 11.5

A coaxial cable has a length of 50 meters and a teflon dielectric. If the radius of the inner conductor is 1.6 mm, and the radius of the outer conductor is 4.8 mm, what is its capacitance?

Solution

The total capacitance is the capacitance per meter multiplied by the length. So

$$\text{for 1 m,} \quad C = \frac{2.41 \times 10^{-5}\, \epsilon_r}{\log_{10}(r_2/r_1)}\ \mu\text{F/m}$$

$$\text{for 50 m,} \quad C = \frac{50\text{ m} \times 2.41 \times 10^{-5} \times 2\ \mu\text{F/m}}{\log_{10}(4.8\text{ mm}/1.6\text{ mm})}$$

$$= \frac{241 \times 10^{-5}}{0.477}\ \mu\text{F}$$

$$C = 5.05 \times 10^{-3}\ \mu\text{F}$$

The algebraic and RPN calculator keystrokes for the solution are:

Quantity	Algebraic	RPN	Display
Capacitance	[5] [0] [×] [2]	[5] [0] [ENTER] [2]	[5.05 – 03]
	[·] [4] [1] [EXP]	[·] [4] [1] [EEX]	
	[5] [+/–] [×] [2]	[5] [CHS] [×] [2]	
	[÷] [[] [[] [4] [·]	[×] [4] [·] [8] [ENTER]	
	[8] [÷] [1] [·] [6]	[1] [·] [6] [÷] [LOG]	
	[]] [LOG] []] [=]	[÷]	

11.6 Dielectric Strength

The electrons in the dielectric do not move from one atom to another, as they do with current. However, as the voltage across the plates is increased, a stress is put on the orbits and the electrons can be torn from the atom. This results in arcing and heating of the dielectric, which produces a condition called dielectric breakdown. The dielectric is no longer an insulator and the unit is no longer a capacitor.

The voltage at which breakdown occurs is called the **breakdown voltage**. Capacitors are usually rated in either dc working voltage or ac root-mean-square (rms) voltage. RMS is the unit in which an alternating voltage or current is measured. The dc voltage that a capacitor can

breakdown voltage

withstand will be greater than its rms voltage rating. This is because the maximum value of the alternating voltage that the capacitor must withstand is greater than the rms value. For a sine wave, V_{max} is 1.414 times the rms value. Thus, a capacitor with a 100-volt rms rating can withstand about 141 volts dc.

dielectric strength

The breakdown voltage is a function of dielectric strength and the spacing. The **dielectric strength** is the number of volts that the unit thickness of dielectric can withstand without breakdown. Typical values for some dielectrics are listed in Table 11.2. Again, impurities, temperature, and other factors can make the actual value vary from these. The breakdown voltage is given by

$$BV = (d)(\text{dielectric strength}) \qquad \textbf{(11.9)}$$

where BV is the breakdown voltage, in volts; d is the spacing, in mm; and dielectric strength is in volts/mm.

EXAMPLE 11.6

A mica capacitor has a spacing of 2 mm. How many volts can it withstand?

Solution

$$BV = (d)(\text{dielectric strength})$$

From Table 11.2, mica has a dielectric strength of 200 kV/mm.

$$BV = (2\text{ mm})(200\text{ kV/mm})$$
$$BV = 400\text{ kV}$$

TABLE 11.2 The Dielectric Strength of Materials

Material	Kilovolts per Millimeter
Air	3
Bakelite	16
Ceramic	3
Glass	40–120
Mica	200
Oil	12–20
Paper	40–79
Plastic film	20
Porcelain	4–10
Rubber	28–47
Teflon	60

EXAMPLE 11.7

A 0.06-microfarad capacitor with a plastic film dielectric must withstand 200 volts without breaking down. What is the area of the plate if the minimum spacing is used?

Solution

$$C = \frac{\epsilon_o \epsilon_r A}{d} \quad \text{so} \quad A = \frac{Cd}{\epsilon_o \epsilon_r}$$

but

$$d = \frac{BV}{\text{dielectric strength}}$$

From Table 11.2, the dielectric strength for film is 20 kV/mm. Therefore,

$$d = \frac{200 \text{ V}}{20{,}000 \text{ V/mm}} = 0.01 \text{ mm}$$

Putting d into Equation 11.7 and arranging the terms gives

$$A = \frac{(0.06 \times 10^{-6} \text{ F})(0.01 \times 10^{-3} \text{ m})}{(8.85 \times 10^{-12} \text{ F/m})(2.5)}$$

$$A = 2.71 \times 10^{-2} \text{ m}^2$$

11.7 Capacitors Connected in Series

series capacitor connection

A typical **series capacitor connection** is shown in Figure 11.9. With no branch points between the capacitors, the current during charging is the same in each part of the circuit, as it was for the series resistor circuit. Since current is Q/t, the charge on each capacitor and that supplied by the source must be the same. Therefore,

$$Q_s = Q_1 = Q_2 = \cdots = Q_n \qquad \textbf{(11.10)}$$

where Q_s is the charge supplied by the source, in coulombs; and Q_1, Q_2, ... , Q_n are the charge on each capacitor, in coulombs.

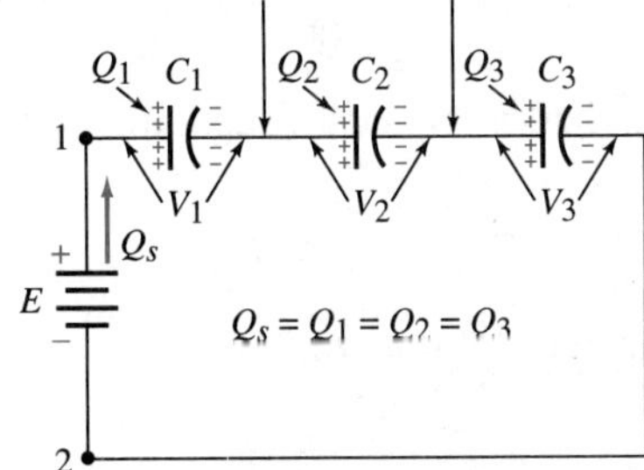

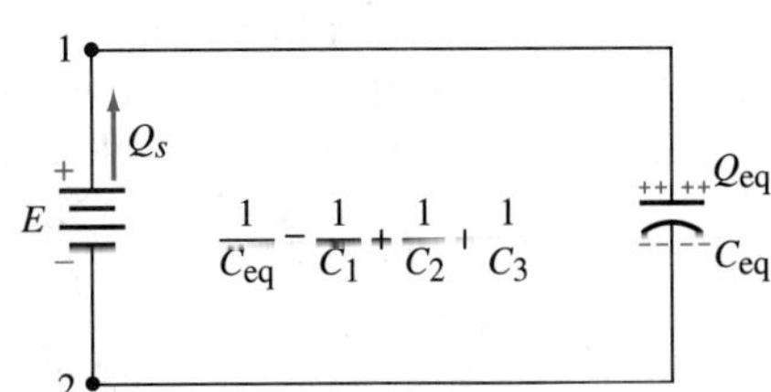

Figure 11.9 The charge supplied by the source must be the same in the equivalent circuit as in the series circuit. (a) Series connection. (b) The relation of the equivalent capacitance to the capacitors in the series circuit.

equivalent capacitance

The series capacitor circuit can be replaced by its equivalent capacitance. The **equivalent capacitance** is the value of capacitance that will store the same amount of charge as the group for which it is equivalent. The use of equivalent capacitance can be helpful in analyzing circuits. Also, several capacitors can be combined to obtain a needed value of capacitance and breakdown voltage.

Consider again the circuit of Figure 11.9(a). The voltage across each capacitor is

$$V_1 = Q_1/C_1$$
$$V_2 = Q_2/C_2$$
$$V_3 = Q_3/C_3$$

Also, by Kirchhoff's Voltage Law,

$$\begin{aligned} E &= V_1 + V_2 + V_3 \\ &= Q_1/C_1 + Q_2/C_2 + Q_3/C_3 \end{aligned}$$

Since $Q_s = Q_1 = Q_2 = Q_3$, we have

$$1/C_{eq} = 1/C_1 + 1/C_2 + 1/C_3$$

The reciprocal of the equivalent capacitance is equal to the sum of the reciprocals of the capacitors in the group. Or, for n capacitors connected in series,

$$1/C_{eq} = 1/C_1 + 1/C_2 + \cdots + 1/C_n \qquad \textbf{(11.11)}$$

Equation 11.11 has the same form as the R_{eq} equation for resistors connected in parallel. Connecting capacitors in series is similar to increasing the thickness of the dielectric, because in each case the capacitance is decreased.

When there are only two capacitors in series, Equation 11.11 takes the form

$$C_{eq} = (C_1C_2)/(C_1 + C_2) \qquad \textbf{(11.12)}$$

Here, the quantities have the same meaning and units as in Equation 11.11.

EXAMPLE 11.8

A 4-microfarad, 5-microfarad, and 20-microfarad capacitor are connected in series across a 25-volt source. Find:

a. the equivalent capacitance.
b. the charge on each capacitor.
c. the voltage across the 5-microfarad capacitor.

Solution

a. To find the equivalent capacitance,

$$\begin{aligned}1/C_{eq} &= 1/C_1 + 1/C_2 + 1/C_3\\ &= (4 \times 10^{-6}\ \text{F})^{-1} + (5 \times 10^{-6}\ \text{F})^{-1} + (20 \times 10^{-6}\ \text{F})^{-1}\\ &= (0.25 \times 10^6 + 0.2 \times 10^6 + 0.05 \times 10^6)\ (\text{F})^{-1}\\ &= 0.5 \times 10^6\ (\text{F})^{-1}\end{aligned}$$

Taking the reciprocal gives

$$C_{eq} = 2.0 \times 10^{-6}\ \text{F}$$
$$C_{eq} = 2\ \mu\text{F}$$

b. To find the charge on each capacitor,

$$\begin{aligned}Q &= EC_{eq}\\ &= (25\ \text{V})(2 \times 10^{-6}\ \text{F})\\ Q &= 50 \times 10^{-6}\ \text{C}\end{aligned}$$

c. To find the voltage,

$$\begin{aligned}V_3 &= Q/C_3\\ &= (50 \times 10^{-6}\ \text{C})/(5 \times 10^{-6}\ \text{F})\\ V_3 &= 10\ \text{V}\end{aligned}$$

EXAMPLE 11.9

A 1-microfarad capacitor is needed, but the only capacitors available are a 2-, 3-, 5-, and 6-microfarad capacitor. Select the combination that will give 1 microfarad.

Solution

If the 2-, 3-, and 6-microfarad capacitors are connected in series, the equivalent capacitance is

$$\begin{aligned}\frac{1}{C_{eq}} &= \frac{1}{2 \times 10^{-6}\ \text{F}} + \frac{1}{3 \times 10^{-6}\ \text{F}} + \frac{1}{6 \times 10^{-6}\ \text{F}}\\ &= \frac{6}{6 \times 10^{-6}\ \text{F}} \quad \text{or} \quad 1 \times 10^6 (\text{F})^{-1}\\ C_{eq} &= 1\ \mu\text{F}\end{aligned}$$

The 2-, 3-, and 6-microfarad capacitor series combination is needed.

The voltage across the capacitors in a series circuit is now considered. For two capacitors connected in series, as shown in Figure 11.10,

$$C_{eq} = \frac{C_1 C_2}{C_1 + C_2}$$

Also,

$$C_{eq} = Q/E \quad \text{but} \quad Q = V_1 C_1$$

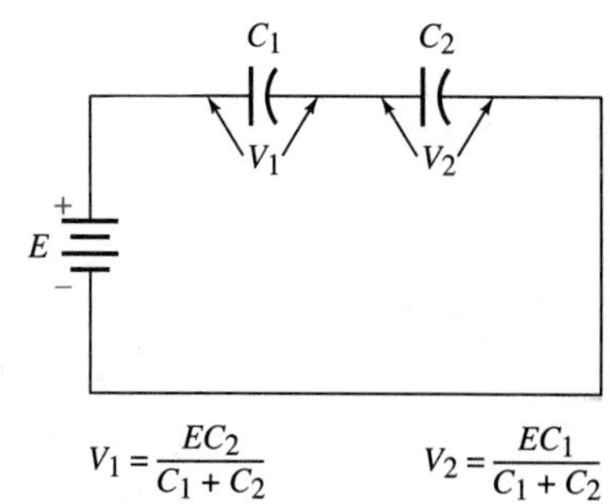

Figure 11.10 The voltage-divider expression for capacitors connected in series.

so

$$C_{eq} = (V_1C_1/E)$$

Putting this into the first expression gives

$$\frac{V_1C_1}{E} = \frac{C_1C_2}{C_1 + C_2}$$

or

$$V_1 = \frac{EC_2}{C_1 + C_2} \qquad \textbf{(11.13)}$$

This is the voltage-divider expression for capacitors connected in series. We can use it to find the voltage across the capacitor without calculating the charge. When there are more than two capacitors, C_2 is the equivalent capacitance of the circuit excluding C_1. An inspection of Equation 11.13 shows that the voltage will be greater across the smaller capacitance. The expression has the same form as the two-branch, current-divider expression.

EXAMPLE 11.10

A 3-microfarad and a 6-microfarad capacitor are connected in series to a power supply. If the voltage across the 3-microfarad capacitor is 12 volts, what is the supply voltage?

Solution

$$V_3 = \frac{EC_6}{C_3 + C_6}$$

or

$$E = \frac{V_3(C_3 + C_6)}{C_6}$$

$$E = \frac{12\text{ V}(3\ \mu\text{F} + 6\ \mu\text{F})}{6\ \mu\text{F}}$$

$$E = 18\text{ V}$$

The voltage across the 3-microfarad capacitor is 2/3 of the source voltage, while that across the 6-microfarad capacitor is 1/3 of the source voltage.

EXAMPLE 11.11

A 4-microfarad capacitor and a 6-microfarad capacitor are connected in series across a 24-volt source. Use Equation 11.13 to find the voltage across the 4-microfarad capacitor.

Solution

$$V_4 = \frac{EC_6}{C_4 + C_6}$$
$$= \frac{(24 \text{ V})(6 \times 10^{-6} \text{ F})}{4 \times 10^{-6} \text{ F} + 6 \times 10^{-6} \text{ F}}$$
$$V_4 = 14.4 \text{ V}$$

The breakdown voltage of n capacitors of equal capacitance connected in series is equal to n times the rating of one capacitor. But if the capacitors have different capacitances, that is not so. For that condition, each capacitor will have a different voltage across it, with the smallest capacitance having the largest voltage. Therefore, the maximum voltage is that which results in any one of the capacitors being at its breakdown voltage.

EXAMPLE 11.12

Find:

a. The breakdown voltage rating of three 5-microfarad, 200-V capacitors connected in series.
b. The breakdown voltage rating of a 4-microfarad, 100-V capacitor and a 5-microfarad, 200-V capacitor series circuit.

Solution

a. $BV = n \times BV_1$
$$= (3)(200 \text{ V})$$
$$BV = 600 \text{ V}$$

b. The 4-microfarad capacitor will have the higher voltage on it. When it is at its rated voltage,

Strategy
The breakdown voltage of the smallest capacitance is used to find the charge.

Charge is used to find the voltage of the other capacitor.

$$Q_4 = C_4V_4$$
$$= (4 \times 10^{-6} \text{ F})\ (100 \text{ V})$$
$$= 400 \times 10^{-6} \text{ C}$$
$$V_5 = Q_5/C_5 = Q_4/C_5$$
$$= (400 \times 10^{-6} \text{ C})/(5 \times 10^{-6} \text{ F})$$
$$= 80 \text{ V}$$
$$V_T = V_4 + V_5$$
$$= 100 \text{ V} + 80 \text{ V}$$
$$V_T = 180 \text{ V}$$

If the total voltage exceeds 180 V, the 4-microfarad capacitor will be damaged. It is apparent that if C_5 had 200 volts across it, V_4 would be much higher than C_4's rating of 100 volts.

11.8 Capacitors Connected in Parallel

parallel capacitor connection

For **parallel capacitor connection**, the capacitors must be connected across the same two points, as shown in Figure 11.11.

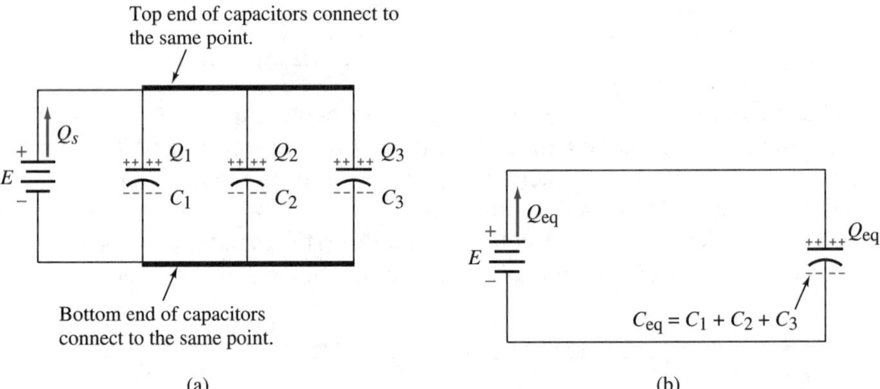

Figure 11.11 The charge supplied by the source must be the same in the equivalent circuit as in the parallel circuit. (a) Parallel connection. (b) The relation of the equivalent capacitance to the capacitors in the parallel circuit.

For this connection, the charge entering the parallel branch must equal the sum of the charge on each capacitor. Or, in general,

$$Q_s = Q_1 + Q_2 + \cdots + Q_n \qquad \textbf{(11.14)}$$

where Q_s is the charge entering the parallel group, in coulombs; and Q_1, Q_2, ... , Q_n are the charges on the respective capacitors, in coulombs.

Equation 11.14 is prŏved from the following:

In Figure 11.11(a),

$$I_s = I_1 + I_2 + I_3$$

since

$$I = Q/t$$
$$I_1 = Q_1/t$$
$$I_2 = Q_2/t$$
$$I_3 = Q_3/t$$

so

$$I_s = (Q_1/t) + (Q_2/t) + (Q_3/t)$$

This reduces to

$$Q_s = Q_1 + Q_2 + Q_3$$

which is the same as Equation 11.14.

The equivalent capacitance of a parallel group is equal to the sum of the capacitances in each branch. This is similar to the form for R_{eq} of a series-connected resistor circuit. Figure 11.11(a) will be used to develop this relation.

In Figure 11.11(a),

$$Q_s = Q_1 + Q_2 + Q_3$$

and

$$Q_1 = C_1E$$
$$Q_2 = C_2E$$
$$Q_3 = C_3E$$

so

$$Q_s = C_1E + C_2E + C_3E$$

For C_{eq} of Figure 11.11(b) to be equivalent to the circuit in Figure 11.11(a), the charge must be the same for both. That is,

$$Q_{eq} = Q_s$$

but

$$Q_{eq} = EC_{eq}$$

so

$$EC_{eq} = C_1E + C_2E + C_3E$$

or

$$C_{eq} = C_1 + C_2 + C_3$$

In general, for n capacitors connected in parallel,

$$C_{eq} = C_1 + C_2 + \cdots + C_n \qquad \textbf{(11.15)}$$

where C_{eq} is the equivalent capacitance, in any unit of capacitance; and $C_1, C_2, \ldots, C_n$ are the capacitance of each capacitor, in the same unit as C_{eq}.

This concept of equivalent capacitance is useful for the same reasons as the series equivalent.

EXAMPLE 11.13

A 5-, 10-, and 3-microfarad capacitor are connected in parallel across a 60-volt source. Find:

a. the equivalent capacitance.
b. the charge on each capacitor.
c. the total charge.

Solution

a. $C_{eq} = C_1 + C_2 + C_3$
$\quad = 5\ \mu F + 10\ \mu F + 3\ \mu F$

$$C_{eq} = 18\ \mu F$$

b. $$Q_1 = V_1C_1 = (60\text{ V})(5 \times 10^{-6}\text{ F})$$

$$Q_1 = 300 \times 10^{-6}\text{ C}$$

$$Q_2 = V_2C_2 = (60\text{ V})(10 \times 10^{-6}\text{ F})$$

$$Q_2 = 600 \times 10^{-6}\text{ C}$$

$$Q_3 = V_3C_3 = (60\text{ V})(3 \times 10^{-6}\text{ F})$$

$$Q_3 = 180 \times 10^{-6}\text{ C}$$

$$Q_s = Q_1 + Q_2 + Q_3 = 300 \times 10^{-6}\text{ C} + 600 \times 10^{-6}\text{ C} + 180 \times 10^{-6}\text{ C}$$

$$Q_s = 1080 \times 10^{-6}\text{ C}$$

Also, as a check

$$Q_s = E \times C_{eq} = (60\text{ V})(18 \times 10^{-6}\text{ F})$$

$$Q_s = 1080 \times 10^{-6}\text{ C}$$

The maximum voltage applied to the parallel group should not exceed the rating of the lowest rated capacitor. Otherwise, that capacitor will be damaged.

EXAMPLE 11.14

A 5-microfarad, 30-volt and a 10-microfarad, 25-volt capacitor are connected in parallel across a 20-volt source. Find the charge on each capacitor and the breakdown voltage of the combination.

Solution

$$Q_s = EC_{eq}$$

and

$$C_{eq} = 5\ \mu F + 10\ \mu F = 15\ \mu F$$

Substituting,

$$Q_s = (20\text{ V})(15 \times 10^{-6}\text{ F}) = 300 \times 10^{-6}\text{ C}$$

Also,

$$Q_5 = EC_5$$

$$= (20\text{ V})(5 \times 10^{-6}\text{ F})$$
$$Q_5 = 100 \times 10^{-6}\text{ C}$$

Likewise,

$$Q_{10} = EC_{10}$$
$$= (20\text{ V})(10 \times 10^{-6}\text{ F})$$
$$Q_{10} = 200 \times 10^{-6}\text{ C}$$

Checking gives

$$Q_5 + Q_{10} = 100 \times 10^{-6}\text{ C} + 200 \times 10^{-6}\text{ C}$$
$$Q_5 + Q_{10} = 300 \times 10^{-6}\text{ C}$$

The breakdown voltage is that of C_2, the lowest rated capacitor.

$$BV = 25\text{ V}$$

11.9 Types of Capacitors

Capacitors can be grouped as variable and fixed. A **variable capacitor** is one that can be adjusted to give values of capacitance between some minimum and maximum limit. The most common variable capacitor is the tuning capacitor used in a radio. It looks similar to that in Figure 11.12. The fixed set of plates is the stator. The movable set of plates is the rotor. The dielectric is air and the capacitance is changed by turning the rotor. This changes the area of the rotor, which meshes with the stator. Fully meshing the plates gives the highest capacitance. Some variable capacitors have two or more sets of plates mounted on one shaft. This permits tuning several circuits at the same time.

variable capacitor

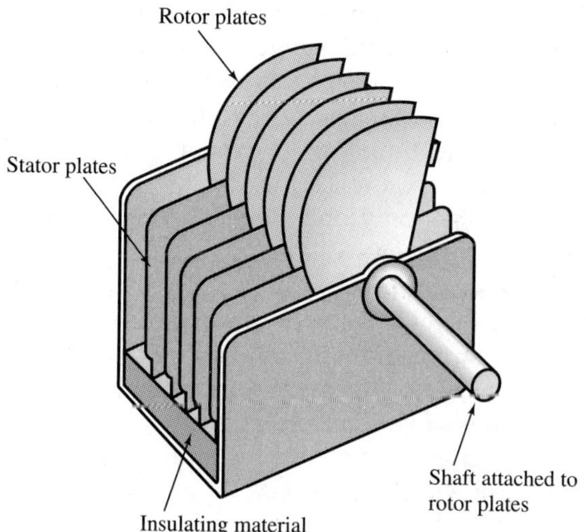

Figure 11.12 A variable air capacitor.

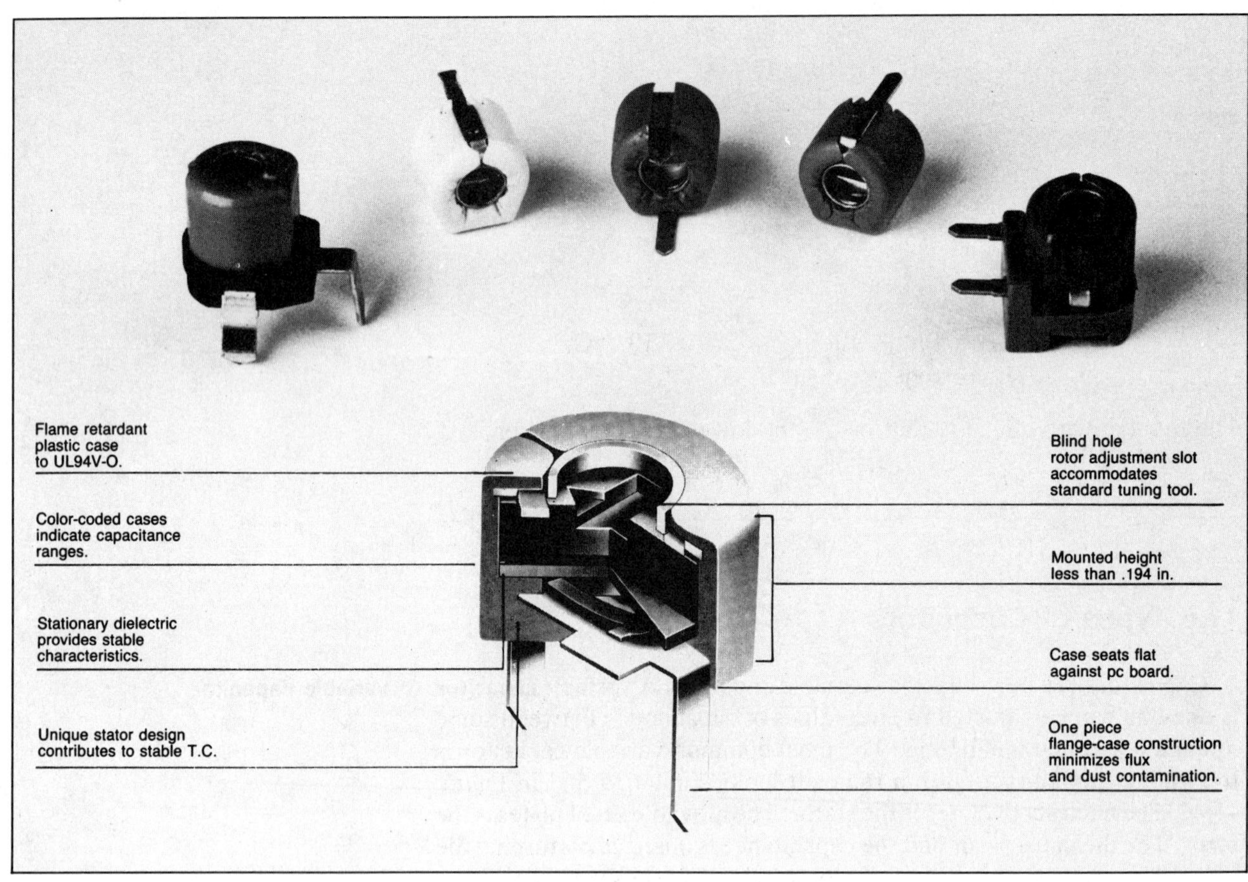

(a)

trimmer capacitor

Another variable capacitor, the **trimmer capacitor**, is shown in Figure 11.13. Trimmers are used where the capacitance is changed infrequently. Turning the slotted shaft changes the relation of the stator to the rotor and changes the capacitance. The capacitance of some trimmers is changed by turning a screw to change the spacing between the plates.

varactor

A third type of variable capacitor is a semiconductor device called a **varactor**. Its capacitance changes as the voltage across it changes. These are used for electronic tuning of television receivers. Some typical varactors are shown in Figure 11.14.

fixed capacitor

A **fixed capacitor** has only one value of capacitance. It is used in applications ranging from low-voltage, low-current electronic circuits to high-voltage, high-current power circuits. Fixed capacitors are usually called by the name of their dielectric. Some of the common ones are paper, mica, ceramic, plastic film, oil, and electrolytic.

The construction of each of these is now described.

paper capacitor

A **paper capacitor** is a fixed capacitor that uses paper for the dielectric. It was one of the first types made, but is now used infrequently. A cross-sectional view is shown in Figure 11.15. Layers of paper and foil are

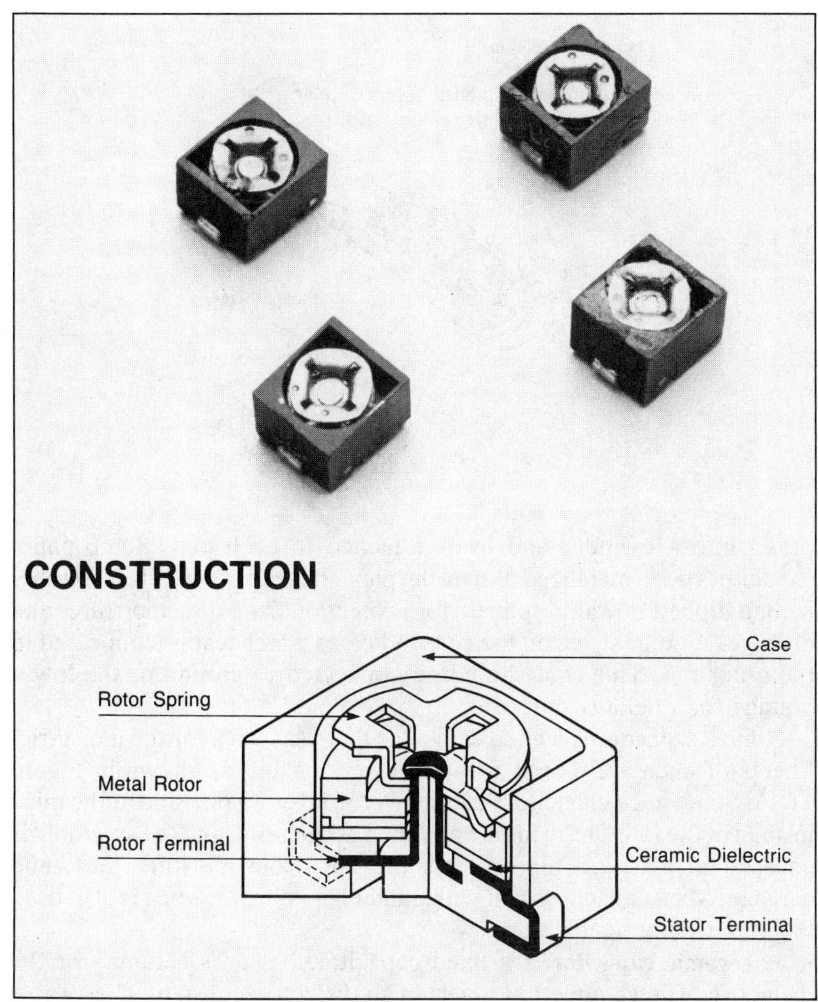

(b)

Figure 11.13 Ceramic trimmer capacitors such as these are used in radios, clocks, VCR systems, and audio equipment. Photos courtesy of MurataErie North America, Inc. (a) Trimcap® Axle-Less capacitor. (b) Chip trimmer capacitor designed for surface mounting.

Figure 11.14 The capacitance of varactors such as these is a function of the voltage applied to them. (Photo courtesy of Electro Ceramic Industries)

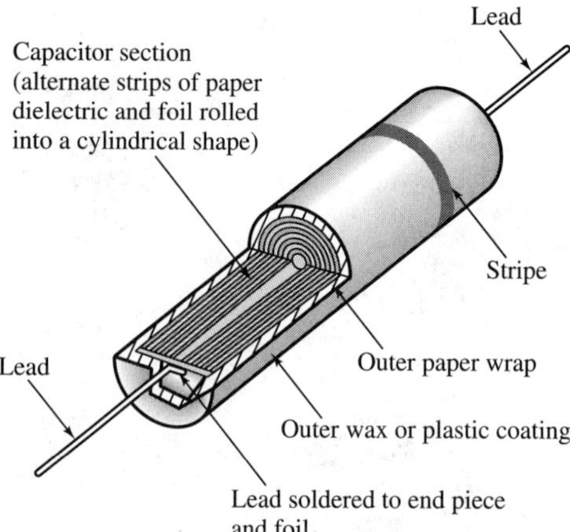

Figure 11.15 A cutaway view of a paper capacitor.

rolled into a cylinder and leads attached to each end. Some paper capacitors use a metalized film on the paper instead of foil. The assembly is then dipped in wax or plastic for protection from dirt, moisture, and damage. A color stripe on the case indicates which lead is connected to the outer foil. This lead should be connected to ground or the lower voltage for shielding purposes.

mica capacitor

A **mica capacitor** is a fixed capacitor that uses mica for the dielectric. Sheets of mica are placed between layers of foil, as shown in Figure 11.16. Some mica capacitors use a silvered layer on one side of the mica instead of the foil. The leads are attached at each end, and the assembly is encased in plastic. This protects the unit from moisture, dirt, and damage. Mica capacitors can withstand high voltages. Mica is also used in some trimmer capacitors.

ceramic capacitor

A **ceramic capacitor** is a fixed capacitor that uses ceramic for the dielectric. The ceramic has a very high dielectric constant. (See Table 11.1.) It is made in many shapes and sizes, including the disk and tubular one shown in Figure 11.17. The conducting plates are a silver or other metal film deposited on the ceramic. The unit is then dipped in a plastic. Ceramic capacitors do not short out as easily as the foil type. This is

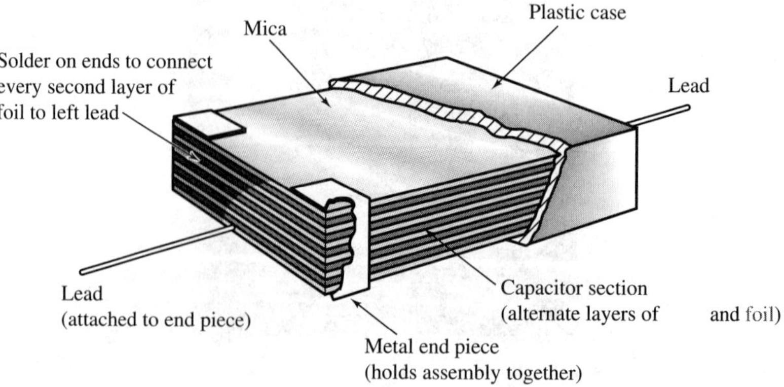

Figure 11.16 A cutaway view of a mica capacitor.

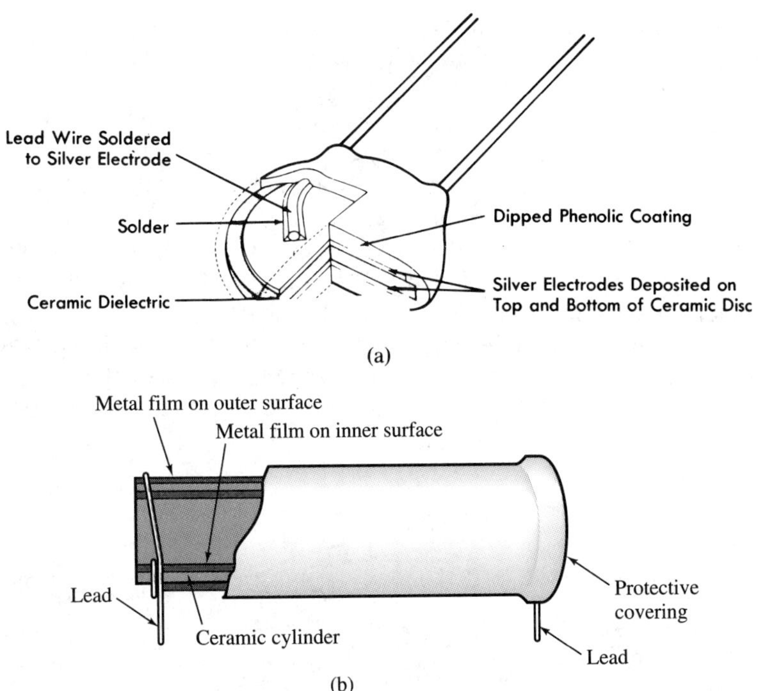

Figure 11.17 Two types of ceramic capacitors. (a) Disk capacitor. (b) Tubular capacitor. (a, courtesy of the Sprague Electric Company)

because the metal film burns away at the breakdown point. In a foil capacitor, the two layers of foil can short out through a hole in the dielectric.

plastic film capacitor

A **plastic film capacitor** is a fixed capacitor that uses plastic film for the dielectric. Two plastic film capacitors are shown in Figure 11.18. Their

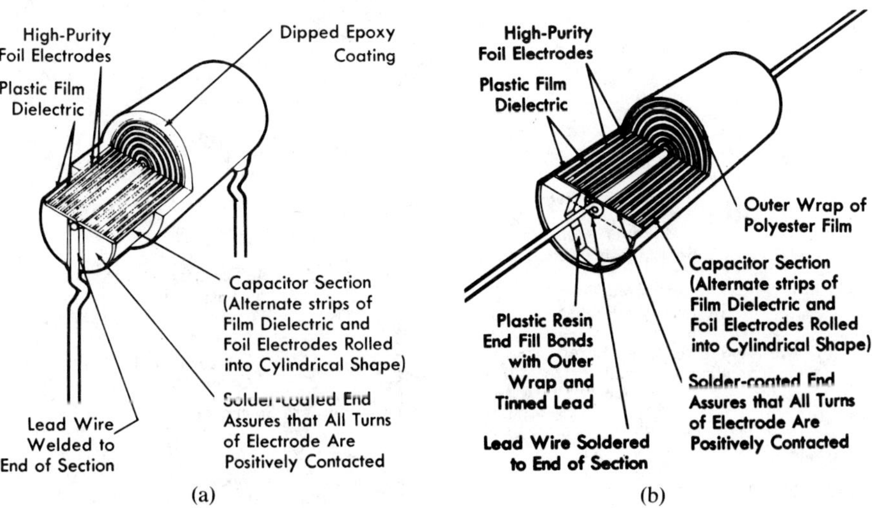

Figure 11.18 Construction of two types of plastic film capacitors. (a) Dipped tubular. (b) Film-wrapped tubular. (Courtesy of the Sprague Electric Company)

construction is much the same as that of the paper capacitor. The thinner film gives higher values of capacitance. Also, these capacitors can be operated at higher temperatures.

oil capacitor

An **oil capacitor** is a fixed capacitor that uses oil or an oil-impregnated material as the dielectric. The unit is hermetically sealed. This type is used mostly in high-voltage and power applications such as correcting the power factor by electric utilities.

electrolytic capacitor

An **electrolytic capacitor** is a tubular or a can type capacitor, which is shown in Figure 11.19. The dielectric is a thin oxide film that forms on the foil electrode when a voltage is applied to the capacitor terminals. The thinness of this film and the large surface area involved give large values of capacitance. Capacitance values thousands of times greater than that from the same size of another type can be obtained. On the negative side,

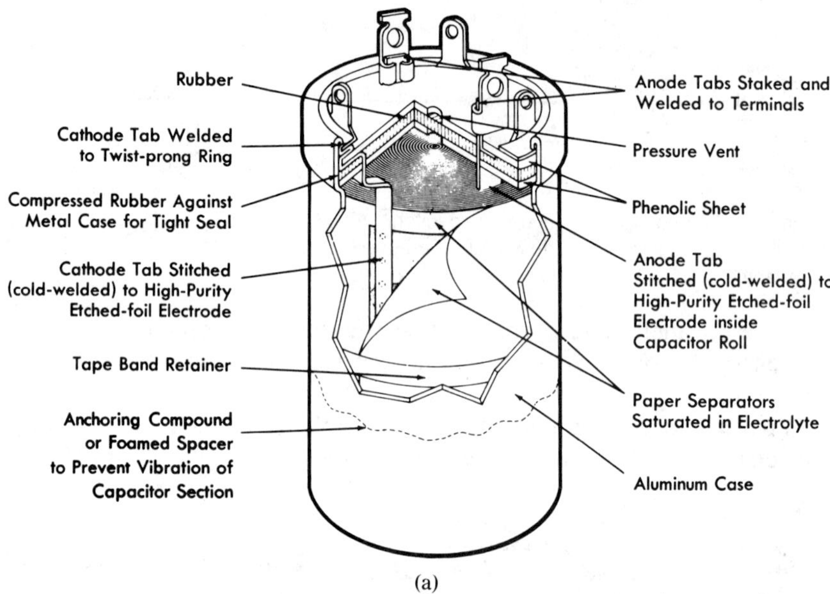

(a)

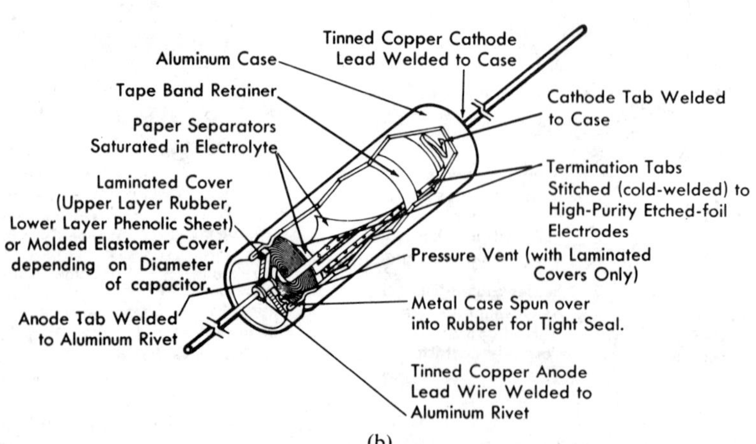

(b)

Figure 11.19 Construction of two wet, eelctrolyte electrolytic capacitors. (a) Can-type electrolytic capacitor. (b) Tubular-type electrolytic capacitor. (Courtesy of the Sprague Electric Company)

electrolytic capacitors have a relatively low working voltage and high leakage current, and a short life compared to the other types.

Most electrolytic capacitors have one terminal marked with a + sign. The capacitor must be connected in the circuit so that the polarity of the terminal is + with respect to the other one. If not, the insulating oxide does not form, and the high capacitor current will cause a buildup of gas that can make the capacitor explode.

Electrolytic capacitors for use in ac circuits are not polarized. These are basically two electrolytic capacitors connected back-to-back. This causes the insulating film to form on one of them, no matter how the terminals are connected. The high capacitance of this type makes it suitable for power supply filtering, and for the starting capacitor on an ac motor.

Electrolytic capacitors are also made with a solid electrolyte such as the tantalum capacitors, as shown in Figure 11.20. These are usually

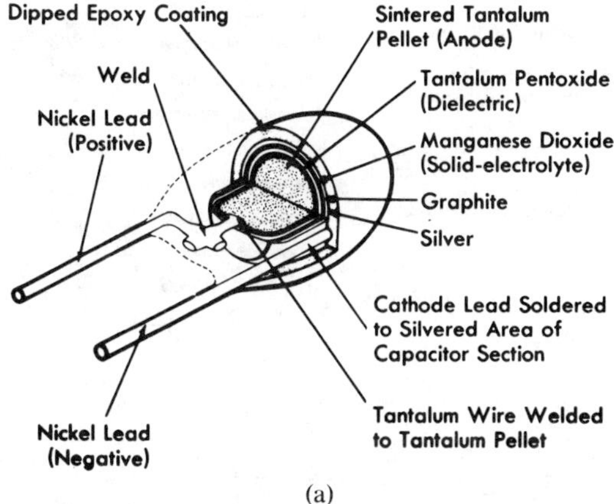

(a)

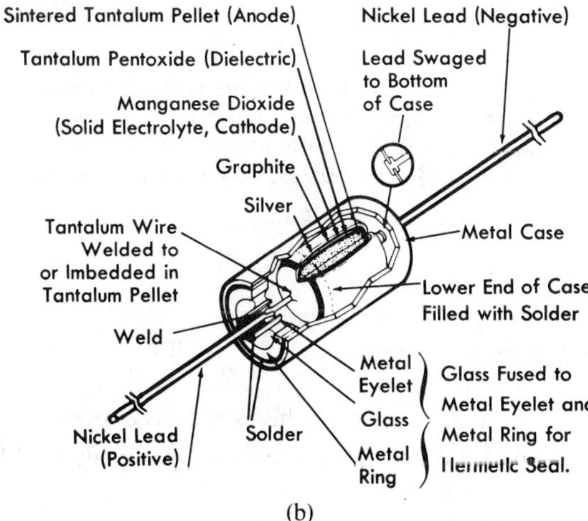

(b)

Figure 11.20 Construction of two solid electrolyte electrolytic capacitors. (a) Epoxy-dipped electrolytic capacitor. (b) Hermetically sealed electrolytic capacitor. (Courtesy of the Sprague Electric Company)

(a)

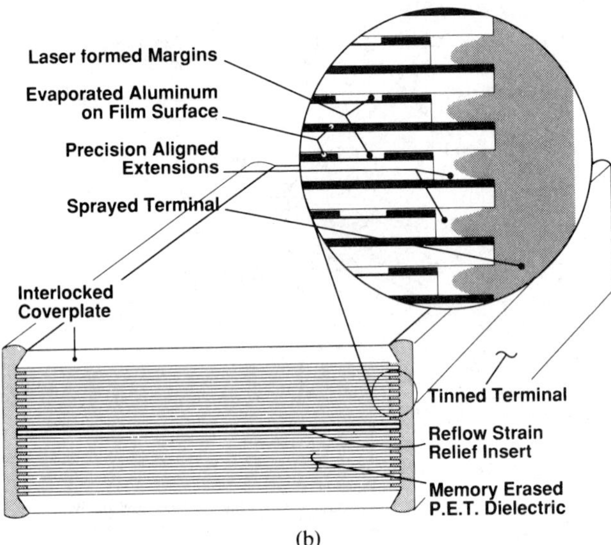

(b)

Figure 11.21 Surface-mount capacitors such as these are small and make use of automation. (a) A comparison of the size of the Surfilm® capacitor. (b) The construction of the Surfilm capacitor. (Photos courtesy of Paktron Division of Illinois Tool Works)

smaller in physical size than the wet type and are used in electronic circuits.

Most capacitors will be one of the types presented here. However, those shown in the illustrations are only a few of the many shapes and sizes in which they are made.

Capacitors are also made in the surface-mount configuration. When mounted, these capacitors look like those in Figure 11.21. They are small units on which the terminals are used both for contact and for attaching to the printed circuit board. They are soldered to the surface of the board, hence the name surface-mount. Surface mounting eliminates the need to drill holes in the board. It is ideal for automation, uses less space on a printed circuit board, and provides a more reliable assembly. Surface mounting was treated in more detail in Chapter 2, Section 2.8.

Picking a capacitor for an application requires knowing the conditions of the application and the characteristics of the capacitor. A capacitor used in a power supply might require high capacitance, while one used in a satellite must be able to function at extreme temperatures. Some characteristics that play a part in the selection of the proper capacitor are:

1. **Capacitance**. This is a measure of how much charge can be stored per volt of potential difference.
2. **Working Voltage**. **working voltage**
   a. dc—This is the highest steady voltage that can be applied across the capacitor terminals without dielectric breakdown.
   b. ac or rms—This is the highest rms voltage that can be applied across the terminals without breakdown. Root-mean-square is the voltage commonly used to measure ac.
3. **Surge Breakdown Voltage**. This is the highest voltage that can be applied across the capacitor terminals for a short time without dielectric breakdown. **surge breakdown voltage**
4. **Leakage Current**. This is the current in the dielectric because of its leakage resistance. Ideally, this should be zero. This current causes power loss in the dielectric and heat buildup.
5. **Temperature Coefficient**. This is the change in capacitance per degree of change in temperature. **temperature coefficient**

A more complete listing of the characteristics is given in Table 11.3. Bear in mind that these are not absolute limits because of the many variations from manufacturer to manufacturer.

Capacitors also have a system of color coding. In fact, there are several different systems. Because of that, and the many shapes and types of capacitors, the use of the codes becomes complex and is not presented here. Basically, the colors have the same numerical values as for the resistor. However, some types use colored dots, others use colored rectangles, and others use colored stripes.

Some capacitors have their value marked on them. Others use a system of color coding. Two of the systems are presented in Tables 11.4 and 11.5. The meaning of the first digit, second digit, multiplier, and tolerance is similar to that for resistors. The meaning of the other terms is explained by the headings in the tables.

TABLE 11.3 Capacitor Application Chart*

Outstanding Characteristic — Limiting Characteristic — Normal Characteristic

CHARACTERISTIC OF CAPACITOR		DIELECTRICS: Electrolytic Aluminum	Tantalum Foil	Tantalum Wet Anode	Tantalum Dry Anode	Paper	Polyester
CAP.	CAPACITANCE Range–Mfd.	.5–1,000,000	0.15–3500 mfd.	1.7–5000	.047–330	.001–200	.001–20
	TOLERANCE Standard %	+50, +100, +150, −10	p/m 20% −15, +30, 50, 75	p/m 20% −15, + 30, 50, 75	±20	±20	±20
	TOLERANCE Minimum %	±20	±15%	±5%	±5	±2	±1
VOLTS	DC OPERATING	2.5–700	3–300	4–125	6–35	50–200,000	50–1,000
	AC 60 Hz OPERATING	≦50 Continuous ≦330 Intermittent	Limited	Limited	Limited	50–75,000	Limited
D.F.	DISSIPATION FACTOR % at 60 Hz	120 Hz 6 and up depending on C/V rating	120 Hz 10–15%	120 Hz varies from <1 to 100 depending on C & V	At 120 Hz 10% max.	.2–.5	.3
	% at 1000 Hz	—	—	—	—	.2–.5	.5
	% at 1 MHz Low Capacitance Values	—	—	—	—	Higher; varies with type	Relatively High
I.R.	INSULATION RESISTANCE Megohm Mfd. at 25°C	Leakage current .1 μa and up depending on C/V rating	Leakage 0.001 μa 0.01 μa/ mfd./volt	Leakage .00040 .0060 μa/ mfd./volt	Leakage at 25°C .02 μa/ mfd./volt	3000–20,000	50,000
	INSULATION RESISTANCE at 85°C compared to 25°C	Leakage current 4 × 25°C value	Leakage 2–10 times 25°C value	Leakage 2–10 times 25°C value	Leakage 10 × 25°C value	$\frac{1}{100}$	$\frac{1}{25}$
TEMP.	OPERATING RANGE °C	−80 +150	−55 +125	−80 +125	−55 +125	−55 +125	−55 +150
STABILITY	CAPACITANCE CHANGE with Temp. Aging	Small to Large	5%	10%	Medium ±10	Medium	Medium
D.A.	% Dielectric Absorption at 25°C	—	—	—	—	.6–3. depending on impreg.	.5
SIZE	Varies as	CV approx.	CV approx.	CV approx.	CV approx.	CV^2	CV^2
	For Equivalent CV Rating	Very Small	Very Small	Very Small	Very Small	Medium Small	Small
	Per KVA 60 Hz	Small for Intermittent duty	Small for Intermittent duty	Small for Intermittent duty	Small for Intermittent duty	Small	Seldom used
	Per KVA 1 MHz	Not used	Not used	Not used	Not used	Not used	Not used
COST	Relative Cost for Equiv. CV Rating	Very Low	Moderate	Moderate	Moderate	Low	Moderately High
	Relative Cost for KVA 60 Hz	Low for Intermittent duty	Not used	Not used	Not used	Low	Seldom used
	Relative Cost for KVA 1 MHz	Not used	Not used	Not used	Not used	Not used	Not used

Values and ranges shown are generally typical or average. Actual limits may vary considerably.

DIELECTRICS							
Teflon†	Paper Metallized	Metallized Polyester	Metallized Paper-Polyester (Comb.)	Multi-Layer Ceramics	Mica Receiving	Mica Transmitting	Reconstituted Mica
.001–4	.001–125	.01–20	.001–12	.000005 to 2.5μf	.000001–.1	.00001–1.0	.01–4
±10	±20	±20	±20	±5 50 GMV	±5	±5	±20
±2	±5	±1	±5	±5	±1	±1	±5
50–1000	50–600	50–600	200–600	20–200	50–2500	200–100,000	200–15,000
Seldom used	25–250	25–250	Seldom used	Seldom used	Seldom used	RF voltage varies with current & freq.	100–7500
<.1	.4–1.0	.2–.3	.4–. 6	Seldom used	Seldom used	Seldom used	Seldom used
.02–.05	.6–1.5	.4–.5	.6–.8	1–2.5	<.1	.04–.07	.5
.04–.07	Relatively High	Relatively High	Relatively High	.1% for NPO	<.1	.03–.06	.7–.9
>100,000	600–1200	5000–50,000	2000	100,000	20,000–100,000 meg./unit	7,500–10,000 meg./unit	10,000
$\frac{1}{10}$	$\frac{1}{60}$	$\frac{1}{40}$	$\frac{1}{12}$	1/20–1/60	$\frac{1}{7}$	$\frac{1}{5}$	$\frac{1}{8}$
−55 +250	−55 +125	−55 +125	−55 +125	−55 +125	−55 +150	−55 +70	−55 +200 −55 +315
Medium	Medium	Medium	Medium	Small	Very Small; Excellent	Very Small; Excellent	Good
.02–.05	—	—	—	—	.3 max.	.3 max.	—
CV^2	CV^2	CV^2	CV^2	CV^2 & K	CV^2	CV^2	CV^2
Large	Small	Small	Small	Small	Large	Large	Large
Seldom used	Seldom used	Seldom used	Seldom used	—	Seldom used	Seldom used	Seldom used
Small	Seldom used	Seldom used	Seldom used	—	Small	Small	Not used
Very high	Moderately High	Moderately High	Moderately High	Low	High	High	High
Seldom used	Not used	Not used	Not used	Low	Seldom used	Seldom used	High
Low	Not used	Not used	Not used	Not used	Low	Low	Medium

*Values taken from Capacitor Application Chart by Cornell Dubilier.
†DU PONT Trademark. ‡Chart courtesy of Cornell Dubilier.

TABLE 11.4 Color Coding of Molded Tubular Capacitors (Picofarads)

Color	Significant Digit	Decimal Multiplier	Tolerance (± %)
Black	0	1	20
Brown	1	10	—
Red	2	100	—
Orange	3	1,000	30
Yellow	4	10,000	40
Green	5	10^5	5
Blue	6	10^6	—
Violet	7	—	—
Gray	8	—	—
White	9	—	—

TABLE 11.5 Color Coding of Molded Mica Capacitors (RETMA and standard MIL specifications)

Color	Significant Figure	Decimal Multiplier	Tolerance (± %)	Class	Temp. Coeff. PPM/°C (not more than)	Cap. Drift (in picofarads; not more than)
Black	0	1	20	A	± 1,000	(5% + 1 pF)
Brown	1	10	—	B	±500	(3% + 1 pF)
Red	2	100	2	C	± 200	(0.5% + 0.5 pF)
Orange	3	1,000	3	D	±100	(0.3% + 0.1 pF)
Yellow	4	10,000	—	E	+100–20	(0.1% + 0.1 pF)
Green	5	—	5	—	—	—
Blue	6	—	—	—	—	—
Violet	7	—	—	—	—	—
Gray	8	—	—	I	+150–50	(0.03% + 0.2 pF)
White	9	—	—	J	+100–50	(0.2% + 0.2 pF)
Gold	—	0.1	—	—	—	—
Silver	—	0.01	10	—	—	—

Note: If both rows of dots are not on one side, rotate the capacitor about the axis, off its leads, to read second row on side or rear.

EXAMPLE 11.15

What are the capacitance and voltage rating of a tubular capacitor having the following colored bands?

Capacitor bands: First: yellow; second: violet; multiplier: red; tolerance: silver; voltage band: blue.

Solution

Referring to Figure 11.22:
The first digit is 4 for yellow.
The second digit is 7 for violet.
The multiplier is 100 for red.
Hence, the capacitance is 47 $\times$ 100 or 4700 picofarads.
The voltage rating is 600 volts.

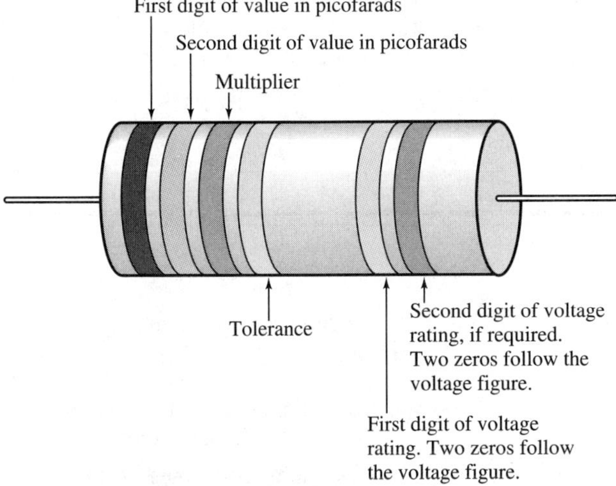

Figure 11.22 The use of colored bands for color coding a tubular capacitor.

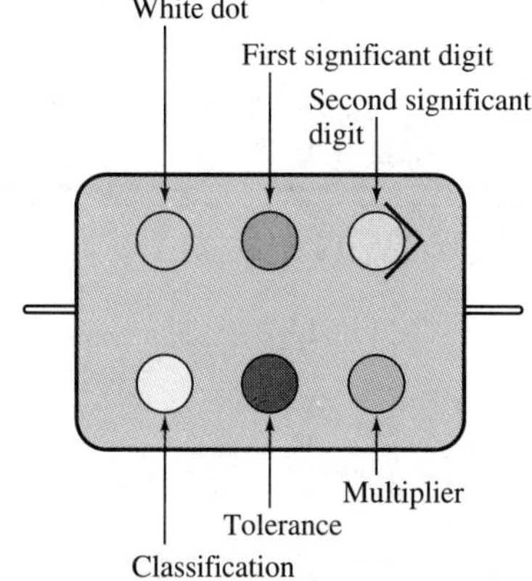

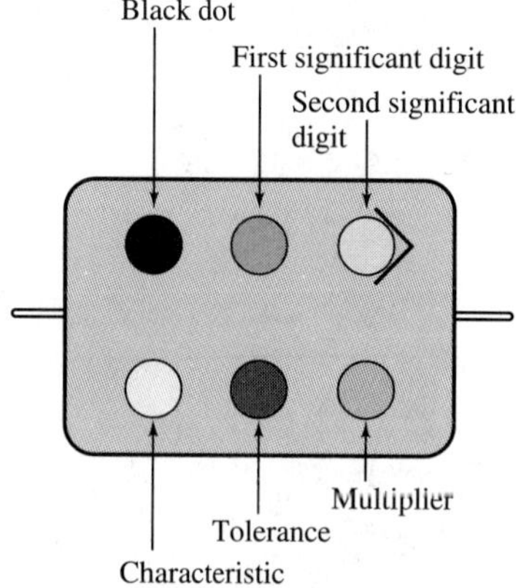

Figure 11.23 The significance of the dots in the capacitor dot color coding system.

SUMMARY

1. Capacitance is the property by which two conducting materials separated by an insulating material can store a charge.
2. The unit of capacitance is the farad. One farad represents a stored charge of one coulomb when one volt of emf is applied.
3. Capacitors, devices designed to have capacitance, are used for time delay, phase shifting, waveshaping, filtering, dc isolation, and ac coupling.
4. A capacitor that has charge stored on its plates is said to be charged. One that has no charge is said to be discharged.
5. Permittivity is a measure of how easily the electrostatic field is established in the dielectric. Relative permittivity is the ratio of a material's permittivity to that of a vacuum.

6. The capacitance depends on the dielectric material; it increases as the spacing is decreased or as the plate area is increased.
7. The breakdown voltage is the voltage that a capacitor can withstand. It depends on the dielectric, and increases as the spacing is increased.
8. Connecting capacitors in series gives a smaller capacitance and a larger breakdown voltage for the combination.
9. Connecting capacitors in parallel gives a larger capacitance. The breakdown voltage is that of the lowest rated capacitor.
10. Leakage current is the current that passes through the dielectric material because of the impurities in it.
11. There are two types of capacitors: fixed and variable.
12. According to the dielectric material, capacitors are air, paper, mica, oil, ceramic, plastic film, and electrolytic.
13. Some characteristics that should be considered when selecting a capacitor are: capacitance, working voltage, leakage current, temperature coefficient, tolerance, surge breakdown voltage, size, and cost.

EQUATIONS

11.1	$C = \dfrac{Q}{V}$	Relation of a capacitance, charge, and voltage.	404
11.2	$\mathscr{E} = \dfrac{V}{d}$	Relation of field intensity, voltage, and distance between the plates.	406
11.3	$\gamma = Q$	The number of lines of flux.	407
11.4	$D = \dfrac{\gamma}{A}$	Relation of flux density to the flux and area.	407
11.5	$\epsilon = \dfrac{D}{\mathscr{E}}$	Relation of permittivity to flux density and field intensity.	409
11.6	$\epsilon = \epsilon_o \epsilon_r$	Relation of permittivity to relative permittivity and the permittivity of a vacuum.	410
11.7	$C = \dfrac{\epsilon_o \epsilon_r A}{d}$	Capacitance of a parallel plate capacitor.	411
11.8	$C = \dfrac{2.4 \times 10^5\, \epsilon_r}{\log_{10}\left(\dfrac{r_2}{r_1}\right)}$	Capacitance of a coaxial cable.	412
11.9	$BV = (d)(\text{dielectric strength})$	Breakdown voltage of a capacitor.	414
11.10	$Q_s = Q_1 = Q_2 = \cdots = Q_n$	Charge on series-connected capacitors.	415

11.11	$\frac{1}{C_{eq}} = \frac{1}{C_1} + \frac{1}{C_2} + \cdots + \frac{1}{C_n}$	Equivalent capacitance of series-connected capacitors.	416
11.12	$C_{eq} = \frac{C_1C_2}{C_1 + C_2}$	Equivalent capacitance of two series-connected capacitors.	416
11.13	$V_1 = \frac{EC_2}{C_1 + C_2}$	Voltage-divider expression for series-connected capacitor.	418
11.14	$Q_s = Q_1 + Q_2 + \cdots + Q_n$	Charge on capacitors connected in parallel.	420
11.15	$C_{eq} = C_1 + C_2 + \cdots + C_n$	Equivalent capacitance of capacitors connected in parallel.	421

QUESTIONS

1. What is capacitance?
2. What is the unit of capacitance?
3. What are the three properties of capacitance that make it useful in electric circuits?
4. What are the basic parts of a capacitor?
5. What is meant by permittivity? Relative permittivity?
6. Which will give more capacitance, a 1-mm sheet of mica or a 1-mm air space?
7. Why is the breakdown voltage important when selecting a capacitor?
8. What two quantities affect the breakdown voltage of a capacitor?
9. What must be done to the plate area to increase capacitance?
10. What will be the capacitance if the two plates are brought in contact with each other?
11. What will happen to the capacitance if the foil material is changed from silver to copper?
12. What two things can cause a capacitor to discharge even though the terminals are not connected to anything?
13. Ideally, should a capacitor have a high or a low leakage current?
14. Can a capacitor with a 400-volt dc working voltage be used in a 400-volt rms application without breakdown?
15. Can a capacitor with a 100-volt rms rating be used in a 100-volt dc circuit without breakdown?
16. If two series-connected capacitors are reconnected in parallel, will the group have more or less capacitance than when they were in series?
17. A 5-microfarad and a 2-microfarad capacitor are connected in series across 100 volts. Which one will have the greater voltage across it?
18. What does one gain by connecting capacitors in parallel?
19. What type of capacitor is suitable for the tuner in a radio, fixed or variable?
20. Which type of capacitor gives the largest capacitance for a given physical size?
21. Which type of capacitor is made from a semiconductor material?
22. What are three types of fixed capacitors?
23. Which type of capacitor might be used for power factor correction?
24. What is the significance of the (+) and (−) markings on the terminals of an electrolytic capacitor?
25. What are some ways in which surface-mount capacitors differ from other types?

PROBLEMS

SECTION 11.1 Capacitance

1. A capacitor of 5 microfarads has 25 volts across its plates. How much charge is stored?

2. What is the capacitance if 24 volts results in a charge of 480×10^{-6} coulombs?

3. How many volts are across two plates when the capacitance is 10 microfarads, and 25×10^{-6} coulombs of charge are stored?

4. The voltage across a 6-microfarad capacitor is reduced from 120 volts to 24 volts.

   (a) How much charge was removed from the plates?

   (b) How much charge is left on the plates?

5. A capacitor of 6 microfarads has 36 volts across its plates. How much charge must be added to increase the voltage to 48 volts?

6. A charge of 240×10^{-6} coulombs is stored when 60 volts are applied to two parallel plates separated by air. What is their capacitance?

7. The charge on two parallel plates connected to a 30-volt source is double the charge on plates having a capacitance of 12 microfarads that are connected to a 20-volt source. What is the capacitance of the first set of plates?

SECTION 11.2 The Electrostatic Field

8. What is the force on 5 microcoulombs of charge placed between two parallel plates 0.04 meter apart that are connected to a 200-volt source?

9. The field intensity between two plates is 50 volts per meter. What voltage is across the plates if they are 0.2 meter apart?

10. How many lines of flux are there between two parallel plates that store a charge of 5×10^{-6} coulombs?

11. A capacitor is connected across a 120-volt source. What is the field intensity if the plate spacing is 0.005 meter?

12. How far apart are two plates if the field intensity is 500 V/m with a potential of 100 volts across the plates?

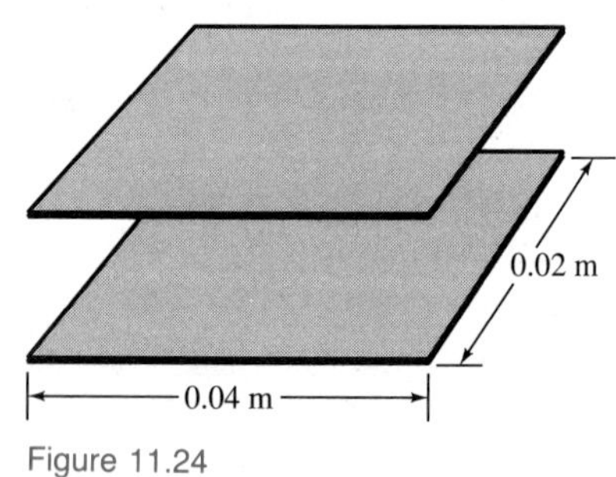

Figure 11.24

13. The flux density in the plates shown in Figure 11.24 is 5×10^{-3} coulombs per square meter.

    (a) How many lines are there?

    (b) What is the capacitance if 20 volts are needed to set up the field?

SECTION 11.4 Permittivity and Relative Permittivity

14. The field intensity between two parallel plates separated by paper is 50 volts per meter. What is the flux density between the plates?

15. Two plates are separated by 0.1 mm and have 120 volts applied across them. What is the field intensity between the plates?

16. What is the permittivity of the following materials?

    (a) Mica. (b) Oil. (c) Paper.

17. A dielectric has a permittivity of 26.55×10^{-12} F/m. What is its relative permittivity?

18. What is the flux density between the plates of a 5-microfarad capacitor connected to a 24-volt source if the plates are 0.01 meter apart and the relative permittivity is 5?

SECTION 11.5 Factors that Affect Capacitance

19. What is the capacitance of a parallel plate capacitor with oil as the dielectric, if the area of one plate is 25×10^{-4} m^2 and the spacing is 0.05 mm?

20. What must the spacing d be for the capacitor in Figure 11.25 to have a capacitance of 12 microfarads?

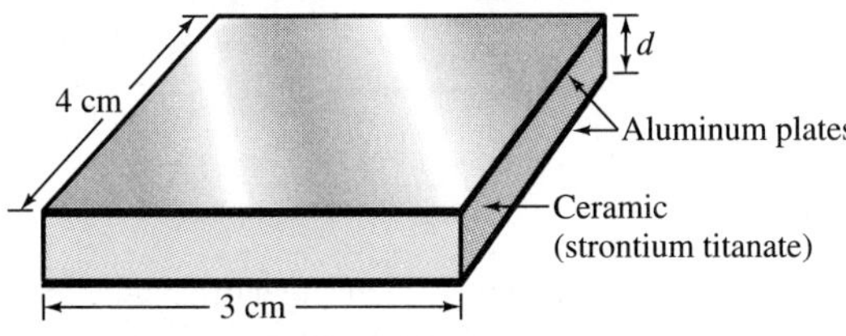

Figure 11.25

21. What is the area of one plate of a 0.2-microfarad, parallel-plate capacitor if the dielectric is plastic film and the spacing is 0.01 mm?

22. A capacitor has 5 microfarads of capacitance when air is used as the dielectric. What will be its capacitance if mica is used?

23. Use Equation 11.7 to calculate the capacitance between the plates in Figure 11.26. Hint: Consider the plates made up of small parallel sections.

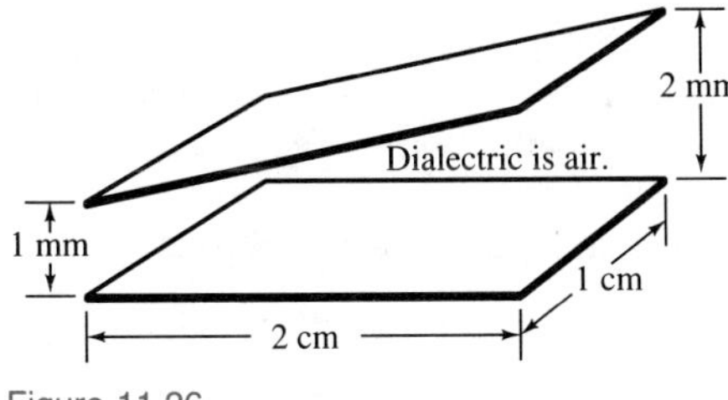

Figure 11.26

24. A capacitor is made of interleaved plates as shown in Figure 11.27. What is the capacitance? Hint: The area is the sum of the areas between any two plate surfaces.

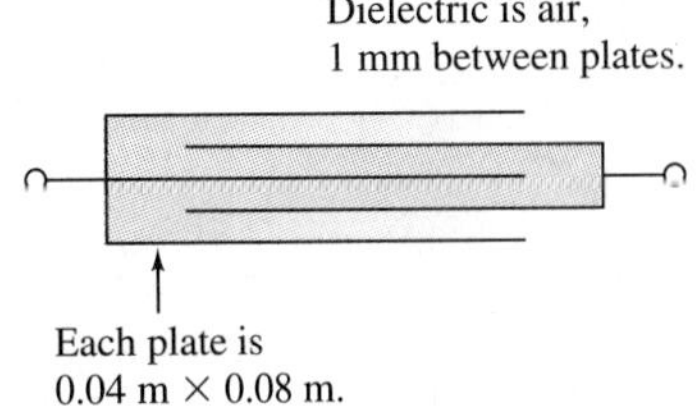

Figure 11.27

25. How much charge will accumulate on one plate of the capacitor in Figure 11.28 when it is connected to a 24-volt source?

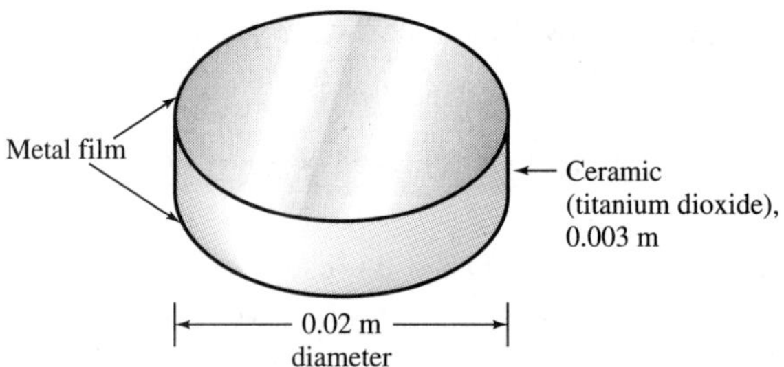

Figure 11.28

26. What is the capacitance of a 20-meter length of coaxial cable if the dielectric is teflon and the outer conductor has double the diameter of the inner one?

SECTION 11.6 Dielectric Strength

27. What is the maximum voltage that can be applied to the capacitor in Figure 11.29 before the dielectric breaks down?

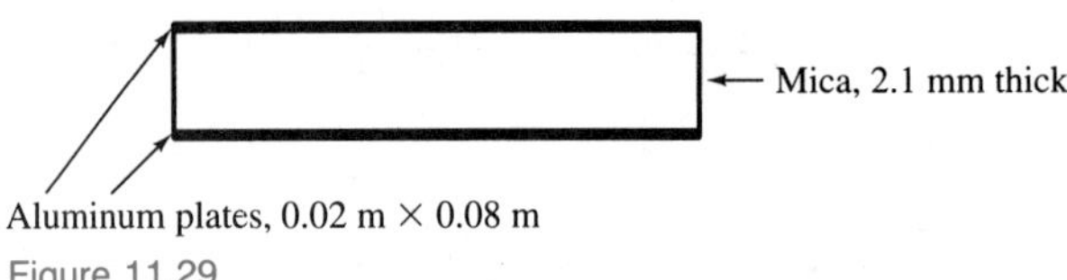

Figure 11.29

28. Two plates are separated by a 0.003-inch sheet of mica. How many volts can be applied across the plates without breakdown?

29. Determine the area of one plate of a 6-microfarad parallel-plate capacitor that has a dc working voltage rating of 500 volts and uses strontium titanate for the dielectric.

30. A 30-microfarad, 600-volt, oil-filled capacitor developed a leak that allowed the oil to seep out. After the oil leaks out

    (a) What is its capacitance?

    (b) What is its breakdown voltage?

    Use 20 kV/mm for the oil.

31. A 10-microfarad parallel-plate capacitor has a breakdown voltage of 250 volts. When the dielectric is replaced with mica, the breakdown voltage is 500 volts. What was the dielectric originally?

SECTION 11.7 Capacitors Connected in Series

32. Find the equivalent capacitance of the circuits in Figure 11.30.

33. A 12-microfarad, 16-microfarad, and 48-microfarad capacitor are connected in series to a 24-volt source. Find:

    (a) The charge on each capacitor.

    (b) The voltage across each capacitor.

    (c) The equivalent capacitance of the group.

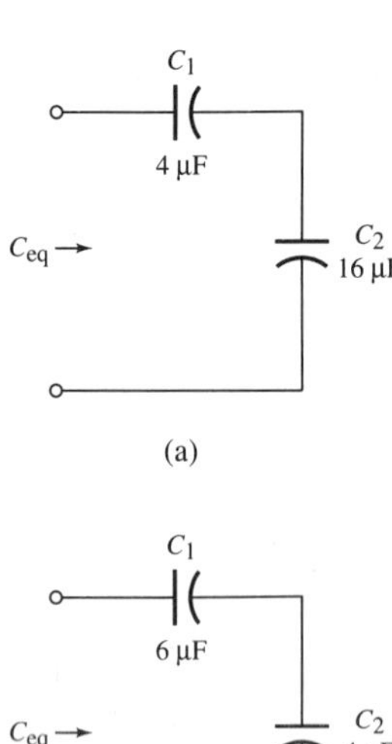

Figure 11.30

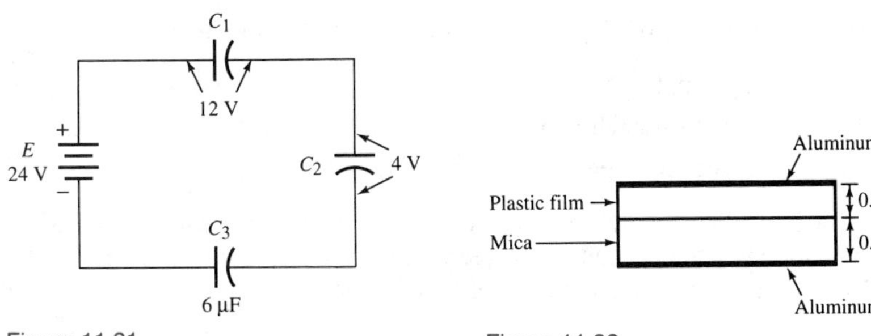

Figure 11.31

Figure 11.32

34. How many 4-microfarad, 200-volt capacitors must be connected in series to withstand 800 volts? What is the capacitance of the group?

35. A series circuit of three capacitors has an equivalent capacitance of 3 microfarads. One capacitor is 6 microfarads and another is 9 microfarads. How many microfarads is the third capacitor?

36. Use Equation 11.12 to find the voltage across each capacitor in Figure 11.30(a) when the circuit is connected to a 120-volt source.

37. Use Equation 11.13 to find the voltage across each capacitor in Figure 11.30(b) when the circuit is connected to a 120-volt source.

38. Given the circuit and conditions shown in Figure 11.31

    (a) What is the capacitance of C_1 and C_2?

    (b) What is the equivalent capacitance of the group?

    (c) What is the voltage across C_3?

39. A 5-microfarad, 200-volt capacitor and a 20-microfarad, 300-volt capacitor are connected in series. What is the maximum voltage that can be impressed across the circuit without damage to any of the capacitors?

40. What is the maximum voltage than can be applied without damage to the capacitor in Figure 11.32? Hint: The two dielectrics form a series circuit.

41. A 3-microfarad capacitor and a 4-microfarad capacitor are connected in series. The breakdown voltage for the combination is 200 volts. If the 3-microfarad capacitor is at its rated voltage, what is its rated voltage?

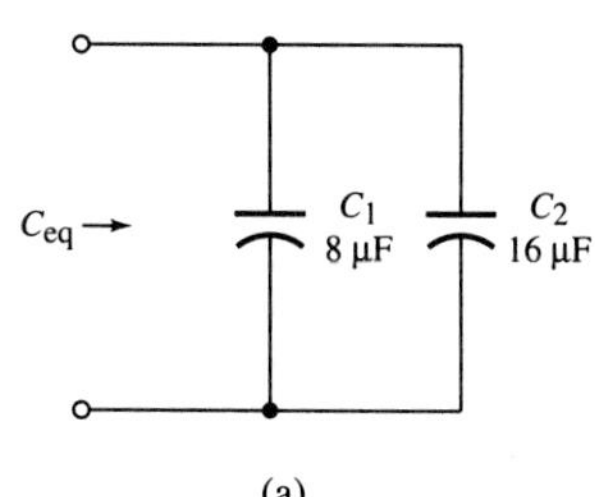

SECTION 11.8 Capacitors Connected in Parallel

42. Find the equivalent capacitance of the circuits in Figure 11.33.

43. A 12-microfarad, 16-microfarad, and 8-microfarad capacitor are connected in parallel across a 24-volt source. Find:

    (a) The equivalent capacitance.

    (b) The charge on each capacitor.

    (c) The charge supplied by the source.

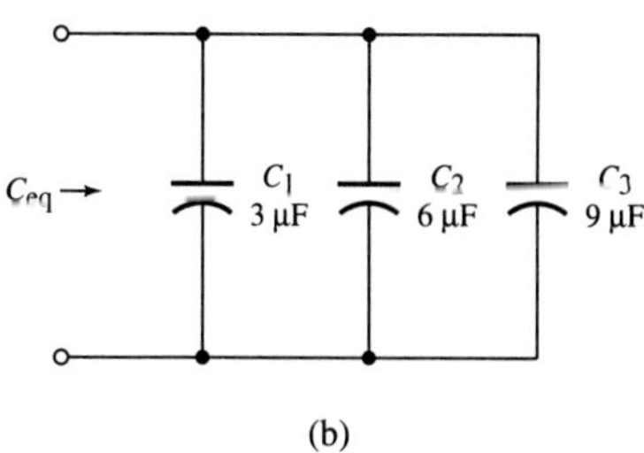

Figure 11.33

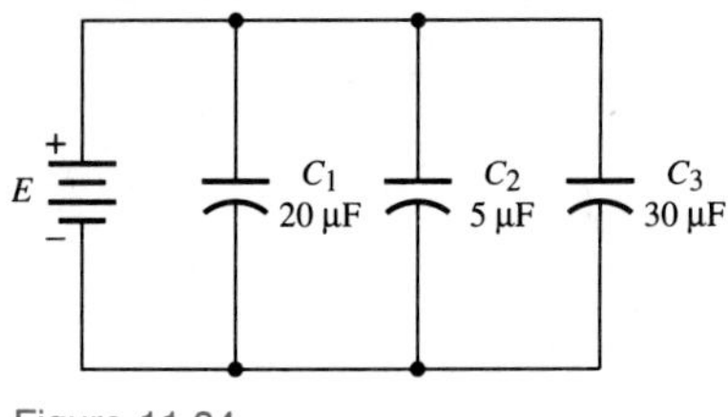

Figure 11.34

44. The charge on C_1 of Figure 11.34 is 250×10^{-6} coulombs. Find:

    (a) The value of E.

    (b) The charge on each capacitor.

    (c) The equivalent capacitance of the circuit.

45. A parallel circuit of three capacitors has an equivalent capacitance of 24 microfarads. One capacitor is 4 microfarads and another is 12 microfarads. How many microfarads is the third capacitor?

46. What is the charge on each capacitor in the circuit of Figure 11.35?

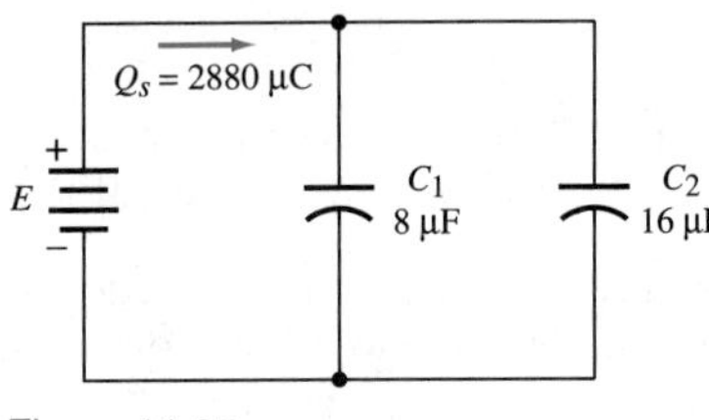

Figure 11.35

47. What is the charge on each capacitor and the source voltage in the circuit of Figure 11.36?

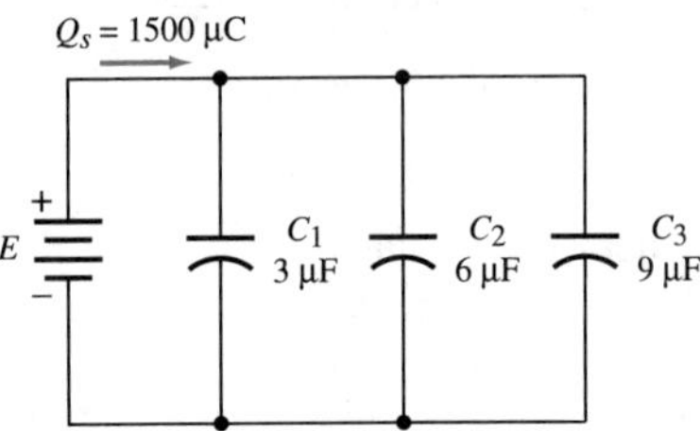

Figure 11.36

48. The switch in the circuit of Figure 11.37 was moved to Position 2 after the capacitor charged to 24 volts. Find:

    (a) The charge on C_1 at 24 volts.

    (b) The voltage across C_1 after the switch is in Position 2.

    (c) The charge on each capacitor after the switch is in Position 2. Hint: The total charge remains the same.

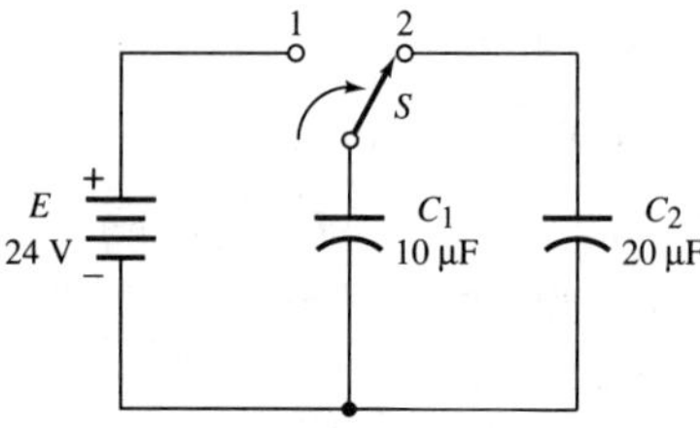

Figure 11.37

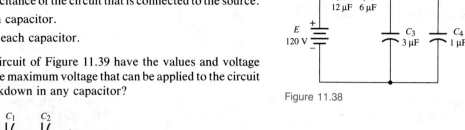

49. In the circuit in Figure 11.38, find:

    (a) The equivalent capacitance of the circuit that is connected to the source.

    (b) The charge on each capacitor.

    (c) The voltage across each capacitor.

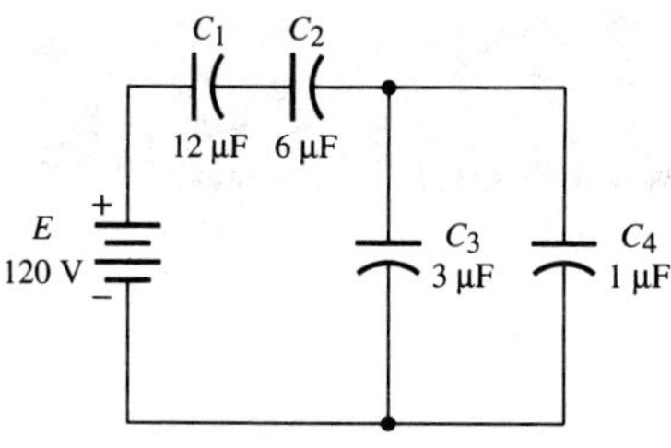

Figure 11.38

50. The capacitors in the circuit of Figure 11.39 have the values and voltage ratings listed. What is the maximum voltage that can be applied to the circuit without expecting breakdown in any capacitor?

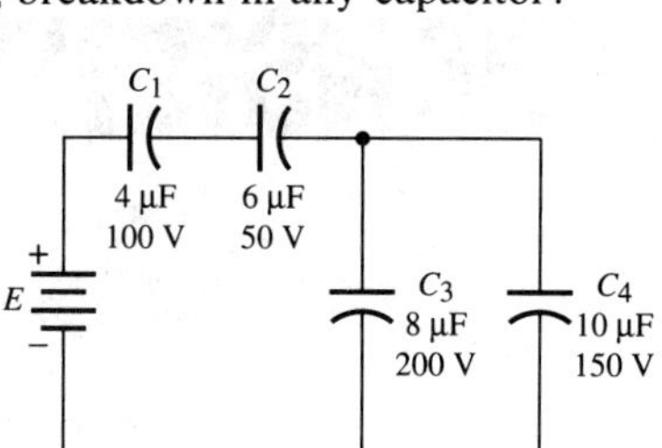

Figure 11.39

CHAPTER 12

THE *R–C* CIRCUIT, TRANSIENT, AND STEADY STATE

CHAPTER OBJECTIVES

After completing this chapter, you should be able to:

1. Describe the voltage and current change during the charging and discharging of a capacitor.
2. List the quantities that affect the rate of change and the initial and final values of current and voltage.
3. Analyze an *R–C* circuit during charging and discharging.
4. Define the time constant and list the quantities that affect it.

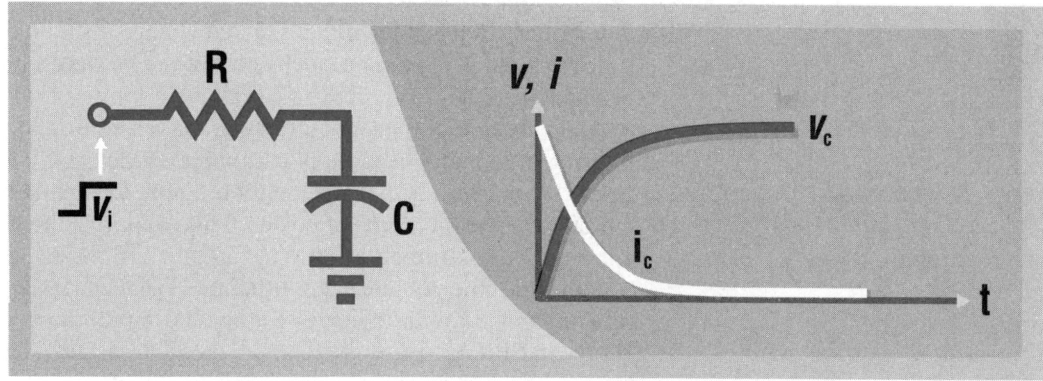

5. Use the Universal Time Constant curve for the analysis of the *R–C* circuit.
6. Calculate the energy that is stored in a capacitor.
7. Analyze the *R–C* circuit for current, voltage, energy, and power during steady state.

KEY TERMS

1. average current
2. charge
3. discharge
4. series *R–C* circuit
5. steady state
6. time constant
7. transient state

INTRODUCTION

The circuits studied thus far had only one type of element in them. It was either resistance or capacitance. In those circuits, the voltage changed instantaneously when a source of emf was connected to the circuit.

In this chapter, circuits that have both resistance and capacitance in them are introduced. These are called *R–C* circuits. Their change is time dependent. That is, the voltages do not change instantaneously, but are a function of time. The time can be controlled by the amount of resistance or capacitance.

The behavior of these *R–C* circuits makes time delays, waveshaping, filtering, and many other applications possible. Many of our electronic circuits would not be possible without them. On the negative side, these circuits change the shape of pulses. This can affect the operation of logic circuits and instrumentation.

How does the voltage vary with time? What can be done to change the rate of change? What effect does the *R–C* circuit have on pulse inputs? How can SPICE be used to analyze the *R–C* circuit? These are some of the questions that are answered in this chapter.

12.1 The Series *R–C* Circuit During Voltage Rise

series *R–C* circuit

A **series *R–C* circuit** is a circuit of resistance and capacitance connected in series as shown in Figure 12.1(a). This circuit can be put in any one of the following three states:

1. Switch open—no current; the voltage across the capacitor is 0 volts.
2. Switch closed—current for a short time; capacitor voltage is rising; the capacitor is charging.
3. Switch closed—no current; the capacitor voltage is *E*; the capacitor is fully charged.

The second state is studied in this section.

charge

When the switch in Figure 12.1(a) is closed, the capacitor voltage increases and the capacitor starts to **charge**. However, the voltage does not rise instantaneously as it does in a purely resistive circuit. It rises as shown by the curve in Figure 12.1(b). The current change during the voltage rise is also shown in Figure 12.1(b). It starts at a maximum and drops to zero. The state during which the voltage changes from one steady state to another steady state is called the **transient state**.

transient state

There is no charge on the capacitor before the switch in Figure 12.1(a) is closed. The electrostatic field is initially zero but becomes stronger as the charge accumulates. Since γ equals Q, initially, the rate of voltage rise is rapid. The steepness of the voltage curve in Figure 12.1(b) at the start shows this. As the field becomes stronger, greater opposition is offered to the charge accumulation. Because of that, the rate of charge accumulation and the voltage rise becomes slower. Again, the voltage curve shows this by the curve flattening out. Finally, there is no further change; the curve is practically horizontal. The capacitor is fully charged.

The time needed for the voltage to build up depends on the resistance and the capacitance. Large resistance slows the transfer of charge. This requires more time to accumulate the amount of charge. Larger capaci-

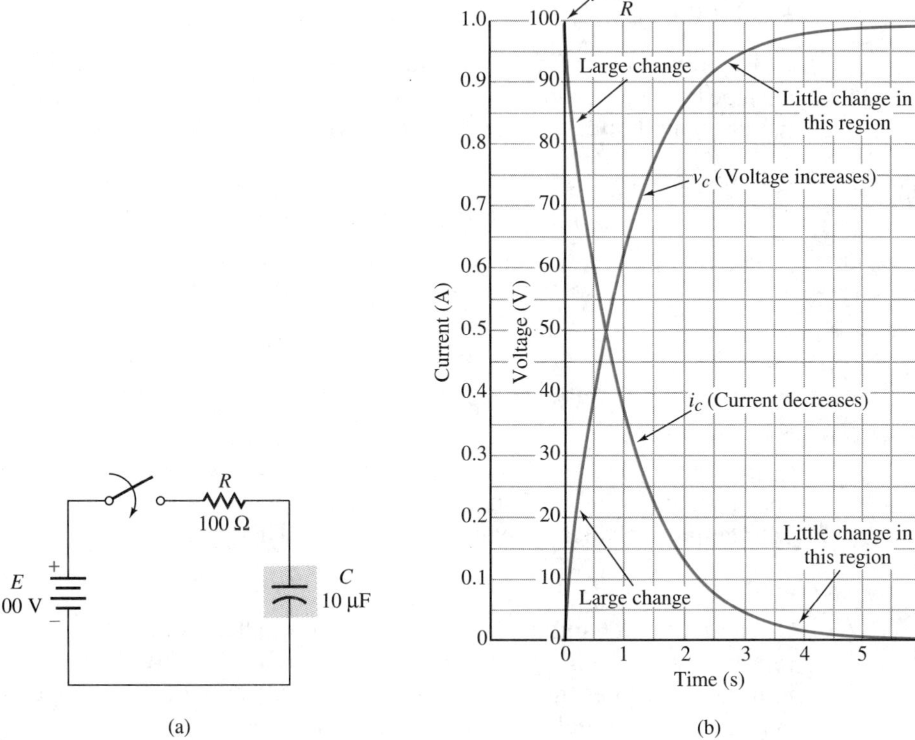

Figure 12.1 The capacitor voltage and current do not change instantaneously in the $R–C$ circuit. (a) A series $R–C$ circuit. (b) The change in voltage and current as a function of time.

tance can store more charge. This means a longer time is needed to fill the capacitor with charge.

As a technician, you might need to know what the exact values of current and voltage will be in the $R–C$ circuit. Some relations for these values are now developed.

Applying Kirchhoff's Voltage Law to the $R–C$ circuit gives

$$E - \upsilon_r - \upsilon_C = 0 \qquad \text{or} \qquad E - iR - V_c = 0$$

Then

$$i = (E - V_c)/R$$

At the start of charging, V_c is 0, so the current at that time (i_o) is

$$I_o = E/R \tag{12.1}$$

where I_o is the current at the start of discharge; $t = 0$, in amperes; E is the source voltage, in volts; and R is the series resistance, in ohms.

Equation 12.1 indicates that the ideal capacitor is initially like a short circuit. Because of that, the current will be very large if the circuit resistance is small. This must be kept in mind when using capacitors. The large current can damage components or blow fuses. One way to reduce the current is to put some resistance in the circuit.

EXAMPLE 12.1

The initial current in the charging circuit of an electronic camera flash must be limited to 2 amperes. How many ohms of resistance are needed if the charging battery has an emf of 45 volts?

Solution

$$I_o = E/R \qquad \text{or} \qquad R = E/i_o$$

Substituting know values gives

$$R = 45\text{ V}/2\text{ A}$$
$$R = 22.5\ \Omega$$

In Chapter 11, you learned that $Q = \upsilon_c C$. Now, as the amount of charge changes, a small amount (dQ) in a short period of time (dt), the voltage must change a small amount ($d\upsilon_c$). This gives

$$dQ/dt = C\ (d\upsilon_c/dt)$$

The terms dQ and $d\upsilon_c$ are calculus terms called differentials. They represent a small change in Q and υ_c for the infinitesimally small change in time. Q/t represents current, so dQ/dt represents the instantaneous current. If the initial current is being considered,

$$dQ/dt = C\ (d\upsilon_c/dt)$$

becomes

$$I_o = C\ (d\upsilon_c/dt)$$

But

$$I_o = E/R$$

so

$$E/R = C\ (d\upsilon_c/dt)$$

or

$$d\upsilon_c/dt = E/(RC) \qquad \textbf{(12.2)}$$

where $d\upsilon_c/dt$ is the rate of voltage change, in V/s; E is the source voltage, in volts; and R is the series resistance, in ohms.

Finally, looking at the charged state, you can see that $\upsilon_c = E$ and $i_c = 0$ when the capacitor is fully charged.

EXAMPLE 12.2

A 5-microfarad capacitor and a 5-ohm resistor are connected in series across a 100-volt source. Find:

a. the initial surge current.

b. the rate of voltage change at the instant the circuit is connected.
c. the final voltage across the capacitor.

Solution

a. $I_o = E/R = 100 \text{ V}/5\ \Omega$
$I_o = 20 \text{ A}$

b. $dv_c/dt = E/RC = \dfrac{100 \text{ V}}{(5\ \Omega)(5 \times 10^{-6} \text{ F})}$
$dv_c/dt = 4 \times 10^6 \text{ V/s}$

c. Final voltage $= E$
Final voltage $= 100$ V

The current i, as found by using $i = C(dv/dt)$, is the current at an instant of time. If Δv and Δt are used, the result will be the **average current**. That is:

average current

$$I_{ave} = C\frac{\Delta v_c}{\Delta t} \qquad \textbf{(12.3)}$$

where $\dfrac{\Delta v_c}{\Delta t}$ is the change in capacitor voltage in a small amount of time, in volts per second; C is the capacitance, in farads; and I_{ave} is the average value of the current in time Δt.

The expressions $i = C(dv/dt)$ and Equation 12.3 show that a current exists only during a change in capacitor voltage. When the capacitor voltage is constant, there is no current. Thus, once steady state is reached, the capacitor acts as an open circuit.

EXAMPLE 12.3

The change in voltage across a 2-microfarad capacitor is shown in Figure 12.2. Find the average current from:

a. 0 s–2 s.
b. 2 s–4 s.
c. 7 s–10 s.

Draw the waveform of the average current.

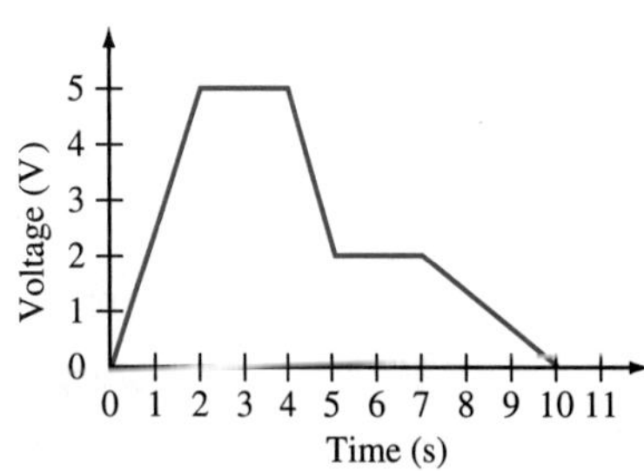

Figure 12.2

Solution

a. $I_{ave} = C\dfrac{\Delta v_c}{\Delta t}$

$= (2 \times 10^{-6} \text{ F})\left(\dfrac{5 \text{ V} - 0 \text{ V}}{2 \text{ s} - 0 \text{ s}}\right)$

$I_{ave} = 5 \times 10^{-6}$ A or 5 microamperes

b. $I_{ave} = C\dfrac{\Delta v_c}{\Delta t}$

$$= (2 \times 10^{-6}\text{ F})\left(\frac{5\text{ V} - 5\text{ V}}{4\text{ s} - 2\text{ s}}\right)$$

$$I_{\text{ave}} = 0\text{ A}$$

c. $$I_{\text{ave}} = C\,\frac{\Delta \upsilon_c}{\Delta t}$$

$$= (2 \times 10^{-6}\text{ F})\left(\frac{0\text{ V} - 2\text{ V}}{10\text{ s} - 7\text{ s}}\right)$$

$$I_{\text{ave}} = -1.334 \times 10^{-6}\text{ A}$$

The (−) sign indicates that the direction of the current is opposite that in a and in b.

The waveform of the average current is shown in Figure 12.3.

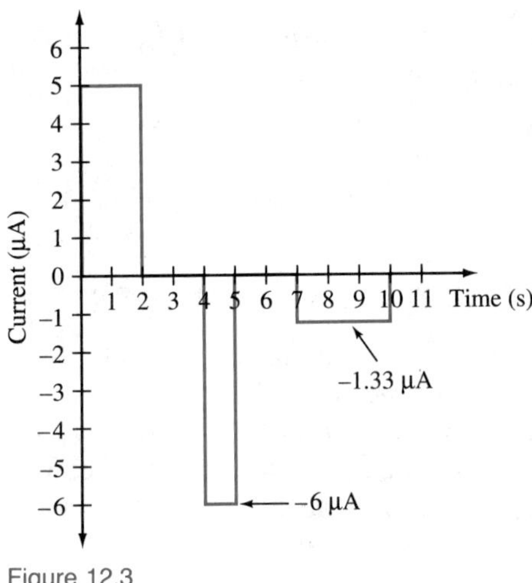

Figure 12.3

Recalling that $i = C(d\upsilon_c/dt)$, and using that in the Kirchhoff's Voltage Law equation for the charging circuit gives

$$E - \upsilon_c - R(C\,d\upsilon_c/dt) = 0$$

or

$$\upsilon_c = E - RC\,d\upsilon_c/dt$$

This equation is known as a differential equation because of the $d\upsilon_c/dt$ term. To obtain υ_c, the method of calculus shown in Appendix C must be used. The result is:

$$\upsilon_c = E(1 - e^{-t/RC}) \qquad \textbf{(12.4)}$$

where υ_c is the capacitor voltage at any time, in volts; E is the source emf, in volts; R is the series resistance, in ohms; C is the capacitance, in farads; t is the length of time that the capacitor has been charging, in seconds; and e is the base for the natural logarithm system and is equal to 2.71828 carried to six digits.

EXAMPLE 12.4

A 5-microfarad capacitor and a 1000-ohm resistor are connected in series across a 24-volt source. Find:

a. the initial current in the circuit.
b. the final voltage across the capacitor.
c. the capacitor voltage at 10 ms after the circuit is connected to the source.
d. the time at which the capacitor voltage is 12 volts.

Solution | **Strategy**

a. $I_o = E/R$

$= 24\text{ V}/1000\ \Omega$

$I_o = 24\text{ mA}$

b. $V_{\text{final}} = E$

$V_{\text{final}} = 24\text{ V}$

c. $v_c = E(1 - e^{-t/RC})$ — Use the exponential equation.

$t/RC = (0.010\text{ s})/(1000\ \Omega \times 5 \times 10^{-6}\text{ F})$ — Find t/RC.

$= 2$

so

$$v_c = 24\text{ V}\ (1 - e^{-2})\text{ volts}$$

$$= 24\text{ V}\ (1 - 0.135)$$ — Use e^x to get e^{-2}.

$$v_c = 20.75\text{ V}$$

d. $12\text{ V} = 24\text{ V}\ (1 - e^{-t/0.005\text{ s}})\ e^{-t/0.005\text{ s}} = 0.5$ — Divide and rearrange.

or

$$-t/0.005\text{ s} = \ln 0.5$$ — Use $\ln x$ to get the exponent of e.

Solving for t,

$$t = (0.005\text{ s}) \ln 0.5$$

$$= (0.005\text{ s})(0.693)$$

$$t = 3.47 \times 10^{-3}\text{ s}$$

The algebraic and RPN calculator keystrokes for the solution are:

Quantity	**Algebraic**	**RPN**	**Display**
E/R	2 4 ÷ 1 EXP 3 =	2 4 ENTER 1 EEX 3 ÷	2.400 −02
t/RC	1 EXP 2 +/− ÷ 1 EXP 3 ÷ 5 EXP 6 +/− =	1 EEX 2 CHS ENTER 1 EEX 3 ÷ 5 EEX 6 CHS ÷	2.000 00
$24(1 - e^{-2})$	2 4 × [1 − 2 +/− e^x] =	1 ENTER 2 CHS e^x − 2 4 × ×	2.075 01
0.005 ln 0.5	5 EXP 3 +/− × · 5 ln × = +/−	5 EEX 3 CHS ENTER · 5 ln × CHS	3.466 −03

The equations for the voltage across the resistor (υ_R), and the current in the circuit (i), are now developed. By Kirchhoff's Voltage Law,

$$E - \upsilon_R - \upsilon_c = 0 \qquad \text{or} \qquad \upsilon_R = E - \upsilon_c$$

since

$$\upsilon_c = E(1 - e^{-t/RC})$$

This reduces to

$$\upsilon_R = Ee^{-t/RC} \tag{12.5}$$

where υ_R is the voltage across the resistance in volts at any time t and E, t, R, C, and e have the same meaning as in Equation 12.4.

Applying Ohm's Law to the resistance gives

$$i = \upsilon_R/R$$

Since

$$\upsilon_R = Ee^{-t/RC}$$

$$i = (E/R)e^{-t/RC} \tag{12.6}$$

where i is the current in the circuit at any time t, in amperes and E, t, R, C, and e have the same meaning as in Equation 12.4.

EXAMPLE 12.5

A series circuit of a 6-microfarad capacitor and a 5000-ohm resistor is connected across a 120-volt source. Find the following at 0.015 seconds after the connection is made:

a. the voltage across the capacitor.
b. the voltage across the resistor.
c. the current in the circuit.

Solution | **Strategy**

a. $\upsilon_c = E(1 - e^{-t/RC})$

$RC = (5000\ \Omega)(6 \times 10^{-6}\ \text{F}) = 0.030\ \text{s}$ — Find RC.

$\upsilon_c = 120\ \text{V}\ (1 - e^{-0.015\ \text{s}/0.030\ \text{s}})$ — Solve for υ_c.

$= 120\ \text{V}\ (1 - e^{-0.5})$

$= 120\ \text{V}\ (1 - 0.607)$

$\upsilon_c = 47.16\ \text{V}$

b. $\upsilon_R = Ee^{-t/RC}$ — Use Equation 12.5 for υ_R.

$= 120\ \text{V}\ e^{-0.5}$

$\upsilon_R = 72.84\ \text{V}$

c. $i = (E/R)e^{-t/RC}$ — Use Equation 12.6 for i.

$= (120\ \text{V}/5000\ \Omega)e^{-0.5}$

$i = 14.57\ \text{mA}$

A term that is commonly used to compare R–C circuits is the **time constant**. The usefulness of the time constant is that it permits comparison of R–C circuits without constructing a complete curve.

time constant

The time constant can be defined in two ways:

1. It is the time in which the voltage rises to 63.21 percent of the final value.
2. It is the time in which the voltage would reach the final value if it rose at the initial rate of rise.

The graphic presentation of these for the circuit of Figure 12.4(a) is shown in Figure 12.4(b). The symbol for the time constant is τ; its unit is the second.

The first definition indicates that τ can be found from the curve or by measurements. One simply notes the time at which υ_c is 63.21 percent of the final voltage. The second definition requires calculus to obtain an exact value of τ. However, it can be used to express τ in terms of R and C of the circuit.

The line 0–A in Figure 12.4(b) represents a rise at the initial rate of E/RC. It reaches E at time t. From the values on the axis,

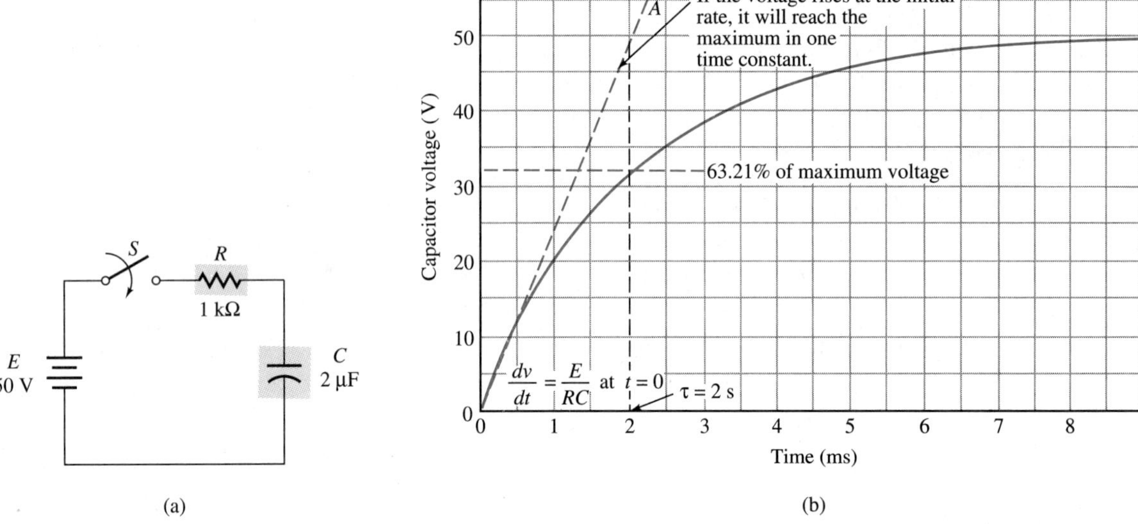

Figure 12.4 The relation of the time constant of an R–C circuit to the voltage rise. (a) A series R–C circuit in which the capacitor is being charged. (b) The voltage rise curve for the circuit and the graphical presentation of the time constant.

$$\Delta v_c/\Delta t = E/t$$

Also, from Equation 12.2,

$$\Delta v_c/\Delta t = E/(RC)$$

so

$$E/t = E/(RC) \qquad \text{or} \qquad t = RC$$

Since, by definition, the time t is the time constant, $\tau = \text{t}$. So

$$\tau = RC \tag{12.7}$$

where τ is the time constant, in seconds; R is the resistance, in ohms; and C is the capacitance, in farads.

TABLE 12.1 The Percent of Final Voltage Reached at 0, 1, 2, 3, 4, and 5 Time Constants During Charging

Number of Time Constants	Percent of Final Voltage
0	0
1	63.21
2	86.47
3	95.02
4	98.17
5	99.33

Equation 12.7 proves that larger values of resistance or capacitance result in a slower voltage rise.

In theory, the exponential curves never reach a final value. But, for practical purposes, the final value is taken as being reached in five time constants. At that time, more than 99.32 percent of the final value has been reached. The percentage of the final voltage at 0, 1, 2, 3, 4, and 5 time constants is listed in Table 12.1. These are useful for checking results, rough plotting of curves, and estimating values at other times.

With the relation in Equation 12.7, we can now design circuits to have certain characteristics. Also, the characteristics of circuits can be compared before they are built.

EXAMPLE 12.6

An R–C circuit has a time constant of 0.5 second. How much resistance is in the circuit if C equals 5 microfarads?

Solution

$$\tau = RC$$

so

$$R = \tau/C$$
$$= 0.5 \text{ s}/5 \times 10^{-6} \text{ F}$$
$$R = 1 \times 10^{5}\ \Omega$$

EXAMPLE 12.7

A 5-microfarad capacitor and a 1000-ohm resistor are connected in series across a 100-volt source. Find:

a. the time needed for the capacitor to reach 63.21 volts.
b. the initial rate of rise of the voltage.
c. the current at the time in a.

Solution

a. 63.21 volts is 63.21 percent of the final value, so the time is one time constant.

$$t = \tau = RC$$
$$= (1000\ \Omega)(5 \times 10^{-6} \text{ F})$$
$$t = 5 \times 10^{-3} \text{ s}$$

b. From the first meaning of the time constant,

$$(\tau)(\text{initial rate of rise}) = 100 \text{ V}$$

So the initial rate of rise $= 100 \text{ V}/(5 \times 10^{-3} \text{ s})$

$$dv/dt = 20 \times 10^{3} \text{ V/s}$$

c. The current will drop to 36.79 percent of its maximum in one time constant.

$$I_{max} = E/R = 100 \text{ V}/1000\ \Omega = 100 \text{ mA}$$

so

$$I = (100 \text{ mA})(0.367\,9)$$
$$I = 36.79 \text{ mA}$$

Some applications result in a capacitor being charged from one level to another. To analyze these, the effect of the initial voltage must be considered. One approach is to add or subtract the initial voltage to Equation 12.4, and use $E \pm V_o$, depending on the polarity, instead of E. Another approach is presented in Section 12.2.

Using the first approach, the expression for υ_c becomes

$$\upsilon_c = \mp V_o + (E \pm V_o)(1 - e^{-t/(RC)})$$

At $t = 0$ seconds, υ_c will equal $\pm V_o$. At $t = \infty$ seconds, υ_c will equal E, which is what it should be. This procedure is now applied to the analysis of a circuit.

EXAMPLE 12.8

Determine the capacitor voltage at 10 ms after the switch in Figure 12.5 is closed.

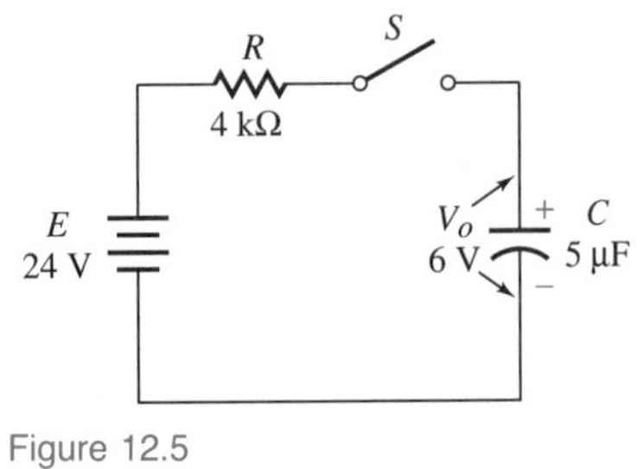

Figure 12.5

Solution

$$\upsilon_c = V_o + (E - V_o)(1 - e^{-t/(RC)})$$
$$RC = (4 \times 10^3\ \Omega)(5 \times 10^{-6} \text{ F}) = 20 \text{ ms}$$

Then

$$\upsilon_c = 6 \text{ V} + (24 \text{ V} - 6 \text{ V})(1 - e^{-10 \times 10^{-3} \text{ s}/20 \times 10^{-3} \text{ s}})$$
$$= 6 \text{ V} + 18 \text{ V}\ (0.393)$$
$$\upsilon_c = 13.07 \text{ V}$$

The first V_o term takes the polarity of the capacitor voltage at $t = 0$ s. The second V_o term takes the opposite polarity.

Some key quantities in the charging transient state are the initial rate of rise, initial current, final voltage, and time constant. Each of these depends on some part of the circuit, as listed in the following.

Quantity	Affected by
Initial rate or rise	E, R, and C
Initial current	E and R
Final voltage	E
Time constant	R and C

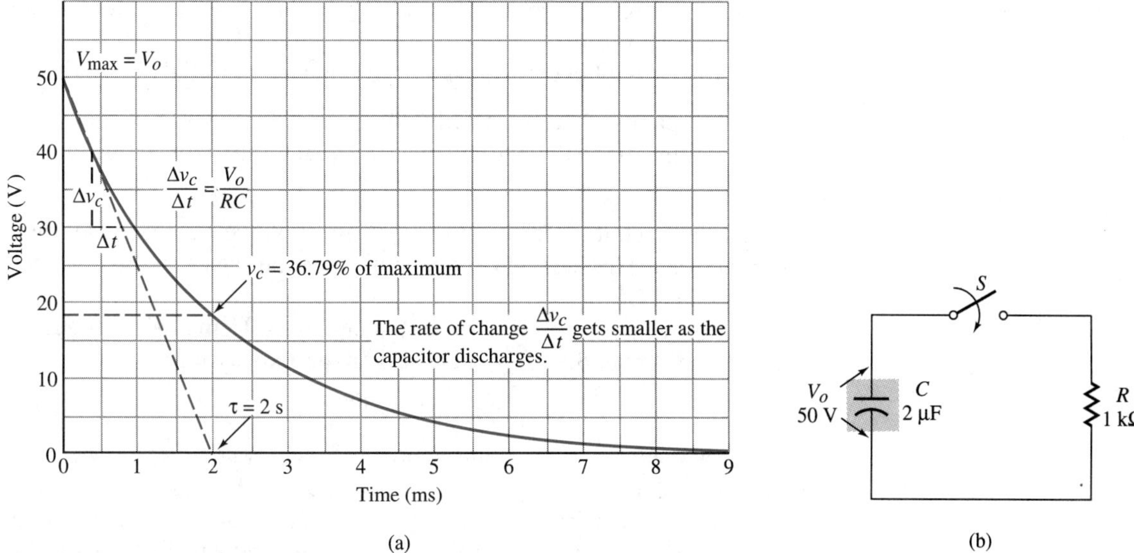

Figure 12.6 The capacitor voltage drops exponentially during the capacitor discharge in an R–C circuit. (a) The discharge curve. (b) The circuit used to get the values on the discharge curve.

12.2 The Series R–C Circuit During Voltage Fall

discharge

The transient state also exists during the capacitor **discharge**. The capacitor voltage does not drop instantaneously, but follows the curve in Figure 12.6(a). Again, the change is opposed by the electrostatic field.

A circuit that can be used to study the discharge transient is shown in Figure 12.6(b). The initial voltage on the capacitor is that to which it was charged, and is called V_o. Applying Kirchhoff's Voltage Law to the circuit gives

$$V_o - I_oR = 0 \quad \text{or} \quad I_o = V_o/R \tag{12.8}$$

where I_o is the capacitor current at the start of the discharge, in amperes; V_o is the voltage to which the capacitor has been charged, in volts; and R is the resistance in the discharge circuit, in ohms.

Both the capacitor voltage and current decrease until they reach zero. The shape of the decrease is shown in Figure 12.6(a).

From Kirchhoff's Voltage Law,

$$V_o - iR = 0$$

Since

$$i = C dv_c/dt,$$
$$V_o = C\,(dv_c/dt)(R)$$

or

$$dv_c/dt = V_o/(RC) \tag{12.9}$$

where dv_c/dt is the initial change in voltage (dv_c) for an extremely small change in time (dt), in V/s; V_o is the initial voltage on the capacitor, in volts; R is the resistance in the discharge path, in ohms; and C is the capacitance, in farads.

The initial rate of decrease will be the same as the initial rate of rise if:

a. the capacitor had been fully charged by the source of emf.
b. the ohms in the discharge path are equal to the ohms in the charge path.

The rate of change for both current and voltage decreases as they approach zero. This is shown by the curves becoming nearer to horizontal in Figure 12.6.

Equation 12.9 is another differential equation. Using calculus again, as shown in Appendix C, yields the voltage expression:

$$\upsilon_c = V_o e^{-t/RC} \qquad \textbf{(12.10)}$$

where υ_c is the capacitor voltage at any instant of time, in volts; V_o is the initial voltage on the capacitor, in volts; R is the resistance in the discharge path, in ohms; C is the capacitance, in farads; t is the time that the capacitor has been discharging, in seconds; and e is the base for the natural logarithm system. It is equal to 2.718 28 carried to six digits.

The time constant for the discharging circuit is defined in the same way as for charging. That is,

1. It is the time in which υ_c will drop to 36.79 percent of the initial value. (A change of 63.21 percent.)
2. It is the time in which υ_c would reach zero if it dropped at the initial rate of drop.

Also, $\tau = RC$ as for charging.

Table 12.2 lists the percent of initial value reached by the voltage and current for 0–5 time constants. From a practical viewpoint, the capacitor is taken as being fully discharged in five time constants.

TABLE 12.2

Number of Time Constant	Percent of Initial Value
0	100
1	36.79
2	13.53
3	4.98
4	1.83
5	0.67

EXAMPLE 12.9

The switch in Figure 12.7 is put in Position 2 after the capacitor is fully charged.

a. What is the voltage across the capacitor at 12 ms after discharge starts?
b. What would be the voltage in a if the switch were in Position 1 for only 12 ms?

Solution

Strategy

a. $\upsilon_c = V_o e^{-t/\tau}$

$V_o = E = 24 \text{ V}$

In Position 2, the discharge path is through R_2

Find the discharge time constant.

$$\tau_2 = R_2 C$$
$$= (2 \times 10^3 \ \Omega)(12 \times 10^{-6} \text{ F})$$
$$\tau_2 = 24 \times 10^{-3} \text{ s}$$
$$\upsilon_c = (24 \text{ V})(e^{(-0.012 \text{ s}/0.024 \text{ s})})$$
$$= (24 \text{ V})(e^{-0.5})$$
$$\upsilon_c = 14.56 \text{ V}$$

b. In position 1,

Check if the capacitor has reached steady state.

$$\tau_1 = R_1 C$$
$$= (1000 \ \Omega)(12 \times 10^{-6} \text{ F})$$
$$\tau_1 = 12 \times 10^{-3} \text{ s}$$

Since the time of 12 ms is less than $5\tau_1$, V_o should not rise to 24 volts.

So

Find the voltage at the start of discharge.

$$V_o = E(1 - e^{-t/\tau})$$
$$= 24 \text{ V}(1 - e^{(-0.012 \text{ s}/0.012 \text{ s})})$$
$$= 24 \text{ V}(1 - e^{-1})$$
$$= 15.17 \text{ V}$$

Then, during discharge,

$$\upsilon_c = V_o \, e^{-t/\tau}$$
$$= 15.17 \, e^{(-0.012/0.024 \text{ s})}$$
$$\upsilon_c = 9.2 \text{ V}$$

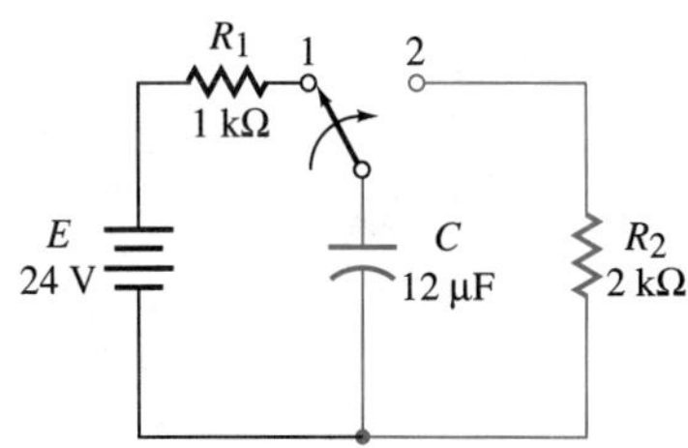

Figure 12.7

We now consider the capacitor polarity and current direction. The capacitor in Figure 12.8 will charge to the voltage E. The direction of the charging current and the terminal polarity are shown in Figure 12.8(a). If the switch is placed in Position 2, the capacitor will discharge through resistor R_2. The discharge current and terminal polarity are shown in Figure 12.8(b). Note that the polarity stays the same, but the direction of the current is reversed. Because of the reversal, the current is given a negative sign. The sequence of charge and discharge can be followed in Figure 12.8(c)–(d).

Now an expression for the current is developed. Applying Kirchhoff's Voltage Law to the circuit that was shown in Figure 12.6(b) yields

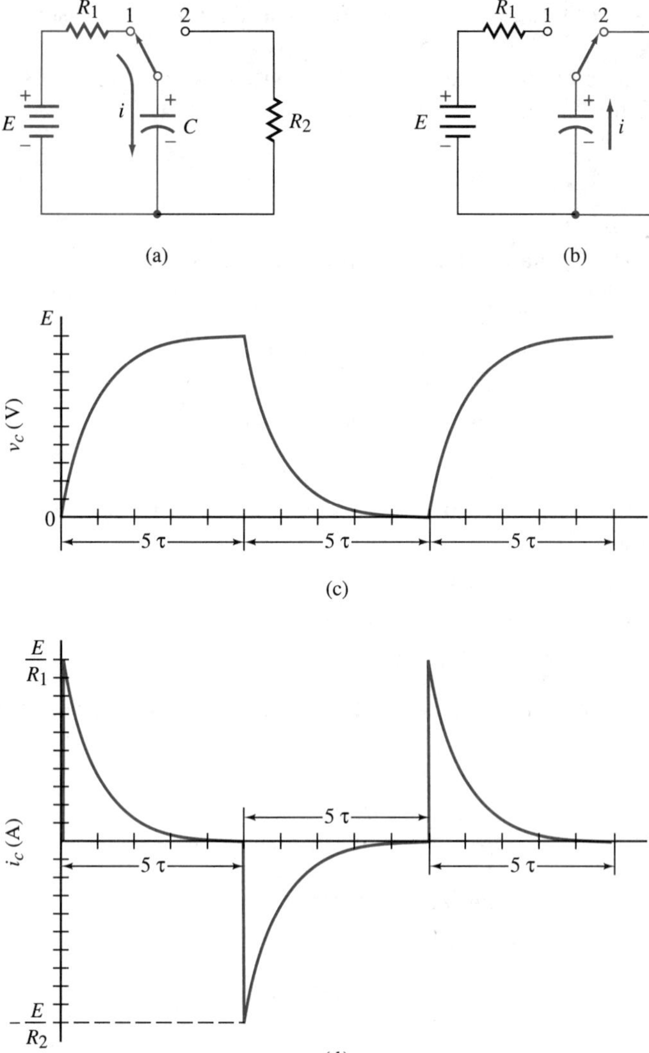

Figure 12.8 The relation of capacitor terminal polarity and current direction during charging and discharging. (a) The charging circuit. (b) The discharging circuit. (c) The terminal polarity is the same during charge and discharge. (d) The current direction reverses during discharge.

$$V_c - iR = 0 \qquad \text{or} \qquad i = \upsilon_c/R$$

Substituting $V_c = V_o\, e^{-t/RC}$ and including the negative sign gives

$$i = -(V_o/R)e^{-t/RC} \tag{12.11}$$

where i is the capacitor current during discharge, in amperes, and V_o, R, t, and C have the same meaning as in Equation 12.9.

An expression for υ_r can now be found.

In Figure 12.6(b), Kirchhoff's Voltage Law yields

$$V_c - \upsilon_r = 0$$

since

$$\upsilon_c = V_o e^{-t/RC},$$
$$\upsilon_R = V_o e^{-t/RC} \qquad \textbf{(12.12)}$$

where υ_R is the voltage across the resistance in the discharge path at any time t, in volts, and V_o, t, R, C, and e have the same meaning as in Equation 12.10.

EXAMPLE 12.10

The capacitor in the circuit of Figure 12.9 has been charged to 120 volts.

a. What is the capacitor voltage and current at 60 ms after the switch is closed?
b. What is the resistor voltage at 60 ms after the switch is closed?
c. Show the direction of the current and mark the polarity of the resistor and capacitor terminals.

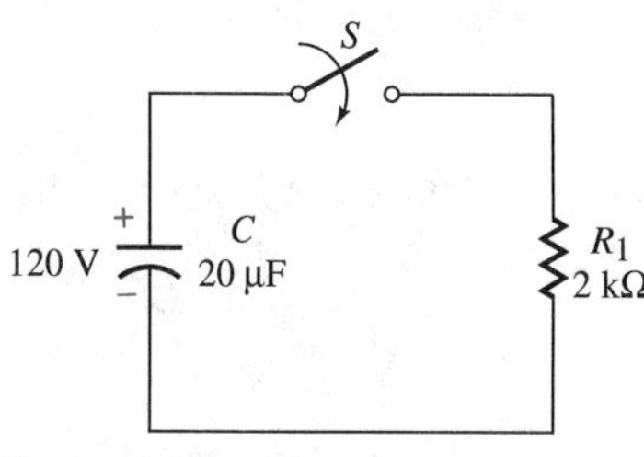

Figure 12.9

Solution

a. From Equation 12.12,

$$\upsilon_c = V_o e^{-t/\tau}$$
$$\tau = RC$$
$$= (2000\ \Omega)(20 \times 10^{-6}\ \text{F})$$
$$\tau = 40\ \text{ms}$$

Then

$$\upsilon_c = 120\ \text{V}\ e^{-0.06\ \text{s}/0.04\ \text{s}}$$
$$= 120\ \text{V}\ e^{-1.5}$$
$$= (120\ \text{V})(0.223)$$
$$\upsilon_c = 26.78\ \text{V}$$
$$i = \upsilon_c/R$$
$$= 26.78\ \text{V}/2000\ \Omega$$
$$i = 13.4\ \text{mA}$$

b. By Equation 12.11,

$$\upsilon_R = \upsilon_c$$

so

$$\upsilon_R = 26.78\ \text{V}$$

c. The current leaves the (+) terminal of the capacitor. The terminal polarities are shown in Figure 12.10.

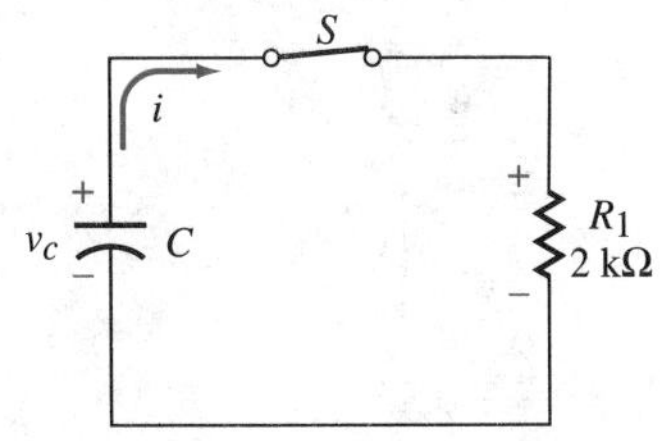

Figure 12.10

EXAMPLE 12.11

The switch in Figure 12.11 was placed in Position 1 for 50 ms, then moved to Position 2. What are the capacitor voltage and the current after 96 ms in Position 2?

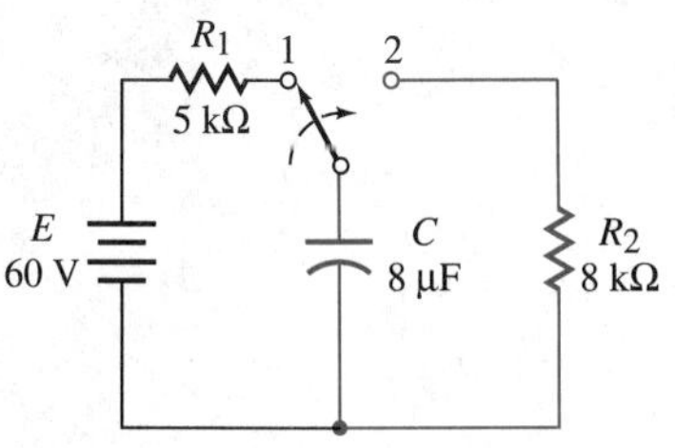

Figure 12.11

Strategy

Solution

The capacitor will discharge in Position 2 so

$$v_c = V_o e^{-t/(RC)}$$

V_o must be obtained from the charging circuit. For it,

Find the charging time constant.

$$\tau = R_1 C$$
$$= (5 \times 10^3\ \Omega)(8 \times 10^{-6}\ \text{F})$$
$$\tau = 40\ \text{ms}$$

Check if the capacitor is charged to E.

$5\ \tau = 200$ ms, so v_c has not reached 60 V.

Find the voltage at the start of discharge.

$$v_c = E(1 - e^{-t/RC})$$
$$v_c = 60\ \text{V}\ (1 - e^{-50\ \text{ms}/40\ \text{ms}})$$
$$v_c = 42.81\ \text{V}$$

Returning to the discharge,

Find the discharge time constant.

$$\tau = R_2 C$$
$$= (8 \times 10^3\ \Omega)(8 \times 10^{-6}\ \text{F})$$
$$\tau = 64\ \text{ms}$$

Find the voltage.

$$v_c = V_o e^{-t/\tau}$$
$$= (42.81\ \text{V})(e^{(-96\ \text{ms}/64\ \text{ms})})$$
$$v_c = 9.55\ \text{V}$$

Use Ohm's Law to find the curent.

$$i_c = v_c / R_2$$
$$= 9.55\ \text{V}/8 \times 10^3\ \Omega$$
$$i_c = 1.19\ \text{mA}$$

Using Equation 12.10 makes it possible to analyze the capacitor charging with an initial voltage by another method. The circuit now can be treated as two parts: the charging of a zero-volt capacitor from the source, and the discharging of the capacitor from its initial voltage. The voltages and currents are then combined using superposition to yield the total voltage and current. Example 12.8 is now analyzed using this procedure.

EXAMPLE 12.12

Work Example 12.8 using the discharge equation.

Solution

The original circuit was shown in Figure 12.5. That can be broken down into the two circuits in Figure 12.12.

For the charging in Figure 12.12(a),

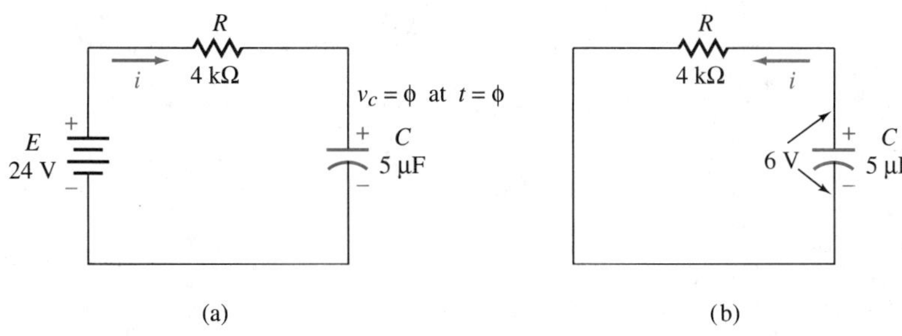

Figure 12.12

$$v_{c1} = E(1 - e^{-t/\tau})$$
$$\tau = RC = (4 \times 10^3\ \Omega)(5 \times 10^{-6}\ \text{F}) = 20\ \text{ms}$$

then

$$v_{c1} = 24\ \text{V}\ (1 - e^{-10\ \text{ms}/20\ \text{ms}})$$
$$v_{c1} = 9.44\ \text{V}$$

For the discharge in Figure 12.12(b),

$$v_{c2} = V_o e^{-t/\tau}$$
$$v_{c2} = 6\ \text{V}\ e^{-10\ \text{ms}/20\ \text{ms}}$$
$$v_{c2} = 3.64\ \text{V}$$

Since both voltages have the same terminal polarity,

$$v_c = v_{1C} + v_{2C}$$
$$= 9.44\ \text{V} + 3.64\ \text{V}$$
$$v_c = 13.08\ \text{V at } t = 10\ \text{ms}$$

Capacitance can present a problem when pulse signals are applied. For example, the rectangular pulse of Figure 12.13(b) can be changed to either the solid wave or the dashed one of Figure 12.13(c). The dashed curve results if the pulse width is less than 5 τ. This change in shape and amplitude can result in faulty operation. To see how the change resulted, we examine the circuit.

The resistance represents the lead resistance and the capacitance represents the shunt capacitance between the leads. Although the time constant should be very small, it can still be large compared to the narrow pulse signals used.

Moving the switch between Position 1 and Position 2 in Figure 12.13(a) generates the pulse in Figure 12.13(b). It t_1 and t_2 are at least five time constants, the output voltage will be the solid line wave in Figure 12.13(c). The square wave has been rounded. If t_1 and t_2 are less than 5 τ, the output will be that of the dashed line. Here, the capacitance cannot charge to the full value of the pulse. Either of these outputs can affect the circuit's operation. The current has the shape shown in Figure 12.13(d).

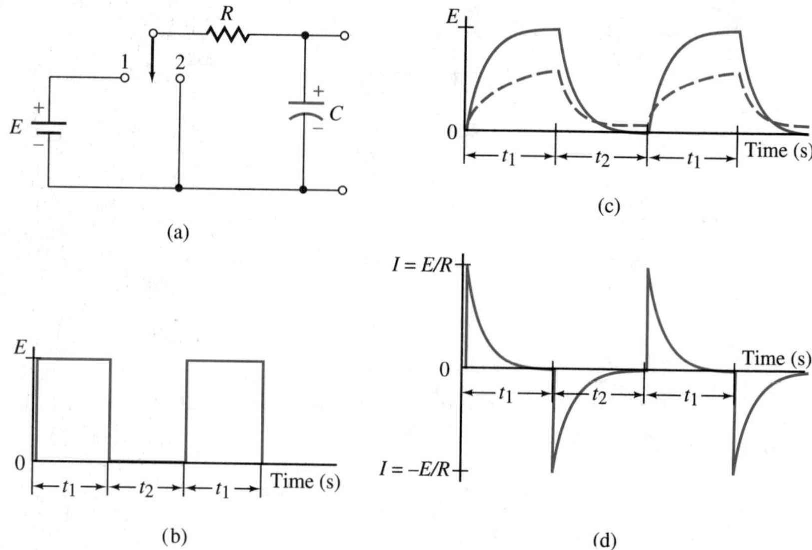

Figure 12.13 The effect of capacitance on the shape of a rectangular pulse. (a) Moving the switch between 1 and 2 generates a rectangular pulse. (b) The pulse that is applied to the R–C circuit. (c) The instantaneous rise and fall are changed to an exponential rise and fall by the capacitor. (d) The current rises instantaneously, but falls exponentially.

The change in waveform becomes critical in measuring instruments such as the oscilloscope. The trace on the display might not be what is in the circuit.

Summarizing the key quantities in the discharge circuit and the quantities that affect each of them, we have:

Quantity	Affected by
Initial voltage	Charging circuit and the length of charging time
Initial current	Initial voltage and the resistance in the discharge circuit
Initial rate of change	Initial voltage, resistance, and capacitance in the discharge circuit
Time constant	Resistance and capacitance in the discharge path

12.3 The Universal Time Constant Curve

The exponential curves of Figure 12.1 provide an alternate method of analyzing the R–C circuit. Values can be found quickly without any

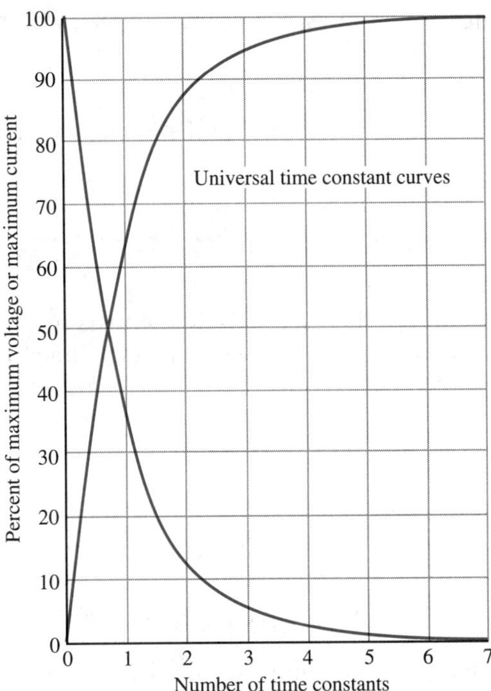

Figure 12.14 The Universal Time Constant Curves plot percent of maximums versus number of time constants and can be used for any circuit values.

exponential mathematics. Also, the curves show the trend at any point on the curve.

On the other hand, any change in circuit values requires the drawing of another curve. Fortunately, that is not necessary if the Universal Time Constant Curves are used. These curves plot the percent of maximums on the *Y*-axis and the number of time constants on the *X*-axis. Such curves are shown in Figure 12.14 and are used in the following manner.

Charging: Use the rising curve for v_c, the falling curve for i_c.

Discharging: Use the falling curve for v_c and i_c.

Finding v_c and i_c:

a. Calculate the number of time constants.
b. At the number of time constants, draw a vertical line to intersect the curve.
c. At the intersection, draw a horizontal line to intersect the "percent of maximum" axis.
d. Read the percent of maximum at the intersection.
e. Calculate the value of v_c and i_c.

Finding R, C, or t:

a. Calculate the percent of maximum for the given v_c and i_c.
b. At the percent of maximum, draw a line across to intersect the curve.

c. At the intersection, draw a line down to intersect the "number of time constants" axis.
d. Calculate the value of R, C, or t.

A few examples illustrate the use of the curves in analyzing the R–C circuit.

EXAMPLE 12.13

A 5-microfarad capacitor and a 3-kilohm resistor are connected in series across a 24-volt source. Use the Universal Time Constant Curves to find the capacitor voltage and current at 30 ms after they are connected.

Strategy
Find the number of time constants.

Solution

$$\tau = RC$$
$$= (3000\ \Omega)(5 \times 10^{-6}\ \text{F})$$
$$\tau = 15\ \text{ms}$$
$$30\ \text{ms}/15\ \text{ms} = 2$$

This is a voltage rise condition so the rising curve must be used.

Determine the percent of maximum.

A vertical line drawn at two time constants crosses the curve at about 86.5 percent.

Determine the voltage.

The maximum voltage is 24 volts, so

$$v_C\ @\ 30\ \text{ms} = (0.865)\ (24\ \text{V})$$
$$v_C = 20.76\ \text{V}$$

For the current, the falling curve must be used. A vertical line drawn at two time constants crosses the curve at about 13.5 percent.

The maximum current is E/R.

Determine the maximum current.

$$i_{max} = 24\ \text{V}/3000\ \Omega = 8\ \text{mA}$$

so

Determine the current.

$$i\ @\ 30\ \text{ms} = (0.135)(8\ \text{mA})$$
$$i = 1.08\ \text{mA}$$

EXAMPLE 12.14

The voltage across C in Figure 12.15 is 60 V when the switch is put into Position 2. Use the Universal Time Constant Curve to find:

a. the length of time that the switch was in Position 1.
b. the capacitor voltage and current at 180 ms after the switch was put into Position 2.

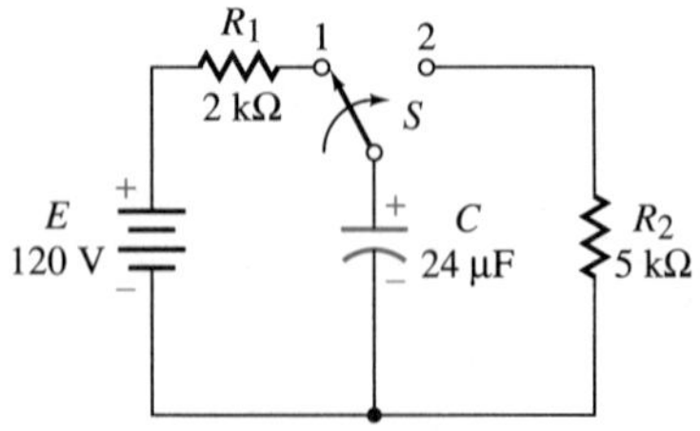

Figure 12.15

Solution

a. This is a charging condition and the maximum voltage is 120 volts.

$$60\ \text{V}/120\ \text{V} = 0.5\ \text{or}\ 50\%$$

A vertical line drawn to cross the rising curve at 50 percent crosses the Universal Time Constant axis at about 0.7 time constants.

$$\tau = R_1C$$
$$= (2000\ \Omega)(24 \times 10^{-6}\ \text{F})$$
$$\tau = 48\ \text{ms}$$
$$t = (0.7)(48\ \text{ms})$$
$$t = 33.6\ \text{ms}$$

This compares to 33.27 ms, calculated analytically.

b. When the switch is in Position 2, the capacitor will discharge. The falling curve must be used. For the discharge,

$$\tau = R_2C$$
$$= (5000\ \Omega)(24 \times 10^{-6}\ \text{F})$$
$$\tau = 120\ \text{ms}$$

180 ms/120 ms = 1.5 time constants

A vertical line drawn at 1.5 time constants crosses the falling curve at about 22 percent.

$$i = 0.22\ I_{max}$$
$$= (0.22)(60\ \text{V}/5000\ \Omega)$$
$$i = 2.64\ \text{mA}$$

This compares to 2.68 mA calculated analytically.

$$v_C = 0.22\ V_{max}$$
$$= (0.22)(60\ \text{V})$$
$$v_C = 13.2\ \text{V}$$

The calculated value is 13.39 V.

12.4 The Analysis of Complex $R-C$ Circuits

Capacitors and resistors are not always in simple series $R-C$ circuits. One need only look at the schematic of a transistor amplifier, power supply filter, or silicon-controlled rectifier circuit for other configurations. A simpler example is shown in Figure 12.16. The charging circuit in Figure 12.16(a) has resistors R_1 and R_2. The discharging circuit in Figure 12.16(b) has resistors R_3 and R_4. Also, $R-C$ circuits can have several sources in them. This section deals with how we analyze these types of circuits.

These circuits are analyzed by changing them to simple series circuits. During charging, the series circuit is obtained by Theveninizing the circuit that is connected to the capacitor. During discharge, the series circuit is obtained by replacing the original circuit connected to the capacitor by its equivalent resistance. Then the capacitor voltage and

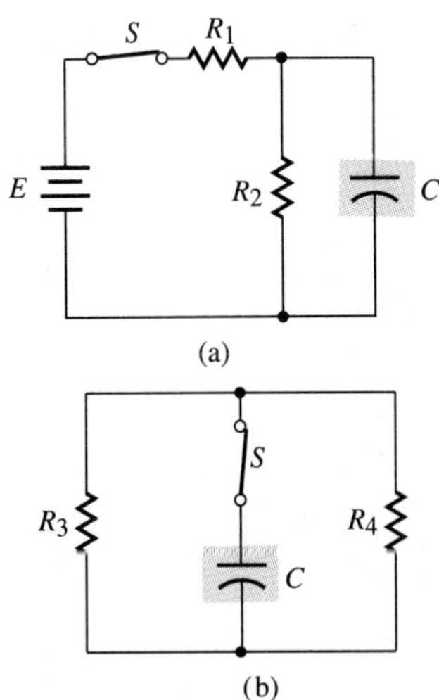

Figure 12.16 Two examples of complex $R-C$ circuits. (a) A charging circuit. (b) A discharging circuit.

current can be found using the material in the Sections 12.1, 12.2, and 12.3. Finally, the basic circuit relations such as Kirchhoff's Voltage Law, Kirchhoff's Current Law, and Ohm's law are used to find other quantities.

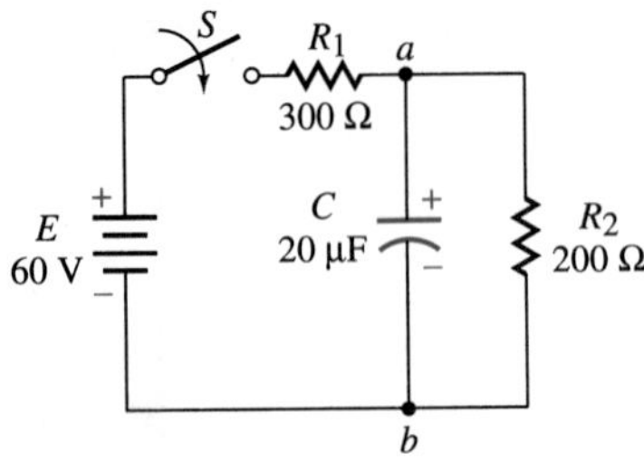

Figure 12.17

EXAMPLE 12.15

Find:

a. the voltage across the capacitor.
b. the current in R_2.
c. the source current at 1.2 ms after the switch is closed in the circuit in Figure 12.17.

Strategy

Solution

Theveninize the circuit.

a. The capacitor is charging, so the circuit connected to it should be Theveninized. For R_{TH}, the circuit is shown in Figure 12.18.

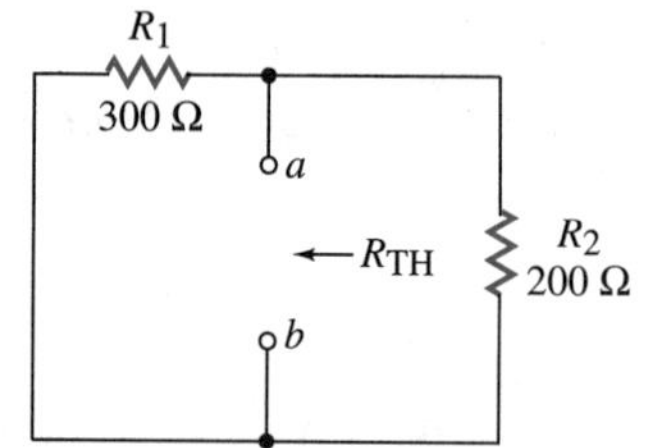

Figure 12.18

$$R_{TH} = R_1 \| R_2$$
$$= \frac{(300\ \Omega)(200\ \Omega)}{300\ \Omega + 200\ \Omega}$$
$$R_{TH} = 120\ \Omega$$

For E_{TH}, the circuit is shown in Figure 12.19.

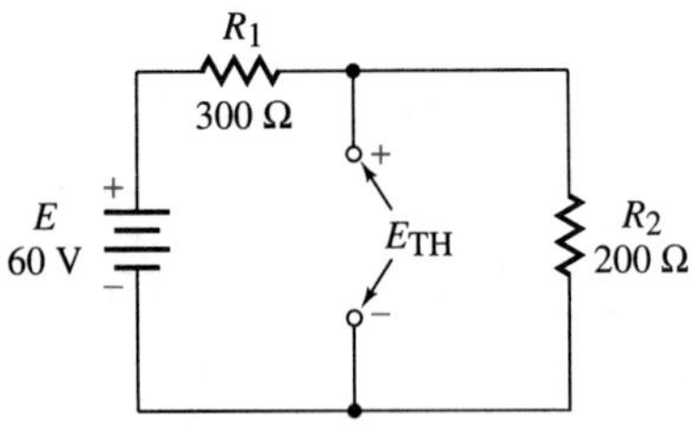

Figure 12.19

$$E_{TH} = V_{ab}$$
$$= \left(\frac{R_2}{R_1 + R_2}\right)E$$
$$= \left(\frac{200\ \Omega}{300\ \Omega + 200\ \Omega}\right)60\ \text{V}$$
$$E_{TH} = 24\ \text{V}$$

The Theveninized circuit is shown in Figure 12.20.

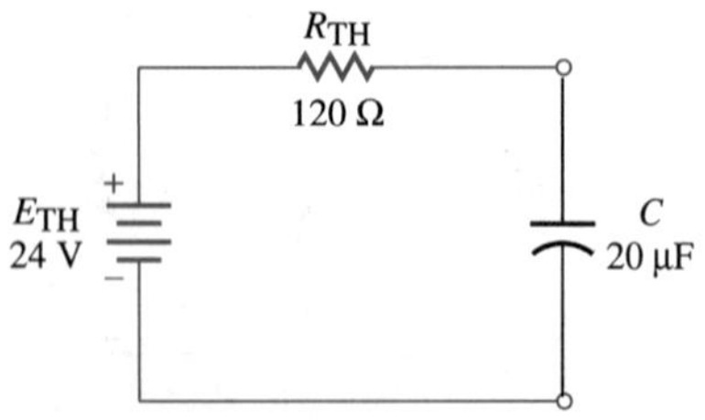

Figure 12.20

$$\tau = R_{TH}C$$
$$= (120\ \Omega)(20 \times 10^{-6}\ \text{F})$$
$$\tau = 2.4\ \text{ms}$$
$$\upsilon_c = E_{TH}(1 - e^{-t/\tau})$$
$$= 24\ \text{V}(1 - e^{-1.2\ \text{ms}/2.4\ \text{ms}})$$
$$= 24\ \text{V}(1 - e^{-0.5})$$
$$\upsilon_c = 9.44\ \text{V}$$

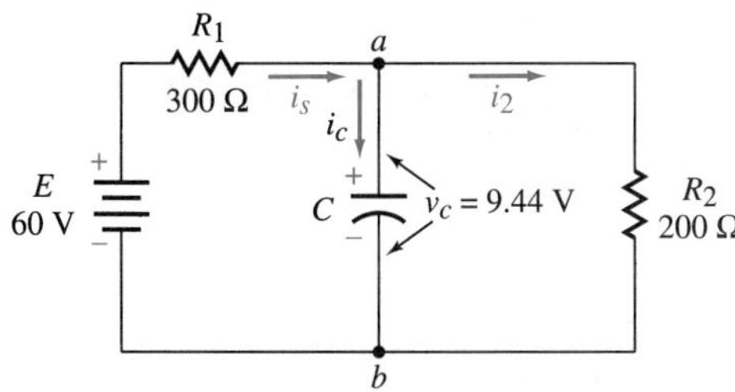

Figure 12.21

b. Putting this into the original circuit, as in Figure 12.21, gives

$$i_2 = v_2/R_2$$

Use Ohm's Law to find i_2.

but

$$v_2 = v_c,$$

so

$$i_2 = v_c/R_2$$
$$i_2 = 9.44\ \text{V}/200\ \Omega$$
$$i_2 = 47.2\ \text{mA}$$

c. By Kirchhoff's Current Law,

$$i_s = i_c + i_2$$
$$i_c = (E_{TH}/R_{TH})\ e^{-t/\tau}$$
$$= (24\ \text{V}/120\ \Omega)\ e^{-0.5}$$
$$i_c = 0.121\ \text{A or } 121\ \text{mA}$$

Relate the known currents by Kirchhoff's Current Law. Find i_c.

so

$$i_s = 47.2\ \text{mA} + 121\ \text{mA}$$
$$i_s = 168.2\ \text{mA}$$

Find I_s.

EXAMPLE 12.16

The capacitor in the circuit in Figure 12.22 is charged to 12 volts when the switch is opened. After it is open for 0.2 ms, find:

a. the capacitor voltage.
b. the capacitor current.
c. the current in R_2.

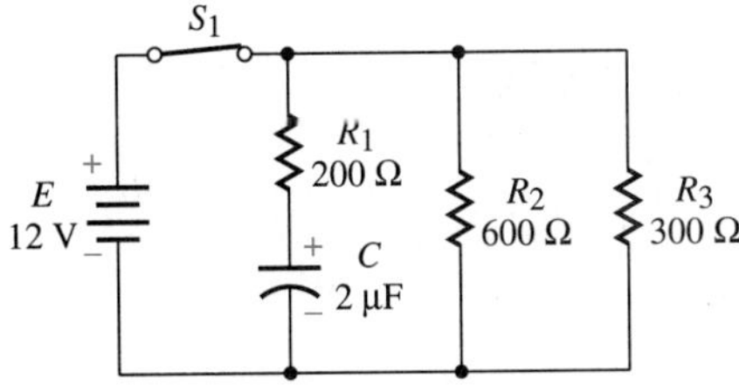

Figure 12.22

Solution

a. With the switch open, the capacitor will discharge. The discharge circuit is shown in Figure 12.23.

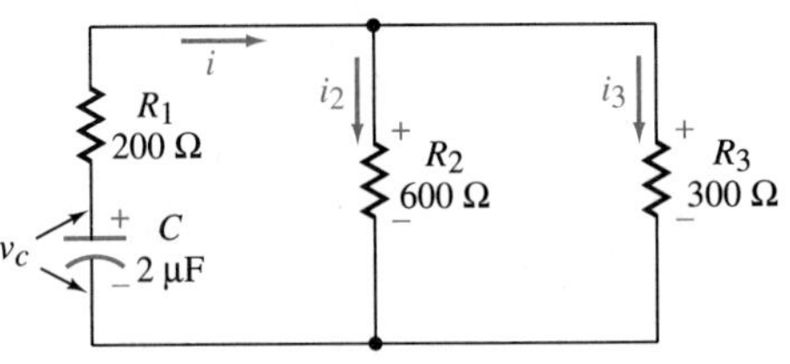

Figure 12.23

The equivalent resistance of the circuit connected to C is:

$$\begin{aligned} R_{eq} &= R_1 + R_2 \| R_3 \\ &= 200\ \Omega + \frac{(600\ \Omega)(300\ \Omega)}{600\ \Omega + 300\ \Omega} \\ &= 200\ \Omega + 200\ \Omega \\ &= 400\ \Omega \end{aligned}$$

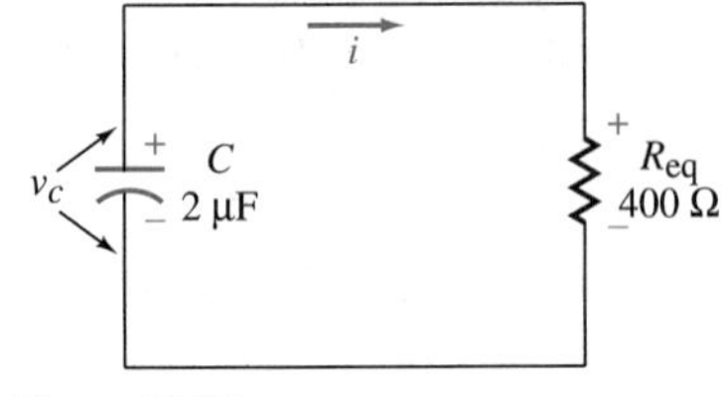

Figure 12.24

The equivalent circuit is shown in Figure 12.24. For the circuit,

$$\begin{aligned} \tau &= R_{eq}C \\ &= (400\ \Omega)(2 \times 10^{-6}\ \text{F}) \\ \tau &= 0.8\ \text{ms} \\ v_c &= V_o e^{-t/RC} \\ &= V_o e^{-0.2\ \text{ms}/0.8\ \text{ms}} \\ &= 12\ \text{V}\ e^{-0.25} \\ v_c &= 9.35\ \text{V} \end{aligned}$$

b.
$$\begin{aligned} i_c &= (V_o/R_{eq})\ e^{-t/\tau} \\ &= (12\ \text{V}/400\ \Omega)e^{-0.25} \\ &= (0.03\ \text{A})(0.779) \\ i_c &= 23.36\ \text{mA} \end{aligned}$$

c. By the current-divider rule,

$$\begin{aligned} i_2 &= \left(\frac{R_3}{R_2 + R_3}\right)i_c \\ &= \left(\frac{300\ \Omega}{600\ \Omega + 300\ \Omega}\right)(23.36\ \text{mA}) \\ i_2 &= 7.79\ \text{mA} \end{aligned}$$

EXAMPLE 12.17

What is the voltage on the capacitor and the source current for E_1 at 8 ms after the switch is closed in Figure 12.25?

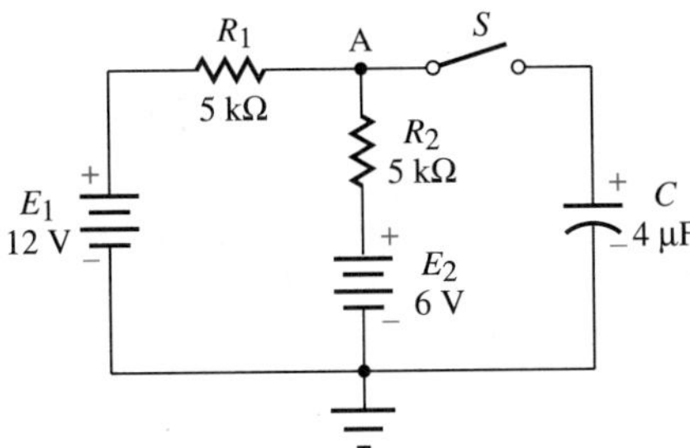

Figure 12.25

Solution

With the switch closed, the circuit is a charging circuit. Its Thevenin equivalent must be found. The circuits for R_{TH} and E_{TH} are shown in Figure 12.26.

$$R_{TH} = R_1 \| R_2$$

$$= \frac{(5000\ \Omega)(5000\ \Omega)}{5000\ \Omega + 5000\ \Omega}$$

$$R_{TH} = 2500\ \Omega$$

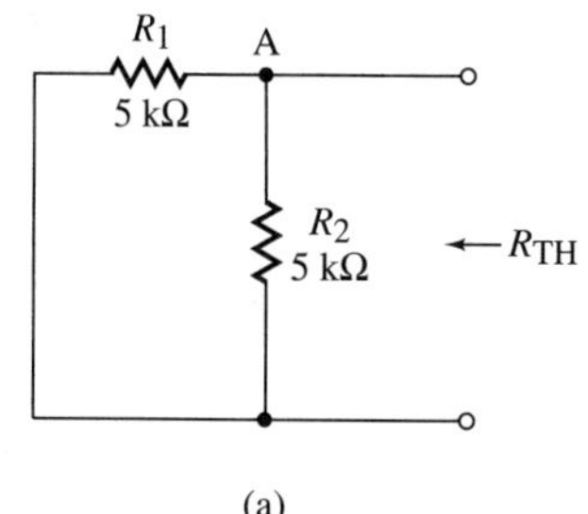

E_{TH} is the same as V_A.
Using Nodal Analysis gives

$$V_A\left(\frac{1}{R_1} + \frac{1}{R_2}\right) = E_1\left(\frac{1}{R_1}\right) + E_2\left(\frac{1}{R_2}\right)$$

$$V_A\left(\frac{1}{5000\ \Omega} + \frac{1}{5000\ \Omega}\right) = 12\ \text{V}\left(\frac{1}{5000\ \Omega}\right) + 6\ \text{V}\left(\frac{1}{5000\ \Omega}\right)$$

or

$$2\ V_A = 18\ \text{V}, \qquad V_A = 9\ \text{V}$$

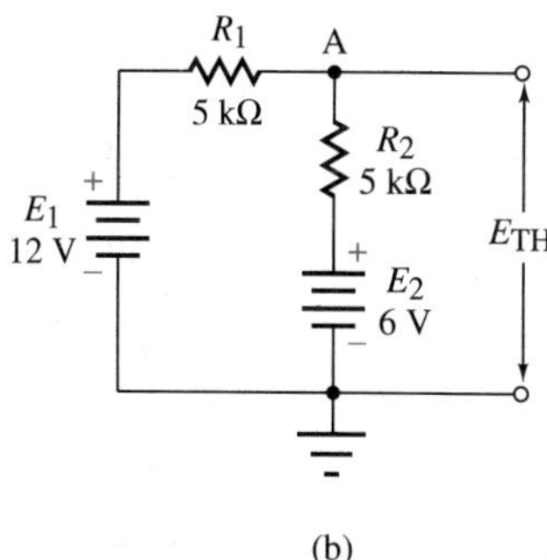

Figure 12.26

The capacitor is now connected to the Thevenin equivalent circuit as in Figure 12.27.

$$\tau = R_{TH}C = (2500\ \Omega)(4 \times 10^{-3}\ \text{F}) = 10\ \text{ms}$$

At $t = 8$ ms,

$$\upsilon_c = E_{TH}(1 - e^{-t/\tau})$$

$$= 9\ \text{V}(1 - e^{-8\ \text{ms}/10\ \text{ms}})$$

$$\upsilon_c = 4.96\ \text{V}$$

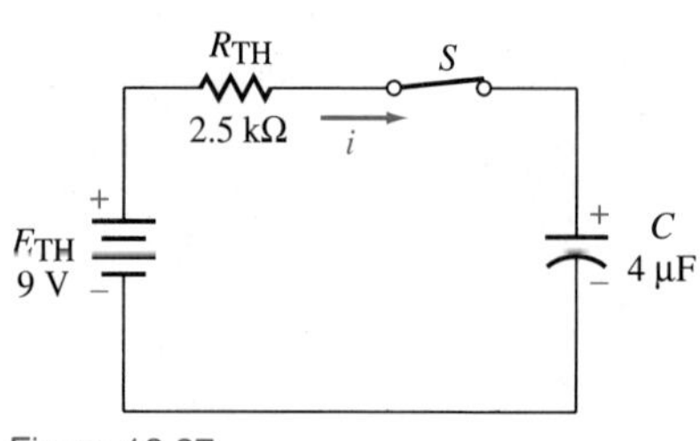

Figure 12.27

Looking at the original circuit, you can see that the voltage on one terminal of R_1, measured from ground, is v_c, while the voltage of the other terminal is 12 volts. Using these voltages in Ohm's Law for R_1 gives

$$E_1 - v_c = i_s R_1$$

or

$$\begin{aligned} i_s &= (E_1 - v_c)/R_1 \\ &= (12\text{ V} - 4.96\text{ V})/5000\ \Omega \\ i_s &= 1.41\text{ mA} \end{aligned}$$

Another solution is to find i_c, then i_2 using V_A, and applying Kirchhoff's Current Law at Node A.

12.5 Energy Stored in the Capacitor

Short circuiting the terminals of a charged capacitor creates a spark. The spark is caused by the release of the energy that was stored in the electrostatic field. During charging, the power builds up to a peak and then drops back down to zero, as shown in Figure 12.28. The curves are

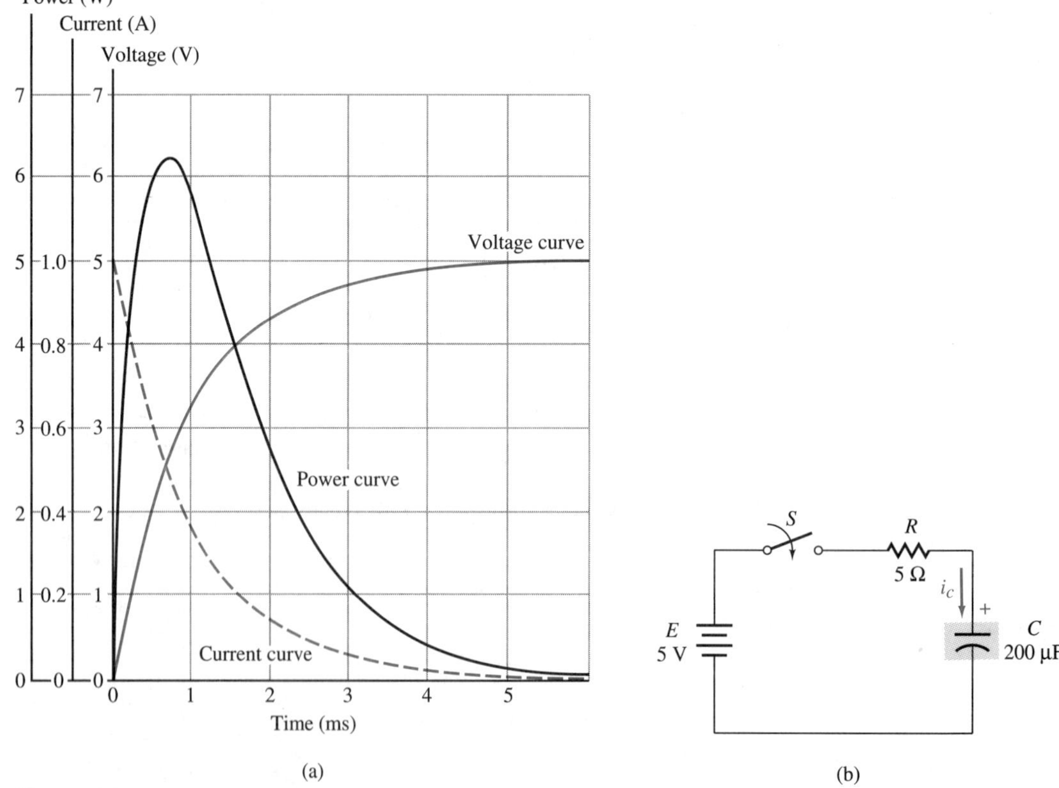

Figure 12.28 Energy is stored in the capacitor during the charging. (a) The stored energy is the area under the power curve. (b) The circuit used to obtain the values for the curve.

for the circuit in Figure 12.28(b). The power at any instant of time is given by $p = v i$. Since energy is the product of power and time, the energy for a small amount of time (Δt) is $\Delta \mathcal{W} = p\Delta t$ or $\Delta \mathcal{W} = v_c i \, \Delta t$. This is the area under that part of the power curve. The stored energy will then be the area under the power curve. Using calculus to get the area under the curve gives

$$\mathcal{W} = (1/2)CV^2 \qquad \textbf{(12.13)}$$

where $\mathcal{W}$ is the energy, in joules; C is the capacitance, in farads; and V is the capacitor voltage, in volts.

Equation 12.13 indicates that the amount of energy depends on C and V. However, for applications of energy such as electronic flashes, both the amount of energy and the charged voltage are critical. If the voltage is not high enough, there will not be any flash even though the amount of energy is large.

EXAMPLE 12.18

How much energy is stored in a 6-microfarad capacitor when it is charged to 120 volts?

Solution

$$\mathcal{W} = (1/2)CV^2$$
$$= (1/2)(6 \times 10^{-6} \text{ F})(120 \text{ V})^2$$
$$\mathcal{W} = 0.043 \text{ J}$$

The algebraic and RPN calculator keystrokes for the solution are:

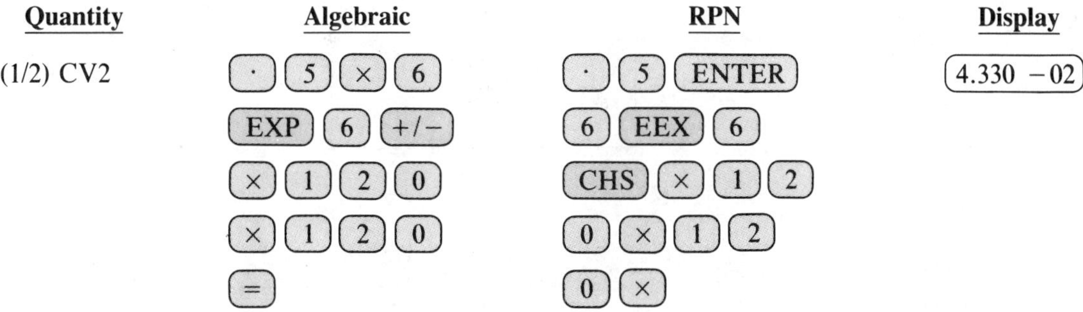

Quantity	Algebraic	RPN	Display
(1/2) CV2	· 5 × 6	· 5 ENTER	4.330 −02
	EXP 6 +/−	6 EEX 6	
	× 1 2 0	CHS × 1 2	
	× 1 2 0	0 × 1 2	
	=	0 ×	

Sometimes the capacitor voltage does not start from 0 volts, but changes from one value to another. The energy increases or decreases, depending on the direction of the change. The change in energy is the difference between the energy at each voltage.

EXAMPLE 12.19

What is the energy change in a 200-microfarad capacitor when the voltage is increased from 30 V to 60 V? Is it an increase or a decrease?

Solution

$$\mathcal{W}_1 = (1/2)CV_1^2$$
$$= (1/2)(200 \times 10^{-6}\ \text{F})(30\ \text{V})^2$$
$$\mathcal{W}_1 = 0.09\ \text{J}$$
$$\mathcal{W}_2 = (1/2)CV_2^2$$
$$= (1/2)(200 \times 10^{-6}\ \text{F})(60\ \text{V})^2$$
$$\mathcal{W}_2 = 0.36\ \text{J}$$

so

$$\mathcal{W}_2 - \mathcal{W}_1 = 0.36\ \text{J} - 0.09\ \text{J}$$

and

$$\mathcal{W} = 0.27\ \text{J}$$

It is an increase because the energy is greater at 60 V.

12.6 The *R–C* Circuit in Steady State

steady state

When the switch in the circuit of Figure 12.29(a) is closed, the capacitors start charging. After some time, each of them is fully charged. The circuit is now in **steady state**. This is a state in which there are no changes in circuit currents and voltages. The capacitors that had appeared as short circuits at the start of the charge now appear as open circuits.

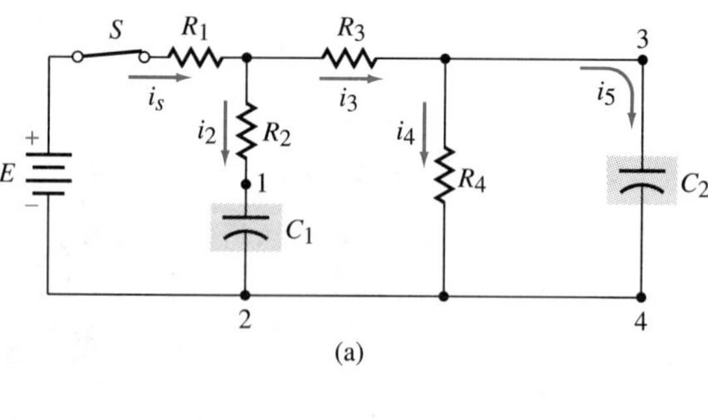

(a)

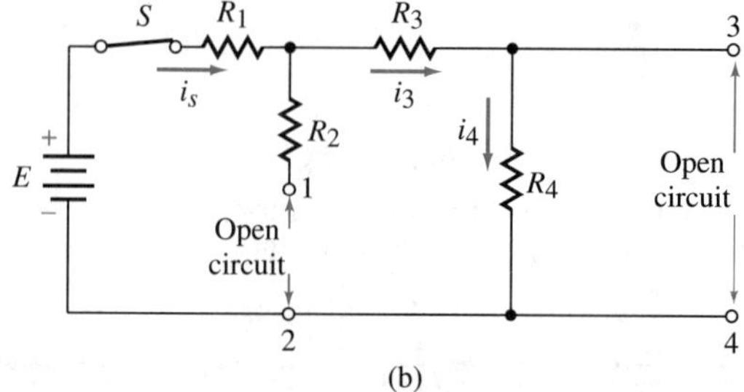

(b)

Figure 12.29 During steady state, the capacitor is an open circuit. (a) A circuit in which the capacitors have reached steady state. (b) The circuit drawn to show the capacitors as open circuits.

The capacitor acting as an open circuit for a constant voltage is used when measuring the ac component of a voltage. It permits the ac part to pass through, but blocks the dc or steady part.

When analyzing the steady-state condition, we must replace the capacitors by their steady-state equivalent, that is, by open circuits, as shown in Figure 12.29(b). This equivalent circuit can then be analyzed using the basic circuit theory.

A few steady-state circuits are now analyzed.

EXAMPLE 12.20

The circuit shown in Figure 12.30 has reached steady state. Find the voltage across each resistor and capacitor and the current in each branch.

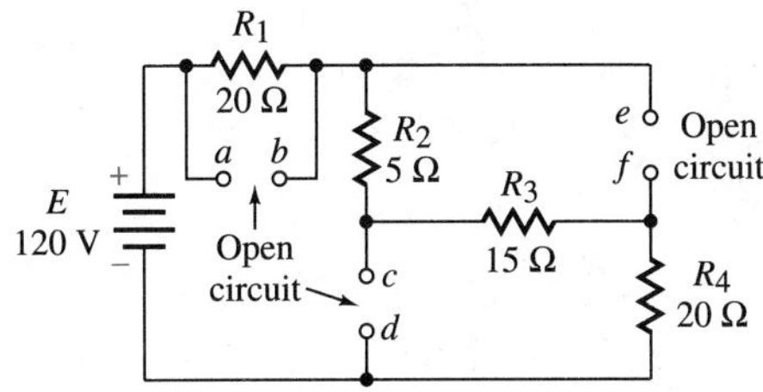

Figure 12.30

Solution

The steady-state equivalent is shown in Figure 12.31.

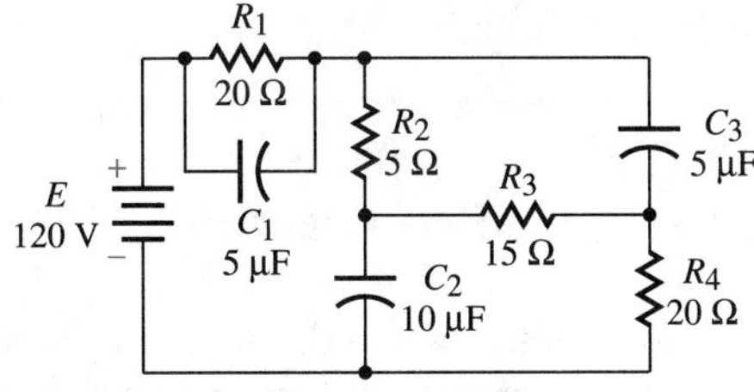

Figure 12.31

The current path is through R_1, R_2, R_3, R_4, and back to the source.

$$\begin{aligned} I &= E/R_{eq} = E/(R_1 + R_2 + R_3 + R_4) \\ &= 120 \text{ V}/(20\ \Omega + 5\ \Omega + 15\ \Omega + 20\ \Omega) \\ &= 2 \text{ A} \end{aligned}$$

$$I = I_1 = I_2 = I_3 = I_4 = 2 \text{ A}$$

$$V_1 = IR_1 = (2 \text{ A})(20\ \Omega), \qquad V_1 = 40 \text{ V}$$

$$V_2 = IR_2 = (2 \text{ A})(5\ \Omega), \qquad V_2 = 10 \text{ V}$$

$$V_3 = IR_3 = (2 \text{ A})(15\ \Omega), \qquad V_3 = 30 \text{ V}$$

$$V_4 = IR_4 = (2 \text{ A})(20\ \Omega), \qquad V_4 = 40 \text{ V}$$

$$V_{c1} = V_1, \qquad V_{c1} = 40\text{ V}$$
$$V_{c2} = V_3 + V_4 = 30\text{ V} + 40\text{ V}, \qquad V_{c2} = 70\text{ V}$$
$$V_{c3} = V_2 + V_3 = 10\text{ V} + 30\text{ V}, \qquad V_{c3} = 40\text{ V}$$

EXAMPLE 12.21

The circuit in Figure 12.32 is in the steady-state condition. Find:
a. the voltage and current for each element.
b. the charge on the capacitor.

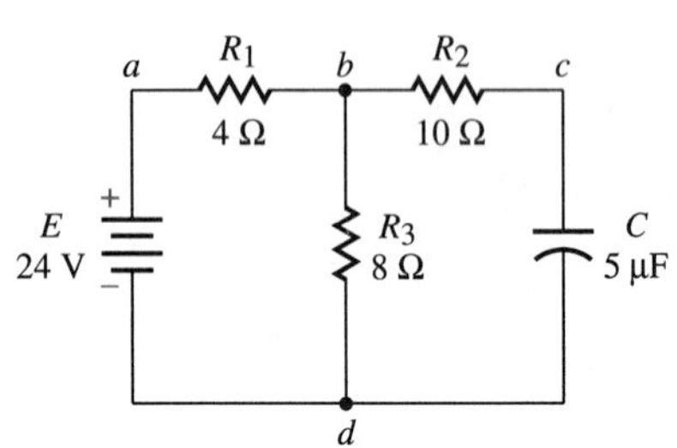

Figure 12.32

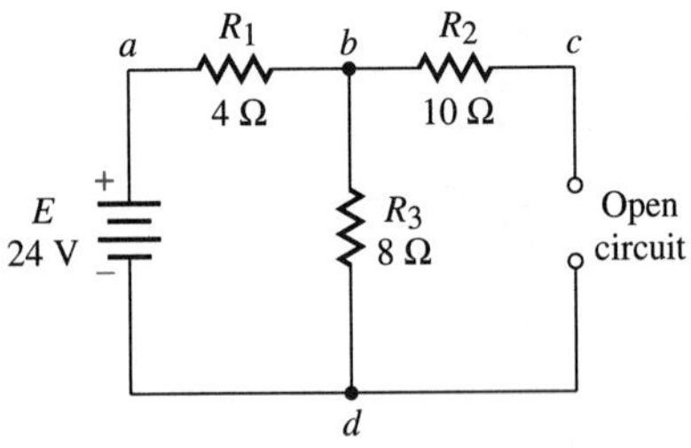

Figure 12.33

Solution

a. The steady-state equivalent is shown in Figure 12.33. The circuit is open, so

$$I_2 = 0$$
$$V_2 = I_2R_2, \qquad V_2 = 0$$
$$I_1 = I_3 = E/(R_1 + R_3)$$
$$= 24\text{ V}/(4\ \Omega + 8\ \Omega)$$
$$I_1 = I_3 = 2\text{ A}$$
$$V_1 = I_1R_1 = (2\text{ A})(4\ \Omega), \qquad V_1 = 8\text{ V}$$
$$V_3 = I_3R_3 = (2\text{ A})(8\ \Omega), \qquad V_3 = 16\text{ V}$$

Since $V_2 = 0$, υ_{cd} must equal V_{bd}, or $V_{cd} = 16$ V. Therefore,

$$\upsilon_c = 16\text{ V}$$
$$Q_c = V_cC$$
$$= (16\text{ V})(5 \times 10^{-6}\text{ F})$$
$$Q_c = 80 \times 10^{-6}\text{ coulombs}$$

S 12.7 SPICE Analysis of R–C Circuits

PSpice can be used to obtain a simulation of the transient response or behavior of *R–C* circuits. The PSpice output is a listing or plot of the voltage and/or current values for any component as a function of time.

PSpice analyzes the transient behavior by computing the value of voltage or current at specific instants of time over a total time period that the user specifies. Insight into the behavior of the circuit and the duration of the transient is needed to choose an appropriate time interval for the simulation. If too short a time interval is specified, output data for the total duration of the transient will not be obtained. On the other hand, if too long a time interval is selected, the transient will be such a small percentage of the total time that the resolution needed will not be obtained.

To use PSpice to simulate the transient behavior of an *R–C* circuit, two things need to be done:

1. The circuit must be described in the same manner we described the interconnection of elements in a purely resistive network.
2. The .TRAN, .PRINT, and .PLOT commands must be used to obtained the desired simulation and results.

To describe a capacitor, the following general form must be used:

```
C<name> <(+)node> <(-)node> [(model)name] <value>
[IC=<(initial)value>]
```

This general form can be shortened to

```
C<name> <(+)node> <(-)node> <value>
```

where `C<name>` is the name of the capacitive element, e.g., Cl, `<(+)node>` is the (+) node identifier, `<(-)node>` is the (−) node identifier, and `<value>` is the component value in farads.

Note: The initial capacitor voltage is assumed to be 0 volts.

EXAMPLE 12.22

Write the PSpice lines to describe R and C in the circuit shown in Figure 12.34.

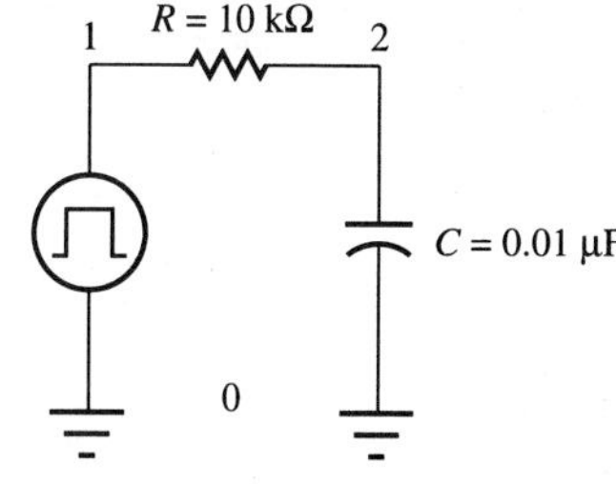

Figure 12.34

Solution

For the resistor,

```
R   1   2   10K
```

For the capacitor,

```
C   2   0   0.01UF
```

Now we need to describe the pulse input signal. The general form used to define a voltage pulse generator is

```
V<name> <(+)node> <(-)node> PULSE (<v1> <v2> <td> <tr>
<tf> <pw> <per>)
```

where `<v1>` is the initial voltage, in volts. No default value. `<v2>` is the pulse voltage, in volts. No default value. `<td>` is the delay time, in seconds. Default value = 0. `<tr>` is the rise time, in seconds. Default value = TSTEP. `<tf>` is the fall time, in seconds. Default value = TSTEP. `<pw>` is the pulse width, in seconds. Default value = TSTOP. `<per>` is the period, in seconds. Default value = TSTOP.

TSTEP and TSTOP are parameters in the .TRAN command that is described in Example 12.24.

The PULSE statement causes the voltage to start at voltage `<v1>` and stay there for `<td>` seconds. Then the voltage goes linearly from voltage `<v1>` to voltage `<v2>` during the next `<tr>` seconds. Then, it goes linearly from voltage <v2> back to voltage <v1> during the next <tf> seconds. It stays at voltage <v1> for the rest of the period, <per> − (<tr> + <pw> + <tf>) seconds, and then the cycle repeats, except for the initial delay of <td> seconds.

EXAMPLE 12.23

Draw the waveform described by the PSpice statement:

```
VS 3 0 PULSE (0 15 0 0.1US 0.1US.5US 2US)
```

Solution

From the statement line, we obtain the following information:

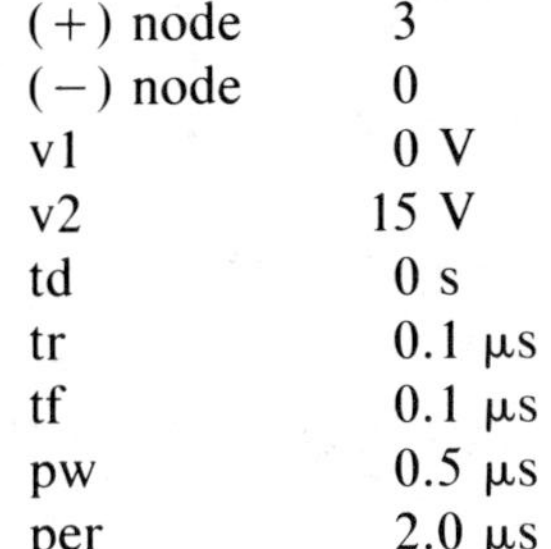

(+) node	3
(−) node	0
v1	0 V
v2	15 V
td	0 s
tr	0.1 μs
tf	0.1 μs
pw	0.5 μs
per	2.0 μs

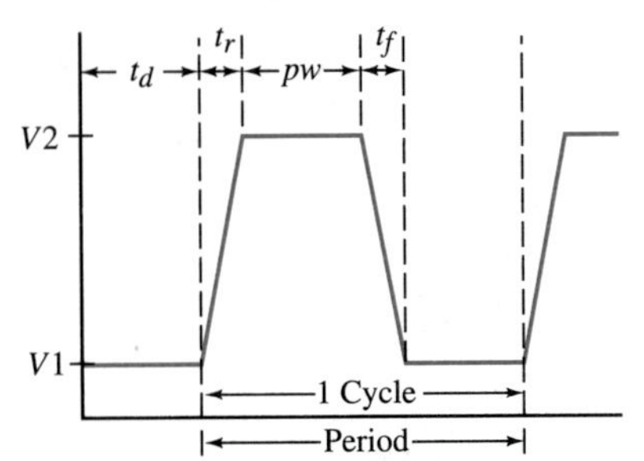

Figure 12.35

Using this information and Figure 12.35, we can graph this waveform. The completed graph is shown in Figure 12.36.

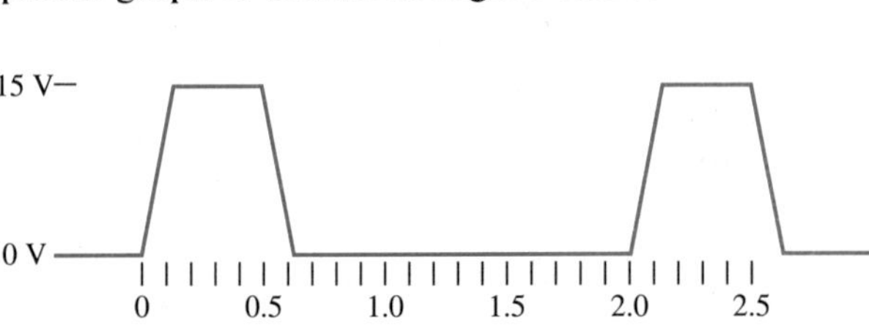

Figure 12.36

EXAMPLE 12.24

Write a PSpice statement to describe the waveform shown in Figure 12.37. Assume the pulse source is connected between Nodes 3 and 2.

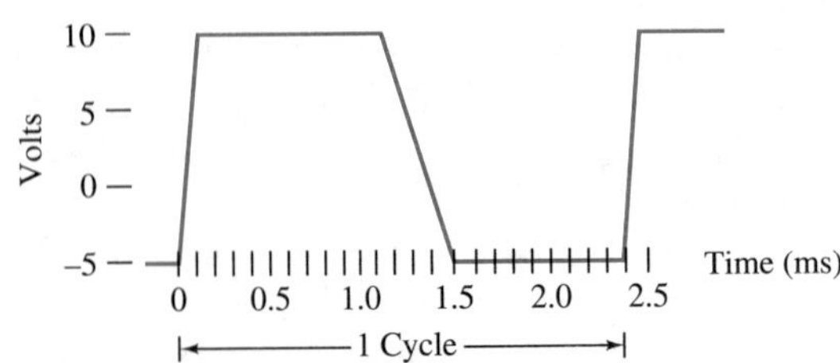

Figure 12.37

Solution

Measuring v1, v2, td, tr, tf, pw, and per on the graph shown in Figure 12.37, we obtain the values shown in Figure 12.38.

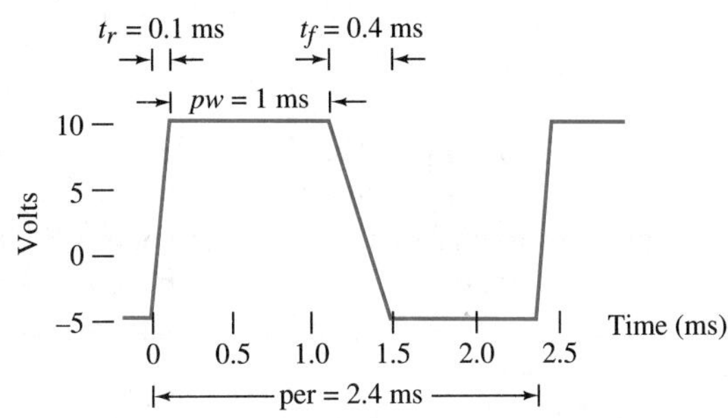

Figure 12.38

Substituting these values into the general format for the PULSE statement yields

```
VS 3 2 PULSE (-5 10 0 0.1MS 0.4MS 1MS 2.4MS)
```

The .TRAN command performs a transient analysis that determines the output variables (current and voltage) as a function of time over a specified time interval. The size of the time step within the interval must be specified. A table of the results may be printed using the .PRINT command and/or the results may be plotted using the .PLOT command.

The basic format for the .TRAN command is

```
.TRAN TSTEP TSTOP
```

where `TSTEP` is the time increment between which values are printed or plotted and `TSTOP` is the total time period over which the transient is computed.

EXAMPLE 12.25

Interpret the PSpice command

```
.TRAN 5MS 50MS
```

Solution

The .TRAN portion calls for an analysis of the transient behavior of the circuit. The 5MS indicates that the step size will be 5 milliseconds. The total time period over which the transient will be computed is 50 milliseconds. Hence, calculations will be performed at the following times: 0 ms, 5 ms, 10 ms, 15 ms, 20 ms, 25 ms, 30 ms, 35 ms, 40 ms, 45 ms, and 50 ms.

The .PLOT command can be used to generate many different types of plots, depending on the needs of the type of analysis being performed. The general format for the .PLOT command is

```
.PLOT PLOTTYPE OUTVAR
```

where `PLOTTYPE` is the type of plot that is being requested and `OUTVAR` is the output variable to be plotted.

Up to eight output variables can be specified in one .PLOT command. However, it is best to restrict the number of variables in one command to two or three. More than this produces a graph that is difficult to read and interpret.

The data for a graph will appear in an output file. The abscissa, or horizontal axis, will be DC volts for DC analysis, frequency for AC analysis, or time for transient analysis.

EXAMPLE 12.26

What does the PSpice command `.PLOT TRAN V(4)` produce?

Solution

This PSpice command plots the data for voltage at Node 4, obtained from a transient analysis, and produces a plot with time on the horizontal axis.

Now, let us analyze some *R–C* circuits.

EXAMPLE 12.27

Use PSpice to compute and print the values of the capacitor voltage v_c in Figure 12.39 after the switch is closed at $t = 0$ seconds.

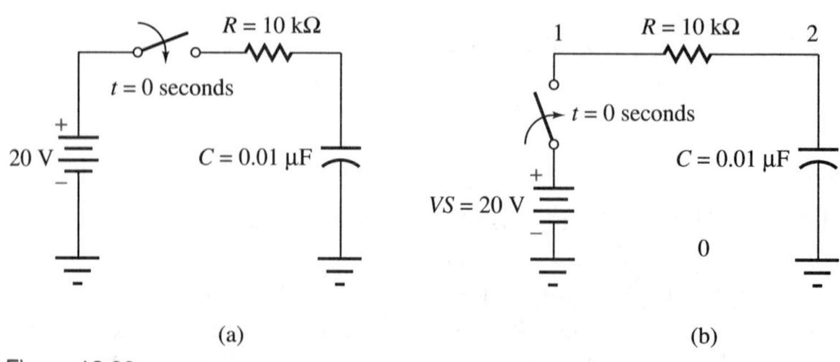

Figure 12.39

Solution

First, let us label the nodes as shown in Figure 12.39(b).

The source file is shown in Figure 12.40.

The output file is shown in Figure 12.41. In this case, only the PRINT command is used.

```
EXAMPLE 11.27
VS     1     0     PULSE (0  20  0  0  0)
R      1     2     10K
C      2     0     0.01UF
.TRAN      25US     0.5MS
.PRINT     TRAN     V(2)
.END
```

Figure 12.40

```
EXAMPLE 11.27
**** TRANSIENT ANALYSIS          TEMPERATURE = 27.000 DEG C
*******************************************************************************
TIME          V(2)
0.000E+00     0.000E+00
2.500E-05     2.296E+00
5.000E-05     6.214E+00
7.500E-05     9.253E+00
1.000E-04     1.164E+01
1.250E-04     1.349E+01
1.500E-04     1.493E+01
1.750E-04     1.605E+01
2.000E-04     1.693E+01
2.250E-04     1.761E+01
2.500E-04     1.814E+01
2.750E-04     1.855E+01
3.000E-04     1.887E+01
3.250E-04     1.912E+01
3.500E-04     1.932E+01
3.750E-04     1.947E+01
4.000E-04     1.959E+01
4.250E-04     1.968E+01
4.500E-04     1.975E+01
4.750E-04     1.980E+01
5.000E-04     1.985E+01
```

Figure 12.41

EXAMPLE 12.28

Modify the source file in Figure 12.40 to obtain a plot of the output data instead of a printout.

Solution

The source file is modified to include a PLOT command. This file is shown in Figure 12.42.

```
EXAMPLE 11.28
VS    1    0    PULSE (0  20  0  0  0)
R     1    2    10K
C     2    3    0.01UF
VM    3    0
.TRAN    25US    0.5MS
.PLOT    TRAN    V(2)
.PLOT    TRAN    I(VM)
.END
```

Figure 12.42

The resulting plots are shown in Figure 12.43(a) and (b).

PSpice is useful tool for analyzing the transient behavior of $R–C$ circuits, and can provide either printed output tables or plots of the output data or both.

```
EXAMPLE 11.28
**** TRANSIENT ANALYSIS          TEMPERATURE = 27.000 DEG C
******************************************************************************
LEGEND:
*: V(2)
  TIME          V(2)
(*)----------       0.0000E+00    5.0000E+00    1.0000E+01    1.5000E+01    2.0000E+
                    - - - - - - - - - - - - - - - - - - - - - - - - - - - - - - - -
 0.000E+00  0.000E+00 *             .             .             .             .
 2.500E-05  2.296E+00 .     *       .             .             .             .
 5.000E-05  6.214E+00 .             .  *          .             .             .
 7.500E-05  9.253E+00 .             .           * .             .             .
 1.000E-04  1.164E+01 .             .             .   *         .             .
 1.250E-04  1.349E+01 .             .             .        *    .             .
 1.500E-04  1.493E+01 .             .             .             *             .
 1.750E-04  1.605E+01 .             .             .             .  *          .
 2.000E-04  1.693E+01 .             .             .             .    *        .
 2.250E-04  1.761E+01 .             .             .             .      *      .
 2.500E-04  1.814E+01 .             .             .             .       *     .
 2.750E-04  1.855E+01 .             .             .             .         *   .
 3.000E-04  1.887E+01 .             .             .             .          *  .
 3.250E-04  1.912E+01 .             .             .             .           * .
 3.500E-04  1.932E+01 .             .             .             .           * .
 3.750E-04  1.947E+01 .             .             .             .            *.
 4.000E-04  1.959E+01 .             .             .             .            *.
 4.250E-04  1.968E+01 .             .             .             .            *.
 4.500E-04  1.975E+01 .             .             .             .            *.
 4.750E-04  1.980E+01 .             .             .             .            *.
 5.000E-04  1.985E+01 .             .             .             .             *
                    - - - - - - - - - - - - - - - - - - - - - - - - - - - - - - - -
```

(a)

```
EXAMPLE 11.28
**** TRANSIENT ANALYSIS          TEMPERATURE = 27.000 DEG C
******************************************************************************
   TIME          I(VM)
(*)----------        0.0000E+00     5.0000E-04     1.0000E-03     1.5000E 03     2.0000E-
                          - - - - - - - - - - - - - - - - - - - - - - - - - - - - - - - -
 0.000E+00   0.000E+00 *               .              .              .              .
 2.500E-05   1.770E-03 .               .              .              .       *      .
 5.000E-05   1.379E-03 .               .              .         *    .              .
 7.500E-05   1.075E-03 .               .              . *            .              .
 1.000E-04   8.357E-04 .               .        *     .              .              .
 1.250E-04   6.515E-04 .               .  *           .              .              .
 1.500E-04   5.066E-04 .               *              .              .              .
 1.750E-04   3.949E-04 .           *   .              .              .              .
 2.000E-04   3.071E-04 .         *     .              .              .              .
 2.250E-04   2.394E-04 .       *       .              .              .              .
 2.500E-04   1.862E-04 .      *        .              .              .              .
 2.750E-04   1.451E-04 .     *         .              .              .              .
 3.000E-04   1.128E-04 .   *           .              .              .              .
 3.250E-04   8.797E-05 .  *            .              .              .              .
 3.500E-04   6.840E-05 .  *            .              .              .              .
 3.750E-04   5.333E-05 .*              .              .              .              .
 4.000E-04   4.147E-05 .*              .              .              .              .
 4.250E-04   3.233E-05 .*              .              .              .              .
 4.500E-04   2.514E-05 .*              .              .              .              .
 4.750E-04   1.960E-05 .*              .              .              .              .
 5.000E-04   1.524E-05 *               .              .              .              .
                          - - - - - - - - - - - - - - - - - - - - - - - - - - - - - - - -
```

(b)

Figure 12.43

EXAMPLE 12.29

Analyze the transient behavior of the circuit shown in Figure 12.44 using PSpice.

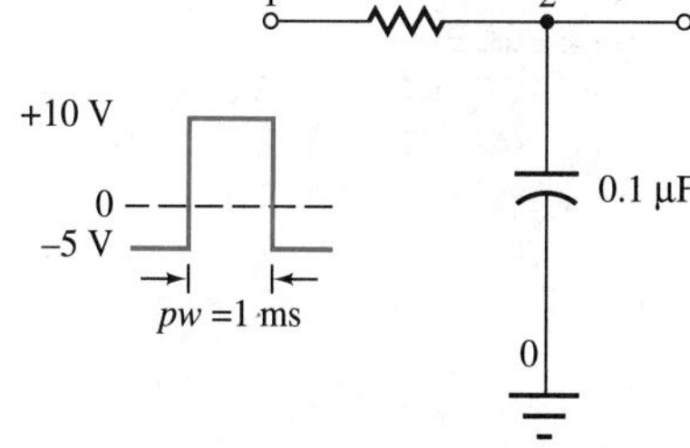

Figure 12.44

Solution

This example shows the effect of an $R-C$ circuit on a pulse input. The circuit time constant is 10 kilohms times 0.1 microfarad or 1 millisecond. Thus, the time constant is equal to the pulse duration. As a result, the capacitor will only charge from -5 volts to a point 63.21% of the way, to $+10$ volts. If we compute this value, the capacitor voltage at the end of the 1 millisecond pulse is 4.45 volts. During the time following the end of the pulse, the capacitor returns to a steady-state level of -5 volts. It takes five time constants to reach this final steady-state value.

The source file and the output are shown in Figure 12.45.

12.8 Applications of the $R-C$ Circuit

$R-C$ circuits have many applications. Each time an electronic flash is used with a camera, an $R-C$ circuit is used. When automatic headlights

```
EXAMPLE 11.29
VS    1    0    PULSE (-5 10 0 0 0 1MS 5MS)
R     1    2    10K
C     2    0    0.1UF
.TRAN     0.2MS    5MS
.PLOT     TRAN     V(2)
.END
```

```
EXAMPLE 11.29
**** TRANSIENT ANALYSIS          TEMPERATURE = 27.000 DEG C
********************************************************************************
  TIME          V(2)
(*)----------      -5.0000E+00    0.0000E+00    5.0000E+00    1.0000E+01    1.5000E+
                            - - - - - - - - - - - - - - - - - - - - - - - - - - - -
 0.000E+00 -5.000E+00 *          .          .          .          .
 2.000E-04 -3.599E+00 .  *       .          .          .          .
 4.000E-04 -1.148E+00 .       *  .          .          .          .
 6.000E-04  8.751E-01 .          . *        .          .          .
 8.000E-04  2.531E+00 .          .     *    .          .          .
 1.000E-03  3.886E+00 .          .        * .          .          .
 1.200E-03  4.995E+00 .          .          *          .          .
 1.400E-03  4.499E+00 .          .         *.          .          .
 1.600E-03  2.786E+00 .          .     *    .          .          .
 1.800E-03  1.373E+00 .          .   *      .          .          .
 2.000E-03  2.170E-01 .          .*         .          .          .
 2.200E-03 -7.297E-01 .        * .          .          .          .
 2.400E-03 -1.505E+00 .      *   .          .          .          .
 2.600E-03 -2.139E+00 .     *    .          .          .          .
 2.800E-03 -2.658E+00 .    *     .          .          .          .
 3.000E-03 -3.083E+00 .   *      .          .          .          .
 3.200E-03 -3.431E+00 .  *       .          .          .          .
 3.400E-03 -3.716E+00 . *        .          .          .          .
 3.600E-03 -3.949E+00 . *        .          .          .          .
 3.800E-03 -4.139E+00 . *        .          .          .          .
 4.000E-03 -4.296E+00 . *        .          .          .          .
 4.200E-03 -4.423E+00 . *        .          .          .          .
 4.400E-03 -4.528E+00 .*         .          .          .          .
 4.600E-03 -4.614E+00 .*         .          .          .          .
 4.800E-03 -4.684E+00 .*         .          .          .          .
 5.000E-03 -4.741E+00 .*         .          .          .          .
                            - - - - - - - - - - - - - - - - - - - - - - - - - - - -
```

Figure 12.45

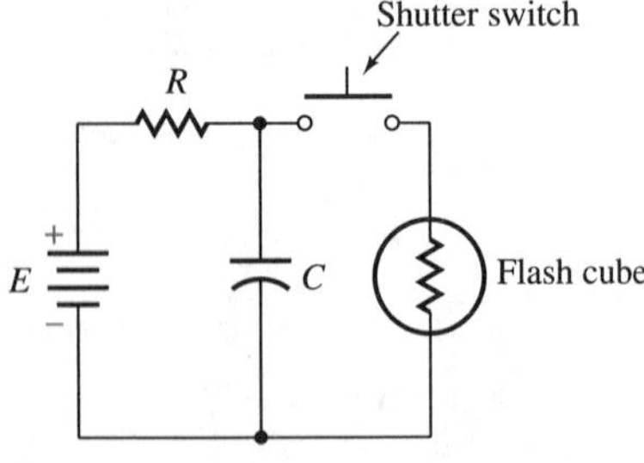

Figure 12.46 A basic R–C circuit for a camera flash.

remain on after they are switched off, an R–C circuit is used to control the time delay. A sawtooth oscillator uses an R–C circuit. The emergency flashers on highway construction use R–C circuits. The filtering circuit that reduces the variation in the output voltage of a dc voltage source uses an R–C circuit. These are just a few of the many R–C circuit applications. Now we will see how some of them work.

A simple capacitor circuit for a camera flash is shown in Figure 12.46. With the shutter open, the capacitor will charge through R. When the shutter is activated, the switch connects the capacitor to the flashcube, and the energy from the capacitor is dissipated in the flashcube material.

This makes it flash. The resistance R is large enough so that very little current is drawn from the battery during the flash. The electronic flash operates on the same principle, but the capacitor is charged to a much higher voltage. This requires additional circuitry. Since the capacitor charges at a slow rate, the current drain on the battery is small, resulting in a longer battery life.

A basic time-delay circuit is shown in Figure 12.47. This circuit keeps relay coils energized after Switch S is opened, and the light is on for a short time after Switch S is opened.

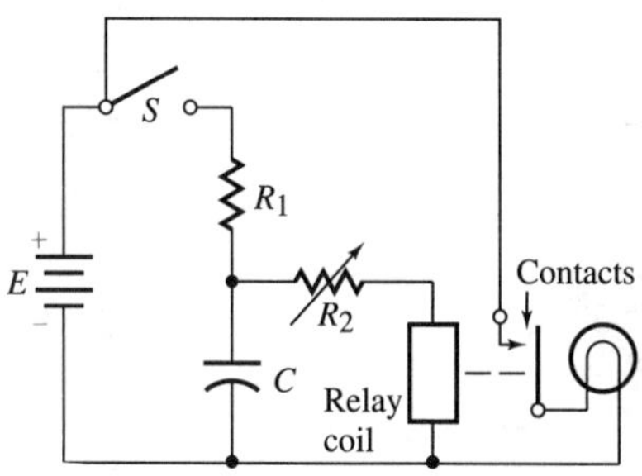

Figure 12.47 The R–C circuit can be used as a time delay circuit.

When Switch S is closed, the relay coil is energized through R_1 and R_2, closing the relay contacts. These connect the lamp to the voltage source. At the same time, capacitor C is charging through R_1. When the switch is opened, the capacitor discharges through R_2 and the relay coil. However, the relay remains energized until the coil voltage drops to the relay's minimum holding voltage. At that time, the relay de-energizes, opening the contact and removing the source voltage from the lamp. The time between the opening of Switch S and the light going out can be adjusted by changing the value of R_2. Long time delays cannot be obtained with this type of circuit because the resistance of the relay coil is not large enough. Replacing the relay with transistors or other semiconductor devices gives longer delays.

A sawtooth oscillator is shown in Figure 12.48. The neon lamp acts as an open circuit until the voltage across it becomes high enough to ionize the gas in it. Then it conducts and becomes a low resistance.

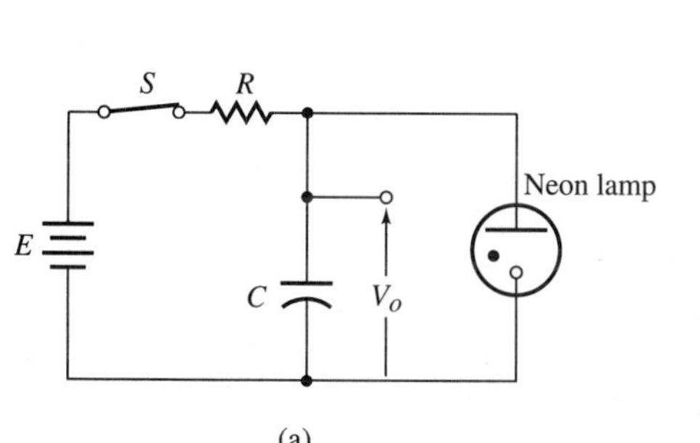

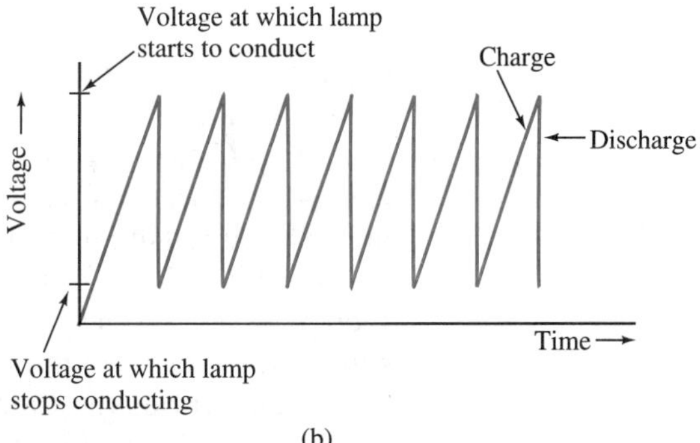

Figure 12.48 The R–C circuit can be used to produce a sawtooth waveform. (a) The sawtooth generator circuit. (b) The waveform of the capacitor voltage.

Closing Switch S–1 starts the capacitor charging. When the capacitor voltage reaches the lamp's breakdown voltage, the lamp starts to conduct and the capacitor discharges through it. The discharge decreases the voltage across the lamp and it stops conducting. The lamp is now an open circuit again and the capacitor starts to charge. The cycle keeps repeating to give the waveform shown in Figure 12.48(b). The frequency of the sawtooth can be varied by changing the time constant of the R–C circuit. The common way is to change the resistance. If the neon lamp is replaced by an electronic flashcube and the source voltage is increased, the result is a strobe light. This is a light that makes rotating or vibrating

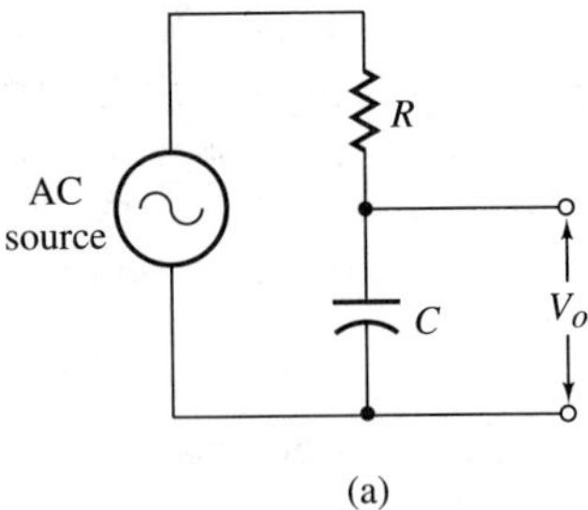

(a)

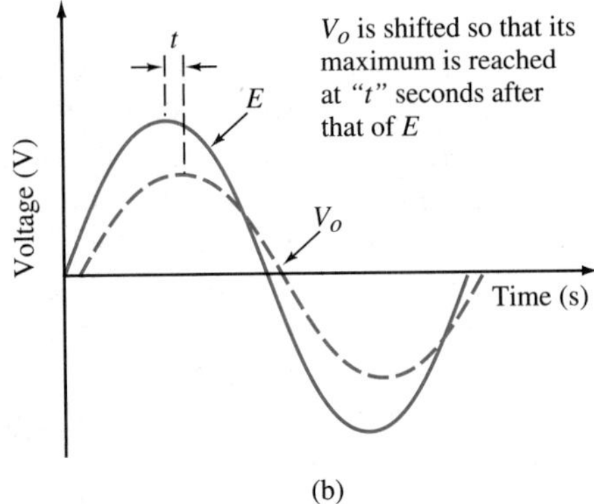

(b)

Figure 12.49 The R–C circuit is used with an ac source for phase shifting. (a) The phase shifting circuit. (b) The relation of the output and the input voltages.

objects seem to be motionless. It is used to measure the revolutions per minute of motors.

The R–C circuit can also be used as a phase shifter. Phase shifting means making one ac voltage happen at a later point in time than another one. The circuit in Figure 12.49(a) will cause the voltage across C when R is varied to shift as shown by the dotted line in Figure 12.49(b). Increasing R causes the capacitor voltage to shift farther away from the source voltage. A common application of this circuit is in light dimmers. The circuit varies the time at which a silicon-controlled rectifier starts conducting. This particular type of circuit (ac source) will be examined in detail in Chapter 16.

These are just a few of the many applications of the R–C circuit. Their low cost, small-size, and reliability make them very attractive for use in electronic circuits.

SUMMARY

1. A capacitor can be in one of three states: steady state—no voltage, transient state—changing voltage, and steady state—some voltage, fully charged.

2. The capacitor voltage cannot change instantaneously. It changes exponentially; the time during the change is the transient state.
3. During charging, the initial rate of rise depends on E, R, and C; the initial current depends on E and R; and the final voltage depends on E.
4. One time constant is the time required for the capacitor to charge to 63.21 percent of the maximum voltage, or drop to 36.79 percent of the maximum during discharge.
5. The time constant is equal to the product of R and C.
6. For practical purposes, a capacitor is considered to be fully charged or fully discharged in five time constants.
7. During discharge, the initial rate of change depends on V_o, R, and C; the initial current depends on V_o and R; and V_o depends on the charging conditions.
8. The Universal Time Constant Curves are curves that plot the percent of maximums versus the number of time constants.
9. The analysis of the voltage rise in a complex R–C circuit requires that the circuit connected to the capacitor be replaced by its equivalent resistance.
10. The analysis of the discharge in a complex R–C circuit requires that the circuit connected to the capacitor be replaced by its equivalent resistance.
11. The amount of energy stored in the field of a capacitor is equal to (1/2) CV^2.
12. At the instant a capacitor starts charging, it is like a short circuit. In the steady state, it is like an open circuit.

EQUATIONS

12.1	$I_o = \dfrac{E}{R}$	The capacitor current at the start of charging.	447
12.2	$\dfrac{dv_c}{dt} = \dfrac{E}{RC}$	The rate of change in capacitor voltage at the start of charging.	448
12.3	$I_{\text{ave}} = C\dfrac{\Delta v_c}{\Delta t}$	The average current during a change in capacitor voltage.	449
12.4	$v_c = E(1 - e^{-t/RC})$	The capacitor voltage during transient charging.	450
12.5	$v_R = Ee^{-t/RC}$	Resistor voltage during transient charging.	451
12.6	$i = \dfrac{E}{R}e^{-t/RC}$	The capacitor current during transient charging.	451

12.7	$\tau = RC$	The time constant of an *R–C* circuit.	454
12.8	$I_o = \dfrac{V_o}{R}$	Initial current during discharge.	457
12.9	$\dfrac{dv_c}{dt} = \dfrac{V_o}{RC}$	The rate of change of the capacitor voltage at the start of discharge.	457
12.10	$v_c = V_o e^{-t/RC}$	Capacitor voltage during the transient discharge.	458
12.11	$i_c = -\dfrac{V_o}{R} e^{-t/RC}$	Current during the transient discharge.	460
12.12	$v_R = V_o e^{-t/RC}$	Resistor voltage during the transient discharge.	461
12.13	$\mathcal{W} = \dfrac{1}{2} CV^2$	Energy stored by the capacitor.	473

QUESTIONS

1. What is the difference between the transient state and the steady state?
2. Can the exponential equations be used for times greater than five time constants?
3. How can the initial surge current in a capacitor be reduced?
4. What are the two meanings of one time constant?
5. A capacitor charges to 20 volts in 50 milliseconds. What can be done to the circuit to increase the time? To decrease the time?
6. At approximately how many time constants is the power at maximum during charging?
7. While the voltage of a charged capacitor was being measured with a voltmeter, the voltage seemed to be decreasing. What do you think is happening? What can be done to prevent the decrease?
8. How do the Universal Time Constant Curves differ from the normal exponential curves?
9. Why are the Universal Time Constant Curves more useful than the normal curves?
10. What type of circuit does the capacitor represent at the start of charge? When in steady state?
11. What are the units that should be used for E, R, C, and t in the exponential equations?
12. How does the capacitor polarity during charging compare to the polarity during the discharge?
13. How does the capacitor current direction during charging compare to the direction during the discharge?
14. One *R–C* circuit has a 5-kilohm resistor and a 4-microfarad capacitor. A second one has a 4-kilohm resistor and a 5-microfarad capacitor. Which characteristics will be the same for both circuits? Which one will be different?

15. A 5-microfarad capacitor is charged to 20 volts in 0.8 second. A second 5-microfarad capacitor is charged to 20 volts in 0.4 second. Is there any difference in the amount of stored energy? Why or why not?
16. A capacitor in a series R–C circuit has been charged to steady state. What effect on the capacitor voltage will reducing R to one-half of its value have?
17. If the charging circuit is similar to that in Example 12.15, what effect will a change in R_2, after steady state is reached, have on the capacitor voltage?
18. Why is it a poor practice to short circuit the terminals of a charged capacitor?
19. Which condition will give the shorter charging time, connecting the R–C circuit to a 20-volt source or connecting it to a 12-volt source?
20. What are two ways by which the amount of stored energy can be increased?
21. A basic sawtooth oscillator has a frequency of 500 Hz. How can the frequency be increased?
22. What would happen to the waveform if a neon lamp were replaced by a higher voltage lamp?
23. What are three applications of the R–C circuit?

PROBLEMS

SECTION 12.1 The Series *R*–*C* Circuit During Voltage Rise

1. A series circuit of a 30-microfarad capacitor and a 500-ohm resistor is connected to a 24-volt source by closing a switch. Find:

   (a) The capacitor current and voltage at the instant the switch is closed.

   (b) The capacitor current and resistor voltage at 6 milliseconds after the switch is closed.

2. A series circuit of a 5-microfarad capacitor and a 2000-ohm resistor is connected across a 120-volt source. Find:

   (a) The time constant of the circuit.

   (b) The initial rate of voltage rise.

   (c) The time needed for the capacitor voltage to reach 60 volts.

3. The voltage across the resistor in the circuit of Figure 12.50 is 20 volts at 30 milliseconds after the switch is closed.

   (a) What is the value of the capacitance in the circuit?

   (b) What is the maximum current in the circuit?

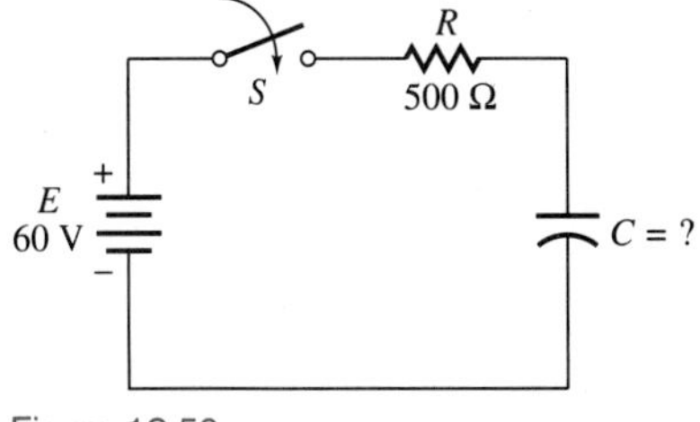

Figure 12.50

4. A series circuit consisting of a 6-microfarad capacitor and a 5-kilohm resistor is connected to a 24-volt source. Find the capacitor voltage, resistor voltage, and circuit current, all at $t = 0.045$ second after the connection.

5. The capacitor in the circuit of Figure 12.51 has a voltage of 6 volts with the polarity shown. What is the circuit current at $t = 0.005$ second after the switch is closed?

6. Repeat Problem 5 for terminal A negative with respect to terminal B. How long will it take for the capacitor voltage to reach 0 volts?

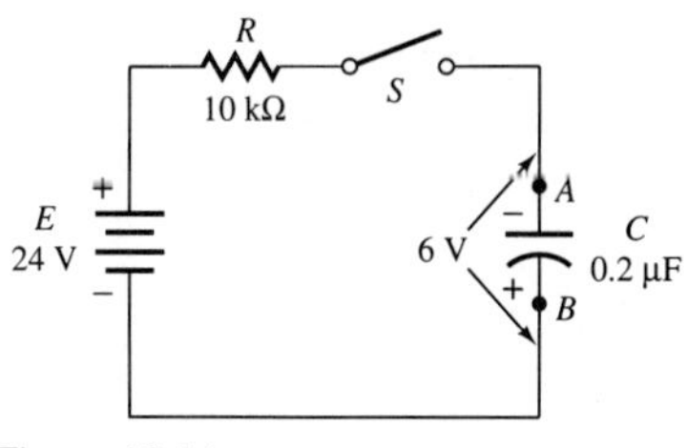

Figure 12.51

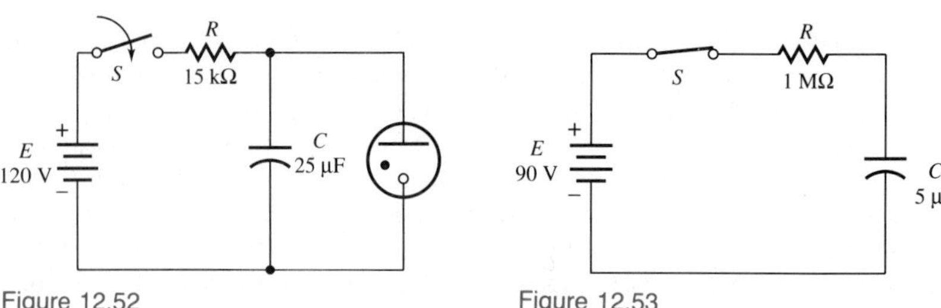

Figure 12.52

Figure 12.53

7. The neon lamp in the circuit of Figure 12.52 starts conducting at 750 milliseconds after the switch is closed.

   (a) What is the time constant of the circuit?

   (b) What should the value of R be if the lamp is to conduct in 500 milliseconds?

8. The switch in the circuit of Figure 12.53 has been closed for 2 seconds.

   (a) What is the time constant of the circuit?

   (b) What are the resistor voltage and current?

9. A series R–C circuit has an initial voltage rise of 200 V/s when connected to a 24-volt source. If the resistance is 6000 ohms, find:

   (a) The capacitance in the circuit.

   (b) The time constant of the circuit.

   (c) The capacitor voltage at $t = 0.080$ second after the connection.

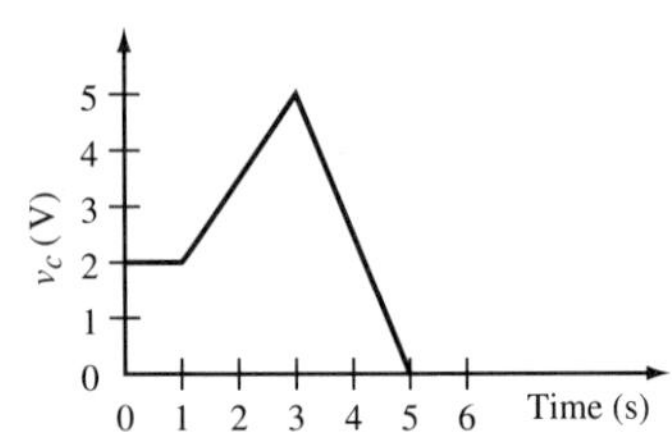

Figure 12.54

10. A timing circuit consists of a switch, a 10-microfarad capacitor, and a 5-megohm resistor connected to a 24-volt source. The switch is closed by the horse leaving the starting gate and opened when it crosses the finish line. If the capacitor voltage is 15 volts, how long does it take for the horse to finish?

11. The voltage across a 5-microfarad capacitor is varied as shown in Figure 12.54. Find the average current from:

   (a) 0–1 second. (b) 1–2 seconds. (c) 3–5 seconds.

SECTION 12.2 The Series *R–C* Circuit During Voltage Fall

12. In the circuit shown in Figure 12.55, S_1 is put into Position 2 after the capacitor is fully charged.

   (a) Place the polarity on the capacitor terminals and show the discharge current's direction.

   (b) Find the time constant of the discharge circuit.

   (c) Find the capacitor current and voltage at 6 milliseconds after the switch is put into Position 2.

   (d) Find the current in R_2 at the instant the switch is put into Position 2.

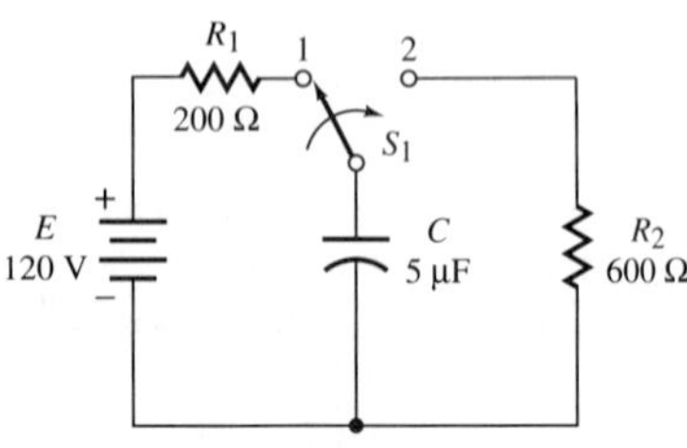

Figure 12.55

13. A 2-ohm resistor is accidentally placed across the terminals of a 30-microfarad capacitor that was charged to 120 volts.

(a) What is the capacitor current at the instant the resistor contacts the terminals?

(b) How many seconds will it take for the capacitor to discharge?

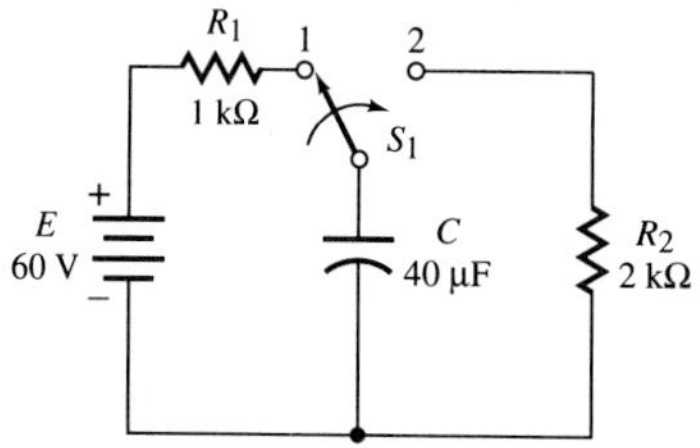

Figure 12.56

14. Switch S_1 in Figure 12.56 is left in Position 1 for five time constants and then put into Position 2 for five time constants.

   (a) What is the time constant while in Position 1?

   (b) What is the time constant while in Position 2?

   (c) Sketch the capacitor current and voltage waveforms.

15. A 50-microfarad capacitor is charged to 60 volts and then connected to a 25-kilohm resistor.

   (a) What is the initial current in the capacitor?

   (b) How many seconds will it take for the capacitor voltage to reach 30 volts?

   (c) What is the initial rate of voltage change?

16. The capacitor voltage in a series R–C circuit drops to 20 volts in 1.5 seconds. If the resistance is 5000 ohms and the capacitance is 200 microfarads, find:

   (a) The initial voltage on the capacitor.

   (b) The initial current.

   (c) The time constant of the circuit.

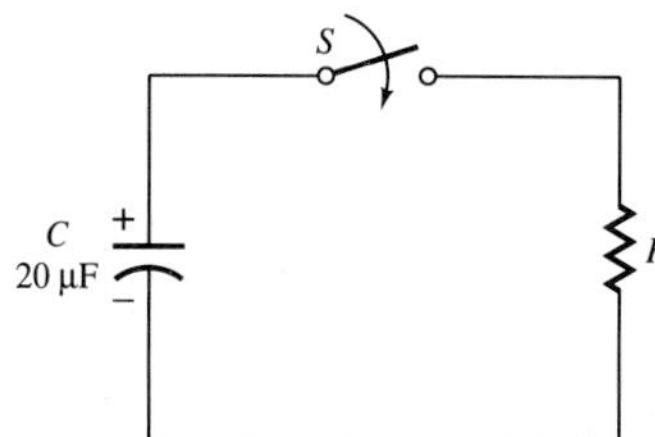

Figure 12.57

17. The capacitor current at the instant the switch is closed in the circuit in Figure 12.57 is 5 milliamperes. If the time constant is 20 milliseconds, find:

   (a) The initial voltage on the capacitor.

   (b) The initial rate of voltage change.

   (c) The capacitor voltage at 5 milliseconds after the switch is closed.

SECTION 12.3 The Universal Time Constant Curve

18. Use the curves of Figure 12.14 to find v_c and i_c at 6 milliseconds after S_1 is closed in Figure 12.58.

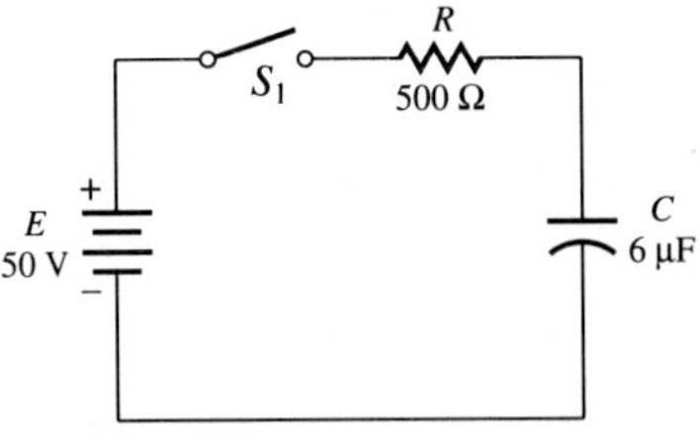

Figure 12.58

19. An R–C series circuit of 20 ohms and 500 microfarads is connected to a 90-volt source. Use the curves of Figure 12.14 to find:

   (a) The capacitor voltage at one time constant, two time constants, and three time constants.

   (b) The resistor voltage at 25 milliseconds after the circuit is connected.

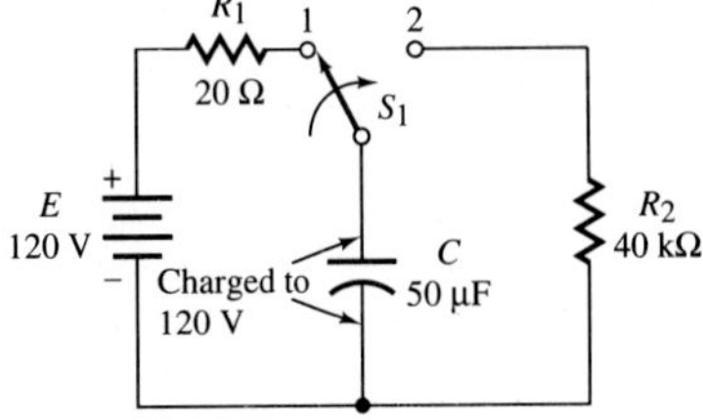

Figure 12.59

20. An R–C series circuit of 2000 ohms and 15 microfarads is connected to a 24-volt source. The capacitor voltage is 12 volts some time after being connected. Use the curves of Figure 12.14 to find:

   (a) The length of time that the circuit was connected.

   (b) The capacitor current.

   (c) The time that it will take for v_c to reach 6 volts.

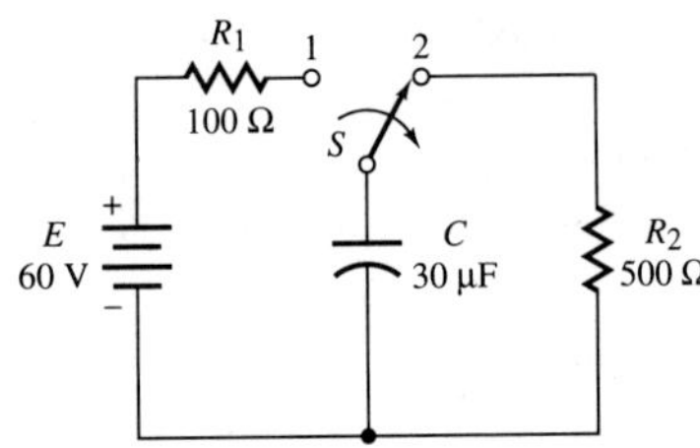

Figure 12.60

21. Use the curves of Figure 12.14 to find the capacitor voltage and current at t = 3 seconds after S_1 is put into Position 2.

22. The switch in the circuit of Figure 12.60 was in Position 1 until the capacitor was fully charged. It was then put into Position 2 where the capacitor current is now 0.06 ampere. Use the curves of Figure 12.14 to find:

    (a) The time that the switch has been in Position 2.

    (b) The voltage across the capacitor.

23. An R–C circuit of 20 ohms and 300 microfarads has an initial current of 2 amperes. Use the curves of Figure 12.14 to find the capacitor current and voltage at 18 milliseconds after the capacitor starts discharging.

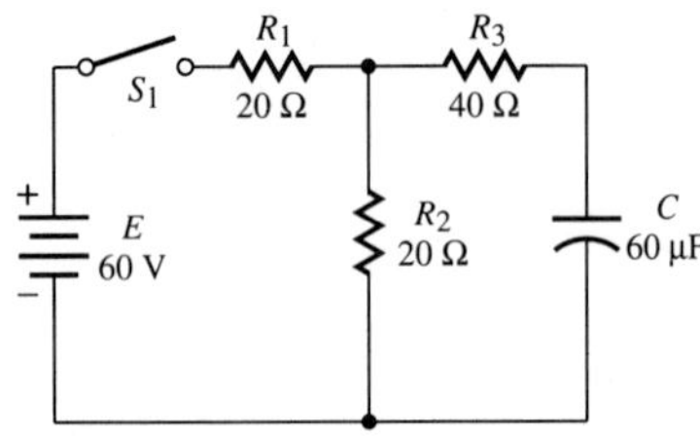

Figure 12.61

SECTION 12.4 The Analysis of Complex *R–C* Circuits

24. For the circuit of Figure 12.61, find:

    (a) The time constant of the circuit.

    (b) v_c, i_c, and i_s at 9 milliseconds after S_1 is closed.

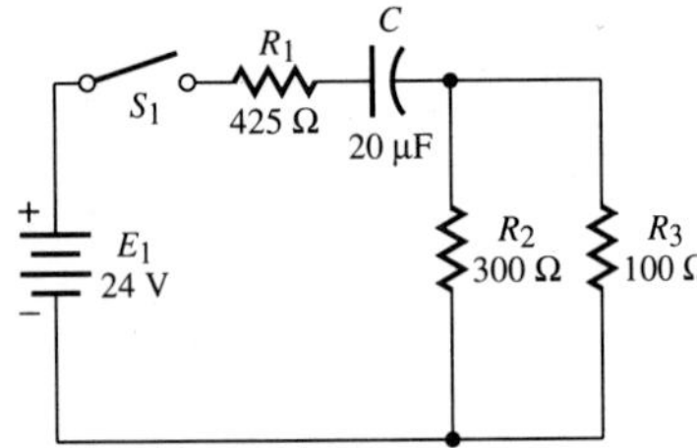

Figure 12.62

25. For the circuit of Figure 12.62, find:

    (a) The time constant of the circuit.

    (b) v_c, i_c, and v_2 at 15 milliseconds after S_1 is closed.

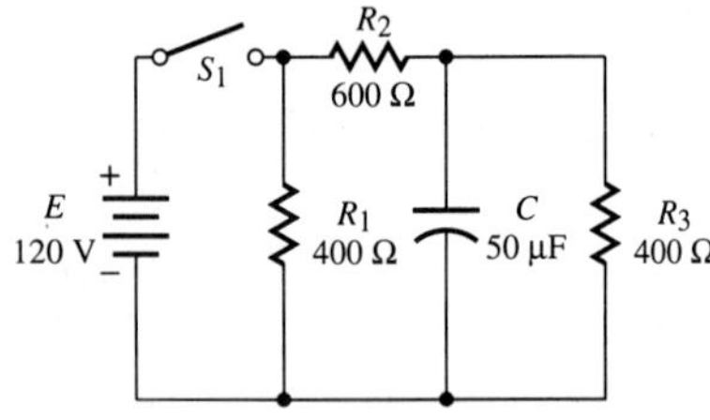

Figure 12.63

26. How long after S_1 is closed will it take for v_c to reach 16 volts in the circuit of Figure 12.63? What is the time constant of the circuit?

27. With S_1 closed, the capacitor voltage has reached its final value in the circuit of Figure 12.64. Find:

    (a) The current in R_2 at 8 milliseconds after S_1 is opened.

    (b) The rate of the capacitor voltage fall at the instant the switch is opened.

28. Capacitor C_1 in Figure 12.65 has been charged to 48 volts. Find:

    (a) v_c after S is closed for 60 microseconds.

    (b) The voltage across R_3 at that time.

29. Determine the capacitor voltage and current in R_2 at t = 45 milliseconds after the switch is Figure 12.66 is closed.

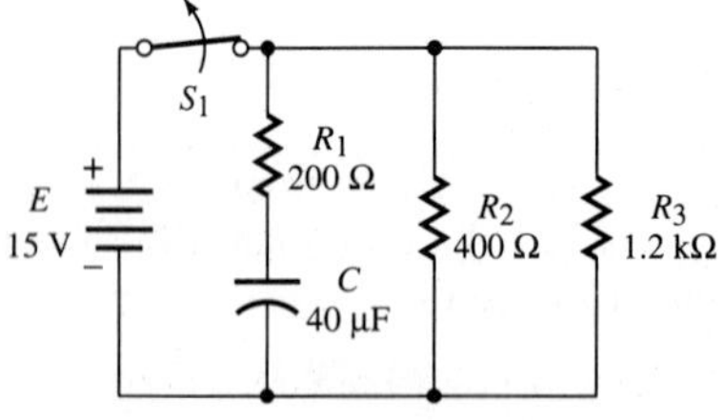

Figure 12.64

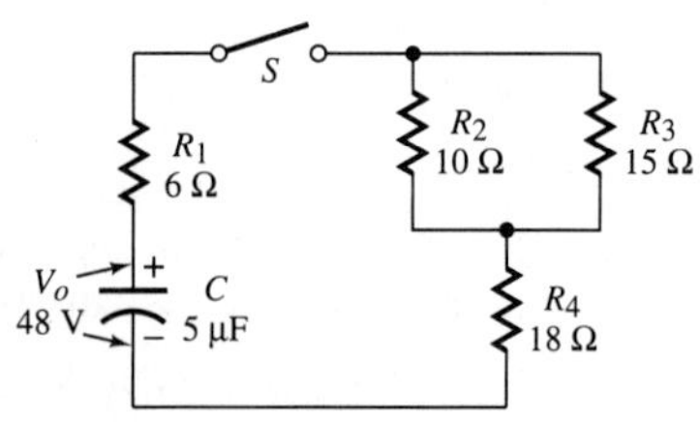

Figure 12.65

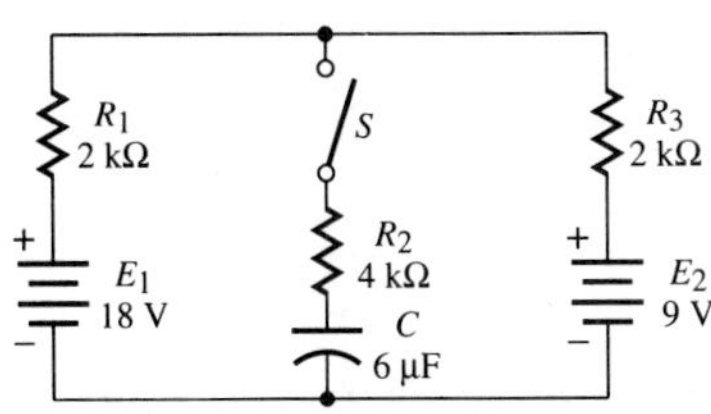

Figure 12.66

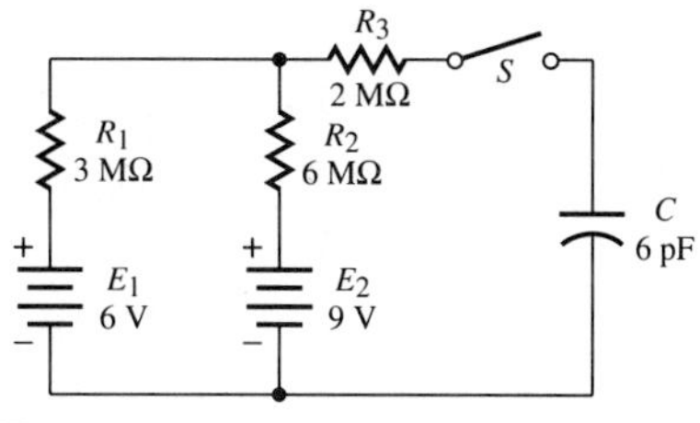

Figure 12.67

30. The voltage across the capacitor in Figure 12.67 is used to trigger an alarm circuit. If the minimum triggering voltage is 4 volts, how long after the switch is closed will the alarm trigger?

31. The capacitor in Figure 12.68 has an initial charge of 12 volts. What will the voltage be at 0.03 second after the switch is closed?

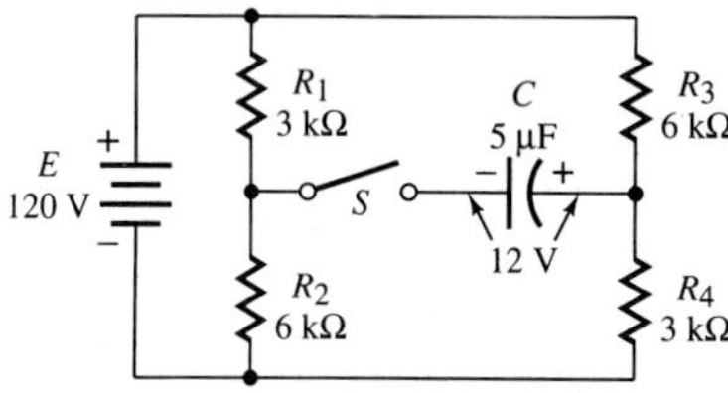

Figure 12.68

SECTION 12.5 Energy Stored in a Capacitor

32. A series R–C circuit of 20 ohms and 5 microfarads is connected across a 24-volt source. How much energy is stored in the capacitor after five time constants?

33. The voltage across a 5-microfarad capacitor is increased from 30 volts to 60 volts.

    (a) How much energy was stored at 30 volts?

    (b) How much energy was added in going to 60 volts?

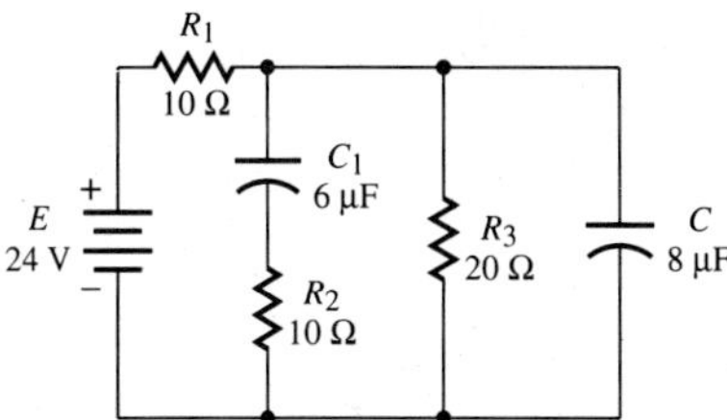

Figure 12.69

34. A series R–C circuit of 500 ohms and 6 microfarads is connected to a 120-volt source. How much energy is stored in the capacitor at 1 millisecond after the circuit is connected?

35. The voltage on a 50-microfarad capacitor, charged to 50 V, decreased to 25 volts due to leakage resistance. How much energy was lost?

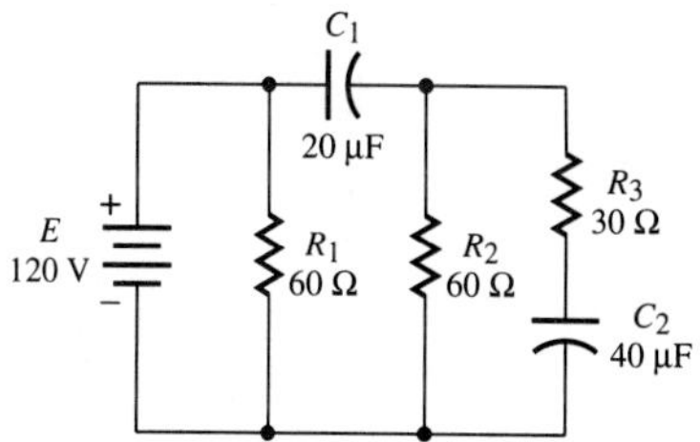

Figure 12.70

SECTION 12.6 The R–C Circuit in Steady State

36. The circuit in Figure 12.69 is in steady state. Find:

    (a) The current in each branch.

    (b) The voltage across each component.

    (c) The charge on each capacitor.

37. The circuit in Figure 12.70 is in steady state. Find:

    (a) The voltage across each component.

    (b) The energy stored in each capacitor.

38. The circuit in Figure 12.71 is in steady state. Find:

    (a) The voltage and current in each component.

    (b) The energy stored in each capacitor.

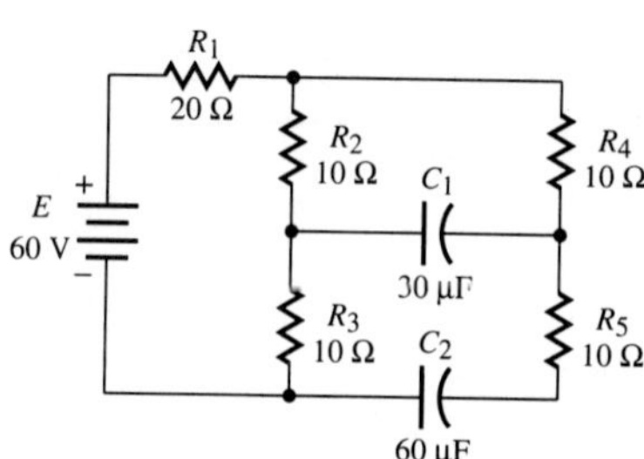

Figure 12.71

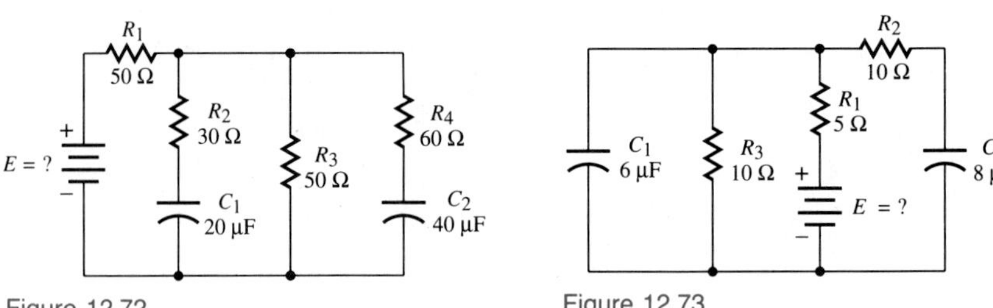

Figure 12.72

Figure 12.73

39. The circuit in Figure 12.72 is in steady state and the energy stored in C_2 is 0.8 J. Find:

 (a) The voltage and current for each component.

 (b) The source emf.

 (c) The charge on each capacitor.

40. The circuit in Figure 12.73 is in steady state and the voltage across C_2 is 20 volts. Find the emf of the source.

SECTION 12.7 SPICE Analysis of *R–C* Circuits

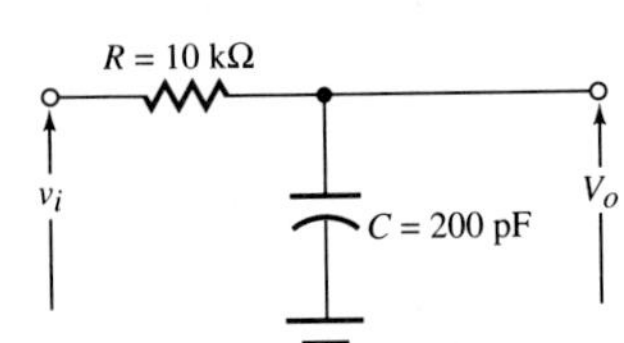

Figure 12.74

Ⓢ 41. Use PSpice to simulate the response of the circuit shown in Figure 12.74 to a 0–15 volt, positive-going pulse, of duration 12 microseconds.

Ⓢ 42. Use PSpice to investigate the effect of connecting the circuit shown in Figure 12.74 to a 50-kilohm load. The load is connected between the output and ground terminals. Compare the new output waveform with the output waveform with no load.

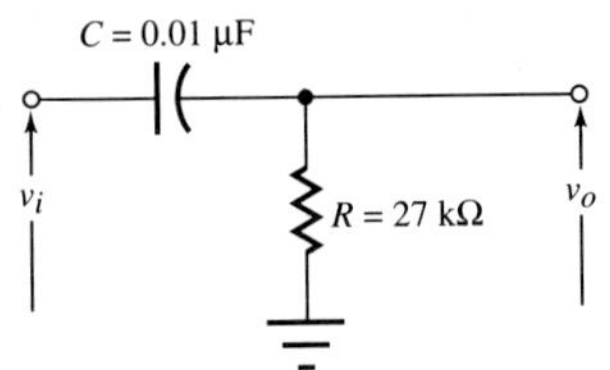

Figure 12.75

Ⓢ 43. Simulate the response of the circuit shown in Figure 12.75 to a 0–10 volt pulse of duration 0.5 millisecond.

Ⓢ 44. Simulate the response of the circuit shown in Figure 12.75 when the circuit is connected to another circuit that has a 100-kilohm, 470-pF input. Compare the response to the no-load response obtained in Problem 43.

Ⓢ 45. Simulate the response of the circuit shown in Figure 12.75 to a +10 volt to −10 volt pulse whose duration is 1 millisecond.

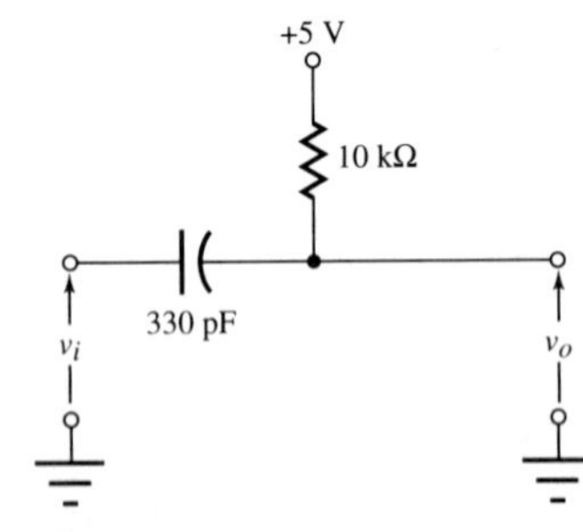

Figure 12.76

Ⓢ 46. Use SPICE to investigate the response of the circuit shown in Figure 12.76 to a 0 to +5 volt input pulse of duration 10 milliseconds.

SECTION 12.8 Applications of the *R–C* Circuit

47. A 500-microfarad capacitor supplies 10 J to flash an electronic flashcube. Find:

 (a) The voltage across the capacitor at the start of the flash.

 (b) The time that it takes to store the energy if the charging resistance is 1000 ohms.

 (c) The maximum current supplied by the source during charging.

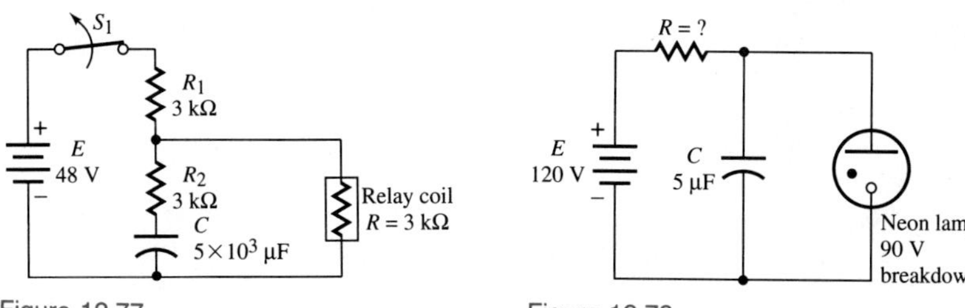

Figure 12.77

Figure 12.78

48. How many seconds after S_1 is opened in Figure 12.77 will the relay de-energize if the relay coil requires at least 8 volts to remain energized? (Assume that the inductance at the coil is negligible and that steady state has been reached with S_1 closed.)

49. The circuit in Figure 12.78 is a simple sawtooth oscillator circuit. What should the value of R be so that the pulse repeats itself every 100 milliseconds?

CHAPTER 13

MAGNETISM AND INDUCTANCE

CHAPTER OBJECTIVES

After completing this chapter, you should be able to:

1. Explain what a magnet is and the molecular theory of magnetism.
2. Describe the magnetic field, and explain the effect of magnetic and nonmagnetic materials on it.
3. Define the quantities flux, flux density, and permeability.
4. List the desirable and undesirable effects of the magnetic field.
5. List some applications that make use of magnets.

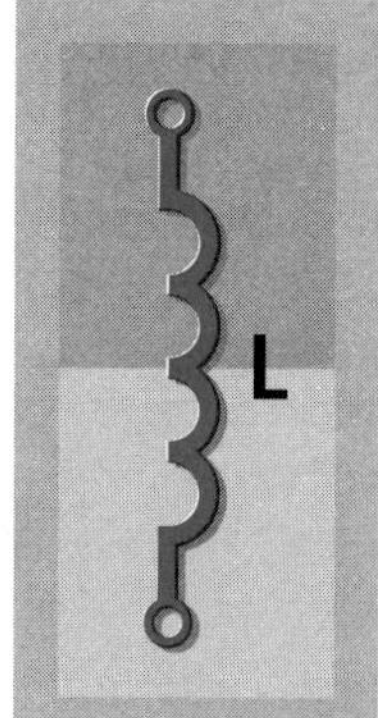

6. Determine the current and turns for the flux in a magnetic circuit.
7. Explain the principle of electromagnetic induction.
8. Explain Faraday's and Lenz's Laws, and use them to determine emf and polarity.
9. Define inductance, give its unit, and explain how the induced emf is related to inductance.
10. List the factors that affect the inductance of a coil, and calculate the inductance of a long coil.
11. Calculate the equivalent inductance of series-connected and parallel-connected inductors.

KEY TERMS

1. average emf
2. coefficient of coupling
3. demagnetizing
4. domains
5. electromagnet
6. electromagnetic induction
7. equivalent inductance
8. Faraday's Law
9. flux density
10. henry
11. inductance
12. inductors
13. lines of force
14. Lenz's Law
15. lodestone
16. magnet
17. magnetic field
18. magnetic flux
19. magnetic materials
20. magnetization curves
21. magnetizing force
22. molecular theory
23. mutual induction
24. nonmagnetic materials
25. north pole
26. parallel inductors
27. permeability
28. permanent magnet
29. relative permeability
30. reluctance

31. right-hand rule
32. series inductors
33. south pole
34. temporary magnet
35. tesla
36. tractive force
37. weber

INTRODUCTION

This chapter introduces the magnetic field and inductance. The magnetic field cannot be seen, but its presence is known by the effect on electronic communication and iron parts.

Magnets have developed from the first-known, natural lodestone magnet to magnets that provide the exact magnitude and shape of magnetic field for the specific application. Some applications that use magnets are motors, door latches, cranes, loudspeakers, microphones, and sensors for burglar alarms.

Magnetism is used to store information on a computer disk. It is also used to store the signal on videotape and audiotape. Transformers use the magnetic field to change voltages from one value to another. AC alternators use the magnetic field to generate the power that is supplied by electric utilities. Without an understanding of magnetism, ac power generation and transmission would either be impossible or very expensive.

Inductance also cannot be seen or felt. It is that property by which a change in current is opposed. It is generally associated with a coil of wire. Inductance is used in applications that include tuning circuits, waveshaping circuits, and even short time delays. Inductance can also create problems because of the high voltage that can be induced by it. This can cause insulation breakdown and damage to parts. A common use of inductance is in the ignition system of an automobile. The interruption of the current in an inductor induces the high voltage for the spark plugs.

After completing this chapter, you will have a better understanding of both magnetism and inductance. This should prepare you better for working in the electronics field.

13.1 The Magnet and the Magnetic Field

magnet
lodestone

A **magnet** is a piece of material that has the property to attract iron. The first known magnetic material was **lodestone**, an oxide of iron (Fe_3O_4). Its magnetic properties were known to the Greeks as early as 600 B.C. In fact, the magnet is probably named after their city of Magnesia, near a large deposit of magnetite where iron ore was mined.

Why learn about magnetism? What can it do for us? Figure 13.1 shows some of the more common applications of magnetism, and as you look around you, others should become apparent to you.

In Figure 13.1(a), the magnet is used to convert electric current to sound. Some earphones and microphones also use magnets. The device in Figure 13.1(b) is a relay. When the magnet coil is energized, the

(a)

(b)

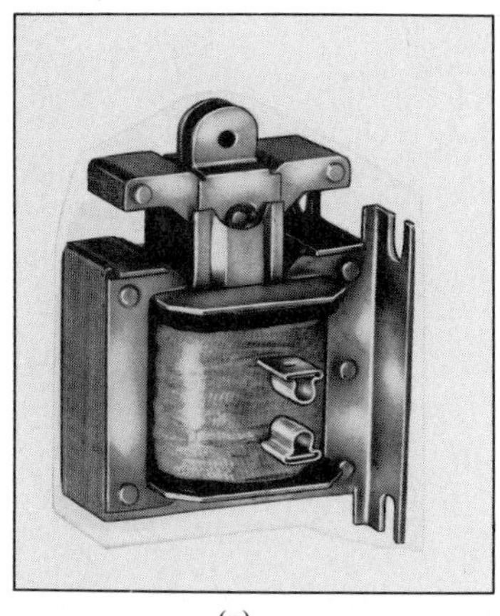
(c)

(d)

(e)

Figure 13.1 Some familiar applications of magnets. (a) A loudspeaker, (b) a relay, (c) a solenoid, (d) transformers, and (e) a motor. (a, Photo courtesy of Electro-Voice, Inc.; b, photo courtesy of Potter & Brumfield A Siemens Company; c, photo courtesy of W.W. Grainger, Inc.; d, photos courtesy of Jensen Transformers, Inc.; (e), photo courtesy of W.W. Grainger, Inc.)

contacts are pulled together. Relays permit remote-controlled power. For instance, outside lights can be turned off from inside the house. Also, relays control large currents with a low-current switch. The low-current light switch in a car energizes the relay that controls the high current to the lights. A relay uses a smaller diameter wire and is more efficient.

Figure 13.1(c) shows a solenoid. When the coil is energized, the rod will move to the left to a position of equilibrium. The movement can be used to actuate a switch, or for mechanical movement. Again, the automobile uses many solenoids. Some are used to lock the doors, others to actuate the hood release, trunk release, and so forth. The starter solenoid is used to engage the gear on the starter motor with the gear on the engine. Solenoids are also used in automatic washers to activate the water valves.

Transformers, such as the one shown in Figure 13.1(d), are used to change the value of an alternating voltage. A door chime transformer changes the voltage from 120 volts to 12 volts. Large transformers are used to provide different voltages to electric utility customers.

A common application of magnetism is the electric motor in Figure 13.1(e). The magnetic field exerts a force on the armature when there is a current in it. This force causes the armature to turn. Generators and alternators also use magnetism to convert mechanical energy to electric energy.

Magnets are also used to separate magnetic materials from nonmagnetic ones. This is an important application in recycling, where iron cans are separated from the aluminum ones.

Electric cranes use magnets to lift large pieces of iron such as car bodies and iron beams. One advantage of the magnet is that the object can be released by reducing the current in the electromagnet. Some other uses for magnetism are in the heads of tape recorders, magnetic stirrers, and magnetic memory applications.

The lodestone was used as a simple compass several thousand years ago by Chinese sailors. A suspended piece would turn until one end pointed to the North Pole of the earth. That end was called the north-seeking or **north pole**. The other end was called the **south pole**. All

north pole
south pole

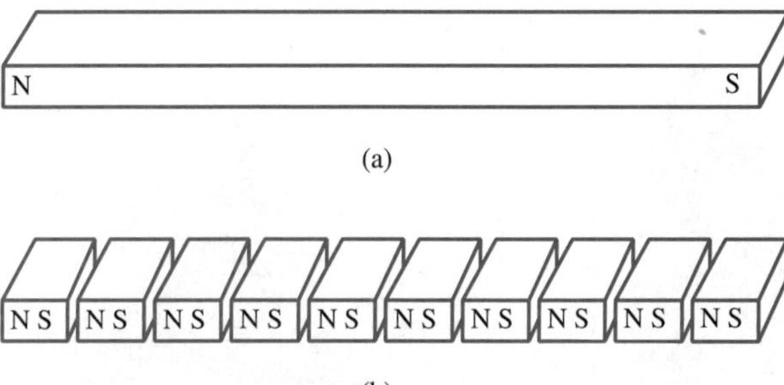

Figure 13.2 Cutting or breaking a magnet makes smaller magnets. (a) The large magnet has a north pole and a south pole. (b) Each small magnet also has a north pole and a south pole.

magnets have a north and a south pole. If a magnet is cut into pieces, each piece will have a north and a south pole. In Figure 13.2, ten magnets are made from the large magnet by cutting it into pieces.

Although a magnet attracts iron, it will either repel or attract another magnet. The action depends on which poles are together. Two north poles or two south poles will repel each other. A north and a south pole will attract each other (see Figure 13.3). This is similar to the force between electric charges.

magnetic field

A **magnetic field** is the space around the magnet in which a force is exerted on iron. In 1600, William Gilbert (1544–1603), a physicist and personal physician to Queen Elizabeth I of England, discovered that there is such a field around the earth. It was this field that made the lodestone point to the North Pole.

lines of force

Iron filings act in an interesting way when sprinkled on a sheet over a magnet. They arrange themselves in the pattern shown in Figure 13.4. The paths formed by the filings are called **lines of force** for the field. A line

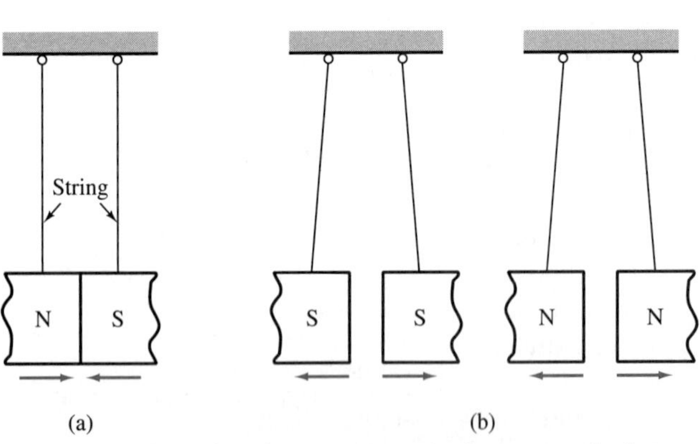

Figure 13.3 The poles of a magnet exert a force on each other. (a) The force is one of attraction for unlike poles. (b) The force is one of repulsion for like poles.

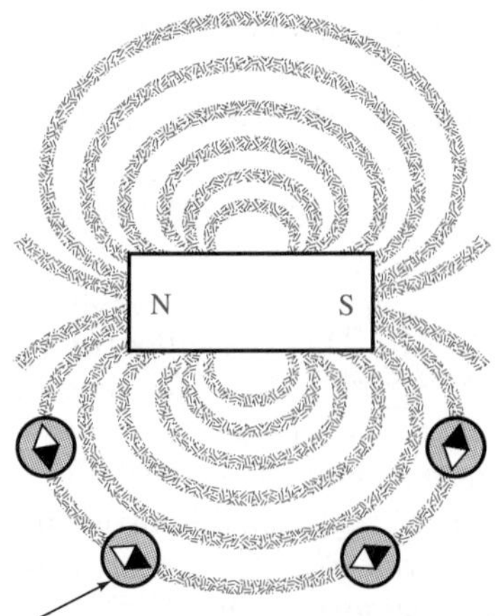

Figure 13.4 The shape of a magnetic field can be outlined using iron filings, as shown here.

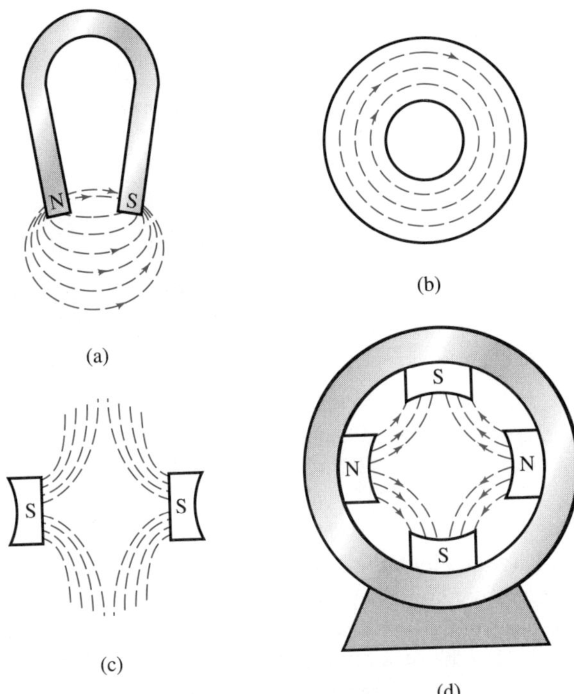

Figure 13.5 Outlines of the magnetic field shape for other magnet configurations. (a) The field of a horseshoe magnet. (b) The field of a circular magnet. (c) The field between two like poles. (d) The field in a four-pole generator.

of force is that path along which an isolated north pole would move if placed in the field. The magnetic field is stronger where the lines are closer together.

The direction of the lines of force is shown by the compass needles in Figure 13.4. This direction is defined as leaving the North Pole and returning into the South Pole. The lines continue through the magnet, forming complete loops. The lines cannot cross each other. The spaces between the lines are caused by the repelling force between them. This force depends on the field strength and the material. The path of the lines for a few other magnets is shown in Figure 13.5.

An electric current in a conductor will set up a magnetic field around the conductor. This was discovered by Hans Christian Oersted (1777–1851), a Danish scientist, in about 1820. This field takes the form of concentric circles, as shown in Figure 13.6(a). Again, the compass needles show the direction of the lines. The direction can also be found by using the **right-hand rule**. This rule states that the fingers of the right hand will point in the direction of the field, if they are wrapped around the conductor so that the thumb points in the direction of the current. When this rule is used with the conductor in Figure 13.6(b), the direction is counterclockwise.

right-hand rule

Winding the conductor into a coil changes the field to that of Figure 13.7. The direction of the lines is found by the right-hand rule. William

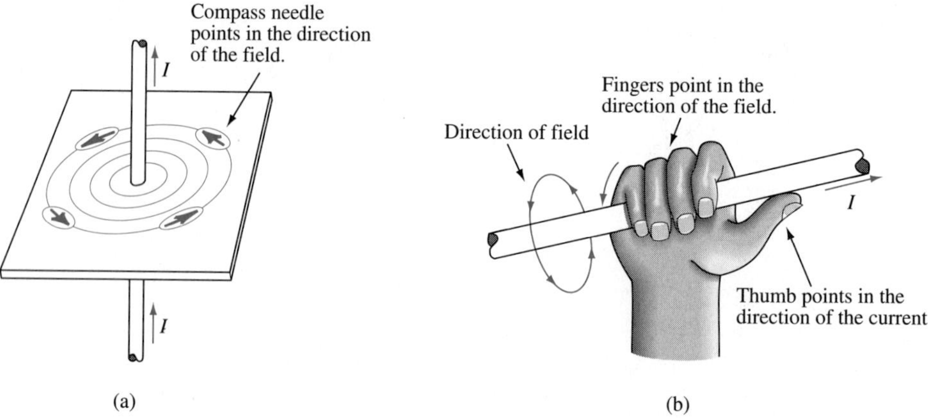
Compass needle
points in the direction
of the field.
I
I
Fingers point in the
direction of the field.
Direction of field
I
Thumb points in the
direction of the current.
(a)
(b)

Figure 13.6 A magnetic field is set up around a conductor by the current in the conductor. (a) The field forms in concentric circles. (b) The right-hand rule can be used to determine the direction of the field.

Sturgeon, an English electrician, discovered that the field in the coil can be made stronger by placing a piece of iron in the coil. The magnetic field around a conductor makes devices such as transformers and electric motors possible. On the other hand, the interference of the magnetic field can affect the operation of electronic equipment. For example, the field around electric transmission lines affects radio and television reception.

molecular theory

A common explanation of magnetism is given by the **molecular theory**.

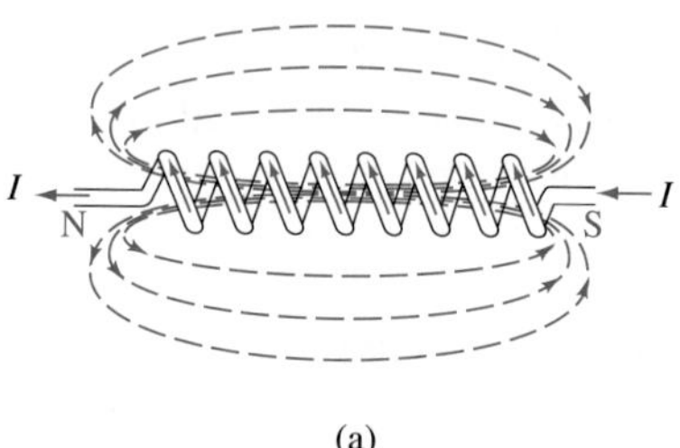
I
N
S
I

(a)

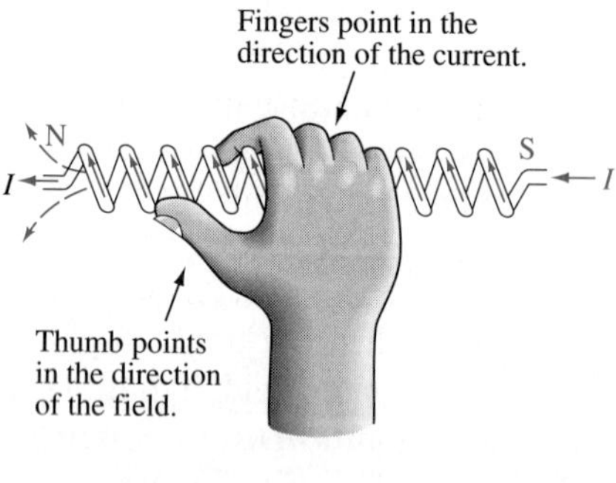
Fingers point in the
direction of the current.
N
S
I
I
Thumb points
in the direction
of the field.

(b)

Figure 13.7 The current in a coil sets up a magnetic field around the coil. (a) The shape of the field around the coil. (b) The right-hand rule can also be used to determine the direction of the field around the coil.

This theory states that the spinning electrons form tiny fields around themselves. These tiny fields in the molecule combine to form small bar magnets or **domains**. When the domains point in different directions, as in Figure 13.8(a), the net field is zero. When the domains become aligned, as in Figure 13.8(b), their fields combine to give a large magnet.

domains

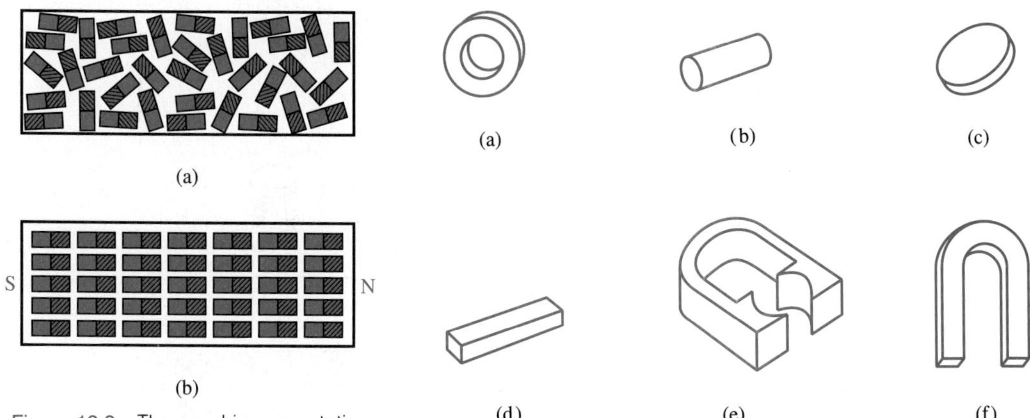

Figure 13.8 The graphic presentation of the domain theory. (a) When unmagnetized, the domains are arranged in random manner, so the fields do not combine. (b) When magnetized, the domains are aligned, so the fields combine.

Figure 13.9 Permanent magnets are made in many shapes to fit the application. (a) A circular magnet. (b) A round bar magnet. (c) A disk magnet. (d) A rectangular bar magnet. (e) A meter movement magnet. (f) A horseshoe magnet.

There are two types of magnets: permanent and temporary. A **permanent magnet** is one that holds its magnetism for a long time. Hard steel is a typical permanent magnet material. A few of the shapes of permanent magnets are shown in Figure 13.9. Some uses of permanent magnets are in dc motors, magnetic mixers, and door latches. The lodestone is a natural permanent magnet. Other permanent magnets must be made by magnetizing a piece of material. This is done by placing the material in a magnetic field. This field is usually from a current-carrying coil of wire. A simple magnetizer can be seen in Figure 13.10. The strength of the magnet depends on the magnetizing field, the type of material, and the amount of magnetism it already has. Commercial magnetizers can provide the exact amount of magnetism for the application.

permanent magnet

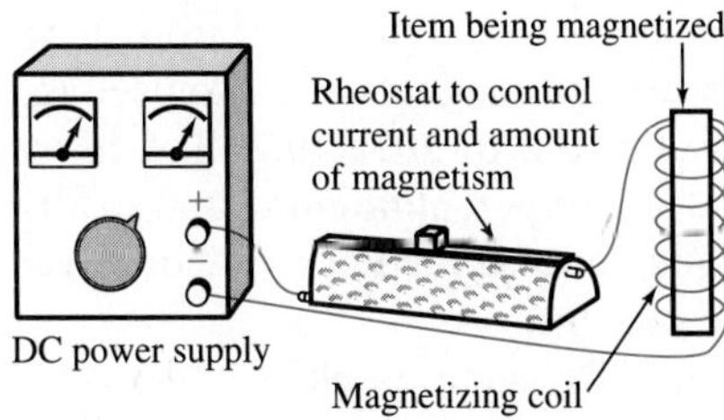

Figure 13.10 A basic magnetizing system consists of a coil and a dc source.

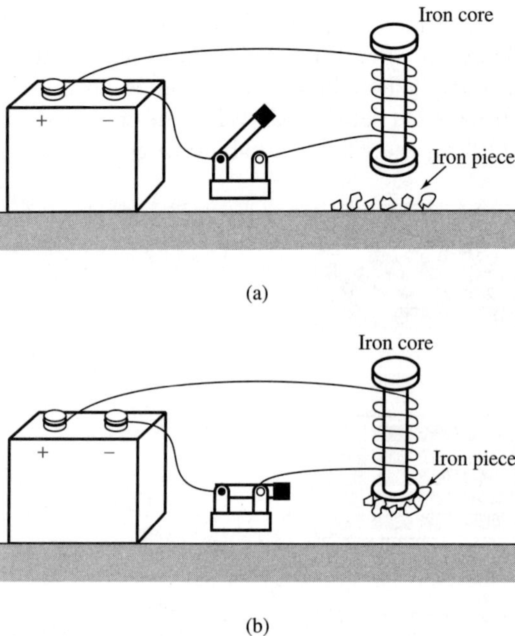

Figure 13.11 The electromagnet is a magnet only when there is current in the coil of the magnet. (a) With the switch open, the core is not magnetic, and the iron particles are not attracted. (b) With the switch closed, the core becomes magnetic, and attracts the iron particles.

demagnetizing

Demagnetizing is a procedure that removes the magnetism from the magnet. Striking a magnet with a sharp blow can demagnetize it. Obviously, this is a crude method. A more controlled way is to place the magnet in a field caused by a gradually decreasing alternating current.

temporary magnet

electromagnet

A **temporary magnet** is one that is a magnet only while acted on by a magnetic field. Soft steels make good temporary magnets. The most common temporary magnet is the **electromagnet**. It is made by winding a coil of wire around a piece of temporary magnet material. The basic construction is shown in Figure 13.11. The material is a magnet as long as there is current in the coil. The first practical electromagnet was made by John Henry, an American physicist. Electromagnets are used in circuit breakers, relays, and solenoids.

Magnetic cranes, such as the one shown in Figure 13.12, use electromagnets to lift large odd-shaped pieces of iron. Turning off the current releases the pieces. Electromagnets are also used to separate iron from other materials. They do this by attracting all of the iron parts but leaving the other parts, such as copper, aluminum, and so forth. The lifter shown in the figure is unique because the part will not be released if there is an interruption of the current.

nonmagnetic materials

Nonmagnetic materials, such as glass and wood, do not affect the magnetic field. But materials such as iron, cobalt, nickel, and ceramic ferrites change the field by giving it an easier path to follow. They are

Figure 13.12 A multimagnet lifter, such as the one shown here, can pick up large assemblies without using any hooks or clamps. (Photo courtesy of PERMADUR Industries, Somerville, New Jersey)

Iron: the path of the lines is affected.

N S

Glass: the path of the lines is not affected.

Figure 13.13 Magnetic materials have a different effect on a magnetic field than nonmagnetic materials. (a) The nonmagnetic glass does not change the shape of the field. (b) The magnetic iron offers an easier path to the field, so the shape changes.

magnetic materials

magnetic materials. This effect is shown in Figure 13.13 where the lines come together to pass through the iron. The path through the glass is just as hard as the lines through the air, so they do not change. Magnetic materials are well suited for transformer cores and shielding sensitive electronic equipment from the magnetic field. This is because they direct the lines along the path of the material.

13.2 Flux, Flux Density, and Other Magnetic Quantities

magnetic flux
weber

This section introduces the terms that are used when analyzing and describing devices that make use of the magnetic field. The magnetic lines of force, introduced in Section 13.1, are also called the **magnetic flux** lines or in shorter form, "flux." The unit for flux in SI is the **weber** (symbol Wb). It is named after Wilhelm E. Weber (1804–1891), a German physicist who advanced the molecular theory of magnetism. The letter symbol for flux is ϕ. The weber is defined in terms of the emf that is induced when a conductor is placed in a changing magnetic field (see Section 13.6). One weber is the amount of flux that will give an emf of 1 volt in a 1-turn coil, when the flux is reduced to 0 in 1 second at a steady rate. In other systems of measurement, flux is measured in maxwells and lines. Each of these is equivalent to 10^{-8} webers.

flux density
tesla

Flux density is the flux per unit area (Wb/m^2). It is measured in **tesla** (T). The unit is named after Nikola Tesla (1856–1943), an electrical engineer who invented the ac induction motor. One tesla is 1 weber of flux per square meter of area. The symbol for flux density is B.

Figure 13.14 shows the flux in two iron cores of the same area. Obviously, there are more lines in Figure 13.14(a) than in Figure 13.14(b). Since the areas are the same, the flux density in Figure (a) is greater than that in (b). The flux density in a magnetic path is given by

$$B = \phi/A \tag{13.1}$$

where B is the flux density, in tesla, $(Wb)/m^2$; ϕ is the flux, in webers; and A is the area perpendicular to the lines of flux, in m^2.

Equation 13.1 shows that the flux density is increased by either a larger flux or a smaller area.

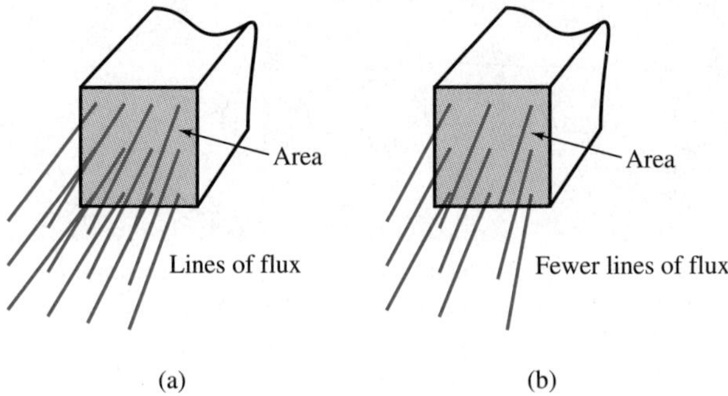

Figure 13.14 Flux density is the number of lines per unit area. (a) More lines in the area gives a larger flux density. (b) Fewer lines in the area gives a smaller flux density.

EXAMPLE 13.1

What is the flux density in a cross section 0.002 m × 0.004 m when the flux is 2×10^{-5} Wb?

Solution

$$B = \phi/A$$

$$A = \text{W} \times \text{h} = (0.004\ \text{m})(0.002\ \text{m}) = 8 \times 10^{-6}\ \text{m}^2$$

$$B = \frac{2 \times 10^{-5}\ \text{Wb}}{8 \times 10^{-6}\ \text{m}^2}$$

$$B = 2.5\ \text{T}$$

permeability

Permeability is a measure of how easily a magnetic flux can be set up in a material. It corresponds to conductivity for electric currents. Materials that offer an easier path to the flux are said to have a higher permeability.

TABLE 13.1 Permeability of Materials

Material	Permeability
Vacuum (free space)	$4\pi \times 10^{-7}$ Wb/A · m
Nonmagnetic Has no effect on the magnetic field. Wood, glass, etc.	$4\pi \times 10^{-7}$ Wb/A · m
Diamagnetic Very slight opposition to the lines of force. Copper, silver, zinc, mercury, gold.	Slightly less than for free space.
Paramagnetic Slightly magnetic. Stainless steels, aluminum, platinum, oxygen, chromium, manganese.	Slightly higher than for free space.
Ferromagnetic Assist the lines of force. Iron, nickel, cobalt, ceramic ferrite magnetite.	Much greater than that of free space value. Depends on the flux density in the material.

Permeability is specified in webers per ampere-meter (Wb/A·m). For the magnetic field, the ampere is the product of the coil current and its number of turns. The old unit for permeability is webers per ampere-turn meter; here, μ will be used as the symbol for permeability.

The permeability of a vacuum and nonmagnetic materials is practically constant for all values of flux density, and is $4\pi \times 10^{-7}$ Wb/A·m. However, the permeability of ferromagnetic materials depends not only on the material but also on the flux density in the material. The permeability decreases as the flux density increases. Eventually, the flux density has very little change for an increase in current. At that point, the material is said to be saturated. A comparison of the permeability of different materials is given in Table 13.1. From a practical sense, most materials are either ferromagnetic or nonmagnetic.

relative permeability

The **relative permeability** (μ_r) compares a material's permeability to that of a vacuum. It is similar to relative conductivity, and has no units. The relative permeability is greater than 100 for ferromagnetic materials, and is 1 for nonmagnetic materials.

$$\mu_r = \mu/\mu_o \tag{13.2}$$

where μ_r is the relative permeability of the material; μ is the permeability of the material, in Wb/A · m; and μ_o is the permeability of a vacuum, $4\pi \times 10^{-7}$ Wb/A · m.

EXAMPLE 13.2

Find the relative permeability of a piece of iron that has a permeability of 2×10^{-3} Wb/A · m.

Solution

$$\mu_r = \frac{\mu}{\mu_o} = \frac{2 \times 10^{-3}\ \text{Wb/A} \cdot \text{m}}{4\pi \times 10^{-7}\ \text{Wb/A} \cdot \text{m}}$$

$$\mu_r = 1592$$

Magnetomotive Force (mmf). This is the force that maintains the flux in a magnetic circuit. Its symbol is $\mathcal{F}$, and its unit is the ampere. Before SI, ampere-turn was used. The mmf is provided by the current in a coil that is wound around the core of the path, as shown in Figure 13.15. There, the right-hand rule shows the flux direction to be clockwise. The force is a function of the current and the turns, and is given by

$$\mathcal{F} = NI \qquad \textbf{(13.3)}$$

where $\mathcal{F}$ is the magnetomotive force, in amperes; N is the number of turns in the coil; and I is the current in amperes.

The fact that $\mathcal{F}$ is a function of current lets us vary the strength of a magnet by changing the current.

reluctance

Reluctance. This is the opposition of a material to the magnetic field. It is analogous to resistance. Wood has a high reluctance while iron has a low reluctance. The effect of **reluctance** on the flux is seen in Figure 13.16. There, the flux in cast iron is compared to the flux in steel.

The symbol for reluctance is $\mathcal{R}$, and its unit is amperes per weber (A/Wb). Ampere-turn per weber is also sometimes used. Nonmagnetic materials have a high reluctance. On the other hand, magnetic materials have a low reluctance. That is one reason why some sort of steel is used for transformer cores.

Reluctance depends on the material, the length of the path, and the area of the cross section at right angles to the flux. That is,

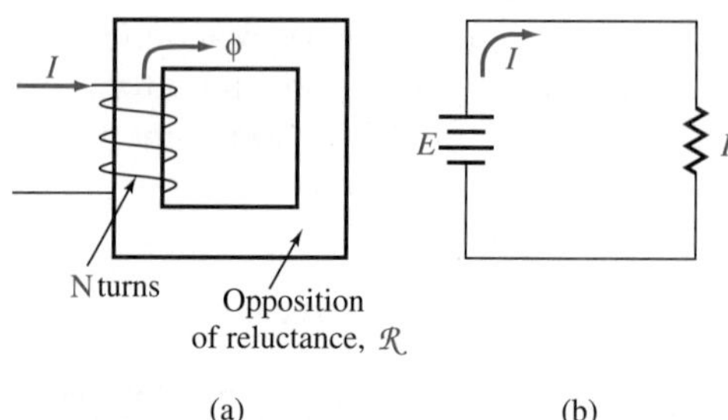

Figure 13.15 The magnetic circuit and its electric circuit parallel. (a) The magnetic circuit: flux ϕ; mmf NI; reluctance, $\mathcal{R}$. (b) The electric circuit: current I; emf E; resistance R.

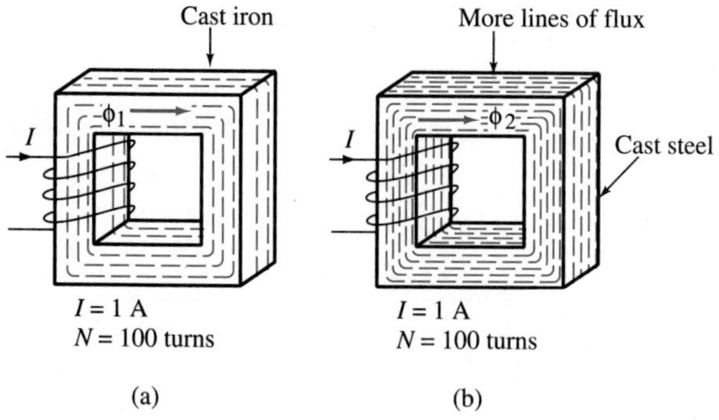

Figure 13.16 A given mmf produces a larger flux in the material having the smaller reluctance. (a) Cast iron, larger reluctance, fewer lines. (b) Cast steel, smaller reluctance, more lines.

$$\mathcal{R} = \frac{\ell}{\mu A} \tag{13.4}$$

where $\mathcal{R}$ is the reluctance in A/Wb; ℓ is the length of the path, in meters; μ is the permeability of the material, in Wb/A · m; and A is the cross-sectional area, in m^2.

EXAMPLE 13.3

What is the reluctance of an air gap that is 0.003 meter long, whose cross section dimensions are 0.02 m × 0.02 m?

Solution

$$\mathcal{R} = \frac{\ell}{\mu A}$$

$$A = (0.02 \text{ m})(0.02 \text{ m}) = 0.0004 \text{ m}^2$$

$$\mathcal{R} = \frac{0.003 \text{ m}}{(4\pi \times 10^{-7} \text{ Wb/A} \cdot \text{m})(0.0004 \text{ m}^2)}$$

$$\mathcal{R} = 5.97 \times 10^6 \text{ A/Wb}$$

13.3 Ohm's Law and Magnetizing Force

Flux, mmf, and reluctance are the parallels of current, emf, and resistance. This might suggest a relation similar to that for E, R, and I. Indeed, the equivalent of Ohm's Law does exist for the magnetic circuit. It states that the mmf is equal to the product of the flux and reluctance. Or

$$\mathcal{F} = \mathcal{R}\phi \tag{13.5}$$

where $\mathcal{F}$ is the mmf, in amperes; $\mathcal{R}$ is the reluctance, in amperes per weber; and ϕ is the flux, in webers.

The relation in Equation 13.5 is now applied in Example 13.4.

EXAMPLE 13.4

Find the reluctance, flux, and flux density for the core in Figure 13.17.

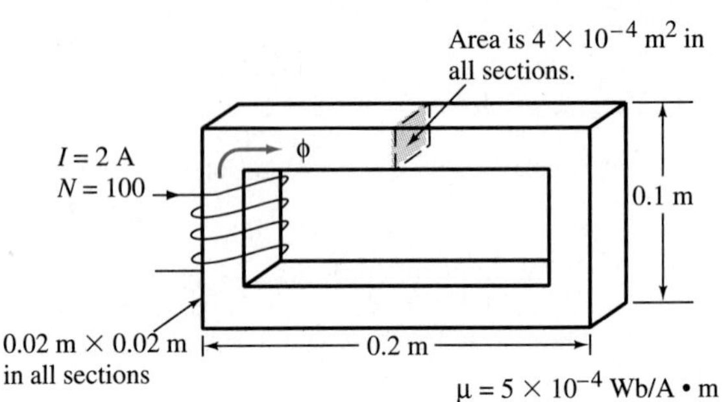

Figure 13.17

Solution

$$\mathcal{R} = \frac{l}{\mu A}$$

An inspection of the coil shows that the length of the path changes from the inner part of the core to the outer part. We will use the length of the mean path. This is a path around the center line of the core, as shown in Figure 13.18.

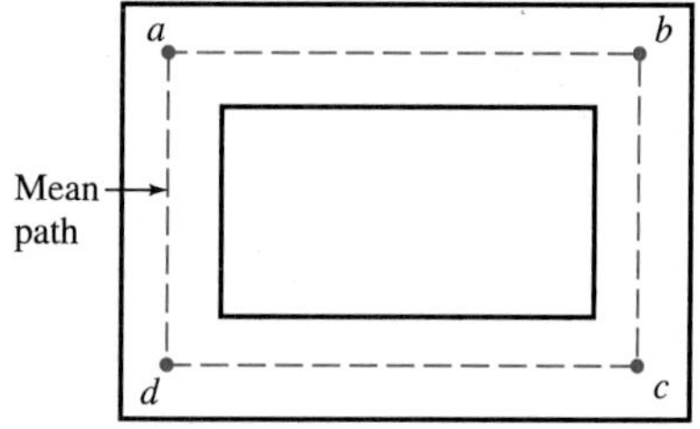

Figure 13.18

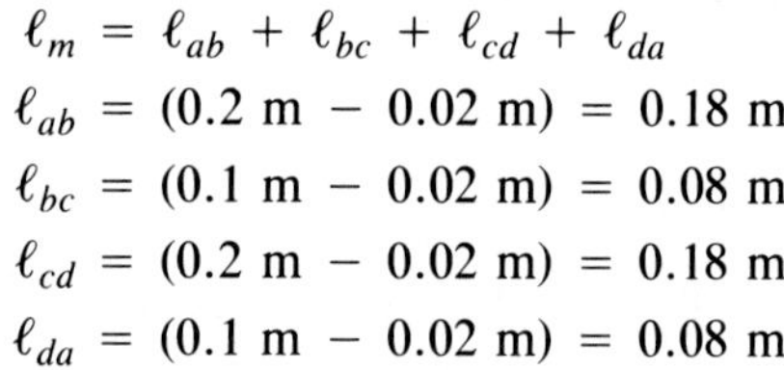

$$\ell_m = \ell_{ab} + \ell_{bc} + \ell_{cd} + \ell_{da}$$
$$\ell_{ab} = (0.2 \text{ m} - 0.02 \text{ m}) = 0.18 \text{ m}$$
$$\ell_{bc} = (0.1 \text{ m} - 0.02 \text{ m}) = 0.08 \text{ m}$$
$$\ell_{cd} = (0.2 \text{ m} - 0.02 \text{ m}) = 0.18 \text{ m}$$
$$\ell_{da} = (0.1 \text{ m} - 0.02 \text{ m}) = 0.08 \text{ m}$$

so

$$\ell_m = 0.18 \text{ m} + 0.08 \text{ m} + 0.18 \text{ m} + 0.08 \text{ m} = 0.52 \text{ m}$$

$$\mathcal{R} = \frac{0.52 \text{ m}}{(5 \times 10^{-4} \text{ WB/(A} \cdot \text{m}))(4 \times 10^{-4} \text{ m}^2)}$$

$$\mathcal{R} = 2.6 \times 10^6 \text{ A/Wb}$$

$$\mathcal{F} = \mathcal{R}\phi \quad \text{so} \quad \phi = \frac{\mathcal{F}}{\mathcal{R}}$$

Also

$$\mathcal{F} = NI$$

then

$$\phi = \frac{NI}{\mathcal{R}} = \frac{(100)(2 \text{ A})}{2.6 \times 10^6 \text{ A/Wb}}$$

$$\phi = 7.69 \times 10^{-5} \text{ Wb}$$

$$B = \frac{\phi}{A} = \frac{7.69 \times 10^{-5}\ \text{Wb}}{4 \times 10^{-4}\ \text{m}^2}$$

$$B = 0.192\ \text{T}$$

Using Equation 13.5 for magnetic materials is not practical because of the change in permeability with flux density. Instead, a quantity called magnetizing force (H) is used. **Magnetizing force** is the mmf needed to set up a flux density in 1 meter of a material (mmf per unit length). Its SI unit is amperes per meter (A/m), but it was once specified in ampere-turn per meter. In terms of circuit quantities, it represents the product of current and coil turns needed to produce an mmf in 1 meter of material. If the magnetizing force is known, the total mmf can be easily found. It is given by

magnetizing force

$$\mathcal{F} = H\ell \qquad \textbf{(13.6)}$$

where $\mathcal{F}$ is the mmf, in amperes; H is the magnetizing force, in amperes per meter; and ℓ is the length of the magnetic path, in meters.

EXAMPLE 13.5

Find the mmf needed to set up a flux in a 0.5-m length of iron if H is 200 A/m.

Solution

$$\mathcal{F} = H\ell$$

$$= (200\ \text{A/m})(0.5\ \text{m})$$

$$\mathcal{F} = 100\ \text{A}$$

People who work with magnetic circuits obtain values of magnetizing force from curves such as those in Figure 13.19.

These curves are called either **magnetization curves**, **B–H** curves, or magnetic saturation curves. They are obtained from experimental results. The curves may vary for different materials. They are also affected

magnetization curves

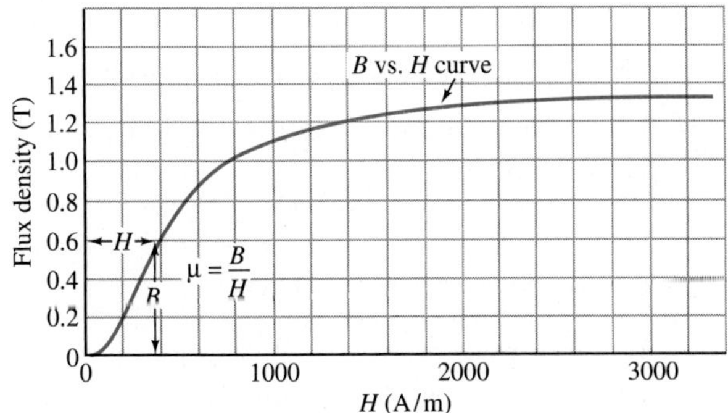

Figure 13.19 A typical B–H magnetizing curve for a magnetic material.

by physical processing, such as heat treatment. But, in general, the shape follows that shown in Figure 13.19.

EXAMPLE 13.6

Using the curves in Figure 13.19, find the magnetomotive force needed to maintain a flux density of 1.2 T in a 0.2-m path.

Solution

A vertical line from $B = 1.2$ T on the magnetization curve gives an H of 1260 A/m. Then

$$\mathcal{F} = H\ell$$
$$= (1260 \text{ A/m})(0.2 \text{ m})$$
$$\mathcal{F} = 252 \text{ A}$$

EXAMPLE 13.7

Find the mmf needed to set up a flux of 4×10^{-4} Wb in the cast steel core of Figure 13.20.

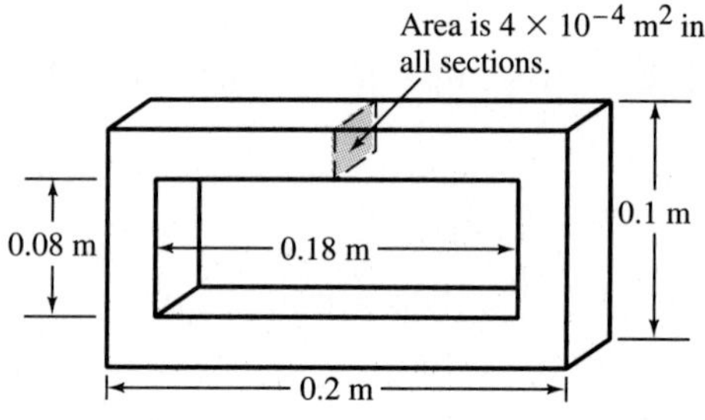

Figure 13.20

Solution

$$\text{mmf} = H\ell$$

ℓ is the length of the mean path. It is

$$\ell = \frac{\ell_{outer} + \ell_{inner}}{2}$$

$$\ell_{outer} = 0.2 \text{ m} + 0.1 \text{ m} + 0.2 \text{ m} + 0.1 \text{ m} = 0.6 \text{ m}$$

$$\ell_{inner} = 0.18 \text{ m} + 0.08 \text{ m} + 0.18 \text{ m} + 0.08 \text{ m} = 0.52 \text{ m}$$

$$\ell = \frac{0.6 \text{ m} + 0.52 \text{ m}}{2} = 0.56 \text{ m}$$

To get H, we must have the flux density.

$$B = \frac{\phi}{A} = \frac{4 \times 10^{-4} \text{ Wb}}{4 \times 10^{-4} \text{ m}^2} = 1 \text{ T}$$

From the *B–H* curve in Figure 13.21, *H* is about 730 A/m.

$$\text{mmf} = H\ell = (730 \text{ A/m})(0.56 \text{ m})$$
$$\text{mmf} = 408.8 \text{ A}$$

Magnetic circuits are designed to make use of the characteristic at a certain point or region on the curve. We now look at each of these regions in Figure 13.19.

Saturation. This is the region where there is very little change in flux density. This region gives the largest flux density. Devices such as power transformers operate here. Smaller cores, with less power loss in them, can be used.

Maximum Incremental Permeability. This is the part of the curve where the slope is the greatest. This region gives the greatest change in flux for a change in current. Devices such as magnetic earphones operate here. The larger change in flux causes the metal diaphragm of the earphone to deflect more. This gives better and louder reproductions.

The Nearly Linear Region. This is the part of the curve where the slope of the curve is constant. The change in flux for a given change in current is almost constant. Audio transformers operate here. The output voltage induced by the changing flux will closely resemble the input for this region.

Maximum Permeability. This is the region where there is maximum flux density for a given mmf. Devices designed for this region will require the least current.

Typical *B–H* curves for some ferromagnetic materials are shown in Figure 13.21. When these curves are used, μ and $\mathcal{R}$ do not have to be calculated.

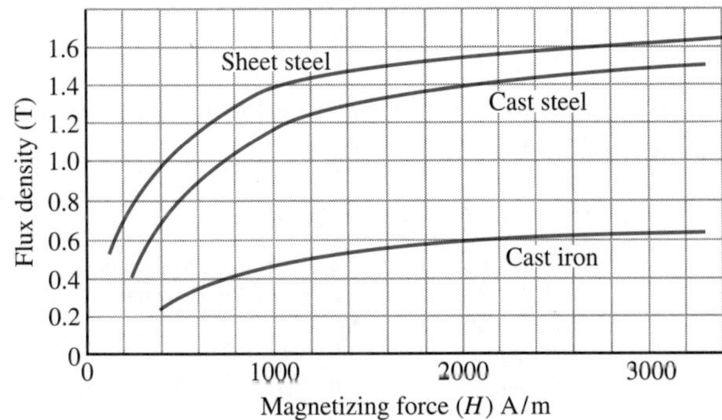

Figure 13.21 The magnetizing *B–H* curves for sheet steel, cast steel, and cast iron.

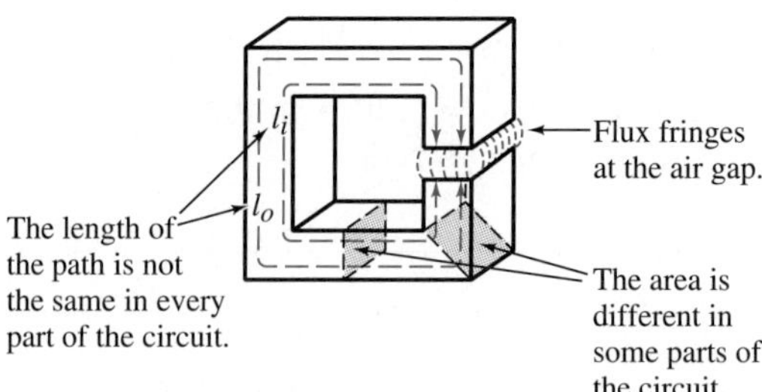

Figure 13.22 Some conditions that cause the difference between the calculated and actual results for the magnetic circuit.

13.4 Magnetic Circuit Analysis

Magnetic circuits can be analyzed using methods similar to those for electric circuit analysis. However, the results will not be as exact because of the reasons shown in Figure 13.22. They are:

1. Fringing of the flux at the edges of the air gap.
2. Differences in the length of the flux path.
3. Variation in the flux density in different parts of the area.

Nevertheless, the analysis provides a reasonable starting point for the design of a circuit.

The analysis of three circuits—the basic circuit, the series circuit, and the series-parallel circuit—is explained in this section. The methods used for these configurations can be applied to other configurations.

The basic circuit has one source of mmf and the entire path has the same permeability. It can be analyzed using the relations presented to this point. The procedure for the analysis is:

1. Find the flux density.
2. Find H from the curves.
3. Determine the mean length.
4. Determine $\mathcal{F}$ using Equation 13.6.
5. Determine N or I using Equation 13.3.

EXAMPLE 13.8

What are the mmf and coil current needed to maintain a flux of 4×10^{-4} webers in the section of Figure 13.23?

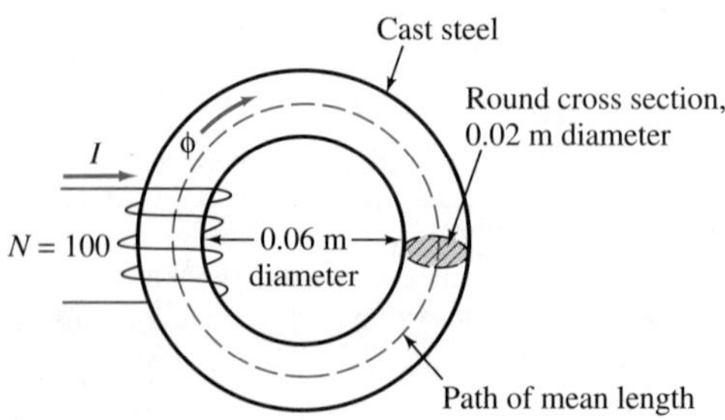

Figure 13.23

Solution

$$B = \frac{\phi}{A}$$

but

$$A = (\pi d^2)/4$$
$$= (\pi)(0.02\ \text{m})^2/4$$
$$= 3.14 \times 10^{-4}\ \text{m}^2$$

so

$$B = \frac{4 \times 10^{-4}\ \text{Wb}}{3.14 \times 10^{-4}\ \text{m}^2}$$
$$= 1.27\ T$$

From the cast steel curve of Figure 13.21, at $B = 1.27\ \text{Wb/m}^2$, H is approximately 1170 A/m.

The mean length is shown by the dashed line. It is the circumference of a circle whose diameter is

$$D = 0.06\ \text{m} + 0.01\ \text{m}$$
$$D = 0.07\ \text{m}$$

Then

$$\ell_m = \pi D$$
$$= (\pi)(0.07\ \text{m})$$
$$\ell_m = 0.22\ \text{m}$$
$$\mathcal{F} = H\ell_m$$
$$= (1700\ \text{A/m})(0.22\ \text{m})$$
$$\mathcal{F} = 374\ \text{A}$$
$$\mathcal{F} = NI$$

so

$$I = \mathcal{F}/N$$
$$= 374\ \text{A} \cdot \text{m}/100\ \text{Turns}$$
$$I = 3.74\ \text{A}$$

A series magnetic circuit also has one path for the flux, but the path includes two or more sections with different permeability. The difference in permeability can be due to different materials or different areas.

The analysis of a series circuit makes use of a loop mmf relation similar to Kirchhoff's Voltage Law. It is

$$\mathcal{F} - H_1\ell_1 - H_2\ell_2 - \cdots - H_n\ell_n = 0 \qquad \textbf{(13.7)}$$

where $\mathcal{F}$ is the magnetizing force in the series circuit, in amperes; H_1, H_2,

... , H_n are the magnetizing forces per unit length for each section, in amperes per meter; and $\ell_1, \ell_2, \ldots, \ell_n$ are the lengths of each section, in meters.

For magnetic materials, the H terms are taken from the B–H curves. For paramagnetic materials, the H terms are calculated using Equation 13.6.

The procedure for analyzing a series circuit when the flux is known is:

1. Identify the sections in the circuit.
2. Find the length of the mean path for each section.
3. Calculate the flux density in each section of the magnetic material.
4. Find B for each magnetic material section using the B–H curve and calculate each H.
5. Calculate $\mathcal{F}$ for each section of nonmagnetic material using Equation 13.5.
6. Use Equation 13.7 to calculate the total mmf for the circuit.

These steps are now used for a series circuit.

EXAMPLE 13.9

Find the mmf needed in the circuit shown in Figure 13.24 and the current in the coil.

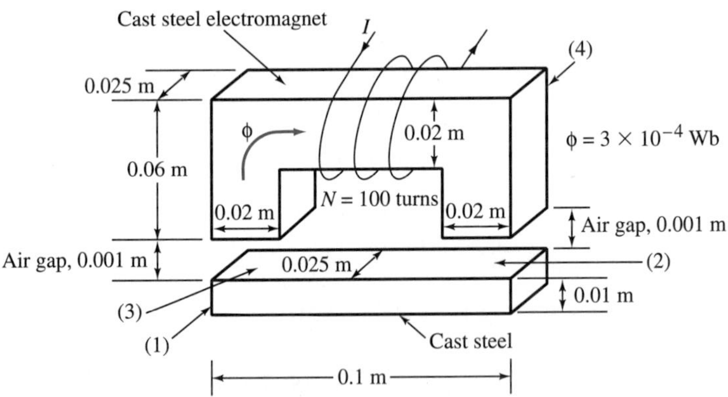

Figure 13.24

Solution

Step 1. The sections are (1) the armature; (2), (3) the air gaps—different material from (1); (4) the electromagnet—different area from (1).

Step 2. The mean path is shown in Figure 13.25 and the length of each section is

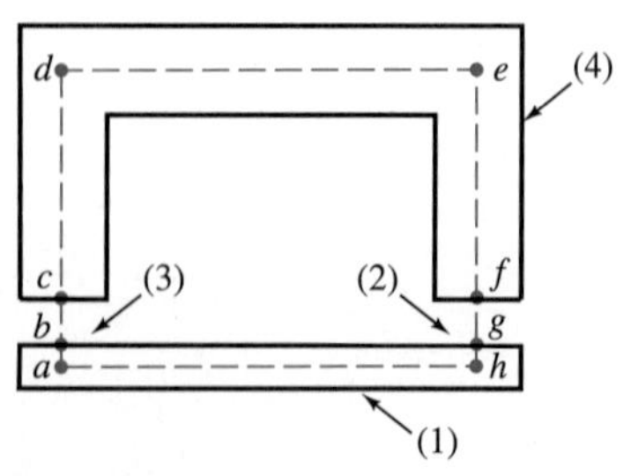

Figure 13.25

$$\ell_1 = \ell_{ab} + \ell_{ah} + \ell_{hg}$$
$$= 0.005 \text{ m} + (0.1 \text{ m} - 0.02 \text{ m}) + 0.005 \text{ m}$$
$$\ell_1 = 0.09 \text{ m}$$

$$\ell_2 = \ell_{fg},\ \ell_3 = \ell_{bc} = 0.001 \text{ m}$$

$$\begin{aligned}\ell_4 &= \ell_{cd} + \ell_{de} + \ell_{ef}\\ &= (0.06 \text{ m} - 0.01 \text{ m}) + (0.1 \text{ m} - 0.02 \text{ m}) + (0.06 \text{ m} - 0.01 \text{ m})\end{aligned}$$

$$\ell_4 = 0.18 \text{ m}$$

Step 3.

$$\begin{aligned}B_1 &= \frac{\phi_1}{A} = \frac{3 \times 10^{-4} \text{ Wb}}{(0.025 \text{ m})(0.01 \text{ m})}\\ &= 1.2 \text{ T}\end{aligned}$$

$$\begin{aligned}B_2 = B_3 &= \frac{\phi_3}{A_3} = \frac{3 \times 10^{-4} \text{ Wb}}{(0.025 \text{ m})(0.02 \text{ m})}\\ &= 0.6 \text{ T}\end{aligned}$$

$$\begin{aligned}B_4 &= \frac{\phi_4}{A_4} = \frac{3 \times 10^{-4} \text{ Wb}}{(0.025 \text{ m})(0.02 \text{ m})}\\ &= 0.6 \text{ T}\end{aligned}$$

Step 4. From the B–H curves for cast steel in Figure 13.21,

$$H_1 \text{ at } 1.2 \text{ T is } 1100 \text{ A/m}$$
$$H_4 \text{ at } 0.6 \text{ T is } 350 \text{ A/m}$$

The mmf drops are

$$H_1\ell_1 = (1100 \text{ A/m})(0.09 \text{ m}) = 99 \text{ A}$$
$$H_4\ell_4 = (350 \text{ A/m})(0.18 \text{ m}) = 63 \text{ A}$$

Step 5. For the air gaps, using Equation 13.5,

$$\begin{aligned}\mathcal{F}_2 &= \phi_2\mathcal{R}_2\\ &= \frac{\phi_2\ell_2}{\mu_o A_2}\\ &= \frac{(3 \times 10^{-4} \text{ Wb})(0.001 \text{ m})}{(4\,\pi \times 10^{-7} \text{ Wb/A} \cdot \text{m})(0.025 \text{ m} \times 0.02 \text{ m})}\end{aligned}$$

$$\mathcal{F}_2 = 477.5 \text{ A}$$

Since the other air gap has the same dimensions and flux,

$$\mathcal{F}_3 = 477.5 \text{ A}$$

Step 6.

$$\begin{aligned}\mathcal{F} &= H_1\ell_1 + H_2\ell_2 + H_3\ell_3 + H_4\ell_4\\ &= 99 \text{ A} + 477.5 \text{ A} + 477.5 \text{ A} + 63 \text{ A}\end{aligned}$$

$$\mathcal{F} = 1117 \text{ A}$$

$$\mathcal{F} = NI$$

so

$$I = \frac{\mathcal{F}}{N}$$

$$= \frac{1117 \text{ A}}{100}$$

$$I = 11.17 \text{ A}$$

Most of the mmf is needed to maintain the flux in the air gap. This can be used as a check on the accuracy of the calculations. In fact, considering only the air gaps gives satisfactory results for many applications.

The series-parallel circuit is the final circuit configuration that is considered. An example of one is shown in Figure 13.26. Notice that the coil flux divides between the two branches of the core. The analysis of these circuits follows the procedure used for electric circuits. For the magnetic circuit, a branch flux relation similar to Kirchhoff's Current Law can be written. It is

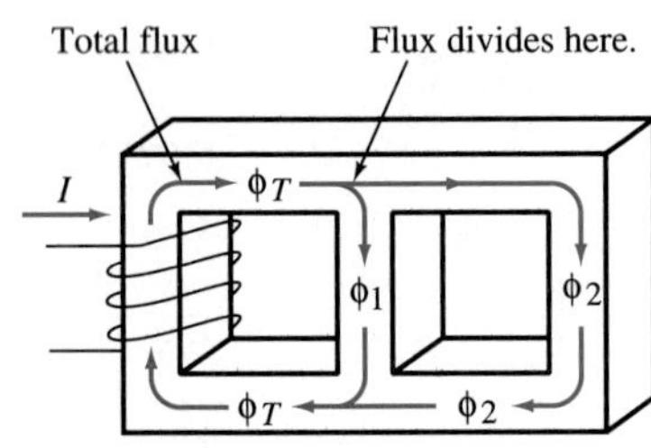

Figure 13.26 An example of a series-parallel magnetic circuit.

$$\phi_T - \phi_1 - \phi_2 - \cdots - \phi_n = 0 \qquad \textbf{(13.8)}$$

where ϕ_T is the flux entering the parallel branches, in webers, and ϕ_1, ϕ_2, ... , ϕ_n are the flux in each parallel branch, in webers.

The procedure for analyzing these circuits is:

1. Use the right-hand rule to find the coil flux direction, and draw the flux in each part of the circuit.
2. Identify the number of loops and the number of junctions of three or more branches.
3. Identify the sections of different permeability and calculate their mean length.
4. Determine the flux density in as many sections as possible. Use the *B–H* curves to determine *H* for the magnetic material sections.
5. Calculate the mmf for as many sections as possible.
6. Write Equation 13.7 for the number of loops less one. Coil mmf is positive if the tracing direction through the coil is in the direction of the flux. $H\ell$ terms are negative if the tracing direction is in the direction of the flux.
7. Insert numerical values in the equations of Step 6 and solve for as many unknown *H*'s as possible.
8. Use the results of Step 7 to determine *B* in other sections. Then calculate the flux in those sections.
9. Write Equation 13.8 to solve for the unknown flux.
10. Calculate the flux density for the flux in Step 9. Then determine the ampere turns from the *B–H* curves or Equation 13.5.
11. Use all known values to solve the other equations.
12. If necessary, repeat Steps 9, 10, and 11 for other fluxes.

These steps are applied to the solution of the next example.

Example 13.10 is a typical series-parallel circuit. The procedure is applied to its analysis.

EXAMPLE 13.10

Find the mmf needed for a flux of 3×10^{-4} Wb in the air gap of Figure 13.27.

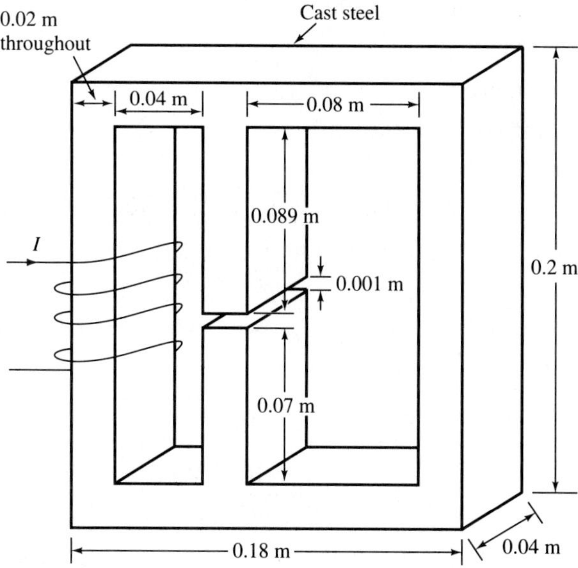

Figure 13.27

Solution

1. The direction of ϕ_T must be upward. At Point C, it branches off into the center leg and the right leg, as shown by ϕ_g and ϕ_1.
2. This is a three-loop circuit, so we have two loop equations and one node equation available.
3. The mean path for the flux is shown in Figure 13.28. The sections are *fabc*, *ch*, *hg*, *gf*, and *cdef*. The mean lengths are

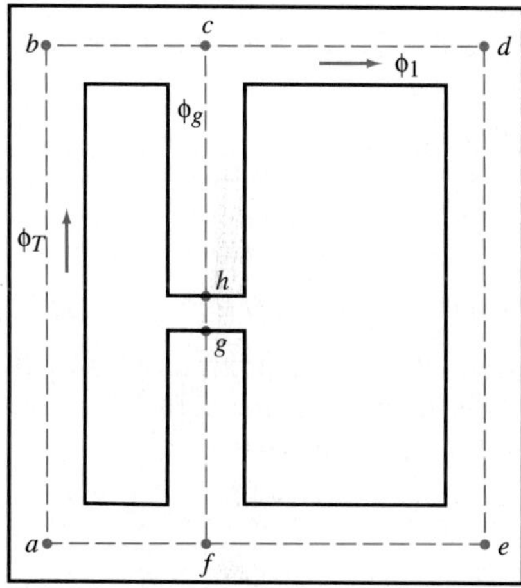

Figure 13.28

$$\ell_{fabc} = (0.01\text{ m} + 0.04\text{ m} + 0.01\text{ m}) + (0.2\text{ m} - 0.02\text{ m}) + (0.01\text{ m} + 0.04\text{ m} + 0.01\text{ m}) = 0.3\text{ m}$$

$$\ell_{ch} = (0.01\text{ m} + 0.089\text{ m}) = 0.099\text{ m}$$

$$\ell_{hg} = 0.001\text{ m}$$

$$\ell_{gf} = (0.07\text{ m} + 0.01\text{ m}) = 0.08\text{ m}$$

$$\ell_{cdef} = (0.01\text{ m} + 0.08\text{ m} + 0.01\text{ m}) + (0.2\text{ m} - 0.02\text{ m}) + (0.01\text{ m} + 0.08\text{ m} + 0.01\text{ m}) = 0.38\text{ m}$$

4. The flux in the center leg is known, so B and H for sections ch, hg, and gf can also be calculated.

$$B_{ch} = B_{hg} = B_{gf} = \frac{\phi_g}{A_{ch}} = \frac{3 \times 10^{-4}\text{ Wb}}{(0.02\text{ m})(0.04\text{ m})}$$

$$B_{ch} = 0.375\text{ T}$$

From the cast steel B–H curve, H is about 220 At/m.

5. $H_{ch}\ell_{ch} = (220\text{ A/m})(0.099\text{ m}) = 21.78\text{ A}$
   $H_{gf}\ell_{gf} = (220\text{ A/m})(0.08\text{ m}) = 17.60\text{ A}$

For the air gap, using Equation 13.5,

$$\mathscr{F}_{gh} = \phi_g \mathscr{R}_{gh} = \frac{\phi_g \ell_{gh}}{\mu_o A_{gh}} = \frac{(3 \times 10^{-4}\text{ Wb})(0.001\text{ m})}{(4\pi \times 10^{-7}\text{ Wb/A}\cdot\text{m})(0.04\text{ m})(0.02\text{ m})}$$

$$\mathscr{F}_{gh} = 298.4\text{ A}$$

6. Writing Equation 13.7 for the two loops, tracing in a clockwise direction, gives

Left loop (clockwise):

$$NI - (H_{fabc})(\ell_{fabc}) - (H_{ch})(\ell_{ch}) - (H_{hg})(\ell_{hg}) - (H_{gf})(\ell_{gf}) = 0$$

Right loop (clockwise):

$$(H_{gf})(\ell_{gf}) + (H_{gh})(\ell_{gh}) + (H_{ch})(\ell_{ch}) - (H_{cdef})(\ell_{cdef}) = 0$$

7. An inspection of these equations shows that every quantity except H_{cdef} is known in the right loop equation. Substituting values into this equation gives

$$17.60\text{ A} + 298.4\text{ A} + 21.78\text{ A} = (H_{cdef})(0.38\text{ m})$$

$$H_{cdef} = \frac{337.78\text{ A}}{0.38\text{ m}} = 888.9\text{ A/m}$$

8. From the B–H curve for cast steel, at 888.9 A/m, B_{cdef} is approximately 1.1 T. Then

$$\begin{aligned}\phi_1 &= (B_{cdef})(A_{cdef})\\ &= (1.1\ \text{Wb/m}^2)(0.02\ \text{m} \times 0.04\ \text{m})\\ &= 8.8 \times 10^{-4}\ \text{Wb}\end{aligned}$$

9. Writing Equation 13.8 gives

$$\begin{aligned}\phi_T &= \phi_g + \phi_1\\ &= 3 \times 10^{-4}\ \text{Wb} + 8.8 \times 10^{-4}\ \text{Wb}\\ \phi_T &= 11.8 \times 10^{-4}\ \text{Wb}\end{aligned}$$

10.

$$\begin{aligned}B_{fabc} &= \frac{\phi_T}{A_{fabc}}\\ &= \frac{11.8 \times 10^{-4}\ \text{Wb}}{(0.02\ \text{m})(0.04\ \text{m})}\\ &= 1.48\ \text{T}\end{aligned}$$

From the B–H curve for cast steel, H is about 3000 A/m at that density.

11. Putting the value of H in the equation for the left loop gives

$$NI - (3000\ \text{A/m})(0.3\ \text{m}) - 21.78\ \text{A} - 298.4\ \text{A} - 17.60\ \text{A} = 0$$

Hence

$$\begin{aligned}NI &= 1237.8\ \text{A}\\ \mathcal{F} &= 1237.8\ \text{A}\end{aligned}$$

Although the size and shape of the circuit can vary, the methods described in these examples will enable you to analyze other circuits.

13.5 The Tractive Force of a Magnet

Some devices use the magnetic field for their operation. Others such as electric cranes, magnetic circuit breakers, relays, and door latches use the **tractive force** of the field. Tractive force is the amount of pull that the magnet exerts on a piece of iron.

tractive force

This force depends on flux density and the area common to both surfaces. The tractive force between two parallel magnetized surfaces in free space is given by

$$F = \frac{B^2 A}{2\mu_o} \tag{13.9}$$

where F is the force, in newtons; B is the flux density, in tesla (Wb/m^2); A is the area common to both surfaces, in m^2; and μ_o is the permeability of free space ($4\pi \times 10^{-7}$) in Wb/A · m.

EXAMPLE 13.11

How much force acts on the armature of the relay shown in Figure 13.29?

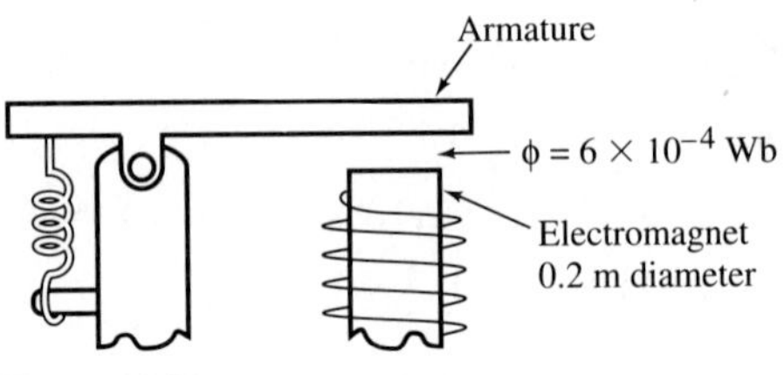

Figure 13.29

Solution

$$F = \frac{B^2A}{2\mu_o}$$

$$B = \frac{\phi}{A} = \frac{6 \times 10^{-4}\ \text{Wb}}{\dfrac{(\pi)(0.2\ \text{m})^2}{4}}$$

$$= \frac{6 \times 10^{-4}\ \text{Wb}}{0.031\ \text{m}^2}$$

$$B = 0.019\ \text{Wb/m}^2$$

$$F = \frac{(0.019\ T)^2(0.031\ \text{m}^2)}{(2)(4\pi \times 10^{-7}\ \text{Wb/A} \cdot \text{m})}$$

$$F = 4.45\ \text{newtons}$$

EXAMPLE 13.12

What flux density is needed in the electromagnet of Figure 13.30 to pick up the bar, if it weighs 20 newtons?

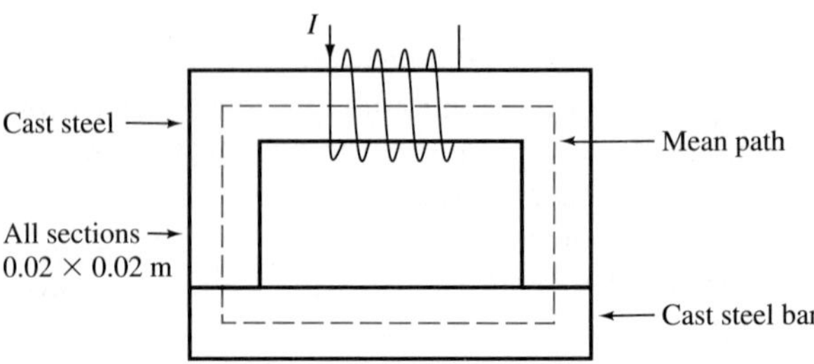

Figure 13.30

Solution

$$F = \frac{B^2A}{2\mu_o}$$

The total common area is (2)(0.02 m × 0.02 m) or 8×10^{-4} m^2.

$$B^2 = \frac{(F)(2\mu_o)}{A}$$

$$= \frac{(20\ N)(2)(4\pi \times 10^{-7}\ \text{Wb/A} \cdot \text{m})}{8 \times 10^{-4}\ \text{m}^2}$$

$$= 62.83 \times 10^{-3}\ \text{T}^2$$

$$B = 0.25\ \text{T}$$

13.6 Electromagnetic Induction

electromagnetic induction

Electromagnetic induction is the generating of an emf in a conductor by exposing it to a change in the lines of a magnetic field. This can be demonstrated by the simple experiment in Figure 13.31. The voltmeter will show a voltage when the conductor is moved up or down. In this example, the change in the lines was caused by the motion of the conductor in the field.

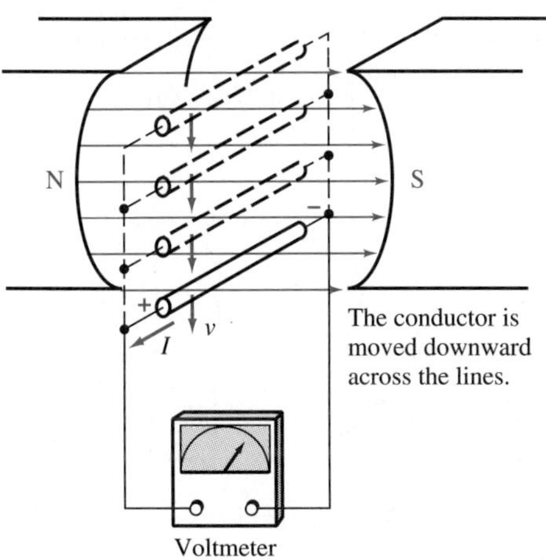

Figure 13.31 An emf is induced in a conductor as it moves across the lines of the field.

It was in the early 1830's that this principle was discovered by Joseph Henry, an American physicist, and Michael Faraday (1791–1867), an English physicist. Before that time, chemical batteries supplied most of the electricity. Now, generators and alternators use electromagnetic induction to produce the electricity for most of the world's needs.

Some characteristics of electromagnetic induction can be observed in the diagram of Figure 13.32. They are:

1. An emf is induced only when the conductor is moved at some angle to the lines. It is greatest when the motion is at right angles to the lines.
2. The emf is larger for faster rates of motion and for larger flux densities.

Figure 13.32 Electromagnetic induction depends on the direction of motion, the rate of motion, and the flux density. (a) The effect of the direction on the emf. (b) The effect of the rate of motion on the emf. (c) The effect of the flux density on the emf.

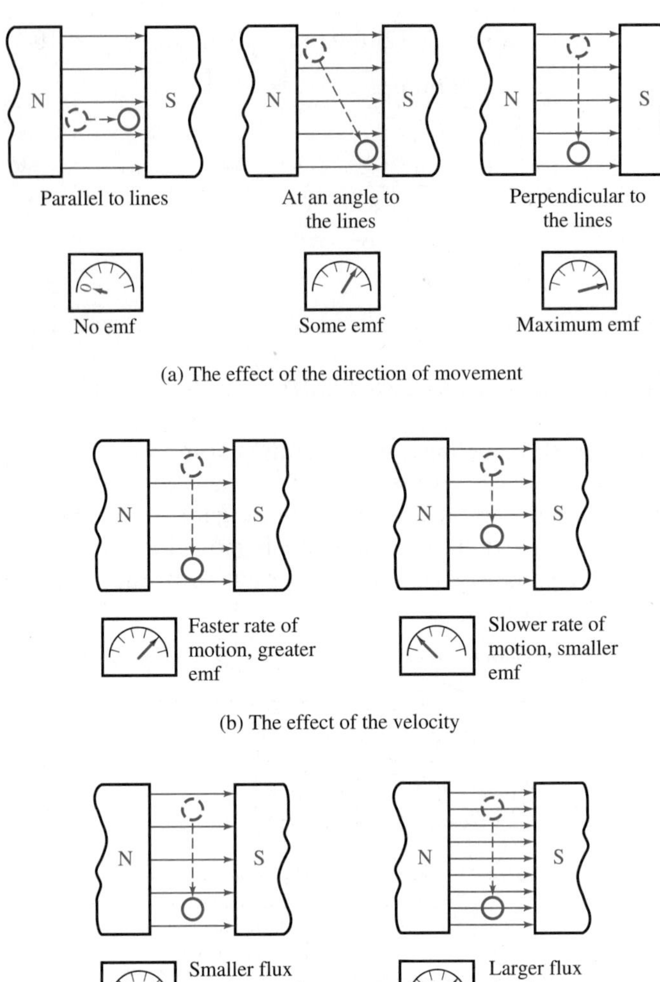

3. The polarity of the emf depends on the direction of the motion. For the magnetic field direction shown, the front end of the conductor is positive when the motion is down. It is negative when the motion is up.

These characteristics are set forth in Faraday's Law, named after M. Faraday, and Lenz's Law, named after Heinrich Lenz (1804–1865), a German physicist.

Faraday's Law

Faraday's Law states that the emf is directly proportional to the rate at which the magnetic field around the conductor changes with respect to time. When the conductor moves parallel to the lines, it does not move across any lines, and the emf is zero. But when it moves perpendicular to the lines, it cuts across the greatest number of lines, and the emf is a maximum. Also, when the flux is larger or the rate of the movement is faster, more lines are cut in a given time.

Lenz's Law

Lenz's Law states that the polarity of the induced emf is such that the induced current will set up a magnetic field opposing the change in the original field. Since the conductor's movement is causing a change in the field, the emf must set up a force that opposes the movement. It does this by forming a magnetic field that adds to the field ahead of the conductor and subtracts from the field behind it. The greater flux density ahead of it exerts an opposing force on the conductor's movement.

The general relation for the induced emf according to Faraday's Law is

$$e = N d\phi/dt \tag{13.10}$$

where e is the instantaneous value of the induced emf, in volts; N is the number of conductors connected in series; and $d\phi/dt$ is the number of lines that are cut by the conductor in Wb/s in the time dt.

The quantities $d\phi$ and dt are called differentials in calculus, and stand for infinitesimally small changes. They were introduced in Chapter 12.

average emf

Most of the time, measuring $d\phi/dt$ is not practical, so the average change for a larger amount of time must be used. This is $\Delta\phi/\Delta t$. The result is an **average emf** of the values induced in the time Δt. It is

$$e_{ave} = N\frac{\Delta\phi}{\Delta t} \tag{13.11}$$

where e_{ave} is the average value of the emf induced in a time Δt, in volts; N is the number of conductors connected in series; and $\Delta\phi$ is the number of lines that are cut by the conductor in a time Δt, in Wb/s.

EXAMPLE 13.13

Find the average emf induced across a 50-turn coil when a flux of 4×10^{-3} Wb is:

a. reduced to 0 in 2 ms.
b. increased to 8×10^{-3} Wb in 2 ms.

Solution

a.
$$e_{ave} = N\frac{\Delta\phi}{\Delta t} = \frac{(50)(4 \times 10^{-3}\text{ Wb})}{2 \times 10^{-3}\text{ s}}$$
$$e_{ave} = 100\text{ V}$$

b.
$$e_{ave} = N\frac{\Delta\phi}{\Delta t} = \frac{(50)(4 \times 10^{-3}\text{ Wb} - 8 \times 10^{-3}\text{ Wb})}{2 \times 10^{-3}\text{ s}}$$
$$e_{ave} = -100\text{ V}$$

Part b is opposite to that of part a because the flux change is opposite to that in part a.

The induced emf in a moving conductor whose length is at right angles to the field can also be expressed as

$$e = B\ell\upsilon \sin \theta \qquad \textbf{(13.12)}$$

where e is the induced emf, in volts; B is the flux density, in T; ℓ is the length of the conductor in the magnetic field, in meters; υ is the velocity of the conductor, in m/s; and θ is the angle between the direction of the velocity and the field.

This relation is more practical than the relation in Equation 13.11. It also shows what was stated in Faraday's Law.

EXAMPLE 13.14

A conductor moves in a circle, as shown in Figure 13.33, at 1200 revolutions per minute (RPM). What is the induced emf:

a. at the position shown?
b. at Point 1?

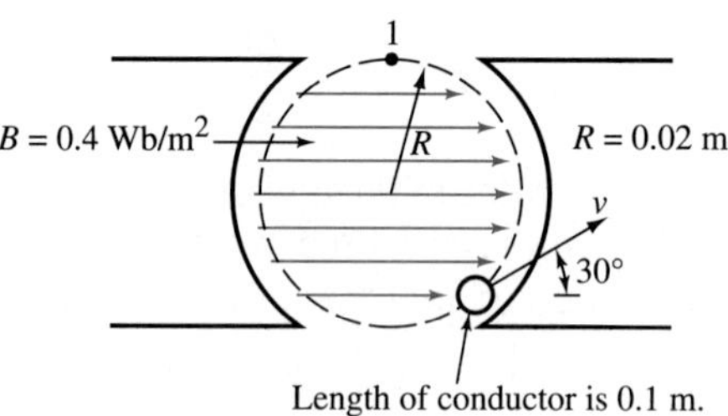

Figure 13.33

Solution

a. $e = B\ell\upsilon \sin \theta$

The RPM must be changed to m/s.

$$(1200 \text{ RPM}) \left(\frac{1 \text{ min}}{60 \text{ s}}\right)(2\pi)(0.02 \text{ m}) = 2.51 \text{ m/s}$$

The angle between the velocity and the lines is 30° so

$$e = (0.4)(0.1)(2.51)(\sin 30°)$$
$$e = 0.05 \text{ V}$$

b. At 1,

$$\sin \theta = \sin 180° = 0$$

so

$$e = 0 \text{ V}$$

This agrees with the fact that there is no emf when the movement is parallel to the lines.

Moving a conductor through a field is one way to obtain a changing flux. Another way is to change the flux's magnitude, direction, or both. A common application of this method is in the transformer. There the changing flux in one transformer winding is used to induce an emf in another transformer winding. This induction is called **mutual induction** because the flux is common to both coils.

mutual induction

There is mutual induction in the basic transformer of Figure 13.34. The flux from the primary coil links, or encircles, the secondary coil. If the current is constant, such as for a dc voltage, $\Delta\phi/\Delta t = 0$ and no emf is induced in the secondary coil. But for a change in flux $(d\phi/dt)$, an emf equal to $N(d\phi/dt)$, $(e_{ave} = Nd\phi/dt)$, will be induced across the secondary coil. According to Lenz's Law, the induced emf produces a current that will counteract the change in flux.

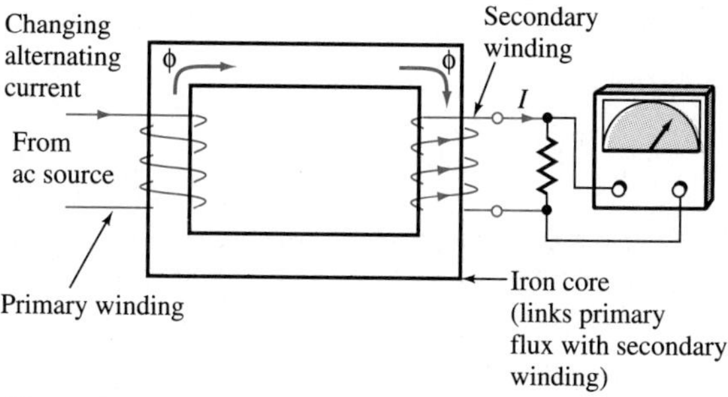

Figure 13.34 The transformer uses mutual induction to induce an emf by the changes in the magnetic field caused by the alternating current.

EXAMPLE 13.15

a. What is the emf induced across Coil 2 of Figure 13.35 if the flux is decreased to 0 webers in 6 milliseconds?
b. What is the polarity of the terminals?

Solution

a. $$e_{ave} = N\frac{\Delta\phi}{\Delta t} = \frac{(200)(6 \times 10^{-3}\text{ Wb} - 0\text{ Wb})}{6 \times 10^{-3}\text{ s}}$$

$$e_{ave} = 200\text{ V}$$

b. The flux is decreasing, so the emf must cause a current that produces a flux which will add to the flux. The flux must be in a downward direction. According to the right-hand rule, the current must be as shown in Figure 13.36.

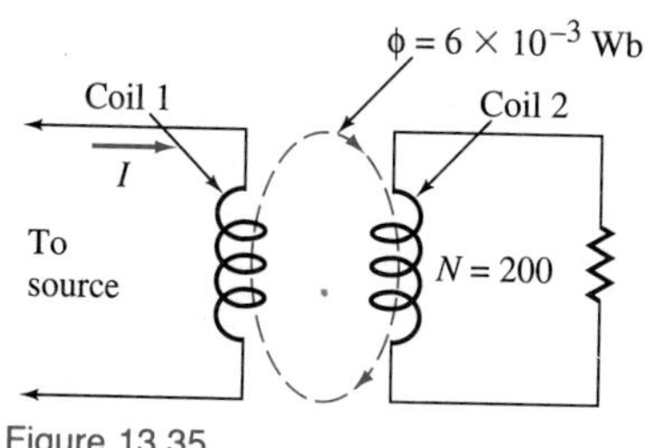

Figure 13.35

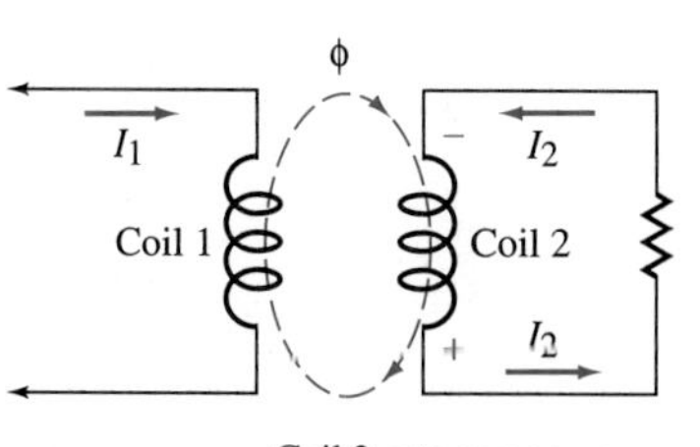

Coil 2 acts as a source so the current leaves the terminal.

Figure 13.36

13.7 Self-Inductance

Self-**inductance** is the property of a device to oppose a change in current. All conductors, even straight ones, have some inductance.

inductance

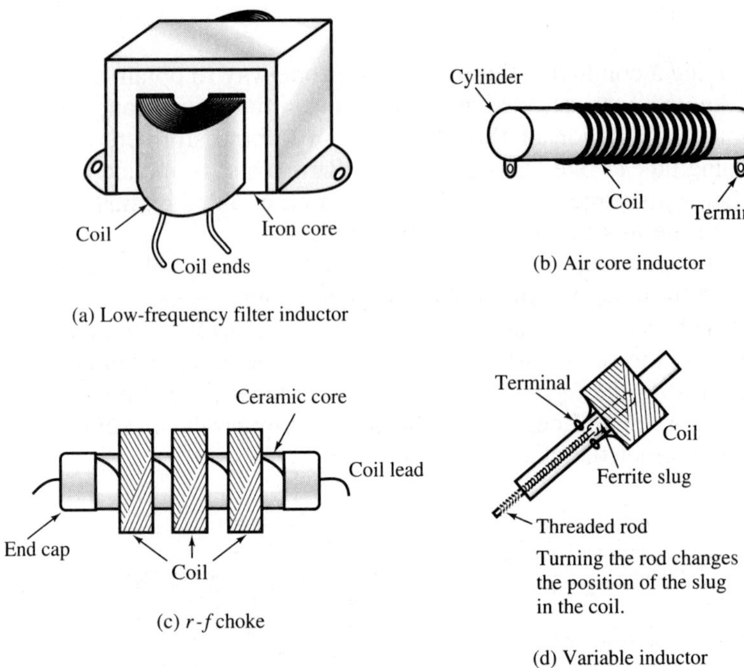

Figure 13.37 Inductors are made in many sizes and shapes. (a) A low-frequency filter inductor. (b) An air core inductor. (c) An *r–f* choke. (d) A variable inductor.

Although of no consequence in low frequency circuits, this small or stray inductance can affect the characteristics and operation of a circuit at higher frequencies.

inductors

Devices that are designed and built to have a specific amount of inductance are called **inductors**. They are wound on a coil form and sometimes use a ferromagnetic core. These cores result in a higher inductance. Some typical inductors are shown in Figure 13.37.

henry

The unit of inductance is the **henry** (H), named after Joseph Henry, and its letter symbol is L. A 1-henry inductor will have an emf of 1 volt induced across its terminals when the current changes at a steady rate of 1 ampere per second. Inductors are used in low and high frequency filter circuits, time delay circuits, tuning circuits, and pulse shaping circuits.

A study of Figure 13.38 will help to understand the action of the inductor. In Figure 13.38(a), a decrease in E will cause a decrease in the current. The changing current results in a change in flux. To oppose the change, the induced emf must add to the emf in the circuit. For this to be so, the induced emf acting as a source must take the polarity shown in Figure 13.38(b). Since the emf acts opposite, or counter, to the change, it is often called a counter emf or "cemf."

If the source emf is increased, causing an increase in current, the cemf must oppose the circuit emf. The induced emf polarity will now be opposite to that for the decreasing current. In each instance, the current from the induced emf tries to make up for the change. The change in current can also be caused by a change in the circuit resistance, with the same results.

Some characteristics of inductance in a circuit are:

1. An emf is induced only when there is a change in current.
2. The emf polarity is such that the current set up by it acts against the change.
3. A faster rate of change in current induces a higher emf.
4. Larger inductance results in a slower rate of change of current.

In terms of the rate of current change and the inductance, the emf is given by

$$e = L\,di/dt \tag{13.13}$$

where e is the instantaneous value of the counter emf induced by the inductance, in volts; L is the inductance, in henries; and di/dt is the change in current in an infinitesimally small amount of time, in amperes per second.

Again, as for $d\phi/dt$, an average emf can be obtained using $\Delta i/\Delta t$. Thus,

$$e_{\text{ave}} = L(\Delta i/\Delta t) \tag{13.14}$$

where e_{ave} is the average value of emf induced by the inductance, in volts; L is the inductance, in henries; and $\Delta i/\Delta t$ is the average change in current in the time Δt, in amperes per second.

EXAMPLE 13.16

What is the average emf induced across a 5-H coil if the current drops from 0.8 A to 0.6 A in 0.01 second?

Solution

$$e_{\text{ave}} = L\frac{\Delta i}{\Delta t} = (5\text{ H})\left(\frac{0.8\text{ A} - 0.6\text{ A}}{0.01\text{ s}}\right)$$

$$e_{\text{ave}} = 100\text{ V}$$

Since $e = L\,di/dt$ and $e = N\,d\phi/dt$, $L = N\,d\phi/di$. Thus, the inductance is related to the amount of flux that links the coils. This linkage is affected by the length, diameter, winding pattern, and permeability of the material in and around the coil. Because of the many possible variations, any equation for the inductance will not give exact results for every coil. But when the length is much greater than the diameter, 10 times or more, and the permeability is constant, a basic relation that gives reasonably close results is

$$L = \frac{\mu N^2 A}{\ell} \tag{13.15}$$

where L is the inductance, in henries; μ is the permeability, in Wb/A · m; A is the area of the coil, in m^2; and ℓ is the length of the coil, in meters.

Equation 13.15 shows that increasing N, μ, or A increases the inductance. Making the coil length longer gives a smaller inductance. Equations for other shapes and winding patterns have been determined empirically, and are available in electrical handbooks.

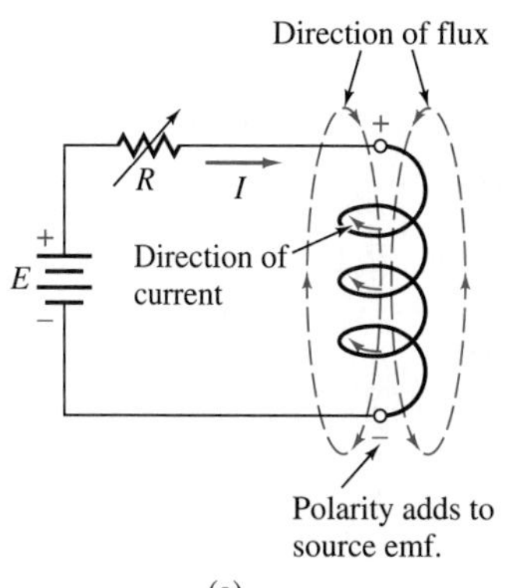

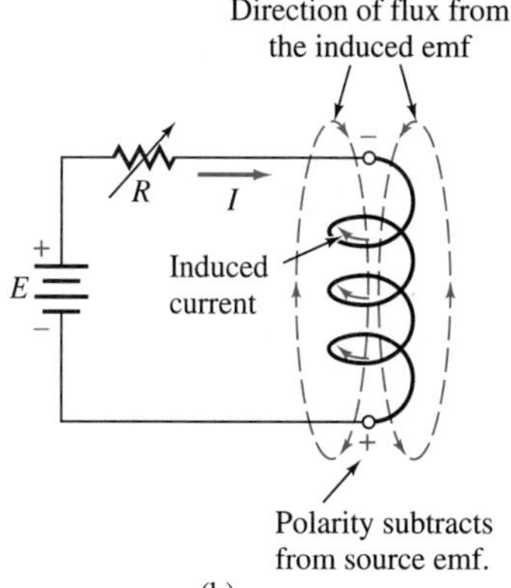

Figure 13.38 The polarity of the induced emf for a change in current and flux. (a) For a decrease in the emf of E, the current will want to decrease. The polarity of the induced emf will add to the emf of E. (b) For an increase in the emf of E, the current will want to increase. The polarity of the induced emf will subtract from the emf of E.

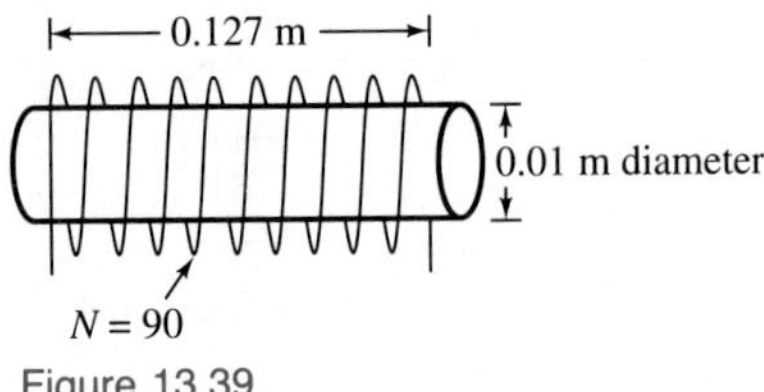

Figure 13.39

EXAMPLE 13.17

Find the inductance of the 90-turn coil shown in Figure 13.39.

Solution

$$L = \frac{\mu N^2 A}{\ell}$$

$$= \left(\frac{\mu N^2}{\ell}\right)\left(\frac{\pi D^2}{4}\right)$$

$$= \left[\frac{(4\pi \times 10^{-7})(90)^2}{0.127 \text{ m}}\right]\left[\frac{(\pi)(0.01 \text{ m})^2}{4}\right]$$

$$L = 6.29 \times 10^{-6} \text{ H}$$

Although the induced emf in generators and transformers serves a useful function, induced emf can also cause damage to equipment. Consider motor windings and relay coils that are very inductive. A sudden change in current can induce a large emf. This emf can break down the winding insulation or cause arcing across the switch contacts. One way of preventing this is to use a "free-wheeling" protective circuit, as shown in Figure 13.40. This circuit provides a path through which the stored energy can be dissipated.

Since the induced emf can be large, even for small currents and source voltage, one should be careful when dealing with inductive circuits. The large emf can result in electric shock.

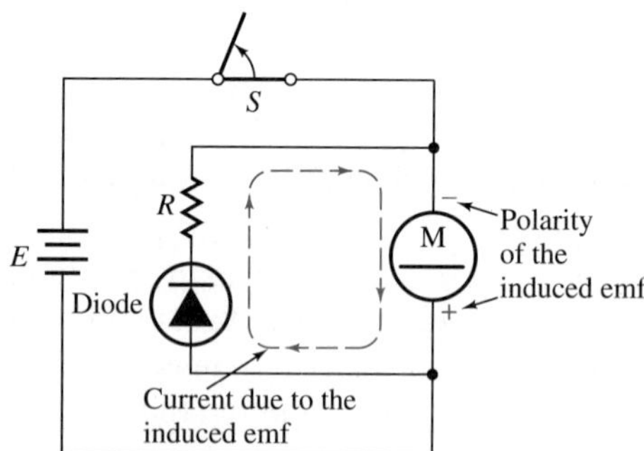

Figure 13.40 The free-wheeling circuit prevents the induced emf from damaging the circuit by providing a path for the energy.

EXAMPLE 13.18

What is the induced emf across a 0.5-H motor winding when the current drops from 10 A to 0 A in 5 milliseconds?

Solution

$$E_{\text{ave}} = L\frac{\Delta i}{\Delta t}$$

$$= \frac{(0.5\text{ H})(10\text{ A} - 0\text{ A})}{5 \times 10^{-3}\text{ s}}$$

$$E_{\text{ave}} = 1000\text{ V}$$

A wire-wound resistor, being coiled wire, has inductance. This inductance can cause problems at high frequencies. One way of reducing the inductance is to wind the turns in opposite directions, as shown in Figure 13.41. The magnetic fields of the turns cancel each other so there is no inductance.

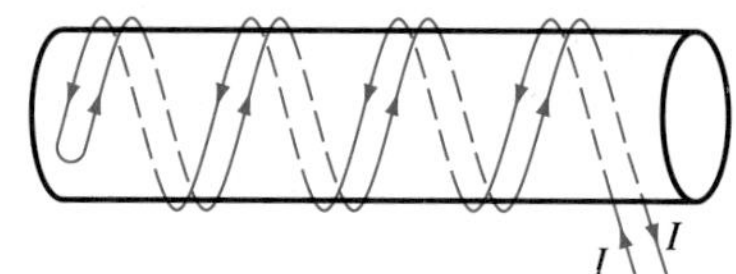

Figure 13.41 A noninductive winding can be made by winding the turns in opposite directions to cancel the magnetic field.

13.8 Equivalent Inductance of Series Inductors

Normally, one would obtain the value of inductance needed, and use only one inductor. But there are times when the desired value is not available, so several inductors may be connected in series to give the needed inductance. Also, some approximations in circuit analysis can result in a series connection of inductors.

series inductors

Series inductors are connected in series when there is no branch point between them. Inductors that are connected in series can be replaced by an **equivalent inductance**. This is an inductance that induces the same emf as the circuit it replaces for the same change in current.

equivalent inductance

In Figure 13.42(a), two ideal inductors are shown connected in series. These are inductors that do not have resistance. Although practical inductors have resistance, it does not affect the equivalent inductance. Considering the circuit of Figure 13.42, a change in current will induce the following emfs.

$$e_1 = L_1\frac{\Delta i}{\Delta t} \quad \text{and} \quad e_2 = L_2\frac{\Delta i}{\Delta t}$$

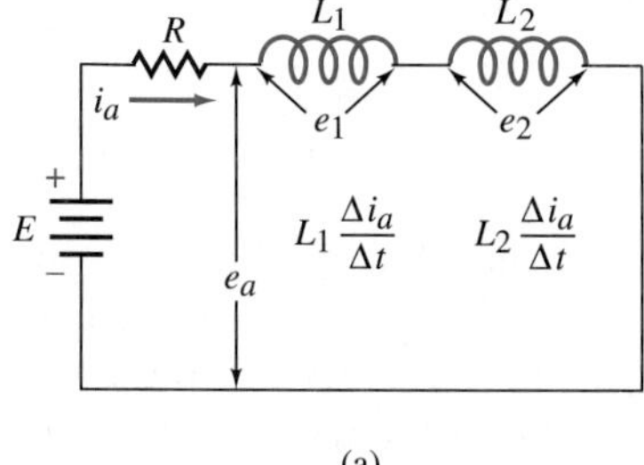

(a)

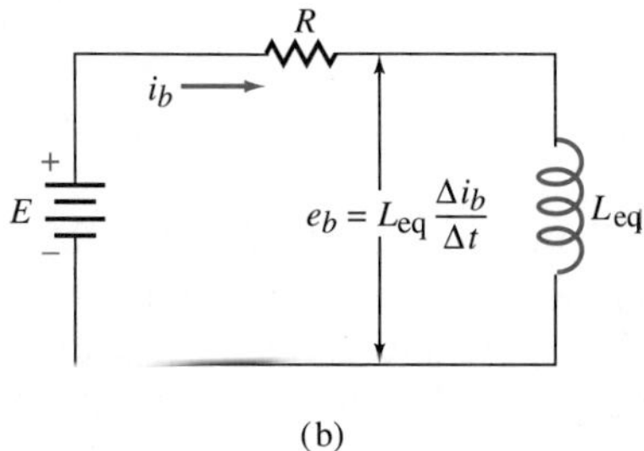

(b)

Figure 13.42 The induced emf for a change in current will be the same for the equivalent inductance as for the original circuit. (a) A series circuit of inductors with no mutual inductance. (b) The equivalent inductance for the series circuit. $e_b = e_a$ and $L_{eq} = L_1 + L_2$.

Since the inductors are in series, the emf polarities are in the same direction. This gives

$$e_a = e_1 + e_2 \qquad \text{or} \qquad L_1 \frac{\Delta i}{\Delta t} + L_2 \frac{\Delta i}{\Delta t}$$

Also

$$e_b = L_{\text{eq}} \frac{\Delta i}{\Delta t}$$

For Figure 13.42(b) and Figure 13.42(a) to be equivalent, e_a must equal e_b as defined by equivalent inductance. So

$$L_{\text{eq}} \left(\frac{\Delta i}{\Delta t}\right) = L_1 \frac{\Delta i}{\Delta t} + L_2 \frac{\Delta i}{\Delta t}$$

but $\frac{\Delta i}{\Delta t}$ is the same for both circuits, so

$$L_{\text{eq}} = L_1 + L_2$$

In general, the equivalent inductance for inductors connected in series with no mutual coupling is given by

$$L_{\text{eq}} = L_1 + L_2 + \cdots + L_n \qquad \textbf{(13.16)}$$

where L_{eq} is the equivalent inductance, in henries, and $L_1, L_2, \ldots, L_n$ are the inductances of each coil, in henries.

This has the same form as the equivalent resistance of a series resistance circuit. L_{eq} will be larger than the largest inductance in the group.

EXAMPLE 13.19

Find the equivalent inductance of a 5-mH, 20-mH, and 15-mH inductor connected in series.

Solution

$$\begin{aligned} L_{\text{eq}} &= L_1 + L_2 + L_3 \\ &= 5 \text{ mH} + 20 \text{ mH} + 15 \text{ mH} \\ L_{\text{eq}} &= 40 \text{ mH} \end{aligned}$$

Equation 13.16 assumes that there is no flux linkage between the inductors. That is, the coupling is zero and there is no mutual inductance. Zero coupling can be obtained by a large distance between the coils, magnetic shielding around them, or orienting them so their fields do not interact.

If the inductors are coupled, the mutual inductance between them will affect their equivalent inductance. When the magnetic fields of the coils aid each other, as shown in Figure 13.43(a), the coils are said to be

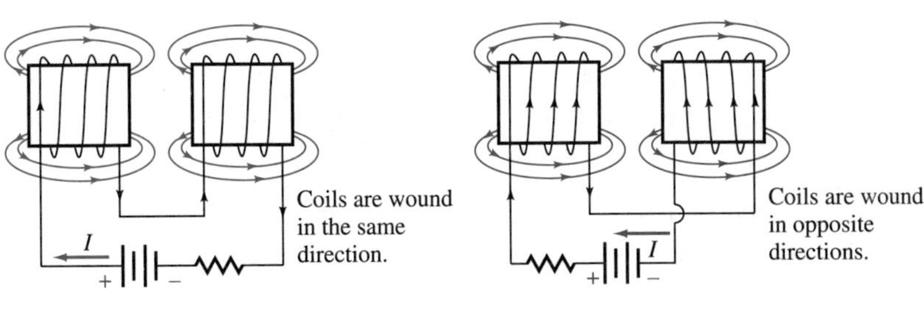

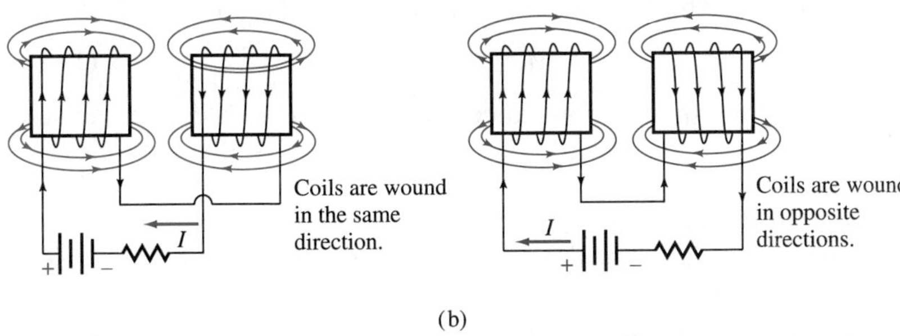

Figure 13.43 Possible coil connections for series-aiding and series-opposing inductors. (a) Series-aiding. (b) Series-opposing.

"series-aiding." The mutual inductance adds to the self-inductance. When the fields oppose each other as shown in Figure 13.43(b), the coils are said to be "series-opposing." The mutual inductance subtracts from the self-inductance.

The symbol for an inductor does not show the direction of the winding. So how can one tell whether the connection of the inductors is series-aiding or series-opposing? One method is to use a dot at one end of the coil as shown in Figure 13.44. The coils are series-aiding when the current enters the same end of the coils. They are series-opposing when the current enters different ends of the coils.

When two coils are connected in series with mutual inductance, the effective inductance of the combination becomes

$$L_{eq} = L_1 + L_2 \pm 2M \tag{13.17}$$

where L_{eq} is the equivalent inductance of the series group, in henries; L_1 and L_2 are the inductances of each coil, in henries; M is the mutual inductance between the coils, in henries; and $2M$ takes a (+) sign for aiding and a (−) sign for opposing.

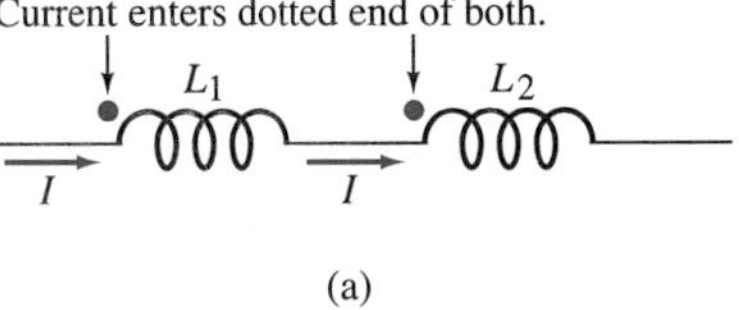

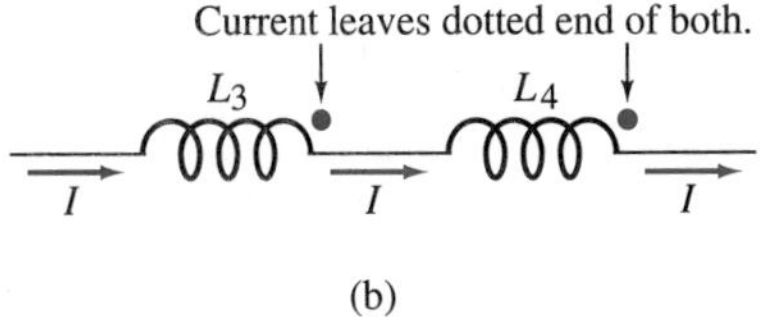

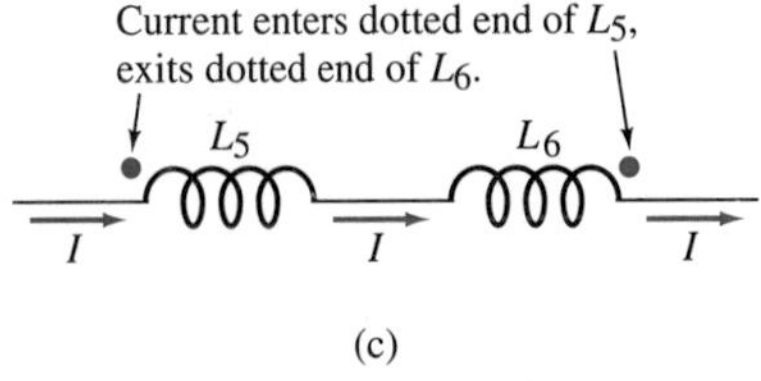

Figure 13.44 The relation of the field to the coil terminals can be shown using a dot. (a) The current enters both dotted ends. L_1 and L_2 are aiding. (b) The current leaves both dotted ends. L_3 and L_4 are aiding. (c) The current enters one dotted end and leaves the other. L_5 and L_6 are opposing.

EXAMPLE 13.20

A 400-mH coil and a 150-mH coil are connected in series. The mutual inductance is 150 mH. Find:

a. the equivalent inductance when series-aiding.
b. the equivalent inductance when series-opposing.

Solution

a. $L_{eq} = L_1 + L_2 + 2M$

$= 400 \text{ mH} + 150 \text{ mH} + 2(150 \text{ mH})$

$L_{eq} = 850 \text{ mH}$

b. $L_{eq} = 400 \text{ mH} + 150 \text{ mH} - 2(150 \text{ mH})$

$L_{eq} = 250 \text{ mH}$

The mutual inductance between two inductors is a function of their self-inductance and the coefficient of coupling. That is,

$$M = k\sqrt{L_1L_2} \qquad \textbf{(13.18)}$$

where M is the mutual inductance, in henries; L_1 and L_2 are the self-inductance of each inductor, in henries; and k is the coefficient of coupling.

coefficient of coupling

The **coefficient of coupling** represents the fraction of the flux from one inductor that is linked to the other. Its value ranges from 0 to 1. Zero represents no linkage, 1 represents total linkage.

Coupling depends on the relative position of the inductors and the distance between them. For instance, the coupling of inductors whose axes are parallel is greater than if they are at right angles.

The mutual inductance and coefficient of coupling can be determined from some simple measurements. Example 13.21 shows one way to determine them.

EXAMPLE 13.21

A 6-mH coil and a 10-mH coil are connected in series. The measured inductance of the combination is 20 mH. Find the mutual inductance and the coefficient of coupling.

Solution

Since L_{eq} is greater than $L_1 + L_2$, the coils must be connected in series-aiding. So

$$L_{eq} = L_1 + L_2 + 2M$$

or

$$20 \text{ mH} = 6 \text{ mH} + 10 \text{ mH} + 2M$$

Solving for M,

$$M = \left(\frac{20 \text{ mH} - 6 \text{ mH} - 10 \text{ mH}}{2}\right)$$

$$M = 2 \text{ mH}$$

$$M = k\sqrt{L_1L_2}$$

so

$$k = \frac{M}{\sqrt{L_1 L_2}}$$

$$= \frac{2 \times 10^{-3}\ \text{H}}{\sqrt{(6 \times 10^{-3}\ \text{H})(10 \times 10^{-3}\ \text{H})}}$$

$$k = 0.258$$

EXAMPLE 13.22

L_{eq} is 20 mH when two coils are connected in series-aiding, and 12 mH when connected in series-opposing. What is the mutual inductance?

Solution

$$\text{Aiding,} \qquad L_{eqa} = L_1 + L_2 + 2M$$

$$\text{Opposing,} \qquad L_{eqo} = L_1 + L_2 - 2M$$

$$L_{eqa} - L_{eqo} = L_1 + L_2 + 2M - (L_1 + L_2 - 2M)$$

or

$$L_{eqa} - L_{eqo} = 4M$$

Solving for M yields

$$M = \left(\frac{20\ \text{mH} - 12\ \text{mH}}{4}\right)$$

$$M = 2\ \text{mH}$$

13.9 Equivalent Inductance of Parallel Inductors

parallel inductors

Parallel inductors are inductors connected across the same two points. The resistance of the wire in an inductor prevents making a true parallel connection. But in many connections this resistance is so small that the connection can be taken as a parallel one.

Two parallel inductors are shown in Figure 13.45(a). The circuit in

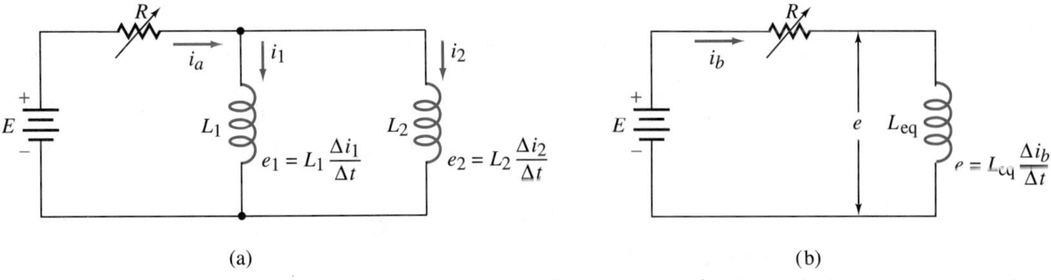

Figure 13.45 The induced emf for a change in current will be the same for the equivalent inductance as for the original parallel circuit. (a) A parallel circuit of inductors with no mutual inductance. (b) The equivalent inductance of the parallel circuit. $e = e_1$ and $1/L_{eq} = 1/L_1 + 1/L_2$.

Figure 13.45(b) will be equivalent to it if the same change in current induces an equal emf in both circuits so that $e = e_1 = e_2$.

Assuming no mutual inductance and no resistance in the circuit of Figure 13.45(a),

$$e_1 = L_1\left(\frac{\Delta i_1}{\Delta t}\right) \qquad \text{so} \qquad \frac{\Delta i_1}{\Delta t} = \frac{e_1}{L_1}$$

$$e_2 = L_2\left(\frac{\Delta i_2}{\Delta t}\right) \qquad \text{so} \qquad \frac{\Delta i_2}{\Delta t} = \frac{e_2}{L_2}$$

For the circuit of Figure 12.45(b),

$$e = L_{\text{eq}}\left(\frac{\Delta i_b}{\Delta t}\right) \qquad \text{so} \qquad \frac{e}{L_{\text{eq}}} = \frac{\Delta i_b}{\Delta t}$$

Since i_a must be the same as i_b,

$$\frac{e}{L_{\text{eq}}} = \frac{\Delta i_a}{\Delta t}$$

Also

$$\Delta i_a = \Delta i_1 + \Delta i_2$$

so

$$\frac{e}{L_{\text{eq}}} = \frac{\Delta i_1}{\Delta t} + \frac{\Delta i_2}{\Delta t}$$

Substituting the expressions from the first equations gives

$$\frac{e}{L_{\text{eq}}} = \frac{e_1}{L_1} + \frac{e_2}{L_2}$$

Since $e = e_1 = e_2$ by the condition that the circuits are equivalent,

$$\frac{1}{L_{\text{eq}}} = \frac{1}{L_1} + \frac{1}{L_2}$$

In general, for n inductors connected in parallel with no mutual coupling,

$$\frac{1}{L_{\text{eq}}} = \frac{1}{L_1} + \frac{1}{L_2} + \cdots + \frac{1}{L_n} \tag{13.19}$$

where L_{eq} is the equivalent inductance, in henries, and $L_1, L_2, \ldots, L_n$ are the inductances, in henries.

Equation 13.19 has the same form as that for resistors in parallel. L_{eq} will always be smaller than the smallest inductance in the group.

EXAMPLE 13.23

Inductors of 16 mH, 20 mH, and 80 mH are connected in parallel. What is the equivalent inductance?

Solution

$$\frac{1}{L_{eq}} = \frac{1}{L_1} + \frac{1}{L_2} + \frac{1}{L_3}$$

$$= \frac{1}{16 \text{ mH}} + \frac{1}{20 \text{ mH}} + \frac{1}{80 \text{ mH}}$$

$$= 0.125 \frac{1}{\text{mH}}$$

$$L_{eq} = \frac{1 \text{ mH}}{0.125}$$

$$L_{eq} = 8 \text{ mH}$$

Deriving the expression for the equivalent inductance of parallel inductors with mutual inductance is rather complex. But for two inductances connected in parallel-aiding, as shown in Figure 13.46(a),

$$L_{eq} = \frac{L_1 L_2 - M^2}{L_1 + L_2 - 2M} \qquad \textbf{(13.20)}$$

For two inductors connected in parallel-opposing, as in Figure 13.46(b), Equation 13.20 becomes

$$L_{eq} = \frac{L_1 L_2 - M^2}{L_1 + L_2 + 2M} \qquad \textbf{(13.21)}$$

where L_{eq} is the equivalent inductance, in henries; L_1 and L_2 are the inductances, in henries; and M is the mutual inductance, in henries.

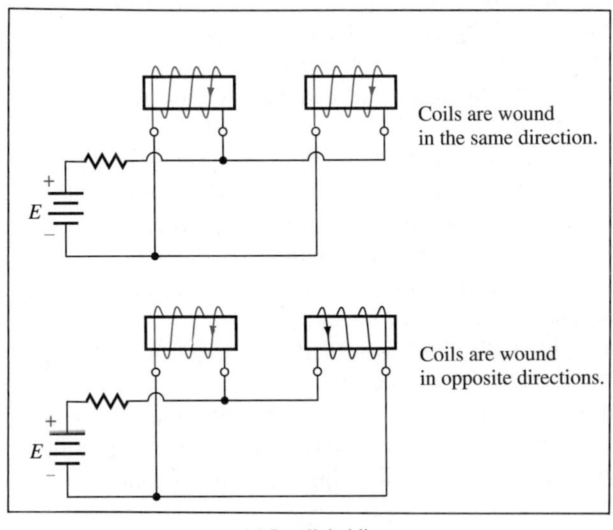

(a) Parallel aiding

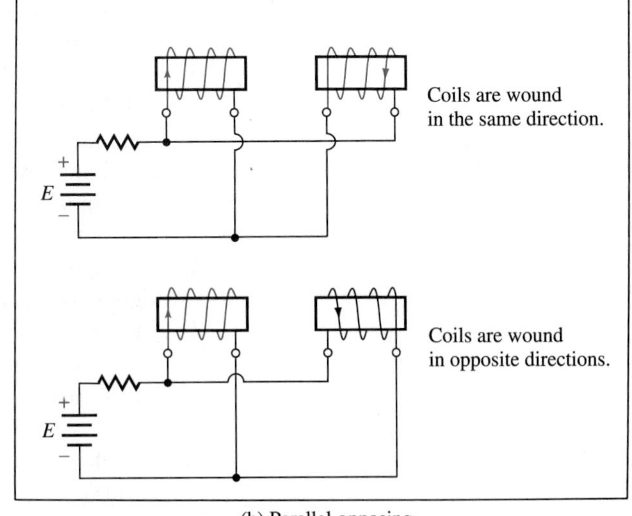

(b) Parallel opposing

Figure 13.46 Possible parallel connections for parallel-aiding and parallel-opposing. (a) Parallel-aiding. (b) Parallel-opposing.

EXAMPLE 13.24

The mutual inductance between a 6-H and an 8-H inductor is 0.5 H. What is the equivalent inductance when they are connected in:

a. parallel-aiding?
b. parallel-opposing?

Solution

a.

$$L_{eq} = \frac{L_1L_2 - M^2}{L_1 + L_2 - 2M}$$

$$= \frac{(6\text{ H})(8\text{ H}) - (0.5\text{ H})^2}{6\text{ H} + 8\text{ H} - 2(0.5\text{ H})}$$

$$L_{eq} = 3.67\text{ H}$$

b.

$$L_{eq} = \frac{L_1L_2 - M^2}{L_1 + L_2 - 2M}$$

$$= \frac{(6\text{ H})(8\text{ H}) - (0.5\text{ H})^2}{6\text{ H} + 8\text{ H} + 2(0.5\text{ H})}$$

$$L_{eq} = 3.18\text{ H}$$

SUMMARY

1. A magnet is a material that has the property to attract iron.
2. Magnets have a north-seeking pole and a south-seeking pole. Like poles will repel each other and unlike poles will attract each other.
3. A magnet has lines of force around it. The lines leave the north pole and enter the south pole.
4. The molecular theory of magnetism holds that a magnet is made by the alignment of small magnetic domains so that their fields add.
5. A permanent magnet is one in which the domains remain aligned even when the magnetizing force is removed.
6. A temporary magnet is one in which the domains do not remain aligned when the magnetizing force is removed.
7. An electric current in a conductor sets up a magnetic field of concentric circles around the conductor.
8. An electromagnet is made up of a temporary magnetic material in a coil of wire. It is magnetic as long as there is current in the coil.
9. The magnetic lines of force are called flux and are measured in webers.
10. Magnetic materials are materials that affect the shape of the magnetic field. Nonmagnetic materials do not affect the shape.

11. Electromagnetic induction is the generating of an emf by the relative motion of a conductor and a magnetic field.
12. Faraday's Law states that the induced emf is directly proportional to the rate at which the flux changes.
13. Lenz's Law states that the current from the induced emf will oppose any change in current.
14. The analysis of a magnetic circuit is analogous to that of the electric circuit. Relations similar to Ohm's Law, Kirchhoff's Current Law, and Kirchhoff's Voltage Law are used.
15. A magnet exerts a tractive force on magnetic materials and other magnets. The force is a function of the flux density squared, the area, and the permeability.
16. Inductance is the property of a device to oppose any change in current. The unit for it is the henry.
17. Mutual inductance is the inductance between two coils that are placed so that their fields interact. The coils are said to be coupled.
18. The equivalent inductance is that inductance that induces the same emf as the original circuit for the same change in current.
19. The equivalent inductance of series-connected inductors is equal to the sum of the inductances of the inductors.
20. The reciprocal of the equivalent inductance of parallel-connected inductors is equal to the sum of the reciprocal of each inductance.
21. The total inductance of coupled coils includes the effect of the mutual inductance. This can add to the total, if aiding, or can subtract from the total, if opposing.

EQUATIONS

13.1	$B = \dfrac{\phi}{A}$	The relation between flux density, flux, and the area of a section.	506
13.2	$\mu_r = \dfrac{\mu}{\mu_o}$	The relation between relative permeability and permeability.	507
13.3	$\mathcal{F} = NI$	The relation of mmf to the coil current and number of turns.	508
13.4	$\mathcal{R} = \dfrac{\ell}{\mu A}$	The relation of reluctance to the dimensions and permeability of the path.	509
13.5	$\mathcal{F} = \mathcal{R}\phi$	The Ohm's Law equivalent for the magnetic path.	509
13.6	$\mathcal{F} = H\ell$	The mmf drop in a magnetic path.	511
13.7	$\mathcal{F} = H_1\ell_1 + H_2\ell_2 + \cdots + H_n\ell_n$	The mmf loop equation for a magnetic circuit.	515

13.8	$\phi_T - \phi_1 - \phi_2 - \cdots - \phi_n = 0$	The parallel branch flux equation for the magnetic circuit.	518
13.9	$F = \dfrac{B^2A}{2\mu_o}$	The general equation for the tractive force of a magnet.	521
13.10	$e = N\dfrac{d\phi}{dt}$	The general equation for the average induced emf.	525
13.11	$e_{\text{ave}} = N\dfrac{\Delta\phi}{\Delta t}$	The general equation for the average induced emf.	525
13.12	$e = B\ell v \sin\theta$	The equation for the emf induced in a conductor rotating in a magnetic field.	526
13.13	$e = L\dfrac{di}{dt}$	The equation for the emf induced in an inductance.	529
13.14	$e_{\text{ave}} = L\dfrac{\Delta i}{\Delta t}$	The equation for the average emf induced in an inductance.	529
13.15	$L = \dfrac{\mu N^2 A}{\ell}$	The equation for the inductance of a coil.	529
13.16	$L_{\text{eq}} = L_1 + L_2 + \cdots + L_n$	The equivalent inductance for series-connected inductors.	532
13.17	$L_{\text{eq}} = L_1 + L_2 \pm 2M$	The equivalent inductance for series-connected inductors with mutual inductance aiding or opposing.	533
13.18	$M = k\sqrt{L_1L_2}$	The relation of mutual inductance to the coefficient of coupling and the inductance of the inductors.	534
13.19	$\dfrac{1}{L_{\text{eq}}} = \dfrac{1}{L_1} + \dfrac{1}{L_2} + \cdots + \dfrac{1}{L_n}$	The equivalent inductance for parallel-connected inductors.	536
13.20	$L_{\text{eq}} = \dfrac{L_1L_2 - M^2}{L_1 + L_2 - 2M}$	The equation for the equivalent inductance of two inductors connected parallel-aiding.	537
13.21	$L_{\text{eq}} = \dfrac{L_1L_2 - M^2}{L_1 + L_2 + 2M}$	The equation for the equivalent inductance of two inductors connected parallel-opposing.	537

QUESTIONS

1. What property of a magnet makes it suitable for use in relays and electric cranes?
2. How does the compass needle indicate the direction of the north pole?
3. Can copper filings be used to show the shape of a magnetic field? Why or why not?
4. How can a compass be used to show if there is a current in a conductor?
5. What is the difference between a permanent magnet and a temporary magnet in terms of the domain theory?
6. Which type of magnet should be used for a door latch, a temporary magnet or a permanent magnet? For use in a door chime?
7. What effect on the operation of a door bell will replacing the coil's iron core with an aluminum core have?
8. How does the permeability of iron compare to that of air?
9. Why are the transformer cores made of iron?
10. Which requires more mmf for the same flux:
    a. a 1 meter length of iron?
    b. a 1 meter length of copper?
11. What are two factors that affect the tractive force of a magnet?
12. Why is the tractive force greater near a magnet than at a distance from it?
13. What is meant by electromagnetic induction?
14. Which law should be used to determine the amount of induced emf, Faraday's or Lenz's?
15. Which law should be used to determine the polarity of the induced emf, Faraday's or Lenz's?
16. The induced emf is greatest when a conductor is moved horizontally in a vertical magnetic field. What will be the induced emf if the movement is vertically?
17. What is the difference between electromagnetic induction and mutual induction?
18. Which type of induction is used in a generator? In a transformer?
19. What is meant by inductance and in what unit is it measured?
20. How does the practical inductor differ from the ideal one?
21. How does inductance cause an arc across the switch contacts when a switch is opened to turn off a motor?
22. Explain the operation of the free-wheeling circuit.
23. Two inductors are connected in series. How will the total inductance change if a third inductor is connected in series?
24. How will the total inductance change if the third inductor is connected in parallel with the two in Question 23?
25. What will happen to the equivalent inductance of two series inductors if the coupling is increased?

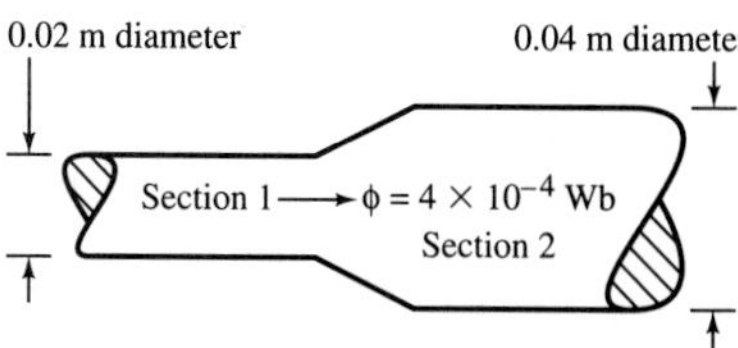

Figure 13.47

Figure 13.48

PROBLEMS

SECTION 13.2 Flux, Flux Density, and Other Magnetic Quantities

1. Find the flux density in Section 1 and Section 2 of the bar in Figure 13.47.
2. Find the flux density in the sections in Figures 13.48 and 13.49.

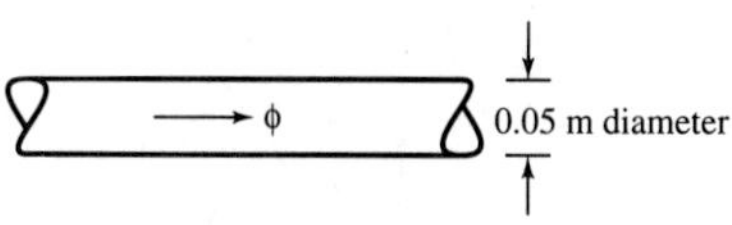

Figure 13.49

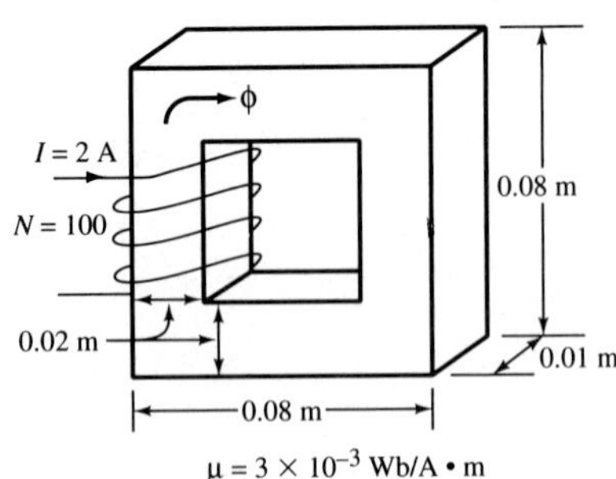

Figure 13.50

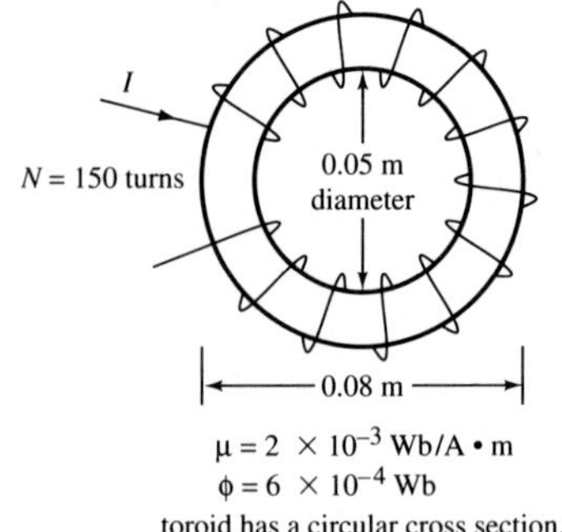

Figure 13.51

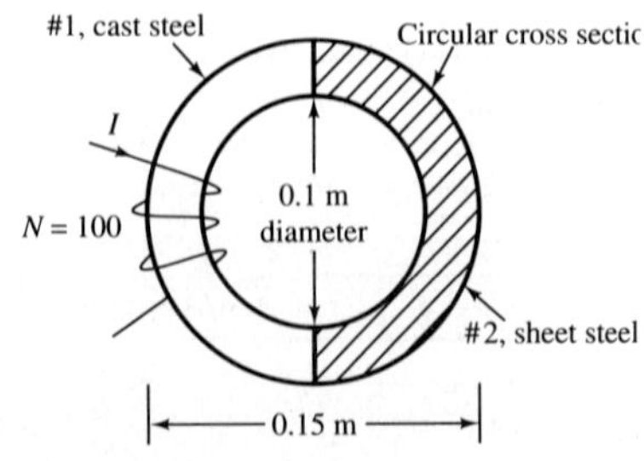

Figure 13.52

3. The flux density in a round iron core of a coil is 0.8 T. How much flux is in the core if its diameter is 1 centimeter?

4. What is the flux in a square bar with sides of 0.05 meter if the flux density is 0.5 T?

5. What is the relative permeability of a piece of iron with a permeability of 3×10^{-3} Wb/A · m?

6. A piece of iron has a relative permeability of 100. What is its permeability?

7. What is the mmf of each of the following?
   (a) 3 amperes, 5 turns.
   (b) 5 amperes, 100 turns.
   (c) 2 amperes, 50 turns.
   (d) 0.5 amperes, 1000 turns.

8. Find the reluctance of an air gap 0.02 meter long between two cylindrical bars of iron 0.02 m in diameter.

9. What is the value of
   (a) ϕ in the core?
   (b) The flux density of the core in Figure 13.50?

10. Find:
    (a) The magnetomotive force needed to set up the flux ϕ in the toroid of Figure 13.51.
    (b) The coil current.

SECTION 13.3 Ohm's Law and Magnetizing Force

11. Two amperes in a 500-turn coil are needed to set up a flux in a path that is 0.5 meter long. What is the magnetizing force?

12. The magnetizing force in a piece of iron is 500 A/m. How many amperes are needed to set up the flux in a 0.02-meter length of iron?

13. Using the curve of Figure 13.19, find the magnetizing force needed for a flux density of:
    (a) 1 T. (b) 0.5 T.

14. What magnetizing force is needed for
    (a) A flux density of 0.8 T?
    (b) 1.2 T for the material of Figure 13.19?

15. What is the flux density in the material of Figure 13.19 if the magnetizing force is
    (a) 900 A/m? (b) 600 A/m?

SECTION 13.4 Magnetic Circuit Analysis

16. In the circuit of Figure 13.52, find:
    (a) The direction of the flux.
    (b) The current needed to set up a flux of 2×10^{-4} Wb.

17. For the circuit of Figure 13.53, find:

    (a) The flux density in the air gap.
    (b) The mmf needed for the flux.
    (c) The number of turns (to the nearest whole number) in the coil.

18. A 1000-turn coil is wound as shown in Figure 13.54.

    (a) Draw the direction of the flux.
    (b) How much current is needed to set up a flux of 1.5×10^{-3} Wb in the core?

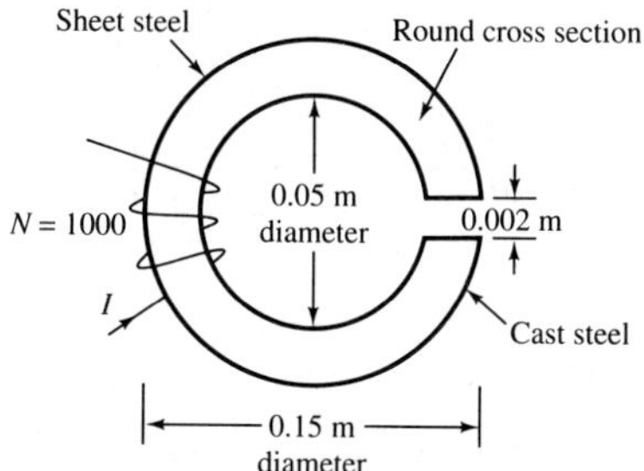

Figure 13.54

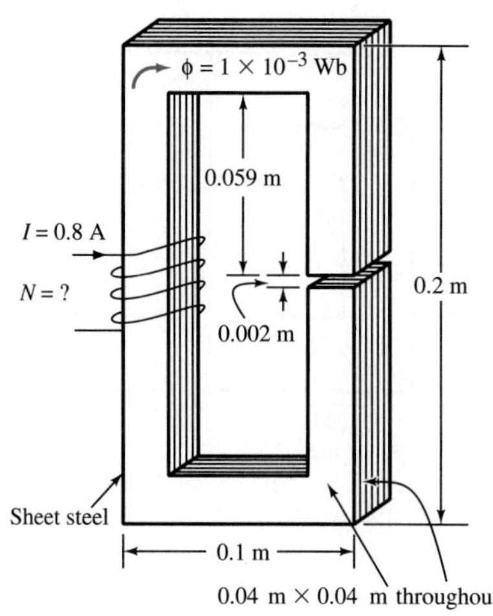

Figure 13.53

19. Find the thickness of the aluminum spacers in Figure 13.55 if the flux is 3×10^{-4} Wb.

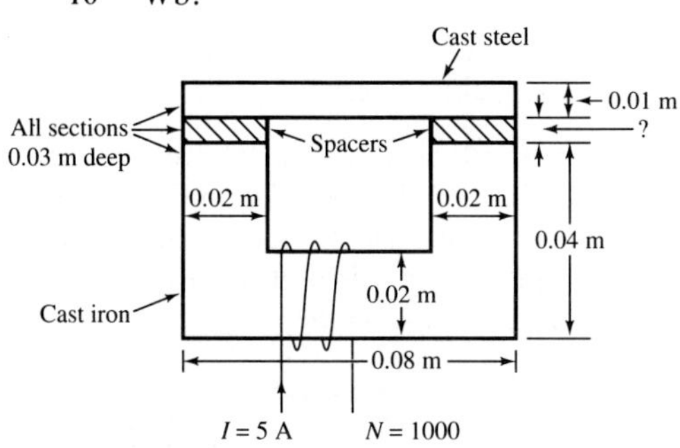

Figure 13.55

Figure 13.56

20. In Figure 13.56, find:

    (a) The mmf needed to maintain a flux of 2×10^{-4} Wb in the right leg.
    (b) The current in the coil.

21. In Figure 13.57, find:

    (a) The mmf needed for a flux of 2×10^{-4} Wb in the air gap.
    (b) The turns in the coil if the current is 2 amperes.

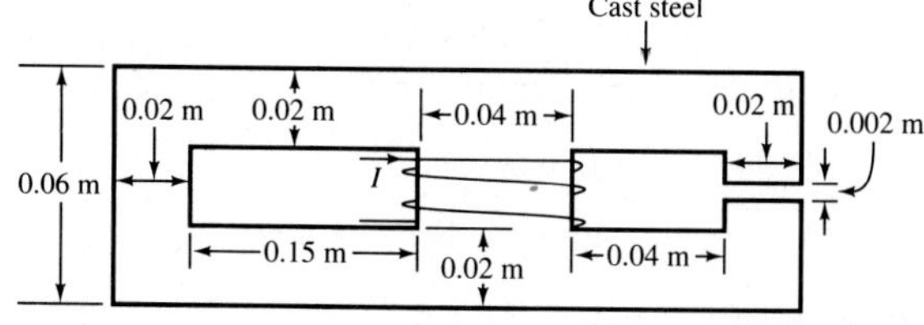

Figure 13.57

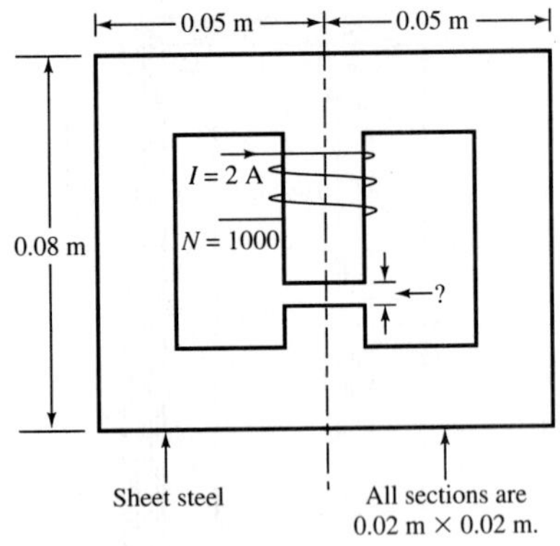

Figure 13.58

22. Find the length of the air gap needed for a flux of 2×10^{-4} Wb in the right leg of Figure 13.58.

SECTION 13.5 The Tractive Force of a Magnet

23. How much current is needed in the coil of the electromagnet in Figure 13.59 to pick up the 60-newton iron bar?

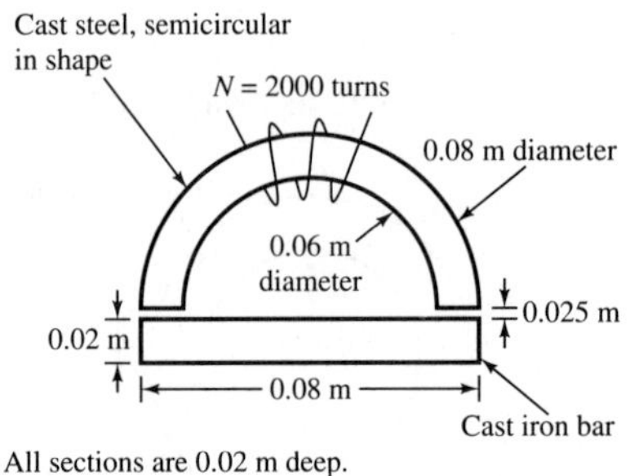

Figure 13.59

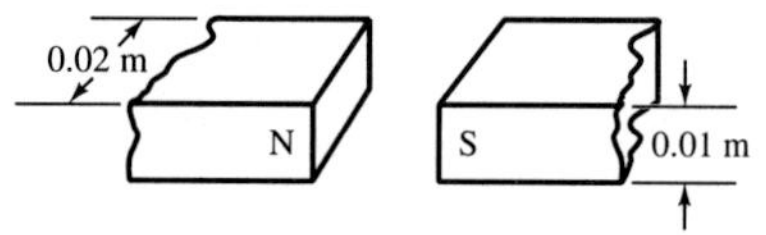

Figure 13.60

24. The force of attraction between the two poles in Figure 13.60 is 50 newtons. What is the flux between the poles?

25. What is the force between the end of the magnet and the relay armature shown in Figure 13.61?

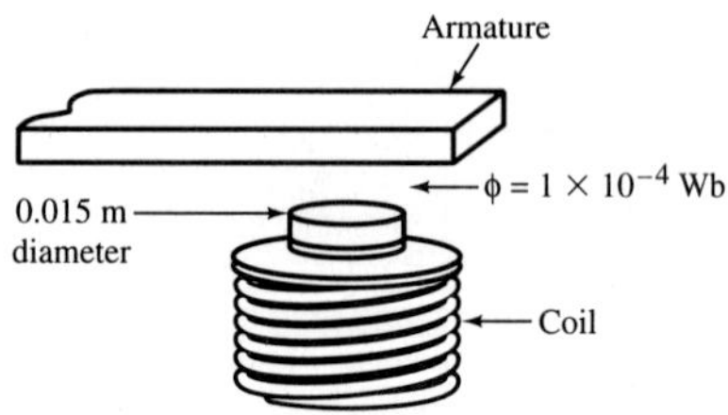

Figure 13.61

26. The force between a magnet and a piece of iron is 50 N. How much flux is needed if the magnet pole piece is a square 19 cm by 10 cm?

27. The space between the two magnets is in Figure 13.62 is 0.01 meter when a latch is open and 0.001 meter when the latch is closed, as in Figure 13.63. How much force is exerted between the magnets in the open condition? How much force is needed to open the latch when it is in the closed position? Assume that the flux is inversely proportional to the spacing.

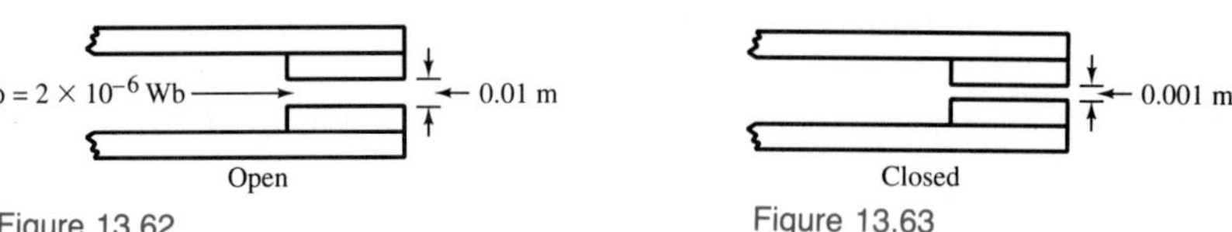

Figure 13.62

Figure 13.63

SECTION 13.6 Electromagnetic Induction

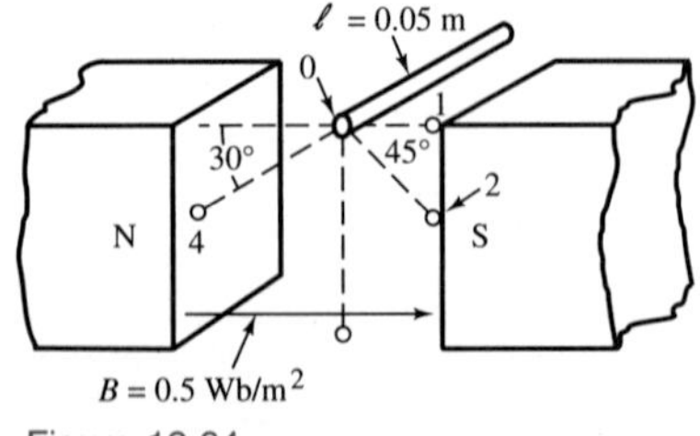

Figure 13.64

28. What is the average emf induced in a 10-conductor coil when it moves across 5×10^{-3} Wb in 8 ms?

29. The induced average emf in a conductor is 0.5 volt when it moves across 3×10^{-4} Wb. How many seconds does it take to move across the field?

30. The conductor in Figure 13.64 moves at a velocity of 0.1 m/s. Find the average emf induced when it moves on a path from:

(a) 0–1. (b) 0–2. (c) 0–3. (d) 0–4.

31. The conductor in Figure 13.65 moves straight down at a rate of 0.2 m/s.

(a) Assuming no fringing, what is the average emf?

(b) What is the polarity of end A?

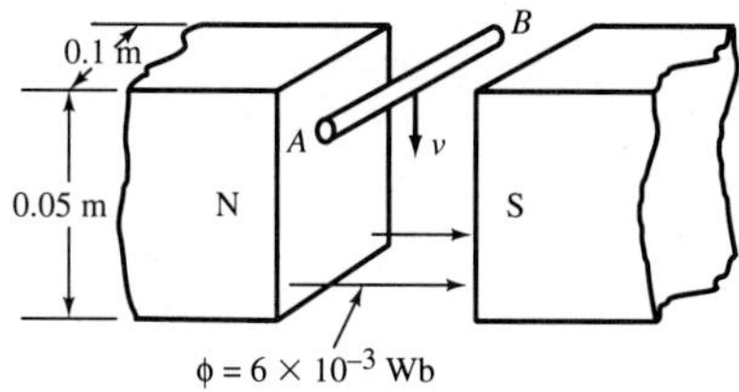

Figure 13.65

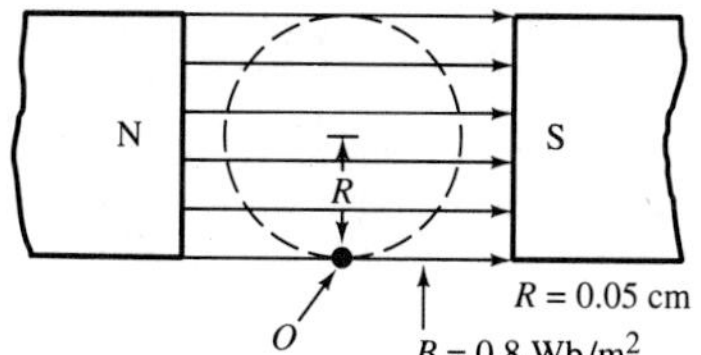

Figure 13.66

32. A 0.1-meter conductor starts at Point "0" and moves in a counterclockwise circular path at 1800 rpm, as shown in Figure 13.66. Plot the curve of induced emf versus time for one revolution.

33. The average emf induced in a coil is 120 volts when the flux changes from 0 Wb to 0.5×10^{-3} Wb in 0.005 second. How many turns are in the coil?

34. The flux in a 200-turn coil varies with time, as shown in Figure 13.67. Find the average emf induced in the coil from 0 to 2×10^{-3} s.

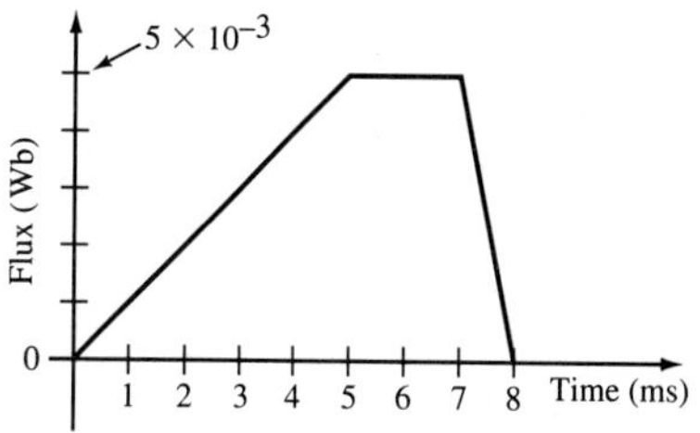

Figure 13.67

SECTION 13.7 Self-Inductance

35. The current in a 250-turn, 0.5-H relay coil is 5 amperes. What average emf is induced across it if the current is reduced to 0 amperes in 0.1 second when the switch is opened?

36. What is the inductance of an inductor if 125 volts is induced when the current changes from 0.5 A to 0.8 A in 5 ms?

37. The current in a 2-H inductor is varied as shown in Figure 13.68. What is the average emf induced in it:

(a) from t_0–t_1. (b) from t_1 to t_2. (c) from t_2–t_3?

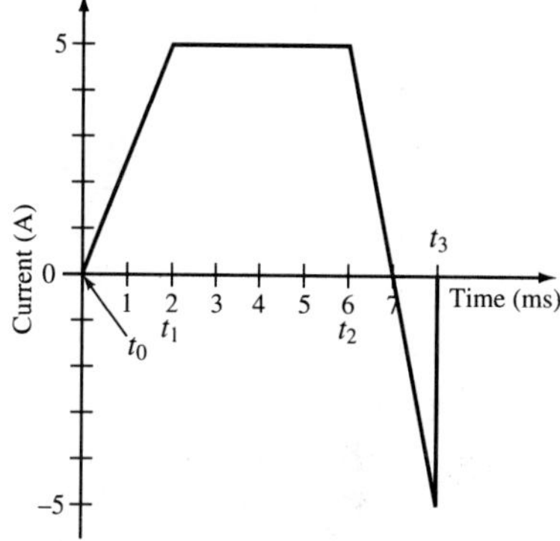

Figure 13.68

38. What is the inductance of the coil shown in Figure 13.69? What would be its inductance if a material with a permeability of 1.5×10^{-3} Wb/A m were inserted in the coil?

39. The field winding of a dc motor has an inductance of 5 H and draws 0.5 A of current. The voltage induced by opening the switch is 600 V. How many seconds does it take for the current to reach 0 amperes?

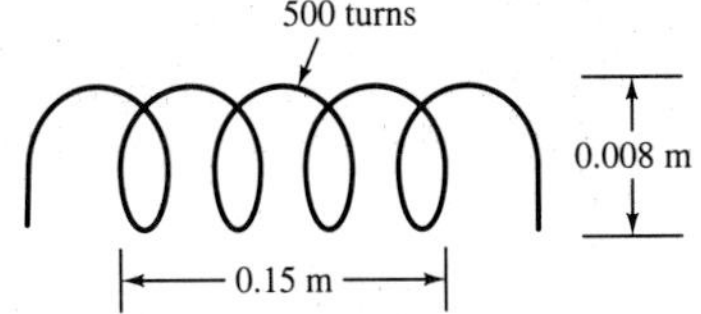

Figure 13.69

SECTION 13.8 Equivalent Inductance of Series Inductors

40. Find the equivalent inductance for the circuits in Figures 13.70 and 13.71 if there is no mutual inductance.

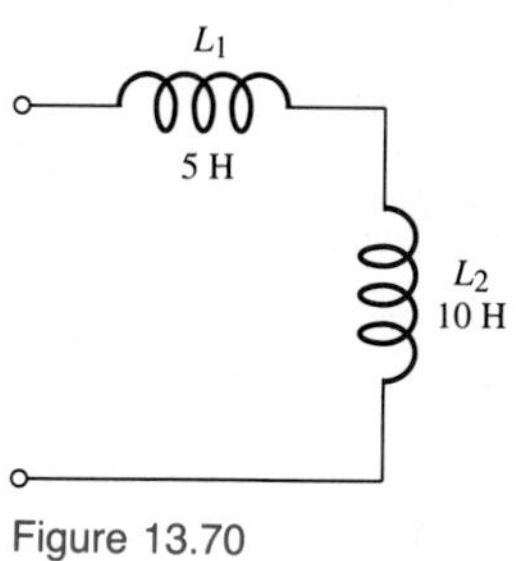

Figure 13.70

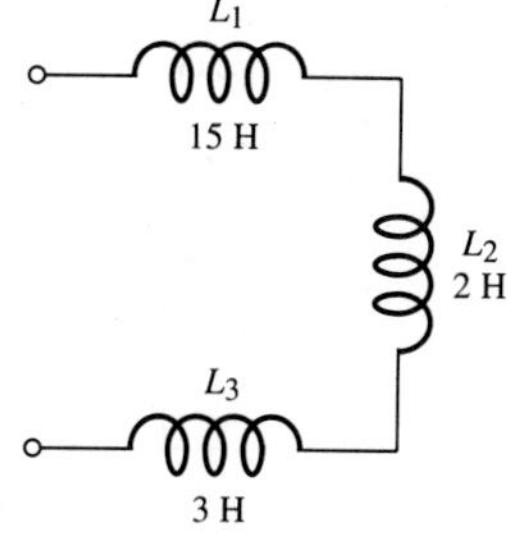

Figure 13.71

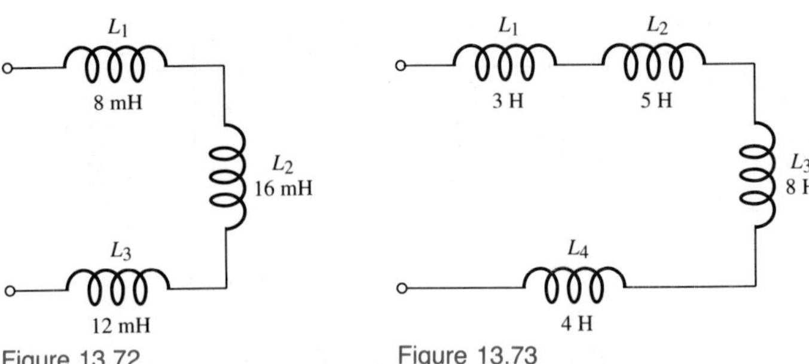

Figure 13.72 Figure 13.73

41. Find the equivalent inductance for the circuits in Figures 13.72 and 13.73 if there is no mutual inductance.

42. The equivalent inductance of three inductors connected in series with no coupling is 24 H. What is the inductance of the third inductor if the first two are 8 H and 10 H?

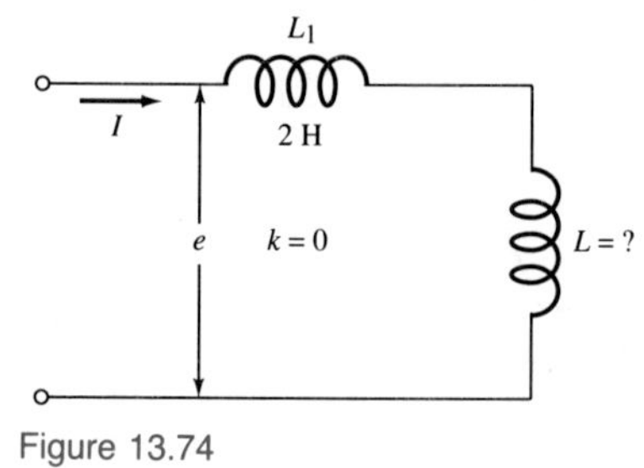

Figure 13.74

43. An emf of 50 volts is induced across the circuit in Figure 13.74 when the current is changed from 1 ampere to 2 amperes in 0.1 second. What is the inductance of L?

44. A 3-H and a 12-H inductor are connected in series. If the connection is series-opposing and the mutual inductance is 5 H, what is the equivalent inductance of the circuit?

45. A 5-H and a 2-H inductor are connected in series-aiding with a mutual inductance of 2 H.

    (a) What is the equivalent inductance of the combination?

    (b) What will the equivalent inductance be when connected in series-opposing?

46. What is the mutual inductance between a 50-mH and a 20-mH inductor if their equivalent inductance is 80 mH when connected in series-aiding?

47. A 3-H and a 6-H inductor have an inductance of 12 H when connected in series-aiding, and 6 H when connected in series-opposing. If a flux of 6×10^{-3} Wb is established in the 3-H inductor, what will the flux in the 6-H inductor be?

SECTION 13.9 Equivalent Inductance of Parallel Inductors

48. Find the equivalent inductance for the circuits in Figures 13.75 and 13.76 if there is no mutual inductance.

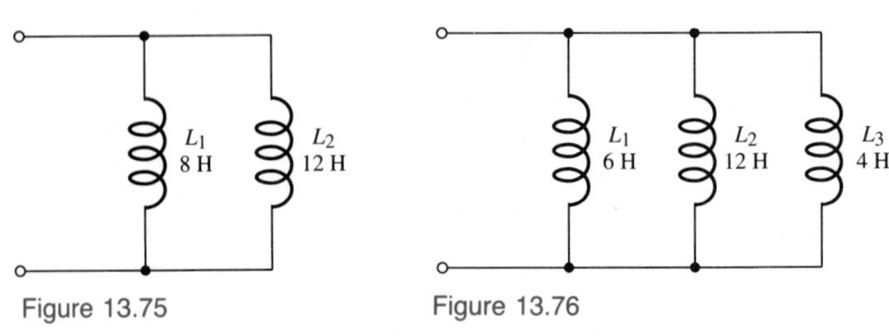

Figure 13.75 Figure 13.76

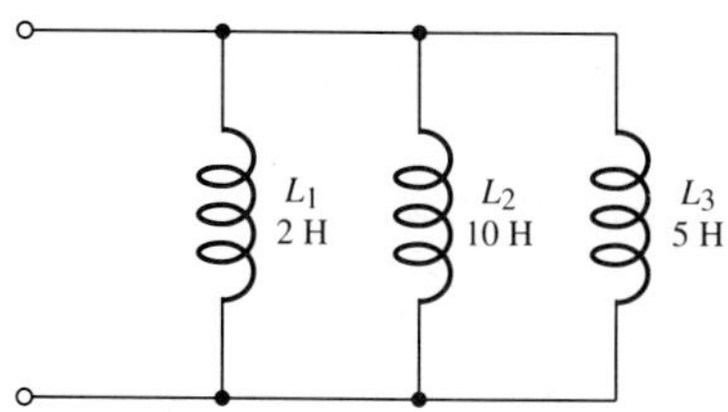

Figure 13.77

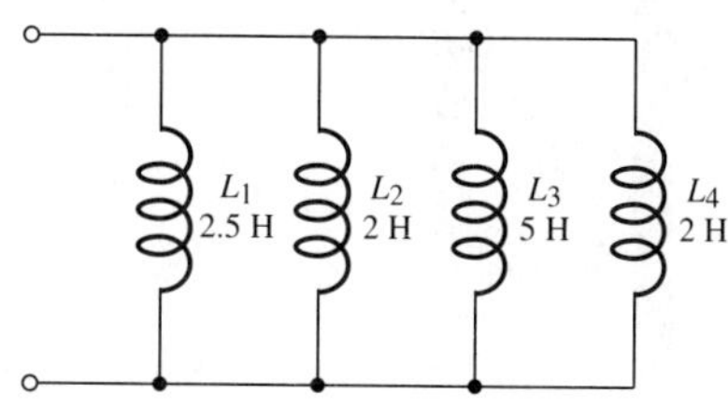

Figure 13.78

49. Find the equivalent inductance for the circuits in Figures 13.77 and 13.78 if there is no mutual inductance.

50. The equivalent inductance of the circuit shown in Figure 13.79 with no coupling is 5 H. What is the inductance of L_3?

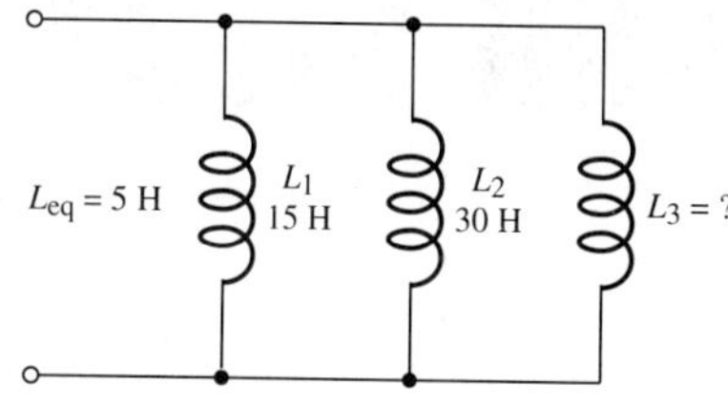

Figure 13.79

51. A 9-H and a 4-H inductor are connected in parallel and have a mutual inductance of 3 H. What is the equivalent inductance if

    (a) They are connected in parallel-aiding?

    (b) Parallel-opposing?

52. A 9-mH and a 21-mH inductor are connected in parallel-aiding to give an equivalent inductance of 8 mH. What is the mutual inductance between them?

53. Two identical coils have an equivalent inductance of 16 H when connected in parallel-aiding, and 8 H when connected in parallel-opposing. What is the mutual inductance between them?

CHAPTER 14

THE *R–L* CIRCUIT, TRANSIENT, AND STEADY-STATE CONDITION

CHAPTER OBJECTIVES

After completing this chapter, you should be able to:

1. Describe the current and voltage change during current rise and fall.
2. List the quantities that affect the transient state condition.
3. Analyze the series *R–L* circuit during current rise and fall.
4. Explain the meaning of the time constant, and give the parts of the circuit that affect it.
5. Use the Universal Time Constant Curve to analyze *R–L* circuits in the transient state.
6. Analyze complex *R–L* circuits during current rise and fall.

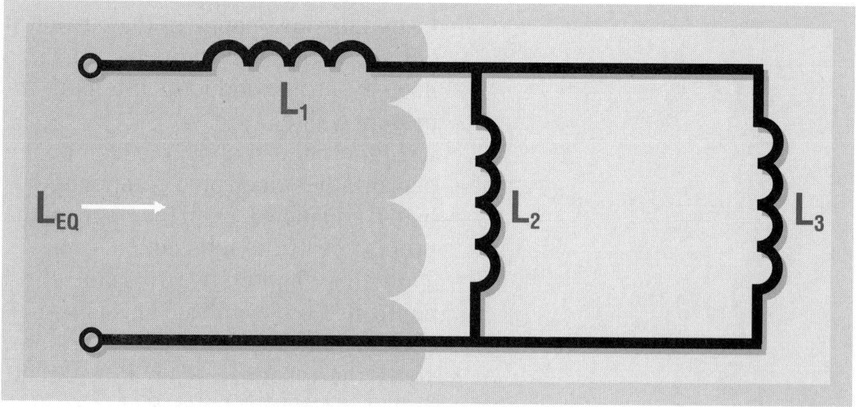

7. Calculate the energy stored by an inductor using the equation and the area under the power curve.
8. Draw the steady-state equivalent of the *R–L* circuit.
9. Analyze the steady-state *R–L* circuit for currents, voltages, energy, and power.
10. Use SPICE to analyze *R–L* circuits for transient behavior.

KEY TERMS

1. complex *R–L* circuit
2. exponential equation
3. inductor stored energy
4. series *R–L* circuit
5. steady state
6. transient state
7. time constant
8. Universal Time Constant Curves

INTRODUCTION

Circuits consisting of resistance and inductance exhibit transient behavior much like circuits consisting of resistance and capacitance do. In the case of *R–L* circuits, one must remember that it is the current that cannot change instantaneously, since that change would require an instantaneous change in the magnetic field around the inductor, and an instantaneous change in the current flowing through the inductor. The voltage across the inductor, on the other hand, will change instantaneously as the inductor attempts to maintain the current flow at its present value.

Inductors can appear as components in the circuit; they can also be part of other electronic components, such as the coil in a relay. At very high frequencies, even the trace on the printed circuit board exhibits the properties of inductance.

In this chapter, we examine the behavior of inductors in circuits controlled by switches. This allows us to separate the rising and falling current time segments. We can then extend our analysis techniques to study the response of an *R–L* circuit to a pulse input.

The analysis of the *R–L* circuit parallels the analysis of the *R–C* circuit. The main difference is that the capacitor opposes a change in voltage whereas the inductor opposes a change in current. Time constant, exponential equation, and universal curves are all items that were introduced in the *R–C* circuit chapter. Finally, we apply our new circuit analysis tool, and analyze several *R–L* circuits using PSpice.

14.1 The Series *R–L* Circuit During Current Rise

series *R–L* circuit

A **series *R–L* circuit** is a circuit of resistance and inductance connected in series, as shown in Figure 14.1(a). The resistance can be the inductor's

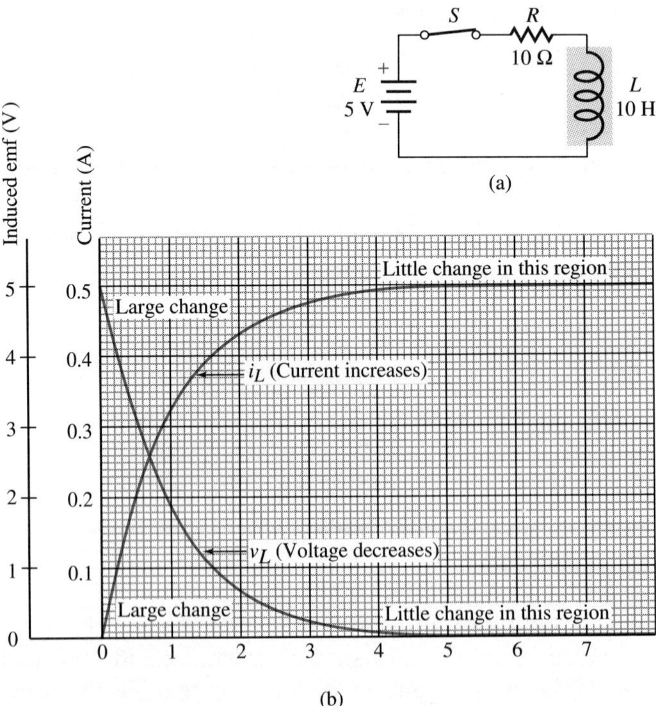

Figure 14.1 The inductor current does not change instantaneously in a series *R–L* circuit. (a) A series *R–L* circuit. (b) The exponential curves for current and voltage.

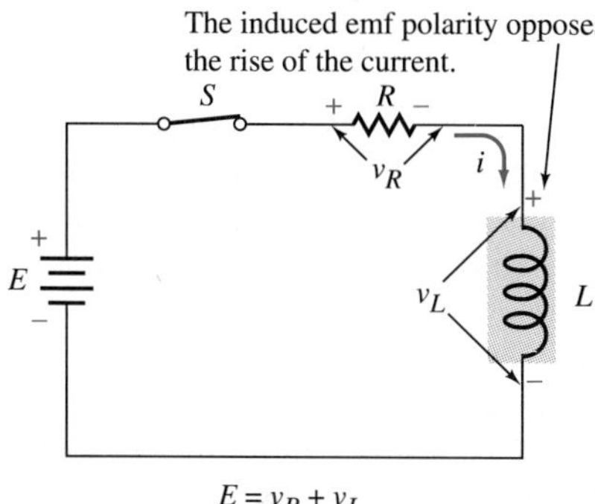

Figure 14.2 The polarity of the induced emf during the current rise.

resistance or other series resistance. In a purely resistive circuit, there are only two states: no current and current; in a series R–L circuit, the current can have three states:

1. **switch open.** No current; the voltage across the inductor is 0 volts.
2. **switch closed.** Current rises from 0 to a maximum value.
3. **switch closed.** Current at the maximum value; the inductor voltage is 0 volts.

The second state, switch closed, is examined in this section.

When the switch in Figure 14.1(a) is closed, the current does not rise at once, as it does in a purely resistive circuit. Instead, some time is needed for it to build up. This is because the inductor's induced emf opposes any change in the current. The result is a response, as shown in Figure 14.1(b). The interval in which the inductor condition changes from one steady state to another steady state is called the **transient state**.

transient state

While the switch in Figure 14.1(a) was open, there was no current flowing in the circuit. Closing the switch completed the circuit and resulted in a current. To oppose the change, an emf is induced in the inductance. It takes the polarity shown in Figure 14.2. Although the current and induced emf are changing with time, Kirchhoff's Voltage Law still applies at any instant of time. So, in Figure 14.2,

$$E = v_R + v_L \qquad \text{or} \qquad E = iR + L\frac{di}{dt}$$

The lowercase letters stand for instantaneous values. Since the current is initially zero, at the instant the switch is closed $t = 0$ seconds,

$$E = (0)(R) + L\frac{di}{dt}$$

or

$$\frac{di}{dt} = \frac{E}{L} \qquad \textbf{(14.1)}$$

where E is the source emf, in volts; L is the inductance, in henries; and

di/dt is the change in current (di) for a small change in time (dt) at the start of the rise, in amperes per second.

As the current increases, di/dt decreases. This is because $di/dt = (E - iR)/L$. Finally, for all practical purposes, there is no further change in current, $di/dt = 0$. This is shown on the curve in Figure 14.1(b) by the slope decreasing. When $di/dt = 0$, $(E - iR)/L$ must equal 0. So

$$i = E/R \tag{14.2}$$

where i is the current after the circuit has reached steady state, in amperes; E is the emf of the source, in volts; and R is the resistance in the circuit, in ohms. On the other hand, v_L, which equals $E - iR$ and also $L(di/dt)$, has its maximum value of E at $t = 0$ seconds and di/dt is a maximum.

The time needed for the current to rise to steady state depends on the inductance and the resistance in the circuit. Large inductances can support more lines of magnetic flux, so more time is needed to build up the lines. Large resistances result in smaller current with less lines of magnetic flux. Thus, less time is needed to build up the weaker field.

The expression $E = iR + L(di/dt)$ is a differential equation, and was also used for the R–C circuit. It is useful for studying the current change in the R–L circuit, but it cannot be used in that form to find the current. Applying the methods of calculus to it, as in Appendix C, results in an expression for the current. It is

$$i = \frac{E}{R}\left(1 - e^{-t/(L/R)}\right) \tag{14.3}$$

where i is the current at t seconds after the switch is closed, in amperes; t is the time in seconds that the switch has been closed; E is the source emf, in volts; R is the resistance in the circuit, in ohms; L is the inductance in the circuit, in henries; and e is the base for the natural logarithm system. It is equal to 2.71828 carried to six digits.

exponential equation

Equation 14.3 is an **exponential equation** because of the term $e^{-t/(L/R)}$. The curve plotted from it is an exponential curve (see Figure 14.1(b)). The rate of change in such curves is great at first, gradually decreases, and finally approaches zero.

An inspection of Equation 14.3 shows that the ideal inductor is an open circuit initially, and a short circuit in steady state. This is just the opposite of the capacitor.

EXAMPLE 14.1

A series circuit consisting of a 5-H inductor and a 2-ohm resistor is connected across a 10-volt source. Find:

a. the maximum rate of change in the current.
b. the final current.
c. the current at 1 second after the switch is closed.
d. the time needed for the current to reach 2 amperes.

Solution

a.

$$\left(\frac{di}{dt}\right)_{\text{max}} = \frac{E}{L} = \frac{10\text{ V}}{5\text{ H}} = 2\text{ A/s}$$

$$di/dt = 2\text{ A/s}$$

b. The final current is

$$I = \frac{E}{R} = \frac{10\text{ V}}{2\ \Omega}$$

$$I = 5\text{ A}$$

c.

$$i = \frac{E}{R}(1 - e^{-t/(L/R)})$$

Since

$$\frac{t}{L/R} = \frac{1\text{ s}}{(5\text{ H})/2\ \Omega} = 0.4$$

$$i = \frac{10\text{ V}}{2\ \Omega}(1 - e^{-0.4})$$

$e^{-0.4}$ is 0.67 so

$$i = (5\text{ A})(1 - 0.67)$$

$$i = 1.65\text{ A}$$

d.

$$i = \frac{E}{R}(1 - e^{-t/(L/R)})$$

t must be found. So, the equation is rearranged to give

$$e^{-t/(L/R)} = \left(1 - \frac{iR}{E}\right)$$

To solve for t, let $\dfrac{t}{L/R} = x$

Then

$$e^{-x} = \left(1 - \frac{Ri}{E}\right) = \left(1 - \frac{(2\ \Omega)(2\text{ A})}{10\text{ V}}\right) = 0.6$$

Then

$$-x = \ln 0.6 = -0.511$$

so

$$\frac{t}{L/R} = 0.511$$

or

$$t = (0.511)\left(\frac{L}{R}\right)$$
$$= 0.511\left(\frac{5 \text{ H}}{2 \ \Omega}\right)$$
$$t = 1.28 \text{ seconds}$$

The equation for the voltage across the resistor and inductance can now be found. By Ohm's Law,

$$v_R = Ri$$
$$= R\frac{E}{R}(1 - e^{-t/(L/R)})$$

or

$$v_R = E(1 - e^{-t/(L/R)}) \tag{14.4}$$

where v_R is the resistor voltage at t seconds after the switch is closed, in volts, and t, E, R, and L and e have the same meaning as in Equation 14.3.

Applying Kirchhoff's Voltage Law gives

$$E = v_R + v_L$$

or

$$v_L = E - v_R$$

so

$$v_L = E - E(1 - e^{-t/(L/R)})$$

This reduces to

$$v_L = Ee^{-t/(L/R)} \tag{14.5}$$

where v_L is the inductance voltage at t seconds after the switch is closed, in volts, and t, E, R, L, and e have the same meaning as in Equation 14.3.

EXAMPLE 14.2

A series circuit of a 2-H inductance and a 400-ohm resistance is connected across a 24-volt source. Find the resistor voltage and inductor voltage at 3 milliseconds after the switch is closed.

Solution

$$v_R = E(1 - e^{-t/(L/R)})$$

Since

$$\frac{t}{L/R} = \frac{3 \times 10^{-3}\ \text{s}}{(2\ \text{H})/(400\ \Omega)} = 0.6$$

so

$$\begin{aligned} v_R &= (24\ \text{V})(1 - e^{-0.6}) \\ &= (24\ \text{V})(1 - 0.549) \\ v_R &= 10.8\ \text{V} \\ v_L &= Ee^{-t/(L/R)} \\ &= (24\ \text{V})e^{-0.6} \\ v_L &= 13.2\ \text{V} \end{aligned}$$

Equation 14.5 does not give the voltage across the terminals of a practical inductor. Since the practical inductor has resistance, the terminal voltage will be equal to $v_L + v_R$.

R–L circuits with sharply rising current curves have a small time constant. Those whose curves build up at a slower rate have a large **time constant**. The symbol for the time constant is τ. The effect of different time constants on the shape of the curve is seen in Figure 14.3. Thus, the time constant can be used to compare the transient state of *R–L* circuits without plotting the curves.

time constant

In theory, the exponential curves never reach a final value. For all practical purposes, the steady state is assumed to be reached in five time constants. At that time, the value is more than 99.33 percent of the final value. Percentages for 0, 1, 2, 3, 4, and 5 time constants are listed in Table 14.1. These values are useful for checking results, rough sketching of curves, and estimating values.

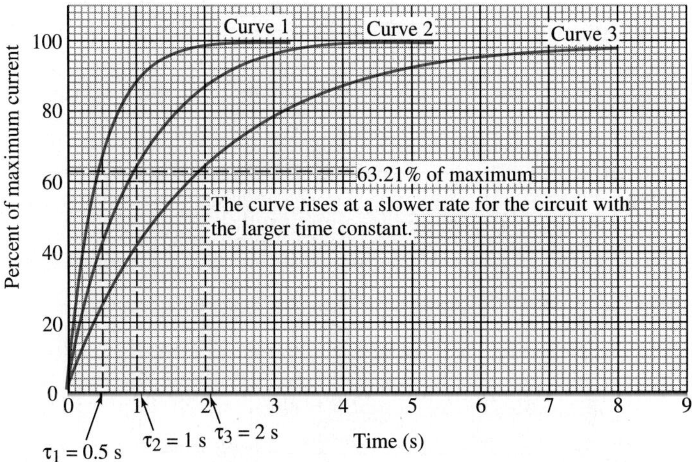

Figure 14.3 The rise of the inductor current is slower for a longer time constant.

TABLE 14.1 Percent of Final Current 0, 1, 2, 3, 4, and 5 Time Constants

Number of Time Constants	Percent of Final Current
0	0
1	63.21
2	86.47
3	95.02
4	98.17
5	99.33

The time constant for the *R–L* circuit has the same meaning as for the *R–C* circuit; that is:

1. It is the time needed for the current to rise to 63.21 percent of its final value. In that time, the voltage will have dropped to 36.79 percent of its initial value. These points are shown in Figure 14.4.

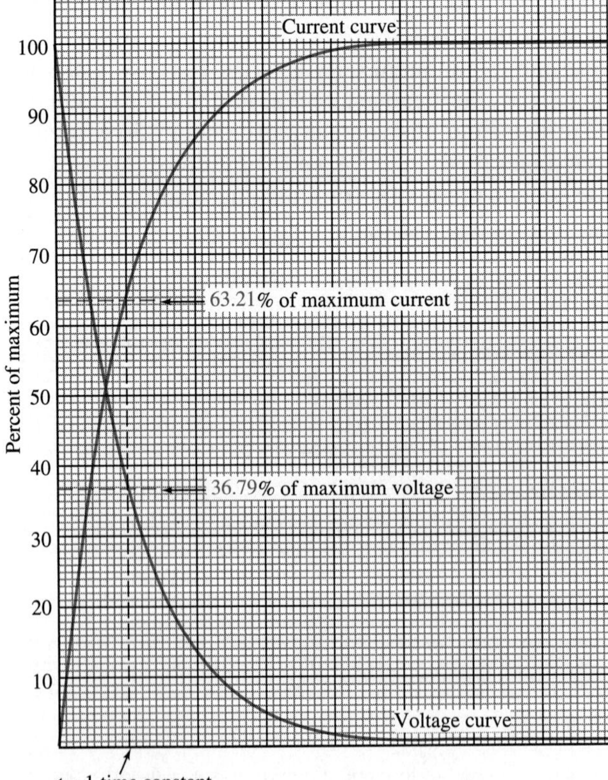

Figure 14.4 Both the inductor current and voltage experience a change of 63.21 percent of the maximum in one time constant.

2. It is the time in which the current would reach the final value if it rose at the initial rate.

The first definition has more practical use, but the second can be used to express the time constant in terms of R and L.

As noted earlier, the time constant depends on the L and R of the circuit. The exact relation is obtained from Figure 14.5. There, line 0 – A represents the rise in current at the initial rate of E/L. The final value is reached in a time of t seconds. From the values on the axis,

$$\frac{\Delta i}{\Delta t} = \frac{E/R}{t}$$

but

$$\frac{\Delta i}{\Delta t} = \frac{E}{L} \qquad \text{since it is the initial rise}$$

so

$$\frac{E}{L} = \frac{E/R}{t}$$

which reduces to

$$t = \frac{L}{R}$$

Since t is the time needed to reach the final value at the initial rate of rise, t is also τ, so

$$\tau = \frac{L}{R} \tag{14.6}$$

Equation 14.6 shows that the buildup takes more time for larger inductance or smaller resistance.

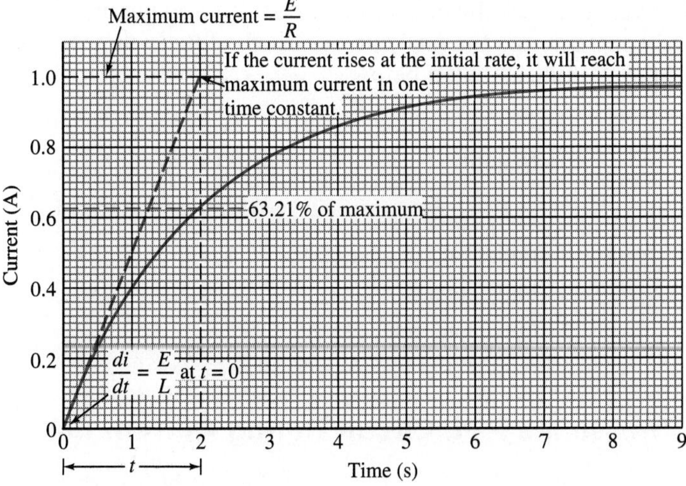

Figure 14.5 One time constant is also the time needed for the current to rise to the maximum at the initial rate of rise.

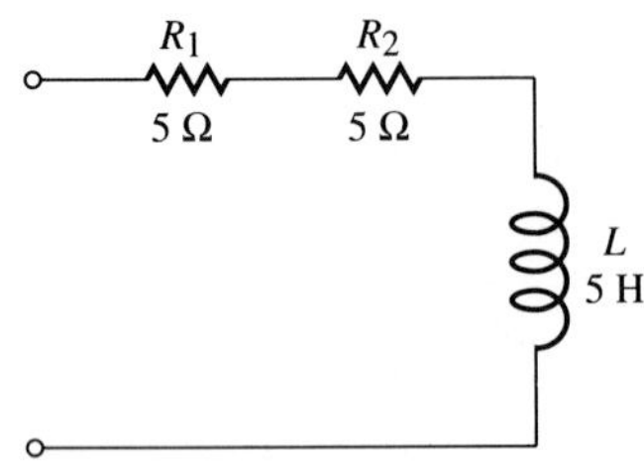

Figure 14.6

EXAMPLE 14.3

Which circuit has a larger time constant: 5 H in series with 10 ohms, or 10 H in series with 5 ohms?

Solution

$$\tau_1 = \frac{L_1}{R_1} = \frac{5\text{ H}}{10\ \Omega} = 0.5\text{ s}$$

$$\tau_2 = \frac{L_2}{R_2} = \frac{10\text{ H}}{5\ \Omega} = 2\text{ s}$$

The second circuit has the larger time constant.

EXAMPLE 14.4

Find the time constant for the circuit shown in Figure 14.6.

Solution

$$\tau = \frac{L}{R}$$

but

$$R = R_1 + R_2 = 5\ \Omega + 5\ \Omega = 10\ \Omega$$

so

$$\tau = \frac{5\text{ H}}{10\ \Omega}$$

$$\tau = 0.5\text{ s}$$

This analysis has shown that some of the key quantities in the transient response are the initial rate of rise, the maximum induced emf, the time constant, and the final current. Each of these depends on some part of the circuit, as listed below.

Quantity	Affected By
Time constant	L and R
Initial rate of change	E and L
Final current	E and R
Maximum induced emf	E

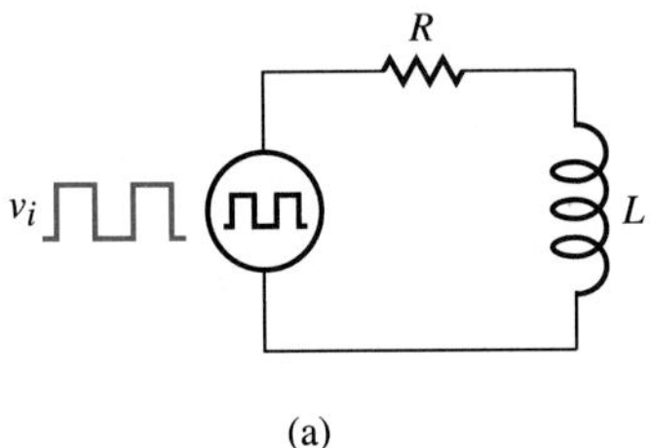

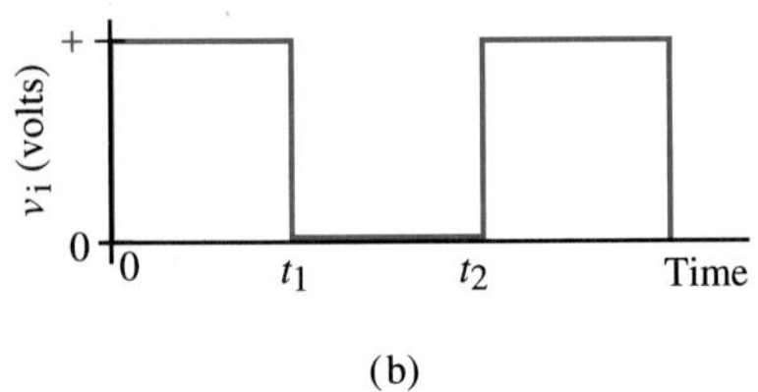

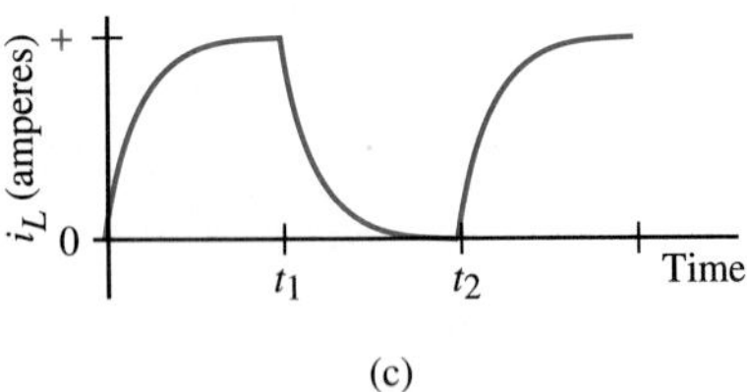

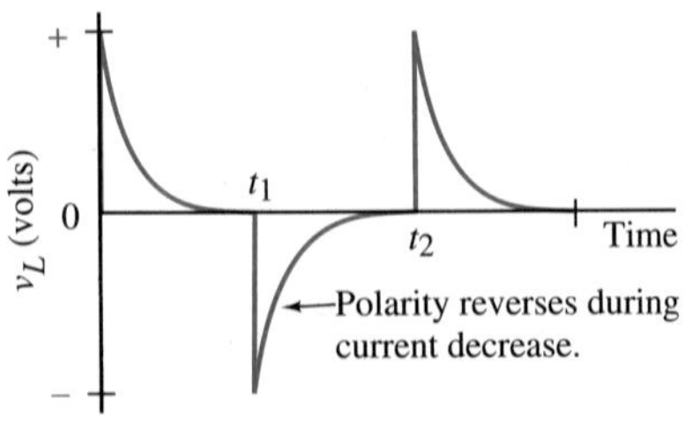

Figure 14.7 The inductor current and voltage during rise and fall. (a) The current direction is the same for rise and fall. (b) The polarity of the induced emf changes for the fall.

14.2 The Series *R–L* Circuit During Current Fall

The transient response also occurs when the current in the R–L circuit decreases. For example, a square wave voltage, as shown in Figure 14.7,

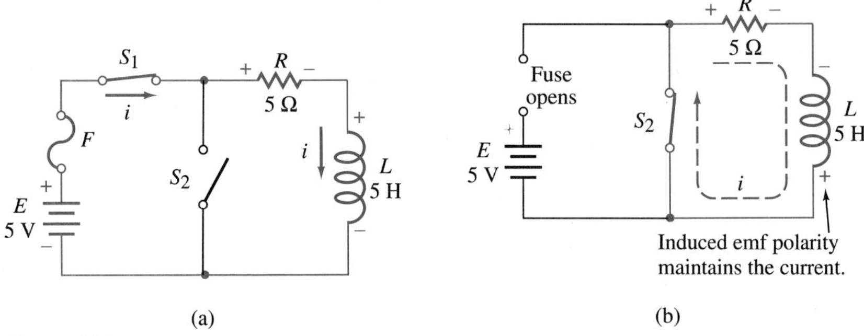

Figure 14.8 The induced emf is such that it opposes the change of the current. (a) During the rise, the induced emf opposes the source emf. (b) During the fall, the induced emf tries to maintain the current.

will cause a current fall from time t_1 to t_2. The current magnitude is the same for the rise and the fall. However, the inductor emf during the fall is opposite to that during the rise. In one case, it opposes the source emf while in the other, it aids the source emf. This is shown in Figure 14.7(d).

A circuit that can be used to study the response during the decrease is shown in Figure 14.8(a). To begin, switch S_1 has been closed long enough for the circuit to reach steady state. In this condition, the source voltage is across the resistance, and the current is E/R amperes.

When switch S_2 is closed, as shown in Figure 14.8(b), the source is shorted out and its emf removed from the circuit. The fuse is there only to protect the source from damage. It opens, leaving a closed path of R, L, and switch S_2. With the source removed, the energy lost in the resistance causes the current to start decreasing. Once again, the current cannot decrease instantaneously because the induced emf opposes a change in current, giving a response as shown in Figure 14.9. The induced emf takes on the polarity shown in Figure 14.8(b). This induced emf main-

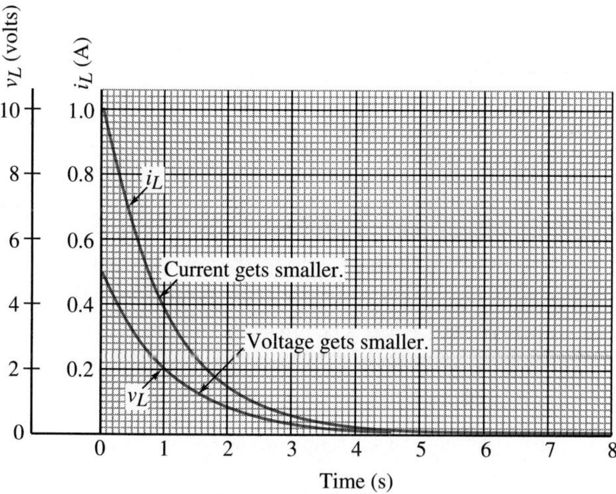

Figure 14.9 The transient response in a series R–L circuit during current fall.

tains some current in the circuit, but eventually both it and the current approach zero. Now we examine this transient response more closely.

Once switch S_2 in Figure 14.8(b) is closed, the series R–L loop is the path with R, L, and switch S_2 in it. Applying Kirchhoff's Voltage Law to this loop, we have

$$-v_R + v_L = 0$$

Since

$$v_R = iR$$

we have

$$-iR + v_L = 0$$

At the instant switch S_2 was closed, the current in the circuit was the steady-state current, and is called I_o. Therefore, at that instant,

$$-I_oR + v_L = 0$$

or

$$v_L = I_oR \tag{14.7}$$

where v_L is the inductor voltage at the start of the fall, in volts; I_o is the current in the inductor at the instant the switch is closed, in amperes; and is equal to E/R; and R is the resistance in the circuit, in ohms.

In general, I_o is the current at the start of the fall. As the energy in the magnetic field is used to move the charge through the resistance, the induced emf will eventually approach 0 volts. The current starts at a value of I_o and also approaches 0 amperes.

Since $v_L = L(di)/(dt)$ just as for the increasing current, and $v_L = I_oR$ at the start of the fall, $L(di)/(dt)$ initially must equal I_oR. The initial value of di/dt is then

$$\frac{di}{dt} = \frac{I_oR}{L} \tag{14.8}$$

where I_o is the current at the start of the current fall, in amperes; R is the resistance in the circuit, in ohms; L is the inductance in the circuit, in henries; and di/dt is the change in current (di) at the start of the fall for a very small change in time (dt), in amperes per second. It is the slope of the curve at $t = 0$ seconds.

As the current decreases, di/dt must decrease until it finally approaches zero. Since di/dt is the slope of the curve, this decrease in di/dt is seen in the slope becoming nearer to a horizontal line.

The expression $iR - L(di/dt)$ is another differential equation. Using calculus, as shown in Appendix C, yields the following equation for the current during the current fall.

$$i_L = I_oe^{-t/(L/R)} \tag{14.9}$$

where i_L is the current at t seconds after the fall has started, in amperes; I_o is the current at the instant the fall begins, in amperes; t is the number of seconds after the fall has started; L is the inductance in the circuit, in henries; R is the resistance in the circuit, in ohms; and e is the base for the natural logarithm system; it is equal to 2.71828 carried to six digits.

Equation 14.9 is also an exponential equation. It has the same form as the inductor voltage equation for the rising current transient. Note that it has the same form as some of the equations for the R–C circuit.

EXAMPLE 14.5

The current in the circuit of Figure 14.10 has reached its steady-state value. Find:

a. the initial rate of current change and the voltage across the inductor at the instant the switch is put into Position 2.
b. the current in the inductor at 5 milliseconds after the switch is put into Position 2 assuming that no energy is lost during the switching.
c. the time that it will take for the current to reach 1 ampere.

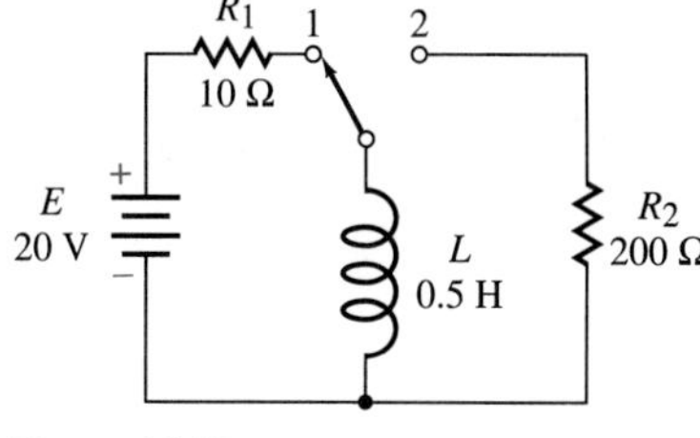

Figure 14.10

Solution

a. The initial rate of change of current equals

$$\frac{di}{dt} = \frac{I_o R_2}{L}$$

Since steady state had been reached,

$$I_o = \frac{E}{R_1} = \frac{20\ \text{V}}{10\ \Omega} = 2\ \text{A}$$

Then

$$\frac{di}{dt} = \frac{(2\ \text{A})(200\ \Omega)}{0.5\ \text{H}}$$

$$\frac{di}{dt} = 800\ \text{A/s}$$

b. The current with S in Position 2 is given by

$$i_L = I_o e^{-t/(L/R)}$$

or

$$\frac{t}{L/R_2} = \frac{0.005\ \text{s}}{(0.5\ \text{H}/200\ \Omega)} = 2$$

so

$$i_L = (2\ \text{A})e^{-2}$$
$$= (2\ \text{A})(0.135)$$
$$i_L = 0.27\ \text{A}$$

c.

$$i_L = I_o e^{-t/(L/R_2)}$$

or

$$1 = 2e^{-t/(L/R_2)}$$

so

$$0.5 = e^{-t/(L/R_2)}$$

Let

$$\frac{t}{L/R_2} = x$$

Then

$$0.5 = e^{-x}$$

and

$$-x = \ln 0.5 = -0.693$$

Hence

$$\frac{t}{L/R_2} = 0.693$$

$$t = (0.693)(0.5 \text{ H})/(200\ \Omega)$$

$$= 0.00173 \text{ s}$$

$$t = 1.73 \text{ ms}$$

Expressions for υ_L and υ_R at any instant of time can now be found. Since $\upsilon_r = iR$,

$$\upsilon_R = RI_o e^{-t/(L/R)} \quad \textbf{(14.10)}$$

where υ_R is the voltage across the resistance at t seconds after the fall has started, in volts, and R, I_o, t, L, and e have the same meaning as in Equation 14.9.

Also, from Kirchhoff's Voltage Law, $\upsilon_L = \upsilon_R$.

So

$$\upsilon_L = RI_o e^{-t/(L/R)} \quad \textbf{(14.11)}$$

where υ_L is the voltage across the inductance at t seconds after the fall has started, in volts, and R, I_o, t, L, and e have the same meaning as in Equation 14.9.

The induced emf υ_L acts as a source while the drop is across the resistor R. Therefore, the magnitudes of υ_L and υ_R are the same.

EXAMPLE 14.6

Switch S_1 in Figure 14.11 has been closed long enough for the inductor current to reach steady state. Find:

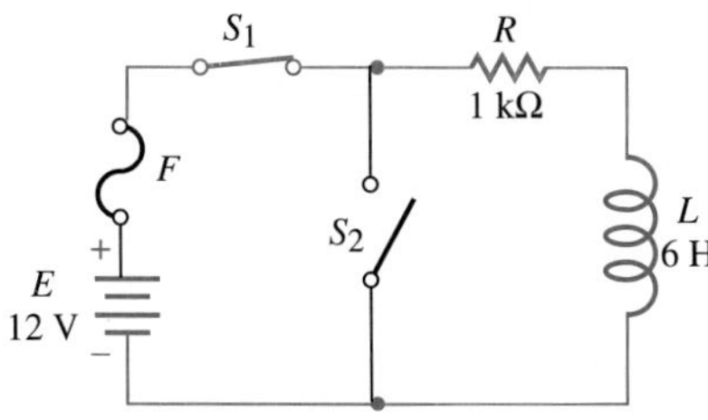

Figure 14.11

a. the voltage and terminal polarity across the inductor at 2 ms after S_2 is closed. Assume the fuse will open.
b. the resistor voltage and terminal polarity at 2 ms after S_2 is closed.
c. the inductor voltage for Item a if S_1 had been closed for only 6 ms.

Solution

a. With S_2 closed, the decay circuit is as in Figure 14.12 and

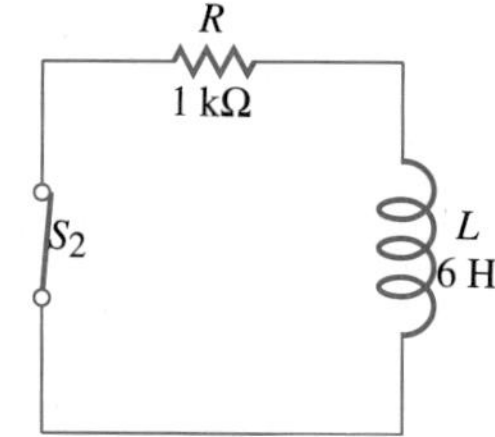

Figure 14.12

$$v_L = I_o R e^{-t/(L/R)}$$

$$I_o = \frac{E}{R} = \frac{12\ \text{V}}{1000\ \Omega} = 0.012\ \text{A}$$

$$\frac{t}{L/R} = \frac{2 \times 10^{-3}\ \text{s}}{6\ \text{H}/(1 \times 10^3\ \Omega)} = 0.333$$

so

$$v_L = (0.012\ \text{A})(1 \times 10^3\ \Omega)\ e^{-0.333}$$
$$= (12\ \text{V})e^{-0.333}$$
$$v_L = 8.6\ \text{V}$$

The bottom terminal is positive.

b. $v_R = V_L$

$v_R = 8.6\ \text{V}$

The left terminal is positive.

c. At 6 ms, the current has not reached steady state because five time constants during rise is $5 \times (L/R)$ or $5(6\ \text{H}/1000\ \Omega) = 30$ ms. Therefore, Equation 14.3 must be used to find I_o.

During the current rise,

$$i_L = \frac{E}{R}(1 - e^{-t/(L/R)})$$

For the values of t, L, and R,

$$\frac{t}{L/R} = \frac{6 \times 10^{-3}\ \text{s}}{(6\ \text{H})/1000\ \Omega} = 1$$

so

$$i_L = (12 \times 10^{-3}\ \text{A})(1 - e^{-1})$$
$$i_L = 7.6 \times 10^{-3}\ \text{A}$$

Then

$$v_L = I_o R e^{-t/(L/R)}$$
$$v_L = (7.6 \times 10^{-3}\ \text{A})(1 \times 10^{3}\ \Omega)e^{-0.333}$$
$$v_L = 5.45\ \text{V}$$

TABLE 14.2 The Percentage of Initial Current Reached at 0, 1, 2, 3, 4, and 5 Time Constants

Time Constants	Percent of Initial Current
0	100
1	36.79
2	13.53
3	4.98
4	1.83
5	0.67

The time constant for the circuit has the same meaning as for the rise. That is, it is the time needed for the current or voltage to reach zero if the rate of change continued as the initial rate; it is also the time needed for the inductor current or voltage to change 63.21 percent of the initial value. Table 14.2 lists the percent of initial value reached at 1, 2, 3, 4, and 5 time constants.

The key characteristics in the current fall circuit analysis are the initial current, initial voltage, time constant, and initial rate of change. Each of these depends on some part of the circuit, which is listed in the following.

Quantity	Affected By
Time constant	L and R
Initial current	E, R, and t for the current rise circuit
Initial rate of change	I_o, L, R
Initial voltage	I_o, R

The terminal polarity of the inductor is now considered for current rise and fall. During the rise, the induced emf opposes the increase in current so the polarity and current direction are as shown in Figure 14.13(a).

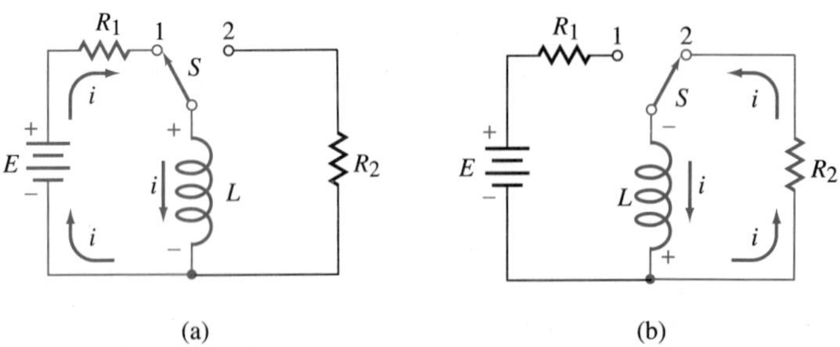

Figure 14.13 The induced emf opposes a change in the current. (a) The current is increasing, so the induced emf subtracts from the source. (b) The current is decreasing, so the induced emf tries to maintain it.

If the switch is put in Position 2, the current will decrease and the induced emf will oppose that decrease. The inductor acts as a source, so the polarity must be as shown in Figure 14.13(b). So the current direction in the inductor remains the same, but the polarity changes. This is the opposite of the polarity and direction for the capacitor.

Inductance can present a problem in circuits with pulse signals. First, the inductance will round the corners of a rectangular wave. Second, if the pulse width is too short, the pulse will not reach its maximum value. The waveforms in Figure 14.14 show some of the possible results. Let us now examine the reason for these results.

The rounding results because the inductor opposes the change in current. As a result, when the current tries to rise instantaneously, the inductance forces an exponential rise, which gives a rounding of the corner. This can result in a delayed triggering of a circuit.

As explained in Sections 14.1 and 14.2, a time of five time constants is needed to reach the maximum current. If the pulse width is less than five time constants, the maximum will not be reached. This can result in a device such as a logic gate not being triggered.

14.3 The Universal Time Constant Curve

The exponential curves give us an alternate method of finding the inductance voltage and current. The values will not be as exact as those from the equations, but they can be found quickly and without involving exponents. Another advantage is that the curves show the trend for that

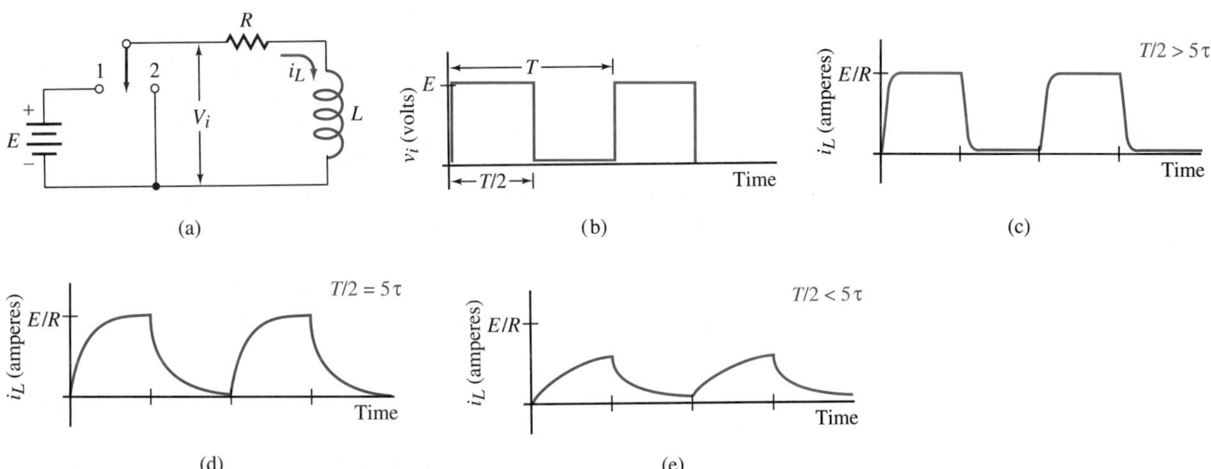

Figure 14.14 The transient response of the R–L circuit can change the shape of a rectangular pulse. (a) Moving the switch between position 1 and 2 will generate a rectangular pulse. (b) The rectangular pulse applied to the R–L circuit. (c) There is a rounding of the pulse, but the maximum is reached in less than T/2 s. (d) There is more rounding, but the maximum is reached in T/2 s. (e) There is rounding, and the maximum is not reached because of the narrow pulse.

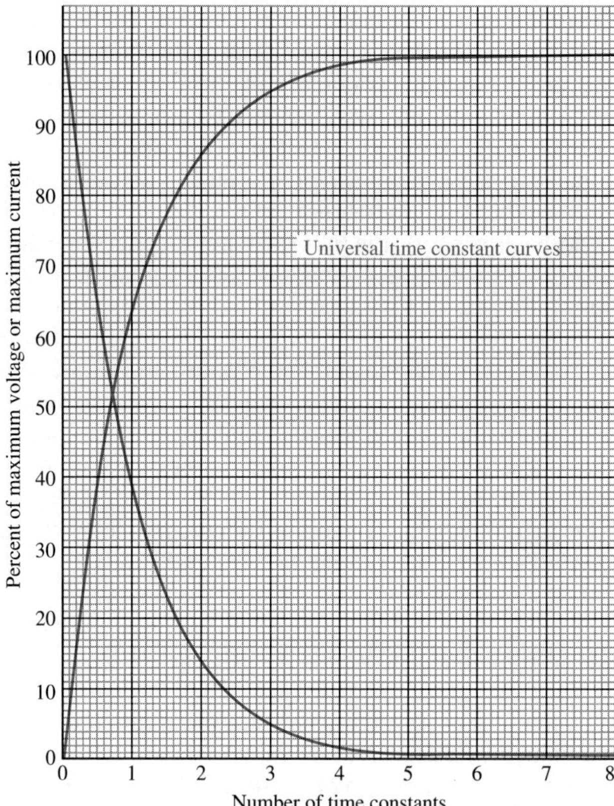

Figure 14.15 The universal time constant curves can be used with any R–L circuit because they plot percent of maximums versus number of time constants.

region on the curve. On the negative side, using the voltage or current versus time curves would require a separate curve for each *R–L* circuit. To avoid this, the **Universal Time Constant Curves** in Figure 14.15 are used. These are curves that plot the percent of maximum versus the number of time constants. Once the number of time constants or percent of maximum is found on the curve, it can be related to the actual time, voltage, or current in the circuit.

Universal Time Constant Curves

The rising curve is used for the current rise. The falling curve is used for voltage during the current rise and fall, and current during the current fall. The procedure for using the curves is as follows:

1. To find i_L and v_L:
   a. Calculate the number of time constants.
   b. Draw a vertical line at the number of time constants to intersect the curve.
   c. At the intersection, draw a horizontal line and note where it intersects the percent axis.
   d. Convert the percentage to current or voltage by multiplying the maximum by the percentage.

2. To find t, R, or L:
   a. Calculate the percent of maximum for the voltage or current.
   b. Draw a horizontal line at the percent to intersect the curve.
   c. At the intersection, draw a vertical line and note where it intersects the number of time constants.
   d. Using the number of time constants, solve for t, L, or R.

EXAMPLE 14.7

The switch in the circuit shown in Figure 14.16 has been closed for 0.08 second. Use the Universal Time Constant Curves to find the inductor current, inductor voltage, and resistor voltage in Figure 14.16 at 0.08 second after the switch is closed.

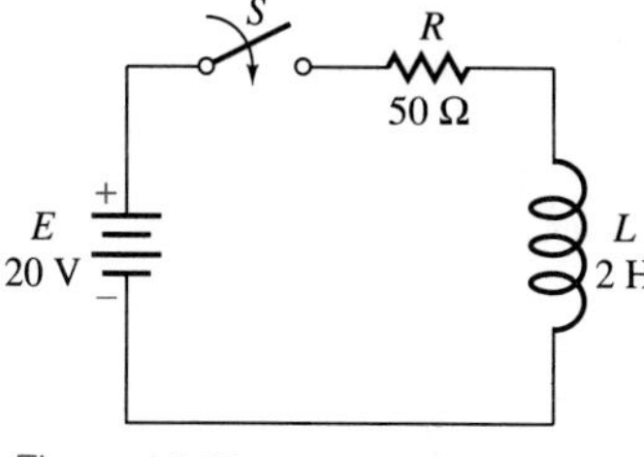

Figure 14.16

Solution

The time must first be converted to time constants.

$$\tau = \frac{L}{R} = \frac{2\text{ H}}{50\ \Omega} = 0.04\text{ s}$$

$$\frac{t}{\tau} = \frac{0.08\text{ s}}{0.04\text{ s}} = 2 \quad \text{or} \quad t = 2\,\tau$$

The current is rising, so the rising curve must be used. A vertical line drawn at 2 τ crosses the rising curve at 86.5 percent of maximum current.

From the circuit, the maximum current is E/R or 20 V/50 Ω = 0.4 A. Therefore, i @ 0.08 second is

$$i \text{ at } t = 0.08\text{ s} = \frac{(86.5\%)(0.4\text{ A})}{(100\%)}$$

$$i = 0.346\text{ A}$$

The voltage is decreasing, so the falling curve must be used. A vertical line drawn at 2 crosses the falling curve at 13.5 percent of maximum. Therefore,

$$v_L = \frac{(13.5\%)(20\text{ V})}{(100\%)}$$

$$v_L = 2.7\text{ V}$$

The resistance voltage cannot be found directly from the curve. But, using Ohm's Law,

$$\begin{aligned} v_R &= iR \\ &= (0.346\text{ A})(50\ \Omega) \\ v_R &= 17.3\text{ V} \end{aligned}$$

An alternative choice can be Kirchhoff's Voltage Law, which gives

$$v_R = E - v_L$$

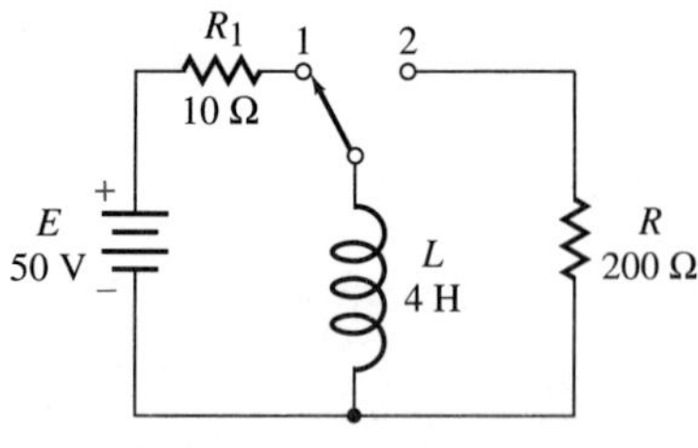

Figure 14.17

EXAMPLE 14.8

The current in the inductor in Figure 14.17 is 5 A when the switch is put into Position 2. Use the Universal Time Constant Curves to find the inductor current and voltage at 0.05 second after the switch is put into Position 2. Assume that no energy is lost in the switching.

Solution

The current and voltage will start decreasing when the switch is put into Position 2. Therefore, the falling curve must be used.

$$\tau = \frac{L}{R} = \frac{4\text{ H}}{200\ \Omega} = 0.02\text{ s}$$

$$\frac{t}{\tau} = \frac{0.05\text{ s}}{0.02\text{ s}} = 2.5 \qquad \text{or} \qquad t = 2.5\ \tau$$

A vertical line drawn at 2.5 τ crosses the falling curve at about 8.3 percent.

The maximum current is the current at the start of the decrease, or 5 amperes.

The maximum voltage is $I_o R$, or (5 A)(200 Ω) = 1000 V. Then

$$i = \frac{(8.3\%)(5\text{ A})}{(100\%)}$$

$$i = 0.415\text{ A}$$

$$v_L = \frac{(8.3\%)(1000\text{ V})}{(100\%)}$$

$$v_L = 83\text{ V}$$

This compares to the calculated value of 82.09 V.

EXAMPLE 14.9

Use the Universal Time Constant curves to determine the time at which the current is 0.2 ampere for the circuit of Figure 14.16.

Solution

$$i_{max} = \frac{E}{R} = \frac{20\text{ V}}{50\ \Omega} = 0.4\text{ A}$$

$$\frac{0.2\text{ A}}{0.4\text{ A}} = 0.5 \qquad \text{or} \qquad 50\%$$

A vertical line drawn to intersect the rising curve at 50 percent intersects the time constant axis at approximately 0.7.

Then

$$t = 0.7\ \tau$$

so

$$t = (0.7)(L/R)$$
$$= (0.7)(2\ \text{H}/50\ \Omega)$$
$$t = 0.028\ \text{second}$$

This compares to the calculated value of 0.0277 s.

14.4 Complex *R–L* Circuits

The series *R–L* circuit provided the basic relations needed to analyze the circuit. Unfortunately, some circuits are not simple series circuits, nor do they all have only one source. The analysis of these **complex *R–L* circuits** is done by converting them to a series circuit and finding the inductor voltage and current. Then, basic circuit relations are used to determine other quantities.

complex *R–L* circuits

When analyzing the current rise, the circuit is replaced by its Thevenin equivalent circuit. For the current fall, the circuit connected to the inductor is replaced by its equivalent resistance.

EXAMPLE 14.10

Assume that the switch in Figure 14.18 is closed at $t = 0$ second. At 0.06 second after the switch is closed, what is:

a. the inductor current.
b. the inductor voltage.
c. the source current.

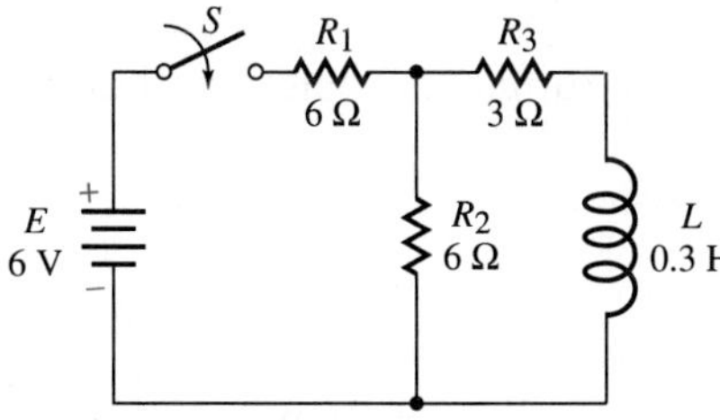

Figure 14.18

Solution

The circuit connected to the inductor must be Theveninized. The circuit used to find R_{TH} is shown in Figure 14.19.

$$R_{TH} = R_3 + R_1 \| R_2$$
$$= R_3 + \frac{R_1 R_2}{R_1 + R_2}$$
$$R_{TH} = 3\ \Omega + \frac{(6\ \Omega)(6\ \Omega)}{6\ \Omega + 6\ \Omega}$$
$$R_{TH} = 6\ \Omega$$

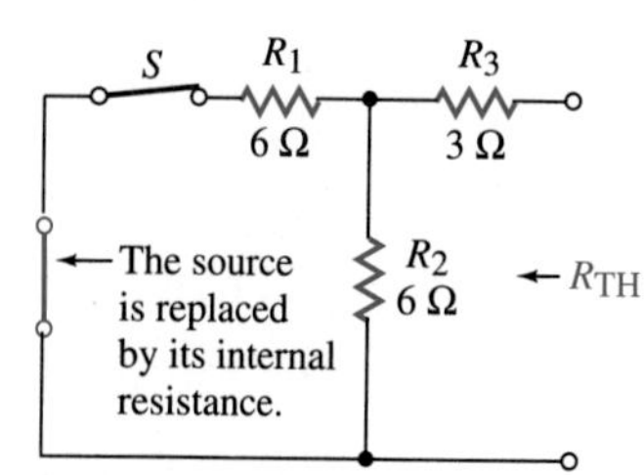

Figure 14.19

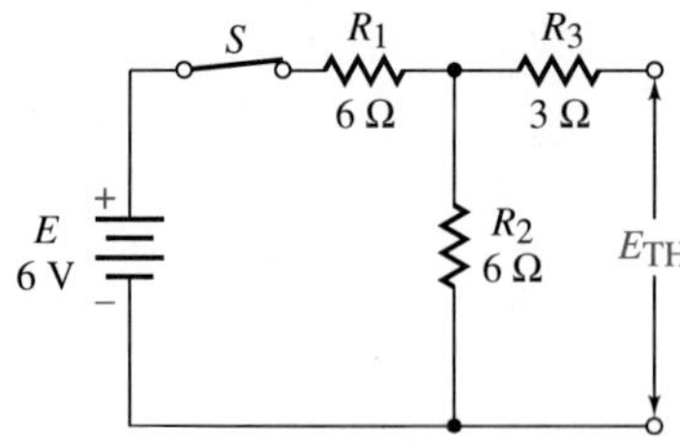

Figure 14.20

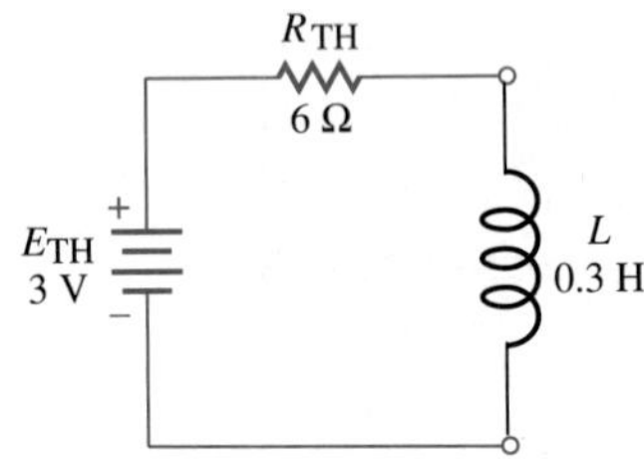

Figure 14.21

The circuit for E_{TH} is shown in Figure 14.20.

$$E_{TH} = \left(\frac{R_2}{R_1 + R_2}\right)E$$

$$= \left(\frac{6\ \Omega}{6\ \Omega + 6\ \Omega}\right)(6\ \text{V})$$

$$E_{TH} = 3\ \text{V}$$

The Theveninized circuit with the switch closed is shown in Figure 14.21.

a.

$$i_L = \frac{E_{TH}}{R_{TH}}(1 - e^{-t/(L/R)})$$

$$= \frac{3\ \text{V}}{6}(1 - e^{-0.06/(0.3\ \text{H}/6\ \Omega)})$$

$$= 0.5(1 - e^{-1.2})$$

$$i_L = 0.35\ \text{A}$$

b.

$$\upsilon_L = E_{TH}(e^{-t/(L/R)})$$

$$= (3\ \text{V})e^{-1.2}$$

$$\upsilon_L = 0.9\ \text{V}$$

Kirchhoff's Voltage Law could also have been used to find υ_L. That is,

$$E_{TH} = \upsilon_R + \upsilon_L$$

or

$$\upsilon_L = E_{TH} - i_L R_{TH}$$

c. Putting these in the original circuit gives the circuit shown in Figure 14.22.

By Kirchhoff's Voltage Law,

$$\upsilon_2 = \upsilon_3 + \upsilon_L$$

$$= i_L R_3 + \upsilon_L$$

$$= (0.35\ \text{A})(3\ \Omega) + 0.9\ \text{V}$$

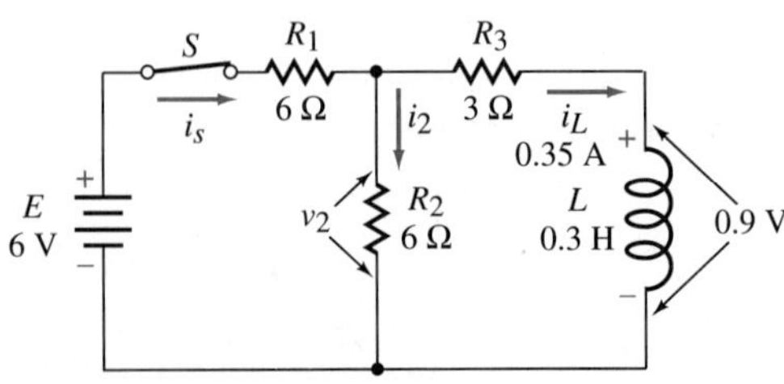

Figure 14.22

$$v_2 = 1.95 \text{ V}$$

Then

$$i_2 = \frac{1.95 \text{ V}}{6 \ \Omega}$$
$$= 0.325 \text{ A}$$

so

$$i_s = 0.325 \text{ A} + 0.35 \text{ A}$$
$$i_s = 0.675 \text{ A}$$

This is only one of several sequences that can be used to relate v_L and i_L to i_s.

The circuit in Figure 14.23(a) is an example of current fall in a complex circuit. When the switch is opened, the energy in the inductance will be dissipated in the combination of resistors. Circuits such as this one can also be analyzed using the exponential equations. But now the resistance combination must be replaced by an R_{eq}, as shown in Figure 14.23(b). Voltage and current are found by using the exponential equations. Finally, basic circuit relations are used to find current, voltage, and other quantities in the original circuit.

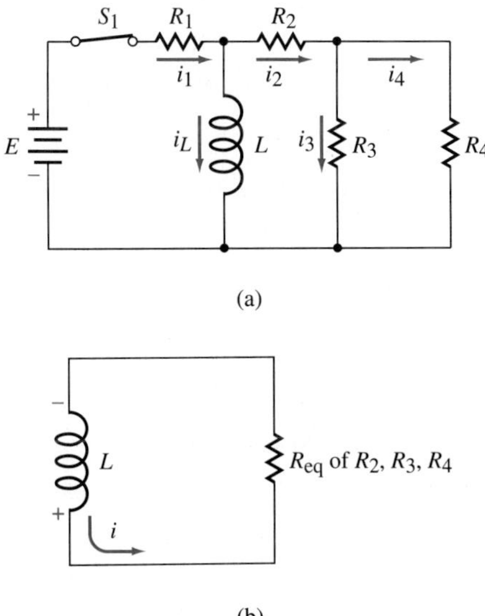

Figure 14.23 During current fall, the complex circuit should be replaced by its equivalent resistance.

EXAMPLE 14.11

The current in the inductor in Figure 14.24 has reached steady state with switch S_1 closed. At 0.5 second after S_1 is opened, what is:

a. the inductor current.
b. the inductor voltage.
c. the current in R_3.

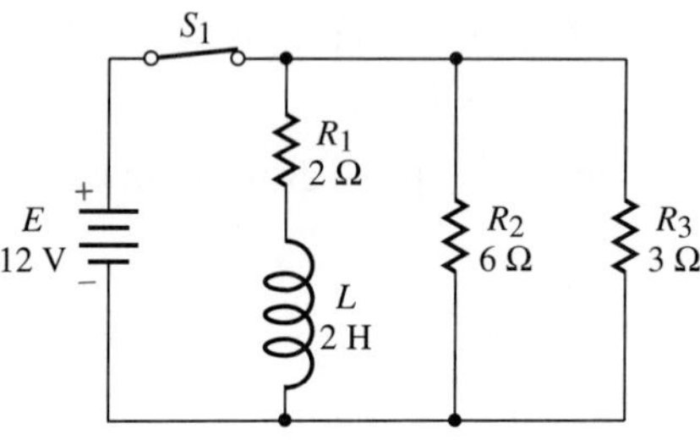

Figure 14.24

Solution

When S_1 is opened, the circuit connected to the inductance is shown in Figure 14.25.

The circuit of R_1, R_2, and R_3 must be replaced by an R_{eq}.

$$R_{eq} = R_1 + R_2 \| R_3$$

$$= R_1 + \frac{R_2 R_3}{R_2 + R_3}$$

$$= 2\ \Omega + \frac{(6\ \Omega)(3\ \Omega)}{6\ \Omega + 3\ \Omega}$$

$$R_{eq} = 4\ \Omega$$

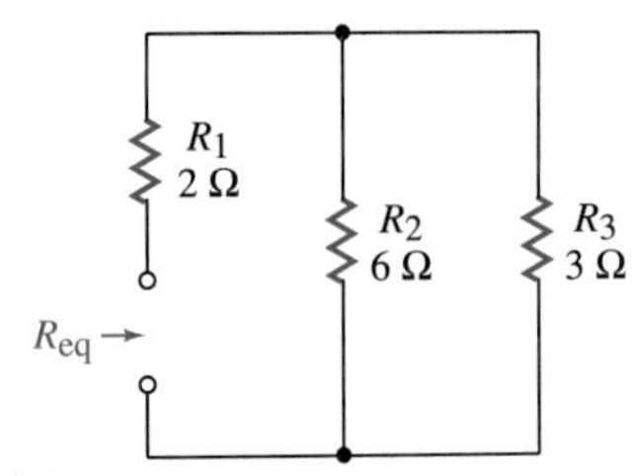

Figure 14.25

The series R–L circuit is now shown in Figure 14.26.

The direction of i_L is the same as it was in L before switch S_1 was opened.

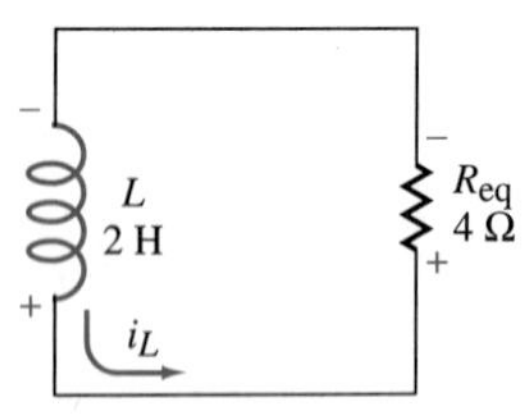

Figure 14.26

a. This is a current-fall condition, so

$$i_L = I_o e^{-t/(L/R)}$$

$$I_o = \frac{E}{R_1} = \frac{12\ \text{V}}{2\ \Omega} = 6\ \text{A}$$

Then

$$i_L = (6\ \text{A})e^{-t/(L/R)}$$

$$= (6\ \text{A})e^{-1}$$

$$i_L = 2.21\ \text{A}$$

b.

$$v_L = I_o R_{eq} e^{-1}$$

$$= (6\text{ A})(4\ \Omega)(0.368)$$
$$v_L = 8.83\text{ V}$$

Putting these in the original circuit gives the circuit shown in Figure 14.27.

c. R_2 and R_3 are in parallel. Hence

$$i_3 = \frac{R_2 i_L}{R_2 + R_3}$$
$$= \frac{(6\ \Omega)(2.21\text{ A})}{6\ \Omega + 3\ \Omega}$$
$$i_3 = 1.47\text{ A}$$

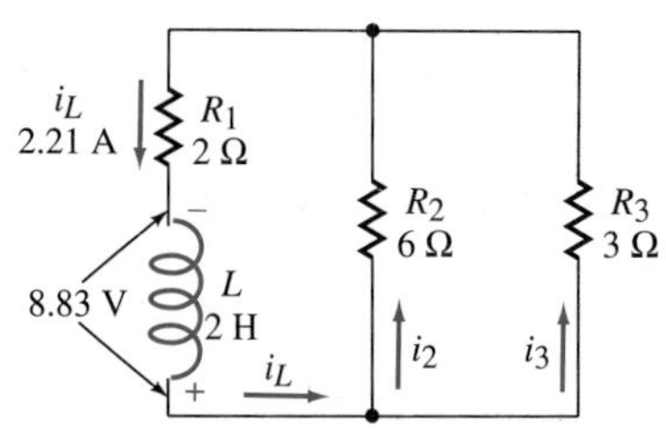

Figure 14.27

Current-fall circuit analysis always involves the initial current. If steady state has not been reached, the initial current is the value that has been reached during the rise of the current.

Next, a circuit with several sources is analyzed.

EXAMPLE 14.12

Determine the inductor voltage and current for the circuit of Figure 14.28 at 0.2 second after the switch is closed.

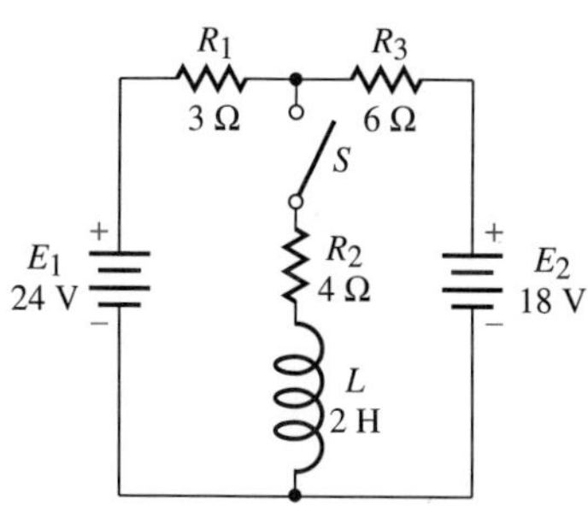

Figure 14.28

Solution

The Thevenin equivalent of the circuit connected to the inductor must be found.

With E_1 and E_2 short-circuited,

$$R_{TH} = R_2 + R_1 \| R_3$$

so

$$R_{TH} = 4\ \Omega + \frac{(3\ \Omega)(6\ \Omega)}{3\ \Omega + 6\ \Omega}$$
$$R_{TH} = 6\ \Omega$$

The Thevenin voltage is the open-circuit voltage of the circuit in Figure 14.29. This is also the voltage of Node A with respect to Node B.

Writing the nodal equation for Node A gives

$$V_A\left(\frac{1}{R_1} + \frac{1}{R_3}\right) = E_1\left(\frac{1}{R_1}\right) + E_2\left(\frac{1}{R_2}\right)$$
$$V_A\left(\frac{1}{3\ \Omega} + \frac{1}{6\ \Omega}\right) = (24\text{ V})\left(\frac{1}{3\ \Omega}\right) + (18\text{ V})\left(\frac{1}{6\ \Omega}\right)$$
$$V_A = 22\text{ V}$$

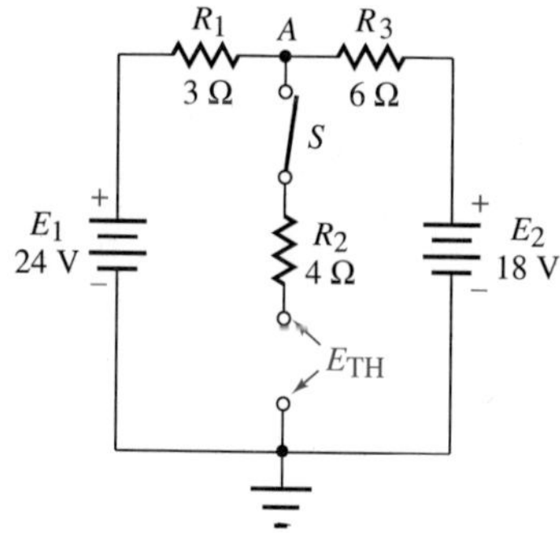

Figure 14.29

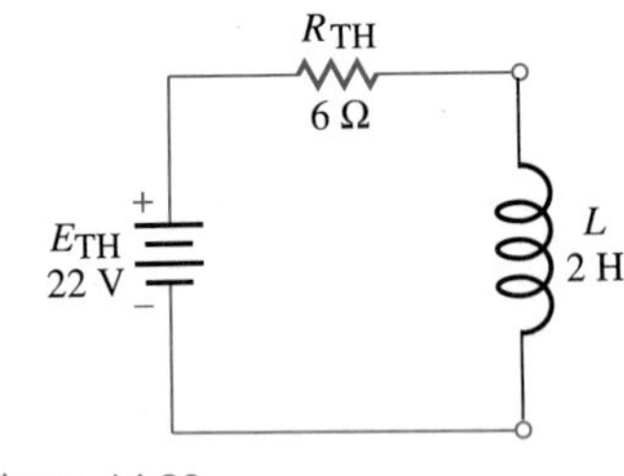

Figure 14.30

The equivalent circuit is shown in Figure 14.30. Then

$$v_L = E_{TH}e^{(-t/(L/R))}$$

At t = 0.2 second,

$$\frac{t}{L/R_{TH}} = \frac{0.2 \text{ s}}{(2 \text{ H})/6\ \Omega} = 0.6$$

so

$$v_L = (22 \text{ V})(e^{-0.6})$$
$$v_L = 12.07 \text{ V}$$
$$i_L = I_o(1 - e^{-t/(L/R)})$$
$$= \frac{E_{TH}}{R_{TH}}(1 - e^{-t/(L/R)})$$
$$= \left(\frac{22 \text{ V}}{6\ \Omega}\right)(1 - e^{-0.6})$$
$$i_L = 1.65 \text{ A}$$

14.5 Energy Stored in an Inductor

As the magnetic field builds up, *energy* is being *stored* in it. The effect of this energy can be seen in things such as the magnetic attraction of iron, and the spark across switch contacts when the switch is opened. An expression for this **inductor stored energy** is now developed.

inductor stored energy

Figure 14.31 shows the inductor current, voltage, and power curves

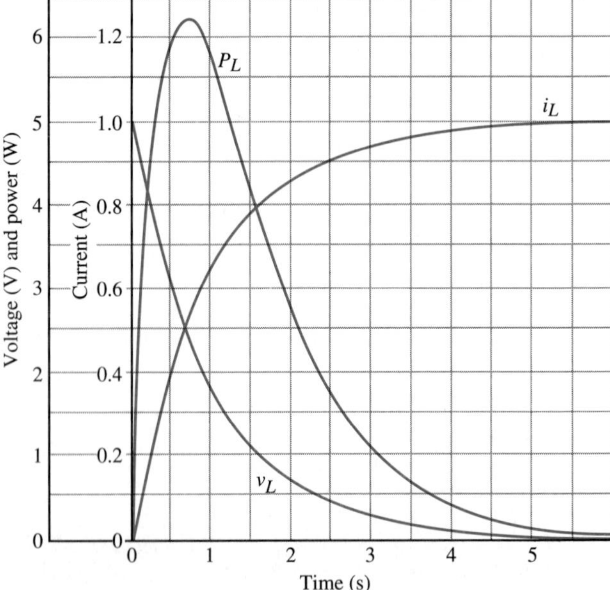

Figure 14.31 Instantaneous voltage, current, and power during the current rise in the inductor.

for the buildup condition. The power at any instant of time is found by multiplying voltage and current at that instant of time. It can be shown that the energy stored in the field of the inductor is the area under the power curve. Since the power is changing with time, calculus must be used, as shown in Appendix C, to obtain the exact area.

When the current starts at 0 amperes, the result is

$$\mathcal{W} = \frac{LI^2}{2} \tag{14.12}$$

where $\mathcal{W}$ is the energy, in joules; L is the inductance, in henries; and I is the current in the inductance, in amperes.

EXAMPLE 14.13

How much energy is stored in a 5-H inductance when the current reaches 2 amperes?

Solution

$$\mathcal{W} = \frac{LI^2}{2} = \frac{(5\text{ H})(2\text{ A})^2}{2}$$

$$\mathcal{W} = 10\text{ J}$$

Sometimes the current does not start at zero, but changes from one value to another. Then the energy either increases or decreases, depending on the direction of change. The change in energy is the difference between the energy at each current.

EXAMPLE 14.14

How much energy is lost in the inductance of Example 14.13 if the current is reduced to 1 ampere?

Solution

$$\mathcal{W}\text{ at 2 A} = \frac{LI_1^2}{2} = \frac{(5\text{ H})(2\text{ A})^2}{2} = 10\text{ J}$$

$$\mathcal{W}\text{ at 1 A} = \frac{LI_2^2}{2} = \frac{(5\text{ H})(1\text{ A})^2}{2} = 2.5\text{ J}$$

$$\mathcal{W}_{\text{lost}} = 10\text{ J} - 2.5\text{ J}$$

$$\mathcal{W}_{\text{lost}} = 7.5\text{ J}$$

14.6 The *R–L* Circuit in Steady State

steady state

Steady state is the condition where there is no change in the inductor current and voltage. Once steady state is reached in an *R–L* circuit, ideal inductances are short circuits, so there is no voltage across them. This can be seen on the v_L versus time curve for times greater than five time constants. At steady state, the inductor current is limited by its resistance. The voltage across its terminals is not zero but equals *IR*. Since there is current in the inductance, there is energy stored in it even though the voltage is zero.

To analyze the steady-state condition, the inductors must be replaced by their steady-state equivalents. These equivalents are short circuit for an ideal inductor, and a resistance for a practical inductor. Then, dc circuit theory can be applied to the circuit. This is done in the following examples.

EXAMPLE 14.15

The circuit shown in Figure 14.32 has reached steady state. Find:
a. the current and voltage for each component.
b. the energy stored in each inductance.

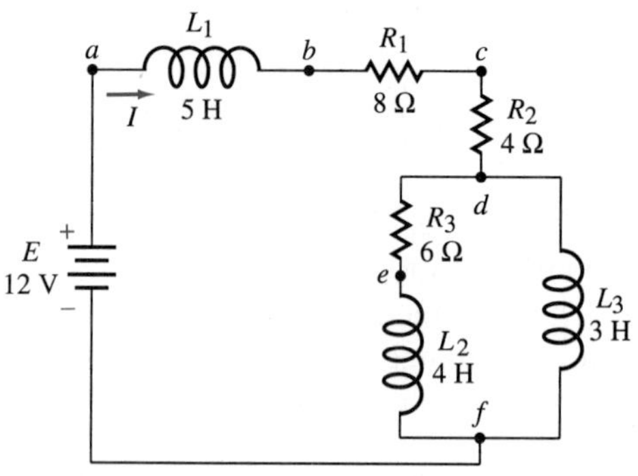

Figure 14.32

Solution

a. The steady-state equivalent circuit is shown in Figure 14.33. The zero resistance of L_3 shorts out R_3 and L_2.

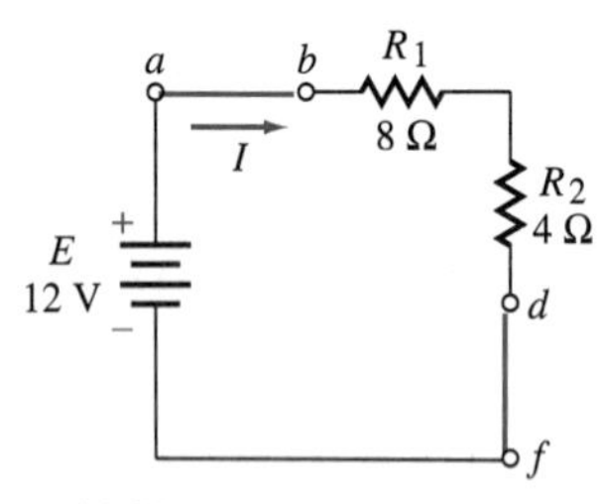

Figure 14.33

$$I = \frac{E}{R_{eq}} = \frac{E}{R_1 + R_2} = \frac{12\text{ V}}{8\ \Omega + 4\ \Omega} = 1\text{ A}$$

I is the current in L_1, R_1, R_2, and L_3 so their current is also 1 ampere. Since R_3 and L_2 are shorted, their current and voltage are zero.

$$V_{L_1} = V_{L_3} = 0\text{ V}$$
$$V_{R_1} = IR_1 = (1\text{ A})(8\ \Omega) = 8\text{ V}$$
$$V_{R_2} = IR_2 = (1\text{ A})(4\ \Omega) = 4\text{ V}$$

b.

$$\mathscr{W}_1 = \frac{L_1 I_1^2}{2} = \frac{(5\text{ H})(1\text{ A})^2}{2}$$
$$\mathscr{W}_1 = 2.5\text{ J}$$
$$\mathscr{W}_2 = \frac{L_2 I_2^2}{2} = \frac{(4\text{ H})(0\text{ A})^2}{2}$$

$$\mathcal{W}_2 = 0 \text{ J}$$

$$\mathcal{W}_3 = \frac{L_3 I_3^2}{2} = \frac{(3 \text{ H})(1 \text{ A})^2}{2}$$

$$\mathcal{W}_3 = 1.5 \text{ J}$$

EXAMPLE 14.16

An inductor with 5-H inductance and 6 ohms of resistance is connected in series with one that has 3-H inductance and 2 ohms of resistance. Find the current in the inductors and the voltage across their terminals when the series group is connected across a 24-volt source and steady state is reached.

Solution

The circuit and its steady-state equivalent are shown in Figures 14.34 and 14.35.

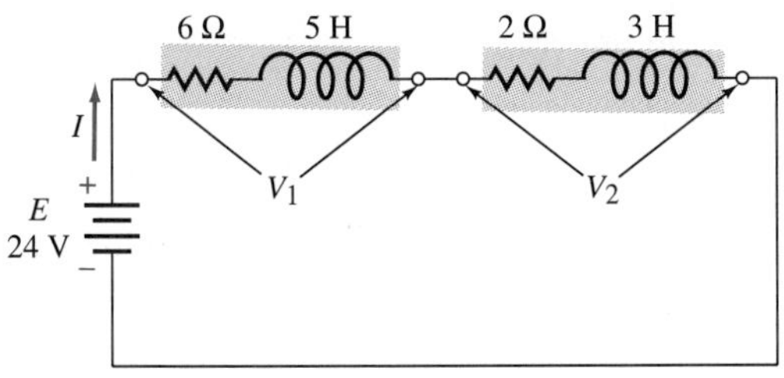

Figure 14.34

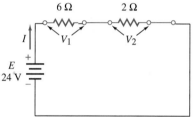

Figure 14.35

At steady state, the inductance parts are shorts circuits so

$$I = \frac{E}{R_{eq}} = \frac{24 \text{ V}}{6\ \Omega + 2\ \Omega}$$

$$I = 3 \text{ A}$$

$$V_1 = (3 \text{ A})(6\ \Omega)$$

$$V_1 = 18 \text{ V}$$

$$V_2 = (3 \text{ A})(2\ \Omega)$$

$$V_2 = 6 \text{ V}$$

S 14.7 SPICE Analysis of *R–L* Circuits

The transient behavior of *R–L* circuits can be simulated using PSpice, just as was done when we studied the transient behavior of *R–C* circuits. The following examples illustrate how this is done.

Consider the *R–L* circuit shown in Figure 14.36(a). The circuit is a series *R–L* circuit consisting of the voltage source, swtich, resistor, and inductor. We assume that the switch is initially open and no current flows in the circuit. We want to study the behavior of the circuit from the time

(a)

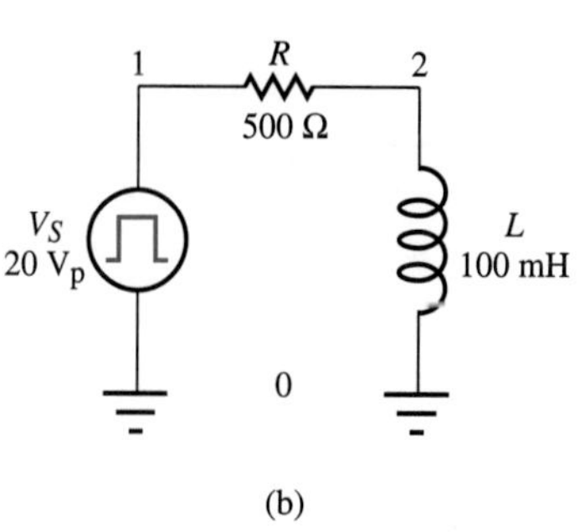

(b)

Figure 14.36 *R–L* circuit. (a) Schematic diagram. (b) Node assignments.

the switch is closed, $t = 0$ seconds, to the point at which the transient behavior of the circuit is complete, a period of time that is approximately equal to five time constants.

The time constant for the circuit is

$$\tau = \frac{L}{R} = \frac{100 \text{ mH}}{500\ \Omega} = 200\ \mu\text{s} \qquad \text{or} \qquad 0.2 \text{ ms}$$

Hence, the transient behavior of the R–L circuit lasts for 5×0.2 ms or 1.0 millisecond.

The first task in developing the PSpice file that will perform the simulation is to label each circuit node and name the node. This is shown in Figure 14.36(b). Remember that the ground node is always assigned as NODE 0.

The SPICE file that we will use is shown in Figure 14.37. Let us examine this file in detail.

Line 1. `ANALYSIS OF AN R–L CIRCUIT USING PSpice`

The first line in a PSpice file is the title line.

Line 2. `R 1 2 500`

This line describes the resistor in the circuit. The resistor is connected between Node 1 and Node 2 and has an ohmic value of 500 ohms.

Line 3. `L 2 0 100M IC = 0`

This line describes the inductor. The inductor is connected between Node 2 and Node 0, ground. The component value is 100 millihenries. Millihenries is indicated by the suffix "M."

Line 4. `VS 1 0 DC 20`

This line describes the dc voltage source. The voltage source is connected between Node 1 and Node 0. The magnitude of the voltage is 20 volts.

This completes the circuit description. Each circuit component has its own line in the PSpice file. The following command lines define the type

```
ANALYSIS OF AN R-L CIRCUIT USING PSPICE
R     1    2   500
L     2    0   100MH   IC=0
VS    1    0   DC  20
.TRAN   50.0U    1.0M 0.0   25.0U   UIC
.PROBE
.END
```

Figure 14.37 PSpice input file.

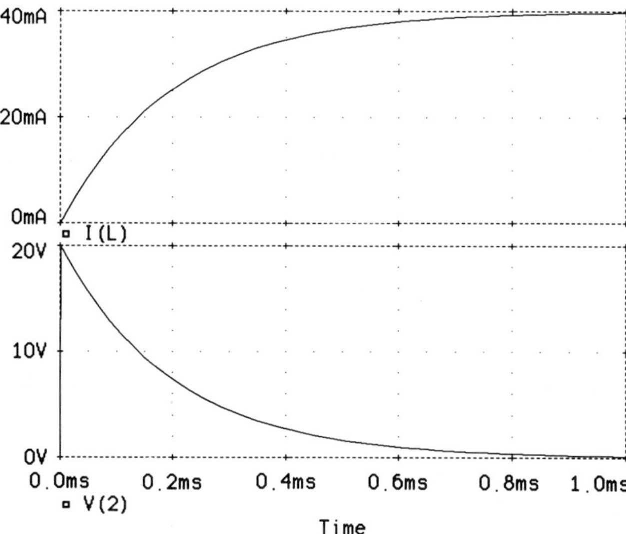

Figure 14.38 Output graphs produced with PROBE.

of analysis to be performed and the format of the simulation output.

Line 5. `.TRAN 50.0U 1.0M 0.0 25.0U UIC`

The .TRAN statement causes a transient analysis to be performed on the circuit. The circuit's behavior is calculated over time, starting at TIME = 0 seconds and going to some final value. In this statement, we have selected a "Print Step Value" of 50 microseconds, a "Final Time Value" of 1.0 millisecond, a "Step Ceiling Value" of 25 microseconds, and "UIC" causes the bias point calculation to be skipped. This will produce 20 plot points and 40 calculation points.

Line 6. `.PROBE`

The .PROBE statement allows the results from the transient analysis to be output in graphical form.

Line 7. `.END`

This is the closing statement for a PSpice file.

Once the simulation has been run, the output file is created, and PROBE can be used to graph current and voltage waveforms. Figure 14.38 shows one graphical output that could be obtained. The bottom graph shows the voltage at Node 2, or V(2), while the top graph shows the current flowing in the inductor. Initially, the entire applied voltage appears across the inductor and the corresponding current is zero. After five time constants, the voltage across the inductor has decreased to zero while the current through the inductor has reached a steady-state value of 40 milliamperes.

Now let us consider two more examples.

EXAMPLE 14.17

Using the PSpice simulator, analyze the transient behavior of the R–L circuit shown in Figure 14.39.

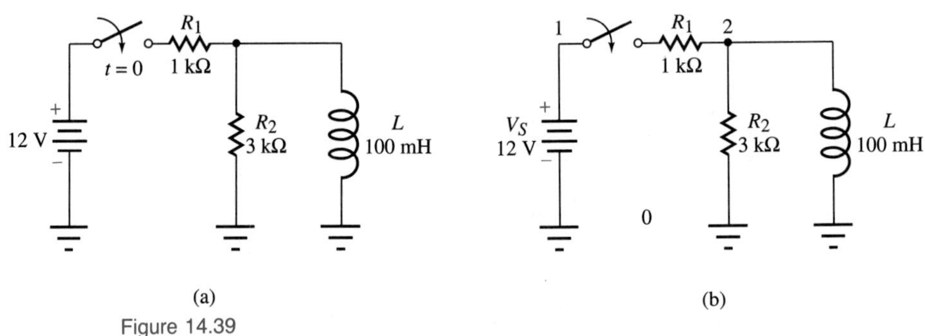

Figure 14.39

Solution

The first step is to label the nodes in the circuit. One labeling scheme is shown in Figure 14.39(b).

The PSpice file is shown in Figure 14.40. In this case, we must Theveninize the circuit connected to the inductor to find the circuit time constant. The Thevenin resistance is 750 ohms and the Thevenin voltage source is +9 volts. Hence, the time constant for the circuit is (100 mH)/(750 Ω) or 0.133 millisecond. The simulation interval specified in the .TRAN statement should be at least 0.66 millisecond or 5 time constants. Hence, the simulation will include the entire transient period.

```
PSPICE ANALYSIS OF THE CIRCUIT IN FIGURE 14.39
VS   1    0   DC   12
R1   1    2   1KOHM
R2   2    0   3KOHM
L    2    0   100MH    IC=0
.TRAN     1US   1.0MS   0.0   25.0U   UIC
.PROBE
.END
```

Figure 14.40 Node assignments and PSpice input file.

Figure 14.41 shows the plot of V(2) and I(L) obtained by using PROBE. Note that the transient appears to be complete by 0.6 millisecond.

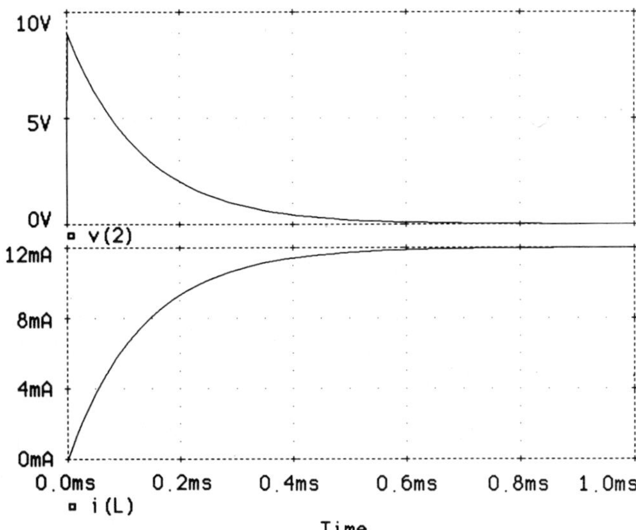

Figure 14.41 Output graphs produced with PROBE.

EXAMPLE 14.18

Simulate the transient behavior of the *R–L* circuit shown in Figure 14.42. The input signal is a pulse whose amplitude is +10 volts, whose duration is 0.4 millisecond, and period is 2 milliseconds. Simulate the behavior of the circuit for a time of 1 millisecond following the rising edge of the input pulse.

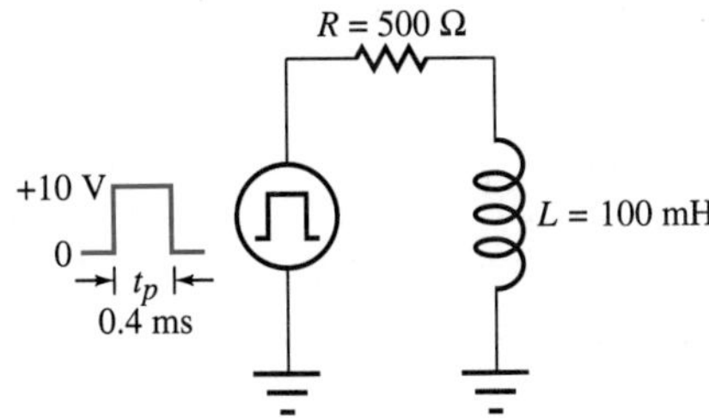

Figure 14.42

Solution

One labeling scheme for the circuit nodes is shown in Figure 14.43(a).

Using this labeling scheme, the PSpice file is created. This file is shown in Figure 14.43(b). Note that the parameters specifying the pulse waveform are inserted into the line for "VS."

The plot of the input voltage, inductor voltage, and inductor current is

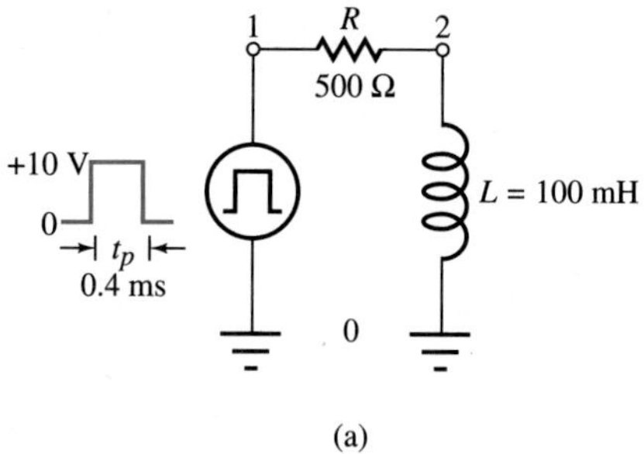


```
PSPICE ANALYSIS OF THE R-L CIRCUIT IN FIGURE 14.42
VS      1     0      PULSE ( 0  10V  0  0  0  0.4MS   2.0MS
R       1     2      500
L       2     0      100MH
.TRAN    1.0US    2.0MS
.PROBE
.END
```

(b)

Figure 14.43 Node assignments and PSpice input file.

shown in Figure 14.44(a). Note that the current begins at 0 amperes and increases toward 20 milliamperes. However, before the current can reach this steady-state value, the input pulse switches back to 0 volts. This causes the circuit current to exponentially decrease toward 0 amperes.

The plot of the inductor voltage shows that the inductor voltage jumps to +10 volts just after the rising edge of the input pulse. The voltage

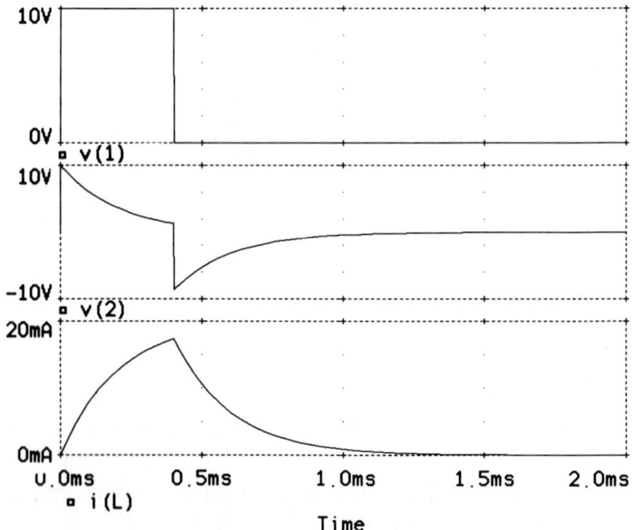

Figure 14.44 Output graphs produced with PROBE.

exponentially decreases toward 0 volts, but before it can reach 0, the input pulse switches from +10 volts to 0 volts, a negative step of −10 volts. This negative step in the voltage waveform is at t = 0.4 millisecond. From this point, the voltage exponentially returns to the zero level.

SUMMARY

1. An inductor can be in one of three states at any instant of time:
   a. steady state with no current.
   b. transient state with current either rising or falling.
   c. steady state with current at the maximum value.
2. The induced emf in an inductor opposes any change in current.
3. The time constant for a R–L circuit is equal to L/R seconds.
4. A transient state in a R–L circuit results when there is a change in the circuit current.
5. The energy stored in an R–L circuit is given by $\mathcal{W} = (LI^2)/2$.
6. When steady state has been reached in a R–L circuit, the ideal inductor is a short circuit. The practical inductor has resistance, so there will be some voltage across its terminals.
7. PSpice analysis can be used to simulate the transient behavior of R–L circuits.

EQUATIONS

14.1	$\frac{di}{dt} = \frac{E}{L}$	The rate of change in current with time at the start of current rise in a series R–L circuit.	551
14.2	$i = \frac{E}{R}$	The current in the inductor after the series R–L circuit has reached steady state.	552
14.3	$i = \frac{E}{R}(1 - e^{-t/(L/R)})$	The exponential equation for the current rise in a series R–L circuit.	552
14.4	$\upsilon_R = E(1 - e^{-t/(L/R)})$	The exponential equation for the resistor voltage during current rise in a series R–L circuit.	554
14.5	$\upsilon_L = Ee^{-t/(L/R)}$	The exponential equation for the inductor voltage during current rise in a series R–L circuit.	554
14.6	$\tau = \frac{L}{R}$	The time constant of the series R–L circuit.	557

14.7	$v_L = I_o R$	The inductor voltage at the start of current fall in a series R–L circuit.	560
14.8	$\dfrac{di}{dt} = \dfrac{I_o R}{L}$	The rate of change of current with time at the start of current fall in a series R–L circuit.	560
14.9	$i_L = I_o e^{-t/(L/R)}$	The exponential equation for the inductor current during current fall in a series R–L circuit.	560
14.10	$v_R = RI_o e^{-t/(L/R)}$	The exponential equation for the resistor voltage during current fall in a series R–L circuit.	562
14.11	$v_L = RI_o e^{-t/(L/R)}$	The exponential equation for the inductor voltage during current fall in a series R–L circuit.	562
14.12	$\mathcal{W} = \dfrac{1}{2} LI^2$	The equation for the energy stored in the magnetic field of the inductor.	575

QUESTIONS

1. What parameter cannot change instantaneously in an R–L circuit?
2. In a series R–L circuit consisting of a resistor, inductor, switch, and dc power source, if the resistor were decreased in value, what would be its effect on the following (increase, decrease, or stay the same)?
   (a) Initial value of current, just after the switch is closed.
   (b) Current at steady state.
   (c) Initial value of di/dt.
   (d) Time to reach steady state.
   (e) Energy stored in the coil.
3. Repeat Question 2 assuming that the inductance value is increased while the resistance value remains constant.
4. Explain why decreasing the source voltage driving a series R–L circuit has no effect on the time it takes the circuit to reach steady state.
5. In response to an applied 20-volt source, which of the following statements refer to an R–C circuit, and which ones refer to a R–L circuit?
   (a) The current is maximum at $t = 0$ seconds.
   (b) The current increases most rapidly at $t = 0$ seconds.
   (c) No current flows after five time constants.
   (d) The current reaches steady state in five time constants.
   (e) The applied voltage, 20 volts, appears across R in steady state.
   (f) The time to reach steady state increases as R is decreased.
6. A 40-mH coil is being discharged through a 4.7-kilohm resistor. The initial coil current is 50 mA. How will each of the following changes affect how much energy will be dissipated in the resistor during the complete discharge?
   (a) Decrease in resistor value.
   (b) Increase in coil inductance.
   (c) Increase in initial current value.

7. Which of the following cases produces the largest coil voltage?
   (a) A 15-mH coil whose current increases from 15 mA to 20 mA in 5 ms.
   (b) A 50-mH coil with a constant current of 200 mA.
   (c) A 20-mH coil whose current changes from 5 mA to 7 mA in 2 microseconds.
   (d) A 3-H coil whose current changes from 1 ampere to 2 amperes in 0.2 second.
8. Which change results in more energy being stored in an inductor?
   (a) doubling the current
   (b) doubling the inductance
9. How does the practical inductor differ from the ideal inductor?
10. How many ohms is the resistance of the ideal inductor in steady state?
11. The voltage across the ideal inductor in steady state is (a) zero volts (b) the source voltage (c) cannot be determined unless the inductance is known.
12. Where is the inductor energy dissipated when a switch is opened to interrupt the current?
13. What is the equation for the inductor current during current rise?
14. What is the equation for the inductor current during current fall?
15. An *R–L* circuit has a time constant of 2 ms. From a practical point, how many seconds are needed for the current to reach steady state?
16. Explain the relation between the slope of the inductor current curve and the induced voltage.

PROBLEMS

SECTION 14.1 The Series *R–L* Circuit During Current Rise

1. A resistance and inductance are connected in series as shown in Figure 14.45.

   (a) What are the inductor current and voltage at the instant the switch is closed?

   (b) What are the inductor current and resistor voltage at 0.8 second after the switch is closed?

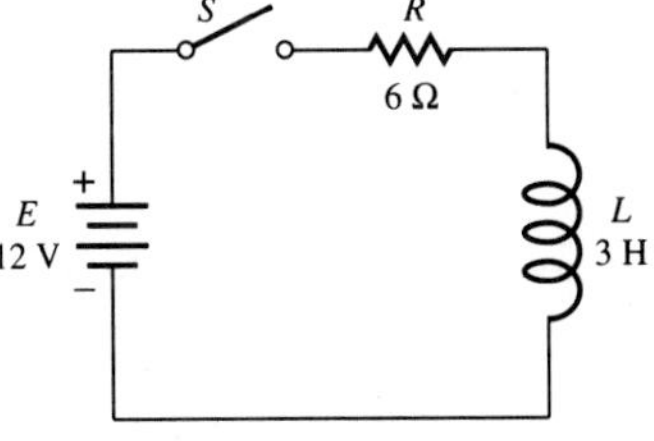

Figure 14.45

2. A series circuit of a 4-H inductance and a 20-ohm resistance is connected to a 24-volt source by closing a switch.

   (a) What is the circuit time constant?

   (b) What is the initial rate of current rise?

   (c) How many seconds are needed for the current to reach 1 ampere after the circuit is connected?

3. To pick up an iron piece, the current in an electromagnet whose coil has an inductance of 6 H and a resistance of 30 ohms must be at least 2 amperes.

   (a) How many seconds after the magnet is connected to a 120-volt source will it be able to pick up the iron?

   (b) What will be the maximum current in the coil?

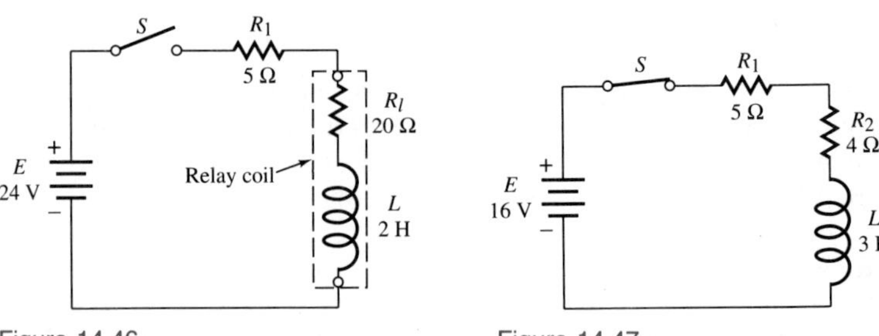

Figure 14.46

Figure 14.47

4. The relay coil shown in Figure 14.46 energizes at 0.1 second after the switch is closed. Find:

   (a) The voltage across the coil's terminals when it energizes.

   (b) The current in the circuit.

   (c) The final current in the relay coil.

5. The switch in the circuit in Figure 14.47 has been closed for 1 second.

   (a) What is the time constant of the circuit?

   (b) What is the voltage across R_1?

   (c) What will be the final current?

6. A series R–L circuit has a time constant of 0.5 second and an initial current rise of 5 A/s when connected to a 15-volt source. Find:

   (a) The resistance in the circuit.

   (b) The inductance in the circuit.

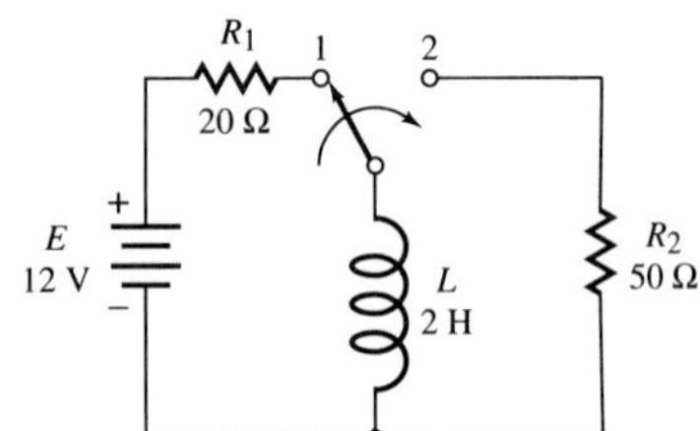

Figure 14.48

SECTION 14.2 The Series R–L Circuit During Current Fall

7. In the circuit shown in Figure 14.48, the switch is put into Position 2 after steady state is reached in L.

   (a) Place the polarity on the inductor terminals.

   (b) Find the time constant of the current fall circuit.

   (c) Find the current in the inductor and voltage across R_2 at 50 ms.

   (d) Find the voltage across R_2 at the instant the switch was put into Position 2. Assume that there is no energy loss during the switching.

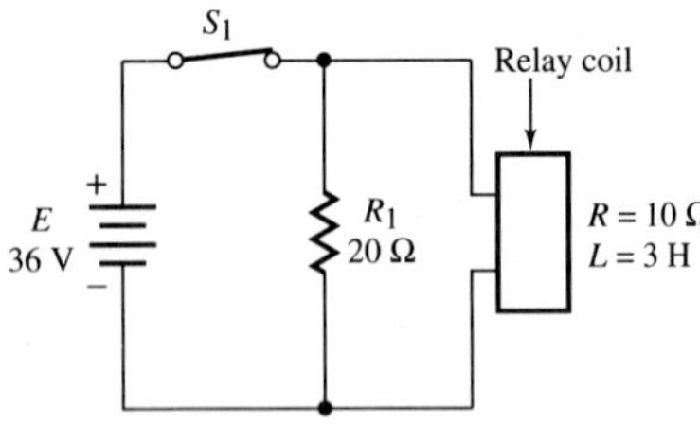

Figure 14.49

8. The current in the relay shown in Figure 14.49 has reached its final value.

   (a) What is the voltage across the relay coil terminals just after S_1 is closed?

   (b) If the relay remains energized as long as the voltage across its terminals is at least 24 volts, for how many seconds will the relay remain energized after S_1 is opened? Assume that there is no energy loss during switching.

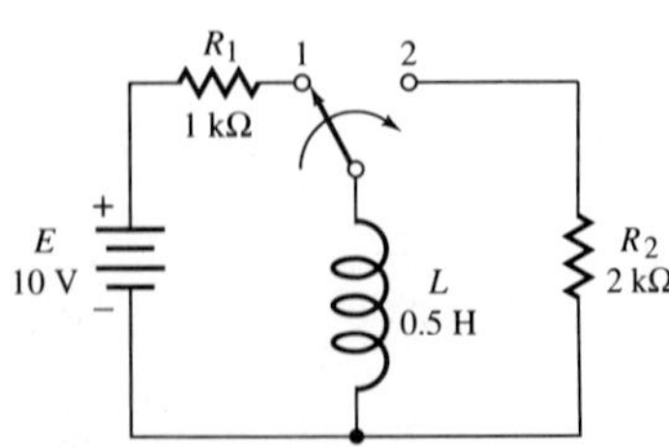

Figure 14.50

9. The switch in Figure 14.50 is left in Position 1 for 5 time constants and then put into Position 2 for 5 time constants.

   (a) What is the time constant while in Position 1?

(b) What is the time constant while in Position 2?

(c) Sketch the inductor current and voltage waveforms for the entire cycle. Assume that there is no energy loss during switching.

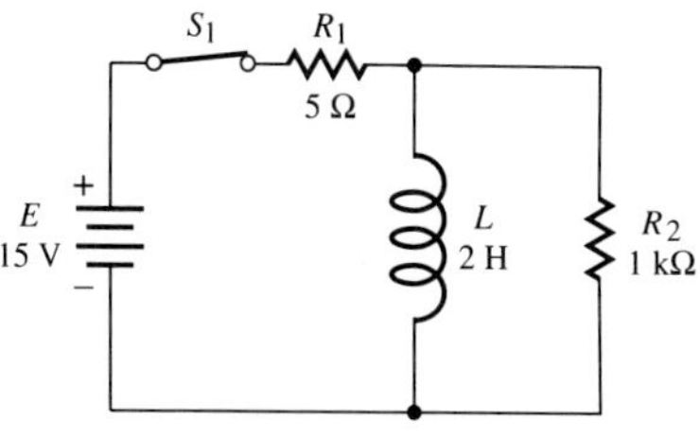

Figure 14.51

10. The current in the inductor shown in Figure 14.51 is 3 amperes when S-1 is opened. Find:

    (a) The time constant of the delay circuit.

    (b) The inductor current at t = 3 time constants.

    (c) The inductor voltage at 2.5 milliseconds after S-1 is opened.

    (d) The final current.

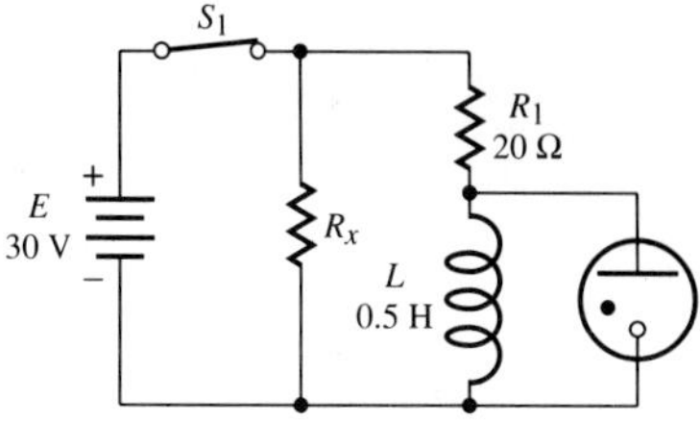

Figure 14.52

11. The neon lamp connected across L in Figure 14.52 acts as an open circuit until 90 volts is across its terminals and then it glows. Find the value of R_x needed to make the lamp glow when S_1 is opened.

12. The current in a series R–L circuit drops to 0.2 ampere in 0.5 second when the source is shorted. If the resistance is 15 ohms and the inductance is 3 H, find:

    (a) The initial current.

    (b) The time constant.

    (c) The source emf to which the circuit had been connected.

SECTION 14.3 The Universal Time Constant Curve

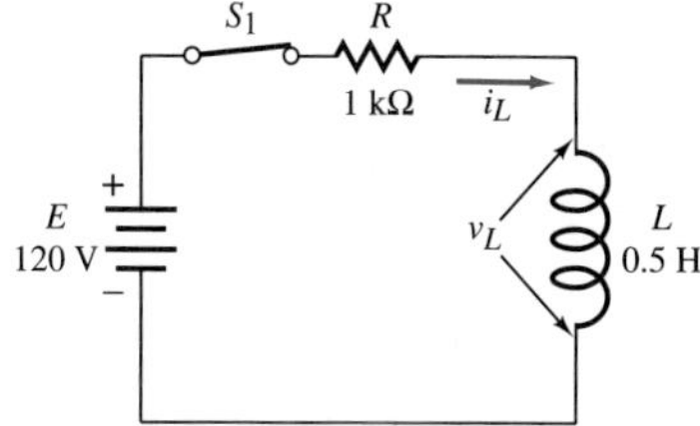

Figure 14.53

13. Use the graph in Figure 14.15 to find v_L and i_L at 0.9 ms after S_1 in Figure 14.53 is closed.

14. A series R–L circuit of 3 H and 6 ohms is connected to a 24-volt source. Use the graph in Figure 14.15 to find:

    (a) The current at τ, 2τ, and 3τ.

    (b) The inductance voltage at 2τ.

    (c) The resistance voltage at τ.

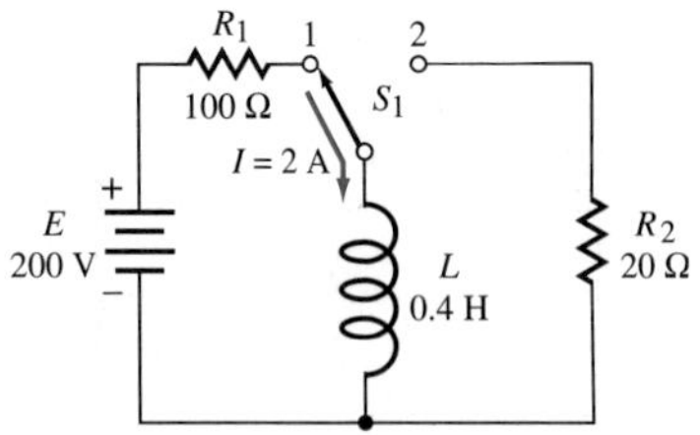

Figure 14.54

15. An R–L series circuit of 4 H and R ohms is connected to a 24-volt source. The inductance voltage is 8 volts at 2.2 seconds after being connected. Use the graph in Figure 14.15 to find:

    (a) The value of R.

    (b) The inductor current.

    (c) The time it will take for v_L to reach 16 volts.

16. Use the graph in Figure 14.15 to find the inductor current and voltage at 0.03 second after S_1 is put into Position 2 in Figure 14.54. Assume that there is no energy loss during the switching.

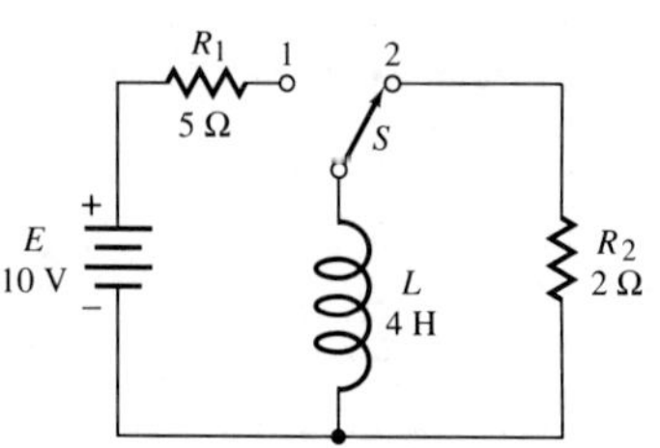

Figure 14.55

17. The switch in the circuit shown in Figure 14.55 was in Position 1 until steady state was reached. It was then put into Position 2 and the current is now 1 ampere. Use the graph in Figure 14.15 to find:

    (a) The time that the switch has been in Position 2.

    (b) The voltage across the inductor.

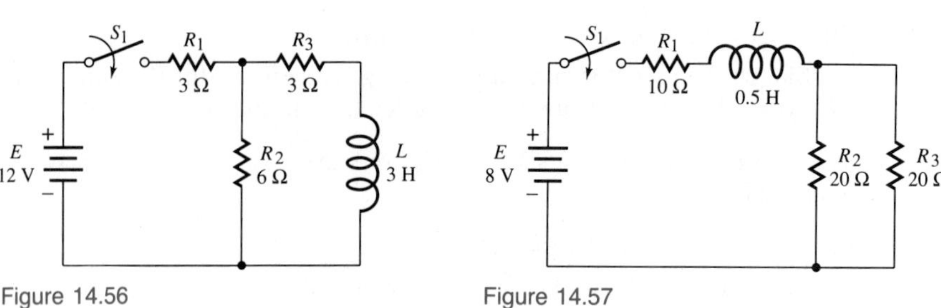

Figure 14.56

Figure 14.57

18. An R–L circuit of 2 H and 8 ohms has an initial current of 4 amperes. Use the graph in Figure 14.15 to find the current and voltage at 0.5 second, and at 0.8 second after the current starts to fall.

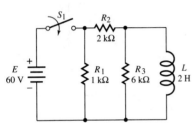

Figure 14.58

SECTION 14.4 Complex R–L Circuits

19. Find:

    (a) The time constant of the circuit shown in Figure 14.56.

    (b) v_L, i_L, and i_s at 1.8 seconds after S_1 is closed.

20. Find:

    (a) The time constant of the circuit shown in Figure 14.57.

    (b) v_L, i_L, and v_2 at 15 milliseconds after S_1 is closed.

21. How long after S_1 is closed will it take for v_L to reach 30 volts in the circuit shown in Figure 14.58?

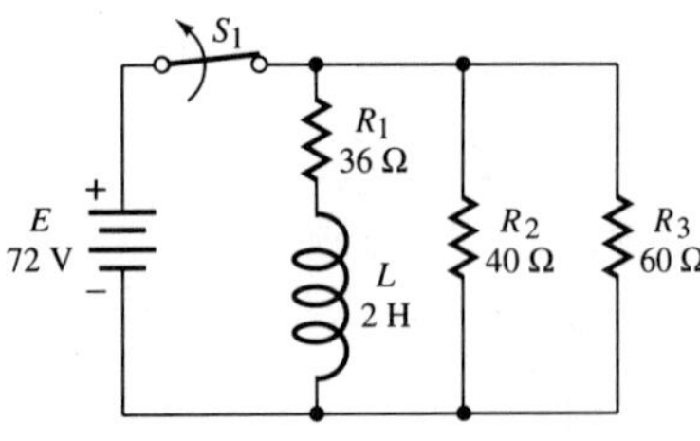

Figure 14.59

22. With S_1 closed in Figure 14.59, the current in the inductor has reached its final value.

    (a) Find the voltage across R_1 at 80 milliseconds after S_1 is opened. Assume that there is no energy loss during the switching.

    (b) Find the circuit time constant.

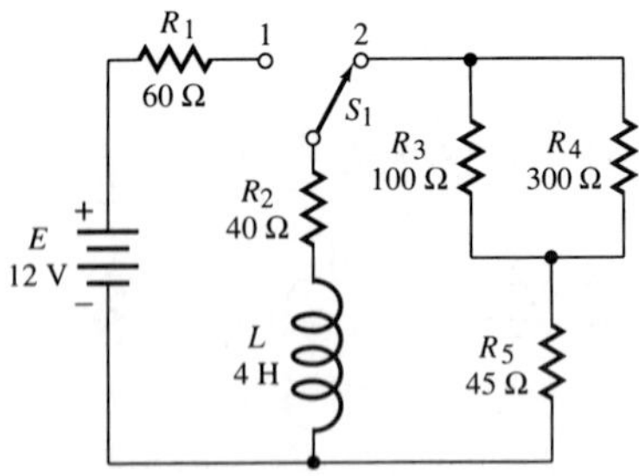

Figure 14.60

23. In Figure 14.60, S_1 was in Position 1 for 60 ms and then put into Position 2. Assuming no energy loss during switching, find:

    (a) i_L and v_L after S_1 is in Position 2 for 50 ms.

    (b) The voltage across R_3 at that time.

24. For the circuit of Figure 14.61, find:

    (a) The inductor current at $t = 0.02$ second after the switch is closed.

    (b) The current in the 40-ohm resistor at $t = 0.02$ second.

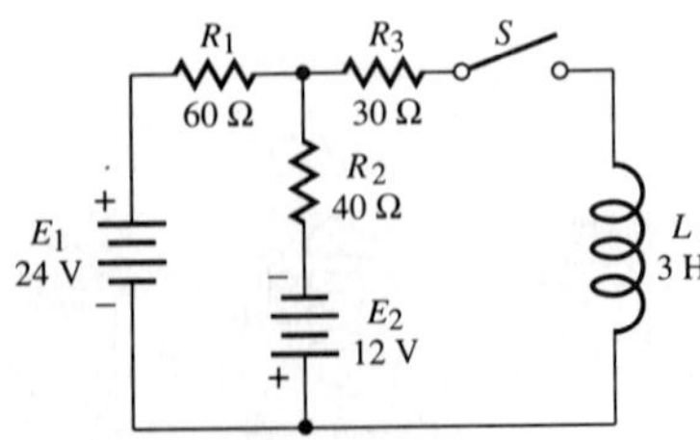

Figure 14.61

25. An inductor is connected to three sources, as shown in Figure 14.62. Find:

    (a) The inductor voltage at $t = 1.5$ seconds after it was connected.

    (b) The inductor current at $t = 1.5$ seconds after it was connected.

26. Determine the number of seconds needed for the inductor current to reach 0.3 ampere after the switch in Figure 14.63 is closed.

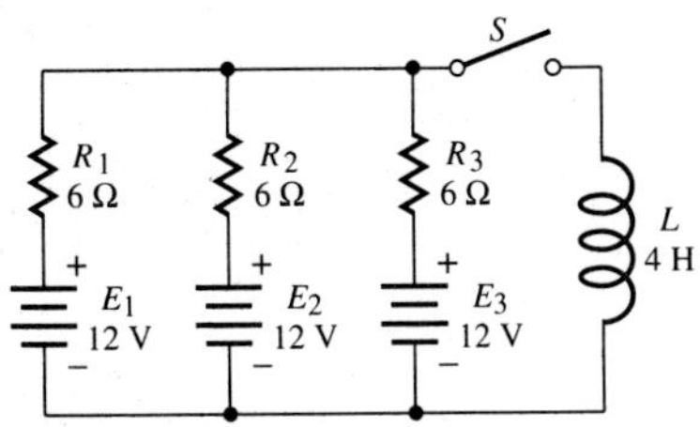

Figure 14.62

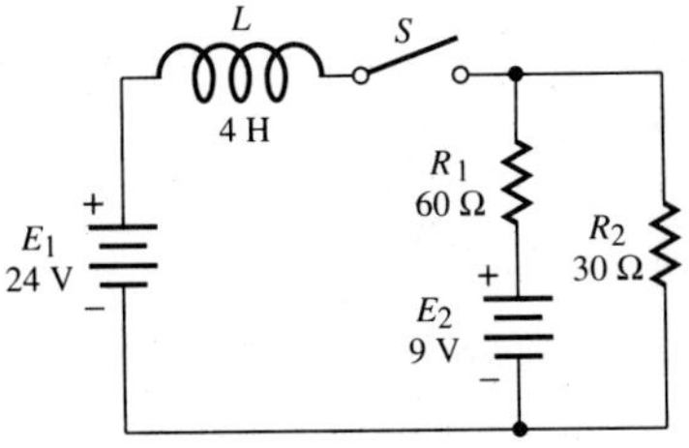

Figure 14.63

Figure 14.64

SECTION 14.5 Energy Stored in an Inductor

27. A series R–L circuit of 5 ohms and 5 H is connected across a 12-volt source. How much energy is stored in the inductance after 5 time constants?

28. The current in a 3-H inductance increased from 2 A to 6 A.

    (a) How much energy was stored in the inductor at 2 A?

    (b) What was the change in energy when the current was increased?

29. The energy in a 4-H inductor at steady state is 8 joules. How much energy does the inductance lose after the current has decreased for one time constant?

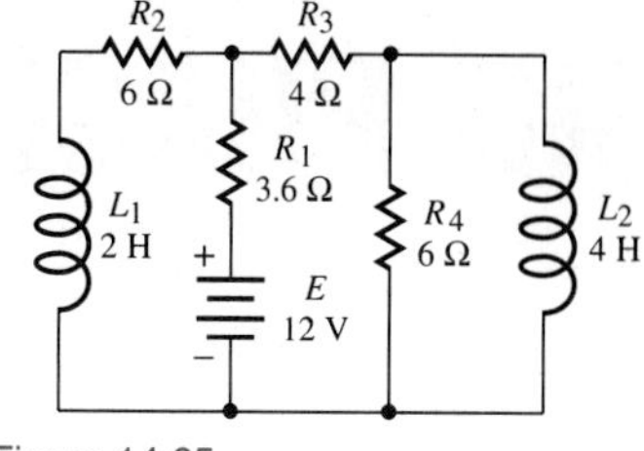

Figure 14.65

SECTION 14.6 The R–L Circuit in Steady State

30. Find the current in each branch in Figure 14.64 and the voltage across each component after steady state has been reached.

31. Repeat Problem 30 for the circuit shown in Figure 14.65.

32. Steady state has been reached in the circuit shown in Figure 14.66. Find:

    (a) The voltage and current for each inductor and resistor.

    (b) The energy stored in each inductor.

Figure 14.66

33. The circuit shown in Figure 14.67 is in the steady state condition and the energy stored in L_2 is 0.5 J. Find:

    (a) The voltage and the current for each inductor and resistor.

    (b) The source emf.

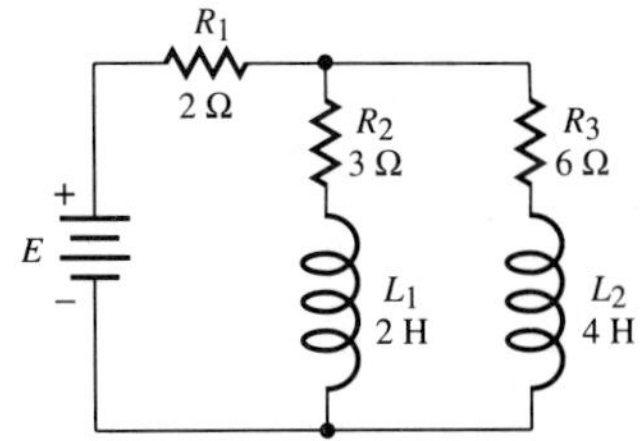

Figure 14.67

SECTION 14.7 SPICE Analysis of R–L Circuits

Ⓢ 34. Perform a PSpice simulation of the transient behavior for the circuits shown in Figure 14.68.

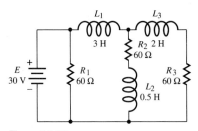

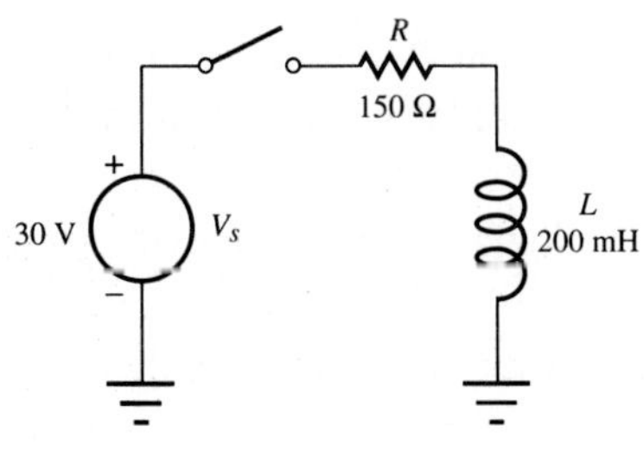

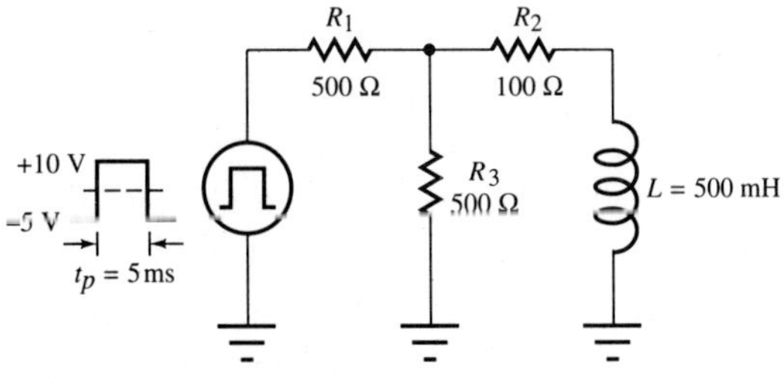

Figure 14.68

CHAPTER 15

ALTERNATING (ac) WAVES AND PHASORS

CHAPTER OBJECTIVES

After completing this chapter, you should be able to:

1. Define an ac waveform and its characteristics.
2. Determine the characteristics of a sinusoidal waveform from the curve.
3. Define angular velocity and its unit; calculate angular velocity for a rotating object.
4. Determine the phase angle of a waveform; determine whether it is leading or lagging.
5. Write the equation of a sinusoidal waveform from the curve, and construct the curve from the equation.

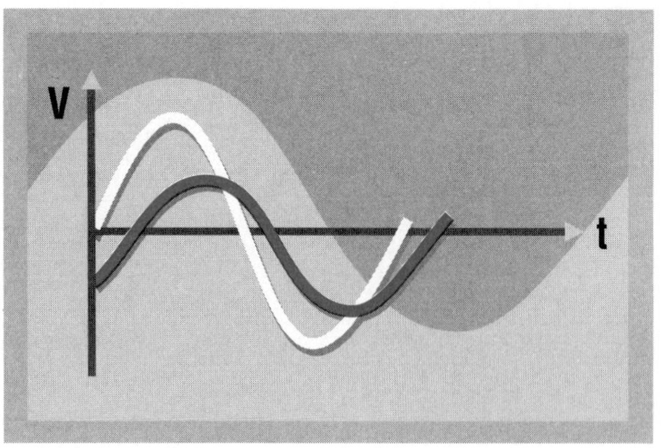

6. Determine the average value and the effective value of the ac component from measurements and from the waveform.
7. Determine the makeup of a waveform by inspection.
8. Define a phasor and explain the advantages of using it.
9. Convert between the sinusoidal and phasor expressions.
10. Determine the phase relationship between phasors.
11. Express the phasor in polar and rectangular form, and convert from one to the other.
12. Explain the phase relation for resistance, capacitance, and inductance.
13. Calculate inductive and capacitive reactance.

KEY TERMS

1. ac
2. amplitude
3. angular velocity
4. average value
5. capacitive reactance
6. division into known shapes
7. cycle
8. effective value
9. frequency
10. hertz
11. inductive reactance
12. instantaneous value
13. lagging wave
14. leading wave
15. peak-to-peak
16. period

17. phase angle
18. phase relation
19. phasor
20. polar form
21. radian
22. rectangular form
23. ripple factor
24. sinusoidal waveform
25. strip method

INTRODUCTION

This chapter introduces the basic concepts of alternating current (ac), and the response of resistors, capacitors, and inductors to it.

Without ac, our present power-generating and distribution system would be impossible or very expensive. The ac permits the use of the transformer to change the levels of voltage in the system, which dc does not permit.

High-frequency ac signals are transmitted through the atmosphere. This gives us radio, television, radar, and other communication systems. Although laser beams can be used in some instances, they are not a practical alternative to ac; dc cannot be transmitted.

Since an alternating waveform varies with time, new quantities must be learned to describe and analyze ac circuits. Some of these quantities are: frequency, amplitude, period, phase shift, average value, and effective value.

How do resistors, capacitors, and inductors behave in an ac circuit? The answer to this question must be learned before one can determine voltages and current in an ac circuit. This chapter provides a way to calculate the amount of opposition, and the relation between the voltage and current for resistors, capacitors, and inductors.

15.1 Alternating (ac) and Sine Wave Characteristics

Many circuits and applications use an alternating voltage instead of dc. Chapter 4 defined an *alternating* quantity as one whose value changes with time, and whose direction reverses at regular intervals. An alternating voltage or current is commonly called **ac**. This follows the same practice as using dc for direct current or voltage.

ac

The most common form of alternating voltage is supplied by electric utilities for our household, industrial, and commercial needs. If viewed on an oscilloscope, the voltage would be similar to that shown in 15.1. It is a sinusoidal waveform. A **sinusoidal waveform** is one that follows the relation given by the sine function. Normally, its variation cannot be

sinusoidal waveform

Figure 15.1 The alternating voltage used in the United States has this sinusoidal shape.

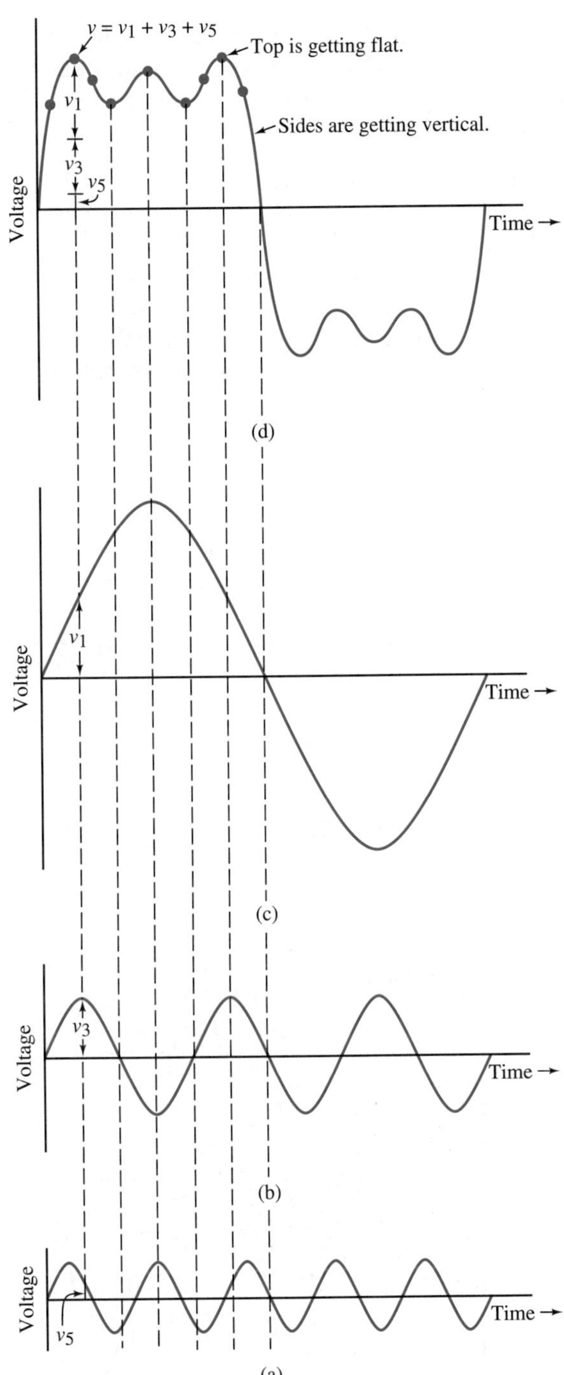

Figure 15.2 Nonsinusoidal waveforms are the combination of sine waves of different frequencies. (a) The fifth harmonic, $f = 5f_1$. (b) The third harmonic, $f = 3f_1$. (c) The fundamental harmonic, $f = f_1$. (d) The sum of f_1, $3f_1$, and $5f_1$ begins to resemble a square wave.

seen because the rate is too fast. However, if the rate is decreased, one can detect a flickering in the light from a lamp as the voltage changes.

Some other forms of ac are pulse, rectangular, and triangular. These forms are used mainly in electronic applications. They are formed by combining different sine waves, as shown in Figure 15.2. All ac is either a sine wave or a group of sine waves.

Since the ac waveform varies, magnitude and direction do not fully describe it. Other characteristics such as amplitude, peak-to-peak value, instantaneous value, cycle, frequency, period, and phase relation are used. Each of these is shown in Figure 15.3, and is now explained.

amplitude

1. **Amplitude** is the largest value of the wave. Figure 15.3(a) has a positive amplitude of +10 volts and a negative amplitude of −10 volts. Amplitude is also called peak value.

peak-to-peak

2. **Peak-to-peak** is the difference between a positive and a negative peak. In Figure 15.3(a), it is 10 − (−10), or 20 volts.

instantaneous value

3. **Instantaneous value** is the value of the waveform at some instant of time. These values are designated by lowercase letters. Some

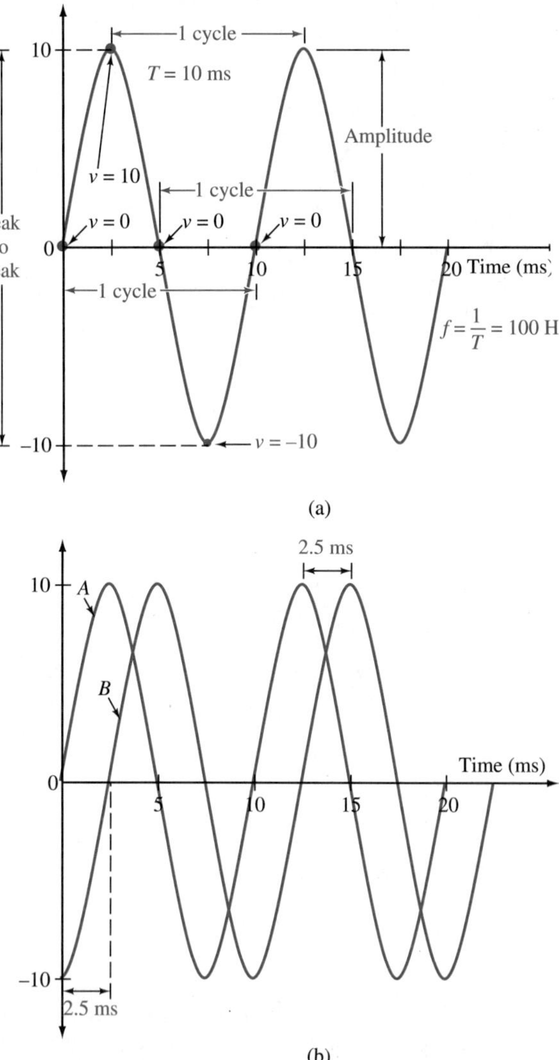

Figure 15.3 The characteristics used to define an alternating waveform. (a) Some common characteristics. (b) The phase relation characteristic.

instantaneous values in Figure 15.3(a) are: $\upsilon = 0$ V at 0 ms; 10 V at 2.5 ms; -10 V at 7.5 ms; and 0 V at 10 ms.

4. **Cycle** is the part of the waveform from one point to the point where it starts repeating. In Figure 15.3(a), one cycle is from 0–10 ms, 2.5 ms–12.5 ms, or any other points separated by 10 ms. **cycle**
5. **Frequency** is the number of cycles completed in 1 second. The unit for it is the **hertz** (Hz) named after the German physicist Heinrich R. Hertz (1857–1894). He demonstrated that ultrahigh frequency waves produced by a spark coil caused electrical oscillations in a wire loop at some distance away. The frequency of our power in the United States is 60 Hz. **frequency** **hertz**
6. **Period** is the length of one cycle. It is 10 ms for Figure 15.3(a). The period and frequency are related by **period**

$$f = 1/T \tag{15.1}$$

where T is the period, in seconds, and f is the frequency, in Hz.

7. **Phase relation** is the displacement between similar parts of two waves. It is often called the "phase shift." In Figure 15.3(b), the phase relation between A and B is $\pi/2$ radians or 90 deg. This quantity is covered in some detail in Section 15.2. **phase relation**

EXAMPLE 15.1

For the wave shown in Figure 15.4, find:

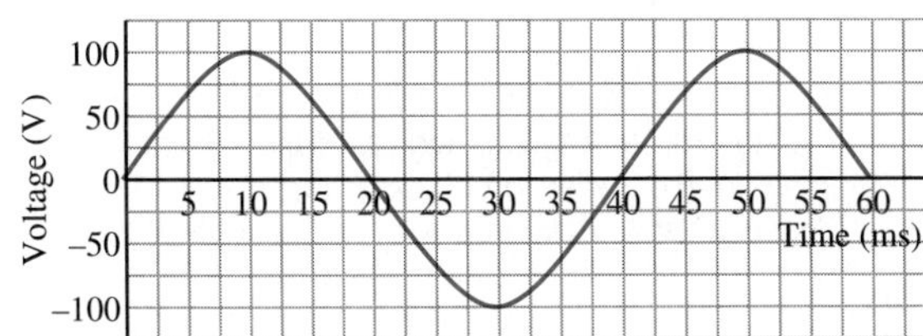

Figure 15.4

a. the peak value.
b. the peak-to-peak value.
c. υ at 25 ms.
d. the period.
e. the frequency.

Solution

a. The peak value is $+100$ V at 10 ms, and -100 V at 30 ms.
b. The peak-to-peak value is 100 V $-$ (-100 V) = 200 V.
c. At 25 ms, the voltage is -70.7 V.
d. One cycle is from 0–40 ms, so the period is 40 ms.
e. The frequency is $1/T = (1/40 \times 10^{-3}$ s) or 25 Hz.

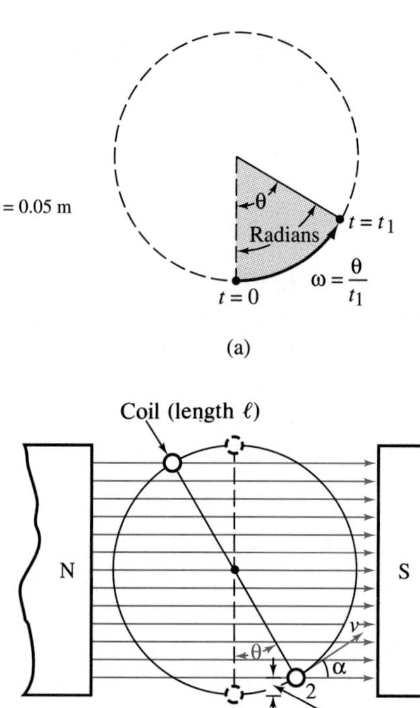

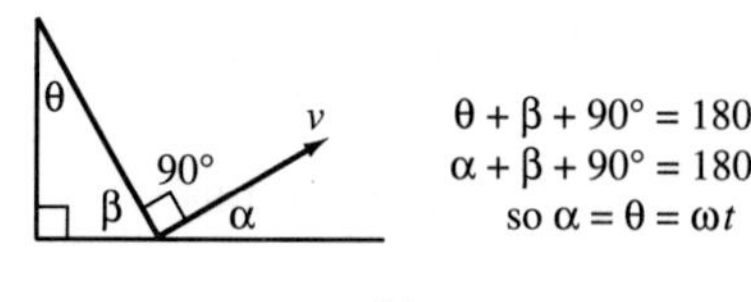

Figure 15.5 An emf is induced in a conductor rotating in a magnetic field. (a) Relation of angular velocity to the rotation. (b) Derivation of the induced emf equation, $e = B\ell v \sin \alpha$.

15.2 Angular Velocity and the Sine Equation

radian

angular velocity

As the point in Figure 15.5(a) rotates, it moves through an angle of 2π radians in one revolution. The **radian** is the SI unit for the plane angle, and is approximately 57.296 degrees. Its letter abbreviation is rad. If the coil makes one revolution in 1 second, its angular velocity will be 2π radians per second. The **angular velocity** is a measure of the radians displaced in 1 second. Its symbol is ω.

Faraday's Law, introduced in Chapter 13, states that the emf induced in the coil of Figure 15.5(b) is given by

$$e = N\frac{d\phi}{dt}$$

By definition, $\phi = B\,dA$, where dA is the area at right angles to the flux. As the conductor in Figure 15.5(b) moves from Point 1 to Point 2,

$dA = h\ell$, so $\phi = Bh\ell$. But h is equal to the component of velocity (υ) along h multiplied by the time, dt.

So $e = N(d\phi/dt)$ becomes

$$e = \frac{(NB\ell\upsilon \sin \alpha)\ dt}{dt}$$

Since $N = 1$ for the coil shown,

$$e = B\ell\upsilon \sin \alpha$$

As the coil rotates at an angular velocity of ω, the angle α through which it moves is equal to ωt. An inspection of Figure 15.5(b) shows that α, the angle between the velocity and the lines of the field, is the same as β. Also, $B\ell\upsilon$ is the maximum induced emf. Therefore, $e = B\ell\upsilon \sin \alpha$ becomes

$$e = E_m \sin \omega t \qquad \textbf{(15.2)}$$

where e is the instantaneous voltage at any time, t; E_m is the peak value of the sine wave, in volts; ω is the angular velocity, in radians per second; and t is the time, in seconds.

Equation 15.2 is the general form for a sine wave, which was shown in Figure 15.3(a). The sine wave is 0 at $\omega t = 0$. The current sine wave would have the same equation, with E_m replaced by I_m, and e by i. That is, $i = I_m \sin \omega t$.

The more common expression for sinusoidal voltages and currents uses frequency instead of angular velocity. This requires ω to be written in terms of frequency. The relation for that is now developed. A coil rotating at ω radians per second takes $2\pi/\omega$ seconds for one revolution. In this case, $2\pi/\omega$ is also the time required to complete one cycle of the sine wave. This gives

$$2\pi/\omega = T$$

Since $f = 1/T$,

$$2\pi/\omega = 1/f$$

or

$$\omega = 2\pi f \qquad \textbf{(15.3)}$$

where ω is the angular velocity, in radians per second, and f is the frequency, in hertz.

Putting $2\pi f$ into Equation 15.2 and into the expression for current gives

$$e - E_m \sin(2\pi f t) \qquad \textbf{(15.4)}$$

$$i = I_m \sin(2\pi f t) \qquad \textbf{(15.5)}$$

where e, i, E_m, I_m, f, and t have the same meaning as in Equations 15.2 and 15.3; $2\pi f t$ is in radians.

The sine wave can also be written with the angle expressed in degrees. Using degrees sometimes makes for easier graphing. The expression then becomes

$$e = E_m \sin \alpha \quad \textbf{(15.6)}$$

$$i = I_m \sin \alpha \quad \textbf{(15.7)}$$

where e, i, E_m, I_m have the same meaning as in Equations 15.2 and 15.3, and α is the angular displacement, in degrees.

The sine wave can be expressed with either radians, degrees, or seconds on the horizontal axis. The choice depends on the application. For example, when using an oscilloscope, the quantity will be seconds, but when making some calculations, it might be radians. The wave will always complete one cycle in 2π radians or 360 degrees. However, the amount of time for one cycle depends on the frequency, as shown in Figure 15.6.

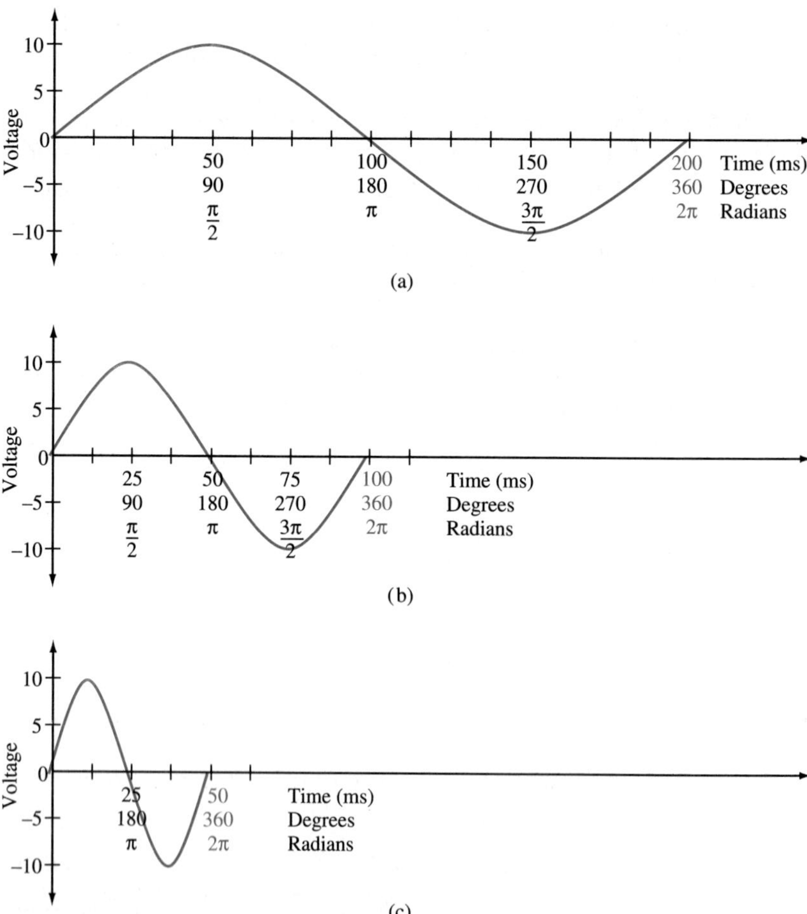

Figure 15.6 A comparison of the period in ms, radians, and degrees for three waveforms. (a) $f = 5$ Hz, $T = 2\pi$ radians and 360°, $T = 200$ ms. (b) $f = 10$ Hz, $T = 2\pi$ radians and 360°, $T = 100$ ms. (c) $f = 20$ Hz, $T = 2\pi$ radians and 360°, $T = 50$ ms.

EXAMPLE 15.2

Given the voltage, $e = 50 \sin(314t)$,

a. find the angular velocity, frequency, and the period in radians, degrees, and seconds.

b. sketch the sine wave with radians on the x-axis; with time on the x-axis.

Solution

a. Comparing the equation to the general expression in Equation 15.2 shows that $\omega = 314$ radians per second. Hence

$$f = \frac{\omega}{2\pi} = \frac{314}{2\pi} \text{ rad/s}$$

$$f = 50 \text{ Hz}$$

T in radians is always 2π radians.

$$T = 2\pi \text{ radians}$$

T in degrees is always 360 degrees.

$$T = 360°$$

$$T \text{ in seconds} = \frac{1}{f} = \frac{1}{50 \text{ Hz}}$$

$$T = 0.02 \text{ s}$$

b. The sine waves are sketched in Figure 15.7.

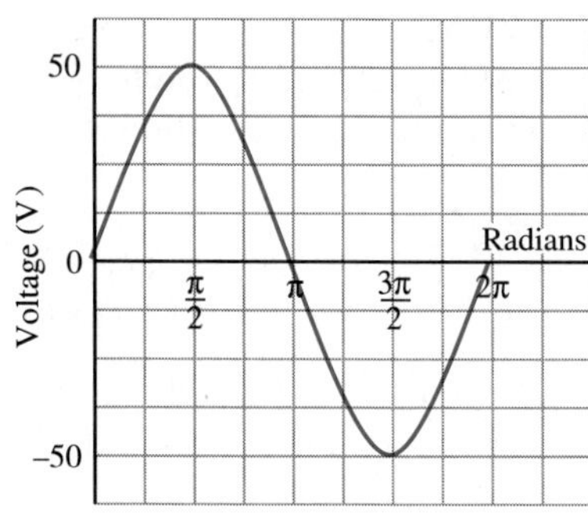

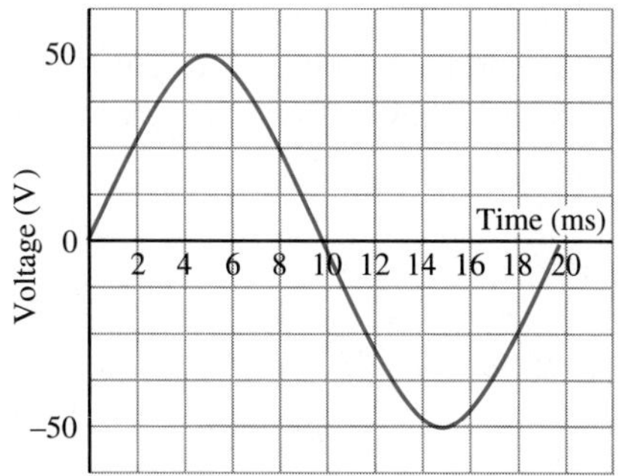

Figure 15.7

15.3 Phase Relation and Phase Angle

Voltages and currents in a circuit having a sinusoidal source of emf also have the shape of a sine wave. But they do not have to be 0 at 0

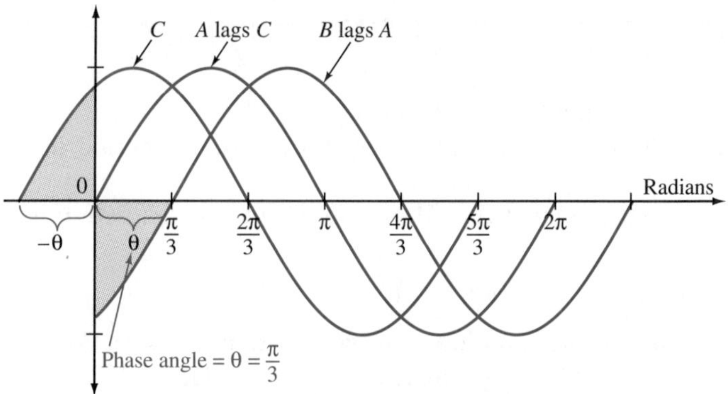

Figure 15.8 The relative position and phase relation for a leading and lagging waveform.

radians and 0 seconds. An example of this is shown in Figure 15.8. There, waves B and C are shifted from the basic sine wave A by the angle θ, which is called the **phase angle**. The phase angle of a wave is the number of radians or degrees between a point on it and a similar point on a sine wave, which is the reference point. For example, in Figure 15.8 it is the angle between the 0 volt point on the sine wave A and the 0 volt points on B and C. This angle is also called ''phase shift'' or ''phase difference.'' Because of the shift, waves B and C are out-of-phase with wave A. An out-of-phase wave can lead or lag the basic sine wave A. A **lagging wave** is one whose peaks, zeros, and other points occur at a larger number of radians or at a greater angle than those on the basic sine wave. This means that they occur at a later time, and the wave is shifted to the right. In Figure 15.8, B lags A by π/3 radians. A **leading wave** is one whose peaks, zeros, and other points occur at a smaller number of radians than those on the basic sine wave. This means they occur at an earlier time or at a smaller number of degrees, and the wave is shifted to the left. In Figure 15.8, C leads A. In general, the forms of the equations for out-of-phase waves are

phase angle

lagging wave

leading wave

$$e = E_m \sin(\omega t \pm \theta) \quad \textbf{(15.8)}$$

$$i = I_m \sin(\omega t \pm \theta) \quad \textbf{(15.9)}$$

where e, i, E_m, I_m, ω, and t have the same meaning as in Equations 15.2 and 15.3, and θ is the phase angle. It takes the negative sign for a lagging wave.

EXAMPLE 15.3

Find the phase angle for the voltage in Figure 15.9. Determine whether it leads or lags the sine wave, and write the equation for it.

Solution

The wave passes through 0 volts at π/6 radians, and has a peak value of 10 volts. Since it passes through 0 at an angle that is greater than 0, it lags the sine wave that is 0 at 0 radians. The amount of lag is $\pi/6$ radians, and the equation for it is

$$e = 10 \sin(\omega t - (\pi/6))$$

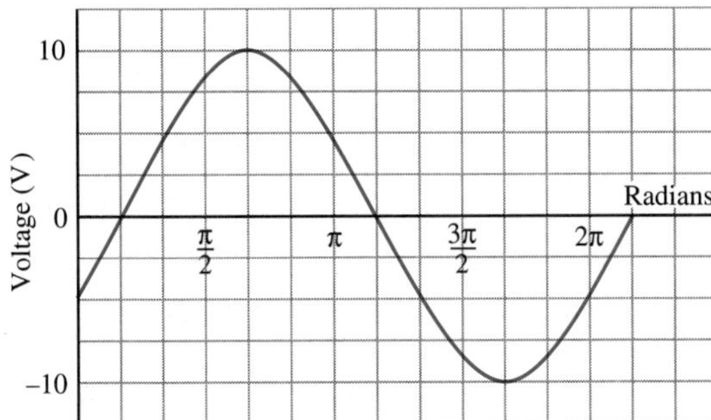

Figure 15.9

EXAMPLE 15.4

Sketch the sine wave described by $i = 5 \sin(\omega t + 45°)$.

Solution

Its amplitude is 5 amperes. θ is $+45$, so i leads by 45° or $\pi/4$ radians.

The wave can be sketched by locating its zero points and peak points. The zeros are located angles where $\sin(\omega t + 45°) = 0$. The peaks are at angles where $\sin(\omega t + 45°) = \pm 1$.

The first zero is at $\omega t = -45°$ or $-\pi/4$ radians because $\sin(-45° + 45°) = 0$. The second and third zeros are at $3\pi/4$ and $7\pi/4$ radians. The peaks of a sine wave fall between the zeros. This puts a positive peak at $\pi/4$ radians and a negative peak at $5\pi/4$ radians. The wave is drawn in Figure 15.10.

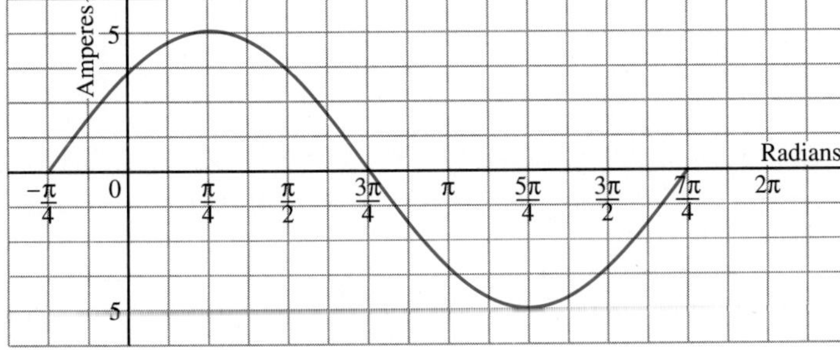

Figure 15.10

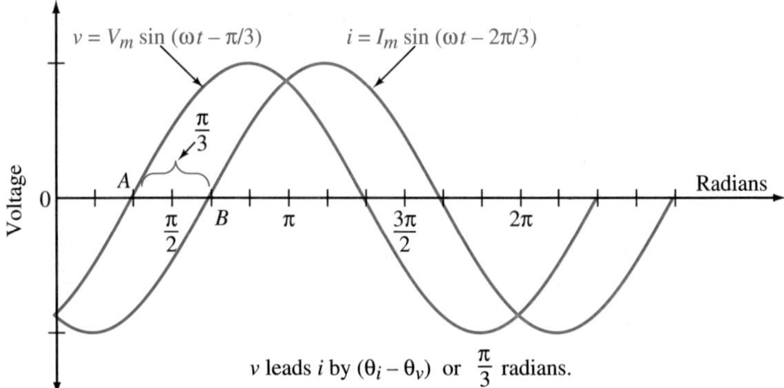

Figure 15.11 The wave with the larger phase angle is the leading wave.

Now we examine the phase relation between two waves, each shifted from 0 degrees (see Figure 15.11). Applying the concept of "which points occur at a later time or angle," one can conclude that v leads i. The phase angle for v is from 0 to A; for i it is from 0 to B. The phase angle between the two is the difference between the two angles. It is $2\pi/6$ radians. The same information can also be obtained from the equations. For the curves in Figure 15.11, $v = V_m \sin(\omega t - (\pi/3))$ and $i = I_m \sin(\omega t - (2\pi/3))$. Both angles have a minus sign because both waves lag the sine wave. The angle $-\pi/3$ radians is more positive than $-2\pi/3$ radians, so v leads i. The phase difference is $(2\pi/3) - (\pi/3) = \pi/3$ radians.

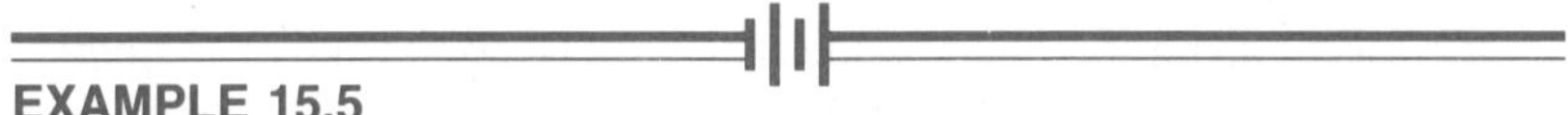

EXAMPLE 15.5

Find the phase angle between the two curves in Figure 15.12. Does i lead or lag?

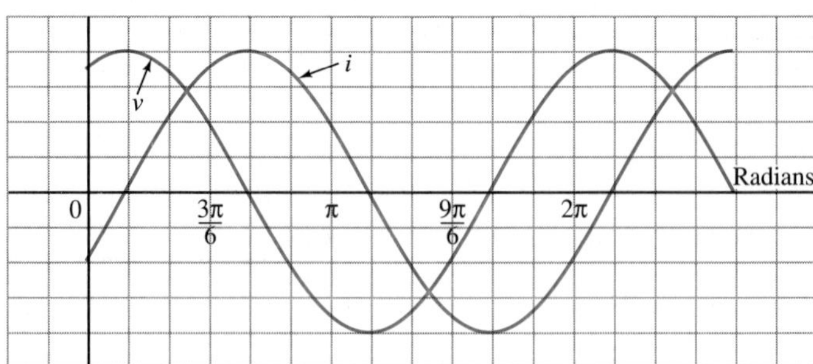

Figure 15.12

Solution

v is 0 at $4\pi/6$ radians and i is at the same point, at $7\pi/6$ radians, so the phase angle is $(7\pi/6) - (4\pi/6)$.

$\theta = \pi/2$ radians.

i lags v since its points come after the v points.

EXAMPLE 15.6

Find the phase angle for each set of equations; determine which quantity is leading.

a. $i = 50 \sin(377t + 60°)$
$\upsilon = 10 \sin(377t + 20°)$

b. $i = 10 \sin(377t - (\pi/3))$
$\upsilon = 20 \sin(377t - (\pi/6))$

Solution

a. i leads υ since its angle is more positive. The phase angle is $(60° - 20°)$

$$\theta = 40°$$

b. υ leads i since its angle is more positive. The phase angle is $(\pi/6 - \pi/3)$

$$\theta = \pi/6 \text{ radians}$$

15.4 The Average Value

The **average value** of a voltage or current is that *constant* voltage or current that causes an equal net amount of charge movement as the varying voltage or current causes. It is more commonly called the dc value.

average value

One way to find the average value is to measure it with a dc meter. A second way is to obtain it from the voltage or current curve. This eliminates the need for constructing a circuit. This second method is now explained.

Suppose the current in a resistance varies, as shown in Figure 15.13(a). Since $Q = I \times t$, the charge moved from time t_0 to t_1 is $I_1 t_1$, and that from t_1 to t_2 is $I_2 (t_2 - t_1)$. However, the direction is reversed for I_2

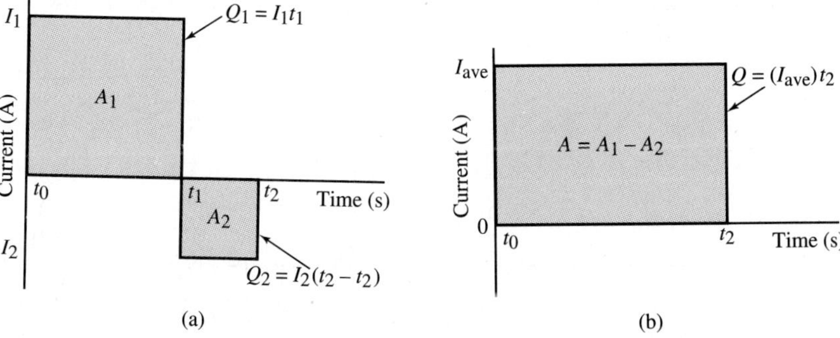

Figure 15.13 The average value will move the same amount of charge in one direction as the changing voltage. (a) Amount of charge that is moved by a changing voltage in a resistor. (b) Amount of charge moved by a steady voltage in the same time.

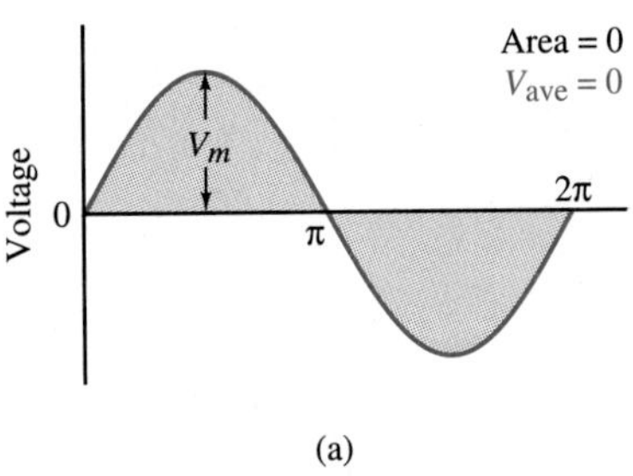

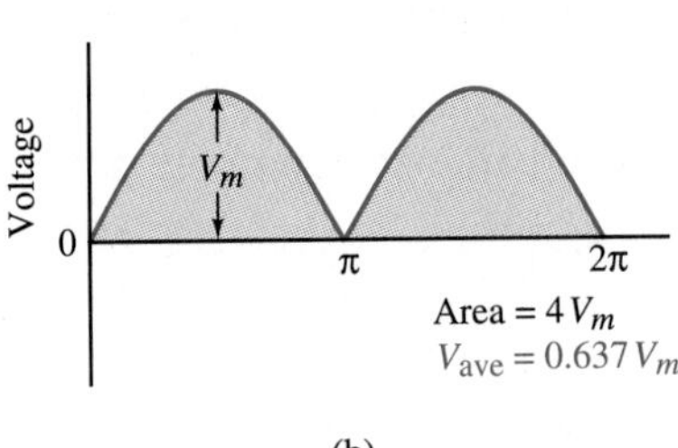

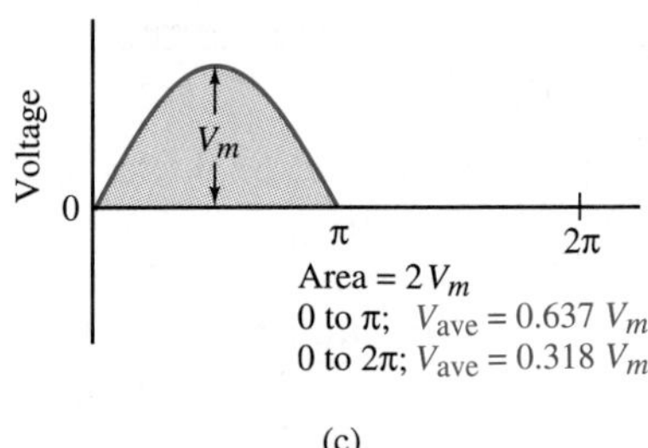

Figure 15.14 The area and average value for some sinusoidal shapes. (a) A sine wave. (b) Full-wave rectification. (c) Half-wave rectification.

because it is negative. The charge moved by a constant I_{ave} in time t_0 to t_2 is $I_{ave}(t_2 - t_0)$. By the definition of the average value, $I_{ave}(t_2 - t_0)$ must equal $I_1(t_1 - t_0) - I_2(t_2 - t_1)$. Then I_{ave} must equal

$$\frac{I_1(t_1 - t_0) - I_2(t_2 - t_1)}{t_2 - t_0}$$

The numerator is the area under the curve (height × base). So, the average current is the net area under the current curve divided by the length of the curve. In general,

$$I_{ave} \quad \text{or} \quad V_{ave} = A/L \qquad \textbf{(15.10)}$$

where I_{ave}, V_{ave} are the average values; L is the length over which the average value is being found (it is usually one cycle); and A is the net area under the curve for the length L.

Equation 15.9 shows that if the net area is zero, the average value must be zero. The sinusoidal voltage falls into this category.

The area under the curve, the average value for the sine wave, and some common modifications of it (as obtained by calculus in Appendix C) are given in Figure 15.14.

division into known shapes

Calculus yields the most exact value for area, but it can be difficult to use for irregularly shaped curves. Two other methods are: **division into known shapes**, and a method of strips. The first method divides the curve into shapes such as triangles, rectangles, and so forth for which the area formulas are known. Rectangles and triangles are used in Figure 15.15. The area of each shape is then calculated and combined to obtain the net

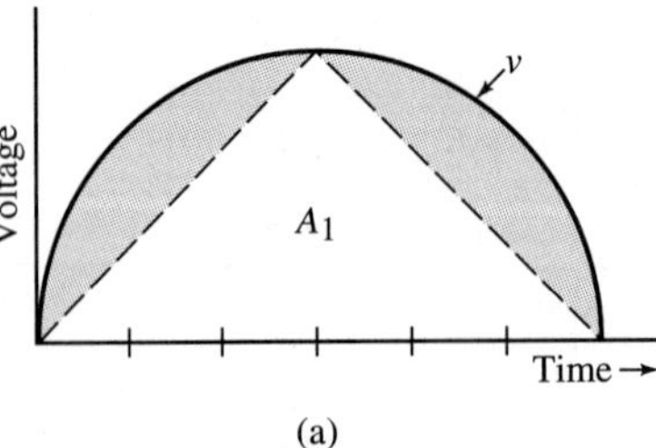

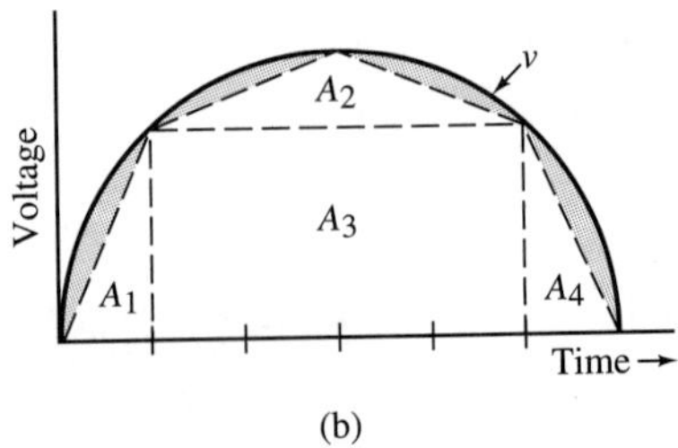

Figure 15.15 Dividing the area into sections of known shapes will give the approximate area of the waveform. (a) Using only one section results in a poor approximation. (b) The approximation becomes more exact when four sections are used.

area. One disadvantage of this method is that the known shapes do not always match the exact area of the curve. Using more sections gives a better match, as seen in Figure 15.15(b). However, the added shapes increase the amount of calculations.

strip method

The **strip method** divides the curve into strips. The areas of the strips are then combined to give the total area. This method is a graphical integration, which gives more exact results when a large number of strips are used. It is suitable for irregularly shaped curves. One advantage of this method is that the curve does not have to be constructed if the h values can be obtained with an equation. When the strip method is used, Equation 15.11 becomes

$$I_{ave} \quad \text{or} \quad V_{ave} = \frac{h_1 + h_2 + \cdots + h_n}{n} \qquad \textbf{(15.11)}$$

where I_{ave}, V_{ave} are the average values; $h_1, h_2, \ldots, h_n$ are the values of current or voltage at the center of each strip; and n is the number of strips.

The two methods are applied in Example 15.7.

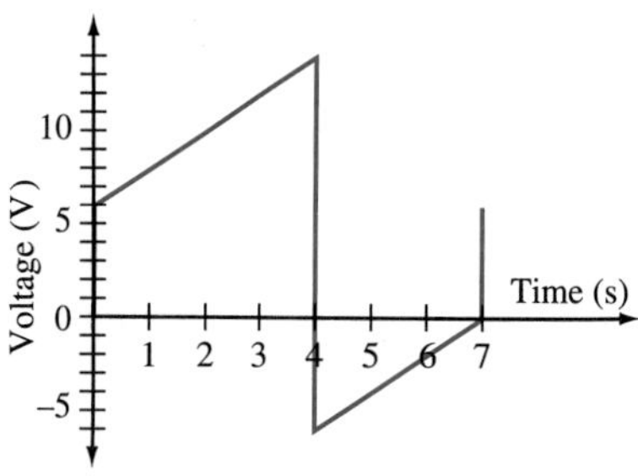

Figure 15.16

EXAMPLE 15.7

Find the average value of the voltage in Figure 15.16 using:
a. known shapes.
b. the strip method.

Solution

a. One cycle is from 0–7 seconds. The curve is divided into the shapes for which the area formulas are known. These are the triangles A_1 and A_3, and the rectangle A_2, in Figure 15.17.

$$A_1 = 1/2\ bh = (0.5)(4 - 0)(14 - 6) = 16\ \text{V} \cdot \text{s}$$
$$A_2 = bh = (4)(6) = 24\ \text{V} \cdot \text{s}$$
$$A_3 = 1/2\ bh = (0.5)(7 - 4)(-6) = -9\ \text{V} \cdot \text{s}$$
$$V_{ave} = \frac{(16 + 24 - 9)\ \text{V} \cdot \text{s}}{7\ \text{s}}$$
$$V_{ave} = 4.43\ \text{V}$$

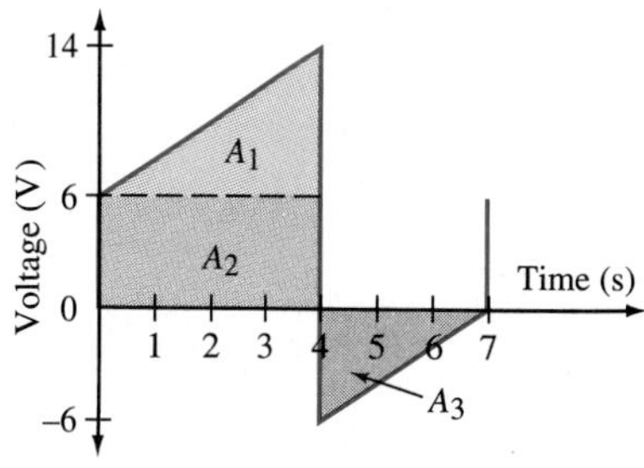

Figure 15.17

b. The curve is divided into strips, as shown in Figure 15.18.

The voltage at the center of each strip is

$V_1 = 7$ V $\quad V_4 = 13$ V $\quad V_7 = -1$ V
$V_2 = 9$ V $\quad V_5 = -5$ V
$V_3 = 11$ V $\quad V_6 = -3$ V

$$V_{ave} = \frac{h_1 + h_2 + \cdots + h_n}{n}$$

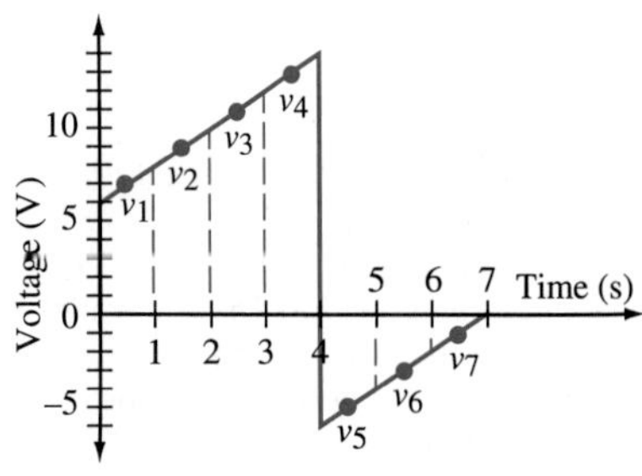

Figure 15.18

$$V_{ave} = \frac{7\text{ V} + 9\text{ V} + 11\text{ V} + 13\text{ V} - 5\text{ V} - 3\text{ V} - 1\text{ V}}{7}$$

$$V_{ave} = 4.43\text{ volts}$$

15.5 The Effective Value

The term used to specify an ac voltage is the effective or root-mean-square (rms) value. For example, the effective value is used as the measure of the sinusoidal voltage and current for our domestic and commercial use.

effective value

The **effective value** of a voltage or current is that *constant* voltage or current that supplies the same energy as the varying voltage or current supplies. An electric heater connected to an ac source of 120 volts will be just as hot as when it is connected to a 120-volt dc source. The difference is that the energy is supplied to the heater at a varying rate with the ac.

When measuring the effective value of a waveform, the proper meter must be used. Although ac meters will measure an effective value, it might not be the correct effective value. The possibilities are as follows:

1. dc movement with a rectifier and some digital meters. These meters are calibrated for a waveform, usually a sine wave. If the shape is different, the reading will be incorrect.
2. ac meters connected so that they are isolated from dc by a transformer or a series capacitor. These meters will measure only the effective value of the ac part.
3. ac meter with an iron vane or dynamometer movement. These meters will measure the effective value of the entire waveform.
4. Digital electronic and true rms meters. These meters measure the effective value of the waveform.

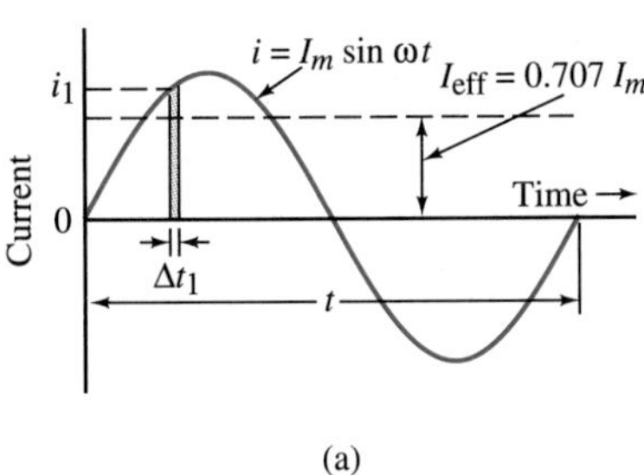

(a)

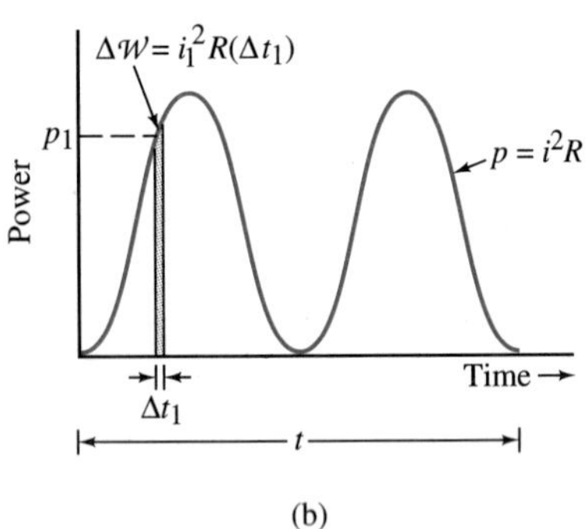

(b)

Figure 15.19 The effective value is the steady value that will supply the same amount of energy as the changing waveform. (a) The sinusoidal current in a resistor. (b) The power curve and energy derivation for the sinusoidal waveform.

The procedure for connecting the meters and making the measurements in a dc circuit was described in Chapter 3. One difference with ac meters is that the pointer will deflect upscale for either terminal polarity connection.

Figure 15.19 shows a sinusoidal current in a resistance along with its power curve, which was obtained by multiplying the resistance by the instantaneous value of the current squared (Ri^2). The energy effect for each strip on the curve will be $i^2R\Delta t$ (i is the current at the center of each strip). The total energy is the sum of the energy in each strip. Now the energy (I_{eff}^2Rt) of the effective value must be the same. Therefore,

$$(I_{eff})^2Rt = i_1^2R\Delta t + i_2^2R\Delta t + \cdots + i_n^2R\Delta t$$

or

$$I_{eff} = \sqrt{\frac{(i_1^2 + i_2^2 + \cdots + i_n^2)\Delta t}{t}}$$

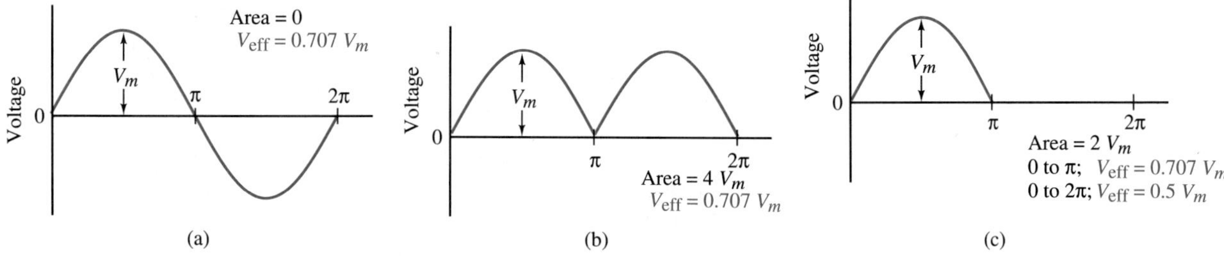

Figure 15.20 The area and effective value for some sinusoidal shapes. (a) A sine wave. (b) Full-wave rectification. (c) Half-wave rectification.

A closer inspection shows that the quantity under the radical is the area under the current squared curve, divided by the length of a cycle. The exact area, as found by calculus, yields $I_{eff} = 0.707 I_m$ and $V_{eff} = 0.707 V_m$ for the sine wave. In general,

$$I_{eff} = \sqrt{\frac{\text{area under the } i^2 \text{ curve}}{\text{length of one cycle}}} \tag{15.12}$$

$$V_{eff} = \sqrt{\frac{\text{area under the } \upsilon^2 \text{ curve}}{\text{length of one cycle}}} \tag{15.13}$$

As with the average value, calculus yields the most exact area value. Figure 15.20 lists the effective values when calculus is used for the sine wave (as in Appendix C). Note that the sine wave has an effective value even though it has no average value.

When strips are used, as in the average value, the equations become

$$I_{eff} \quad \text{or} \quad V_{eff} = \sqrt{\frac{h_1^2 + h_2^2 + \cdots + h_n^2}{n}} \tag{15.14}$$

where I_{eff}, V_{eff} are the effective values; $h_1, h_2, \ldots, h_n$ are the current or voltage values at the center of the strip; and n is the number of strips.

Some advantages of the strip method are: an i^2 or υ^2 power curve is not needed; no curve is needed if the h values can be calculated with an equation; the same h values can be used for both average and effective value calculations; and results are more exact for irregular curves. This contrasts with the method of known shapes where the υ^2 and i^2 curves must be available.

Equations 15.11–15.13 show that the effective value is the square root of the average or mean value of the current or voltage squared curve. This results in it being called the rms value.

EXAMPLE 15.8

A 60-Hz sine wave has a peak of 170 volts. Use the strip method to find the effective value.

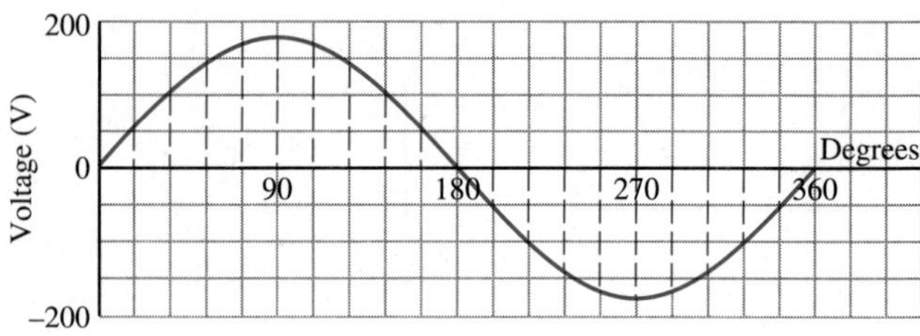

Figure 15.21

Solution

The sinusoidal voltage is drawn and divided into 20 strips, as shown in Figure 15.21. Each strip is 18° wide and the centers are 18° apart starting at 9°.

The voltage at each point will be given by $\upsilon = 170 \sin \theta$. For example, $\upsilon_1 = 170 \sin 9° = 26.59$ V. The values for the other strips are

Strip	υ	υ^2	Strip	υ	υ^2
1	26.59	707	11	−26.59	707
2	77.18	5,956.7	12	−77.18	5,956.7
3	120.21	14,450.4	13	−120.21	14,450.4
4	151.47	22,943.2	14	−151.47	22,943.2
5	167.9	28,190.4	15	−167.9	28,190.4
6	167.9	28,190.4	16	−167.9	28,190.4
7	151.47	22,943.2	17	−151.47	22,943.2
8	120.21	14,450.4	18	−120.21	14,450.4
9	77.18	5,956.7	19	−77.18	5,956.7
10	26.59	707	20	−26.59	707

From Equation 15.13,

$$\upsilon_{\text{eff}} = \sqrt{\frac{\upsilon_1^2 + \upsilon_2^2 + \cdots + \upsilon_{20}^2}{20}}$$

$$= \upsilon_1^2 + \upsilon_2^2 + \cdots + \upsilon_{20}^2 = 288{,}990.8 \text{ V}^2$$

so

$$V_{\text{eff}} = \sqrt{\frac{288{,}990.8 \text{ V}^2}{20}}$$

$$V_{\text{eff}} = 120.21 \text{ V}$$

This compares with 120.19 volts, as obtained by the calculus relation $V_{\text{eff}} = (0.707)V_{\text{pk}}$. It is worth noting that construction of the curve was not needed to find the h values.

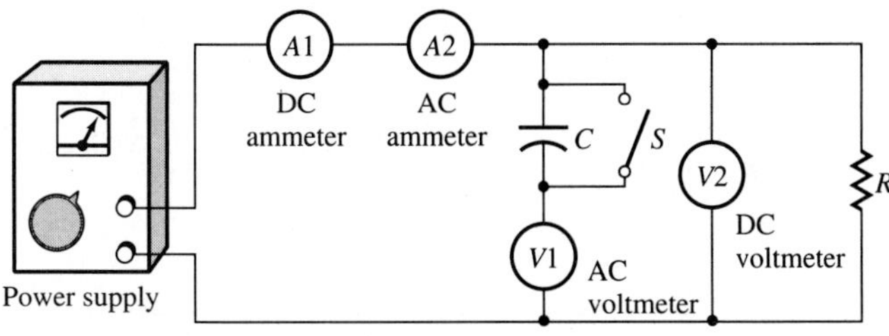

Figure 15.22 A practical method of measuring the dc and effective value of the ac component of a voltage.

Many voltages and currents contain both dc and ac parts. To prevent problems or improve the circuit operation, it might be necessary to know the exact content of the wave. Figure 15.22 shows how this can be done by measurements. The dc meter measures the dc or average value, and the ac meter, with the switch closed, measures the effective or rms value. This includes ac and dc. When the switch is open, the ac meter measures *only* the effective value of the ac component—since the capacitor blocks out the dc part. The effective value of the ac component can also be calculated using Equation 15.14.

$$V_{ac} = \sqrt{V_{eff}^2 - V_{dc}^2} \qquad \textbf{(15.15)}$$

where V_{ac} is the effective value of the ac part of the voltage or current; V_{eff} is the effective value of the voltage or current; and V_{dc} is the average or dc value of the voltage or current.

Although the dc meters do not measure ac, the ac current still passes through their windings and causes an I^2R loss. Therefore, the dc meter windings should be rated for the effective value of the current—not just for the dc value. Otherwise, damage can result.

A term that is often used to indicate the amount of ac in a waveform is the **ripple factor** (γ). The ripple factor is the ratio of the effective value of the ac part to the dc value.

ripple factor

$$\gamma = \frac{\text{effective value of ac}}{\text{dc value}} \qquad \textbf{(15.16)}$$

EXAMPLE 15.9

For the waveform shown in Figure 15.23, find:

a. the dc voltage in the output.
b. the effective value of the output.
c. the effective value of the ac voltage in the output.
d. the ripple factor.

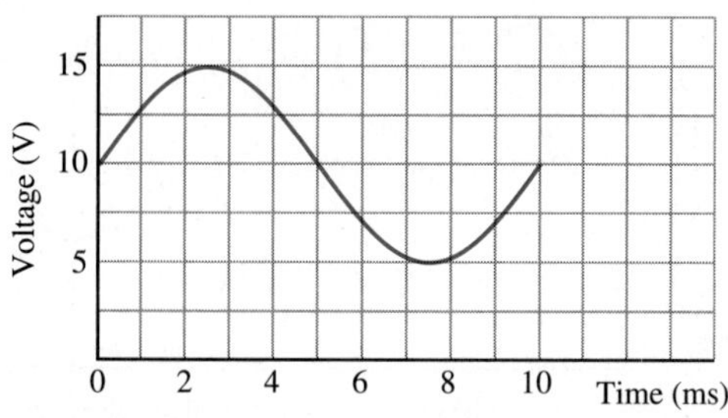

Figure 15.23

Solution

One cycle is divided into 10 strips, as shown in Figure 15.24.

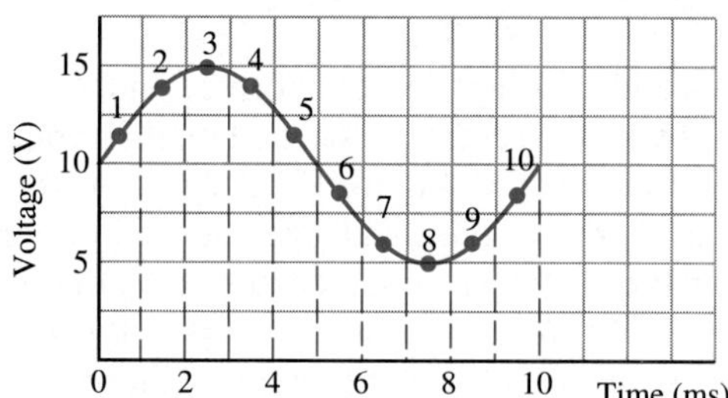

Figure 15.24

a.

$$V_{dc} = \frac{\upsilon_1 + \upsilon_2 + \cdots + \upsilon_{10}}{10}$$

Using the calculated values for υ_1, υ_2, and so forth gives

$$V_{dc} = (11.54 + 14.04 + 15 + 14.04 + 11.54 + 8.45 + 5.96 + 5 + 5.96 + 8.45)(1/10)$$

$$V_{dc} = 9.998 \text{ V}$$

b.

$$V_{eff} = \sqrt{\frac{\upsilon_1^2 + \upsilon_2^2 + \cdots + \upsilon_{10}^2}{10}}$$

$$= \sqrt{\frac{1124.42 \text{ V}^2}{10}}$$

$$V_{eff} = 10.60 \text{ V}$$

c. From Equation 15.16,

$$V_{ac} = \sqrt{V_{eff}^2 - V_{dc}^2}$$

$$V_{ac} = \sqrt{(10.6)^2 - (10)^2}$$

$$V_{ac} = 3.52 \text{ V}$$

d.

$$\gamma = V_{ac}/V_{dc}$$

$$= 3.52 \text{ V}/9.998 \text{ V}$$

$$\gamma = 0.352$$

Table 15.1 presents a summary of the content of various waveforms. It can be used to check results and to determine the content by inspection.

TABLE 15.1 Makeup of Waveforms

Characteristics of Waveform	Average or dc Value	Effective or rms value of the Waveform	Effective or rms Value of the ac Part
1. No variation with time (battery)	$\frac{h_1 + h_2 + \cdots + h_n}{n}$		
(a) net area is positive See Figure 15.25	dc value is positive	equals the dc value	zero
(b) net area is negative See Figure 15.26	dc value is negative	equals the dc value	zero
2. Net area is zero (sine wave) See Figure 15.27	zero	$\sqrt{\frac{h_1^2 + h_2^2 + \cdots + h_n^2}{n}}$	equals the effective value of the waveform
3. Variation with time and net area (full-wave, half-wave rectification) See Figure 15.28	$\frac{h_1 + h_2 + \cdots + h_n}{n}$	$\sqrt{\frac{h_1^2 + h_2^2 + \cdots + h_n^2}{n}}$	$\sqrt{(\text{eff})^2 - (\text{dc})^2}$

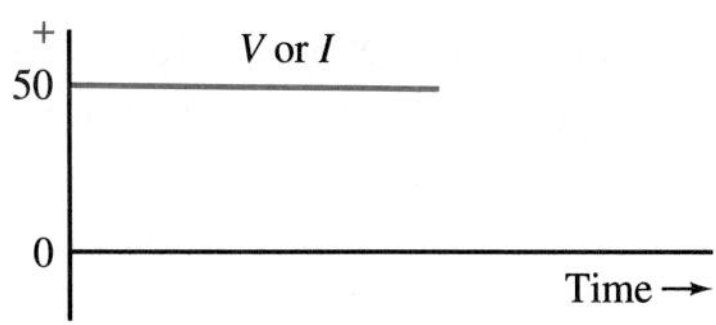

Figure 15.25 A positive dc voltage.

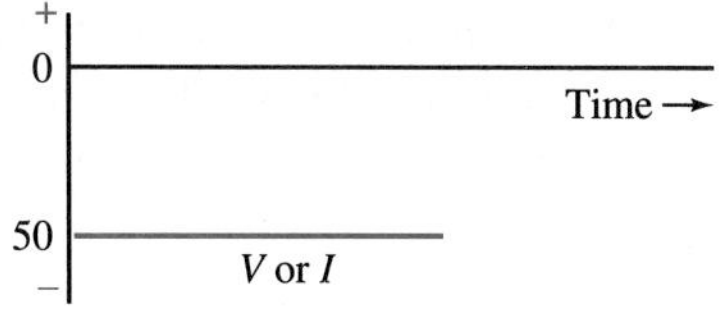

Figure 15.26 A negative dc voltage.

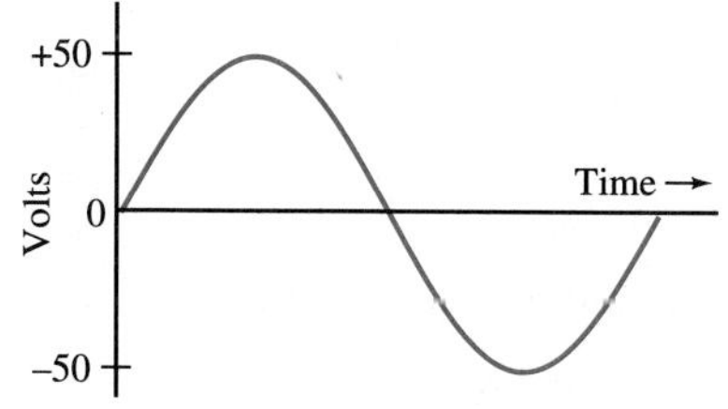

Figure 15.27 A sinusoidal ac voltage.

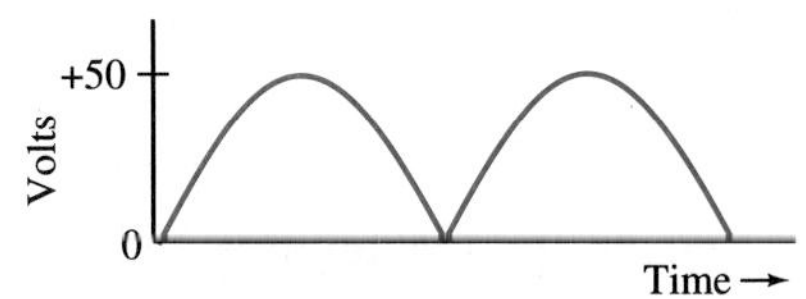

Figure 15.28 A sinusoidal full-wave rectified voltage.

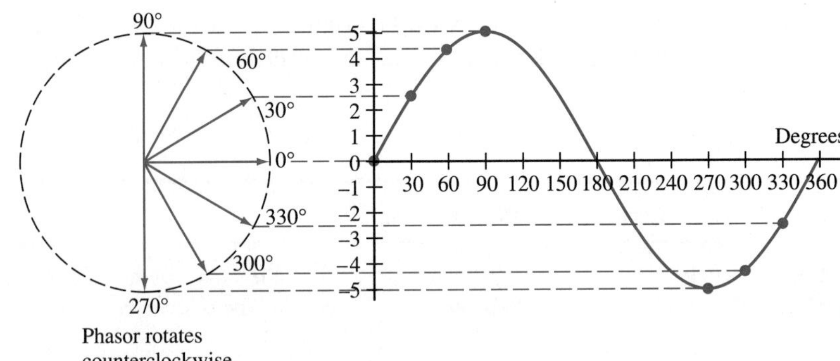

Figure 15.29 The relation between the phasor and the sine wave.

15.6 Phasors

phasor

A sine wave can be represented as a phasor. A **phasor** is the electrical circuit's equivalent of a vector, which has magnitude and direction. The vector has a direction in space. The phasor's direction is the phase angle. The use of phasors eliminates the need to combine sine waves or use calculus in the analysis of ac circuits. Use of phasors results in simpler and shorter calculations, more exact results, and less difficulty in drawing. On the other hand, phasors do not show those characteristics of the waveform such as frequency, instantaneous value, and others.

The relation between a phasor and a sine wave can be seen in Figure 15.29. Each point on the sine wave is the vertical projection of the phasor at that angle, as it is rotated counterclockwise. For example, the height at 30° is 5 sin 30°, or 2.5. The projection will be 0 at 180°, followed by negative values. Thus, a 360° rotation of the phasor generates one cycle of the sine wave.

polar form
rectangular form

Phasors can be written in either **polar form** or **rectangular form**. A detailed explanation of these is given in Appendix E, but it might be helpful to review it. As noted in the Appendix, the relation between the two is:

Polar	Rectangular	
$C \angle \pm\theta$	$A \pm jB$	
$C = \sqrt{A^2 + B^2}$	$A = C \cos\theta,$	$B = C \sin\theta$

Using the magnitudes and signs of A and B, the relation for θ is:

If A is positive, $\theta = \tan^{-1}(B/A)$
If A is negative, $\theta = \pm 180° + \tan^{-1}(B/A)$

Most electronic calculators automatically give the correct angle in all quadrants, so the foregoing relations do not have to be used with them.

EXAMPLE 15.10

Express the phasor 40 $\angle 30°$ in rectangular form.

Solution

$$A = C \cos \theta = 40 \cos 30°$$
$$A = (40)(0.866) = 34.64$$
$$B = C \sin \theta = 40 \sin 30°$$
$$= (40)(0.5) = 20$$
$$40 \angle 30° = 34.6 + j20$$

The algebraic and RPN calculator keystrokes for the conversion are:

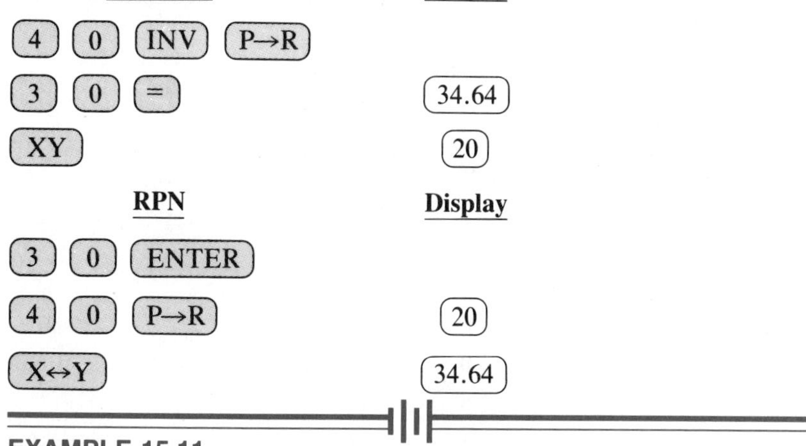

EXAMPLE 15.11

Express the phasor 20 $\angle 200°$ in rectangular form.

Solution

$$A = 20 \cos 200° = (20)(-0.94) = -18.8$$
$$B = 20 \sin 200° = (20)(-0.34) = -6.8$$
$$20 \angle 200° = -18.8 - j6.8$$

EXAMPLE 15.12

Express the phasor $-3 - j4$ in polar form.

Solution

$$C = \sqrt{3^2 + 4^2} = \sqrt{25} = 5$$

Since A and B are negative, the angle is in the third quadrant, and

$$\theta = 180° + \tan^{-1} (B/A)$$
$$= 180° + \tan^{-1} (4/3)$$
$$= 180° + 53.13°$$
$$\theta = 233.13°$$
$$-3 - j4 = 5 \angle 233.13°$$

If $-180°$ is used, θ will be $-126.87°$. This gives the same location as measured in a clockwise direction.

The phasor can use either the peak or the effective value of the sine wave. Since ac meters measure the effective value, and circuit analysis uses the effective value, the effective value is usually used. Unless otherwise noted, the magnitude of any voltage or current phasor will be the effective value.

EXAMPLE 15.13

Express the phasor $4 + j3$ in polar form.

Solution

$$C = \sqrt{A^2 + B^2}$$
$$= \sqrt{4^2 + 3^2}$$
$$= \sqrt{25}$$
$$C = 5$$
$$\tan \theta = 3/4 = 0.75$$
$$\theta = \tan^{-1} 0.75$$
$$\theta = 36.78°$$
$$4 + j3 = 5 \angle 36.87°$$

The algebraic and RPN calculator keystrokes for the conversion are:

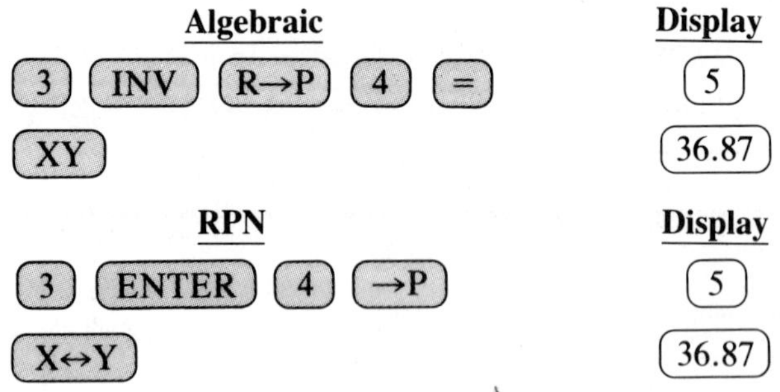

The magnitude of the phasor is the effective value of the sine wave. The angle is the phase angle. A phasor for a leading wave will have a positive angle, whereas one for a lagging wave will have a negative angle. Figure 15.30 shows the relationship among the equation, polar form, rectangular form, and the diagrams for some phasors.

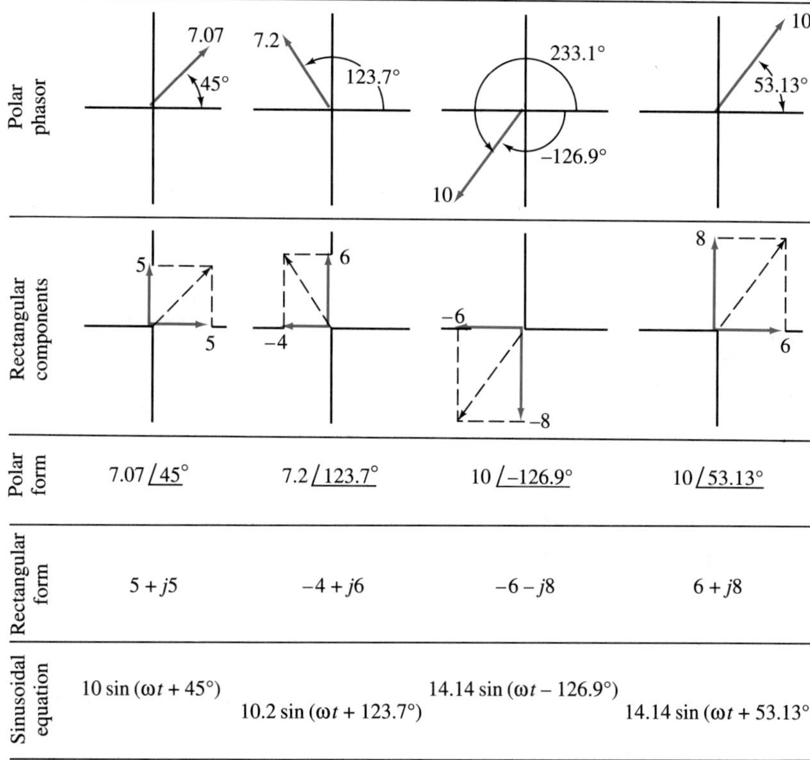

Figure 15.30 The graphic presentation and mathematical expressions for some phasors.

EXAMPLE 15.14

Write the phasors for the following in polar and rectangular form.

a. $\upsilon = 70.7 \sin(\omega t + 20°)$
b. $i = 100 \sin(\omega t)$

Solution

a. $C = (70.7)(0.707) = 50$
   $\theta = 20°$
   The polar form for the phasor is $50 \angle 20°$.
   The rectangular form is $46.98 + j17.10$.
b. $C = (100)(0.707) = 70.7$
   $\theta = 0°$
   The polar form for the phasor is $70.7 \angle 0°$.
   The rectangular form is $70.7 + j0$.

EXAMPLE 15.15

Write $\upsilon = 50 \sin(\omega t - 60°)$ in polar and rectangular form.

Solution

Polar form:

$$C = (50)(0.707) = 35.35$$

$$\theta = -60°$$
$$V = 35.35 \angle -60°$$

Rectangular form:

$$A = C \cos\theta = (35.35)\cos(-60°) = 17.68$$
$$B = C \sin\theta = (35.35)\sin(-60°) = -30.61$$
$$V = 17.68 - j30.61$$

EXAMPLE 15.16

Write the sinusoidal expression for $5 \angle 60°$ if ω is 377 rad/sec.

Solution

The amplitude is $5/0.707 = 7.07$.
The phase angle is 60°.
$5 \angle 60° = 7.07 \sin(377t + 60°)$

15.7 Resistance, Capacitance, and Inductance in ac

When a resistance is connected to a sinusoidal source, as in Figure 15.31(a), the current remains in phase with the voltage, as in Figure

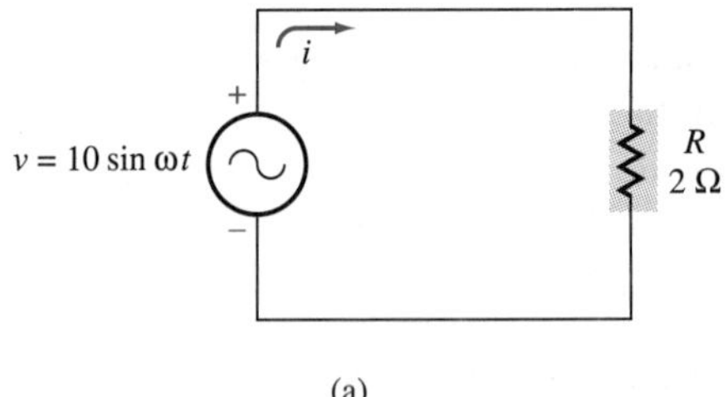

(a)

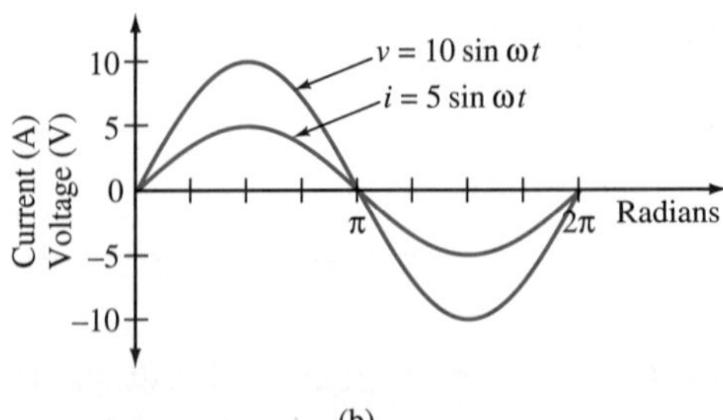

(b)

Figure 15.31 The resistor voltage and current are in phase. (a) A resistor in an ac circuit. (b) The resistor voltage and current waveforms.

15.31(b). The voltage is given by $R \times i$, so if $i = I_m \sin \omega t$, $v_r = RI_m \sin \omega t$. Comparing this to the general expression in Equation 14.8 shows that RI_m must be V_m. So

$$v_R = V_m \sin \omega t \tag{15.17}$$

where v_R is the voltage across the resistance at any time t, in volts; V_m is the peak value of the voltage, in volts; ω is the angular velocity, in radians per second; and t is the time, in seconds.

Equation 15.17 is a sine equation in phase with the current. The opposition to the current is the resistance R, just as for dc. If $V_m = RI_m$,

$$R = V_m/I_m, \quad \text{also} \quad V/I \tag{15.18}$$

where R is the resistance, in ohms; V_m, I_m are the peak values; and V, I are the effective values of voltage and current in volts and amperes.

EXAMPLE 15.17

The current in a 10-ohm resistor is given by $i = 20 \sin 377t$. What is the sinusoidal expression for the voltage?

Solution

$$v_R = V_m \sin 377t$$

$$R = V_m/I_m \quad \text{so} \quad V_m = I_mR = (20\text{ V})(10) = 200\text{ V}$$

$$v_R = 200 \sin 377t$$

The current in a capacitor leads the voltage by $\pi/2$ radians, as shown in Figure 15.32. This is because $i_c = C\,(dv/dt)$ where dv/dt is the slope of the voltage curve at any time t (see Chapter 12). A look at the voltage curve shows that dv/dt is 0 at the wave peaks, and maximum at the zero points. It is maximum positive at 0 radians and maximum negative at π radians. Therefore, i will be 0 at the peaks of the voltage curve, and maximum at

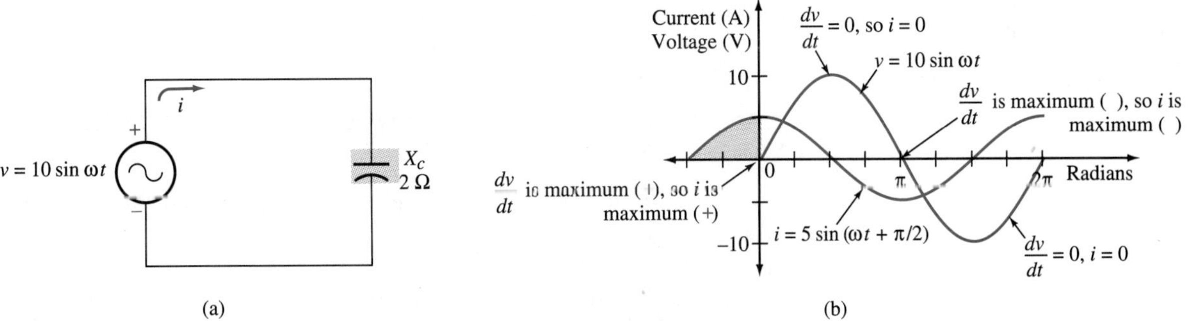

Figure 15.32 The capacitor voltage lags the current by $\pi/2$ radians (90°). (a) A capacitor in an ac circuit. (b) The capacitor voltage and current waveforms.

its zeros. Plotting these points shows that the current curve is shifted $\pi/2$ radians with a leading phase angle.

Constructing the current curve by drawing and calculating the slope is tedious and inaccurate. So, once again, calculus must be used to obtain an equation for dv/dt of the voltage curve. If $v = V_m \sin \omega t$, $dv/dt = \omega V_m \sin(\omega t + \pi/2)$ and

$$i_c = \omega C V_m \sin(\omega t + \pi/2) \tag{15.19}$$

where i_c is the capacitor current at any time t, in amperes; ω is the angular velocity, in radians per second; C is the capacitance, in farads; V_m is the amplitude of the voltage, in volts; and t is the time, in seconds.

Comparing Equation 15.19 to Equation 15.9 shows that $\omega C\, V_m$ must be I_m. Then

$$i_c = I_m \sin(\omega t + \pi/2) \tag{15.20}$$

where i_c, ω, and t have the same meaning as in Equation 15.19, and I_m is the amplitude of the current wave, in amperes.

Since $\omega C V_m$ equals I_m, V_m/I_m must equal $1/(\omega C)$. The quantity $1/(\omega C)$ is defined as the **capacitive reactance**.

Capacitive reactance is the opposition of capacitance to the ac current. Its symbol is X_C, and its unit is the ohm. When 1 volt (effective value) across the capacitor results in a current of 1 ampere (effective value), X_C is 1 ohm. Although the charge does not pass through the dielectric, the electrostatic field rises, decays, and reverses with the sinusoidal voltage. The capacitance opposes this change, and by doing so, opposes the current. Then

$$X_C = 1/(\omega C) \tag{15.21}$$

Also

$$X_C = V/I \tag{15.22}$$

where X_C, ω, and C have the same meaning and units as in Equation 15.19, and V, I are the capacitor voltage and current, either effective or peak, in volts and amperes.

X_C changes with frequency and capacitance, as seen in Figure 15.33. At high frequencies, it can be very small and offer little opposition to the current. This can cause interaction between parts of a circuit even though the parts are not physically connected.

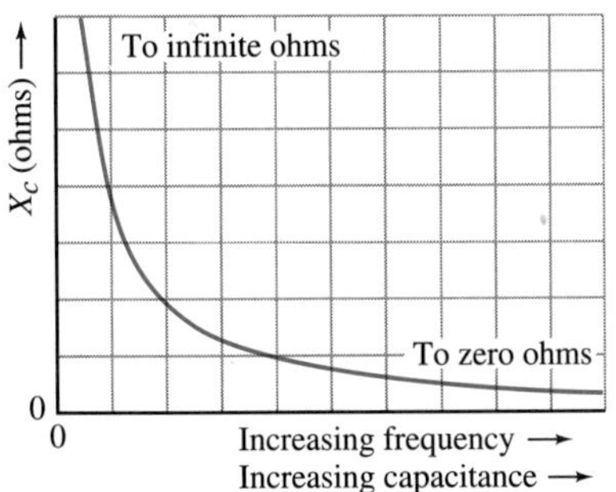

Figure 15.33 The variation of capacitor reactance with capacitance and frequency.

EXAMPLE 15.18

A 5-microfarad capacitor is connected across a source whose voltage is 100 sin(1000t).

a. Find the value of X_C.
b. Write the equation for the current.
c. Find the effective value of the current.

Solution

a. $$X_C = 1/(\omega C) = \frac{1}{(1000 \text{ rad/s})(5 \times 10^{-6} \text{ F})}$$
$$X_C = 200\ \Omega$$

b. $$I_m = V_m/X_C = \frac{100 \text{ V}}{200\ \Omega}$$
$$I_m = 0.5 \text{ A}$$

The current must lead v by $\pi/2$ radians. So

$$i = I_m \sin(1000t + \pi/2)$$
$$i = 0.5 \sin(1000t + \pi/2)$$

c. $$I_{\text{eff}} = 0.707 I_m = (0.707)(0.5 \text{ A})$$
$$I_{\text{eff}} = 0.354 \text{ A}$$

The current in an ideal inductor lags the voltage by $\pi/2$ radians, as shown in Figure 15.34(b). This is because $v_L = L\,(di/dt)$ where di/dt is the slope of the current curve at any time t (see Chapter 13). A look at the

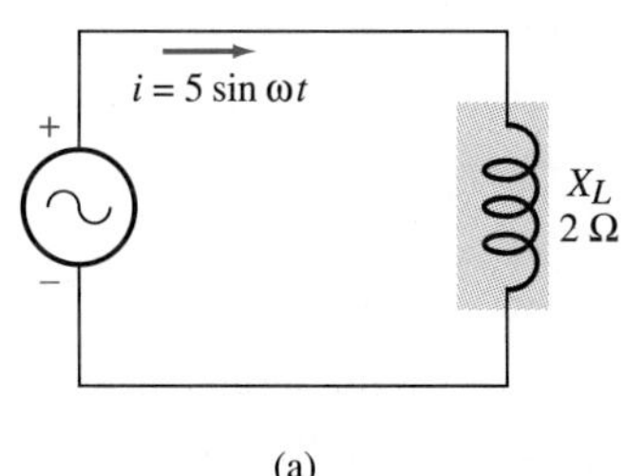

(a)

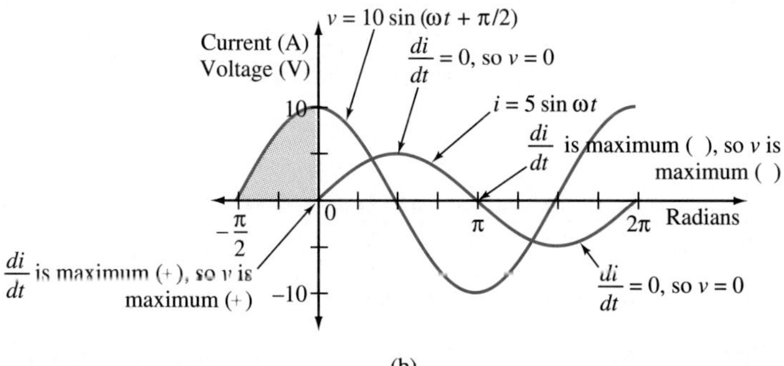

(b)

Figure 15.34 The inductor current lags the voltage by $\pi/2$ radians (90°). (a) An inductor in an ac circuit. (b) The inductor voltage and current waveforms.

current curve shows that di/dt is 0 at its peaks, and maximum at its zero points. Therefore, υ_L will be 0 at the peaks of the current curve, and maximum at its zeros. This is similar to the relation for i_c and υ_c. Plotting these points shows that the voltage curve is shifted by $\pi/2$ radians, and is leading the current.

As with the capacitor, calculus gives the most exact expression for di/dt. If $i = I_m \sin \omega t$, $di/dt = \omega I_m \sin(\omega t + \pi/2)$. Calculus and integration gives

$$\upsilon_L = \omega L I_m \sin(\omega t + \pi/2) \qquad \textbf{(15.23)}$$

where υ_L is the voltage across the inductor at any time t, in volts; ω is the angular velocity, in radians; L is the inductance, in henries; I_m is the amplitude of the current, in amperes; and t is the time, in seconds.

Comparing Equation 15.23 to Equation 15.8 shows that $\omega L I_m$ must be V_m. Then

$$\upsilon_L = V_m \sin(\omega t + \pi/2) \qquad \textbf{(15.24)}$$

where υ_L, ω, and t have the same meaning as in Equation 15.23, and V_m is the amplitude of the voltage wave, in volts.

Since $\omega L I_m = V_m$, V_m/I_m must equal ωL. The quantity ωL is defined as the inductive reactance. **Inductive reactance** is the opposition of inductance to an ac current. Its symbol is X_L, and its unit is the ohm. When 1 volt (effective value) across an inductor results in 1 ampere (effective value) of current, X_L is 1 ohm.

The change and reversal of the current changes and reverses the magnetic field. The inductance opposes this change, thus causing an opposition to the current. Then

$$X_L = \omega L \qquad \textbf{(15.25)}$$

Also

$$X_L = V/I \qquad \textbf{(15.26)}$$

where X_L, ω, and L have the same meaning and units as before, and V, I are the inductor voltage and current, either effective or peak, in volts and amperes.

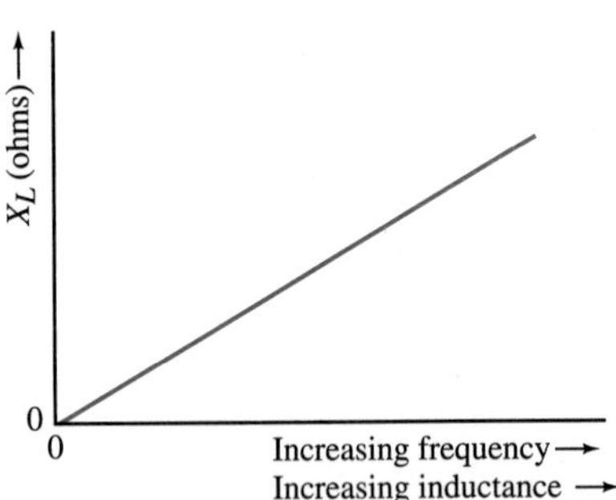

Figure 15.35 The variation of inductor reactance with inductance and frequency.

X_L is directly proportional to the frequency and inductance, as shown in Figure 15.35. Thus, small stray inductance can have a large X_L at high frequencies and cause a large drop in voltage.

EXAMPLE 15.19

A 2-henry inductor has a current given by $i = 0.3 \sin 200t$.

a. Find the value of X_L and the effective value of the current.
b. Write the equation for the voltage.

Solution

a. $X_L = \omega L$

$= (200 \text{ radians per second})(2 \text{ H})$

$X_L = 400\ \Omega$

$I_{\text{eff}} = (0.707)(I_m) = (0.707)(0.3 \text{ A})$

$I_{\text{eff}} = 0.212 \text{ A}$

b.

$$\begin{aligned} V_m &= I_m X_L \\ &= (0.3 \text{ A})(400\ \Omega) \\ &= 120 \text{ V} \end{aligned}$$

υ must lead i by $\pi/2$ radians so

$$\upsilon = 120 \sin(200t + \pi/2)$$

Chapters 12 and 14 explained that capacitance and inductance change the shape of waveforms. This is not so for the sine wave. The voltage and current remain sinusoidal, but are shifted by $\pi/2$ radians. The sine wave is the only waveform whose shape is *not* changed by capacitance or inductance.

SUMMARY

1. An alternating waveform is one that varies in magnitude and changes signs at regular intervals of time. The sinusoidal voltage generated by electric utilities is the most common alternating electrical waveform.
2. Some characteristics that are used to describe ac are amplitude, peak-to-peak value, instantaneous value, cycle, frequency, period, and phase relation.
3. The sinusoidal expression can be in terms of angular velocity, frequency, or degrees. Angular velocity is a measure of the number of radians displaced per second by a rotating object.
4. A sinusoidal wave can be in phase with, lag, or lead another one. If it passes through its zeros, and peaks at the same time as the first wave, it is in phase. If it does so at a later time, it is lagging. If it does so at an earlier time, it is leading the first wave.
5. If there is a net area under the curve for one cycle, an alternating waveform has an average value. A dc meter measures the average value, which can be 0 for some waveforms.
6. The amount of an alternating voltage or current is usually given in terms of its effective value. Effective value is related to the energy supplied by the waveform.

7. The average value of a sine wave is 0. The effective value is 0.707 of the peak value.
8. In circuit analysis, sinusoidal expressions are replaced by phasors. These express the wave by its effective value and phase angle.
9. The opposition of a resistance to an ac current is also its resistance, as it is for dc. The current in a resistor is in phase with the resistor voltage.
10. The opposition of a capacitance to an ac current is capacitive reactance (X_c). It is equal to $1/(2\pi fC)$.
11. The current in a capacitor leads the voltage by $\pi/2$ radians.
12. The opposition of an inductor to an ac current is inductive reactance (X_L). It is equal to $2\pi fL$.
13. The current in an ideal inductor lags the voltage by $\pi/2$ radians.

EQUATIONS

15.1	$f = \dfrac{1}{T}$	The relation of frequency and period.	595
15.2	$e = E_m \sin \omega t$	The induced emf in a rotating conductor.	597
15.3	$\omega = 2\pi f$	The relation between angular velocity and frequency.	597
15.4	$e = E_m \sin(2\pi f t)$	The induced emf in terms of frequency.	597
15.5	$i = I_m \sin(2\pi f t)$	The current in a rotating conductor in terms of frequency.	597
15.6	$e = E_m \sin \alpha$	The induced emf in terms of the angle displaced by a rotating conductor.	598
15.7	$i = I_m \sin \alpha$	The current in a rotating conductor in terms of the angle displaced by a rotating conductor.	598
15.8	$e = E_m \sin(\omega t \pm \theta)$	The sinusoidal expression for the voltage wave with a phase shift.	600
15.9	$i = I_m \sin(\omega t \pm \theta)$	The sinusoidal expression for the current wave with a phase shift.	600
15.10	$I_{ave} \quad \text{or} \quad V_{ave} = \dfrac{A}{L}$	The relation of the average value to the area and cycle length.	604
15.11	$I_{ave} \quad \text{or} \quad V_{ave} = \dfrac{h_1 + h_2 + \cdots + h_n}{n}$	The expression for the average value in the strip method.	605
15.12	$I_{eff} = \sqrt{\dfrac{\text{area under the } i^2 \text{ curve}}{\text{length of one cycle}}}$	The general expression for the effective value of a current wave.	607

15.13	$V_{\text{eff}} = \sqrt{\dfrac{\text{area under the } \upsilon^2 \text{ curve}}{\text{length of one cycle}}}$	The general expression for the effective value of a voltage wave.	607
15.14	$I_{\text{eff}} \quad \text{or} \quad V_{\text{eff}} = \sqrt{\dfrac{h_1^2 + h_2^2 + \cdots + h_n^2}{n}}$	The expression for the effective value in the strip method.	607
15.15	$V_{\text{ac}} = \sqrt{V_{\text{eff}}^2 - V_{\text{dc}}^2}$	The equation for the effective value of the ac component.	609
15.16	$\gamma = \dfrac{\text{effective value of ac}}{\text{dc value}}$	The ripple factor of a voltage.	609
15.17	$V_R = V_m \sin \omega t$	The voltage across a resistor with a sinusoidal source.	617
15.18	$R = \dfrac{V_m}{I_m} \quad \text{also} \quad V/I$	The relation of resistance to the peak voltage and current in the resistor.	617
15.19	$i_c = \omega C V_m \sin\left(\omega t + \dfrac{\pi}{2}\right)$	The sinusoidal expression for the current in a capacitor with a sinusoidal source.	618
15.20	$i_c = I_m \sin\left(\omega t + \dfrac{\pi}{2}\right)$	The capacitor current in terms of the peak value of current.	618
15.21	$X_C = \dfrac{1}{\omega C}$	Capacitive reactance in terms of angular frequency and capacitance.	618
15.22	$X_C = \dfrac{V}{I}$	Capacitive reactance in terms of voltage and current.	618
15.23	$\upsilon_L = \omega L I_m \sin\left(\omega t + \dfrac{\pi}{2}\right)$	The sinusoidal expression for the inductor voltage with a sinusoidal source.	620
15.24	$\upsilon_L = V_m \sin\left(\omega t + \dfrac{\pi}{2}\right)$	The inductor voltage in terms of the peak value of voltage.	620
15.25	$X_L = \omega L$	Inductive reactance in terms of angular frequency and inductance.	620
15.26	$X_L = \dfrac{V}{I}$	Inductive reactance in terms of voltage and current.	620

QUESTIONS

1. What are two ways in which an alternating waveform differs from dc?
2. What is meant by the instantaneous value of a voltage?
3. What effect does increasing the frequency of a waveform have on the period?
4. What is meant by frequency?

5. What will be the positive amplitude of an ac voltage sine wave if the negative amplitude is -5 V?
6. What are the three quantities that can be used on the x-axis when plotting an alternating voltage or current?
7. In the equation, $\upsilon = 10 \sin 377t$, which quantity represents the angular velocity? The amplitude?
8. What is the meaning of a lagging wave? A leading wave?
9. How can one determine whether a wave is leading or lagging from the picture of the wave?
10. How can one determine whether a wave is leading or lagging from the equation for the wave?
11. What is the phase angle between the waves $\upsilon = 5 \sin(377t - 30°)$ and $i = 1 \sin(377t - 10°)$?
12. What is meant by the average value of a voltage or current?
13. What is meant by the effective value of a voltage or a current?
14. What is the average value of a 1 1/2-V flashlight cell? What is the effective value?
15. What does rms stand for, and how is the meaning related to the expression for the effective value?
16. Does a sinusoidal waveform have an average value? Why?
17. What will happen to the average value of a sine wave if the wave is shifted up by 2 volts?
18. What effect does changing the frequency have on the average and effective values of a sinusoidal voltage?
19. What effect does changing the amplitude of a sine wave have on the average and effective values?
20. How can one measure the average value with a voltmeter? The effective value?
21. The average value of a voltage is 10 V and the effective value is 12 V. What accounts for the difference?
22. How can one measure the effective value of the ac component in a voltage?
23. How is the effective value related to the energy supplied by an ac voltage?
24. Which light bulb will be brighter: a 100-W bulb connected to a 120-V dc source, or a 100-W bulb connected to a 120-V (effective value) ac source?
25. When does the method of known shapes give poor results for the average and effective values?
26. What can be done to obtain more exact results from the method of strips?
27. What is the phase relation between current and voltage for a resistor?
28. What is the phase relation between current and voltage for a capacitor?
29. What is the phase relation between current and voltage for an inductor?
30. What effect will increasing the frequency have on the phase angle for the capacitor? For the inductor?
31. What is meant by capacitive reactance? By inductive reactance?
32. What will happen to the current in a capacitor if the source frequency is doubled? Halved?
33. What will happen to the current in an inductor if the frequency is doubled? Halved?
34. Which of the following is correct?
    a. $X_C = 2\pi fC$
    b. $X_C = C/(2\pi f)$
    c. $X_C = 1/(2\pi fC)$

35. Which of the following is correct?
   a. $X_L = 2\pi f L$
   b. $X_L = L/(2\pi f)$
   c. $X_L = 1/(2\pi f L)$

PROBLEMS

SECTION 15.1 Alternating (ac) and Sine Wave Characteristics

1. Determine the amplitude, period, frequency, and instantaneous value of the voltage at 5 ms for the wave shown in Figure 15.36.
2. Repeat Problem 1 for the wave shown in Figure 15.37.
3. An ac voltage has a frequency of 4×10^3 Hz. How many seconds does it take to complete one cycle?
4. The voltage during charging and discharging of a capacitor varies, as shown in Figure 15.38.
   (a) What is the period of the wave?
   (b) What is the instantaneous value of 2 ms?
   (c) What is the amplitude?
5. How many cycles of the waveform are shown in Figure 15.39? What is the period of one cycle?
6. Given the square wave in Figure 15.40, find its:
   (a) Peak-to-peak value.
   (b) Frequency.
   (c) Voltage at 5 ms.

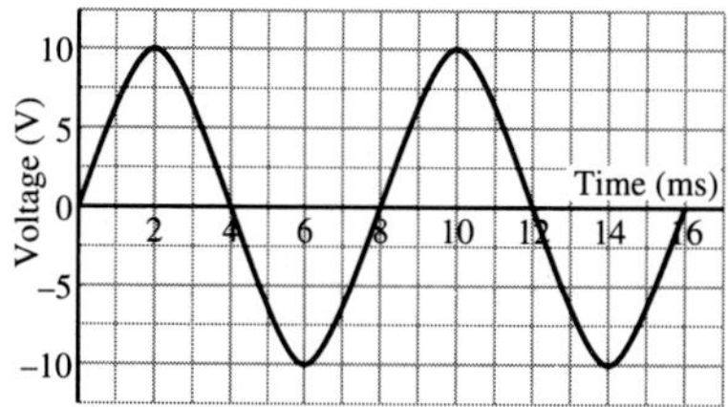

Figure 15.36

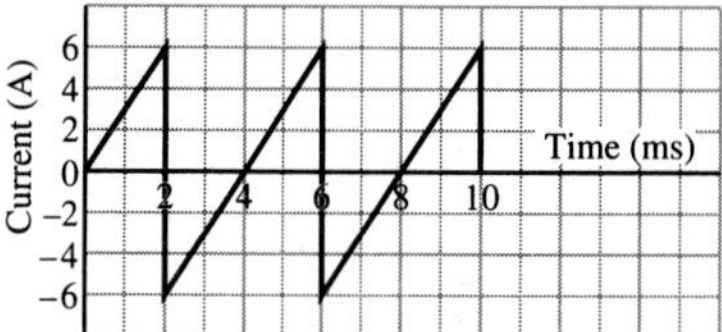

Figure 15.37

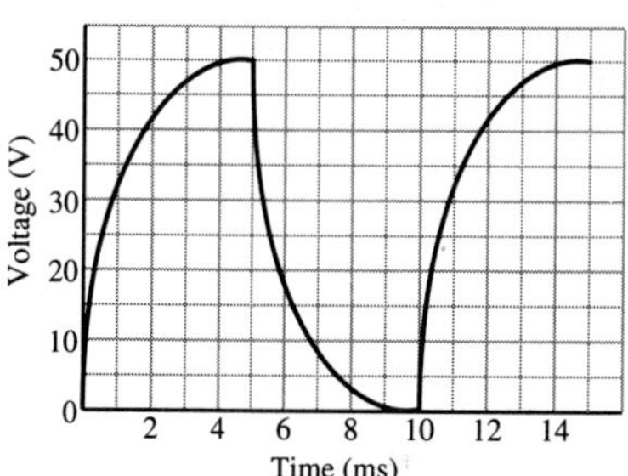

Figure 15.38

SECTION 15.2 Angular Velocity and the Sine Equation

7. A point on a wheel turns through 270 deg in 0.5 second. What is the angular velocity of the point?
8. What is the angular velocity of a wheel that rotates at 1800 RPM?
9. Given the sine wave, $v = 20 \sin 1256t$:
   (a) Find the amplitude of the wave, the frequency in Hz, and the voltage at 1 ms.
   (b) Sketch the waveform.
10. A sine wave has an amplitude of 50 volts and a frequency of 380 Hz.
   (a) Write the equation for the waveform.
   (b) Sketch the waveform.
   (c) At what time is the voltage 30 volts?

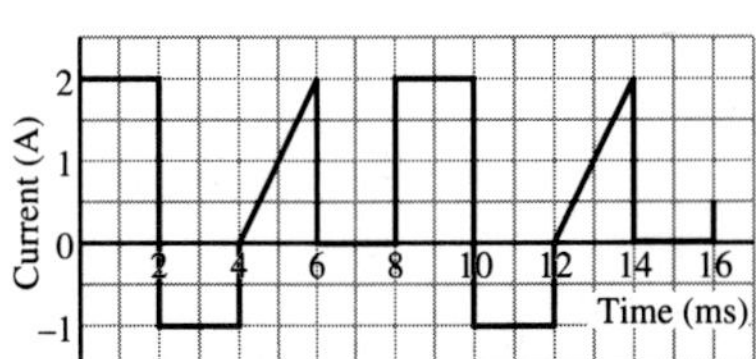

Figure 15.39

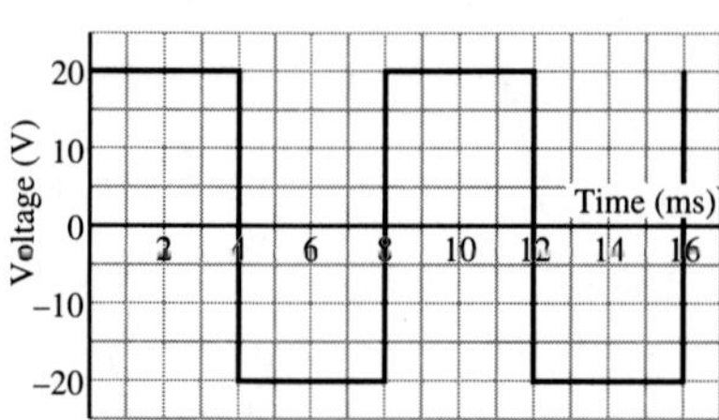

Figure 15.40

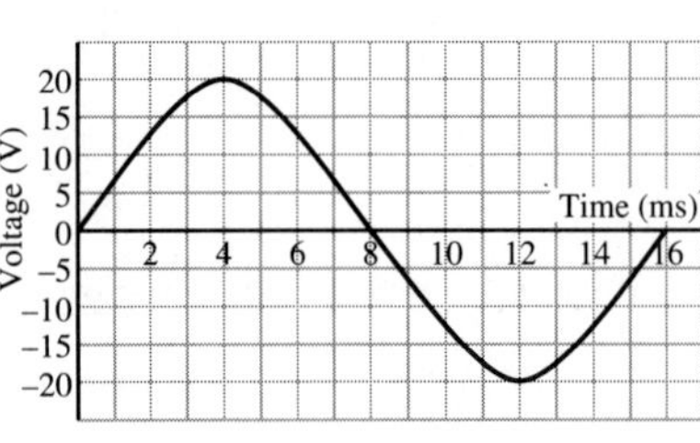

Figure 15.41

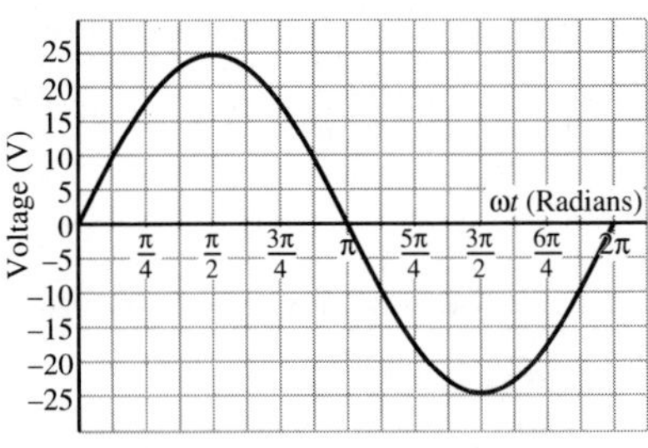

Figure 15.42

11. The output of a transformer is shown in Figure 15.41.

    (a) Write the equation for the waveform.
    (b) What is the frequency, in Hz?
    (c) What is the peak-to-peak value?

12. For the waveform shown in Figure 15.42:

    (a) Write the equation for the sine wave.
    (b) What is the voltage at $9\pi/8$ radians?
    (c) What is its amplitude?

13. A sine wave with an amplitude of 50 volts has a value of 20 volts at 6.5 ms.

    (a) Write the equation for the waveform.
    (b) Sketch the waveform.

SECTION 15.3 Phase Relation and Phase Angle

14. Determine the phase angle for each wave and tell whether it is leading or lagging the basic sine wave.

    (a) $\upsilon = 50 \sin(377t + \pi/2)$
    (b) $\upsilon = 10 \sin(377t + \pi/4)$
    (c) $i = 0.5 \sin(1000t - 30°)$
    (d) $i = 0.3 \sin(800t + 20°)$

15. Repeat Problem 14 for the following signals:

    (a) $\upsilon = 170 \sin(188t - 15°)$
    (b) $\upsilon = 25 \sin(66t + \pi/4)$
    (c) $i = 0.1 \sin(377t - \pi/6)$
    (d) $i = 5 \times 10^{-3} \sin(1131t + \pi/4)$

16. What is the phase angle of υ in Figure 15.43? Is it leading or lagging?

17. What is the phase angle of i in Figure 15.44? Is it leading or lagging?

18. A sine wave has a leading phase angle of $(3\pi)/4$ radians and an amplitude of 10. Sketch the waveform using radians on the horizontal axis.

19. Repeat Problem 18 for a wave with a lagging angle of 30 deg and an amplitude of 20.

20. Determine whether i leads or lags υ, and by how much for each set.

    (a) $i = 3 \sin(377t + \pi/2)$
    $\upsilon = 120 \sin(377t + \pi/4)$

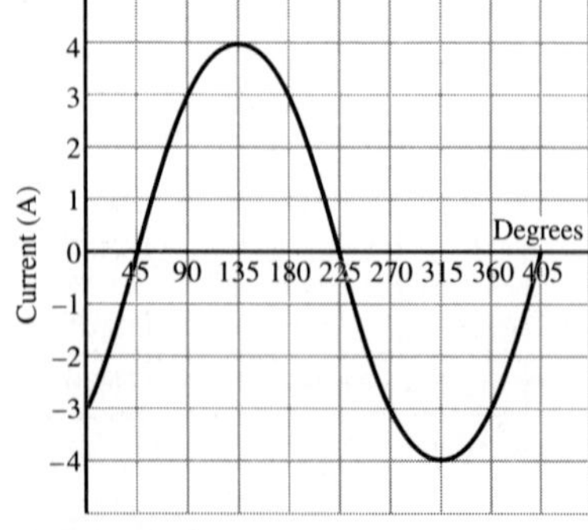

Figure 15.43

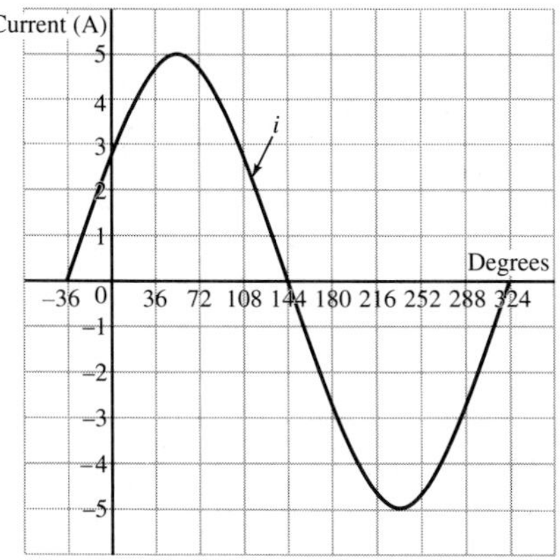

Figure 15.44

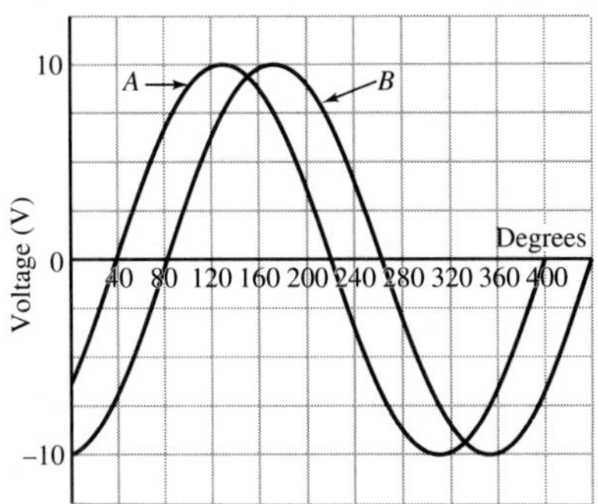

Figure 15.45

(b) $i = 2 \sin(900t - \pi/3)$
$\upsilon = 100 \sin(900t - \pi/6)$

(c) $i = 0.5 \sin(1131t - 20°)$
$\upsilon = 20 \sin(1131t + 30°)$

21. Work Problem 20 for the following values:

(a) $\upsilon = 15 \sin(628t + (3\pi/4))$
$i = 0.2 \sin(628t - (3\pi/4))$

(b) $\upsilon = 50 \sin(314t - 30°)$
$i = 2 \times 10^{-3} \sin(314t - 45°)$

(c) $\upsilon = 120 \sin(471t - \pi/8)$
$i = 2 \sin(471t + (3\pi/8))$

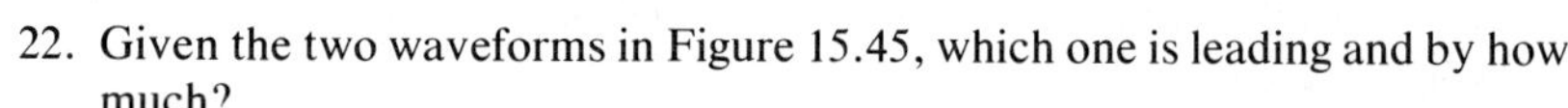

22. Given the two waveforms in Figure 15.45, which one is leading and by how much?

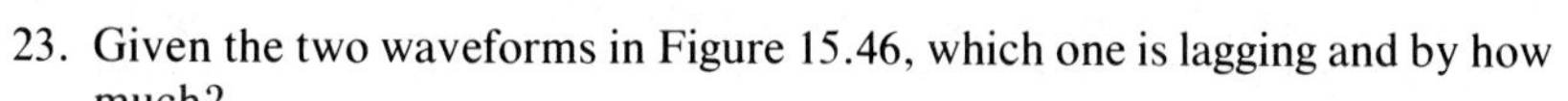

23. Given the two waveforms in Figure 15.46, which one is lagging and by how much?

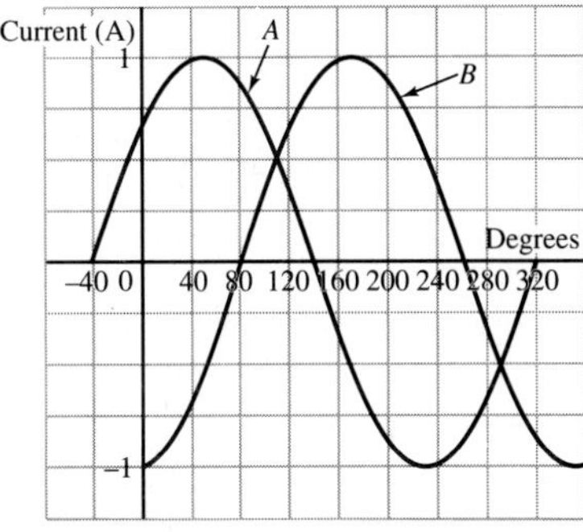

Figure 15.46

SECTION 15.4 The Average Value

24. By inspection, determine which of the waveforms in Figure 15.47 has an average value for one cycle. Is it negative or positive?

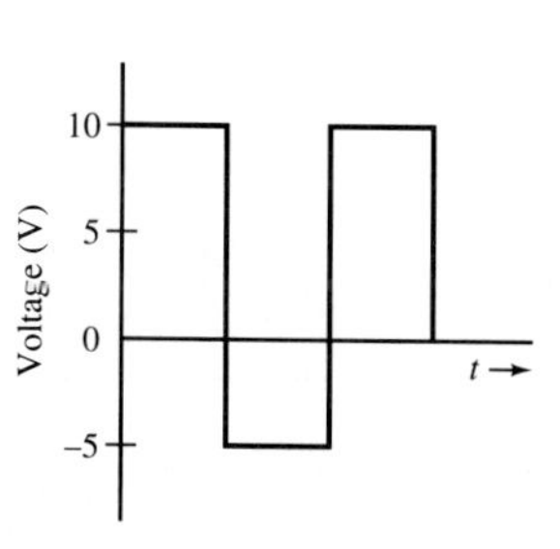

(a)

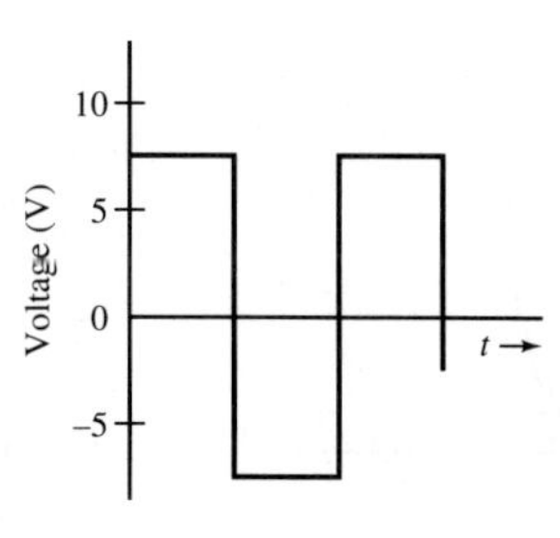

(b)

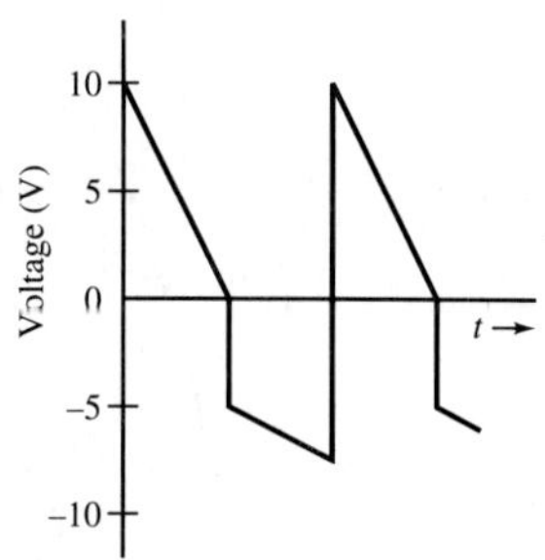

(c)

Figure 15.47

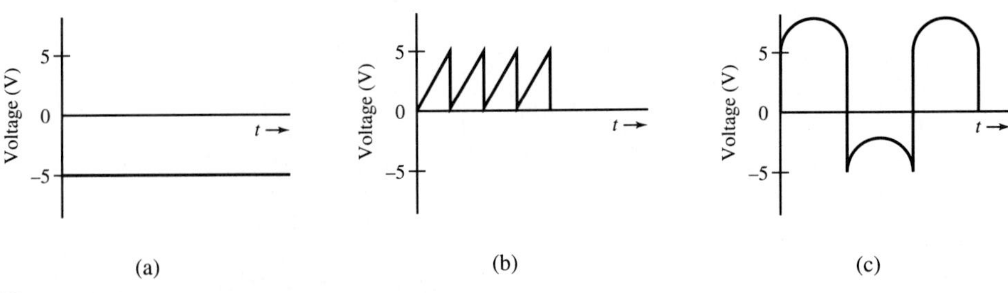

Figure 15.48

25. Repeat Problem 24 for the waveforms in Figure 15.48.

26. Find the average value for one cycle of the current in Figure 15.49 by:
(a) Known shapes. (b) Strips.

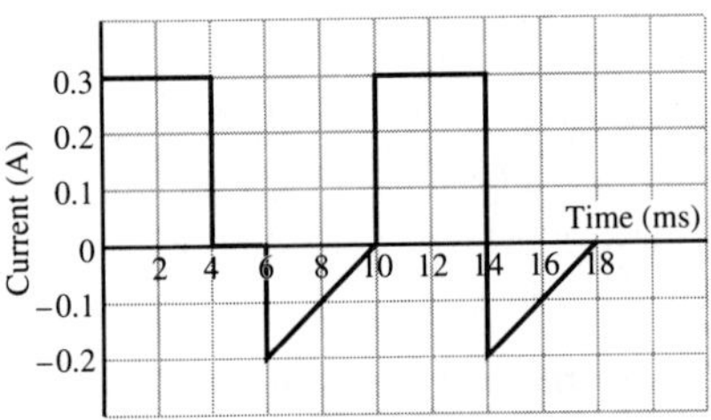

Figure 15.49

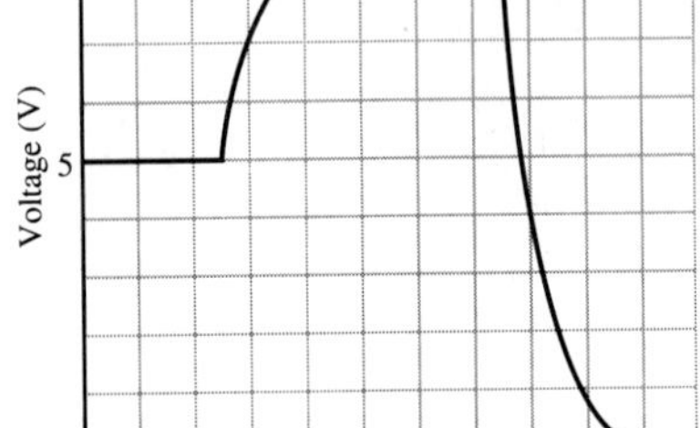

Figure 15.50

27. The waveform in Figure 15.50 is for a capacitor charging from one voltage to another and then discharging. Find its average value by
(a) Known shapes. (b) Strips.

28. The curve in Figure 15.51 is the charging curve for a 5-microfarad capacitor. How much charge, in coulombs, is stored in the capacitor after 5 time constants?

29. A dc voltage of 40 volts is needed from a voltage source whose output looks like Figure 15.52. Use the strip method to find whether the source will be satisfactory.

SECTION 15.5 The Effective Value

30. A sinusoidal voltage from a transformer delivers 1200 joules of energy to a 50-ohm resistor in 10 minutes. What is the effective value of the voltage?

31. Use the method of known shapes to find the effective (rms) value of the voltage in Figure 15.53 for one cycle.

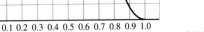

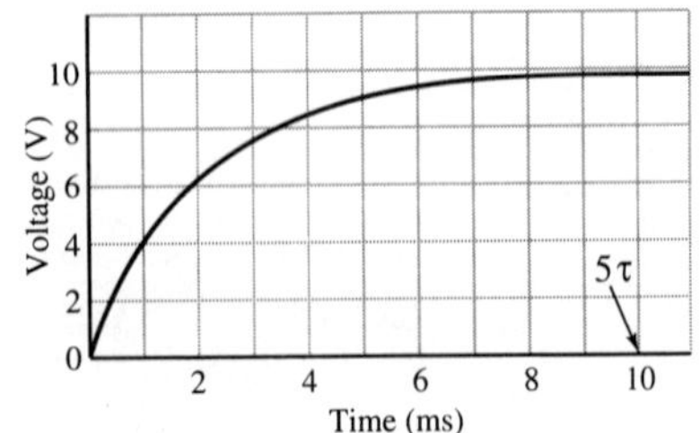

Figure 15.51

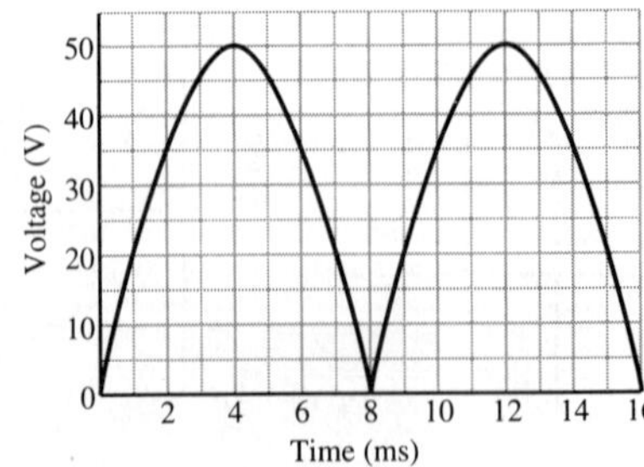

Figure 15.52

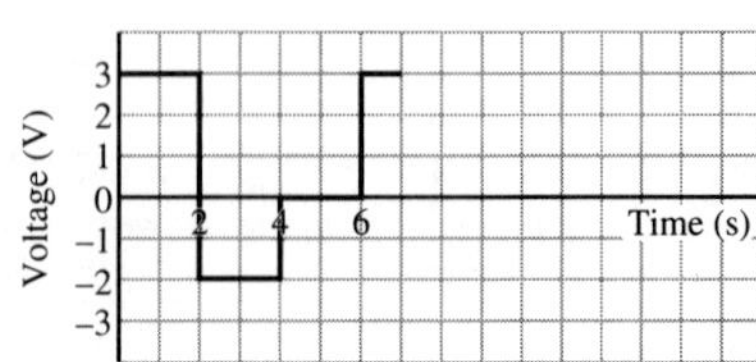

Figure 15.53

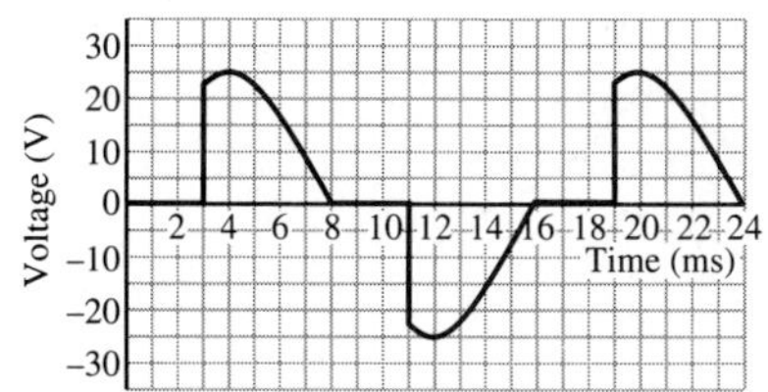

Figure 15.54

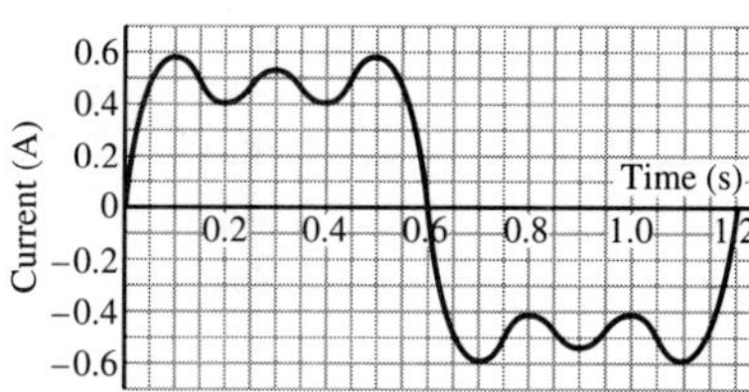

Figure 15.55

32. A semiconductor TRIAC controls the conduction in a circuit so that the voltage is as shown in Figure 15.54. Use the method of strips to find its effective value over one cycle.

33. Use the method of strips to find the effective value of the current in Figure 15.55 for one cycle.

34. Use the method of strips to find the effective value of the voltage in Figure 15.56.

35. Tell whether the waves in Figure 15.57 are made of dc only, ac only, or ac and dc.

36. Repeat Problem 35 for the waves in Figure 15.58.

37. Determine the average value and the effective value of the ac component for one cycle of the voltage in Figure 15.59.

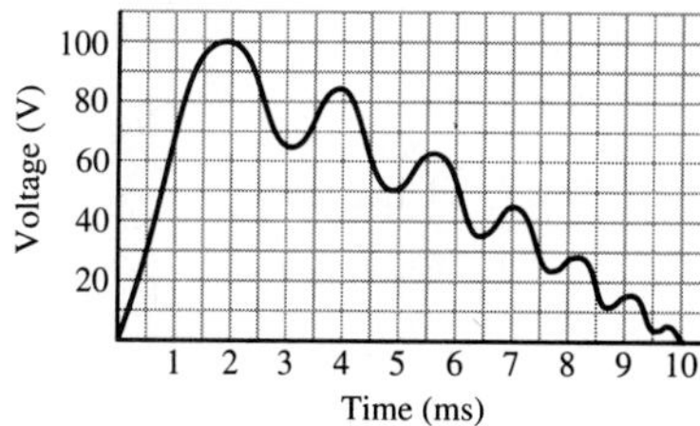

Figure 15.56

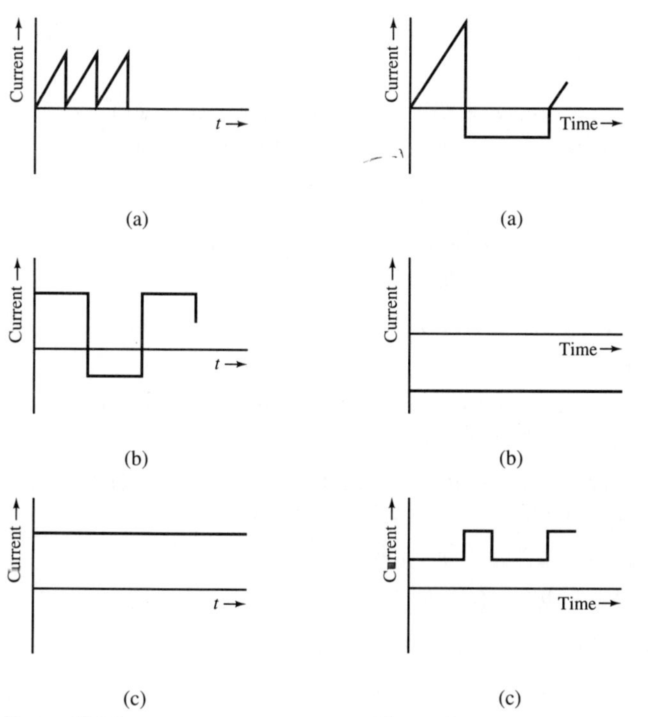

Figure 15.57

Figure 15.58

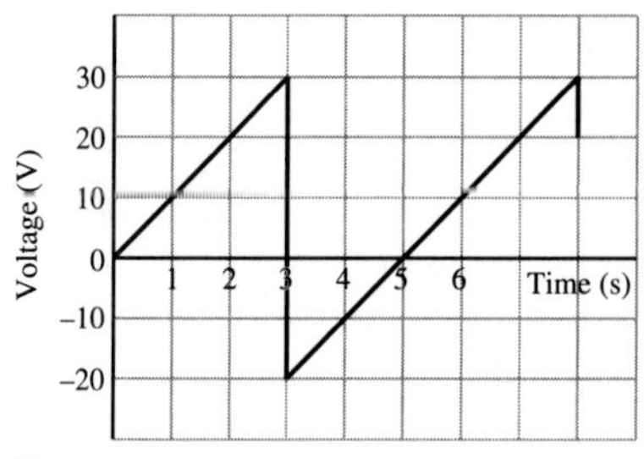

Figure 15.59

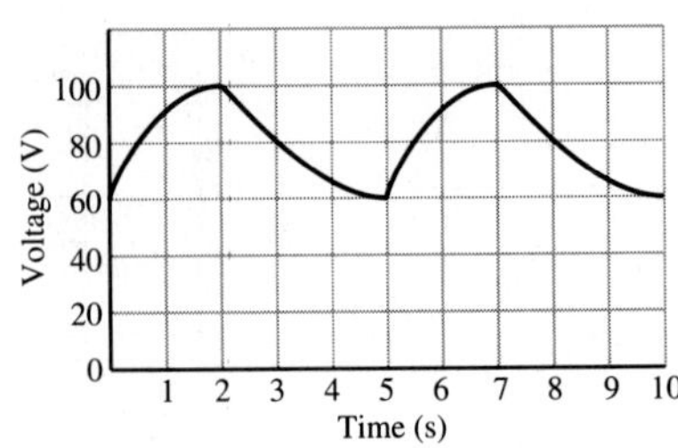

Figure 15.60

38. The output voltage of a dc source is shown in Figure 15.60. Use the method of strips to find:

    (a) The dc voltage.

    (b) The effective value.

    (c) The effective value of the ac voltage for one cycle.

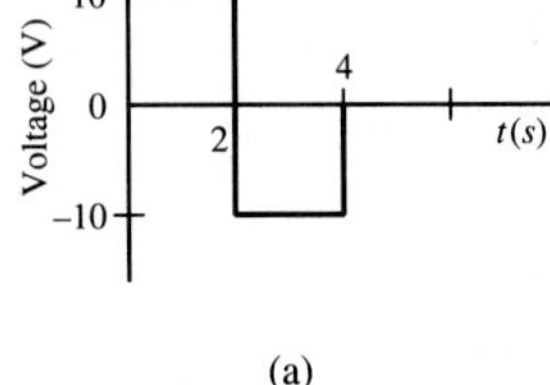

(a)

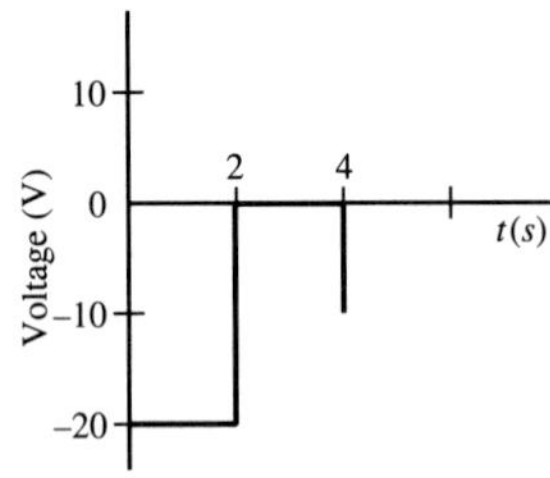

(b)

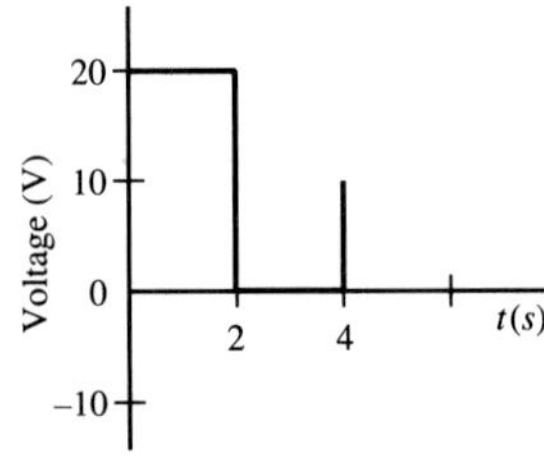

(c)

Figure 15.61

39. Shown in Figure 15.61 are three voltages, each the same shape but shifted on the vertical axis. Use the method of known shapes to find:

    (a) The dc value.

    (b) The effective value.

    (c) The effective value of the ac part for one cycle.

    Compare the results.

SECTION 15.6 Phasors

40. Draw the phasors for the following sine waves. (Use the amplitude for their magnitude.)

    (a) $\upsilon = 20 \sin(\omega t + 30°)$ (c) $\upsilon = 10 \sin(\omega t - 20°)$

    (b) $i = 30 \sin(\omega t - 60°)$ (d) $i = 10 \sin(\omega t + 90°)$

41. Repeat Problem 40 for the following:

    (a) $\upsilon = 5 \sin(\omega t + 30°)$ (c) $\upsilon = 15 \sin(\omega t + 120°)$

    (b) $\upsilon = 10 \sin(\omega t - 90°)$ (d) $\upsilon = 20 \sin(\omega t - 150°)$

42. Write equations (a)–(d) of Problem 40 in phasor form using the effective value of the waveform.

43. Write equations (a)–(d) of Problem 41 in phasor form using the effective value of the waveforms.

44. Write the sinusoidal equations for the following phasors using a frequency of 60 Hz. The phasor values are effective values.

    (a) $5 \angle 60°$ (c) $8 \angle 120°$

    (b) $20 \angle -30°$ (d) $16 \angle 0°$

45. Repeat Problem 44 for the following phasors:

    (a) $12 \angle -30°$ (c) $6 \angle 0°$

    (b) $120 \angle 135°$ (d) $15 \angle 200°$

46. Draw the phasor for the following:

    (a) $5 \angle 30°$ (b) $10 \angle 240°$ (c) $10 \angle -120°$

47. Draw the phasor for the following:

    (a) $16 \angle -30°$ (b) $10 \angle 150°$ (c) $20 \angle 220°$

48. Write the polar form for each phasor in Figure 15.62.

49. Write the polar form for each phasor in Figure 15.63.

50. Write the rectangular form for each of the phasors in Figure 15.64.

51. Write the rectangular form for each of the phasors in Figure 15.65.

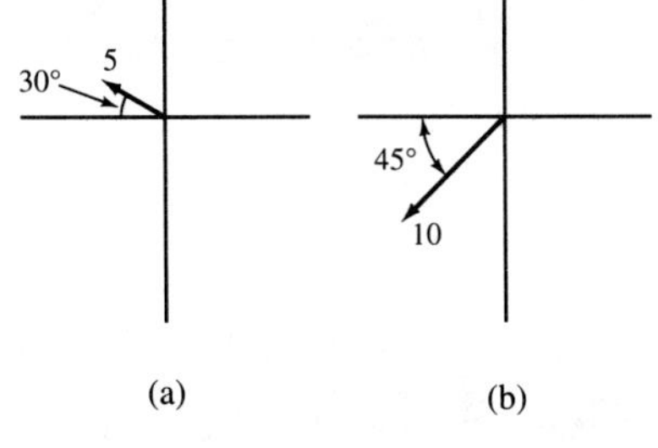

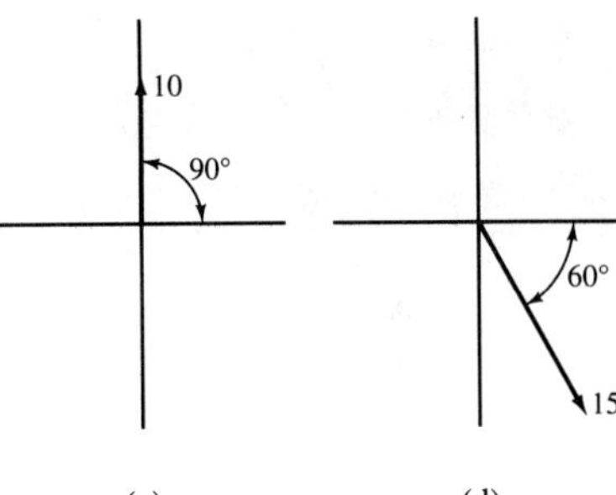

Figure 15.62

SECTION 15.7 Resistance, Capacitance, and Inductance in ac

52. A 20-ohm resistor is connected to a source whose voltage is $20 \sin 377t$.

    (a) Write the equation for the current.

    (b) What is the effective value of the current?

53. What is the equation for the voltage across a 50-ohm resistor if the current is $0.5 \sin 754t$?

54. A 5-microfarad capacitor is connected across a voltage given by $170 \sin 377t$.

    (a) Find the capacitive reactance.

    (b) Write the equation for the current.

55. Find the reactance of a 5-microfarad capacitor to 60 Hz, 500 Hz, 3 kHz, and 1 MHz.

56. A capacitor has a reactance of 30 ohms at 300 Hz. What is its reactance at 600 Hz?

57. The current in a capacitor is $0.5 \sin 754t$. If the reactance is 50 ohms, what is the equation for the voltage across the capacitor?

58. A 3-henry inductor is connected across a voltage given by $170 \sin 754t$.

    (a) Find the inductive reactance.

    (b) Write the equation for the current.

59. An inductance has an inductive reactance of 200 ohms at 300 Hz. What is the frequency when its reactance is 300 ohms?

60. Find the reactance of a 3-henry inductor at 60 Hz, 500 Hz, 3 kHz, and 1 MHz.

61. The current in an inductor is $0.6 \sin 377t$. If its reactance is 30 ohms, what is the equation for the voltage across the inductor?

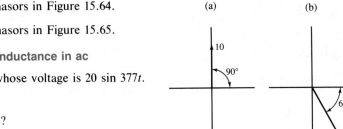

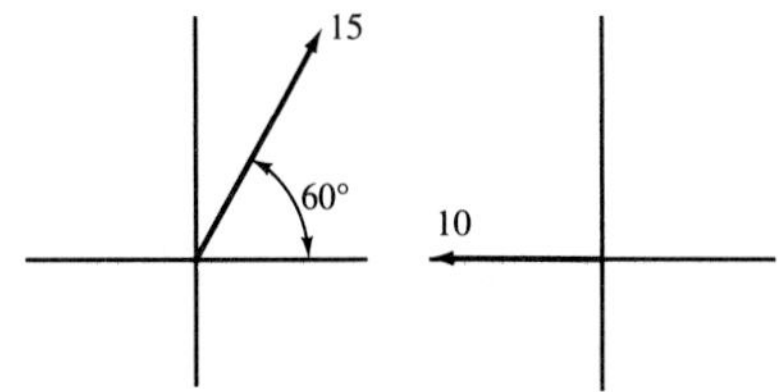

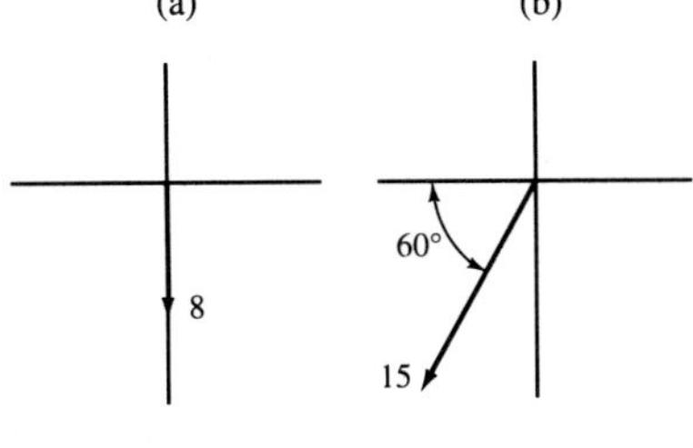

Figure 15.63

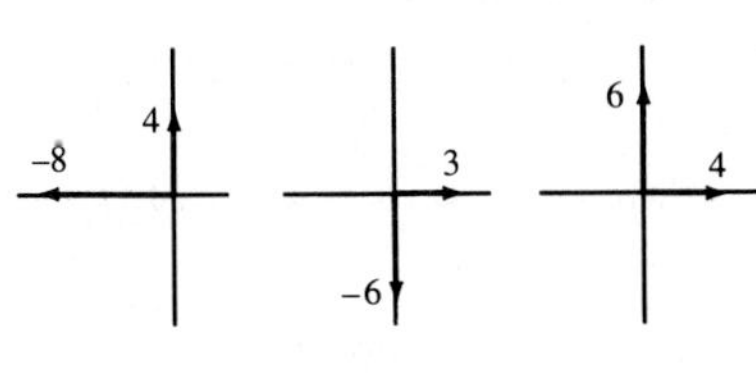

Figure 15.64

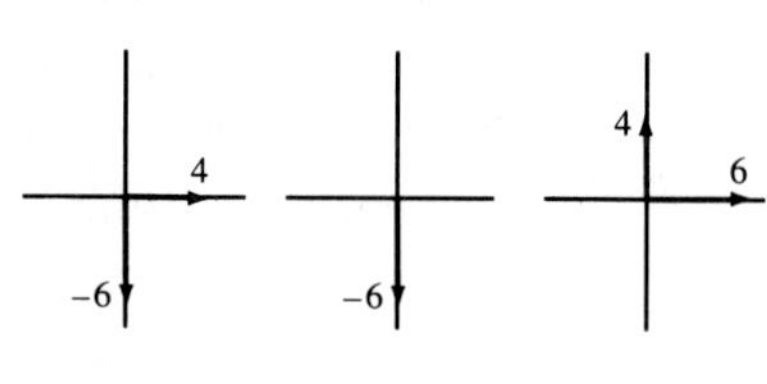

Figure 15.65

CHAPTER 16

ALTERNATING CURRENT (ac) CIRCUITS WITH A SINGLE FREQUENCY SOURCE

CHAPTER OBJECTIVES

After completing this chapter, you should be able to:

1. Determine whether parts of an ac circuit are connected in series, parallel, or series-parallel.
2. Define impedance and calculate the equivalent impedance of series, parallel, and series-parallel circuits.
3. Draw the phasor diagram for the series, parallel, and series-parallel circuit.

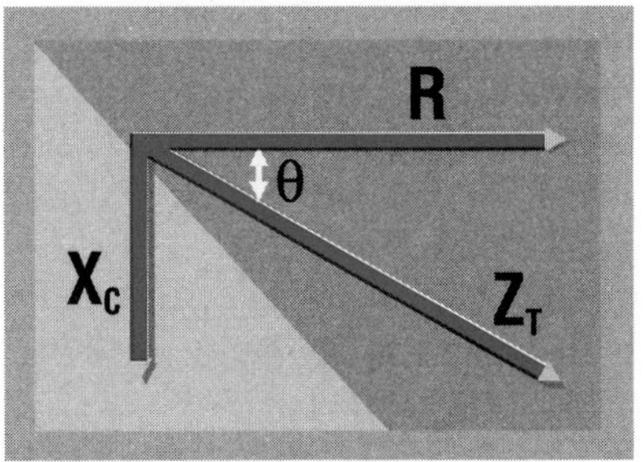

4. Effectively apply Kirchhoff's Voltage Law and the voltage divider rule to series circuits.
5. Effectively apply Kirchhoff's Current Law and the current divider rule to parallel circuits.
6. Construct the phasor diagram for the series, parallel, and series-parallel circuits from the calculated voltages and currents.
7. Calculate the power and power factor of series, parallel, and series-parallel circuits.
8. From the equivalent impedance or current and voltage phasors, determine whether a circuit has a leading or lagging power factor.
9. Relate the power and energy in parts of a circuit to the power and energy supplied by the source.
10. Perform the analysis and synthesis of series, parallel, and basic series-parallel circuits.

KEY TERMS

1. admittance
2. analysis
3. average power
4. conductance
5. current divider rule
6. equivalent impedance
7. impedance
8. impedance triangle
9. Kirchhoff's Current Law
10. Kirchhoff's Voltage Law

11. lagging power factor
12. leading power factor
13. parallel ac circuit
14. phase angle
15. phasor diagram
16. power factor
17. reactance
18. series ac circuit
19. series-parallel ac circuits
20. siemens
21. susceptance
22. synthesis
23. voltage divider rule

INTRODUCTION

Before one can understand the operation of complex electronic and electrical circuits, one must understand the basic circuits. Whether you will be designing, analyzing, or troubleshooting, an understanding of these concepts is important.

Most of the concepts and relations used for ac circuit analysis are similar to those used for dc circuits. Kirchhoff's Voltage and Current Laws, equivalent impedance, and Ohm's Law are just a few of the similar concepts. The major difference between ac and dc circuits is in the use of phasors. Phasors are necessary in ac circuits because the voltage and current are time-varying and can also be out of phase.

Surprisingly, current and voltage in some components of an ac circuit can be greater than the source current and voltage. Also, the power in a circuit can be zero, even though there is current and voltage. Again, this is because of the phase relation.

This chapter first examines the basic series and parallel circuits; then it presents the more complex series-parallel circuit. For success in future electronic and polyphase courses, as well as your work as technicians, the concepts in this chapter must be understood and mastered.

To make the calculations in this chapter, the rules for phasor arithmetic must be understood. They are presented in Appendix E; you should review them before proceeding with this chapter.

16.1 Series Circuit Characteristics

series ac circuit

A **series ac circuit** connection is a group of components connected so that there are no branch points between any of them, as in Figure 16.1. The components can be resistors, capacitors, inductors, or any combination of these components. Because there are no branch points, the current must be the same in each part of the circuit. Also, a change in any part of the circuit affects the current and voltage in the other parts of the circuit. A common example of the series ac circuit is a switch connected in series with the fuse and lamp in a house. If the switch or fuse opens, the lamp will not light.

impedance

The opposition to the current in an ac circuit is known as **impedance**. Its symbol is Z, and its unit is the ohm. Impedance corresponds to resistance in the dc circuit. In the ac circuit, impedance can be resistance, reactance, or both. **Reactance** is the opposition of inductance or

reactance

capacitance to an ac current. There is 1 ohm of impedance when an effective value of the voltage of 1 volt results in an effective value of 1 ampere of current. Impedance is a phasor quantity and has magnitude and direction. The resistance phasor R is at 0°, and the current is in phase with the voltage. The inductive reactance phasor X_L is at 90° or $+j$, and the current lags the voltage by 90°. The capacitive reactance phasor X_C is at $-90°$ or $-j$, and the current leads the voltage by 90°. These phasors are written as R, $X_L \angle 90°$ or jX_L, and $X_C \angle -90°$ or $-jX_C$. The angle for any combination depends on the resistance and reactance in the combination.

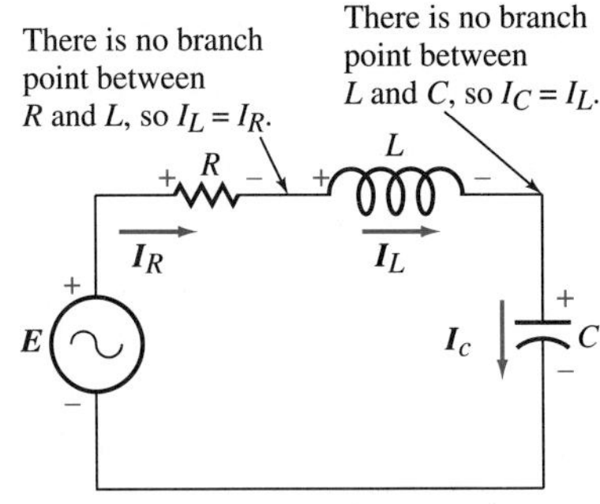

Figure 16.1 The series connection does not have any branch points between elements; therefore, the current is the same in all parts of the circuit, $I_R = I_L = I_C$.

An instrument that can measure ac impedance, phase angle, and the Q of a component, resistance, inductance, and capacitance is the impedance bridge. Many bridges can measure these quantities at different frequencies. Some bridges require adjusting a knob control until the bridge is balanced, as indicated by nulling a sensitive meter. Others, such as that shown in Figure 16.2, vary the control automatically and indicate the value on a digital display. The unit shown can determine a component's value in less than 100 ms.

Equivalent Impedance. Just as a series dc circuit can be replaced by an equivalent resistance, a series ac circuit can also be replaced by an **equivalent impedance**. This is an impedance whose overall characteristics are the same as those of the circuit. It offers the same opposition to the current as the original circuit.

equivalent impedance

The equivalent impedance of a series ac circuit is equal to the phasor sum of the impedance of each part of the series circuit. In general,

$$\mathbf{Z}_{eq} = \mathbf{Z}_1 + \mathbf{Z}_2 + \cdots + \mathbf{Z}_n \tag{16.1}$$

where $\mathbf{Z}_{eq}$ is the equivalent impedance, in ohms, and $\mathbf{Z}_1, \mathbf{Z}_2, \ldots, \mathbf{Z}_n$ are the impedances of each part of the circuit, in ohms.

Note that the form of Equation 16.1 is the same as that for equivalent resistance. However, the quantities are phasors. Because X_C and X_L are functions of frequency, the impedance of an ac circuit varies with frequency.

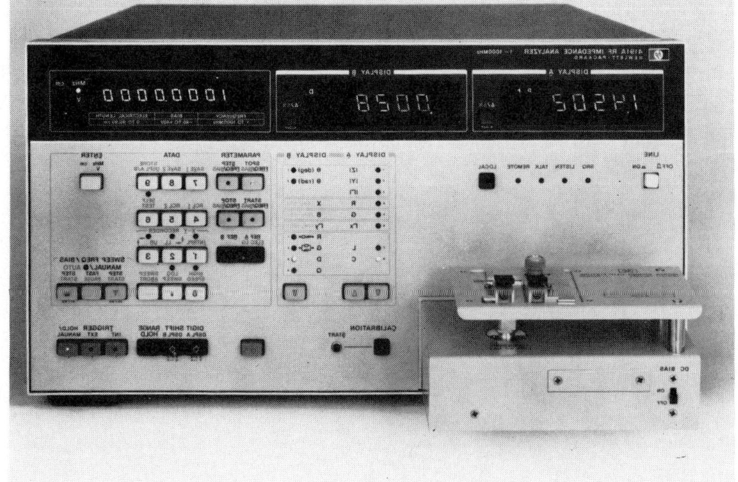

Figure 16.2 Impedance bridges such as this HP4191A can measure an impedance in less than 0.1 second. Photo Courtesy of Hewlett-Packard Company

Before continuing with equivalent impedance, the basic rules of phasor arithmetic are reviewed. Refer to Appendix E for a more detailed explanation.

a. **Addition and subtraction.** The real terms and the j terms are combined separately.

$$(5 + j2) + (3 - j6) = (5 + 3) + j(2 - 6) \quad \text{or} \quad 8 - j4$$

b. **Multiplication.** The coefficients are multiplied. The angles are added.

$$(5 \angle 30^\circ) \times (4 \angle -60^\circ) = (5 \times 4) \angle 30^\circ + (-60^\circ)$$
$$\text{or} \quad 20 \angle -30^\circ$$

c. **Division.** The divisor coefficient is divided into the dividend coefficient. The divisor angle is subtracted from the dividend angle.

$$\frac{8 \angle 30^\circ}{4 \angle -60^\circ} = \frac{8}{4} \angle 30^\circ - (-60^\circ) \quad \text{or} \quad 2 \angle 90^\circ$$

The equivalent impedance of a circuit made up of the same type of components will be greater than the largest impedance in the circuit. But if the components are different, $\mathbf{Z}_{eq}$ can be smaller, equal to, or larger than the largest impedance. The following are examples of the $\mathbf{Z}_{eq}$ of some circuits.

EXAMPLE 16.1

Find the equivalent impedance of each of the circuits in Figure 16.3.

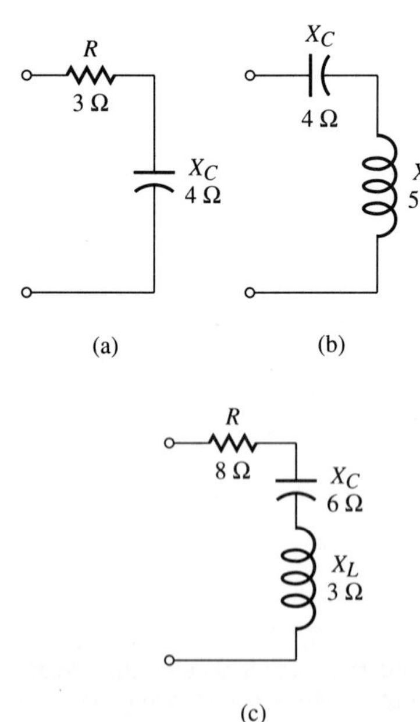

Figure 16.3

Solution

a.
$$\begin{aligned} \mathbf{Z}_{eq} &= \mathbf{Z}_R + \mathbf{Z}_C \\ &= R \angle 0^\circ - jX_C \\ &= (3 - j4)\ \Omega \\ \mathbf{Z}_{eq} &= 5 \angle -53.13^\circ\ \Omega \end{aligned}$$

b.
$$\begin{aligned} \mathbf{Z}_{eq} &= \mathbf{Z}_C + \mathbf{Z}_L \\ &= jX_C + jX_L \\ &= (-j4 + j5)\ \Omega \\ &= j1\ \Omega \\ \mathbf{Z}_{eq} &= 1 \angle 90^\circ\ \Omega \end{aligned}$$

c.
$$\begin{aligned} \mathbf{Z}_{eq} &= \mathbf{Z}_R + \mathbf{Z}_C + \mathbf{Z}_L \\ &= R \angle 0^\circ - jX_C + jX_L \\ &= (8 - j6 + j3)\ \Omega \\ \mathbf{Z}_{eq} &= (8 - j3)\ \Omega \\ \mathbf{Z}_{eq} &= 8.54 \angle -20.56^\circ \end{aligned}$$

$\mathbf{Z}_{eq}$ can be found using a hand calculator. The magnitude is $\sqrt{R^2 + X^2}$ and the angle is $\tan^{-1}(X/R)$. R is the sum of the resistances and X is the algebraic sum of the reactances $(+X_L)$, $(-X_C)$. The solution for part c of Example 16.1 is:

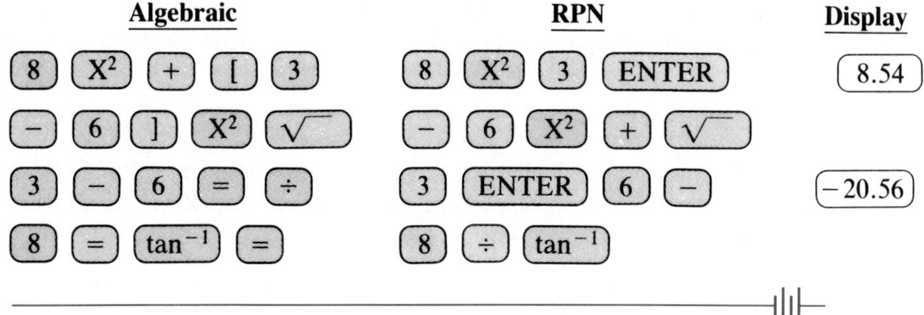

The impedances can also be added graphically. Doing this yields an impedance triangle, as shown in Figure 16.4. The **impedance triangle** is a triangle made up of resistance on the X-axis, reactance on the j-axis, and the equivalent impedance as the hypotenuse. The angle between the resistance and the impedance gives the phase angle of the circuit. The **phase angle** is the angle between the voltage across an impedance and the current in the impedance. The addition is done by using the tip-to-tail method. That is, the tail of one phasor is drawn at the tip of the other phasor until all phasors are drawn. $\mathbf{Z}_{eq}$ is the resultant, and is drawn from the tail of the first phasor to the tip of the last. Thus, the hypotenuse in Figure 16.4 is the resultant of the resistance and the reactance.

impedance triangle

phase angle

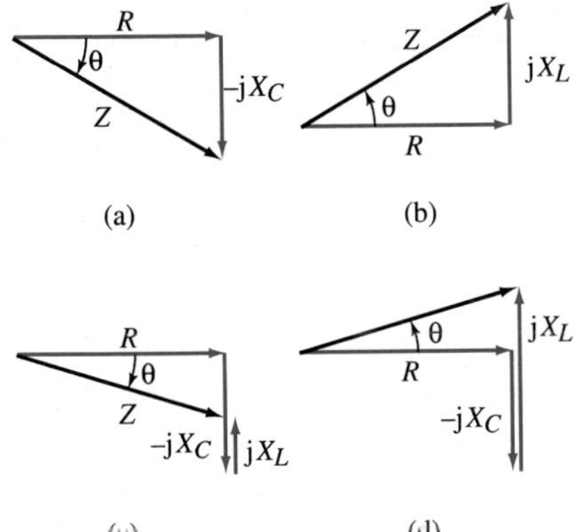

Figure 16.4 The impedance triangle for some series circuits. (a) An R–C circuit. (b) An R–L circuit. (c) An R–L–C circuit with X_C greater than X_L. (c) An R–L–C circuit with X_C smaller than X_L.

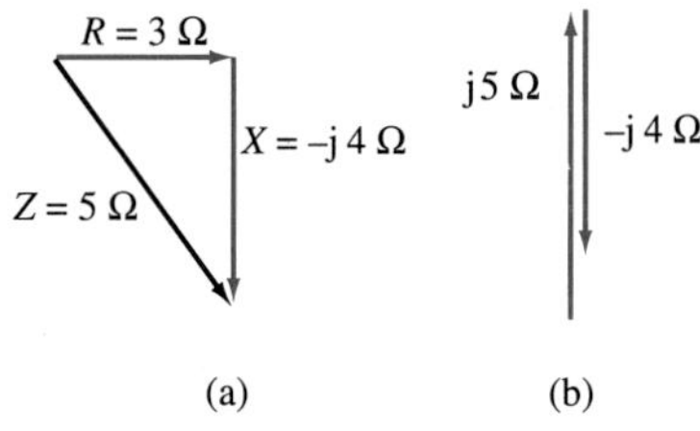

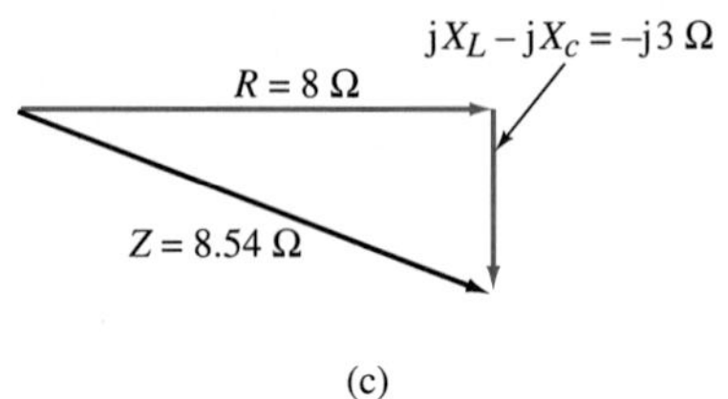

Figure 16.5

EXAMPLE 16.2

Draw the impedance triangles for the circuits of Example 16.1.

Solution

The triangles are shown in Figure 16.5.

The current in each component of the series circuit results in a voltage drop across the component. Ohm's Law also holds true for the ac circuit so that

$$\boldsymbol{V} = \boldsymbol{IZ} \qquad \textbf{(16.2)}$$

where $\boldsymbol{V}$ is the phasor voltage across the component, in volts; $\boldsymbol{I}$ is the phasor current in the component, in amperes; and $\boldsymbol{Z}$ is the phasor impedance of the component, in ohms.

EXAMPLE 16.3

For the circuit of Figure 16.6, find:

a. the voltage across each component.
b. the voltage across the combination of R and L.

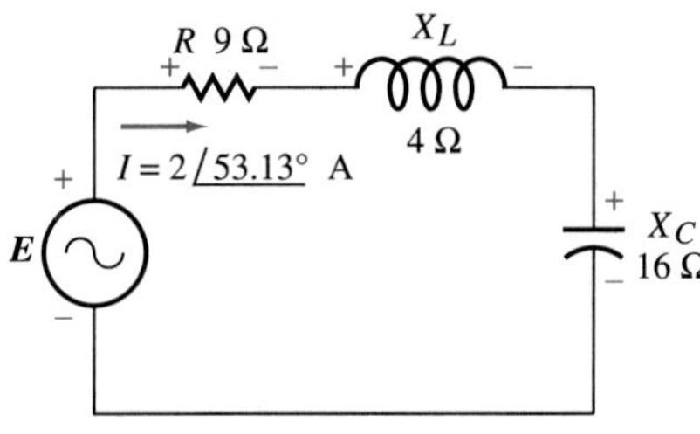

Figure 16.6

Solution

a.

$$\boldsymbol{V}_R = \boldsymbol{IZ}_R = \boldsymbol{I}R \angle 0° = (2 \angle 53.13° \text{ A})(9 \angle 0° \ \Omega)$$
$$\boldsymbol{V}_R = 18 \angle 53.13° \text{ V}$$
$$\boldsymbol{V}_L = \boldsymbol{IZ}_L = \boldsymbol{I}X_L \angle 90° = (2 \angle 53.13° \text{ A})(4 \angle 90° \ \Omega)$$
$$\boldsymbol{V}_L = 8 \angle 143.13° \text{ V}$$
$$\boldsymbol{V}_C = \boldsymbol{IZ}_C = \boldsymbol{I}X_C \angle -90° = (2 \angle 53.13° \text{ A})(16 \angle -90° \ \Omega)$$
$$\boldsymbol{V}_C = 32 \angle -36.87° \text{ V}$$

The capacitor voltage lags the current by 90° while the inductor voltage leads the current by 90°.

b.

$$\boldsymbol{V}_{R-L} = \boldsymbol{IZ}_{R-L}$$
$$\boldsymbol{Z}_{R-L} = \boldsymbol{R} + jX_L = (9 + j4) \ \Omega \quad \text{or} \quad 9.85 \angle 23.96° \ \Omega$$
$$\boldsymbol{V}_{R-L} = (2 \angle 53.13° \text{ A})(9.85 \angle 23.96° \ \Omega)$$
$$\boldsymbol{V}_{R-L} = 19.7 \angle 77.09° \text{ V}$$

Before continuing, let us consider the (+) and (−) markings on the terminals of an ac source, as in Figure 16.6. That polarity and the polarities shown on all terminals exist for only one-half of a cycle. When the source polarity changes, the current direction and all element polarities will also change. This concept of polarity relation is used to obtain the correct relative polarities and current directions.

Kirchhoff's Voltage Law. The phasor sum of the voltages around a closed loop must equal zero. This relation was first introduced in dc circuits, and is known as **Kirchhoff's Voltage Law**. In general, for a series circuit with one source,

Kirchhoff's Voltage Law

$$\mathbf{E} - \mathbf{V}_1 - \mathbf{V}_2 - \cdots - \mathbf{V}_n = 0 \qquad \textbf{(16.3)}$$

where $\mathbf{E}$ is the source voltage, in volts, and $\mathbf{V}_1, \mathbf{V}_2, \ldots \mathbf{V}_n$ are the voltages across each part of the circuit.

Equation 16.3 can also be used for parts of a series circuit if E is replaced by the voltage across the part of the circuit.

EXAMPLE 16.4

In the circuit of Figure 16.7, use Equation 16.3 to find:

a. the source voltage.
b. the voltage $\mathbf{V}_x$.

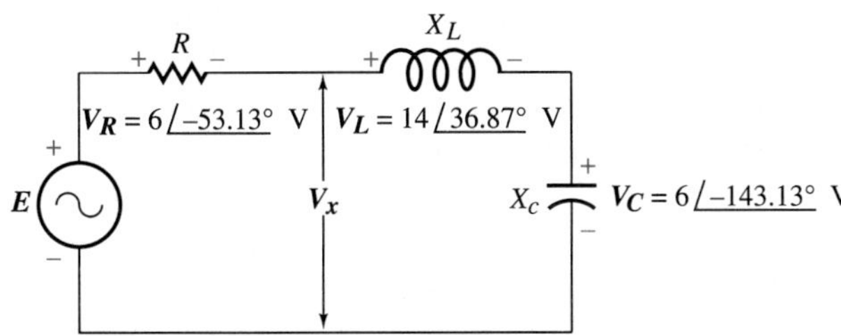

Figure 16.7

Solution

a.

$$\mathbf{E} - \mathbf{V}_R - \mathbf{V}_L - \mathbf{V}_C = 0$$

so

$$\begin{aligned}\mathbf{E} &= 6\angle{-53.13°}\text{ V} + 14\angle 36.87°\text{ V} + 6\angle{-143.13°}\text{ V}\\ &= (3.6 - j4.8 + 11.2 + j8.4 - 4.8 - j3.6)\text{ V}\\ &= (10 + j0)\text{ V}\\ \mathbf{E} &= 10\angle 0°\text{ V}\end{aligned}$$

b.

$$\begin{aligned}\mathbf{V}_x &= \mathbf{V}_L + \mathbf{V}_C\\ &= 14\angle 36.87°\text{ V} + 6\angle{-143.13°}\text{ V}\\ &= (11.2 + j8.4 - 4.8 - j3.6)\text{ V}\\ &= (6.4 + j4.8)\text{ V}\\ \mathbf{V}_x &= 8\angle 36.87°\text{ V}\end{aligned}$$

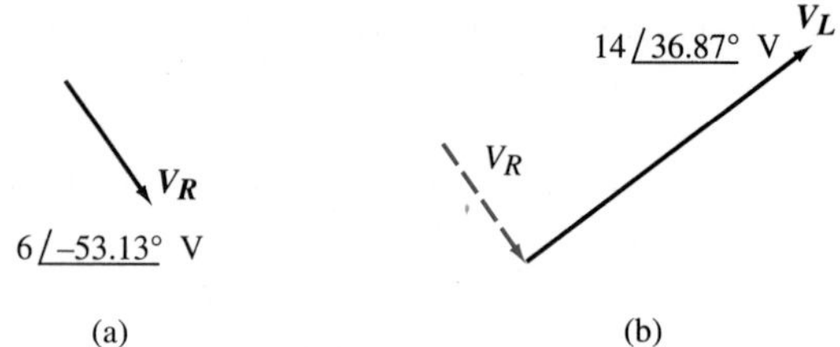

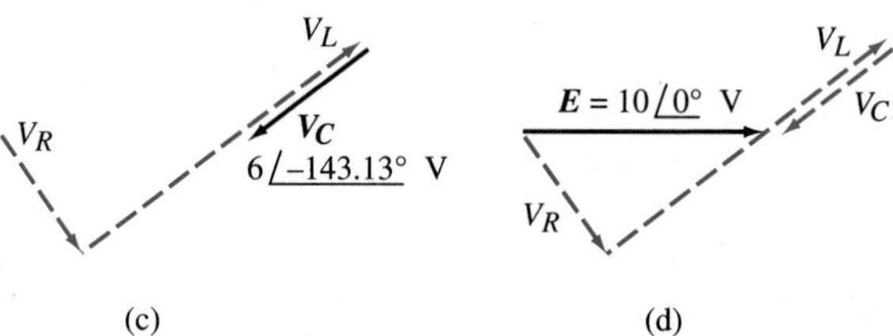

Figure 16.8 The steps in the graphic addition of phasor voltages for Example 16.4. (a) Resistor voltage. (b) Resistor voltage plus the inductor voltage. (c) Resistor voltage, plus inductor voltage, plus capacitor voltage. (d) The source voltage is the graphic sum of the *R*, *L*, and *C* voltages.

The addition of the phasors can also be done graphically. Although the result is not as exact, it provides a quick check of an answer. The graphical addition uses the tip-to-tail method, as shown in Figure 16.8. The tail of $\mathbf{V}_L$ is drawn at the tip of $\mathbf{V}_R$. Then the tail of $\mathbf{V}_C$ is drawn at the tip of $\mathbf{V}_L$. $\mathbf{E}$ is the resultant. It is drawn from the tail of the first phasor to the tip of the last. The example used the $\mathbf{V}_R$, $\mathbf{V}_L$, $\mathbf{V}_C$ sequence, but they can be drawn in any order.

voltage divider rule

Voltage Divider. The **voltage divider rule** states that the voltage across a part of a series circuit is equal to the voltage across the series circuit, multiplied by the impedance of the part, and divided by the equivalent impedance of the series circuit. In general, for a circuit such as that in Figure 16.9(a),

$$\mathbf{V}_x = \mathbf{E}(\mathbf{Z}_x/\mathbf{Z}_{eq}) \qquad \textbf{(16.4)}$$

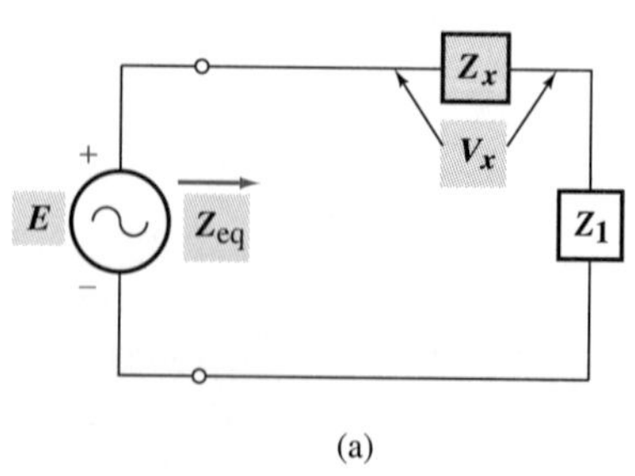

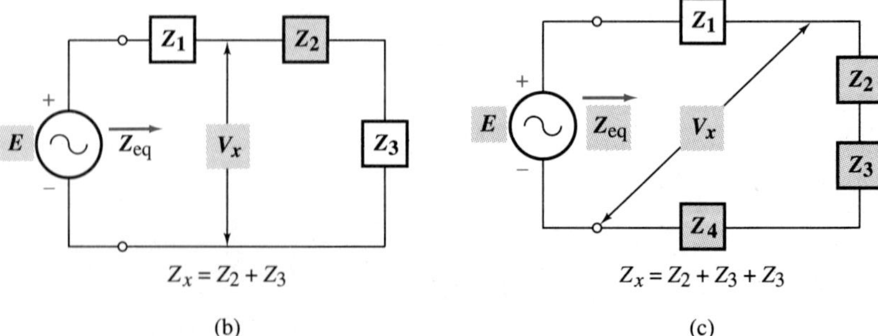

Figure 16.9 The relation of circuit components in some series circuits to the quantities used in the voltage divider equation.

where $\mathbf{V}_x$ is the voltage across $\mathbf{Z}_x$, in volts; $\mathbf{E}$ is the source voltage, in volts; and $\mathbf{Z}_{eq}$ is the impedance of the circuit, in ohms.

Figure 16.9(b) and (c) show how this relation is applied to some other circuits. In each case, the impedance $\mathbf{Z}_x$ is the impedance of the part for which the voltage is being found. The equivalent impedance is for the series circuit across which the total voltage is known. The advantage of using Equation 16.4 is that the current does not have to be calculated to obtain the voltage. This also holds true for dc circuits.

EXAMPLE 16.5

Use Equation 16.4 to find the voltage across X_L in Figure 16.10.

Solution

$$\mathbf{V}_L = \frac{\mathbf{E}\mathbf{Z}_L}{\mathbf{Z}_R + \mathbf{Z}_L}$$

$$\mathbf{V}_L = \frac{\mathbf{E}X_L \angle 90^\circ}{R + jX_L}$$

$$= \frac{(120 \angle 0^\circ \text{ V})(8.66 \angle 90^\circ \ \Omega)}{(5 + j8.66)\ \Omega}$$

$$= \frac{1039.2 \angle 90^\circ \text{ V} \cdot \Omega}{10 \angle 60^\circ \ \Omega}$$

$$\mathbf{V}_L = 103.92 \angle 30^\circ \text{ V}$$

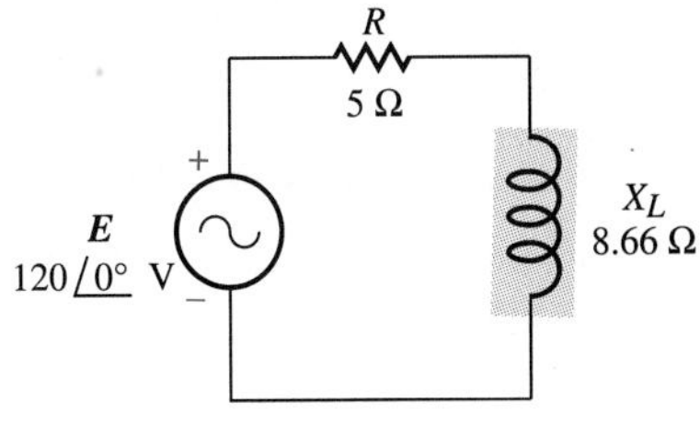

Figure 16.10

EXAMPLE 16.6

Use Equation 16.4 to find the voltage $\mathbf{V}$ in the circuit of Figure 16.11.

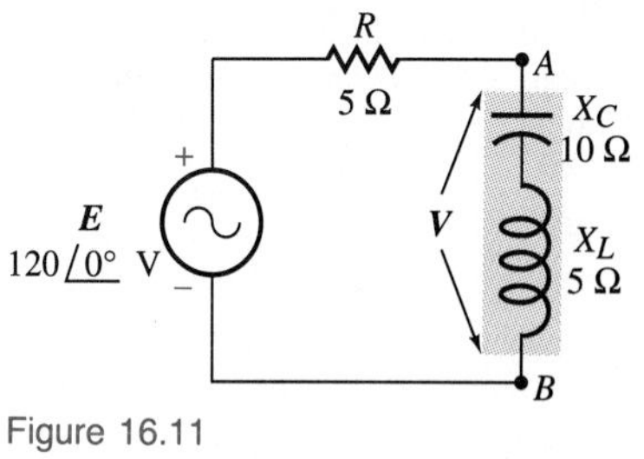

Solution

The impedance between Points A and B is

$$\mathbf{Z}_{AB} = \mathbf{Z}_C + \mathbf{Z}_L$$

$$\mathbf{Z}_{AB} = -jX_C + jX_L$$

$$= (-j10 + j5)\ \Omega$$

$$= -j5\ \Omega$$

Then

$$\mathbf{V} = \mathbf{E}\frac{\mathbf{Z}_{AB}}{\mathbf{Z}_{eq}} = \frac{(120 \angle 0^\circ \text{ V})(-j5\ \Omega)}{(5 - j10 + j5)\ \Omega}$$

$$= \frac{600 \angle -90^\circ \text{ V} \cdot \Omega}{7.07 \angle -45^\circ \ \Omega}$$

$$\mathbf{V} = 84.87 \angle -45^\circ \text{ V}$$

Figure 16.11

An open circuit in an ac circuit has the same effect as it has in a dc circuit: the current is 0 amperes and the source voltage is across the terminals. However, the effect of a short circuit depends on what part is short-circuited. Short-circuiting a resistor causes an increase in current, but short-circuiting a capacitor or inductor either increases or decreases the current. Why? Because the short-circuited reactance could have been canceling some reactance of the other component. For example, suppose $Z = 4 + j6 - j3$, or $5 \angle 36.87°$ ohms. If the 3-ohm capacitor is shorted, $Z = 4 + j6$, or $7.2 \angle 56.3°$ ohms. The larger impedance results in a lower current. Short-circuiting the inductor gives no change in current, but gives a change in the phase angle.

16.2 Parallel Circuit Characteristics

parallel ac circuit

A **parallel ac circuit** is a group of components connected across the same two electrical points, as shown in Figure 16.12. The components can be resistors, inductors, capacitors, or combinations of them.

The most common example of a parallel connection is the distribution system of electric utilities. The lamps, motors, and other loads on the system are connected in parallel across the two lines. In a parallel circuit with no resistance in the connecting lines, a change in one parallel component does not affect the other components, as long as the source voltage does not change. However, in the real-life circuit, there is a slight effect on the branches. The effect is the result of the resistance in the lines. An extreme change, such as a short circuit in one branch, results in nearly 0 volts across the other branches.

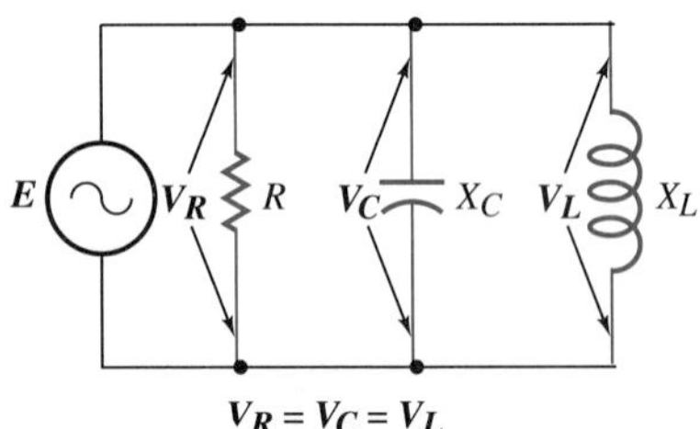

$V_R = V_C = V_L$

Figure 16.12 The parallel connection has the branches connected across the same pair of points; therefore, the voltage is the same across each branch.

A parallel ac circuit also can be replaced by an equivalent impedance. This impedance offers the same opposition to the current and has the same electrical characteristics as the original circuit. The reciprocal of the equivalent impedance is equal to the sum of the reciprocals of the impedance in each branch, which is the same as the equivalent resistance in a parallel dc circuit. In general,

$$\frac{1}{\mathbf{Z}_{eq}} = \frac{1}{\mathbf{Z}_1} + \frac{1}{\mathbf{Z}_2} + \cdots + \frac{1}{\mathbf{Z}_n} \qquad \textbf{(16.5)}$$

where $\mathbf{Z}_{eq}$ is the equivalent impedance of a parallel circuit, in ohms, and $\mathbf{Z}_1, \mathbf{Z}_2, \ldots \mathbf{Z}_n$ are the impedances in the branches, in ohms. When there are only two branches, Equation 16.5 simplifies to:

$$\mathbf{Z}_{eq} = \frac{\mathbf{Z}_1\mathbf{Z}_2}{\mathbf{Z}_1 + \mathbf{Z}_2} \qquad \textbf{(16.6)}$$

EXAMPLE 16.7

A 25-ohm resistor, a 10-ohm capacitive reactance, and a 20-ohm inductive reactance are connected in parallel. What is the equivalent impedance of the group?

Solution

$$\frac{1}{\mathbf{Z}_{\text{eq}}} = \frac{1}{\mathbf{Z}_R} + \frac{1}{\mathbf{Z}_C} + \frac{1}{\mathbf{Z}_L}$$

$$\frac{1}{\mathbf{Z}_{\text{eq}}} = \frac{1}{R} + \frac{1}{X_C \angle -90^\circ} + \frac{1}{X_L \angle 90^\circ}$$

$$= \frac{1}{25\ \Omega} + \frac{1}{10 \angle -90^\circ\ \Omega} + \frac{1}{20 \angle 90^\circ\ \Omega}$$

$$= (0.04 + 0.1 \angle 90^\circ + 0.05 \angle 90^\circ)\ \frac{1}{\Omega}$$

$$= (0.04 + j0.1 - j0.05)\ \frac{1}{\Omega}$$

$$= (0.04 + j0.05)\ \frac{1}{\Omega} \quad \text{or} \quad 0.064 \angle 51.34^\circ\ \frac{1}{\Omega}$$

$$\mathbf{Z}_{\text{eq}} = \frac{1}{0.064 \angle 51.34^\circ}\ \Omega$$

$$\mathbf{Z}_{\text{eq}} = 15.63 \angle -51.34^\circ\ \Omega \quad \text{or} \quad (9.76 - j12.20)\ \Omega$$

An inspection of $\mathbf{Z}_{\text{eq}}$ in Example 16.7 shows that it consists of a resistance and a reactance in series. Thus, $\mathbf{Z}_{\text{eq}}$ converts the parallel circuit into a series equivalent. The parallel circuit for Example 16.7 and its series equivalent is shown in Figure 16.13(a).

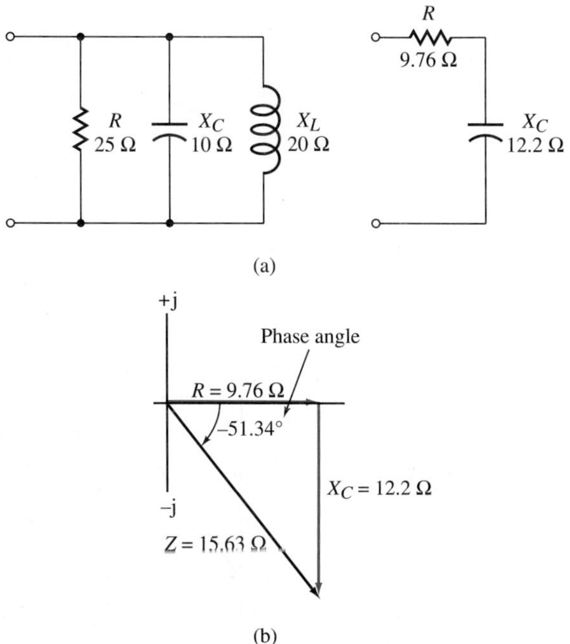

Figure 16.13 A parallel circuit can be replaced by a series equivalent. (a) The parallel circuit and its equivalent. (b) The impedance triangle for the series equivalent.

The impedance triangle for the parallel circuit is the triangle of its series equivalent. The triangle for the example is shown in Figure 16.13(b). Again, the phase angle is the angle between the resistance part and the equivalent impedance. It is important to recognize that this equivalent impedance changes with frequency because X_C and X_L are frequency-dependent.

admittance

Admittance. An alternate method for parallel circuit analysis is to use admittance instead of impedance. **Admittance** is an indicator of how well a circuit admits the ac current. Admittance (Y) is the reciprocal of impedance. Thus, if $\mathbf{Z} = 5\angle 45°$, $\mathbf{Y}$ will equal $1/\mathbf{Z}$ or $0.2\angle -45°$. The unit of admittance is the **siemens** (S), named in honor of the German scientist Ernst Werner Von Siemens (1816–1892).

siemens

conductance
susceptance

Admittance can also have two parts, conductance (G) and susceptance (B). **Conductance** is the reciprocal of resistance, and **susceptance** is the reciprocal of reactance. The unit for these is also the siemens.

The capacitive susceptance has a plus j, and the inductive susceptance has a minus j—the opposite of the capacitive and inductive reactance. This results because $\mathbf{Y} = 1/\mathbf{Z}$, so the angle for $\mathbf{Y}$ will be the negative of the angle for $\mathbf{Z}$. Thus, if $\mathbf{Z} = R + jX$, $\mathbf{Y}$ will equal $G - jB$.

Some admittance triangles are shown in Figure 16.14. Conductance is

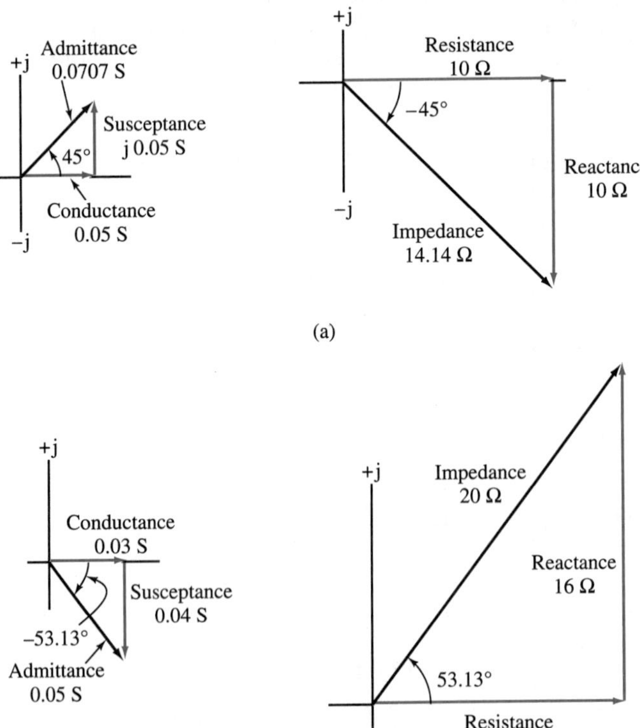

Figure 16.14 The admittance triangle consists of susceptance, conductance, and admittance. (a) The admittance and impedance triangles for a parallel R–C circuit. (b) The admittance and impedance triangles for a parallel R–L circuit.

on the real axis just as resistance, and susceptance is on the j axis. The admittance is the hypotenuse of the triangle. The phase angle of the circuit is the negative of the angle between the conductance and the admittance. Figure 16.14(a) is for a capacitive impedance while Figure 16.14(b) is for an inductive impedance.

EXAMPLE 16.8

Find the admittance, conductance, and susceptance of the following impedances:

a. $50 \angle 30° \ \Omega$.
b. $10 \angle -60° \ \Omega$.
c. $5 \angle 0° \ \Omega$.
d. $2 \angle -90° \ \Omega$.

Solution

a. $$\boldsymbol{Y} = \frac{1}{\boldsymbol{Z}} = \frac{1}{50 \angle 30° \ \Omega} = 0.02 \angle -30° \text{ S}$$

$$Y = (0.017 - j0.01) \text{ S}$$

The conductance is 0.017 S.
The susceptance is 0.01 S, inductive.

b. $$\boldsymbol{Y} = \frac{1}{\boldsymbol{Z}} = \frac{1}{10 \angle -60° \ \Omega} = 0.1 \angle 60° \text{ S}$$

$$Y = (0.05 + j0.09) \text{ S}$$

The conductance is 0.05 S.
The susceptance is 0.09 S, capacitive.

c. $$\boldsymbol{Y} = \frac{1}{\boldsymbol{Z}} = \frac{1}{5 \angle 0° \ \Omega}$$

$$\boldsymbol{Y} = 0.2 \text{ S}$$

The conductance is 0.2 S.
Since there is no j term, the susceptance is 0 S.

d. $$\boldsymbol{Y} = \frac{1}{\boldsymbol{Z}} = \frac{1}{2 \angle -90° \ \Omega} = 0.5 \angle 90° \text{ S}$$

$$\boldsymbol{Y} = (0 + j0.5) \text{ S}$$

The conductance is 0 S.
The susceptance is 0.5 S, capacitive.
Substituting admittance into Equation 16.5 gives

$$\boldsymbol{Y}_{eq} = \boldsymbol{Y}_1 + \boldsymbol{Y}_2 + \cdots + \boldsymbol{Y}_n \qquad \textbf{(16.7)}$$

where: $\boldsymbol{Y}_{eq}$ is the equivalent admittance of the circuit in siemens, and $\boldsymbol{Y}_1$, $\boldsymbol{Y}_2$, ..., $\boldsymbol{Y}_n$ are the admittances of each branch, in siemens.

Also, putting admittance into the Ohm's Law relation $V = IZ$ gives

$$V = \frac{I}{Y} \tag{16.8}$$

where V is the voltage across a part of a circuit, in volts; I is the current, in amperes; and Y is the admittance of the circuit, in siemens.

Having learned the meaning of the terms and their relation to impedance, resistance, and reactance, the following examples show how they are used in circuit analysis.

EXAMPLE 16.9

Find the equivalent admittance and the current in the circuit of Figure 16.15. Construct the admittance diagram and admittance triangle.

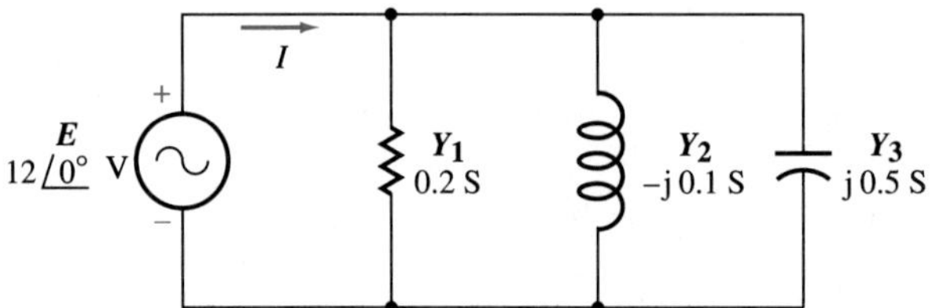

Figure 16.15

Solution

$$\begin{aligned} Y_{eq} &= Y_1 + Y_2 + Y_3 \\ &= (0.2 - j0.1 + j0.5)\text{ S} \\ &= (0.2 + j0.4)\text{ S} \\ Y_{eq} &= 0.45\ \angle 63.43^\circ\text{ S} \end{aligned}$$

$$\begin{aligned} I &= EY_{eq} \\ &= (12\ \angle 0^\circ\text{ V})(0.45\ \angle 63.43^\circ\text{ S}) \\ I &= 5.4\ \angle 63.43^\circ\text{ A} \end{aligned}$$

The admittance diagram and admittance triangle are shown in Figure 16.16.

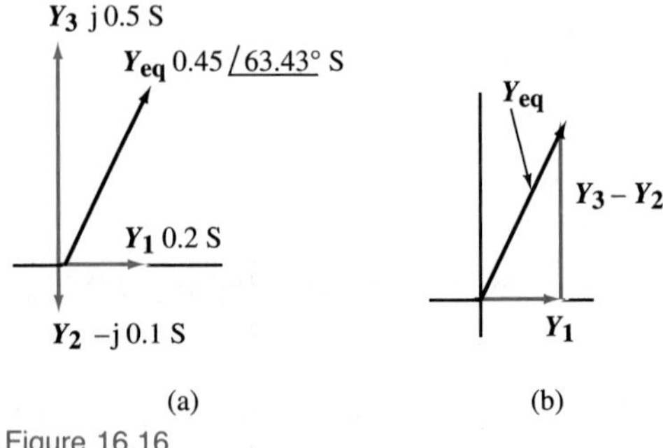

Figure 16.16

EXAMPLE 16.10

Find the source current in the circuit in Figure 16.17.

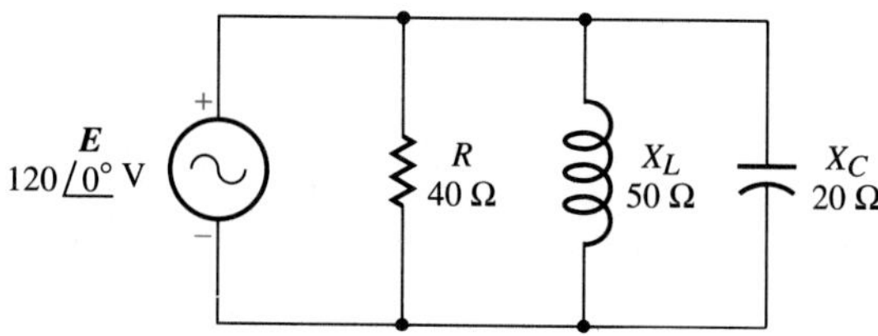

Figure 16.17

Solution

$$I = EY_{eq}$$
$$Y_{eq} = Y_R + Y_L + Y_C$$
$$Y_R = \frac{1}{Z_R} = \frac{1}{40\ \Omega} = 0.025\ \text{S}$$
$$Y_L = \frac{1}{Z_L} = \frac{1}{50\ \angle 90°\ \Omega} = 0.02\ \angle -90°\ \text{S}$$
$$Y_C = \frac{1}{Z_C} = \frac{1}{20\ \angle -90°\ \Omega} = 0.05\ \angle 90°\ \text{S}$$

Then

$$Y_{eq} = (0.025 - j0.02 + j0.05)\ \text{S}$$
$$Y_{eq} = (0.025 + j0.03)\ \text{S}$$

or

$$Y_{eq} = (0.039\ \angle 50.19°)\ \text{S}$$
$$I = (120\ \angle 0°\ \text{V})(0.039\ \angle 50.19°\ \text{S})$$
$$I = 4.68\ \angle 50.19°\ \text{A}$$

Kirchhoff's Current Law. In a parallel circuit, the voltage must be the same across each branch. Because of this, a branch with a small impedance has a larger current than a branch with a large impedance.

Since the current in the branches must come from the line to which they are connected, the phasor sum of the current must be equal to zero. This is **Kirchhoff's Current Law** as applied to the ac circuit. In general,

Kirchhoff's Current Law

$$I_s - I_1 - I_2 - \cdots - I_n = 0 \qquad \textbf{(16.9)}$$

where I_s is the current in the line entering the branches, in any unit of current, and $I_1, I_2, \ldots, I_n$ are the currents in the branches, in the same unit as I_s.

EXAMPLE 16.11

Use Kirchhoff's Current Law to find the source current in Figure 16.18.

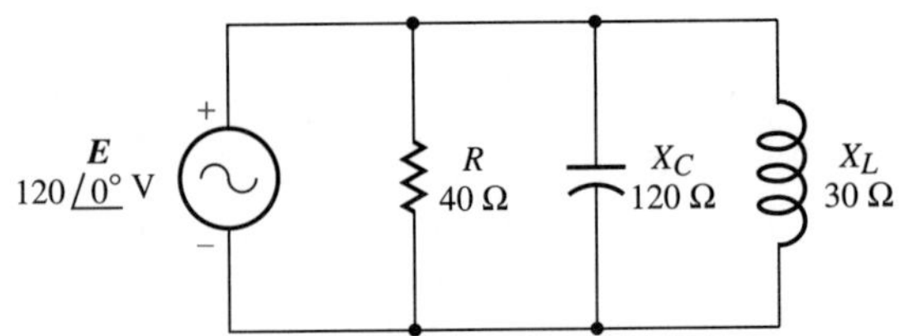

Figure 16.18

Solution

$$I_s - I_R - I_C - I_L = 0$$

Using Ohm's Law,

$$I_R = \frac{E}{Z_R} = \frac{E}{R\ \angle 0^\circ} = \frac{120\ \angle 0^\circ\ \text{V}}{40\ \angle 0^\circ\ \Omega} = 3\ \text{A}$$

$$I_C = \frac{E}{Z_C} = \frac{E}{X_C\ \angle -90^\circ} = \frac{120\ \angle 0^\circ\ \text{V}}{120\ \angle -90^\circ\ \Omega} = 1\ \angle 90^\circ\ \text{A}$$

$$I_L = \frac{E}{Z_L} = \frac{E}{X_L\ \angle 90^\circ} = \frac{120\ \angle 0^\circ\ \text{V}}{30\ \angle 90^\circ\ \Omega} = 4\ \angle -90^\circ\ \text{A}$$

Then

$$I_s = (3 + 1\ \angle 90^\circ + 4\ \angle -90^\circ)\ \text{A}$$

$$= (3 + j1 - j4)\ \text{A} \quad \text{or} \quad (3 - j3)\ \text{A}$$

$$I_s = 4.24\ \angle -45^\circ\ \text{A}$$

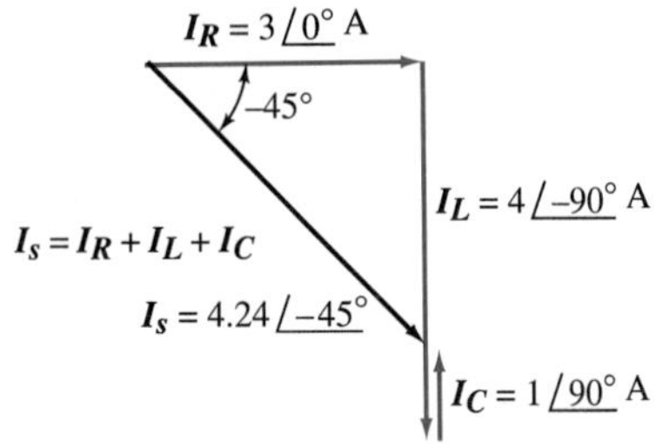

Figure 16.19 The graphic addition of current phasors for the circuit of Example 16.11 gives the source current.

Current phasors also can be added graphically, just as the voltages were. Applying the tip-to-tail method to the currents in Example 16.11 gives the current I_s, as shown in Figure 16.19.

current divider rule

Current Divider Rule. A current divider expression can be derived for the parallel ac circuit, as was done for the parallel dc circuit. The **current divider rule** states that the current in one branch of a parallel ac circuit is equal to the current entering the branches, multiplied by the equivalent impedance of the circuit, and divided by the impedance of the branch. In general, for a circuit such as the one shown in Figure 16.20(a),

$$I_x = \frac{IZ_{eq}}{Z_x} \tag{16.10}$$

where I_x is the branch current, in any unit of current; I is the current entering the circuit, in the same unit as I_x; Z_x is the impedance of the branch in which the current is being found, in ohms; and Z_{eq} is the equivalent impedance of the parallel circuit.

If the circuit has only two branches, as in Figure 16.20(b), Equation 16.10 becomes

$$I_x = \frac{IZ}{Z + Z_x} \quad \textbf{(16.11)}$$

where I_x, I, and Z_x have the same meaning as in Equation 16.10, and Z is the impedance of the other branch.

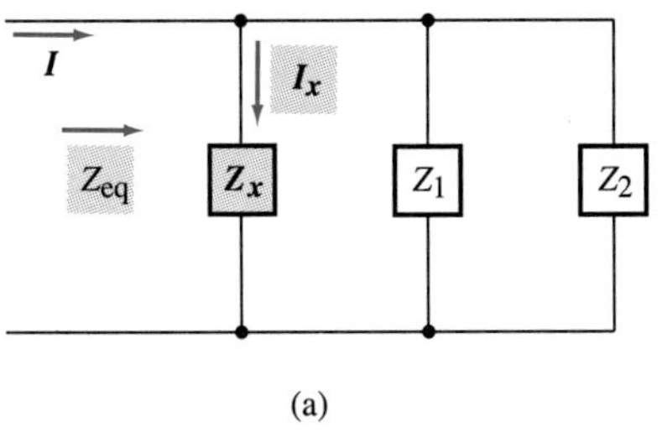

(a)

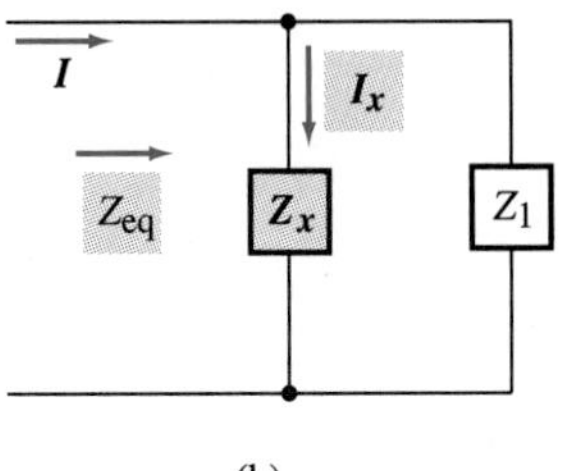

(b)

Figure 16.20 The relation of circuit components in the parallel circuit to the quantities in the current divider equations.

EXAMPLE 16.12

Use Equation 16.10 to find the current in the capacitor of Figure 16.21.

Solution

First, we must find the Z_{eq} of the circuit.

$$\frac{1}{Z_{eq}} = \frac{1}{3\ \Omega} + \frac{1}{3\ \angle 90^\circ\ \Omega} + \frac{1}{4\ \angle -90^\circ\ \Omega} = 0.343\ \angle -14^\circ\ \text{S}$$

so

$$Z_{eq} = 2.92\ \angle 14^\circ\ \Omega$$

Then

$$I_C = \frac{IZ_{eq}}{Z_C}$$

$$I_C = \frac{IZ_{eq}}{X_c\ \angle -90^\circ} = \frac{(2\ \angle 0^\circ\ \text{A})(2.92\ \angle 14^\circ\ \Omega)}{4\ \angle -90^\circ\ \Omega}$$

$$I_C = 1.46\ \angle 104^\circ\ \text{A}$$

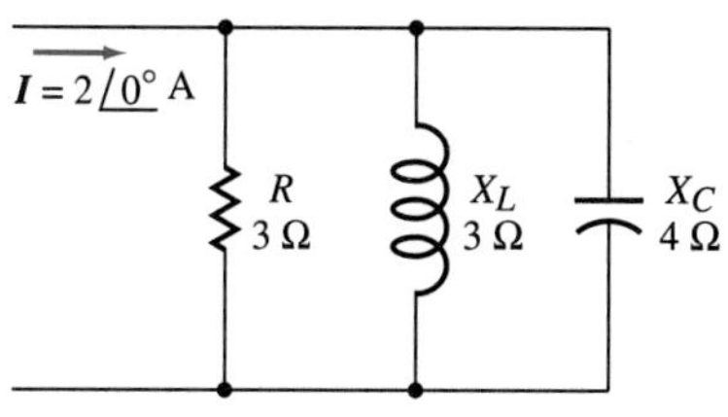

Figure 16.21

EXAMPLE 16.13

Use Equation 16.11 to find the current in the resistor of Figure 16.22.

Solution

$$I_R = \frac{IZ_L}{Z_R + Z_L}$$

$$I_R = \frac{(I)(X_L\ \angle 90^\circ)}{R + jX_L} = \frac{(5\ \angle 30^\circ\ \text{A})(40\ \angle 90^\circ\ \Omega)}{(20 + j\,40)\ \Omega}$$

$$= \frac{200\ \angle 120^\circ\ \text{V}}{44.72\ \angle 63.43^\circ\ \Omega}$$

$$I_R = 4.47\ \angle 56.57^\circ\ \text{A}$$

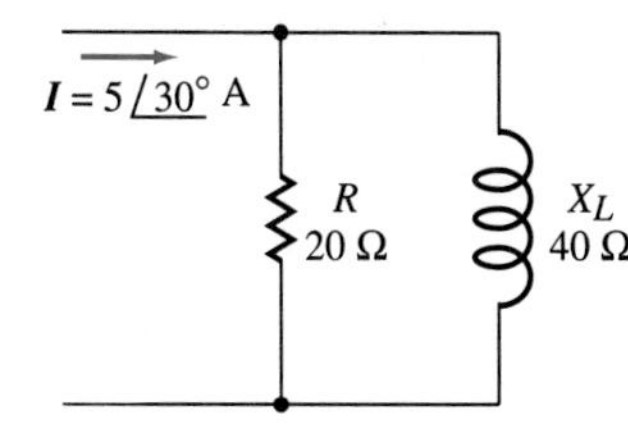

Figure 16.22

An open circuit in a branch of a parallel circuit results in no current in that branch. The effect on circuits connected to the parallel circuit depends on the impedance in the branches. This is because of the phasor quantities.

A short circuit results in 0 volts across every branch of the parallel circuit. Again, what happens to circuits that are connected to the parallel circuit depends on how the impedance in the total circuit is changed because of the short circuit.

16.3 Phasor Diagrams

Although numerical values can be useful, they do not present a visual picture of the relative magnitudes and angles. Waveforms can be used, but their construction is time-consuming and difficult. An alternative solution is to use a phasor diagram.

A **phasor diagram** is a diagram that shows the phasors for the current and voltages in a circuit. These quantities are drawn to some scale at their phase angle. They are simple to construct, requiring only a scale and a protractor. The diagrams for the series and parallel *R–C*, *R–L*, and *R–L–C* circuits are constructed in this section.

The calculated values for the series *R–C* circuit in Figure 16.23 are:

$$\mathbf{Z}_{eq} = 50 \angle -53.13^\circ\ \Omega$$
$$\mathbf{I} = 1.41 \angle 53.13^\circ \text{ A} \quad \text{or} \quad 2 \sin(\alpha + 53.13^\circ)$$
$$\mathbf{V}_R = 42.43 \angle 53.13^\circ \text{ V} \quad \text{or} \quad 60 \sin(\alpha + 53.13^\circ)$$
$$\mathbf{V}_C = 56.57 \angle -36.87^\circ \text{ V} \quad \text{or} \quad 80 \sin(\alpha - 36.87^\circ)$$
$$\mathbf{E} = 70.7 \angle 0^\circ \text{ V} \quad \text{or} \quad 100 \sin \alpha$$

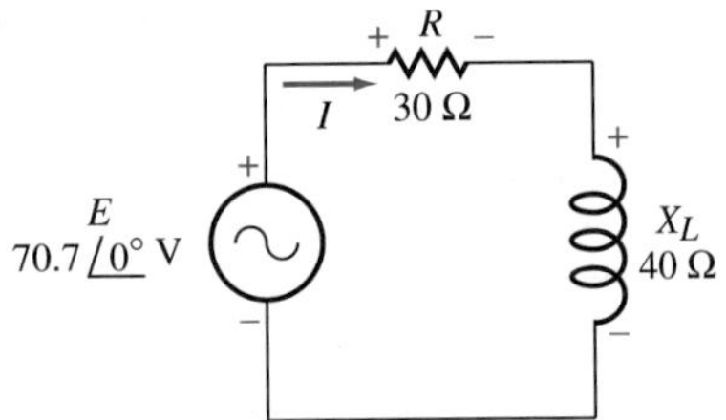

Figure 16.23 This series circuit is capacitive, so the current leads the source voltage.

If these quantities are drawn to some scale, the phasor diagram is as shown in Figure 16.24(a). The leading phasor is the one that first passes a reference point as the phasors rotate clockwise. $\mathbf{V}_R$ leads $\mathbf{V}_C$, but is in

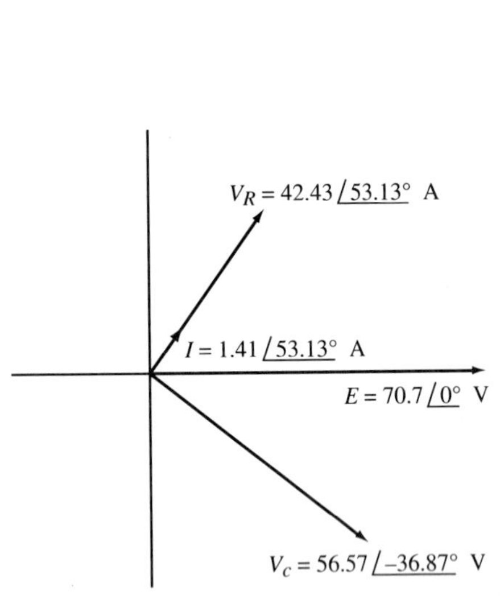

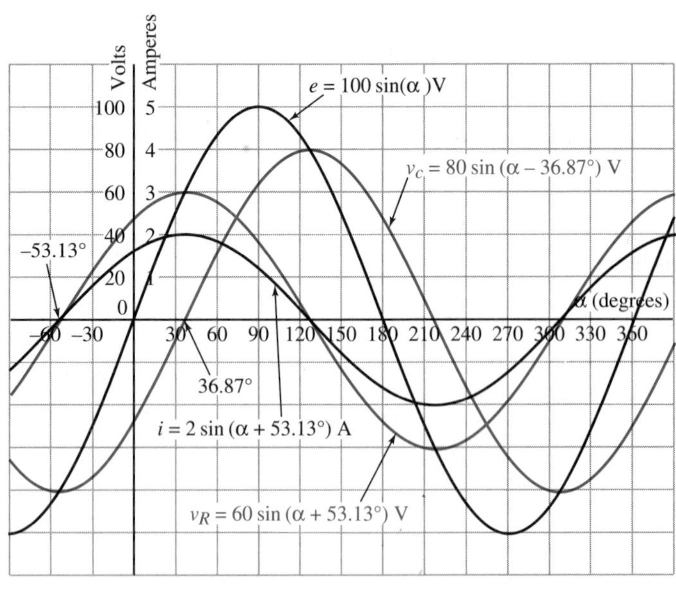

Figure 16.24 The phasor and wave diagrams for the capacitive circuit of Figure 16.23. (a) The phasor diagram is simple, but shows only magnitudes and phase angles. (b) The wave diagrams are more complex, but also show the variation with time.

phase with $\boldsymbol{I}$. Since the circuit is capacitive, the current leads the source voltage.

The wave diagram for the series circuit is shown in Figure 16.24(b). Constructing this diagram requires more time than the phasor diagram. Also, interpreting these curves is a little more difficult because of the many curves.

If the capacitor in Figure 16.23 is replaced by an inductor, as in Figure 16.25, the calculated quantities become:

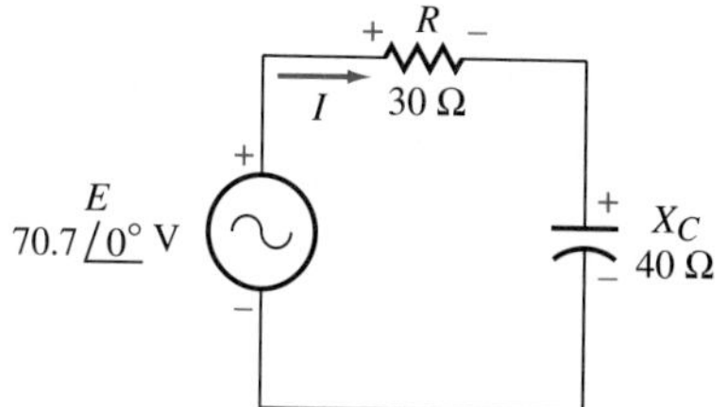

Figure 16.25 This series circuit is inductive, so the current lags the source voltage.

$$\begin{aligned} \boldsymbol{Z}_{eq} &= 50 \angle 53.13^\circ\ \Omega \\ \boldsymbol{I} &= 1.41 \angle -53.13^\circ\ \text{A} \quad &\text{or} \quad & 2\sin(\alpha - 53.13^\circ) \\ \boldsymbol{V}_R &= 42.43 \angle -53.13^\circ\ \text{V} \quad &\text{or} \quad & 60\sin(\alpha - 53.13^\circ) \\ \boldsymbol{V}_L &= 56.57 \angle 36.87^\circ\ \text{V} \quad &\text{or} \quad & 80\sin(\alpha + 36.87^\circ) \\ \boldsymbol{E} &= 70.7 \angle 0^\circ\ \text{V} \quad &\text{or} \quad & 100\sin\alpha \end{aligned}$$

The current now lags the source voltage because the circuit is inductive.

The phasor diagram and wave diagram are drawn in Figure 16.26. Both diagrams show the resistor voltage in phase with the current and the inductor voltage leading the current by 90°. Again the phasor diagram is easier to interpret.

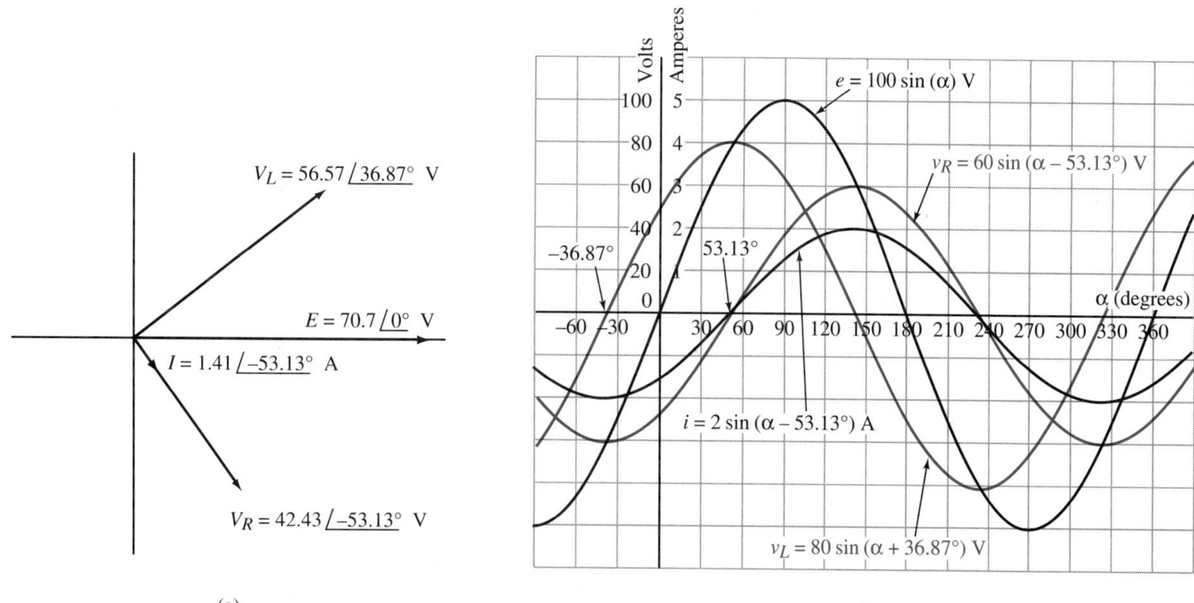

Figure 16.26 Phasor and wave diagrams for the inductive circuit of Figure 16.25. (a) The phasor diagram. (b) The wave diagram.

Finally, the series R–L–C circuit in Figure 16.27 is considered. The calculated quantities for this circuit are:

$$\begin{aligned} \boldsymbol{Z}_{eq} &= 50 \angle -36.87^\circ\ \Omega \\ \boldsymbol{I} &= 1.41 \angle 36.87^\circ\ \text{A} \quad &\text{or} \quad & 2\sin(\alpha + 36.87^\circ) \\ \boldsymbol{V}_R &= 56.57 \angle 36.87^\circ\ \text{V} \quad &\text{or} \quad & 80\sin(\alpha + 36.87^\circ) \end{aligned}$$

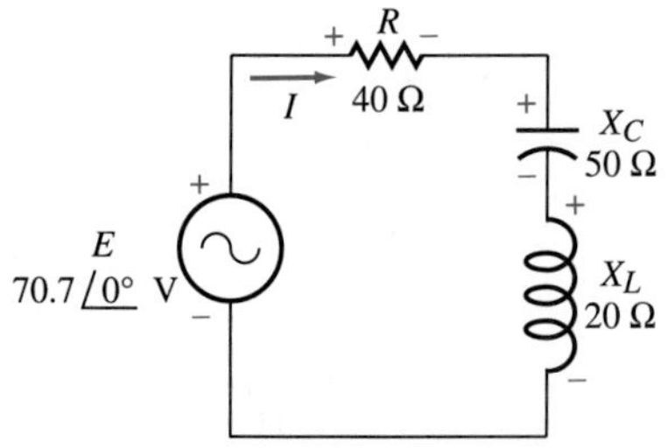

Figure 16.27 In the general series circuit, when X_C is greater than X_L, the current leads the source voltage.

$$\boldsymbol{V}_C = 70.7 \angle -53.13° \text{ V} \quad \text{or} \quad 100 \sin(\alpha - 53.13°)$$
$$\boldsymbol{V}_L = 28.28 \angle 126.87° \text{ V} \quad \text{or} \quad 40 \sin(\alpha + 126.87°)$$
$$\boldsymbol{E} = 70.7 \angle 0° \text{ V} \quad \text{or} \quad 100 \sin \alpha$$

If each phasor is drawn at its angle and to some scale, the phasor diagram in Figure 16.28(a) results. The diagram shows $\boldsymbol{V}_R$ in phase with $\boldsymbol{I}$, $\boldsymbol{V}_C$ lagging $\boldsymbol{I}$, and $\boldsymbol{V}_L$ leading $\boldsymbol{I}$.

The wave diagram for the circuit is shown in Figure 16.28(b). Because of the many waveforms, the diagram is rather cluttered.

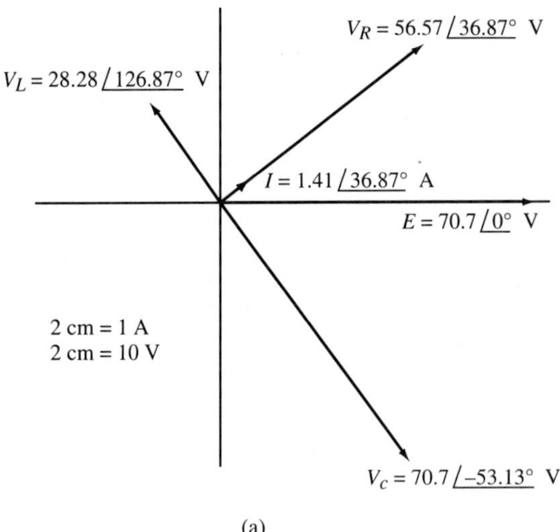

(a)

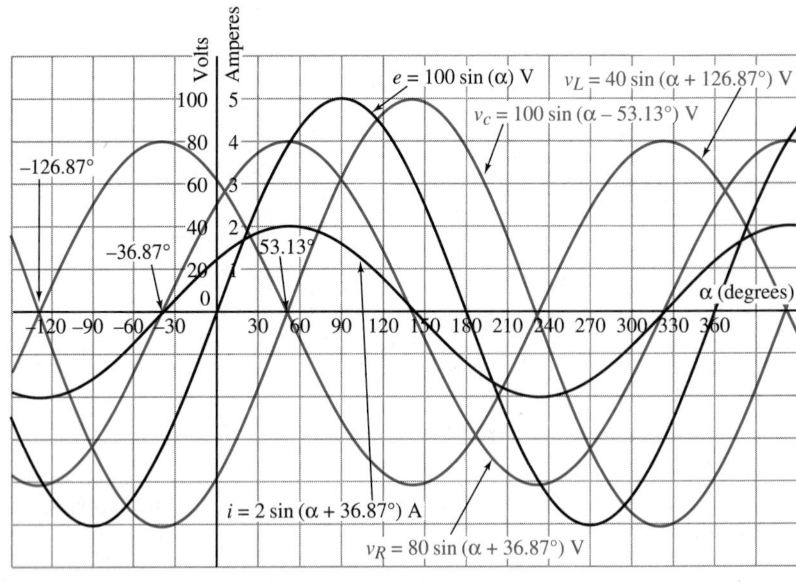

(b)

Figure 16.28 Phasor and wave diagrams for the circuit of Figure 16.27. (a) The phasor diagram. (b) The wave diagram.

Next, the phasor diagrams for some parallel configurations are constructed. The calculated quantities for the parallel R–C circuit in Figure 16.29 are:

$$
\begin{aligned}
\mathbf{Z}_{eq} &= 24 \angle -53.13^\circ\ \Omega \\
\mathbf{I}_s &= 3.54 \angle 53.13^\circ\ \text{A} &\quad \text{or} \quad i_s &= 5 \sin(\alpha + 53.13^\circ)\ \text{A} \\
\mathbf{I}_R &= 2.13 \angle 0^\circ\ \text{A} &\quad \text{or} \quad i_R &= 3 \sin \alpha\ \text{A} \\
\mathbf{I}_C &= 2.83 \angle 90^\circ\ \text{A} &\quad \text{or} \quad i_C &= 4 \sin(\alpha + 90^\circ)\ \text{A} \\
\mathbf{E} &= 85 \angle 0^\circ\ \text{V} &\quad \text{or} \quad e &= 120 \sin \alpha\ \text{V}
\end{aligned}
$$

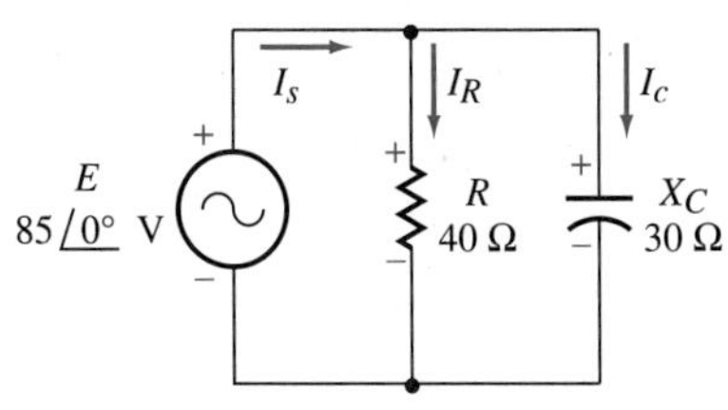

Figure 16.29 This parallel circuit is capacitive, so the current leads the source voltage.

The phasors are shown in Figure 16.30(a). $\mathbf{I}_R$ and $\mathbf{E}$ are in phase. $\mathbf{I}_C$ leads $\mathbf{E}$ by 90°. Also, $\mathbf{I}_S$ will equal $\mathbf{I}_R + \mathbf{I}_C$. The wave diagram for the circuit is in Figure 16.30(b). Here also, at any degree point, the value of i_S will be equal to $i_R + i_C$.

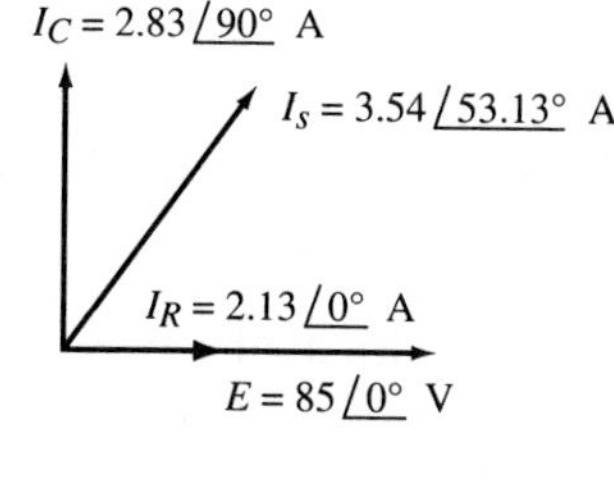

(a)

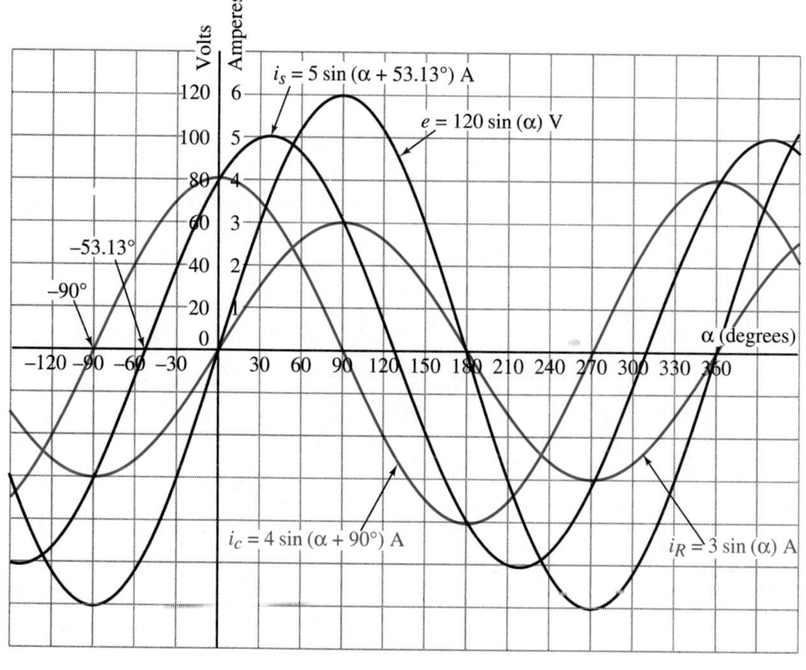

(b)

Figure 16.30 Phasor and wave diagrams for the parallel circuit of Figure 16.29. (a) The phasor diagram. (b) The wave diagram.

Replacing the capacitor in Figure 16.29 with an inductor, as in Figure 16.31, gives the following quantities:

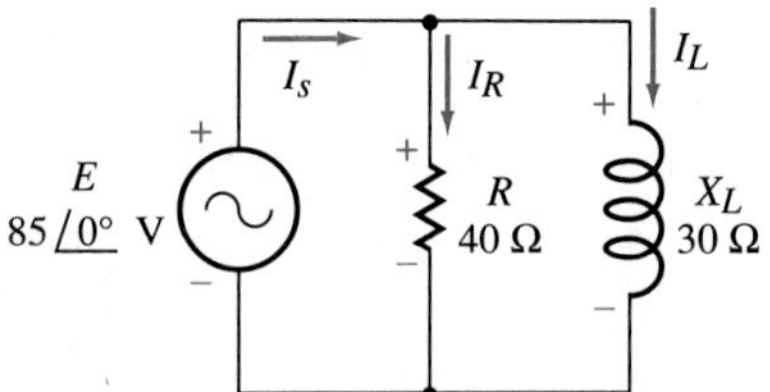

Figure 16.31 This parallel circuit is inductive, so the current lags the voltage.

$\mathbf{Z}_{eq} = 24 \angle 53.13° \ \Omega$			
$\mathbf{I}_s = 3.54 \angle -53.13° \text{ A}$	or	$i_s = 5 \sin(\alpha - 53.13°) \text{ A}$
$\mathbf{I}_R = 2.13 \angle 0° \text{ A}$	or	$i_R = 3 \sin \alpha \text{ A}$
$\mathbf{I}_L = 2.83 \angle -90° \text{ A}$	or	$i_L = 4 \sin(\alpha - 90°) \text{ A}$
$\mathbf{E} = 85 \angle 0° \text{ V}$	or	$e = 120 \sin \alpha \text{ V}$

Now the resistor current is still in phase with the source voltage, but the inductance causes the source current to lag the voltage. The phase and wave diagrams are shown in Figure 16.32.

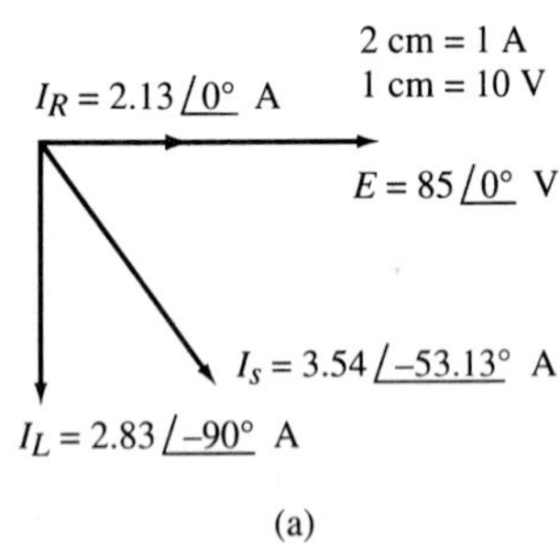

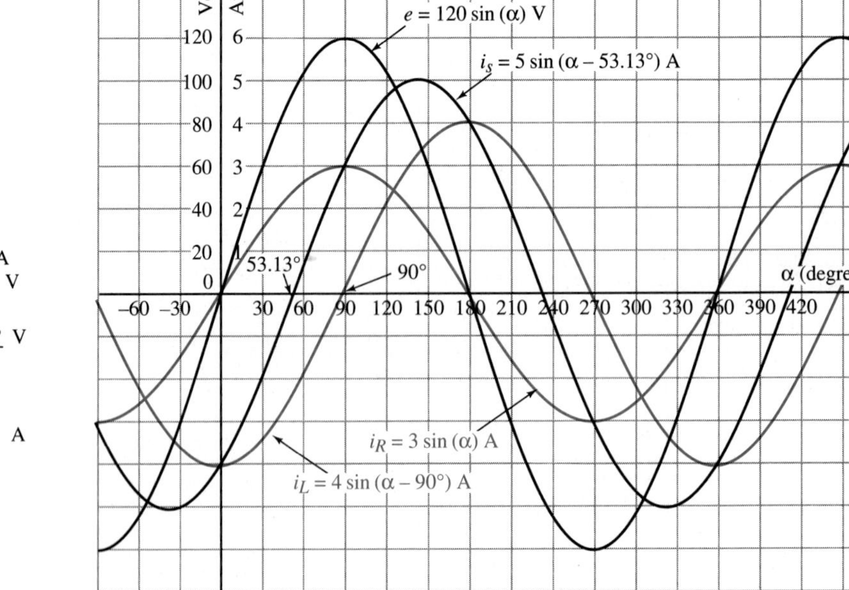

Figure 16.32 Phasor and wave diagrams for the inductive circuit of Figure 16.31. (a) The phasor diagram. (b) The wave diagram.

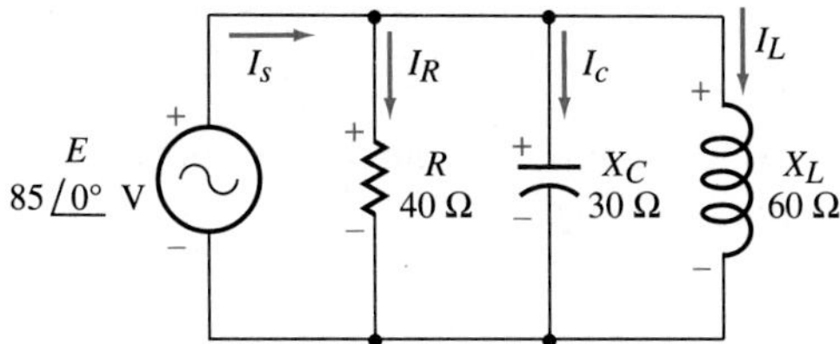

Figure 16.33 In the general parallel circuit, when X_C is smaller than X_L the source current leads the source voltage.

Finally, the parallel *R–L–C* circuit in Figure 16.33 are considered. Using ac analysis, the resulting quantities are:

$$\begin{aligned}
\mathbf{Z}_{eq} &= 33.28 \angle -33.7^\circ\ \Omega \\
\mathbf{I}_R &= 2.13 \angle 0^\circ \text{ A} &\quad \text{or} \quad i_R &= 3 \sin \alpha \text{ A} \\
\mathbf{I}_L &= 1.42 \angle -90^\circ \text{ A} &\quad \text{or} \quad i_L &= 2 \sin(\alpha - 90^\circ) \text{ A} \\
\mathbf{I}_C &= 2.83 \angle 90^\circ \text{ A} &\quad \text{or} \quad i_C &= 4 \sin(\alpha + 90^\circ) \text{ A} \\
\mathbf{I}_s &= 2.55 \angle 33.7^\circ \text{ A} &\quad \text{or} \quad i_s &= 3.61 \sin(\alpha + 33.7^\circ) \text{ A} \\
\mathbf{E} &= 85 \angle 0^\circ &\quad \text{or} \quad e &= 120 \sin \alpha \text{ V}
\end{aligned}$$

The phasor diagram includes all currents and the source voltage. Both the phasor and wave diagrams are shown in Figure 16.34.

In these circuits, the diagrams were used to show the currents and voltages in the circuits. But the phasor diagrams can be used to determine an unknown phasor. This is done by using the tip-to-tail method to combine the phasors.

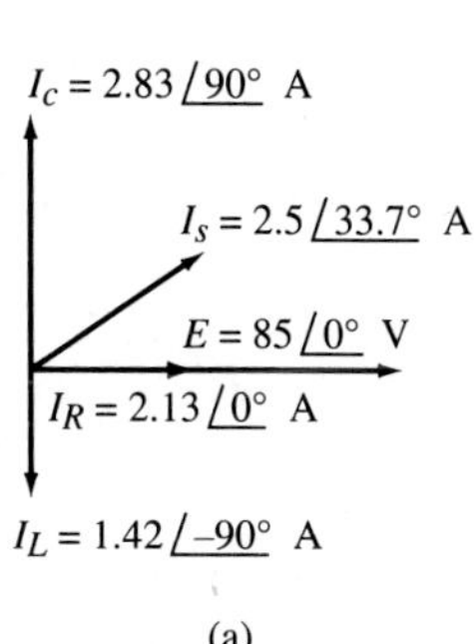

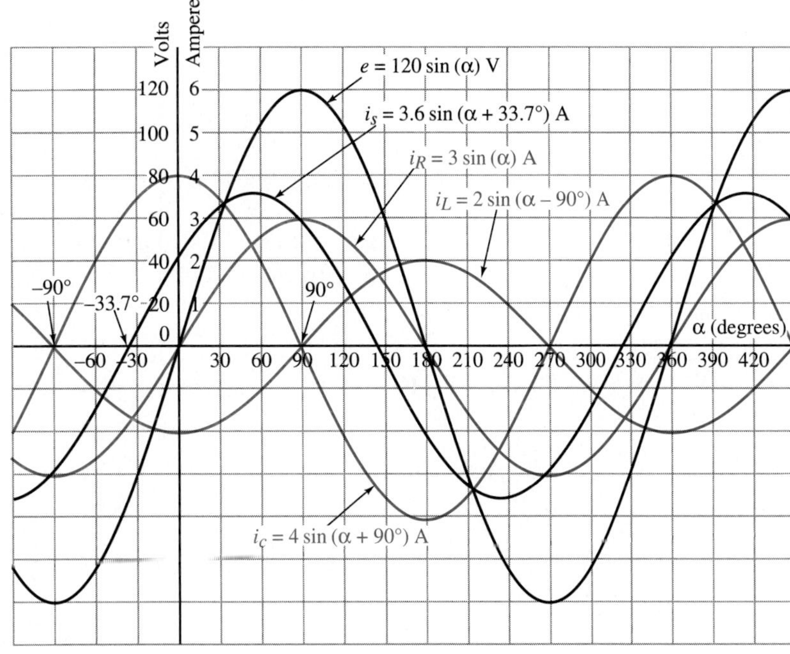

Figure 16.34 Phasor and wave diagrams for the circuit of Figure 16.33. (a) The phasor diagram. (b) The wave diagram.

EXAMPLE 16.14

Given the phasors for $\boldsymbol{V_R}$, $\boldsymbol{V_L}$, and $\boldsymbol{V_C}$ in the series R–L–C circuit, determine the source voltage.

Solution

The solution is shown in Figure 16.35. It is obtained as follows:

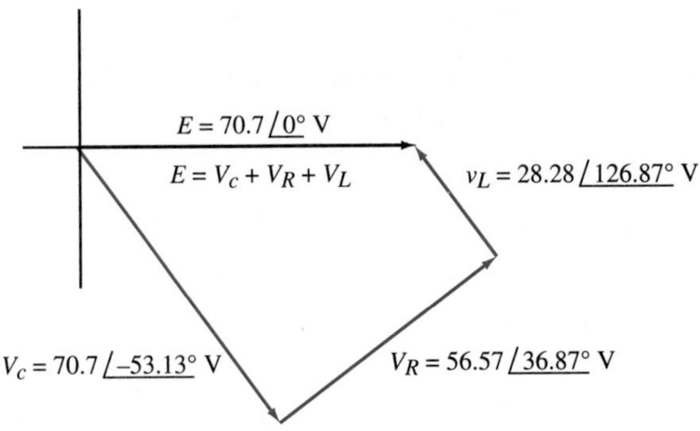

Figure 16.35

Starting at the origin, draw $\boldsymbol{V_C}$ at its angle, $-53.13°$.
Starting at the tip (arrowhead) of $\boldsymbol{V_C}$, draw $\boldsymbol{V_R}$ at its angle 36.87°.
Starting at the tip of $\boldsymbol{V_R}$, draw $\boldsymbol{V_L}$ at its angle, 126.87°.
Since $\boldsymbol{E} = \boldsymbol{V_R} + \boldsymbol{V_C} + \boldsymbol{V_L}$, $\boldsymbol{E}$ is the phasor drawn from the origin to the tip of $\boldsymbol{V_L}$. It is approximately 70.7 volts at an angle of 0°.

This solution used the order $\boldsymbol{V_C}$, $\boldsymbol{V_R}$, $\boldsymbol{V_L}$, but any order could have been used.

If we knew $\boldsymbol{E}$, $\boldsymbol{V_R}$, and $\boldsymbol{V_C}$ and wanted to find $\boldsymbol{V_L}$, the graphical operation of $\boldsymbol{E} - \boldsymbol{V_R} - \boldsymbol{V_C}$ would have to be performed. $\boldsymbol{E}$ would be drawn, then $-\boldsymbol{V_R}$, and finally $-\boldsymbol{V_C}$. $\boldsymbol{V_L}$ would be the phasor from the start of $\boldsymbol{E}$ to the end of $\boldsymbol{V_C}$. The diagram is shown in Figure 16.36.

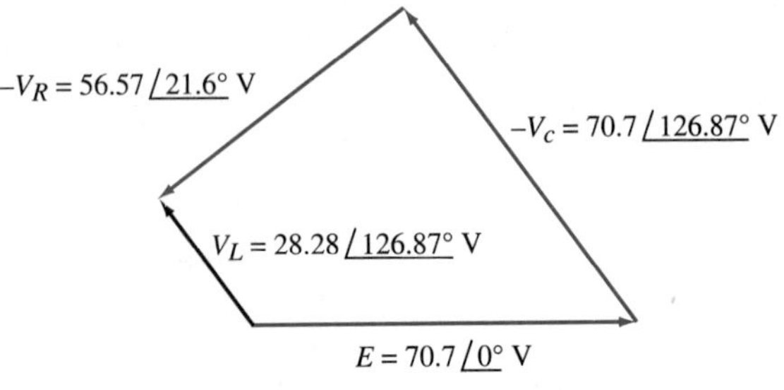

$V_L = E - V_c - V_R$

Figure 16.36 The graphical addition of voltages is used to determine the voltage across the inductor in Example 16.14.

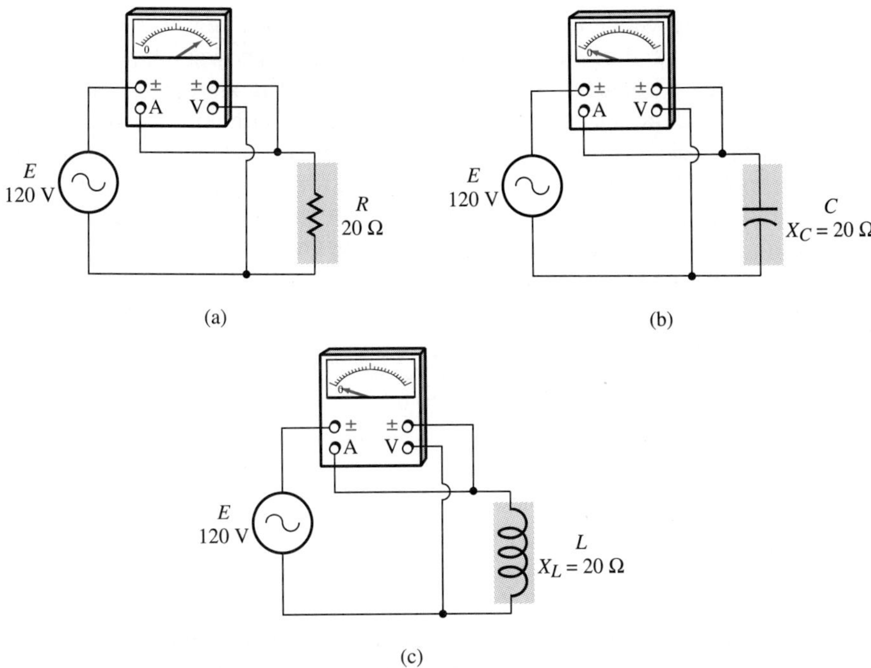

Figure 16.37 AC circuits connected across the same voltage do not always have equal power. (a) The wattmeter indicates power for the resistor. (b) The wattmeter indicates zero power for the capacitor. (c) The wattmeter indicates zero power for the inductor.

16.4 Power in the ac Circuit

Power in the dc circuit was given by VI, I^2R, and V^2/R, but these relations are not necessarily true for the ac circuit. This can be seen in Figure 16.37 where the wattmeters indicate power for the resistance but not for the capacitance and inductance. To understand why this is so, each one is considered individually.

The power dissipated in a resistance varies from zero to a peak, as seen in Figure 16.38. The instantaneous power at each point is obtained

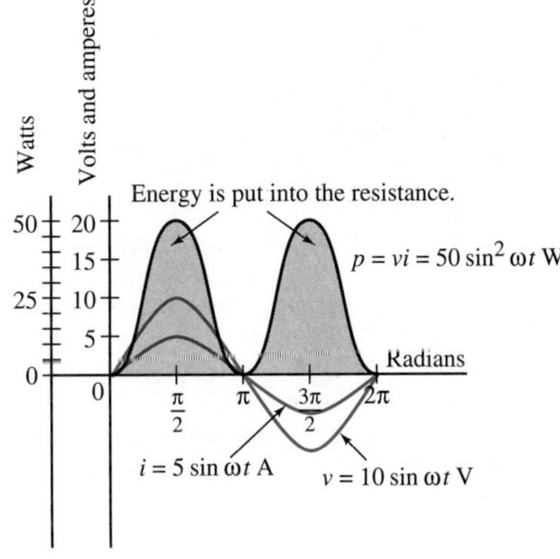

Figure 16.38 Energy is supplied to the resistor during both halves of the cycle so there is an average power.

from $p = vi$. Since $v = V_M \sin(\omega t)$ and $i = I_M \sin(\omega t)$, $vi = V_M I_M \sin^2 (\omega t)$. Although the peak power and instantaneous power are important in some applications, here we are concerned with the **average power**. This is the constant power that results in the same area under the power curve for one cycle as under the power curve for the varying waveform. The area under the curve represents the energy for the cycle. This area can be found by the strip method or method of known shapes, but a more exact value is found by calculus integration. The area, as derived in Appendix C, is equal to $\pi V_m I_m$. Since the average power equals the area divided by the length of the cycle,

average power

$$P = \frac{V_m I_m (\pi)}{2\pi} = \frac{V_m I_m}{2}$$

But

$$V_m = V_{\text{eff}} \sqrt{2} \quad \text{and} \quad I_m = I_{\text{eff}} \sqrt{2},$$

so for a resistance

$$P = VI \tag{16.12}$$

where P is the average power dissipated in the resistance, in watts; V is the effective value of the voltage across the resistance, in volts; and I is the effective value of the resistor current, in amperes.

Equation 16.12 can be put into two other forms with the use of Ohm's Law. They are:

$$P = I^2 R \tag{16.13}$$

$$P = V^2/R \tag{16.14}$$

These equations show that the power relation for resistance in an ac circuit is the same as the power relation for the dc circuit.

Next, the power in the capacitor is considered. In Figure 16.39, the power curve for the capacitor differs from the power curve for the

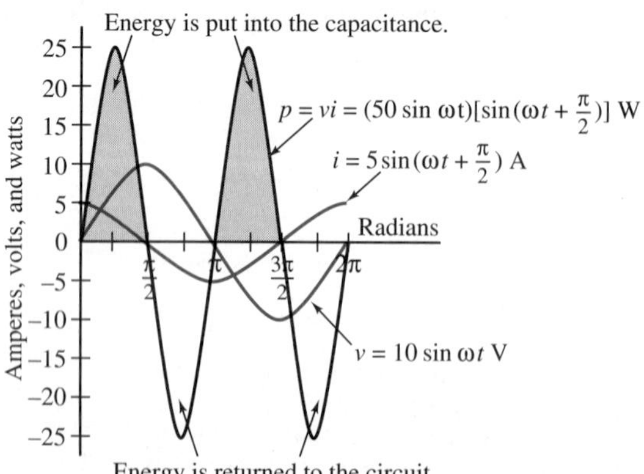

Figure 16.39 The ideal capacitor returns energy to the circuit during half of the cycle so the average power is zero watts.

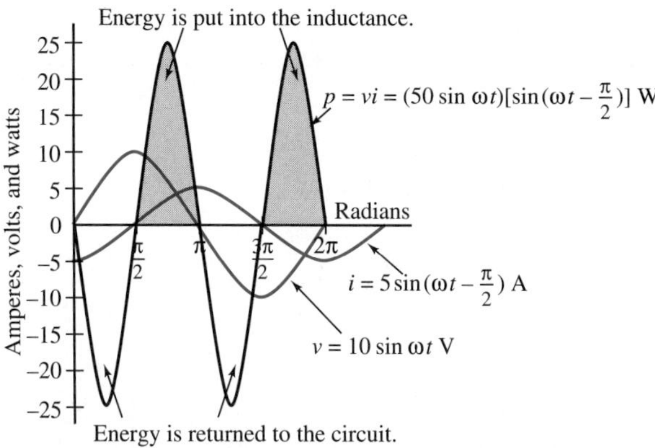

Figure 16.40 The ideal inductor returns energy to the circuit during half of the cycle so the average power is zero watts.

resistor. This curve has positive and negative areas. Also, because the v and i sine waves are 90° out of phase, the positive area equals the negative area. Thus, the average power is equal to zero. During the time of positive power values, energy is being put into the electrostatic field. Then during the time of negative power values, the energy is returned to the circuit. If the capacitor has no resistance, the energy returned equals the energy put in, and the average power is zero. Practical capacitors do have some leakage resistance, so there will be a small amount of power dissipated in them.

The v and i curves for the inductor in Figure 16.40 are also 90° out of phase. This yields the same condition as for the capacitor, except the power curve is shifted 90° from that of the capacitor. The ideal inductor, having no resistance, does not dissipate any power. Instead, during the positive time of power values, energy is put into the magnetic field. Then during the negative time of power values, the field is collapsing and the energy is returned to the circuit. The practical inductor has resistance. This results in power being dissipated in it.

The relations in the previous sections can now be used to find the power in the general series ac circuit, which is shown in Figure 16.41.

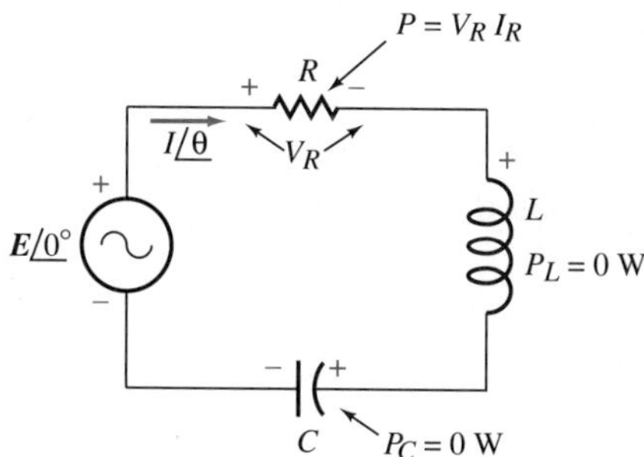

Figure 16.41 The general relation for power in the ac circuit is $P = EI \cos \theta$.

Since the power in the capacitance and inductance must be zero, the power in the circuit must also be the power in the resistance.

Then

$$P = V_R I$$

But, by the voltage divider relation,

$$\mathbf{V}_R = \frac{\mathbf{E}\mathbf{Z}_R}{\mathbf{Z}_{eq}}$$

$$\mathbf{V}_R = \frac{(\mathbf{E})(R)}{\mathbf{Z}_{eq}}$$

so the magnitude of $\mathbf{V}_R$ is $(ER)/|Z|_{eq}$ and $P = (ERI)/Z_{eq}$.

In the impedance triangle $(R/|\mathbf{Z}_{eq}|) = \cos\theta$, where θ is the phase angle for the circuit and also the angle for $\mathbf{Z}_{eq}$. Therefore, $P = (EIR)/Z_{eq}$ becomes $P = EI\cos\theta$. In general, the power in a complete series circuit or parts of it is given by

$$P = VI\cos\theta \qquad \textbf{(16.15)}$$

where P is the power dissipated in one cycle, in watts; V is the effective value of the voltage across the series circuit, in volts; I is the effective value of the circuit's current, in amperes; and θ is the angle between the circuit's voltage and current. It is also the angle of Z_{eq}.

power factor

Cos θ is also known as the **power factor**, F_p. The value of the power factor ranges from zero for capacitance and inductance, to unity for resistance. If the angle of Z_{eq} is negative, the circuit is capacitive and the current leads the voltage. Therefore, it is said to be a **leading power factor**. If the angle of Z_{eq} is positive, the circuit is inductive and the current lags the voltage. Therefore, it is said to be a **lagging power factor**.

leading power factor

lagging power factor

According to Equation 16.15, the current needed to provide a given amount of power is inversely proportional to the power factor. That is, $I = P/(V\cos\theta)$. Smaller power factors mean more current and a larger power loss in the conductors. These larger currents require larger generating capacities. Also, a lagging power factor causes wide voltage variation. Because of this, electric power companies require their commercial customers to keep their load power factor above a set lagging value.

EXAMPLE 16.15

In the circuit shown in Figure 16.42, find:

a. the power factor, and determine if it is a leading or lagging one.
b. the power dissipated.

Solution

a.
$$\begin{aligned}
F_p &= \cos\theta \\
\mathbf{Z}_{eq} &= \mathbf{Z}_R + \mathbf{Z}_L - \mathbf{Z}_C \\
\mathbf{Z}_{eq} &= R + jX_L - jX_C \\
&= (60 + j20 - j100)\ \Omega \\
&= (60 - j80)\ \Omega \quad \text{or} \quad 100\angle -53.13^\circ\ \Omega
\end{aligned}$$

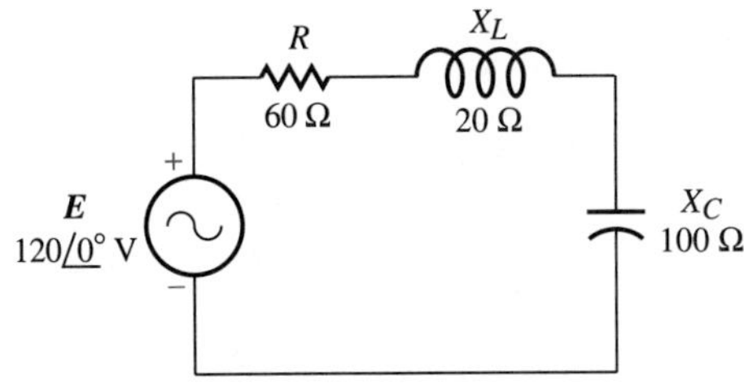

Figure 16.42

$$F_p = \cos(-53.13°)$$
$$F_p = 0.6$$

Since the angle of $\mathbf{Z}_{eq}$ is (−), the circuit is capacitive. So, F_p is a leading power factor.

b. $P = EI \cos \theta$

$$\mathbf{I} = \frac{\mathbf{E}}{\mathbf{Z}_{eq}} = \frac{120 \angle 0° \text{ V}}{100 \angle -53.13° \ \Omega} = 1.2 \angle 53.13° \text{ A}$$

Then

$$P = (120 \text{ V})(1.2 \text{ A})(0.6)$$
$$P = 86.4 \text{ W}$$

Since all the power is in the resistance, P also equals $I^2R = (1.2 \text{ A})^2(60 \ \Omega) = 86.4$ watts.

The total power in an ac series circuit is also equal to the sum of the power dissipated in each part of the circuit. That is,

$$P = P_1 + P_2 + \cdots + P_n \qquad \textbf{(16.16)}$$

where P is the power dissipated in the circuit, in watts, and $P_1, P_2, \ldots, P_n$ is the power dissipated in each part of the circuit, in watts.

EXAMPLE 16.16

Find the power dissipated in the circuit of Figure 16.43 using:
a. Equation 16.15.
b. Equation 16.16.

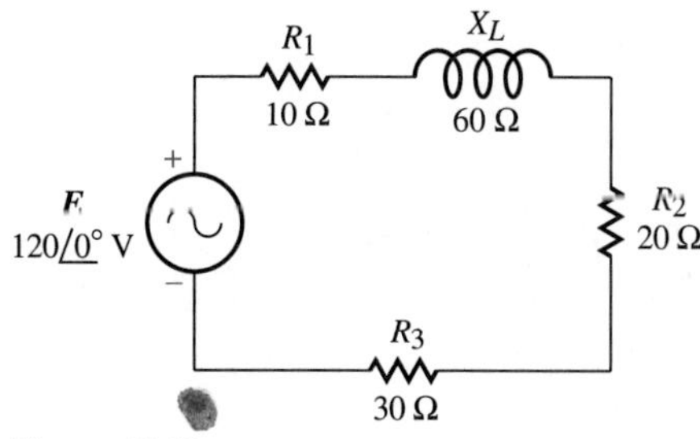

Figure 16.43

Solution

a. $P = EI \cos \theta$

$$\mathbf{Z}_{eq} = \mathbf{Z}_1 + \mathbf{Z}_2 + \mathbf{Z}_3 + \mathbf{Z}_L$$

$$\mathbf{Z}_{eq} = R_1 + R_2 + R_3 + jX_L$$

$$= 10\ \Omega + 20\ \Omega + 30\ \Omega + j60\ \Omega$$

$$= (60 + j60)\ \Omega \quad \text{or} \quad 84.853\ \angle 45^\circ\ \Omega$$

$$\mathbf{I} = \frac{\mathbf{E}}{\mathbf{Z}_{eq}} = \frac{120\ \angle 0^\circ\ \text{V}}{84.853\ \angle 45^\circ\ \Omega} = 1.414\ \angle -45^\circ\ \text{A}$$

Then

$$P = (120\ \text{V})(1.414\ \text{A})(\cos 45^\circ)$$

$$P = 119.98\ \text{W}$$

b. $P = P_1 + P_2 + P_3$

$$= I^2R_1 + I^2R_2 + I^2R_3$$

$$= (1.414\ \text{A})^2(10\ \Omega) + (1.414\ \text{A})^2(20\ \Omega) + (1.414\ \text{A})^2(30\ \Omega)$$

$$= 19.99\ \text{W} + 39.99\ \text{W} + 59.98\ \text{W}$$

$$P = 119.96\ \text{W}$$

The answers agree within the accuracy of the digits used in the calculations.

Although Equations 16.15 and 16.16 were derived for the series circuit, they also hold for any circuit configuration. A parallel circuit is now considered in Example 16.17.

EXAMPLE 16.17

For the circuit in Figure 16.44, find:

a. the power factor and power.
b. whether F_p is leading or lagging.

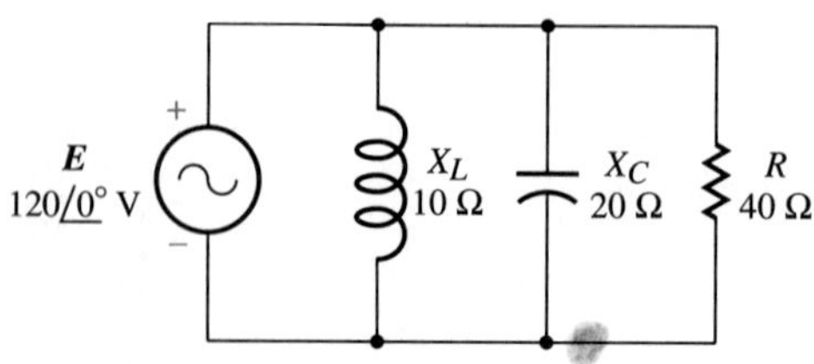

Figure 16.44

Solution

a. Power factor $= \cos\theta$

$$\theta = \text{the angle of } \boldsymbol{Z}_{eq}$$

$$\frac{1}{\boldsymbol{Z}_{eq}} = \frac{1}{\boldsymbol{Z}_L} + \frac{1}{\boldsymbol{Z}_C} + \frac{1}{\boldsymbol{Z}_R}$$

$$\frac{1}{\boldsymbol{Z}_{eq}} = \frac{1}{X_L \angle 90^\circ} + \frac{1}{X_C \angle -90^\circ} + \frac{1}{R}$$

$$= \frac{1}{10 \angle 90^\circ\ \Omega} + \frac{1}{20 \angle -90^\circ\ \Omega} + \frac{1}{40\ \Omega}$$

$$= 0.1 \angle -90^\circ\ \text{S} + 0.05 \angle 90^\circ\ \text{S} + 0.025 \angle 0^\circ\ \text{S}$$

$$= (-j0.1 + j0.05 + 0.025)\ \text{S}$$

$$\text{or } (0.025 - j0.05)\ \text{S} \quad \text{or} \quad 0.0559 \angle -63.44^\circ\ \text{S}$$

$$\boldsymbol{Z}_{eq} = 17.89 \angle 63.44^\circ\ \Omega$$

$$\boldsymbol{I} = \frac{\boldsymbol{E}}{\boldsymbol{Z}_{eq}} = \frac{120 \angle 0^\circ\ \text{V}}{17.89 \angle 63.44^\circ\ \Omega} = 6.71 \angle -63.44^\circ\ \text{A}$$

$$F_p = \cos(63.44^\circ)$$

$$F_p = 0.45$$

Power can be found in two ways. Since the power is dissipated in R,

$$P = \frac{V^2}{R} = \frac{E^2}{R}$$

$$= \frac{(120\ \text{V})^2}{40\ \Omega}$$

$$P = 360\ \text{W}$$

Also, using the general power relation,

$$P = EI\cos\theta$$

$$P = (120\ \text{V})(6.71\ \text{A})\cos(63.44^\circ)$$

$$P = 360.57\ \text{W}$$

b. E is at 0° and I is at −63.44°, so F_p is a lagging power factor.

EXAMPLE 16.18

Find the power dissipated in the parallel circuit of Figure 16.45 using Equation 16.16.

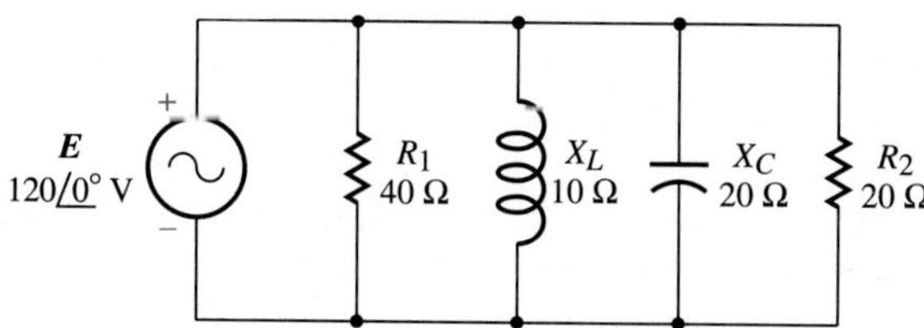

Figure 16.45

Solution

$$P = P_{R1} + P_{R2}$$
$$P_{R1} = \frac{E^2}{R_1} = \frac{(120\text{ V})^2}{40\ \Omega} = 360\text{ W}$$
$$P_{R2} = \frac{E^2}{R_2} = \frac{(120\text{ V})^2}{20\ \Omega} = 720\text{ W}$$
$$P = 360\text{ W} + 720\text{ W}$$
$$P = 1080\text{ W}$$

Energy and power in an ac circuit are related in the same way as they are in the dc circuit. Energy is the product of power and time. Multiplying the power equation by time gives

$$\mathcal{W} = Pt \tag{16.17}$$
$$\mathcal{W} = \mathcal{W}_1 + \mathcal{W}_2 + \cdots + \mathcal{W}_n \tag{16.18}$$

where $\mathcal{W}$ is the energy in a circuit, in joules; $\mathcal{W}_1, \mathcal{W}_2, \ldots, \mathcal{W}_n$ is the energy in each part of a circuit, in joules; t is the time, in seconds; V and I are the effective value of voltage and current, in volts and amperes; and P is the power, in watts.

The energy will be in kWh if time is in hours, and power is in kilowatts.

EXAMPLE 16.19

How much energy in joules and kWh is supplied to the circuit of Example 16.16 in 10 minutes?

Solution

$$\mathcal{W} = Pt = (119.98\text{ W})(10\text{ min})(60\text{ s/min})$$
$$\mathcal{W} = 71{,}988\text{ J}$$
$$\mathcal{W} = \left(\frac{119.98\text{ W}}{1000\text{ W/kW}}\right)\left(\frac{10\text{ min}}{60\text{ min/hr}}\right)$$
$$\mathcal{W} = 0.02\text{ kWh}$$

16.5 Series Circuit Analysis

The analysis of an ac circuit follows the same procedure as was used for dc analysis. One or more concepts of equivalent impedance, current, voltage division, the voltage divider rule, and power must be selected to provide a path to the answer. If the path is not apparent, the unknowns in the equation can be used as a guide. Since the best way to learn this is from some examples, some are now worked.

EXAMPLE 16.20

In the circuit of Figure 16.46, find:

a. the power dissipated in the relay coil.
b. the current in the circuit.
c. the power supplied by the source.

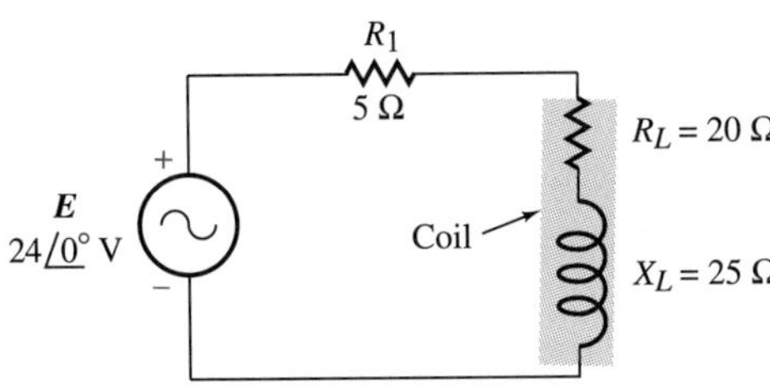

Figure 16.46

Solution

Strategy

One procedure is to find the current using $\mathbf{E}$ and $\mathbf{Z}_{eq}$. This can be used to calculate the coil power. Then, $P = EI \cos \theta$ gives the power in the circuit. Now let us see how the equations can lead us to a solution.

a.

$$P_{coil} = I^2 R_L$$

Begin with a power relation.

$$I = \frac{E}{Z_{eq}}$$

The power relation includes current, so a current relation follows.

$$\mathbf{Z}_{eq} = \mathbf{Z}_{R1} + \mathbf{Z}_L$$
$$\mathbf{Z}_{eq} = R_1 + R_L + jX_L$$
$$= (5 + 20 + j25)\ \Omega$$
$$= (25 + j25)\ \Omega \quad \text{or} \quad 35.36 \angle 45°\ \Omega$$

The current equation indicates that $\mathbf{Z}_{eq}$ is needed.

$$\mathbf{I} = \frac{24 \angle 0°\ \text{V}}{35.36 \angle 45°\ \Omega} = 0.68 \angle -45°\ \text{A}$$

$\mathbf{Z}_{eq}$ is put in the equation for I.

Then

$$P_{coil} = (0.68\ \text{A})^2 (20\ \Omega)$$
$$P_{coil} = 9.248\ \text{W}$$

The first equation gives power.

b. From Part a,

$$I = 0.68 \angle -45°\ \text{A}$$

c. $P_s = EI \cos \theta$

$$= (24\ \text{V})(0.68\ \text{A})(\cos 45°)$$
$$= (24\ \text{V})(0.68\ \text{A})(0.707)$$
$$P_s = 11.54\ \text{W}$$

E, I, and $\cos\theta$ are used to find source power.

EXAMPLE 16.21

For the circuit of Figure 16.47:

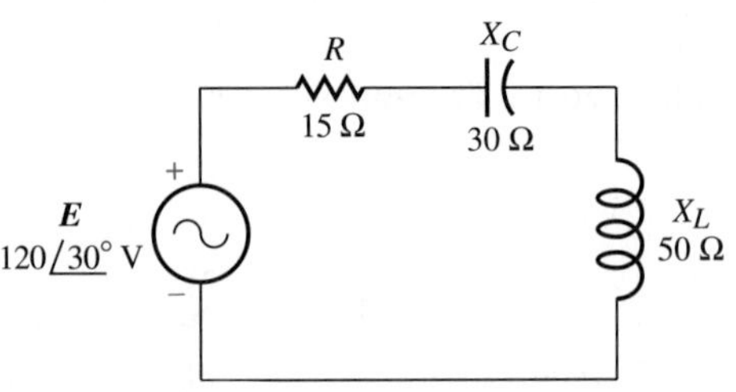

Figure 16.47

a. find the current, the voltage across the coil, and the power dissipated.

b. draw the impedance triangle.

Strategy

Solution

The current can be found using Ohm's Law for the circuit. Then the voltage divider equation gives the coil voltage. Finally, power is obtained from $P = EI\cos\theta$.

a.

Begin with a relation for current.

$$\mathbf{I} = \frac{\mathbf{E}}{\mathbf{Z}_{eq}}$$

Solve for $\mathbf{Z}_{eq}$.

$$\mathbf{Z}_{eq} = \mathbf{Z}_R + \mathbf{Z}_C + \mathbf{Z}_L$$

$$\mathbf{Z}_{eq} = R - jX_C + jX_L$$

$$= (15 - j30 + j50)\ \Omega$$

$$\mathbf{Z}_{eq} = 25\ \angle 53.13^\circ\ \Omega$$

$$\mathbf{I} = \frac{120\ \angle 30^\circ\ \text{V}}{25\ \angle 53.13^\circ\ \Omega}$$

$$\mathbf{I} = 4.8\ \angle -23.13^\circ\ \text{A}$$

The voltage divider expression includes the coil voltage.

$$\mathbf{V}_{coil} = \frac{\mathbf{Z}_L\mathbf{E}}{\mathbf{Z}_{eq}}$$

$$\mathbf{V}_{coil} = \frac{(X_L\ \angle 90^\circ)\ \mathbf{E}}{\mathbf{Z}_{eq}}$$

$$= \frac{(50\ \angle 90^\circ\ \Omega)(120\ \angle 30^\circ\ \text{V})}{25\ \angle 53.13^\circ\ \Omega}$$

$$\mathbf{V}_{coil} = 240\ \angle 66.87^\circ\ \text{V}$$

Note that the voltage across a part of the circuit is greater than that of the source. This is possible because of the phase difference.

$$P = EI\cos\,\theta$$

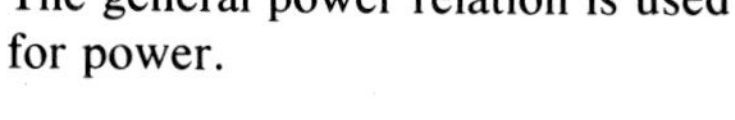

The general power relation is used for power.

Since θ is the angle of $\mathbf{Z}_{eq}$, or between $\boldsymbol{E}$ and $\boldsymbol{I}$, $\theta = 53.13°$. So

$$P = (120\text{ V})(4.8\text{ A})\cos 53.13°$$

$$P = 345.6\text{ W}$$

b. The impedance diagram and the triangle are shown in Figure 16.48.

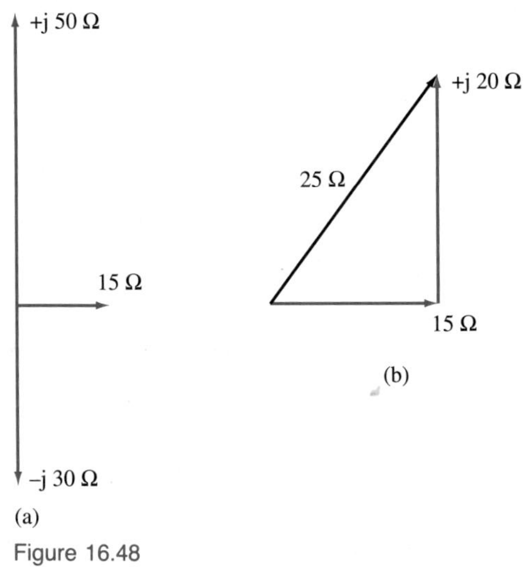

Figure 16.48

EXAMPLE 16.22

The power in a series R–L–C circuit is 90 watts. If $\boldsymbol{I} = 3\angle 0°$ A, $\boldsymbol{X}_L = 20\angle 90°\ \Omega$, and $X_C = 30\angle -90°$, find:

a. the resistance value.
b. the voltage across each element.
c. the source voltage.

Solution

Strategy

Begin with a resistance relation that includes the knowns.

$$P = I^2R$$

so

$$R = \frac{P}{I^2} = \frac{90\text{ W}}{(3\text{ A})^2}$$

$$R = 10\ \Omega$$

Since $\boldsymbol{I}$, R, X_C, X_L are known, Ohm's Law is used.

$$\boldsymbol{V_R} = \boldsymbol{IZ_R}$$
$$\boldsymbol{V_R} = \boldsymbol{I}R\ \angle 0° = (3\ \angle 0°\ \text{A})(10\ \angle 0°\ \Omega)$$
$$\boldsymbol{V_R} = 30\ \angle 0°\ \text{V}$$
$$\boldsymbol{V_C} = \boldsymbol{IZ_C} = \boldsymbol{I}X_C\ \angle -90° = (3\ \angle 0°\ \text{A})(30\ \angle -90°\ \Omega)$$
$$\boldsymbol{V_C} = 90\ \angle -90°\ \text{V}$$
$$\boldsymbol{V_L} = \boldsymbol{IZ_L} = \boldsymbol{I}X_L\ \angle 90° = (3\ \angle 0°\ \text{A})(20\ \angle 90°\ \Omega)$$
$$\boldsymbol{V_L} = 60\ \angle 90°\ \text{V}$$

Kirchhoff's Voltage Law includes $\boldsymbol{E}$, so it is used.

By Kirchhoff's Voltage Law,

$$\boldsymbol{E} - \boldsymbol{V_R} - \boldsymbol{V_L} - \boldsymbol{V_C} = 0$$

so

$$\boldsymbol{E} = 30\ \angle 0°\ \text{V} + 60\ \angle 90°\ \text{V} + 90\ \angle -90°\ \text{V}$$

or

$$\boldsymbol{E} = (30 - j30)\ \text{V}$$
$$\boldsymbol{E} = 42.43\ \angle -45°\ \text{V}$$

EXAMPLE 16.23

The power input to an ac series R–L load is 450 watts when connected to a 120-volt source. If the current is 5 amperes, find:

a. the power factor of the load.
b. the resistance of the load.
c. the inductive reactance of the load.

Strategy

Solution

Begin with the relation for power factor.

$$F_p = \frac{P}{EI} = \frac{450\ \text{W}}{(120\ \text{V})(5\ \text{A})}$$
$$F_p = 0.75$$

The power relation gives us R.

$$P = I^2R \qquad \text{so} \qquad R = \frac{P}{I^2}$$

$$R = \frac{450\ \text{W}}{(5\ \text{A})^2}$$
$$R = 18\ \Omega$$

The impedance triangle gives the relation for X.

From the impedance triangle,

$$\tan\theta = \frac{X_L}{R} \qquad \text{or} \qquad X_L = R\tan\theta$$

Since the power factor is $\cos\theta$,

The relation includes θ, so it is found from the power factor.

$$\theta = \cos^{-1}(F_P)$$
$$= \cos^{-1}(0.75)$$
$$\theta = 41.41°$$

Then

$$X_L = (18\ \Omega)\tan(41.41°)$$
$$X_L = 15.87\ \Omega$$

16.6 Parallel Circuit Analysis

The individual concepts for the parallel circuit are used to determine the conditions in a parallel circuit. Several examples show how this is done.

EXAMPLE 16.24

In the circuit of Figure 16.49, find:

a. the source current.
b. the current in each branch.
c. the power dissipated in the circuit.

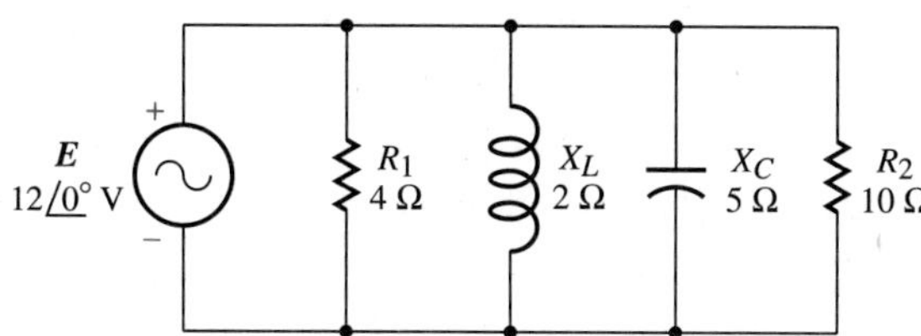

Figure 16.49

Solution

Using Ohm's Law for the circuit gives the source current. Using it for each branch gives the current in R_1, L, C, and R_2. Then, $P = EI\cos\theta$ gives the circuit power.

a.

$$\boldsymbol{I} = \frac{\boldsymbol{E}}{\boldsymbol{Z}_{\text{eq}}}$$

$$\frac{1}{\boldsymbol{Z}_{\text{eq}}} = \frac{1}{\boldsymbol{Z}_{R1}} + \frac{1}{\boldsymbol{Z}_L} + \frac{1}{\boldsymbol{Z}_C} + \frac{1}{\boldsymbol{Z}_{R2}}$$

$$\frac{1}{\boldsymbol{Z}_{\text{eq}}} = \frac{1}{R_1} + \frac{1}{X_L \angle 90°} + \frac{1}{X_C \angle -90°} + \frac{1}{R_2}$$

$$= \frac{1}{4\ \Omega} + \frac{1}{2 \angle 90°\ \Omega} + \frac{1}{5 \angle -90°\ \Omega} + \frac{1}{10\ \Omega}$$

$$= 0.25\ \text{S} + 0.5 \angle -90°\ \text{S} + 0.2 \angle 90°\ \text{S} + 0.1\ \text{S}$$

$$= 0.35 - j0.3\ \text{S} \quad \text{or} \quad 0.46 \angle -40.6°\ \text{S}$$

and

$$Z_{eq} = \frac{1}{0.46 \angle -40.6^\circ \text{ S}} = 2.17 \angle 40.6^\circ \ \Omega$$

$$I = \frac{12 \angle 0^\circ \text{ V}}{2.17 \angle 40.6^\circ \ \Omega}$$

$$I = 5.53 \angle -40.6^\circ \text{ A}$$

b.

$$I_{R1} = \frac{E}{Z_1} = \frac{E}{R_1} = \frac{12 \angle 0^\circ \text{ V}}{4 \ \Omega}$$

$$I_{R1} = 3 \angle 0^\circ \text{ A}$$

$$I_L = \frac{E}{Z_L} = \frac{E}{X_L \angle 90^\circ} = \frac{12 \angle 0^\circ \text{ V}}{2 \angle 90^\circ \ \Omega}$$

$$I_L = 6 \angle -90^\circ \text{ A}$$

$$I_C = \frac{E}{Z_C} = \frac{E}{X_C \angle -90^\circ} = \frac{12 \angle 0^\circ \text{ V}}{5 \angle -90^\circ \ \Omega}$$

$$I_C = 2.4 \angle 90^\circ \text{ A}$$

$$I_{R2} = \frac{E}{R_2} = \frac{E}{R_2} = \frac{12 \angle 0^\circ \text{ V}}{10 \ \Omega}$$

$$I_{R2} = 1.2 \angle 0^\circ \text{ A}$$

c. $P = EI\cos\theta = (12 \text{ V})(5.53 \text{ A}) \cos 40.6^\circ$

$P = 50.39 \text{ W}$

In this example, the branch current, I_L, is greater than the source current. This is possible because of the phase relation.

EXAMPLE 16.25

In the circuit of Figure 16.50:

a. find the total current and branch currents.
b. draw the phasor diagram.

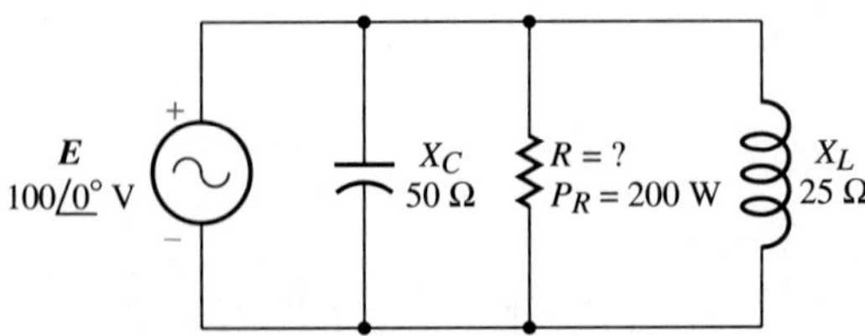

Figure 16.50

Solution

a.

$$\boldsymbol{I} = \frac{\boldsymbol{E}}{\boldsymbol{Z}_{eq}}$$

$$\frac{1}{\boldsymbol{Z}_{eq}} = \frac{1}{\boldsymbol{Z}_C} + \frac{1}{R} + \frac{1}{\boldsymbol{Z}_L}$$

$$\frac{1}{\boldsymbol{Z}_{eq}} = \frac{1}{X_C \angle -90°} + \frac{1}{R} + \frac{1}{X_L \angle 90°}$$

$$P_R = \frac{E^2}{R} \quad \text{so} \quad R = \frac{E^2}{P_R}$$

$$R = \frac{(100\text{ V})^2}{200\text{ W}} = 50\ \Omega$$

so

$$\frac{1}{\boldsymbol{Z}_{eq}} = \frac{1}{50 \angle -90°\ \Omega} + \frac{1}{50\ \Omega} + \frac{1}{25 \angle 90°\ \Omega}$$

$$= 0.02 \angle 90°\text{ S} + 0.02\text{ S} + 0.04 \angle -90°\text{ S}$$

$$= (0.02 - j0.02)\text{ S}$$

$$= 0.028 \angle -45°\text{ S}$$

$$\boldsymbol{Z}_{eq} = \frac{1}{0.028 \angle -45°\text{ S}} = 35.71 \angle 45°\ \Omega$$

Then

$$\boldsymbol{I} = \frac{100\text{ V}}{35.71 \angle 45°\ \Omega}$$

$$\boldsymbol{I} = 2.80 \angle -45°\text{ A}$$

$$\boldsymbol{I}_C = \frac{\boldsymbol{E}}{\boldsymbol{Z}_C} = \frac{\boldsymbol{E}}{X_C \angle -90°} = \frac{100 \angle 0°\text{ V}}{50 \angle -90°\ \Omega}$$

$$\boldsymbol{I}_C = 2 \angle 90°\text{ A}$$

$$\boldsymbol{I}_R = \frac{\boldsymbol{E}}{\boldsymbol{Z}_R} = \frac{\boldsymbol{E}}{R} = \frac{100 \angle 0°\text{ V}}{50\ \Omega}$$

$$\boldsymbol{I}_R = 2 \angle 0°\text{ A}$$

$$\boldsymbol{I}_L = \frac{\boldsymbol{E}}{\boldsymbol{Z}_L} = \frac{\boldsymbol{E}}{X_L \angle 90°} = \frac{100 \angle 0°\text{ V}}{25 \angle 90°\ \Omega}$$

$$\boldsymbol{I}_L = 4 \angle -90°\text{ A}$$

b. The phasor diagram is shown in Figure 16.51.

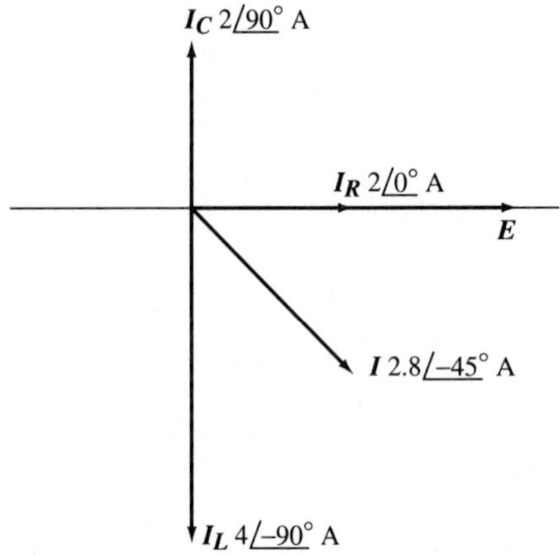

Figure 16.51

EXAMPLE 16.26

With the capacitor out of the circuit of Figure 16.52, the current is 1.7 $\angle -45°$ amperes. With the capacitor in the circuit, the current is 1.2 $\angle 0°$ amperes. Find:

a. the value of X_L.
b. the power factor of the circuit with the capacitor in the circuit.

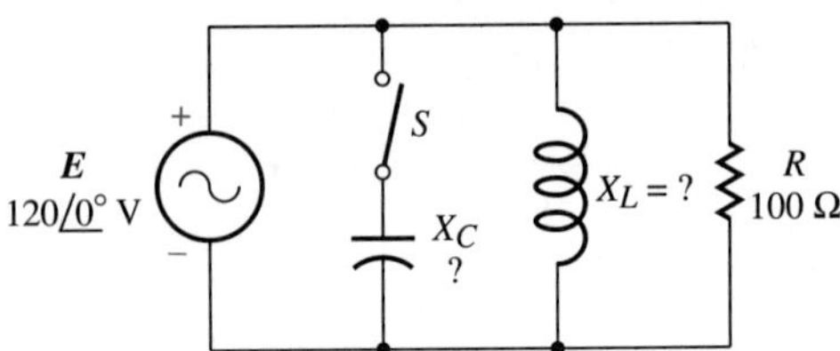

Figure 16.52

Solution

a. With the switch open,

$$Z_L = \frac{E}{I_L} \quad \text{but} \quad I_L = I - I_R = I - \frac{E}{R}$$

$$I_L = 1.7 \angle -45° \text{ A} - 1.2 \angle 0° \text{ A} \quad \text{or} \quad (1.2 - j1.2 - 1.2) \text{ A}$$

$$I_L = -j1.2 \text{ A} \quad \text{or} \quad 1.2 \angle -90° \text{ A}$$

Then

$$Z_L = \frac{E}{I_L} = \frac{120 \angle 0° \text{ V}}{1.2 \angle -90° \text{ A}}$$

So

$$X_L = 100 \ \Omega$$

b. $P = EI\cos\theta \quad \text{or} \quad \cos\theta = \dfrac{P}{EI}$

but

$$P = \frac{E^2}{R}$$

so

$$\cos\theta = \frac{E}{RI} = \frac{120\text{ V}}{(100\ \Omega)(1.2\text{ A})}$$
$$F_p = \cos\theta$$
$$F_p = 1$$

EXAMPLE 16.27

Find the value and the type of the component represented by Y_X in Figure 16.53.

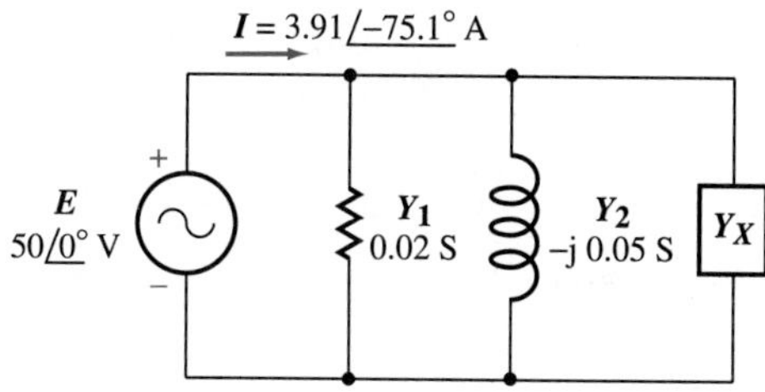

Figure 16.53

Solution

$$I = EY_{eq}$$

or

$$Y_{eq} = \frac{I}{E} = \frac{3.91\ \angle -75.1^\circ\text{ A}}{50\ \angle 0^\circ\text{ V}} = 0.078\ \angle -75.1^\circ\text{ S}$$
$$Y_{eq} = Y_R + Y_L + Y_X$$

or

$$Y_x = Y_{eq} - Y_R - Y_L$$
$$Y_x = (0.078\ \angle -75.1^\circ - 0.02 - (-j0.05))\text{ S}$$
$$Y_x = (0.02 - j0.075 - 0.02 + j0.05)\text{ S}$$
$$= -j0.025\text{ S}$$
$$Z_x = \frac{1}{Y_x} = \frac{1}{-j0.025\text{ S}} = j40\ \Omega \quad \text{or} \quad 40\ \angle 90^\circ\ \Omega$$

These are only a few of the many parallel circuit configurations possible. The analysis and synthesis of each one makes use of the basic parallel circuit relations.

16.7 Series-Parallel Circuit Analysis

series-parallel ac circuits

Many ac circuits include groups of series circuits, parallel circuits, and individual impedances. These are called **series-parallel ac circuits**. A few examples are shown in Figure 16.54.

Actually, many practical circuits that are considered to be parallel circuits are really series-parallel circuits. This is because the conductors that connect the branches have resistance. Usually this resistance is small compared to the impedance of the parts of the circuit, so that circuit is treated as a parallel circuit.

How are the conditions in these circuits determined? How does one find the current in a resistor, or the voltage across a capacitor in these circuits? The same approach as used for the dc series parallel circuit is used. The circuit is divided into series and parallel groups. Then the relations for series and parallel circuits are used. A summary of these relations for the ac circuit is given in Table 16.1 as a refresher.

The phasor diagram for the series-parallel circuit usually includes more quantities than for the series and parallel circuits. However, it is

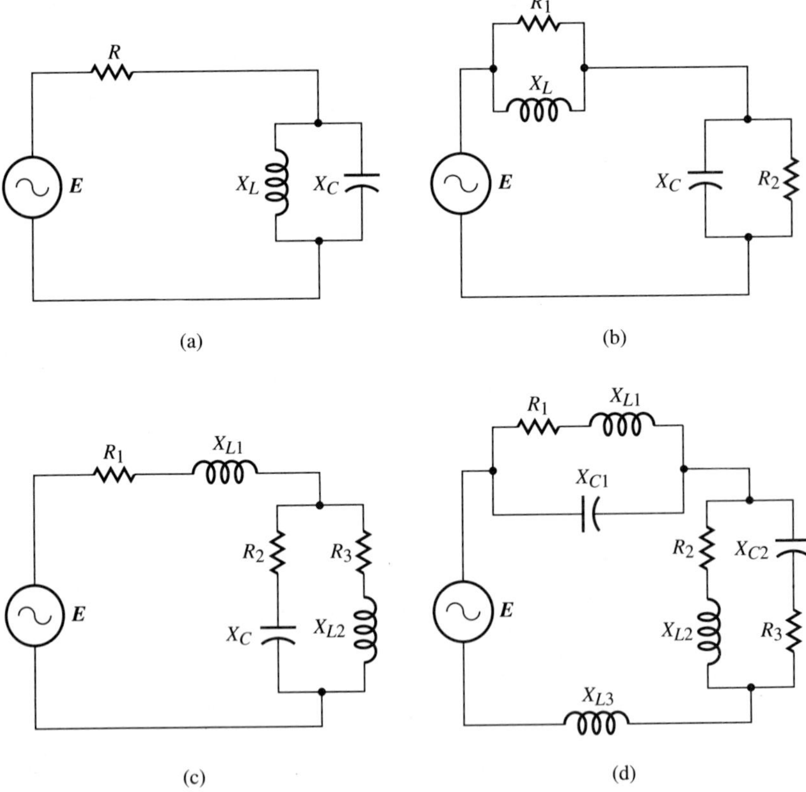

Figure 16.54 Series-parallel ac circuits can have many configurations. (a) R is in series with the parallel group of X_L and X_C. (b) The R_1,X_L parallel group is in series with the R_2,X_C parallel group. (c) The R_1,X_{L1} series group is in series with the series-parallel R_2,X_C, R_3,X_{L2} group. (d) X_{L3} is in series with the series-parallel R_1,X_{L1},X_{C1} group, and the series-parallel R_2,X_{L2},X_{C2},R_3 group.

TABLE 16.1 Series and Parallel Circuit Characteristics

Series Circuit	Parallel Circuit
$I_1 = I_2 = \cdots = I_n$	$V_1 = V_2 = \cdots = V_n$
$V_T = V_1 + V_2 + \cdots + V_n$	$I_T = I_1 + I_2 + \cdots + I_n$
	$Y_{eq} = Y_1 + Y_2 + \cdots + Y_n$
$Z_{eq} = Z_1 + Z_2 + \cdots + Z_n$	$\frac{1}{Z_{eq}} = \frac{1}{Z_1} + \frac{1}{Z_2} + \cdots + \frac{1}{Z_n}$
$E = IZ_{eq}$	$E = IZ_{eq}$
$V_x = \frac{EZ_x}{Z_{eq}}$	$I_x = \frac{IZ_{eq}}{Z_x}$
$P = EI\cos\theta$	$P = EI\cos\theta$
$\cos\theta = \frac{R_{eq}}{Z_{eq}}$	$\cos\theta = \frac{R_{eq}}{Z_{eq}}$
$\tan\theta = \frac{X_{eq}}{R_{eq}}$	$\tan\theta = \frac{X_{eq}}{R_{eq}}$
$P_T = P_1 + P_2 + \cdots + P_n$	$P_T = P_1 + P_2 + \cdots + P_n$
$\mathcal{W}_T = \mathcal{W}_1 + \mathcal{W}_2 + \cdots + \mathcal{W}_n$	$\mathcal{W}_T = \mathcal{W}_1 + \mathcal{W}_2 + \cdots + \mathcal{W}_n$

constructed in the same manner. That is, once the magnitude and angles are found, the phasor is drawn at the angle and to that magnitude. The phasor voltages for parts in series can be added graphically to give the total voltage. Also, the phasor currents for parts in parallel can be added graphically to give the total current.

Although a series-parallel ac circuit can also be replaced by an equivalent impedance, a standard equation for the impedance does not exist. This is because the circuits vary. The impedance can be determined by breaking down the circuit into its parallel and series groups. This can be done by inspection, or by a mechanical sequence. The first method is shorter, but requires the ability to recognize the arrangement of the parts. The second method is longer, but follows a set procedure. The procedure for the second method is as follows:

1. Replace the series and parallel groups with their equivalent impedance and redraw the circuit.
2. Repeat Step 1 until the circuit is reduced to no more than one resistance and one reactance.

EXAMPLE 16.28

Find the equivalent impedance of the circuit in Figure 16.55 by:

a. inspection.
b. series and parallel groups.

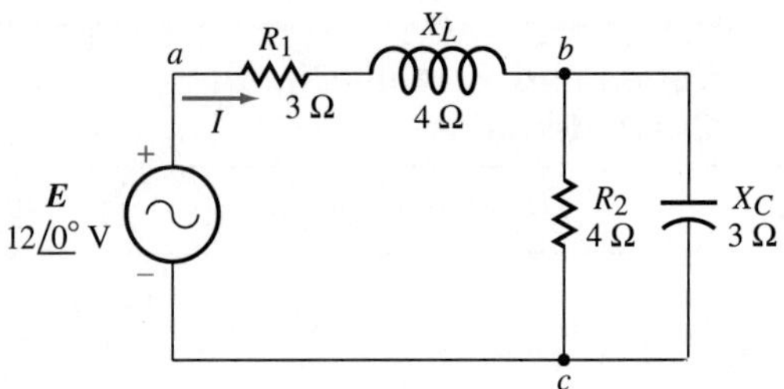

Figure 16.55

Solution

a. R_1, X_L, and the parallel group of R_2 and X_C are in series. The equivalent impedance is

$$\mathbf{Z}_{\mathrm{eq}} = \mathbf{Z}_{R1} + \mathbf{Z}_L + \left(\frac{\mathbf{Z}_{R2}\mathbf{Z}_C}{\mathbf{Z}_{R2} + \mathbf{Z}_C}\right)$$

$$\mathbf{Z}_{\mathrm{eq}} = R_1 + jX_L + \left(\frac{R_2 X_C \angle -90^\circ}{R_2 - jX_C}\right)$$

$$= 3\ \Omega + j4\ \Omega + \frac{(4 \angle 0^\circ\ \Omega)(3 \angle -90^\circ\ \Omega)}{(4 - j3)\ \Omega}$$

$$= (4.44 + j2.08)\ \Omega$$

$$\mathbf{Z}_{\mathrm{eq}} = 4.90 \angle 25.10^\circ\ \Omega$$

b. R_1 and X_L are in series

$$\mathbf{Z}_{ab} = \mathbf{Z}_{R1} + \mathbf{Z}_L$$

so

$$\mathbf{Z}_{ab} = R_1 + jX_L = (3 + j4)\ \Omega$$

The circuit is now shown in Figure 16.56.

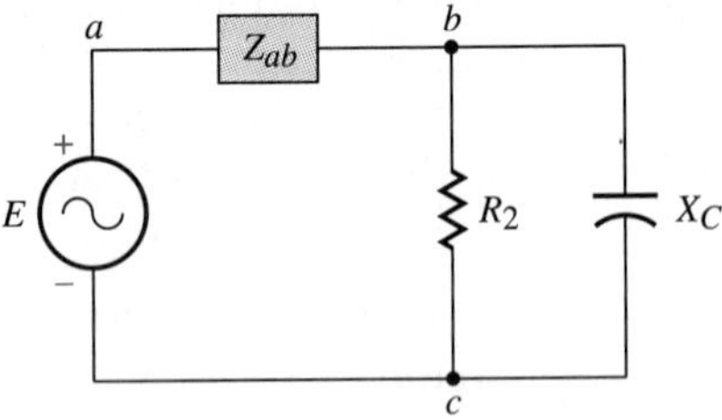

Figure 16.56

R_2 and X_C are in parallel

$$\mathbf{Z}_{bc} = \frac{\mathbf{Z}_{R2}\mathbf{Z}_C}{\mathbf{Z}_{R2} + \mathbf{Z}_C}$$

so

$$\mathbf{Z}_{bc} = \frac{R_2 X_C \angle -90°}{R_2 - jX_C}$$
$$= \frac{(4\ \Omega)(3\ \angle -90°\ \Omega)}{(4 - j3)\ \Omega}$$
$$= 2.4\ \angle -53.13°\ \Omega$$

Now the circuit is shown in Figure 16.57.

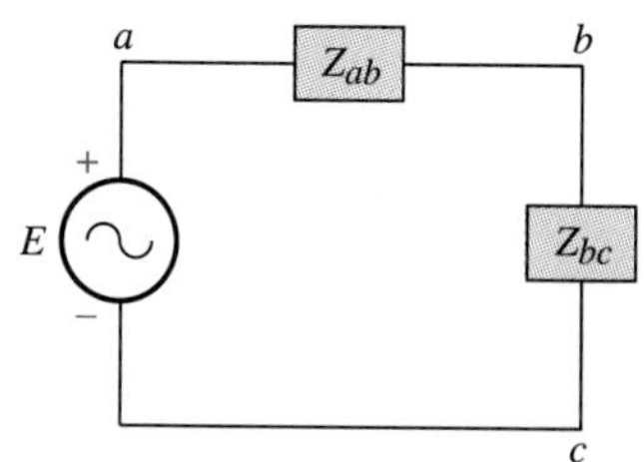

Figure 16.57

c. $\mathbf{Z}_{ab}$ and $\mathbf{Z}_{bc}$ are in series so

$$\mathbf{Z}_{ac} = \mathbf{Z}_{ab} + \mathbf{Z}_{bc}$$
$$= 3\ \Omega + j4\ \Omega + 2.4\ \angle -53.13°\ \Omega$$
$$\mathbf{Z}_{ac} = (3 + j4 + 1.44 - j1.92)\ \Omega$$
$$= (4.44 + j2.08)\ \Omega$$

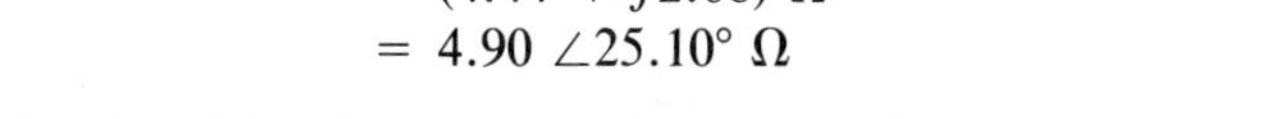

$$= 4.90\ \angle 25.10°\ \Omega$$

The circuit has been reduced to one resistance and one reactance so

$$\mathbf{Z}_{eq} = 4.90\ \angle 25.10°\ \Omega$$

The next three examples deal with the analysis of some series-parallel circuits. **Analysis** is the procedure in which the conditions are found when the parts of the circuit are known.

analysis

Recall, from dc series-parallel circuits, that the key was to develop a sequence that relates the known quantities to the wanted ones. Since the circuit configuration can vary, the sequence can vary. One approach is to use the unknowns in the concepts and relations as a guide—that is—the unknown quantities in each equation steer you to the next one. How this is done is shown in the following examples.

EXAMPLE 16.29

a. Find the current and voltage for each component in the circuit of Example 16.28.
b. Find the power supplied by the source.
c. Draw the phasor diagram.

Solution

Strategy

Ohm's Law is used to find the current.

a. From Example 16.28

$$\mathbf{Z}_{eq} = 4.9\ \angle 25.1°\ \Omega$$
$$\mathbf{Z}_{bc} = 2.4\ \angle -53.13°\ \Omega$$

$$I = \frac{E}{Z_{eq}} = \frac{12 \angle 0° \text{ V}}{4.9 \angle 25.1° \ \Omega} = 2.45 \angle -25.1° \text{ A}$$

The first equation indicates a need for one with V_{R2} in it. The equation for V_{R2} directs us to one with V_{bc} in it.

$$I_{R2} = \frac{V_{R2}}{R_2} \quad \text{and} \quad V_{R2} \text{ is } V_{bc}$$

so

$$I_{R2} = \frac{V_{bc}}{R_2}$$

From the circuit of Figure 16.55, $V_{bc} = IZ_{bc}$ so

$$V_{bc} = (2.45 \angle -25.1° \text{ A})(2.4 \angle -53.13° \ \Omega) = 5.88 \angle -78.23° \text{ V}$$

Solve for I_{R2}.

Then

$$I_{R2} = \frac{5.88 \angle -78.23° \text{ V}}{4 \ \Omega}$$

$$I_{R2} = 1.47 \angle -78.23° \text{ A}$$

$$V_{R1} = I_s Z_1$$

so

Ohm's Law is used for the voltages.

$$V_{R1} = I_s R_1 = (2.45 \angle -25.1° \text{ A})(3 \ \Omega)$$

$$V_{R1} = 7.35 \angle -25.1° \text{ V}$$

$$V_L = I_s Z_L$$

so

$$V_L = I_s X_L \angle 90° = (2.45 \angle -25.1° \text{ A})(4 \angle 90° \ \Omega)$$

$$V_L = 9.8 \angle 64.9° \text{ V}$$

$$V_C = V_{bc} = 5.88 \angle -78.23° \text{ V}$$

$$V_C = 5.88 \angle -78.23° \text{ V}$$

$$I_C = \frac{V_{bc}}{Z_C}$$

so

$$I_C = \frac{V_{bc}}{X_C \angle -90°}$$

$$I_C = \frac{(5.88 \angle -78.23° \text{ V})}{3 \angle -90° \ \Omega}$$

$$I_C = 1.96 \angle 11.77° \text{ A}$$

$$V_{R2} = V_C$$

$$I_L = I$$

The general power relation has only P as the unknown, so P can be determined.

b. $P = EI \cos \theta$

$E = 12$ V, $I = 2.45$ A, and θ is the angle of Z_{eq}, which is 25.10°.

So

$$P = (12\text{ V})(2.45\text{ A})(0.91)$$
$$P = 26.62\text{ W}$$

c. The phasor diagram for the circuit is shown in Figure 16.58(a). Figure 16.58(b) shows the phasor addition used to obtain the source voltage and current.

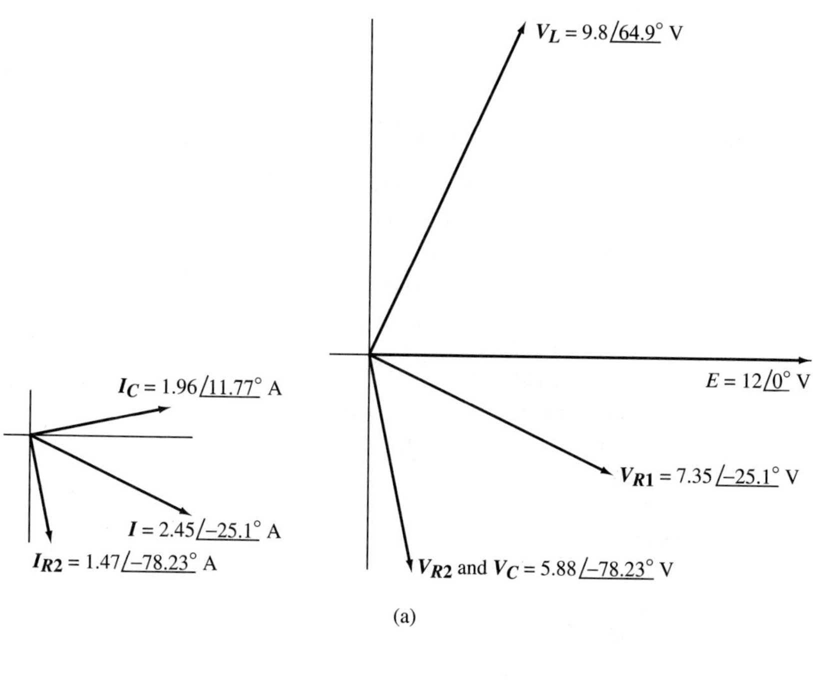

(a)

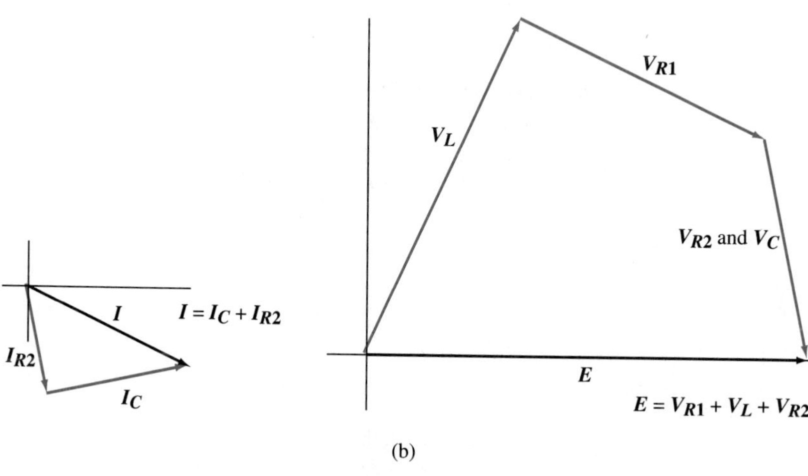

(b)

Figure 16.58

EXAMPLE 16.30

Find the voltage across the resistor R_2 and the power dissipated in the circuit of Figure 16.59 between a and c.

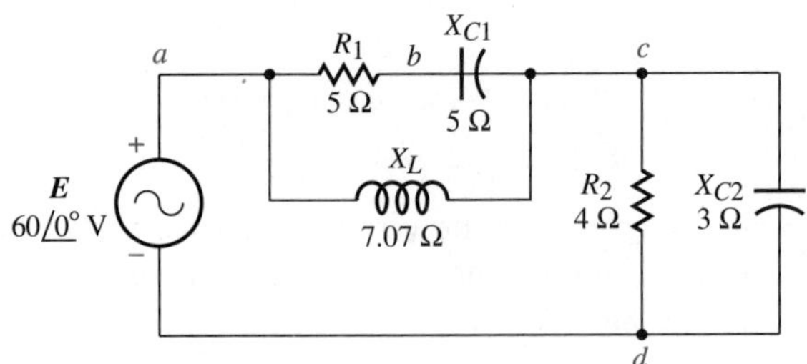

Figure 16.59

Strategy

Solution

In this example, we start with the voltage divider relation for V_{R2}. This gives

This equation indicates that Z_{cd} and Z_{eq} are needed.

$$\mathbf{V}_{R2} = \frac{\mathbf{Z}_{cd}\mathbf{E}}{\mathbf{Z}_{eq}}$$

$$\mathbf{Z}_C = \mathbf{Z}_{R2} \| \mathbf{Z}_{C2}$$

$$\mathbf{Z}_{cd} = \mathbf{R}_2 \| X_{C2} \angle -90°$$

$$\mathbf{Z}_{cd} = \frac{(R_2)(X_{C2} \angle -90°)}{R_2 - jX_{C2}}$$

$$= \frac{(4\ \Omega)(3 \angle -90°\ \Omega)}{(4 - j3)\ \Omega}$$

$$= 2.4 \angle -53.13°\ \Omega$$

$$\mathbf{Z}_{eq} = \mathbf{Z}_{cd} + \mathbf{Z}_{ac}$$

$\mathbf{Z}_{ac}$ is the series branch of R_1 and X_{C1}, in parallel with X_L, so

$$\mathbf{Z}_{ac} = \frac{(\mathbf{Z}_{R1} + \mathbf{Z}_{C1})\mathbf{Z}_L}{\mathbf{Z}_{R1} + \mathbf{Z}_{C1} + \mathbf{Z}_L}$$

$$\mathbf{Z}_{ac} = \frac{(R_1 - jX_{C1})(X_L \angle 90°)}{R_1 - jX_{C1} + jX_L}$$

$$= \frac{((5 - j5)\ \Omega)(7.07 \angle 90°\ \Omega)}{(5 - j5 + j7.07)\ \Omega}$$

which gives

$$\mathbf{Z}_{ac} = \frac{50 \angle 45°\ \Omega^2}{5.41 \angle 22.49°\ \Omega}$$

$$= 9.24 \angle 22.51°\ \Omega$$

Therefore,

$$\mathbf{Z}_{eq} = 2.4 \angle -53.13°\ \Omega + 9.24 \angle 22.51°\ \Omega$$

$$= (1.44 - j1.92 + 8.54 + j3.54)\ \Omega$$

$$= (9.98 + j1.62)\ \Omega$$

or

$$\mathbf{Z}_{eq} = 10.11 \angle 9.22°\ \Omega$$

Putting this into the expression for V_{R2} gives

$$V_{R2} = \frac{(2.4 \angle -53.13° \ \Omega)(60 \angle 0° \ \text{V})}{10.11 \angle 9.22° \ \Omega}$$

$$V_{R2} = 14.24 \angle -62.35° \ \text{V}$$

V_{R2} can now be determined.

The circuit power is the power in resistor R_1 because there is no power lost in the capacitor and the ideal inductor. Writing an equation for the power in R_1 gives

The circuit power is in the resistance.

$$P_{R1} = \frac{V_{R1}^2}{R_1}$$

But, by the voltage divider rule for section ac,

$$V_{R1} = \frac{Z_{R1}V_{ac}}{Z_{R1} - Z_C} = \frac{R_1 V_{ac}}{R_1 - jX_{C1}}$$

This relation indicates that V_{ac} is needed.

and by Kirchhoff's Voltage Law,

$$V_{ac} = E - V_{R2}$$

Solving the three previous equations gives

$$V_{ac} = 60 \angle 0° \ \text{V} - 14.24 \angle -62.35° \ \text{V}$$
$$= 54.86 \angle 13.28° \ \text{V}$$

Kirchhoff's Voltage Law is used.

Then

$$V_{R1} = \frac{(5 \ \Omega)(54.86 \angle 13.28° \ \text{V})}{(5 - j5)}$$
$$= 38.8 \angle 58.28° \ \text{V}$$

so

$$P = \frac{(38.8 \ \text{V})^2}{5 \ \Omega}$$

$$P_{ac} = 301.1 \ \text{W}$$

This sequence is only one of the several sequences that can be used to solve this problem.

EXAMPLE 16.31

Find the total current I, the voltage across R_2, and the power dissipated in the circuit of Figure 16.60.

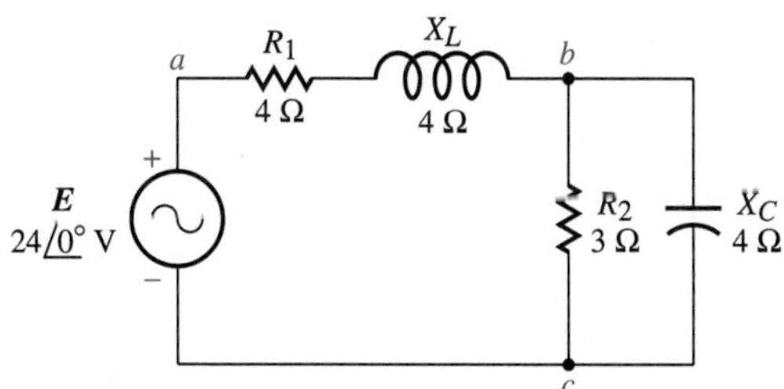

Figure 16.60

Solution

$$\mathbf{I} = \frac{\mathbf{V}_{bc}}{\mathbf{Z}_{bc}}$$

$$\mathbf{V}_{bc} = \frac{(\mathbf{Z}_{bc})(\mathbf{E})}{\mathbf{Z}_{bc} + \mathbf{Z}_{ab}}$$

$$\mathbf{Z}_{bc} = \mathbf{Z}_{R2} \| \mathbf{Z}_C$$

$$= \frac{(\mathbf{Z}_{R2})(\mathbf{Z}_C)}{\mathbf{Z}_{R2} + \mathbf{Z}_C}$$

$$= \frac{(R_2 \angle 0°)(X_C \angle -90°)}{R_2 - jX_C}$$

$$\mathbf{Z}_{ab} = \mathbf{Z}_{R1} + \mathbf{Z}_L$$

so

$$\mathbf{Z}_{ab} = R_1 + jX_L$$

At this point, the number of equations equals the number of unknowns, so we can begin to solve them.

$$\mathbf{Z}_{ab} = R_1 + jX_L = (4 + j4)\ \Omega \quad \text{or} \quad 5.66 \angle 45°\ \Omega$$

$$\mathbf{Z}_{bc} = \frac{(R_2)(X_C \angle -90°)}{R_2 - jX_C} = \frac{(3\ \Omega)(4 \angle -90°\ \Omega)}{(3 - j4)\ \Omega}$$

$$= 2.4 \angle -36.87°\ \Omega$$

Then

$$\mathbf{V}_{bc} = \frac{(\mathbf{Z}_{bc})(\mathbf{E})}{\mathbf{Z}_{ab} + \mathbf{Z}_{bc}} = \frac{(2.4 \angle -36.87°\ \Omega)(24 \angle 0°\ \text{V})}{5.66 \angle 45°\ \Omega + 2.4 \angle -36.87°\ \Omega}$$

$$= \frac{57.6 \angle -36.87°\ \text{V}}{6.45 \angle 23.39°\ \Omega} = 8.93 \angle -60.26°\ \text{V}$$

$$\mathbf{I} = \frac{\mathbf{V}_{bc}}{\mathbf{Z}_{bc}} = \frac{8.93 \angle -60.26°\ \text{V}}{2.4 \angle -36.87°\ \Omega}$$

$$\mathbf{I} = 3.72 \angle -23.39°\ \text{A}$$

$\mathbf{V}_{R2}$ is the voltage across b–c so

$$V_{R2} = 8.93 \angle -60.26°\ \text{V}$$

The first group of equations did not include one for power. That equation must be written now. One such relation is

$$P = P_{R1} + P_{R2}$$

$$P_{R1} = I^2 R_1$$

$$P_{R2} = \frac{V_{bc}^2}{R_2}$$

$$= \frac{(8.93\ \text{V})^2}{3\ \Omega}$$

$$= 26.58\ \text{W}$$

$$P_{R1} = (3.72\ \text{A})^2(4\ \Omega) = 55.35\ \text{W}$$
$$P = 55.35\ \text{W} + 26.58\ \text{W}$$
$$P = 81.93\ \text{W}$$

These examples are meant to show the steps in the solution. They do not necessarily represent the shortest sequence. As one becomes more proficient, the solution to the equation would be done without first listing them. This reduces the steps, and makes the solution shorter.

It is often necessary to determine what parts are needed to obtain certain conditions. One place where this is done is during the design of a circuit. This procedure is called the **synthesis** of a circuit. The approach for synthesis is the same as for the analysis, except that the starting point depends on which quantity is wanted. As in dc, it simply means that we must solve for a different unknown.

synthesis

The next examples illustrate the synthesis of some circuits. These examples include series circuits, parallel circuits, and combination circuits.

EXAMPLE 16.32

A 120-volt motor is to be connected to a 240-volt source, as shown in Figure 16.61. The motor has a resistance of 10 ohms and an inductive reactance of 30 ohms. Find:

a. the value of the series resistance R_x needed to drop the voltage to 120 volts.
b. the wattage rating of the resistor.

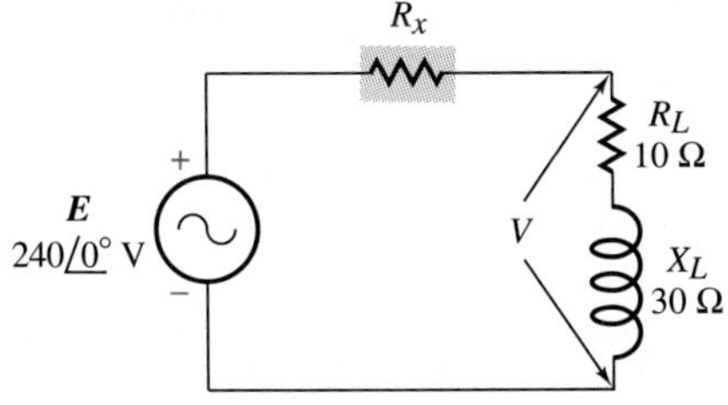

Figure 16.61

Solution

The solution needs some relation that includes R_x in it. The voltage divider relation includes R_x.

$$V = \left(\frac{R_L + jX_L}{R_x + R_L + jX_L}\right)E = \frac{(10\ \Omega + j30\ \Omega)(240\ \angle 0^\circ\ \text{V})}{(R_x + 10 + j30)\ \Omega}$$
$$= \frac{7589.5\ \angle 71.57^\circ\ \Omega \cdot \text{V}}{(R_x + 10 + j30)\ \Omega}$$

The magnitude of V is

$$\frac{7589.5\ \Omega \cdot \text{V}}{\sqrt{(R_x + 10\ \Omega)^2 + (30\ \Omega)^2}}$$

But the magnitude of V is 120 volts so

$$120\ \text{V} = \frac{7589.5\ \text{V}}{\sqrt{(R_x + 10\ \Omega)^2 + (30\ \Omega)^2}}$$

This reduces to

$$(R_x + 10\ \Omega)^2 = \frac{(7589.5\ \Omega \cdot \text{V})^2}{(120\ \text{V})^2} - (30\ \Omega)^2$$

$$R_x = 45.68\ \Omega$$

The power rating of the resistor must be greater than the power dissipated in it.

$$P_x = I^2 R_x \quad \text{and} \quad I = \frac{V_M}{Z_M}$$

The magnitude of I is given by

$$I = \frac{120\ \text{V}}{\sqrt{(R^2 + X_L^2)}} = \frac{120\ \text{V}}{\sqrt{(10\ \Omega)^2 + (30\ \Omega)^2}} = 3.79\ \text{A}$$

so

$$P_x = (3.79\ \text{A})^2(45.68\ \Omega) = 656.2\ \text{W}$$

$$P_{\text{rating}} = 656\ \text{W minimum}$$

EXAMPLE 16.33

For the circuit in Figure 16.62 to have a power factor of 0.8 leading, what value of X_C is needed?

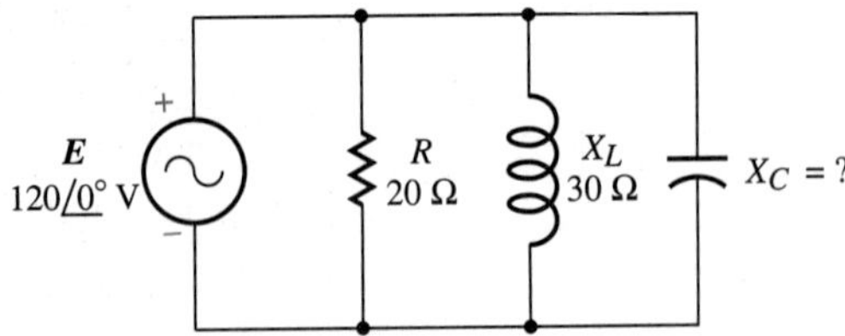

Figure 16.62

Solution

The solution starts with a relation that has X_C in it.

$$X_C \angle -90° = \frac{\mathbf{E}}{\mathbf{I}_C}$$

$$\mathbf{I} = \mathbf{I}_R + \mathbf{I}_L + \mathbf{I}_C$$

and

$$I_R = \frac{E}{R} = \frac{120\ \angle 0°\ \text{V}}{20\ \Omega} = 6\ \angle 0°\ \text{A}$$

also

$$I_L = \frac{E}{X_L\ \angle 90°} = \frac{120\ \angle 0°\ \text{V}}{30\ \angle 90°\ \Omega} = 4\ \angle -90°\ \text{A}$$

An equation with I in it is needed.

$$I = \frac{P}{E \cos \theta}$$

and

$$P = \frac{E^2}{R} = \frac{(120\ \text{V})^2}{20\ \Omega} = 720\ \text{W}$$

so

$$\text{I} = \frac{720\ \text{W}}{(120\ \text{V})(0.8)} = 7.5\ \text{A}$$

The angle of I is arc cos 0.8, or 36.87°. Since the power factor is a leading one, the angle is positive. So

$$I = 7.5\ \angle 36.87°\ \text{A}$$

Putting this into $I = I_R + I_L + I_C$ gives

$$7.5\ \angle 36.87°\ \text{A} = I_R + I_L + I_C$$

or

$$\begin{aligned} I_C &= 7.5\ \angle 36.87°\ \text{A} - 6\ \angle 0°\ \text{A} - 4\ \angle -90°\ \text{A} \\ &= (6 + j4.5 - 6 + j4)\ \text{A} \\ &= j8.5\ \text{A} \qquad \text{or} \qquad 8.5\ \angle 90°\ \text{A} \end{aligned}$$

Then

$$X_C\ \angle -90° = \frac{E}{I_c} = \frac{120\ \angle 0°\ \text{V}}{8.5\ \angle 90°\ \text{A}} = 14.12\ \angle -90°\ \Omega$$

$$X_C = 14.12\ \Omega$$

EXAMPLE 16.34

The inductor and capacitor in Figure 16.63 form the filtering circuit of a dc power supply. What must the value of X_L be if the voltage across R is 3 volts?

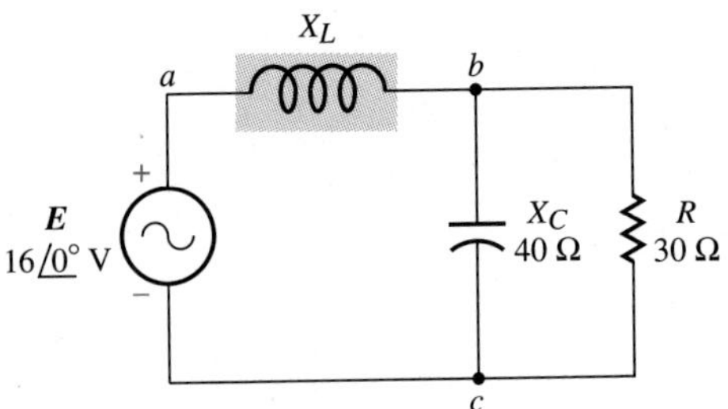

Figure 16.63

Solution

Again starting with the voltage divider relation, which has X_L in it,

$$\mathbf{V}_{bc} = \frac{\mathbf{Z}_{bc}\mathbf{E}}{\mathbf{Z}_{bc} + \mathbf{Z}_L}$$

$$\mathbf{Z}_{bc} = \mathbf{Z}_R \| \mathbf{Z}_c$$

so

$$\mathbf{Z}_{bc} = \frac{(\mathbf{Z}_R)(\mathbf{Z}_C)}{\mathbf{Z}_R + \mathbf{Z}_c}$$

$$= \frac{(R \angle 0°)(X_C \angle -90°)}{R - jX_C}$$

$$= \frac{(30\ \Omega)(40 \angle -90°\ \Omega)}{(30 - j40)\ \Omega}$$

$$= 24 \angle -36.87°\ \Omega$$

Then

$$\mathbf{V}_{bc} = \frac{(24 \angle -36.87°\ \Omega)(16 \angle 0°\ \text{V})}{(24 \angle -36.87° + jX_L)\ \Omega}$$

$$= \frac{384 \angle -36.87°\ \Omega \cdot \text{V}}{(19.2 + j(X_L - 14.40))\ \Omega}$$

The magnitude of $\mathbf{V}_{bc}$ is given by

$$\frac{384\ \Omega \cdot \text{V}}{\sqrt{(19.2\ \Omega)^2 + (X_L - 14.40\ \Omega)^2}}$$

This must be equal to 3 V so

$$\frac{384\ \Omega \cdot \text{V}}{\sqrt{(19.2\ \Omega)^2 + (X_L - 14.40\ \Omega)^2}} = 3\ \text{V}$$

Then

$$\sqrt{(19.2\ \Omega)^2 + (X_L - 14.40\ \Omega)^2} = \frac{384\ \Omega \cdot \text{V}}{3\ \text{V}}$$

Squaring both sides and rearranging it gives

$$(X_L - 14.40\ \Omega)^2 = \left(\frac{384\ \Omega \cdot \mathrm{V}}{3}\right)^2 - (19.2\ \Omega)^2$$

so

$$X_L = 140.95\ \Omega$$

S 16.8 SPICE analysis of Single Source ac Circuits

SPICE analysis of circuits driven by a single-frequency ac source is a simple extension of the dc analysis technique that has already been covered. Looking ahead, in Chapter 19, the use of SPICE will be extended to cover a range of input frequencies where SPICE analyses produce amplitude and phase frequency response plots.

Independent ac voltage and current sources can be described in SPICE. The key parameters that you must provide include the name of the source, the nodes to which it is connected, and the magnitude and phase of the voltage or current. The following format must be used to specify these sources.

Independent ac Voltage Source

The general format for specifying an ac voltage source is:

```
V<name> <(+) node> <(-) node> [AC<(magnitude) value>
                  [(phase) value]]
```

where: V<name> is the name of the ac voltage source, (+) node is the positive node, (−) node is the reference node, <(magnitude) value> is the amplitude of the voltage in volts, and <(phase) value> is the phase angle in degrees.

EXAMPLE 16.35

Interpret the following PSpice statement:

```
VS   1   0   AC   10V   0DEG
```

Draw a schematic symbol for this source.

Solution

The name of the circuit component is VS. The first letter V in the component name tells the simulator that the component is an independent voltage source. The voltage source is connected between nodes 1 and 0. Recall that node 0 is the ground reference node in the circuit. The magnitude of the ac voltage source is 10 volts and is the peak value of the voltage waveform. The phase angle is 0°.

The schematic symbol can then be drawn, which is shown in Figure 16.64.

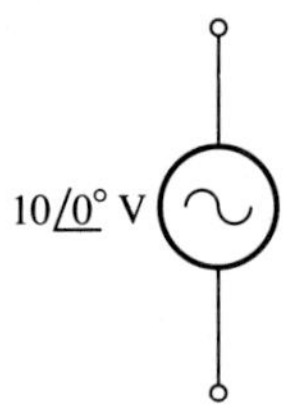

Figure 16.64 Schematic symbol of an ac voltage source.

Independent ac Current Source

Likewise, an independent ac current source can be specified in PSpice. The general format for an independent current source is:

```
I<name> <(+)node> <(-)node> [AC<(magnitude) value>
                              [(phase)]]
```

where the magnitude is in peak amperes and the phase is in degrees. The default values are zero for ac values.

EXAMPLE 16.36

Interpret the following PSpice statement:

```
IAC   2   3   AC   5MA   90DEG
```

Solution

The component name begins with the letter I, which indicates that the circuit component is an independent current source. The current source is connected between nodes 2 and 3. The magnitude of the current is 5 milliamperes and the phase angle is 90°.

The .AC statement is used to calculate the frequency response over a range of frequencies. The general form of the statement is:

```
.AC [type of sweep] <(points) value> <(start frequency)
         value> <(end frequency) value>
```

where `[type of sweep]` can be a linear sweep, LIN, a sweep by octaves, OCT, or a sweep by decades, DEC; `<(points) value>` is the number of data points in the sweep; `<(start frequency) value>` is the starting frequency value; and `<(end frequency) value>` is the ending frequency value. Note that the end frequency value must be greater or equal to the start frequency value, and both values must be greater than zero.

If the .AC command is to specify a single frequency value, the points value should be set to 1 and the start frequency and end frequency values should be the same.

Now let us perform an ac analysis on a circuit and print out the voltage and current values. The circuit is shown in Figure 16.65. The nodes are numbered, as shown in the figure, with ground assigned the node number of 0.

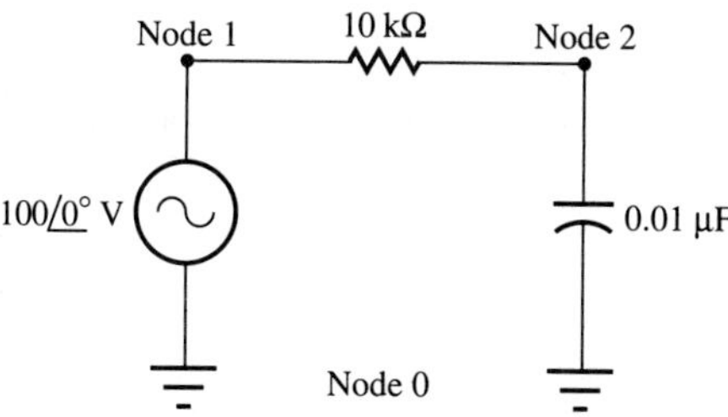

Figure 16.65 Series *R–C* circuit driven with an ac source.

The first line in the PSpice file is the title line:

```
AC Analysis of a Simple R-C Circuit
```

The next three lines describe the connection of components and their component values:

```
VS   1   0   AC      100V    0DEG
R1   1   2   10K
C1   2   0   0.01UF
```

The command lines are as follows. The first command line is the .AC statement, which is used to specify the frequency range. In this case, only one frequency is specified, 1500 Hz. Therefore, a linear sweep can be chosen with one data point. The start frequency and end frequency are both set to 1500 Hz.

```
.AC   LIN   1    1500Hz    1500HZ
```

The next two command lines specify the results that are to be printed out. Both the magnitude and the phase angle for the voltage and current are output variables.

```
.PRINT AC    VM(C1)    VP(C1)
.PRINT AC    IM(C1)    IP(C1)
```

The .OPTIONS NOPAGE is used to save paper and does not paginate the output listing. And, of course, the last line is the .END command.

Figure 16.66 shows the source file listing and the results.

Now you need to ask whether the results obtained are reasonable. Mentally, compute the capacitive reactance of C_1 and then estimate the voltage and current values. Then compare them to the values returned by the PSpice program. The two sets of values should be close to each other.

Another capability of PSpice that is useful in the analysis of ac circuits is the graphical display of voltage and current waveforms using PROBE. In Chapter 12, the .TRAN command was used to display voltage and current waveforms in *R–C* circuits. Now the .TRAN command to ac circuits is applied.

```
AC ANALYSIS OF A SIMPLE R-C CIRCUIT
****     CIRCUIT DESCRIPTION
*****************************************************************************

VS      1      0     AC            100V    0DEG
R1      1      2     10K
C1      2      0     0.01UF
.AC     LIN    1     1500HZ                1500HZ
.OPTIONS NOPAGE
.PRINT AC VM(C1)     VP(C1)
.PRINT AC IM(C1)     IP(C1)
.END

****     AC ANALYSIS                                    TEMPERATURE =    27.000 DEG C
 FREQ            VM(C1)          VP(C1)
  1.500E+03     7.277E+01       -4.330E+01

****     AC ANALYSIS                                    TEMPERATURE =    27.000 DEG C
 FREQ            IM(C1)          IP(C1)
  1.500E+03     6.859E+03       4.670E+01
```

Figure 16.66 PSpice output file.

The circuit to be analyzed is shown in Figure 16.67. The independent voltage source is a 10-volt peak sine wave of frequency 1000 Hz with zero dc offset. The PSpice statement that describes the voltage source is:

```
VS    1    0    SIN(0    10V    1000HZ)
```

The parameters for the sine wave are as follows:

```
SIN(<voff> <vampl> <freq> <td> <tf> <phase>)
```

where <voff> is the offset voltage, in volts; <vampl> is the peak amplitude of the voltage, in volts; <freq> is the frequency, in Hertz; <td> is the delay, in seconds; <tf> is the damping factor, in seconds^{-1}; and <phase> is the phase angle in deg.

If td, df, and phase are not specified, the default values for these parameters are zero.

In our example, the offset voltage is 0 volts, the peak amplitude of the voltage is 10 volts, and the frequency is 1000 Hz.

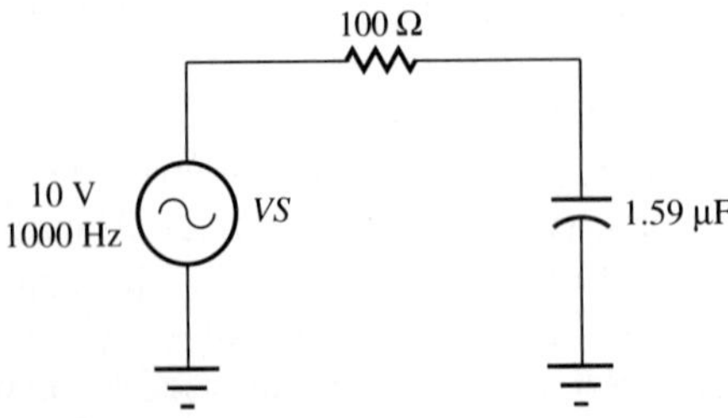

Figure 16.67 A series *R–C* circuit driven by an ac source.

```
TRANSIENT ANALYSIS OF A R-C CIRCUIT
VS    1    0    SIN( 0 10V  1000HZ )
R1    1    2    100
C1    2    0    1.59UF
.TRAN    5US    2MS
.PROBE
.END
```

Figure 16.68 The PSpice source file for the circuit in Figure 16.67.

The statements for the resistor and capacitor are:

```
R1    1    2    100
C1    2    0    1.59UF
```

The .TRAN command needs to specify the print step value and the final value of time. The statement has the general form:

```
.TRAN <(print step) value> <(final time) value>
```

In this analysis, let us select two cycles as the time interval over which the voltage waveforms are graphed. Let us also select 5 microseconds as the print step value. Hence, the .TRAN command can be written as:

```
.TRAN  5US  2MS
```

The complete file is shown in Figure 16.68.

Once the analysis is complete, .PROBE is called. Within .PROBE, the waveforms for V(1), V(2), and V(1,2) can be drawn. The resulting graphs are plotted on the same axes, and the printout is shown in Figure 16.69.

The last example in this section illustrates how PSpice can be used to graph current waveforms in an ac circuit.

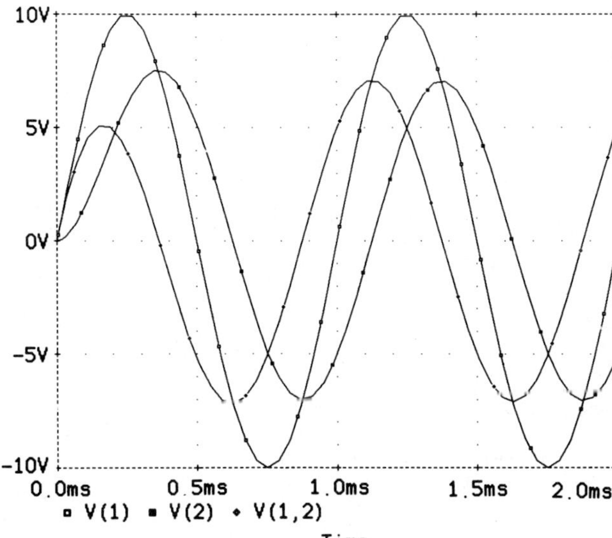

Figure 16.69

EXAMPLE 16.37

For the circuit shown in Figure 16.70, use PSpice to draw the current waveforms for the resistor, inductor, and capacitor.

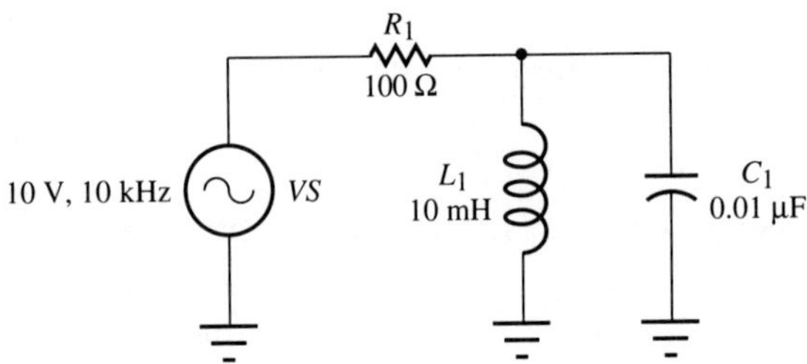

Figure 16.70

Solution

The PSpice file for the analysis is:

```
EXAMPLE 16.37
VS    1    0    SIN(0 10V 10KHZ)
R1    1    2    100
L1    2    0    10MH
C1    2    0    0.01UF
.TRAN    0.1US    0.3MS
.PROBE
.END
```

The graph produced by .PROBE is shown in Figure 16.71.

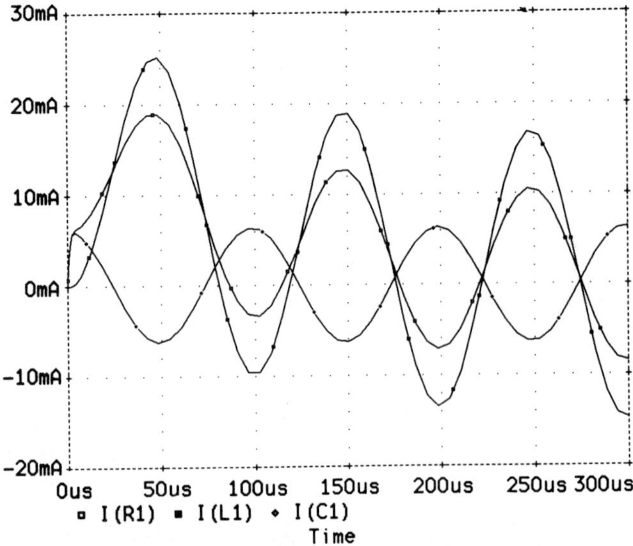

Figure 16.71

Examining the output waveforms, the following observations can be made:

1. There is a transient portion to the output.
2. The inductor and capacitor currents are 180° out of phase.
3. The current through the resistor equals the inductor current plus the capacitor current at any instant of time.

The ability to use PSpice to perform single frequency analyses and graph voltage and current waveforms for ac circuits is a useful tool in analyzing circuits. Simulation allows the technician and engineer to investigate the operational properties of circuits before they are built, and to experiment with different circuit configurations and component values without the time-consuming computations that were heretofore done by hand.

SUMMARY

1. Two basic ac circuit configurations are the series connection and the parallel connection.
2. The series circuit has the elements connected so that the same current is in each one of them.
3. The parallel circuit has the elements connected across the same points so the same voltage is across each of them.
4. Impedance is the opposition of a circuit to an ac current. It can include the opposition of resistance and the opposition of capacitance and inductance.
5. Equivalent impedance is a single impedance that has the same characteristics as the circuit to which it is equivalent. There is no specific relation between the magnitude of Z_{eq} and the parts of the circuit as for dc.
6. The impedance triangle is a triangle formed by the phasors for resistance, reactance, and impedance.
7. Kirchhoff's Voltage Law and the voltage divider expression relate the voltage in parts of a series circuit to the total voltage.
8. Kirchhoff's Current Law and the current divider expression relate the current in parts of a parallel circuit to the total current.
9. Admittance is the reciprocal of impedance, and its use sometimes simplifies the analysis of a parallel circuit.
10. A phasor diagram is a diagram of the circuit voltage and current phasors drawn on a set of axes. The diagram shows the magnitude and angular relation of the quantities.
11. Power in an ac circuit is a function of voltage, current, and the phase angle. $P = VI \cos \theta$.
12. The power factor is the cosine of the phase angle. It ranges from zero to unity. Lower power factors require larger currents to provide an equal amount of power.
13. Power is dissipated only in the resistance of a circuit. Ideal capacitors and inductors do not dissipate any power over a cycle in an ac circuit because their phase angle is 90°.

14. A series-parallel circuit is made up of series and parallel groups of impedances. The series circuit and parallel circuit relations are used to analyze the series-parallel circuit.
15. The key to the analysis of any circuit is to develop a sequence or path between the known quantities and the wanted quantities. This sequence should begin with a relation that includes one of the wanted quantities. The unknowns in each equation can be used as a guide to the next equation.

EQUATIONS

16.1	$Z_{eq} = Z_1 + Z_2 + \cdots + Z_n$	Equivalent impedance of a series circuit.	635
16.2	$V = IZ$	Ohm's Law for an ac impedance.	638
16.3	$E - V_1 - V_2 - \cdots - V_n = 0$	Kirchhoff's Voltage Law for an ac series circuit.	639
16.4	$V_x = E(Z_x/Z_{eq})$	Voltage divider equation for an ac series circuit.	640
16.5	$\frac{1}{Z_{eq}} = \frac{1}{Z_1} + \frac{1}{Z_2} + \cdots + \frac{1}{Z_n}$	Equivalent impedance of a parallel circuit.	642
16.6	$Z_{eq} = \frac{Z_1Z_2}{Z_1 + Z_2}$	Equivalent impedance of two parallel branches.	642
16.7	$Y_{eq} = Y_1 + Y_2 + \cdots + Y_n$	Equivalent admittance of a parallel circuit.	645
16.8	$V = \frac{I}{Y}$	Ohm's Law for admittance.	646
16.9	$I_s - I_1 - I_2 - \cdots - I_n = 0$	Kirchhoff's Current Law for an ac parallel circuit.	647
16.10	$I_X = \frac{IZ_{eq}}{Z_X}$	Current divider equation for an ac parallel circuit.	648
16.11	$I_X = \frac{IZ}{Z + Z_X}$	Current divider for two parallel branches.	649
16.12	$P = VI$	Average power in a resistor in an ac circuit.	658
16.13	$P = I^2R$	Average power in a resistor in an ac circuit.	658
16.14	$P = \frac{V^2}{R}$	Average power in a resistor in an ac circuit.	658

16.15	$P = VI \cos \theta$	Power in an ac circuit of any elements and configuration.	660
16.16	$P_T = P_1 + P_2 + \cdots + P_n$	Relation of the total average power in a circuit to the power in each part of the circuit.	661
16.17	$\mathcal{W} = Pt$	Relation of energy to power and time.	664
16.18	$\mathcal{W} = \mathcal{W}_1 + \mathcal{W}_2 + \cdots + \mathcal{W}_n$	Relation of the total energy in a circuit to the energy in each part of the circuit.	664

QUESTIONS

1. A 5-ohm resistor is connected to an ac source. What would happen to the source current if the resistor were replaced with a 5-ohm capacitive reactance?
2. Which series combination will have the greater equivalent impedance: two 5-ohm resistors or a 5-ohm resistor and a 5-ohm capacitive reactance?
3. A series circuit has an equivalent impedance of $4 \angle 45° \ \Omega$. What change in the impedance will result if the source frequency is doubled?
4. What effect will doubling the source frequency have on an equivalent impedance of $4 \angle -45° \ \Omega$?
5. Voltage measurements in a series R–L–C circuit give $V_R = 12$ V, $V_C = 30$ V, and V across the coil = 30 V. The voltage across the group is only 12 volts. Explain why the total does not equal the sum.
6. Which parallel combination will have the greater equivalent impedance: two 4-ohm resistors or a 4-ohm resistor and a 4-ohm capacitive reactance?
7. A parallel circuit has an equivalent impedance of $10 \angle 45°$. What change in impedance will result if the source frequency is doubled?
8. What effect will doubling the source frequency have on an equivalent impedance of $10 \angle -45° \ \Omega$?
9. What information can be obtained from a phasor diagram?
10. How does one obtain the angle from the phasor diagram?
11. How does one obtain the magnitude of voltage or current from the phasor diagram?
12. What is the difference between the phasor diagram and the impedance triangle?
13. What is meant by the power factor of a circuit?
14. Why do electric power companies want the loads in their systems to have a power factor near unity?
15. What is the value of the power factor when V and I are in phase? When V and I are out of phase by 45°. When V and I are out of phase by 90°?
16. For what type of circuit element does the relation $P = VI$ hold in an ac circuit?
17. The power in a series circuit of a 30-ohm resistor and a 40-ohm capacitor is 120 watts. What change will result in the power if the capacitor is replaced by a 40-ohm inductive reactance? By a 40-ohm resistor?
18. The power factor of a parallel R–C circuit is 0.7. What effect will connecting another capacitor in parallel with the circuit have on the power factor?

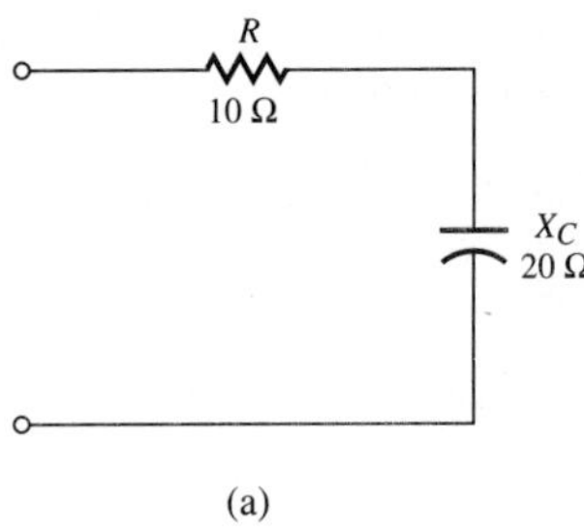

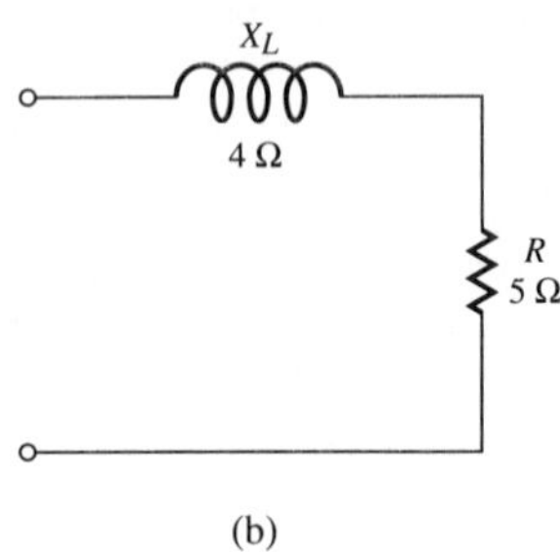

Figure 16.72

19. What effect will connecting an inductance in parallel with the original circuit of Question 18 have on the power factor?
20. What is the difference between analysis and synthesis, as related to ac circuits?

PROBLEMS

SECTION 16.1 Series Circuit Characteristics

1. Determine the equivalent impedance of the circuits shown in Figure 16.72.

2. Determine the equivalent impedance of the circuits shown in Figure 16.73.

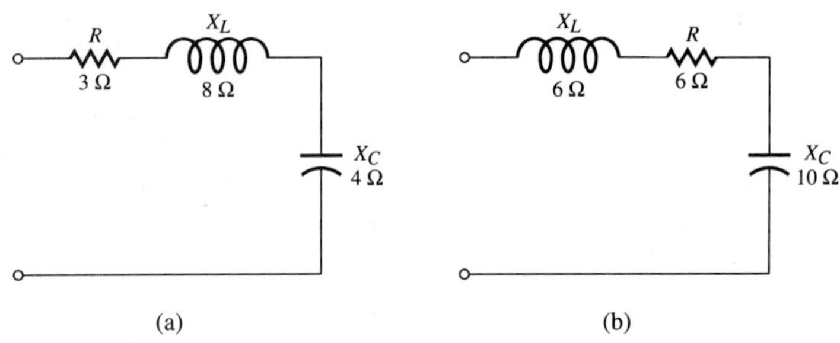

Figure 16.73

3. A 6-ohm resistor, an inductive reactance of 10 ohms, and a capacitive reactance are connected in series. Find the value of the capacitive w09reactance if the equivalen0034impedance of the circuit is $10\ \angle -53.13°\ \Omega$.

4. Draw the impedance triangle and determine the phase angle for a series circuit made up of $R = 10\ \Omega$, $X_c = 15\ \Omega$, and $X_L = 20\ \Omega$.

5. An impedance triangle has the values shown in Figure 16.74.

   (a) How many ohms is R?

   (b) Is the circuit inductive or capacitive?

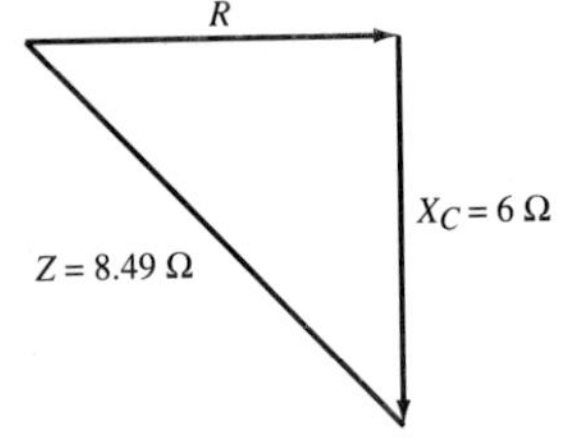

Figure 16.74

6. What is the current in a circuit that has an impedance of $5\ \angle 45°\ \Omega$ and is connected to a voltage of $25\ \angle 0°$ volts?

7. The current in an impedance of $10\ \angle 60°\ \Omega$ is $2\ \angle 30°$ amperes. What is the voltage across the impedance?

8. What is the source voltage E if $V_L = 18\ \angle 53.13°$ volts and $V_R = 24\ \angle -36.87°$ volts for a series R–L circuit?

9. The voltage across C in Figure 16.75 is $150\ \angle -45°$ volts, and that across L is $50\ \angle 135°$ volts. What is the voltage across R?

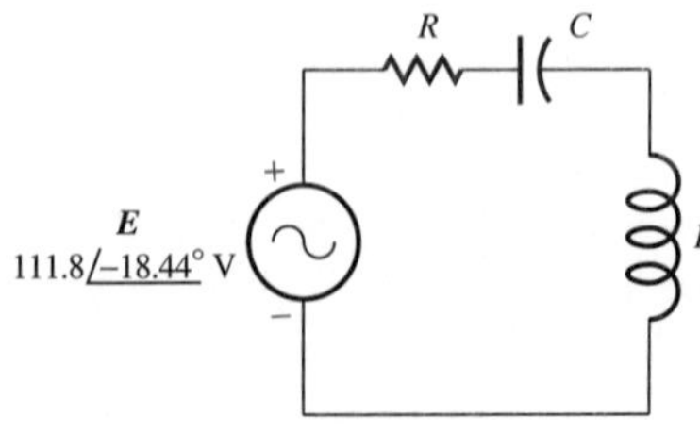

Figure 16.75

10. A series circuit has the following voltages in it. $V_R = 10\ \angle 30°$ volts, $V_L = 20\ \angle 120°$ volts, and $V_c = 30\ \angle -60°$ volts. Determine the source voltage graphically.

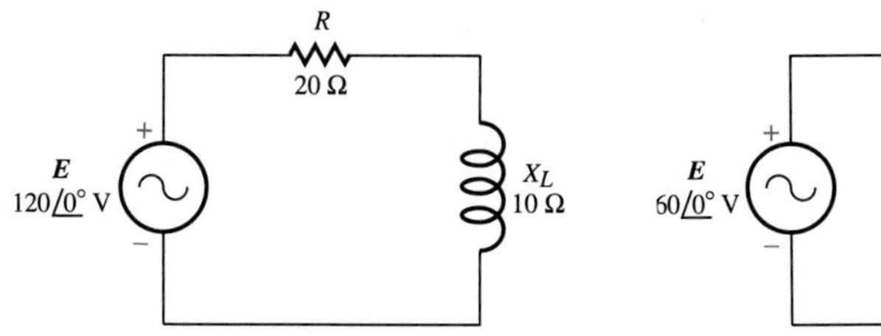

Figure 16.76

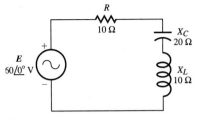

Figure 16.77

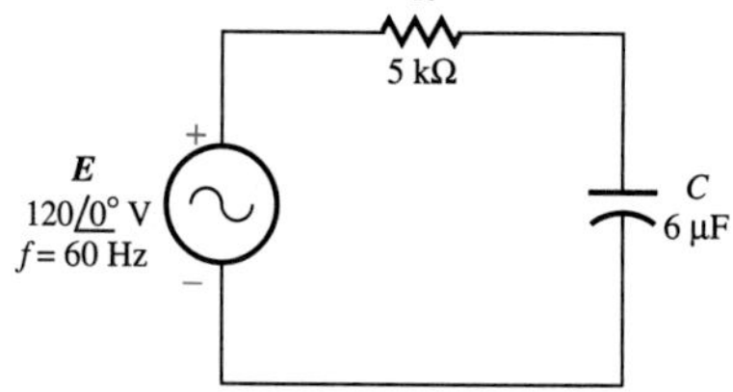

Figure 16.78

11. The source voltage in a series circiut is 25 $\angle 0°$ V. The voltage across the resistance is 15 $\angle -53.13°$ volts, and that across the capacitance is 30 $\angle -143.13°$ volts. Determine the inductor voltage graphically.

12. Use the voltage divider expression to find the voltage across X_L in Figure 16.76.

13. Use the voltage divider expression to find the voltage across R in Figure 16.77.

14. The circuit in Figure 16.78 is used to trigger an SCR (Silicon Controlled Rectifier). Use the voltage divider expression to find the voltage across the capacitor and the phase shift between it and the source.

15. Use the voltage divider expression to find the voltage across points A–B in Figure 16.79.

16. The voltage across A–B is 20 $\angle 30°$ V. Use the voltage divider expression to find the source voltage in Figure 16.80.

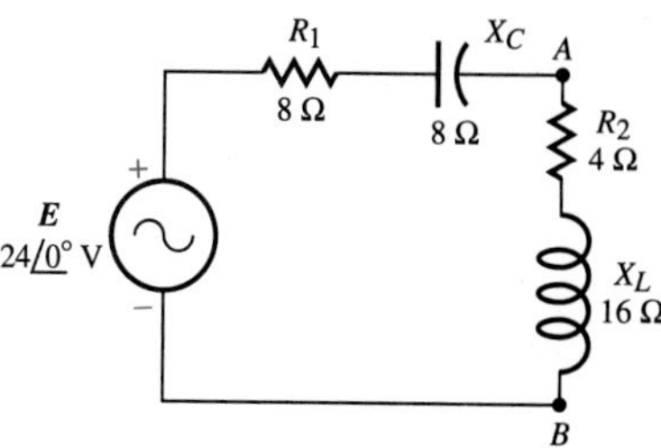

Figure 16.79

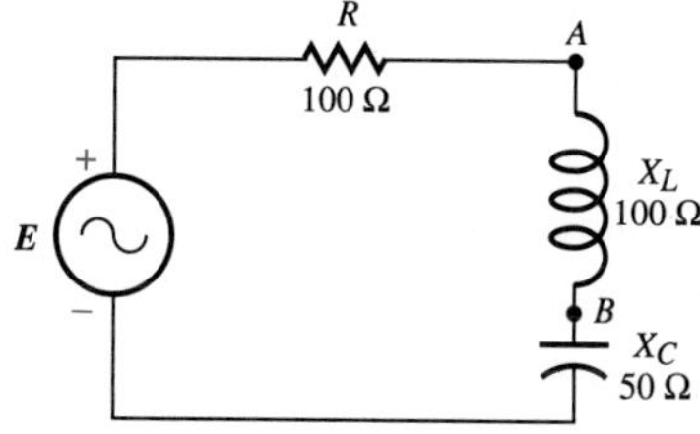

Figure 16.80

SECTION 16.2 Parallel Circuit Characteristics

17. Determine the equivalent impedance and the phase angle for the circuit. Draw the equivalent series circuit for the circuit in Figure 16.81.

18. Determine the equivalent impedance of the circuit in Figure 16.82, and draw the series equivalent of it.

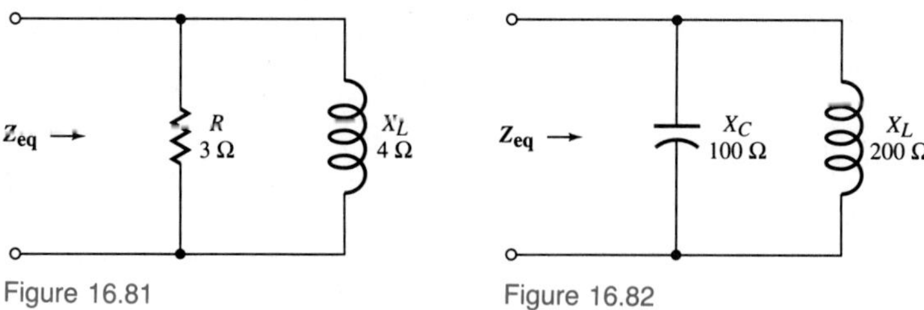

Figure 16.81

Figure 16.82

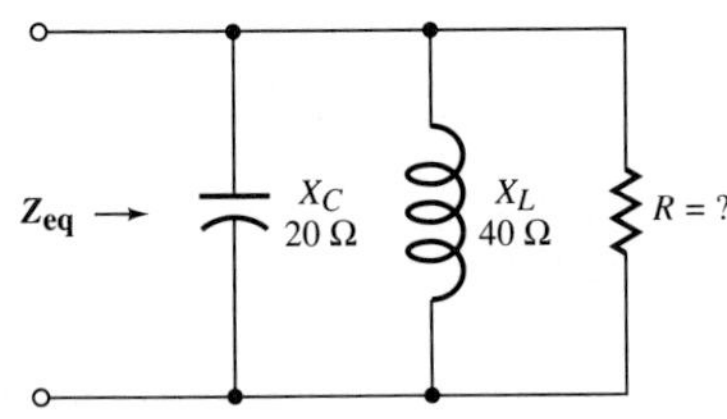

Figure 16.83

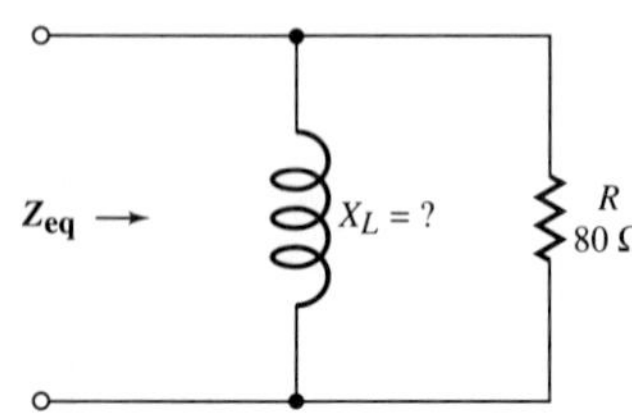

Figure 16.84

19. The circuit in Figure 16.83 has an equivalent impedance of 28.28 $\angle -45°$. What is the ohmic value of R?

20. What is the value of X_L if the equivalent impedance of the circuit in Figure 16.84 is 48 $\angle 53.13$ Ω? Draw the series equivalent of the circuit.

21. Determine the admittance, conductance, and susceptance of the following impedances.

    (a) 5 $\angle 30°$ Ω.
    (b) 10 $\angle 45°$ Ω.
    (c) 10 $\angle -53.13°$ Ω.
    (d) 5 $\angle 0°$ Ω.

22. Repeat Problem 21 for the following:

    (a) 20 $\angle -60°$ Ω.
    (b) 7.07 $\angle -45°$ Ω.
    (c) 50 $\angle 36.87°$ Ω.
    (d) 8 $\angle -90°$ Ω.

23. Determine the equivalent admittance, the source current, and the current in each branch of the circuit in Figure 16.85.

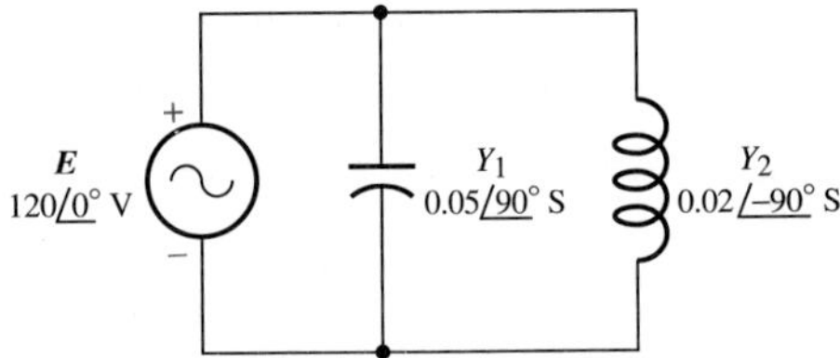

Figure 16.85

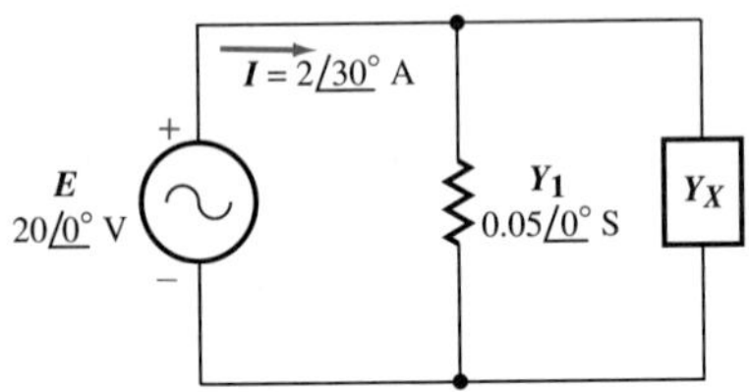

Figure 16.86

24. What is the admittance Y_X if the source current is 2 $\angle 30°$ A, in Figure 16.86?

25. Determine the equivalent admittance, its conductance, and susceptance, and draw the admittance diagram for the circuit in Figure 16.87.

26. How many amperes are supplied by a 120 $\angle 0°$ V ac source when connected to a parallel circuit of a 100-ohm resistor and a 50-ohm inductive reactance?

27. How many amperes of current are in the resistor branch of the circuit in Figure 16.88?

28. What is the current in the capacitor of the circuit in Figure 16.89?

29. A resistor, capacitor, and inductor are connected in parallel. The current in the resistor is 2 $\angle 0°$ A, that in the inductor is 3 $\angle -90°$ A, and that in the capacitor is 1 $\angle 90°$ A. Graphically determine the total current.

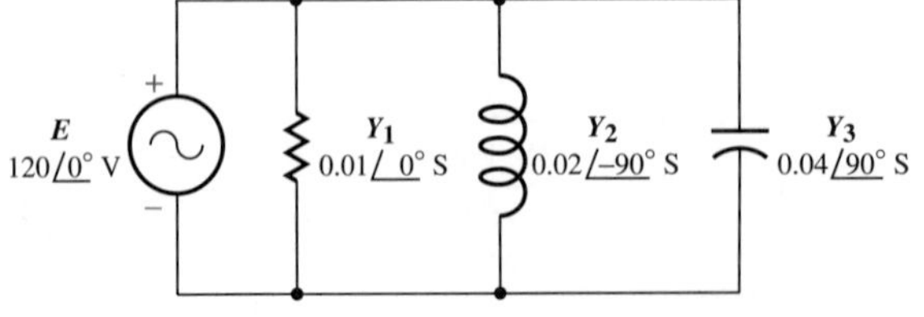

Figure 16.87

I = 4/30° A
+
E
−
IL
2/60° A
R
I ?

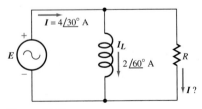

Figure 16.88

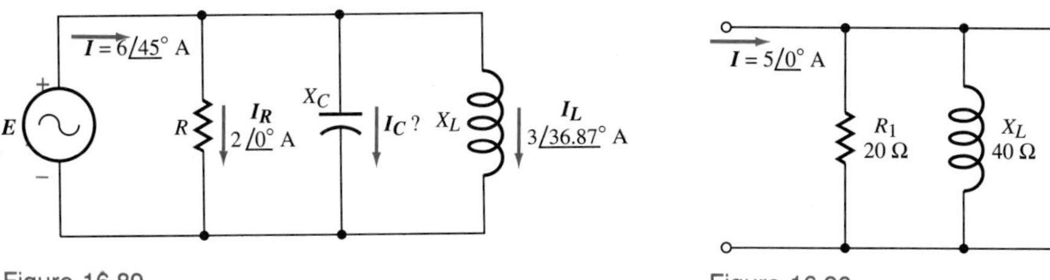

Figure 16.89

Figure 16.90

30. A parallel circuit of a 25-ohm resistance, 20 ohms X_L, and a capacitive reactance are connected to an ac source. Graphically find $\boldsymbol{I}_C$ if $\boldsymbol{I}_R$ is 4 $\angle 0°$ A, $\boldsymbol{I}_L$ is 4 $\angle -90°$ A, and $\boldsymbol{I}_{\text{TOTAL}}$ is 5 $\angle -36.87°$ A.

31. The total current in a parallel circuit of a 30-ohm resistor and a 40-ohm inductive reactance is 2 $\angle -36.87°$ amperes. Use the current divider equation to find the current in the inductance.

32. A 50-ohm capacitive reactance is connected in parallel with a 100-ohm inductive reactance. The current in the capacitance is 2 $\angle 60°$ amperes. Use the current divider equation to find the total curent.

33. Use the current divider rule to find the current in R_1 of the circuit in Figure 16.90.

34. A 30-ohm capacitive reactance is connected in parallel with another impedance. What is the impedance of the other branch if the capacitor current is 2 $\angle 90°$ amperes when the total current is 2.24 $\angle 63.43°$ amperes?

SECTION 16.3 Phasor Diagrams

35. Draw the phasor diagram for the circuit in Figure 16.91.

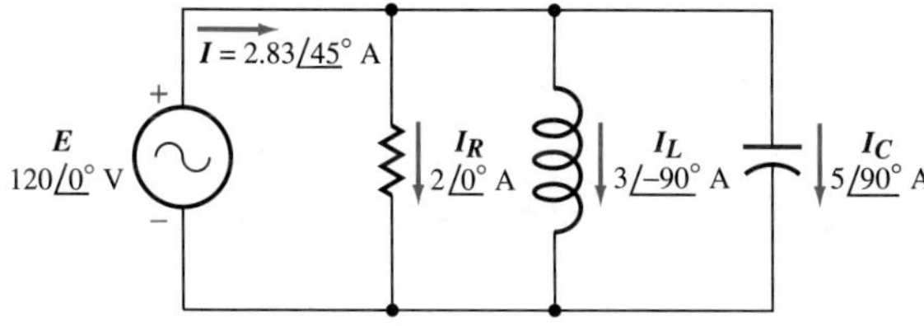

Figure 16.91

36. Draw the phasor diagram of the series circuit shown in Figure 16.92. Determine the impedance in each branch.

37. Find the current in the circuit in Figure 16.93, find the voltage across each component, and draw the phasor diagram.

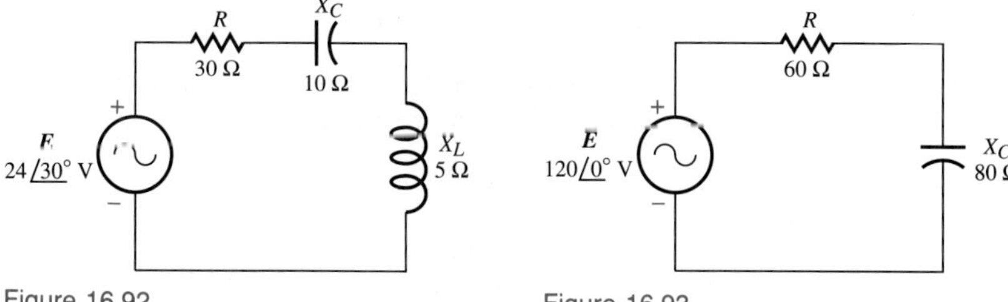

Figure 16.92

Figure 16.93

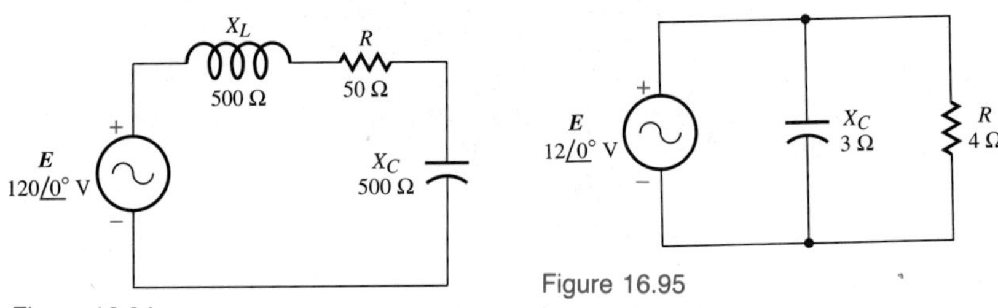

Figure 16.94

Figure 16.95

38. Find the current in the circuit of Figure 16.94 and the voltage across each component. Draw the phasor diagram showing all voltages.

39. Repeat Problem 38 for the circuit in Figure 16.95.

40. Determine the currents in the circuit in Figure 16.96 and draw the phasor diagram. Is the circuit inductive or capacitive?

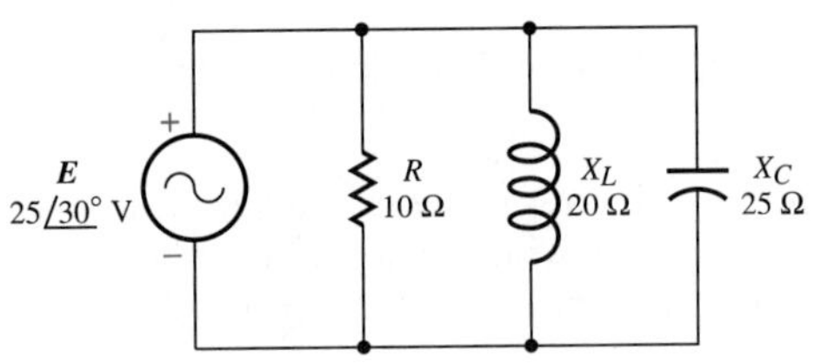

Figure 16.96

41. Repeat Problem 40 for the circuit in Figure 16.97.

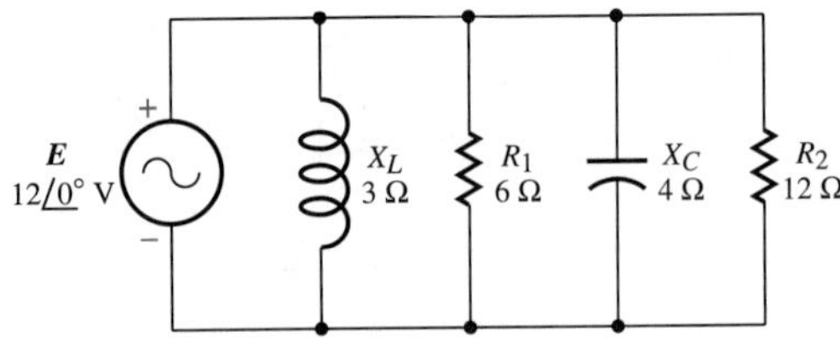

Figure 16.97

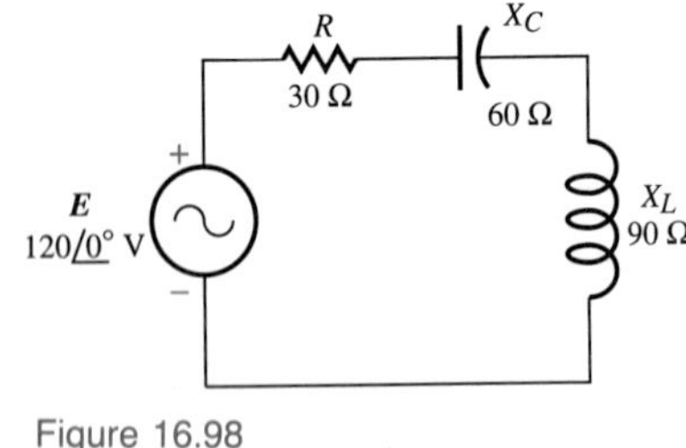

Figure 16.98

42. Repeat Problem 40 for the circuit in Figure 16.98.

SECTION 16.4 Power in the ac Circuit

43. A 60-Hz, 120-volt sinusoidal voltage results in 2 joules of energy being converted to heat in a resistance in one cycle. Find:

    (a) The average power for this waveform.

    (b) The current in the resistor.

44. What is the power factor of the circuit in Figure 16.99? Is it a leading or lagging power factor?

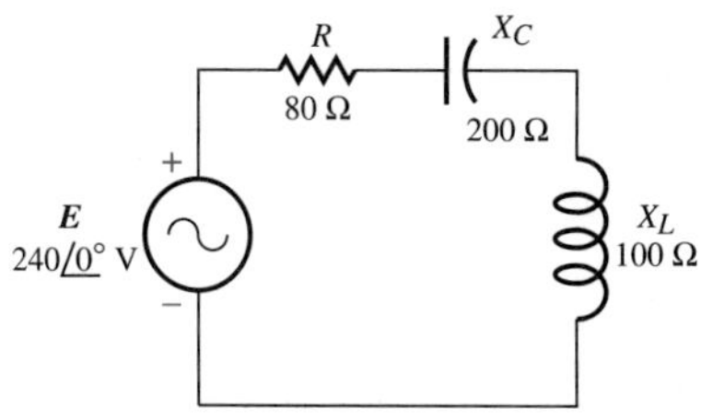

Figure 16.99

45. How much power is dissipated in the circuit that has the phasor diagram in Figure 16.100? What is its power factor? Is it leading or lagging?

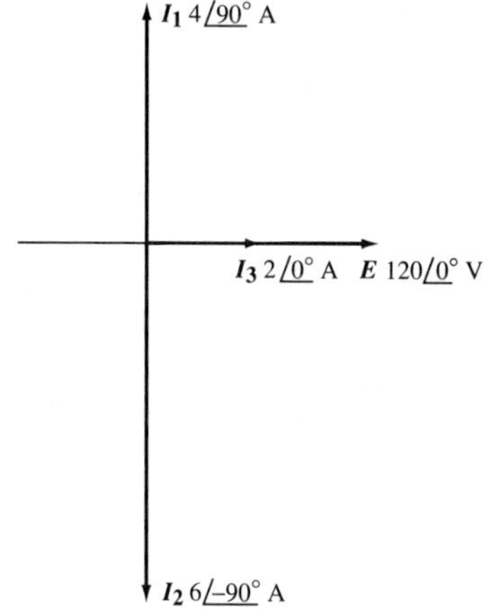

Figure 16.100

46. An electric power company provides 5 kW of power at 120 volts to a customer whose load has a power factor of 0.7.

    (a) If the power factor is improved to 0.9, how much less current would be required?

    (b) If the lines have a resistance of 0.5 ohm, how much power is lost in the lines for both conditions?

47. A load has a leading power factor of 0.8. It dissipates 1000 W when connected to a 120-volt source.

    (a) What is the current in the load?

    (b) What are the magnitude and phase angle of the load?

48. The power dissipated in the circuit in Figure 16.101 is 150 watts. If 50 W are dissipated in R_1, how many watts are dissipated in R_2? How many joules of energy are used in the circuit in one hour?

49. A 50-ohm resistor and a 100-ohm inductive reactance are connected in parallel across a source of 120 $\angle 30°$ V. Find:

    (a) The power factor of the parallel circuit.

    (b) The power dissipated in the circuit.

50. For the circuit in Figure 16.102, find:

    (a) The power factor.

    (b) The power dissipated.

    (c) The circuit current.

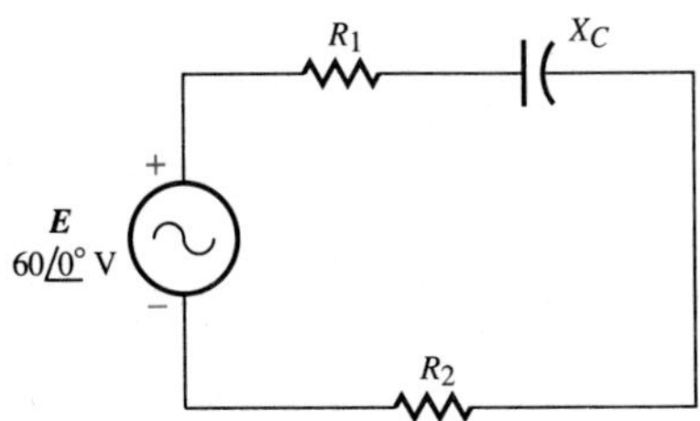

Figure 16.101

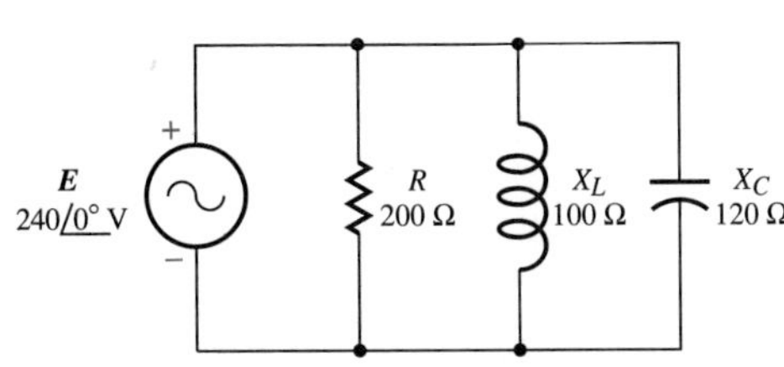

Figure 16.102

51. The power dissipated in a two-branch parallel circuit is 240 watts when connected to a 120 $\angle 0°$ V source. If the power factor is 0.8 leading, determine:

    (a) The total current.

    (b) The components in each branch.

52. The power factor of the circuit in Figure 16.103 with S open is 0.6 lagging.

    (a) What is the power factor with S closed?

    (b) How much power is dissipated with S open and closed?

    (c) What is the change in source current caused by closing S?

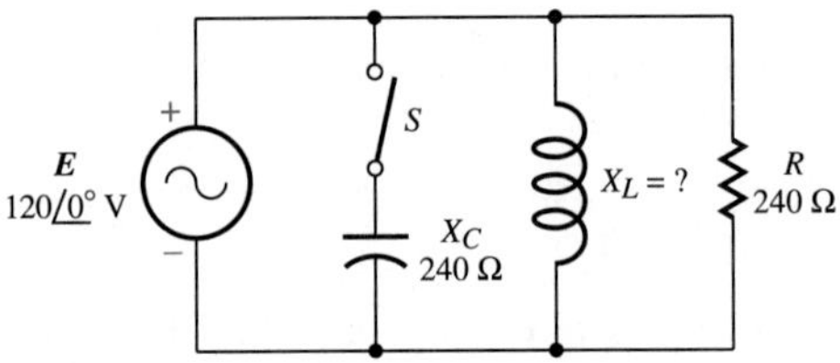

Figure 16.103

53. Use Equation 16.11 to find the power supplied by the source in the circuit in Figure 16.104.

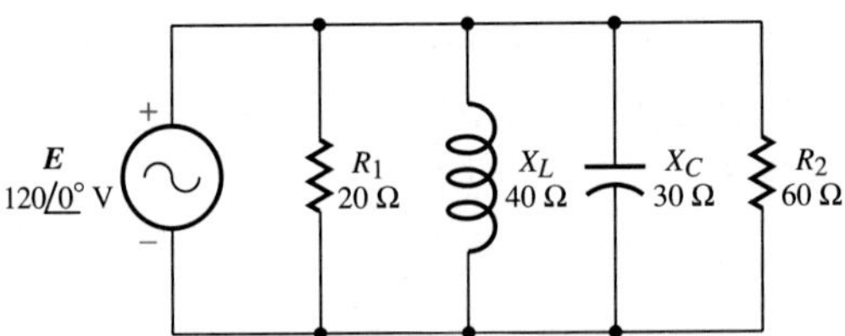

Figure 16.104

54. The power dissipated in the circuit in Figure 16.105 is 240 watts. If 72 watts is dissipated in R_2:

    (a) How much power is dissipated in R_1?

    (b) What is the current in R_1?

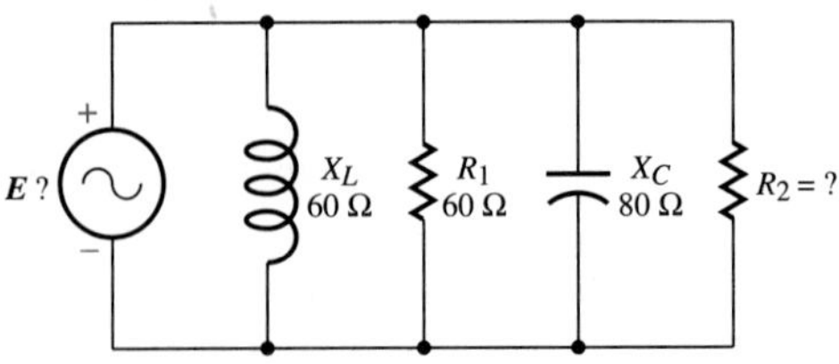

Figure 16.105

SECTION 16.5 Series Circuit Analysis

55. For the circuit in Figure 16.106:

    (a) Find the current.

    (b) Find the power dissipated.

    (c) Find the power factor.

    (d) Draw the impedance triangle and the phasor diagram.

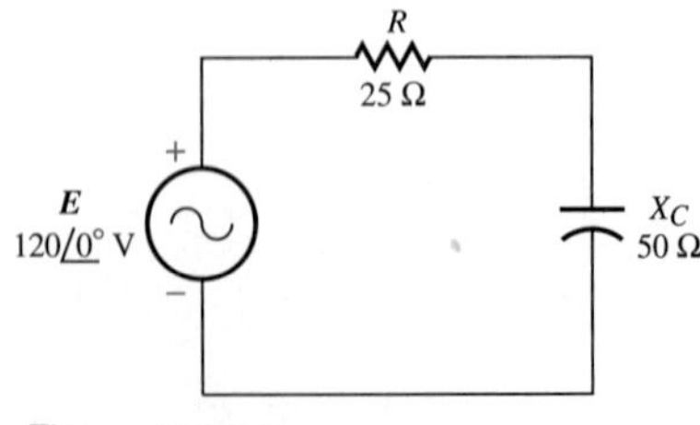

Figure 16.106

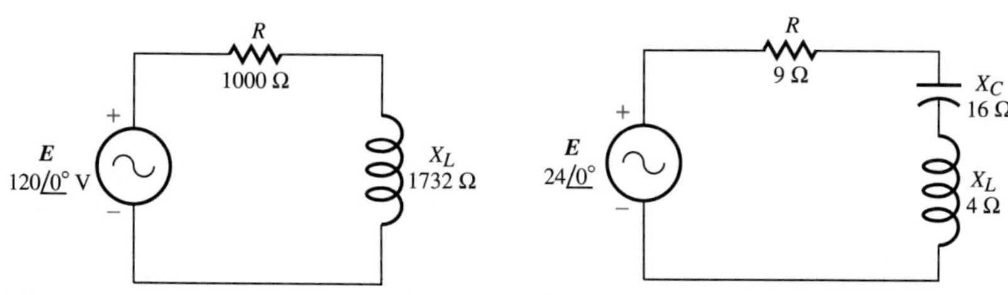

Figure 16.107

Figure 16.108

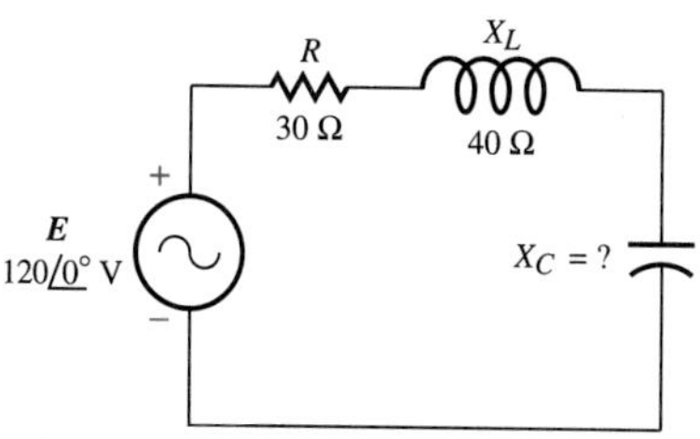

Figure 16.109

56. Repeat Problem 55 for the circuit in Figure 16.107.

57. Repeat Problem 55 for the circuit in Figure 16.108.

58. In the circuit of Figure 16.109, determine the capacitve reactance that must be connected in series to raise the power factor to 0.8 lagging.

59. The voltage across R is 30 $\angle 30°$ volts and the power factor is 0.6 leading for the circuit in Figure 16.110. Determine the value of X_C, and the source voltage, E.

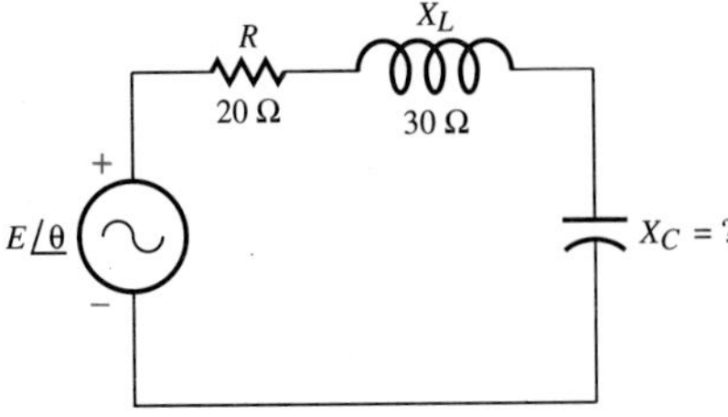

Figure 16.110

SECTION 16.6 Parallel Circuit Analysis

60. For the circuit in Figure 16.111:
    (a) Find the source current.
    (b) Find the power factor.
    (c) Find the power dissipated in it.
    (d) Draw the phasor diagram.

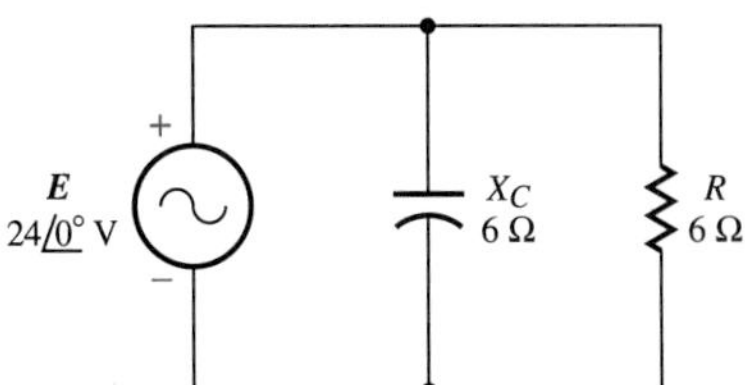

Figure 16.111

61. Repeat Problem 60 for the circuit shown in Figure 16.112.

62. The admittance of a circuit is 0.0057 $\angle -45°$ S.
    (a) What is the component in the third branch if the admittances in the first and second branches are 0.004 $\angle 0°$ S and 0.004 $\angle 90°$ S?
    (b) What is the circuit current when connected to a 120 $\angle 0°$ volt source?
    (c) What is the equivalent impedance of the circuit?

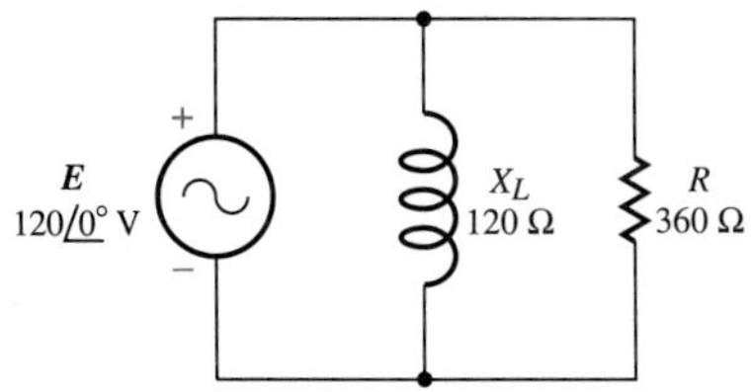

Figure 16.112

63. In the circuit in Figure 16.113:
    (a) Find the source voltage.
    (b) Find the power factor.
    (c) Find the power dissipated.
    (d) Draw the phasor diagram.

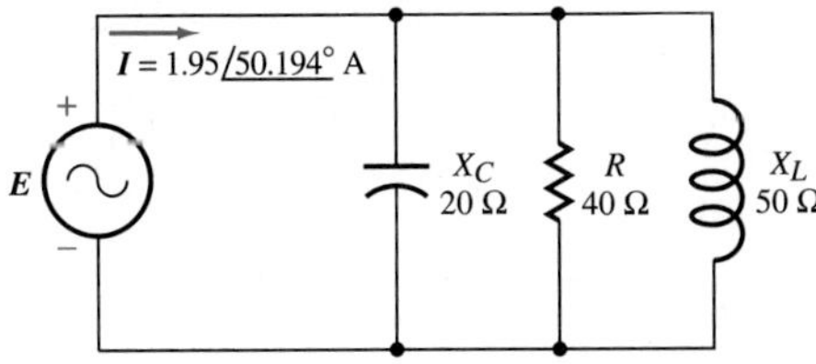

Figure 16.113

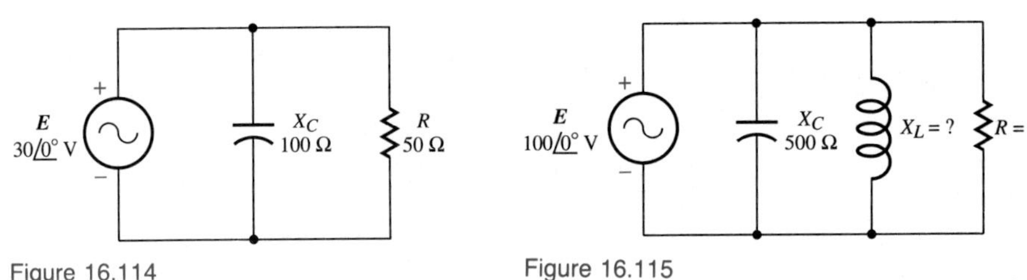

Figure 16.114

Figure 16.115

64. Use the equivalent admittance method to find the current in the circuit in Figure 16.114. How much power is dissipated in the circuit?

65. The equivalent impedance of the circuit in Figure 16.115 is 354 ∠45°.

    (a) Determine the value of X_L and R if the power dissipated is 20 watts.

    (b) What is the current in each branch?

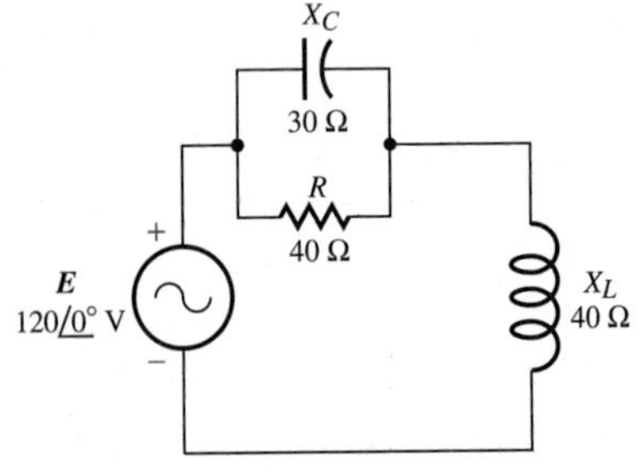

Figure 16.116

SECTION 16.7 Series-Parallel Circuit Analysis

66. Find the equivalent impedance for the circuit shown in Figure 16.116.

67. Work Problem 66 for the circuit shown in Figure 16.117.

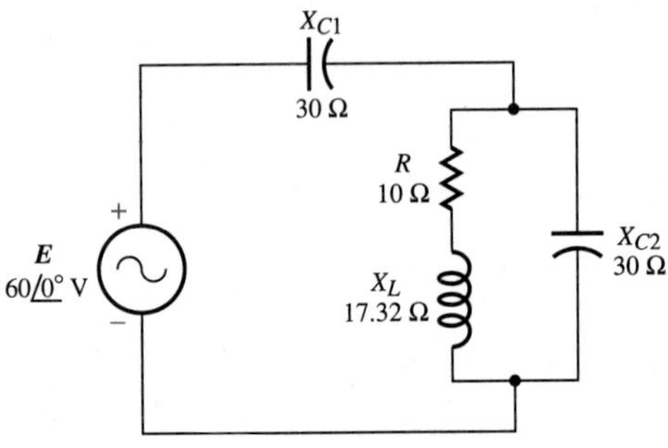

Figure 16.117

68. Work Problem 66 for the circuit shown in Figure 16.118.

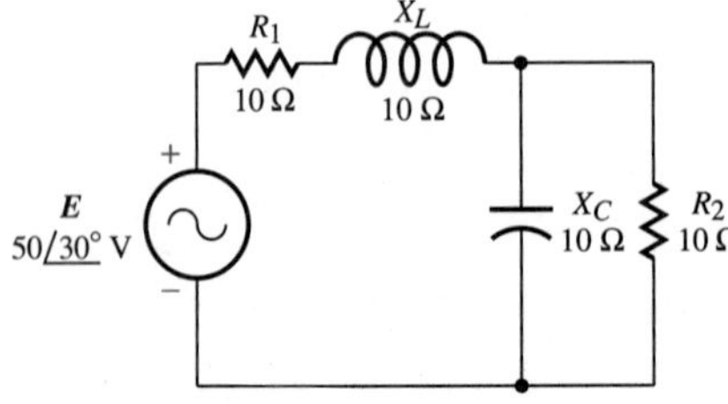

Figure 16.118

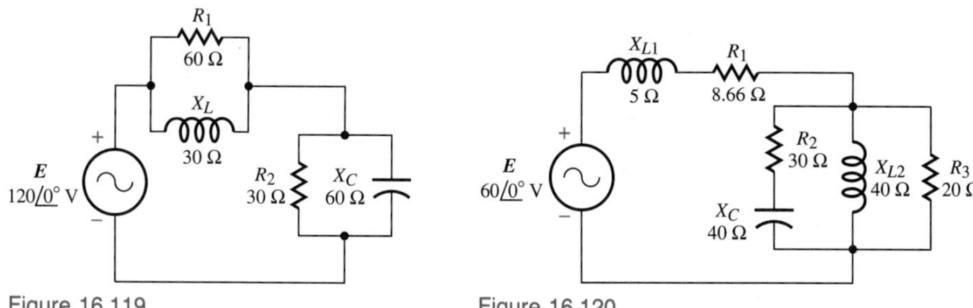

Figure 16.119

Figure 16.120

69. Find the current in R_2 and the voltage across X_L in the circuit of Figure 16.119.

70. Find the current in X_{L2} and the power dissipated in the parallel group shown in Figure 16.120.

71. What is the source voltage, the current in R_1, and the voltage across X_{c2} in the circuit of Figure 16.121?

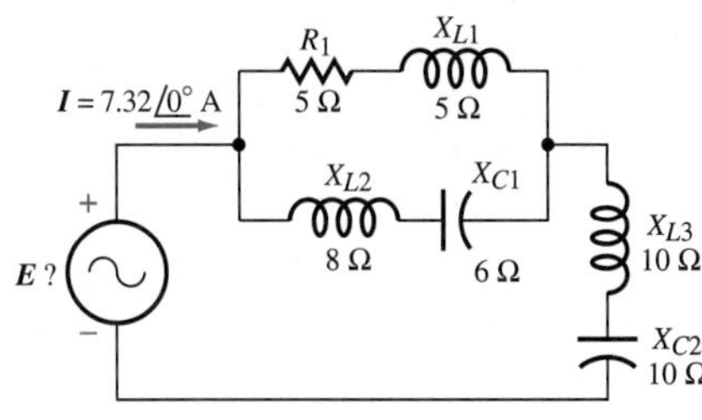

Figure 16.121

72. How many ohms must X_L be if the current in the circuit of Figure 16.122 is 3.03 $\angle 8.13°$ amperes?

73. For the relay to energize, the current in the relay coil in Figure 16.123 must be 1 ampere. How many ohms should R be? What is the power factor of the circuit?

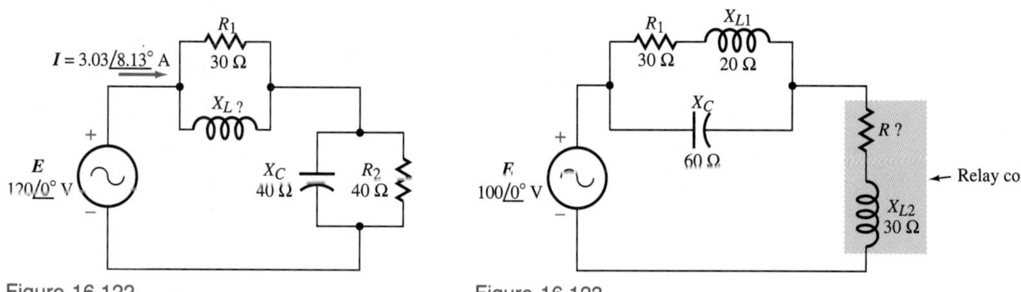

Figure 16.122

Figure 16.123

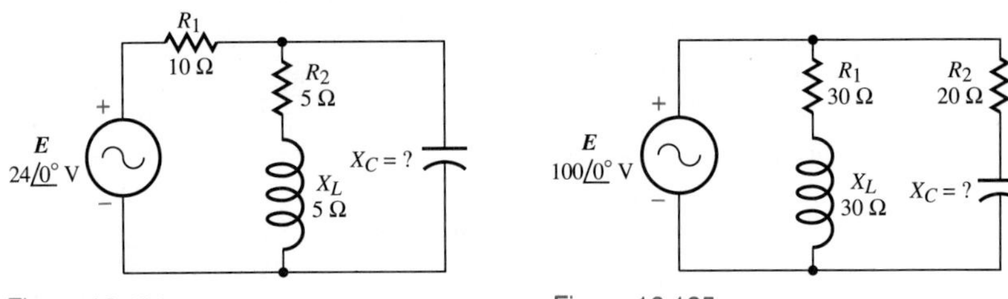

Figure 16.124

Figure 16.125

74. Find the value of X_C needed to make the power factor of the circuit in Figure 16.124 be equal to one.

75. Find the value of X_C needed to make the circuit in Figure 16.125 have a power factor of unity. Hint: The *j* part of one branch current must cancel out the *j* part of the other.

SECTION 16.8 SPICE Analysis of Single Source ac Circuits

Ⓢ 76. Use PSpice to analyze the circuit shown in Figure 16.126.

(a) Determine the magnitude and phase angle of V(OUT).

(b) Graph VS and V(OUT).

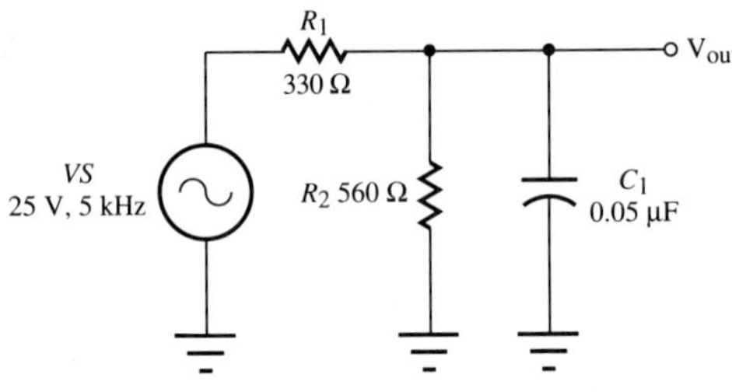

Figure 16.126

Ⓢ 77. Use PSpice to analyze the circuit shown in Figure 16.127.

(a) Determine the magnitude and phase angle of V(OUT).

(b) Graph VS and V(OUT).

Ⓢ 78. Use PSpice to determine the voltage drop across the resistor and the voltage drop across the capacitor in the circuit in Figure 16.92. Assume a source frequency of 1 kHz.

Ⓢ 79. Use PSpice to graph the voltage across each circuit element in the circuit in Figure 16.95. Assume a source frequency of 1 kHz.

Ⓢ 80. Use PSpice to graph the current waveforms of each parallel branch in the circuit in Figure 16.96. Assume a source frequency of 1 kHz.

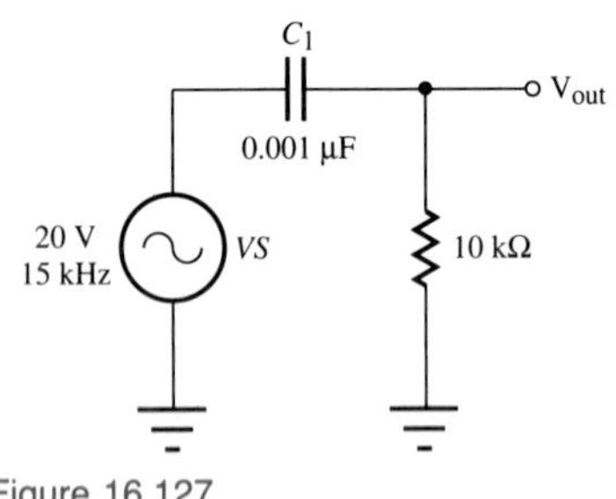

Figure 16.127

CHAPTER 17

THE POWER TRIANGLE AND ALTERNATING CURRENT (ac) RESISTANCE

CHAPTER OBJECTIVES

After completing this chapter, you should be able to:

1. Define reactive power and apparent power, and give the units for each.
2. Draw a power triangle and relate it to the impedance.
3. Combine power triangles and use them to analyze an ac circuit.
4. Calculate the number of vars needed to change the power factor of a circuit.
5. Define ac resistance and the items that are included in it.
6. Recognize at which frequency each part of the ac resistance is significant.
7. Explain what items affect the hysteresis and eddy current losses.

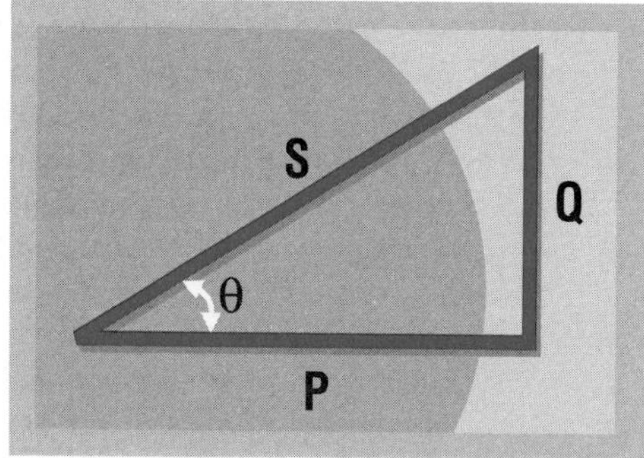

KEY TERMS

1. ac resistance
2. apparent power
3. dc resistance
4. eddy current loss
5. electromagnetic radiation
6. hysteresis loss
7. power factor correction
8. power triangle
9. reactive power
10. skin effect
11. var
12. volt ampere
13. watts

INTRODUCTION

Why do electric companies connect banks of capacitors across the lines that supply power to a customer? What causes the resistance of a conductor to appear to be greater in an ac circuit than in a dc circuit? These are some of the questions that are answered in this chapter.

First, the concept of reactive power and apparent power is explained. Then, it is applied to the analysis of a circuit. Electric power companies use this approach in planning and checking the loading of their distribution systems. Calculations are simplified, so complex arithmetic is not needed.

The effect of frequency on the losses in a conductor or component is also presented. Electromagnetic radiation, skin effect, hysteresis, and eddy currents are all factors that contribute to what is known as the ac resistance. Because of these losses, the ac resistance is greater than the dc resistance. In this chapter you will learn what factors affect the losses, and how they are reduced.

17.1 Reactive Power and Apparent Power

Thus far, our concept of the ac circuit has been in terms of the impedance in the circuit. Another way to look at the circuit is from the point of the power in the impedance. This approach is common in applications such as power transmission and distribution. It can also be used when it is easier to measure power, current, and voltage rather than to measure impedance. To understand this approach, you must first understand the meaning of reactive power and apparent power.

Previous to this chapter, you learned that energy is converted to heat in the resistance of a complex impedance. The rate at which the conversion takes place was defined as power, and is measured in **watts**. This power is measured with a wattmeter (see Chapter 3). Along with that conversion, some energy is also exchanged between the source and the field of the reactance. Recall the discussion of the ideal capacitor and inductor in Section 16.4. The rate at which this energy is delivered to the field is known as **reactive power** (symbol Q). Its unit is the **var** (volt ampere reactive), which can be inductive or capacitive.

watts

reactive power

var

The reactive power is the product of the voltage and the component of current that is 90° out of phase with the voltage. That is,

$$Q = VI \sin \theta \tag{17.1}$$

where Q is the reactive power, in vars; V is the voltage across the impedance, in volts; I is the current in the impedance, in amperes; and θ is the phase angle of the impedance, in degrees or radians.

For a pure reactance, where θ is 90°, Equation 17.1 becomes

$$Q = I^2X \tag{17.2}$$

$$Q = \frac{V^2}{X} \tag{17.3}$$

$$Q = VI \tag{17.4}$$

where Q is the reactance, in vars; I and V have the same meaning as in Equation 17.1; and X is the reactance, in ohms.

Since the reactive power equations have the same form as the real power equations, they also let us determine the reactance from the vars, just as resistance is determined from power.

EXAMPLE 17.1

A load draws 2 amperes when connected to a 120-volt source. The power factor is 0.6 leading. Find:

a. the real power dissipated in the load.
b. the reactive power.
c. the reactance and resistance of the load, assuming it is a series circuit.

Solution

a. $P = EI \cos \theta$

$= (120\text{ V})(2\text{ A})(0.6)$

$P = 144\text{ W}$

b. $Q = EI \sin \theta$

$= (120\text{ V})(2\text{ A})(0.8)$

$Q = 192\text{ vars}$

c. $Q = I^2X \quad \text{so} \quad X = \dfrac{Q}{I^2} \quad \text{or} \quad X = \dfrac{192\text{ vars}}{(2\text{ A})^2}$

$X = 48\ \Omega$

It is capacitive since the power factor is a leading one. I leads V_c.

$$P = I^2R \quad \text{so} \quad R = \frac{P}{I^2}$$

$$R = 144\text{ W}/(2\text{ A})^2$$

$$R = 36\ \Omega$$

apparent power
volt ampere

Apparent power (symbol S) is the product of voltage and current. Its unit is the **volt ampere** (VA), or kilovolt ampere (kVA) for larger amounts. In equation form,

$$S = VI \tag{17.5}$$

where S is the apparent power, in volt-amperes, and V and I have the same meaning as in Equation 17.1.

Transformers and large machines are usually rated in apparent power. For example, a transformer can be rated at 1 kVA at 220 volts. One advantage of this rating is that it can be used to find the item's real power capacity at any power factor. Since the apparent power rating is for a power factor of unity, the power capacity at any power factor will be

$$P_m = S_mF_p \tag{17.6}$$

where P_m is the power capacity at the power factor, and S_m is the volt-ampere rating of the unit.

TABLE 17.1 Power and Component Relation

Component	Power (W)	Reactive Power (vars)	Apparent Power (VA)
Resistance	$VI,\ I^2R,\ \frac{V^2}{R}$	0	$VI,\ I^2R,\ \frac{V^2}{R}$
Reactance	0	$VI,\ I^2X,\ \frac{V^2}{X}$	$VI,\ I^2X,\ \frac{V^2}{X}$
Impedance	$VI \cos \theta$	$VI \sin \theta$	VI

EXAMPLE 17.2

A transformer is rated at 5 kVA at 440 volts.

a. How much power can it supply to a load having a power factor of 0.7?

b. What is the maximum current rating of the transformer?

Solution

a.
$$\begin{aligned} P_m &= S_m\,(F_p) \\ &= (5 \times 10^3\ \text{VA})(0.7) \\ &= 3.5 \times 10^3\ \text{W} \\ P_m &= 3.5\ \text{kW} \end{aligned}$$

b.
$$S = VI \quad \text{so} \quad I = \frac{S}{V} = \frac{5000\ \text{VA}}{440\ \text{V}}$$
$$I = 11.36\ \text{A}$$

A summary of the power relations for each component is given in Table 17.1. Note that there is no reactive power in a resistor, no real power in a reactance, and apparent power in all impedances.

EXAMPLE 17.3

Determine the values of the components in each load of Figure 17.1.

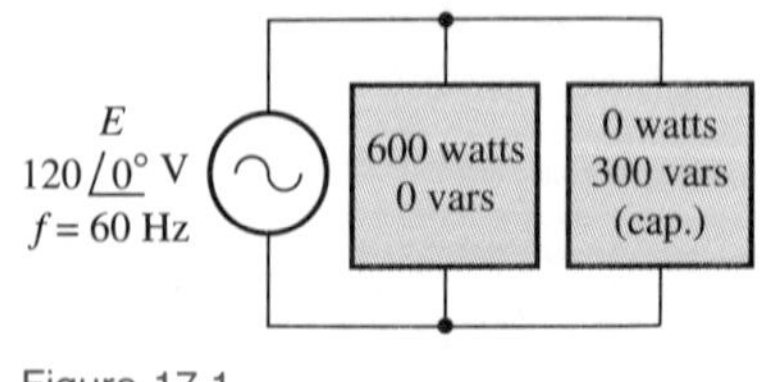

Figure 17.1

Solution

Load 1 has no reactance since there are 0 vars. Since it is connected across the source voltage E,

$$P = \frac{E^2}{R} \quad \text{or} \quad R = \frac{E^2}{P}$$

$$R = \frac{(120\ \text{V})^2}{600\ \text{W}}$$

$$R = 24\ \Omega$$

Load 2 has no resistance since there are 0 watts. Then

$$Q = \frac{E^2}{X_C} \quad \text{or} \quad Xc = \frac{E^2}{Q}$$

$$X_C = \frac{(120\ \text{V})^2}{300\ \text{vars}} = 48\ \Omega$$

$$X_C = \frac{1}{2\pi fC} \quad \text{so} \quad C = \frac{1}{2\pi fX_C}$$

$$C = \frac{1}{(2\pi)(60\ \text{Hz})(48\ \Omega)}$$

$$C = 55.26\ \mu\text{F}$$

17.2 The Power Triangle

power triangle

The **power triangle** is a triangle formed using the reactive power as the leg, real power as the base, and apparent power as the hypotenuse. It is the impedance triangle with all terms multiplied by I^2.

Since $P = VI \cos\theta$ and $Q = VI \sin\theta$, they form the base and leg of the triangle. The hypotenuse is the apparent power, VI. Power triangles for an R–C and R–L impedance are shown in Figure 17.2. Some characteristics of these triangles are:

1. Power is the base of the triangle.
2. Reactive power is the leg of the triangle. In this text, inductive vars will be drawn downward and capacitive vars upward. Thus, a lagging current will have a negative angle in the triangle.
3. The angle between S and P is the phase angle of the impedance.
4. S, P, and Q are related by the relations for a right triangle.

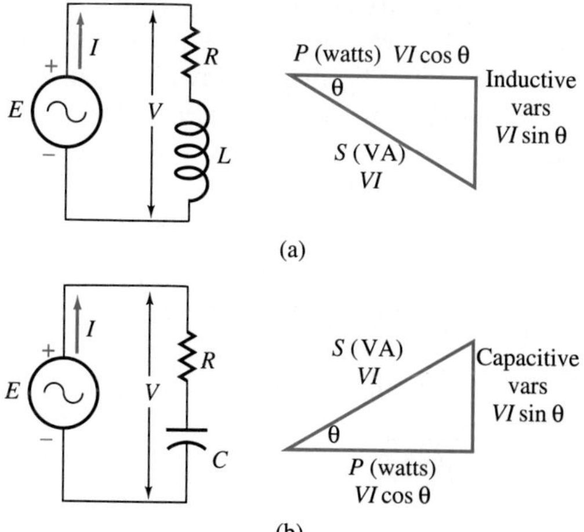

Figure 17.2 The power triangle is made up of real power, reactive power, and apparent power. (a) The vars for the R–C series circuit are drawn upward. (b) The vars for the R–L series circuit are drawn downward.

EXAMPLE 17.4

a. Draw the power triangle for an impedance of 20 $\angle 36.87°$ ohms when connected across a voltage of 120 $\angle 0°$ V.
b. Repeat for an impedance of 20 $\angle -36.87°$ ohms.

Solution

a. $Q = VI \sin \theta$
but

$$I = \frac{V}{Z} = \frac{120 \angle 0° \text{ V}}{20 \angle 36.87° \ \Omega} = 6 \angle -36.87° \text{ A}$$

So

$$Q = (120 \text{ V})(6 \text{ A})(\sin 36.87°)$$
$$Q = 432 \text{ vars}$$
$$P = VI \cos \theta = (120 \text{ V})(6 \text{ A})(\cos 36.87°)$$
$$P = 576 \text{ W}$$

In the triangle,

$$\cos \theta = \frac{P}{S}$$

so

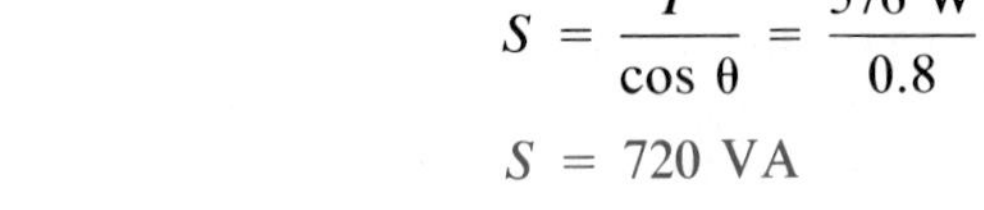

$$S = \frac{P}{\cos \theta} = \frac{576 \text{ W}}{0.8}$$
$$S = 720 \text{ VA}$$

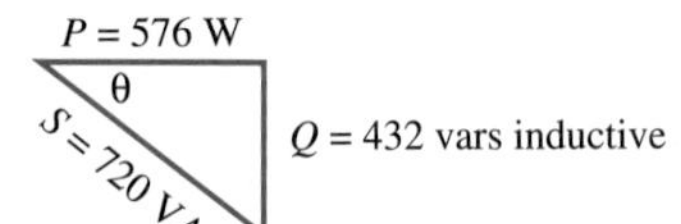

Figure 17.3

Since the angle of Z is positive, the impedance is inductive, and the vars are drawn downward as in Figure 17.3.

b. P, Q, and S will have the same values, but the vars will be capacitive. They must be drawn upward as in Figure 17.4.

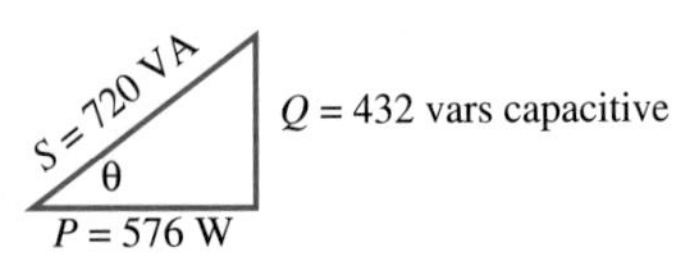

Figure 17.4

When P and Q are known, S and θ can be found using the polar-rectangular conversion on the electronic calculator. The calculator sequence for Example 17.3 a is:

Algebraic	RPN	DISPLAY
5 7 6 INV R→P	4 3 2 ENTER 5	
4 3 2 =	7 6 P	720
X Y	X↔Y	36.87°

17.3 Circuit Analysis Using the Power Triangle

The real and reactive power in the parts of a circuit can be combined, respectively, to give the total quantities in the circuit. Then these can be used to obtain the power triangle for the circuit.

The procedure is similar to using the equivalent impedance of the circuit. An advantage of this method is that it does not require the use of complex numbers. The combining is done as follows:

1. The power for the circuit triangle is equal to the sum of the power in each part. That is,

$$P_T = P_1 + P_2 + \cdots + P_n$$

2. The reactive power is equal to the algebraic sum of the reactive power in each part. Inductive vars are negative, capacitive vars are positive.
3. The apparent power is found using the power and reactive power from Steps 1 and 2, and right triangle relations. Because of the phase difference, apparent power cannot be added directly.

EXAMPLE 17.5

Construct the power triangle for a circuit made up of the following three impedances: 200 W, 100 vars inductive; 300 W, 400 vars capacitive; 400 W, 0 vars.

Solution

$$P_T = P_1 + P_2 + P_3$$
$$= 200\text{ W} + 300\text{ W} + 400\text{ W}$$
$$P_T = 900\text{ W}$$
$$Q_T = Q_1 + Q_2 + Q_3$$
$$= -100\text{ vars} + 400\text{ vars} + 0\text{ vars}$$
$$Q_T = +300\text{ vars capacitive}$$
$$S_T = \sqrt{P_T^2 + Q_T^2}$$
$$= \sqrt{(900\text{ W})^2 + (300\text{ vars})^2}$$
$$S_T = 948.68\text{ VA}$$
$$\tan\theta_T = \frac{300\text{ vars}}{900\text{ W}} = 0.333$$
$$\theta_T = 18.42°$$

The triangle is shown in Figure 17.5. Since the vars are capacitive and the angle θ is positive, I leads V.

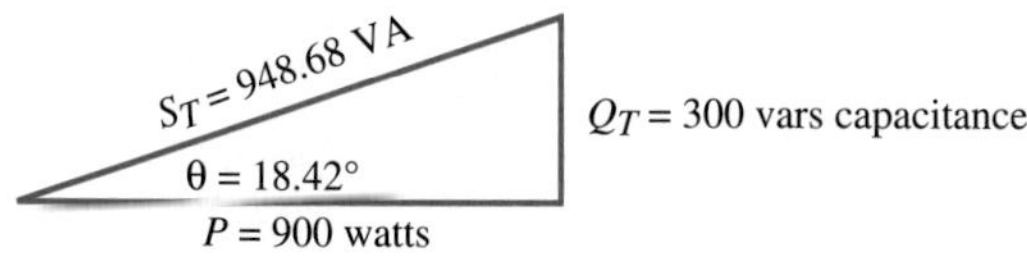

Figure 17.5

The steps in the application of the power triangle to circuit analysis are:

1. Construct the power triangle for the circuit, combining the watts and vars, as in Example 17.5.
2. Using the source voltage and apparent power, calculate the magnitude of the source current.
3. Determine the circuit's phase angle from the power triangle. If the circuit vars are inductive, the current angle is the voltage angle minus the triangle angle. If the vars are capacitive, the current angle is the voltage angle plus the triangle angle.
4. Note whether the circuit current or voltage is the current or voltage for any other part of the circuit. Then repeat Step 3 for that part of the circuit using the current or voltage.
5. Repeat Step 4 for other parts until all wanted quantities are found.
6. If needed, determine reactance and resistance using the relations in Table 17.1.

The procedure is applied in Example 17.6.

EXAMPLE 17.6

Find the source current, branch currents, voltage across each impedance, real power dissipated in the circuit, the power factor for the circuit in Figure 17.6, and the R and X part of $\mathbf{Z}_1$, assuming they are in series.

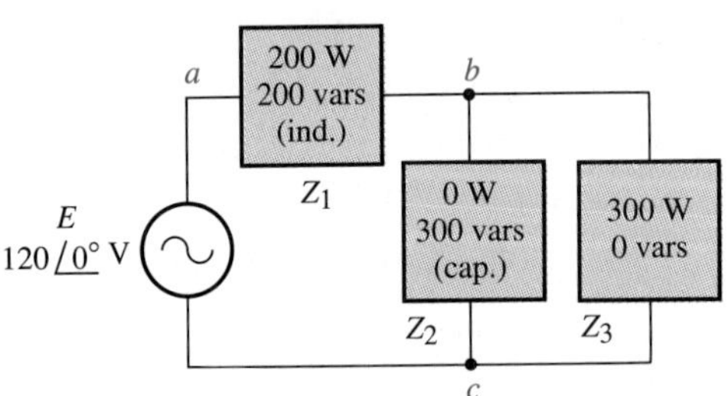

Figure 17.6

Strategy

Construct the circuit power triangle, and find the current.

Solution

The source voltage is known, so the source current will be found from the circuit power triangle.

$$P_T = P_1 + P_2 + P_3$$
$$= 200 \text{ W} + 0 \text{ W} + 300 \text{ W}$$
$$P_T = 500 \text{ W}$$
$$Q_T = -Q_1 + Q_2 + Q_3$$
$$= -200 \text{ vars} + 300 \text{ vars} + 0 \text{ vars}$$
$$= 100 \text{ vars capacitive}$$
$$S_T = \sqrt{P_T^2 + Q_T^2}$$

$$= \sqrt{(500 \text{ W})^2 + (100 \text{ vars})^2}$$

$$S_T = 509.9 \text{ VA}$$

$$\tan \theta_T = \frac{100 \text{ vars}}{500 \text{ W}} = 0.2$$

$$\theta_T = 11.31°$$

$$S_T = V_s I_s \quad \text{so} \quad I_s = \frac{S_T}{V_s} = \frac{509.9 \text{ VA}}{120 \text{ V}} = 4.25 \text{ A}$$

The circuit is capacitive and E is at 0° so

$$I_s = 4.25 \angle 11.31° \text{ A}$$

The total real power dissipated in the circuit is P_T or 500 W.

$$F_p = \cos(11.31°)$$

$$F_p = 0.981$$

Inspection of the circuit shows that I_s is also the current in Z_1. Therefore, the power triangle for Z_1 will be used to find V_1.

Relate the current to another impedance. Then find the voltage from current and apparent power.

$$S_1 = \sqrt{P_1^2 + Q_1^2}$$

$$= \sqrt{(200 \text{ W})^2 + (-200 \text{ vars})^2}$$

$$S_1 = 282.84 \text{ VA}$$

$$\tan \theta_1 = \frac{-200 \text{ vars}}{200 \text{ W}} = 1, \qquad \theta_1 = -45°$$

$$S_1 = V_1 I_1 \quad \text{so} \quad V_1 = \frac{S_1}{I_1} = \frac{282.84 \text{ VA}}{4.25 \text{ A}}$$

$$V_1 = 66.55 \text{ V}$$

The negative vars means that z_1 is inductive so V_1 leads I_1 by θ_1. I_1 is at 11.31° so V_1 will be at 11.31° + 45° or

$$V_1 = 66.55 \angle 56.31° \text{ V}$$

The source current is the current in the parallel combination of Z_2 and Z_3. Therefore, the power triangle for $Z_2 \parallel Z_3$ will be used to find the voltage across Z_2 and Z_3,

Use the current in the parallel group and its triangle to find voltage.

$$P_{bc} = P_2 + P_3$$

$$= 0 \text{ W} + 300 \text{ W}$$

$$= 300 \text{ W}$$

$$Q_{bc} = Q_2 + Q_3$$

$$= +300 \text{ vars} + 0 \text{ vars}$$

$$= 300 \text{ vars capacitance}$$

$$S_{bc} = \sqrt{P_{bc}^2 + Q_{bc}^2}$$

$$= \sqrt{(300 \text{ W})^2 + (300 \text{ vars})^2}$$

$$= 424.3 \text{ VA}$$

$$\tan \theta_{bc} = \frac{+300 \text{ vars}}{300 \text{ W}} = +1$$

$$\theta_{bc} = +45°$$

Now the power triangle can be drawn. It is shown in Figure 17.7.

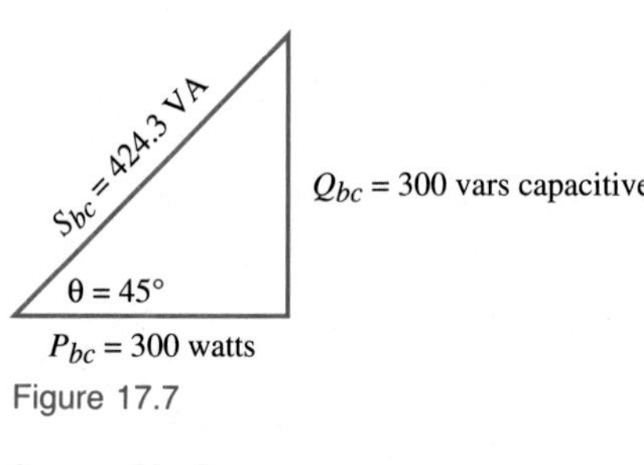

Figure 17.7

$$S_{bc} = V_{bc} I_{bc}$$

so

$$V_{bc} = \frac{S_{bc}}{I_{bc}} = \frac{424.3 \text{ VA}}{4.25 \text{ A}} = 99.84 \text{ V}$$

Knowing the voltage, find the current.

Now the combination is capacitive so V_{bc} lags I_{bc} by 45°. I_{bc} is at 11.1° so V_{bc} is at (11.31° − 45°) or

$$V_{bc} = 99.84 \angle -33.69° \text{ V}$$

Now that V_{bc} is known, the Z_2 and Z_3 power triangles can be used to find the current in them.

$$S_2 = \sqrt{(0 \text{ W})^2 + (300 \text{ vars})^2}$$

$$= 300 \text{ VA}$$

$$\theta_2 = 90° \text{ since} \qquad P_2 = 0 \text{ W}$$

$$S_2 = V_2 I_2 = V_{bc} I_2$$

so

$$I_2 = \frac{S_2}{V_{bc}} = \frac{300 \text{ VA}}{99.84 \text{ V}} = 3 \text{ A}$$

The branch is capacitive so I_2 leads V_2 by 90°. V_2 is at −33.69° so I_2 is at −33.69° + 90°.

$$I_2 = 3 \angle 56.31° \text{ A}$$

$$S_3 = \sqrt{(300 \text{ W})^2 + (0 \text{ vars})^2}$$

$$= 300 \text{ VA}$$

$$\theta_3 = 0° \text{ since } Q_T = 0 \text{ vars}$$

$$S_3 = V_3 I_3 = V_{bc} I_3$$

so

$$I_3 = \frac{S_3}{V_{bc}} = \frac{300 \text{ VA}}{99.84 \text{ V}} = 3 \text{ A}$$

I_3 is in phase with V_{bc}, so $I_3 = 3 \angle -33.69°$ A; $I_3 = 3 \angle -33.69°$ A.
For the R and X parts of Z_1, assuming a series circuit,

Find the value of R and X from the current and triangle.

$$P_1 = I_s^2 R_1$$

or

$$R_1 = \frac{P_1}{I_s^2}$$

$$= \frac{200 \text{ W}}{(4.25 \text{ A})^2}$$

$$R_1 = 11.07\ \Omega$$

$$Q_1 = I_s^2 X_1$$

$$X_1 = \frac{Q_1}{I_s^2}$$

$$X_1 = \frac{200 \text{ vars}}{(4.25 \text{ A})^2}$$

$$X_1 = 11.07\ \Omega \text{ inductive reactance}$$

The analysis of a series or series-parallel circuit using the power triangle can be done as an exercise, but its application to actual circuits is not practical. The reason is that if the vars and watts are known, the voltage or current must already be known; the vars and watts are for only one voltage. If the voltage and current are known, there is no need for the analysis.

On the other hand, the method is useful for parallel impedances where specific voltage are used. A prime application is in electric power distribution. There, the voltages are 120 V, 220 V, 240 V, 440 V, and a few others. For these voltages, loads and reactances can be specified in terms of vars, watts, and apparent power. The next example considers such a load.

EXAMPLE 17.7

The following loads are connected across the 220-V lines of an electric utility: 500 W, 200 vars inductive; 300 W, 500 vars capacitive; 400 W, 0 vars; 1000 VA at a lagging power factor of 0.8. Find:

a. the total current.
b. the total real power.
c. the power factor of the total.

Solution

The total real power is

$$P_T = P_1 + P_2 + P_3 + P_4$$

$$P_4 = S_4 \cos \theta_4$$

$$= (1000 \text{ VA})(0.8) = 800 \text{ W}$$

so

$$P_T = 500\text{ W} + 300\text{ W} + 400\text{ W} + 800\text{ W}$$
$$P_T = 2000\text{ W}$$
$$Q_T = -Q_1 + Q_2 + Q_3 - Q_4$$
$$Q_4 = S_4 \sin \theta_4$$
$$\theta_4 = \cos^{-1}(0.8) = 36.87°$$

so

$$Q_4 = (1000\text{ VA})(\sin 36.87°) = 600\text{ vars inductive}$$

Then

$$Q_T = -200\text{ vars} + 500\text{ vars} + 0\text{ vars} - 600\text{ vars}$$
$$= -300\text{ vars}$$
$$S_T = \sqrt{P_T^2 + Q_r^2}$$
$$= \sqrt{(2000\text{ W})^2 + (-300)^2}$$
$$= 2022\text{ VA}$$
$$S_T = EI_T \quad \text{so} \quad I_T = S_T/E$$
$$I_T = \frac{2022\text{ VA}}{220\text{ V}}$$
$$I_T = 9.19\text{ A}$$
$$\cos\theta_T = \frac{P_T}{S_T} = \frac{2000\text{ W}}{2022\text{ VA}} = 0.989$$
$$F_p = \cos\theta_T$$
$$F_p = 0.989$$

17.4 Power Factor Correction

Residential customers of electric utilities use loads that are mostly resistive and have a power factor of unity. These are loads such as electric lights, electric ranges, toasters, and heating units. However, industrial customers use loads that are inductive, such as transformers, motors, welders, and fluorescent lights. These loads can have a power factor much smaller than one. Because $P = VI\cos\theta$, this lower power factor requires more current to supply a given amount of energy. This additional current requires larger alternators and increases the utility's cost. In addition, the current can cause a low voltage at the customer's site. Because of this increased expense, the customer must either correct the power factor or pay a penalty charge. **Power factor correction** is a procedure in which vars are added to increase the power factor by canceling some of the vars in the original load.

power factor correction

The customer's load usually cannot be changed, so the power factor correction must be done by some other means. Since most industrial

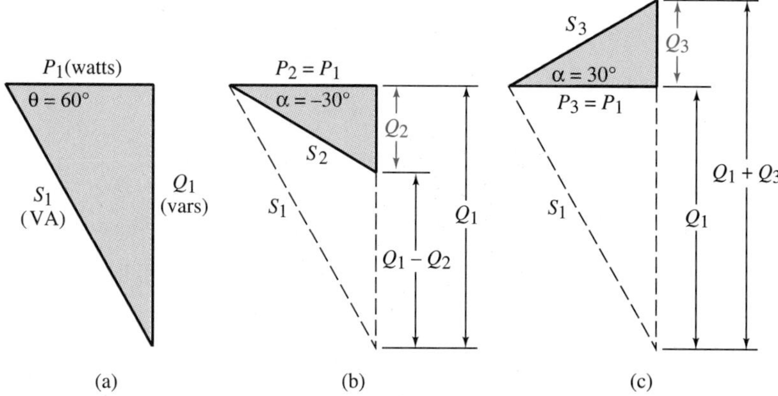

Figure 17.8 The power factor is corrected by adding vars to the circuit vars. (a) This load has a 60° lagging current and a 0.5 power factor. (b) Adding $Q_1 - Q_2$ capacitive vars gives a 0.866 lagging power factor. (c) Adding $Q_1 + Q_3$ capacitive vars gives a 0.866 leading power factor.

loads are inductive, the correction is usually made by connecting capacitors across the lines that supply power to the customer. The capacitive vars cancel some of the inductive vars. It is important to recognize that these capacitors do not change the customer's load. They change the load that is connected to the source.

An understanding of power factor correction can be obtained by studying Figure 17.8. Q_1 is the number of vars at a power factor of cos θ. Q_2 is the number of vars at a power factor of cos α. The difference in vars is $Q_1 - Q_2$. Therefore enough capacitive vars must be added to numerically equal $Q_1 - Q_2$. The power factor is still lagging but is now cos α. Adding an amount of capacitive vars greater than Q_1 will produce a leading power factor. The resulting triangle is shown in Figure 17.8(c).

The sequence in calculations involving power factor correction is:

1. Determine the vars in the present load.
2. Determine the vars needed for the desired power factor.
3. The vars that must be connected in the circuit are the difference between those in Step 2 and those in Step 1.

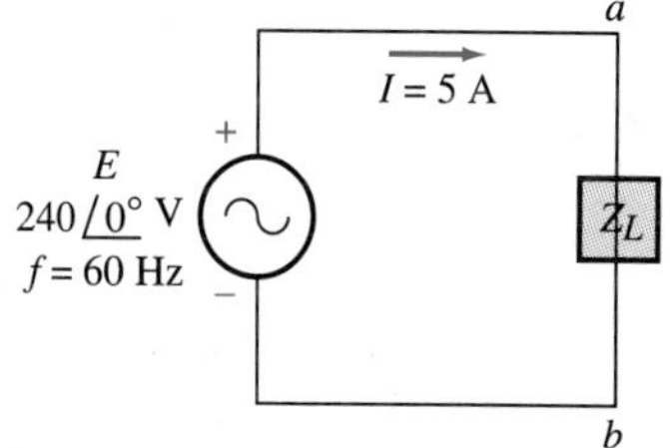

Figure 17.9

EXAMPLE 17.8

An electric utility company supplies 720 W of power to the inductive load Z_L, as in Figure 17.9.

a. How many capacitive vars must be connected across a–b to change the power factor to 0.8 lagging?
b. What value of capacitance will give the required vars?

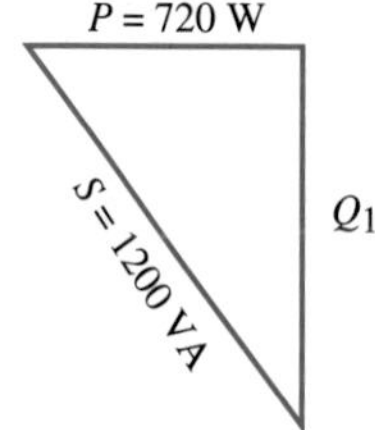

Figure 17.10

Solution

Strategy

$$S = VI = (240 \text{ V})(5 \text{ A}) = 1200 \text{ VA}$$

The power triangle is drawn in Figure 17.10.

$$Q_1 = P \tan \theta_1$$

Find the vars in the present circuit.

$$\cos \theta_1 = \frac{P}{S} = \frac{720 \text{ W}}{1200 \text{ VA}} = 0.6$$

so

$$|\theta_1| = 53.13°$$

$$Q_1 = (720 \text{ W}) \tan 53.13° = 960 \text{ vars inductive or } -960 \text{ vars}$$

At a power factor of 0.8,

Find the vars at a power factor of 0.8 lagging.

$$\theta_2 = \text{arc cos } 0.8 \text{ or } 36.87°$$

so

$$\begin{aligned} Q_2 &= P \tan \theta_2 \\ &= (720 \text{ W}) \tan 36.87° \\ &= 540 \text{ vars inductive or } -540 \text{ vars} \end{aligned}$$

Calculate the difference.

The difference is $Q_1 - Q_2 = -960 - (-540)$ or -420 vars.

Since 420 inductive vars must be cancelled, 420 capacitive vars must be added.

The voltage across the capacitor will be 240 volts.

Find the value of C using Q and $X_c = 1/(2\pi f C)$

$$Q_c = \frac{V^2}{X_C} \quad \text{so} \quad X_C = \frac{V^2}{Q_C} = \frac{(240 \text{ V})^2}{420 \text{ vars}} \quad \text{or} \quad 137.1\ \Omega$$

Since

$$X_c = \frac{1}{2\pi f C} \qquad C = \frac{1}{2\pi f X_C}$$

$$C = \frac{1}{(2\pi)(60 \text{ Hz})(137.1\ \Omega)}$$

$$C = 19.3\ \mu\text{F}$$

17.5 Alternating Current (ac) Resistance

dc resistance

For a dc current in a conductor, the only energy loss is that needed to overcome the resistance in the conductor. This is the resistance that was calculated using $R = (\rho l)/A$, which is the **dc resistance**.

On the other hand, for an ac current, there are also losses caused by the reversal of the magnetic field around the conductor. These losses are greater if the conductor is wound around an iron core.

ac resistance

The **ac resistance** is a value of resistance that accounts for the losses due to the opposition to the ac current and the changing magnetic field around the conductor. It is also called the effective resistance. This resistance will be greater than the dc resistance because it accounts for the additional losses. The ac resistance can be determined by measuring the power and current when in the ac circuit. Then

$$P = I^2R_{ac}$$

or

$$R_{ac} = \frac{P}{I^2} \quad \textbf{(17.7)}$$

where R_{ac} is the resistance, in ohms; P is the power, in watts; and I is the current, in amperes.

The additional energy losses are due to skin effect, electromagnetic radiation, hysteresis, and eddy currents. The first two can usually be ignored at frequencies below radio frequencies. The last two can be considerable, even at low frequencies.

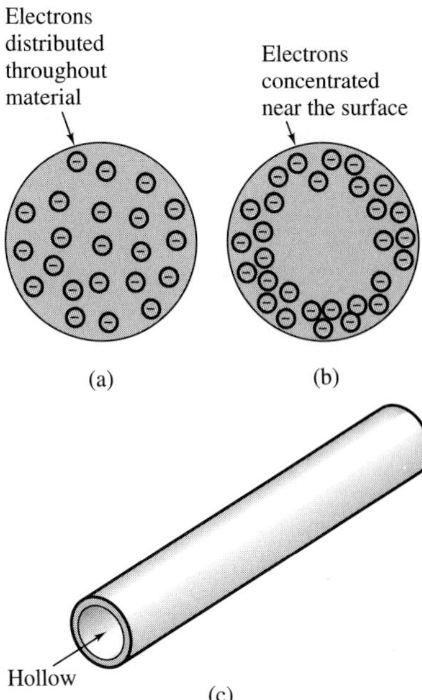

Figure 17.11 The skin effect causes the charge to concentrate near the outer surface of the conductor. (a) The distribution of electrons at low frequency. (b) The distribution of electrons at a high frequency. (c) A hollow conductor such as this one is often used at high frequencies.

EXAMPLE 17.9

Find the ac resistance of an iron core coil if the power is 16 W at a current of 2 amperes.

Solution

$$R_{ac} = \frac{P}{I^2} = \frac{16\text{ W}}{(2\text{ A})^2}$$

$$R_{ac} = 4\ \Omega$$

skin effect

Skin effect is a condition where the current tries to concentrate near the surface of the conductor, as seen in Figure 17.11. The magnetic field is stronger near the center of the conductor, so a change in the field induces a larger emf at the center. This emf, which is larger at the center, keeps the current smaller in that region. The skin effect increases as the frequency is increased. Because of this, many high-frequency conductors are hollow instead of solid. Since the current is concentrated near the surface, the area is effectively reduced, thus increasing the resistance.

A useful application of skin effect is to heat-treat metals. The surface of the part can be heat-treated without affecting the deeper parts. The depth of the heating is controlled by the frequency of the current. Higher frequencies result in less depth.

electromagnetic radiation

Electromagnetic radiation is the radiation of energy into the atmosphere as the magnetic field alternates. An extreme example of this is the radiation of energy from an antenna. Since the energy is not returned, it is lost. This loss also increases with frequency.

hysteresis loss

Hysteresis loss is the loss of energy in a magnetic material as the field reverses. As the material is magnetized and demagnetized, energy is used to realign the domains. The hysteresis loss can be reduced by adding elements such as silicon to the iron. The hysteresis loss varies with frequency and the flux density, and is given by

$$P_h = K_h f B^n \tag{17.8}$$

where P_h is the power loss due to hysteresis, in watts; K_h is a number whose value depends on such things as material, structure, and shape; f is the frequency, in hertz; B is the flux density, in tesla; and n is an exponent whose value is approximately 1.5–2.5.

eddy current loss

Eddy current loss is the loss caused by circulating currents in the magnetic material. The currents are a result of the emf induced by the changing magnetic field. Making the core from laminations instead of one solid piece reduces these losses. The laminations are coated with an insulating material. This effectively provides a high resistance path for the eddy current, as seen in Figure 17.12.

The eddy current loss is given by

$$P_e = K_e f^2 B^2 \tag{17.9}$$

where P_e is the power loss due to eddy currents, in watts; K_e is a number whose value depends on such things as the shape, structure, and material; f is the frequency, in hertz; and B is the flux density, in tesla.

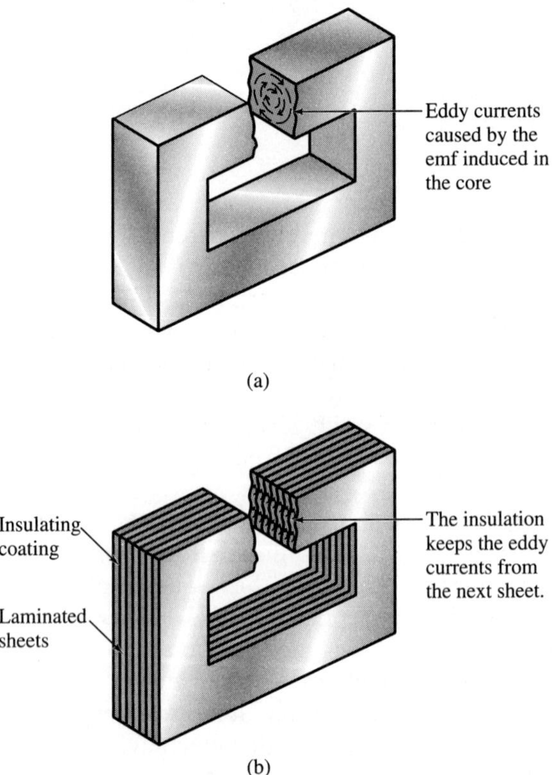

Figure 17.12 Eddy current losses in an iron core can be reduced by using laminations. (a) The solid core offers little resistance to the eddy currents. (b) The insulation between the laminations prevents the eddy currents from going from one sheet to the other.

In transformers and machinery, hysteresis and eddy current losses result in a rise in the temperature of the units.

EXAMPLE 17.10

The hysteresis loss in a transformer is 3 W and the eddy current loss is 2 W at a flux density of 1.2 tesla and a frequency of 60 Hz. What will these losses be at $B = 0.6$ T and $f = 30$ Hz? Assume $n = 2$.

Solution

For the hysteresis losses,

$$P_h = K_h f B^n$$
$$P_{30} = (K_h)(30\text{ Hz})(0.6\text{ T})^2$$
$$P_{60} = (K_h)(60\text{ Hz})(1.2\text{ T})^2$$

or

$$K_h = \frac{P_{60}}{(60\text{ Hz})(1.2\text{ T})^2}$$

Putting this in the expression for P_{30} gives

$$P_{30} = \frac{(30\text{ Hz})(0.6\text{ T})^2(P_{60})}{(60\text{ Hz})(1.2\text{ T})^2}$$
$$= \frac{(30\text{ Hz})(0.6\text{ T})^2(3)}{(60\text{ Hz})(1.2\text{ T})^2}$$
$$P_{30} = 0.375\text{ W}$$

For the eddy current losses,

$$P_e = K_e f^2 B^2$$
$$P_{30} = (K_e)(30\text{ Hz})^2(0.6\text{ T})^2$$
$$P_{60} = (K_e)(60\text{ Hz})^2(1.2\text{ T})^2$$

or

$$K_e = \frac{P_{60}}{(60\text{ Hz})^2(1.2\text{ T})^2}$$

Putting this in the expression for P_{30} gives

$$P_{30} = \frac{(30\text{ Hz})^2(0.6\text{ T})^2(P_{60})}{(60\text{ Hz})^2(1.2\text{ T})^2}$$
$$= \frac{(30\text{ Hz})^2(0.6\text{ T})^2(2)}{(60\text{ Hz})^2(1.2\text{ T})^2}$$
$$P_{30} = 0.125\text{ W}$$

An examination of the four additional losses shows that only one, the skin effect, represents an actual change in the resistance of the conductor. The others are related to the material surrounding the conductor.

SUMMARY

1. Reactive power is the rate at which energy is exchanged in the field of a reactance. Its unit is the var.
2. Apparent power is the product of the voltage and current in an impedance. Its unit is the volt-ampere.
3. Reactive power is equal to $VI \sin \theta$, and apparent power is equal to VI.
4. There is power and apparent power in a resistance, reactive power and apparent power in a reactance, and all three in a complex impedance.
5. The power triangle has watts as its base, vars as the leg, and apparent power as the hypotenuse.
6. Making the power factor larger by adding vars to cancel some of the vars in the circuit is known as power factor correction.
7. The ac resistance is a resistance value that accounts for the resistance of the conductor and losses due to skin effect, electromagnetic radiation, hysteresis, and eddy currents.
8. Skin effect is the effect where the current tries to concentrate near the surface of the conductor.
9. Electromagnetic radiation is the radiation of energy into the atmosphere as the magnetic and electric fields alternate.
10. Hysteresis loss is the loss of energy in a magnetic material as the field reverses and the domains are realigned.
11. Eddy current loss is the loss caused by circulating currents in the magnetic material.

EQUATIONS

17.1	$Q = VI \sin \theta$	Equation for reactive vars in any impedance.	710
17.2	$Q = I^2X$	Equation for reactive vars in a reactance in terms of current and reactance.	710
17.3	$Q = \dfrac{V^2}{X}$	Equation for reactive vars in a reactance in terms of voltage and reactance.	710
17.4	$Q = VI$	Equation for reactive vars in a reactance in terms of voltage and current.	710
17.5	$S = VI$	Equation for apparent power.	711
17.6	$P_m = S_m F_p$	Equation for the power rating of a transformer.	711

17.7	$R_{ac} = \frac{P}{I^2}$	Equation for the ac resistance of a conductor.	723
17.8	$P_h = K_h f B^n$	Equation for the hysteresis loss in a magnetic material.	724
17.9	$P_e = K_e f^2 B^2$	Equation for the eddy current loss in a magnetic material.	724

QUESTIONS

1. How can one determine the apparent power when the vars and watts in an impedance are known?
2. How can the power factor of a circuit be determined from the power triangle?
3. For what type of load is the apparent power equal to the reactive power?
4. What type of vars must be added to increase the power factor of an inductive load?
5. For what type of load is the apparent power equal to the average power?
6. What is one advantage of rating a transformer by apparent power?
7. A load is made up of a 5-microfarad capacitor and a 50-ohm resistor connected in parallel across a 120-volt source. What effect will increasing the source frequency have on the reactive power? On the power factor?
8. What effect will decreasing the frequency of the source in Question 7 have on the power factor and reactive power?
9. What type of vars must be added to increase the power factor of an impedance with a leading power factor?
10. When correcting the power factor, why are the added vars connected across the lines instead of in series with the load?
11. Why is the power factor of a residential load generally higher than that of an industrial one?
12. What type of power is measured with a wattmeter: average, reactive, or apparent?
13. A wattmeter connected correctly in a circuit indicates 0 watts, but there is current and voltage. Why might this be so?
14. What effect will making a transformer core solid have on the eddy current losses?
15. Can two circuits have the same apparent power, even though the average power and reactive power are not the same?
16. When determining the ac resistance of an iron core coil, using ac voltage and current measurements, what characteristics of the voltage must be the same as those in the actual application in order to get correct results?
17. The surface of a round rod is being hardened by the skin effect to a depth of 0.01 mm. What effect will decreasing the frequency have on the process?
18. Why do transformers used in high-frequency electronic circuits generally have an air core instead of iron?
19. A 5-kVA transformer overheats when supplying 5 kW of power. What might be the cause of the overheating?
20. Why can conductors for high frequency signals be hollow instead of solid?
21. What type of losses are generally found in the core of a transformer?

PROBLEMS

SECTION 17.1 Reactive Power and Apparent Power

1. What are the watts, vars, and volt-amperes for a motor that draws 5 amperes of current from a 120-volt source? It has a lagging power factor of 0.8.

2. A 120-volt source supplies 3 amperes to a load, which dissipates 300 watts of real power and has a lagging power factor. Find:

   (a) The vars.

   (b) The volt-amperes.

   (c) The power factor of the load.

3. The apparent power in a load is 500 volt-amperes. The reactive power is 400 vars capacitive. Find:

   (a) The real power dissipated in the load.

   (b) The power factor of the load.

4. How many capacitive vars are introduced when a 200-microfarad capacitor is connected across a 120-volt, 60-Hz source?

5. A 0.5-H inductor has a resistance of 150 ohms. What are the watts, vars, and volt-amperes in the inductor when the current is 0.4 ampere at a frequency of 60 Hz?

6. A transformer has a kVA rating of 15 kVA at 440 volts at its output.

   (a) What is the maximum current that it can supply?

   (b) How many watts of power can it supply at a power factor of 0.6?

7. How many microfarads must a capacitor be to have the same reactive power as a 5-H inductor at 60 Hz?

8. What must be the kVA rating of an alternator if its load characteristics are 440 V and 750 kW, at a power factor of 0.75?

SECTION 17.2 The Power Triangle

9. Construct the power triangles for the following loads, and find the phase angle:

   (a) 300 W, 400 vars inductive.

   (b) 500 W, 707 VA, Q is capacitive.

10. Construct the power triangle for the following loads:

    (a) 600 vars capacitive, 900 VA.

    (b) 800 W, 800 vars inductive.

11. Construct the power triangle for an impedance of $20 \angle 53.13°$ ohms when its current is $5 \angle 0°$ A.

12. A motor has a lagging power factor of 0.85. It draws 5 A from a 120-volt line. Construct the power triangle.

13. A load of 500 W and 800 vars inductive is connected across a 120-volt line. Find its resistance and reactance:

    (a) If they are considered to be connected in series.

    (b) If they are considered to be connected in parallel.

14. A load dissipates 200 W and has 100 vars capacitive when connected across a 220-volt line. What are the resistance and reactance of the load's series equivalent circuit?

15. The power triangle for a series R–C circuit is shown in Figure 17.13. Find:

    (a) The circuit resistance.

    (b) The circuit reactance.

    (c) The magnitude and angle of the current if E is $120 \angle 0°$ V.

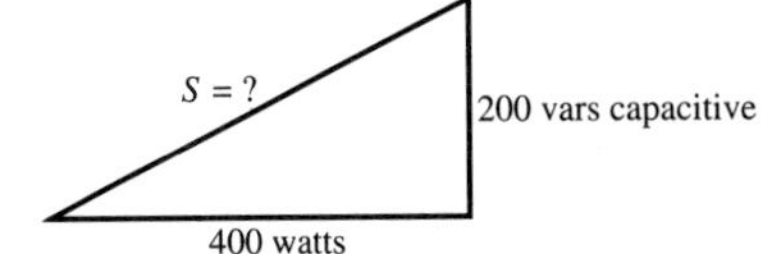

Figure 17.13

16. Determine the power factor and current for a circuit in which Q is 800 capacitive vars, S is 1600 volt-amperes, and R is 50 ohms.

SECTION 17.3 Circuit Analysis Using the Power Triangle

17. Use the power triangle method for the following problems.

    (a) Find the watts, vars, and volt-amperes for a circuit made up of the following loads:
        500 W, 500 wars inductive
        200 W, 300 vars capacitive
        100 W, 100 vars inductive

    (b) Construct the power triangle and find the power factor.

18. Three loads of a circuit have the following characteristics:
        Z_1, 50 W, 60 VA, leading F_p
        Z_2, 40 W, 40 vars inductive
        Z_3, 10 W, 30 vars capacitive

    (a) Construct the power triangle for the circuit.

    (b) Find the power dissipated in the circuit, the power factor, and the source current if the voltage is $120 \angle 0°$ V.

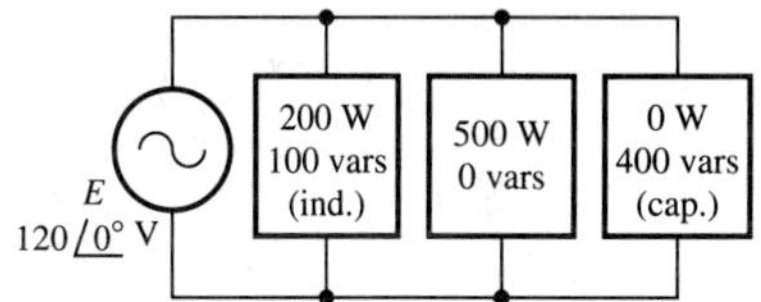

Figure 17.14

19. The loads in a house are shown in Figure 17.14. Find:

    (a) The current, including the phase angle, in each branch.

    (b) The source current.

    (c) The power factor of the circuit.

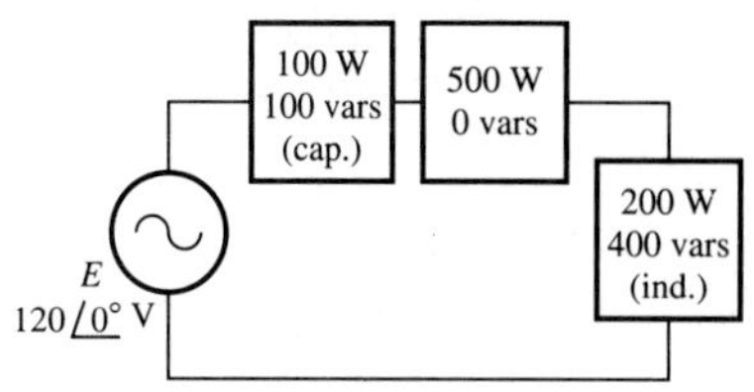

Figure 17.15

20. For the circuit of Figure 17.15, find:

    (a) The voltage across each impedance.

    (b) The source current.

    (c) The power factor of the circuit.

21. The load shown in Figure 17.16 will be connected to an ac alternator of $220 \angle 0°$ V, 60 Hz. Find:

    (a) The circuit current and circuit power factor.

    (b) The power dissipated in the circuit.

    (c) The current and voltage for each load.

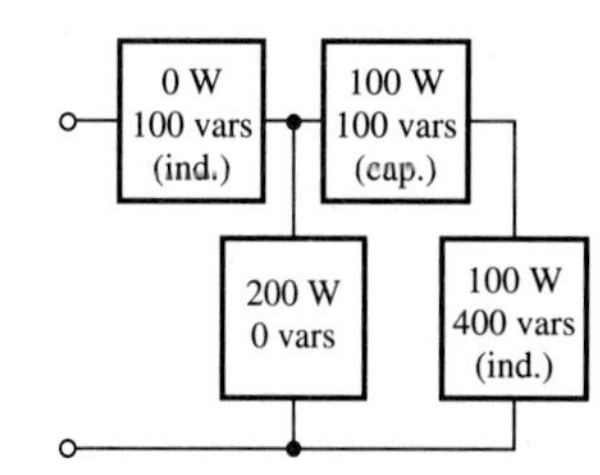

Figure 17.16

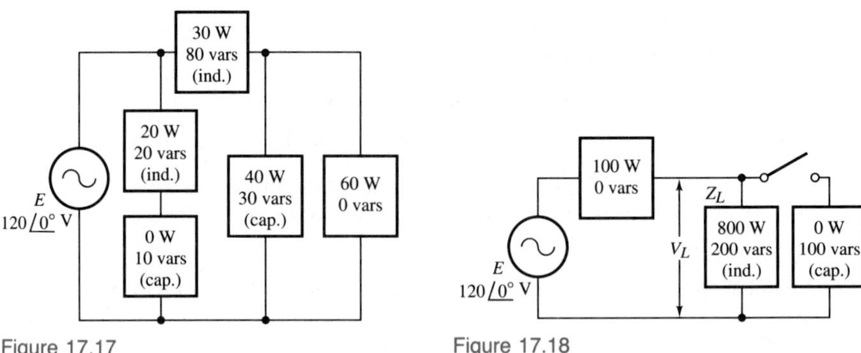

Figure 17.17

Figure 17.18

22. For the circuit in Figure 17.17, find:
    (a) The source current and power factor.
    (b) The resistance and reactance in each load, assuming a series combination.

SECTION 17.4 Power Factor Correction

Assume that the watts and vars do not change in a load when the circuit currents and voltage change.

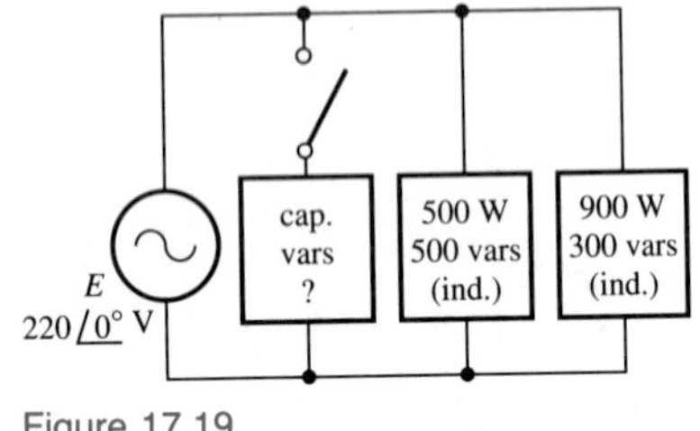

Figure 17.19

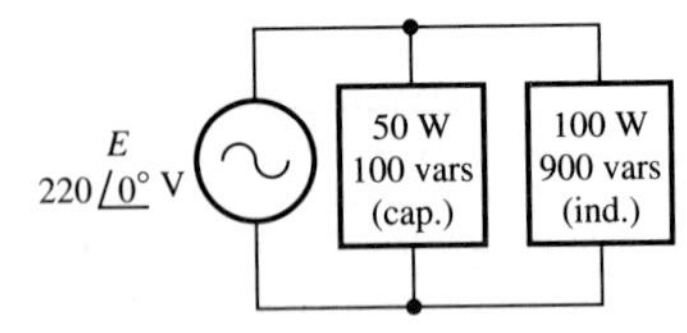

Figure 17.20

23. In the circuit of Figure 17.18, what will the decrease in the effective value of V_L be when 100 vars capacitive are connected in parallel with Z_L?

24. In the circuit of Figure 17.19, how many capacitive vars must be connected across the line to correct the power factor to 0.9 lagging?

25. How much capacitance must be connected across a 220-volt, 60-Hz line to change the power factor to 0.9 lagging, if the power factor is 0.7 lagging and the current is 2 amperes?

26. Work Problem 24 to correct the power factor to 0.8 leading.

27. In the circuit of Figure 17.20, how many capacitive vars must be connected across the line to reduce the current to 3 amperes? What capacitance is needed if the frequency is 60 Hz?

28. At a certain industrial plant, loads 3 and 4 of Figure 17.21 are disconnected after 3 p.m. How much does the power factor improve when they are disconnected?

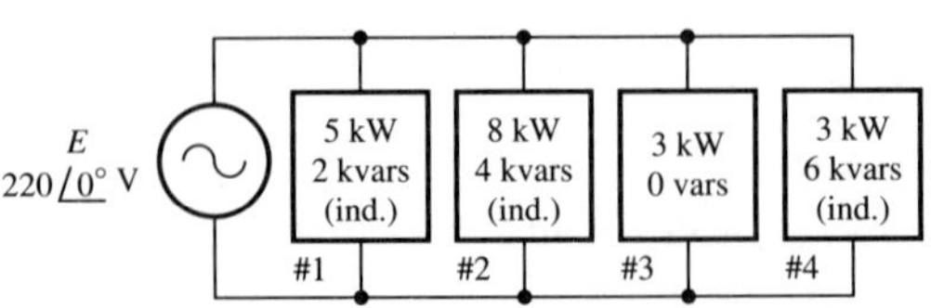

Figure 17.21

SECTION 17.5 Alternating Current (ac) Resistance

29. What is the ac resistance of a coil in which 20 W of power is dissipated when the current is 2 A?

30. The dc resistance of a coil wound around an iron core coil is 80 ohms. What is its ac resistance if it is 1.25 times the dc resistance?

31. The hysteresis loss in a transformer is 10 watts at its rated current of 5 A and a frequency of 60 Hz. If the flux density is assumed to be proportional to the current, and $n = 2$, what is the hysteresis loss at 3 A?

32. Tests on a 220-V, 10-A, 60-Hz transformer show the hysteresis loss as 80 W and the eddy current loss to be 30 W at rated conditions. If n is 1.8 and $B = 1.4$ T at rated conditions, find the total loss when B is 0.8 T.

CHAPTER 18

ALTERNATING CURRENT (ac) NETWORK THEOREMS AND ANALYSIS

CHAPTER OBJECTIVES

After completing this chapter, you should be able to:

1. Analyze ac circuits using the loop current method, branch current method, nodal method, and superposition.
2. Construct the Thevenin equivalent of an ac circuit and use it in circuit analysis.
3. Construct the Norton equivalent of an ac circuit and use it in circuit analysis.
4. Convert from one equivalent to the other.

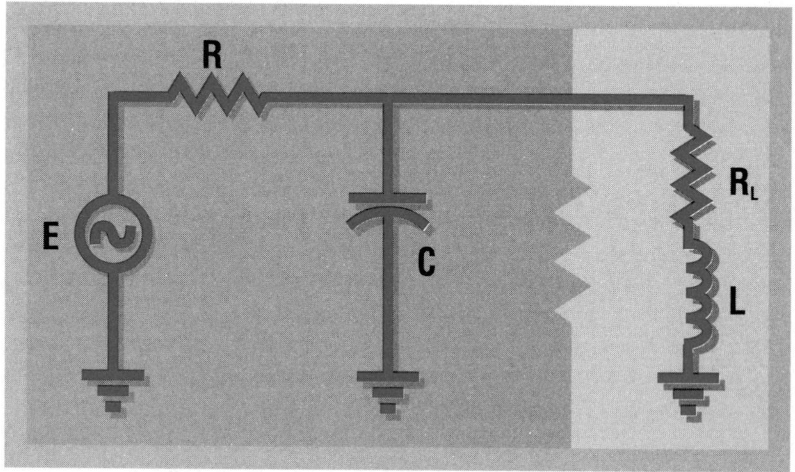

5. Apply the conditions of maximum power transfer to a circuit.
6. Analyze a bridge circuit to determine the value of an unknown impedance.

KEY TERMS

1. balanced bridge
2. branch current method
3. bridge circuit
4. conjugate
5. Hay Bridge
6. loop
7. loop current method
8. maximum power transfer
9. Maxwell-Wein Bridge
10. nodal method
11. Norton's Theorem
12. Schering Bridge
13. superposition
14. Thevenin's Theorem

INTRODUCTION

The method of equivalent impedance developed in Chapter 16 works well for single-source and series-parallel circuits. Unfortunately, many practical applications do not use such circuits. For instance, electronic amplifier circuits usually have a dc source and an ac signal. Three-phase power distribution systems are three-source circuits. Also, their loads are not series-parallel loads. Circuits used in impedance bridges have only one course, but are not a series-parallel connection. Methods for the analysis of these circuits must be available and understood.

This chapter presents several methods for analysis, as well as some electrical theorems. The methods are loop current, branch current, nodal, and superposition. They were first introduced for dc circuits. For ac circuits, the results of the analysis are true only for the frequency at which they were calculated. A change in frequency requires another analysis. The application of the analysis to ac circuits is now explained.

Analysis of multisource and nonseries-parallel circuits involves the solution of simultaneous equations. Because of the phasors, this can be lengthy and tedious. Fortunately, alternatives are available. Some electronic calculators can be programmed for the solution. Also, equation-solving programs are available for the personal computer. These alternatives can be used to solve the equations or to check results.

In most instances, the material in this chapter parallels the same topics for the dc circuits. The only difference is the phasor arithmetic. Therefore, reviewing the dc material can be helpful in your study of this chapter.

18.1 The Loop Current Method

loop current method
loop

The **loop current method** is studied first. This is a method based on the use of an assumed current for each loop. Recall that a **loop** is formed by tracing a path from one point, through the circuit, and back to the point. In this method, fewer equations have to be solved simultaneously. Kirchhoff's Voltage Law is applied to the loops to develop a series of equations that are used to determine the loop currents. These are then used to find the branch currents. Finally, Ohm's Law and other relations are used to find other quantities. A procedure for using the mesh method for an ac circuit is as follows:

1. Identify the loops in the circuit and let their number be N.
2. Draw a loop current clockwise in any $(N - 1)$ loop. If one views the circuit as a window, $N - 1$ is the number of panes. The circuit has $N - 1$ independent loops.
3. Mark the terminal polarity of each element in the circuit as follows:
   a. Sources: (+) or (−), according to the terminal markings.
   b. Impedances: terminal where the current enters (+); terminal where the current leaves (−). If an impedance has two loop currents in it, mark the polarity for each current.
4. Apply Kirchhoff's Voltage Law in a clockwise direction to $N - 1$ loops as follows:
   a. Write the voltage as the product of the current and impedance. Where there are two loop currents, write the voltage for each current.
   b. Assign a (−) sign to the term if the potential difference across the element is a drop. Otherwise, assign a (+) sign.
   c. Set the phasor sum of the voltages around the loop equal to zero.
5. Solve the set of equations in Step 4 for the loop currents.

6. Determine the branch currents. When only one loop current is in the branch, a branch current is the loop current. For a branch with two loop currents in the same direction, the branch current is their phasor sum. For a branch with two loop currents in opposite directions, the branch current is their phasor difference. The directions is that of the phasor that is given the positive sign in the subtraction.
7. Use the basic circuit concepts to find other quantities.

The procedure is applied to Example 18.1.

Find the current in each impedance, the voltage across each impedance, and the power dissipated in the circuit of Figure 18.1.

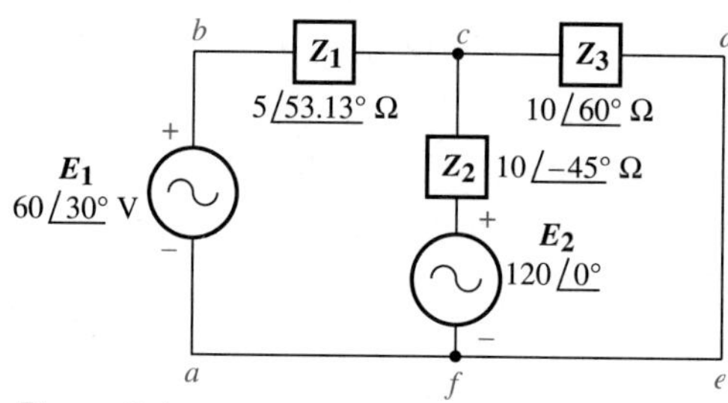

Figure 18.1

Solution

Strategy

1. This circuit has three loops. They are *abcfa*, *fcdef*, and *abcdefa*.
2. The clockwise loop currents in $(N - 1)$ loops are shown as I_A and I_B in Figure 18.2.
3. The terminal polarity is shown in Figure 18.2.

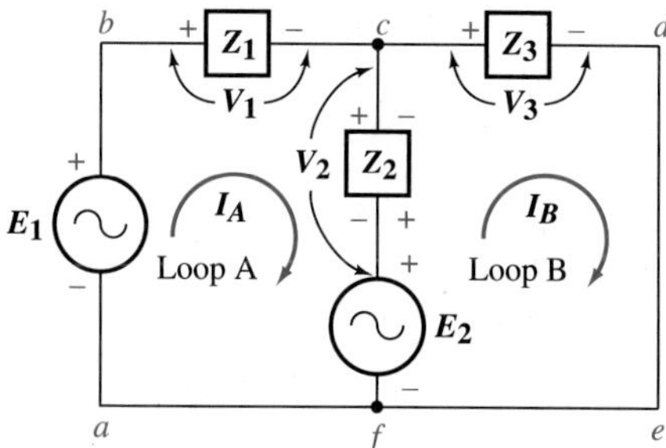

Figure 18.2

4. Writing Kirchhoff's Voltage Law in a clockwise direction for Loop A gives

Write the equations as directed in Step 3.

$$-(Z_1 + Z_2)I_A + Z_2 I_B + E_1 - E_2 = 0$$

Writing Kirchhoff's Voltage Law in a clockwise direction for Loop B gives

$$+ \mathbf{Z}_2\mathbf{I}_A - (\mathbf{Z}_2 + \mathbf{Z}_3)\mathbf{I}_B + \mathbf{E}_2 = 0$$

Impedance and source values will be added to the equations later. It is simpler to work with the letters at this time.

Solve the equations.

5. Since there are two equations with two unknowns, $\mathbf{I}_A$ and $\mathbf{I}_B$, they can be solved. Any valid method can be used. Since we have only two equations, the method of substitution, as explained in Appendix A, is used here. It might be helpful to refer to that explanation. Rearranging the equation for Loop A gives

$$\mathbf{I}_A = \frac{\mathbf{E}_1 - \mathbf{E}_2 + \mathbf{Z}_2\mathbf{I}_B}{\mathbf{Z}_1 + \mathbf{Z}_2}$$

or

$$\mathbf{I}_A = \frac{60 \angle 30^\circ \text{ V} - 120 \text{ V} + (10 \angle -45^\circ \ \Omega)(\mathbf{I}_B)}{5 \angle 53.13^\circ \ \Omega + 10 \angle -45^\circ \ \Omega}$$

$$= \frac{74.36 \angle 156.21^\circ \text{ V} + (10 \angle -45^\circ \ \Omega)(\mathbf{I}_B)}{10.53 \angle -16.95^\circ \ \Omega}$$

This reduces to

$$\mathbf{I}_A = 7.06 \angle 173.16^\circ \text{ A} + (0.95 \angle -28.05^\circ)\ \mathbf{I}_B.$$

Substituting this into the Loop B equation

$$+\mathbf{Z}_2\mathbf{I}_A - (\mathbf{Z}_2 + \mathbf{Z}_3)\mathbf{I}_B + \mathbf{E}_2 = 0$$

gives

$$(10 \angle -45^\circ \ \Omega)[7.06 \angle 173.16^\circ \text{ A} + (0.95 \angle -28.05^\circ \ \Omega)\mathbf{I}_B] - (12.17 \angle 7.5^\circ \ \Omega)\ \mathbf{I}_B + 120 \text{ V} = 0$$

or

$$\mathbf{I}_B(-9.3 \ \Omega - j10.68 \ \Omega) = (-76.38 - j55.51) \text{ V}$$

so

$$\mathbf{I}_B = \frac{94.42 \angle -143.99^\circ \text{ V}}{14.16 \angle -131.05^\circ \ \Omega}$$

$$\mathbf{I}_B = 6.67 \angle -12.94^\circ \text{ A}$$

$\mathbf{I}_A$ is found by putting this value into the equation for loop A.

$$\mathbf{I}_A = 7.06 \angle 173.16^\circ \text{ A} + (0.95 \angle -28.05^\circ)(6.67 \angle -12.94^\circ) \text{ A} = (-2.22 - j3.32) \text{ A}$$

$$\mathbf{I}_A = 3.99 \angle -123.77^\circ \text{ A}$$

Determine the branch currents.

6. An inspection of the circuit shows that the current is $\mathbf{I}_A$ in $\mathbf{Z}_1$, and $\mathbf{I}_B$ in $\mathbf{Z}_3$. So, $\mathbf{I}_1 = 3.99 \angle -123.77^\circ$ A, $\mathbf{I}_3 = 6.67 \angle -12.94^\circ$ A. The currents are in opposite directions in $\mathbf{Z}_2$, so the branch current is the phasor difference of the mesh currents. Therefore,

$I_2 = I_B - I_A$

$I_2 = 6.67 \angle -12.94° \text{ A} - 3.99 \angle -123.77° \text{ A}$

$I_2 = 8.91 \angle 11.85° \text{ A}$ in the direction of I_B because I_B was given the (+) sign.

7. Ohm's Law can be used for the voltages. Then

Determine the voltages.

$$V_1 = I_1Z_1 = (3.99 \angle -123.77° \text{ A})(5 \angle 53.13° \ \Omega)$$
$$V_1 = 19.95 \angle -70.64° \text{ V}$$
$$V_2 = I_2Z_2 = (8.91 \angle 11.85° \text{ A})(10 \angle -45° \ \Omega)$$
$$V_2 = 89.1 \angle -33.15° \text{ V}$$
$$V_3 = I_3Z_3 = (6.67 \angle -12.94° \text{ A})(10 \angle 60° \ \Omega)$$
$$V_3 = 66.7 \angle 47.06° \text{ V}$$

The last step is to calculate the power. This can be done by summing the power that is dissipated in each resistor. This gives

Determine the power.

$$P_T = P_1 + P_2 + P_3$$

Since the resistive part of an impedance is $Z \cos \theta$,

$$P_1 = I_1^2R_1 = I_1^2Z_1 \cos \theta_1 = (3.99 \text{ A})^2(5 \ \Omega) \cos 53.13° = 47.86 \text{ W}$$
$$P_2 = I_2^2R_2 = I_2^2Z_2 \cos \theta_2$$
$$= (8.91 \text{ A})^2(10 \ \Omega) \cos 45°$$
$$= 561.3 \text{ W}$$
$$P_3 = I_3^2R_3 = I_3^2Z_3 \cos \theta_3$$
$$= (6.67 \text{ A})^2(10 \ \Omega) \cos 60°$$
$$= 222.4 \text{ W}$$
$$P_T = 831.56 \text{ W}$$

In dc analysis, an incorrect assumed direction gave a minus sign before the value. In ac analysis, the negative sign appears as a 180° shift in the angle. Thus, if I_A is assumed counterclockwise in Example 18.1, the result will be $I_A = 3.99 \angle 56.23°$. At first glance, the answers seem to be different, but they describe the same phasor. This can cause some confusion if not understood and might lead one to think that the answer is incorrect.

A second example is used to further show the steps in the procedure.

EXAMPLE 18.2

Write the loop equations for the circuit in Figure 18.3 and determine the branch currents.

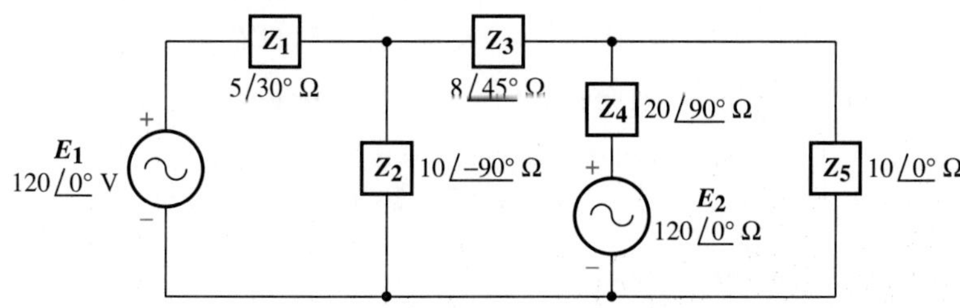

Figure 18.3

Strategy

Identify the loops and draw the loop currents.

Solution

1,2. There are four loops. The loop currents for three of them are as shown in Figure 18.4.

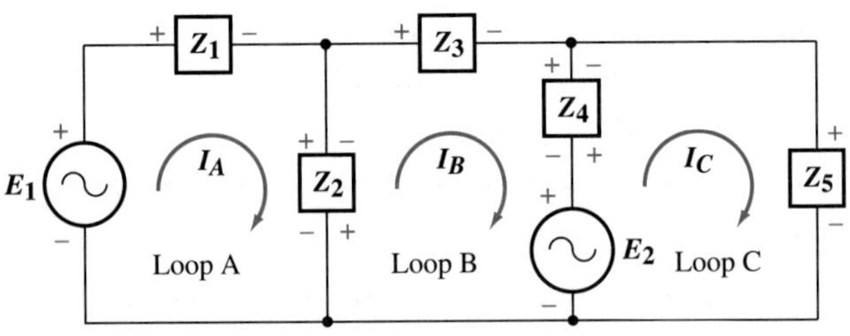

Figure 18.4

3. The terminal polarity is marked in Figure 18.4.

Write the equations as directed in Step 3.

4. Writing Kirchhoff's Voltage Law in a clockwise direction for Loop A gives

$$-I_A(Z_1 + Z_2) + I_B Z_2 + E_1 = 0$$

or

$$-(8.66 \angle -60° \ \Omega)I_A + (10 \angle -90° \ \Omega)I_B + 120 \angle 0° \text{ V} = 0$$

Writing Kirchhoff's Voltage Law in a clockwise direction for Loop B gives

$$-I_B(Z_2 + Z_3 + Z_4) + I_A Z_2 + I_C Z_4 - E_2 = 0$$

or

$$(10 \angle -90° \ \Omega)I_A - (16.65 \angle 70.13° \ \Omega)I_B + (20 \angle 90° \ \Omega)I_c - 120 \angle 0° \text{ V} = 0$$

Writing Kirchhoff's Voltage Law in a clockwise direction for Loop C gives

$$-I_C(Z_4 + Z_5) + I_B Z_4 + E_2 = 0$$

or

$$+ (20 \angle 90° \ \Omega)I_B - (22.36 \angle 63.43° \ \Omega)I_C + 120 \angle 0° \text{ V} = 0$$

Equations are arranged for determinants.

5. Since there are three equations, the method of determinants is better suited for the solution. First the equations are arranged to put similar terms in the columns. This gives

$$\begin{aligned} -(8.66 \angle -60° \ \Omega)I_A + (10 \angle -90° \ \Omega)I_B &= -120 \angle 0° \text{ V} \\ (+10 \angle -90° \ \Omega)I_A - (16.65 \angle 70.13° \ \Omega)I_B + (20 \angle 90° \ \Omega)I_C &= +120 \angle 0° \text{ V} \\ +(20 \angle 90° \ \Omega)I_B - (22.36 \angle 63.43° \ \Omega)I_C &= -120 \angle 0° \text{ V} \end{aligned}$$

Set up and evaluate each matrix.

The determinant matrixes and their values are:

$$D = \begin{vmatrix} -8.66 \angle -60° & +10 \angle -90° & 0 \\ +10 \angle 90° & -16.65 \angle 70.13° & +20 \angle 90° \\ 0 & +20 \angle 90° & -22.36 \angle 63.43° \end{vmatrix}$$

$$= 4202.30 \angle 209.86°$$

$$D_A = \begin{vmatrix} -120\ \angle 0^\circ & +10\ \angle -90^\circ & 0 \\ +120\ \angle 0^\circ & -16.65\ \angle 70.13^\circ & +20\ \angle 90^\circ \\ -120\ \angle 0^\circ & +20\ \angle 90^\circ & -22.36\ \angle 63.43^\circ \end{vmatrix}$$
$$= 47{,}598.11\ \angle 248.8^\circ$$

$$D_B = \begin{vmatrix} -8.66\ \angle -60^\circ & -120\ \angle 0^\circ & 0 \\ +10\ \angle -90^\circ & +120\ \angle 0^\circ & +20\ \angle 90^\circ \\ 0 & -120\ \angle 0^\circ & -22.36\ \angle 63.43^\circ \end{vmatrix}$$
$$= 19{,}040.63\ \angle 170.06^\circ$$

$$D_C = \begin{vmatrix} -8.66\ \angle -60^\circ & +10\ \angle -90^\circ & -120\ \angle 0^\circ \\ +10\ \angle -90^\circ & -16.65\ \angle 70.13^\circ & +120\ \angle 0^\circ \\ 0 & +\ 20\ \angle 90^\circ & -120\ \angle 0^\circ \end{vmatrix}$$
$$= 35{,}795.94\ \angle 168.15^\circ$$

Then the loop currents are

Solve for the loop currents.

$$I_A = \frac{D_A}{D} = \frac{47{,}598.11\ \angle 248.8^\circ}{4202.30\ \angle 209.86^\circ}$$
$$I_A = 11.33\ \angle 38.94^\circ\ \text{A}$$
$$I_B = \frac{D_B}{D} = \frac{19{,}040.63\ \angle 170.94^\circ}{4202.30\ \angle 209.86^\circ}$$
$$I_B = 4.53\ \angle -38.92^\circ\ \text{A}$$
$$I_C = \frac{D_C}{D} = \frac{35{,}795.94\ \angle 168.15^\circ}{4202.30\ \angle 209.86^\circ}$$
$$I_C = 8.52\ \angle -41.71^\circ\ \text{A}$$

The branch currents are

Solve for the branch currents.

In Z_1, $I_1 = I_A$, $I_1 = 11.33\ \angle 38.94^\circ$ A
In Z_3, $I_3 = I_B$, $I_3 = 4.53\ \angle -38.92^\circ$ A
In Z_5, $I_5 = I_C$, $I_5 = 8.52\ \angle -41.71^\circ$ A

There are two loop currents of opposite direction in Z_2 and in Z_4 so

$$I_2 = I_B - I_A = 4.53\ \angle -38.92^\circ\ \text{A} - 11.33\ \angle 38.94^\circ\ \text{A}$$
$I_2 = 11.28\ \angle -117.91^\circ$ A in the direction of I_B
$$I_4 = I_C - I_B = 8.52\ \angle -41.71^\circ\ \text{A} - 4.53\ \angle -38.92^\circ\ \text{A}$$
$I_4 = 4.01\ \angle -44.9^\circ$ A in the direction of I_C

For the ac circuit, the loop equations can also be written without marking the terminal polarity. The first step in the procedure is to draw the loop currents in a clockwise direction. Then, the following terms are obtained:

1. **Loop Current Terms**
   + (loop current) × (phasor sum of the impedances in the loop)

2. **Mutual Loop Current Terms**
   – (mutual loop current) × (phasor sum of the impedances in the mutual branch)

3. **Loop Source Terms**
   Value of the source emf with the sign of the terminal at which the loop current enters. If the source is a current source, it can be converted to a voltage source. The procedure was explained in Chapter 8, Section 8.6, and is also shown in Section 18.6.

The loop equation is formed by setting the phasor sum of the terms in Steps 1–3 equal to zero.

This procedure gives the correct relation between rises and drops, but it is important to recognize that it assigns a (+) to a rise and a (−) to a drop in a counterclockwise summation.

EXAMPLE 18.3

Write the loop equations for the circuit in Figure 18.5.

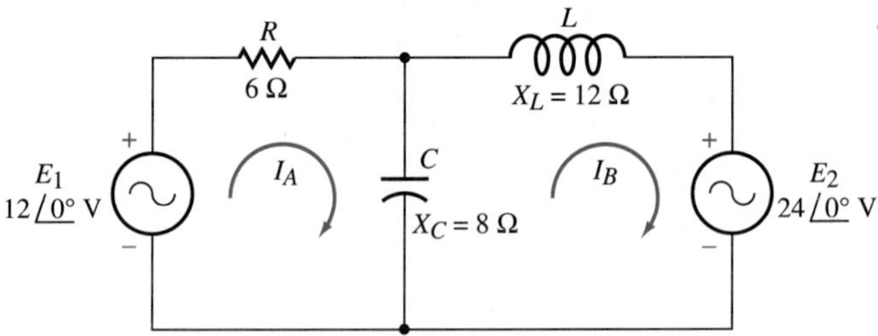

Figure 18.5

Solution

The loop equation are already shown in the circuit. The terms are

Loop	Loop Currents	Mutual Loop Current Terms	Source Terms
A	$+\boldsymbol{I}_A(R - jX_C)$	$-\boldsymbol{I}_B(-jX_C)$	$-E_1$
B	$\boldsymbol{I}_B(-jX_C + jX_L)$	$-\boldsymbol{I}_A(-jX_C)$	$+E_2$

The equation for Loop A is

$$\boldsymbol{I}_A(R - jX_C) - \boldsymbol{I}_B(-jX_C) - \boldsymbol{E}_1 = 0$$
$$\boldsymbol{I}_A(6\ \Omega - j8\ \Omega) - \boldsymbol{I}_B(-j8\ \Omega) - 12\ \angle 0°\ \text{V} = 0$$

or

$$(10\ \angle 53.13°\ \Omega)\boldsymbol{I}_A - (8\ \angle -90°\ \Omega)\boldsymbol{I}_B - 12\ \angle 0°\ \text{V} = 0$$

The equation for Loop B is

$$I_B(-jX_C + jX_L) - I_A(-jX_C) + E_2 = 0$$

$$I_B(-j8\ \Omega + j12\ \Omega) - I_A(-j8\ \Omega) + 24\ \angle 0°\ \text{V} = 0$$

or

$$(-8\ \angle -90°\ \Omega)\, I_A + (4\ \angle 90°\ \Omega) I_B + 24\ \angle 0°\ \text{V} = 0$$

EXAMPLE 18.4

Write the loop equations for the circuit of Figure 18.6.

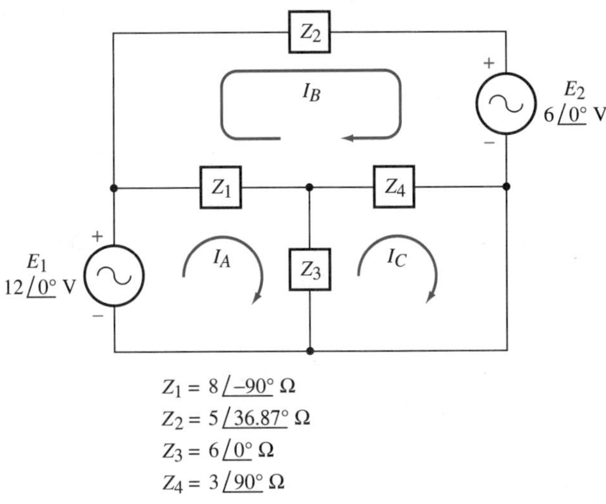

Figure 18.6

Solution

There are three independent loops, so three loop equations are needed. The loop currents I_A, I_B, and I_C are shown in Figure 18.6. The terms for the equations are listed in the following.

Loop	Loop Current Terms	Mutual Loop Current Terms		Source Terms
A	$+I_A(Z_1 + Z_3)$	$-I_BZ_1,$	$-I_CZ_3$	$-E_1$
B	$+I_B(Z_1 + Z_2 + Z_4)$	$-I_AZ_1,$	$-I_CZ_4$	$+E_2$
C	$+I_C(Z_3 + Z_4)$	$-I_AZ_3,$	$-I_BZ_4$	0

Putting the terms in the equation for Loop A gives

$$I_A(Z_1 + Z_3) - I_BZ_1 - I_CZ_3 - E_1 = 0$$

$$I_A(-j8\ \Omega + 6\ \Omega) - I_B(8\ \angle -90°\ \Omega) - (6\ \Omega)I_C - 12\ \angle 0°\ \text{V} = 0$$

or

$$(10\ \angle -53.13°\ \Omega)I_A - (8\ \angle -90°\ \Omega)I_B - (6\ \Omega)I_C - 12\ \angle 0°\ \text{V} = 0$$

For Loop B

$$I_B(Z_1 + Z_2 + Z_4) - I_A Z_1 - I_C Z_4 + E_2 = 0$$

$$I_B(-j8\ \Omega + 4\ \Omega + j3\ \Omega + j3\ \Omega) - I_A(8\ \angle -90°\ \Omega) - I_C(3\ \angle 90°\ \Omega) + 6\ \angle 0°\ \text{V} = 0$$

or

$$(-8\ \angle -90°\ \Omega)I_A + (4.47\ \angle -26.57°\ \Omega)I_B - (3\ \angle 90°\ \Omega)I_C + 6\ \angle 0°\ \text{V} = 0$$

For Loop C

$$I_C(Z_3 + Z_4) - I_A Z_3 - I_B Z_4 = 0$$

$$I_C(6\ \Omega + j3\ \Omega) - I_A(6\ \Omega) - I_B(3\ \angle 90°\ \Omega) = 0$$

or

$$(-6\ \Omega)I_A - (3\ \angle 90°\ \Omega)I_B + (6.71\ \angle 26.57°\ \Omega)I_C = 0$$

Each method presented in this section has advantages and disadvantages. You should select the one that lets you analyze a circuit with the least difficulty.

18.2 The Branch Current Method

If one prefers to work with the actual currents in the branches, the branch current method can be used. This method is suited to an analysis where only one or two branch currents are needed.

branch current method

In the **branch current method** the loop equations are written using the actual currents in the branches instead of loop currents. The method was introduced for dc circuit analysis in Chapter 9.

One disadvantage of this method is that more equations must be written before any algebraic manipulation can begin. In Example 18.1 for instance, three equations are needed before the solution can start. On the positive side, currents do not have to be combined in a mutual branch.

The procedure for using this method is similar to that used for the dc analysis. Again, the main difference is the phasor arithmetic. The basic steps are:

1. Identify the loops, the nodes, and the branches. Also, note the number of each. These numbers will be referred to as N.
2. Draw a current in each branch, arbitrarily assigning the direction.
3. Mark the polarity of the impedances and sources as follows:

a. Impedances: the (+) terminal is the one at which the current enters.
b. Sources: as designated for the problem or by other given conditions.

4. Apply Kirchhoff's Voltage Law in a clockwise direction to $N - 1$ loops, writing the potential difference across each impedance in terms of the branch current and the impedance. Give each term the sign of the polarity on the terminal first encountered. This automatically gives the correct relation for rises and drops. However, recognize that it assigns a (+) to a drop and a (−) to a rise.
5. Apply Kirchhoff's Current Law to $N - 1$ nodes. A node is a junction of three or more branches. Currents entering a node are taken as positive. Currents leaving a node are taken as negative.
6. Solve the equations from Steps 4 and 5 for the branch currents.
7. Find other quantities by using basic circuit concepts.

The procedure is now applied to find the currents and voltages in a three-loop, multisource circuit.

EXAMPLE 18.5

Use the branch current method to find the current in each impedance, and the voltage across each impedance, in the circuit of Figure 18.7.

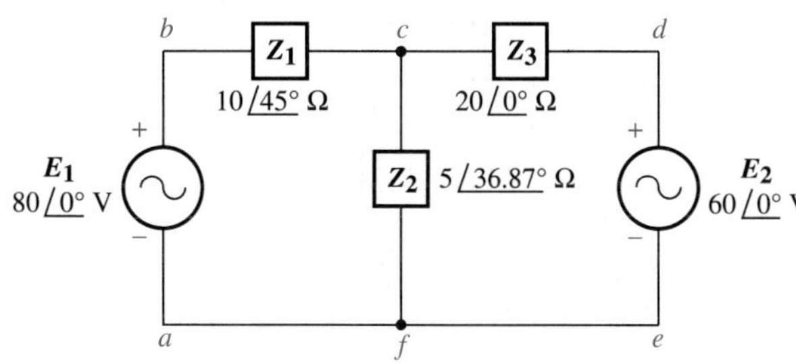

Figure 18.7

Solution

Strategy

Prepare the circuit for the solution.

1. This circuit has three loops. They are *abcfa*, *fcdef*, and *abcdefa*. It has two nodes, one at *c* and one at *f*.
2. The branch currents are shown as I_1, I_2, and I_3 in Figure 18.8.

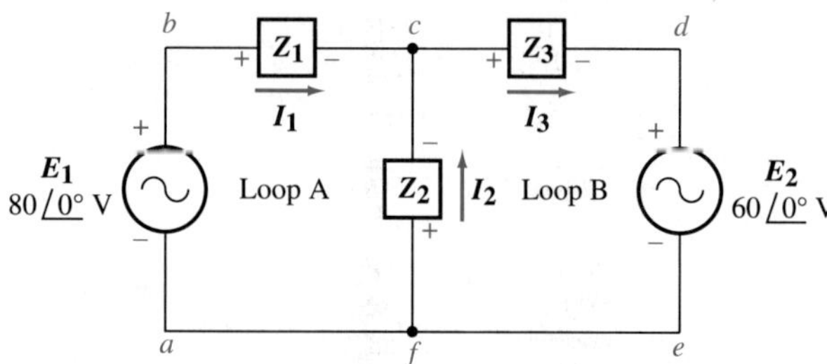

Figure 18.8

3. The polarity of the source and impedance terminals are marked in Figure 18.8.
4. Applying Kirchhoff's Voltage Law to the loops gives

Write the loop equations as directed in Step 4.

Starting at point a in loop A and traveling clockwise gives

$$-\mathbf{E}_1 + \mathbf{I}_1\mathbf{Z}_1 - \mathbf{I}_2\mathbf{Z}_2 = 0$$

Starting at point f in loop B and traveling clockwise gives

$$\mathbf{I}_2\mathbf{Z}_2 + \mathbf{I}_3\mathbf{Z}_3 + \mathbf{E}_2 = 0$$

There are now two equations and three unknowns. One more equation is needed.

Write the node equations for one node.

5. Applying Kirchhoff's Current Law at node C gives

$$\mathbf{I}_1 + \mathbf{I}_2 - \mathbf{I}_3 = 0$$

The three equations are

a. $-\mathbf{E}_1 + \mathbf{I}_1\mathbf{Z}_1 - \mathbf{I}_2\mathbf{Z}_2 = 0$

b. $\mathbf{E}_2 + \mathbf{I}_2\mathbf{Z}_2 + \mathbf{I}_3\mathbf{Z}_3 = 0$

c. $\mathbf{I}_1 + \mathbf{I}_2 - \mathbf{I}_3 = 0$

Put numbers in the equations and arrange for determinants.

6. Putting numbers in the equations and arranging them to get similar terms in each column gives

$$\begin{aligned}(10\angle 45^\circ\ \Omega)\mathbf{I}_1 - (5\angle 36.87^\circ\ \Omega)\mathbf{I}_2 &= +80\angle 0^\circ\\ (5\angle 36.87^\circ\ \Omega)\mathbf{I}_2 + (20\angle 0^\circ\ \Omega)\mathbf{I}_3 &= -60\angle 0^\circ\ \text{V}\\ \mathbf{I}_1 + \mathbf{I}_2 - \mathbf{I}_3 &= 0\end{aligned}$$

Set up and evaluate each matrix.

The determinant matrixes and their values are

$$D = \begin{vmatrix} 10\angle 45^\circ & -5\angle 36.87^\circ & 0\\ 0 & 5\angle 36.87^\circ & 20\angle 0^\circ\\ 1 & 1 & -1\end{vmatrix}$$

$$= 339.37\angle 227.68^\circ$$

$$D_1 = \begin{vmatrix} 80\angle 0^\circ & -5\angle 36.87^\circ & 0\\ -60\angle 0^\circ & 5\angle 36.87^\circ & 20\angle 0^\circ\\ 0 & 1 & -1\end{vmatrix}$$

$$= 1681.07\angle 182.04^\circ$$

$$D_2 = \begin{vmatrix} 10\angle 45^\circ & 80\angle 0^\circ & 0\\ 0 & -60\angle 0^\circ & 20\angle 0^\circ\\ 1 & 0 & -1\end{vmatrix}$$

$$= 2068.25\angle 11.84^\circ$$

$$D_3 = \begin{vmatrix} 10\angle 45^\circ & -5\angle 36.87^\circ & 80\angle 0^\circ\\ 0 & 5\angle 36.87^\circ & -60\angle 0^\circ\\ 1 & 1 & 0\end{vmatrix}$$

$$D_3 = 501.20\angle 46.62^\circ$$

Solve for the currents.

Then

$$\mathbf{I}_1 = \frac{D_1}{D} = \frac{1681.07\angle 182.04^\circ}{339.37\angle 227.68^\circ}$$

$$I_1 = 4.95 \angle -45.64° \text{ A}$$

$$I_2 = \frac{D_2}{D} = \frac{2068.25 \angle 11.84°}{339.37 \angle 227.68°}$$

$$I_2 = 6.09 \angle -215.84° \text{ A}$$

$$I_3 = \frac{D_3}{D} = \frac{501.20 \angle 46.62°}{339.37 \angle 227.68°}$$

$$I_3 = 1.48 \angle -181.06° \text{ A}$$

7. The voltages are obtained using Ohm's Law:

Solve for the voltages.

$$V_1 = I_1 Z_1 = (4.95 \angle -45.64° \text{ A})(10 \angle 45° \ \Omega)$$

$$V_1 = 49.5 \angle -0.64° \text{ V}$$

$$V_2 = I_2 Z_2 = (6.09 \angle -215.84° \text{ A})(5 \angle 36.87° \ \Omega)$$

$$V_2 = 30.45 \angle -178.97° \text{ V}$$

$$V_3 = I_3 Z_3 = (1.48 \ -181.06° \text{ A})(20 \angle 0° \ \Omega)$$

$$V_3 = 29.60 \angle -181.06° \text{ V}$$

The next example compares the branch current method to the mesh method.

EXAMPLE 18.6

Use the branch current method to determine the currents in the circuit of Figure 18.9. The circuit is the same as in Example 18.2.

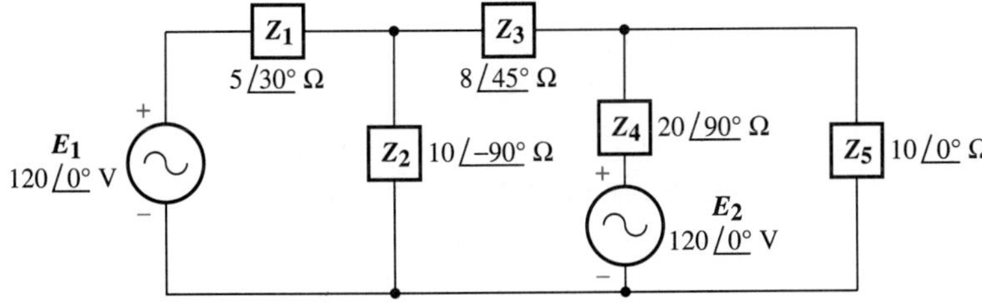

The circuit, showing the branch currents and the polarities, is shown in Figure 18.10. The circuit has four loops, so only three loop equations are needed.

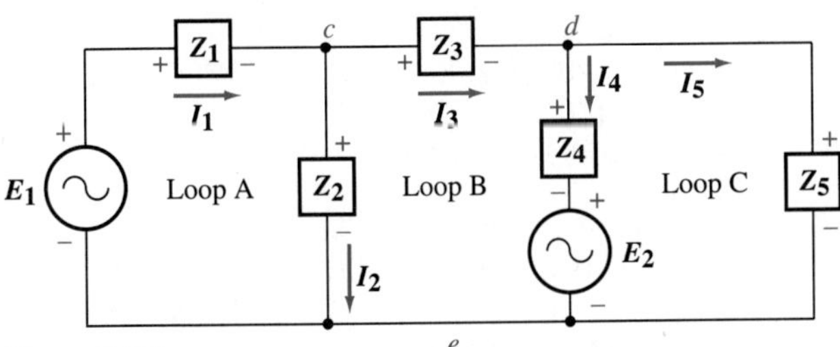

Figure 18.10

Writing Kirchhoff's Voltage Law in a clockwise direction gives

$$\begin{aligned} &\text{For A:} && -\boldsymbol{E}_1 + \boldsymbol{I}_1\boldsymbol{Z}_1 + \boldsymbol{I}_2\boldsymbol{Z}_2 = 0 \\ &\text{For B:} && -\boldsymbol{I}_2\boldsymbol{Z}_2 + \boldsymbol{I}_3\boldsymbol{Z}_3 + \boldsymbol{I}_4\boldsymbol{Z}_4 + \boldsymbol{E}_2 = 0 \\ &\text{For C:} && -\boldsymbol{E}_2 - \boldsymbol{I}_4\boldsymbol{Z}_4 + \boldsymbol{I}_5\boldsymbol{Z}_5 = 0 \end{aligned}$$

There is one node at c, one at d, and one at the bottom where four branches meet. Kirchhoff's Current Law must be written for two of these nodes.

$$\begin{aligned} &\text{At } c, && \boldsymbol{I}_1 - \boldsymbol{I}_2 - \boldsymbol{I}_3 = 0 \\ &\text{At } d, && \boldsymbol{I}_3 - \boldsymbol{I}_4 - \boldsymbol{I}_5 = 0 \end{aligned}$$

There are now five equations. Putting the values in the equation gives

$$\begin{aligned} &\text{For A,} && -120\angle 0^\circ\text{ V} + (5\angle 30^\circ\ \Omega)\boldsymbol{I}_1 + (10\angle -90^\circ\ \Omega)\boldsymbol{I}_2 = 0 \\ &\text{For B,} && -(10\angle -90^\circ\ \Omega)\boldsymbol{I}_2 + (8\angle 45^\circ\ \Omega)\boldsymbol{I}_3 + (20\angle 90^\circ\ \Omega)\boldsymbol{I}_4 \\ & && \quad + 120\angle 0^\circ\text{ V} = 0 \\ &\text{For C,} && -120\angle 0^\circ\text{ V} - (20\angle 90^\circ\ \Omega)\boldsymbol{I}_4 + (10\ \Omega)\boldsymbol{I}_5 = 0 \\ &\text{At node } c, && \boldsymbol{I}_1 - \boldsymbol{I}_2 - \boldsymbol{I}_3 = 0 \\ &\text{At node } d, && \boldsymbol{I}_3 - \boldsymbol{I}_4 - \boldsymbol{I}_5 = 0 \end{aligned}$$

The five equations are now reduced to three so that the method of determinants can be used.

From the last two equations,

$$\boldsymbol{I}_1 = \boldsymbol{I}_2 + \boldsymbol{I}_3 \quad \text{and} \quad \boldsymbol{I}_5 = \boldsymbol{I}_3 - \boldsymbol{I}_4$$

Putting the first in the equation for Loop A and the second in the equation for Loop C gives three equations.

They are

$$\begin{aligned} (8.66\angle -60^\circ\ \Omega)\boldsymbol{I}_2 + (5\angle 30^\circ\ \Omega)\boldsymbol{I}_3 &= 120\angle 0^\circ\text{ V} \\ -(10\angle -90^\circ\ \Omega)\boldsymbol{I}_2 + (8\angle 45^\circ\ \Omega)\boldsymbol{I}_3 + (20\angle 90^\circ\ \Omega)\boldsymbol{I}_4 &= -120\angle 0^\circ\text{ V} \\ (10\angle 0^\circ\ \Omega)\boldsymbol{I}_3 - (22.36\angle 63.43^\circ\ \Omega)\boldsymbol{I}_4 &= 120\angle 0^\circ\text{ V} \end{aligned}$$

The determinant matrixes and their values are

$$D = \begin{vmatrix} 8.66\angle -60^\circ & 5\angle 30^\circ & 0 \\ -10\angle -90^\circ & 8\angle 45^\circ & 20\angle 90^\circ \\ 0 & 10\angle 0^\circ & -22.36\angle 63.43^\circ \end{vmatrix}$$
$$= 4201.59\angle 209.86^\circ$$

$$D_2 = \begin{vmatrix} 120 & 5\angle 30^\circ & 0 \\ -120 & 8\angle 45^\circ & 20\angle 90^\circ \\ 120 & 10\angle 0^\circ & -22.36\angle 63.43^\circ \end{vmatrix}$$
$$= 47{,}390.95\angle 271.92^\circ$$

$$D_3 = \begin{vmatrix} 8.66\angle -60^\circ & 120 & 0 \\ -10\angle -90^\circ & -120 & 20\angle 90^\circ \\ 0 & 120 & -22.36\angle 63.43^\circ \end{vmatrix}$$
$$= 19{,}040.63\angle 170.94^\circ$$

$$D_4 = \begin{vmatrix} 8.66 \angle -60° & 5 \angle 30° & 120 \\ -10 \angle -90° & 8 \angle 45° & -120 \\ 0 & 10 \angle 0° & 120 \end{vmatrix}$$
$$= 16{,}798.66 \angle 345°$$

Then

$$I_2 = \frac{D_2}{D} = \frac{47{,}390.95 \angle 271.92°}{4201.59 \angle 209.86°}, \qquad I_2 = 11.28 \angle 62.06° \text{ A}$$

$$I_3 = \frac{D_3}{D} = \frac{19{,}040.63 \angle +170.94°}{4201.59 \angle 209.86°}, \qquad I_3 = 4.53 \angle -38.92° \text{ A}$$

$$I_4 = \frac{D_4}{D} = \frac{16{,}798.66 \angle 345.00°}{4201.59 \angle 209.86°}, \qquad I_4 = 4.00 \angle 135.14° \text{ A}$$

From the node equation for node C,

$$I_1 = I_2 + I_3 = 11.28 \angle 62.06° \text{ A} + 4.53 \angle -38.92° \text{ A}$$
$$I_1 = 11.33 \angle 38.94° \text{ A in the direction shown in Figure 18.10}$$

From the node equation for node D,

$$I_5 = I_3 - I_4 = 4.53 \angle -38.92 \text{ A} - 4.00 \angle 135.14° \text{ A}$$
$$I_5 = 8.51 \angle -41.76° \text{ A in the direction shown in Figure 18.10}$$

Note that the angles of I_2 and I_4 are 180° out of phase with those in Example 18.2. This is because their chosen direction is opposite from that in Example 18.2.

18.3 The Nodal Method

nodal method

The third method of ac circuit analysis is the **nodal method**. In this method, the branch currents are written in terms of the difference between node voltages at the ends of the branch and the parts in the branch. Node voltage means the potential difference between the node and the reference node. Kirchhoff's Current Law is used to find the node voltages, which are then used to find the branch currents. Ohm's Law and other circuit concepts can then be used to find other quantities.

The method presented here combines the branch current equations and Kirchhoff's Current Law in one step. It also eliminates the need to convert a voltage source to a current source. Finally, the branch currents are obtained directly without combining a source current and a branch current.

The procedure for using the nodal method is:

1. Identify the nodes and branches.
2. Select any node as a reference node and mark it with the ground symbol. All node voltages will be with respect to this node. Mark the other node voltages using a letter and subscript, V_A, V_B, and so forth.

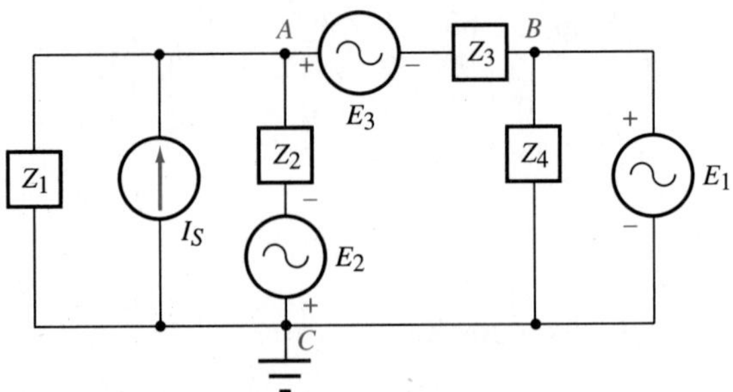

Figure 18.11 A typical ac circuit showing the nodes labeled for nodal analysis.

3. Write an expression for each node, except the reference node, to include the following terms, as described for Node A of Figure 18.11.

   a. Node Term

   + (node voltage) × (phasor sum of the reciprocals of the impedances in each branch connected to the node)

For Node A:

$$+\mathbf{V}_A\left(\frac{1}{\mathbf{Z}_1}+\frac{1}{\mathbf{Z}_2}+\frac{1}{\mathbf{Z}_3}\right)$$

   b. Adjacent Node Terms

   − (adjacent node voltage) × (reciprocal of the impedances in the branch between the nodes)

For Node A:

$$-\mathbf{V}_B(1/\mathbf{Z}_3)$$

   c. Source Terms

   (source emf) × (reciprocal of the impedance in the branch)

   This term is given the sign of the terminal facing the node.

For Node A:

$$-\mathbf{E}_2(1/\mathbf{Z}_2), \qquad +\mathbf{E}_3(1/\mathbf{Z}_3)$$

For a current source, the value of the source current is used. Current into the node is given a (+) sign. Current away from the node is given a (−) sign.

For Node A:

$$+\mathbf{I}_s$$

4. Set the phasor sum of the terms in a and b equal to the phasor sum of the terms in c. For Node A, the result is

$$\mathbf{V}_A\left(\frac{1}{\mathbf{Z}_1}+\frac{1}{\mathbf{Z}_2}+\frac{1}{\mathbf{Z}_3}\right)-\mathbf{V}_B\left(\frac{1}{\mathbf{Z}_3}\right)=-\mathbf{E}_2\left(\frac{1}{\mathbf{Z}_2}\right)+\mathbf{E}_3\left(\frac{1}{\mathbf{Z}_3}\right)+\mathbf{I}_s$$

5. Solve the equations of Step 4 for the node voltage.
6. Determine the current in each branch as follows:

a. Arbitrarily assign a direction for the current in a branch and mark the terminal polarity.
b. Starting with the voltage at one end of the branch, add the rises and drops while moving toward the other end. Set the sum equal to the voltage at the other end.
c. Arrange the equation in b to give the current.

For the $\mathbf{Z}_2$ branch of Figure 18.11 with $\mathbf{I}_2$ downward, the equation is

$$\mathbf{V}_A - \mathbf{I}_2\mathbf{Z}_2 + \mathbf{E}_2 = \mathbf{V}_C$$

But $\mathbf{V}_C$ is 0 volts, so

$$\mathbf{V}_A - \mathbf{I}_2\mathbf{Z}_2 + \mathbf{E}_2 = 0$$

or

$$\mathbf{I}_2 = \frac{\mathbf{V}_A + \mathbf{E}_2}{\mathbf{Z}_2}$$

The phase angle is for the assigned direction of the current. Had the direction been assigned as upward, there would be a 180° difference in the angle.

7. Use basic circuit relations to find other quantities.

We now apply the procedure to the solution of Example 18.7.

EXAMPLE 18.7

Use the nodal method to find the current in each branch of the circuit in Figure 18.12.

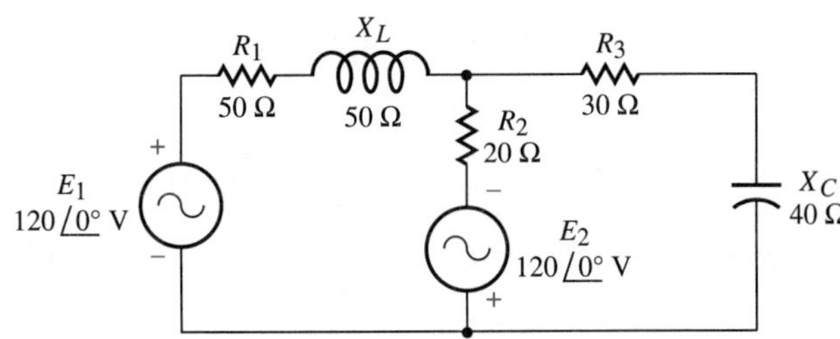

Figure 18.12

Solution

Strategy
Label the circuit.

1. The circuit, labeled and ready for analysis, is shown in Figure 18.13. There are two nodes, a and b, and three branches.

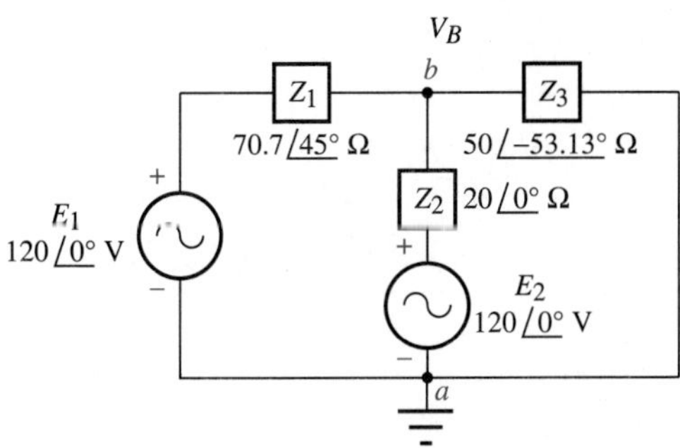

Figure 18.13

Mark the nodes.

2. Node a is the reference node. The voltage at b is marked V_B. Since a is the reference node, $V_A = 0$ volts.
3. Since there are two nodes, the nodal equation is needed for only one node, node b.

Combine the impedances into one and write the terms.

4. The impedances in the branches connected to Node b are

$$\mathbf{Z}_1 = R_1 + jX_L = (50 + j50)\ \Omega \quad \text{or} \quad 70.7\ \angle 45°\ \Omega$$
$$\mathbf{Z}_2 = R_2 = 20\ \angle 0°\ \Omega$$
$$\mathbf{Z}_3 = R_3 = jX_C = (30 - j40)\ \Omega \quad \text{or} \quad 50\ \angle -53.13°\ \Omega$$

The node term is

$$+\ \mathbf{V}_B\left(\frac{1}{\mathbf{Z}_1} + \frac{1}{\mathbf{Z}_2} + \frac{1}{\mathbf{Z}_3}\right)$$

The only other node is the reference node, so there are no adjacent node terms.

The source terms are

$$+\mathbf{E}_1(1/\mathbf{Z}_1) \quad \text{and} \quad -\mathbf{E}_2(1/\mathbf{Z}_2)$$

Then

Write the equation. E_2 is assigned a (−) sign because its (+) terminal is away from the node.

$$\mathbf{V}_B\left(\frac{1}{\mathbf{Z}_1} + \frac{1}{\mathbf{Z}_2} + \frac{1}{\mathbf{Z}_3}\right) = \frac{\mathbf{E}_1}{\mathbf{Z}_1} - \frac{\mathbf{E}_2}{\mathbf{Z}_2}$$

$$\mathbf{V}_B\left(\frac{1}{70.7\ \angle 45°\ \Omega} + \frac{1}{20\ \Omega} + \frac{1}{50\ \angle -53.13°\ \Omega}\right) = \frac{120\ \text{V}}{70.7\ \angle 45°\ \Omega} - \frac{120\ \text{V}}{20\ \Omega}$$

or

$$(0.072\ \angle 4.76°\ \text{S})\mathbf{V}_B = 4.93\ \angle -165.91°\ \text{A}$$

Solve the equation.

5. Since there is one unknown and one equation, $\mathbf{V}_B$ can be found.

$$\mathbf{V}_B = 68.47\ \angle -170.67°\ \text{V}$$

Start at the node with the larger voltage.

6. The currents will be found by writing the expressions for them:

For $\mathbf{I}_1$, starting at Node B and taking the current as being out of Node B

$$\mathbf{V}_B - \mathbf{I}_1\mathbf{Z}_1 - \mathbf{E}_1 = 0$$

or

$$\mathbf{I}_1 = \frac{\mathbf{V}_B - \mathbf{E}_1}{\mathbf{Z}_1}$$

$$\mathbf{I}_1 = \frac{68.47\ \angle -170.67°\ \text{V} - 120\ \text{V}}{70.7\ \angle 45°\ \Omega}$$

$$I_1 = 2.66\ \angle -221.6°\ \text{A}$$

For I_2, starting at Node B and taking the current as being out of Node B

$$V_B - I_2Z_2 + E_2 = 0$$

$$I_2 = \frac{V_B + E_2}{Z_2}$$

$$= \frac{68.47 \angle -170.67^\circ \text{ V} + 120 \text{ V}}{20 \angle 0^\circ \ \Omega}$$

$$I_2 = 2.68 \angle -11.95^\circ \text{ A}$$

For I_3, starting at Node B and taking the current as being out of Node B

$$V_B - I_3Z_3 = 0$$

$$I_3 = \frac{V_B}{Z_3}$$

$$= \frac{68.47 \angle -170.67^\circ \text{ V}}{50 \angle -53.13^\circ \ \Omega}$$

$$I_3 = 1.37 \angle -117.54^\circ \text{ A}$$

Had the currents been toward Node B, the angles would change by 180°.

Next the method is applied to a more complex circuit in Example 18.8.

EXAMPLE 18.8

Use the nodal method to find the current in each branch of Figure 18.14.

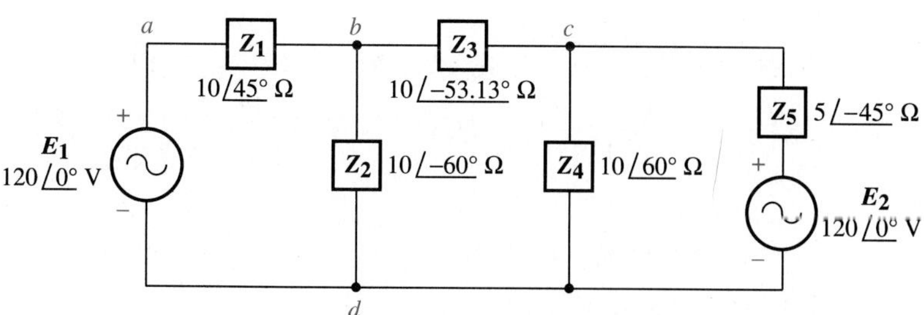

Figure 18.14

Strategy

Prepare the circuit for the solution.

Solution

The circuit, labeled and ready for analysis, is shown in Figure 18.15.

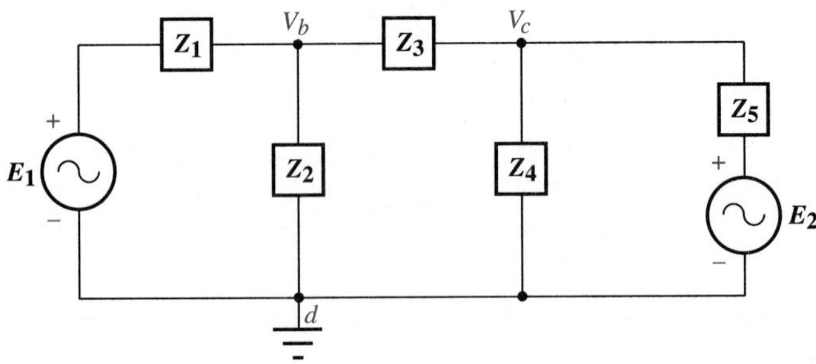

Figure 18.15

1. There are three nodes, one at b, one at c, and one at d. There are five branches, one for each impedance.
2. Node d is selected as the reference node. The other node voltages are marked as $\mathbf{V}_b$ and $\mathbf{V}_c$. These represent the voltages between the node and the reference node.

Write the nodal equation for Node b.

3,4. Writing the terms and equations for Node b gives

$$\underbrace{\mathbf{V}_b\left(\frac{1}{\mathbf{Z}_1}+\frac{1}{\mathbf{Z}_2}+\frac{1}{\mathbf{Z}_3}\right)}_{\text{Node Term}} - \underbrace{\mathbf{V}_c\left(\frac{1}{\mathbf{Z}_3}\right)}_{\text{Adjacent Node Term}} = \underbrace{\mathbf{E}_1\left(\frac{1}{\mathbf{Z}_1}\right)}_{\text{Source Term}}$$

$$\mathbf{V}_b\left(\frac{1}{10\ \angle 45^\circ\ \Omega}+\frac{1}{10\ \angle -60^\circ\ \Omega}+\frac{1}{10\ \angle -53.13^\circ\ \Omega}\right) - \mathbf{V}_c\left(\frac{1}{10\ \angle -53.13^\circ\ \Omega}\right) = (120\ \angle 0^\circ\ \text{V})\left(\frac{1}{10\ \angle 45^\circ\ \Omega}\right)$$

or

$$(0.205\ \angle 27.95^\circ)\mathbf{V}_b - (0.1\ \angle 53.13^\circ)\mathbf{V}_c = 12\ \angle -45^\circ\ \text{V}$$

Write the node equation for Node c.

Writing the terms and equation for Node c gives

$$\underbrace{\mathbf{V}_c\left(\frac{1}{\mathbf{Z}_3}+\frac{1}{\mathbf{Z}_4}+\frac{1}{\mathbf{Z}_5}\right)}_{\text{Node Term}} - \underbrace{\mathbf{V}_b\left(\frac{1}{\mathbf{Z}_3}\right)}_{\text{Adjacent Node Term}} = \underbrace{\mathbf{E}_2\left(\frac{1}{\mathbf{Z}_5}\right)}_{\text{Source Term}}$$

$$\mathbf{V}_c\left(\frac{1}{10\ \angle -53.13^\circ\ \Omega}+\frac{1}{10\ \angle 60^\circ\ \Omega}+\frac{1}{5\ \angle -45^\circ\ \Omega}\right) - \mathbf{V}_b\left(\frac{1}{10\ \angle -53.13^\circ\ \Omega}\right) = (120\ \angle 0^\circ\ \text{V})\left(\frac{1}{5\ \angle -45^\circ\ \Omega}\right)$$

or

$$-(0.1\ \angle 53.13^\circ)\mathbf{V}_b + (0.285\ \angle 28.2^\circ)\mathbf{V}_c = 24\ \angle 45^\circ\ \text{V}$$

Solve for the node voltages.

5. Solving the two equations gives

$$\mathbf{V}_b = 61.86\ \angle -22.49^\circ\ \text{V}$$
$$\mathbf{V}_c = 105.29\ \angle 13.88^\circ\ \text{V}$$

These values mean that the effective value of voltage is 61.86 V between b and the reference, and 105.29 V between c and the reference.

6. For the current in $\mathbf{Z}_1$, starting at b and taking the current as being out of Node b.

Solve for the currents as directed in Step 5.

$$\mathbf{V}_b - \mathbf{I}_1\mathbf{Z}_1 - \mathbf{E}_1 = 0$$

so

$$\mathbf{I}_1 = \frac{\mathbf{V}_b - \mathbf{E}_1}{\mathbf{Z}_1}$$

$$\mathbf{I}_1 = \frac{61.86 \angle -22.49° \text{ V} - 120 \angle 0° \text{ V}}{10 \angle 45° \ \Omega}$$

$$\mathbf{I}_1 = 6.71 \angle -204.4° \text{ A}$$

For the current in $\mathbf{Z}_2$, starting at b and taking the current as being out of Node b,

$$\mathbf{V}_b = \mathbf{I}_2\mathbf{Z}_2 = 0$$

so

$$\mathbf{I}_2 = \frac{\mathbf{V}_b}{\mathbf{Z}_2}$$

$$\mathbf{I}_2 = \frac{61.86 \angle -22.49° \text{ V}}{10 \angle -60° \ \Omega}$$

$$\mathbf{I}_2 = 6.19 \angle 37.51° \text{ A}$$

For the current in $\mathbf{Z}_3$, starting at c and taking the current as being out of Node c,

$$\mathbf{V}_c - \mathbf{I}_3\mathbf{Z}_3 = \mathbf{V}_b$$

so

$$\mathbf{I}_3 = \frac{\mathbf{V}_c - \mathbf{V}_b}{\mathbf{Z}_3}$$

$$= \frac{105.29 \angle 13.88° \text{ V} - 61.86 \angle -22.49° \text{ V}}{10 \angle -53.13° \ \Omega}$$

$$\mathbf{I}_3 = 6.65 \angle 100.48° \text{ A}$$

For the current in $\mathbf{Z}_4$, starting at c and taking the current as being out of Node c,

$$\mathbf{V}_c - \mathbf{I}_4\mathbf{Z}_4 = 0$$

so

$$\mathbf{I}_4 = \frac{\mathbf{V}_c}{\mathbf{Z}_4}$$

$$\mathbf{I}_4 = \frac{105.29 \angle 13.88° \text{ V}}{10 \angle 60° \ \Omega}$$

$$\mathbf{I}_4 = 10.53 \angle -46.12° \text{ A}$$

For the current in $\mathbf{Z}_5$, starting at c and taking the current as being out of Node c,

$$\mathbf{V}_c - \mathbf{E}_2 - \mathbf{I}_5\mathbf{Z}_5 = 0$$

so

$$\mathbf{I}_5 = \frac{\mathbf{V}_c - \mathbf{E}_2}{\mathbf{Z}_5}$$

$$= \frac{105.29 \angle 13.88° \text{ V} - 120 \angle 0° \text{ V}}{5 \angle -45° \ \Omega}$$

$$\mathbf{I}_5 = 6.18 \angle 170.14° \text{ A}$$

18.4 Superposition

superposition

> The **superposition** theorem states: The current and voltage in any part of an ac circuit of linear bilateral components is equal to the phasor combination of the currents and voltages from each source, taken one at a time, while all other sources are replaced by their internal impedance.

A linear impedance is one that does not change with current variation. If its impedance is the same for either direction of current, it is also bilateral.

The superposition of voltages is present in many practical applications. An electronic amplifier uses the superposition of the dc bias voltages and the ac signal voltage. Also, the wave transmitted by a radio station is the superposition of the high-frequency carrier and the low-frequency audio signal.

The graphical superposition of waveforms can be a useful tool for determining wave shapes of sources having different frequencies and shapes. The wave is obtained by adding the instantaneous values and plotting the resulting wave.

One advantage of using superposition for circuit analysis is that it does not involve solving simultaneous equations. On the other hand, the need to analyze the circuit for each source results in a large number of calculations.

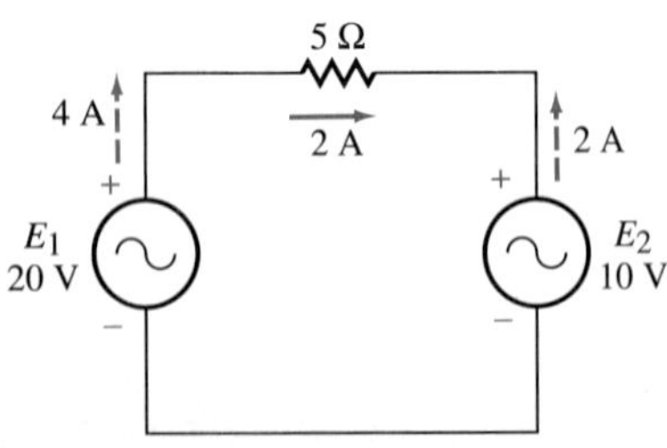

Figure 18.16 The principle of superposition is demonstrated in this simple two-source circuit.

Although superposition can be used for current and voltage, it cannot be applied to power in a circuit. If two sources result in currents of I_1 and I_2 in a resistance, the current is $(I_1 + I_2)$. The power is $(I_1 + I_2)^2R$. Combining power by superposition gives $(I_1^2 + I_2^2)R$. Obviously, this is not the same as $(I_1 + I_2)^2R$. Therefore, the power must be obtained by combining the currents or voltages first, and using the combined value to calculate the power.

The concept of superposition is illustrated in the simple circuit of Figure 18.16. The current from the 20-volt source is 4 amperes clockwise. The current from the 10-volt source is 2 amperes counterclockwise. The actual current is 4 − 2 or 2 amperes clockwise. The same principles that apply for this simple circuit apply for more complex circuits being analyzed.

The procedure for using the superposition theorem for ac circuits is:

1. Replace all sources except one by their internal impedance.
2. Calculate the currents and voltages for that one source using any of the methods studied.
3. Repeat Steps 1 and 2 for each source.
4. Find the currents by combining the currents due to each source. The phasor currents in one direction are subtracted from those in the other direction.
5. Find the voltages by using the total current and Ohm's Law.
6. Use basic circuit concepts to find other quantities.

A comparison of superposition to the loop method can be made by solving Example 18.1 by superposition as shown in Example 18.9.

EXAMPLE 18.9

Use superposition to find the current in each impedance, the voltage across each impedance, and the power dissipated in the circuit of Figure 18.17.

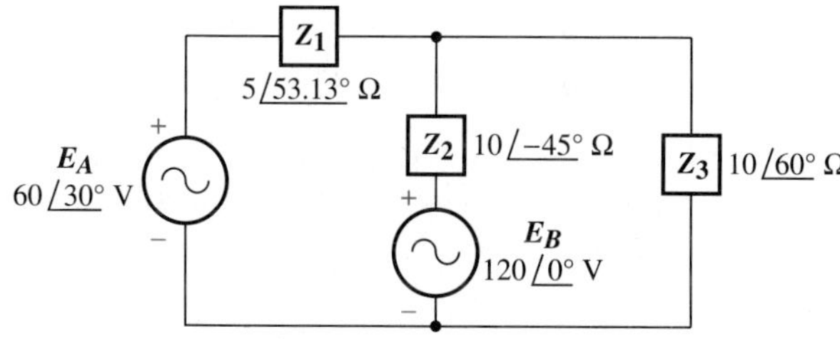

Figure 18.17

Solution

Strategy
Determine the currents from Source A.

1. Source E_B is replaced by its internal impedance, a short circuit, while the currents from E_A are found. The circuit is shown in Figure 18.18.

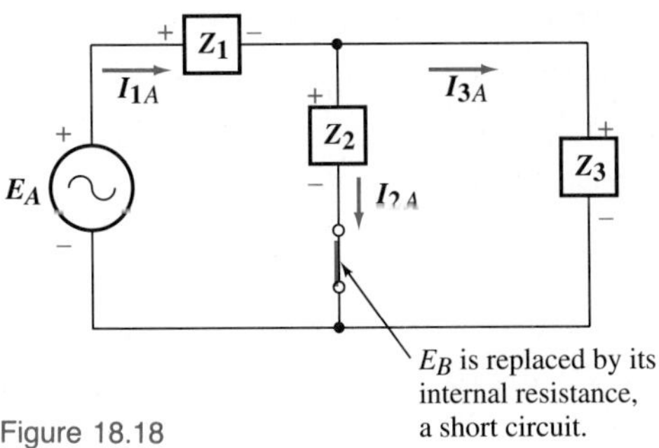

Figure 18.18

2. Then

$$I_{1A} = \frac{E_A}{Z_{eqA}}$$

$$Z_{eqA} = Z_1 + Z_2 \| Z_3$$

$$= 5 \angle 53.13° \ \Omega + \frac{(10 \angle -45° \ \Omega)(10 \angle 60° \ \Omega)}{10 \angle -45° \ \Omega + 10 \angle 60° \ \Omega}$$

$$= 12.24 \angle 24.47° \ \Omega$$

$$I_{1A} = \frac{60 \angle 30° \ \text{V}}{12.24 \angle 24.47° \ \Omega}$$

$$= 4.9 \angle 5.53° \ \text{A}$$

Using the current divider rule,

$$I_{2A} = \frac{(Z_3)I_{1A}}{Z_2 + Z_3}$$

$$= \frac{(10 \angle 60° \ \Omega)(4.9 \angle 5.53° \ \text{A})}{10 \angle -45° \ \Omega + 10 \angle 60° \ \Omega}$$

$$= 4.02 \angle +58.03° \ \text{A}$$

$$I_{3A} = \frac{Z_2 I_{1A}}{Z_2 + Z_3}$$

$$= \frac{(10 \angle -45° \ \Omega)(4.9 \angle 5.53° \ \text{A})}{10 \angle -45° \ \Omega + 10 \angle 60° \ \Omega}$$

$$= 4.02 \angle -46.97° \ \text{A}$$

Determine the currents from Source B.

3. The currents from E_B are now found. E_A is replaced by its internal impedance, a short circuit. The circuit becomes as shown in Figure 18.19.

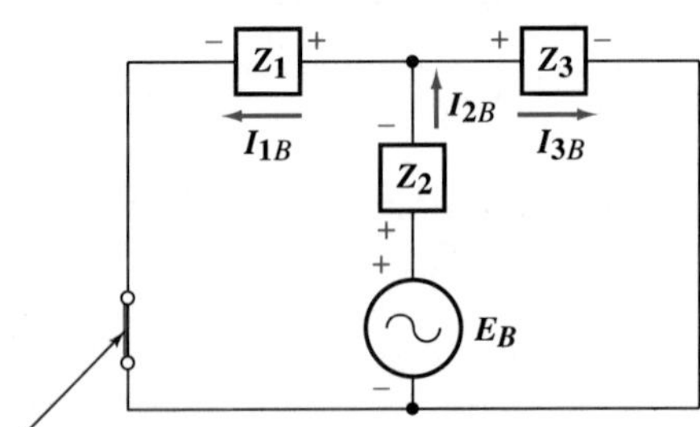

Figure 18.19

$$I_{2B} = \frac{E_B}{Z_{eqB}}$$

$$Z_{eqB} = Z_2 + Z_1 \| Z_3$$

$$= 10 \angle -45° \ \Omega + \frac{(5 \angle 53.13° \ \Omega)(10 \angle 60° \ \Omega)}{5 \angle 53.13° \ \Omega + 10 \angle 60° \ \Omega}$$

$$= 9.96 \angle -25.72° \ \Omega$$

$$\boldsymbol{I}_{2B} = \frac{120 \angle 0° \text{ V}}{9.96 \angle -25.72° \ \Omega}$$

$$= 12.05 \angle 25.72° \text{ A}$$

Using the current divider rule,

$$\boldsymbol{I}_{1B} = \frac{\boldsymbol{Z}_3 \boldsymbol{I}_{2B}}{\boldsymbol{Z}_1 + \boldsymbol{Z}_3}$$

$$= \frac{(10 \angle 60° \ \Omega)(12.05 \angle 25.72° \text{ A})}{5 \angle 53.13° \ \Omega + 10 \angle 60° \ \Omega}$$

$$\boldsymbol{I}_{1B} = 8.04 \angle 28.01° \text{ A}$$

$$\boldsymbol{I}_{3B} = \frac{\boldsymbol{Z}_1 \boldsymbol{I}_{2B}}{\boldsymbol{Z}_1 + \boldsymbol{Z}_3}$$

$$= \frac{(5 \angle 53.13° \ \Omega)(12.05 \angle 25.72° \text{ A})}{5 \angle 53.13° \ \Omega + 10 \angle 60° \ \Omega}$$

$$\boldsymbol{I}_{3B} = 4.02 \angle 21.14° \text{ A}$$

4. All sources have been considered, so the currents are now combined. $\boldsymbol{I}_{1A}$ and $\boldsymbol{I}_{1B}$ are in different directions so

Combine the currents to obtain currents with both sources.

$$\boldsymbol{I}_1 = \boldsymbol{I}_{1A} - \boldsymbol{I}_{1B}$$

$$= 4.9 \angle 5.53° \text{ A} - 8.04 \angle 28.01° \text{ A}$$

$$= 4.88 + j0.47 - 7.10 - j3.78$$

$$\boldsymbol{I}_1 = 3.99 \angle -123.85° \text{ A in the direction of } \boldsymbol{I}_{1A}$$

$\boldsymbol{I}_{2A}$ and $\boldsymbol{I}_{2B}$ are in different directions so

$$\boldsymbol{I}_2 = \boldsymbol{I}_{2B} - \boldsymbol{I}_{2A}$$

$$= 12.05 \angle 25.72° \text{ A} - 4.02 \angle 58.03° \text{ A}$$

$$\boldsymbol{I}_2 = 8.92 \angle 11.78° \text{ A in the direction of } \boldsymbol{I}_{2B}$$

$\boldsymbol{I}_{3A}$ and $\boldsymbol{I}_{3B}$ are in the same direction so

$$\boldsymbol{I}_3 = \boldsymbol{I}_{3A} + \boldsymbol{I}_{3B}$$

$$= 4.02 \angle -46.97° \text{ A} + 4.02 \angle 21.14° \text{ A}$$

$$\boldsymbol{I}_3 = 6.66 \angle -12.93° \text{ A in the direction of } \boldsymbol{I}_{3A} \text{ and } \boldsymbol{I}_{3B}$$

The slight differences between these values and those in Example 18.1 are due to the rounding off of the numbers in the procedure.

5. The voltages across an impedance from each source can also be combined by superposition. However, a simpler way is to use the total currents and Ohm's Law. Then

Solve for the voltages using the total currents and impedances.

$$\boldsymbol{V}_1 = \boldsymbol{I}_1 \boldsymbol{Z}_1$$

$$= (3.99 \angle -123.85° \text{ A})(5 \angle 53.13° \ \Omega)$$

$$V_1 = 19.95 \angle -70.72° \text{ V}$$
$$\mathbf{V_2} = \mathbf{I_2 Z_2}$$
$$= (8.92 \angle 11.78° \text{ A})(10 \angle -45° \ \Omega)$$
$$V_2 = 89.2 \angle -33.22° \text{ V}$$
$$\mathbf{V_3} = \mathbf{I_3 Z_3}$$
$$= (6.66 \angle -12.93° \text{ A})(10 \angle 60° \ \Omega)$$
$$V_3 = 66.6 \angle 47.57° \text{ V}$$

Using the same procedure for power as in Example 18.1,

$$P_T = P_1 + P_2 + P_3$$
$$= I_1^2 Z_1 \cos\theta_1 + I_2^2 Z_2 \cos\theta_2 + I_3^2 Z_3 \cos\theta_3$$
$$= (3.99 \text{ A})^2(5 \ \Omega) \cos 53.13° + (8.92 \text{ A})^2(10 \ \Omega) \cos 45° + (6.66 \text{ A})^2(10 \ \Omega) \cos 60°$$
$$= 47.76 \text{ W} + 562.62 \text{ W} + 221.78 \text{ W}$$
$$P_T = 832.16 \text{ W}$$

A second example helps to further explain the procedure.

EXAMPLE 18.10

Use superposition to find:

a. The current in Z_1 of the circuit shown in Figure 18.20.

b. The minimum power rating of the resistance in Z_1.

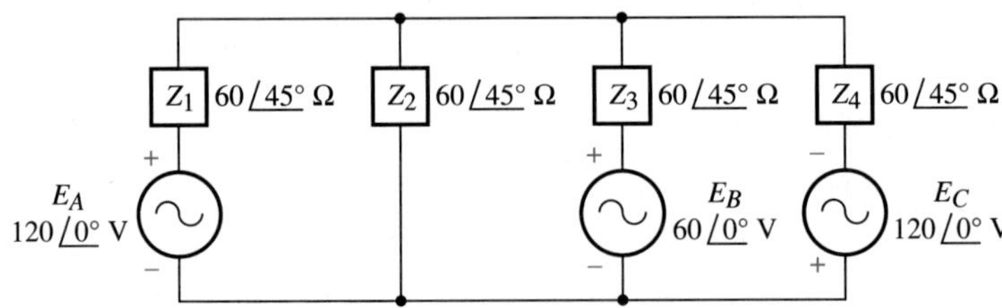

Figure 18.20

Solution

1. E_B and E_C are replaced by short circuits while the current from E_A is calculated. The circuit and current directions are shown in Figure 18.21.

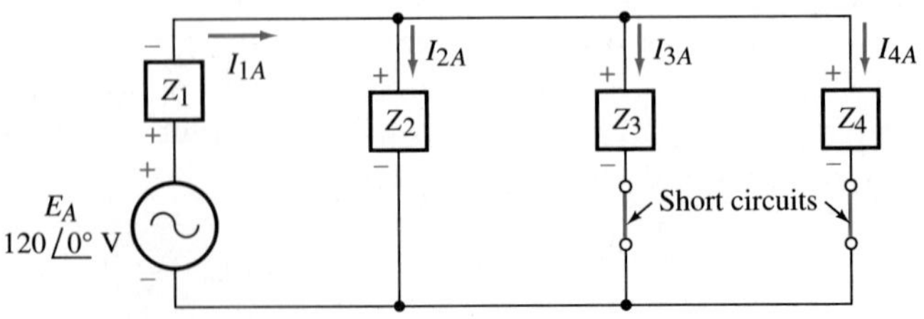

Figure 18.21

2.

$$I_{1A} = \frac{E_A}{Z_{eqA}}$$

$$\begin{aligned} Z_{eqA} &= Z_1 + Z_2 \| Z_3 \| Z_4 \\ &= 60 \angle 45^\circ\ \Omega + 20 \angle 45^\circ\ \Omega \\ &= 80 \angle 45^\circ\ \Omega \end{aligned}$$

Then

$$\begin{aligned} I_{1A} &= \frac{120 \angle 0^\circ \text{ V}}{80 \angle 45^\circ\ \Omega} \\ &= 1.5 \angle -45^\circ \text{ A} \end{aligned}$$

Next, E_A and E_B are replaced by short circuits while the current from E_C is found. The circuit and current directions are as shown in Figure 18.22.

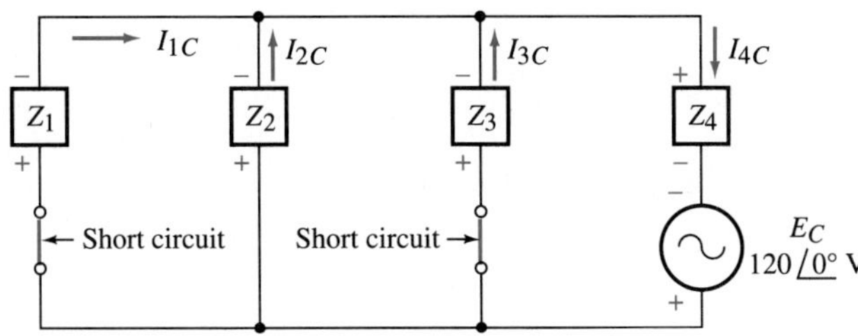

Figure 18.22

Since Z_1, Z_2, and Z_3 are of equal magnitude and angle, the current I_{4C} will divide equally. Then

$$I_{1C} = \frac{I_{4C}}{3}$$

Solving for I_{4C} gives

$$I_{4C} = \frac{E_C}{Z_{eqC}}$$

$$\begin{aligned} Z_{eqC} &= Z_4 + Z_1 \| Z_2 \| Z_3 \\ &= 60 \angle 45^\circ\ \Omega + 20 \angle 45^\circ\ \Omega \\ &= 80 \angle 45^\circ\ \Omega \end{aligned}$$

Then

$$\begin{aligned} I_{4C} &= \frac{120 \angle 0^\circ \text{ V}}{80 \angle 45^\circ\ \Omega} \\ &= 1.5 \angle -45^\circ \text{ A} \end{aligned}$$

and

$$\begin{aligned} I_{1C} &= \frac{1.5 \angle -45^\circ \text{ A}}{3} \\ &= 0.5 \angle -45^\circ \text{ A} \end{aligned}$$

3. Next, E_A and E_C are replaced by short circuits while the current from E_B is calculated. The circuit and current directions are as shown in Figure 18.23.

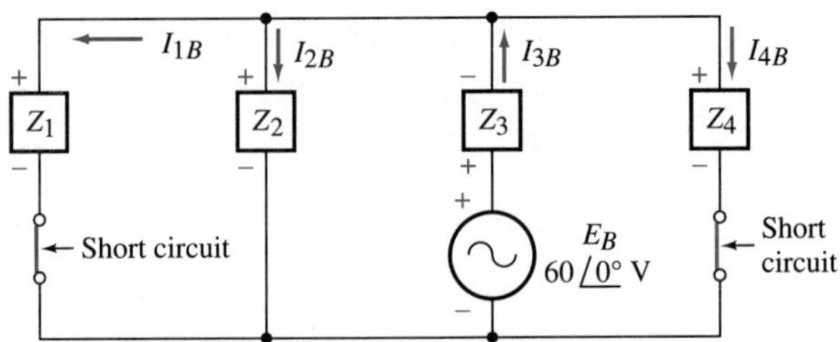

Figure 18.23

Since $\mathbf{Z}_1$, $\mathbf{Z}_2$, and $\mathbf{Z}_4$ are of equal magnitude and angle, the current $\boldsymbol{I}_{3B}$ will divide equally. Then

$$\boldsymbol{I}_{1B} = \frac{\boldsymbol{I}_{3B}}{3}$$

$$\boldsymbol{I}_{3B} = \frac{\boldsymbol{E}_B}{\mathbf{Z}_{\text{eq}B}}$$

$$\begin{aligned}\mathbf{Z}_{\text{eq}B} &= \mathbf{Z}_3 + \mathbf{Z}_1 \| \mathbf{Z}_2 \| \mathbf{Z}_4 \\ &= 60 \angle 45^\circ\ \Omega + 20 \angle 45^\circ\ \Omega \\ &= 80 \angle 45^\circ\ \Omega\end{aligned}$$

Then

$$\begin{aligned}\boldsymbol{I}_{3B} &= \frac{60 \angle 0^\circ\ \text{V}}{80 \angle 45} \\ &= 0.75 \angle -45^\circ\ \text{A}\end{aligned}$$

and

$$\begin{aligned}\boldsymbol{I}_{1B} &= \frac{0.75 \angle -45^\circ\ \text{A}}{3} \\ &= 0.25 \angle -45^\circ\ \text{A}\end{aligned}$$

4. Since every source has been considered, the currents can now be combined to obtain $\boldsymbol{I}_1$.

$$\begin{aligned}\boldsymbol{I}_1 &= \boldsymbol{I}_{1A} - \boldsymbol{I}_{1B} + \boldsymbol{I}_{1C} \\ &= 1.5 \angle -45^\circ\ \text{A} - 0.25 \angle -45^\circ\ \text{A} + 0.5 \angle -45^\circ\ \text{A} \\ I_1 &= 1.75 \angle -45^\circ\ \text{A}\end{aligned}$$

5. The power dissipated in Z_1 is

$$\begin{aligned}P_1 &= I_1^2 R_1 \\ &= I_1^2 Z_1 \cos \theta_1 \\ P_1 &= (1.75\ \text{A})^2 (60 \cos 45^\circ) \\ P_1 &= 129.93\ \text{W}\end{aligned}$$

The rating of the resistance must be at least 129.93 watts.

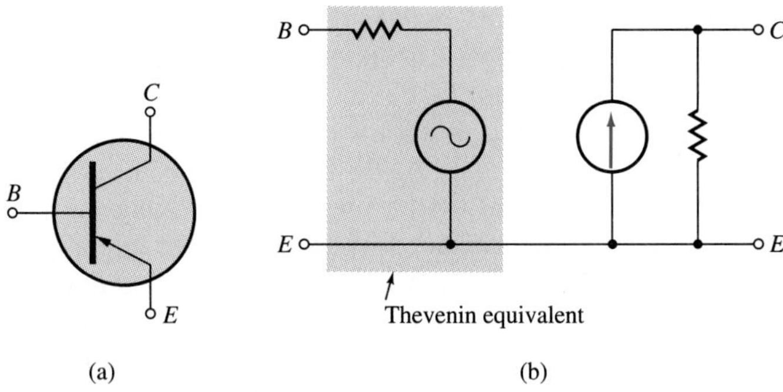

Figure 18.24 A Thevenin circuit is used to replace components by a circuit. (a) The symbol for a PNP transistor. (b) The ac equivalent circuit includes the Thevenin circuit.

18.5 Thevenin's Theorem for Alternating Current (ac) Circuits

Many ac circuits and devices can be replaced by an equivalent circuit of one source of emf and a series impedance. This is in accordance with **Thevenin's Theorem** for ac circuits. The theorem states that a two-terminal, linear, bilateral ac circuit of fixed impedances and sources can be replaced by one source of emf and a series impedance. The Thevenin equivalent of the ac circuit holds for only one frequency. This is because X_C and X_L change with frequency.

Thevenin's Theorem

The use of Thevenin's equivalent can reduce the work in repetitive calculations. Also, it permits replacing components by an equivalent ac circuit that can be analyzed using circuit theory. For example, a transistor can be replaced by a circuit such as that in Figure 18.24(b). It includes a Thevenin circuit on the left side.

A comparison of an actual circuit to its Thevenin equivalent in Figure 18.25 helps to clarify the conversion. The circuits are equivalent if the

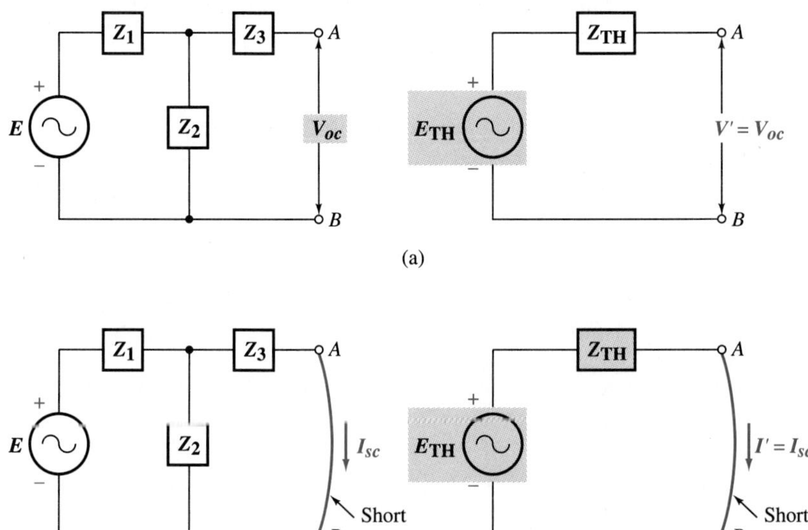

Figure 18.25 The relation between the Thevenin circuit and the circuit that it replaces. (a) The open-circuit voltage must be the same for both. (b) The short-circuit current must be the same for both circuits.

voltage across their terminals A–B is the same for all loads. Since the circuit is linear, making both the open-circuit and the short-circuit conditions equivalent makes the circuits equivalent for all conditions. The open-circuit condition is met if the emf of E_{TH} is equal to the open-circuit voltage across terminals A–B. The short-circuit condition will then be met if Z_{TH} is equal to the impedance looking back into the terminals with all sources replaced by their internal impedances.

An examination of the Thevenin equivalent shows that $E_{TH} = Z_{TH}I_{sc}$, where I_{sc} is the short-circuit current. Thus, the equivalent can also be obtained by either measuring or calculating the open-circuit voltage (E_{TH}) and the short-circuit current (I_{sc}). This was presented in more detail in Chapter 10. This relationship becomes useful when only the terminals of the circuit are accessible, or when calculating the short-circuit current is easier than calculating the open-circuit voltage.

The procedure for finding the Thevenin equivalent of the ac circuit from its components is:

1. Identify the circuit that is to be Theveninized and the load that is connected to it.
2. Disconnect the load from the circuit that is to be Theveninized.
3. Find the impedance looking into the two terminals with all sources replaced by their internal impedance. This is Z_{TH}.
4. Use circuit concepts to find the voltage across the two open-circuited terminals. This is E_{TH}.
5. Reconnect the items of Step 2 to the Thevenin equivalent and make any required analysis.

EXAMPLE 18.11

Determine the Thevenin equivalent for the circuit connected to R_2 in Figure 18.26.

Solution

1. The circuit to be Theveninized is that connected to R_2.
2. The sources are replaced by their internal impedance, a short circuit in this case. The circuit for Z_{TH} is shown in Figure 18.27.

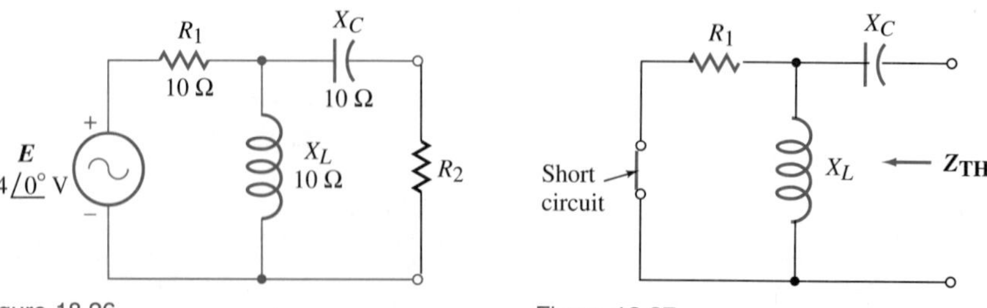

Figure 18.26

Figure 18.27

3. $$\mathbf{Z}_{TH} = -jX_C + \frac{(R)(X_L \angle 90^\circ)}{R + jX_L}$$
$$= -j10\ \Omega + \frac{(10\ \Omega)(10 \angle 90^\circ\ \Omega)}{(10 + j10)\ \Omega}$$
$$\mathbf{Z}_{TH} = 7.07 \angle -45^\circ\ \Omega$$

4. Putting the source back in the circuit gives the circuit of Figure 18.28.

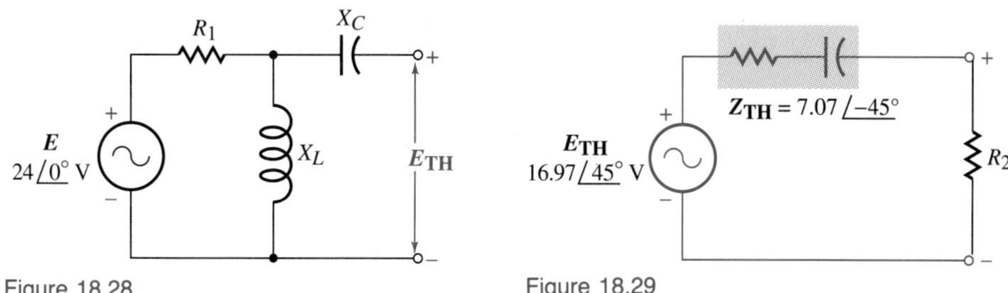

Figure 18.28

Figure 18.29

E_{TH} is the same as the voltage across X_L. Since X_L and R_1 are in series across E, we have

$$E_{TH} = \frac{(E)(X_L \angle 90^\circ)}{R_1 + jX_L}$$
$$= \frac{(24 \angle 0^\circ\ \text{V})(10 \angle 90^\circ\ \Omega)}{(10 + j10)\ \Omega}$$
$$E_{TH} = 16.97 \angle 45^\circ\ \text{V}$$

5. The circuit connected to R_2 is shown in Figure 18.29.

EXAMPLE 18.12

Determine the Thevenin equivalent for the circuit connected to R_2 in Figure 18.30.

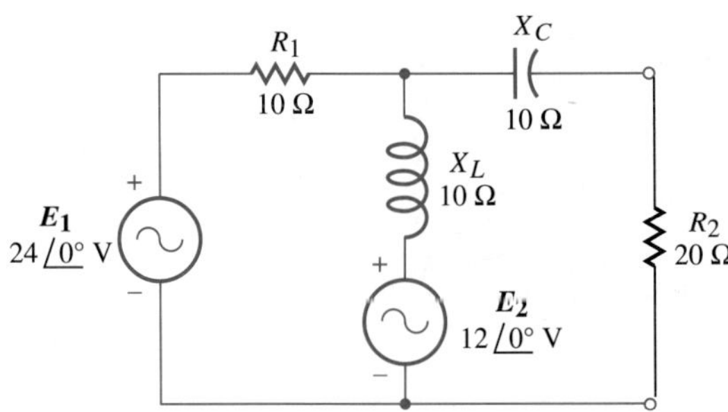

Figure 18.30

Solution

With the sources removed, the circuit is the same as in Example 18.11. Therefore,

$$Z_{TH} = 7.07 \angle -45° \ \Omega$$

Putting the sources back into the circuit gives the circuit of Figure 18.31.

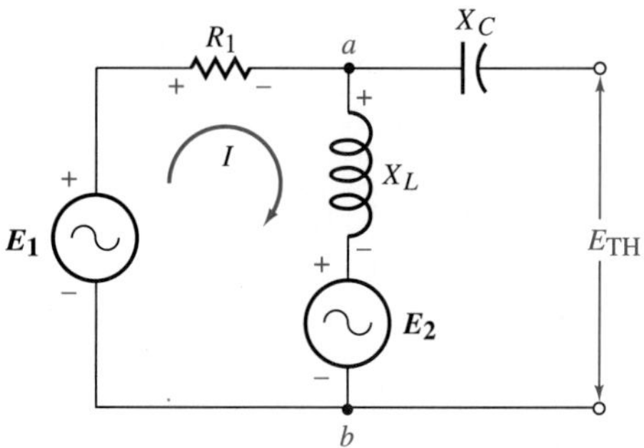

Figure 18.31

E_{TH} is the same as the voltage across Points a and b. This is made up of E_2 and the voltage across X_L.

$$E_1 - IR_1 - IX_L \angle 90° - E_2 = 0$$

so

$$I = \frac{E_1 - E_2}{R_1 + X_L \angle 90°}$$
$$= \frac{24 \text{ V} - 12 \text{ V}}{(10 + j10) \ \Omega}$$
$$I = 0.849 \angle -45° \text{ A}$$

Then

$$E_{TH} = E_2 + (I)(jX_L)$$
$$= 12 \text{ V} + (0.849 \angle -45° \text{ A})(10 \angle 90° \ \Omega)$$
$$E_{TH} = 18.97 \angle 18.43° \text{ V}$$

This is the voltage of the upper terminal with respect to the lower one. The circuit connected to R_2 is shown in Figure 18.32.

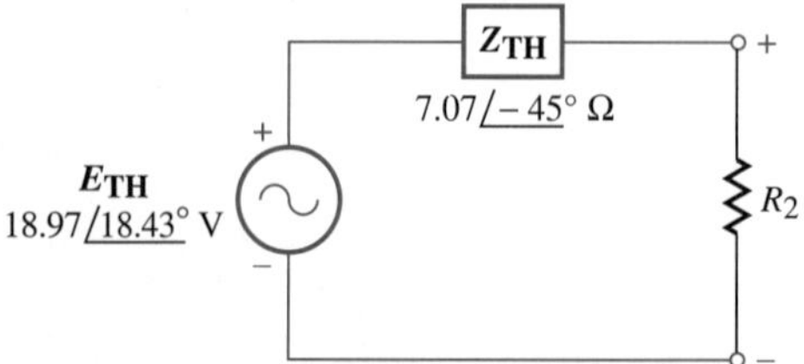

Figure 18.32

One can easily see that making repetitive calculations using the Thevenin equivalent is much shorter than having to analyze the entire circuit each time.

EXAMPLE 18.13

Determine the Thevenin equivalent of the circuit in Figure 18.33.

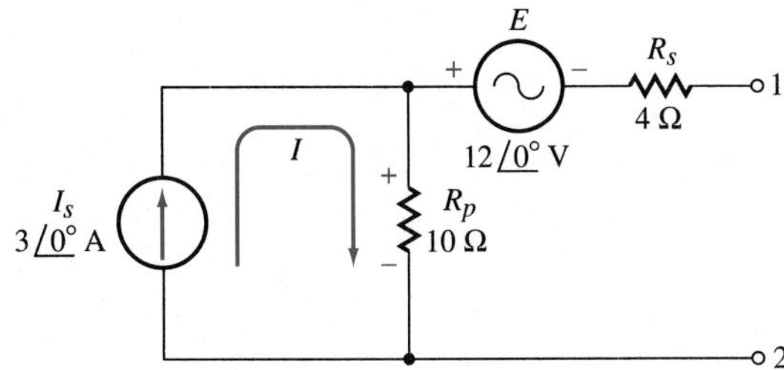

Figure 18.33

Solution

The current source is an open circuit, so the circuit for Z_{TH} is as shown in Figure 18.34.

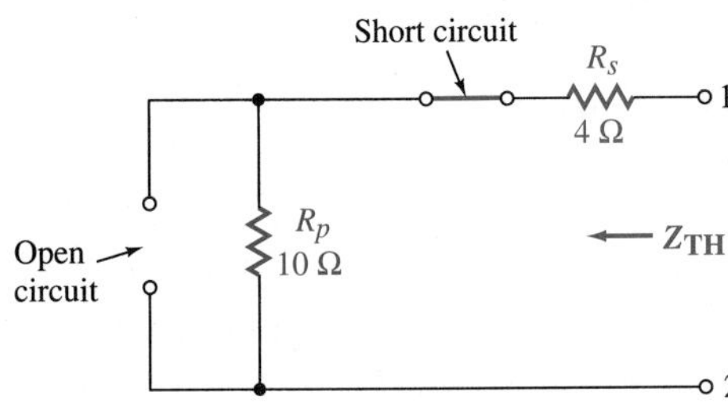

Figure 18.34

$$Z_{TH} = R_p + R_s$$
$$Z_{TH} = 10\ \angle 0^\circ\ \Omega + 4\ \angle 0^\circ\ \Omega$$
$$Z_{TH} = 14\ \angle 0^\circ\ \Omega$$

E_{TH} is the voltage across terminals 1 and 2. The voltage of terminal 1 with respect to terminal 2 is

$$V_{Rp} - E$$

Now with terminals 1–2 open-circuited, the current in R_p is I_s with the top terminal positive. Therefore,

$$V_{1-2} = +I_sR_p - E$$
$$= (3\ \angle 0^\circ\ \text{A})(10\ \Omega) - 12\ \angle 0^\circ\ \text{V} = 18\ \angle 0^\circ\ \text{V}$$
$$E_{TH} = 18\ \angle 0^\circ\ \text{V}$$

The Thevenin equivalent is shown in Figure 18.35.

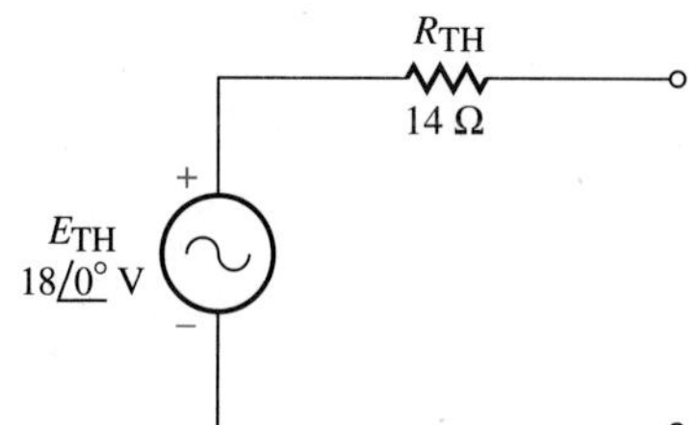

Figure 18.35

This solution could also have been obtained by converting the current source to a voltage source. The procedure for that was given in Chapter 8, and is also explained in Section 18.6.

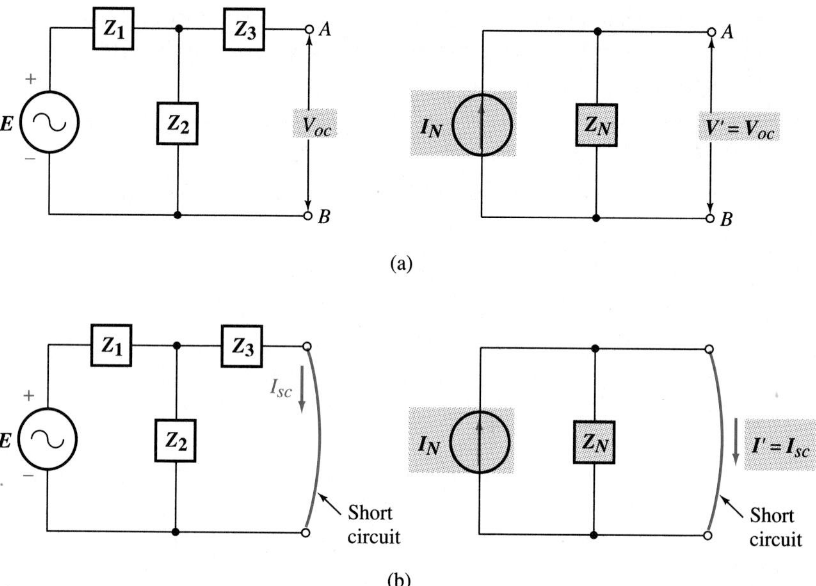

Figure 18.36 The relation between the Norton circuit and the circuit that it replaces. (a) The open-circuit voltage must be the same for both. (b) The short-circuit current must be the same for both.

18.6 Norton's Theorem for Alternating Current (ac) Circuits

Norton's Theorem

Norton's Theorem is the current analog of Thevenin's Theorem. It states that a two-terminal, linear, bilateral ac circuit of fixed impedances and sources can be replaced by a current source and a parallel impedance.

A study of the Norton equivalent and the circuit it replaces in Figure 18.36 helps to understand the conversion. The circuits are equivalent if the load voltage is the same for all currents. Again, meeting the open-circuit and short-circuit conditions will ensure this.

For the short-circuit condition, the current source must have a current equal to the current through the terminals when they are short-circuited. Then, if the impedance $\mathbf{Z}_N$ is equal to the impedance looking into the circuit, with the sources replaced by their internal impedance, the open-circuit conditions are met.

An examination of this equivalent shows that

$$\mathbf{I}_N = \frac{\mathbf{E}_{oc}}{\mathbf{Z}_N}$$

where $\mathbf{E}_{oc}$ is the open-circuit voltage at the terminals.

As in the Thevenin equivalent, the values of $\mathbf{I}_N$ and $\mathbf{Z}_N$ change with frequency.

The procedure for finding the Norton equivalent of a circuit is:

1. Identify the circuit that is to be Nortonized, and the load that is connected to it.

2. Disconnect the load from the circuit that is to be Nortonized.
3. Short circuit the terminals and use circuit concepts to find the short-circuit current. This is I_N.
4. Open the terminals, replace the sources by their internal impedance, and find the impedance looking into the terminals. This is Z_N.
5. Reconnect the items of Step 2 and make any required analysis.

EXAMPLE 18.14

Determine the Norton equivalent for the circuit of Example 18.12.

Solution

Z_N is found in the same way as it was in Example 18.12.

$$Z_N = 7.07 \angle -45° \ \Omega$$

Putting the sources back into the circuit and shorting out terminals 1–2 gives the circuit of Figure 18.37. The short-circuit current is I_B and can be found using the mesh method.

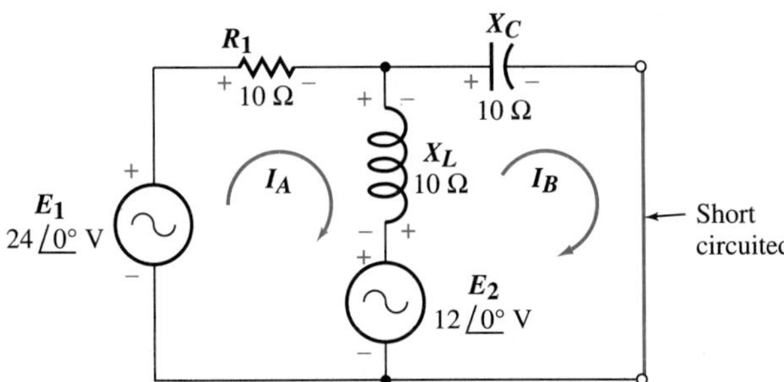

Figure 18.37

Writing Kirchhoff's Voltage Law in a clockwise direction for Loop A gives

$$-E_1 + I_A R_1 + I_A X_L \angle 90° - I_B X_L \angle 90° + E_2 = 0$$
$$-24 \text{ V} + I_A(10 \ \Omega + j10 \ \Omega) - I_B(10 \angle 90° \ \Omega) + 12 \text{ V} = 0$$

or

$$-12 \text{ V} + (14.14 \angle 45° \ \Omega)I_A - (10 \angle 90° \ \Omega)I_B = 0$$

And for Loop B:

$$-E_2 + I_B X_L \angle 90° - I_A X_l \angle 90° + I_B X_c \angle -90° = 0$$
$$-12 \text{ V} + I_B(j10 \ \Omega - j10 \ \Omega) - I_A(10 \angle 90° \ \Omega) = 0$$

or

$$-12 \text{ V} + I_B(0 \ \Omega) - I_A(10 \angle 90° \ \Omega) = 0$$
$$I_A = 1.2 \angle 90° \text{ A}$$

Putting this into the equation for Loop A gives

$$I_B = 2.68 \angle 63.43^\circ \text{ A}$$

Since the short-circuit current is I_B,

$$I_N = 2.68 \angle 63.43^\circ \text{ A}$$

This is the Norton equivalent current.

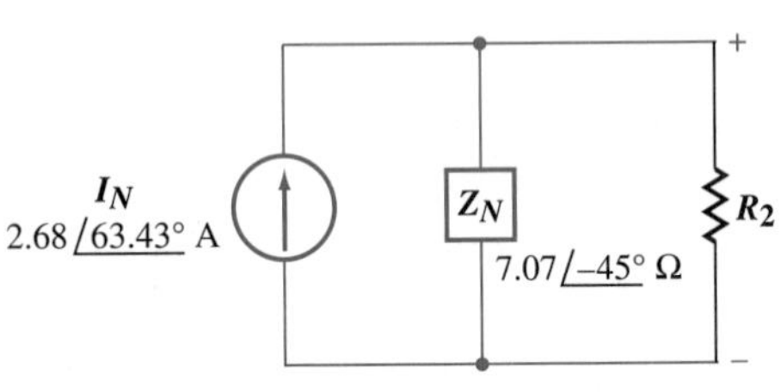

Figure 18.38

The Norton equivalent connected to R_2 is shown in Figure 18.38. The direction of the source current is such that the direction through R_2 is the same as for the original circuit.

Conversions between the Thevenin and Norton ac equivalents are done just as they were for dc equivalents. This lets us change ac voltage sources to ac current sources. These conversions can simplify the analysis of some circuits. The relation in the conversion between the sources is shown in the following equations.

Thevenin to Norton:

$$I_N = \frac{E_{TH}}{Z_{TH}} \tag{18.1}$$

$$Z_N = Z_{TH} \tag{18.2}$$

Norton to Thevenin:

$$E_{TH} = I_N Z_N \tag{18.3}$$

$$Z_{TH} = Z_N \tag{18.4}$$

EXAMPLE 18.15

Convert the voltage source in Figure 18.39(a) to a current source, and the current source of Figure 18.39(b) to a voltage source.

Solution

To make the conversion, we must find the Norton equivalent of Figure 18.39(a) and the Thevenin equivalent of Figure 18.39(b).

a.

$$I_N = \frac{E_{TH}}{Z_{TH}}$$

$$= \frac{10 \angle 0^\circ \text{ V}}{5 \angle 53.13^\circ \ \Omega}$$

$$I_N = 2 \angle -53.13^\circ \text{ A}$$

$$Z_N = Z_{TH}$$

$$Z_N = 5 \angle 53.13^\circ \ \Omega$$

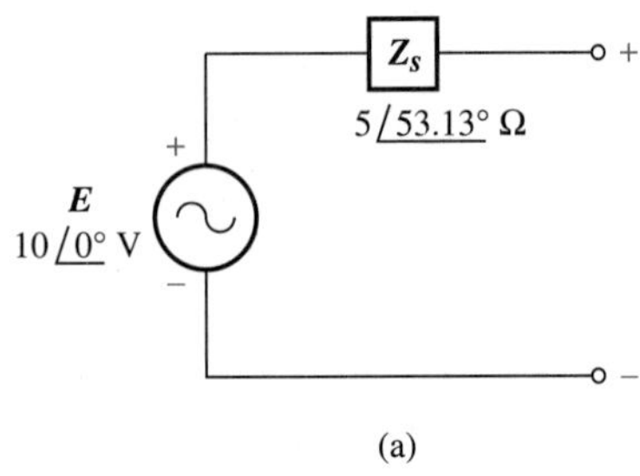

(a)

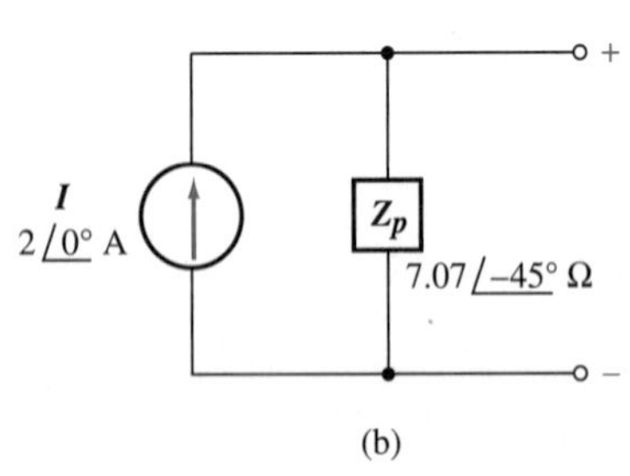

(b)

Figure 18.39

The equivalent is shown in Figure 18.40.

Note that the direction of I_N is the same as the direction of the current in E_{TH}.

b.

$$E_{TH} = I_N Z_N$$
$$= (2 \angle 0° \text{ A})(7.07 \angle -45° \ \Omega)$$
$$E_{TH} = 14.14 \angle -45° \text{ V}$$
$$Z_{TH} = Z_N$$
$$Z_{TH} = 7.07 \angle -45° \ \Omega$$

The equivalent is shown in Figure 18.41.

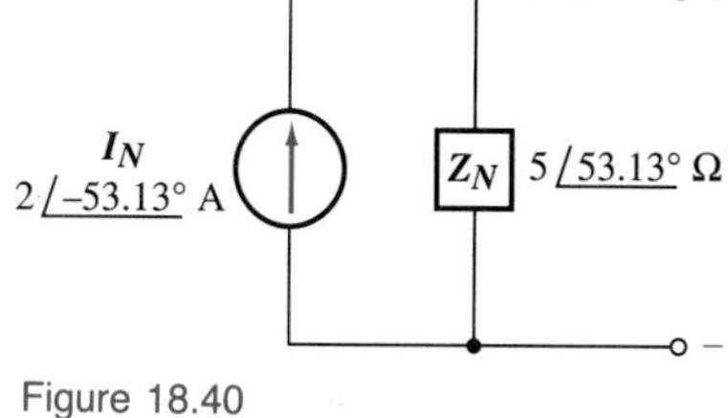

Figure 18.40

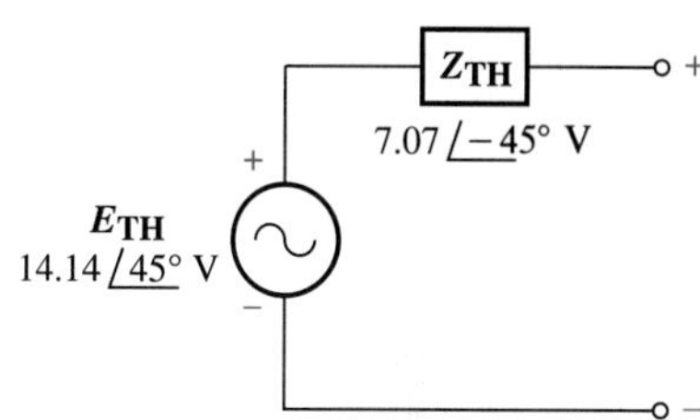

Figure 18.41

18.7 Maximum Power Transfer

An alternator connected to its load and an antenna connected to its receiver share one thing in common. Each one transfers power to a load. For the alternator, a main consideration is obtaining a high efficiency, but for the antenna, a main consideration is the transfer of as much power as possible. The reason for these different objectives is that the alternator involves large amounts of power. Therefore, small inefficiencies mean large loss and high costs. On the other hand, the power level in the antenna is low; therefore, we must transfer as much power to the receiver as possible. How this is done is now considered.

maximum power transfer

Maximum power transfer is the condition when the most power is being delivered to a load by a source. When this condition is met, the load is also said to be matched.

Chapter 10 explained that the maximum power transfer in a dc circuit is obtained when the load resistance is equal to the resistance looking back into the source terminals. Now we examine the transfer of power in an ac circuit. This is done for two conditions. They are:

1. Z_s and Z_L are both resistive.
2. Z_L and Z_s are complex, and the angle of Z_L can be varied.

Case 1: Z_L and Z_s are both resistive.

Since Z_L is $R_L + j0$, the power in Z_L is

$$P_L = I^2 R_L$$

but

$$I = \frac{E}{R_s + R_L}$$

so

$$P_L = \frac{E^2 R_L}{(R_s + R_L)^2}$$

This is the same form used for the dc circuit in Chapter 9. There, we saw that P_L was maximum when $R_L = R_s$. The power transferred for that condition is

$$P_{max} = \frac{E^2}{4R_s} \tag{18.5}$$

Also

$$P_{max} = \frac{E^2}{4R_L} \tag{18.6}$$

where P_{max} is the maximum power transferred, in watts; E is the source voltage, in volts; and R_s and R_L are the source and load resistance, in ohms.

The efficiency of the system is only 50 percent at maximum power transfer. One-half of the power is lost in the internal resistance of the source. This helps explain why electric utilities do not strive for maximum power transfer. If they obtained it, 50 percent of their power would be lost.

EXAMPLE 18.16

The voltage across the terminals of an antenna connected to a receiver is 5 mV. Its impedance is a resistance of 75 ohms. Find:

a. the impedance of the receiver for maximum power transfer.
b. the amount of power delivered to the receiver.
c. the efficiency of the system.

Solution

a. $\mathbf{Z}$ of the receiver must be the same as that of the antenna.

$$\mathbf{Z} = 75 \angle 0° \ \Omega$$

b. $$P = \frac{E^2}{4R_s} = \frac{(5 \times 10^{-3}\ \text{V})^2}{(4)(75\ \Omega)}$$

$$P = 0.083\ \mu\text{W}$$

c. % Efficiency is 50 percent.

Case 2: $\mathbf{Z}_s$ and $\mathbf{Z}_L$ are complex, with a variable angle.

The power in $\mathbf{Z}_L$ is again given by $P_L = I^2R_L$. But now

$$|I| = \frac{|E|}{|Z_{eq}|} = \frac{E}{\sqrt{(R_s + R_L)^2 + (X_s + X_L)^2}}$$

so

$$P_L = \frac{E^2R_L}{(R_s + R_L)^2 + (X_s + X_L)^2}$$

Making X_L equal to and the opposite of X_s cancels out the $X_s + X_L$ term, giving

$$P_L = \frac{E^2 R_L}{(R_s + R_L)^2}$$

This is the same as Case 1, where making $R_L = R_s$ further maximizes P_L.

Based on these results, P_L is a maximum when Z_L is the **conjugate** of Z_s. That is, if $Z_s = R + jX$, $Z_L = R - jX$. If $Z_s = R - jX$, $Z_L = R + jX$. The conjugate of an impedance is one whose reactance is of the opposite sign, and whose resistance has the same value. The conjugate cancels X_s and lets $R_L = R_s$.

conjugate

The amount of power transferred and the efficiency are the same as for Case 1, as given by Equations 18.5 and 18.6.

EXAMPLE 18.17

For the circuit of Figure 18.42, determine:

a. $\mathbf{Z}_L$ for maximum power transfer.
b. the power transferred to $\mathbf{Z}_L$.
c. the power lost in $\mathbf{Z}_s$.

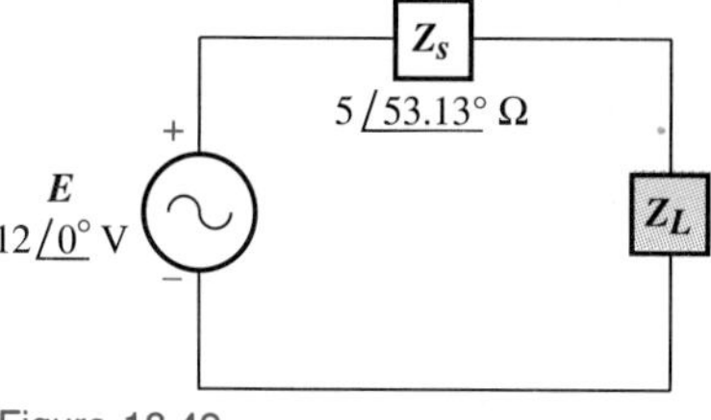

Figure 18.42

Solution

a. For maximum transfer, $\mathbf{Z}_L$ must be the conjugate of $\mathbf{Z}_s$.

$$\mathbf{Z}_s = (3 + j4)$$

so

$$\mathbf{Z}_L = (3 - j4)$$
$$\mathbf{Z}_L = 5 \angle -53.13°\ \Omega$$

b.

$$P_L = \frac{E^2}{4R_L} = \frac{(12\text{ V})^2}{(4)(3\ \Omega)}$$
$$P_L = 12\text{ W}$$

c. Since the efficiency is 50 percent, the power lost is equal to the power in $\mathbf{Z}_L$.

$$P_{\text{lost}} = 12\text{ W}$$

The discussion and the examples used a single source with one impedance. In practical applications such as power transfer from an amplifier to a speaker, the source can be a network rather than a single source. The power transfer in these circuits can be found by Theveninizing the circuit. This gives one source and one impedance. Then, $\mathbf{E}_{\text{TH}}$ is considered to be $\mathbf{E}$ and $\mathbf{Z}_{\text{TH}}$ to be $\mathbf{Z}_s$.

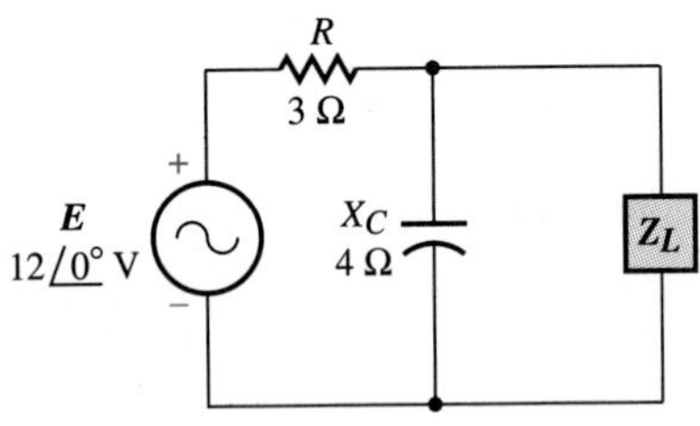

Figure 18.43

EXAMPLE 18.18

Determine $\mathbf{Z}_L$ for maximum power transfer and the power transferred to $\mathbf{Z}_L$ in Figure 18.43.

Solution

$$\mathbf{Z}_{\text{TH}} = R \| X_C \angle -90^\circ = \frac{(3\ \Omega)(4\ \angle -90^\circ\ \Omega)}{(3 - j4)\ \Omega} = 2.4\ \angle -36.87^\circ\ \Omega$$

$$\mathbf{E}_{\text{TH}} = \frac{(X_C \angle -90^\circ)\mathbf{E}}{R - jX_C} = \frac{(4\ \angle -90^\circ\ \Omega)(12\ \angle 0^\circ\ \text{V})}{(3 - j4)\ \Omega} = 9.6\ \angle -36.87^\circ\ \text{V}$$

The circuit is now shown in Figure 18.44.

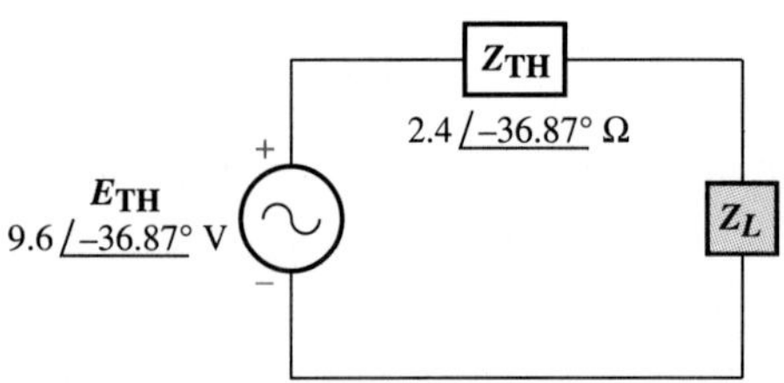

Figure 18.44

$\mathbf{Z}_L$ must be the conjugate of $\mathbf{Z}_{\text{TH}}$.

$$\mathbf{Z}_L = 2.4\ \angle +36.87^\circ\ \Omega \qquad \text{or} \qquad (1.92 + j1.44)\ \Omega$$

The power transferred to $\mathbf{Z}_L$ is

$$P = \frac{E_{\text{TH}}^2}{4R_L} = \frac{(9.6\ \text{V})^2}{(4)(1.92\ \Omega)}$$

$$P = 12\ \text{W}$$

18.8 The Bridge Circuit

bridge circuit

A **bridge circuit** is a circuit that is used to accurately determine the value of an unknown impedance. The configuration of the circuit is shown in Figure 18.45.

balanced bridge

In operation, one or more of the impedances is varied until the current in the meter is zero or a minimum. Then it is said to be a **balanced bridge**.

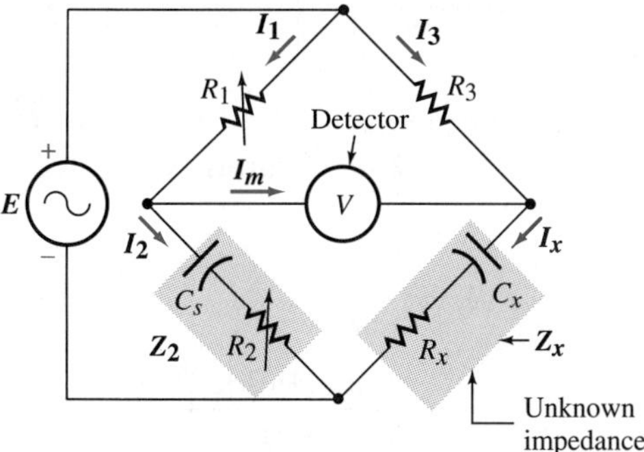

Figure 18.45 The circuit of a capacitance comparison bridge, which is used to measure capacitance.

A pair of headphones or an oscilloscope can also be used in place of the meter. The device should have a high impedance so that no damage occurs before balance is obtained.

When balance is reached, the following conditions exist.

$$\begin{aligned} I_m &= 0 \\ V_m &= 0 \\ I_1R_1 &= I_3R_3 \\ I_2Z_2 &= I_xZ_x \\ I_1 &= I_2 \\ I_3 &= I_x \end{aligned}$$

Combining the last four equations gives

$$\frac{R_1}{Z_2} = \frac{R_3}{Z_x}$$

or

$$Z_x = \frac{R_3Z_2}{R_1} \tag{18.7}$$

For this equality to hold, both the magnitudes and the angles must be equal. This means that the real part of Z_x equals the real part of $(R_3Z_2)/R_1$, and the reactive part of Z_x equals the reactive part of $(R_3Z_2)/R_1$. Since $Z_x = R_x - jX_{cx}$ and $Z_2 = R_2 - jX_{cs}$,

$$R_x = \frac{R_2R_3}{R_1} \tag{18.8}$$

and

$$C_x = \frac{C_sR_1}{R_3} \tag{18.9}$$

If the quantities to the right of the equation sign are known, R_x and C_x can be calculated. In practice, it is not necessary to calculate these values. They are shown on a dial or by a digital readout.

EXAMPLE 18.19

Determine the value of C_x and R_x if $R_1 = 5$ kilohms, $R_2 = 10$ kilohms, $R_3 = 2$ kilohms, and $C_s = 0.2$ microfarad at balance.

Solution

$$R_x = \frac{R_2 R_3}{R_1}$$

$$= \frac{(10\ \text{k}\Omega)(2\ \text{k}\Omega)}{5\ \text{k}\Omega}$$

$$R_x = 4\ \text{k}\Omega$$

$$C_x = \frac{C_s R_1}{R_3}$$

$$= \frac{(0.2\ \mu\text{F})(5\ \text{k}\Omega)}{2\ \text{k}\Omega}$$

$$C_x = 0.5\ \mu\text{F}$$

The circuit of Figure 18.45 can be used to measure inductance if C_s is replaced by an inductance L_s, as seen in Figure 18.46(a). An undesirable

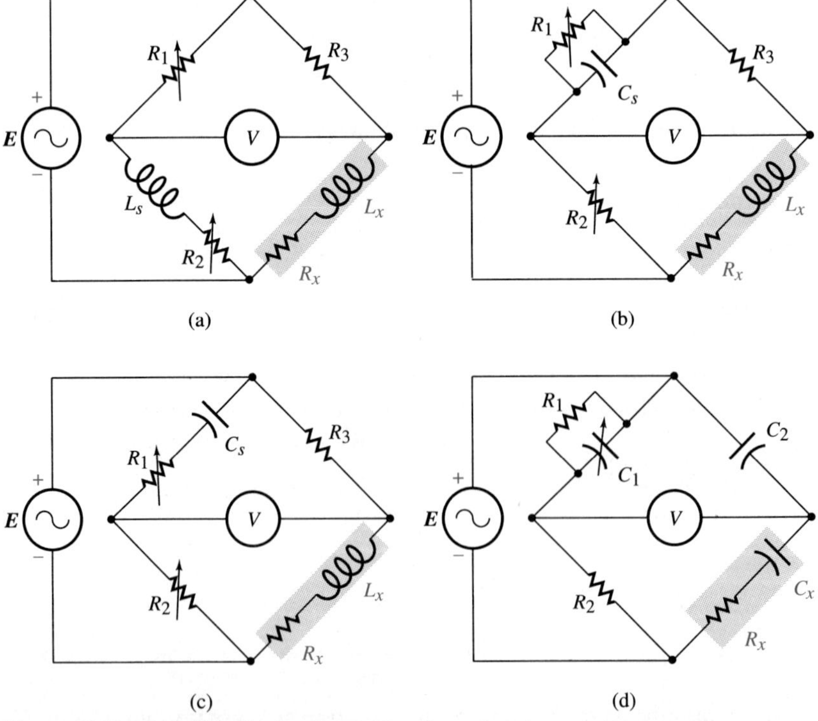

Figure 18.46 Bridges with circuits such as these are also used to measure capacitance or inductance. (a) The inductance comparison bridge measures inductance. (b) The Maxwell-Wein bridge also measures inductance. (c) The Hay Bridge. (d) The Schering Bridge.

feature of that arrangement is that inductances are bulky, expensive, and more sensitive to the fields around them. The **Maxwell-Wein Bridge**, shown in Figure 18.46(b), uses capacitors instead of inductors to avoid these disadvantages.

Maxwell-Wein Bridge

When the Maxwell-Wein Bridge is balanced,

$$R_x = \frac{R_2 R_3}{R_1} \tag{18.10}$$

$$L_x = C_s R_2 R_3 \tag{18.11}$$

This bridge is suitable for coils where the coil's resistance is a large fraction of X_L. When the resistance is a small fraction of X_L, the **Hay Bridge**, shown in Figure 18.46(c), is more suitable. At balance,

Hay Bridge

$$R_x = \frac{\omega^2 C_s R_1 R_2 R_3}{1 + \omega^2 C_s^2 R_1^2} \tag{18.12}$$

$$L_x = \frac{C_s R_2 R_3}{1 + \omega^2 C_s^2 R_1^2} \tag{18.13}$$

Another bridge circuit that measures capacitance is the **Schering Bridge**, shown in Figure 18.46(d). When this bridge is balanced,

Schering Bridge

$$C_x = \frac{C_2 R_1}{R_2} \tag{18.14}$$

$$R_x = \frac{R_2 C_1}{C_2} \tag{18.15}$$

These are a few of the common bridge circuits. Although their components are different, the relations for R_x and L_x or C_x are based on the fact that $(Z_1/Z_2) = (Z_3/Z_4)$ at the balance condition.

SUMMARY

1. The method of equivalent impedance cannot be used to analyze ac circuits with several sources or those that are not series-parallel. Some methods that can be used are loop current, branch current, nodal, and superposition.
2. In the loop current method, Kirchhoff's Voltage Law is applied to the loops using loop currents. Solving the equations yields the loop currents. These can then be used to find the branch currents and other quantities.
3. In the branch current method, Kirchhoff's Voltage Law is applied to the loops using the branch currents. Kirchhoff's Current Law is then applied to the nodes. Solving the equations yields the branch currents.
4. In the nodal method, the branch currents are expressed in terms of the voltage across the branch and the elements in the branch. Solving the equations yields the node voltages. These can be used to find branch currents and other quantities.

5. Superposition is a method in which the currents are found for one source at a time. The currents are then combined to yield the currents with all of the sources in the circuit. The equations are simpler, but there are more of them.
6. An ac circuit of bilateral and linear impedances can be replaced by its Thevenin equivalent. The equivalent circuit consists of a constant voltage generator and a series impedance. The conditions at the output terminals of the equivalent are the same as the conditions for the original circuit.
7. An ac circuit of bilateral and linear impedances can be replaced by its Norton equivalent. The equivalent consists of a constant current generator and a shunt impedance. The conditions at the output terminals of the equivalent are the same as the conditions for the original circuit.
8. The amount of power transferred from a source to a load is affected by the relation of the load and source impedances. For two resistances, maximum power is transferred when $R_L = R_s$. For a complex impedance, maximum power is transferred when Z_L is the conjugate of Z_s.
9. Bridge circuits are used to accurately determine the value of an impedance. When he voltage across the output terminals is zero, the bridge is balanced. At balance, there is a set relation between the unknown impedance and the impedances in the bridge circuit.

EQUATIONS

18.1	$I_N = \dfrac{E_{TH}}{Z_{TH}}$	The relation between the Norton circuit current and Thevenin circuit values.	768
18.2	$Z_N = Z_{TH}$	The relation between the Norton circuit impedance and the Thevenin circuit impedance.	768
18.3	$E_{TH} = I_N Z_N$	The relation between the Thevenin circuit voltage and the Norton circuit values.	768
18.4	$Z_{TH} = Z_N$	The relation between the Thevenin circuit impedance and Norton circuit impedance.	768
18.5	$P_{max} = \dfrac{E^2}{4R_s}$	The power in a resistive load at maximum power transfer in terms of R_s.	770
18.6	$P_{max} = \dfrac{E^2}{4R_L}$	The power in a resistive load at maximum power transfer in terms of R_L.	770

18.7	$Z_x = \frac{R_3 Z_2}{R_1}$	The relation of the unknown impedance to the impedances in the bridge circuit at balanced condition.	773
18.8	$R_x = \frac{R_2 R_3}{R_1}$	Equation that gives the resistance of a capacitor on the bridge.	773
18.9	$C_x = \frac{C_s R_1}{R_3}$	Equation that gives the capacitance on the bridge.	773
18.10	$R_x = \frac{R_2 R_3}{R_1}$	Equation that gives the resistance of an inductor on the Maxwell-Wein bridge.	775
18.11	$L_x = C_s R_2 R_3$	Equation that gives the inductance value on the Maxwell-Wein bridge.	775
18.12	$R_x = \frac{\omega^2 C_s R_1 R_2 R_3}{1 + \omega^2 C_s^2 R_1^2}$	Equation that gives the resistance of an inductor on the Hay Bridge.	775
18.13	$L_x = \frac{C_s R_2 R_3}{1 + \omega^2 C_s^2 R_1^2}$	Equation that gives the inductance of an inductor on the Hay Bridge.	775
18.14	$C_x = \frac{C_2 R_1}{R_2}$	Equation that gives the capacitance on the Schering Bridge.	775
18.15	$R_x = \frac{R_2 C_1}{C_2}$	Equation that gives the resistance on the Schering Bridge.	775

QUESTIONS

1. Can the equivalent impedance method be used for circuits that have several sources in them? Why?
2. How many loop equations are needed in the loop method for a circuit of three loops?
3. For what combination of loop currents will a branch current be 0 amperes in the loop method?
4. What changes should be made in Step 3 of the loop method if the loop currents are drawn counterclockwise?
5. How many loop equations are needed in the branch current method for a circuit of four loops?
6. What equations are needed in addition to the loop equations in the branch current method?
7. A branch current is calculated as $5 \angle 60°$ A by the branch current method. What would the current be if the direction were taken as opposite to that for the analysis?
8. How many nodal equations are needed for a circuit of four nodes?
9. Node voltages are calculated as $V_A = 5 \angle 20°$ V, $V_B = 10 \angle 30°$ V with node C as the reference. What would be the voltages of nodes B and C if A were the reference?

10. How can the equivalent impedance of a single-source, three-loop circuit that is not a series-parallel circuit be calculated?
11. Can the loop method, branch current method, and nodal method be used for single-source circuits?
12. Will superposition provide correct results for a circuit that has a nonlinear element in it? Why?
13. What is one disadvantage of superposition?
14. Is it possible for two different circuits to have the same Thevenin equivalent? Norton equivalent?
15. Will the value of Z_{TH}, as obtained from open-circuit and short-circuit measurement, be greater or smaller than the actual value?
16. Is the Thevenin equivalent of a circuit for one frequency still the Thevenin equivalent at another frequency?
17. How does the Norton equivalent circuit differ from the Thevenin equivalent circuit?
18. Maximum power is transferred from a source to a complex impedance. What effect will a change in the source frequency have on the power transfer?
19. What is meant by the conjugate of an impedance?
20. Does the bridge relation in Equation 18.7 hold when the bridge is not balanced? Why?
21. What effect would using an ammeter as an indicating device have on the operation of the bridge circuit?

PROBLEMS

SECTION 18.1 The Loop Current Method

Use the loop current method for the following problems.

1. Find the current in each branch, the voltage across each impedance, and the power dissipated in the circuit of Figure 18.47.

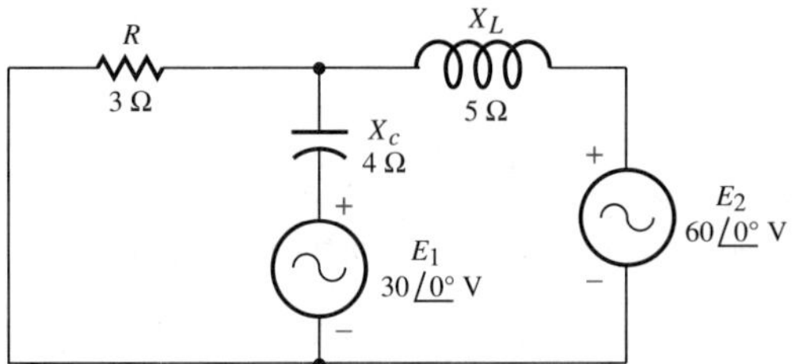

Figure 18.47

2. Work Problem 1 for the circuit in Figure 18.48.

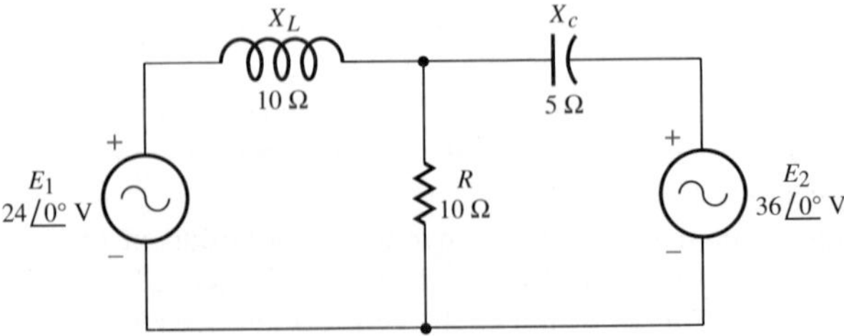

Figure 18.48

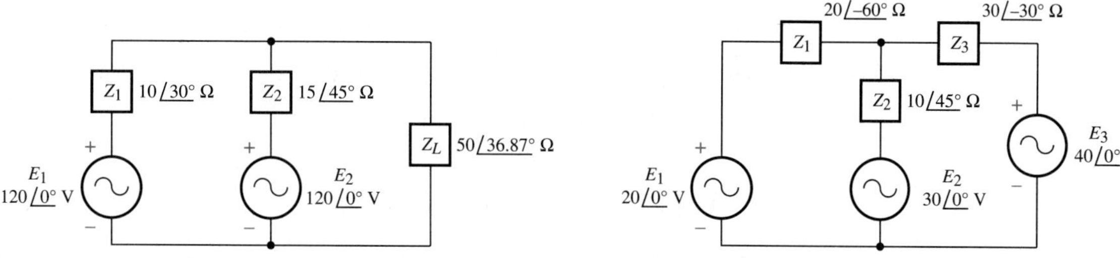

Figure 18.49

Figure 18.50

3. Two alternators are connected, as shown in Figure 18.49, to supply current to Z_L. Determine:

   (a) The current supplied by each alternator.

   (b) The power in Z_L.

4. Find the current in each branch, the voltage across each impedance, and the power dissipated in the circuit of Figure 18.50.

5. Find the current in each branch, the voltage across each impedance, and the power in Z_4 of the circuit in Figure 18.51.

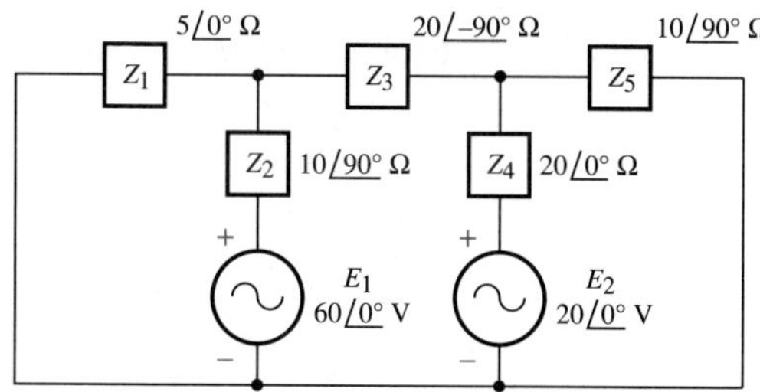

Figure 18.51

6. Determine Z_s in Figure 18.52 for a source current of $0.2\angle -50.78°$ A upward.

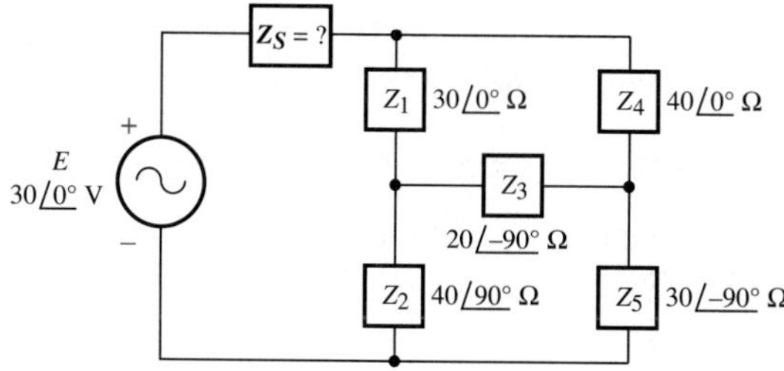

Figure 18.52

SECTION 18.2 The Branch Current Method

7. Work Problem 1 using the branch current method.

8. Work Problem 2 using the branch current method.

9. Work Problem 3 using the branch current method.

10. Work Problem 4 using the branch current method.

11. Work Problem 5 using the branch current method.

12. Work Problem 6 using the branch current method.

SECTION 18.3 The Nodal Method

13. Work Problem 1 using the nodal method.

14. Work Problem 2 using the nodal method.

15. Work Problem 3 using the nodal method.

16. Work Problem 4 using the nodal method.

17. Work Problem 5 using the nodal method.

18. Work Problem 6 using the nodal method.

SECTION 18.4 Superposition

19. Work Problem 1 using superposition.

20. Work Problem 2 using superposition.

21. Work Problem 3 using superposition.

22. Work Problem 5 using superposition.

23. Determine the current in Z_3 of Figure 18.53 and the voltage across it, using superposition.

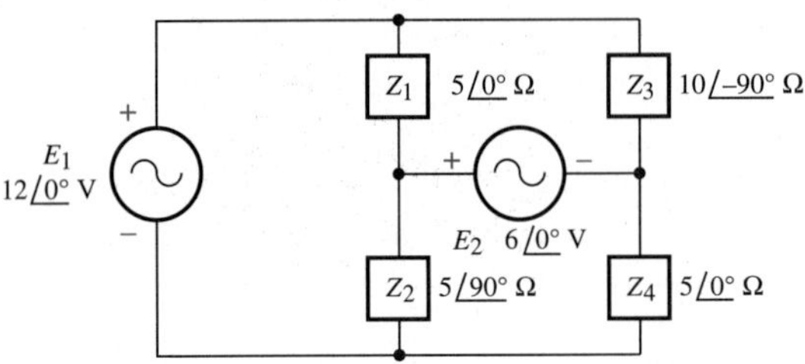

Figure 18.53

Figure 18.54

24. The output of source E_1 in Figure 18.54 is 5 sin (377t) and that of E_2 is 5 volts dc. Use superposition to plot the voltage across R_L for one cycle of E_1.

SECTION 18.5 Thevenin's Theorem for Alternating Current (ac) Circuits

25. Use Thevenin's Theorem to find the voltage V_4 and the current I_4 in the circuit of Figure 18.55.

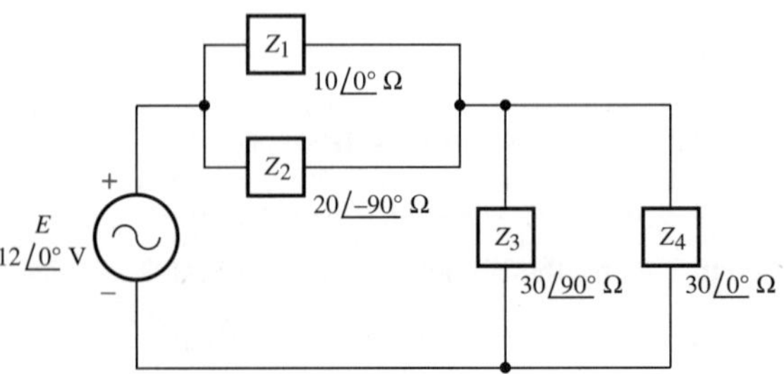

Figure 18.55

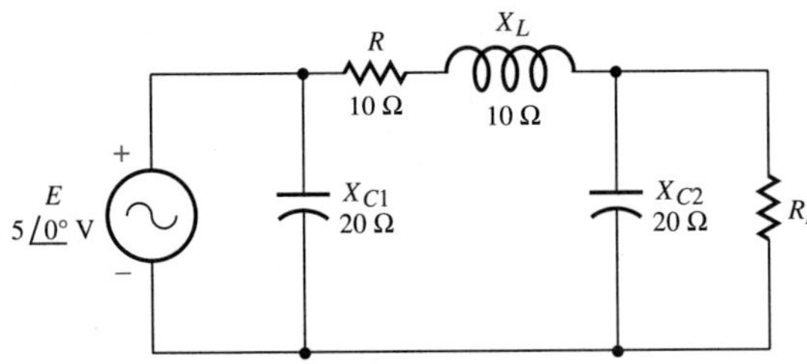

Figure 18.56

26. The circuit in Figure 18.56 is the filter circuit of a power supply. Use Thevenin's Theorem to find the ac voltage across R_L for R_L = 30, 20, and 10 ohms.

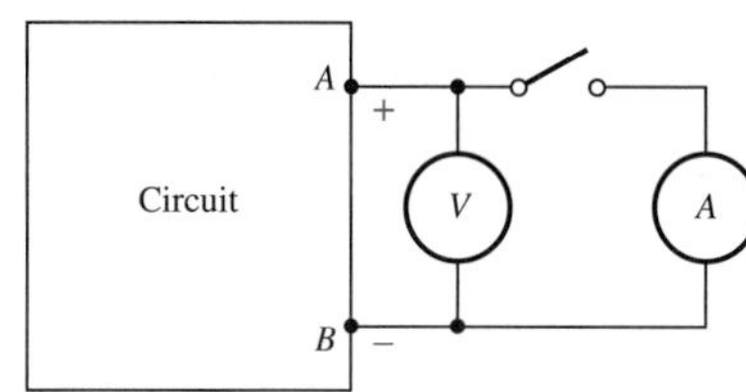

Figure 18.57

27. With S open in Figure 18.57, the voltage measured across A − B is 120 $\angle 0°$ V. With S closed, the current is 5 $\angle 30°$ A. Replace the circuit by its Thevenin equivalent.

28. In the circuit of Figure 18.58, use Thevenin's Theorem to find:

(a) The current in Z_2.

(b) The voltage across Z_2.

(c) The power dissipated in Z_2.

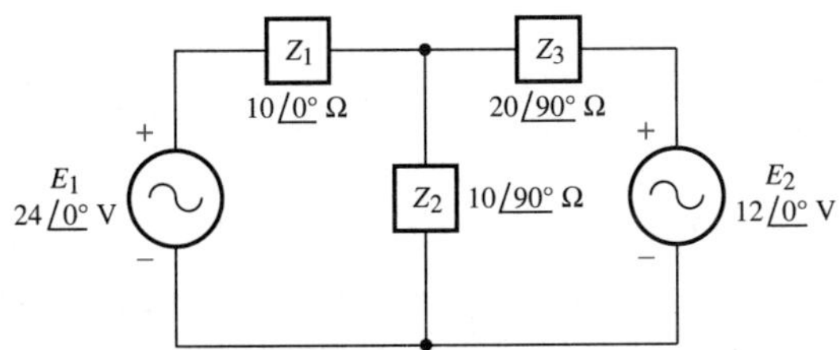

Figure 18.58

29. Use Thevenin's Theorem to find the voltage across Z_1 and the power in Z_1 of Figure 18.59. The top terminal of each source is (+).

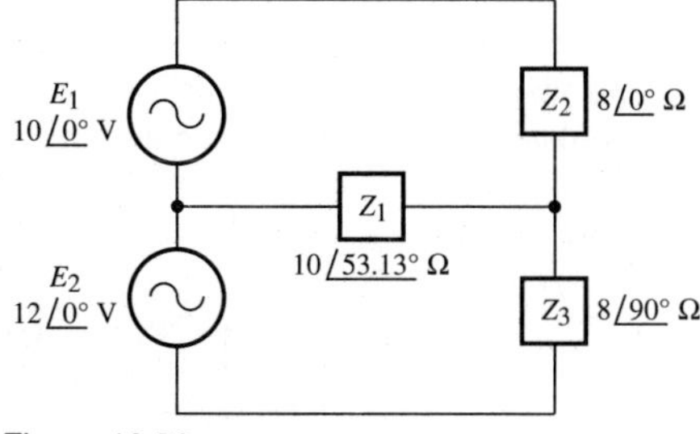

Figure 18.59

30. Develop a proof to show that the Thevenin equivalent will have the same short-circuit current as the circuit in Figure 18.60, if E_{TH} is the open-circuit voltage and Z_{TH} is the impedance measured across 1 − 2, with E short-circuited.

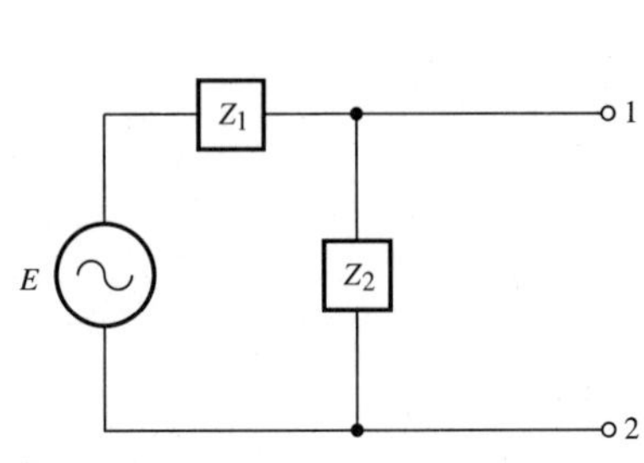

Figure 18.60

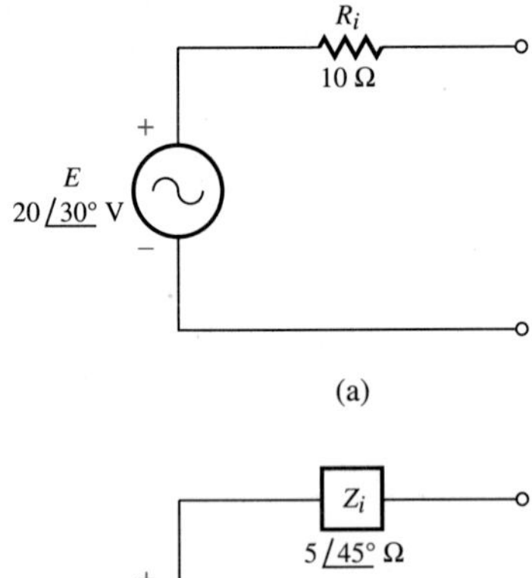

Figure 18.61

SECTION 18.6 Norton's Theorem for Alternating Current (ac) Circuits

31. Work Problem 25 using Norton's Theorem.
32. Work Problem 26 using Norton's Theorem.
33. Find the Norton equivalent for Problem 27.
34. Work Problem 28 using Norton's Theorem.
35. Work Problem 29 using Norton's Theorem.
36. Convert the voltage sources shown in Figure 18.61 to current sources.
37. Convert the current sources shown in Figure 18.62 to voltage sources.

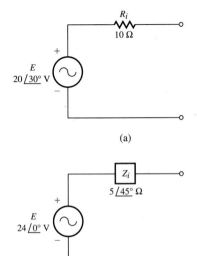

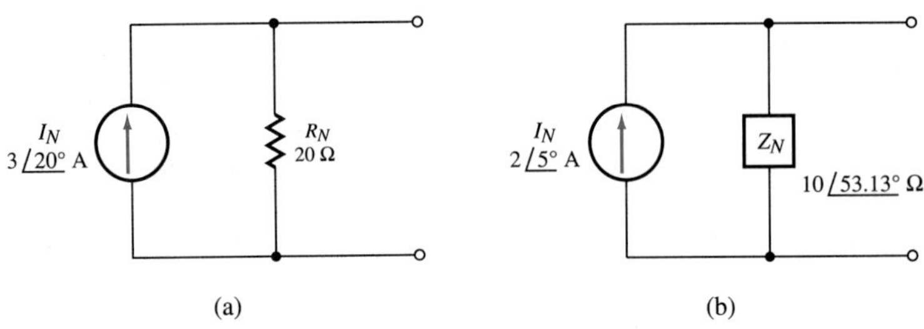

Figure 18.62

38. Develop a proof to show that the Norton equivalent will have the same open-circuit voltage as the circuit in Figure 18.63, if I_N is the short-circuit current and Z_N is the impedance measured across 1 − 2, with E short-circuited.

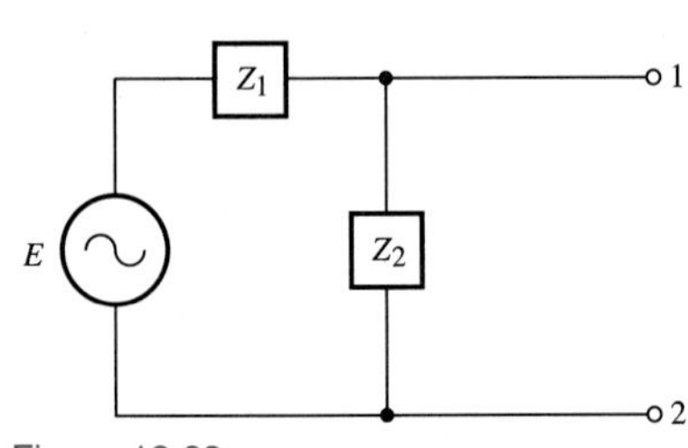

Figure 18.63

SECTION 18.7 Maximum Power Transfer

39. Maximum power is being transferred to R_L in Figure 18.64.
    (a) How many ohms is R_L?
    (b) How much power is transferred?
    (c) What is the voltage across R_L?
40. What must the value of R_L be in Figure 18.65 for it to receive maximum power? How much power is dissipated in it?
41. The source E is connected through the filter network of Figure 18.66 to Z_L. If the phase angle and magnitude of Z_L can be varied, what must Z_L be for maximum power transfer? What are the current and power in Z_L?

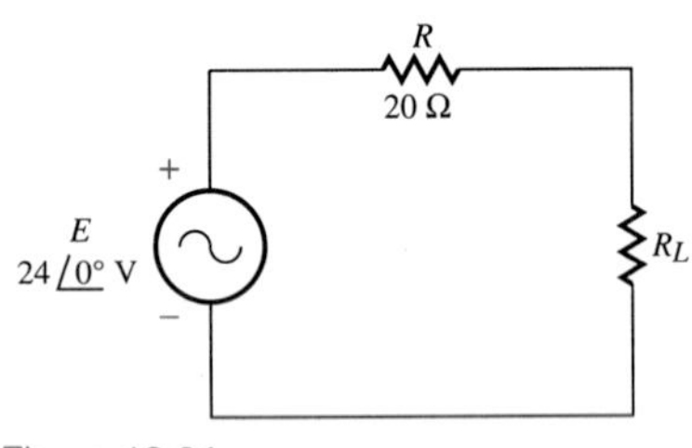

Figure 18.64

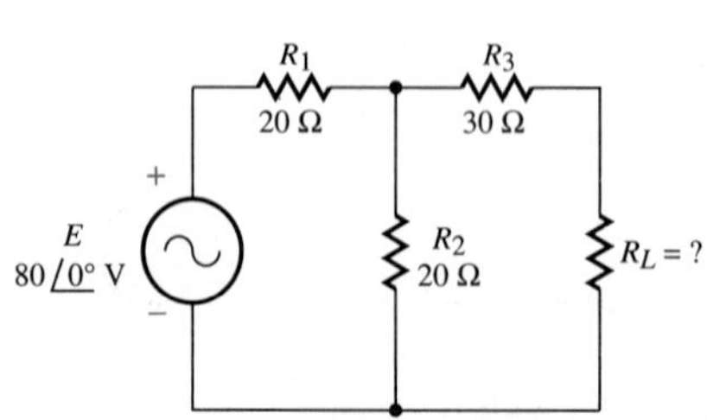

Figure 18.65

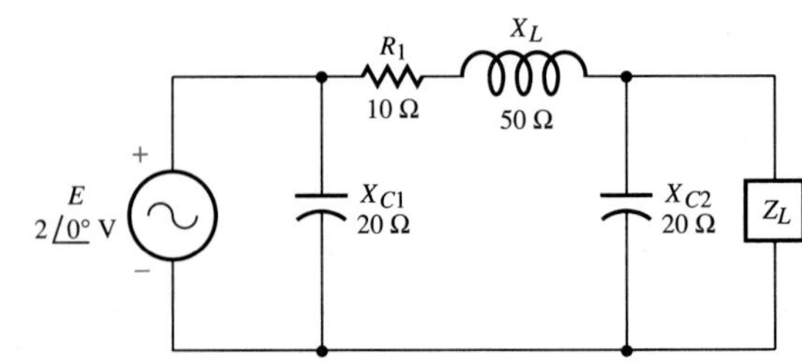

Figure 18.66

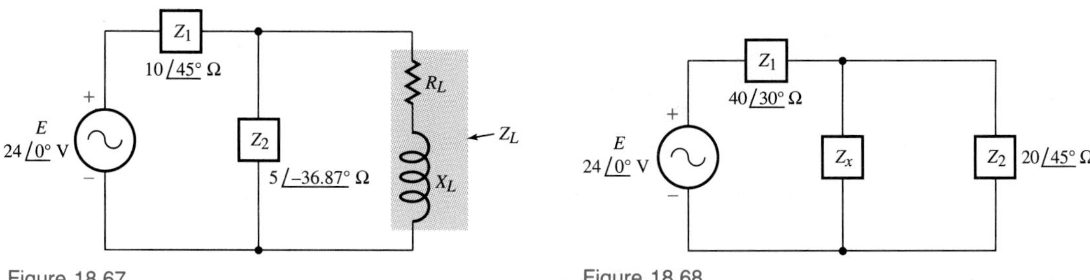

Figure 18.67

Figure 18.68

42. The impedance Z_L in Figure 18.67 can be varied until maximum power is transferred to Z_L. Determine:

    (a) The value of R_L.

    (b) The phase angle of Z_L.

    (c) The power dissipated in Z_L and the power supplied by E.

43. If the phase angle of Z_L in the circuit of Figure 18.67 is kept constant at 30°, show that the maximum power is transferred when the magnitude of Z_L is equal to the magnitude of Z_{TH}.

44. Z_x is used to match Z_2 to the source in Figure 18.68 to obtain maximum power transfer. Determine the magnitude and angle of Z_x.

45. Maximum power of 2 watts is being transferred to an 8-ohm loudspeaker connected to an amplifier. Assuming no changes in the amplifier, how much power will be transferred to each loudspeaker if another 8-ohm loudspeaker is connected in parallel with it?

SECTION 18.8 The Bridge Circuit

46. Determine the impedance Z_4 in Figure 18.69 if the bridge is balanced with the values of Z_1, Z_2, and Z_3 as shown.

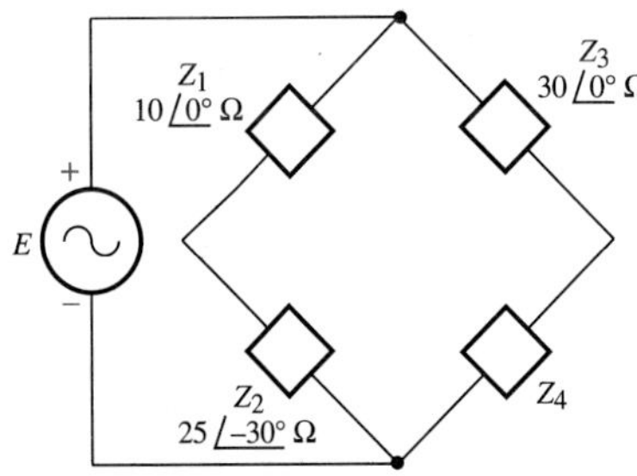

Figure 18.69

47. An inductance is measured in a Maxwell-Wein Bridge. Balance is obtained when $R_2 = 50$ ohms, $R_3 = 80$ ohms, $R_1 = 20$ ohms, and $C_s = 10$ microfarads. What is the value of L_x and R_x?

48. A capacitor is measured on a Schering Bridge, and balance is obtained at $C_1 = 12$ microfarads, $C_2 = 6$ microfarads, and $R_2 = 90$ ohms. What is the value of C_x and R_x?

49. Derive the expression for R_x and L_x on the Maxwell-Wein Bridge.

50. Work Problem 48 for the Schering Bridge.

CHAPTER 19

FREQUENCY RESPONSE ANALYSIS

CHAPTER OBJECTIVES

After completing this chapter, you should be able to:

1. Identify low-pass, high-pass, bandpass, and band-reject filters and describe their frequency response characteristics.
2. Derive the transfer function for an R–C, R–L, or R–L–C circuit.
3. Separate the transfer function into its magnitude and phase components.
4. Determine the bandwidth of a filter circuit.
5. Graph a Bode plot given the transfer function.
6. Use PSpice to analyze the frequency response characteristics of a passive circuit.
7. Determine the bandwidth and Q of a resonant circuit.

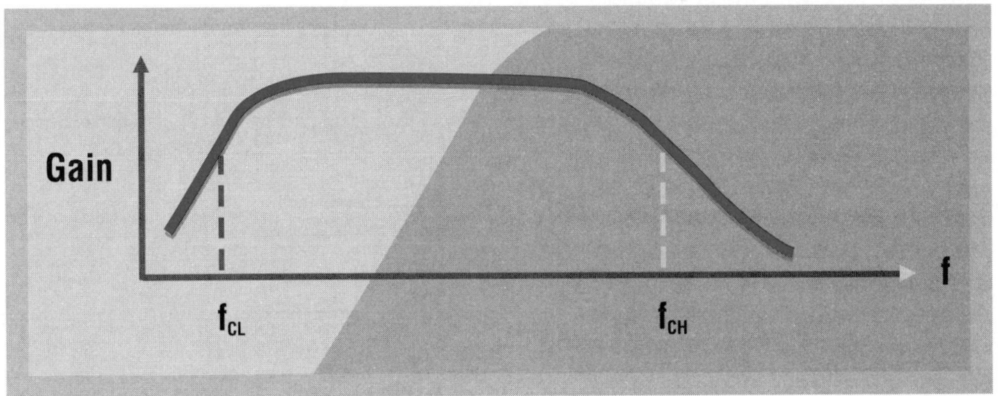

KEY TERMS

1. bandpass
2. band-reject
3. bandwidth
4. Bode plot
5. high-pass
6. low-pass
7. phase shift
8. resonant frequency
9. −3-dB frequency
10. transfer function

INTRODUCTION

The frequency response of a circuit, instrument, or electronic system is an important measure of its performance. Each circuit, instrument, or system is designed with a specific operating range.

A stereo sound system has an operating range of 40 Hz–20,000 Hz. Within this range of frequencies, the components of the stereo system must detect and amplify signal frequencies that make up the music which we will hear. Likewise, an electronic measuring instrument such as an oscilloscope must amplify signal frequencies within its operating range. Oscilloscopes have operating frequency ranges from dc to a few MHz, to frequencies into the gigahertz range.

The frequency response of systems can be complex. In this chapter, however, we investigate four basic types: low-pass, high-pass,

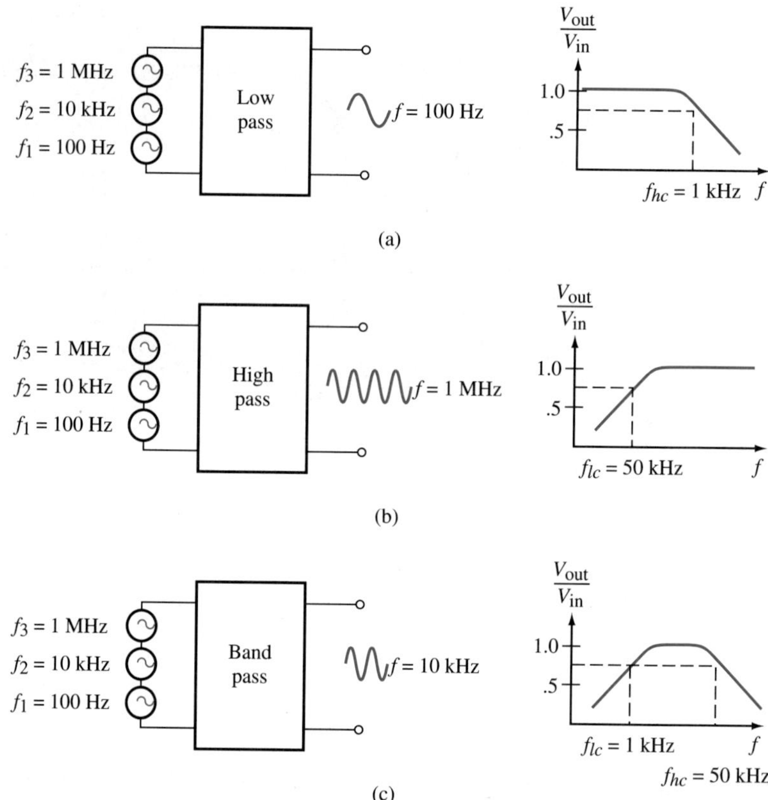

Figure 19.1 The frequency response of systems to complex signals. (a) Low-pass. (b) High-pass. (c) Band-pass.

bandpass, and band-reject. Figure 19.1 illustrates the characteristics of the first three types. The input signal is composed of three sinusoidal components—100 Hz, 10 kHz, and 1 MHz.

low-pass

The **low-pass** characteristic shown in Figure 19.1(a) has an upper frequency limit of 1 kHz. The circuit will pass signals up to this upper limit with little attenuation or change in waveshape. Hence, the output signal is composed of only the 100-Hz sinusoidal component.

high-pass

The **high-pass** circuit shown in Figure 19.1(b) allows all signal components above a certain frequency to pass through the circuit without attenuation or change in phase. The low-frequency limit in this example is 50 kHz. The only input signal component above 50 kHz is the 1-MHz component; hence, the output signal will contain this single 1-MHz component, and the other two input signals will be greatly attenuated.

The third type of characteristic shown in Figure 19.1(c) is the bandpass characteristic. The bandpass characteristic has two frequency limits: a lower limit and an upper limit. The bandpass circuit will only allow those signal frequencies between these two limits to pass through the circuit without attenuation. Hence, the output only contains the 10-kHz component. The 100-Hz and the 1-MHz components are outside the passband and are greatly attenuated.

In this chapter, we learn how to express the frequency response characteristics of these basic circuit types in mathematical form using transfer functions, and in graphical form using Bode plots.

Circuit simulation is also used to generate frequency response information, and we learn how to interpret the results of PSpice simulations.

19.1 Transfer Functions

In Chapter 16, we studied the response of a circuit to a driving signal of a single frequency. In this chapter, we expand the analysis to include input signals that contain a range of frequency components. In Chapter 20, we use these concepts to study the response of circuits to nonsinusoidal signals by considering these nonsinusoidal signals to be merely a composite of many sinusoidal signals of different amplitudes, frequencies, and phase.

Here, our concern in the study of circuits is to determine some convenient mathematical description of a circuit or system from an overall external viewpoint. This description allows us to determine the response of the circuit or system from a knowledge of the input signal, without necessarily knowing the details of the hardware inside the circuit or system.

For example, our stereo system has a frequency response of 40 Hz–20,000 Hz. This indicates that any signal in this frequency range will be amplified with the same amplification and sent on to the speakers. Any signal outside of this range will not be amplified, and will not be present in the sound reproduced by the speakers. Hence, we can predict the circuit's behavior without knowing the specific details of the circuitry inside the stereo amplifier.

For our study, we restrict our discussion to linear systems, although the concepts of transfer functions and frequency response apply to nonlinear systems as well. A system is said to be linear with respect to an input signal $x(t)$ if the following properties are satisfied.

First, if an input signal $x(t)$ produces an output signal $y(t)$, then an input signal $K\,x(t)$ should produce an output signal $K\,y(t)$ for any value of K, where K is a constant. For example, if an amplifier produces an output signal of 1 volt in response to an input signal of 10 mV, then the same amplifier should produce an output signal of 2 volts if the input signal is increased to 20 mV, or an output signal of 0.5 volt if the input signal is decreased to 5 mV.

Second, if a linear circuit has more than one input signal, the superposition property must be satisfied. That is, if the circuit has an input $x_1(t)$ that produces an output $y_1(t)$ and another input $x_2(t)$ that produces an output $y_2(t)$, then the input signal $x_1(t) + x_2(t)$ should produce an output signal $y_1(t) + y_2(t)$ for any arbitrary signals $x_1(t)$ and $x_2(t)$. For example, if we apply a 10-volt peak-to-peak sine wave with a 10-volt dc level to a resistive voltage divider consisting of two 1-kilohm resistors, the output signal will be composed of the response of the voltage divider to the sinusoidal component, plus the response of the voltage divider to the dc component, both of which are attenuated by a factor of 0.5.

Now let us turn our attention to finding a means of representing the behavior of linear circuits consisting of R, L, and C from an external point of view. Consider the block diagram in Figure 19.2. The input signal $v_{in}(t)$ is processed by the circuit inside the box. The response of the circuit to the input signal $v_{in}(t)$ is described by a mathematical function called the **transfer function** of the circuit. The result is the output signal $v_{out}(t)$.

transfer function

We can write an equation to describe the relationship of $x(t)$, TF, and $v_{out}(t)$ as

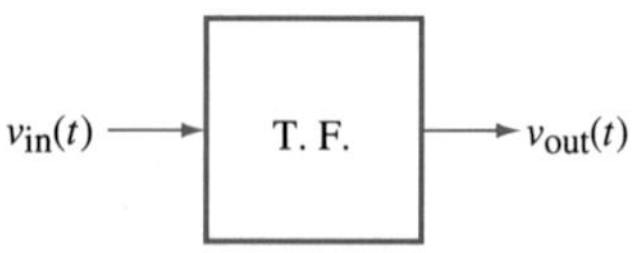

Figure 19.2 Block diagram of a system component.

$$v_{out}(t) = \text{TF}\ v_{in}(t) \tag{19.1}$$

where $x(t)$ is the input signal, TF is the transfer function of the circuit, and $y(t)$ is the output signal.

Dividing Equation 19.1 by the input signal $x(t)$ yields

$$\frac{v_{out}(t)}{v_{in}(t)} = \text{TF} \tag{19.2}$$

The transfer function can be divided into two parts, one part describing the magnitude response of the circuit and the other part describing the phase response of the circuit. The magnitude response means the ratio of the magnitude of the output signal to the magnitude of the input signal. The phase response means the difference in phase between the output signal and the input signal.

Let us look at some familiar circuits and determine their transfer functions.

EXAMPLE 19.1

Given the circuit shown in Figure 19.3, determine the transfer function for the circuit.

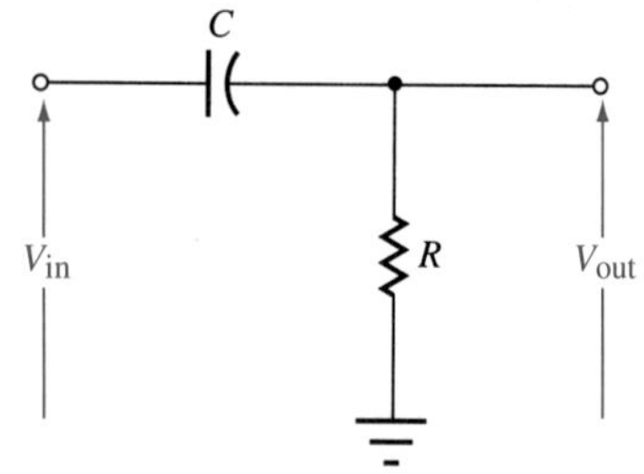

Figure 19.3

Solution

The impedance of the capacitor is $-jX_c$ ohms and the impedance of the resistor is R ohms. The circuit is a voltage divider circuit so

$$v_{out} = \left(\frac{R}{R - jX_c}\right) v_{in}$$

Dividing both sides of the equation by v_i yields the transfer function

$$\text{TF} = \frac{v_{out}}{v_{in}} = \frac{R}{R - jX_c}$$

EXAMPLE 19.2

Given the R–L circuit in Figure 19.4, determine the transfer function for the circuit.

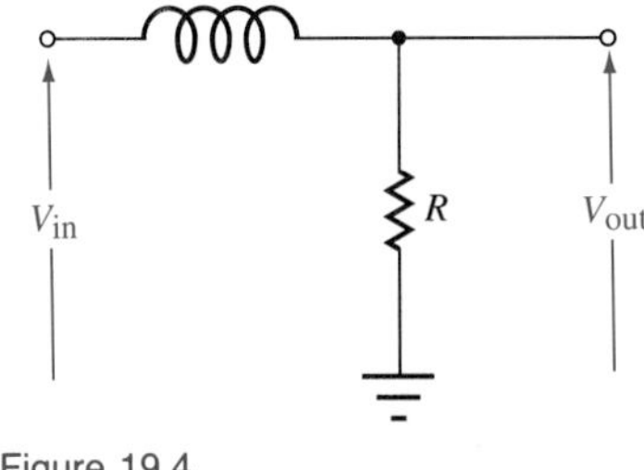

Figure 19.4

Solution

The impedance of the inductor is jX_L. Again, applying the divider principle,

$$v_{out} = \left(\frac{R}{R + jX_L}\right)v_{in}$$

Dividing both sides of the equation by v_{in} yields the transfer function

$$\text{TF} = \frac{v_{out}}{v_{in}} = \frac{R}{R + jX_L}$$

Now, let us try a more complex circuit. In Example 19.3, we combine the impedance of R_L, L, and C into a total or equivalent impedance, Z_{eq} for the series combination. Our goal is to reduce the circuit to an equivalent divider circuit consisting of two impedances.

EXAMPLE 19.3

Determine the transfer function for the circuit shown in Figure

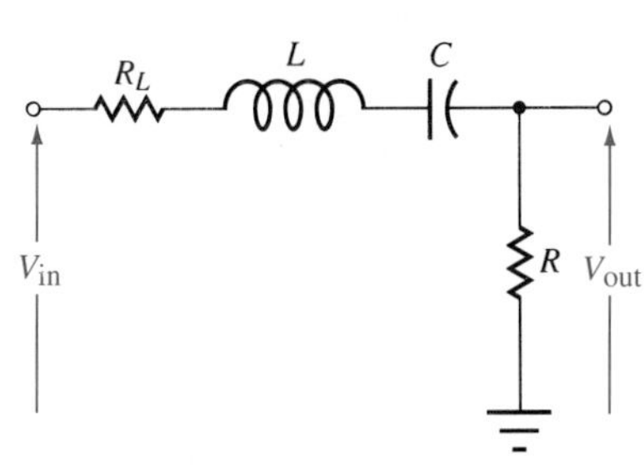

Figure 19.5

Solution

The equivalent impedance of the series combination of R_L, L, an

$$\begin{aligned} Z_{eq} &= R_L + jX_L - jX_C \\ &= R_L + j(X_L - X_C) \end{aligned}$$

Applying the voltage divider principle yields

$$v_{out} = \left(\frac{R}{(R_L + R) + j(X_L - X_C)}\right)v_{in}$$

Dividing both sides of the equation by v_{in} yields the transfer fun

$$\text{TF} = \frac{v_{out}}{v_{in}} = \frac{R}{(R_L + R) + j(X_L - X_C)}$$

Transfer functions describe both the magnitude response and the phase response of the circuit. By separating the transfer function into its magnitude and phase components, we can easily analyze the operating characteristics of the circuit. The following examples show you how this is performed.

EXAMPLE 19.4

Determine the magnitude and phase components of the transfer function derived in Example 19.1.

Solution

The transfer function derived in Example 19.1 is

$$\text{TF} = \frac{v_{out}}{v_{in}} = \frac{R}{R - jX_C}$$

To perform the division on the right-hand side of the equation, we must convert the two impedances to their polar form.

R can be represented as $R\angle 0°$

$R - jX_C$ can be represented as $\sqrt{R^2 + X_c^2}\ \angle -\tan^{-1}(X_C/R)$

Substituting the polar form of each impedance into the transfer function yields

$$\frac{v_{out}}{v_{in}} = \frac{R\angle 0°}{\sqrt{R^2 + X_C^2}\ \angle -\tan^{-1}(X_C/R)}$$

$$= \left(\frac{R}{\sqrt{R^2 + X_C^2}}\right)\angle \tan^{-1}(X_C/R)$$

The magnitude response equation is described by

$$\frac{v_{out}}{v_{in}} = \frac{R}{\sqrt{R^2 + X_C^2}}$$

Examining the magnitude response equation, we note: At $f = 0$ Hz, X_C is infinity. Hence,

$$\frac{v_{out}}{v_{in}} = \frac{R}{\sqrt{R^2 + \infty^2}} = \frac{1}{\sqrt{\infty^2}} \approx 0$$

and $v_{out} = 0$ volts. The circuit does not allow a dc signal to pass through to the output.

At $f = \infty$ Hz, $X_C = 0$ ohms. Substituting 0 ohms for X_C in the transfer function equation gives

$$\frac{v_{out}}{v_{in}} = \frac{R}{\sqrt{R^2 + 0^2}} = \frac{R}{\sqrt{R^2}} = 1$$

and $v_{out} = v_{in}$.

The phase response is described by

$$\theta = \tan^{-1}\left(\frac{X_C}{R}\right)$$

Analyzing this function at $f = 0$ Hz and $f = \infty$ Hz yields the following results.

At $f = 0$ Hz, X_C is infinite. Hence,

$$\theta = \tan^{-1}\left(\frac{\infty}{R}\right)$$

$$\theta = +90°$$

At $f = \infty$ Hz, $X_C = 0$ ohms. Hence,

$$\theta = \tan^{-1}\frac{0}{R}$$

$$\theta = 0°$$

EXAMPLE 19.5

From the transfer function derived in Example 19.2, determine the magnitude and phase response.

Solution

The transfer function derived in Example 19.2 is

$$\text{TF} = \frac{v_{\text{out}}}{v_{\text{in}}} = \frac{R}{R + jX_L}$$

Converting R and $R + jX_L$ to polar form yields $R\angle 0°$ and $\sqrt{R^2 + X_L^2}$ $\angle\tan^{-1}(X_L/R)$. Substituting the polar values into the transfer function gives

$$\frac{v_{\text{out}}}{v_{\text{in}}} = \frac{R}{\sqrt{R^2 + X_L^2}\angle\tan^{-1}(X_L/R)}$$

$$= \frac{R}{\sqrt{R^2 + X_L^2}}\angle -\tan^{-1}(X_L/R)$$

Hence, the magnitude response is

$$\frac{v_{\text{out}}}{v_{\text{in}}} = \frac{R}{\sqrt{R^2 + X_L^2}}$$

and the phase response is

$$\theta = -\tan^{-1}\frac{X_L}{R}$$

Let us check the phase and magnitude response equations at $f = 0$ Hz and $f = \infty$ Hz.

At $f = 0$ Hz, $X_L = 0$ ohms. Hence, the value of the magnitude response equation is

$$\frac{v_{out}}{v_{in}} = \frac{R}{\sqrt{R^2 + 0^2}} = \frac{R}{\sqrt{R^2}} = 1$$

and the value of the phase response equation is

$$\theta = -\tan^{-1}\left(\frac{0}{R}\right) = 0°$$

At $f = \infty$ Hz, $X_L = \infty$ ohms. Hence, the value of the magnitude response equations is

$$\frac{v_{out}}{v_{in}} = \frac{R}{\sqrt{R^2 + \infty^2}} = \frac{R}{\infty} = 0$$

and the value of the phase response equation is

$$\theta = -\tan^{-1}\left(\frac{\infty}{R}\right) = -90°$$

Between $f = 0$ Hz and $f = \infty$ Hz, the transfer function changes from 1.0 to 0.0, and the phase angle between v_{in} and v_{out} changes from 0° to −90°.

19.2 Bandwidth and Phase Shift

Bandwidth is an important characteristic of many circuits and systems. For example, a general-purpose oscilloscope is commonly described in terms of its bandwidth. A Tektronix 2235 dual-trace oscilloscope has a bandwidth of 100 MHz.

bandwidth

Bandwidth is defined as the range of frequencies within which the circuit or system performance, with respect to some characteristic, falls within specified limits. Bandwidth is commonly defined at the frequencies where the response is 3 decibels less than the reference level. Consider the graph in Figure 19.6. The characteristic plotted is the gain of a circuit. The reference level is 10 decibels. The 3-dB points are the two frequencies where the gain is 3 dB less than the reference level, that is, 7 dB. The two frequencies are 40 Hz and 28 kHz. The bandwidth is calculated by taking the difference of the two frequencies

$$\text{BW} = 28 \text{ kHz} - 40 \text{ Hz} \simeq 28 \text{ kHz}$$

In some cases like the oscilloscope, the lower limit is dc. In this case, the bandwidth is equal to the upper 3-dB frequency. For the Tektronix 2235 oscilloscope, the operating range is dc to 100 MHz and the bandwidth is said to be 100 MHz.

−3-dB frequency

What is the significance of the −3-dB frequency on the frequency response graph? The **−3-dB frequency** is the frequency at which the

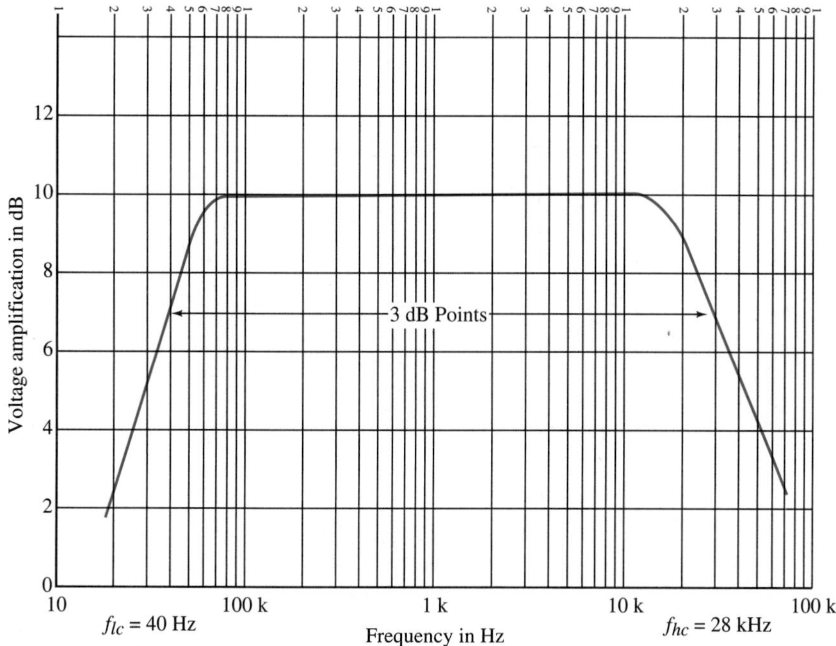

Figure 19.6 Frequency response graph showing bandwidth measurement.

power in the output signal is one-half the power in the pass band. By definition,

$$\text{dB} = 10 \log_{10}\left(\frac{P_{\text{out}}}{P_{\text{in}}}\right) = 20 \log_{10}\left(\frac{V_{\text{out}}}{V_{\text{in}}}\right)$$

Setting dB equal to -3 dB gives

$$\text{dB} = -3 \text{ dB} = 10 \log_{10}\left(\frac{P_{\text{out}}}{P_{\text{in}}}\right)$$

Dividing both sides of the equation by 10,

$$-0.3 = \log_{10}\left(\frac{P_{\text{out}}}{P_{\text{in}}}\right)$$

Raising both sides of the equation to a power of 10,

$$10^{-0.3} = 10^{(\log_{10}(P_{\text{out}}/P_{\text{in}}))}$$

$$0.5 = \frac{P_{\text{out}}}{P_{\text{in}}}$$

or

$$P_{\text{out}} = 0.5\, P_{\text{in}}$$

Hence, the -3-dB frequency on the frequency response curve represents the half-power frequency. There are normally two half-power frequencies, one low and one high.

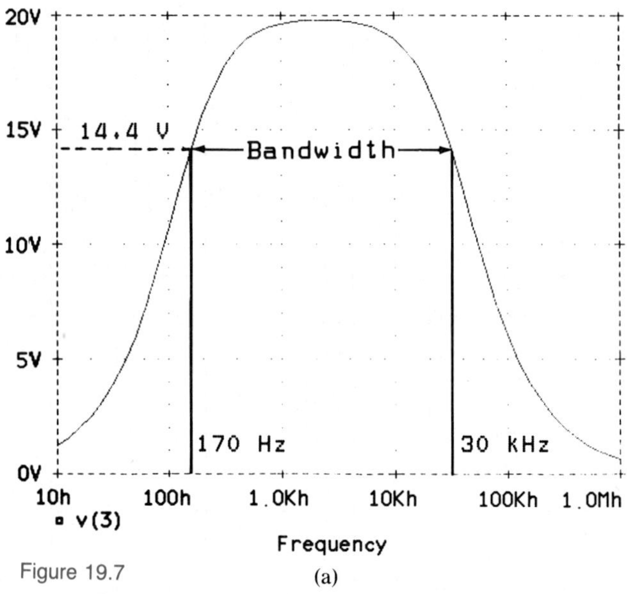

Figure 19.7 (a)

In terms of voltage,

$$\text{dB} = 20 \log_{10}\left(\frac{V_{out}}{V_{in}}\right)$$

At the -3-dB frequency,

$$-3 \text{ dB} = 20 \log_{10}\left(\frac{V_{out}}{V_{in}}\right)$$

Dividing both sides of the equation by 20 yields

$$-0.15 = \log_{10}\left(\frac{V_{out}}{V_{in}}\right)$$

and raising both sides of the equation to a power of 10 gives

$$0.707 = \frac{V_{out}}{V_{in}}$$

or

$$V_{out} = 0.707\ V_{in}$$

Hence, the output voltage at the -3-dB point is 0.707 of the input voltage.

Now let us apply the bandwidth definitions in the following examples.

EXAMPLE 19.6

Given the frequency response curve shown in Figure 19.7(a), determine the bandwidth. Assume that the input signal has an amplitude of 20 volts.

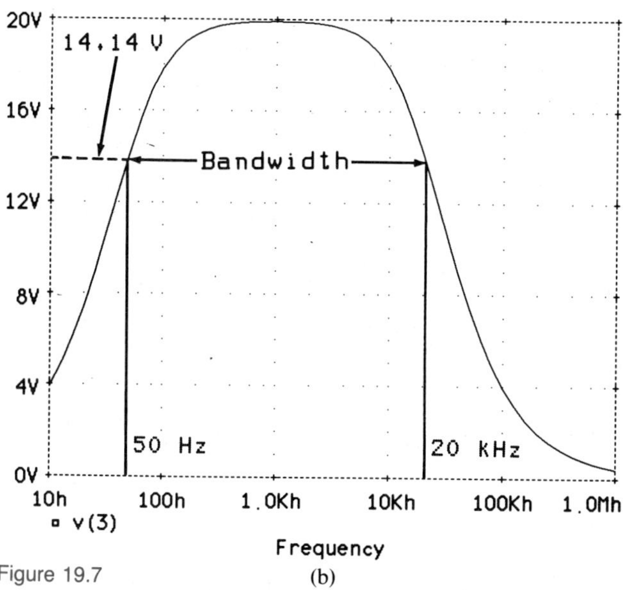

Figure 19.7 (b)

Solution

Since the graph is given in volts, we look for the frequencies where the output voltage is 0.707 of the input voltage, 0.707 of 20 volts, or 14.14 volts.

The two points are shown in Figure 19.7(b).

The bandwidth is

$$\text{BW} = 20 \text{ kHz} - 50 \text{ Hz} \simeq 20 \text{ kHz}$$

EXAMPLE 19.7

Given the frequency response curve shown in Figure 19.8, determine the bandwidth.

Solution

In this case, the gain is given in dB. Hence, we look for the frequencies where the response is 3 dB less than the gain in the pass band, or 21 dB.

Figure 19.8(b) shows the two −3-dB frequencies. The lower 3-dB frequency is 2 kHz and the upper 3-dB frequency is 300 kHz. Hence the bandwidth is

$$\text{BW} = 300 \text{ kHz} - 2 \text{ kHz} = 298 \text{ kHz}$$

The other parameter that is of interest when describing the frequency response of a circuit or systems is the **phase shift** between the input and output. Let us return to the high-pass *R–C* circuit and take a look at the phase change as a function of frequency. The phase angle for a high-pass *R–C* circuit is

phase shift

$$\theta = +\tan^{-1}(X_c/R)$$

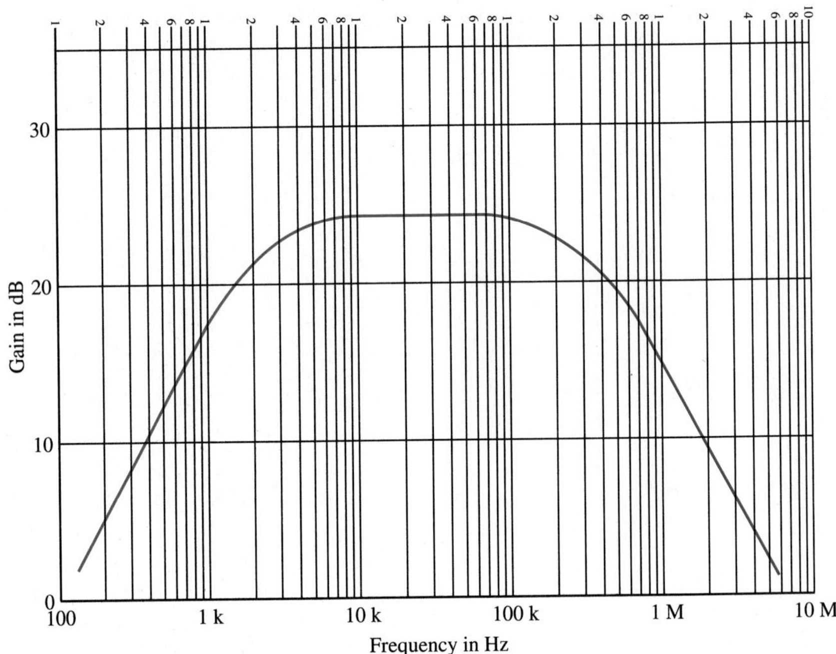

Figure 19.8

If $R = 10$ kilohms and $C = 0.01$ microfarad, calculating the phase angle at various points between 100 Hz and 100 kHz yields the data in Table 19.1.

TABLE 19.1 Phase Angle versus Frequency

High-Pass *R–C* Circuit	
Frequency	***Phase Angle***
100 Hz	86°
200 Hz	83°
500 Hz	73°
1 kHz	58°
2 kHz	38°
5 kHz	18°
10 kHz	9°
20 kHz	4.5°
50 kHz	1.8°
100 kHz	0.9°

Plotting these data produces the graph shown in Figure 19.9.

As the frequency increases, the phase angle decreases from 90° to 0°. At the 3-dB frequency or corner frequency, the phase angle is 45°.

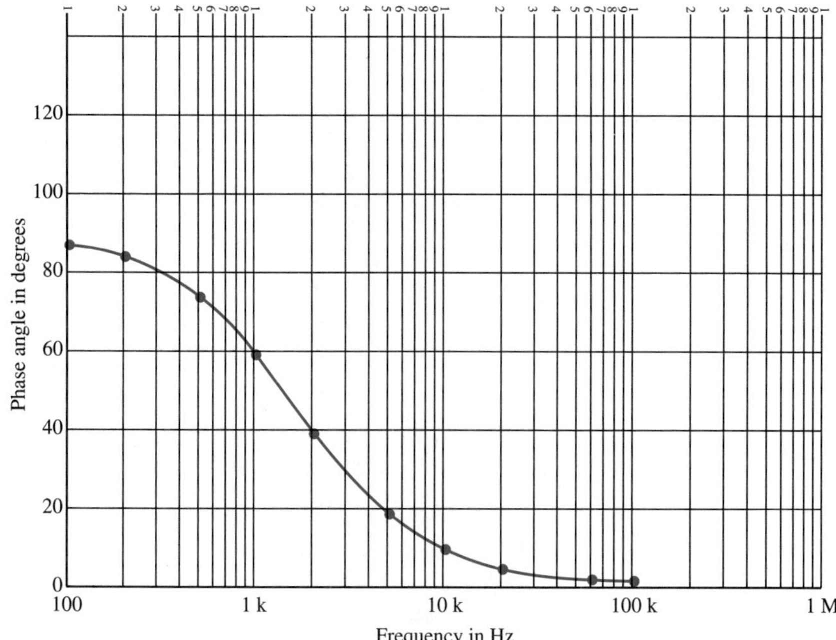

Figure 19.9 Graph of phase angle versus frequency.

19.3 Bode Plots

Rather than plotting the ratio of v_{out} and v_{in}, if the gain or attenuation of a circuit is plotted in dB versus frequency on a semilog plot, a special graph called a **Bode plot** is obtained. The advantage of the Bode plot is the ability to compress a wide range of values onto a linear vertical axis.

Bode plot

Consider the value listed in Table 19.2. The ratio, v_{out}/v_{in}, is the gain of a circuit or system. Note that the gain column varies from 1–100,000 while the dB column only ranges from 0–100 dB.

Also note that a factor of 10 increase in gain results in an increase of 20 dB. Likewise, a decrease in gain by a factor of 10 results in a decrease of 20 dB. A doubling of gain results in an increase of 6 dB, and a reduction in gain of 50 percent results in a decrease of 6 dB.

A Bode plot is actually a straight-line approximation of the actual frequency response curve. It is composed of the passband portion of the response, and the portion outside the passband.

To construct a Bode plot, use the following procedure:

1. Graph the constant dB line within the passband.
2. Graph the $20\log_{10} (f_c)/(f)$ line, where f_c is the corner or 3-dB frequency. The slope will be positive for the lower 3-dB frequency

TABLE 19.2

Gain, V_{out}/V_{in}	Gain in dB
1	0 dB
2	6 dB
5	14 dB
10	20 dB
20	26 dB
50	34 dB
100	40 dB
200	46 dB
500	54 dB
1,000	60 dB
2,000	66 dB
5,000	74 dB
10,000	80 dB
20,000	86 dB
50,000	94 dB
100,000	100 dB

and negative for the upper 3-dB frequency. The straight-line approximation goes through the point (f_c, 0 dB) with a slope of +20 dB per decade or −20 dB per decade.

3. Correct for the difference between the straight-line approximation and the actual frequency response curve using the following correction factors:

Correction Factors	
$f_c/2$	−1 dB
f_c	−3 dB
$2\ f_c$	−1 dB

4. Draw the complete frequency response curve.

Let us use our high-pass *R–C* circuit as an example. The corner frequency of the circuit is f_c. In the passband of the circuit, v_{out} equals v_{in} and the gain is 1. Converting this gain of 1 into dB yields 0 dB. This value is plotted for frequencies greater than f_c. The *R–C* circuit response then rolls off for frequencies less than f_c and the gain is less than 1. The approximation for the response for $f < f_c$ is a line passing through f_c on the *x*-axis with a slope of 20 dB per decade or 6 dB per octave. Hence, we can plot either −6 dB at $f_c/2$ or −20 dB at $f_c/10$. Since two points define a line, we can draw a straight line through the two points. This is shown in Figure 19.10.

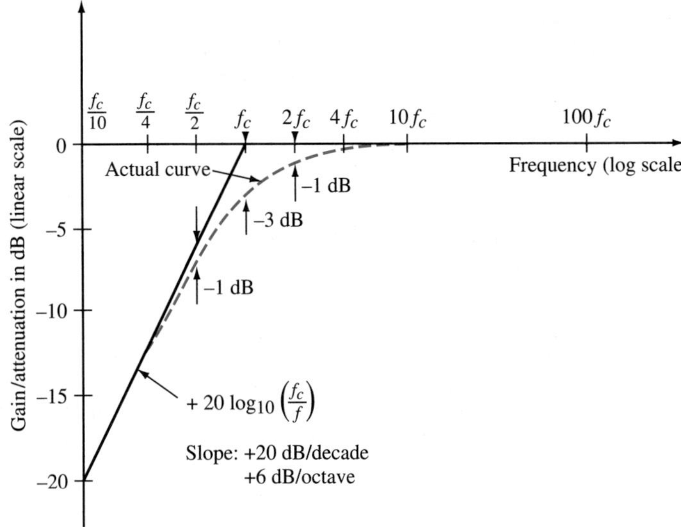

Figure 19.10 A Bode plot of a high-pass circuit.

To apply the correction, at $1/2\,f_c$ and at $2\,f_c$, we plot a point 1 dB below the straight-line approximation yielding -1 dB at $1/2\,f_c$ and -7 dB at $2\,f_c$. At the corner frequency, the correction factor is -3 dB. Since the straight-line approximation is 0 dB, subtracting 3 dB yields a point at -3 dB.

The actual response follows the straight-line approximation except for the region around the corner frequency.

Let us apply these techniques to determine the Bode plot for the following circuit examples.

EXAMPLE 19.8

Draw a Bode plot to describe the frequency response for the low-pass filter shown in Figure 19.11.

Solution

First, we need to find the corner frequency, f_c.

$$f_c = \frac{1}{2\pi RC}$$

$$= \frac{1}{2\pi(15.9\ \text{k}\Omega)(0.01\ \mu\text{F})}$$

$$f_c = 1000\ \text{Hz}$$

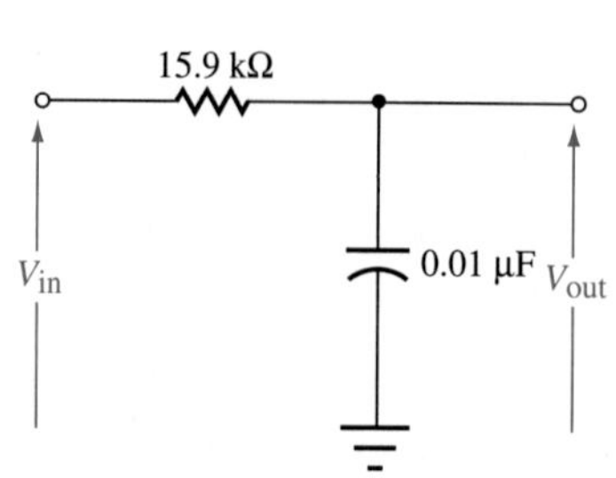

Figure 19.11

Step 1. In the passband for the low-pass circuit, v_{out} equals v_{in} and the gain is unity. Hence, the gain can be expressed as 0 dB. The straight-line approximation for frequencies less than f_c is a horizontal line representing 0 dB.

Step 2. For frequencies greater than f_c, a line is drawn through (f_c, 0 dB) with a slope of −20 dB per decade. This line should pass through the point (10 kHz, −20 dB).

Step 3. Now we can apply the correction factors for frequencies around the corner frequency.

Frequency		Correction	Actual dB
$f_c/2$	= 500 Hz	−1 dB	−1 dB
f_c	= 1000 Hz	−3 dB	−3 dB
$2\,f_c$	= 2000 Hz	−1 dB	−7 dB

Step 4. Drawing the complete response curve is a matter of following the straight-line approximations except for the region around the corner frequency where the curve passes through the correction points. The complete graph is shown in Figure 19.12.

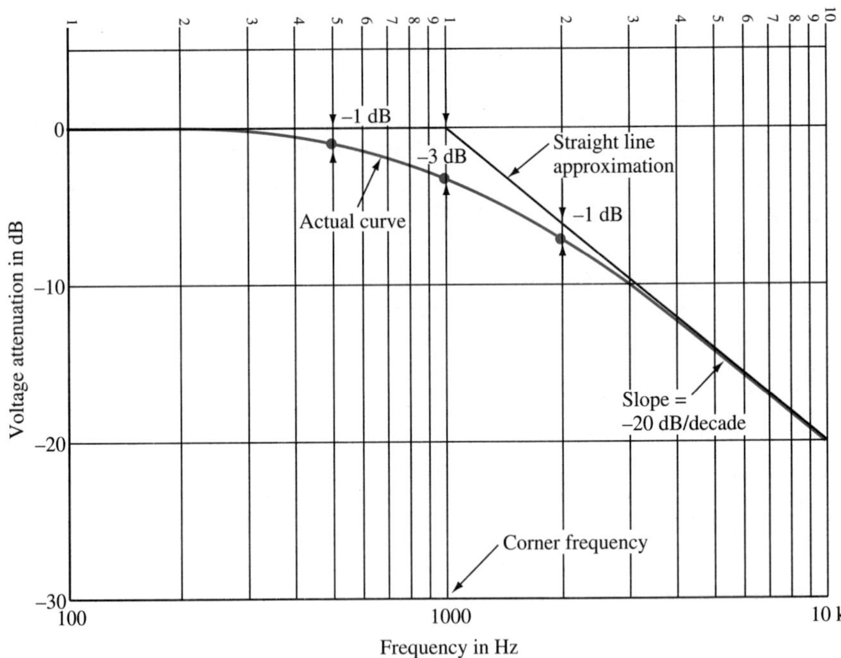

Figure 19.12 A Bode plot for the low-pass filter.

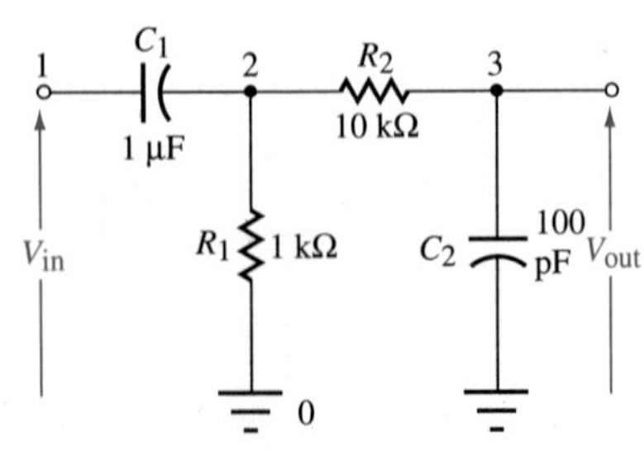

Figure 19.13

EXAMPLE 19.9

Draw the Bode plot for the circuit shown in Figure 19.13.

Solution

We separate the circuit and the analysis into two parts—the low-frequency response and the high-frequency response. This is a reasonable step because each circuit's components are such that they have a minimum effect on the other circuit's performance.

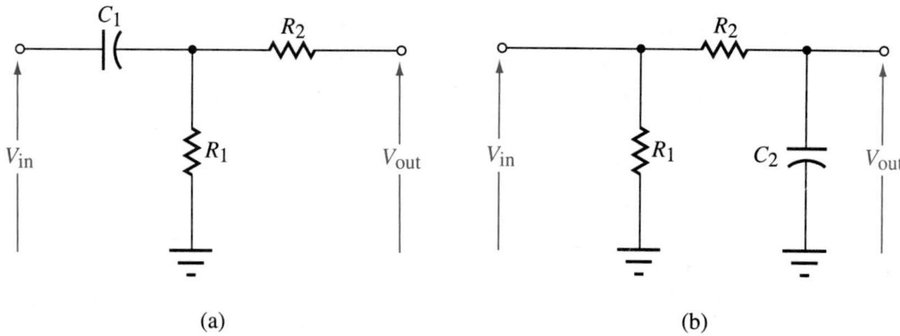

Figure 19.14 Equivalent circuits. (a) Low-frequency equivalent circuit. (b) High-frequency equivalent circuit.

The low-frequency equivalent circuit is shown in Figure 19.14(a). Capacitor C_2 is replaced by an open circuit since it is much smaller in value than C_1, and has little effect at low frequencies.

The circuit is a high-pass circuit with the corner frequency determined by R_1 and C_1. The transfer function is

$$\frac{v_{out}}{v_{in}} = \frac{R_1}{R_1 - jX_{C1}}$$

The corner frequency can be found by equating R_1 and X_{C1} and solving for f_C, the corner frequency.

$$f_{C1} = \frac{1}{2\pi R_1 C_1}$$

The high-frequency equivalent circuit is shown in Figure 19.14(b). Capacitor C_1 is essentially a short circuit at high frequencies.

The transfer function representing the high-frequency response is

$$\frac{v_{out}}{v_{in}} = \frac{-jX_{C2}}{R_2 - jX_{C2}}$$

and the corner frequency for the high-frequency roll-off is

$$f_{C2} = \frac{1}{2\pi R_2 C_2}$$

Substituting the circuit values into the f_{C1} and f_{C2} equations yields

$$f_{C1} = \frac{1}{2\pi(1\ \text{k}\Omega)(1\ \mu\text{F})} = 159\ \text{Hz}$$

$$f_{C2} = \frac{1}{2\pi(10\ \text{k}\Omega)(100\ \text{pF})} = 159\ \text{kHz}$$

To graph the complete response, we need five-cycle semilog graph paper, since we want to plot frequency on the horizontal axis and the frequency range will extend from 10 Hz–1 MHz.

In the passband, $v_{out} = v_{in}$ so the output will be represented by a 0-dB level. Above f_{C2} and below f_{C1}, the response curve will roll off at -20 dB/decade.

We begin by drawing the straight-line approximation, as shown in Figure 19.15. Then, we can apply the corrections around each corner frequency to obtain the actual response, as shown in Figure 19.15.

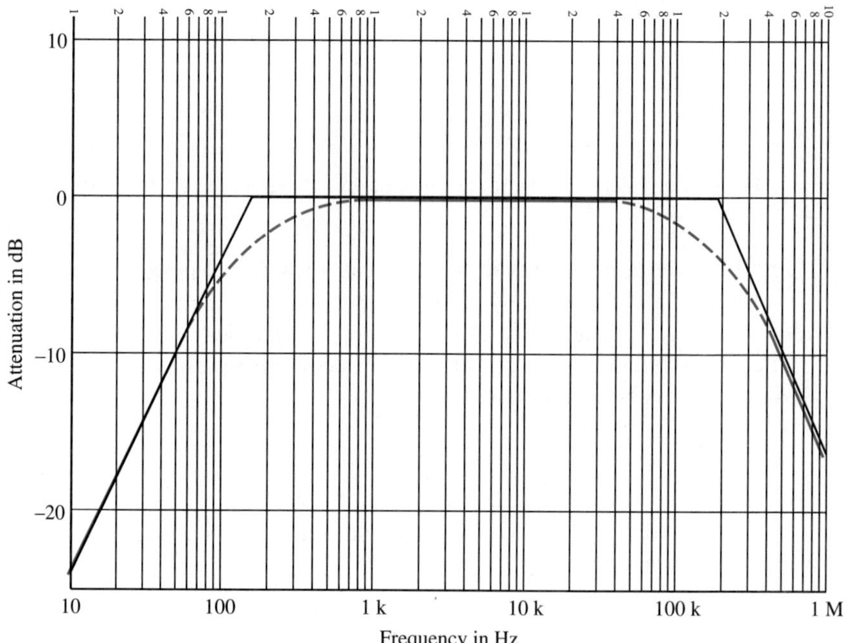

Figure 19.15 Frequency response curve.

The ability to draw, read, and interpret frequency response curves is an important skill for the technician and engineer. Many device data sheets include frequency response curves that describe the operating characteristics of the device. Instruments and systems will also use frequency response curves to convey operating characteristics.

In the following section, we learn how to use SPICE to perform circuit simulations and obtain Bode plots using the computer to perform the necessary computations.

S 19.4 Using SPICE for Frequency Response Analysis

Using PSpice to obtain the magnitude and phase response graphs for a circuit is a four-step process.

Step 1. The first step is to describe the circuit connections as we have done in previous examples. Again, remember that GROUND is Node 0. The circuit must contain at least one ac source.

Step 2. The .AC statement must be used to specify the type of sweep desired, the number of points that will be calculated, and the start and stop frequencies.

Step 3. A PSpice simulation is run.
Step 4. To view the results, .PROBE is used to format the displays of the output waveforms.

Your PSpice file should look like this:

```
TITLE line
Device lines
. . .
.AC . . . .
.PROBE
.END
```

The .AC statement causes PSpice to calculate the frequency response of a circuit over a range of frequencies. The general form of the .AC statement is

```
.AC <sweep type> <(points) value> <(start frequency)
        value> <(stop frequency) value>
```

There are three types of sweeps that can be used. They are linear sweep, sweep by octaves, and sweep by decades. A brief description of these three sweeps is:

- LIN — Linear sweep. The frequency is swept linearly from the starting frequency to the ending frequency. <(points value> is the total number of points in the sweep.
- OCT — Sweep by octaves. The frequency is swept logarithmically by octaves. <(points) value> is the number of points per octave.
- DEC — Sweep by decades. The frequency is swept logarithmically by decades. <(points) value> is the number of points per decade.

There are several things to note when using the .AC statement. First, one of the foregoing three sweep types must be specified. Second, the start frequency must be less than the ending frequency and both must be greater than 0. The start and end frequency, however, can be the same, producing a sweep that consists of only one point.

Let us now consider a couple of examples to illustrate the power that PSpice brings to circuit analysis.

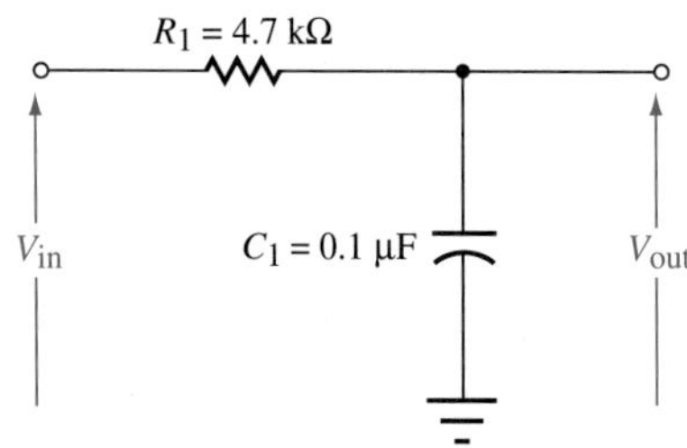

Figure 19.16

EXAMPLE 19.10

Use PSpice to determine the frequency response of the circuit shown in Figure 19.16.

Solution

We begin by numbering the nodes in the circuit, as shown in Figure 19.17, making sure that GROUND is assigned the node value of 0. We will title our file "Example 19.10 R-C circuit."

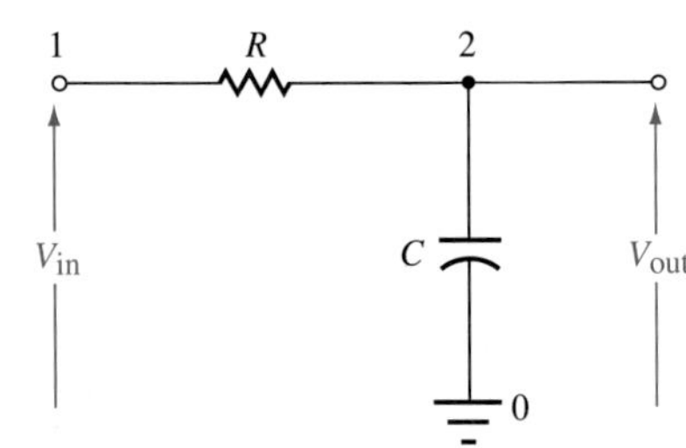

Figure 19.17

The circuit description portion of our file will look like the following:

```
EXAMPLE 19.10 R-C CIRCUIT
R1     1   2    4.7K
C1     2   0    0.1U
VIN    1   0    AC    10.0    0.0
```

The .AC statement will have the following parameters:

```
<sweep type>                         DEC
<(points) value>                      10
<(start frequency) value>             10    Hz
<(stop frequency) value>            1000    Hz
```

Hence, the .AC state will look like:

```
.AC    DEC    10    10    1000
```

The last type statements will be:

```
.PROBE
.END
```

Now, let us enter this file and run it through the PSpice simulator.

The output graphs are shown in Figure 19.18. The upper graph shows the gain versus frequency characteristic and the lower graph shows the phase angle versus frequency characteristic.

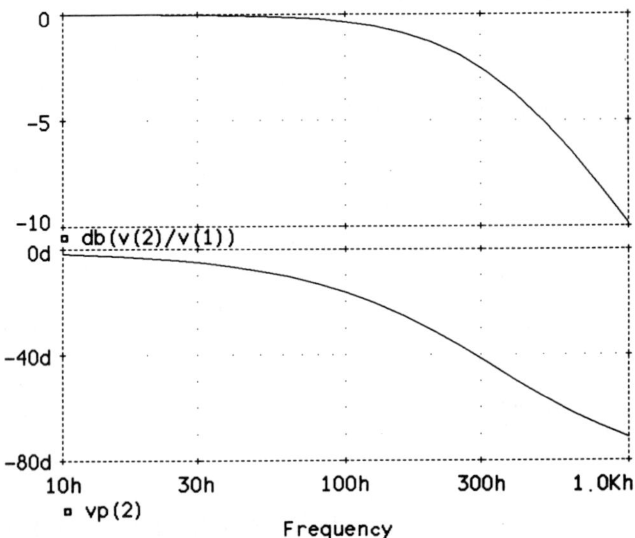

Figure 19.18 PSpice output using PROBE.

EXAMPLE 19.11

Use PSpice to determine the frequency response of the circuit shown in Figure 19.19.

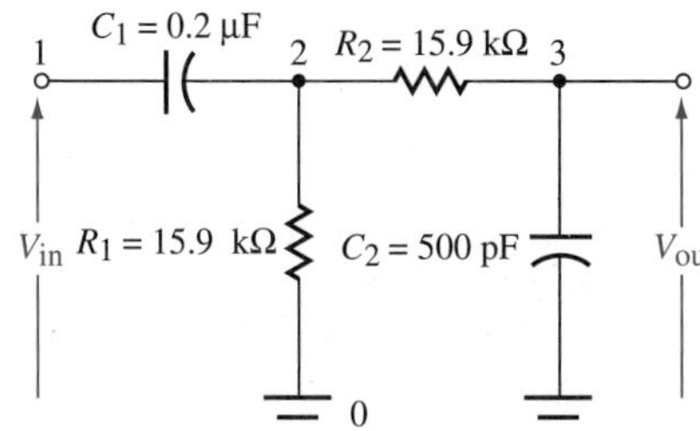

Figure 19.19

Solution

The circuit contains four nodes, which are numbered as shown in Figure 19.19. The file can then be written to perform an ac analysis, and graphical outputs can be obtained using .PROBE.

The PSpice file is:

```
EXAMPLE 19.11--FIGURE 19.19
VS    1    0    AC      20.0    0.0
R1    2    0    15.9K
R2    2    3    15.9K
C1    1    2    0.2U
C2    3    0    500P
.AC   DEC  10   10      1MEG
.PROBE
.END
```

The Bode plot produced by PSpice is shown in Figure 19.20. The low-

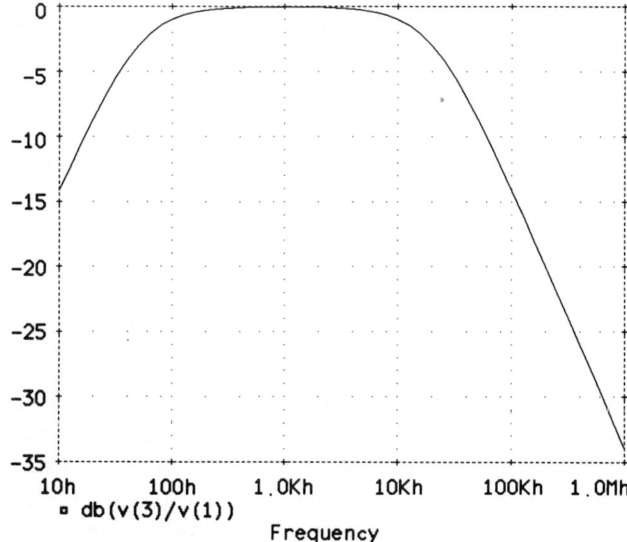

Figure 19.20 Bode plot.

frequency 3-dB point, as read from the graph, is 50 Hz and the upper-frequency 3-dB point is 20 kHz.

Checking these values with our calculations,

$$f_{C1} = \frac{1}{2\pi(15.9\ \text{k}\Omega)(0.2\ \mu\text{F})} = 50\ \text{Hz}$$

$$f_{C2} = \frac{1}{2\pi(15.9\ \text{k}\Omega)(500\ \text{pF})} = 20\ \text{kHz}$$

These calculated values agree with the corner frequencies determined from the graphs generated by PSpice.

Figure 19.21 was also generated by PSpice. This graph shows the phase shift between v_{in} and v_{out} as a function of frequency. The phase angle changes from $+90°$ at low frequencies to zero in the pass band, to $-90°$ above the high-frequency roll-off point.

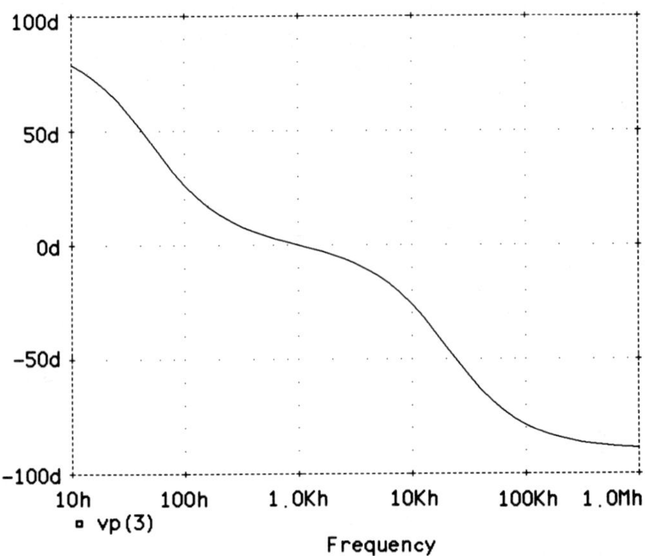

Figure 19.21 Phase plot.

PSpice files are easy to create, and with the graphical output capability built into PSpice, the frequency response of circuits can be obtained quickly and efficiently.

19.5 Resonance

So far in this chapter, we have only considered R–C and R–L circuits. In this section, we investigate the behavior of circuits containing all three circuit elements. By combining R, L, and C in different configurations, we can create a variety of **bandpass** and **band-reject** circuits.

bandpass

band-reject

There are three different regions of operation. Since the R–L–C circuit contains both inductance and capacitance, there will be a range of frequencies where the inductance will dominate, and the impedance of the circuit is said to be inductive. There will also be a range of frequencies where the capacitance will dominate and the circuit is said to be capacitive. There is also a frequency where the inductive and capacitive reactances exactly cancel each other, and the circuit appears totally resistive. This frequency where the circuit appears totally resistive is called the resonance frequency.

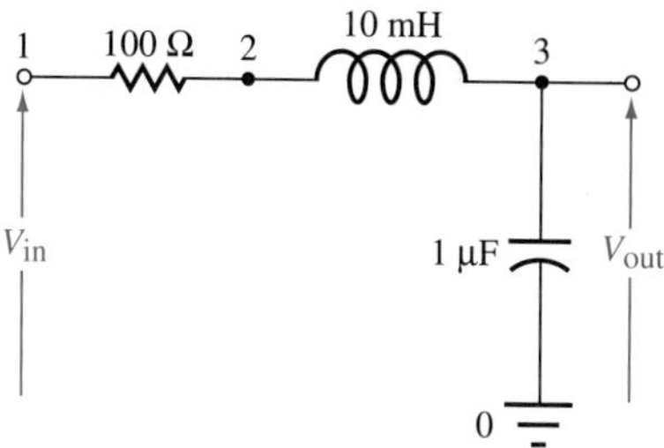

Figure 19.22 A series resonant circuit.

Let us examine the series *R–L–C* circuit shown in Figure 19.22. We will assume that the inductor and capacitor are ideal components. As frequency varies, the impedance of the inductor and capacitor varies. The impedance of the resistor is constant over the range where resonance occurs. The inductive reactance, jX_L, increases linearly with frequency. The curve for the capacitive reactance, on the other hand, is an equilateral parabola, asymptotic to the horizontal frequency axis at high frequencies, and asymptotic to the vertical reactance at low frequencies. Plotting these impedance relationships on the same set of axes gives the results shown in Figure 19.23.

Note that for frequencies below the resonant frequency, the total impedance is capacitive. Above the resonant frequency, the total impedance of the circuit is inductive. At the resonant frequency, the total impedance is totally resistive.

resonant frequency

The **resonant frequency** is that frequency where the inductive reactance equals the capacitive reactance. Equating the two impedances gives

$$X_L = X_C$$

$$2\pi f_r L = \frac{1}{2\pi f_r C}$$

Solving for the resonant frequency, f_r, gives

$$f_r^2 = \frac{1}{4\pi^2 LC}$$

or

$$f_r = \frac{1}{2\pi\sqrt{LC}} \qquad \textbf{(19.3)}$$

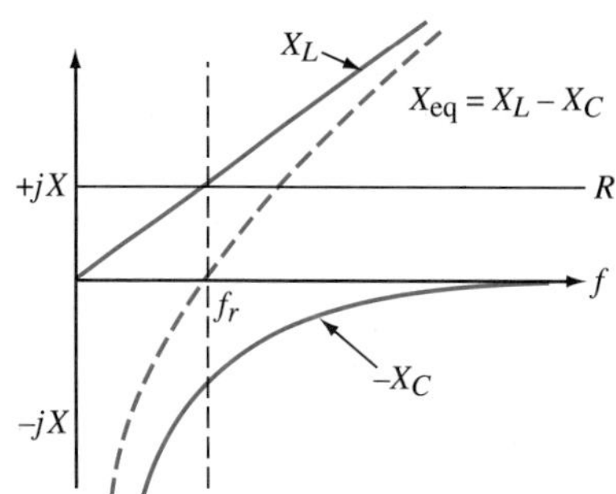

Figure 19.23 An impedance graph.

The impedance of the series *R–L–C* circuit is described mathematically as

$$Z = R + jX_L - jX_C$$

The magnitude of the impedance can be computed using the formula

$$|Z| = \sqrt{R^2 + (X_L - X_C)^2}$$

Inspecting this formula for Z yields the following observations:

1. The impedance is minimum at the resonant frequency, when the reactance term goes to zero.
2. The graph of Z versus frequency is asymmetrical since X_L is linear while X_C is nonlinear. Below f_r, the capacitive reactance dominates, producing an asymptotic, while above f_r, the curve follows the straight line of X_L.

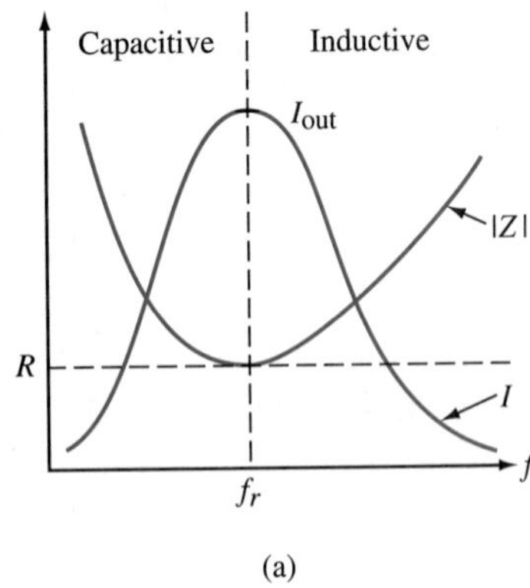

(a)

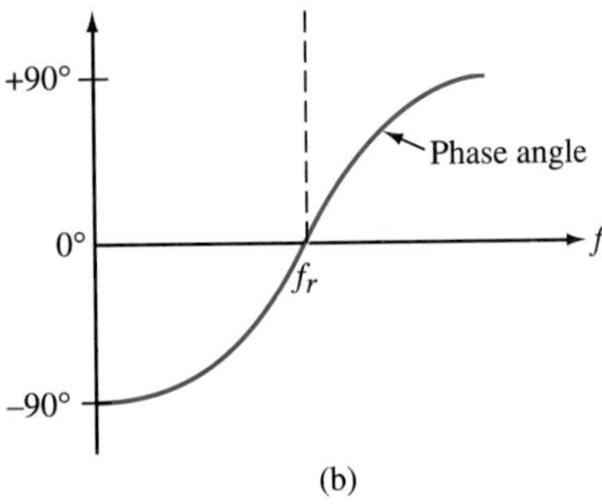

(b)

Figure 19.24 Impedance and phase relationships.

Since the impedance is minimum at the resonant frequency and the current is inversely related to impedance, the current reaches a maximum at f_r, the resonant frequency. These relationships are shown in Figure 19.24.

At the resonant frequency, the current is in phase with the supply voltage. Below the resonant frequency, the capacitive nature of the circuit causes the current to lead the supply voltage, and above f_r the inductive nature of the circuit causes the current to lag behind the supply voltage.

The phase angle must vary from $-90°$ to $+90°$ as frequency increases. This is based on the following observations:

1. At very low frequencies, the capacitive reactance is almost infinity, and the inductive reactance is very small. The capacitive reactance dominates the impedance equation and produces a phase angle of $-90°$.
2. As frequency increases, the capacitive reactance decreases, and the inductive reactance increases. When X_C decreases to the point where it equals R, the phase angle (defined by $\tan^{-1}(X_r/R)$ equals $-45°$.
3. When $f = f_r$, X_r is zero, and the phase angle is 0°.
4. Above f_r, the circuit is inductive, resulting in a positive phase angle.
5. At very high frequencies above f_r, the inductive reactance dominates and the phase angle is $+90°$.

The variation of phase angle with frequency is shown in Figure 19.24(b).

Since R–L–C circuits can form bandpass or band-reject circuits, the selectivity of the circuit is of interest. The selectivity is related to the steepness of the sides of the peak in a bandpass circuit or the valley in a band-reject circuit. A measure of the selectivity is the quality factor, Q_r, of the circuit. Quality factor is defined as the ratio of the reactive power in the inductance or capacitance to the real power at resonance. The higher the Q, the more selective the circuit.

$$Q_r = \frac{Q_L}{P} \tag{19.4}$$

where Q_L is the reactive power stored in either the coil or the capacitor, in vars; P is the power dissipated in the resistor, in watts, at resonance; and Q_r is the Q of the circuit at f_r.

Do not be confused by the use of Q for two different quantities in Equation 19.4. Q_r is a ratio (reactive power over real power) where Q_L and Q_C are reactive power and have the units of volt-ampere reactive (vars).

Af f_r, $X_L = X_C$, so

$$Q_r = \frac{I^2X_L}{I^2R} = \frac{\omega_r L}{R} = \frac{1}{R(\omega_r C)}$$

Solving the first equality for ω_r yields

$$\omega_r = \frac{Q_r R}{L}$$

Substituting this expression into the previous equation gives

$$Q_r = \frac{1}{R\left(\frac{Q_r R}{L}\right)C}$$

$$Q_r^2 = \frac{1}{R^2}\frac{L}{C}$$

$$Q_r = \frac{1}{R}\sqrt{\frac{L}{C}} \tag{19.5}$$

Equation 19.5 indicates the following properties of series R–L–C circuits:

1. The higher the resistance, the lower the resonant Q_r, and vice versa.
2. The higher the L/C ratio, the higher the circuit resonant Q_r, and the more selective the circuit.

These properties are illustrated in Figure 19.25.

And finally, the bandwidth of a series R–L–C circuit is found from the relation

$$\text{bandwidth} = f_2 - f_1 = \frac{f_r}{Q_r}$$

where f_1 is the lower 3-dB frequency, in hertz; f_2 is the upper 3-dB frequency, in hertz; f_r is the resonant frequency, in hertz; and Q_r is the Q of the circuit.

Since the Q of the circuit is defined as X_L/R,

$$\text{bandwidth} = \frac{f_r R}{X_L} = \frac{R}{2\pi L}$$

For essentially all resonant circuits, the upper and lower bandwidth frequencies (f_1 and f_2) are symmetrical around the resonant frequency (f_r).

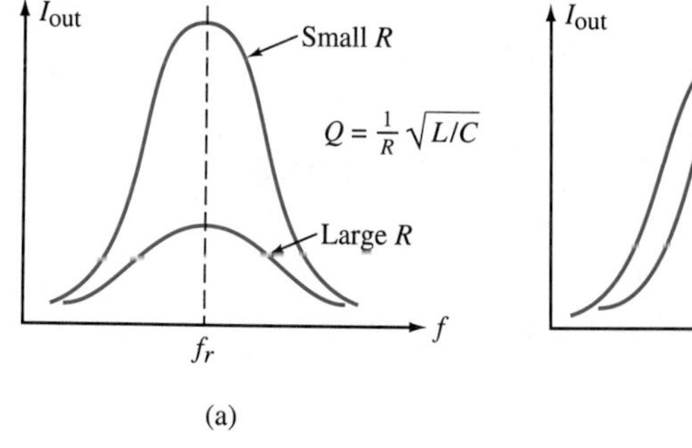

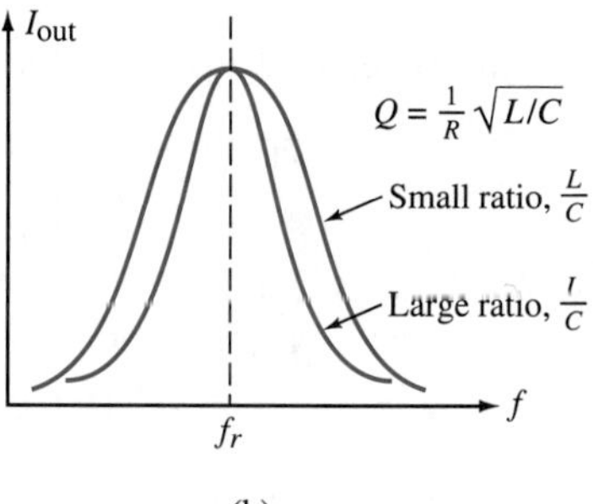

Figure 19.25 Current as a function of frequency.

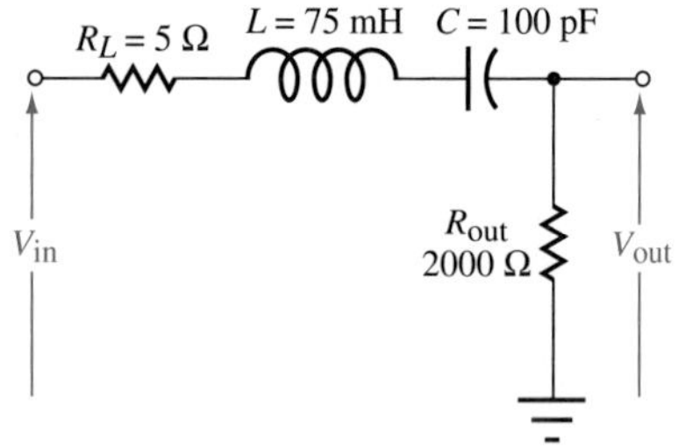

Figure 19.26

EXAMPLE 19.12

For the circuit shown in Figure 19.26, determine the following quantities:

a. the resonant frequency, f_r.
b. the Q of the circuit at the resonant frequency.
c. the bandwidth of the circuit.
d. the output voltage, v_{out}, at the resonant frequency.
e. use PSpice to graph the frequency response, both magnitude and phase, for the circuit.

Solution

a. The resonant frequency occurs when X_L equals X_C. This occurs at

$$f_r = \frac{1}{2\pi\sqrt{LC}}$$

$$= \frac{1}{2\pi\sqrt{(75\text{ mH})(100\text{ pF})}}$$

$$f_r = 58.1\text{ kHz}$$

b. The Q of the circuit is

$$Q_r = \frac{2\pi f_r L}{R_T}$$

$$= \frac{2\pi(58.1\text{ kHz})(75\text{ mH})}{(5\ \Omega + 2000\ \Omega)}$$

$$Q = 13.7$$

c. The bandwidth is calculated from the resonant frequency and the Q of the circuit.

$$\text{BW} = \frac{f_r}{Q_r}$$

$$= \frac{58.1\text{ kHz}}{13.7}$$

$$\text{BW} = 4.24\text{ kHz}$$

The lower and upper 3-dB frequencies are

$$f_{lower} = f_r - (\text{BW}/2)$$

$$= 58.1\text{ kHz} - (4.24\text{ kHz}/2)$$

$$f_{lower} = 55.98\text{ kHz}$$

$$f_{upper} = f_r + (\text{BW}/2)$$

$$= 58.1\text{ kHz} + (4.24\text{ kHz}/2)$$

$$f_{upper} = 60.22\text{ kHz}$$

d. The output voltage at the resonant frequency is

$$\upsilon_{out} = \left(\frac{R_{out}}{R_L + R_{out}}\right)V_{in}$$

$$= \left(\frac{2000\ \Omega}{5\ \Omega + 2000\ \Omega}\right)(10\ \angle 0°\ \text{V})$$

$$\upsilon_{out} = 9.97\ \text{V}\ \angle 0°$$

e. The circuit node assignments are shown in Figure 19.27(a).

Using the node assignments shown in Figure 19.27(a), the PSpice circuit description shown in Figure 19.27(b) is written and is used to perform the PSpice analysis that includes both magnitude and phase response.

1 2 3 4

$L_1 = 75$ mH

$R_L = 5\ \Omega$

$C_1 = 100$ pF

$V_{in} = 10\angle 0°$ V

$R_{out} = 2$ kΩ

$V_{out} = V(4)$

0

(a)

```
EXAMPLE 19.12 - SERIES RLC CIRCUIT
****      CIRCUIT DESCRIPTION
******************************************************

VIN   1   0   AC    10.0   0.0
RL    1   2   5
L1    2   3   75M
C1    3   4   100P
RO    4   0   2K
.AC DEC 200   40K   80K
.PRINT AC     VM(4)  VP(4)
.PROBE
.END
```

(b)

Figure 19.27 (a) Circuit node assignments. (b) PSpice circuit description.

The magnitude response is shown in Figure 19.28(a). The response curve peaks at 58 kHz, and the magnitude of the output voltage is almost 10 volts. One of the limitations of PSpice is that one of the calculation

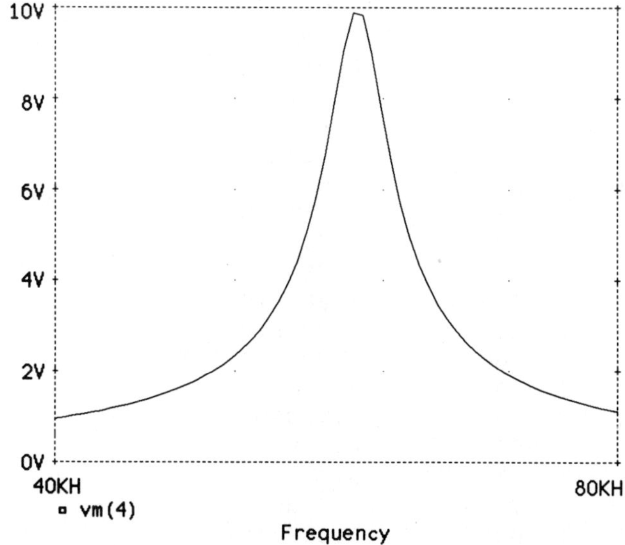

(a)

Figure 19.28 Magnitude and phase response curve produced by PROBE. (*Continued* on p. 812)

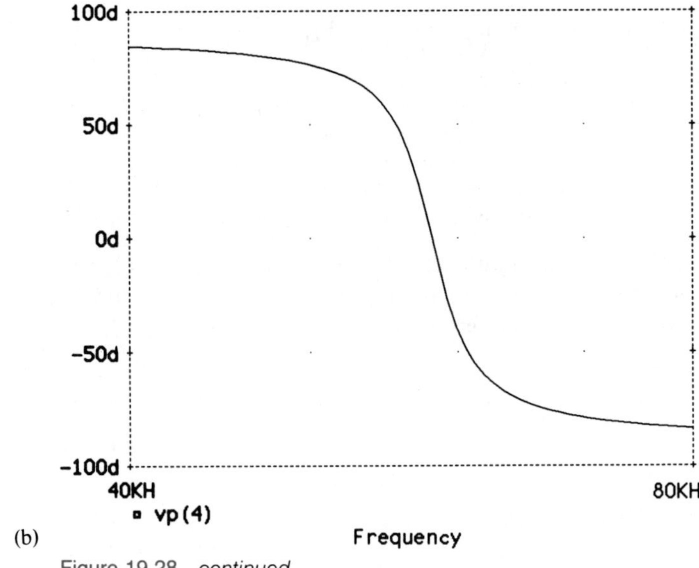

(b)

Figure 19.28 *continued*

points may not fall exactly on the resonant frequency. Examining the .OUT file, a portion of the listing is given in Figure 19.29, which shows that calculations were performed at 57.82 kHz and at 58.49 kHz. The voltages at these frequencies are 9.878 and 9.827 volts, respectively. Hence, the graph produced by PSpice does not capture the peak voltage of the resonance curve.

```
EXAMPLE 19.12--SERIES RLC CIRCUIT
**** AC ANALYSIS          TEMPERATURE = 27.000 DEG C
*******************************************************

FREQ          VM(4)          VP(4)
4.000E+04     9.508E-01      8.453E+01
4.046E+04     9.821E-01      8.435E+01
4.093E+04     1.015E+00      8.416E+01
4.141E+04     1.051E+00      8.395E+01
4.189E+04     1.089E+00      8.373E+01
4.237E+04     1.129E+00      8.350E+01
4.286E+04     1.173E+00      8.325E+01
4.336E+04     1.220E+00      8.298E+01
4.386E+04     1.270E+00      8.269E+01
4.437E+04     1.325E+00      8.237E+01
4.488E+04     1.384E+00      8.203E+01
4.540E+04     1.448E+00      8.165E+01
4.593E+04     1.519E+00      8.124E+01
4.646E+04     1.597E+00      8.079E+01
4.700E+04     1.682E+00      8.029E+01
4.754E+04     1.777E+00      7.974E+01
4.809E+04     1.883E+00      7.912E+01
4.865E+04     2.001E+00      7.843E+01
4.921E+04     2.135E+00      7.764E+01
```

Figure 19.29 Output file listing.

```
4.978E+04    2.287E+00     7.675E+01
5.036E+04    2.461E+00     7.572E+01
5.094E+04    2.663E+00     7.452E+01
5.153E+04    2.898E+00     7.311E+01
5.213E+04    3.177E+00     7.143E+01
5.273E+04    3.510E+00     6.940E+01
5.334E+04    3.913E+00     6.690E+01
5.396E+04    4.410E+00     6.377E+01
5.458E+04    5.027E+00     5.974E+01
5.522E+04    5.801E+00     5.444E+01
5.585E+04    6.763E+00     4.731E+01
5.650E+04    7.906E+00     3.757E+01
5.716E+04    9.080E+00     2.445E+01
5.782E+04    9.878E+00     7.983E+00
5.849E+04    9.827E+00    -9.886E+00
5.916E+04    8.962E+00    -2.605E+01
5.985E+04    7.776E+00    -3.878E+01
6.054E+04    6.650E+00    -4.819E+01
6.124E+04    5.709E+00    -5.509E+01
6.195E+04    4.953E+00    -6.023E+01
6.267E+04    4.351E+00    -6.414E+01
6.340E+04    3.866E+00    -6.720E+01
6.413E+04    3.470E+00    -6.964E+01
6.487E+04    3.144E+00    -7.163E+01
6.562E+04    2.871E+00    -7.327E+01
6.638E+04    2.639E+00    -7.466E+01
6.715E+04    2.441E+00    -7.584E+01
6.793E+04    2.269E+00    -7.605E+01
6.872E+04    2.120E+00    -7.773E+01
6.951E+04    1.988E+00    -7.851E+01
7.032E+04    1.871E+00    -7.919E+01
7.113E+04    1.766E+00    -7.980E+01
7.195E+04    1.672E+00    -8.035E+01
7.279E+04    1.588E+00    -8.084E+01
7.363E+04    1.511E+00    -8.129E+01
7.448E+04    1.441E+00    -8.169E+01
7.535E+04    1.377E+00    -8.206E+01
7.622E+04    1.319E+00    -8.240E+01
7.710E+04    1.264E+00    -8.272E+01
7.799E+04    1.214E+00    -8.301E+01
7.890E+04    1.168E+00    -8.328E+01
7.981E+04    1.125E+00    -8.353E+01
8.073E+04    1.085E+00    -8.376E+01
```

The 3-dB frequencies are approximately 56 kHz and 60 kHz. This agrees with our calculations in part c.

The frequency response curve produced by PSpice is shown in Figure 19.28(b). At low frequencies, the phase angle is $+84°$. The phase angle decreases to zero at the resonant frequency. At frequencies greater than f_r, the phase angle is negative and approaches $-90°$.

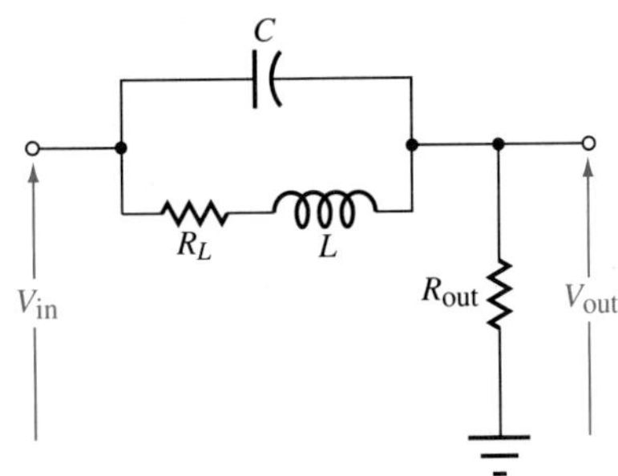

Figure 19.30 A band-reject circuit.

Bandpass and band-reject circuits can also be constructed using parallel R–L–C configurations. Consider the parallel R–L–C circuit shown in Figure 19.30.

Applying our knowledge of resistors, inductors, and capacitors, we know that the circuit will pass low-frequency signals, since the inductive branch of the parallel circuit appears as a short in low-frequency signals.

At high frequencies, the capacitive branch of the parallel circuit appears as a low-impedance path for the input signal. Hence, high-frequency signals pass through the circuit without attenuation.

In between these two extremes in frequency, the impedance of the parallel circuit is greater. The impedance can be calculated using the following relationship:

$$\mathbf{Z}_p = \frac{\mathbf{Z}_L \mathbf{Z}_C}{\mathbf{Z}_L + Z_C}$$

where $\mathbf{Z}_L = R_L + jX_L$ and $\mathbf{Z}_C = -jX_C$.

For given values of R, L, and C, the impedance varies with frequency, as shown in Figure 19.31. The impedance of the parallel circuit begins at R_L, rises to a peak, and then decreases to zero at high frequencies.

The peak in Figure 19.31 occurs at the resonant frequency. The resonant frequency for parallel circuits using high Q coils ($Q > 10$) can be calculated using the following relationship:

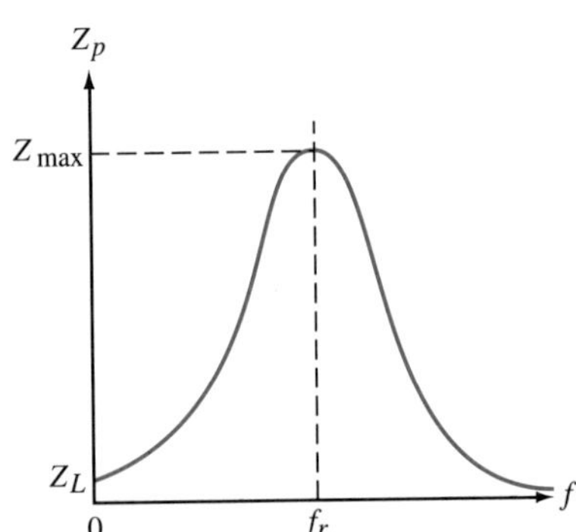

Figure 19.31 Impedance as a function of frequency.

$$f_r = \frac{1}{2\pi\sqrt{LC}}$$

If Z_p varies with frequency, as shown in Figure 19.31, then the frequency response curve must exhibit the following characteristics:

1. A low frequencies, $v_{out}/v_{in} = 1$ since

$$\frac{v_{out}}{v_{in}} = \frac{R_{out}}{R_{out} + Z_p} = \frac{R_{out}}{R_{out} + 0} = 1$$

2. At high frequencies, $v_{out}/v_{in} = 1$ since

$$\frac{v_{out}}{v_{in}} = \frac{R_{out}}{R_{out} + Z_p} = \frac{R_{out}}{R_{out} + 0} = 1$$

3. At f_r, $X_L = X_C$ and

$$Z_p = \frac{(jX_L)(-jX_C)}{R_L} = \frac{|X_L|\,|X_C|}{R_L}$$

$$\frac{v_{out}}{v_{in}} = \frac{R_{out}}{R_{out} + Z_p\ @\ f_r}$$

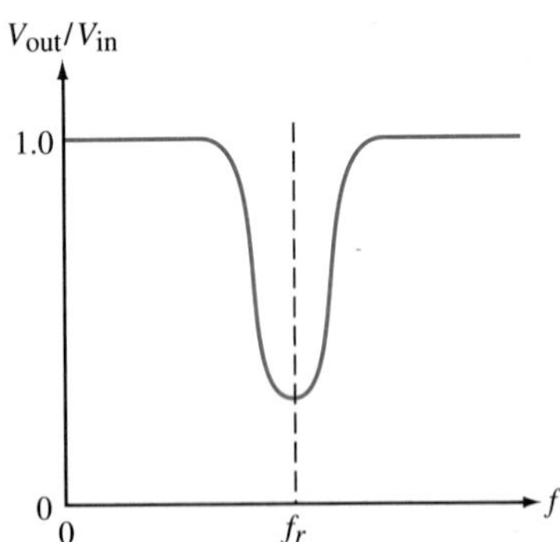

Figure 19.32 Transfer characteristics.

The transfer function says that the ratio of v_{out} to v_{in} will be less than 1 around the resonant frequency. The frequency response curve will exhibit band-reject characteristics. A band-reject response curve is shown in Figure 19.32.

The sharpness of the curve and the bandwidth can be related to the Q of the parallel circuit. For parallel resonant circuits using high Q coils ($Q > 10$), the Q of the circuit is

$$Q_r = \frac{X_L}{R_L}$$

and the bandwidth is

$$\text{BW} = \frac{f_r}{Q_r}$$

Since

$$Z_p = \frac{|X_L|\,|X_C|}{R_L}$$

at the resonant frequency $X_L = X_C$

$$Z_p = \frac{X_L^2}{R_L}$$

$$= \frac{X_L^2}{R_L^2} R_L$$

$$Z_p = Q_r^2 R_L$$

Now, let us use these relationships to analyze a circuit of the type shown in Figure 19.30.

EXAMPLE 19.13

Given the parallel-resonant, band-reject circuit shown in Figure 19.33, determine:

a. the resonant frequency, f_r
b. the Q of the circuit
c. the bandwidth of the circuit
d. the output voltage at f_r
e. the magnitude and phase response using PSpice

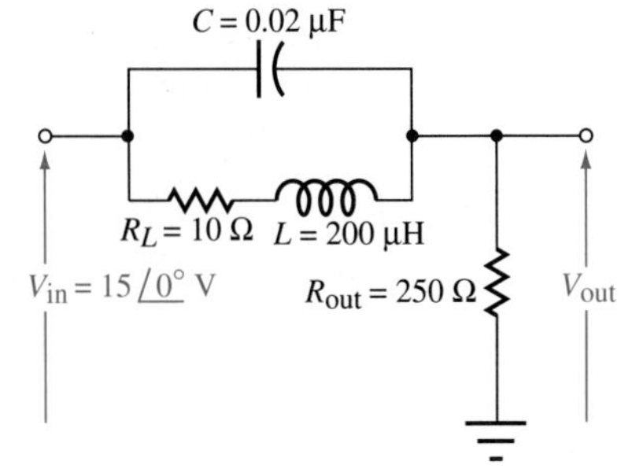

Figure 19.33

Solution

a.

$$f_r = \frac{1}{2\pi\sqrt{(200\ \mu\text{H})(0.02\ \mu\text{F})}}$$

$$f_r = 79.58\ \text{kHz}$$

b.

$$Q_r = \frac{X_L}{R_L}$$

$$= \frac{2\pi f_r L}{R_L}$$

$$= \frac{2\pi(79.58 \text{ kHz})(200 \ \mu\text{H})}{10 \ \Omega}$$

$$Q_r = 10$$

c.

$$\text{BW} = \frac{f_r}{Q_r}$$

$$= \frac{79.58 \text{ kHz}}{10}$$

$$\text{BW} = 7.96 \text{ kHz}$$

d.

$$v_{out} = \frac{R_{out}}{R_{out} + Z_p \ @ \ f_r}$$

where $Z_p \ @ \ f_r = Q^2 R_L = (10)^2(10 \ \Omega) = 1\text{-k}\Omega$

$$v_{out} = \frac{250 \ \Omega}{250 \ \Omega + 1 \text{ k}\Omega} v_{in}$$

$$= (0.2)(15 \text{ V})$$

$$v_{out} = 3.0 \text{ V}$$

e. The node assignments are shown in Figure 19.34(a) and the PSpice circuit description is listed in Figure 19.34(b). The magnitude

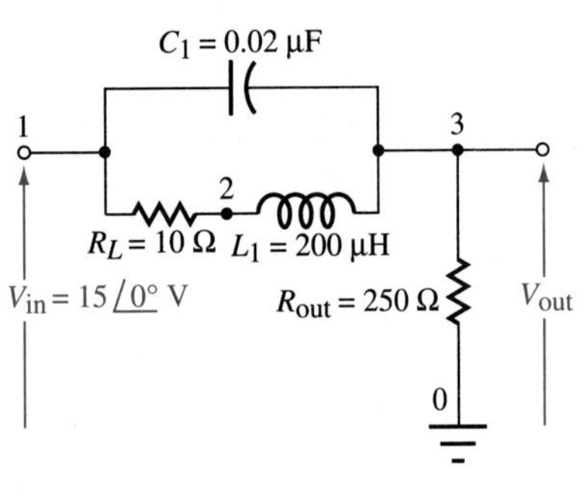

(a)

```
EXAMPLE 19.13--BAND-REJECT CIRCUIT
**** CIRCUIT DESCRIPTION
******************************************

VIN      1     0      AC      15.0     0.0
RL       1     2      10
L1       2     3      200U
C1       1     3      0.02U
RO       3     0      250
.AC      DEC   50     1K      1MEG
.PRINT   AC    VM(3)          VP(3)
.PROBE
.END
```

(b)

Figure 19.34 Node assignments and PSpice file.

response curve shown in Figure 19.35 shows a dip at just before 80 kHz. The output magnitude drops to 3.0 volts. Both of these values match our calculated results.

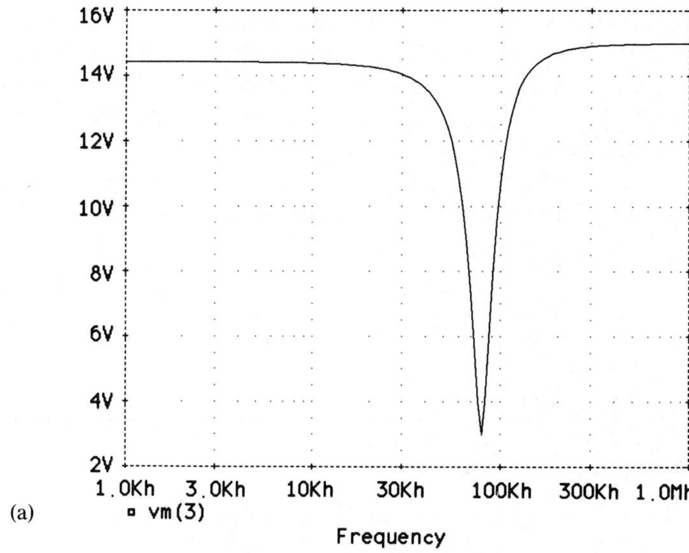

(a)

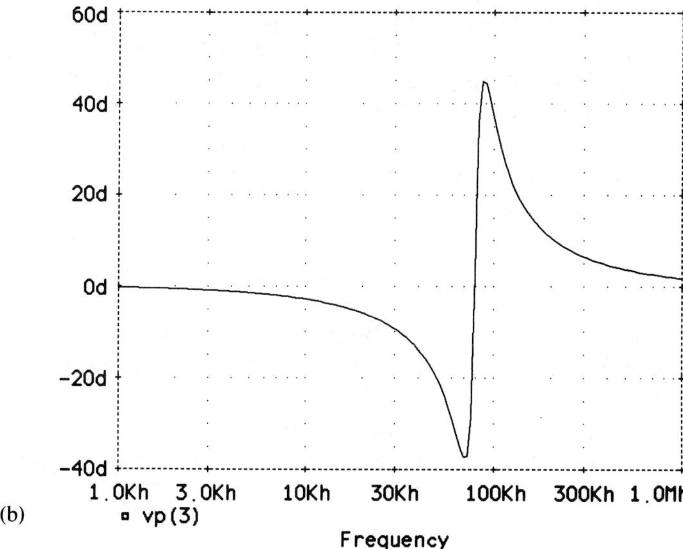

(b)

Figure 19.35 Frequency response curves: (a) magnitude response and (b) phase response.

The phase response of the circuit is shown in Figure 19.35(b). At low frequencies, the phase angle is negative and reaches a negative peak at 70 kHz. The phase angle is zero at f_r. At frequencies greater than f_r, the phase angle is positive, reaching a peak at 90 kHz and then decreasing to zero.

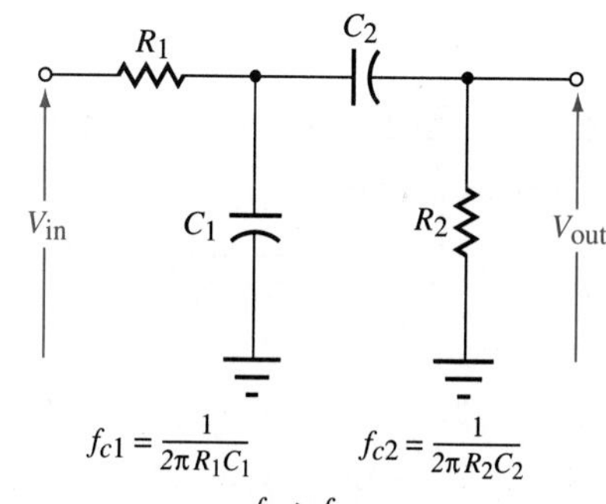

Figure 19.36 Phase response curve.

In this chapter we have examined the frequency response characteristics of a number of circuits. We have learned how to write mathematical statements and transfer functions that describe the signal-processing capabilities of these circuits. We have calculated 3-dB frequencies, bandwidths, and phase angles.

We have also used PSpice to simulate circuit behavior, obtaining graphs of the magnitude and phase response of the circuit. Being able to visualize the frequency response characteristics of circuits is an important skill for the technician and engineer. Cultivate this skill as you progress in your study of electronics.

SUMMARY

1. The relationship between the input and output of a circuit can be defined in terms of a transfer function. The transfer function describes both the magnitude and the phase response of the circuit.
2. The bandwidth of a circuit is measured between the upper and lower half-power or 3-dB frequencies. The low-frequency response of some circuits extends down to dc, and the bandwidth is measured from dc to the upper half-power or 3-dB frequency.
3. The frequency response of a circuit or system is often displayed in graphical form, with gain or attenuation plotted on the y-axis and frequency on the x-axis, using semilog graph paper.
4. A Bode plot is a special plot of the frequency response of a circuit or system. Straight-line approximations are used to represent the frequency response on a graph of gain/attenuation versus frequency.
5. PSpice can be used to obtain both the magnitude and the phase response information for a circuit. Graphical outputs can be produced by .PROBE.
6. Resonant circuits are circuits that contain resistance, capacitance, and inductance in such a way that bandpass and band-reject responses can be obtained.

EQUATIONS

19.1	$V_{out}(t) = \text{TF}\ V_{in}(t)$	The output signal in terms of the transfer function and the input signal.	788
19.2	$\dfrac{V_{out}(t)}{V_{in}(t)} = \text{TF}$	The relation that defines the transfer function.	788
19.3	$f_r = \dfrac{1}{2\pi\sqrt{LC}}$	The equation for the resonant frequency of a series *R-L-C* circuit.	807

19.4	$Q_r = \frac{Q_L}{P}$	The equation for the Q of a series R-L-C circuit.	808
19.5	$Q_r = \frac{1}{R}\sqrt{\frac{L}{C}}$	The relation between the Q of a series R-L-C circuit and the values of R, L, and C.	809

QUESTIONS

1. What name is given to a circuit that only allows frequency components above a certain frequency to pass through the circuit?
2. What name is given to a circuit that only allows a narrow range of frequencies to pass through the circuit?
3. What is the name given to the frequency that represents the lowest frequency that will pass through a high-pass filter with less than 3-dB attenuation?
4. Define the term bandwidth.
5. Describe the phase shift characteristics of the following circuits:
   (a) Low-pass R–C circuit.
   (b) High-pass R–C circuit.
   (c) Low-pass R–L circuit.
   (d) High-pass R–L circuit.
6. What name is given to the mathematical function that describes the relationship between the input and output signals for a circuit?
7. What is a Bode plot and what information is plotted?
8. When constructing a Bode plot, what are the correction factors one octave above and below the corner frequency and at the corner frequency that are used to obtain the actual frequency response?
9. On a Bode plot of gain in dB versus frequency, what is the slope of the graph at frequencies greater than the corner frequency for a simple R–C low-pass circuit?
10. For a simple low-pass R–C circuit, what effect will increasing the series resistance have on the 3-dB frequency?
11. For a simple high-pass R–C circuit, what effect will decreasing the series capacitance have on the 3-dB frequency?
12. For a simple series R–L–C circuit, what is the phase angle difference between the input voltage and the current flowing in the series circuit at the resonant frequency?
13. Draw a series R–L–C bandpass circuit.
14. In a series R–L–C bandpass circuit, what effect will the following changes in circuit component values have on the bandwidth of the circuit?
    (a) Increasing the series resistance.
    (b) Decreasing the series inductance.
    (c) Doubling the series capacitance.
15. For a series R–L–C bandpass circuit, what effect will the following circuit changes have on the peak output at the resonant frequency?
    (a) Doubling the inductance value.
    (b) Adding a load resistor across the output of the circuit.
16. Write a PSpice statement that will perform an ac analysis over the frequency range of 1 kHz–10 MHz with 20 calculation points per decade.
17. Write a PSpice statement that will use PROBE to plot the outputs of the ac analysis.

18. A PSpice analysis is performed on a bandpass $R–L–C$ circuit that has a very narrow peak. The graph of the PSpice simulation shows a peak that is only one-third of the calculated peak output voltage. What could have caused the discrepancy between the calculated value and the simulation value?
19. Sketch the general shape of the frequency response curve for a band-reject $R–L–C$ circuit.
20. If we can create a bandpass circuit by cascading a high-pass $R–C$ circuit and a low-pass $R–C$ circuit, why can't we create a band-reject circuit by reversing the position of the two stages, as shown in Figure 19.36?

PROBLEMS

SECTION 19.1 Transfer Functions

1. Write the transfer function for the circuit shown in Figure 19.37.
2. Write the transfer function for the circuit shown in Figure 19.38.

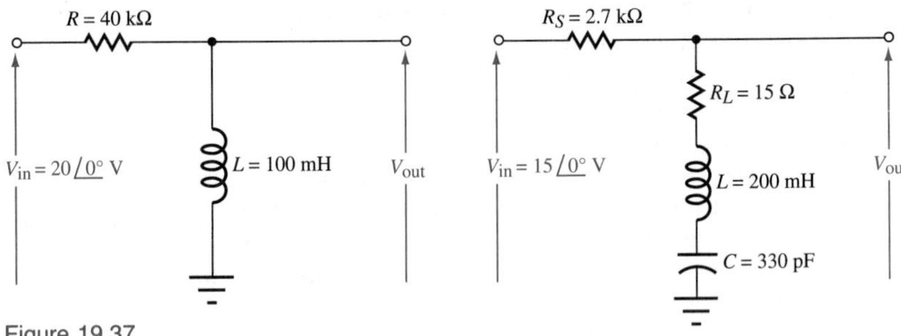

Figure 19.37

Figure 19.38

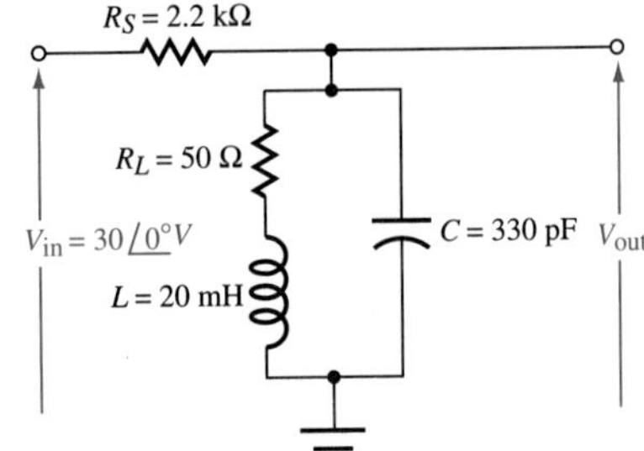

Figure 19.39

3. Write the transfer function for the circuit shown in Figure 19.39.
4. Write the transfer function for the circuit shown in Figure 19.40.
5. Using the transfer function derived in Problem 1, write the expression for the magnitude and the phase response of the circuit.
6. Using the transfer function derived in Problem 2, write the expression for the magnitude and the phase response of the circuit.

SECTION 19.2 Bandwidth and Phase Shift

7. Determine the 3-dB point and the passband frequencies for the circuit shown in Figure 19.38.
8. Determine the 3-dB point and the passband frequencies for the circuit shown in Figure 19.39.
9. Determine the bandwidth for the circuit shown in Figure 19.40.

SECTION 19.3 Bode Plots

10. Draw a Bode plot for the circuit shown in Figure 19.37.
11. Draw a Bode plot for the circuit shown in Figure 19.38.

12. Draw a Bode plot for the circuit shown in Figure 19.39.

13. Draw a graph showing the phase response of the circuit shown in Figure 19.37.

14. Draw a graph showing the phase response of the circuit shown in Figure 19.38.

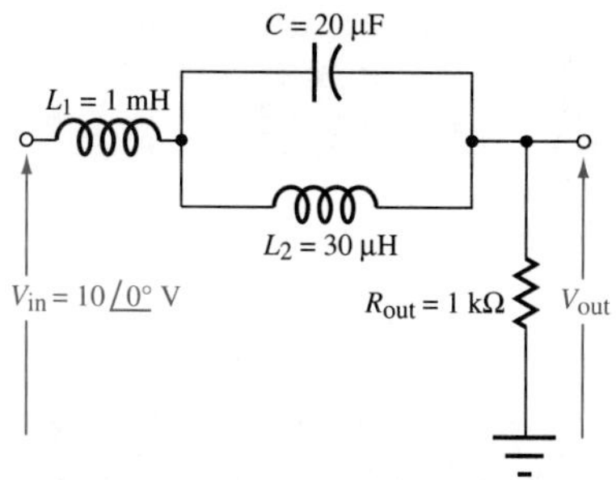

Figure 19.40

SECTION 19.4 Using SPICE for Frequency Response Analysis

Ⓢ 15. Use PSpice to obtain a Bode plot for the circuit in Figure 19.37.

Ⓢ 16. Use PSpice to obtain a graph of attenuation versus frequency for the circuit shown in Figure 19.38.

Ⓢ 17. Use PSpice to obtain a graph of the phase response for the circuit shown in Figure 19.39.

Ⓢ 18. Use PSpice to determine the frequency response (magnitude and phase) of the circuit shown in Figure 19.40.

Ⓢ 19. Use PSpice to determine the frequency response (magnitude and phase) of the circuit shown in Figure 19.41.

Ⓢ 20. Use PSpice to determine the frequency response (magnitude and phase) of the circuit shown in Figure 19.42.

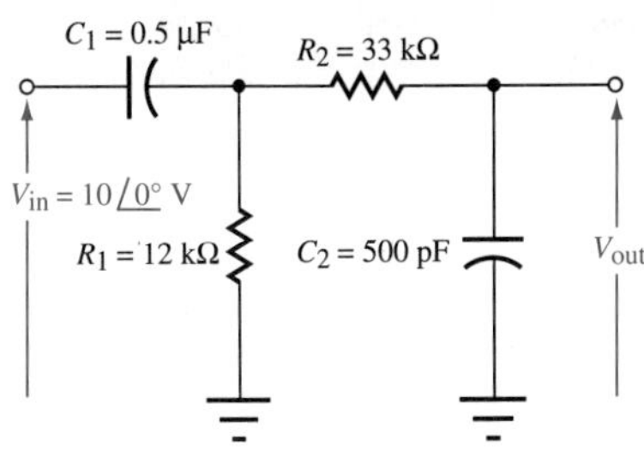

Figure 19.41

SECTION 19.5 Resonance

21. Find the resonant frequency and Q_r for the R–L–C circuit shown in Figure 19.43.

22. Find the resonant frequency and Q_r for the R–L–C circuit shown in Figure 19.44.

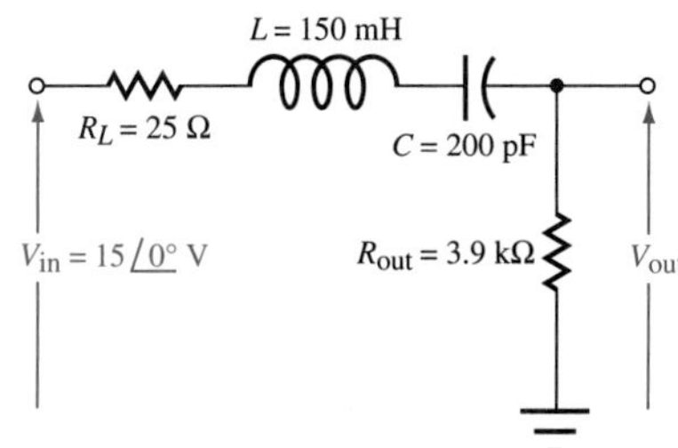

Figure 19.42

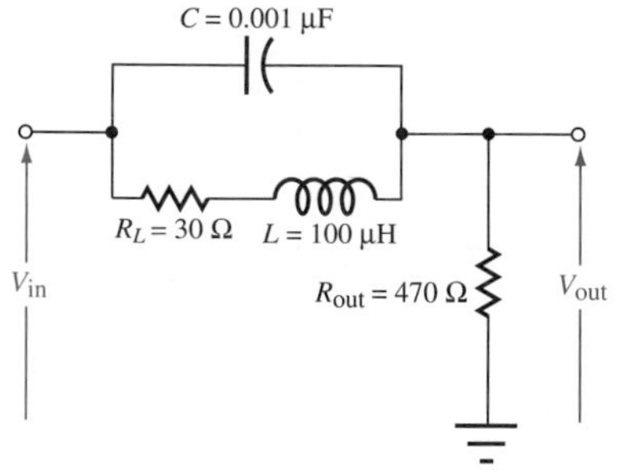

Figure 19.43

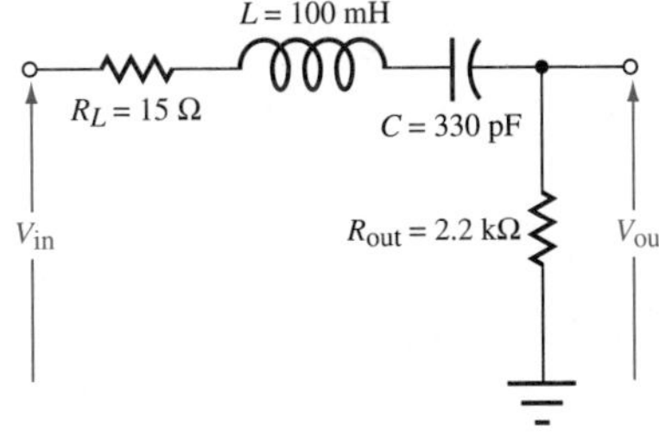

Figure 19.44

CHAPTER 20

NONSINUSOIDAL SOURCES

CHAPTER OBJECTIVES

After completing this chapter, you should be able to:

1. Identify a nonsinusoidal waveform.
2. Write the general form of the Fourier expression.
3. Graphically construct a wave from the terms in the Fourier series.
4. Identify the dc component, the fundamental, and the harmonics in the Fourier expression.
5. Determine the coefficients for the Fourier expression using the computer program.
6. Analyze a waveform for the conditions of Table 20.1.
7. Determine the effective value and the power in a circuit with a nonsinusoidal source.

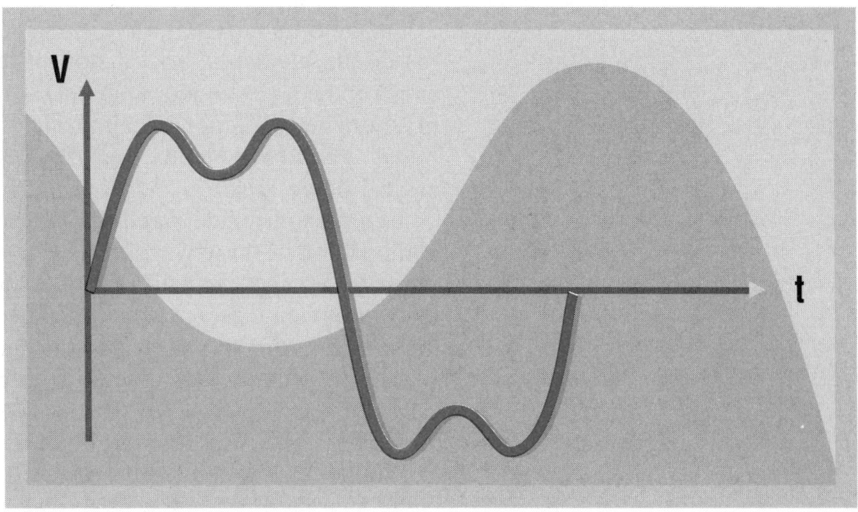

8. Analyze a circuit with a nonsinusoidal source.
9. Explain what a spectrum analyzer measures and displays.

KEY TERMS

1. distortion
2. even function wave
3. exponential wave
4. Fourier analysis
5. Fourier analyzer
6. Fourier series
7. fundamental
8. harmonics
9. nonsinusoidal waveforms
10. odd function wave
11. periodic waves
12. rectification
13. sawtooth wave
14. signal analyzers
15. square wave
16. spectrum analyzer
17. triangular wave
18. wave analyzer

INTRODUCTION

Nonsinusoidal voltages and currents are used in many electronic applications. The square wave is used in the logic circuits of computers; an exponential wave is generated when a capacitor charges and discharges; and the sawtooth wave is used to sweep the electron beam across the face of the oscilloscope screen.

Unlike the sinusoidal wave, nonsinusoidal waves can be distorted by the circuit elements. This distortion can affect the characteristics, or even render a circuit inoperable. Distortion in an audio amplifier results in a poor reproduction of the music or voice. Distortion in the navigational circuit of a satellite can affect its course, transmission, and other critical operations.

To have some understanding of these effects, one must learn about the composition of these waveforms. Thus far, they have been treated as a single frequency. Actually, they are made up of many sinusoidal waves of different amplitudes and frequencies. This chapter introduces you to this concept. First, you learn how to determine what components are in a waveform. Finally, some circuits with nonsinusoidal sources are analyzed.

As you study this chapter, you should understand why capacitance and inductance round the corners of the square wave. Also, you learn how the relation for the effective value of the ac component was developed.

20.1 Nonsinusoidal Waveforms

nonsinusoidal waveforms

Nonsinusoidal waveforms are those waveforms that do not have a sinusoidal shape. These waveforms include (but are not limited to) square waves, triangular waves, and sawtooth waves. These types of waveforms are generally used for control purposes rather than for power transmission; they are found mostly in electronic circuits. Figure 20.1 shows the shapes of the more common nonsinusoidal waveforms. They are all **periodic waves**. Periodic means that they repeat themselves at regular periods of time. Let us now take a closer look at these nonsinusoidal waveforms.

periodic waves

Figure 20.1(a) and (b) show the waves for full-wave and half-wave rectification. **Rectification** is a procedure that changes an alternating voltage to a unidirectional one, with dc in it. Although the input is sinusoidal, the output is not. These waves provide a dc output from the sinusoidal ac input.

rectification

sawtooth wave

Another common wave, the **sawtooth wave**, is shown in Figure 20.1(c). This waveform increases linearly and decreases instantaneously. One use for this type of wave is to move the electron beam across the screen of the oscilloscope. The voltage is applied to the horizontal deflection plates. The voltage increase causes the electron beam to deflect more, thus moving the beam across the screen. A simple method for generating a sawtooth is by partially charging a capacitor through a resistor, then discharging it instantaneously.

triangular wave

Figure 20.1(d) is a **triangular wave**. It is similar to the sawtooth, but it also decreases linearly. It can also be generated by charging and discharging a capacitor. However, the discharge must be controlled to give the linear discharge. This can be done by changing the resistance in the discharge path.

exponential wave

An **exponential wave**, shown in Figure 20.1(e), is simply the complete charging and discharging of a capacitor. If the source is a pulse, the

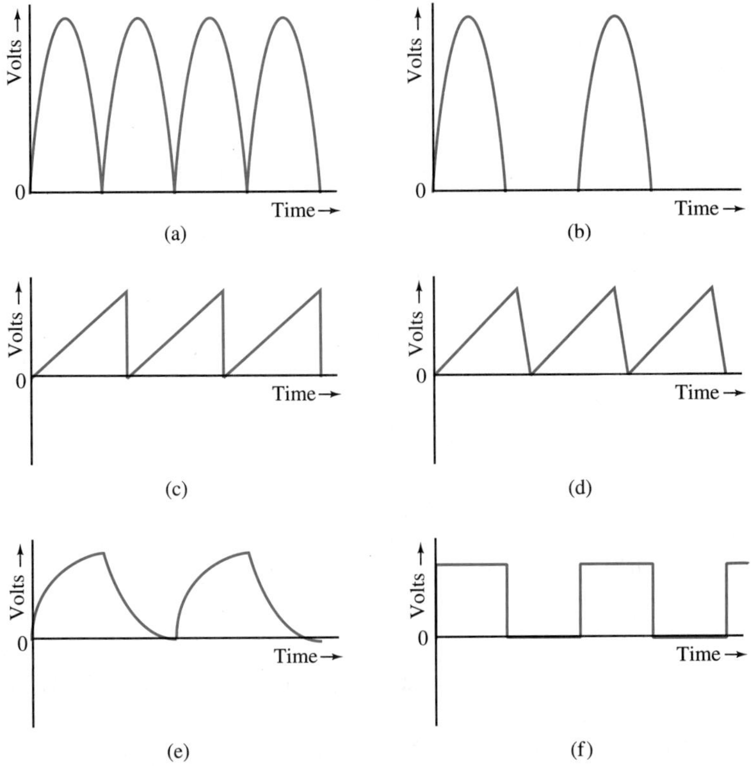

Figure 20.1 The shapes of some common nonsinusoidal waveforms. (a) Full-wave rectification. (b) Half-wave rectification. (c) Sawtooth. (d) Triangular. (e) Exponential. (f) Square.

frequency, charge, and discharge times are controlled by the frequency of the pulse.

square wave

The wave in Figure 20.1(f) is a **square wave**. It alternates between two constant values. A common use for this wave is in logic circuits. The waveform can be generated by switching a dc source on and off. One circuit that does this is an astable multivibrator. In this circuit, the charging of capacitors is used to turn transistors on and off, thus switching the dc voltage. An alternate method for generating a reasonable approximation of a square wave is to clip the top of a sine wave, as shown in Figure 20.2(b). The zener diode circuit, shown in Figure 20.2(a), performs this clipping automatically. The diode starts to conduct

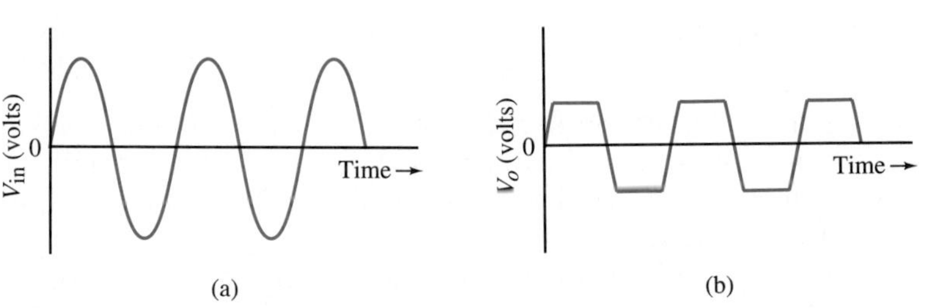

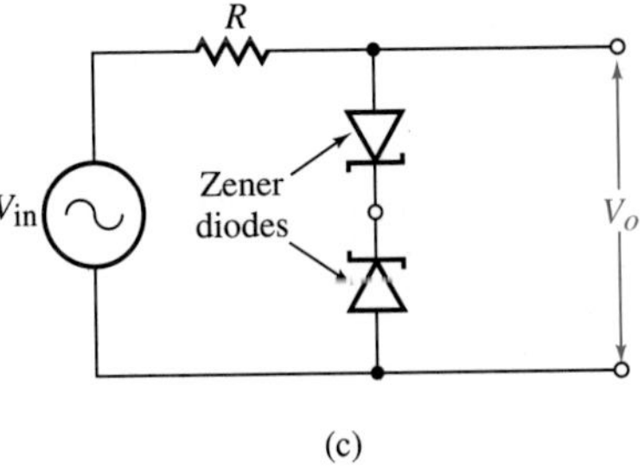

Figure 20.2 The shaping of an approximate square wave from a sine wave by a zener diode circuit.

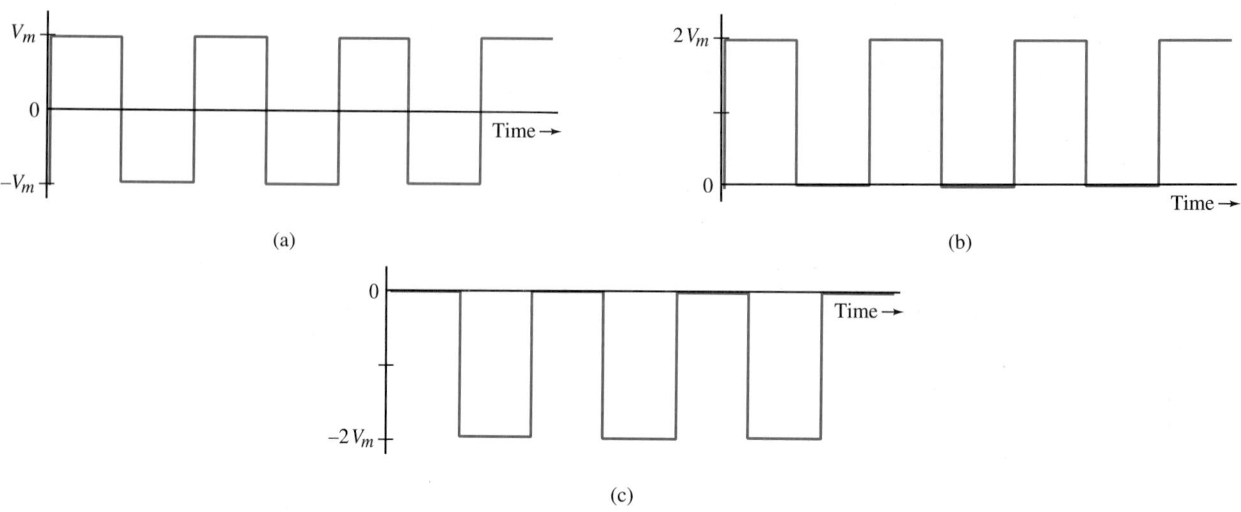

(c)

Figure 20.3 The addition of dc changes the voltage level of the wave. (a) A square wave without any dc. (b) Adding a positive dc shifts the wave up. (c) Adding a negative dc shifts the wave down.

when the voltage across it reaches a certain level; the diode then maintains that voltage.

There is a dc component in the waves in Figure 20.1, since the net area under the curve is positive. If no dc component were in the waves, they would be shifted down. The effect of dc is shown in Figure 20.3.

20.2 Fourier Analysis

Fourier analysis

Nonsinusoidal waveshapes are made up of sine and cosine waves of different frequencies and amplitudes. **Fourier analysis** was first proposed by Jean Baptiste Fourier (1768–1830), a French physicist and mathematician, in his text, *Analytical Theory of Heat*. In effect, he proposed that a periodic function can be expressed by a mathematical series of sine and cosine terms. This series is known as the **Fourier series**, which is seen in the construction of the square wave in Figure 20.4. Only four components are included, but the wave is already approaching a square shape. Fourier proposed the following general expression for the waveform:

Fourier series

$$y = A_0 + (A_1 \sin \omega t + B_1 \cos \omega t) + (A_2 \sin 2\omega t + B_2 \cos 2\omega t) + \cdots + (A_n \sin n\omega t + B_n \cos n\omega t) \qquad \textbf{(20.1)}$$

where y is the value of the function at ωt radians; ω is the angular frequency, in radians per second; ωt is the distance along the axis, in radians; A_1, B_2, and so forth, are the amplitudes of the waves; and A_0 is the dc part of the wave.

The general wave of Equation 20.1 is made up of a dc component, a fundamental component that has the angular frequency ω, and the harmonics of frequencies 2ω, 3ω, . . . , $n\omega$. The **fundamental** is the term

fundamental

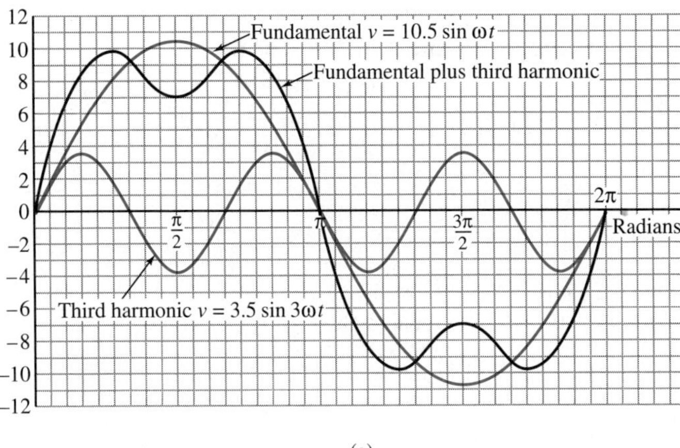

(a)

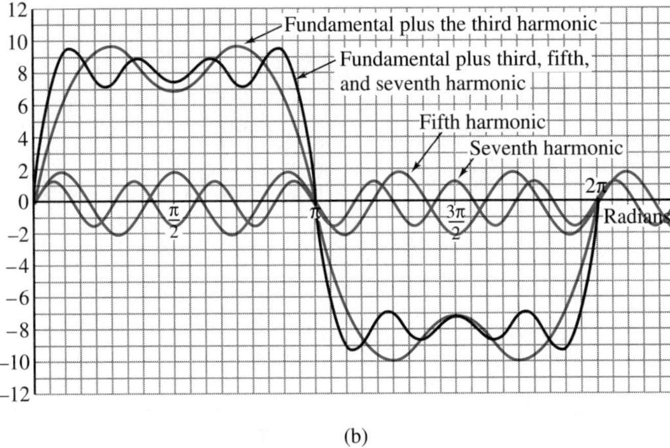

(b)

Figure 20.4 The graphical construction of a square wave from sinusoidal voltages. (a) The fundamental plus the third harmonic. (b) The fundamental plus the third, fifth, and seventh harmonic.

that has the same frequency as the original repeating waveform. The **harmonics** are the terms whose frequency is a multiple of the fundamental. In theory, an infinite number of terms are needed in the series to give the wave. In actuality, only a few yield a reasonably close shape. Again, this can be seen in the shaping of the square wave in Figure 20.4. Periodic waves must have the fundamental. But whether they have a dc component and all harmonics depends on the waveshape. A guide to the characteristics that affect the content is given in Table 20.1.

harmonics

The theory of a wave being made up of different frequency sine and cosine waves is not just a mathematical concept. These terms actually exist in the wave, and can lead to distortion and other undesirable effects. For instance, a filter that does not pass any frequencies above the third harmonic changes the square wave to what is shown in Figure 20.4(a). Obviously, this can cause problems in some electronic applications.

TABLE 20.1 Effect of Wave Shape on the Fourier Terms

1. A_0 is zero when the area above the axis is equal to the area below the axis.

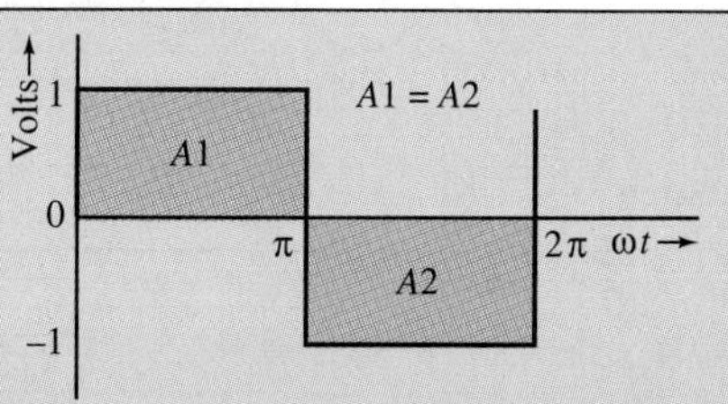

Figure 20.5 The negative area is equal to the positive area, so the dc term is zero.

2. The cosine terms are zero when the value of the wave at $+\alpha$ is the negative of the value at $-\alpha$, as measured from the A_0 axis. An example is the sine wave. This type of wave is an **odd function wave**.

odd function wave

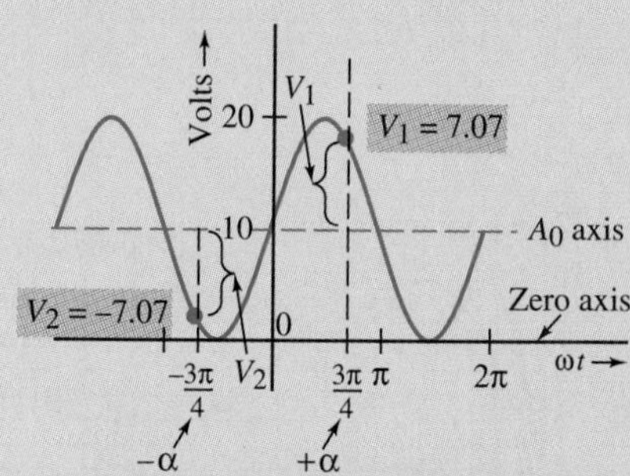

Figure 20.6 $V_2 = -V_1$ so the cosine terms are zero.

3. The sine terms are zero when the value of the wave at $+\alpha$ is the same as at $-\alpha$. An example is the cosine wave. This type of wave is an **even function wave**.

even function wave

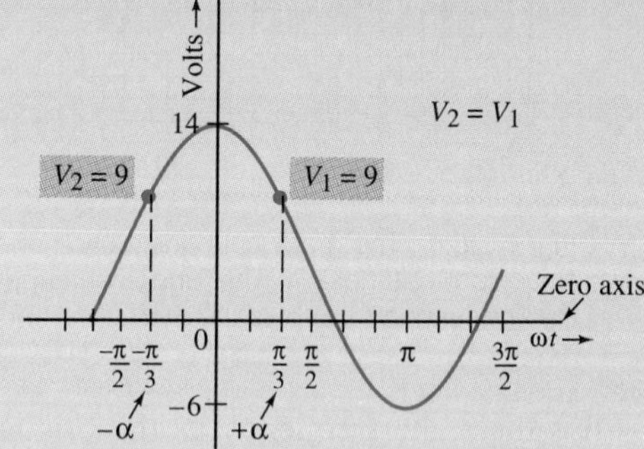

Figure 20.7 $V_2 = V_1$ so the sine terms are zero.

4. The even harmonics are zero when the value of the wave at α is the negative of the value at $(\alpha + \pi)$, as measured from the A_0 axis.

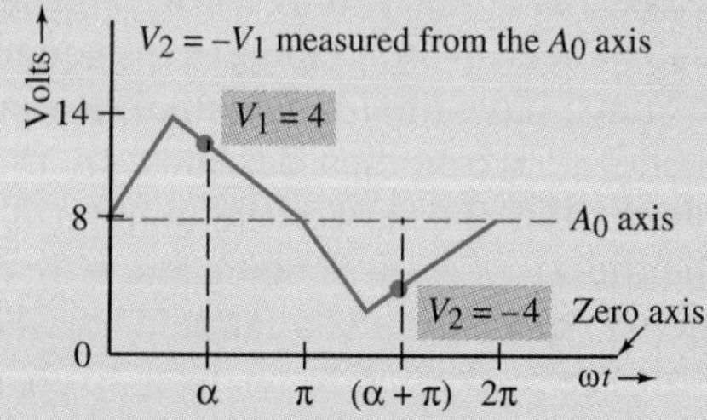

Figure 20.8 The voltage at $(\alpha + \pi)$ is equal to the voltage at α, so the even harmonics are zero.

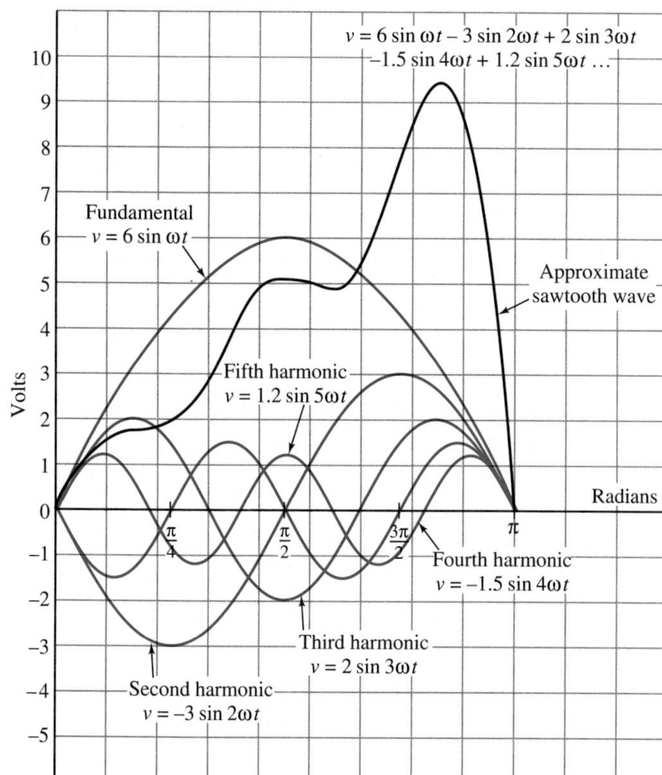

Figure 20.9 The graphical construction of the first half of a sawtooth wave from sinusoidal waves.

The Fourier series for the square wave of Figure 20.4 is

$$v = \frac{4V_m}{\pi}\left(\sin \omega t + \frac{1}{3}\sin 3\omega t + \frac{1}{5}\sin 5\omega t + \frac{1}{7}\sin 7\omega t + \ldots\right) \tag{20.2}$$

where V_m is the maximum value of the wave; ω is the angular frequency of the wave, in radians per second; and t is the time, in seconds.

Equation 20.2 indicates that the square wave does not have a dc value or any even harmonics.

The construction of the first half of a sawtooth waveform without any dc is as shown in Figure 20.9. The Fourier expression for this waveform is

$$v = \frac{2V_m}{\pi}\left(\sin \omega t - \frac{1}{2}\sin 2\omega t + \frac{1}{3}\sin 3\omega t - \frac{1}{4}\sin 4\omega t + \frac{1}{5}\sin 5\omega t - \ldots\right) \tag{20.3}$$

Here, all harmonics are present, but there are no dc component or cosine terms. This agrees with the summary in Table 20.1.

Some other waveforms and their Fourier expressions are shown in Figure 20.10. The expressions can be adapted to variations of the waves as follows:

1. A wave with a dc component must have the amount of dc added to the expression as the A_0 term.

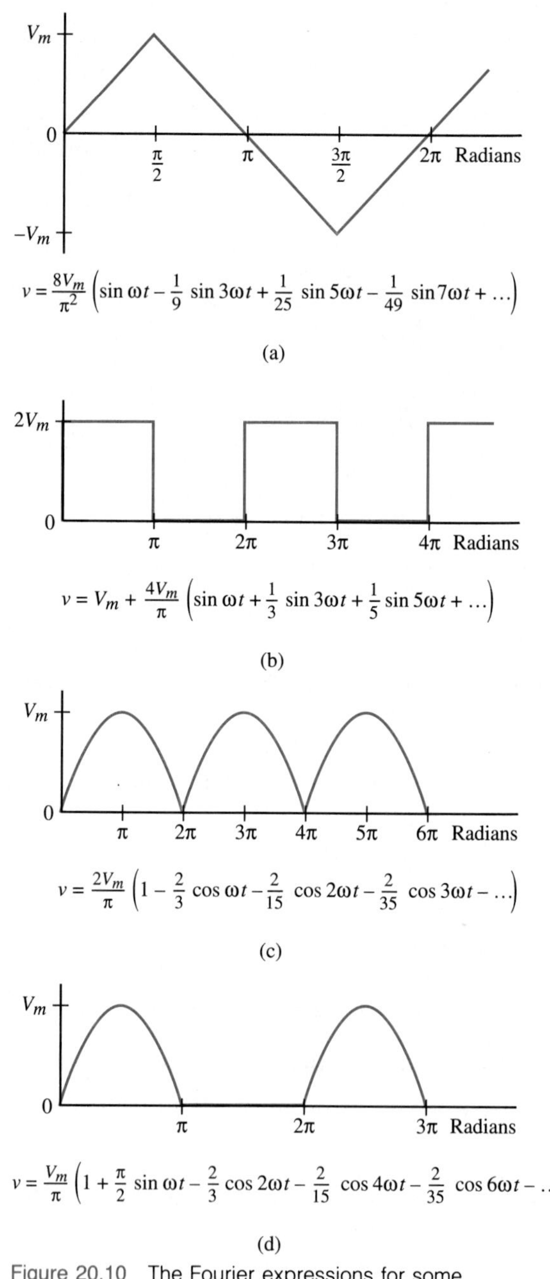

Figure 20.10 The Fourier expressions for some nonsinusoidal waves. (a) A triangular wave. (b) A square wave with positive dc. (c) Full-wave rectification. (d) Half-wave rectification.

2. A wave shifted θ radians to the right lags the original wave, so the angle must be $\omega t - \theta$ instead of ωt.
3. A wave shifted θ radians to the left leads the original wave, so that angle must be $\omega t + \theta$ instead of ωt.

20.3 Computer Determination of the Fourier Series

The following integrals give the general forms for A_0, A_1, B_1, and so forth, in Equation 20.1. $f(\alpha)$ is the equation for the waveshape. For example, if $v = 4 \sin \alpha$, $f(\alpha)$ is $4 \sin \alpha$.

The derivation of these integrals is not shown here. It is beyond the scope of this text.

$$A_0 = \frac{1}{2\pi} \int_0^{2\pi} f(\alpha)\, d\alpha$$

$$A_n = \frac{1}{\pi} \int_0^{2\pi} f(\alpha) \sin(n\alpha)\, d\alpha$$

$$B_n = \frac{1}{\pi} \int_0^{2\pi} f(\alpha) \cos(n\alpha)\, d\alpha$$

Integration, as you might remember from average and effective values, is a procedure for finding the area under a curve. It gives the exact value of the coefficients, but using it has two disadvantages. First, the equation for the waveform is needed. Second, one must be able to perform the integration. An alternate method is explained here. Its results are less exact, and the procedure is longer. However, the equation of the curve is not required, so it can be used with any waveform.

From the definition of integration, the expressions for A_0, A_n, and B_n can be written as

$$A_0 = \frac{\text{area under the } f(\alpha) \text{ curve}}{2\pi} \tag{20.4}$$

$$A_n = \frac{\text{area under the } f(\alpha) \sin n\alpha \text{ curve}}{\pi} \tag{20.5}$$

$$B_n = \frac{\text{area under the } f(\alpha) \cos n\alpha \text{ curve}}{\pi} \tag{20.6}$$

An inspection of Equations 20.4 to 20.6 shows that:

1. A_0 is the average or dc value of the waveform over 360°, one cycle of the waveform.
2. A_n is double the average value of the $f(\alpha) \sin n\alpha$ curve over 360°, one cycle of the waveform.
3. B_n is double the average value of the $f(\alpha) \cos n\alpha$ curve over 360°, one cycle of the waveform.

Writing the average values of the curves in terms of the value at the center of the strips, as was done for the effective value in Chapter 15, Section 15.5, gives

$$A_0 = \frac{h_1 + h_2 + \cdots + h_m}{m} \tag{20.7}$$

$$A_n = 2\left(\frac{h_1 \sin n\alpha_1 + h_2 \sin n\alpha_2 + \cdots + h_m \sin n\alpha_m}{m}\right) \tag{20.8}$$

$$B_n = 2\left(\frac{h_1 \cos n\alpha_1 + h_2 \cos n\alpha_2 + \cdots + h_m \cos n\alpha_m}{m}\right) \quad \textbf{(20.9)}$$

where m is the number of strips of equal width into which the waveform is divided; $\alpha_1, \alpha_2, \ldots \alpha_m$ are the angles at the center of each strip on the degrees axis; $h_1, h_2, \ldots, h_m$ are the y values of the wave at the center of each strip; and n represents the fundamental and harmonics. It has values of 1, 2, 3, 4, ...

The procedure for finding the terms in the Fourier expression with these equations is:

1. Mark off the x-axis in degrees so that 360° corresponds to one cycle of the waveform.
2. Divide one cycle into m strips of equal width. Increasing the number of strips gives more exact results.
3. Find the value of the curve (h) and the angle (α) at the center of each strip. Tabulate these values. The values of h can also be obtained from the equation, if it is available. This eliminates the need for the curve.
4. For $n = 1$, calculate $h_m \sin n\alpha_m$ and $h_m \cos n\alpha_m$ for each value of h and α. Tabulate these values.
5. Repeat Step 4 for as many harmonics as are needed.
6. Calculate A_0 using Equation 20.7.
7. Calculate A_n and B_n for each value of n using Equations 20.8 and 20.9.

The procedure is better understood by working an example.

EXAMPLE 20.1

Determine the Fourier series up to the fourth harmonic for the waveform shown in Figure 20.11.

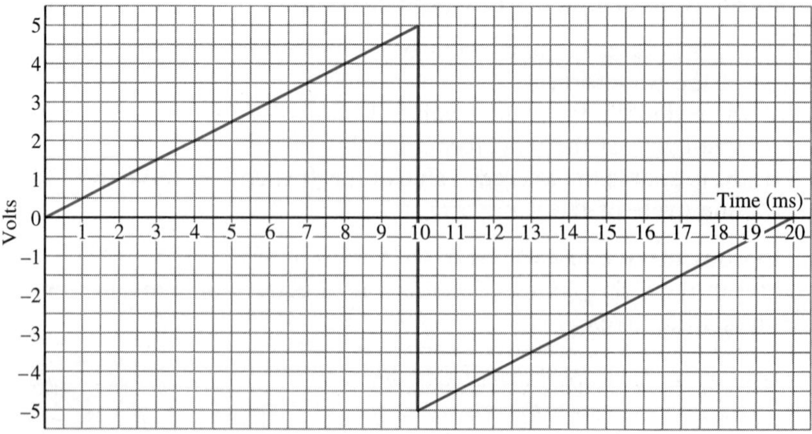

Figure 20.11

Solution

1. One cycle is 20 milliseconds, so that point is marked as 360°.
2. The cycle is divided into 20 strips, each 18° wide, as shown in Figure 20.12.

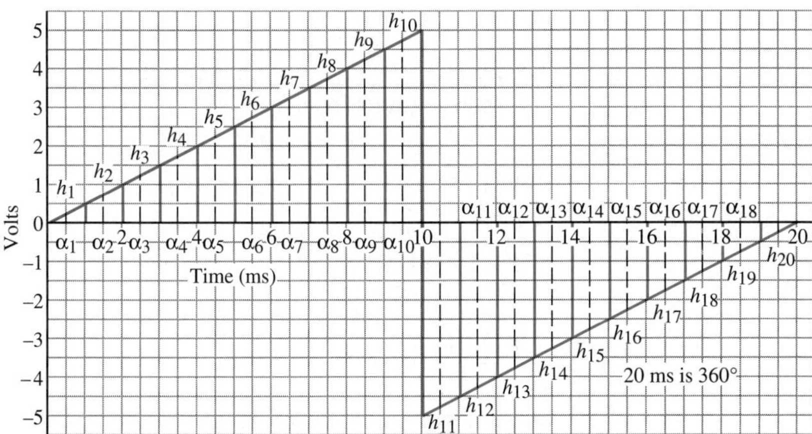

Figure 20.12

3. The values of h and α at the center of some strips are:

$$h_1 = 0.25 \qquad \alpha_1 = 9^\circ$$
$$h_2 = 0.75 \qquad \alpha_2 = 27^\circ$$
$$h_{11} = -4.75 \qquad \alpha_{11} = 189^\circ$$

These and the other values are tabulated in Table 20.2.

4. For $n = 1$ (fundamental)

$$h_1 \sin \alpha_1 = 0.25 \sin(9^\circ) = 0.039$$
$$h_2 \sin \alpha_2 = 0.75 \sin(27^\circ) = 0.340$$
$$h_{11} \sin \alpha_{11} = (-4.75) \sin(189^\circ) = (-4.75)(-0.156) = 0.743$$
$$h_1 \cos \alpha_1 = 0.25 \cos(9^\circ) = 0.247$$
$$h_2 \cos \alpha_2 = 0.75 \cos(27^\circ) = 0.668$$
$$h_{11} \cos \alpha_{11} = (-4.75) \cos(189^\circ) = 4.692$$

These and the other values are tabulated in Table 20.2.

5. For $n = 2$ (second harmonic)

$$h_1 \sin(2\alpha_1) = 0.25 \sin[(2)(9^\circ)] = 0.077$$
$$h_1 \cos(2\alpha_1) = 0.25 \cos[(2)(9^\circ)] = 0.238$$

For $n = 3$ (third harmonic)

$$h_1 \sin(3\alpha_1) = 0.25 \sin[(3)(9^\circ)] = 0.113$$
$$h_1 \cos(3\alpha_1) = 0.25 \cos[(3)(9^\circ)] = 0.223$$

For $n = 4$ (fourth harmonic)

$$h_1 \sin(4\alpha_1) = 0.25 \sin[(4)\,(9°)] = 0.147$$
$$h_1 \cos(4\alpha_1) = 0.25 \cos[(4)\,(9°)] = 0.202$$

These values and the values for h_2 through h_{20} are tabulated in Table 20.2.

6. $A_0 = \left(\dfrac{h_1 + h_2 + \cdots + h_m}{m}\right)$

From Table 20.2, $h_1 + h_2 + \cdots + h_m = 0$, so $A_0 = 0$.

7. Using Equation 20.8 and the totals from Table 20.2 for each value of n gives

$$A_1 = \frac{(2)(31.96)}{20} \qquad A_1 = 3.196$$

$$A_2 = \frac{(2)(-16.18)}{20} \qquad A_2 = -1.618$$

$$A_3 = \frac{(2)(11.01)}{20} \qquad A_3 = 1.101$$

$$A_4 = \frac{(2)(-8.51)}{20} \qquad A_4 = -0.851$$

The B_n terms are equal to zero because the sum of the cosine terms in Table 20.2 is equal to zero. The general expression for the waveform becomes

$$y = A_1 \sin \omega t + A_2 \sin 2\omega t + A_3 \sin 3\omega t + A_4 \sin 4\omega t + \ldots$$

The frequency of the waveform in Figure 20.11 is 50 Hz, so $\omega = 314.1$ radians per second. Putting the value of ω and A_1, A_2, A_3, and A_4 into the general expression gives

$$\upsilon = 3.196 \sin(341.1t) - 1.618 \sin(628.2t) + 1.101 \sin(942.3t) - 0.851 \sin(1256.4t)$$

As a point of interest, the exact terms obtained using Equation 20.3 are:

$$A_1 = \frac{2V_m}{\pi} = \frac{(2)(5)}{\pi} = 3.18$$

$$A_2 = \left(\frac{2V_m}{\pi}\right)\left(\frac{-1}{2}\right) = \frac{-(2)(5)}{2\pi} = -1.59$$

$$A_3 = \left(\frac{2V_m}{\pi}\right)\left(\frac{1}{3}\right) = \frac{(2)(5)}{3\pi} = 1.06$$

$$A_4 = \left(\frac{2V_m}{\pi}\right)\left(\frac{-1}{4}\right) = -\frac{(2)(5)}{4\pi} = -0.796$$

Increasing the number of strips would reduce the difference between these terms and those obtained with the strips.

The expression is more useful if written in terms of V_m and ω, as in Equations 20.2 and 20.3. It can then be used with any sawtooth wave by

TABLE 20.2 Values for Example 20.1

strip	α degrees	h	$h \sin \alpha$	$h \cos \alpha$	$h \sin 2\alpha$	$h \cos 2\alpha$	$h \sin 3\alpha$	$h \cos 3\alpha$	$h \sin 4\alpha$	$h \cos 4\alpha$
1	9	0.25	0.039	0.247	0.077	0.238	0.113	0.223	0.147	0.202
2	27	0.75	0.340	0.668	0.607	0.441	0.741	0.117	0.713	−0.232
3	45	1.25	0.884	0.884	1.250	0	0.884	−0.884	0	−1.250
4	63	1.75	1.559	0.794	1.416	−1.029	−0.274	−1.728	−1.664	−0.541
5	81	2.25	2.222	0.352	0.695	−2.140	−2.005	−1.021	−1.323	1.820
6	99	2.75	2.716	−0.430	−0.850	−2.615	−2.450	1.248	1.616	2.225
7	117	3.25	2.896	−1.475	−2.629	−1.910	−0.508	3.210	3.091	−1.004
8	135	3.75	2.652	−2.652	−3.750	0	2.652	2.652	0	−3.750
9	153	4.25	1.929	−3.787	−3.438	2.498	4.198	−0.665	−4.042	−1.313
10	171	4.75	0.743	−4.692	−1.468	4.517	2.156	−4.232	−2.792	3.843
11	189	−4.75	0.743	4.692	−1.468	−4.517	2.156	4.232	−2.792	−3.843
12	207	−4.25	1.929	3.787	−3.438	−2.498	4.197	−0.665	−4.042	1.313
13	225	−3.75	2.652	2.652	−3.750	0	2.652	−2.652	0	3.750
14	243	−3.25	2.896	1.475	−2.629	1.910	−0.508	−3.210	3.091	1.004
15	261	−2.75	2.716	0.430	−0.850	2.615	−2.450	−1.248	1.616	−2.225
16	279	−2.25	2.222	−0.352	0.695	2.140	−2.004	1.021	−1.323	−1.820
17	297	−1.75	1.559	−0.794	1.416	1.029	−0.274	1.728	−1.664	0.541
18	315	−1.25	0.884	−0.884	1.250	0	0.884	0.884	0	1.250
19	333	−0.75	0.340	−0.668	0.607	−0.441	0.741	−0.117	0.713	0.232
20	351	−0.25	0.039	−0.247	0.077	−0.238	0.113	−0.223	0.147	−0.202
	Total =	0	31.96	0	−16.18	0	11.01	0	−8.51	0

putting the V_m and ω of the wave into the expression. This form can be obtained by multiplying the general expression by V_m, and dividing each coefficient by the value of V_m. Example 20.1, where $V_m = 5$, gives

$$v = V_m \left[\frac{3.196}{5} \sin(\omega t) - \frac{1.618}{5} \sin(2\omega t) + \frac{1.101}{5} \sin(3\omega t) - \frac{0.851}{5} \sin(4\omega t) \right]$$

$$v = V_m [0.639 \sin(\omega t) - 0.324 \sin(2\omega t) + 0.22 \sin(3\omega t) - 0.17 \sin(4\omega t)]$$

The values that were calculated for Example 20.1 are shown in Table 20.2. The calculations can be lengthy and tedious, even when done with an electronic calculator. Fortunately, computer programs, including

PSpice, are available to do the analysis. Some of the programs also plot the harmonics and the total waveform. A simple program in BASIC that determines the terms and effective value is listed in Table 20.3. This program can also be used to calculate the average and effective values of waveforms. The program lets you enter the curve as linear sections, or using values at the center of strips, or both. Again, using more strips gives more exact results.

TABLE 20.3 Fourier Series Program

```
10 CLS
20 PRINT "     ************************************"
30 PRINT "     ***** FOURIER ANALYSIS PROGRAM ***** "
40 PRINT "     ************************************"
50 PRINT
60 PRINT "THIS PROGRAM GIVES THE VALUES OF Ao, An TERMS, AND Bn TERMS IN THE GENERAL"
70 PRINT "EXPRESSION y = Ao + A1 sinwt + B1 sinwt + ...Ansin(nwt) + Bncos(nwt)"
80 PRINT
90 PRINT "TO USE THE PROGRAM, DIVIDE THE CYCLE INTO SECTIONS. EACH SECTION CAN BE A LINEAR PART, A CURVED PART, OR A LINEAR APPROXIMATION OF A CURVED PART"
100 PRINT
110 DIM AA(100), AB(100), X1(100), X2(100), Y1(100), Y2(100), W(100), SN(100), SW(100), L(100), H(100), X(100), Y(100), AR(100), AE(100), R(100)
120 PRINT
130 PRINT "
140 PRINT "
150 PRINT "
160 PRINT "
170 PRINT "
180 PRINT "
190 PRINT "
200 PRINT
210 PRINT "THE SECTIONS OF THE CURVE CAN BE ENTERED AS LINEAR OR BY VALUES AT CENTER OF THE STRIP. USING VALUES AT THE CENTER GIVES MORE CORRECT RESULTS FOR NONLINEAR SECTIONS."
220 PRINT
230 INPUT "HOW MANY SECTIONS ARE THERE IN ONE CYCLE" ;P
240 INPUT "WHAT IS THE TOTAL NUMBER OF STRIPS THAT YOU WANT TO USE FOR ONE CYCLE" ;MT
250 INPUT "HOW MANY HARMONIC TERMS DO YOU WANT" ;F
```

```
260 PRINT
270 FOR S=1 TO P
280 PRINT
290 PRINT "FOR SECTION" ;S;
300 INPUT "START: DEGREES, VALUE";X1(S), Y1(S)
310 INPUT "END: DEGREES, VALUE";X2(S), Y2(S)
320 INPUT "ANALYZE SECTION AS LINEAR (Y/N) CAPS LOCK
OFF!";T$(S)
330 IF T$(S) = "Y" THEN 320
340 NEXT S
350 S=1
360 FOR N=1 TO P
370 L(N)=X2(N)-X1(N)
380 IF T$(N) = "y" THEN 520
390 SN(N)=MT
400 IF P=1 THEN 440
410 CLS
420 PRINT "HOW MANY STRIPS IN SECTION";N;"? SHOULD
NOT BE MORE THAN";MT-TNS;
430 INPUT SN(N)
440 FOR Q=1 TO SN(N)
450 PRINT "FOR SECTION";N;": VALUE AT CENTER OF
STRIP";Q;
460 INPUT Y(S)
470 S=S+1
480 NEXT Q
490 SW(N) = L(N)/SN(N)
500 TND = TND + L(N)
510 TNS = TNS + SN(N)
520 NEXT N
530 M = MT - TNS
540 TLD = 360 -TND
550 FOR N= 1 TO P
560 IF T$(N) <> "y" THEN 610
570 SN(N) = INT(.5+L(N)*M/TLD)
580 IF SN(N) <> 0 THEN 600
590 SN(N) = 1
600 SW(N) = L(N)/SN(N)
610 NEXT N
620 D=1
630 FOR N=1 TO P
640 FOR J=1 TO SN(N)
650 W(D) = X1(N)+L(N)*(J-.5)/SN(N)
```

```
660 IF T$(N) <> "y" THEN 690
670 H(D) = Y1(N)+(W(D)-X1(N))*(Y2(N)-Y1(N))/L(N)
680 GOTO 710
690 E=E+1
700 H(D) = Y(E)
710 AR(D) = H(D)*SW(N)+AR(D-1)
720 AE(D) = (H(D))^2*SW(N)+AE(D-1)
730 D=D+1
740 NEXT J
750 NEXT N
760 A0=AR(D-1)/360
770 EFF = (AE(D-1)/360)^.5
780 CLS
790 PRINT "IN THE GENERAL EXPRESSION y = Ao + A1sinwt +
B1coswt + .....+ Ansin nwt + Bncos nwt, the terms are;
800 PRINT
810 PRINT "A0=";A0," EFF=";EFF
820 FOR T = 1 TO F
830 D=1
840 FOR N=1 TO P
850 FOR U = 1 TO SN(N)
860 R(D) = W(D) * .01745293233
870 AA(T) = AA(T) + H(D)*SW(N)*SIN(T*R(D))
880 AB(T) = AB(T) + H(D)*SW(N)*COS(T*R(D))
890 D=D+1
900 NEXT U
910 NEXT N
920 PRINT "A";T;"=";AA(T)/180," B";T;"=";AB(T)/180
930 NEXT T
940 END
```

The following steps should be completed before starting the program:

1. Assign one cycle a length of 360°.
2. Select the total number of strips into which you want the curve divided. Using more strips gives more exact results, but takes longer to run.
3. Divide the curve into sections. These can be linear lengths, curved lengths, or both. A length of a section should not include an instantaneous change in the *y* value.
4. Note the degrees and *y* value at the start and end of each section.
5. Divide each curved section into a number of equal width strips. Note the *y* value for the degrees at the center of each strip.
6. Decide how many harmonics terms are wanted in the analysis.

EXAMPLE 20.2

Find the effective value and the Fourier series terms to the fifth harmonic for the square wave in Figure 20.13.

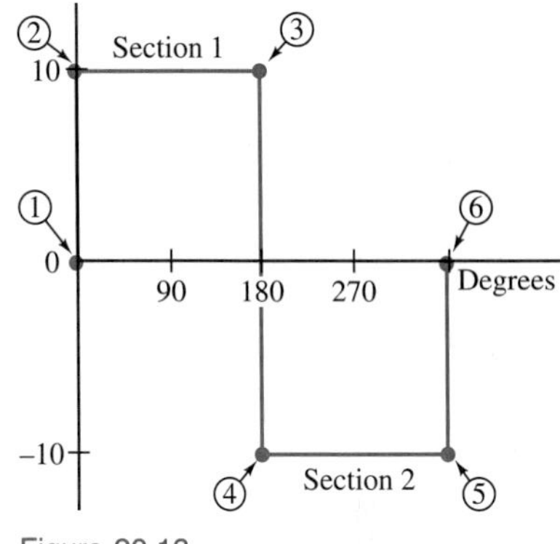

Figure 20.13

Solution

Forty strips are used for the solution. There are two sections, both linear. Five harmonics terms are included. The values for each section are:

Section 1. Start 0, 10 End 180, 10
Section 2. Start 180, −10 End 360, −10

Load the program and type RUN to start it. The displays and entries are as follows:

```
HOW MANY SECTIONS ARE THERE IN ONE CYCLE?
ENTER; 2
WHAT IS THE TOTAL NUMBER OF STRIPS THAT YOU WANT TO
USE FOR ONE CYCLE?
ENTER; 40
HOW MANY HARMONICS DO YOU WANT?
ENTER; 5
FOR SECTION 1 START: DEGREES, VALUE?
ENTER; 0, 10
END: DEGREES, VALUE?
ENTER; 180, 10
ANALYZE SECTION AS LINEAR (Y/N)?
ENTER; Y
FOR SECTION 2 START: DEGREES, VALUE?
ENTER; 180, -10
END: DEGREES, VALUE?
ENTER; 360, -10
ANALYZE SECTION AS LINEAR (Y/N) CAPS LOCK OFF!?
ENTER; Y
```

The analysis is performed and the following results displayed.

IN THE GENERAL EXPRESSION $y = A_0 + A_1 \sin \omega t + B_1 \cos \omega t + \cdots + A_n \sin(n\omega t) + B_n \cos(n\omega t)$

```
A0 = 0                    EFF = 10
A1 = 12.74576              B1 = 8.262635E-04
A2 = -2.278222E-07         B2 = -1.356337E-06
A3 = 4.283748              B3 = 8.344439E-04
A4 = 1.37753E-07           B4 = 1.695421E-06
A5 = 2.613176              B5 = 8.474562E- 04
```

Note: Values less than 1E-4 times the largest term are generally zero.

These values can be compared to the exact values, which are

$$A_1 = \frac{(4)(10)}{\pi} = 12.73$$

$$A_3 = \frac{(4)(10)}{3\pi} = 4.24$$

$$A_5 = \frac{(4)(10)}{5\pi} = 2.55$$

Since $V_m = 10$, the general expression is

$$\upsilon = V_m \left[\frac{12.746}{10} \sin(\omega t) + \frac{4.284}{10} \sin(3\omega t) + \frac{2.613}{10} \sin(5\omega t) \right]$$

or

$$\upsilon = V_m \, [1.27 \sin(\omega t) + 4.28 \sin(3\omega t) + 2.61 \sin(5\omega t)]$$

Table 20.4 shows the quantities that would have to be calculated for only 20 strips.

TABLE 20.4 Values for Example 20.2

strip	(α) degrees	*h*	*h* sin α	*h* sin 3α	*h* sin 5α
1	9	10	1.564	4.540	7.071
2	27	10	4.540	9.877	7.071
3	45	10	7.071	7.071	−7.071
4	63	10	8.910	−1.564	−7.071
5	81	10	9.877	−8.910	7.071
6	99	10	9.877	−8.910	7.071
7	117	10	8.910	−1.564	−7.071
8	135	10	7.071	7.071	−7.071
9	153	10	4.540	9.877	7.071
10	171	10	1.564	4.540	7.071
11	189	−10	1.564	4.540	7.071
12	207	−10	4.540	9.877	7.071
13	225	−10	7.071	7.071	−7.071
14	243	−10	8.910	−1.564	−7.071
15	261	−10	9.877	−8.910	7.071
16	279	−10	9.877	−8.910	7.071
17	297	−10	8.910	−1.564	−7.071
18	315	−10	7.071	7.071	−7.071
19	333	−10	4.540	9.877	7.071
20	351	−10	1.564	4.540	7.071

It might be interesting for you to analyze the same waveform with the sections entered as curved. The height at the center of each strip should be entered as 10 for Section 1 and −10 for Section 2.

PSpice can also be used to determine the amplitudes of the dc component and the first nine frequency components, the fundamental and the second through eighth harmonics. However, when using PSpice, a harmonic analysis can only be performed in conjunction with a transient analysis. Hence, the file must contain a .TRAN command statement.

The general form of the .FOUR control statement is

```
.FOUR <(frequency)value> <output variable>
```

where <(frequency)value> is the fundamental frequency, and <output variables> is the voltage or current in the circuit.

There is one requirement when using the .FOUR and .TRAN control statements. To insure that PSpice has enough data points to perform the Fourier calculations, the transient analysis must be at least 1/⟨(frequency)value⟩ seconds long.

To illustrate the use of the .FOUR command statement, let us use PSpice to compute the Fourier coefficients for the waveform in Figure 20.11; then we compare our PSpice results to the values found in Example 20.1. Since we must perform a transient analysis, we apply the input voltage to a circuit consisting of a single 1-ohm resistor. The circuit is shown in Figure 20.14.

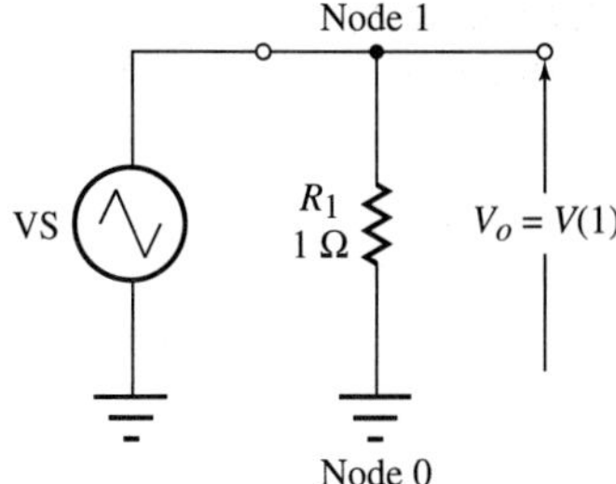

Figure 20.14 A resistive circuit driven by a repetitive ramp waveform.

To specify the nonsinusoidal input, we must use the piecewise-linear waveform (PWL) specification. The general form of the PWL statement is

```
PWL (<t1> <v1> <t2> <v2> . . . <tn> <vn>)
```

where <tn> is the time at a corner, in seconds, and <vn> is the voltage at a corner, in volts.

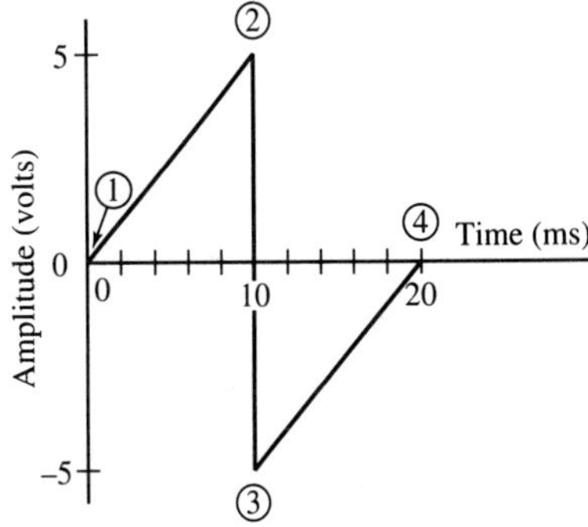

Figure 20.15 A repetitive ramp waveform.

Figure 20.15 shows the input waveform. Piecewise-linear means that the waveform can be approximated by line segments. In Figure 20.15, the waveform shown can be approximated by three line segments. Four corner points, labeled 1, 2, 3, and 4, are used to specify the three line segments. Each corner point has a pair of time-voltage values.

TABLE 20.5 Time-Voltage Values for Figure 20.15

Point *n*	*tn*	*vn*
1	0 seconds	0 volts
2	9.95 msec.	5 volts
3	10.05 msec.	−5 volts
4	20.00 msec.	0 volts

```
PWL EXAMPLE - FIGURE 20.15
****     CIRCUIT DESCRIPTION
*********************************************************


VS    1    0    PWL(0 0V 9.95M 5V 10.05M -5V 20M 0.0)
R1    1    0    1
.TRAN 0.1M20.0M
.FOUR 50 V(1)
.END
```

Figure 20.16 PSpice source file.

The voltages at times between the corners is the linear interpolation of the voltages at the corners.

The PSpice file is listed in Figure 20.16. Let us examine each statement.

Line 1. Specifies the input voltage source, VS, using the PWL specification. VS is connected between Nodes 1 and 0.

Line 2. Specifies a resistor connected between Nodes 1 and 0. The value of the resistor is 1 ohm.

Line 3. Causes a transient analysis to be performed for the interval 0–20 milliseconds. The time increment selected is 0.1 millisecond.

Line 4. Causes a Fourier analysis to be performed. The fundamental frequency selected is 50 Hz, which matches the period of the waveform specified using PWL in Line 1. The output variable selected is the voltage at Node 1, V(1).

Line 5. The .END statement that closes the file.

The results of the Fourier analysis are automatically written to the output file, eliminating the need for a .PRINT statement. The results are shown in Figure 20.17.

Let us compare these results to the Fourier coefficients obtained in Example 20.1. The dc value returned by PSpice is 2.8×10^{-9}, or 0, for all practical purposes, which agrees with our previous results. The amplitude of the fundamental is 3.199, which agrees with 3.196 in Example 20.1.

The second harmonic at a frequency of 100 Hz is listed in Figure 20.17 as 1.600, but in Example 20.1 the value calculated was negative, -1.618. PSpice give the Fourier component as a positive value plus a phase angle. The phase angle listed in Figure 20.17 is $+180°$. Taking the phase angle into account yields

$$1.600 \angle 180° = -1.600 \angle 0° \approx -1.618$$

Our results again agree. Note that the same situation arises for the fourth harmonic. The phase angle is listed in Figure 20.17 as $-180°$, and we observe that

$$8.003 \times 10^{-1} \angle -180° = -0.8003 \approx -0.851$$

```
PWL EXAMPLE - FIGURE 20.15
****     FOURIER ANALYSIS              TEMPERATURE =    27.000 DEG C
*****************************************************************


FOURIER COMPONENTS OF TRANSIENT RESPONSE V(1)


DC COMPONENT =     2.865769E-09

HARMONIC  FREQUENCY     FOURIER    NORMALIZED      PHASE      NORMALIZED
   NO        (HZ)      COMPONENT   COMPONENT       (DEG)      PHASE (DEG)

    1      5.000E+01   3.199E+00   1.000E+00  -1.545E-07    0.000E+00
    2      1.000E+02   1.600E+00   5.001E-01   1.800E+02    1.800E+02
    3      1.500E+02   1.067E+00   3.334E-01  -2.343E-07   -7.985E-08
    4      2.000E+02   8.003E-01   2.502E-01  -1.800E+02   -1.800E+02
    5      2.500E+02   6.405E-01   2.002E-01  -1.673E-07   -1.283E-08
    6      3.000E+02   5.340E-01   1.669E-01  -1.800E+02   -1.800E+02
    7      3.500E+02   4.579E-01   1.431E-01   2.029E-06    2.184E-06
    8      4.000E+02   4.010E-01   1.253E-01   1.800E+02    1.800E+02
    9      4.500E+02   3.567E-01   1.115E-01  -2.830E-06   -2.675E-06

TOTAL HARMONIC DISTORTION =     7.351113E+01 PERCENT
```

Figure 20.17 The output file generated by PSpice.

20.4 Effective Value and Power for a Nonsinusoidal Source

The effective value of a nonsinusoidal source is given by

$$E_{\text{eff}} = \sqrt{E_0^2 + E_1^2 + E_2^2 + \cdots + E_n^2} \qquad \textbf{(20.10)}$$

where E_0 is the dc component, and E_1, E_2, . . . , E_n are the effective values of each term in the series.

A few lines are used to show how Equation 20.10 was obtained. Recall that the effective value is the square root of the area under the squared curve, divided by the length of a cycle. The area under the e^2 curve is now considered.

If

$$e = E_0 + E_{m1}\sin(\alpha) + E_{m2}\sin(2\alpha) + E_{m3}\sin(3\alpha)$$

$$e^2 = [E_0 + E_{m1}\sin(\alpha) + E_{m2}\sin(2\alpha) + E_{m3}\sin(3\alpha)]^2$$

When squared, the quantity on the right yields

$$\begin{aligned} e^2 = {} & [E_0^2 + E_{m1}^2\sin^2(\alpha) + E_{m2}^2\sin^2(2\alpha) + E_{m3}^2\sin^2(3\alpha)] \\ & + E_0\,[E_{m1}\sin(\alpha) + E_{m2}\sin(2\alpha) + E_{m3}\sin(3\alpha)] \\ & + [E_{m1}E_{m2}\sin(\alpha)\sin(2\alpha) + E_{m1}E_{m3}\sin(\alpha)\sin(3\alpha) \\ & + E_{m2}E_{m3}\sin(2\alpha)\sin(3\alpha)] \end{aligned}$$

The area under the terms in the e^2 curve is now examined.

1. The second set of terms does not contribute any area, since the area under a sine wave is zero.
2. Sin (α) sin(2α) can be written as 0.5 [cos(α − 2α) − cos(α + 2α)]. Since the area under a cosine wave is zero, the sin(α)sin(2α) term also has a zero area. This also holds true for the sin(α) sin(3α) and sin(2α)sin(3α) terms.
3. The E_0^2 and $\sin^2 n\alpha$ terms are the only terms that have a net area. The area under the E_0^2 curve is $2\pi E_0^2$. The area under the $\sin^2 n\alpha$ curve is $\alpha(E_{mn})^2$. Since, $E_{\text{eff}} = \sqrt{\text{Area}/(2\pi)}$

$$E_{\text{eff}} = \sqrt{\frac{2E_0^2 + E_{m1}^2 + E_{m2}^2 + E_{m3}^2}{2}}$$

or

$$E_{\text{eff}} = \sqrt{E_0^2 + \frac{E_{m1}^2}{2} + \frac{E_{m2}^2}{2} + \frac{E_{m3}^2}{2}}$$

But $E_m/\sqrt{2}$ is the effective value of a sine wave, so $E_m^2/2$ is the effective value squared. Writing E_{eff} in terms of the effective value gives

$$E_{\text{eff}} = \sqrt{E_0^2 + E_1^2 + E_3^2}$$

which is the same as Equation 20.10.

As a point of interest, Equation 20.10 was used to find the effective value of the ac component in Chapter 15, Section 15.5.

Likewise, the effective value for the current is given by

$$I_{\text{eff}} = \sqrt{I_0^2 + I_1^2 + I_2^2 + \cdots + I_n^2} \qquad \textbf{(20.11)}$$

where I_0 is the dc component, and $I_1, I_2, \ldots, I_n$ are the effective values of each term in the series.

EXAMPLE 20.3

Find the effective value of

a. $e = 5 + 3\sin(\alpha) + 2\sin(2\alpha) + 1\sin(3\alpha)$

b. $i = 6 + 4\sin(\alpha) + 3\cos(\alpha) + 2\sin(2\alpha) + 1\cos(2\alpha)$

Solution

a.

$$E_{\text{eff}} = \sqrt{E_0^2 + E_1^2 + E_2^2 + E_3^2}$$

$$E_{\text{eff}} = \sqrt{(5)^2 + \left(\frac{3}{\sqrt{2}}\right)^2 + \left(\frac{2}{\sqrt{2}}\right)^2 + \left(\frac{1}{\sqrt{2}}\right)^2}$$

$$E_{\text{eff}} = 5.66 \text{ V}$$

b.

$$I_{\text{eff}} = \sqrt{I_0^2 + I_1^2 + I_2^2 + I_3^2}$$

$$I_{\text{eff}} = \sqrt{(6)^2 + \left(\frac{4}{\sqrt{2}}\right)^2 + \left(\frac{3}{\sqrt{2}}\right)^2 + \left(\frac{2}{\sqrt{2}}\right)^2 + \left(\frac{1}{\sqrt{2}}\right)^2}$$

$$I_{\text{eff}} = 7.14 \text{ A}$$

The power from a nonsinusoidal source connected to a resistor R is

$$P = \frac{E_0^2}{R} + \frac{E_1^2}{R} + \frac{E_2^2}{R} + \cdots + \frac{E_n^2}{R} \quad \textbf{(20.12)}$$

where E_0 is the dc component in the series, and $E_1, E_2, \ldots, E_n$ are the effective values of each term in the series.

The power expression can also be obtained in terms of the current. The result is

$$P = I_0^2R + I_1^2R + I_2^2R + \cdots + I_n^2R \quad \textbf{(20.13)}$$

where I_0 is the dc component of the current, and $I_1, I_2, \ldots, I_n$ are the effective values of each term in the series.

EXAMPLE 20.4

The voltage across a 50-ohm resistor is given by

$$e = 10 + 10\sin(\omega t) + 5\sin(3\omega t).$$

Find the power in the resistor.

Solution

$$\begin{aligned} P &= \frac{E_0^2}{R} + \frac{E_1^2}{R} + \frac{E_3^2}{R} \\ &= \frac{(10\text{ V})^2}{50\ \Omega} + \frac{(10\text{ V})^2}{(\sqrt{2})^2(50\ \Omega)} + \frac{(5\text{ V})^2}{(\sqrt{2})^2(50\ \Omega)} \\ &= \frac{100\text{ V}^2}{50\ \Omega} + \frac{100\text{ V}^2}{100\ \Omega} + \frac{25\text{ V}^2}{100\ \Omega} \\ P &= 3.25\text{ W} \end{aligned}$$

The power from a nonsinusoidal wave results only from the voltage and current of the same frequency. For example, there is no power interaction between the second harmonic voltage and the third harmonic current. Thus, the total power is also given by

$$P = E_0I_0 + E_1I_1\cos(\theta_1) + E_2I_2\cos(\theta_2) + \cdots + E_nI_n\cos(\theta_n) \quad \textbf{(20.14)}$$

where E_0, I_0 are the dc components of the voltage and current; $E_1, E_2, \ldots, E_n$ and $I_1, I_2, \ldots, I_n$ are the effective values of the voltage and current for each term in the series; and $\theta_1, \theta_2, \ldots, \theta_n$ are the angles between the voltage and current in the fundamental and each harmonic.

EXAMPLE 20.5

The voltage and current in a circuit are given by

$$e = 120\sin(\omega t) - 45\cos(3\omega t) + 15\sin(5\omega t)$$

$$i = 2.4\sin(\omega t - 60°) - 1.56\cos(3\omega t - 30°) + 0.424\sin(5\omega t - 45°)$$

Calculate the power using Equation 20.14.

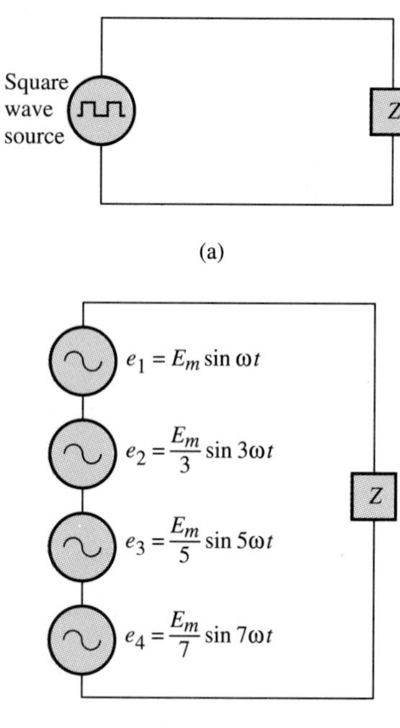

Figure 20.18 Analysis of a circuit with a nonsinusoidal source. (a) The circuit and its nonsinusoidal source. (b) Each source in the equivalent circuit represents a term in the Fourier series.

Solution

There are no dc terms in the equations, so

$$P = E_1 I_1 \cos(\theta_1) + E_3 I_3 \cos(\theta_3) + E_5 I_5 \cos(\theta_5)$$

$$= \left(\frac{120 \text{ V}}{\sqrt{2}}\right)\left(\frac{2.4 \text{ A}}{\sqrt{2}}\right)\cos(60°) + \left(\frac{45 \text{ V}}{\sqrt{2}}\right)\left(\frac{+1.56\text{A}}{\sqrt{2}}\right)\cos(30°)$$

$$+ \left(\frac{15 \text{ V}}{\sqrt{2}}\right)\left(\frac{0.424 \text{ A}}{\sqrt{2}}\right)\cos(45°)$$

$$= (144)(0.5) + (35.1)(0.866) + (3.18)(0.707)$$

$$P = 104.65 \text{ W}$$

20.5 Analysis of Circuits with Nonsinusoidal Sources

The analysis of these circuits is more complex than the analysis of a circuit with a sinusoidal source. The procedure that is used is:

1. Let each term in the series represent an individual source in series, as in Figure 20.18.
2. Use superposition to analyze the circuit for each term and to obtain the totals.

A series circuit and a parallel circuit are now analyzed.

EXAMPLE 20.6

A series circuit of a 60-ohm resistor, a 2-H inductor, and a 3-microfarad capacitor is connected across a nonsinusoidal source. The emf of the source is $e = 20 \sin(377t) + 10 \sin(754t)$. Find:

a. the sinusoidal expression for the current.
b. the sinusoidal expression for the voltage across each component.
c. the effective value of the source voltage and current.
d. the power dissipated in the circuit.

Solution

a. The source is considered to be two sources

$$e_1 = 20 \sin(377t)$$
$$e_2 = 10 \sin(754t)$$

For e_1,

$$X_{L1} = \omega_1 L = (377 \text{ rad/s})(2 \text{ H}) = 754 \ \Omega$$

$$X_{C1} = \frac{1}{\omega_1 C} = \frac{1}{(377 \text{ rad/s})(3 \times 10^{-6} \text{ F})} = 884.2 \ \Omega$$

$$Z_1 = R + jX_{L1} - jX_{C1}$$
$$= 60 \ \Omega + j754 \ \Omega - j884.2 \ \Omega$$
$$Z_1 = 143.36 \angle -65.26° \ \Omega$$

Then

$$I_{m1} = \frac{E_{m1}}{Z_1} = \frac{20 \text{ V}}{143.36\ \Omega} = 0.14 \text{ A}$$

and

$$\theta_1 = 65.26°$$

Since the circuit is capacitive, i_1 must lead e_1 by 65.26°. So

$$i_1 = 0.14 \sin(377t + 65.26°) \text{ A}$$

For e_2,

$$\begin{aligned}
X_{L2} &= \omega_2 L = (754 \text{ rad/s})(2 \text{ H}) = 1508\ \Omega \\
X_{C2} &= \frac{1}{\omega_2 C} = \frac{1}{(754 \text{ rad/s})(3 \times 10^{-6} \text{ F})} = 442.1\ \Omega \\
Z_2 &= R + jX_L - jX_C \\
&= 60\ \Omega + j1508\ \Omega - j442.1\ \Omega \\
Z_2 &= 1067.6 \angle 86.78°\ \Omega
\end{aligned}$$

Then

$$I_{m2} = \frac{E_{m2}}{Z_2} = \frac{10 \text{ V}}{1067.6\ \Omega} = 0.0094 \text{ A}$$

and

$$\theta_2 = 86.78°$$

Since the circuit is inductive, i_2 must lag e_2 by 86.78°. So

$$i_2 = 0.0094 \sin(754t - 86.78°) \text{ A}$$

By the Superposition Theorem,

$$\begin{aligned}
i &= i_1 + i_2 \\
i &= 0.14 \sin(377t + 65.26°) \text{ A} + 0.0094 \sin(754t - 86.78°) \text{ A}
\end{aligned}$$

b. The voltage across R is in phase with I and is equal to iR. So

$$\upsilon_R = (60\ \Omega)[0.14 \sin(377t + 65.26°) \text{ A}] + 60\ \Omega\ [0.0094 \sin(754t - 86.78°) \text{ A}]$$

or

$$\upsilon_R = 8.4 \sin(377t + 65.26°) \text{ V} + 0.564 \sin(754t - 86.78°) \text{ V}$$

The voltage across L leads the current by 90° and is equal to iX_L. For e_1,

$$\upsilon_{L1} = X_{L1} i_1$$

so

$$v_{L1} = (754\ \Omega)[0.14 \sin(377t + 65.26° + 90°)\ \text{A}]$$

or

$$v_{L1} = 105.56 \sin(377t + 155.26°)\ \text{V}$$

For e_2,

$$v_{L2} = X_{L2} i_2$$

and must lead i_2 by 90°, so

$$v_{L2} = (1508\ \Omega)[0.0094 \sin(754t - 86.78° + 90°)\ \text{A}]$$

or

$$v_{L2} = 14.18 \sin(754t + 3.22°)\ \text{V}$$

By the Superposition Theorem,

$$v_L = v_{L1} + v_{L2}$$

or

$$v_L = 105.56 \sin(377t + 155.26°)\ \text{V} + 14.18 \sin(754t + 3.22°)\ \text{V}$$

The voltage across C lags the current by 90° and is equal to iX_C. For e_1,

$$v_{C1} = X_{C1} i_1$$

So

$$v_{C1} = (884.2\ \Omega)[0.14 \sin(377t + 65.26° - 90°)\ \text{A}]$$

or

$$v_{C1} = 123.8 \sin(377t - 24.74°)\ \text{V}$$

For e_2,

$$v_{C2} = X_{C2} i_2$$

So

$$v_{C2} = (442.1\ \Omega)\ [0.0094 \sin(754t - 86.78° - 90°)\ \text{A}]$$

or

$$v_{C2} = 4.16 \sin(754t - 176.78°)\ \text{V}$$

By the Superposition Theorem,

$$v_C = v_{C1} + v_{C2}$$

or

$$v_C = 123.8 \sin(377t - 24.74°)\ \text{V} + 4.16 \sin(754t - 176.78°)\ \text{V}$$

The analysis can be summarized by stating that the quantities must be calculated for each term and combined by superposition.

c. The effective value of the source is

$$E_{eff} = \sqrt{E_1^2 + E_2^2}$$

$$= \sqrt{\left(\frac{20\ V}{\sqrt{2}}\right)^2 + \left(\frac{10\ V}{\sqrt{2}}\right)^2}$$

$$E_{eff} = 15.8\ V$$

$$I_{eff} = \sqrt{I_1^2 + I_2^2}$$

$$= \sqrt{\left(\frac{0.14\ A}{\sqrt{2}}\right)^2 + \left(\frac{0.0094\ A}{\sqrt{2}}\right)^2}$$

$$I_{eff} = 0.099\ A$$

d. The power equals I^2R. So

$$P = (0.099\ A)^2(60\ \Omega)$$

$$P = 0.588\ W$$

Also

$$P = E_1I_1\cos(\theta_1) + E_2I_2\cos(\theta_2)$$

$$P = \left(\frac{20\ V}{\sqrt{2}}\right)\left(\frac{0.14\ A}{\sqrt{2}}\right)\cos(65.26°) + \left(\frac{10\ V}{\sqrt{2}}\right)\left(\frac{0.0094\ A}{\sqrt{2}}\right)\cos(86.78°)$$

$$P = (1.4\ W)(0.419) + (0.047\ W)(0.0561)$$

$$P = 0.589\ W$$

A dc component did not enter into the calculations because there was no dc component. Had there been a dc voltage, then dc current would have been zero because of the capacitor in the circuit.

EXAMPLE 20.7

The parallel circuit shown in Figure 20.19 is connected to a source of $e = 40\sin(377t)\ V + 20\sin(754t)\ V$. Find:

a. the sinusoidal expression for the current in each branch.
b. the effective value of the source voltage and current.
c. the power dissipated in the circuit.

Solution

a. The source is considered to be two sources

$$e_1 = 40\sin(377t)\ V \text{ and } e_2 = 20\sin(754t)\ V$$

The current in R is in-phase with e_1 and e_2 and is

$$i_{R1} = \frac{e_1}{R} = \frac{40\ V}{50\ \Omega}\sin(377t)$$

$$i_{R1} = 0.8\sin(377t)$$

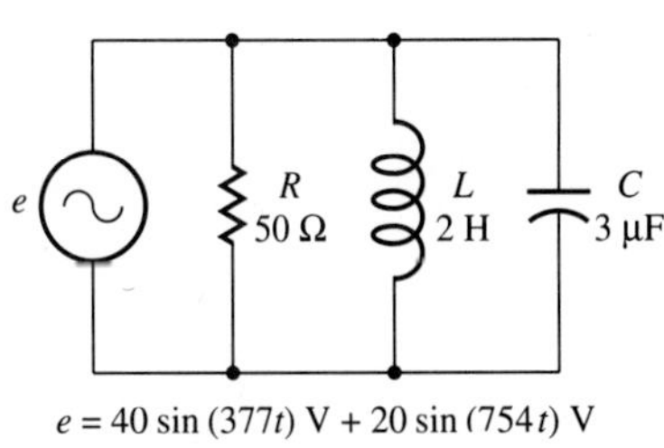

$e = 40\sin(377t)\ V + 20\sin(754t)\ V$

Figure 20.19

$$i_{R2} = \frac{e_2}{R} = \frac{20 \text{ V}}{50 \ \Omega} \sin(754t)$$

$$i_{R2} = 0.4 \sin(754t) \text{ A}$$

By the Superposition Theorem,

$$i_R = i_{R1} + i_{R2}$$

so

$$i_R = 0.8 \sin(377t) \text{ A} + 0.4 \sin(754t) \text{ A}$$

The currents in L lag each of the voltages e_1 and e_2 by 90°.

$$i_{L1} = \frac{e_1}{X_{L1}} \quad \text{and} \quad i_{L2} = \frac{e_2}{X_{L2}}$$

$$X_{L1} = \omega_1 L = (377 \text{ rad/s})(2 \text{ H}) = 754 \ \Omega$$

$$X_{L2} = \omega_2 L = (754 \text{ rad/s})(2 \text{ H}) = 1508 \ \Omega$$

Then

$$i_{L1} = \frac{40 \text{ V}}{754 \ \Omega} \sin(377t - 90°) = 0.053 \sin(377t - 90°) \text{ A}$$

and

$$i_{L2} = \frac{20 \text{ V}}{1508 \ \Omega} \sin(754t - 90°) = 0.013 \sin(754t - 90°) \text{ A}$$

By the Superposition Theorem,

$$i_L = i_{L1} + i_{L2}$$

so

$$i_L = 0.053 \sin(377t - 90°) \text{ A} + 0.013 \sin(754t - 90°) \text{ A}$$

The currents in C lead each of the voltages e_1 and e_2 by 90°.

$$i_{C1} = \frac{e_1}{X_{C1}} \quad \text{and} \quad i_{C2} = \frac{e_2}{X_{C2}}$$

$$X_{C1} = \frac{1}{\omega_1 C} = \frac{1}{(377)(3 \times 10^{-6} \text{ F})} = 884.2 \ \Omega$$

$$X_{C2} = \frac{1}{\omega_2 C} = \frac{1}{(754)(3 \times 10^{-6} \text{ F})} = 442.1 \ \Omega$$

Then

$$i_{C1} = \frac{40 \text{ V}}{884.2 \ \Omega} \sin(377t + 90°)$$

$$= 0.045 \sin(377t + 90°) \text{ A}$$

and

$$i_{C2} = \frac{20 \text{ V}}{442.1 \ \Omega} \sin(754t + 90°)$$

$$= 0.045 \sin(754t + 90°) \text{ A}$$

By the Superposition Theorem,

$$i_C = i_{C1} + i_{C2}$$

so

$$i_C = 0.045 \sin(377t + 90°) \text{ A} + 0.045 \sin(754t + 90°) \text{ A}$$

b. The effective value of the source is

$$E_{\text{eff}} = \sqrt{E_1^2 + E_2^2}$$

$$E_{\text{eff}} = \sqrt{\left(\frac{40 \text{ V}}{\sqrt{2}}\right)^2 + \left(\frac{20 \text{ V}}{\sqrt{2}}\right)^2}$$

$$E_{\text{eff}} = 31.62 \text{ V}$$

The effective value of the current is

$$I_{\text{eff}} = \sqrt{I_1^2 + I_2^2 + \cdots + I_n^2}$$

$$= \sqrt{I_{R1}^2 + I_{R2}^2 + I_{L1}^2 + I_{L2}^2 + I_{C1}^2 + I_{C2}^2}$$

$$I_{\text{eff}} = 0.635 \text{ A}$$

c. The power is given by E^2/R.

so

$$P = \frac{(31.62 \text{ V})^2}{50 \ \Omega}$$

$$P = 20 \text{ W}$$

Also

$$P = I_R^2 R$$

$$= \left[\left(\frac{0.8 \text{ A}}{\sqrt{2}}\right)^2 + \left(\frac{0.4 \text{ A}}{\sqrt{2}}\right)^2\right] 50 \ \Omega$$

$$P = 20 \text{ W}$$

Also

$$P = E_1 I_1 \cos(\theta_1) + \cdots + E_n I_n \cos(\theta_n)$$

$$P = \left(\frac{40 \text{ V}}{\sqrt{2}}\right)\left(\frac{0.8 \text{ A}}{\sqrt{2}}\right)\cos(0) + \left(\frac{20 \text{ V}}{\sqrt{2}}\right)\left(\frac{0.4 \text{ A}}{\sqrt{2}}\right)\cos(0)$$

$$+ \left(\frac{40 \text{ V}}{\sqrt{2}}\right)\left(\frac{0.053 \text{ A}}{\sqrt{2}}\right)\cos(90°) + \left(\frac{20 \text{ V}}{\sqrt{2}}\right)\left(\frac{0.013 \text{ A}}{\sqrt{2}}\right)\cos(90°)$$

$$+ \left(\frac{40 \text{ V}}{\sqrt{2}}\right)\left(\frac{0.045 \text{ A}}{\sqrt{2}}\right)\cos(90°) + \left(\frac{20 \text{ V}}{\sqrt{2}}\right)\left(\frac{0.045 \text{ A}}{\sqrt{2}}\right)\cos(90°)$$

$$P = 16 \text{ W} + 4 \text{ W} + 0 \text{ W} + 0 \text{ W} + 0 \text{ W} + 0 \text{ W}$$

$$P = 20 \text{ W}$$

In Example 20.7, the ratio of the second harmonic's amplitude to that of the fundamental is 0.5 for the source voltage, 0.5 for i_R, 0.245 for i_L, and 1 for i_C. Since the ratio for i_L and i_C is not the same as the ratio for the source voltage, the shape of the i_L and i_C current is not the same as that of the voltage. In other words, **distortion** occurs. This is because the

distortion

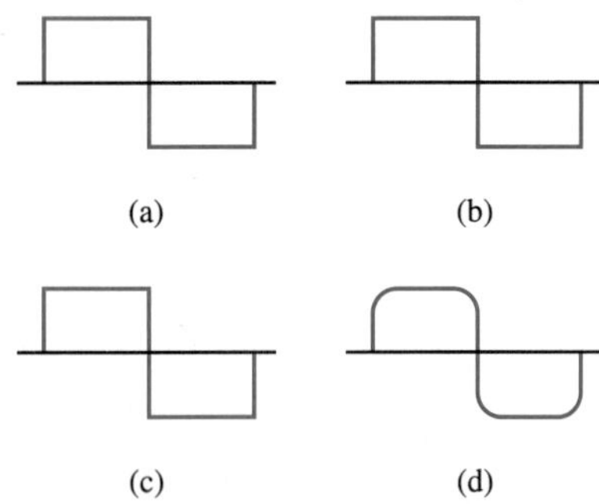

Figure 20.20 The high-frequency components are attenuated by the inductive reactance, resulting in a rounding of the square wave. (a) Voltage across a resistor. (b) The resistor current has no rounding. (c) Voltage across an inductor. (d) The high-frequency components are attenuated, so the wave cannot rise instantaneously.

reactance is different for each harmonic frequency. The low-frequency harmonics will be predominant in the inductive current, and the high-frequency ones will be predominant in the capacitive circuit.

Figure 20.20 shows the effect that resistance and inductance have on the current for a square wave voltage. The resistor current has the same shape as the voltage. However, the reduction in the high-frequency harmonics for the inductor results in a rounding of the pulse. This type of distortion can cause serious problems in some circuits.

Thus, the understanding of the Fourier series gives you a better idea of why stray capacitance and inductance should be avoided in electronic circuits.

20.6 Signal Analyzers

signal analyzers
spectrum analyzer
Fourier analyzer
wave analyzer

Various instruments are available for measuring and studying the frequency content of a waveform. They are generally known as **signal analyzers**. Three specific analyzers are the **spectrum analyzer**, **Fourier analyzer**, and the **wave analyzer**. These instruments are very useful for waveform analysis, vibration analysis, structural analysis, and other areas.

A spectrum analyzer can easily be mistaken for an oscilloscope, as seen by looking at Figure 20.21. However, the oscilloscope is a time-domain instrument, whereas the analyzer is a frequency-domain instrument.

A signal input to the spectrum analyzer results in a vertical line on the CRT at that frequency. The amplitude of the signal is related to the length of the line. A sample display is shown in Figure 20.22.

Some areas where spectrum analyzers are used are broadcasting, manufacturing, harmonic distortion, satellite communications, radar, pulse analysis, and cellular radio. Many spectrum analyzers can be interfaced with a computer, printer, and other peripherals, as shown in

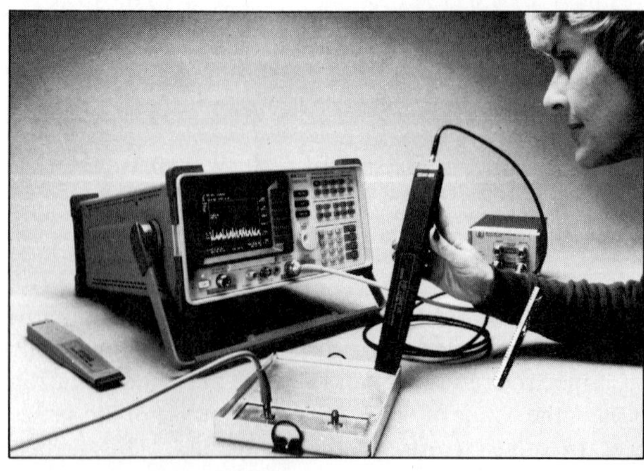

Figure 20.21 This spectrum analyzer resembles an oscilloscope, but displays amplitude versus frequency. (Photo courtesy of Hewlett-Packard, Signal Analysis Division)

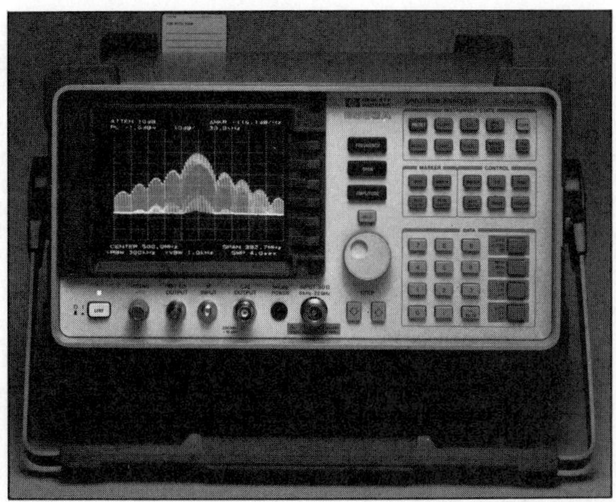

Figure 20.22 The height of the spikes in the display of this spectrum analyzer represents the amplitude of the frequency component. (Photo courtesy of Hewlett-Packard, Signal Analysis Division)

Figure 20.23. This allows automated operation and obtaining a hard copy of the display. For instance, an analyzer can be controlled by a computer to test the frequency content of the magnetrons that are used in microwave ovens. The program would direct the testing and rejecting of any units that are out of limits.

The number of controls and their location differ between models and manufacturers, so a specific explanation of making a measurement is meaningless. In general, one must input the signal and set the sweep range. Then other controls such as sensitivity, position, and brightness must be set, as with an oscilloscope.

How does the analyzer work? Basically, a superheterodyne receiver is swept-tuned over the frequency range. The sweeping is done automatically. As a frequency is tuned, any signal of that frequency will pass through and be displayed on the cathode ray tube. The analyzer in Figure 20.21 can be tuned to study signals from 50 kHz–22 GHz in five steps.

Figure 20.23 Interfacing the analyzer with peripherals permits automatic testing and recording of data. (Photo courtesy of Hewlett-Packard, Signal Analysis Division)

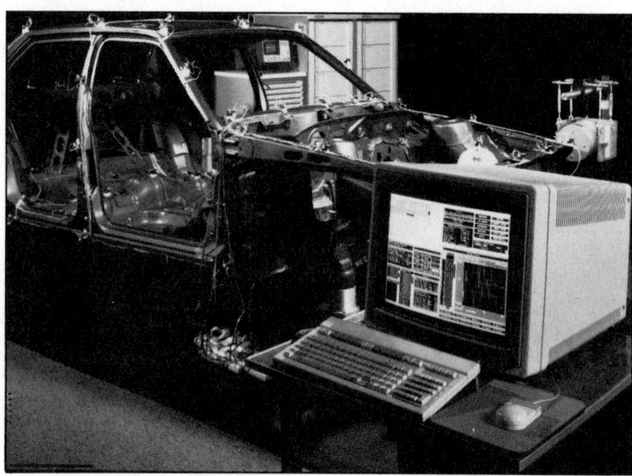

Figure 20.24 Systems such as this HP3565S are used for structural and vibration analysis. (Photo courtesy of Hewlett-Packard, Lake Stevens Instrument Division)

Although the spectrum analyzer provides frequency and amplitude information, it does not measure phase angles.

Some of the measurements that can be made with a spectrum analyzer are actual and relative frequency; actual and relative magnitude; noise level; AM, FM, and pulse RF modulation; and frequency response.

The Fourier analyzer is essentially a spectrum analyzer that uses parallel-tuned filters. Using parallel filters lets it measure all of the frequencies at one time. This is unlike the spectrum analyzer, which was swept-tuned through one frequency at a time. The Fourier analyzer also measures phase relations.

Systems such as that shown in Figure 20.24 are used in structural testing, environmental testing, and vibration and acoustic testing.

The third instrument that can be used is the wave analyzer. This instrument also provides amplitude and frequency information. The frequency must be swept manually and the amplitude is shown on the analog meter. The unit shown in Figure 20.25 has a frequency range of 15 Hz–50 kHz. The frequency is shown by the digital read-out at the top of the unit.

Figure 20.25 This wave analyzer displays the amplitude of the frequency component as the frequency is manually changed. (Photo courtesy of Hewlett-Packard Company)

The wave analyzer is essentially a voltmeter with a tuned filter whose pass-frequency can be varied. The meter measures the strength of the signal passed by the filter. The meter that was shown in Figure 20.25 can be connected to an X–Y recorder to obtain a hard copy of the amplitude-frequency response. As with the spectrum analyzer, phase relation is not measured.

The manual sweeping makes this instrument less suited for fast operations. The lack of a CRT display is a handicap when the general value of the signal frequency is not known. Wave analyzers are more commonly used in communication systems. They are generally used to measure the amplitude of a single component in a complex wave, measure a signal amplitude in the presence of noise and interfering signals, and measure the signal energy in a specific well-defined bandwidth.

Instruments such as these are invaluable when the effect of frequency must be considered. Whether it be the frequency response of a network, the harmonic content in a vibrating body, or the effect of forces on a structure, these instruments provide the information needed by the engineer.

SUMMARY

1. The circuits in many electronic applications use sources that produce nonsinusoidal waveforms such as square, sawtooth, and triangular.
2. A nonsinusoidal waveform is composed of the following:
   (a) A dc waveform (in some cases).
   (b) A fundamental sinusoidal waveform. This has the same frequency as the nonsinusoidal waveform.
   (c) Harmonic sinusoidal waveforms. These have frequencies that are integer multiples of the fundamental.
3. The series of terms that are used to define the nonsinusoidal waveform is called the Fourier series.
4. The A_0 term in the series is the dc value. The A_n terms are double the area under the $f(\alpha)$ sin $(n\alpha)$ curve, divided by a cycle-length of the fundamental. The B_n terms are double the area under the $f(\alpha)$ cos $(n\alpha)$ curve, divided by the cycle-length of the fundamental.
5. Nonsinusoidal waveforms also transfer energy. The effective value of the waveform is equal to the square root of the sum of the squares of the effective value of each term. The effective value is also called the rms value.
6. The power delivered by a nonsinusoidal source is equal to the sum of the power from each part.
7. Circuits with nonsinusoidal sources can be analyzed by considering each term as a source and using superposition.
8. The dependence of reactance on frequency causes distortion of the waveform. Capacitance rounds the corners of a voltage. Inductance rounds the corner of a current.
9. An instrument that is used to study the frequency content and amplitudes of the harmonics is the spectrum analyzer.

EQUATIONS

Eq.	Equation	Description	Page
20.1	$y = A_0 + (A_1 \sin \omega t + B_1 \cos \omega t) + (A_2 \sin 2\omega t + B_2 \cos 2\omega t) + \cdots + (A_n \sin n\omega t + B_n \cos n\omega t)$	The Fourier series for the terms in a nonsinusoidal waveform.	826
20.2	$\upsilon = \frac{4V_m}{\pi}\left(\sin \omega t + \frac{1}{3}\sin 3\omega t + \frac{1}{5}\sin 5\omega t + \frac{1}{7}\sin 7\omega t + \cdots\right)$	The Fourier expression for a square wave with no dc, which goes from $+V_m$ to $-V_m$.	829
20.3	$\upsilon = \frac{2V_m}{\pi}\left(\sin \omega t - \frac{1}{2}\sin 2\omega t + \frac{1}{3}\sin 3\omega t - \frac{1}{4}\sin 4\omega t + \frac{1}{5}\sin 5\omega t - \cdots\right)$	The Fourier expression for a sawtooth waveform with no dc and a peak of V_m.	829
20.4	$A_0 = \frac{\text{area under the } f(\alpha) \text{ curve}}{2\pi}$	The basic relation for the value of A_0 in the Fourier series.	831
20.5	$A_n = \frac{\text{area under the } f(\alpha) \sin n\alpha \text{ curve}}{\pi}$	The basic relation for the value of the A_n terms in the Fourier series.	831
20.6	$B_n = \frac{\text{area under the } f(\alpha) \cos n\alpha \text{ curve}}{\pi}$	The basic relation for the value of the B_n terms in the Fourier series.	831
20.7	$A_0 = \frac{h_1 + h_2 + \cdots + h_m}{m}$	The equation for determining A_0 from points on the waveform and the number of strips.	831
20.8	$A_n = 2\left(\frac{h_1 \sin n\alpha_1 + h_2 \sin n\alpha_2 \cdots + h_m \sin n\alpha_m}{m}\right)$	The equation for determining A_n from points on the waveform and $\sin(n\alpha)$ curve, and the number of strips.	831
20.9	$B_n = 2\left(\frac{h_1 \cos n\alpha_1 + h_2 \cos n\alpha_2 + \cdots + h_m \cos n\alpha_m}{m}\right)$	The equation for determining B_n from points on the waveform and $h \cos(n\alpha)$ curve, and the number of strips.	832
20.10	$E_{\text{eff}} = \sqrt{E_0^2 + E_1^2 + E_2^2 + \cdots + E_n^2}$	The equation for the effective value of a nonsinusoidal voltage waveform.	843
20.11	$I_{\text{eff}} = \sqrt{I_0^2 + I_1^2 + I_2^2 + \cdots + I_n^2}$	The equation for the effective value of a nonsinusoidal current waveform.	844
20.12	$P = \frac{E_0^2}{R} + \frac{E_1^2}{R} + \frac{E_2^2}{R} + \cdots + \frac{E_n^2}{R}$	The equation for power in terms of the voltage of each term in the series and the resistance.	845
20.13	$P = I_0^2R + I_1^2R + I_2^2R + \cdots + I_n^2 R$	The equation for power in terms of the current from each term in the series and the resistance.	845
20.14	$P = E_0I_0 + E_1I_1 \cos(\theta_1) + E_2I_2 \cos(\theta_2) + \cdots + E_nI_n \cos(\theta_n)$	The equation for power in terms of the voltage, current, and power factor for each term in the series.	845

QUESTIONS

1. What are some examples of nonsinusoidal waveform?
2. What is the name of the series of terms used to express a nonsinusoidal waveform?
3. What is a harmonic?
4. Which term in the series cannot be zero?
5. Which terms in the series are zero for an even function?
6. Which terms in the series are zero for an odd function?
7. Which term in the series represents the dc value?
8. Which terms in the series are zero if the series is used to define a 60-Hz sine wave?
9. What will happen to the square wave if the higher frequency terms are omitted?
10. Can two nonsinusoidal waves have the same effective value even though the terms are different?
11. Can two nonsinusoidal waves have the same dc value even though all terms are different?
12. Which terms in the series, high or low frequency, will be attenuated more by a capacitor in series with a resistive load?
13. Which terms in the series, high or low frequency, will be attenuated more by a capacitor in parallel with a resistive load?
14. What will happen to the effective value of the nonsinusoidal waveform if the values of the harmonics are increased? Decreased?
15. How does one find the total instantaneous value from the terms in the Fourier series for a nonsinusoidal waveform?
16. Which will give more correct results in determining the Fourier series, 10 strips or 40 strips?
17. How many strips are used when finding the terms by integration?
18. Which theorem is used for the analysis of circuits with nonsinusoidal sources?
19. The expression for a nonsinusoidal voltage is $e = \sin(\omega t) - 0.11\sin(3\omega t) - 0.04\sin(5\omega t) - 0.25\sin(7\omega t)$. What will be the expression if 5 V dc is added?
20. Which element will cause greater distortion in a nonsinusoidal voltage, a 5-ohm resistor or a 5-ohm capacitor?
21. Which instrument is used to study the frequency content of a waveform?
22. What information does a Fourier analyzer give that is not given by the spectrum analyzer?

PROBLEMS

SECTION 20.1 Nonsinusoidal Waveforms

1. Find the value of υ at $\omega t = \pi/6$ radians for each of the Fourier expressions below:

   (a) $\upsilon = 10 - 6.37\sin(\omega t) - 3.185\sin(2\omega t) - 2.12\sin(3\omega t)$

   (b) $\upsilon = 5 + 15\sin(\omega t) + 5\sin(3\omega t) + 3\sin(5\omega t)$

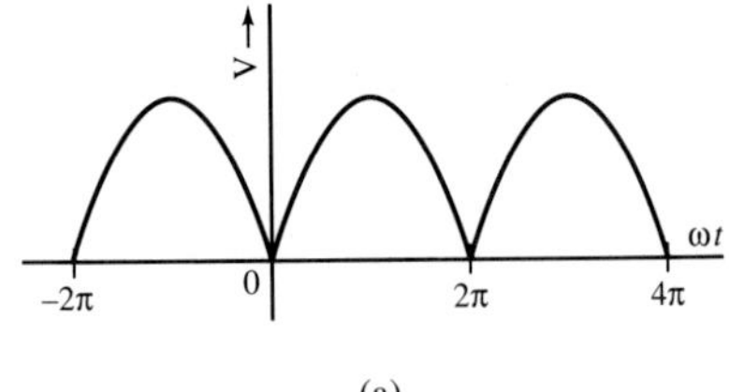

(a)

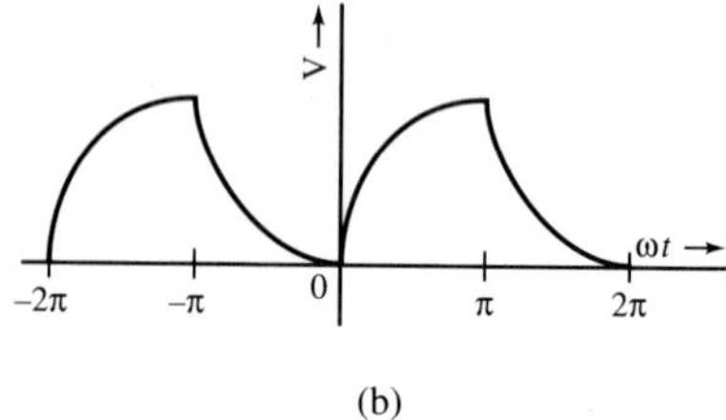

(b)

Figure 20.26

2. Find the value of υ at $\omega t = \pi/3$ radians for each of the following Fourier expressions:

   (a) $\upsilon = 5 - 4.05\cos(\omega t) - 0.45\cos(3\omega t) + 6.37\sin(\omega t) - 3.18\sin(2\omega t) + 2.12\sin(3\omega t)$

   (b) $\upsilon = 10 - 15.7\sin(\omega t) - 6.67\cos(2\omega t) - 1.33\cos(4\omega t) - 0.57\cos(6\omega t)$

3. Determine the dc value and the frequency in Hz of the fundamental for each of the following:

   (a) $\upsilon = 5 - 3.14\sin(377t) - 1.59\sin(754t) - 1.06\sin(1131t)$

   (b) $\upsilon = -2 - 3.14\sin(251.3t) - 1.33\cos(502.6t) - 0.27\cos(1005.2t)$

4. Work Problem 3 for the following:

   (a) $i = 0.52 + 1\cos(62.8t) + 0.866\cos(125.6t) + 0.667\cos(188.4t) + 0.443\cos(251.2t)$

   (b) $\upsilon = 25 + 25\sin(503t) + 15\sin(1006t) + 10\sin(2012t)$

SECTION 20.2 Fourier Analysis

5. Determine by inspection whether there is a dc term, even harmonics, cosine terms, and sine terms in each of the waves of Figure 20.26.

6. Work Problem 5 for the waveform of Figure 20.27.

7. Work Problem 5 for the waveform of Figure 20.28.

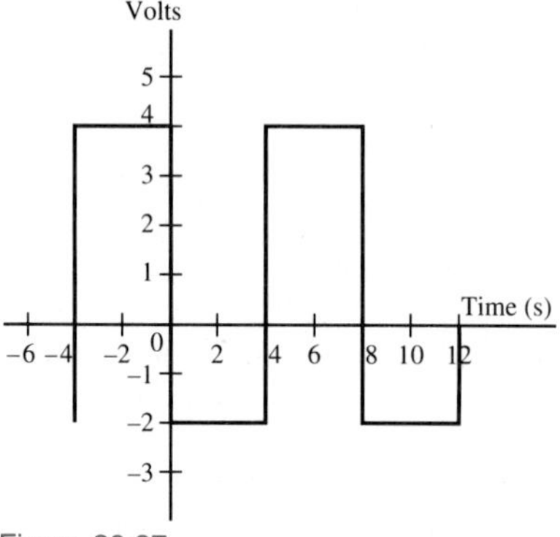

Figure 20.27

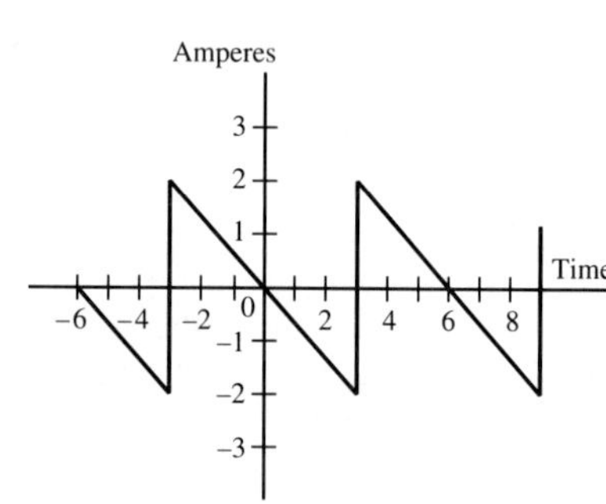

Figure 20.28

8. Work Problem 5 for the waves in Figure 20.29.

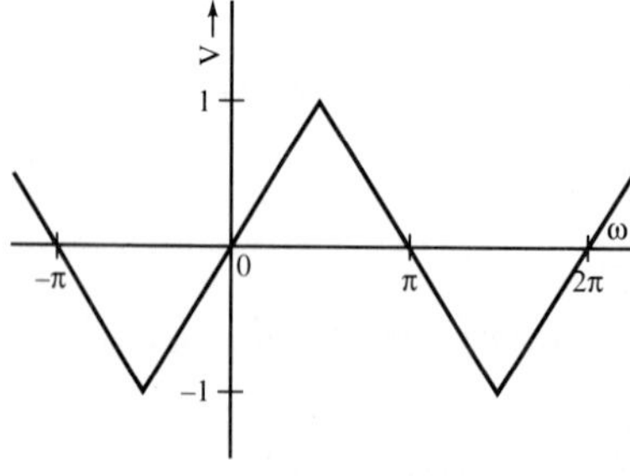

(a)

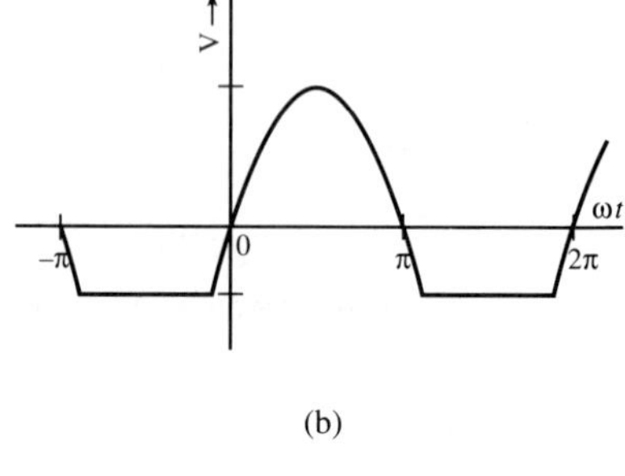

(b)

Figure 20.29

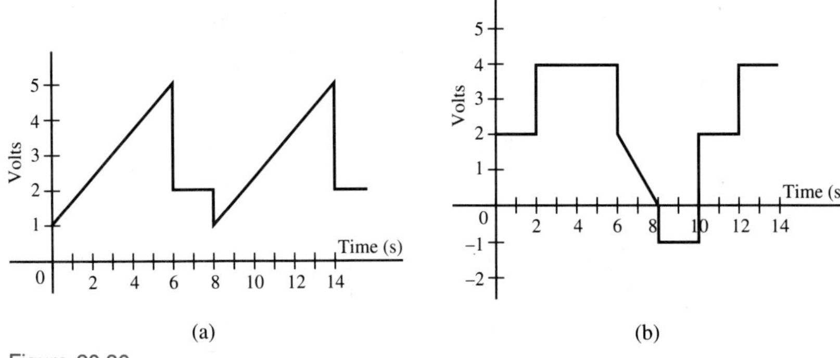

Figure 20.30

SECTION 20.3 Computer Determination of the Fourier Series

9. Divide one cycle of each waveform in Figure 20.30 into its linear sections, and give the degrees and value of the ends of each section.

10. Repeat Problem 9 for the waveforms in Figure 20.31.

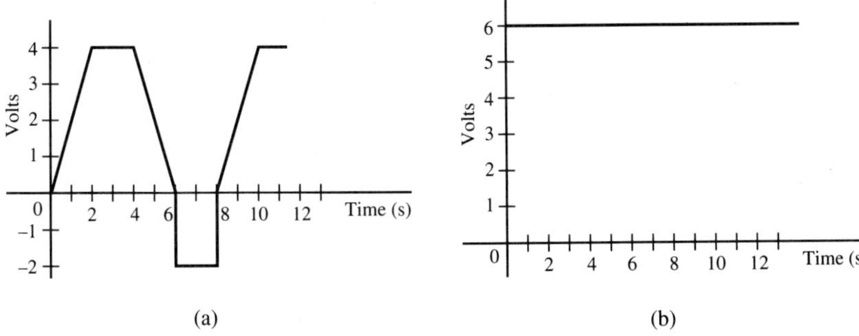

Figure 20.31

11. Approximate the shape of one cycle of the waveform in Figure 20.32 by dividing each curved section into at least four linear sections. Give the degrees and value of the ends of each section.

12. Repeat Problem 11 for the waveform in Figure 20.33.

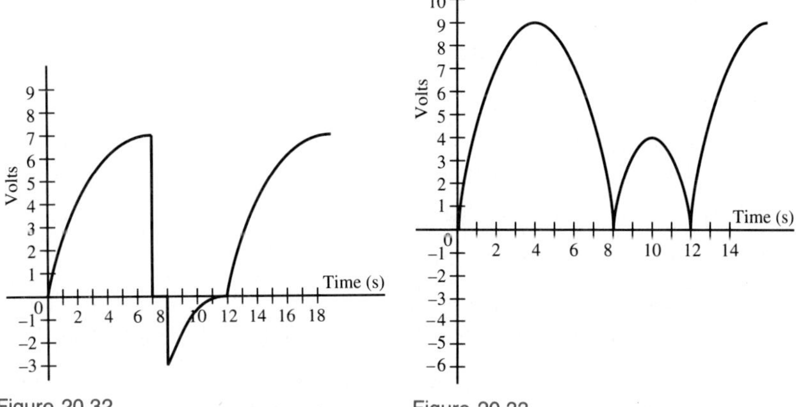

Figure 20.32

Figure 20.33

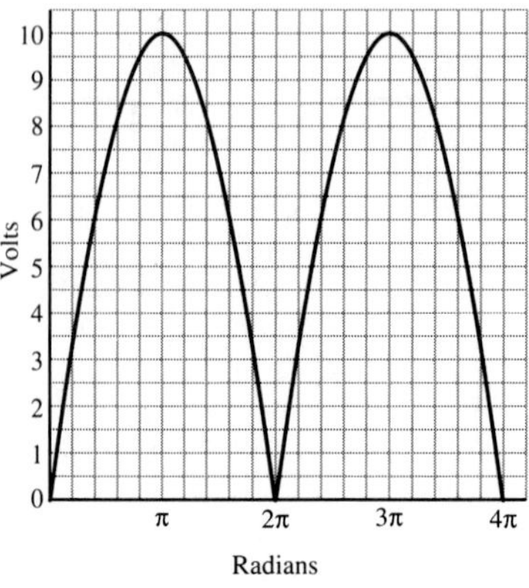

Figure 20.34

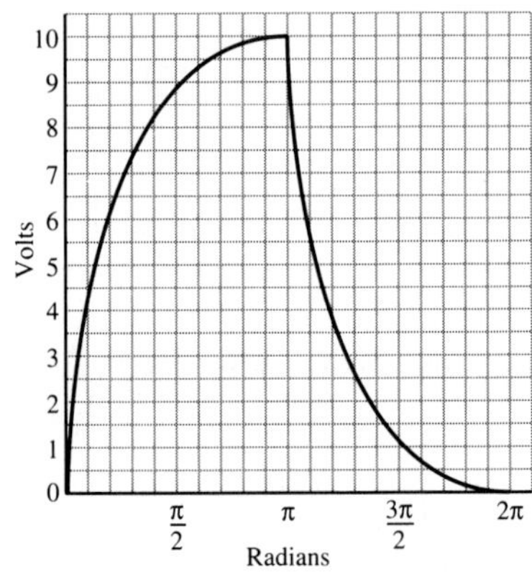

Figure 20.35

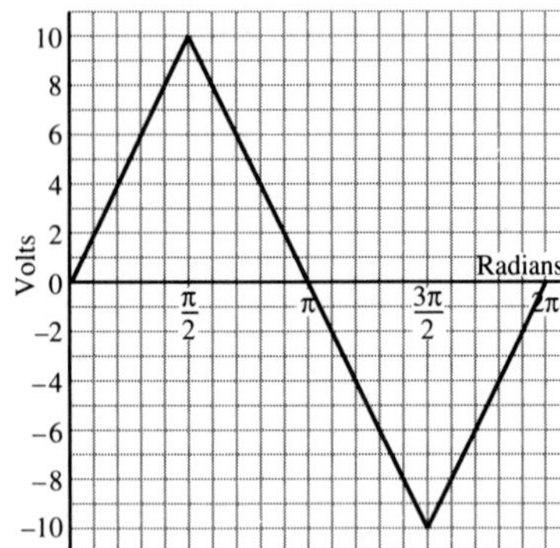

Figure 20.36

13. Use the computer program to determine the Fourier expression in terms of V_m to the fifth harmonic for the wave of Figure 20.34 for one cycle.

14. Use the computer program to determine the Fourier expression in terms of V_m to the fifth harmonic for the wave of Figure 20.35.

15. Use the computer program to determine the Fourier expression in terms of V_m for the fifth harmonic for the wave in Figure 20.36.

16. Use the computer program to determine the Fourier expression to the fifth harmonic for the wave shown in Figure 20.37.

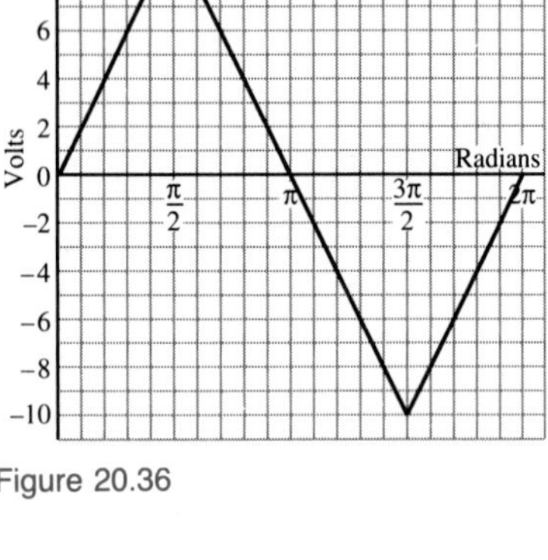

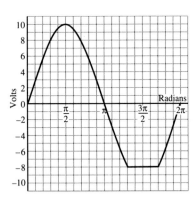

Figure 20.37

SECTION 20.4 Effective Value and Power for a Nonsinusoidal Source

17. A 20-ohm resistor is connected to a nonsinusoidal source of $e = 10 - 6.37 \sin(\omega t) - 3.18 \sin(2\omega t) - 2.12 \sin(3\omega t)$. Find:

    (a) The effective value of the voltage and current.

    (b) The power dissipated in the resistor.

18. Work Problem 17 for a 50-ohm resistor and a source voltage of $e = 25 + 25 \sin(\omega t) + 15 \sin(2\omega t) + 10 \sin(4\omega t)$.

19. The current in a 30-ohm resistor is $i = 1 + 5 \sin(200t) + 2 \sin(600t) + 1 \sin(1000t)$. Find:

    (a) The effective value of the voltage and current.

    (b) The power dissipated in the resistor.

20. The current in a circuit is $i = 20 \sin (100t + 30° + 1.27 \sin(300t - 60°)$ when the voltage is $e = 100 \sin(100t) + 11 \sin(330t)$. Find:

    (a) The power supplied to the circuit.

    (b) The power factor of the circuit.

21. Work Problem 20 for a voltage of $e = 250 \sin(377t + 53.2°) + 310 \sin(1131t + 136°)$ and a current of $i = 50 \sin(377t) + 25 \sin(1131t + 60°)$.

22. Work Problem 20 for a voltage of $e = 200 \sin(377t) + 100 \sin(1131t)$ and a current of $i = 20 \sin(377t + 36.8°) + 11.54 \sin(1131t - 22.6°)$.

SECTION 20.5 Analysis of Circuits with Nonsinusoidal Sources

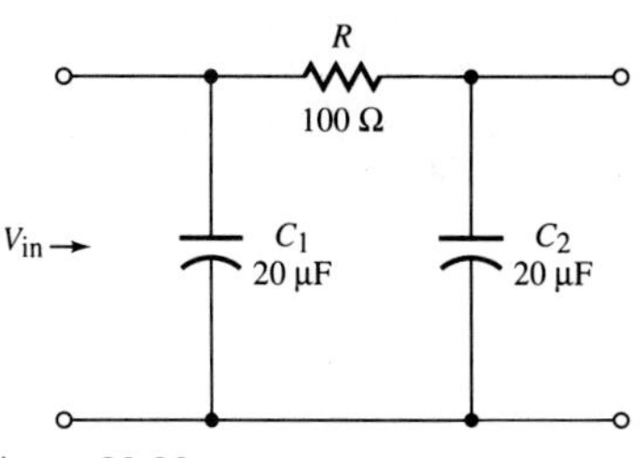

Figure 20.38

23. A pulse, as in Figure 20.10(b), is connected to the filter circuit of Figure 20.38. V_m is 20 V and the fundamental frequency is 60 Hz.

    (a) Determine the effective value of the voltage across C_2.

    (b) Determine the ripple factor for the voltage across C_2.

24. Determine the effective value of the pulse in Figure 20.10(c) for a V_m of 100 V using Equation 20.10. Compare this to $V_{eff} = 0.707\ V_m$.

25. A source pulse having the shape in Figure 20.10(a) with a V_m of 20 V is connected to a 50-ohm resistor.

    (a) Determine the effective value of current and voltage.

    (b) Determine the power dissipated in the resistor.

26. A series circuit of a 12-ohm resistor and an inductive reactance of 16 ohms at the fundamental frequency is connected to a source of $e = 3.18 \sin(377t) + 1.59 \sin(754t) + 1.06 \sin(1131t)$. Find:

    (a) The sinusoidal expression for the current.

    (b) The effective value of the voltage and current.

    (c) The power dissipated in the circuit.

27. A series circuit of a 16-ohm resistor and an inductor whose reactance is 12 ohms at the fundamental is connected to a source of $e = 24 + 25 \sin(314t) + 15 \sin(628t) + 10 \sin(1256t)$. Find:

    (a) The sinusoidal expression for the current.

    (b) The effective value of the current and the voltage.

    (c) The power dissipated in the circuit.

28. Work problem 26 for a series circuit of a 16-ohm resistor and a capacitor whose reactance is 12 ohms at the fundamental.

29. The source of Problem 27 is connected across a parallel circuit of a 12-ohm resistance and a capacitor whose reactance is 16 ohms at the fundamental.

    (a) Determine the effective value of the source voltage.

    (b) Determine the power in the circuit.

30. A voltage source of $e = 40 \sin(375t) + 30 \sin(750t)$ is connected across a parallel circuit of a 20-ohm resistor, an inductive reactance of 20 ohms at the fundamental, and a capacitive reactance of 40 ohms at the fundamental. Find:

    (a) The effective value of the source voltage.

    (b) The power dissipated in the circuit.

CHAPTER 21

THE TRANSFORMER

CHAPTER OBJECTIVES

After completing this chapter, you should be able to:

1. Connect and use a transformer.
2. Explain the function of a transformer and describe its construction.

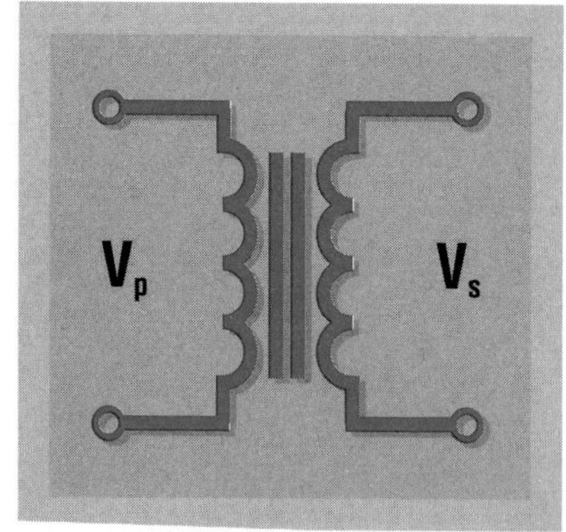

3. Explain the difference between the ideal transformer and the practical one.
4. Apply the voltage, current, and impedance relations to a transformer.
5. Explain what step-up and step-down transformers do.
6. List the three functions for which transformers are used.
7. Determine the terminal polarity from the terminal markings or testing.
8. Determine whether a transformer is defective, and isolate the defect.
9. Determine efficiency and voltage regulation using the equivalent circuits.
10. Determine efficiency and voltage regulation using the short-circuit, open-circuit test data.
11. Explain why a transformer should be operated at the rated frequency.
12. Perform the analysis of an autotransformer.
13. List the types of transformers and explain the function of each one.
14. Differentiate between an air core transformer and an iron core one; determine Z_i and the voltage output of the air core transformer.

KEY TERMS

1. air core transformer
2. audio transformer
3. autotransformers
4. buck-boost transformer
5. center-tapped transformer
6. conductively transferred power
7. constant voltage transformer
8. control transformers
9. copper losses
10. core losses
11. current transformer
12. distribution transformer
13. equivalent circuit
14. impedance matching
15. impedance matching transformer
16. inductively transferred power
17. iron core transformers
18. isolation
19. isolation transformer
20. open-circuit test
21. potential transformer
22. power transformer
23. primary winding
24. radio frequency transformer
25. secondary winding
26. short-circuit test
27. step-down transformer
28. step-up transformer
29. three-phase transformer
30. transformer
31. transformer impedance
32. tuned circuit
33. turns ratio
34. volt-ampere rating

INTRODUCTION

The many benefits of alternating current could not be realized without the transformer. Power distribution would be limited to short distances and single voltages. Electronic applications would be inefficient or impossible. This is because resistors would have to be used to vary the voltage. Increasing the voltage would be impossible.

The basic principle of the transformer is simple: two coils of wire linked by a magnetic field. From this simplicity come many specialized units such as audio transformers, distribution transformers, and isolation transformers.

The two coils can be wound on an iron core or an air core. Each of these has different characteristics, some of which are examined in this chapter. Transformer ratings and the effect of frequency are also introduced.

What is a transformer? How does it operate? And how can we analyze its operation? These questions, along with others, are answered in this chapter. After completing this chapter, you should be able to appreciate the importance of the transformer. In addition to introducing transformer terminology, this chapter presents the material needed to select, connect, test, analyze, and troubleshoot the transformer.

21.1 Transformer Construction and Terminology

transformer

A **transformer** is a device that transfers energy from one coil to another through the magnetic field that couples the coils. This energy can be transferred at the same voltage, a lower voltage, or a higher voltage. This ability to change voltage was a major factor in the adoption of ac instead of dc for our power use. Power is generated at several thousand volts, and transformers increase the voltage to several hundred thousand volts. This permits transmission at low current with a low power loss. Finally, the voltage is reduced with transformers to suit the customer's needs.

Transformers are shown on schematics by the symbols in Figure 21.1. When the coupling of the windings can be varied, an arrow is added, as in Figure 21.1(d). The number of turns in these symbols is not related to the number of turns in the winding.

iron core transformers

Most transformers are either **iron core transformers** or air core transformers. The coils of an iron core transformer are wound around an iron core. The iron provides a high permeability path for the magnetic flux. This results in practically 100 percent coupling between the coils.

air core transformer

The **air core transformer** has the coils wound on a nonmagnetic core, as shown in Figure 21.2(c). Air core transformers are used in radio and high-frequency applications. They have no core losses and their frequency characteristics are superior to the iron core transformer. However, the strength of the field will not be as great as for the iron core transformer. This is because of the air's smaller permeability. The coefficient of coupling ranges from low values of about 0.002 to 0.7.

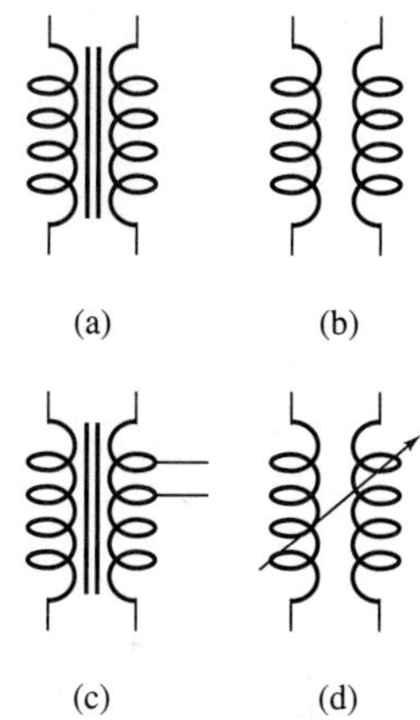

Figure 21.1 Transformer symbols. (a) Fixed iron core transformer. (b) Air core tranformer. (c) Tapped iron core transformer. (d) Variable air core transformer.

Two iron core transformers are shown in Figure 21.2. The windings of the shell type of transformer are wound on the center section and are surrounded by the outer shell. The windings are normally wound one on top of the other. The core type has the windings around one of the iron legs. The iron core in both transformers is made from thin sheets or

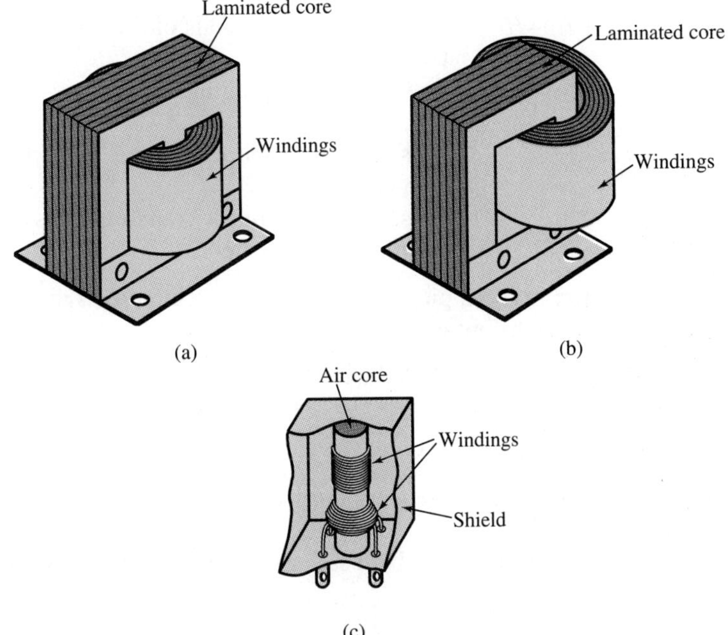

Figure 21.2 Examples of iron core and air core transformers. (a) The windings of the core-type iron core transformer surround the core. (b) The windings of the shell-type iron core transformer are surrounded by the core. (c) A variable air core transformer.

laminations of iron. These are coated with insulating material to reduce eddy current losses. A small amount of silicon is added to the iron to reduce the hysteresis loss, which is due to the reversal of the flux in the iron.

primary winding
secondary winding

A transformer has at least one primary winding and one secondary winding. The **primary winding** is the one that connects to a source of emf. The **secondary winding** is the one that connects to a load. Either winding can be used as the primary or secondary as long as the ratings of the winding are not exceeded.

step-up transformer
step-down transformer

If a transformer is connected to increase the voltage, it is a **step-up transformer**. If it is connected to decrease the voltage, it is a **step-down transformer**.

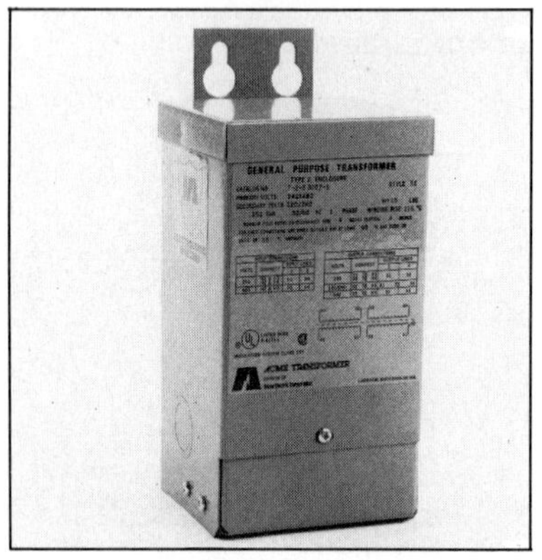

Figure 21.3 Transformer ratings and terminal designation are some of the information provided on the nameplate. (Photo courtesy of ACME Transformer, A Division of Acme Electric Corporation)

The nameplate data give the information needed to connect and operate the transformer, and to prevent damage to it. Transformers are rated in volt-amperes, with large ones being rated in kilovolt-amperes (kVA). Some other quantities are shown in the nameplate data of Figure 21.3.

The meanings of the terms are:

volt-ampere rating

240×480	These are the voltages that can be applied to the primary.
120/240	These are the rated secondary voltages. They will be 120 V for the 240 V input, and 240 V for the 480 V input.
$X1$–$X2$ and $X3$–$X4$	These are the terminals of the low-voltage windings.
$H1$–$H2$ and $H3$–$H4$	These are the terminals of the high-voltage windings.
50/60 Hz	These are the operating frequencies of the transformer.
0.250 kVA	This is the **volt-ampere rating** of the transformer.
1 phase	This specifies that the transformer is a single-phase transformer.

The information in the input and output connection tables gives the winding connections to obtain the voltages.

Another example of a voltage rating specification is:

Primary volts	240/480/600
Secondary volts	120/100

This represents the possible secondary voltages from the three primary voltages.

The volt-ampere rating can be used to determine the secondary current rating as follows:

$$I_s = \frac{\text{kVA}}{kV_s} \tag{21.1}$$

where I_s is the maximum secondary current, in amperes; kVA is the kilovolt-ampere rating of the transformer; and kV_s is the kilovolt rating of the secondary.

The maximum power that a transformer can safely supply is given by

$$P_{max} = (\text{VA rating}) \times (F_p) \quad \textbf{(21.2)}$$

EXAMPLE 21.1

How much current can a 220/110 volt, 2-kVA step-down transformer safely supply to a load?

Solution

$$I_s = \frac{\text{kVA}}{kV_s} = \frac{2\text{ kVA}}{0.110\text{ kV}}$$

$$I_s = 18.18\text{ A}$$

EXAMPLE 21.2

How much power can the transformer in Example 21.1 supply if the load has a power factor of 0.8?

Solution

$$P = (\text{VA})(F_p)$$

$$= (2000\text{ kVA})(0.8)$$

$$P = 1600\text{ W}$$

A parameter of importance, but not shown on the nameplate, is impedance. **Transformer impedance** is the current-limiting characteristic of a transformer. It is used to determine the capacity of a fuse or circuit breaker, which is used to protect the primary from the high inrush current if the secondary is short-circuited. This inrush current is much greater than the load current rating.

transformer impedance

$$\%\text{ impedance} = \frac{(\text{rated primary current}) \times 100\%}{\text{short-circuit primary current}} \quad \textbf{(21.3)}$$

EXAMPLE 21.3

A 5-kVA transformer has a primary rating of 480 volts and a 5 percent impedance. Determine the fuse capacity to protect the primary from the instantaneous inrush current.

Solution

$$\text{Rated primary current} = \frac{\text{VA}}{V_p} = \frac{5000\ \text{VA}}{480\ \text{V}} = 10.42\ \text{A}$$

From Equation 21.3,

$$\text{Short-circuit current} = \frac{\text{rated current}}{\%\ \text{impedance}} = \frac{10.42\ \text{A}}{0.05}$$

$$I_{\text{sc}} = 208.4\ \text{A}$$

The fuse should interrupt at a minimum inrush current of 208.4 amperes. This value represents the instantaneous inrush current, and lasts for only about one-tenth of a second. For steady currents, the fuse or breaker characteristics are such that they protect against currents greater than full load.

When the voltage applied exceeds the rated primary voltage, saturation can result. You might recall from Chapter 13, Section 13.3, that this is the condition where there is very little change in the magnetic field. If this happens, the magnetizing current becomes dangerously high. Saturation can be caused by a dc component in the primary voltage or the secondary current.

21.2 The Ideal Transformer

The ideal transformer is used as a model to explain the transformer relations. This is a transformer in which permeability is infinite, losses are zero, and there is 100 percent coupling of the flux. In many cases, the ideal analysis is acceptable for the practical transformer. A case where it is not acceptable is when the losses or efficiency of the transformer are being determined.

Consider the ideal transformer in Figure 21.4. Based on the material in Chapter 13, the voltage applied to the primary winding of the transformer is given by

$$e_p = N_p \frac{d\phi_p}{dt}$$

The change in flux caused by this voltage induces a secondary voltage of

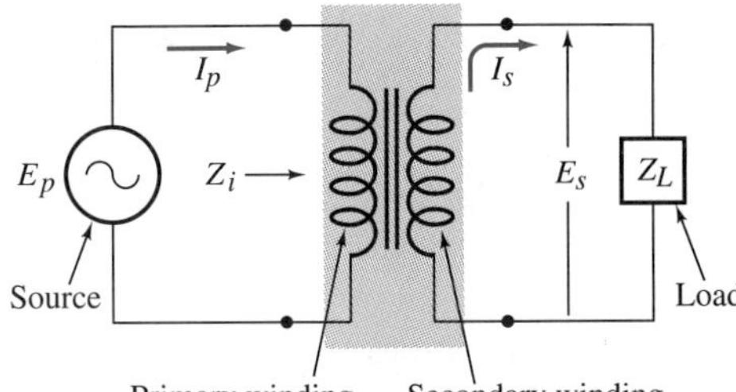

Figure 21.4 The connection of the primary and secondary windings of the ideal transformer to the source and the load.

$$e_s = kN_s \frac{d\phi_s}{dt}$$

Since there are no losses, e_s is also the voltage at the secondary output.

Since the coupling is 100 percent, $k = 1$, $\phi_p = \phi_s$, and $(d\phi_p/dt) = (d\phi_s/dt)$. Dividing e_s by e_p and using the foregoing equalities gives

$$\frac{e_s}{e_p} = \frac{N_s}{N_p}$$

That is, the voltage ratio is equal to the ratio of the turns. In terms of rms values,

$$\frac{E_s}{E_p} = \frac{N_s}{N_p} = \frac{1}{\alpha} \tag{21.4}$$

where E_s and E_p are the secondary and primary voltage, N_s and N_p are the number of turns on the secondary and primary, and α is called the **turns ratio**, which equals (N_p/N_s).

turns ratio

Connecting a load across the secondary results in a secondary current. This current causes a magnetomotive force (mmf) that opposes the magnetizing current in the primary. The core flux would decrease but the primary current increases and produces an mmf that cancels the secondary mmf. Since mmf is the product of current and turns, the following relation holds true.

$$N_pI_p = N_sI_s$$

Therefore, any increase in I_s results in an increase in I_p to supply the additional current. Likewise, a decrease in I_s results in a decrease in I_p. Rearranging $N_pI_p = N_sI_s$ gives

$$\frac{I_s}{I_p} = \frac{N_p}{N_s} = \alpha \tag{21.5}$$

where I_s and I_p are the secondary and primary currents, and N_s, N_p, and α have the same meaning as in Equation 21.4.

Since $I_s/I_p = \alpha$ and $E_s/E_p = 1/\alpha$, a voltage step-up transformer is a current step-down. Conversely, a voltage step-down transformer is a current step-up.

EXAMPLE 21.4

An ideal step-down transformer with a turns ratio (α) of 10 is connected to a 120 volt-source. It has a 24-ohm resistance connected across its secondary. Find:

a. the resistor voltage, current, and power.
b. the primary current.
c. the primary turns if the secondary has 80 turns.

Solution

a.

$$\frac{E_s}{E_p} = \frac{1}{\alpha}, \qquad E_s = \frac{E_p}{\alpha}$$

$$E_s = \frac{120 \text{ V}}{10}$$

$$E_s = 12 \text{ V}$$

$$I_s = \frac{E_s}{R_L} = \frac{12 \text{ V}}{24\ \Omega}$$

$$I_s = 0.5 \text{ A}$$

Since the load is resistive,

$$P_L = E_s I_s = (12 \text{ V})(0.5 \text{ A})$$

$$P_L = 6 \text{ W}$$

b.

$$\frac{I_s}{I_p} = \alpha, \qquad I_p = \frac{I_s}{\alpha} = \frac{0.5 \text{ A}}{10}$$

$$I_p = 0.05 \text{ A}$$

c.

$$N_p = \alpha N_s = (10)(80 \text{ turns})$$

$$N_p = 800 \text{ turns}$$

Combining Equations 21.4 and 21.5 shows that the volt-ampere rating of the primary equals the rating of the secondary. This relation provides a method of calculating one of these values when the others are known. In equation form,

$$(\text{VA})_{\text{primary}} = (\text{VA})_{\text{secondary}} \tag{21.6}$$

EXAMPLE 21.5

The rating of a transformer secondary is 540 VA. What is the rated primary current if the primary is rated at 120 volts?

Solution

$$(\text{VA})_{\text{primary}} = (\text{VA})_{\text{secondary}}$$

or

$$I_p = \frac{(\text{VA})_{\text{sec}}}{V_p} = \frac{540 \text{ VA}}{120 \text{ V}}$$

$$I_p = 4.5 \text{ A}$$

Some transformers have taps on the secondary winding to supply more than one load. They are called tapped transformers. Example 21.6 provides some insight into the analysis of an ideal tapped transformer.

EXAMPLE 21.6

Determine the voltage across each load and the primary current for the transformer in Figure 21.5.

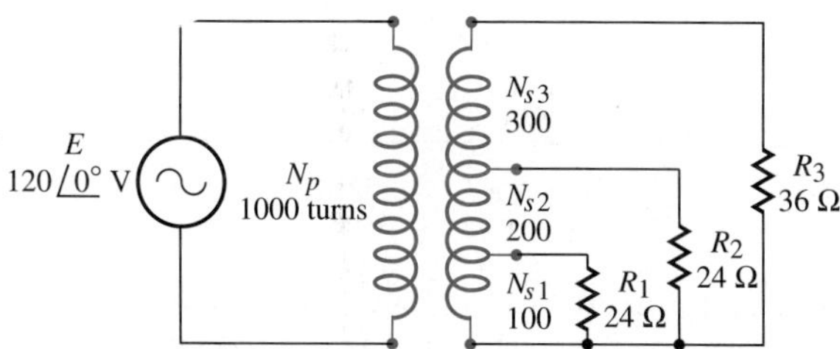

Figure 21.5

Solution

The relation of Equation 21.4 applies for each section so

$$V_1 = \frac{N_{S1}E_p}{N_p} = \left(\frac{100}{1000}\right)(120 \text{ V})$$

$$V_1 = 12 \text{ V}$$

$$V_2 = \frac{(N_{S1} + N_{S2})E_p}{N_p} = \frac{(100 + 200)(120 \text{ V})}{1000}$$

$$V_2 = 36 \text{ V}$$

$$V_3 = \frac{(N_{S1} + N_{S2} + N_{S3})E_p}{N_p} = \frac{(100 + 200 + 300)(120 \text{ V})}{1000}$$

$$V_3 = 72 \text{ V}$$

$$I_1 = \frac{V_1}{R_1} = \left(\frac{12 \text{ V}}{24 \text{ }\Omega}\right)$$

$$I_1 = 0.5 \text{ A}$$

$$I_2 = \frac{V_2}{R_2} = \left(\frac{36 \text{ V}}{24 \text{ }\Omega}\right)$$

$$I_2 = 1.5 \text{ A}$$

$$I_3 = \frac{V_3}{R_3} = \left(\frac{72 \text{ V}}{36 \text{ }\Omega}\right)$$

$$I_3 = 2 \text{ A}$$

Since the transformer is ideal and there are no losses, the power out must be equal to the power in.

$$\begin{aligned} P_{\text{out}} &= P_1 + P_2 + P_3 \\ &= I_1V_1 + I_2V_2 + I_3V_3 \\ &= (0.5 \text{ A})(12 \text{ V}) + (1.5 \text{ A})(36 \text{ V}) + (2 \text{ A})(72 \text{ V}) \end{aligned}$$

$$P_{\text{out}} = 204 \text{ W}$$

$$P_{\text{in}} = I_pE_p \quad \text{so} \quad I_p = P_{\text{in}}/E_p$$

Since $P_{\text{in}} = P_o$,

$$I_p = \frac{204 \text{ W}}{120 \text{ V}}$$

$$I_p = 1.7 \text{ A}$$

---

impedance matching

An impedance connected to the secondary of a transformer is made to appear as a different value in the primary. This permits the transformer to be used for **impedance matching** to match a load to a source. Consider the connection of an 8-ohm speaker to a transistor whose output resistance is several thousand ohms. Connecting the speaker directly would almost short circuit the transistor. A transformer can be used to make the 8-ohm speaker appear to be a high impedance to the transistor. The exact relationship of the impedance change is now examined.

The impedance $\boldsymbol{Z}_i$ looking into the primary of the transformer in Figure 21.4 is $\boldsymbol{Z}_i = \boldsymbol{E}_p/\boldsymbol{I}_p$. The load impedance is $\boldsymbol{Z}_L = \boldsymbol{E}_s/\boldsymbol{I}_s$. Dividing one by the other gives

$$\boldsymbol{Z}_i = \alpha^2 \boldsymbol{Z}_L \tag{21.7}$$

where $\boldsymbol{Z}_L$ is the impedance connected across the secondary, in ohms; $\boldsymbol{Z}_i$ is the impedance $\boldsymbol{Z}_L$ reflected into the primary side, in ohms; and α is the same as for Equation 21.4.

According to Equation 21.7, the transformer can be used to make an impedance look smaller, larger, or the same. If α were 50, the 8-ohm speaker across the secondary would appear to be 20,000 ohms in the primary.

EXAMPLE 21.7

A 16-ohm loudspeaker is to be matched to a 6400-ohm amplifier. Find:

a. the turn ratio needed to match the two.

b. the turns in the secondary if the primary has 200 turns.

Solution

a.

$$Z_i = \alpha^2 Z_L$$

$$\alpha^2 = \frac{Z_i}{Z_L} = \frac{6400\ \Omega}{16\ \Omega} = 400$$

so

$$\alpha = 20$$

b.

$$\alpha = \frac{N_p}{N_s}, \qquad N_s = \frac{N_p}{\alpha}$$

$$N_s = \frac{(200\ \text{turns})}{20}$$

$$N_s = 10\ \text{turns}$$

There is no connection between the primary and secondary, so the transformer provides electrical isolation. **Isolation** can prevent damage to equipment and electrical shock. Isolation also permits connecting dc power supplies in series.

isolation

This isolation can also be used to block a dc voltage. The transformers in the amplifier circuit of Figure 21.6 transfer the ac but not the dc. Since

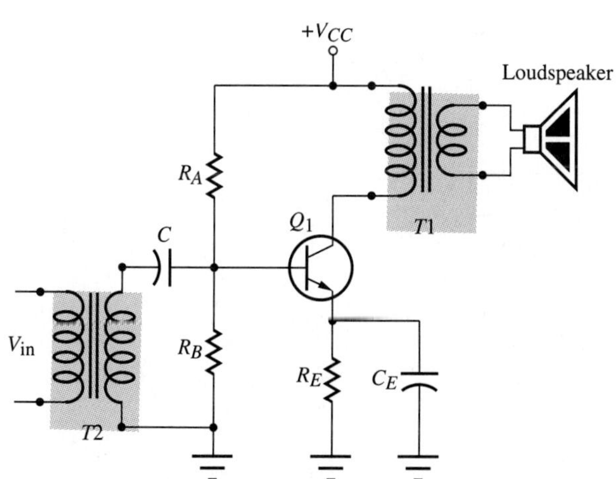

Figure 21.6 The transformers in this transistor amplifier are used to couple the ac signal and block the dc.

the primary and secondary are not connected, the dc is blocked. Thus, the dc bias circuit of each section can be designed independently of the other section.

21.3 Terminal Polarity and Troubleshooting

The phase relation of transformer outputs is important for some applications, such as when transformers are connected in parallel or in series; when single-phase transformers are connected as a three-phase unit; and when connecting speakers of a stereo amplifier. In these applications, the result can range from a short circuit to poor sound reproduction. To avoid these conditions, the schematic must show the instantaneous terminal polarity. This can be done by placing a dot at the terminal of each winding that has the same instantaneous polarity. You might recall that this was introduced in Chapter 13, Section 13.8 for the inductor.

The dots on the windings in Figure 21.7 show that the top ends of N_p

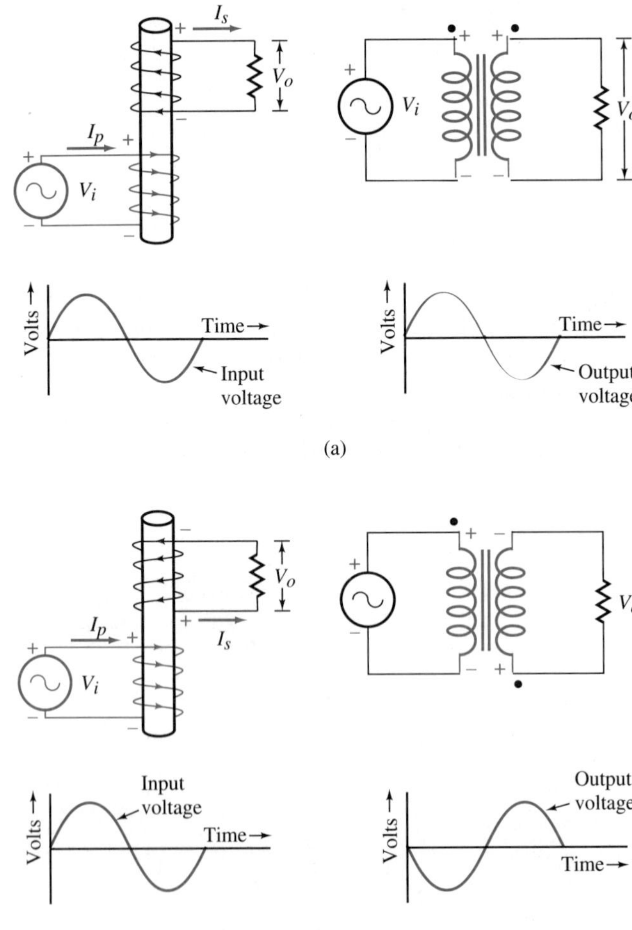

Figure 21.7 The effect of the terminal polarity on the phase relation between the input and the output voltages. (a) The winding direction, terminal polarity, and waveforms for the in-phase connection. (b) The winding direction, terminal polarity, and waveforms for the out-of-phase connection.

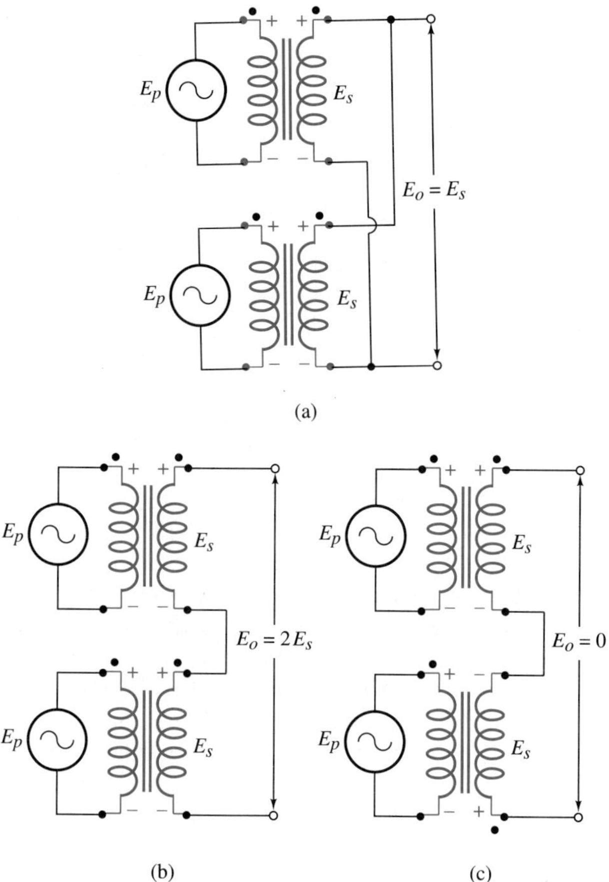

Figure 21.8 Possible output voltages obtained by interconnecting transformers. (a) Parallel connection, like terminals connected together, $E_o = E_s$. (b) Series connection, unlike terminals connected together, $E_o = 2E_s$. (c) Series connection, like terminals connnected together, $E_o = 0$.

and N_s have the same instantaneous polarity. That is, when one is negative, so is the other. When one is positive, the other is also positive. The output will be in phase with the input. The dots are on the opposite ends in Figure 21.7(b), so the output is 180° out-of-phase with the input. Polarity is set by the relative direction of the windings of the transformer. It is not related to the turns on the symbol. This can also be seen in Figure 21.7.

The effect of the terminal polarity for the series and parallel connection of two transformers is shown in Figure 21.8. These effects assume that the voltages have the same frequency, are in phase, and the resistance of each coil is the same.

To connect the transformer as shown on a schematic, the relative terminal polarity must be marked on the transformer. Figure 21.9 shows a method of doing that. The letter *H* is used for the high voltage side. This is the side with more turns. The letter *X* is used for the low voltage side. The transformer in Figure 21.9 has both the high and low voltage side

Figure 21.9 This nameplate shows the use of *H* and *X* for voltage and odd and even numbers for terminal polarity. (Photo courtesy of ACME Transformer, A Division of Acme Electric Corporation)

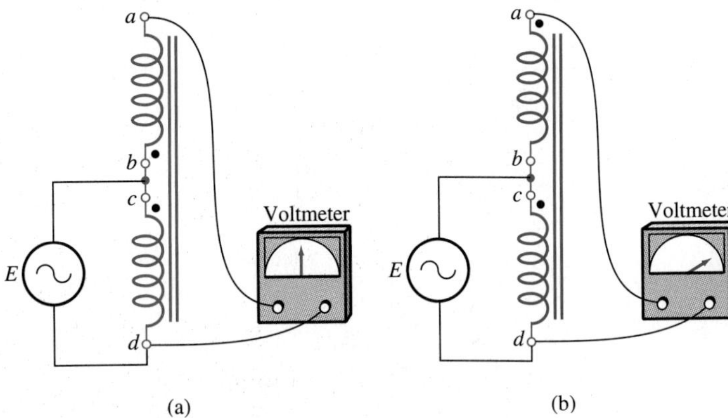

Figure 21.10 Voltage measurements can be used to determine terminal polarity. (a) The voltmeter reading is low because the voltages are subtractive. (b) The voltmeter reading is higher, the voltages are additive, so the unlike terminals must be connnected together.

made of several coils. The terminal polarity is related to the numbers. The lower numbered terminals 1 and 3 have the same polarity, and the higher numbered ones have the same polarity. Thus, the groups of terminals with the same polarity are $H1, H3, X1, X3$, and $H2, H4, X2, X4$.

A simple test such as that in Figure 21.10 can be used to determine terminal polarity. The connection that gives the higher voltmeter reading is noted. The voltages must be adding for this connection. The polarity is as shown for the higher reading connection.

EXAMPLE 21.8

A transformer has two windings on the secondary, as shown in Figure 21.11. Find:

a. the voltage across $H1$–$H4$ if $H2$ and $H3$ are connected.
b. the voltage across $H1$–$H3$ if $H2$ and $H4$ are connected.

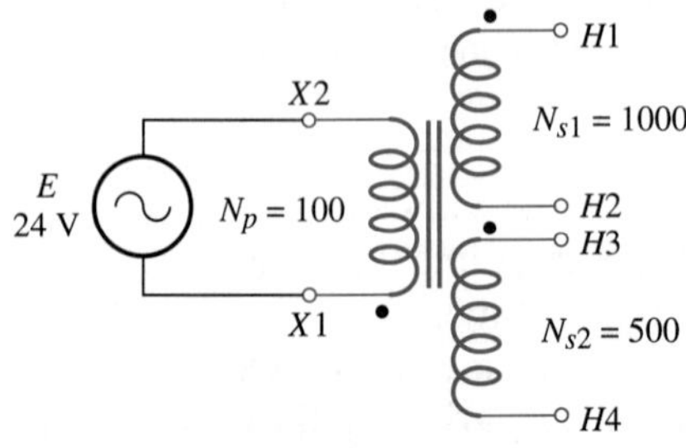

Figure 21.11

Solution

a. The relative polarities are as shown in Figure 21.12.

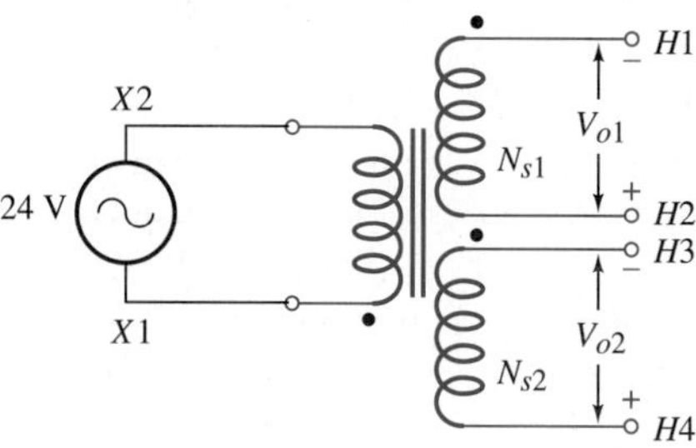

Figure 21.12

$$V_{o1} = \frac{N_{s1}}{N_p} V_i$$

$$= \frac{1000}{100}(24 \text{ V})$$

$$V_{o1} = 240 \text{ V}$$

$$V_{o2} = \frac{N_{s2}}{N_p} V_i$$

$$= \frac{500}{100}(24 \text{ V})$$

$$V_{o2} = 120 \text{ V}$$

When $H2$ and $H3$ are connected, the voltages are additive. The voltage across the combination is 240 volts + 120 volts, or 360 volts, with $H4$ positive (+) with respect to $H1$.

b. If $H2$ and $H4$ are connected, the voltages are subtractive. The voltage across the combination is 240 V − 120 V, or 120 V with $H3$ positive (+) with respect to $H1$.

shows another test for determining the terminal polarity. With Terminal 1 connected to the positive (+) of the source, the voltmeter pointer will momentarily move when the switch is closed. For

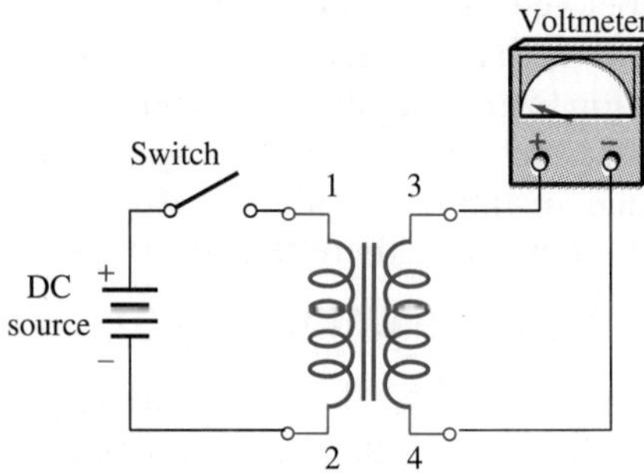

Figure 21.13 This test uses the instantaneous deflection caused by a dc voltage to determine the terminal polarity.

TABLE 21.1

Symptom	Possible Cause
No output.	Open winding.
Output voltage is greater than rated.	Short circuit between turns of the primary winding. Short circuit between windings.
Output voltage is smaller than rated.	Short circuit between turns of the secondary. Short circuit between windings.
Transformer overheats.	Short circuit between turns of a winding. Short circuit between windings. Frequency is too high or low.

up-scale movement, Terminals 1 and 3 have the same polarity. If it is down-scale, terminals 1 and 4 have the same polarity.

A note of caution: Since the dc winding resistance can be small, a low voltage should be used so as not to exceed the current rating.

Before leaving this section, some simple troubleshooting procedures are presented. Some symptoms and their possible causes are listed in Table 21.1. Whether these causes exist can be determined by the following tests.

Open winding. Check for continuity between the terminals of the winding with an ohmmeter.

Short circuit in a winding.

a. If the original resistance is known, measure the resistance and compare it to the original. If it is lower, there is a short circuit.
b. If the original resistance is not known, determine whether the primary turns or secondary turns are short-circuited from the output voltage of the transformer as listed under symptom and cause.

Short circuit between windings.

a. If a source or an insulation tester is not available, check the continuity between a terminal of one winding and a terminal of the other winding with an ohmmeter. There should not be continuity.
b. If an insulation tester or a source is available, use them to check the continuity between the terminals with a voltage applied. If a source is used, place an ammeter and a resistor in the circuit to measure and limit the current if there is a short circuit.

21.4 The Exact Equivalent Circuit

The practical transformer has losses and leakage, and the coupling is not 100 percent. Thus, the voltage does change with load, and some power is lost in the transformer.

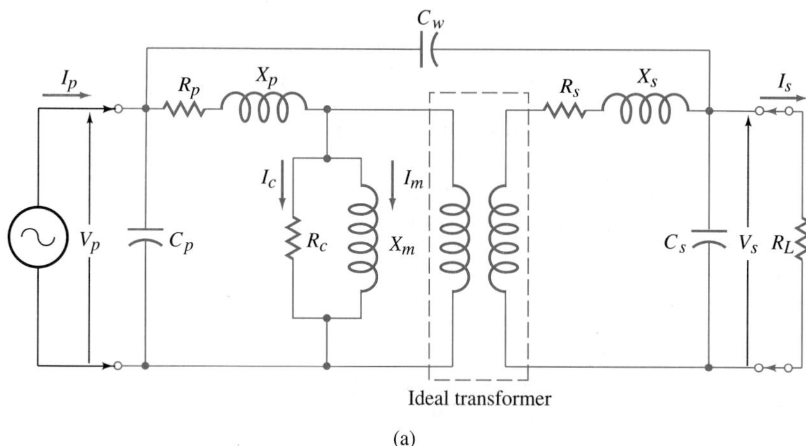

(a)

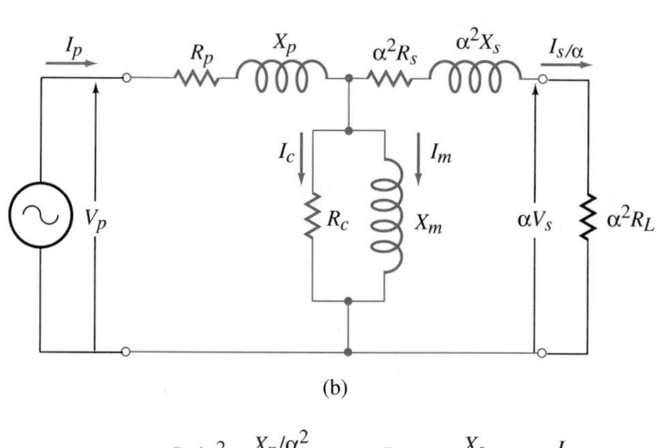

(b)

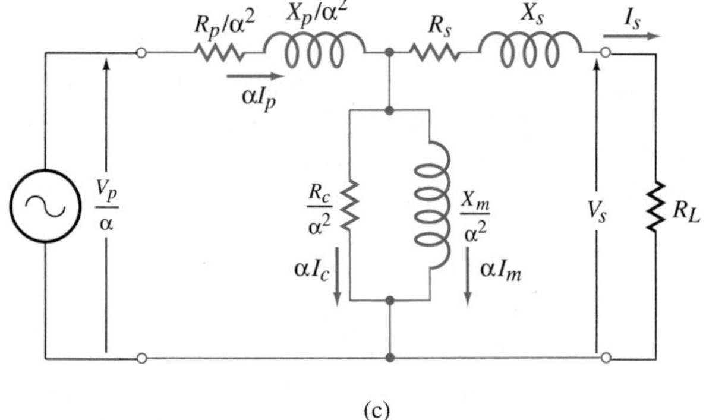

(c)

Figure 21.14 Alternating current (ac) equivalent circuits for an iron core transformer. (a) The exact equivalent circuit with quantities referred to the primary. (b) The equivalent circuit ignoring the effect of C_p and C_w, and C_s with quantities referred to the primary. (c) The equivalent circuit of Part b with quantities referred to the secondary.

The losses in the transformer are core losses, copper losses, and leakage and magnetizing current losses. The **core losses** are hysteresis and eddy currents, which were explained in Chapter 16. The **copper losses** are caused by the resistance of the primary and secondary windings. They are proportional to the current squared. Well-designed iron core transformers have an efficiency of greater than 95 percent.

core losses

copper losses

The practical transformer can be represented by the exact **equivalent circuit** of Figure 21.14. It consists of the ideal transformer plus the

equivalent circuit

components that account for the losses and the voltage drops. The values of the quantities can be obtained from design data or test data.

The meanings of the symbols in the diagram are:

V_p	terminal voltage of the primary
V_s	terminal voltage of the secondary
E_p	primary induced voltage
E_s	secondary induced voltage
I_p	primary current
I_m	magnetizing current
I_c	current that accounts for the core loss
I_s	secondary current
R_p	resistance of the primary winding
X_p	leakage reactance of the primary winding
R_c	resistance to account for core losses
X_m	magnetizing reactance
R_s	resistance of the secondary winding
X_s	leakage reactance of the secondary winding
R_L	load resistance
C_p	capacitance between primary turns
C_s	capacitance between secondary turns
C_w	capacitance between primary and secondary turns

C_w, C_p, and C_s are usually very small, so removing them does not have much effect on the result at power frequencies.

If all quantities are referred to the primary, as was done for impedance in the ideal transformer, the circuit is as shown in Figure 21.14(b). The circuit can also be drawn with all quantities referred to the secondary side, as in Figure 21.14(c).

It can be seen that the practical transformer's input current includes the no-load excitation current plus the load current. Also, the load voltage will be less than the induced secondary voltage because of the drop in X_s and R_s.

As a point of interest, the phasor diagrams of the voltages and currents for a resistance, capacitance, and inductance load are drawn in Figure 21.15.

These diagrams present the general relationships. The actual value will depend on the parameters in the equivalent circuit. At no load, V_s is approximately equal to V_p. The inductor load voltage is smaller than the no-load voltage. But the capacitance diagram shows that the load voltage can be greater than the no-load voltage for a capacitive load. The capacitive reactance cancels some of the inductive reactance of the transformer.

Since the equivalent circuit has an ac source, resistors, and reactances, it can be analyzed using the methods of ac analysis. Also, programs for transformer analysis, including PSpice, are available for the personal computer. These can allow one to quickly determine the characteristics of a proposed design.

The exact equivalent circuit is now used to determine the efficiency and other characteristics of a transformer.

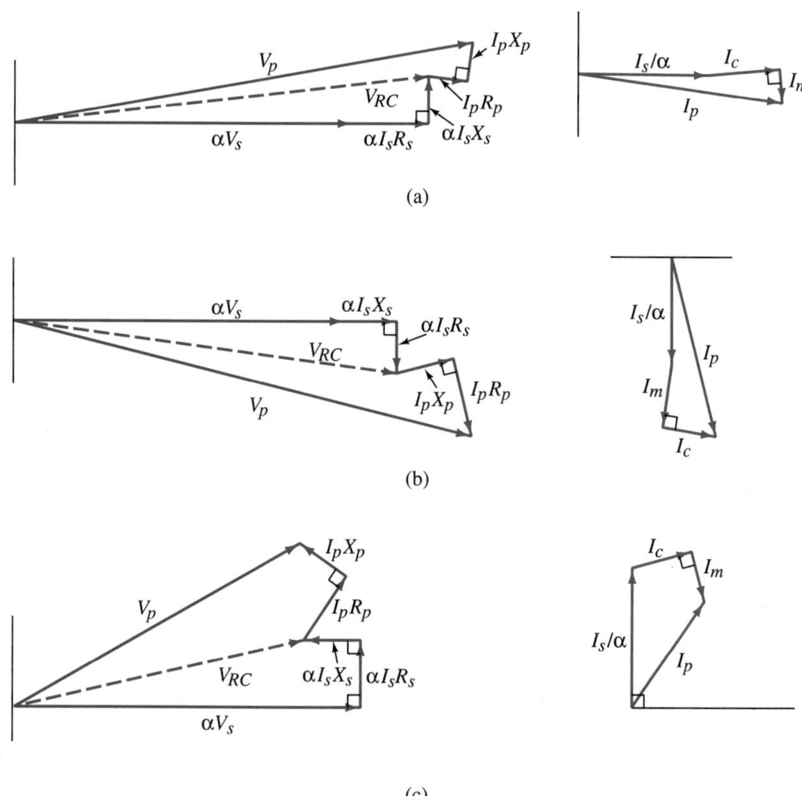

Figure 21.15 The phasor diagrams for three types of loads. (a) Resistance load. (b) Inductance load. (c) Capacitance load.

EXAMPLE 21.9

A 240/120-V, 5-kVA transformer supplies rated current to a resistance load. Find the present voltage regulation, the percent efficiency, and the Z_i if the components in the equivalent circuit have the following values:

$$R_p = 0.2\ \Omega \qquad X_p = 1.0\ \Omega \qquad R_c = 15\text{-k}\ \Omega$$
$$X_m = 2\text{-k}\ \Omega \qquad R_s = 0.01\ \Omega \qquad X_s = 0.05\ \Omega$$

Solution

From the transformer rating,

$$\alpha = \frac{240\ \text{V}}{120\ \text{V}} = 2$$

$$I_s = \frac{\text{kVA}}{V_2} = \frac{5000\ \text{VA}}{120\ \text{V}} = 41.67\ \text{A}$$

Using the relations of Section 21.2 to refer the secondary quantities to the primary gives

$$\alpha^2 R_s = (2^2)(0.01\ \Omega) = 0.04\ \Omega$$
$$\alpha^2 X_s = (2^2)(0.05\ \Omega) = 0.2\ \Omega$$
$$\alpha V_s = (2)(120\ \text{V}) = 240\ \text{V}$$

The equivalent circuit with quantities referred to the primary is shown in Figure 21.16.

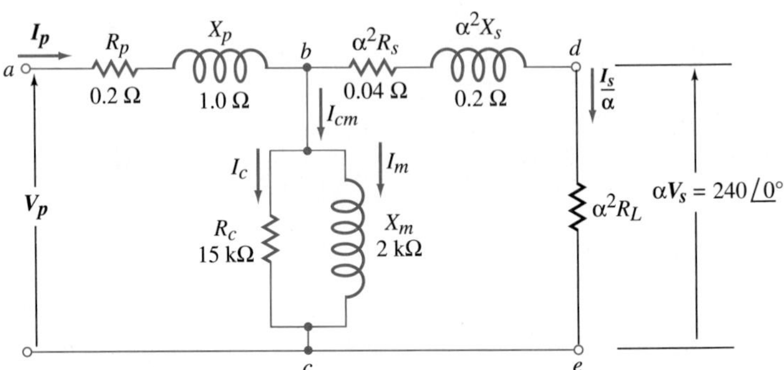

Figure 21.16

For the circuit,

$$\%\text{VR} = \left(\frac{V_{\text{NL}} - V_{\text{L}}}{V_{\text{L}}}\right) \times 100\%$$
$$= \left(\frac{\alpha V_{\text{SNL}} - \alpha V_{\text{SL}}}{\alpha V_{\text{SL}}}\right) \times 100\%$$

where V_{SL} is the secondary output at rated load. It is 120 V, so αV_{SL} equals 240 V.

The analysis proceeds as follows:

a. Since V_S at no load equals V_{bc}, the voltage divider equation is used to write V_{bc} in terms of V_p.
b. Kirchhoff's Voltage Law for the primary is used to determine V_p with the load connected.
c. Since V_p is the same with and without load, the value from part b is used in part a.

The secondary voltage at no load is now determined from the voltage divider expression for the open circuit.

$$\mathbf{V}_{bc} = \frac{(\mathbf{V}_p)(\mathbf{Z}_{bc})}{R_p + jX_p + \mathbf{Z}_{bc}}$$

but

$$\mathbf{Z}_{bc} = R_c \parallel X_m = \frac{(15 \times 10^3\ \angle 0^\circ\ \Omega)(2 \times 10^3\ \angle 90^\circ\ \Omega)}{(15 \times 10^3 + j2 \times 10^3)\ \Omega}$$
$$= 1.98 \times 10^3\ \angle 82.41^\circ\ \Omega$$

so at no load

$$V_{bc} = \frac{(V_p)(1.98 \times 10^3 \angle 82.41^\circ)\ \Omega}{(0.2 + j1.0 + 1.98 \times 10^3 \angle 82.41^\circ)\ \Omega} \approx V_p$$

V_p is the primary input voltage, which is the same at load or no load. Using the given data, it can be found at load. Applying Kirchhoff's Voltage Law to Path a, b, c gives

$$V_p = I_p(R_p + jX_p) + V_{bc}$$

V_{bc} and I_p must be determined with the load connected. Taking V_s at 0°,

$$\begin{aligned} V_{bc} &= \frac{I_s \angle 0^\circ}{\alpha}(\alpha^2 R_s + j\alpha^2 X_s) + \alpha V_s \angle 0^\circ \\ &= 20.84\text{ A }(0.04 + j0.2)\ \Omega + 240 \angle 0^\circ\text{ V} \\ &= 240.87 \angle 1^\circ\text{ V} \end{aligned}$$

Note: I_s is in phase with V_s for the resistance load. By Kirchhoff's Current Law,

$$I_p = I_{cm} + \frac{I_s}{\alpha}$$

but

$$\begin{aligned} I_{cm} &= \frac{V_{bc}}{Z_{bc}} \\ &= \frac{240.87 \angle 1^\circ\text{ V}}{1.98 \times 10^3 \angle 82.41^\circ\ \Omega} \\ &= 0.122 \angle -81.41^\circ\text{ A} \end{aligned}$$

so

$$\begin{aligned} I_p &= 0.122 \angle -81.41^\circ\text{ A} + 20.84 \angle 0^\circ\text{ A} \\ &= 20.86 \angle -0.33^\circ\text{ A} \end{aligned}$$

Putting V_{bc} and I_p in the expression for V_p gives

$$\begin{aligned} V_p &= (20.86 \angle -0.33^\circ\text{ A})(0.2 + j1.0)\ \Omega + 240.87 \angle 1^\circ\text{ V} \\ &= (245.12 + j25.04)\text{ V} \quad \text{or} \quad 246.4 \angle 5.83^\circ\text{ V} \end{aligned}$$

Since $\alpha V_{SNL} = V_{bc}$ at no load, and $V_{bc} \approx V_p$,

$$\alpha V_{SNL} = 246.4 \angle 5.83^\circ\text{ V}$$

The voltage at rated load, referred to the primary, is 240 V.

$$\text{Using } \%VR = \left(\frac{V_{NL} - V_L}{V_L}\right) \times 100\%$$

gives

$$\%VR = \left(\frac{246.4\text{ V} - 240\text{ V}}{240\text{ V}}\right) \times 100\%$$

$$\%\text{VR} = 2.67\%$$

$$\%\ \text{efficiency} = \left(\frac{P_{\text{out}}}{P_{\text{in}}}\right) \times 100\%$$

For the circuit of Figure 21.16

$$P_{\text{out}} = (\alpha V_{\text{SL}})\left(\frac{I_s}{\alpha}\right)$$

$$P_{\text{out}} = (240\ \text{V})(20.84\ \text{A}) = 5002\ \text{W}$$

$$P_{\text{in}} = V_p I_p \cos\theta_p$$

$$\theta_p = 5.83° - (-0.33°) = 6.16°$$

$$P_{\text{in}} = (246.4\ \text{V})(20.86\ \text{A})(\cos 6.16°) = 5110.2\ \text{W}$$

$$\%\ \text{efficiency} = \left(\frac{5002\ \text{W}}{5110.2\ \text{W}}\right) \times 100\%$$

$$\%\ \text{efficiency} = 97.88\%$$

$$Z_i = \frac{V_p}{I_p} = \frac{246.4\ \angle 5.83°\ \text{V}}{20.86\ \angle -0.33°\ \text{A}}$$

$$Z_i = 11.81\ \angle 6.16°\ \Omega$$

The circuit could also have been analyzed using any of the other methods of analysis. For instance, $\mathbf{V}_{bc}$ could have been obtained using nodal analysis instead of the voltage divider.

EXAMPLE 21.10

An iron core transformer has the equivalent circuit shown in Figure 21.17. Determine $\mathbf{Z}_i$ when alpha is 4 and:

a. $\mathbf{Z}_L$ is an 8-ohm resistor.
b. $\mathbf{Z}_L$ is an 8-ohm inductive reactance.
c. $\mathbf{Z}_L$ is an 8-ohm capacitive reactance.

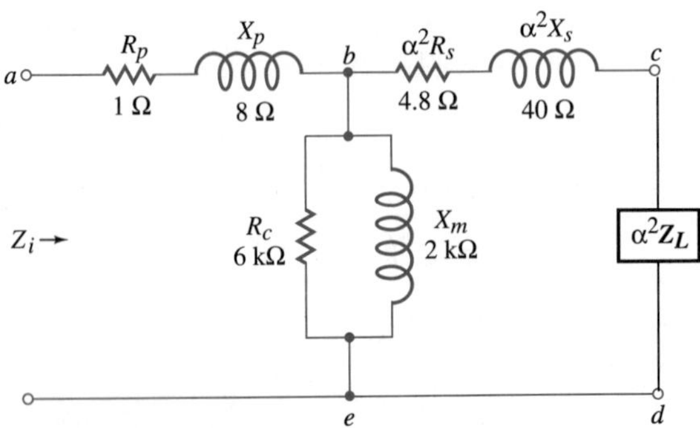

Figure 21.17

Solution

a. For $\mathbf{Z}_L = 8 \angle 0° \ \Omega$

$$\begin{aligned}
\mathbf{Z}_i &= \mathbf{Z}_{ab} + \mathbf{Z}_{be} \parallel \mathbf{Z}_{bcde} \\
\mathbf{Z}_{ab} &= R_p + jX_p \\
&= (1 + j8)\ \Omega \qquad \text{or} \qquad 8.06 \angle 82.87° \ \Omega \\
\mathbf{Z}_{be} &= R_c \parallel jX_m \\
&= \frac{(6 \times 10^3\ \Omega)(2 \times 10^3 \angle 90° \ \Omega)}{(6 \times 10^3 + j2 \times 10^3)\Omega} \\
&= 1.9 \times 10^3 \angle 71.57° \ \Omega \\
\mathbf{Z}_{bcde} &= \alpha^2 R_s + j\alpha^2 X_s + \alpha^2 Z_L \\
&= 4.8\ \Omega + j40\ \Omega + (16)(8\ \Omega) \\
&= 132.8\ \Omega + j40\ \Omega \\
&= 138.7 \angle 16.76° \ \Omega
\end{aligned}$$

so

$$\begin{aligned}
\mathbf{Z}_i &= 8.06 \angle 82.87° \ \Omega + \frac{(1.9 \times 10^3 \angle 71.57° \ \Omega)(138.7 \angle 16.76° \ \Omega)}{1.9 \times 10^3 \angle 71.57° \ \Omega + 138.7 \angle 16.76° \ \Omega} \\
&= (125.84 + j53.53)\ \Omega \\
\mathbf{Z}_i &= 136.75 \angle 23.04° \ \Omega
\end{aligned}$$

b. For $\mathbf{Z}_L = 8 \angle 90° \ \Omega$

$$\begin{aligned}
\mathbf{Z}_i &= \mathbf{Z}_{ab} + \mathbf{Z}_{be} \parallel \mathbf{Z}_{bcde} \\
\mathbf{Z}_{ab} &= 8.06 \angle 82.87° \ \Omega \text{ as in part a} \\
\mathbf{Z}_{be} &= 1.9 \times 10^3 \angle 71.57° \ \Omega \\
\mathbf{Z}_{bcde} &= \alpha^2 R_s + j\alpha^2 X_s + \alpha^2 Z_L \\
&= 4.8\ \Omega + j40\ \Omega + j128\ \Omega \\
&= 4.8\ \Omega + j168\ \Omega \qquad \text{or} \qquad 168.06 \angle 88.36° \ \Omega \\
\mathbf{Z}_{be} \parallel \mathbf{Z}_{bcde} &= \frac{(1.9 \times 10^3 \angle 71.57° \ \Omega)(168.06 \angle 88.36° \ \Omega)}{(1.9 \times 10^3 \angle 71.57° \ \Omega) + (168.06 \angle 88.36° \ \Omega)} \\
&= 154.9 \angle 87.01° \ \Omega
\end{aligned}$$

so

$$\begin{aligned}
\mathbf{Z}_i &= 1\ \Omega + j8\ \Omega + 154.9 \angle 87.01° \ \Omega \\
&= (9.08 + j162.69)\ \Omega \\
\mathbf{Z}_i &= 162.94 \angle 86.81° \ \Omega
\end{aligned}$$

c. For $\mathbf{Z}_L = 8 \angle -90° \ \Omega$

$$\begin{aligned}
\mathbf{Z}_i &= \mathbf{Z}_{ab} + \mathbf{Z}_{be} \parallel \mathbf{Z}_{bcde} \\
\mathbf{Z}_{ab} &= 8.06 \angle 82.87° \ \Omega \text{ as in part a} \\
\mathbf{Z}_{be} &= 1.9 \times 10^3 \angle 71.57° \ \Omega \text{ as in part a} \\
\mathbf{Z}_{bcde} &= \alpha^2 R_s + j\alpha^2 X_s + \alpha^2 Z_L
\end{aligned}$$

$$= 4.8\ \Omega + j40\ \Omega - j128\ \Omega$$

$$= 4.8\ \Omega - j88\ \Omega \quad \text{or} \quad 88.13\ \angle -86.88^\circ\ \Omega$$

$$\mathbf{Z}_{be} \parallel \mathbf{Z}_{bcde} = \frac{(1.9 \times 10^3\ \angle 71.57^\circ\ \Omega)(88.13\ \angle -86.88^\circ\ \Omega)}{(1.9 \times 10^3\ \angle 71.57^\circ\ \Omega) + (88.13\ \angle -86.88^\circ\ \Omega)}$$

$$= 92.09\ \angle -85.86^\circ\ \Omega$$

so

$$\mathbf{Z}_i = 1\ \Omega + j8\ \Omega + 92.09\ \angle -85.86^\circ\ \Omega$$

$$= (7.65 - j83.85)\ \Omega$$

$$Z_i = 84.20\ \angle -84.79^\circ\ \Omega$$

These values compare to those found using $Z_i = \alpha^2 Z_L$ as shown in the following.

Z_L	$\alpha^2 Z_L$	Actual Z_L
$8\ \angle 0^\circ\ \Omega$	$128\ \angle 0^\circ\ \Omega$	$136.75\ \angle 23.04^\circ\ \Omega$
$8\ \angle 90^\circ\ \Omega$	$128\ \angle 90^\circ\ \Omega$	$162.94\ \angle 86.81^\circ\ \Omega$
$8\ \angle -90^\circ\ \Omega$	$128\ \angle -90^\circ\ \Omega$	$84.2\ \angle -84.79^\circ\ \Omega$

Thus, the relation $\mathbf{Z}_i = \alpha^2 Z_L$ does not hold exactly for the practical transformer. However, the relation does give results that are acceptable in many cases.

21.5 The Approximate Equivalent Circuit

Many times, the need does not justify using the exact equivalent circuit. Since the excitation current in a well-designed transformer is 0.5–5 percent of the total current, the equivalent circuit in Figure 21.18 can be used without too much error. However, that approximation could not be used to determine leakage losses. Why?

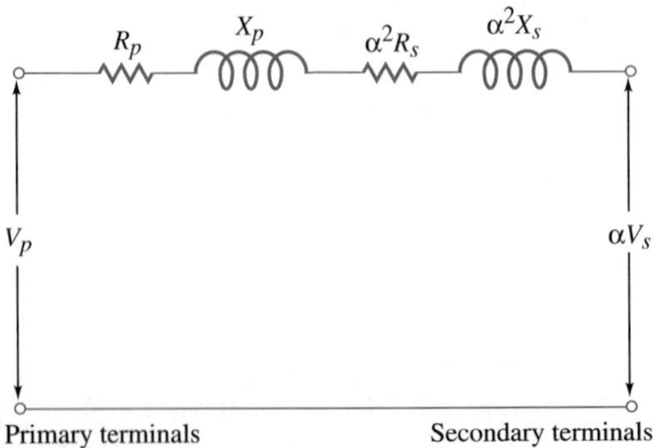

Figure 21.18 The approximate transformer equivalent circuit ignoring excitation losses.

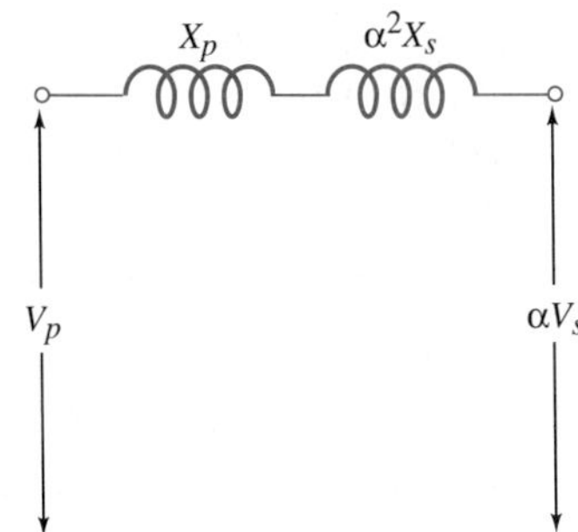

Primary terminals Secondary terminals

Figure 21.19 The approximate transformer equivalent circuit ignoring excitation losses and winding resistance.

An even simpler equivalent circuit is shown in Figure 21.19. This circuit is based on the fact that $X_p \gg R_p$ and $X_s \gg R_s$. This circuit cannot be used for efficiency analysis because there is no resistance in the circuit.

Finally, the simplest approximation is that of the ideal transformer in Section 21.20. The choice of the circuit depends on factors such as accuracy, time, quantity being studied, and parameters that are known.

EXAMPLE 21.11

Use the approximation of Figure 21.18 to determine the primary voltage of the transformer in Example 21.9.

Solution

For the circuit of Figure 21.18

$$\mathbf{V}_p = \frac{\mathbf{I}_s}{\alpha}(R_p + jX_p + \alpha^2 R_s + j\alpha^2 X_s + \alpha \mathbf{V}_s)$$

Inserting the values from Example 21.9 gives

$$\mathbf{V}_p = \left(\frac{41.67 \text{ A}}{2}\right)\bigg(0.2\ \Omega + j1\ \Omega + (2)^2(0.01\ \Omega) + j(2)^2(0.05\ \Omega)\bigg) + (2)(120 \text{ V})$$

$$\mathbf{V}_p = (20.84 \text{ A})(1.22\ \angle 78.69^\circ\ \Omega) + 240 \text{ V}$$

$$V_p = 246.25\ \angle 5.81^\circ \text{ V}$$

This answer is almost the same as that obtained by the exact solution.

Before leaving equivalent circuits, we once again consider transformer impedances, this time in terms of transformer parameters.

In general, the short-circuit current will be V_p/Z_i. If the rated current and voltage are I_p and V_p,

$$\%\text{ impedance} = \left(\frac{I_p Z_i}{V_p}\right) \times 100\% \tag{21.8}$$

where Z_i is the magnitude of the impedance looking into the primary with the secondary short-circuited, in ohms, and V_p and I_p are the rated primary voltage and current, in volts and amperes.

EXAMPLE 21.12

Determine the percent impedance of a 120/240-V, 1-kVA transformer using Figure 21.18 if $R_p = 0.01$ ohm, $X_p = 1$ ohm, $R_s = 0.1$ ohm, and $X_s = 2$ ohms.

Solution

$$\%\ \text{impedance} = \frac{I_p Z_i \times 100\%}{V_p}$$

$$I_p = \frac{\text{VA}}{V_p} = \frac{1000\ \text{VA}}{120\ \text{V}} = 8.33\ \text{A}$$

$$Z_i = R_p + jX_p + \alpha^2 R_s + \alpha^2 jX_s$$

$$\alpha = \frac{V_p}{V_s} = \frac{120\ \text{V}}{240\ \text{V}} = 0.5$$

so

$$Z_i = 0.01\ \Omega + j1\ \Omega + (0.5)^2(0.1\ \Omega) + (0.5)^2(j2\ \Omega)$$

$$= 1.5\ \angle 88.66^\circ\ \Omega$$

$$\%\ \text{impedance} = \frac{(8.33\ \text{A})(1.5\ \Omega)}{120\ \text{V}}$$

$$\%\ \text{impedance} = 10.4\%$$

21.6 Short-Circuit and Open-Circuit Tests

A practical method that can be used to determine the efficiency of a transformer is the short-circuit and open-circuit test. One advantage of this method is that equipment to operate the transformers at rated load is not needed. However, the transformers must be available, so the method cannot be used at the design stage.

short-circuit test

The connection for the tests is shown in Figure 21.20. For the **short-circuit test**, the secondary is short-circuited. The source is set to give the current that is in the winding at full load. If this is not known, then the current, as calculated from the kVA rating, is used. The voltage will be small because of the short circuit, so the excitation losses will be negligible. The power is being dissipated in the resistance of the windings. This loss is also called the I^2R loss, and is proportional to the current squared.

open-circuit test

For the **open-circuit test**, the secondary winding is open-circuited, and rated voltage is applied to the primary. The current is now very small because it is only the excitation current. The small current gives a

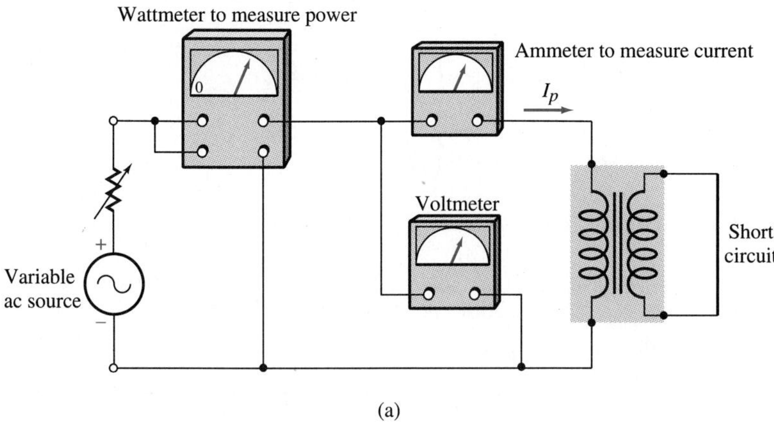

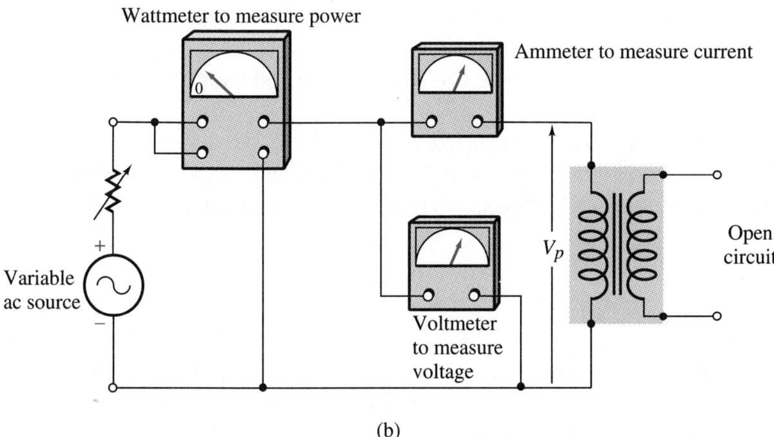

Figure 21.20 Transformer connections for determining losses. (a) The short-circuit test connection. (b) The open-circuit test connection.

negligible I^2R loss. The power is being dissipated in the core and other excitation losses. Thus, the short-circuit test is used for the winding losses, and the open-circuit test is used for the excitation losses.

The source can be connected to either winding of the transformer. The choice usually depends on the equipment that is available. If it is connected to the low-voltage side, exercise caution. The high-voltage terminals can have a potentially dangerous voltage across them. The efficiency of a transformer is now calculated from the test data.

EXAMPLE 21.13

Measurements on a 3-kVA, 120/240-V transformer give the following:

Short-circuit: $P = 100$ W, $V_p = 10$ V, $I_p = 25$ A.
Open-circuit: $P = 24$ W, $V_p = 120$ V, $I_p = 0.25$ A.

Find the percent efficiency when operating at rated output for a resistance load.

Solution

$$\% \text{ efficiency} = \left(\frac{P_{\text{out}}}{P_{\text{in}}}\right) \times 100\%$$

$$= \left(\frac{P_{\text{out}}}{(P_{\text{out}} + P_{\text{lost}})}\right) \times 100\%$$

For a resistance load, rated P_{out} is 3 kW so

$$\% \text{ efficiency} = \left(\frac{3000 \text{ W}}{3000 \text{ W} + 100 \text{ W} + 24 \text{ W}}\right) \times 100\%$$

$$\% \text{ efficiency} = 96\%$$

The short-circuit and open-circuit test data can also be used to determine the voltage regulation of the transformer at any load current. The procedure is explained for the approximate equivalent circuit of Figure 21.18.

For this approximation, $(I_s/\alpha) = I_p$. If V_s is taken at 0° for the reference, (I_s/α) must lead or lag V_s by the phase angle (θ_L) of the load. Then, the secondary voltage, referred to the primary side, is αV_s and is given by

$$\alpha V_s \angle 0° = V_p \angle \theta_p - \left(\frac{I_s \angle \theta_L}{\alpha}\right)(Z_{\text{eq}} \angle \theta_z) \qquad \textbf{(21.9)}$$

θ_L is (+) for a capacitive load and (−) for an inductive load.

Z_{eq} and θ_z are found using the magnitudes of V_p, I_p, and P from the short-circuit test. This gives

$$Z_{\text{eq}} = \frac{V_p}{I_p} \qquad \textbf{(21.10)}$$

$$\theta_z = \text{arc cos}\left(\frac{V_p I_p}{P}\right) \qquad \textbf{(21.11)}$$

The last quantity, θ_p, is found by setting the j part of the right-hand side of Equation 21.9 to 0. This gives

$$\theta_p = \text{arc sin}\left[\frac{I_s Z_{\text{eq}}}{\alpha V_p} \sin(\theta_z + \theta_L)\right] \qquad \textbf{(21.12)}$$

θ_L in Equation 21.12 takes the same sign as in Equation 21.9.

Now that all of the relations are known, we apply them to a problem.

EXAMPLE 21.14

A transformer with an α of 2 is connected to a 240-volt source. It supplies 10 A to a load with a power factor of 0.7 lagging. The short-circuit test data are $V_p = 20$ V, $I_p = 5$ A, $P = 80$ W. Determine the % VR.

Solution

$$V_p = 240 \text{ V}$$

From the short-circuit test data,

$$Z_{\text{eq}} = \frac{20 \text{ V}}{5 \text{ A}} = 4 \ \Omega$$

$$\theta_z = \text{arc cos}\left(\frac{80 \text{ W}}{(20 \text{ V})(5 \text{ A})}\right)$$

$$= 36.87°$$

so

$$\mathbf{Z}_{\text{eq}} = 4 \angle 36.87° \ \Omega$$

The load is inductive because of the lagging power factor, so

$$\theta_L = -\text{arc cos}(0.7) = -45.57°$$

Then, using Equation 21.12 gives

$$\theta_p = \text{arc sin}\left[\frac{(10 \text{ A})(4)}{(2)(240 \text{ V})} \sin(36.87° - 45.57°)\right]$$

$$= -0.72°$$

Putting these values in Equation 21.9 gives

$$\alpha \mathbf{V}_s = 240 \angle -0.72° \text{ V} - \left[\left(\frac{10}{2} \angle -45.57° \text{ A}\right)(4 \angle 36.87° \ \Omega)\right]$$

$$= (220.21 + j0.009) \text{ V}$$

$$= 220.21 \angle +0.002° \text{ V}$$

$$\%\text{VR} = \left(\frac{240 \text{ V} - 220.21 \text{ V}}{220.21 \text{ V}}\right) \times 100\%$$

$$\%\text{VR} = 8.99\%$$

The short-circuit and open-circuit tests can also be used to obtain the transformer parameters.

The relations of the parameters to the test results are listed in Equations 21.13 to 21.15. Their derivation is left as an exercise for the student. First, R_p must be measured with a bridge or ohmmeter with the secondary open-circuited. Then, from the short-circuit test,

$$R_s = \frac{1}{\alpha^2}\left(\frac{P}{I_p^2} - R_p\right) \quad \textbf{(21.13)}$$

$$X_p = \frac{V_p \sin \theta_p}{2I_p} \quad \textbf{(21.14)}$$

$$X_s = \frac{V_p \sin \theta_p}{\alpha^2 2I_p} \quad \textbf{(21.15)}$$

where P is the short-circuit power, in watts; R_p is the primary resistance, in ohms; I_p is the short-circuit primary current, in amperes; V_p is the short-circuit primary voltage, in volts; and θ_p is equal to arc cos $[(P)/(V_pI_p)]$, and is positive.

From the open-circuit test,

$$R_c = \frac{V_c^2}{P - I_p^2 R_p} \tag{21.16}$$

$$X_m = \frac{V_c^2}{V_p I_p \sin\theta_p - I_p^2 X_p} \tag{21.17}$$

where P is the open-circuit power, in watts; I_p is the open-circuit primary current, in amperes; V_p is the open-circuit primary voltage, in volts; X_p and R_p are obtained from the short-circuited test; θ_p is equal to arc $\cos(P)/(V_pI_p)$, and is positive; and V_c is equal to the magnitude of $V_p - I_p \angle -\theta_p\ (R_p + jX_p)$. θ_p is negative since I_p lags V_p.

An example helps to show how those relations are applied.

EXAMPLE 21.15

Use the short-circuit and open-circuit data from Example 21.13 to determine the equivalent circuit parameters for the transformer. R_p was measured as 0.1 ohm.

Solution

Using the short-circuit conditions first

$$R_s = \frac{1}{\alpha^2}\left(\frac{P}{I^2} - R_p\right)$$

$$\alpha = \frac{120\ \text{V}}{240\ \text{V}} = 0.5$$

so

$$R_s = \frac{1}{(0.5)^2}\left(\frac{100\ \text{W}}{(25\ \text{A})^2} - 0.1\ \Omega\right)$$

$$R_s = 0.24\ \Omega$$

$$X_p = \frac{V_p \sin\theta_p}{2I_p}$$

and

$$\theta_p = \text{arc}\cos\left(\frac{P}{VI}\right)$$

$$= \text{arc}\cos\left(\frac{100\ \text{W}}{(10\ \text{V})(25\ \text{A})}\right)$$

$$\theta_p = 66.42^\circ$$

Then

$$X_p = \frac{(10\ \text{V})(\sin 66.42°)}{(2)(25\ \text{A})}$$

$$X_p = 0.183$$

$$X_s = \frac{V_p \sin \theta_p}{\alpha^2 2 I_p}$$

or

$$X_s = \frac{X_p}{\alpha^2}$$

$$= \frac{0.183\ \Omega}{(0.5)^2}$$

$$X_s = 0.73\ \Omega$$

Using the open-circuit conditions and the relation for power,

$$R_c = \frac{V_c^2}{P - I_p^2 R_p}$$

but

$$V_c = V_p - I_p \angle \theta_p\ (R_p + jX_p)$$

and

$$\theta_p = \text{arc cos}\left(\frac{P}{V_p I_p}\right)$$

$$= \text{arc cos}\left(\frac{24\ \text{W}}{(120\ \text{V})(0.25\ \text{A})}\right)$$

$$= 36.87°$$

so

$$V_c = 120 \angle 0°\ \text{V} - (0.25 \angle -36.87°\ \text{A})(0.1 + j0.183)\ \Omega$$

$$= 119.95 \angle -0.01°\ \text{V}$$

Then

$$R_c = \frac{(119.95\ \text{V})^2}{24\ \text{V} - (0.25\ \text{A})^2(0.1\ \Omega)^2}$$

$$R_c = 599.66\ \Omega$$

$$X_m = \frac{V_c^2}{V_p I_p \sin \theta_p - I_p^2 X_p}$$

$$= \frac{(119.95\ \text{V})^2}{(120\ \text{V})(0.25\ \text{A}) \sin (36.87°) - (0.25\ \text{A})^2\ (0.183)\ \Omega}$$

$$X_m = 799.8\ \Omega$$

When the input for the test is on the secondary side of the transformer, the relation can be stated as

$$R_p = \alpha^2 \left(\frac{P}{I_s^2} - R_s \right) \tag{21.18}$$

$$X_s = \frac{V_s \sin \theta_s}{2I_s} \tag{21.19}$$

where

$$\theta_s = \text{arc cos}\left(\frac{P}{V_s I_s} \right)$$

$$R_c = \frac{V_c^2}{P - I_s^2 R_s} \tag{21.20}$$

$$X_m = \frac{V_c^2}{V_s I_s \sin \theta_s - I_s^2 X_s} \tag{21.21}$$

where

$$V_c = V_s - I_s \angle \theta_s (R_s + jX_s)$$

and

$$\theta_s = -\text{arc cos}\left(\frac{P}{V_s I_s} \right)$$

21.7 Frequency Effects

Transformers are designed to operate at a specific frequency or range of frequencies. Most power transformers operate at 50 or 60 Hz, while transformers used in aircraft equipment operate at 400 Hz. When considering the effect of frequency, we must consider its effect on the voltage and on the losses. First, the voltage is considered.

At lower frequencies, the shunting effect of X_m will cause a drop in the output, since $X_m = 2\pi f L_m$. The drop is shown in curve 1 in Figure 21.21. At higher frequencies, since $X_c = 1/(2\pi f C)$, the shunting effect of C_s and C_p will likewise cause a drop, as shown in curve 2 of Figure 21.21. At some frequency, the capacitive and inductive series resonance causes a peak, as shown in curve 3 of Figure 21.21.

The shape of the response curve is not too critical for transformers that operate at a single frequency. On the other hand, audio transformers

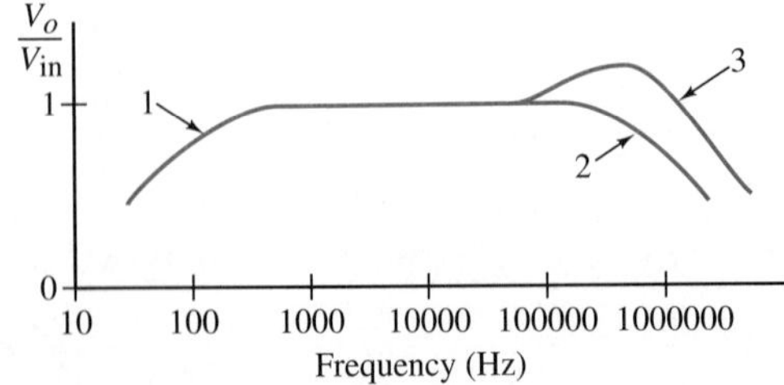

Figure 21.21 The general effect of frequency on the output of the transformer.

or others that operate over a range of frequencies should have a flat response over the operating range, as shown by the flat section in Figure 21.21.

Next, we consider the losses. At higher frequencies, eddy current and hysteresis losses are greater. Doubling the frequency doubles the hysteresis losses and quadruples the eddy current losses. These losses cause a rise in temperature and can damage the transformer. They also reduce the efficiency. At lower frequencies, the inductive reactances are smaller, so the current will be greater. This results in a larger I^2R loss and a temperature rise. Again, damage can result and the efficiency will be reduced. Although a transformer can operate at higher and lower frequencies, its efficiency and characteristics will not be the same and its life can be shortened.

21.8 The Autotransformer

Some transformers, such as that in Figure 21.22, transfer power by both transformer action and directly. They are called **autotransformers**. The power that is transferred by transformer action is **inductively transferred power**. The power that is transferred directly is **conductively transferred power**. The connections for a step-down autotransformer are shown in Figure 21.23. Commercial units are made with one winding on one core, but a regular transformer also can be connected as an autotransformer. The tap can be fixed or variable, and there can be more than one tap. A variable tap provides an ac voltage from zero to the voltage at the tap. Varying ac with an autotransformer is more efficient than using a potentiometer or rheostat.

autotransformers

inductively transferred power

conductively transferred power

Figure 21.22 This variable autotransformer provides 0 to 140 volts from a 120 volt input. (Photo courtesy of Superior Electric, Bristol, CT)

Autotransformers are usually more efficient and have a smaller physical size than a two-winding transformer of the same volt-ampere output. They also have a smaller percent voltage regulation. Two reasons for this are the single-winding and smaller iron core. The disadvantage of the autotransformer is its lack of electrical isolation. Notice that the input and output are connected at Point c. This can result in damage to equipment and presents a shock hazard. Also, any dc in the source will be conducted to the load.

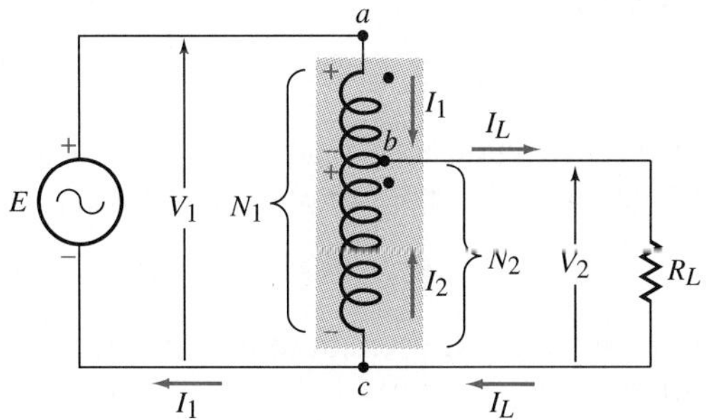

Figure 21.23 The winding connection for a step-down autotransformer.

A voltage applied across coil N_1 of Figure 21.23 induces a voltage in N_2. For an ideal two-winding autotransformer, such as the one in Figure 21.23,

$$\frac{V_2}{V_1} = \frac{N_2}{N_1} = \frac{1}{\alpha} \tag{21.22}$$

The voltage across N_2 is applied to R_L, causing the current I_L. This current is made up of the current I_1 in the primary N_1, and the current I_2 in the secondary. That is, $I_L = I_1 + I_2$. The directions of these currents are shown in Figure 21.23. Being an ideal transformer, the volt-amperes in the load must equal the volt-amperes supplied by the source. That is,

$$I_L V_2 = I_1 V_1$$

or

$$\frac{I_L}{I_1} = \frac{V_1}{V_2}$$

but

$$\frac{V_1}{V_2} = \frac{N_1}{N_2}$$

so

$$\frac{I_L}{I_1} = \frac{N_1}{N_2} = \alpha \tag{21.23}$$

This is the same relation as for the two-winding transformer. The power in R_L is

$$P_L = V_2 I_L$$

but

$$I_L = I_1 + I_2$$

so

$$P_L = V_2 I_1 + V_2 I_2$$

I_2 is the current transferred to the secondary of the transformer. So the inductively transferred power is

$$P_i = V_2 I_2 \tag{21.24}$$

I_1 is the current passing directly from the source to the load through the upper winding. So the conductively transferred power is

$$P_c = V_2 I_1 \tag{21.25}$$

The next example applies these relations to find the characteristics of an autotransformer.

EXAMPLE 21.16

A 400-VA, 220/110-V transformer is to be connected as a step-down autotransformer to supply 220 V from a 330-V source. Determine the current in each part, the maximum power that can be supplied to a resistive load, the conductively transferred power, the inductively transferred power, and compare the power supplied to the transformer rating.

Solution

The step-down connection with the currents labeled is shown in Figure 21.24.

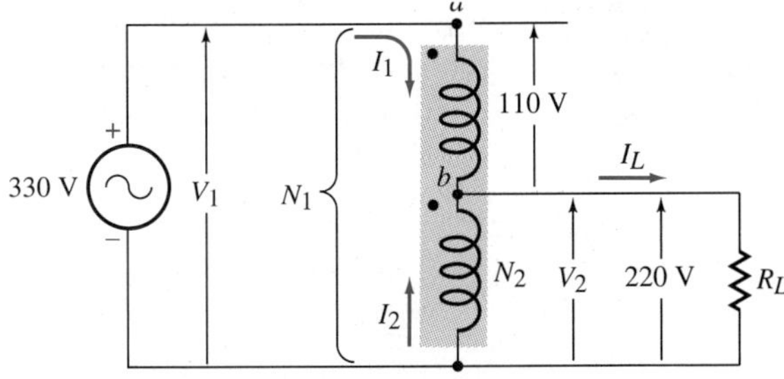

Figure 21.24

The maximum current in N_2 is

$$I_2 = \frac{\text{VA rating}}{V_2} = \frac{400 \text{ VA}}{220 \text{ V}}$$

$$I_2 = 1.82 \text{ A}$$

The maximum current that can be in winding *ab* is

$$I_1 = \frac{\text{VA rating of } ab}{V_{ab}} = \frac{400 \text{ VA}}{110 \text{ V}}$$

$$I_1 = 3.64 \text{ A}$$

$$I_L = I_1 + I_2 = 3.64 \text{ A} + 1.82 \text{ A}$$

$$I_L = 5.46 \text{ A}$$

The maximum load power is

$$P_L = V_L I_L = (220 \text{ V})(5.46 \text{ A})$$

$$P_L = 1201.2 \text{ W}$$

The conductive power is

$$P_c = V_2 I_1 = (220 \text{ V})(3.64 \text{ A})$$

$$P_c = 800.8 \text{ W}$$

The inductively transferred power is

$$P_i = V_2 I_2 = (220 \text{ V})(1.82 \text{ A})$$

$$P_i = 400.4 \text{ W}$$

The ratio of the power handled to the transformer rating is

$$\frac{1201 \text{ W}}{400.4 \text{ W}}$$

$$\text{Ratio} = 3.0$$

As an autotransformer, it can handle three times the power capacity of the two-winding transformer.

The transformed power is equal to the transformer rating because the two windings are at their rated voltage. The additional power is supplied directly through the upper winding.

A step-up autotransformer connection is shown in Figure 21.25. This type of autotransformer can provide a voltage higher than the source voltage. It can be used to connect low-voltage conditions. With a variable tap, ac voltage from zero to some voltage greater than the source is possible.

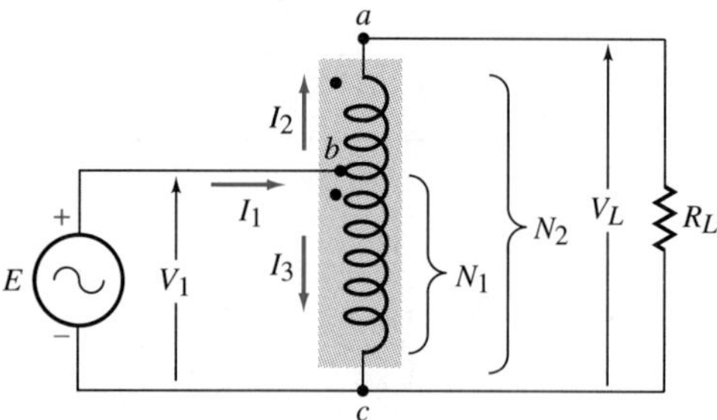

Figure 21.25 The winding connection for a step-up autotransformer.

The relations for the step-up transformer are derived similarly to those for the step-down unit. They are

$$\frac{V_L}{V_1} = \frac{N_2}{N_1} = \frac{1}{\alpha} \tag{21.26}$$

$$\frac{I_2}{I_1} = \frac{N_1}{N_2} = \alpha \tag{21.27}$$

$$P_c = V_1 I_2 \tag{21.28}$$

$$P_i = V_1 I_3 \tag{21.29}$$

The transformer of Example 21.16 is now used as a step-up transformer to illustrate these relations and the transformer's characteristics.

EXAMPLE 21.17

The transformer of Example 21.16 is connected as a step-up autotransformer to supply 330 V from a 220-V source. Determine the current in each part, the maximum power that can be supplied to a resistive load, the conductively transferred power, the inductively transferred power, and compare the power supplied to the transformer rating.

Solution

The step-up connection with the currents labeled is shown in Figure 21.26.

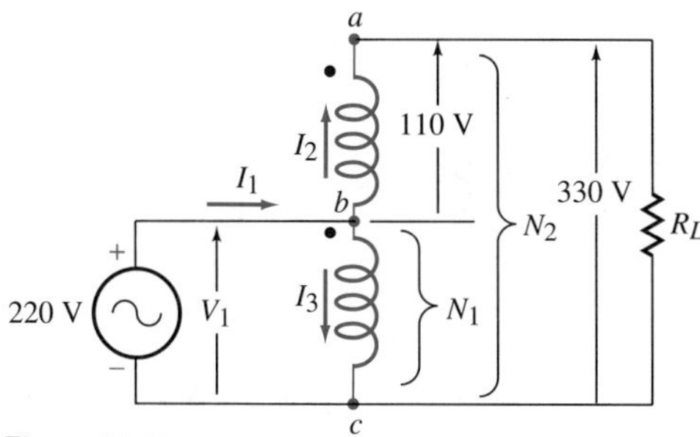

Figure 21.26

The maximum current in N_1 is

$$I_3 = \frac{\text{VA rating}}{V_1} = \frac{400\text{ VA}}{220\text{ V}}$$
$$I_3 = 1.82\text{ A}$$

The maximum current in winding *ab* is

$$I_2 = \frac{\text{VA rating of }_{ab}}{V_{ab}} = \frac{400\text{ VA}}{110\text{ V}}$$
$$I_2 = 3.64\text{ A}$$
$$P_L = V_L I_2 = (330\text{ V})(3.64\text{ A})$$
$$P_L = 1201.2\text{ W}$$
$$P_c = V_1 I_2 = (220\text{ V})(3.64\text{ A})$$
$$P_c = 800.8\text{ W}$$
$$P_i = V_1 I_3 = (220\text{ V})(1.82\text{ A})$$
$$P_i = 400.4\text{ W}$$

Again, the inductively transferred power is equal to the rating of the transformer.

$$\text{Power ratio} = \frac{1201.2\text{ W}}{400.4\text{ W}}$$
$$\text{Power ratio} = 3.0$$

21.9 The Air Core Transformer

The air core transformer does not have any iron in it. Coupling is obtained through the nearness of the coils and is less than one.

Air core transformers are used mainly in electronic applications. Some reasons for their use are that there are no hysteresis and eddy

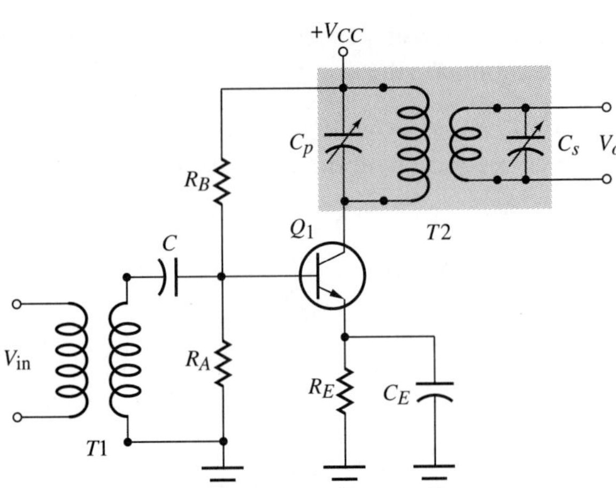

Figure 21.27 A double-tuned coupling circuit is used in the output of this transistor amplifier.

current losses and they are less costly. Also, they are usually used as part of a **tuned circuit**. This is a circuit in which a capacitor is connected across the winding and either the capacitance or transformer inductance is adjusted to obtain a resonant condition. This condition develops a maximum voltage across the transformer.

tuned circuit

A common application is in the Intermediate Frequency (IF) section of television and radio receivers. They are tuned to obtain maximum voltage at the I.F. frequencies, 455 kHz for A.M. and 10.7 MHz for F.M.

A basic double-tuned coupling circuit is shown in Figure 21.27. It has a capacitor across each winding.

In operation, the capacitance is varied to obtain resonance at the desired frequency. Resonance is the condition where Z_i is maximum. The maximum impedance results in maximum voltage across the primary and secondary. Tuning can also be done by varying the position of a ferrite rod in the core of the transformer.

Since the transformer is used for coupling, our concern is with the input impedance and the output voltage. Because there is no iron core, the coupling is much less than 100 percent, so Equations 21.4 to 21.6 do not hold for the air core transformer in Figure 21.28.

For that transformer, the secondary impedance will be transferred to the primary as $(\omega M)^2/(Z_s + Z_L)$.

Also:

A resistance will reflect to the primary as a resistance.
An inductance will reflect to the primary as a capacitance.
A capacitance will reflect to the primary as an inductance.

The reflected impedance will be in series with $\mathbf{Z}_p$, so

$$\mathbf{Z}_i = \mathbf{Z}_p + \frac{(\omega M)^2}{\mathbf{Z}_s + \mathbf{Z}_L} \tag{21.30}$$

where $\mathbf{Z}_i$ is the impedance looking into the primary winding, in ohms; ω is

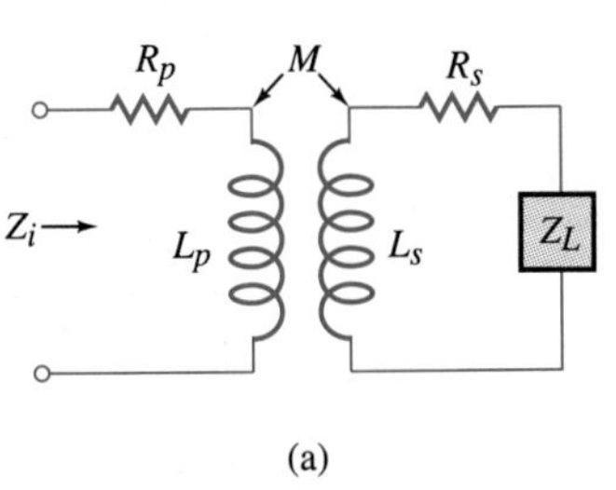

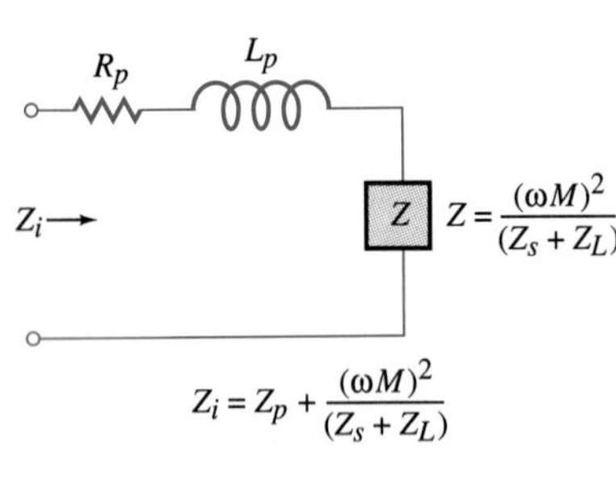

Figure 21.28 Determination of the input impedance for the air core transformer. (a) The ac equivalent circuit with the load connected. (b) The ac equivalent circuit with the load transferred to the primary.

the angular frequency, in radians per second; M is the mutual inductance, in henrys; $\mathbf{Z}_L$ is the impedance connected to the secondary, in ohms; $\mathbf{Z}_p$ is $R_p + jX_p$; and $\mathbf{Z}_s$ is $R_s + jX_s$.

At this point, it is helpful to recall from Chapter 13, Section 13.8, that $M = k\sqrt{L_p L_s}$, where k is the coefficient of coupling.

EXAMPLE 21.18

An air core transformer has the following parameters:

$R_p = 1.6$ ohms $L_p = 0.05$ H $R_s = 1$ ohm
$L_s = 0.03$ H $k = 0.3$

What is the input impedance for a frequency of 4 kHz and Z_L of:

a. 50 ∠0° ohms.
b. 50 ∠90° ohms.
c. 50 ∠−90° ohms.

Solution

a. From Equation 21.30,

$$\mathbf{Z}_i = \mathbf{Z}_p + \frac{(\omega M)^2}{\mathbf{Z}_s + \mathbf{Z}_L}$$

$$X_p = \omega L_p = 2\pi f L_p$$
$$X_p = (2)(\pi)(4000 \text{ Hz})(0.05 \text{ H}) = 1256.6\ \Omega$$
$$X_s = \omega L_s = 2\pi f L_s$$
$$X_s = (2)(\pi)(4000 \text{ Hz})(0.03 \text{ H}) = 753.98\ \Omega$$
$$M = k\sqrt{L_p L_s} = 0.3\sqrt{(0.05 \text{ H})(0.03 \text{ H})} = 0.012 \text{ H}$$

so

$$\mathbf{Z}_i = 1.6\ \Omega + j1256.6\ \Omega + \frac{[(2)(\pi)(4000 \text{ Hz})(0.012 \text{ H})]^2}{(1 + j753.98 + 50)\ \Omega}$$
$$= 1.6\ \Omega + j1256.6\ \Omega + 120.36\ \angle -86.13°\ \Omega$$
$$\mathbf{Z}_i = 1136.56\ \angle 89.51°\ \Omega$$

b.

$$\mathbf{Z}_i = 1.6\ \Omega + j1256.6\ \Omega + \frac{[(2)(\pi)(4000 \text{ Hz})(0.012 \text{ H})]^2}{1\ \Omega + j753.98\ \Omega + j50\ \Omega}$$
$$= 1.6\ \Omega + j1256.6\ \Omega + 113.13\ \angle -89.93°\ \Omega$$
$$\mathbf{Z}_i = 1143.47\ \angle 89.91°\ \Omega$$

c.

$$\mathbf{Z}_i = 1.6\ \Omega + j1256.6\ \Omega + \frac{[(2)(\pi)(4000 \text{ Hz})(0.012 \text{ H})]^2}{1\ \Omega + j753.98\ \Omega - j50\ \Omega}$$
$$= 1.6\ \Omega + 1256.6\ \Omega + 129.21\ \angle -89.9°\ \Omega$$
$$\mathbf{Z}_i = 1127.39\ \angle 89.91°\ \Omega$$

The magnitude of the output voltage at the tuned frequency for the double-tuned circuit is given by

$$V_L = V_{in}\sqrt{\frac{L_s}{L_p}}\left(\frac{k}{k^2 + (1/(Q_pQ_s))}\right) \tag{21.31}$$

where V_L is the output voltage, in volts; V_{in} is the voltage applied to the primary, in volts; L_p and L_s are the inductance of the primary and secondary windings, in henrys; k is the coefficient of coupling; Q_p and Q_s are the quality factors of the primary and secondary circuits; and $Q_p = X_p/R_p$ and $Q_s = X_s/R_s$ or X_c/R_s if the secondary has no load connected to it.

Equation 21.31 also gives acceptable results for high-Q single-tuned circuits.

EXAMPLE 21.19

A double-tuned coupling circuit with a Q_p of 40 and a Q_s of 50 has 5 volts applied to the primary. What is the secondary voltage if $L_p = 20$ mH, $L_s = 30$ mH, and $k = 0.5$?

Solution

$$V_L = E_1\sqrt{\frac{L_s}{L_p}}\left(\frac{k}{k^2 + (1/Q_pQ_s)}\right)$$

$$V_L = (5\text{ V})\sqrt{\frac{0.030\text{ H}}{0.020\text{ H}}}\left(\frac{0.5}{(0.5)^2 + (1)/(40)(50)}\right)$$

$$V_L = 12.22\text{ V}$$

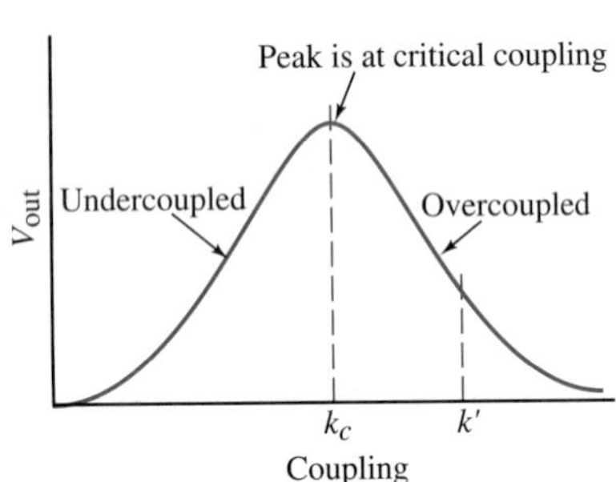

Figure 21.29 Undercoupling and overcoupling result in a lower output voltage than at critical coupling.

Connecting a resistance load across the secondary reduces Q_s and also the output voltage. However, practical applications use a large impedance load so the results do not change much.

The value of coupling at which V_L peaks is the critical coupling (k_c), and is shown in Figure 21.29. Circuits coupled at $k < k_c$ are undercoupled. Those coupled at $k > k_c$ are overcoupled. The voltage drops off for both of these conditions.

A comparison of the response curves for the three degrees of coupling is shown in Figure 21.30. Critical coupling gives the maximum output at the resonant frequency. Undercoupling also gives a maximum, but it is less than that for critical coupling. Overcoupling gives a dip at the resonant frequency but is resonant at two other frequencies. Thus, it effectively provides an increased bandwidth.

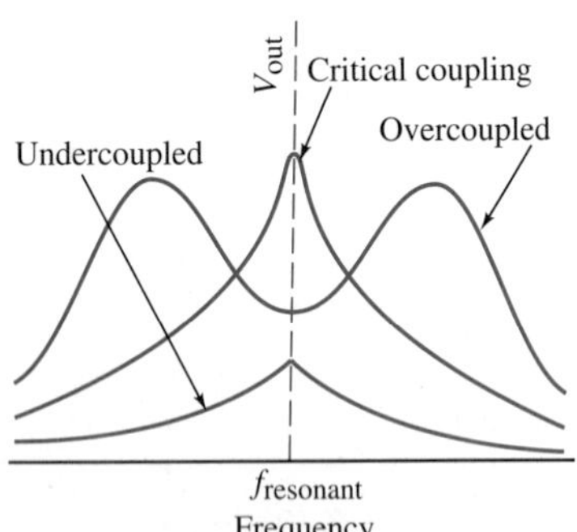

Figure 21.30 The frequency response curves for undercoupling, critical coupling, and overcoupling.

21.10 Transformer Types

Transformers are often named according to their application. Before concluding this chapter, it might be helpful to learn what some of these transformers are. The description of the more common ones is given in the following sections. Although the appearance of some might be very similar, the design is selected to provide the required characteristics.

Figure 21.31 This distribution transformer is used to provide power to many electrical loads. (Photo courtesy of ACME Transformer, a Division of ACME Electric Corporation)

Distribution Transformer

distribution transformer

The **distribution transformer** is used to step down the high voltage to a level that is needed by the customer. Such a transformer is shown in Figure 21.31. They are generally rated in the 100-kVA range and step the voltage down to 2400, 600, 480, 240, and 120 volts. They are available as single-phase and three-phase units. They also are made as dry and oil-filled units.

Potential Transformer

potential transformer

Connecting a voltmeter across a high voltage, such as 33,000 volts, is not practical and can be dangerous. Therefore, a **potential transformer** is connected across the high-voltage line to reduce the voltage to a level safe for meters. It is a step-down transformer. The meter scale is either marked to include the effect of the transformer or a multiplier is used. Since the primary is connected across a high voltage, it must be well insulated from the secondary. As an added safety feature, the secondary is grounded.

Current Transformer

current transformer

The **current transformer** is used to step down the high current to a level that can be measured on an ammeter. The primary is connected in series in the line. The primary has a few turns of heavy wire and the secondary has many turns of fine wire. The meter scale is either marked to include the effect of the transformer, or a multiplier is used.

Audio Transformer

audio transformer

The **audio transformer** is used to connect the output of an amplifier to a loudspeaker. Speakers usually have a small impedance such as 8 or 16 ohms, and the amplifier output usually has a large impedance. The audio transformer matches the speaker's impedance to the amplifier's. The transformer should have a flat response over the audio frequency range. That is, approximately 50 Hz–20 kHz. If it does not, distortion can result because the different frequency voltages are not transformed at the same ratio.

Power Transformer

power transformer

A **power transformer** is used to supply the power in electronic equipment such as a radio, oscilloscope, or signal generator. It usually has several taps including a center tap on the secondary winding. The transformer must be able to handle the total power used by the system.

Center-Tapped Transformer

center-tapped transformer

The **center-tapped transformer** has a tap located at the midpoint of the secondary. This tap provides two outputs of equal magnitude. These transformers are generally used for two-diode, full-wave rectifier circuits.

Isolation Transformer

isolation transformer

The **isolation transformer** is used to provide electrical isolation between circuits. All transformers provide isolation, but the isolation transformer has a 1–1 ratio so that there is no change in voltage value. The input voltage and current will have the same value as the output voltage and current if the ratio is 1–1.

Impedance Matching Transformer

impedance matching transformer

The **impedance matching transformer** is used to match a load impedance to the output of a circuit. The audio transformer is one of these. Others are used at the input to an amplifier or other section. This transformer matches the input or output impedances to that required for the application.

Radio Frequency (RF) Transformer

radio frequency transformer

The **radio frequency transformer** is usually used at radio frequencies and higher. Some have air cores. The IF (intermediate frequency) transformer has a ferrite core whose position can be adjusted. The transformer is usually used with a capacitor. The core's position is adjusted to tune the combination to the desired frequency. Some RF transformers are shielded to prevent interference from magnetic fields.

Figure 21.32 Control transformers such as this one provide voltages for the operation of contactors, relays, and solenoids. (Photo courtesy of ACME Transformer, a Division of ACME Electric Corporation)

Buck-boost Transformer

buck-boost transformer

The **buck-boost transformer** is connected as a step-down or step-up autotransformer. It is used where the voltage is constantly high or low by a fixed amount to bring the voltage to the desired level.

Constant Voltage Transformer

constant voltage transformer

The **constant voltage transformer** is used to maintain a constant ouptut voltage, even though the load current or input voltage changes. Some transformers use an electric motor drive to change the setting of an autotransformer. Others use some means to change the flux or coupling, which in turn changes the output voltage.

Control Transformers

control transformers

Control transformers are used for the operation of devices such as relays, solenoids, and other electromagnetic devices. These transformers must be able to handle the high momentary current when the device is energized. A control transformer is shown in Figure 21.32.

Three-Phase Transformer

three-phase transformer

A **three-phase transformer** is designed to operate on three-phase systems. These systems are described in Chapter 22. Basically, the transformer is three transformers in one.

The transformer is a very useful and versatile device. It is also reliable and efficient. Because of these and other characteristics, it has many applications in our daily lives.

SUMMARY

1. A transformer transforms energy from the primary coil to the secondary coil through the magnetic field that couples the coils. This results in a voltage across the secondary.
2. A transformer that increases the voltage is a step-up transformer. One that reduces the voltage is a step-down transformer.
3. Transformers are used to change the voltage value, match impedances, and obtain electrical isolation.
4. The relations for an ideal transformer are $E_s/E_p = N_s/N_p$, $I_s/I_p = N_p/N_s$, and $\mathbf{Z}_i = \alpha^2\mathbf{Z}_L$. These are approximate relations for the practical transformer.
5. When transformers are connected in groups, the terminal polarity must be considered in order to obtain the proper voltage or prevent short circuits.
6. The practical transformer has losses and can be represented by an equivalent circuit. This circuit accounts for the losses and can be analyzed as an ac circuit.
7. Short-circuit and open-circuit tests can be used to determine the losses in a transformer. The data can be used to determine the efficiency and voltage regulation.
8. At frequencies above and below the rated frequency, the losses will be greater and, in general, the output voltage will be smaller than at the rated frequency.
9. An autotransformer is one in which energy is transferred both inductively and directly. This enables it to handle more power at a higher efficiency.
10. The air core transformer has no iron losses, and the relations for the iron core transformer do not hold. It is used mainly in tuned circuits for coupling of sections.
11. Transformers are named according to their application. Some examples are: distribution, current, isolation, and audio transformers.

EQUATIONS

	Equation	Description	Page
21.1	$I_s = \dfrac{\text{kVA}}{kV_s}$	Relation of the rated transformer current to the voltage and kVA ratings.	866
21.2	$P_{\text{max}} = (\text{VA rating}) \times (F_p)$	The equation for finding the power capacity from the kVA rating and the power factor.	867
21.3	$\%\ \text{impedance} = \dfrac{(\text{rated primary current}) \times 100\%}{\text{short-circuit primary current}}$	The % impedance in terms of the primary current.	867

21.4	$\dfrac{E_s}{E_p} = \dfrac{N_s}{N_p} = \dfrac{1}{\alpha}$	The equation relating the ideal transformer voltages to the turns ratio.	869
21.5	$\dfrac{I_s}{I_p} = \dfrac{N_p}{N_s} = \alpha$	The equation relating ideal transformer currents to the turns ratio.	869
21.6	$(\text{VA})_{\text{primary}} = (\text{VA})_{\text{secondary}}$	The relation of the output and input volt-amperes for the ideal transformer.	870
21.7	$\mathbf{Z}_i = \alpha^2\mathbf{Z}_L$	The equation that refers the load impedance to the primary side.	872
21.8	$\%\ \text{impedance} = \left(\dfrac{I_pZ_i}{V_p}\right) \times 100\%$	The relation between the % impedance and transformer parameters.	887
21.9	$V_s \angle 0° = V_p \angle \theta_p - \dfrac{I_s \angle \theta_L}{\alpha}(Z_{\text{eq}} \angle \theta_z)$	The equation for obtaining the secondary voltage from the short-circuit test data.	890
21.10	$Z_{\text{eq}} = \dfrac{V_p}{I_p}$	The equation for finding the magnitude of Z_{eq} from the short-circuit test data.	890
21.11	$\theta_z = \text{arc cos}\left(\dfrac{V_pI_p}{P}\right)$	The equation for finding the angle of Z_{eq} from the short-circuit test data.	890
21.12	$\theta_p = \text{arc sin}\left[\dfrac{I_s Z_{\text{eq}}}{V_p}\sin(\theta_Z + \theta_L)\right]$	The equation for finding the input phase angle from the short-circuit test.	890
21.13	$R_s = \dfrac{1}{\alpha^2}\left(\dfrac{P}{I_p^2} - R_p\right)$	The equation for finding R_s of the equivalent circuit from the short-circuit test data.	891
21.14	$X_p = \dfrac{V_p \sin \theta_p}{2I_p}$	The equation for finding X_p of the equivalent circuit from the short-circuit test data.	891
21.15	$X_s = \dfrac{V_p \sin \theta_p}{\alpha^2 2I_p}$	The equation for finding X_s of the equivalent circuit from the short-circuit test data.	891
21.16	$R_c = \dfrac{V_c^2}{P - I_p^2R_p}$	The equation for finding R_c of the equivalent circuit from the open-circuit test data.	891
21.17	$X_m = \dfrac{V_c^2}{V_pI_p \sin \theta_p - I_p^2X_p}$	The equation for finding X_m of the equivalent circuit from the open-circuit test data.	891

21.18	$R_p = \alpha^2\left(\dfrac{P}{I_s^2} - R_s\right)$	The equation for R_p for the circuit referred to the secondary.	894
21.19	$X_s = \dfrac{V_s \sin\theta_s}{2I_s}$	The equation for X_s for the circuit referred to the secondary.	894
21.20	$R_c = \dfrac{V_c^2}{P - I_s^2R_s}$	The equation for R_c for the circuit referred to the secondary.	894
21.21	$X_m = \dfrac{V_c^2}{V_sI_s\sin\theta_s - I_s^2X_s}$	The equation for X_m for the circuit referred to the secondary.	896
21.22	$\dfrac{V_2}{V_1} = \dfrac{N_2}{N_1} = \dfrac{1}{\alpha}$	The relation between the voltages and turns ratio of a step-down autotransformer.	896
21.23	$\dfrac{I_L}{I_1} = \dfrac{N_1}{N_2} = \alpha$	The relation between the currents and turns ratio for a step-down autotransformer.	896
21.24	$P_i = V_2I_2$	The equation for the inductively transferred power in a step-down autotransformer.	896
21.25	$P_c = V_2I_1$	The equation for the conductively transferred power in a step-down autotransformer.	896
21.26	$\dfrac{V_L}{V_1} = \dfrac{N_2}{N_1} = \dfrac{1}{\alpha}$	The relation between the voltages and turns ratio of a step-up autotransformer.	898
21.27	$\dfrac{I_2}{I_1} = \dfrac{N_1}{N_2} = \alpha$	The relation between currents and turns ratio in a step-up autotransformer.	898
21.28	$P_c = V_1I_2$	The equation for the conductively transferred power in a step-up autotransformer.	898
21.29	$P_i = V_1I_3$	The equation for the inductively transferred power in a step-up autotransformer.	898
21.30	$Z_i = Z_p + \dfrac{(\omega M)^2}{Z_s + Z_L}$	The equation for the input impedance of an air core transformer.	900
21.31	$V_L = V_{in}\sqrt{\dfrac{L_s}{L_p}}\left(\dfrac{k}{k^2 + \dfrac{1}{Q_p\,Q_s}}\right)$	The equation for the open-circuit secondary voltage of a double-tuned circuit.	902

QUESTIONS

1. What is one advantage of using an iron core in a transformer? One disadvantage?
2. How can the kVA rating of a transformer be used to determine the current rating of the secondary?
3. A 5-kVA transformer can supply 50 amperes of current at a unity power factor. How many amperes can it supply at a power factor of 0.8?
4. Which transformer will have a higher short-circuit inrush current if each has the same rated primary current?
   (a) 120/240 V, 5% impedance
   (b) 120/240 V, 8% impedance
5. The output of a transformer is 50 volts. What effect on the output voltage will shorting some of the secondary turns have? Shorting some of the primary turns?
6. The 100 turns of a transformer secondary winding are replaced by a smaller diameter wire. What effect will this have on the output voltage at no load? What effect will it have when there is current in the secondary?
7. Why can a test with an ohmmeter not show a short circuit between windings, while a test with an isolation tester might show a short circuit?
8. What will the output voltage be of a 1–1 isolation transformer if a 12-volt dc source is connected to the primary winding? Will there be any primary current? Will there be any secondary current if a 5-ohm resistor is connected across the secondary?
9. What will happen to the primary current of a transformer if the secondary is short-circuited?
10. What are the markings for the high-voltage terminals of a transformer? For the low-voltage side?
11. A transformer has two secondary windings connected in series. A voltmeter, which is connected across the free ends, indicates 0 volts. Assuming that the windings are not open and the meter is not defective, what could be the cause of the zero voltage?
12. Which test is used to determine the copper losses of a transformer?
13. Which test is used to determine the core losses of a transformer?
14. What is the form of the equation used to calculate the efficiency from the output power and losses?
15. What is the form of the equation used to calculate the efficiency from the input power and losses?
16. The copper losses in a transformer at 6 amperes of current are 10 watts. What effect will decreasing the current to 3 amperes have on the losses? How many watts will the losses be at 3 amperes?
17. The core loss in a transformer at rated voltage and current is 2 watts. What effect will reducing the voltage to 50 percent of rated voltage have on the core loss?
18. What will happen to the core loss in a transformer if the insulation between the laminations is damaged?
19. Why aren't there any hysteresis and eddy current losses in an air core transformer?
20. Explain how turning the screw at the top of an I.F. transformer results in a change in the volume of the received signal.

21. What are two reasons for using an autotransformer instead of a regular transformer?
22. What are the two ways by which the power is transferred in an autotransformer?
23. What effect does increasing the frequency of the input voltage have on the core loss of a transformer?
24. Why does the potential transformer have many turns on the primary winding while the current transformer has many turns on the secondary?
25. What is one difference between the transformer used for isolation and that used for impedance matching?

PROBLEMS

SECTION 21.1 Transformer Construction and Terminology

1. What is the kVa rating of a 2200/440-V transformer if the rated secondary current is 500 amperes?

2. A 5.5-kVA transformer has a rated secondary current of 50 amperes. What is its rated secondary voltage?

3. How much power can the transformer in Problem 1 safely supply to:

   (a) A load with a 0.8 power factor?
   (b) A load with a unity power factor?

4. What must be the kVA rating of a transformer that is to supply 5 kW at a power factor of 0.6?

5. What is the short-circuit inrush current in a 220/440-V, 10-kVA transformer that has a 4% impedance?

6. The short-circuit inrush current in a 240/120-V transformer is 50 amperes. What is the rated primary current if its impedance is 8%?

SECTION 21.2 The Ideal Transformer

7. An ideal transformer has an α of 8 and a secondary of 200 turns. If the secondary voltage is 120 volts, find

   (a) The primary turns.
   (b) The primary voltage.
   (c) The secondary current if the primary current is 0.5 ampere.

8. A transformer for a doorbell steps down the voltage from 120 volts to 12 volts. Its secondary has 50 turns. The doorbell draws 5 amperes of current. Find:

   (a) The primary current.
   (b) The power supplied to the transformer.
   (c) The α of the transformer.

9. A 220/110-volt transformer has a secondary current of 20 amperes. Determine:

   (a) The turns ratio.
   (b) The primary current.
   (c) The resistance of the load, assuming it is resistance.
   (d) The resistance looking back into the primary terminals.

10. An iron core transformer has 500 turns on the primary and has two secondary windings. One has 250 turns and one has 100 turns. If the input is 120 volts, determine:

    (a) The voltage across each secondary.
    (b) The current in each secondary winding if a 20-ohm resistance is across each one.
    (c) The power input to the transformer.

11. An impedance matching transformer must match a 75-ohm load to a 675-ohm source. What turns ratio is needed?

12. An audio transformer is to be used to match an 8-ohm loudspeaker to a 7200-ohm amplifier. Determine:

    (a) The turn ratio, α.
    (b) The primary turns if the secondary has 20 turns.

13. A 220/110-V iron core transformer has a 220-ohm resistance connected across it. Find:

    (a) The primary current.
    (b) The resistance looking into the primary.
    (c) The power input to the transformer.

14. A center-tapped transformer supplies two resistance loads of 50 ohms each. The primary has 400 turns and the secondary has a total of 100 turns. If 120 volts is applied to the primary, what are the voltage and current for each load, and what is the primary current?

15. A center-tapped transformer has a primary of 1000 turns, and the voltage at each load is 120 volts. How many turns are in the secondary if there are 240 volts across the primary?

SECTION 21.3 Terminal Polarity and Troubleshooting

16. The secondary coils of a transformer are connected as shown in Figure 21.33. What is the voltage across 1–2? What will be the voltage if the 120-volt coil is reversed?

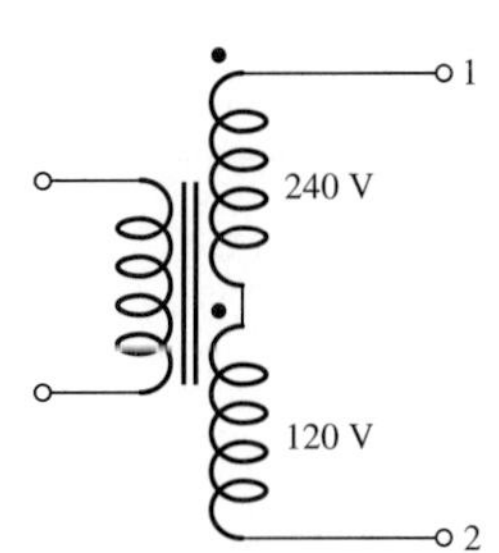

Figure 21.33

17. A transformer has a primary winding with 1000 turns. One secondary winding has 600 turns and the other has 300 turns. If the input is 110 volts, find the output when the windings are connected so their voltages are:

    (a) Additive.
    (b) Subtractive.

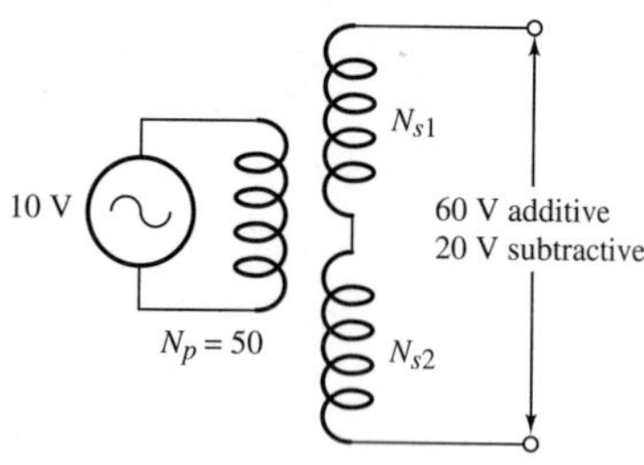

Figure 21.34

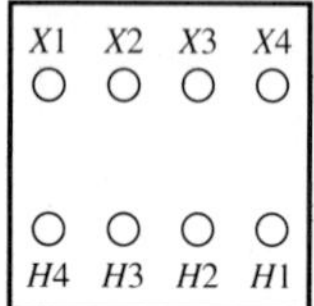

Figure 21.35

18. The transformer shown in Figure 21.34 has a secondary voltage of 60 V when N_{s1} and N_{s2} are additive, and 20 V when they are subtractive. How many turns are in each secondary winding?

19. The terminals on a transformer having two primary and two secondary windings are marked as shown in Figure 21.35. Show the proper connection for:

    (a) Connecting the primary in series-additive and the secondary in series-additive.

    (b) Connecting the two primary winding in parallel and the two secondary windings in parallel.

SECTION 21.4 The Exact Equivalent Circuit

20. A 4400/220-V transformer is rated at 10 kVA. Find the maximum power output that it can safely supply if the load power factor is:

    (a) 1.

    (b) 0.8 Lagging.

    (c) 0.8 Leading.

21. A transformer with an α of 3 supplies a 40-ohm resistance load at 120 volts. Determine:

    (a) The secondary current.

    (b) The primary current and voltage.

    (c) The power input to the transformer.

    The parameters are: $R_p = 0.4$ ohm, $X_p = 8$ ohms, $R_s = 0.004$ ohm, $X_s = 6$ ohms, $R_c =$ 3-kilohms, and $X_m =$ 4-kilohms.

22. An iron core transformer has the following parameters: $R_p = 0.3$ ohm, $X_p = 3$ ohms, $R_s = 3$ ohms, $X_s = 40$ ohms, $R_c =$ 8-kilohms, $X_m =$ 4-kilohms, and $\alpha = 0.5$. Find the % efficiency and % voltage regulation if the input voltage is $240 \angle 0°$ volts when connected to a load of:

    (a) $100 \angle 0°$.

    (b) $100 \angle 60°$.

23. Work Problem 22 for a load of $100 \angle -30°$.

24. An iron core 240/720-V transformer has the following parameters: $R_p = 0.04$ ohm, $X_p = 0.06$ ohm, $R_s = 1.8$ ohms, $X_s = 2.7$ ohms, $R_c =$ 1-kilohm, and $X_m =$ 1.5-kilohms. Find the % efficiency and % voltage regulation when supplying 13.02 amperes at 675 volts to a load with a 0.8 lagging power factor.

25. Work Problem 24 for a resistance load with 13.02 amperes.

26. The circuit shown in Figure 21.36 has all quantities referred to the primary side. Determine the value of each item for the circuit referred to the secondary side.

27. An iron core transformer has the following parameters: $L_p = 0.8$ mH, $R_p = 0.8\ \Omega$, $L_s = 7$ mH, $R_s = 7\ \Omega$, $\alpha = 0.5$, $R_c = 3000\ \Omega$, and $X_m = 5000\ \Omega$. The load is $80 \angle 0°\ \Omega$ and the frequency is 500 Hz. What is the input impedance?

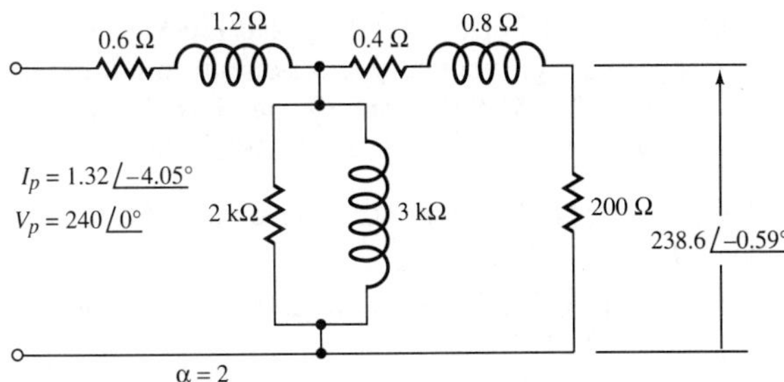

Figure 21.36

28. An audio-frequency transformer has the following parameters: $R_p = 60\ \Omega$, $X_p = 250\ \Omega$, $R_s = 600\ \Omega$, $X_s = 1250\ \Omega$, $R_c = 2000\ \Omega$, $X_m = 4000\ \Omega$, and $\alpha = 0.2$. The voltage applied to the primary is 2 volts. What are the output voltage and current for a load of 6000 ohms resistance?

SECTION 21.5 The Approximate Equivalent Circuit

29. Work Problem 21 using the circuit of Figure 21.18.

30. Work Problem 22 using the circuit of Figure 21.18.

31. Work Problem 23 using the circuit of Figure 21.18.

32. Work Problem 24 using the circuit of Figure 21.18.

33. Work Problem 25 using the circuit of Figure 21.18.

SECTION 21.6 Short-Circuit and Open-Circuit Tests

34. The power loss in a 220/110-V transformer is 50 watts for the short-circuit test and 10 watts for the open-circuit test. Find the % efficiency of the transformer when the transformer is operated at rated output with a load of:

    (a) $22\ \angle 0°\ \Omega$.

    (b) $22\ \angle 53.13°\ \Omega$.

35. Short-circuit and open-circuit test data on a transformer gave the following data: short circuit: $V_p = 50$ V, $P = 100$ W, $I_p = 7$ A; open circuit: $V_p = 240$ V, $P = 50$ W, $I_p = 0.9$ A. Calculate the % efficiency, the power out, and the secondary current if the load has a resistance of 30 ohms and the power input is 1650 W at rated primary current of 7 amperes.

36. A 2.5-kVA, 120/240-V transformer has the following test data with the input to the primary: short circuit: $V_p = 22$ V, $I_p = 20.8$ A, $P = 120$ W; open circuit: $V_p = 120$ V, $I_p = 0.06$ A, $P = 60$ W. Calculate the % efficiency and % voltage regulation when operating at $I_s = 10.4$ A and $V_s = 240$ V with a resistance load.

37. Calculate the % efficiency and % voltage regulation for the transformer of Problem 36 if V_p is 120 V, I_s is 10.4 A, and the load has a leading power factor of 0.6.

38. Determine the values of the equivalent circuit parameters referred to the primary for the transformer of Problem 36 if the R_p is 0.1 ohm.

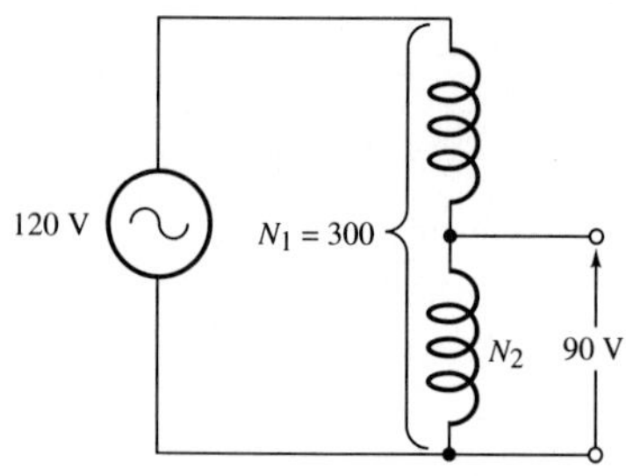

Figure 21.37

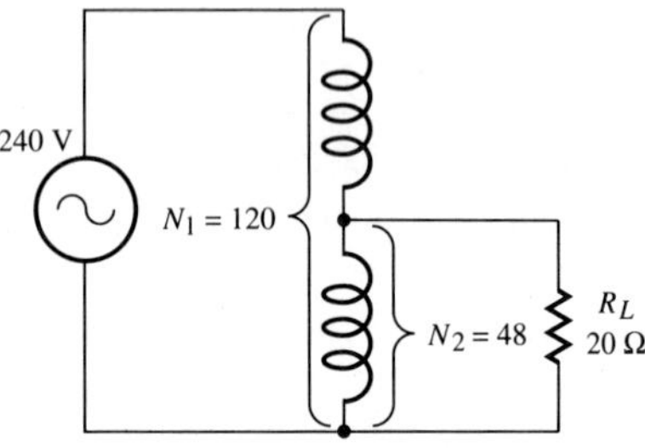

Figure 21.38

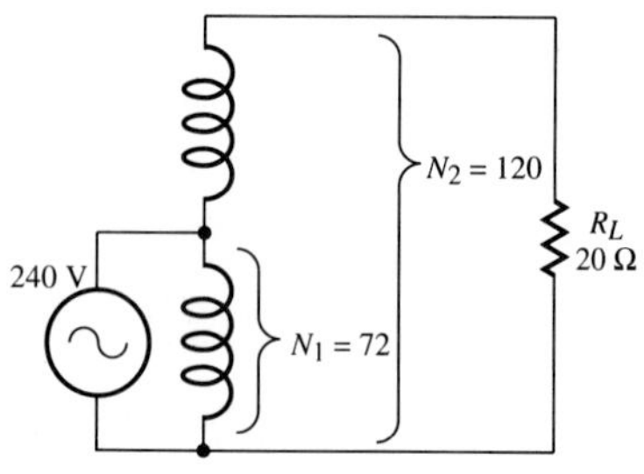

Figure 21.39

39. The following data were taken on a 1.5-kVA 240/120-V iron core transformer used by an electric utility: secondary short-circuit, input to primary: V_p = 15.44 V, I_p = 6.25 A, P = 50.8 W; primary open-circuit, input to secondary: V_s = 120 V, I_s = 0.268 A, P = 14.38 W. Determine:

    (a) The % efficiency of the transformer when operated at rated output with a resistance load.

    (b) The equivalent circuit parameters referred to the primary if R_p = 0.5 ohm.

40. Determine the % efficiency of the transformer in Problem 28 for a load current of 6 A at 120 V. Assume the excitation losses are constant.

SECTION 21.8 The Autotransformer

41. There are 300 turns on the autotransformer winding in Figure 21.37. How many turns should be in N_2?

42. For the autotransformer shown in Figure 21.38, calculate:

    (a) The voltage across R_L.

    (b) The current in R_L, N_1, and N_2.

    (c) The power supplied by the source.

43. A step-down autotransformer connected to a 240-volt line is used to reduce a motor's starting voltage to 200 volts. Find the current drawn from the line if the motor's starting current is 5 amperes. Assume a power factor of one.

44. A step-down autotransformer supplies a 2-kW resistance load at 120 V. If the transformer is connected to a 240-V source, what is:

    (a) The load current.

    (b) The source current.

    (c) The inductively transferrred power.

    (d) The conductively transferred power.

45. In Figure 21.39, calculate the:

    (a) Voltage across R_L.

    (b) Current in R_L, N_1, and N_2.

    (c) Power supplied by the source in the circuit.

46. A 2200/220-V transformer is connected to a 2200-V source as an autotransformer to provide 25 amperes at 2420 volts to a resistance load. Find:

    (a) The source current.

    (b) The load current.

    (c) The power transferred inductively.

    (d) The power transferred conductively.

    (e) The kVA rating that the two-winding transformer should have.

47. A 220/330-V autotransformer supplies 10 kW of power to a load that has a power factor of 0.8. Find:

    (a) The current in each section.

    (b) The kVA rating of the transformer.

SECTION 21.9 The Air Core Transformer

48. The transformer parameters in a double-tuned circuit are: $R_p = 3\ \Omega$, $L_p = 6$ mH, $R_s = 2.5\ \Omega$, $L_s = 6$ mH, and $k = 0.04$. What is the input impedance at 100 kHz if a 1000-ohm resistive load is connected to the secondary?

49. What will be the input impedance of the circuit in Problem 48 if the coefficient of coupling is 0.2?

50. An air core transformer is designed to reflect a 1000-ohm resistance in the secondary as an impedance with a magnitude of 3200 ohms in the primary. If $R_s = 4$ ohms, $X_s = 20{,}000$ ohms, and the frequency is 50 kHz, what must the coefficient of coupling be? Note: The reflected secondary does not include Z_p and $X_p = X_s$

51. The parameters for an air core transformer in a double-tuned circuit are: $Q_p = 80$, $Q_s = 80$, $L_p = 400\ \mu$H, $L_s = 500\ \mu$H, and $M = 90\ \mu$H. How many volts will the output voltage be if $V_p = 0.5$ V?

52. Determine the value of k that will give the maximum output voltage for the circuit in Problem 51.

53. A transformer in a double-tuned circuit has the following parameters: $L_p = 0.04$ H, $L_s = 0.08$ H, $Q_p = 40$, $Q_s = 50$, and $k = 0.4$. What must the input voltage be for an output of 10 volts?

54. The coils of the transformer in Problem 53 were accidentally moved to give a k of 0.01. What is the output voltage with the determined input?

CHAPTER 22

THREE-PHASE SYSTEMS

CHAPTER OBJECTIVES

After completing this chapter, you should be able to:

1. Explain what a three-phase source is and give some of its advantages.
2. Express voltages using the double subscript notation.
3. Connect the three coils of an alternator to form a three-phase Wye and Delta source.
4. Give the relation between the line and phase quantities for Wye and Delta sources.

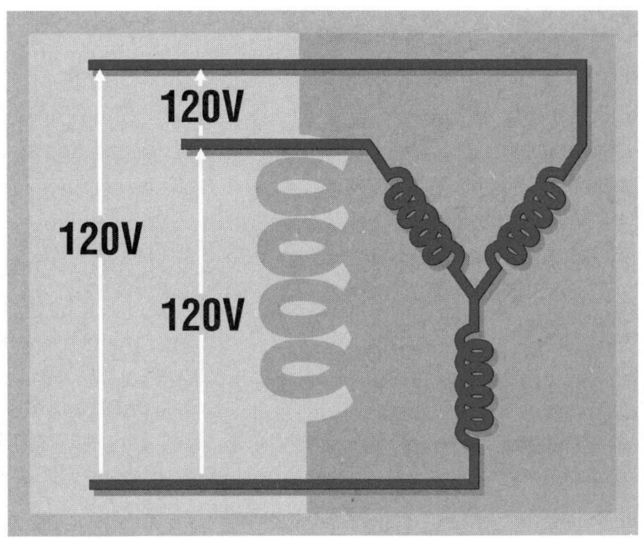

5. Perform the analysis of the three-wire and four-wire systems with balanced and unbalanced loads.
6. Perform the analysis of a three-phase system that includes line impedance.
7. Make the connection to measure three-phase power with two wattmeters.
8. Determine the power factor from the wattmeter readings.

KEY TERMS

1. balanced load
2. double subscript notation
3. four-wire, three-phase (3–ϕ) Wye load
4. line current
5. line impedance
6. line voltage
7. neutral current
8. neutral line
9. phase current
10. phase sequence
11. phase sequence indicator
12. phase voltage
13. three-phase
14. three-phase Delta Δ source
15. three-wire, three-phase Delta load
16. three-wire, three-phase Wye load
17. three-phase Wye (Y) source
18. two wattmeter method
19. unbalanced load

INTRODUCTION

As the applications of electricity increased, motors, lighting loads, and other loads also became larger. Steel mills and other industrial plants required motors of hundreds of horsepower instead of tens.

Single-phase ac and dc were not able to provide such amounts of power without large losses in the lines and larger alternators. As a result, the three-phase system was developed. This is a system that has three voltages each displaced by 120°. The modern distribution system uses one of the three voltages for single-phase loads, and all three voltages for three-phase loads such as motors and large heating units. It does this using only four lines, compared to six for three single-phase sources. The three-phase lines provide power distribution with a smaller voltage drop and less loss in the lines than for an equivalent single-phase system.

This chapter deals with the types of three-phase sources, three-phase loads, and their analysis. After a study of the theory, it presents the procedures for power measurement. The chapter provides the preparation for working with power distribution. For those not working in power distribution, it provides a general understanding of three-phase systems.

22.1 Three-Phase Source

three-phase

As the use of electricity increased, it was found that its generation, distribution, and use were more efficient and economical if a **three-phase** ($3-\phi$) source was used. What is a three-phase source? It is a system of three emf's displaced by a fixed number of degrees. The common three-phase system used in the United States has a phase difference of 120°. The amplitude and frequency of each emf is usually the same. There are some specific applications that use more than three emf's, but they are not as efficient as the three-phase system. The variation of the phase voltages for a three-phase system is shown in Figure 22.1. The emf's in

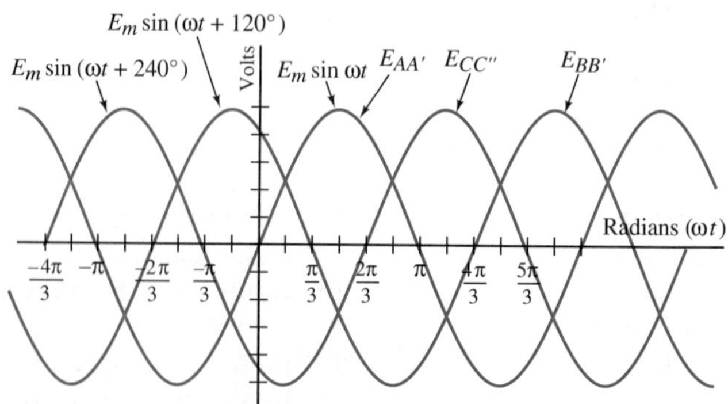

Figure 22.1 The three-phase alternator generates three sinusoidal emf's that are 120° out of phase.

the system are $E_m \sin \omega t$, $E_m \sin(\omega t + 120°)$, and $E_m \sin(\omega t + 240°)$. Their phasor form is $E \angle 0°$, $E \angle 120°$, and $E \angle 240°$.

Some specific reasons for using three-phase instead of single-phase are:

1. There is less power loss in the line for a given amount of load power. This permits use of a smaller wire size.
2. Three-phase can be generated more economically and efficiently.
3. The three-phase voltage has less pulsation. This results in less vibration of the alternators.
4. Three-phase motors are more efficient and smaller in size and weight and do not require special starting means.
5. Three-phase motors produce a constant shaft torque resulting in less variation and vibration.
6. Rectified three-phase does not require as much filtering as rectified single-phase.
7. The three-phase source can supply three-phase loads, single-phase loads, and combinations of both at different voltages.

As a point of interest, the alternator of the modern automobile is a three-phase source. It replaced the dc generator because of some of the advantages listed in the foregoing.

Most of the three-phase power used in the world is generated by three-phase alternators. An alternator uses electromagnetic induction to generate an emf by the relative motion between a conductor and a magnetic field. Small three-phase alternators have three sets of coils that turn in a stationary magnetic field. Larger alternators have stationary coils, which turn the magnet, as shown in Figure 22.2. This eliminates problems caused by the large mass of the coils. Also, it greatly reduces the amount of current to be carried from the rotor to the stator by the slip rings.

Because the rotor moves past each winding at the same angular velocity and the magnetic field is the same, the emf for each winding has the same amplitude and frequency. Also, because the windings are

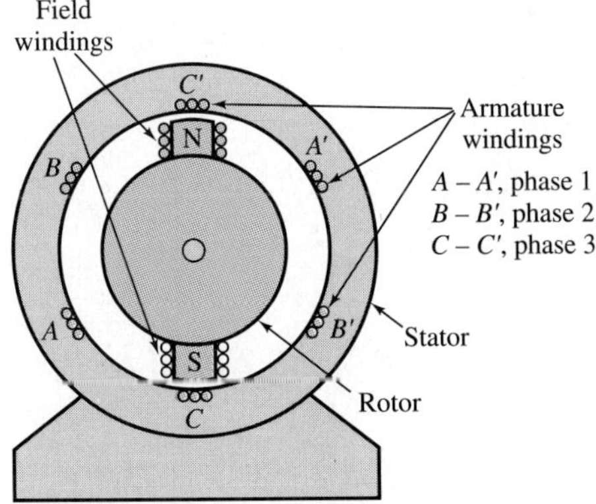

Figure 22.2 This sectional view shows the parts of a basic three-phase alternator.

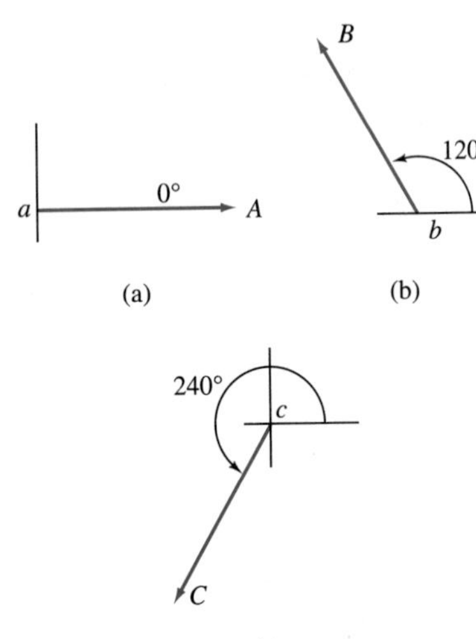

Figure 22.3 The double subscript notation and its relation to the phasor (a) $\mathbf{E}_{Aa} = 100 \angle 0°$. (b) $\mathbf{E}_{Bb} = 100 \angle 120°$. (c) $\mathbf{E}_{Cc} = 100 \angle 240°$.

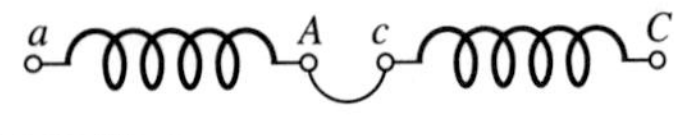

Figure 22.4

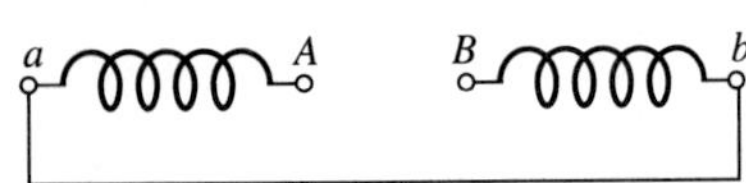

Figure 22.5

permanently fixed, the displacement of each emf will be 120° apart. The resulting output for a counterclockwise rotation was shown in Figure 22.1.

To connect the alternator coils properly to form a three-phase system, some method of showing the voltage difference and instantaneous relative polarity is needed. One method is to use **double subscript notation**. This notation was first introduced in Chapter 5, Section 5.6. It is a notation that gives the voltage of the point at the first subscript with respect to the point of the second one. Thus, $\mathbf{E}_{Aa} \angle 0°$ means that the voltage of Terminal A as measured from Terminal a is the phasor $E \angle 0°$. The other voltages can be written as $\mathbf{E}_{Bb} = E \angle 120°$, and $\mathbf{E}_{Cc} = E \angle 240°$, where B,b and C,c are the terminals of the other coils. It might appear confusing, but actually it is the same as stating that the voltage of the $(-)$ terminal of a battery is $-1\ 1/2$ volts when measured from the $(+)$ terminal. The phasors that represent $\mathbf{E}_{Aa}$, $\mathbf{E}_{Bb}$, and $\mathbf{E}_{Cc}$ are shown in Figure 22.3. $\mathbf{E}_{aA}$ is $-\mathbf{E}_{Aa}$, $\mathbf{E}_{bB}$ is $-\mathbf{E}_{Bb}$, and $\mathbf{E}_{cC}$ is $-\mathbf{E}_{Cc}$.

EXAMPLE 22.1

The emf's for three coils are

$$\mathbf{E}_{Aa} = 208 \angle 30° \text{ V}$$
$$\mathbf{E}_{Bb} = 208 \angle 150° \text{ V}$$
$$\mathbf{E}_{Cc} = 208 \angle -90° \text{ V}$$

a. For the connection in Figure 22.4, find $\mathbf{E}_{Ca}$ and $\mathbf{E}_{aC}$.
b. For the connection in Figure 22.5, find $\mathbf{E}_{AB}$ and $\mathbf{E}_{BA}$.

Solution

When adding the phasors, we start at the reference subscript and end at the first one.

a. $\mathbf{E}_{Ca}$ equals $\mathbf{E}_A$ as measured from a, plus $\mathbf{E}_C$ as measured from c. So

$$\mathbf{E}_{Ca} = \mathbf{E}_{Aa} + \mathbf{E}_{Cc} = 208 \angle 30° \text{ V} + 208 \angle -90° \text{ V}$$
$$\mathbf{E}_{Ca} = 208 \angle -30° \text{ V}$$
$$\mathbf{E}_{aC} = -\mathbf{E}_{Ca}$$
$$\mathbf{E}_{aC} = 208 \angle 150° \text{ V}$$

b. $\mathbf{E}_{AB} = \mathbf{E}_b$ as measured from B, plus $\mathbf{E}_A$ as measured from a. So

$$\mathbf{E}_{AB} = \mathbf{E}_{bB} + \mathbf{E}_{Aa}$$

But

$$\mathbf{E}_{bB} = -\mathbf{E}_{Bb}$$

so

$$\mathbf{E}_{AB} = -208 \angle 150° \text{ V} + 208 \angle 30° \text{ V}$$

$$E_{AB} = 360 \angle 0° \text{ V}$$
$$\boldsymbol{E_{BA}} = -\boldsymbol{E_{AB}}$$
$$E_{BA} = 360 \angle 180° \text{ V}$$

22.2 Wye and Delta Source Connections

three-phase Wye (Y) source
three-phase Delta (Δ) source

The two common 3–ϕ connections are the **three-phase Wye (Y) source** and the **three-phase Delta (Δ) source**. The three-phase Wye source has the three coils connected in the shape of a wye (see Figure 22.6). The load is

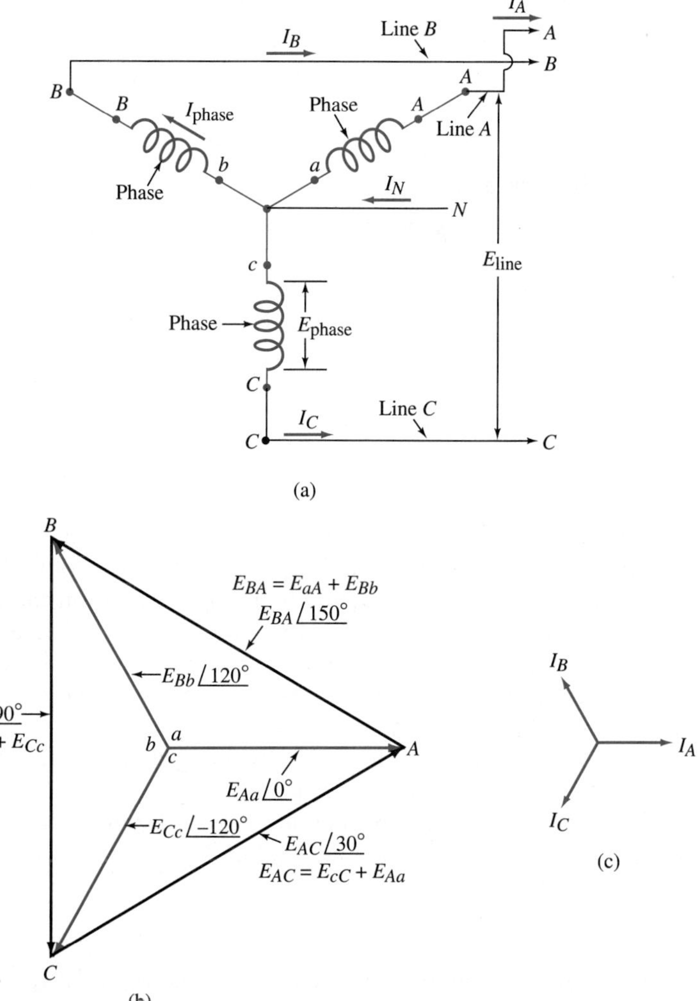

Figure 22.6 The four-wire, three-phase Wye connection and its phasor diagrams. (a) The connection of the phase windings. (b) The phasor diagram of the phase and line voltages. (c) The phasor diagram of the line currents.

connected to the legs of the wye. A four-wire Wye source has a line brought out from the common point of the wye. This line is called the **neutral line**. A three-wire Wye does not have a neutral line. The four-wire Wye connection is more popular because it can provide two voltages at 3–ϕ and at single-phase. The three-wire Wye provides only one voltage at three-phase, and one at single-phase.

neutral line

phase voltage
line voltage

The **phase voltage** is the voltage across the winding. $\boldsymbol{E}_{Aa} \angle 0°$, $\boldsymbol{E}_{Bb} \angle 120°$, and $\boldsymbol{E}_{Cc} \angle -120°$ are phase voltages. The **line voltage** is the voltage across any two lines. These are $\boldsymbol{E}_{AB}$, $\boldsymbol{E}_{BC}$, and $\boldsymbol{E}_{CA}$. When the windings are connected with the proper polarity relation, as in Figure 22.6(a), the phase and line voltage is related by the $\sqrt{3}$, because the three-phase voltages are out of phase by 120 degrees.

In Figure 22.6,

$$\boldsymbol{E}_{AC} = -E_{Cc} \angle -120° + E_{Aa} \angle 0°$$

or

$$\boldsymbol{E}_{\text{line}} = -E_{\phi} \angle -120° + E_{\phi} \angle 0°$$

This can be written as

$$\boldsymbol{E}_{\text{line}} = E_{\phi} \angle 60° + E_{\phi} \angle 0°$$

or

$$\boldsymbol{E}_{\text{line}} = E_{\phi}(\cos 60° + j \sin 60° + 1)$$

The magnitude of the quantity in the parentheses is equal to the $\sqrt{3}$, so

$$E_L = \sqrt{3}E_{\phi} \quad \textbf{(22.1)}$$

Since $E_L = \sqrt{3}\,E_{\phi}$, the voltage output is greater than the voltage rating of the winding. Thus, a 100-volt winding can safely provide 173 volts.

If the polarity is not correct, the line voltage will not be equal, Equation 22.1 does not hold, and the line voltage will not be 120° apart.

The line and phase voltages and their angular relation can be seen in the phasor diagram of Figure 22.6(b). The voltages are equal and 120° apart. Therefore, they bisect the angle formed by each pair of line voltages. Also, their reference ends must meet at a point equidistant from the points of the phasor triangle.

neutral current

The phase current is the current in a winding of the alternator. The line current is the current in the line of the three-phase system. The current in the neutral line is the **neutral current**. The Wye connection shows that each line is connected to only one winding, so the line current is also the phase current. That is,

$$\boldsymbol{I}_L = \boldsymbol{I}_{\phi} \quad \textbf{(22.2)}$$

This limits the maximum line current to the rating of the alternator winding.

Applying Kirchhoff's Current Law to the node in the connection of Figure 22.6(a) yields

$$\boldsymbol{I}_A + \boldsymbol{I}_B + \boldsymbol{I}_C - \boldsymbol{I}_N = 0 \quad \textbf{(22.3)}$$

If the load is balanced, $I_A = I_B = I_C$, so $I_N = 0$. Also, the line currents must be 120° apart, as shown in Figure 22.6(c). A **balanced load** is one in which the impedance is the same in each phase of the load.

balanced load

EXAMPLE 22.2

Find the voltages, E_{AC}, E_{BA}, and E_{CB} in Figure 22.7, if $E_{Aa} = 120 \angle 30°$ V, $E_{Bb} = 120 \angle 150°$ V, and $E_{Cc} = 120 \angle -90°$ V.

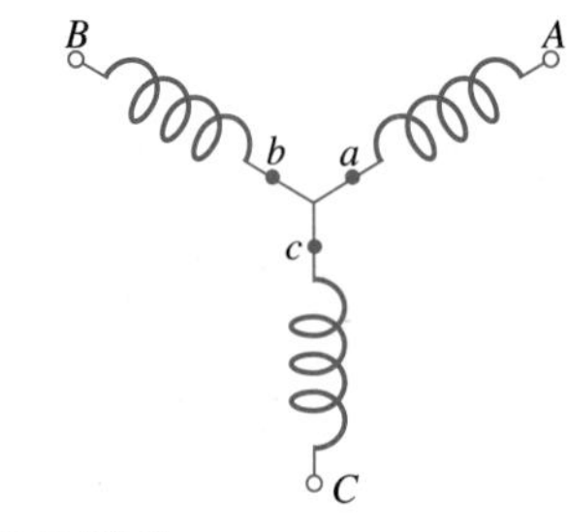

Figure 22.7

Solution

$$E_{AC} = E_{cC} + E_{Aa} = -E_{Cc} + E_{Aa}$$

so

$$E_{AC} = -120 \angle -90° \text{ V} + 120 \angle 30° \text{ V}$$
$$E_{AC} = 208 \angle 60° \text{ V}$$
$$E_{BA} = E_{aA} + E_{Bb} = -E_{Aa} + E_{Bb}$$

so

$$E_{BA} = -120 \angle 30° \text{ V} + 120 \angle 150°$$
$$E_{BA} = 208 \angle 180° \text{ V}$$
$$E_{CB} = E_{bB} + E_{Cc} = -E_{Bb} + E_{Cc}$$

so

$$E_{CB} = -120 \angle 150° \text{ V} + 120 \angle -90°$$
$$E_{CB} = 208 \angle -60° \text{ V}$$

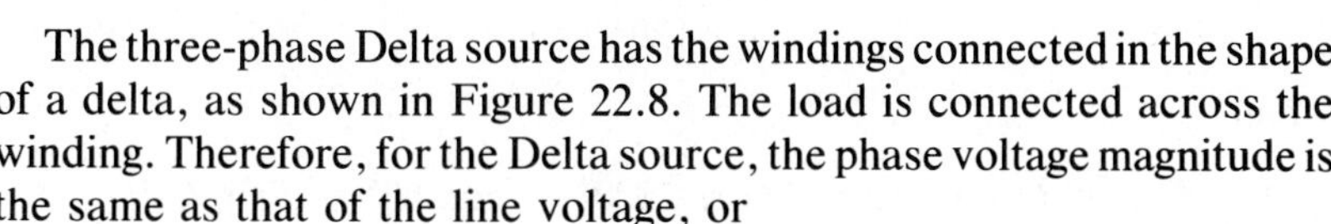

The three-phase Delta source has the windings connected in the shape of a delta, as shown in Figure 22.8. The load is connected across the winding. Therefore, for the Delta source, the phase voltage magnitude is the same as that of the line voltage, or

$$E_L = E_\phi \tag{22.4}$$

The phasor sum of the source voltages in the closed Delta circuit of Figure 22.8(a) is $E_{Aa} + E_{Bb} + E_{Cc}$, or $E_\phi \angle 120°$ V + $E_\phi \angle -120°$ V + $E_\phi \angle 0°$ V. This turns out to be zero volts. Because the net voltage in the Delta is zero, the circulating current in it must be zero amperes. If a balanced load is connected to the Delta, the magnitudes of I_{AB}, I_{BC}, and I_{CA} are equal. The relation between the line and phase currents can be obtained from Figure 22.8. There

$$I_A = I_{CA} \angle 120° - I_{AB} \angle -120°$$

or

$$I_{\text{line}} = I_\phi \angle 120° - I_\phi \angle -120$$

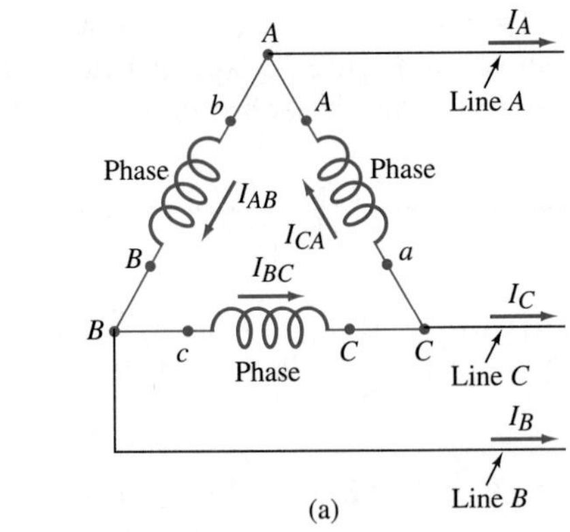

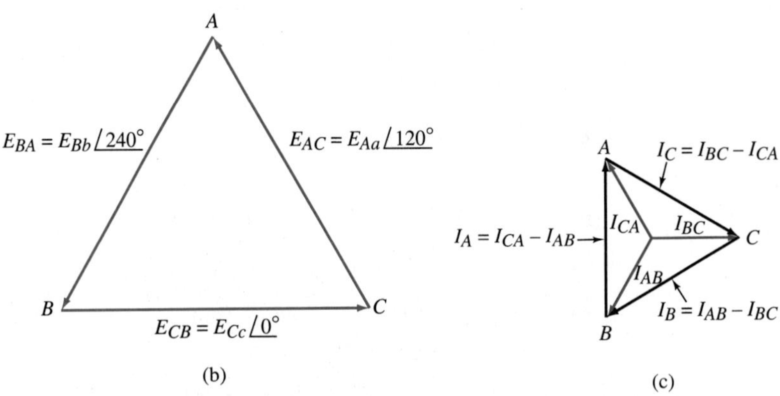

Figure 22.8 The three-phase Delta connection and its phasor diagrams. (a) The connection of the phase windings. (b) The voltage phasor diagram. (c) The current phasor diagram for a balanced resistance load.

This can be written as

$$\boldsymbol{I}_{\text{line}} = I_\phi \angle 120° + I_\phi \angle 60°$$

Then

$$\boldsymbol{I}_{\text{line}} = I_\phi (\cos 120° + j \sin 120° + \cos 60° + j \sin 60°)$$

The magnitude of the quantity in the parentheses is equal to the $\sqrt{3}$, so

$$I_{\text{line}} = \sqrt{3} I_\phi \tag{22.5}$$

Since $I_L = \sqrt{3} I_\phi$, the Delta connection can provide a higher current than the rating of the coils.

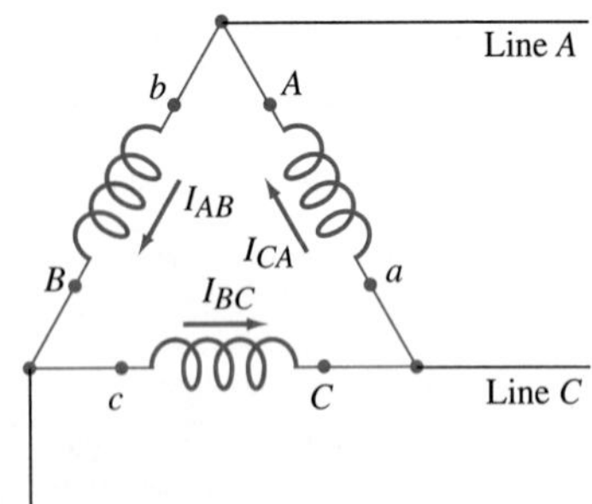

Figure 22.9

EXAMPLE 22.3

In Figure 22.9, $\boldsymbol{E}_{Aa} = 120 \angle 120°$ V, $\boldsymbol{E}_{Bb} = 120 \angle -120°$ V, and $\boldsymbol{E}_{Cc} = 120 \angle 0°$ V. Find

a. the voltages $\boldsymbol{E}_{AB}$, $\boldsymbol{E}_{BC}$, and $\boldsymbol{E}_{CA}$.
b. the magnitude of the current in each winding, if $I_{\text{line}} = 5$ amperes.

Solution

$$\mathbf{E}_{AB} = \mathbf{E}_{bB} = -\mathbf{E}_{Bb}$$
$$\mathbf{E}_{AB} = 120 \angle 60° \text{ V}$$
$$\mathbf{E}_{BC} = \mathbf{E}_{cC} = -\mathbf{E}_{Cc}$$
$$\mathbf{E}_{BC} = 120 \angle 180° \text{ V}$$
$$\mathbf{E}_{CA} = \mathbf{E}_{aA} = -\mathbf{E}_{Aa}$$
$$\mathbf{E}_{CA} = 120 \angle -60° \text{ V}$$

The current in each winding is

$$I_{\phi} = \frac{I_L}{\sqrt{3}} = \frac{5 \text{ A}}{\sqrt{3}}$$
$$I_{\phi} = 2.89 \text{ A}$$

$$\mathbf{I}_A = \mathbf{I}_{CA} - \mathbf{I}_{AB}$$
$$\mathbf{I}_B = \mathbf{I}_{AB} - \mathbf{I}_{BC}$$
$$\mathbf{I}_C = \mathbf{I}_{BC} - \mathbf{I}_{CA}$$

These reduce to

$$\mathbf{I}_A + \mathbf{I}_B + \mathbf{I}_C = 0 \quad \textbf{(22.6)}$$

Thus, for the Delta source, the phasor sum of the line currents must equal zero. With a balanced load, $I_A = I_B = I_C$. The only way that the phasor sum can equal zero is for the line currents to be displaced by 120° between them. The angular relation between the voltages and currents in the Delta for a resistive load is shown in Figure 22.8.

Incorrect winding polarity in the connection can have a more serious effect than in the Wye connection. Connecting with the incorrect polarity can result in arcing and damage to the windings when the last two terminals are connected.

For instance, suppose winding *cC* of Figure 22.10(a) is reversed so that terminals *a* and *c* are the last two connected, as in Figure 22.10(a). Then

$$\mathbf{E}_{ca} = \mathbf{E}_{Aa} + \mathbf{E}_{Bb} + \mathbf{E}_{cC}$$

but

$$\mathbf{E}_{cC} = -\mathbf{E}_{Cc}$$

so

$$\mathbf{E}_{ca} = \mathbf{E}_{Aa} + \mathbf{E}_{Bb} - \mathbf{E}_{Cc}$$
$$= E_{\phi} \angle 120° \text{ V} + E_{\phi} \angle -120° \text{ V} - E_{\phi} \angle 0° \text{ V}$$
$$\mathbf{E}_{ca} = 2E_{\phi} \angle 180° \text{ V}$$

E_{ϕ} is the phase voltage. If it were 110 V, the potential difference between *c* and *a* would be 220 volts. Any connection would cause severe

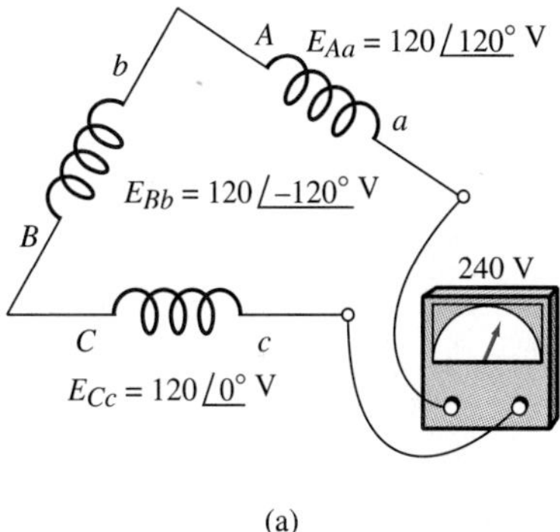

(a)

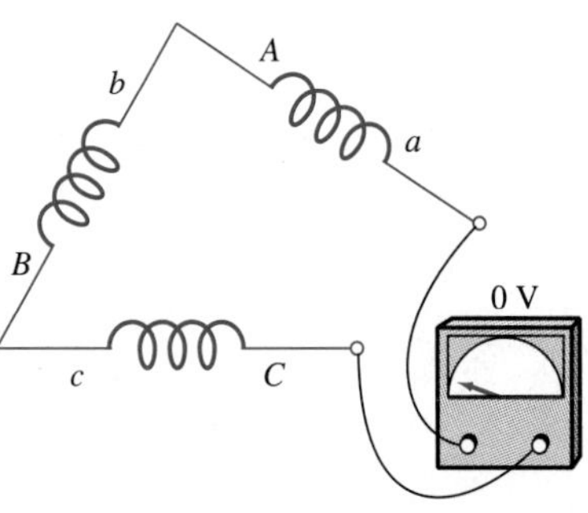

(b)

Figure 22.10 A practical method for determining the correct phase relation when closing the Delta. (a) Incorrect phase relation; the voltmeter measures 240 V. (b) The correct phase relation; the voltmeter measures 0 V.

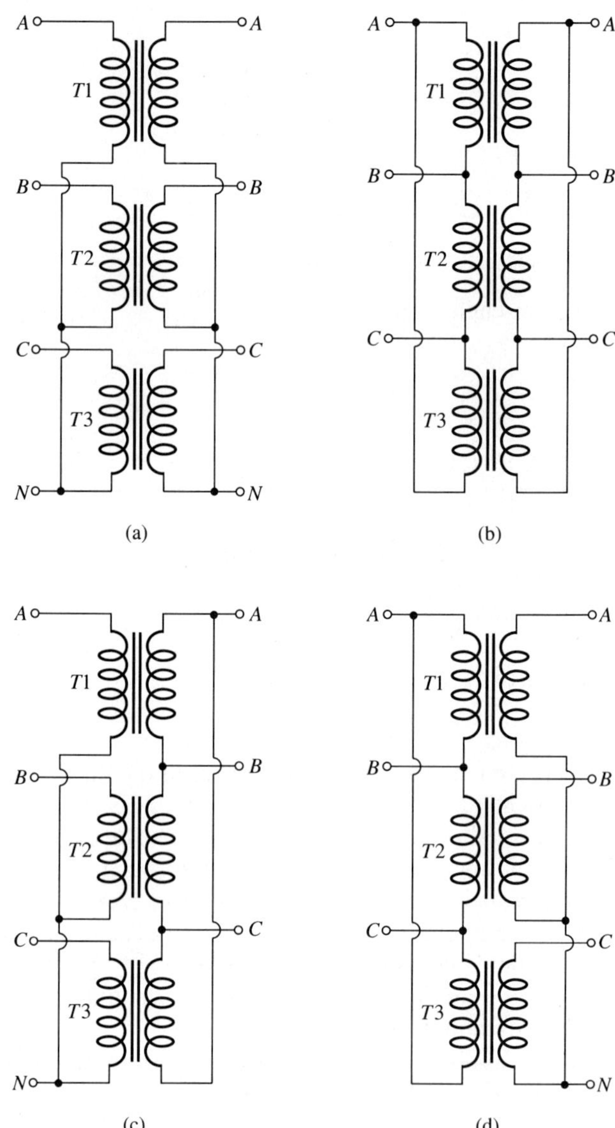

Figure 22.11 Transformers are also connected in Wye and Delta. (a) The Wye-Wye connection. (b) The Delta-Delta connection. (c) The Wye-Delta connection. (d) The Delta-Wye connection.

arcing and high circulating current. Differences between the magnitude or angle of the phase voltages also result in a voltage across the last two terminals. However, it should not be as great as that caused by the incorrect phase relation.

A method of determining when the polarities are correct for closure is shown in Figure 22.10(b). For the correct phase relation, the voltmeter should indicate a few volts. It generally will not indicate zero because of the slight differences in the windings.

This presentation is centered on the coils of an alternator, but transformers are also connected in Delta and Wye connections. Some common connections are shown in Figure 22.11. The Wye-Delta provides a voltage step-down with increased current capacity. The Delta-Wye provides a voltage step-up with a lower current capacity. It also provides terminals for a four-wire Wye system. The Wye-Wye and Delta-Delta provide a one-to-one ratio when the transformer windings have an equal number of turns.

22.3 Phase Sequence

As the rotor of the alternator turns, the emf's build up and fall in a sequence, one after the other. The order in which the phase voltages build up is called the **phase sequence**. If the phases are termed *A*, *B*, and *C*, the phase sequence can be either *ABC* or *ACB*, depending on the direction of rotation of the rotor. The phase sequence at the alternator can be reversed by changing the direction of rotation. The phase sequence at the load can be changed by interchanging two lines coming to the load. In the latter case, the phase sequence of the lines remains unchanged.

phase sequence

Phase sequence is also another way of indicating which wave or phasor leads. In Figure 22.12(a), $\mathbf{E}_{Aa}$ leads $\mathbf{E}_{Bb}$ and $\mathbf{E}_{Bb}$ leads $\mathbf{E}_{Cc}$. Therefore, the sequence of the waves is $\mathbf{E}_{Aa}$, $\mathbf{E}_{Bb}$, and $\mathbf{E}_{Cc}$. Writing the first subscript in that order gives the sequence *ABC*.

The order of the leads is found from the phasor diagram by rotating the phasors counterclockwise past a point. Doing this for Figure 22.12(b)

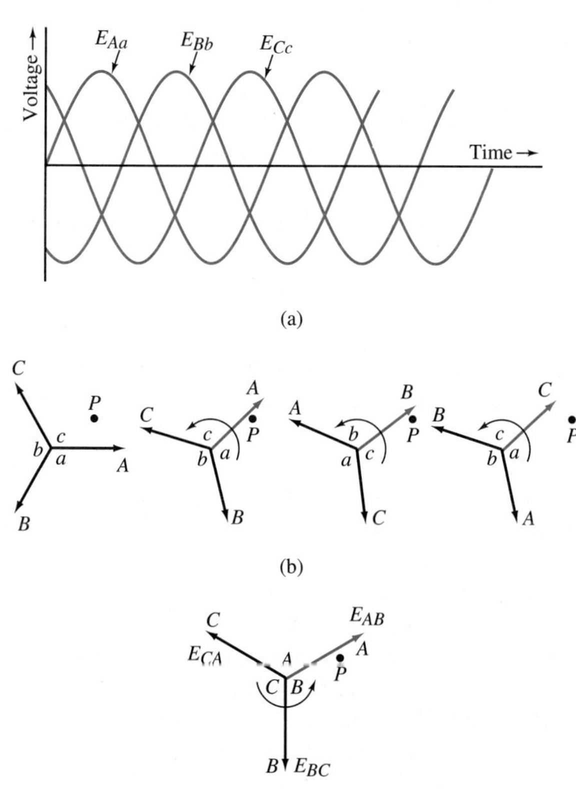

Figure 22.12 The phase sequence is the order in which the waveforms reach their maximum point. (a) Listing the first subscript in the order that the maximum point is reached gives *ABC*. (b) Listing the first subscript in the order that the phase quantities pass a point when rotated *CCW* gives *ABC*. (c) Listing the first subscript in the order that the line voltages pass a point when rotated *CCW* gives *ABC*.

results in E_{Aa} passing Point P first, then E_{Bb}, and finally E_{Cc}. The sequence again is ABC. Figure 22.12(c) shows the line voltages for the phase voltages of Figure 22.12(b). Rotating these past Point P gives the sequence E_{AB}, $E_{BC}E_{CA}$. Writing the first subscript gives the sequence $BCABC$ or ABC.

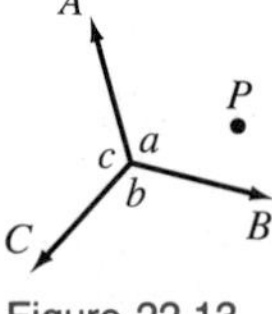

Figure 22.13

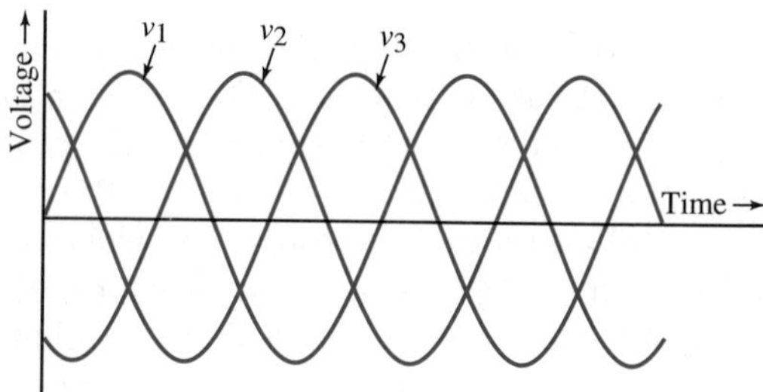

Figure 22.14

EXAMPLE 22.4

Find the phase sequence of:
a. the phasors in Figure 22.13.
b. the waves in Figure 22.14.

Solution

a. When the phasors are rotated counterclockwise past a point on the zero degree axis, B passes first, then C, and then A. The phase sequence is $BCABCA$, which is also ABC.
b. v_1 is the leading wave. v_2 leads v_3, and v_3 lags v_1 by a greater time. The phase sequence is v_1, v_2, v_3.

The phase sequence sets the direction of rotation of a three-phase motor because the magnetic field in the motor rotates with the phase sequence. Switching any two of the connections to the motor reverses its direction.

Electric utilities connect alternators in parallel to meet high load demands. When alternator outputs are connected in parallel, they must have the same phase sequence, frequency, magnitude, and phase angle. Otherwise, a large potential difference can exist between terminals, resulting in possible damage when the outputs are paralleled. The manner in which the phase sequence affects the voltage between the alternator terminals is shown in Figure 22.15. The difference in potential can be quite large, as shown by the calculation in Figure 22.15.

unbalanced load

Changing the phase sequence of a balanced load does not affect the magnitude of the currents. However, if it is an **unbalanced load**, there will be some change in the currents in the load.

To alternator A — C $V_{CC'}$ C' — To alternator B

ACB sequence — ABC sequence

$E_{AB} = 120\angle 0°$ V — B B' — $E_{A'B'} = 120\angle 0°$ V

$E_{BC} = 120\angle 120°$ V — Connection of terminals BB' — $E_{B'C'} = 120\angle -120°$ V

$E_{CA} = 120\angle -120°$ V — $E_{C'A'} = 120\angle 120°$ V

A A'

$$V_{CC'} = E_{B'C'} + E_{CB} = 120\angle -120° \text{ V} - 120\angle 120° \text{ V} = 208\angle -90° \text{ V}$$

Figure 22.15 Three-phase sources must have the same voltage, frequency, and phase sequence when connected in parallel.

Two practical devices that can be used to determine the phase sequence are the three-phase motor and the phase sequence indicator. A brief explanation shows how each one is used.

To use the three-phase motor, the direction of rotation must first be observed when connected to a source of known sequence. The motor is then connected to the unknown sequence and the direction noted. If the direction is the same as for the known sequence, so is the sequence. If the direction is opposite, the sequence is reversed.

phase sequence indicator

A **phase sequence indicator** indicates the sequence by a bulb or digital readout. The circuit for one such indicator is shown in Figure 22.16. If the sequence is ABC, Lamp B is brighter than Lamp C. If the sequence is ACB, Lamp C is brighter than Lamp B. More sophisticated indicators use digital readouts to indicate the sequence. The reason for the difference in lamp brightness can be seen by the analysis of the circuit in Example 22.5.

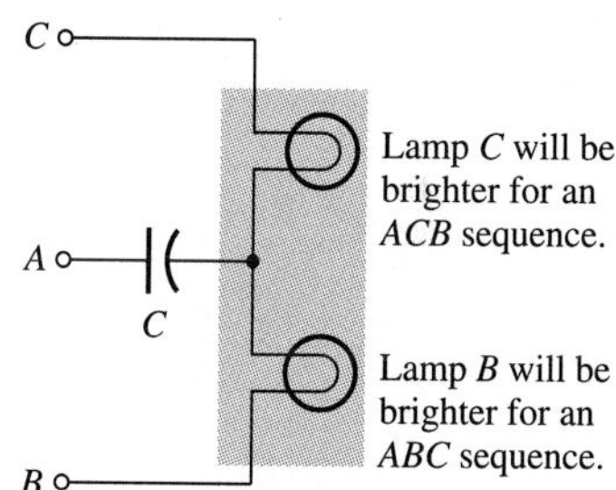

Figure 22.16 This circuit uses the brightness of a lamp to indicate the phase sequence.

EXAMPLE 22.5

The phase voltages in the circuit of Figure 22.17 are $\boldsymbol{E}_{Aa} = 100\angle 0°$ V, $\boldsymbol{E}_{Bb} = 100\angle -120°$ V, and $\boldsymbol{E}_{Cc} = 100\angle 120°$ V.

a. Draw the phasor diagram of the phase voltages and determine the phase sequence.
b. Find the currents: $\boldsymbol{I}_A$, $\boldsymbol{I}_B$, and $\boldsymbol{I}_C$.

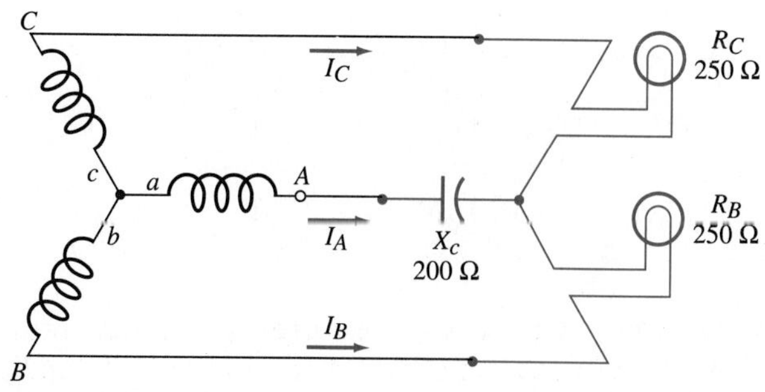

Figure 22.17

Solution

a. The phase sequence, as found from Figure 22.18, is *ABC*.
b. The windings are replaced by three ac sources to give the more familiar form of an ac circuit in Figure 22.19.

C P A b c a B The sequence is *ABC*.

Figure 22.18

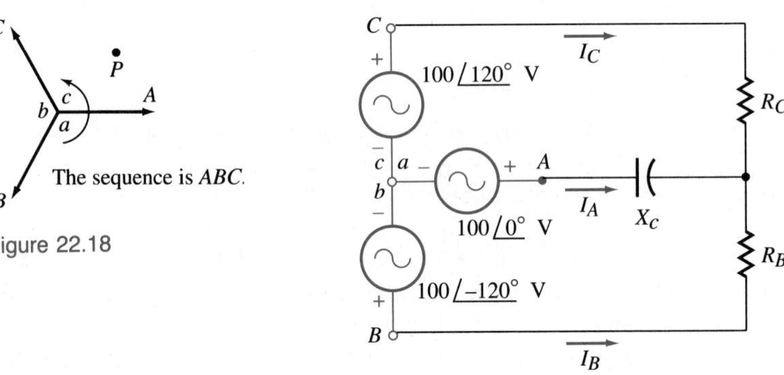

Figure 22.19

In the top loop

$$E_{Cc} - I_C R_C + I_A X_C \angle -90^\circ + E_{aA} = 0$$

$$100 \angle 120^\circ \text{ V} - I_C(250\ \Omega) + (I_A)(200 \angle -90^\circ\ \Omega) - 100 \angle 0^\circ \text{ V} = 0$$

or

$$-(I_A)(200 \angle -90^\circ\ \Omega) + I_C(250\ \Omega) = 173.2 \angle 150^\circ \text{ V}$$

In the bottom loop

$$E_{bB} + E_{Aa} - I_A X_C \angle -90^\circ + I_B R_B = 0$$

$$-100 \angle -120^\circ \text{ V} + 100 \angle 0^\circ \text{ V} - I_A(200 \angle -90^\circ\ \Omega) + I_B(250\ \Omega) = 0$$

or

$$I_A(200 \angle -90^\circ\ \Omega) - I_B(250\ \Omega) = 173.2 \angle 30^\circ \text{ V}$$

At the junction of R_B, R_C, and X_C,

$$I_A + I_B + I_C = 0$$

Solving the equation gives

$$I_A = 0.636 \angle 58^\circ \text{ A}$$
$$I_B = 0.639 \angle 254.7^\circ \text{ A}$$

and

$$I_C = 0.185 \angle 155.5^\circ \text{ A}$$

Lamp *B* will be brighter because the current in it is greater than that in Lamp *C*. Therefore, when Terminals *A*, *B*, and *C* of the indicator are connected to a three-phase source and Lamp B is brighter, the sequence is ABC. It can also be shown that for the ACB sequence, Lamp C will be brighter.

22.4 The Four-Wire, Three-Phase Load

Having examined the three-phase system from the source end, we now look at the three-phase load. The load can be a three- or four-wire Wye load or a three-wire Delta load. It can be balanced or unbalanced. The four-wire, balanced-load is considered first.

four-wire, three-phase (3–ϕ) Wye load

A **four-wire, three-phase (3–ϕ) Wye load** is shown in Figure 22.20. It consists of three impedances with one terminal of each impedance connected together and a neutral line connected to the junction. This load can be used only with a four-wire Wye source. The neutral point of the load is connected to the neutral point of the source. The load phase impedance is connected across a phase of the source. So, $V_L = \sqrt{3}V_\phi$ for the source. Since the load is balanced, the line currents have the same magnitude.

line current

We generally need to know only the magnitude of voltage and current when using three-phase systems. Also, the values usually listed are line values. For instance, a motor nameplate might list values of 208 V and 5 A. These represent the line voltage and **line current**. Bearing this in mind, the analysis of 3–ϕ systems still includes angles in many cases. But for practical purposes, the angles might not be needed.

The current directions in Figure 22.20 are arbitrarily selected. It should be obvious that all currents cannot have the same direction at one time. However, some direction must be assumed so that the calculated current phasors can be related to that direction. The power supplied by

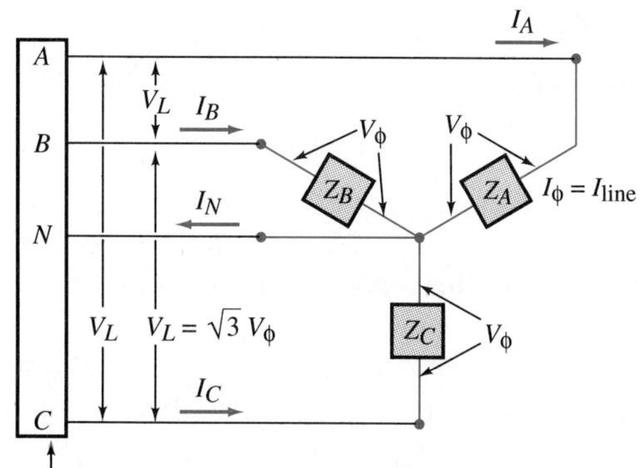

Figure 22.20 The four-wire, three-phase Wye load connection has a neutral line and $I_\phi = I_L$, $V_L = \sqrt{3}\, V_\phi$.

the 3–ϕ source must equal the power dissipated in the load. Therefore, for balanced and unbalanced loads

$$P_T = V_\phi I_{\phi A} \cos \theta_{\phi A} + V_{\phi B} I_{\phi B} \cos \theta_{\phi B} + V_{\phi C} I_{\phi C} \cos \theta_{\phi C} \quad \textbf{(22.7)}$$

where P_T is the power supplied by the source, V_ϕ is the phase voltage, I_ϕ is the phase current, and θ_ϕ is the phase angle of the phase impedance.

For a balanced load, $\mathbf{Z}_A = \mathbf{Z}_B = \mathbf{Z}_C$. Since the phase voltages are equal, the phase currents are also equal. Therefore, the power is the same in each phase, and Equation 22.7 becomes

$$P_T = 3V_\phi I_\phi \cos \theta_\phi \quad \textbf{(22.8)}$$

where V_ϕ, I_ϕ, and θ_ϕ are the voltage, current, and impedance angle for any phase.

Because practical data usually lists line values, a power expression in terms of line values is more convenient. This can be obtained with a few substitutions.

Putting $V_\phi = V_L/\sqrt{3}$ and $I_\phi = I_L$ into Equation 22.8 gives

$$P_T = (3V_L/\sqrt{3})I_L \cos \theta_\phi$$

$$P_T = \sqrt{3}V_L I_L \cos \theta_\phi \quad \textbf{(22.9)}$$

Also, the reactive and apparent power are given by

$$Q_T = \sqrt{3}V_L I_L \sin \theta_\phi \quad \textbf{(22.10)}$$

$$S_T = \sqrt{P_T^2 + Q_T^2} \quad \textbf{(22.11)}$$

where Q_T is the total number of vars and S_T is the total number of volt-amperes.

If the load is balanced,

$$\mathbf{Z}_A = \mathbf{Z}_B = \mathbf{Z}_C = \mathbf{Z}_\phi$$

So

$$\mathbf{I}_N = \frac{1}{\mathbf{Z}_\phi}(V_\phi \angle 0° + V_\phi \angle -120° + V_\phi \angle 120°)$$

The term in the parentheses is equal to zero for a balanced load, so

$$\mathbf{I}_N = 0 \quad \textbf{(22.12)}$$

When the load is balanced, the neutral line can be disconnected without affecting the circuit. This leaves a three-wire, 3–ϕ system. This implies that the balanced three-wire system can be treated the same as the balanced four-wire system.

What purpose does the neutral line serve? To find out, consider the neutral line disconnected, and $\mathbf{Z}_A$ made smaller in Figure 22.20. The voltages now are $V_A \approx 0$ V, $V_B \approx 208$ V, and $V_C \approx 208$ V. The neutral prevents this change in phase voltage by clamping Point N to the same potential as Point n at the source. As a result, V_L also equals $\sqrt{3}V_\phi$ for an unbalanced load, as long as the phase voltages of the source do not change.

Having examined the characteristics of the 3–ϕ, four-wire circuit for a balanced load, it is time to apply the relations to some practical examples.

EXAMPLE 22.6

In a circuit similar to that of Figure 22.20,

$$\mathbf{Z}_A = \mathbf{Z}_B = \mathbf{Z}_C = 100 \text{ ohms}$$
$$\mathbf{V}_{AN} = 120 \angle 0° \text{ V}$$
$$\mathbf{V}_{BN} = 120 \angle -120° \text{ V}$$
$$\mathbf{V}_{CN} = 120 \angle 120° \text{ V}$$

Calculate:
a. the line voltages, line and phase currents.
b. the voltage across $\mathbf{Z}_A$, $\mathbf{Z}_B$, $\mathbf{Z}_C$, the power and reactive power.
c. sketch the phasor diagram for a resistive load, capacitive load, and an inductive load.

Solution

a. $\mathbf{V}_{CA} = \mathbf{V}_{CN} + \mathbf{V}_{NA} = \mathbf{V}_{CN} - \mathbf{V}_{AN}$

$\mathbf{V}_{CA} = 120 \angle 120° \text{ V} - 120 \angle 0° \text{ V}$

$\mathbf{V}_{CA} = 208 \angle 150° \text{ V}$

$\mathbf{V}_{AB} = \mathbf{V}_{AN} + \mathbf{V}_{NB} = \mathbf{V}_{AN} - \mathbf{V}_{BN}$

$\mathbf{V}_{AB} = 120 \angle 0° \text{ V} - 120 \angle -120° \text{ V}$

$\mathbf{V}_{AB} = 208 \angle 30° \text{ V}$

$\mathbf{V}_{BC} = \mathbf{V}_{BN} + \mathbf{V}_{NC} = \mathbf{V}_{BN} - \mathbf{V}_{CN}$

$\mathbf{V}_{BC} = 120 \angle -120° \text{ V} - 120 \angle 120°$

$\mathbf{V}_{BC} = 208 \angle -90° \text{ V}$

For $\mathbf{R}_A$,

$$\mathbf{V}_{AN} = \mathbf{I}_A\mathbf{R}_A$$

Then

$$\mathbf{I}_A = \frac{\mathbf{V}_{AN}}{\mathbf{R}_A} = \frac{120 \angle 0° \text{ V}}{100\ \Omega}$$
$$\mathbf{I}_A = 1.2 \angle 0° \text{ A}$$

For $\mathbf{R}_B$, $\mathbf{V}_{BN} = \mathbf{I}_B\mathbf{R}_B$
Then

$$\mathbf{I}_B = \frac{\mathbf{V}_{BN}}{\mathbf{R}_B} = \frac{120 \angle -120° \text{V}}{100\ \Omega}$$
$$\mathbf{I}_B = 1.2 \angle -120° \text{ A}$$

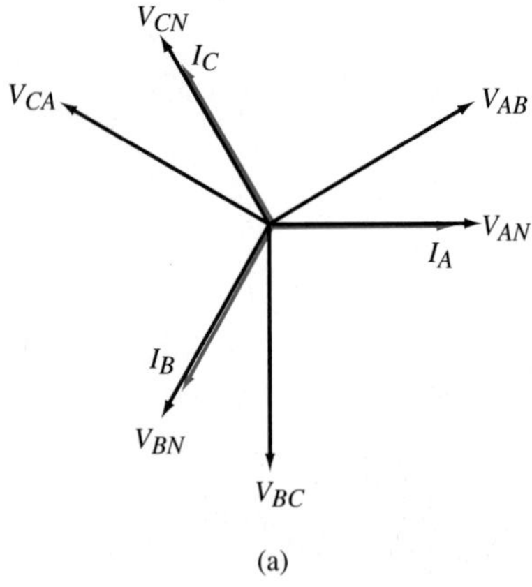

(a)

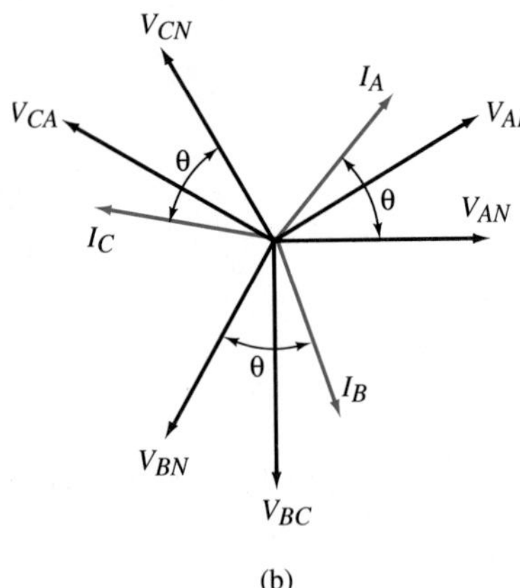

(b)

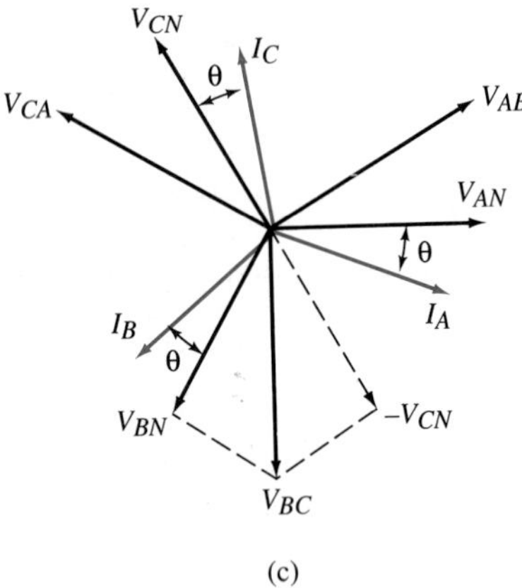

(c)

Figure 22.21

For R_C, $\mathbf{V}_{CN} = \mathbf{I}_C R_C$
Then

$$\mathbf{I}_C = \frac{\mathbf{V}_{CN}}{R_C} = \frac{120 \angle 120° \text{ V}}{100 \ \Omega}$$

$$\mathbf{I}_C = 1.2 \angle 120° \text{ A}$$

The phase currents are the same as the line currents.
b.

$$V_A = |\mathbf{V}_{AN}|$$

$$V_A = 120 \text{ V}$$

V_B and V_C also equal $|\mathbf{V}_{AN}|$

$$V_B = V_C = 120 \text{ V}$$

At Node N,

$$\mathbf{I}_A + \mathbf{I}_B + \mathbf{I}_C - \mathbf{I}_N = 0$$

Then

$$\mathbf{I}_N = \mathbf{I}_A + \mathbf{I}_B + \mathbf{I}_C$$

$$= 1.2 \angle 0° \text{ A} + 1.2 \angle -120° \text{ A} + 1.2 \angle 120° \text{ A}$$

$$\mathbf{I}_N = 0 \text{ A}$$

$$P_T = \sqrt{3} V_L I_L \cos \theta$$

$$P_T = (\sqrt{3})(208 \text{ V})(1.2 \text{ A}) \cos 0°$$

$$P_T = 432.32 \text{ W}$$

$$Q_T = (\sqrt{3}) V_L I_L \sin \theta$$

$$Q_T = (\sqrt{3})(208 \text{ V})(1.2 \text{ A}) \sin 0°$$

$$Q_T = 0 \text{ vars}$$

c. The phasor diagrams are shown in Figure 22.21. Note that the current leads the voltage for the capacitive load and lags it for the inductive load.

Next, an unbalanced four-wire Wye load is considered.

EXAMPLE 22.7

Calculate the phase currents and line currents for Example 22.6 if $\mathbf{Z}_A = 100 \angle 0° \ \Omega$, $\mathbf{Z}_B = 50 \angle 0° \ \Omega$, and $\mathbf{Z}_C = 25 \angle 0° \ \Omega$; $\mathbf{V}_{AN} = 120 \angle 0°$ V, $\mathbf{V}_{BN} = 120 \angle -120°$ V, and $\mathbf{V}_{CN} = 120 \angle 120°$ V.

Solution

The circuit is shown in Figure 22.22. Since the neutral wire is connected, the phase voltages must be equal. Ohm's Law can be used to find the currents.

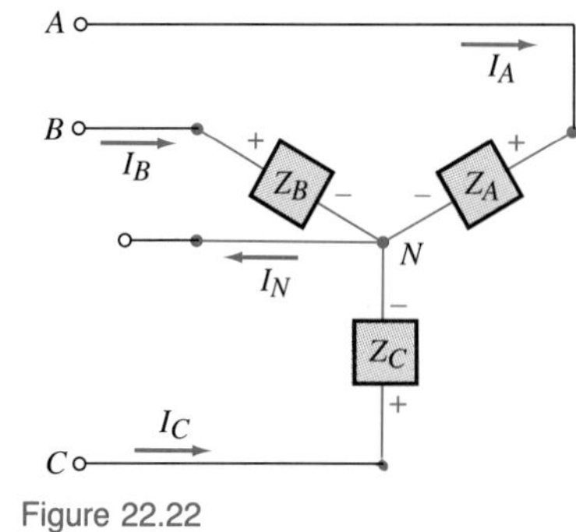

Figure 22.22

$$\boldsymbol{I_A} = \frac{\boldsymbol{V_{AN}}}{\boldsymbol{Z_A}} = \frac{120\ \angle 0^\circ\ \text{V}}{100\ \angle 0^\circ\ \Omega}$$

$$I_A = 1.2\ \angle 0^\circ\ \text{A}$$

$$\boldsymbol{I_B} = \frac{\boldsymbol{V_{BN}}}{\boldsymbol{Z_B}} = \frac{120\ \angle -120^\circ\ \text{V}}{50\ \angle 0^\circ\ \Omega} = 2.4\ \angle -120^\circ\ \text{A}$$

$$I_B = 2.4\ \angle -120^\circ\ \text{A}$$

$$\boldsymbol{I_C} = \frac{\boldsymbol{V_{CN}}}{\boldsymbol{Z_C}} = \frac{120\ \angle 120^\circ\ \text{V}}{25\ \angle 0^\circ\ \Omega} = 4.8\ \angle 120^\circ\ \text{A}$$

$$I_C = 4.8\ \angle 120^\circ\ \text{A}$$

$$\boldsymbol{I_N} = \boldsymbol{I_A} + \boldsymbol{I_B} + \boldsymbol{I_C}$$

$$\boldsymbol{I_N} = 1.2\ \angle 0^\circ\ \text{A} + 2.4\ \angle -120^\circ\ \text{A} + 4.8\ \angle 120^\circ\ \text{A}$$

$$I_N = 3.18\ \angle 139^\circ\ \text{A}$$

The phase currents are the same as the line currents.

With the neutral line connected, the unbalance in the phase loads resulted in a neutral current. Had the neutral line been disconnected, the voltage values across each phase of the load would shift and there would not be any neutral current. That is, the phasor sum of the phase currents would equal zero.

Distribution systems use the neutral line to keep the voltage across the customer's loads from changing. Since the neutral line normally does not carry as much current as the other lines, it can have a smaller diameter.

22.5 Three-Wire, Three-Phase Balanced Loads

Three-wire, three-phase loads can be connected either in Wye or Delta. A **three-wire, three-phase Wye load** is shown in Figure 22.23. It consists of three impedances having one terminal of each connected

three-wire, three-phase Wye load

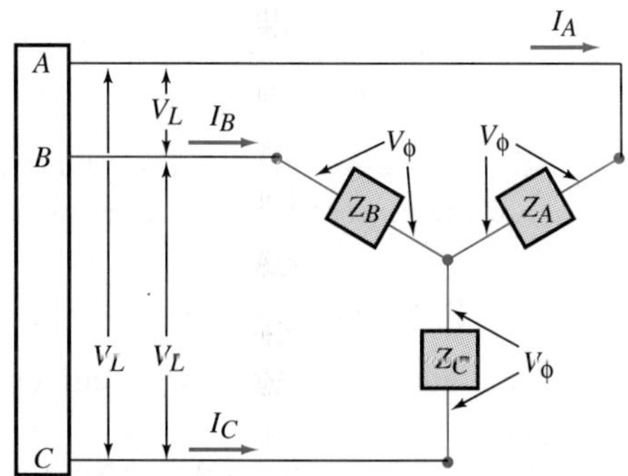

Figure 22.23 The three-wire, three-phase Wye load, $I_\phi = I_L$, $V_L = \sqrt{3}\, V_\phi$ if $Z_A = Z_B = Z_C$.

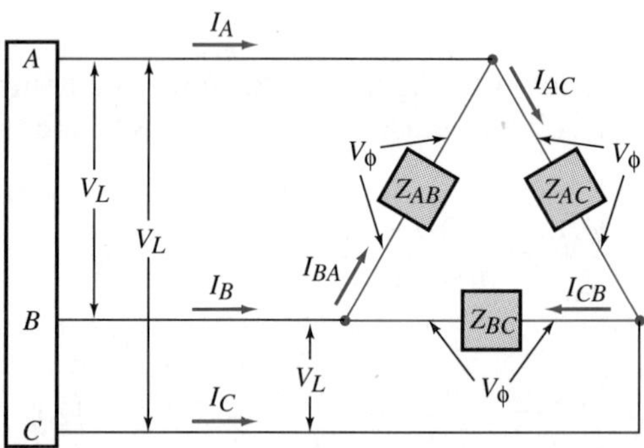

Figure 22.24 The three-phase Delta load connection. $V_L = V_\phi$, and $I_L = \sqrt{3}\, I_\phi$ if $Z_{AB} = Z_{BC} = Z_{AC}$.

together, and the other connected to the lines of a 3–ϕ source. If $\mathbf{Z}_A = \mathbf{Z}_B = \mathbf{Z}_C$, the load is balanced. For that condition, it is simply a balanced four-wire, three-phase load without the neutral. As such, current, voltage, and other characteristics can be found using the same procedure as for the four-wire load.

three-wire, three-phase Delta load

A **three-wire, three-phase Delta load** is shown in Figure 22.24. It has three impedances connected in the form of a delta. The points are connected to the lines of a 3–ϕ source. If $\mathbf{Z}_{AB} = \mathbf{Z}_{BC} = \mathbf{Z}_{AC}$, the load is balanced. The conditions and relations for this load parallel the conditions for the Delta source. $V_L = V\phi$, and if the load is balanced, $I_L = \sqrt{3} I_\phi$.

The derivation is not included but it can be shown that for both balanced and unbalanced Delta loads, the phasor sum of the line currents must equal zero. That is,

$$\mathbf{I}_A + \mathbf{I}_B + \mathbf{I}_C = 0 \qquad \textbf{(22.13)}$$

When the load is balanced, the magnitudes are equal, and the phase difference is 120°. For an unbalanced load, the magnitudes are not equal, and the displacement varies.

The total power in the load must equal the sum of the power in each phase impedance. So, in general, the relation is the same as for the Y load.

$$P_T = V_{AB}I_{AB} \cos \theta_{AB} + V_{BC}I_{BC} \cos \theta_{BC} + V_{CA}I_{CA} \cos \theta_{CA} \qquad \textbf{(22.14)}$$

For a balanced load, the magnitudes of the phase currents are the same, and the impedance angles are the same. Therefore, Equation 22.14 becomes

$$P_T = 3V_LI_\phi \cos \theta_\phi \qquad \textbf{(22.15)}$$

where V_L is the line voltage, I_ϕ is the phase current, and θ_ϕ is the impedance phase angle.

Also, for a balanced load, $I_\phi = I_L/\sqrt{3}$. Putting this into Equation 22.15 gives the power in terms of the line quantities

$$P_T = \sqrt{3}V_LI_L \cos \theta_\phi \qquad \textbf{(22.16)}$$

Comparing these power relations to those for the Y load shows that both loads have the same relations. In addition, the relations of Equations 22.10 and 22.11 also hold true.

EXAMPLE 22.8

$\mathbf{Z}_{AB} = \mathbf{Z}_{BC} = \mathbf{Z}_{AC} = 100\angle 0°\ \Omega$ in Figure 22.25, find:

a. the current in each impedance if $\mathbf{V}_{AC} = 208\ \angle 0°\ \text{V}$, $\mathbf{V}_{BA} = 208\ \angle 120°\ \text{V}$, $\mathbf{V}_{CB} = 208\ \angle -120°\ \text{V}$.

b. the line current.

c. the power dissipated in the Delta load of Figure 22.25.

d. sketch the phasor diagram for a resistive load, inductive load, and capacitive load.

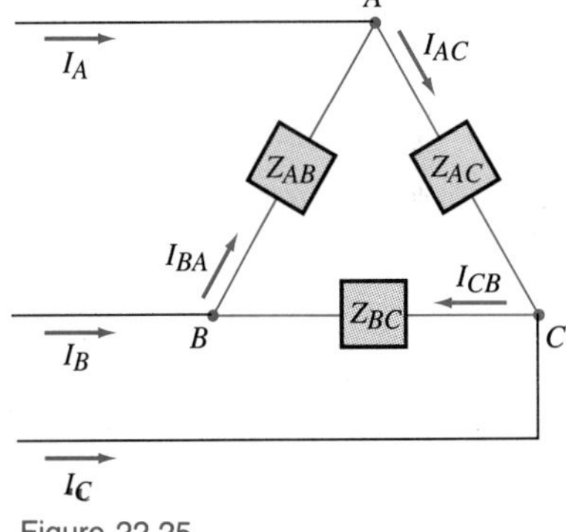

Figure 22.25

Solution

a.

$$\mathbf{I}_{BA} = \frac{\mathbf{V}_{BA}}{\mathbf{Z}_{AB}} = \frac{208\ \angle 120°\ \text{V}}{100\ \Omega}$$

$$\mathbf{I}_{BA} = 2.08\ \angle 120°\ \text{A}$$

$$\mathbf{I}_{CB} = \frac{\mathbf{V}_{CB}}{\mathbf{Z}_{BC}} = \frac{208\ \angle -120°\ \text{V}}{100\ \Omega}$$

$$\mathbf{I}_{CB} = 2.08\ \angle -120°\ \text{A}$$

$$\mathbf{I}_{AC} = \frac{\mathbf{V}_{AC}}{\mathbf{Z}_{AC}} = \frac{208\ \angle 0°\ \text{V}}{100\ \Omega}$$

$$\mathbf{I}_{AC} = 2.08\ \angle 0°\ \text{A}$$

b.

$$\mathbf{I}_A = \mathbf{I}_{AC} - \mathbf{I}_{BA} = 2.08\ \angle 0°\ \text{A} - 2.08\ \angle 120°\ \text{A}$$

$$\mathbf{I}_A = 3.6\ \angle -30°\ \text{A}$$

$$\mathbf{I}_B = \mathbf{I}_{BA} - \mathbf{I}_{CB} = 2.08\ \angle 120°\ \text{A} - 2.08\ \angle -120°\ \text{A}$$

$$\mathbf{I}_B = 3.6\ \angle 90°\ \text{A}$$

$$\mathbf{I}_C = \mathbf{I}_{CB} - \mathbf{I}_{AC} = 2.08\ \angle -120°\ \text{A} - 2.08\ \angle 0°\ \text{A}$$

$$\mathbf{I}_C = 3.6\ \angle -150°\ \text{A}$$

c.

$$P = \sqrt{3}V_L I_L \cos\theta_\phi$$

$$= \sqrt{3}(208\ \text{V})(3.6\ \text{A})(1)$$

$$P = 1297\ \text{W}$$

d. The phasor diagram in Figure 22.26 shows the currents in phase for the resistive load, lagging for the inductive load, and leading for the capacitive load.

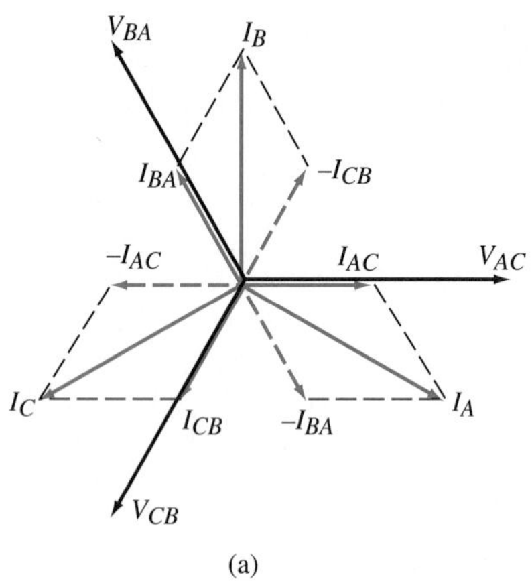

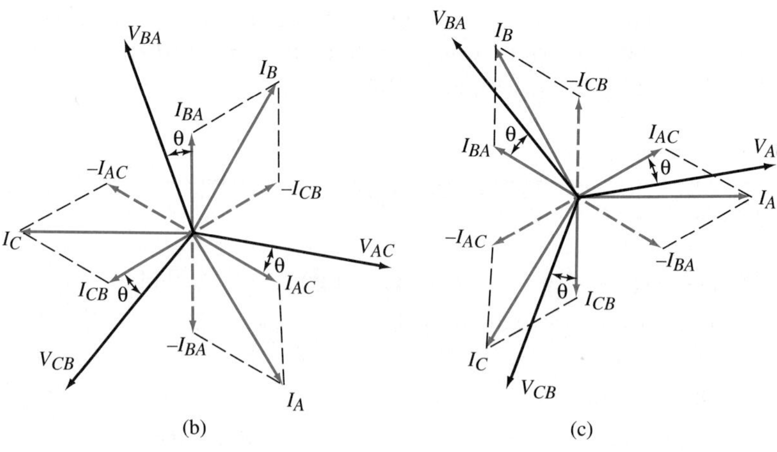

Figure 22.26

22.6 Unbalanced Three-Phase Delta and Wye Loads

Analyzing a four-wire, unbalanced load is not difficult since the neutral wire keeps the load phase voltages at the same value as the source phase voltages. This allows us to use Ohm's Law instead of the more complex loop analysis. The three-wire, three-phase Delta load, balanced or unbalanced, is also not difficult to analyze since the loads are connected across the lines, and the line voltage is known. If the phase impedance is known, the **phase current** can be found using Ohm's Law. If the phase current is known, the phase impedance can be found using Ohm's Law. It is important to remember that the $\sqrt{3}$ relations do not hold true for the unbalanced loads. Let us now consider an unbalanced Delta load.

phase current

EXAMPLE 22.9

An unbalanced Delta load is connected across a 3–ϕ system in which $\boldsymbol{V}_{BA} = 208 \angle 120°$ V, $\boldsymbol{V}_{CB} = 208 \angle -120°$ V, and $\boldsymbol{V}_{AC} = 208 \angle 0°$ V; $\boldsymbol{Z}_{AB} = 50 \angle 0°$ ohms, $\boldsymbol{Z}_{BC} = 100 \angle 0°$ ohms, and $\boldsymbol{Z}_{AC} = 25 \angle 30°$ ohms.

a. Find the phase currents, line currents, power, and reactive power for the circuit.
b. Draw the phasor diagram.

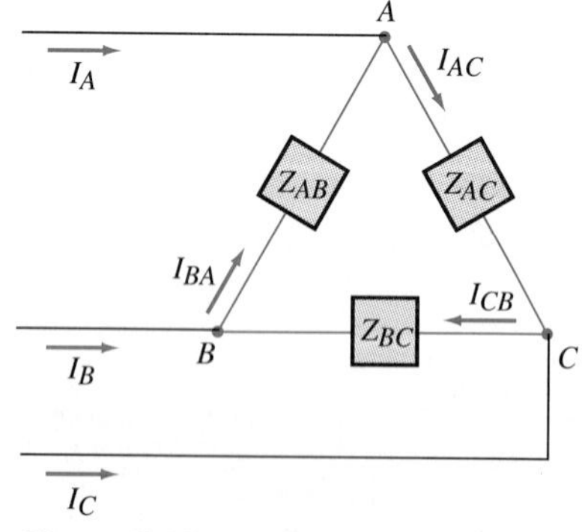

Figure 22.27

Solution

a. The circuit is drawn in Figure 22.27.
Applying Ohm's Law to the phases gives

$$\boldsymbol{I}_{AC} = \frac{\boldsymbol{V}_{AC}}{\boldsymbol{Z}_{AC}} = \frac{208 \angle 0° \text{ V}}{25 \angle 30° \ \Omega}$$

$$\boldsymbol{I}_{AC} = 8.32 \angle -30° \text{ A}$$

$$\boldsymbol{I}_{BA} = \frac{\boldsymbol{V}_{BA}}{\boldsymbol{Z}_{AB}} = \frac{208 \angle 120° \text{ V}}{50 \angle 0° \ \Omega}$$

$$\boldsymbol{I}_{BA} = 4.16 \angle 120° \text{ A}$$

$$\mathbf{I}_{CB} = \frac{\mathbf{V}_{CB}}{\mathbf{Z}_{BC}} = \frac{208\ \angle -120°\ \text{V}}{100\ \angle 0°\ \Omega}$$

$$\mathbf{I}_{CB} = 2.08\ \angle -120°\ \text{A}$$

Next, Kirchhoff's Current Law will be used to find the line current.

$$\mathbf{I}_A = \mathbf{I}_{AC} - \mathbf{I}_{BA} = 8.32\ \angle -30°\ \text{A} - 4.16\ \angle 120°\ \text{A}$$

$$\mathbf{I}_A = 12.1\ \angle -39.9°\ \text{A}$$

$$\mathbf{I}_B = \mathbf{I}_{BA} - \mathbf{I}_{CB} = 4.16\ \angle 120°\ \text{A} - 2.08\ \angle -120°\ \text{A}$$

$$\mathbf{I}_B = 5.5\ \angle 100.9°\ \text{A}$$

$$\mathbf{I}_C = \mathbf{I}_{CB} - \mathbf{I}_{AC} = 2.08\ \angle -120°\ \text{A} - 8.32\ \angle -30°\ \text{A}$$

$$\mathbf{I}_C = 8.58\ \angle 164°\ \text{A}$$

The load is unbalanced, so the general relations for power must be used.

$$P_T = V_{BA}I_{BA} \cos \theta_{AB} + V_{CB}I_{CB} \cos \theta_{BC} + V_{AC}I_{AC} \cos \theta_{CA}$$

$$P_T = (208\ \text{V})(4.16\ \text{A})\cos 0° + (208\ \text{V})(2.08\ \text{A}) \cos 0° + (208\ \text{V})(8.32\ \text{A}) \cos 30°$$

$$P_T = 865.28\ \text{W} + 432.64\ \text{W} + 1498.7\ \text{W}$$

$$P_T = 2796.62\ \text{W}$$

Finally, the reactive power is given by

$$Q_T = V_{BA}I_{BA} \sin \theta_{AB} + V_{CB}I_{CB} \sin \theta_{BC} + V_{AC}I_{AC} \sin \theta_{CA}$$

$$Q_T = (208\ \text{V})(4.16\ \text{A}) \sin 0° + (208\ \text{V})(2.08\ \text{A}) \sin 0° + (208\ \text{V})(8.32\ \text{A}) \sin 30°$$

$$Q_T = 865.28\ \text{vars}$$

b. The phasor diagram is shown in Figure 22.28.

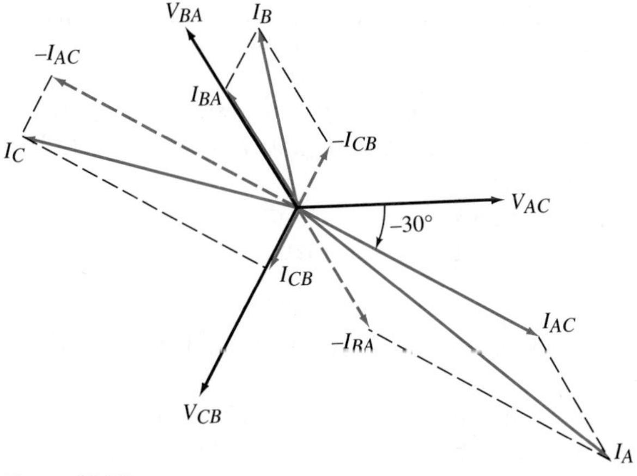

Figure 22.28

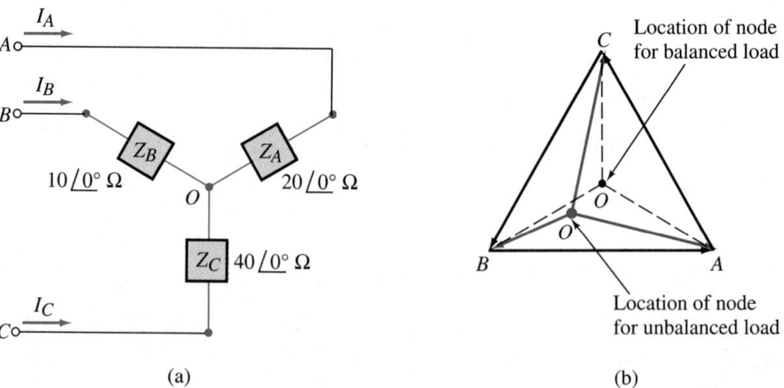

Figure 22.29 The effect of an unbalanced load on the phasor diagram of the three-wire Wye load. (a) An unbalanced load. (b) The unbalance causes point O to shift since the phase voltages are not equal.

The unbalanced three-wire Wye load does present some problems because of the floating neutral. Now the voltage across the phase impedance is no longer $E_L/\sqrt{3}$, but takes on different values, as stated in Section 22.4. The common point in the phasor diagram is moved, as in Figure 22.29(b).

Since V_ϕ is not known, Ohm's Law cannot be used to provide a simple solution as with the four-wire Wye and the three-wire Delta.

One option is to replace the Wye with its equivalent Delta. The Delta will be unbalanced, but that does not present any problem. However, the conversion process and relating the delta quantities to the Wye quantities can be long, and result in errors. A more direct method is to use the nodal method and to write the currents in terms of the voltage at each end of the impedance. In other words, Ohm's Law is applied to each impedance. Then, Kirchhoff's Current Law is applied at the node of the Wye to calculate the phase voltages. These can be used with the impedances to find the currents. Example 22.10 illustrates this procedure.

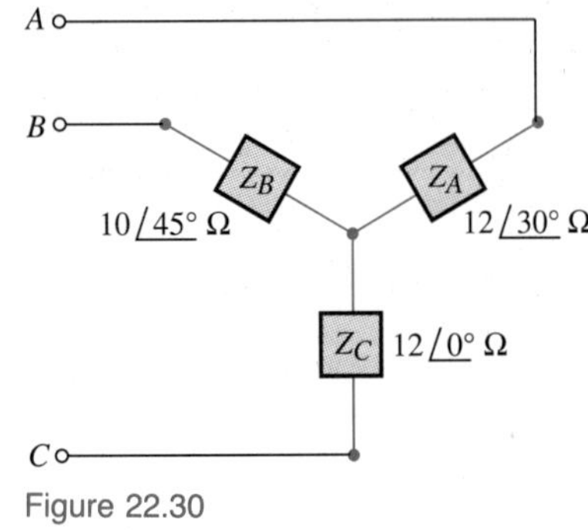

Figure 22.30

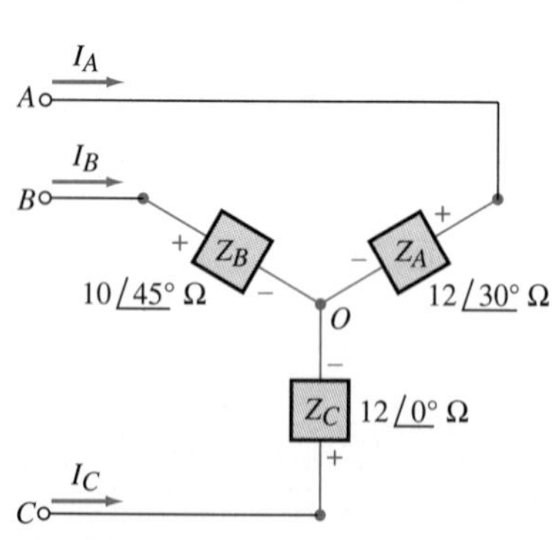

Figure 22.31

EXAMPLE 22.10

A three-phase load is connected to a source, as in Figure 22.30. $V_{CA} = 208\ \angle -120°$ V, $V_{AB} = 208\ \angle 0°$ V, and $V_{BC} = 208\ \angle 120°$ V. Find:

a. the line currents.
b. the phase currents.
c. the power dissipated in the load.

Solution

The system with current direction and polarity is shown in Figure 22.31.

a. The voltage at the terminals of each impedance is written with respect to one of the lines as a reference. Line C will be used here. The voltage at the (+) end of Z_B, as measured from C, is V_{BC} and that on the (−) end is V_{OC}. Then

1. $V_{BC} - V_{OC} = I_B Z_B$

and

$$I_B = \frac{V_{BC} - V_{OC}}{Z_B}$$

Likewise, for Z_C and Z_A,

2. $V_{AC} - V_{OC} = I_A Z_A$

and

$$I_A = \frac{V_{AC} - V_{OC}}{Z_A}$$

3. $0 - V_{OC} = I_C Z_C$

and

4. $$I_C = \frac{-V_{OC}}{Z_C}$$

The zero (0) in Equation 3 represents V_{CC}.

At Point O,

$$I_A + I_B + I_C = 0$$

Putting 1–3 into 4 gives

$$\frac{-V_{OC}}{Z_C} + \frac{V_{BC} - V_{OC}}{Z_B} + \frac{V_{AC} - V_{OC}}{Z_A} = 0$$

which can be reduced to

$$V_{OC}\left(\frac{1}{Z_A} + \frac{1}{Z_B} + \frac{1}{Z_C}\right) = \frac{V_{BC}}{Z_B} + \frac{V_{AC}}{Z_A} \tag{22.17}$$

Since $V_{AC} = -V_{CA}$, we have

$$V_{OC}\left(\frac{1}{12\ \angle 30°\ \Omega} + \frac{1}{10\ \angle 45°\ \Omega} + \frac{1}{12\ \angle 0°\ \Omega}\right) = \frac{208\ \angle 120°\ \text{V}}{10\ \angle 45°\ \Omega} - \frac{208\ \angle -120°\ \text{V}}{12\ \angle 30°\ \Omega}$$

which results in $V_{OC} = 139.58\ \angle 81.07°\ \text{V}$.

Now that V_{OC} is known, the three currents can be calculated using Equations 1–3.

$$I_C = \frac{-V_{OC}}{Z_C} = \frac{-139.39\ \angle 81.07°\ \text{V}}{12\ \angle 0°\ \Omega}$$

$$I_C = 11.62\ \angle 261.07°\ \text{A}$$

$$I_B = \frac{V_{BC} - V_{OC}}{Z_B} = \frac{208\ \angle 120°\ \text{V} - 139.39\ \angle 81.07°\ \text{V}}{10\ \angle 45°\ \Omega}$$

$$I_B = 13.26\ \angle 116.34°\ \text{A}$$

$$I_A = \frac{V_{AC} - V_{OC}}{Z_A} = \frac{-208 \angle -120° \text{ V} - 139.39 \angle 81.07° \text{ V}}{12 \angle 30° \ \Omega}$$

$$I_A = 7.72 \angle -2.74° \text{ A}$$

b. The load is a Wye load so the line currents are the same as the phase currents.
c. Using the general relations for power gives

$$P_T = V_A I_A \cos \theta_A + V_B I_B \cos \theta_B + V_C I_C \cos \theta_C$$
$$V_C = V_{OC} = 139.39 \text{ V}$$
$$V_B = I_B Z_B = (13.26 \text{ A})(10 \ \Omega) = 132.6 \text{ V}$$
$$V_A = I_A Z_A = (7.72 \text{ A})(12 \ \Omega) = 92.64 \text{ V}$$

so

$$P_T = (92.64 \text{ V})(7.72 \text{ A}) \cos 30° + (132.61 \text{ V})(13.26 \text{ A}) \cos 45° + (139.39 \text{ V})(11.63 \text{ A}) \cos 0°$$
$$P_T = 619.36 \text{ W} + 1243.29 \text{ W} + 1621.11 \text{ W}$$
$$P_T = 3483.76 \text{ W}$$

Equation 22.17 can also be written for V_{OB} and V_{OA} in terms of the impedances and the other line voltages. The result is

$$V_{OB}\left(\frac{1}{Z_A} + \frac{1}{Z_B} + \frac{1}{Z_C}\right) = \frac{V_{CB}}{Z_C} + \frac{V_{AB}}{Z_A} \quad \textbf{(22.18)}$$

$$V_{OA}\left(\frac{1}{Z_A} + \frac{1}{Z_B} + \frac{1}{Z_C}\right) = \frac{V_{CA}}{Z_C} + \frac{V_{BA}}{Z_B} \quad \textbf{(22.19)}$$

22.7 Three-Phase Loads and Line Impedance

The examples considered thus far gave the voltage at the load, neglecting any effect of the line impedance. In the electric distribution systems, the transmission lines have a **line impedance** of resistance, inductance, and capacitance. These can result in a drop between the source and load, unequal line voltages, and the phase angles not being 120 degrees apart. This section considers the effect that line impedance has on what occurs at the load.

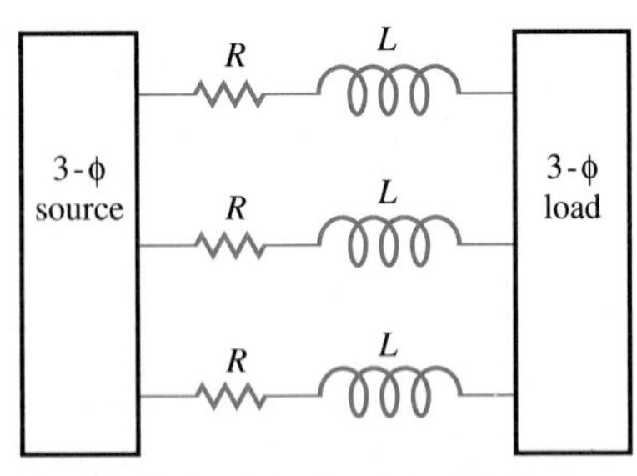

Figure 22.32 The line impedance causes a voltage drop and a power loss in the line.

The analysis includes the effect of resistance and inductance, but not capacitance. From a practical view, X_C will be very large at 60 Hz and will not have much effect unless the line is very long.

The representation of a practical line is shown in Figure 22.32. R is the ac resistance and L is the inductance. The ac resistance at 60 Hz will be approximately 4–10 percent greater than the dc resistance. The inductance changes with line length and spacing between the lines, but is not directly proportional to them.

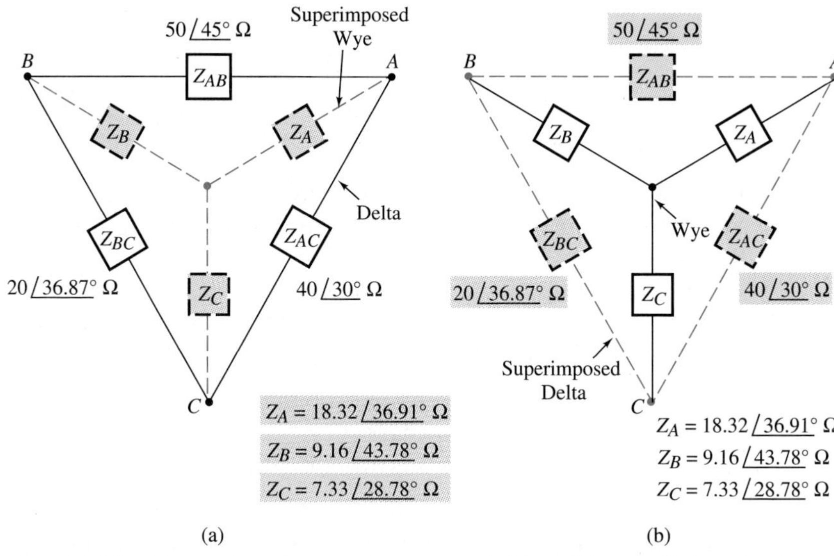

Figure 22.33 A three-phase load can be represented by a Wye or a Delta network configuration. (a) The three-phase Delta and its equivalent Wye. (b) The three-phase Wye and its equivalent Delta.

When including the line impedance, it is convenient to convert the load to an equivalent Wye. Then the line impedance can be added to the impedance in the leg of the Wye. The result is a Wye load connected directly to the source.

The physical relation between the Wye and Delta networks is shown in Figure 22.33. Equations 22.20–22.27 are used to convert from one network to the other. These equations are similar to the ones introduced for resistance in Chapter 10, Section 10.5.

They are

Delta to Wye

$$\mathbf{Z}_A = \frac{\mathbf{Z}_{AB}\mathbf{Z}_{AC}}{\mathbf{Z}_{AB} + \mathbf{Z}_{AC} + \mathbf{Z}_{BC}} \tag{22.20}$$

$$\mathbf{Z}_B = \frac{\mathbf{Z}_{AB}\mathbf{Z}_{BC}}{\mathbf{Z}_{AB} + \mathbf{Z}_{AC} + \mathbf{Z}_{BC}} \tag{22.21}$$

$$\mathbf{Z}_C = \frac{\mathbf{Z}_{BC}\mathbf{Z}_{AC}}{\mathbf{Z}_{AB} + \mathbf{Z}_{AC} + \mathbf{Z}_{BC}} \tag{22.22}$$

If $\mathbf{Z}_{AB} = \mathbf{Z}_{BC} = \mathbf{Z}_{AC}$ then $\mathbf{Z}_A = \mathbf{Z}_B = \mathbf{Z}_C = \mathbf{Z}_{AB}/3$ **(22.23)**

Wye to Delta

$$\mathbf{Z}_{AB} = \frac{\mathbf{Z}_A\mathbf{Z}_B + \mathbf{Z}_B\mathbf{Z}_C + \mathbf{Z}_A\mathbf{Z}_C}{\mathbf{Z}_C} \tag{22.24}$$

$$\mathbf{Z}_{BC} = \frac{\mathbf{Z}_A\mathbf{Z}_B + \mathbf{Z}_B\mathbf{Z}_C + \mathbf{Z}_A\mathbf{Z}_C}{\mathbf{Z}_A} \tag{22.25}$$

$$Z_{AC} = \frac{Z_A Z_B + Z_B Z_C + Z_A Z_C}{Z_B} \tag{22.26}$$

If $Z_A = Z_B = Z_C$, then $Z_{AB} = Z_{AC} = Z_{BC} = 3Z_A$ **(22.27)**

The conversion can be made as follows:

1. Identify the Wye or Delta and mark the points A, B, C.
2. Superimpose the desired network on the actual one, as shown by the dashed lines in Figure 22.33.
3. Use the equations to calculate the values of the impedance.
4. Remove the original network and leave the equivalent in its place, connected to A, B, and C.

An important part in the conversion is obtaining the proper relation between the positions of the impedances.

An inspection of the equations for the conversion shows the general relation to be:

Delta to Wye. The leg impedance equals the product of the adjacent impedances in the Delta, divided by the phasor sum of the three Delta impedances.

Wye to Delta. The Delta impedance equals the phasor sum of the products of all pairs of Delta impedances, divided by the opposite Wye impedance.

Although the loads on a three-phase system might not always be balanced, for the most part, they are kept at or near balance. Therefore, our examples are limited to the balanced condition. The solution to these examples shows how the circuit concepts are applied. They are not meant to be memorized.

EXAMPLE 22.11

Determine the % voltage regulation and the % efficiency of the three-phase system in Figure 22.34(a).

Solution

The load can be taken as a Delta or a Wye. The Wye will be used and is shown in Figure 22.34(b).

$$\%\ \text{VR} = \frac{(V_{\text{no load}} - V_{\text{load}})}{V_{\text{load}}} \times 100\%$$

The voltage at load is 220 kV, as given in Figure 22.34.

The voltage at no load is the source voltage. This is found with the load connected.

Since the load is balanced,

$$V_s = \sqrt{3}\,(V_\phi \text{ at the source})$$
$$V_s = \sqrt{3} I_L (R_L + jX_L + Z_\phi \text{ of the load})$$

X_L and Z_ϕ must be determined.

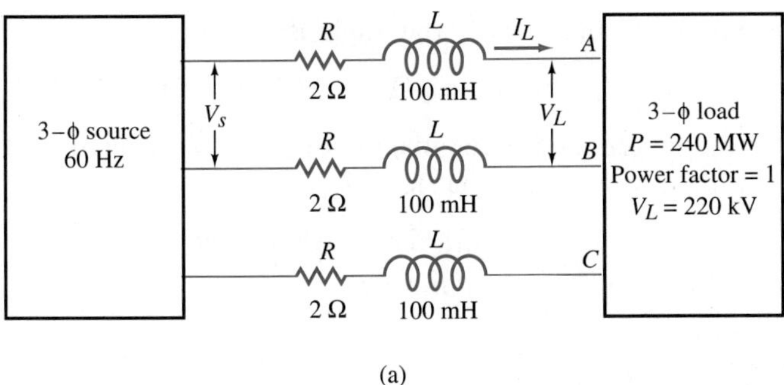

(a)

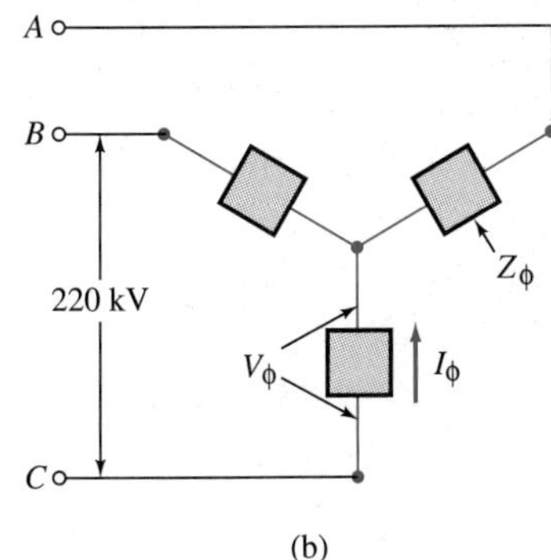

(b)

Figure 22.34

$$X_L = 2\pi fL = 2\pi(60 \text{ Hz})(100 \times 10^{-3} \text{ H})$$
$$X_L = 37.7 \ \Omega$$

The magnitude of the phase impedance in one leg is

$$Z_\phi = \frac{V_\phi}{I_\phi} = \frac{V_\phi}{I_L}$$

I_L is found from the power relation. It is

$$P_T = \sqrt{3} I_L V_L \cos\theta$$

or

$$I_L = \frac{P_T}{\sqrt{3} V_L \cos\theta}$$

cos θ is 1. So

$$I_L = \frac{240 \times 10^6 \text{ W}}{(\sqrt{3})(220 \times 10^3 \text{ V})(1)} = 629.84 \text{ A}$$

Substituting I_L in the equation for Z_ϕ gives

$$Z_\phi = \frac{220 \times 10^3 \text{ V}}{\sqrt{3} \times 629.84 \text{ A}} = 201.67$$

Since the power factor is 1, the angle is 0°. Then

$$V_s = (\sqrt{3})(629.84\ \text{A})(2\ \Omega + j37.7\ \Omega + 201.67\ \Omega)$$

We are only interested in the magnitude of V_s. The magnitude of the quantity in parentheses is equal to 207.13. So

$$V_s = (\sqrt{3})(629.84\ \text{A})(207.13)$$

$$V_s = 225.96\ \text{kV}$$

Then

$$\%\ \text{VR} = \frac{(225.96\ \text{kV} - 220\ \text{kV})(100\%)}{220\ \text{kV}}$$

$$\%\ \text{VR} = 2.7\%$$

$$\text{efficiency} = \frac{P_{\text{load}}}{P_{\text{source}}} \times 100\%$$

$$P_{\text{source}} = P_{\text{load}} + P_{\text{lines}}$$

$$P_{\text{lines}} = 3I_L^2R$$

$$P_{\text{lines}} = (3)(629.84\ \text{A})^2(2\ \Omega)$$

so

$$P_{\text{source}} = 242.38 \times 10^6\ \text{W}$$

$$\%\ \text{efficiency} = \frac{(240 \times 10^6\ \text{W}) \times 100\%}{(242.38 \times 10^6\ \text{W})}$$

$$\%\ \text{efficiency} = 99\%$$

EXAMPLE 22.12

A three-phase distribution system supplies three loads, as shown in Figure 22.35. The line impedance is $R = 0.3$ ohm, $L = 0.3$ mH. Use the power triangle method to find the power supplied by the system, the current in each line, the total KVA supplied by the source, and the power factor.

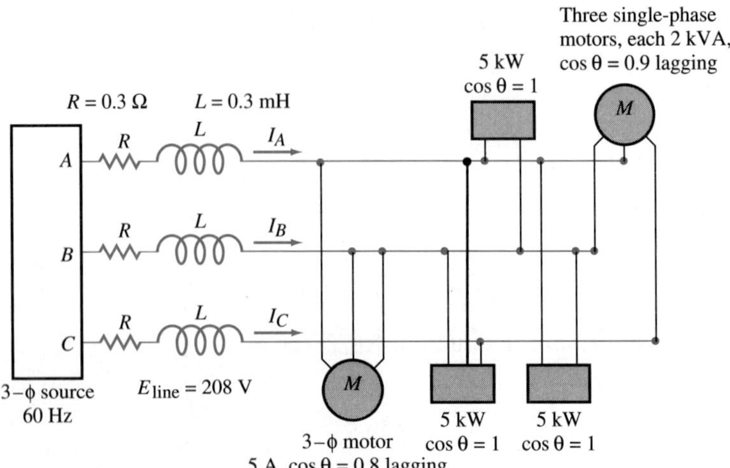

Figure 22.35

Solution

The loads are the three-phase motor in Delta across the lines, a three-phase Y load made up of the three lighting loads, and a three-phase Y load made up of three single-phase motors. The power triangle is found for each load.

For the lighting load,

$$P = 15\text{ kW} \quad \text{and} \quad Q = 0\text{ vars}$$
$$S = \sqrt{P^2 + Q^2} = 15\text{ kVa}$$

The triangle is a horizontal line, as in Figure 22.36.

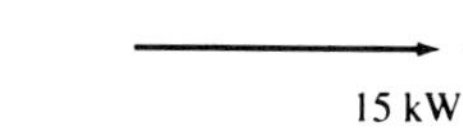

Figure 22.36

The 3–ϕ motor line current is 5 A and the line voltage is 208 V. So

$$P = \sqrt{3}V_L I_L \cos\theta$$
$$= (\sqrt{3})(208\text{ V})(5\text{ A})(0.8)$$
$$P = 1441.1\text{ W}$$
$$\cos\theta = (P/S) \quad \text{so} \quad S = P/\cos\theta$$
$$S = \frac{1441.1\text{ W}}{0.8} = 1801.4\text{ VA}$$

Then

$$Q = \sqrt{S^2 - P^2}$$
$$Q = \sqrt{(1801.4\text{ VA})^2 - (1441.1\text{ W})^2}$$
$$Q = 1080.9\text{ vars}$$

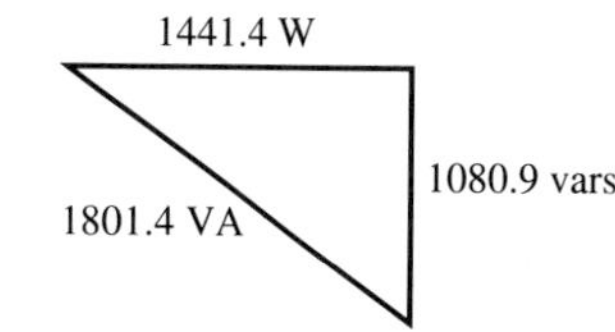

Figure 22.37

The motor power triangle is shown in Figure 22.37.

The total motor power of the three single-phase motors can be calculated using the kVa rating and power factor.

$$P = (3)(2\text{ kVA})(0.9) = 5.4\text{ kW}$$
$$S = P/\cos\theta = 5400\text{ W}/0.9$$
$$S = 6\text{ kVA}$$

Then

$$Q = \sqrt{S^2 - P^2}$$
$$= \sqrt{(6000\text{ VA})^2 - (5400\text{ W})^2}$$
$$= 2615.3\text{ vars}$$

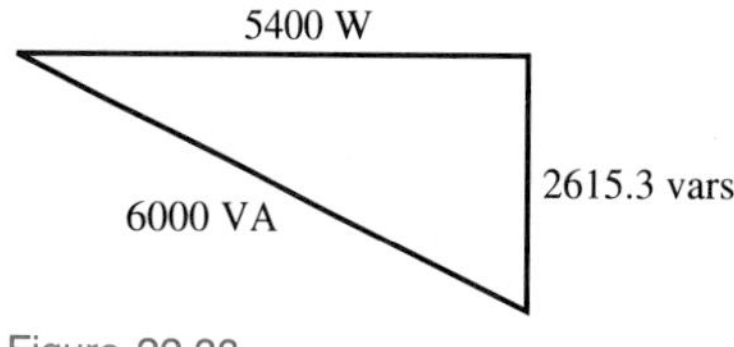

Figure 22.38

The power triangle is shown in Figure 22.38 for the single-phase motors.

The total load power on the lines is the sum of the power in each load.

$$P_{\text{load}} = 15{,}000\text{ W} + 1441.1\text{ W} + 5400\text{ W}$$
$$P_{\text{load}} = 21{,}841.1\text{ W}$$

The total vars is the sum of the vars in each load:

$$Q_{\text{load}} = 0\text{ vars} + 1080.9\text{ vars} + 2615.3\text{ vars}$$
$$Q_{\text{load}} = 3696.2\text{ vars}$$

The total volt-amperes is

$$S_T = \sqrt{P_{\text{load}}^2 + Q_{\text{load}}^2}$$
$$S_T = \sqrt{(21{,}841.1 \text{ W})^2 + (3696.2 \text{ vars})^2}$$
$$S_T = 22.15 \text{ kVA}$$

The load power factor is

$$F_p = \cos\theta = P_{\text{load}}/S_{\text{load}}$$
$$F_p = (21{,}841.1 \text{ W})/(22{,}150 \text{ VA})$$
$$F_p = 0.986$$

The power relation is used to obtain I_L. At the load,

$$P_T = \sqrt{3}V_L I_L \cos\theta$$

but

$$\cos\theta = F_p = 0.986$$

So

$$I_L = \frac{21{,}841.1 \text{ W}}{\sqrt{3}\,(208 \text{ V})(0.986)}$$
$$I_L = 61.49 \text{ A}$$

The load power and line power are added to obtain the source power.

$$P_s = P_{\text{load}} + P_{\text{lines}}$$
$$P_{\text{lines}} = (3)(I_L)^2(R)$$
$$P_{\text{lines}} = (3)(61.49 \text{ A})^2(0.3\ \Omega)$$
$$P_{\text{lines}} = 3402.9 \text{ W}$$

So

$$P_s = 21{,}841.1 \text{ W} + 3{,}402.9 \text{ W}$$
$$P_s = 25{,}244 \text{ W}$$

The load vars and line vars are combined to obtain the source vars.

$$Q_s = Q_{\text{load}} + Q_{\text{lines}}$$
$$Q_{\text{lines}} = 3I_L^2 X_L$$
$$X_L = 2\pi f L = (2\pi)(60 \text{ Hz})(0.3 \times 10^{-3} \text{ H})$$
$$X_L = 0.11\ \Omega$$
$$Q_{\text{lines}} = (3)(61.49 \text{ A})^2(0.11\ \Omega)$$
$$Q_{\text{lines}} = 1247.74 \text{ vars}$$

so

$$Q_s = 3696.2 \text{ vars} + 1247.74 \text{ vars}$$
$$Q_s = 4943.94 \text{ vars}$$

The source volt-amperes is

$$S_T = \sqrt{(25{,}244)^2 + (4943.94)^2}$$
$$S_T = 25.72 \text{ kVA}$$

The concept of power factor correction also applies to three-phase systems. A lagging power factor can be improved by connecting capacitance across the lines. A leading power factor can be improved by connecting inductance.

An example that makes use of the Delta-Wye conversion is considered next.

EXAMPLE 22.13

Two three-phase loads are connected to a three-phase system, as shown in Figure 22.39. Determine the line current, load power, and % voltage regulation

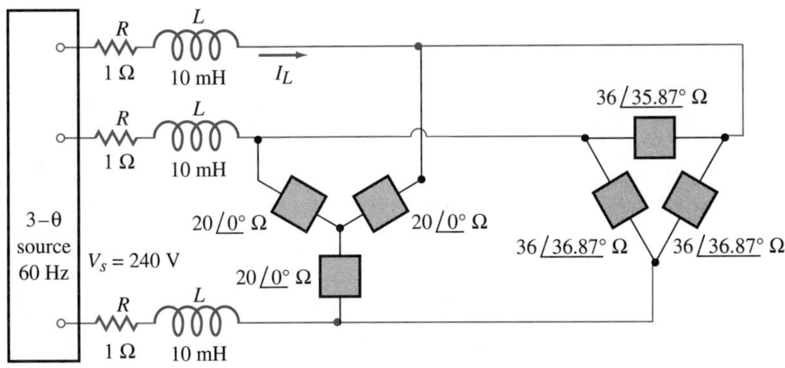

Figure 22.39

Solution

The Delta is balanced, so each impedance in the equivalent Wye is

$$\mathbf{Z}_{\text{Wye}} = (1/3)(\mathbf{Z}_{\text{Delta}})$$
$$\mathbf{Z}_{\text{Wye}} = \frac{36 \angle 36.87^\circ \ \Omega}{3}$$
$$\mathbf{Z}_{\text{Wye}} = 12 \angle 36.87^\circ \ \Omega$$

Since the loads are both balanced, the neutral point of each is at the same potential. For this condition, the respective legs of the two Wye's can be treated as being in parallel. The impedance in each leg of the equivalent wye is

$$Z_{Aeq} = (20 \angle 0^\circ \ \Omega)\|(12 \angle 36.87^\circ \ \Omega$$

or

$$Z_{Aeq} = \frac{(20 \angle 0° \ \Omega)(12 \angle 36.87° \ \Omega)}{20 \ \Omega + 12 \angle 36.87° \ \Omega}$$
$$Z_{Aeq} = 7.88 \angle 23.2° \ \Omega$$

Z_{Beq} and Z_{Ceq} also equal 7.88 ∠23.2° Ω since the loads are balanced.
The system, with the equivalent Wye connected, will look like Figure 22.40.

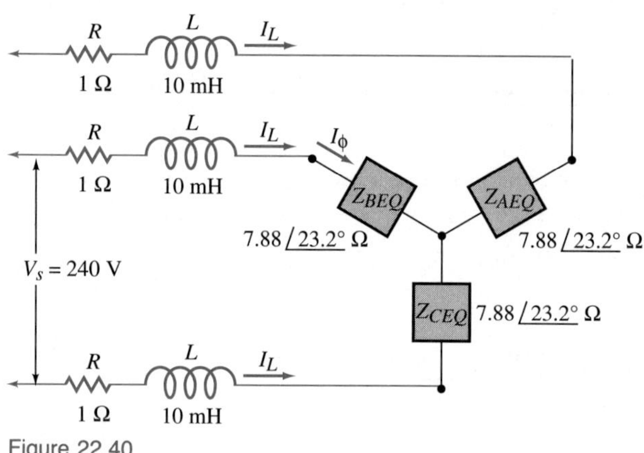

Figure 22.40

The impedance of each phase connected to the source is

$$Z'_\phi = R + jX_L + Z_\phi$$
$$X_L = 2\pi fL = (2\ \pi)(60 \text{ Hz})(10 \times 10^{-3} \text{ H})$$
$$= 3.77 \ \Omega$$

so

$$Z'_\phi = 1 \ \Omega + j3.77 \ \Omega + 7.88 \angle 23.2° \ \Omega$$
$$Z'_\phi = (8.24 + j6.87) \ \Omega \qquad \text{or} \qquad 10.73 \angle 39.82° \ \Omega$$
$$I_L = I\phi = V_{s\phi}/Z'_\phi$$

but

$$V_{s\phi} = V_s/\sqrt{3}$$

so

$$I_\phi = \frac{V_s}{\sqrt{3}\, Z'_\phi} = \frac{240 \text{ V}}{(\sqrt{3})(10.73 \ \Omega)} = 12.91 \text{ A}$$

so

$$I_L = 12.91 \text{ A}$$
$$P_{\text{load}} = 3P_\phi \text{ but } P_\phi = I_\phi^2 R_\phi$$

so

$$P_{\text{load}} = (3)(I_\phi)^2(Z_\phi \cos \theta)$$

$$P_{load} = (3)(12.91\ A)^2(7.88\ \Omega)\cos 23.2°$$
$$P_{load} = 3621.42\ W$$
$$\%VR = \left(\frac{V_{no\ load} - V_{load}}{V_{load}}\right) \times 100\%$$

If line voltages are used,

$$V_{load} = \sqrt{3}V_{\phi} = \sqrt{3}I_{\phi}Z_{\phi}$$
$$V_{load} = (\sqrt{3})(12.91\ A)(7.88\ \Omega)$$
$$V_{load} = 176.20\ V$$
$$\%\ VR = \left(\frac{240\ V - 176.2\ V}{176.2\ V}\right) \times 100\%$$
$$\%\ VR = 36\%$$

Such a large drop in voltage would not be acceptable. It might be remedied by installing a transformer or changing the lines.

Next, the condition where the source quantities are known is considered.

EXAMPLE 22.14

For the system in Figure 22.41, determine the power lost in the lines and the line voltage at the load, for a 7011 VA source.

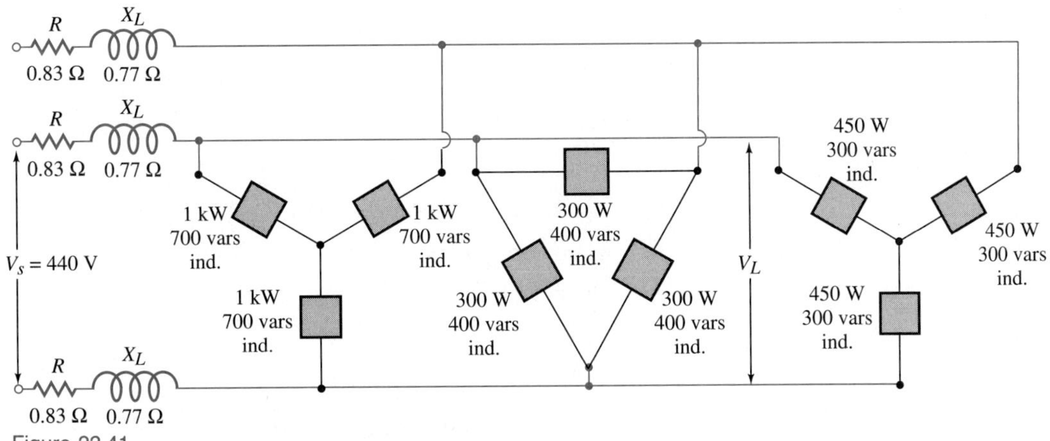

Figure 22.41

Solution

$$P_{lines} = (3)(I_L)^2(R)$$
$$S_{source} = \sqrt{3}V_sI_L$$

so

$$I_L = \frac{S_{source}}{\sqrt{3}V_s}$$

$$I_L = \frac{7011\text{ VA}}{(\sqrt{3})(440\text{ V})}$$
$$I_L = 9.2\text{ A}$$

so

$$P_{\text{lines}} = (3)(9.2\text{ A})^2(0.83\ \Omega)$$
$$P_{\text{lines}} = 210.75\text{ W}$$

For the load,

$$P = \sqrt{3}V_L I_L \cos\theta$$

so

$$V_L = \frac{P}{\sqrt{3}I_L \cos\theta}$$

The load is balanced, so

$$\theta = \tan^{-1}\left(\frac{Q}{P}\right)$$

For the total load,

$$\begin{aligned} P &= P_1 + P_2 + P_3 \\ &= 3000\text{ W} + 900\text{ W} + 1350\text{ W} \\ &= 5250\text{ W} \end{aligned}$$

and

$$\begin{aligned} Q &= Q_1 + Q_2 + Q_3 \\ &= 2100\text{ vars} + 1200\text{ vars} + 900\text{ vars} \\ &= 4200\text{ vars} \end{aligned}$$

so

$$\begin{aligned} \theta &= \tan^{-1}(4200\text{ vars}/5250\text{ vars}) \\ &= \tan^{-1}(0.8) \\ \theta &= 38.66^\circ \end{aligned}$$

so

$$V_L = \frac{5250\text{ W}}{\sqrt{3}(9.2\text{ A})(\cos 38.66^\circ)}$$
$$V_L = 421.92\text{ V}$$

The analysis of the three-phase systems can also be done using loop, branch, and superposition. However, one must be careful to include the correct angles with the quantities. This can be confusing and result in errors. If phase angles are not needed in the results, using watts, vars, volt-amperes, and magnitudes is much simpler.

Unbalanced loads were not considered in this section. In general, their analysis requires finding the impedance or watts and vars in each phase. The line impedance is then added, as in the balanced loads. Then the method of Example 22.10 must be used because the $\sqrt{3}$ relations do not hold.

22.8 Power Measurement in Three-Phase Circuits

The preceding sections introduced the three-phase system and the methods of analyzing it. Now, we examine the procedure for measuring power in an operating system or load. Although the most convenient method of measuring power is with a polyphase wattmeter, single-phase wattmeters can be used.

The first method that comes to mind is to use three single-phase wattmeters, one for each phase. The three wattmeters in Figure 22.42 are connected to measure the power of the load. The power equals the sum of the meter readings.

The three wattmeter connection is limited to loads that have both terminals of the phases available. These are mostly loads on a four-wire

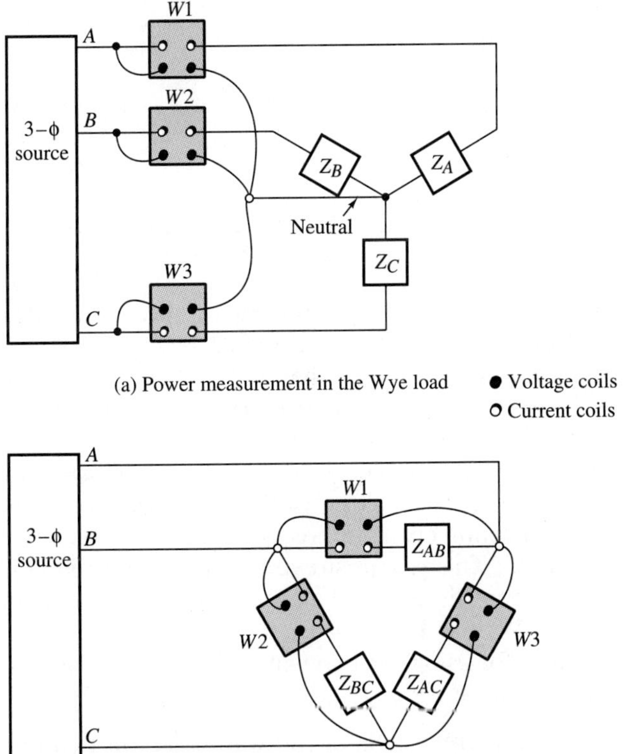

Figure 22.42 Measuring three-phase power with a wattmeter in each phase. (a) The meter connection for a Wye load. (b) The meter connection for a Delta load.

TABLE 22.1 Wattmeter Connections for 3–ϕ Power Measurements

Method	Meters Needed	Balanced Load	Unbalanced Load	Remarks
1. Three wattmeter method	3	Yes	Yes	Both phase terminals must be accessible.
2. One wattmeter method	1	Yes	No	Both phase terminals must be accessible.
3. Neutral floating	3	Yes	Yes, if there is a neutral wire	Both phase terminals do not need to be accessible.
4. Voltage divider	1	Yes	No	Two resistors of equal value are needed.
5. Simulated neutral	1	Yes	No	Three resistors of equal value are needed.
6. Two wattmeter method	2	Yes	Yes	Inconvenient for load with a changing power factor. Neutral terminal not needed.

Wye system. Other connections for making the measurement are possible and some of them are listed in Table 22.1.

two wattmeter method

Detailed descriptions of these other methods can be found in power handbooks and are not included here. However, the **two wattmeter method** is the most popular one, and is explained. Some reasons for its use is that it does not require access to both phase terminals and does not require any additional resistance networks. One disadvantage is that the meter pointer deflection changes direction when the load power factor passes through 0.5. This requires reversing the connection of the two leads. The method can be used for balanced and unbalanced loads. It can also be used for Wye or Delta loads.

The procedure for using the two wattmeter method for power measurement is as follows:

1. Connect the current coils of the wattmeters in two lines, as in Figure 22.43.
2. Connect the voltage coils of the wattmeters across the two lines, as in Figure 22.43. Reverse the connections if needed to get an upscale deflection on both meters.

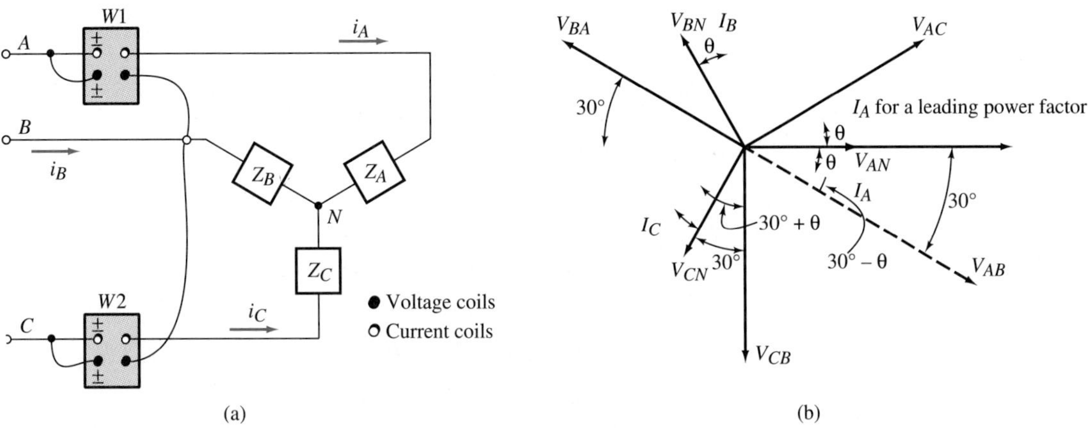

Figure 22.43 Measuring three-phase power with only two wattmeters. (a) The connection of the wattmeters in the system. (b) The phasor diagram for a lagging power factor load.

3. Disconnect the voltage coil lead of the lower reading wattmeter from the line that has no meter in it, and touch the lead to the line with the other wattmeter in it.
4. If the lower reading wattmeter pointer deflects up-scale in Step 3, the meter readings must be added. If it deflects down-scale, the lower reading must be subtracted from the higher one.

The power factor of the load can be determined from the meter readings by using the relation in Equation 22.28. It is

$$F_p = \cos\left[\arc \tan \sqrt{3}\left(\frac{P_1 - P_2}{P_1 + P_2}\right)\right] \tag{22.28}$$

where F_p is the power factor, and P_1 and P_2 are the readings of the wattmeters. If the pointer deflected down-scale in Step 3 of the procedure, a negative value must be entered.

The reactive power can also be obtained from the wattmeter readings by the following relation:

$$Q = \sqrt{3}\text{ (higher reading} \pm \text{lower reading)} \tag{22.29}$$

(+) if the deflection in Step 3 is down-scale

(−) if the deflection in Step 3 is up-scale

EXAMPLE 22.15

Power was measured by the two wattmeter method for three different loads. The readings are as follows, with the (−) sign indicating that the deflection was down-scale in Step 4 of the procedure.

a. Load *A* $P_2 = 333$ W, $P_1 = 667$ W
b. Load *B* $P_2 = 1294$ W, $P_1 = 294$ W (−)
c. Load *C* $P_2 = 294$ W (−), $P_1 = 1294$ W

Find the power factor, power, and vars for each load.

Solution

a.

$$\sqrt{3}\left(\frac{P_1 - P_2}{P_1 + P_2}\right) = \sqrt{3}\left(\frac{667 \text{ W} - 333 \text{ W}}{667 \text{ W} + 333 \text{ W}}\right) = 0.578$$

$$\arctan(0.578) = 30°$$

$$F_p = \cos 30°$$

$$F_p = 0.866$$

$$P = P_1 + P_2 = 667 \text{ W} + 333 \text{ W}$$

$$P = 1000 \text{ W}$$

$$Q = \sqrt{3}(P_1 - P_2) = \sqrt{3}(667 \text{ W} - 333 \text{ W})$$

$$Q = 578.5 \text{ vars}$$

b.

$$\sqrt{3}\left(\frac{P_1 - P_2}{P_1 + P_2}\right) = \sqrt{3}\left(\frac{-294 \text{ W} - 1294 \text{ W}}{-294 \text{ W} + 1294 \text{ W}}\right) = -2.75$$

$$\arctan(-2.75) = -70°$$

$$F_p = \cos(-70°) = 0.342$$

$$F_p = 0.342$$

$$P = P_2 - P_1 = 1294 \text{ W} - 294 \text{ W}$$

$$P = 1000 \text{ W}$$

$$Q = \sqrt{3}(P_2 + P_1) = \sqrt{3}(1294 \text{ W} + 294 \text{ W})$$

$$Q = 2750.50 \text{ vars}$$

c.

$$\sqrt{3}\left(\frac{P_1 - P_2}{P_1 + P_2}\right) = \sqrt{3}\left(\frac{1294 \text{ W} - (-294 \text{ W})}{1294 \text{ W} + (-294 \text{ W})}\right) = 2.75$$

$$\arctan(2.75) = 70°$$

$$F_p = \cos 70°$$

$$F_p = 0.342$$

$$P = P_1 - P_2 = 1294 \text{ W} - 294 \text{ W}$$

$$P = 1000 \text{ W}$$

$$Q = \sqrt{3}(P_1 + P_2) = \sqrt{3}(1294 \text{ W} + 294 \text{ W})$$

$$Q = 2750.50 \text{ vars}$$

S 22.9 SPICE Applications to Three-Phase Circuits

Three-phase systems can be simulated using PSpice. The first examples shows how a balanced three-phase system can be analyzed.

EXAMPLE 22.16

Use PSpice to analyze the balanced, three-phase system shown in Figure 22.44. Determine the magnitude and phase of each branch current.

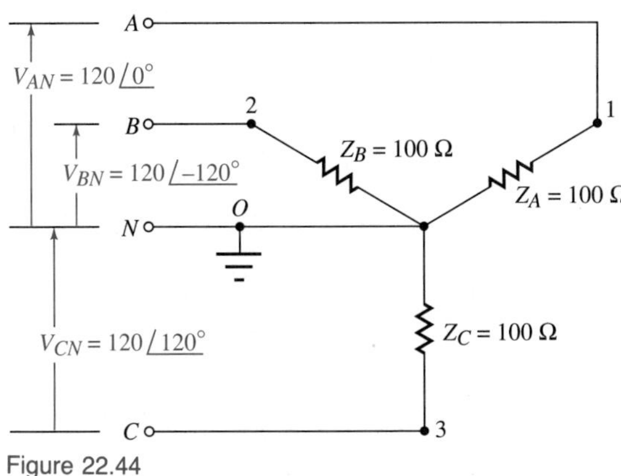

Figure 22.44

Solution

To create a PSpice file, the first line is the title line. A suitable title line is

```
THREE PHASE EXAMPLE 22.16
```

The three voltages can then be described:

```
VAN   1   0   AC   120V       0DEG
VBN   2   0   AC   120V   -120DEG
VCN   3   0   AC   120V    120DEG
```

The three branch resistances are then specified:

```
RA   1   0   100
RB   2   0   100
RC   3   0   100
```

The .AC command line specifies the single-frequency for the analysis:

```
.AC   LIN   1   60HZ   60HZ
```

Note: Only one data point is requested.

The .PRINT statement ask for the magnitude and phase angle of the three branch currents:

```
.PRINT   AC   IM(RA)   IP(RA)
.PRINT   AC   IM(RB)   IP(RB)
.PRINT   AC   IM(RC)   IP(RC)
```

And finally, the .END command is used to close the file. The entire file is shown in Figure 22.45(a).

After running the file through the PSpice simulator and examining the OUTPUT file, the results shown in Figure 22.45(b) are obtained. The

```
THREE PHASE EXAMPLE 22.16
**** CIRCUIT DESCRIPTION
**********************************************

VAN     1    0    AC    120V      0DEG
VBN     2    0    AC    120V   -120DEG
VCN     3    0    AC    120V    120DEG
RA    1    0    100
RB    2    0    100
RC    3    0    100
.AC   LIN    1    60HZ    60HZ
.PRINT AC IM(RA)  IP(RA)
.PRINT AC IM(RB)  IP(RB)
.PRINT AC IM(RC)  IP(RC)
.OPTIONS NOPAGE
.END
```

(a)

```
**** AC ANALYSIS
 FREQ           IM(RA)          IP(RA)
  6.000E+01      1.200E+00       0.000E+00

**** AC ANALYSIS
 FREQ           IM(RB)          IP(RB)
  0.000E+01      1.200E+00      -1.200E+02

**** AC ANALYSIS
 FREQ           IM(RC)          IP(RC)
  6.000E+01      1.200E+00       1.200E+02
```

(b)

Figure 22.45 PSpice output. (a) Circuit description. (b) ac analysis results.

magnitude of the current through resistor RA is 1.2 amperes and the phase angle is 0 degrees. The magnitude of the current through resistor RB is 1.2 amperes and the phase angle is − 120 degrees. And finally, the current through RC is 1.2 amperes with a phase angle of + 120 degrees.

The next example shows the analysis of an unbalanced, three-phase system. One caution that needs to be exercised in simulating a three-phase system is to avoid voltage loops that contain no impedance. This can occur when specifying the three-source voltage sources. The solution is to include a small resistance in series with one of the voltage sources.

EXAMPLE 22.17

Analyze the unbalanced, three-phase system shown in Figure 22.46. Determine the magnitude and phase angle of each branch current.

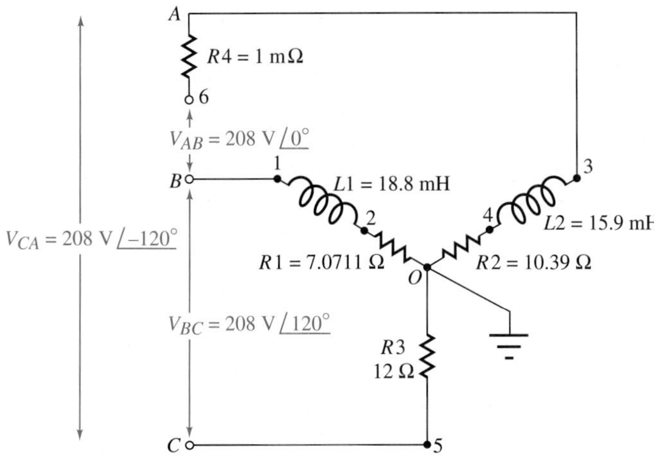

Figure 22.46

Solution

The source file is shown in Figure 22.47(a). Note that a resistor, R4, is inserted in series with source VAB, and is given a small value of 0.001 ohm.

```
UNBALANCED WYE LOAD
**** CIRCUIT DESCRIPTION
****************************

VAB   6   1   AC   208V      0DEG
VBC   1   5   AC   208V    120DEG
VCA   5   3   AC   208V   -120DEG
R1    2   0   7.0711
R2    4   0   10.39
R3    5   0   12
R4    6   3   1M
L1    1   2   18.8MH
L2    3   4   15.9MH
.AC   LIN   1   60HZ   60HZ
.PRINT AC   IM(R1)   IP(R1)
.PRINT AC   IM(R2)   IP(R2)
.PRINT AC   IM(R3)   IP(R3)
.OPTIONS  NOPAGE
.END
```

(a)

```
**** AC ANALYSIS
 FREQ          IM(R1)         IP(R1)
  6.000E+01     1.325E+01      1.164E+02

**** AC ANALYSIS
 FREQ          IM(R2)         IP(R2)
  6.000E+01     7.708E+00     -2.809E+00

***** AC ANALYSIS
 FREQ          IM(R3)         IP(R3)
  6.000E+01     1.163E+01     -9.896E+01
```

(b)

Figure 22.47 PSpice output file for the unbalanced Wye load.

The three currents are given in their magnitude–phase angle form. The results of the simulation are

Current	Magnitude	Phase Angle
I(R1)	13.25 A	116.4 degrees
I(R2)	7.71 A	−2.8 degrees
I(R3)	11.63 A	−99.0 degrees

The results of the simulation match the results that were obtained in Example 22.10. Note that the simulation required component values for the inductors. Hence, the impedances given in Example 22.10 were converted to equivalent inductance values at the frequency the analysis was performed.

And finally, if graphs of the current waveforms are desired for the unbalanced, three-phase system of Example 22.18, the source file can be modified as shown in Example 22.18.

EXAMPLE 22.18

Graph the current waveforms for the unbalanced, three-phase system shown in Figure 22.46.

Solution

The source file given in Figure 22.47 is modified in the following manner:

The source voltage statements must use the SIN function as follows:

```
V<name> <(+)node> <(−)node> SIN(Voff Vampl Freq Delay
                  Damping-Factor Phase)
```

where `Voff` is the Offset voltage, in volts; `Vampl` is the peak amplitude of voltage, in volts; `Freq` is the frequency, in hertz; `Delay` is the delay, in seconds; `Damping-Factor` is the damping factor, in seconds^{-1}; and `Phase` is the phase angle, in degrees.

For example,

```
VAB   6   1   SIN(0   208V   60HZ   0   0   0DEG)
```

where offset voltage = 0 volts, peak amplitude = 208 volts, frequency = 60 Hz, delay = 0 seconds, damping factor = 0 seconds^{-1}, and phase angle = 0 degrees.

To obtain a graphical output, the .TRAN and .PROBE commands are used. Since a 60-Hz signal has a period of approximately 17 milliseconds, the .TRN command specifies a time duration of 40 milliseconds and a step size of 50 microseconds. The file is shown in Figure 22.48(a).

The graphical output is shown in Figure 22.48(b).

```
UNBALANCED WYE LOAD
**** CIRCUIT DESCRIPTION
*************************************************************

VAB  6  1  SIN( 0  208V  60HZ  0  0      0DEG )
VBC  1  5  SIN( 0  208V  60HZ  0  0    120DEG )
VCA  5  3  SIN( 0  208V  60HZ  0  0   -120DEG )
R1   2  0  7.0711OHM
R2   4  0  10.39OHM
R3   5  0  12OHM
R4   6  3  1MOHM
L1   1  2  18.8MH
L2   3  4  15.9MH
.TRAN    50US    40MS
.PROBE
.OPTIONS NOPAGE
.END

**** INITIAL TRANSIENT SOLUTION
 NODE VOLTAGE        NODE VOLTAGE
(   1)    46.7640 (    2) 46.7640
(   3) 46.7640 (    4) 46.7640
(   5) -133.3700 (    6) 46.7640

  VOLTAGE SOURCE CURRENTS
  NAME         CURRENT

  VAB           5.030E-12
  VBC          -6.613E+00
  VCA           4.501E+00

  TOTAL POWER DISSIPATION 2.00E+03 WATTS
```

(a)

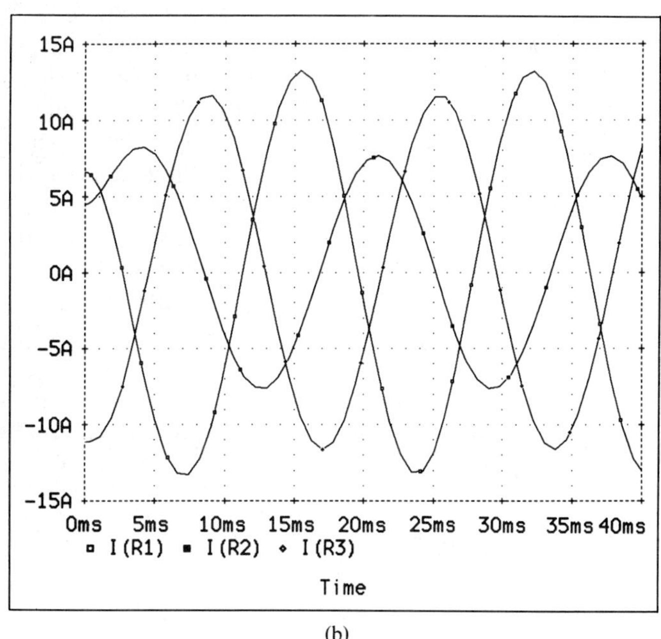

(b)

Figure 22.48 PSpice output. (a) Circuit description and initial transient solution. (b) Graph of the current waveforms produced by PROBE.

SUMMARY

1. A three-phase alternator has three sets of windings. These induce three voltages that are displaced by 120 degrees.
2. The coils of the alternator can be connected in Wye or Delta. The Wye connection has one end of each coil connected together and the

three other ends connected to the load. The Delta connection has the three coils connected in series and the two free ends connected together. The load is connected to the junction between each pair of coils.

3. In the Wye connection, the line voltage is $\sqrt{3} \times$ (the phase voltage), and the line current is the phase current.
4. In the Delta connection, the line current is $\sqrt{3} \times$ (the phase current), and the line voltage is the phase voltage.
5. The three voltages build up in a sequence, one leading the other. This sequence is known as the phase sequence.
6. Three-phase loads are also connected in Wye and Delta. The configuration is the same as for the source connection.
7. A balanced three-phase load has like impedances in each phase.
8. For a balanced Wye load and a balanced Delta load, the line and phase relations are the same as for the source in Items 3 and 4.
9. A four-wire Wye load has a neutral line connecting the junction of the Wye load to that of the Wye source. It keeps the load phase voltages at the same value as the source phase voltages.
10. The impedance of the lines can result in voltage drops and power loss. One procedure for including the lines in the analysis is to reduce the loads to a single Wye and add the line impedance to the phase impedance of the Wye.
11. Power in a three-phase load can be measured with a polyphase wattmeter. Two single-phase wattmeters can be used in the two wattmeter method.

EQUATIONS

22.1	$E_L = \sqrt{3}E_\phi$	The relation between the line and phase voltage for a three-phase Wye source.	922
22.2	$I_L = I_\phi$	The relation between the line and phase currents for a three-phase Wye source.	922
22.3	$\mathbf{I}_A + \mathbf{I}_B + \mathbf{I}_C - \mathbf{I}_N = 0$	Kirchhoff's current equation for the currents of a three- phase Wye source.	922
22.4	$\mathbf{E}_L = \mathbf{E}_\phi$	The relation between the line and phase voltages for a three-phase Delta source.	923
22.5	$I_{\text{line}} = \sqrt{3}I_\phi$	The relation between the line and phase currents for a three-phase Delta source.	924

	Equation	Description	Page
22.6	$I_A + I_B + I_C = 0$	Kirchhoff's current law for the line currents in a Delta source.	925
22.7	$P_T = V_{\phi A} I_{\phi A} \cos(\theta_{\phi A}) + V_{\phi B} I_{\phi B} \cos(\theta_{\phi B}) + V_{\phi C} I_{\phi C} \cos(\theta_{\phi C})$	The general equations for power in a three-phase Wye load.	932
22.8	$P_T = 3V_\phi I_\phi \cos \theta_\phi$	The expression for power in a balanced three-phase Wye load.	932
22.9	$P_T = \sqrt{3} V_L I_L \cos \theta_\phi$	The power in a three-phase balanced Wye load in terms of line voltage and current.	932
22.10	$Q_T = \sqrt{3} V_L I_L \sin \theta_\phi$	The reactive power in a balanced three-phase Wye load in terms of line voltage and current.	932
22.11	$S_T = \sqrt{P_T^2 + Q_T^2}$	The relation of apparent power, power, and reactive power in a three-phase load.	932
22.12	$I_N = 0$	The neutral current in a balanced three-phase, four-wire load.	932
22.13	$I_A + I_B + I_C = 0$	Kirchhoff's current equation for the line currents in a Delta load.	936
22.14	$P_T = V_{AB} I_{AB} \cos \theta_{AB} + V_{BC} I_{BC} \cos \theta_{BC} + V_{CA} I_{CA} \cos \theta_{CA}$	The general expression for power in a three-phase balanced Delta load.	936
22.15	$P_T = 3V_L I_\phi \cos \theta_\phi$	The expression for power in a three-phase balanced Delta load.	936
22.16	$P_T = \sqrt{3} V_L I_L \cos \theta_\phi$	The power in a three-phase balanced Delta load in terms of line voltage and current.	936
22.17	$V_{OC}\left(\frac{1}{Z_A} + \frac{1}{Z_B} + \frac{1}{Z_C}\right) = \frac{V_{BC}}{Z_B} + \frac{V_{AC}}{Z_A}$	The expression for V_{OC} in an unbalanced three-wire, three-phase load.	941
22.18	$V_{OB}\left(\frac{1}{Z_A} + \frac{1}{Z_B} + \frac{1}{Z_C}\right) = \frac{V_{CB}}{Z_C} + \frac{V_{AB}}{Z_A}$	The expression for V_{OB} in an unbalanced three-wire, three-phase load.	942
22.19	$V_{OA}\left(\frac{1}{Z_A} + \frac{1}{Z_B} + \frac{1}{Z_C}\right) = \frac{V_{CA}}{Z_C} + \frac{V_{BA}}{Z_B}$	The expression for V_{OA} in an unbalanced three-wire, three-phase load.	942
22.20	$Z_A = \frac{Z_{AB} Z_{AC}}{Z_{AB} + Z_{AC} + Z_{BC}}$	The equation to determine one leg in the equivalent Wye from the Delta impedances.	943

22.21	$Z_B = \dfrac{Z_{AB}Z_{BC}}{Z_{AB} + Z_{AC} + Z_{BC}}$	The equation to determine the second leg in the equivalent Wye from the Delta impedances.	943
22.22	$Z_C = \dfrac{Z_{BC}Z_{AC}}{Z_{AB} + Z_{AC} + Z_{BC}}$	The equation to determine the third leg in the equivalent Wye from the Delta impedances.	943
22.23	$Z_A = Z_B = Z_C = Z_{AB}/3$	The equation for the impedances in the equivalent Wye when the Delta impedances are the same.	943
22.24	$Z_{AB} = \dfrac{Z_AZ_B + Z_BZ_C + Z_AZ_C}{Z_C}$	The equation to determine one side of the equivalent Delta from the Wye impedances.	943
22.25	$Z_{BC} = \dfrac{Z_AZ_B + Z_BZ_C + Z_AZ_C}{Z_A}$	The equation to determine the second side of the equivalent Delta from the Wye impedances.	943
22.26	$Z_{AC} = \dfrac{Z_AZ_B + Z_BZ_C + Z_AZ_C}{Z_B}$	The equation to determine the third side of the equivalent Delta from the Wye impedances.	944
22.27	$Z_{AB} = Z_{AC} = Z_{BC} = 3Z_A$	The equation for the impedances in the equivalent Delta when the Wye impedances are the same.	944
22.28	$F_p = \cos\left[\text{arc tan}\,\sqrt{3}\left(\dfrac{P_1 - P_2}{P_1 + P_2}\right)\right]$	The equation for obtaining the power factor from the two wattmeter method.	955
22.29	$Q = \sqrt{3}(\text{higher reading} \pm \text{lower reading})$	The equation for obtaining vars by the two wattmeter method.	955

QUESTIONS

1. Which three-phase source connection should be used to give a line voltage greater than the alternator coil voltage?
2. Which three-phase source connection should be used to give a line current greater than the alternator coil current capacity?
3. What is the meaning of phase sequence in terms of the leading or lagging relation of the voltages?
4. What effect will reversing two leads of a three-phase motor have on the direction of rotation?
5. How can one tell if the proper polarity exists when the open ends of the Delta are closed?
6. One source coil is connected with reverse polarity in the Wye connection. If each coil voltage is 120 volts, what is the voltage across each pair of terminals?

7. Connecting two three-phase alternator outputs in parallel results in much arcing. What are some possibilities that could have caused the arcing?
8. Why isn't a neutral line needed if the load is a balanced load?
9. Why is a neutral line needed if the load is an unbalanced one?
10. A three-phase load is connected as a four-wire Wye load. If allowed only one measurement, what would you measure to determine if the load is balanced?
11. What are some load characteristics that can be affected by the line impedance?
12. A three-phase Wye resistive load is connected to a source by a line that has an impedance of $2 + j5$ ohms. Explain how putting some capacitance in the load can increase the voltage across the resistance.
13. In the two wattmeter method, if the meters were placed in the two lower lines of Figure 22.39, where would the voltage leads be connected?

PROBLEMS

SECTION 22.1 Three-Phase Source

1. The voltages for two windings are $\mathbf{E}_{Aa} = 120 \angle 0°$ V, and $\mathbf{E}_{Cc} = 120 \angle -120°$ V. Find $\mathbf{E}_{ca}$ for Figure 22.49(a) and $\mathbf{E}_{Ac}$ for Figure 22.49(b).

2. Draw the phasors for $\mathbf{E}_{ca}$ and $\mathbf{E}_{Ac}$ in Problem 1.

3. The voltages for two windings are $\mathbf{E}_{Bb} = 120 \angle 120°$ V and $\mathbf{E}_{Cc} = 120 \angle -120°$ V. Find $\mathbf{E}_{BC}$ for Figure 22.50(a) and $\mathbf{E}_{Cb}$ for Figure 22.50(b).

4. Draw the phasors for $\mathbf{E}_{BC}$ and $\mathbf{E}_{Cb}$ in Problem 3.

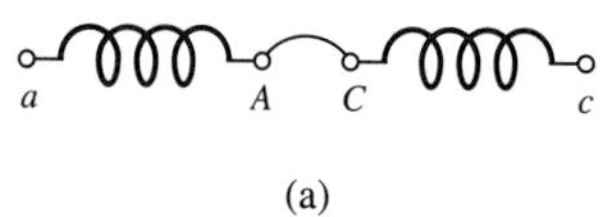

(a)

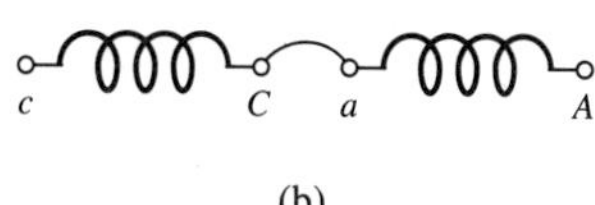

(b)

Figure 22.49

SECTION 22.2 Wye and Delta Source Connections

5. A three-phase Wye source has a line voltage of 440 volts magnitude.

   (a) What is the magnitude of the phase voltages?

   (b) What is the magnitude of the phase current if the line current is 10 A?

6. Work Problem 5 for a line voltage of 600 volts and a line current of 5 A.

7. The phase voltages in a three-phase Wye source are $\mathbf{E}_{Aa} = 208 \angle 0°$ V, $\mathbf{E}_{Bb} = 208 \angle 120°$ V, and $\mathbf{E}_{Cc} = 208 \angle -120°$ V. Find the line voltages and draw the phasor diagram.

8. Work Problem 7 for $\mathbf{E}_{Aa} = 440 \angle -120$ V, $\mathbf{E}_{Cc} = 440 \angle 0°$ V, and $\mathbf{E}_{Bb} = 440 \angle 120°$ V.

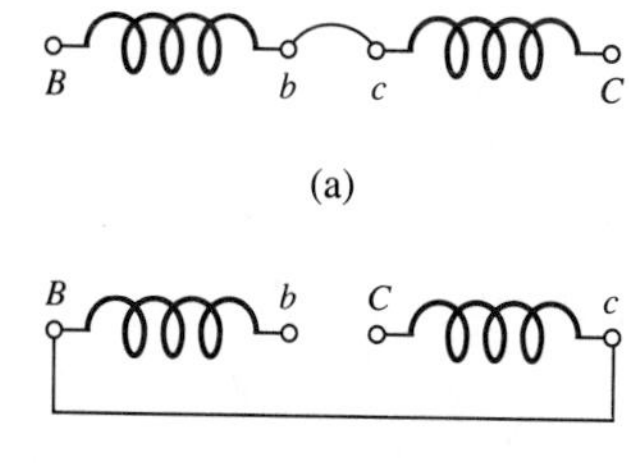

(b)

Figure 22.50

9. A three-phase Delta source has a line voltage of 220 V and a line current of 6 A.

   (a) What is the magnitude of the phase voltage?

   (b) What is the magnitude of the phase current, if the source supplies a balanced load?

10. A Delta source supplies a line current of 8 A to a balanced load.

    (a) What is the magnitude of the phase voltage if the line voltage is 120 volts for a resistance load?

    (b) Draw the phasor diagram if $\mathbf{E}_{AB} = 120 \angle 0°$ V, $\mathbf{E}_{BC} = 120 \angle 120°$ V, and $\mathbf{E}_{CA} = 120 \angle -120°$ V.

11. Work Problem 10 for a line current of 12 A and $\boldsymbol{E}_{AB} = 120\angle 0°$ V, $\boldsymbol{E}_{BC} = 120\angle -120°$ V, and $\boldsymbol{E}_{CA} = 120\angle 120°$ V.

12. The phase currents in a Delta source are $\boldsymbol{I}_{AB} = 3\angle 0°$ A, $\boldsymbol{I}_{CA} = 3\angle 120°$ A, and $\boldsymbol{I}_{BC} = 3\angle -120°$ A. Find the line currents: $\boldsymbol{I}_A$, $\boldsymbol{I}_B$, and $\boldsymbol{I}_C$.

SECTION 22.3 Phase Sequence

13. What is the phase sequence for each of the phasor sets shown in Figure 22.51?

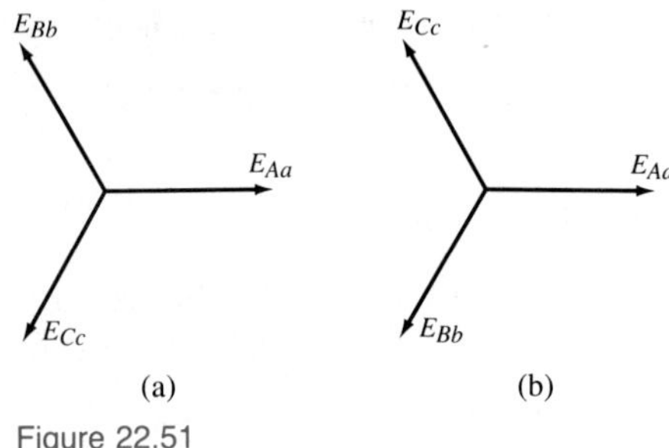

Figure 22.51

14. Work Problem 13 for the sets shown in Figure 22.52.

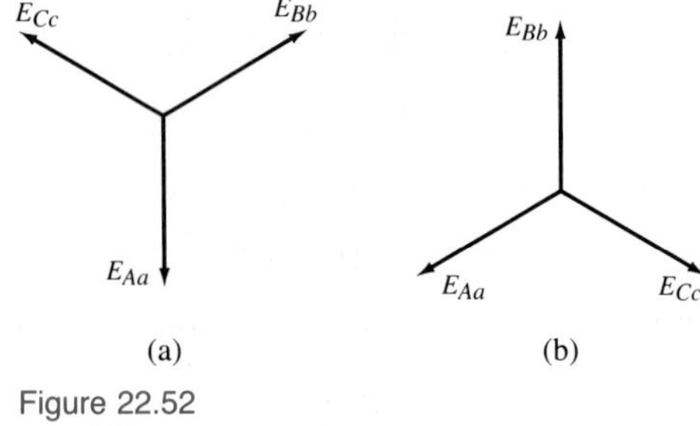

Figure 22.52

15. What is the phase sequence for the waves shown in Figure 22.53?

16. What is the sequence for the following set of equations?

(a) $v_A = 5 \sin(\omega t + 30°)$.
(b) $v_B = 5 \sin(\omega t + 150°)$.
(c) $v_C = 5 \sin(\omega t - 90°)$.

17. Work Problem 16 for the following sets:

(a) $v_A = 120 \sin(\omega t + 150°)$.
(b) $v_B = 120 \sin(\omega t + 30°)$.
(c) $v_C = 120 \sin(\omega t - 90°)$.

18. Work Example 22.5 for an ACB sequence to show that Lamp C will be brighter.

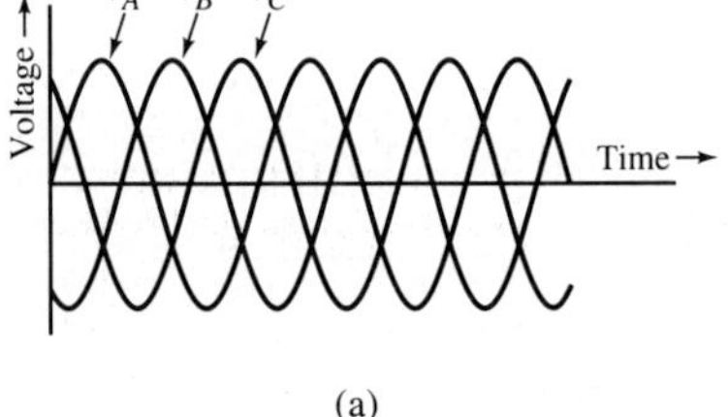

(a)

v_B v_A v_C
Voltage
Time
(b)

Figure 22.53

SECTION 22.4 The Four-Wire, Three-Phase Load

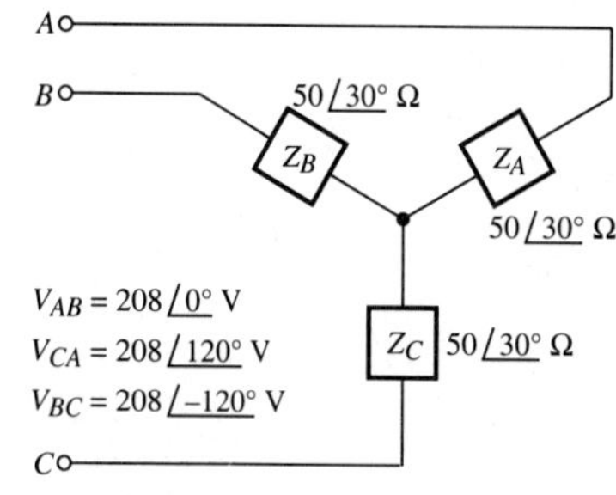

Figure 22.54

19. For the circuit in Figure 22.54, find:

    (a) The magnitude of the current in each line.
    (b) The magnitude of the voltage across each impedance.
    (c) The power dissipated in the three-phase load.
    (d) Sketch the phasor diagram.

20. Work Problem 19 for $\mathbf{Z}_A = 50\ \angle 30°\ \Omega$, $\mathbf{Z}_B = 20\ \angle -45°\ \Omega$, and $\mathbf{Z}_C = 100\ \angle 60°\ \Omega$ when there is a neutral line. Omit the phasor diagram.

21. Work Problem 19 for $\mathbf{Z}_A = 50\ \angle 30°\ \Omega$, $\mathbf{Z}_B = 50\ \angle 0°\ \Omega$, and $\mathbf{Z}_C = 50\ \angle -30°\ \Omega$ when there is a neutral line. Omit the phasor diagram.

22. Work Problem 21 for an ACB phase sequence where $\mathbf{V}_{AB} = 208\ \angle 0°$ V, $\mathbf{V}_{CA} = 208\ \angle -120°$ V, and $\mathbf{V}_{BC} = 208\ \angle 120°$ V. Omit the phasor diagram. What was the effect of changing the phase sequence?

23. The magnitude of each impedance and the power dissipated in each one are listed in Figure 22.55. Find the line current's magnitude and angle.

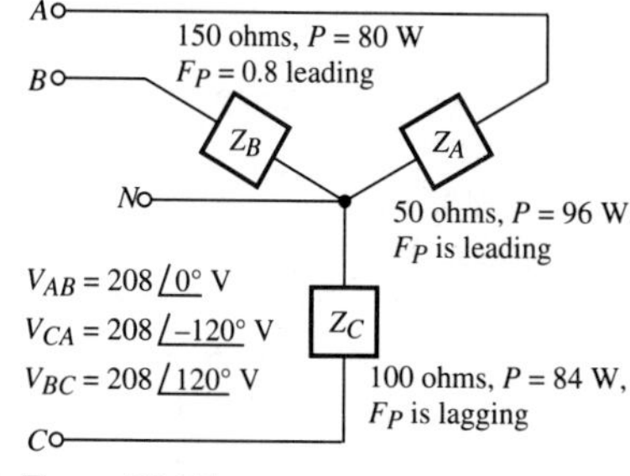

Figure 22.55

SECTION 22.5 Three-Wire, Three-Phase Balanced Loads

24. A three-phase source supplies 5 kW of power to a balanced Wye load whose power factor is 0.6 leading. Find:

    (a) The magnitude of the line current if the line voltage is 220 V.
    (b) The power in each phase.
    (c) The current in each phase.

25. A 240-V three-phase system supplies current to a balanced three-wire Wye load. If each impedance is $50\ \angle 45°\ \Omega$, find:

    (a) The magnitude of the voltage across each impedance.
    (b) The line current.
    (c) The power supplied by the source.
    (d) The kVA rating that the source must have.

26. The phase voltage in a balanced Wye load is 208 volts. Each impedance in the Wye has a magnitude of 200 ohms. Find:

    (a) The magnitude of the line voltage.
    (b) The magnitude of the line current.
    (c) The power factor of the load if the total power is 500 W.

27. The line voltages in a three-phase Delta balanced load are $\mathbf{V}_{AB} = 220\ \angle 0°$ V, $\mathbf{V}_{CA} = 220\ \angle 120°$ V, and $\mathbf{V}_{BC} = 220\ \angle -120°$ V. Each impedance in the Delta is $40\ \angle 60°$ ohms. Find:

    (a) The phase currents, magnitude, and angle.
    (b) The line currents, magnitude, and phase.
    (c) The power dissipated in the load.

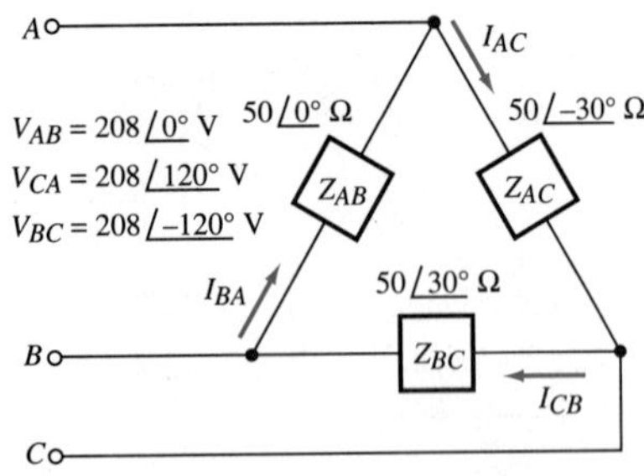

Figure 22.56

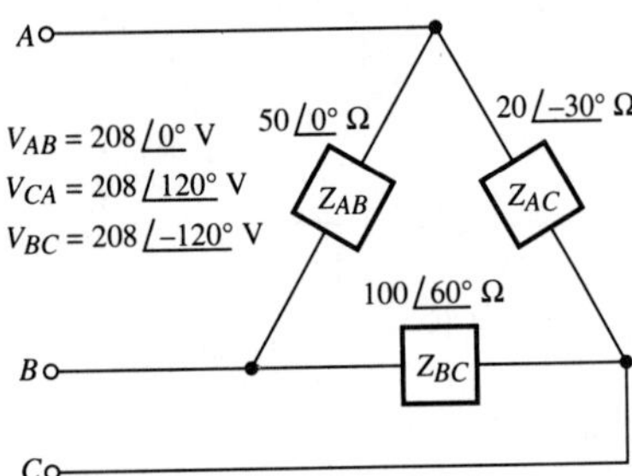

Figure 22.57

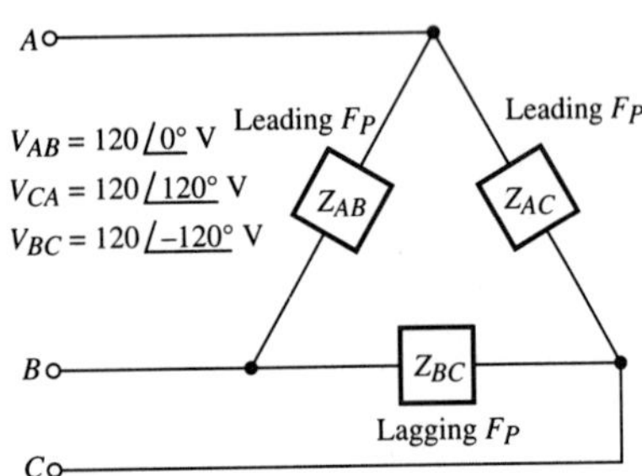

Figure 22.58

28. A three-phase Delta load has the following characteristics. $V_L = 440$ V, $\cos\theta = 0.8$ lagging, $P_T = 12$ kW. Find:

    (a) The magnitude of the line current.

    (b) The magnitude of the phase currents.

    (c) The impedance in each phase.

29. What is:

    (a) The line current and power in a 2-kVA, three-phase motor if its power factor is 0.8 lagging when connected to a 220-V, three-phase source?

    (b) The magnitude of the phase current?

SECTION 22.6 Unbalanced Three-Phase Delta and Wye Loads

30. For the circuit in Figure 22.56, find:

    (a) The phase currents, magnitude, and angle.

    (b) The line currents, magnitude, and angle.

    (c) The power dissipated in the circuit.

    (d) Sketch the phasor diagram.

31. Work Problem 30 for the circuit in Figure 22.57. Omit the phasor diagram.

32. In the circuit of Figure 22.58, $P_{AB} = 125$ W, $I_{BA} = 1.2$ A; $P_{AC} = 72$ W, $I_{AC} = 0.8$ A; and $B_{BC} = 225$ W, $I_{CB} = 4$ A. Find:

    (a) The impedances $\mathbf{Z}_{BC}$, $\mathbf{Z}_{Ac}$, and $\mathbf{Z}_{AC}$.

    (b) The line currents.

33. For the circuit in Figure 22.59, find:

    (a) The current in each line.

    (b) The voltage across each impedance.

    (c) The power dissipated in the load.

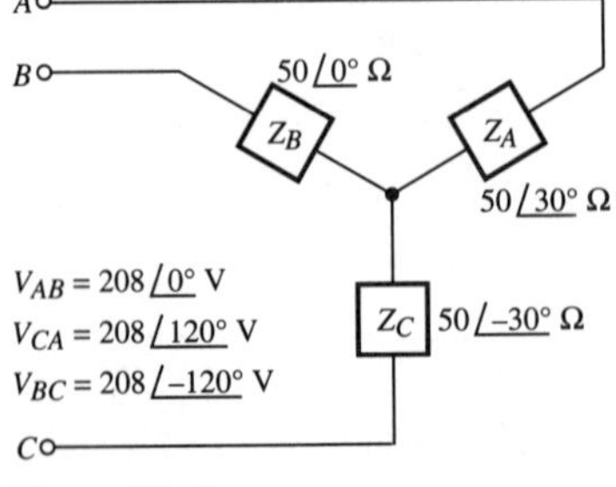

Figure 22.59

34. Work Problem 33 for an *ACB* phase sequence where $\mathbf{V}_{AB} = 208\angle 0°$ V, $\mathbf{V}_{CA} = 208\angle -120°$ V, $\mathbf{V}_{BC} = 208\angle 120°$ V. What was the effect of changing the phase sequence?

35. In a circuit similar to that of Problem 33, the current is $2.3\angle -16°$ A in Z_A, and $2.25\ \angle 208°$ A in Z_B. $Z_A = 30\ \angle 60°$ ohms. Find:

(a) The impedance Z_B and Z_C.

(b) The voltage across each impedance.

(c) The power dissipated in the circuit.

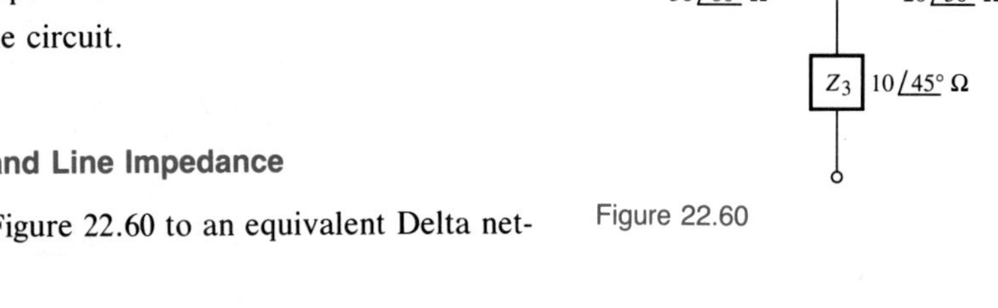

Figure 22.60

SECTION 22.7 Three-Phase Loads and Line Impedance

36. Convert the Wye network in Figure 22.60 to an equivalent Delta network.

37. Convert the Delta network in Figure 22.61 to an equivalent Wye network.

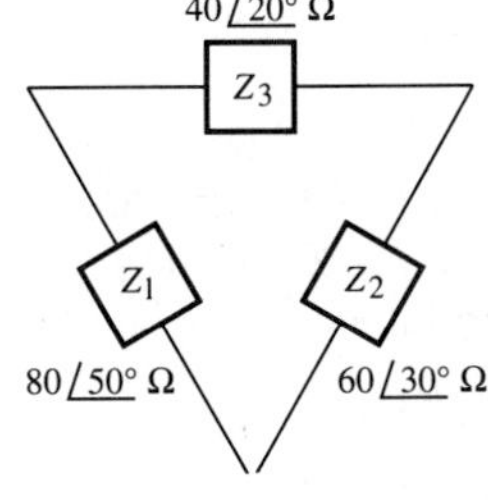

Figure 22.61

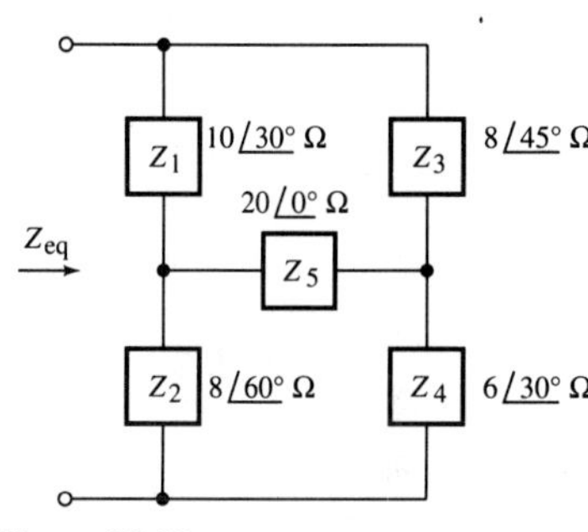

Figure 22.62

38. Change the network of Figure 22.62 to a series-parallel network and calculate its equivalent impedance.

39. A 220-volt, three-phase, 60-Hz transformer supplies the two loads in Figure 22.63. Find:

(a) The magnitude of the line currents.

(b) The power supplied by the source.

(c) The kVA rating of the source.

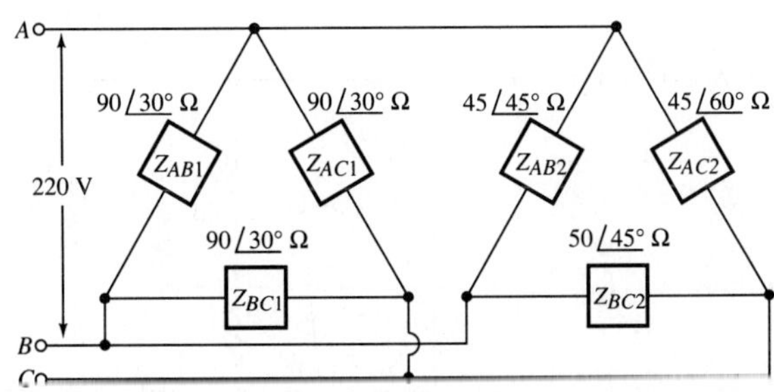

Figure 22.63

40. Determine the line voltages at the load and the power lost in the lines for the circuit of Problem 39 if each line has an impedance of $0.15 + j0.18$ ohm.

41. The source of Problem 39 is connected to the load in Figure 22.64. Find the magnitude of the line currents, the power supplied by the source, and the kVA of the source.

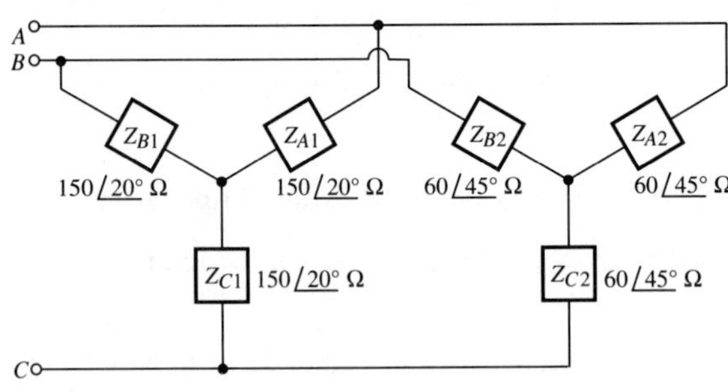

Figure 22.64

42. A three-phase, four-wire 208/120-V source supplies the following balanced loads:
one 10 kW, 208 V, three-phase motor at $F_p = 0.8$;
three single-phase 120-V lighting loads of 5 kW, each at $F_p = 1$;
three single-phase 2 A, 208-V motors at $F_p = 0.8$ lagging.
If each line has an impedance of $0.11 + j0.05$ ohm. Find:

(a) The magnitude of the line current.
(b) The total power supplied to the loads.
(c) The line voltage at the load.

43. Find:

(a) The line currents.
(b) The line voltages at the load.
(c) The power factor.
(d) The power lost in the lines for the circuit of Figure 22.65. Each line has a resistance of 0.07 ohm and an inductive reactance at 0.07 ohm.

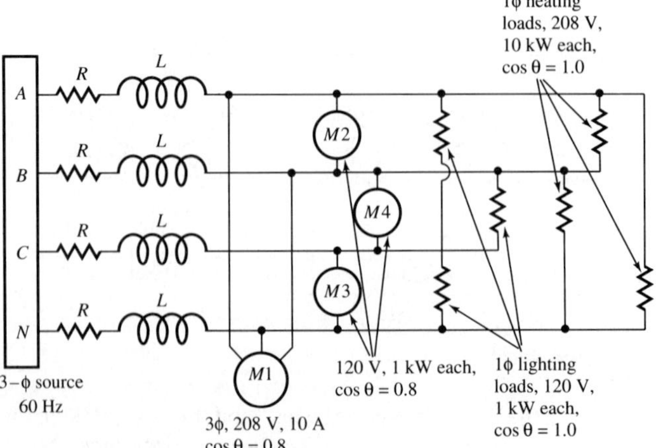

Figure 22.65

44. Two unbalanced loads are supplied by a three-phase, three-wire system, as in Figure 22.66. Find:

(a) The line currents.

(b) The power supplied by the source.

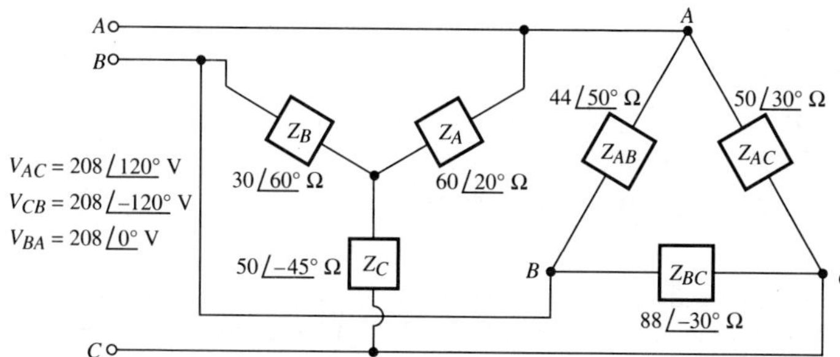

Figure 22.66

45. Work Problem 44 for a line impedance of $0.8 + j0.6$ ohm to find:

(a) The line voltages at the load.

(b) The power lost in the lines.

46. A three-phase distribution system supplies two unbalanced loads, as in Figure 22.67. Find:

(a) The line currents.

(b) The magnitude of the phase voltage at the load.

(c) The total power supplied by the source.

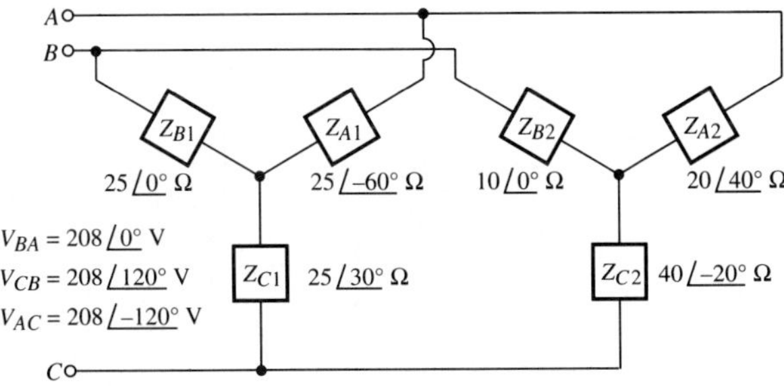

Figure 22.67

47. Work Problem 46 for a line impedance of $0.8 + j0.8$ ohm to find:

(a) The magnitude of the line currents.

(b) The magnitude of the line voltages at the load.

SECTION 22.8 Power Measurement in Three-Phase Circuits

48. The two wattmeter method is used to measure power in a three-phase balanced load. What will the meter readings be for each of the following loads?

(a) $V = 220$ V, $I = 5$ A, power factor = 0.6 lagging

(b) $V = 220$ V, $I = 5$ A, power factor = 0.3 lagging

49. Work Problem 48 for the following loads:

(a) $V = 220$ V, $I = 5$ A, power factor $= 0.9$ leading
(b) $V = 220$ V, $I = 5$ A, power factor $= 0.2$ leading

50. Calculate the power factor, power, and reactive power for each set of wattmeter readings:

(a) $P_1 = 5000$ W, $P_2 = -500$ W
(b) $P_1 = 800$ W, $P_2 = 600$ W
(c) $P_2 = 400$ W, $P_1 = -50$ W

51. The two wattmeter method is used to measure power in a balanced three-phase load. One meter indicates 750 W and the total power is 600 W. Find:

(a) The reading on the other meter.
(b) The power factor of the load.

52. A three-phase source supplies 15 kW of power to a balanced three-phase load whose power factor is 0.8 lagging. The line voltage is 208 V. If the two wattmeter method is used, what is the reading on each meter?

SECTION 22.9 SPICE Applications to Three-Phase Circuits

Ⓢ 53. Use PSpice to determine the branch currents in the three-phase system shown in Figure 22.68.

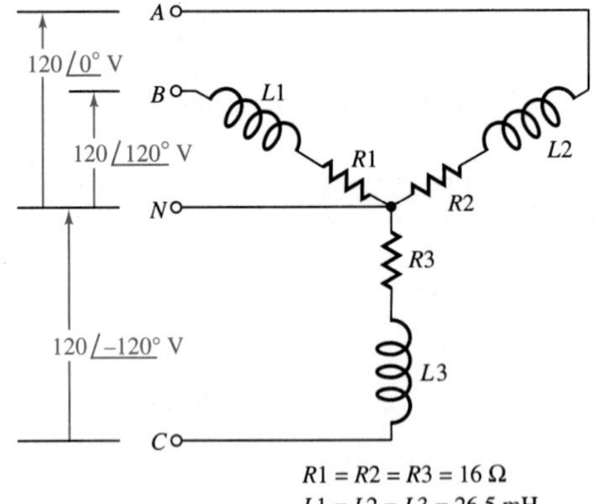

Figure 22.68

Ⓢ 54. Use PSpice to determine the branch current in the three-phase system in Figure 22.69.

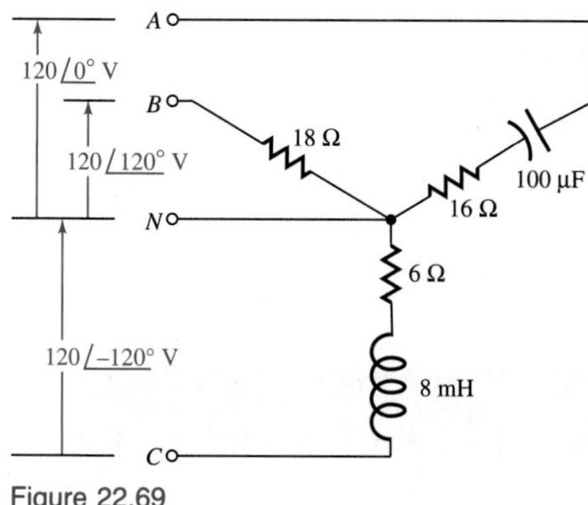

Figure 22.69

Ⓢ 55. Use PSpice to graph the current waveforms in the three-phase system in Figure 22.69.

APPENDIX

EQUATION-SOLVING METHODS

The three basic methods for determining the value of the unknowns in a set of equations are: determinants, substitution, and addition or subtraction. Each of these methods is now explained.

A.1 Determinants

The advantage of this method is that the same format is used for every set of equations. A disadvantage is that many numbers must be manipulated in a set of three equations. This can result in arithmetic errors.

In this method, a matrix that is made up of the terms in the equations must be evaluated. The matrix for two equations has four terms in it, as shown in (a). The matrix for three equations has nine terms in it, as shown in (b).

$$\begin{vmatrix} 5 & -2 \\ 4 & 3 \end{vmatrix} \qquad \begin{vmatrix} 4 & -5 & 3 \\ -2 & 15 & 6 \\ 0 & 4 & -2 \end{vmatrix}$$

(a) Matrix for two equations (b) Matrix for three equations

The steps in evaluating a two-equation matrix are:

1. Draw the matrix diagonals, as shown by lines a–a and b–b.

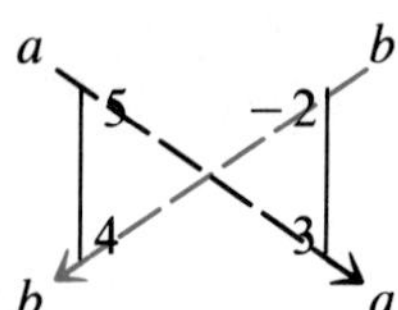

2. Multipy the terms on the a–a diagonal.

$$(5)(3) = 15$$

3. Multiply the terms on the b–b diagonal.

$$(-2)(4) = -8$$

4. The value of the matrix is equal to the product in Step 2 minus the product in Step 3. This value is called D. Then, $D = 15 - (-8) = 23$.

The steps in evaluating a three-equation matrix are:

1. Draw the matrix, repeating the first two columns. Then draw the diagonals a–a, b–b, c–c, d–d, e–e, and f–f.

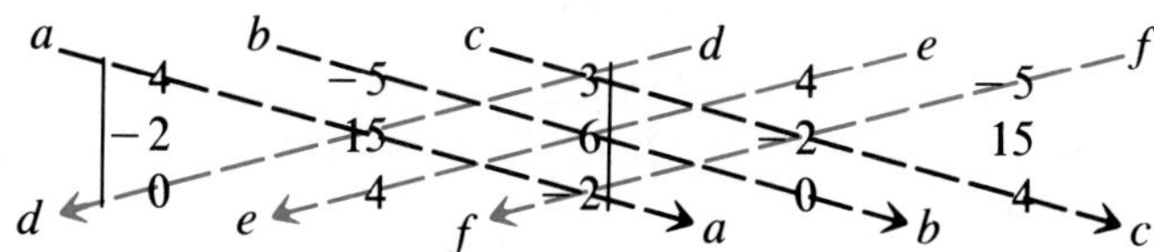

2. Multiply the terms on diagonal a–a.

$$(4)(15)(-2) = -120$$

3. Multiply the terms on diagonal b–b.

$$(-5)(6)(0) = 0$$

4. Multiply the terms on diagonal c–c.

$$(3)(-2)(4) = -24$$

5. Add the products in Steps 2, 3, and 4.

$$-120 + 0 + (-24) = -144$$

6. Multiply the terms on diagonal d–d.

$$(3)(15)(0) = 0$$

7. Multiply the terms on diagonal e–e.

$$(4)(6)(4) = 96$$

8. Multiply the terms on diagonal f–f.

$$(-5)(-2)(-2) = -20$$

9. Add the products in Steps 6, 7, and 8.

$$0 + 96 + (-20) = 76$$

10. D is equal to the sum of the products in Step 5 minus the sum of the products in Step 9.

$$D = -144 - 76 = -220$$

Now that the method of evaluating the matrix is known, some equations are solved.

Two Equations with Two Unknowns

The equations are

$$3X + 2Y - 13 = 0 \qquad \text{and} \qquad -6Y + 2X + 6 = 0$$

a. The equations must be arranged so that each unknown is in one column, and the numbers are on the right-hand side of the equation.

$$3X + 2Y = 13$$
$$2X - 6Y = -6$$

b. The D matrix is formed by writing the coefficients of X and Y.

$$D = \begin{vmatrix} 3 & 2 \\ 2 & -6 \end{vmatrix}$$

$$\text{Then } D = (3)(-6) - (2)(2) = -22$$

c. The D_X matrix is formed by replacing the coefficients of X in the D matrix with the numbers from the right-hand side of the equation. This gives

$$D_X = \begin{vmatrix} 13 & 2 \\ -6 & -6 \end{vmatrix}$$

$$\text{Then } D_X = (13)(-6) - (2)(-6) = -66$$

To find the value of X, we must find the ratio of D_X/D. So

$$X = D_X/D = -66/-22 = 3$$

d. The D_Y matrix is formed by replacing the coefficients of Y in the D matrix with the numbers from the right-hand side of the equation. This gives

$$D_Y = \begin{vmatrix} 3 & 13 \\ 2 & -6 \end{vmatrix}$$

$$\text{Then } D_Y = (3)(-6) - (2)(13) = -44$$

The value of Y is equal to the ratio of D_Y/D. So

$$Y = D_Y/D = -44/-22 = 2$$

Three Equations with Three Unknowns

The equations are

$$2X + 4Y - 4Z + 12 = 0$$
$$2Z + 6X - 3Y = 23$$
$$-8Z + 6Y + 26 = 0$$

a. The equations must be arranged so that each unknown is in one column:

$$\begin{array}{rrrcr} 2X & +4Y & -4Z & = & -12 \\ 6X & -3Y & +2Z & = & 23 \\ 0X & +6Y & -8Z & = & -26 \end{array}$$

b. The D matrix is formed by writing the coefficients of X, Y, and Z in columns. This gives

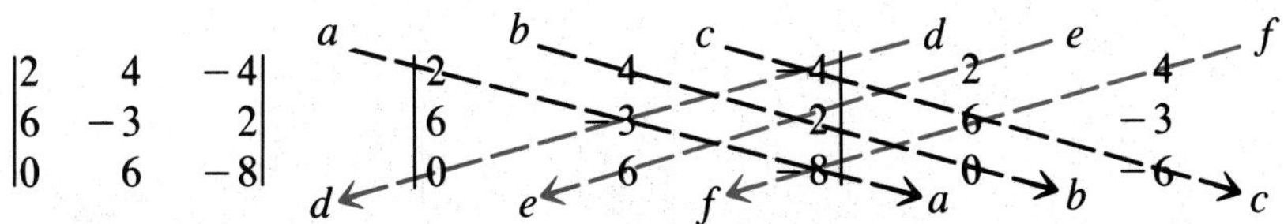

Matrix Matrix for evaluation

Using the method for evaluating a three-equation matrix gives

$$\begin{aligned} D &= [(2)(-3)(-8) + (4)(2)(0) + (-4)(6)(6)] \\ &\quad - [(-4)(-3)(0) + (2)(2)(6) + (4)(6)(-8)] \\ &= (48 + 0 - 144) - (0 + 24 - 192) = 72 \end{aligned}$$

c. The D_X matrix is formed by replacing the coefficients of X in the D matrix with the numbers at the right-hand side of the equation. This gives

$$D_X = \begin{vmatrix} -12 & 4 & -4 \\ 23 & -3 & 2 \\ -26 & 6 & -8 \end{vmatrix} \qquad \begin{array}{|ccc|cc} -12 & 4 & -4 & -12 & 4 \\ 23 & -3 & 2 & 23 & -3 \\ -26 & 6 & -8 & -26 & 6 \end{array}$$

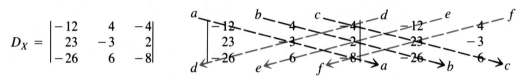

Matrix Matrix for evaluation

$$\begin{aligned} D_X &= [(-12)(-3)(-8) + (4)(2)(-26) + (-4)(23)(6)] \\ &\quad - [(-4)(-3)(-26) + (-12)(2)(6) + (4)(23)(-8)] \\ &= (-288 - 208 - 552) - (-312 - 144 - 736) = 144 \end{aligned}$$

The value of X is equal to D_X/D. So

$$X = D_X/D = 144/72 = 2$$

d. The D_Y matrix is formed by replacing the coefficients of Y in the D matrix with the numbers at the right-hand side of the equation. This gives

$$\begin{vmatrix} 2 & -12 & -4 \\ 6 & 23 & 2 \\ 0 & -26 & -8 \end{vmatrix} \qquad \begin{array}{|ccc|cc} 2 & -12 & -4 & 2 & -12 \\ 6 & 23 & 2 & 6 & 23 \\ 0 & -26 & -8 & 0 & -26 \end{array}$$

Matrix Matrix for evaluation

$$\begin{aligned} D_Y &= [(2)(23)(-8) + (-12)(2)(0) + (-4)(6)(-26) \\ &\quad - [(-4)(23)(0) + (2)(2)(-26) + (-12)(6)(-8) \\ &= (-368 + 0 + 624) - (0 - 104 + 576) \\ &= -216 \end{aligned}$$

The value of Y is equal to D_Y/D. So

$$Y = D_Y/D = -216/72 = -3$$

e. The D_Z matrix is formed by replacing the coefficients of Z in the D matrix with the numbers at the right-hand side of the equation. This gives

$$D_Z = \begin{vmatrix} 2 & 4 & -12 \\ 6 & -3 & 23 \\ 0 & 6 & -26 \end{vmatrix} \qquad \begin{array}{|ccc|cc} 2 & 4 & -12 & 2 & 4 \\ 6 & -3 & 23 & 6 & -3 \\ 0 & 6 & -26 & 0 & 6 \end{array}$$

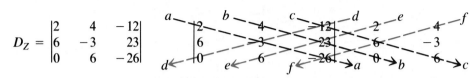

Matrix — Matrix for evaluation (diagonals a, b, c, d, e, f)

$$\begin{aligned} D_Z &= [(2)(-3)(-26) + (4)(23)(0) + (-12)(6)(6)] \\ &\quad - [(-12)(-3)(0) + (2)(23)(6) + (4)(6)(-26)] \\ &= (156 + 0 - 432) - (0 + 276 - 624) = 72 \end{aligned}$$

The value of Z is equal to D_Z/D. So

$$Z = D_Z/D = 72/72 = 1$$

Two Equations with Phasor Quantities

The equations are

$$(5 \angle 40°)X + (3 \angle 60°)Y = 18.75 \angle 76.1°$$
$$(6 \angle 30°)Y + (2 \angle 30°)X - (17 \angle 76.9°) = 0$$

The procedure for equations with phasor quantities is the same as for those without phasors. However, the arithmetic manipulation must be done according to the rules of phasor algebra.

a. The equations are arranged to place each unknown in the same column and the number terms on the right-hand side of the equation. This gives

$$(5 \angle 40°)X + (3 \angle 60°)Y = 18.75 \angle 76.1°$$
$$(2 \angle 30°)X + (6 \angle 30°)Y = 17 \angle 76.9°$$

b. The D matrix is formed by writing the coefficients of X and Y. This gives

$$D = \begin{vmatrix} 5 \angle 40° & 3 \angle 60° \\ 2 \angle 30° & 6 \angle 30° \end{vmatrix}$$

$$\begin{aligned} D &= (5 \angle 40°)(6 \angle 30°) - (3 \angle 60°)(2 \angle 30°) \\ &= 30 \angle 70° - 6 \angle 90° = 24.46 \angle 65.2° \end{aligned}$$

c. The D_X matrix is formed by replacing the coefficients of X in the D matrix by the terms from the right-hand side of the equation. This gives

$$D_X = \begin{vmatrix} 18.75 \angle 76.1° & 3 \angle 60° \\ 17 \angle 76.9° & 6 \angle 30° \end{vmatrix}$$

$$\begin{aligned} D_X &= (18.75 \angle 76.1°)(6 \angle 30°) - (3 \angle 60°)(17 \angle 76.9°) \\ &= 112.5 \angle 106.1° - 51 \angle 136.9° = 73.49 \angle 85.2° \end{aligned}$$

d. As for nonphasor equations, the value of X is equal to D_X/D. Therefore,

$$X = D_X/D = (73.49 \angle 85.2^\circ)/(24.46 \angle 65.2^\circ) = 3.0 \angle 20^\circ$$

e. The D_Y matrix is formed by replacing the coefficients of Y in the D matrix by the terms from the right-hand side of the equation. This gives

$$D_Y = \begin{vmatrix} 5 \angle 40^\circ & 18.75 \angle 76.1^\circ \\ 2 \angle 30^\circ & 17 \angle 76.9^\circ \end{vmatrix}$$

$$\begin{aligned} D_Y &= (5 \angle 40^\circ)(17 \angle 76.9^\circ) - (18.75 \angle 76.1^\circ)(2 \angle 30^\circ) \\ &= 85 \angle 116.9^\circ - 37.5 \angle 106.1^\circ = 48.67 \angle 125.2^\circ \end{aligned}$$

f. The value of Y is equal to D_Y/D. Therefore,

$$Y = D_Y/D = (48.67 \angle 125.2^\circ)/(24.46 \angle 65.2^\circ) = 1.99 \angle 60^\circ$$

Three Equations with Phasor Quantities

The solution of three equations with phasors follows the same procedure that is used for three equations without phasors. The difference is in the use of phasor arithmetic because of the angles.

Solving equations with phasors can be lengthy and tedious. This is so whether determinants, substitution, or addition or subtraction is used. Many programs are available for solving the equations on a computer or a programmable hand calculator. These can be used to reduce the time spent on the solution, or for checking the answers.

A.2 Substitution

This method is suited for use with two equations. It can become lengthy when used for three equations, since double substitution must be used. The first substitution reduces the three equations to two, and the second substitution reduces the two equations to one. Disadvantages of this method are that the sequence of substitution must be determined for each set of equations, and the substitution often requires clearing of fractions. Its main advantage is that it is relatively simple for use with two equations.

In this method, one unknown in one of the equations is written in terms of the other unknowns. This expression is then substituted into the other equations. The procedure is repeated until there is only one equation with one unknown. Now let us look at a few examples.

Two Equations with Two Unknowns

The equations are

$$(1)\ 6X + 2Y = 20 \qquad \text{and} \qquad (2)\ 3X - 5Y = 22$$

One of the unknowns must be written in terms of the other. Arbitrarily selecting the first equation, we express X in terms of Y. This gives

$$X = \frac{20 - 2Y}{6}$$

This expression must be substituted into the second equation. Doing so gives

$$3\,\frac{(20 - 2Y)}{6} - 5Y = 22$$

or

$$10 - Y - 5Y = 22$$

which reduces to

$$Y = -2$$

X can be found in the same manner. However, it is usually shorter to substitute the value of Y into the expression for X. This gives

$$X = \frac{(20 - 2Y)}{6} = \frac{20 - (2)(-2)}{6} = 4$$

These values can be checked by putting them into the second equation.

$$3(4) - 5(-2) = 22$$

Three Equations with Three Unknowns

The equations are

$$\begin{aligned}
&(1) \qquad 2X + 3Y + 4Z = 17\\
&(2) \qquad 4X - 3Y - 2Z = 21\\
&(3) \qquad -3X + 4Y + 6Z = -3
\end{aligned}$$

One of the unknowns in one of the equations must be expressed in terms of the others. The first equation is selected and X is written in terms of Y and Z. The result is

$$X = \frac{17 - 3Y - 4Z}{2}$$

This expression must now be substituted into the other two equations. Doing so gives

$$\frac{4(17 - 3Y - 4Z)}{2} - 3Y - 2Z = 21$$

and

$$\frac{-3(17 - 3Y - 4Z)}{2} + 4Y + 6Z = -3$$

These reduce to

$$(4) \qquad -9Y - 10Z = -13$$

$$(5) \qquad 17Y + 24Z = 45$$

We now have two equations with two unknowns. The procedure for solving them was explained in the preceding example.

From Equation (4),

$$Y = \frac{13 - 10Z}{9}$$

Substituting this expression into Equation (5) gives

$$\frac{17(13 - 10Z)}{9} + 24Z = 45 \qquad \text{or} \qquad Z = 4$$

Substitution can also be used to find X and Y. For X, the first step is to write an expression for Y. This is substituted into the other two equations. Then substitution is used to eliminate Z. For Y, the first step is to write an expression for Z. This is substituted into the other two equations. Then substitution is used to eliminate X. However, a shorter way is to use the value of Z and put it into the expression for Y. This gives

$$Y = \frac{13 - 10(4)}{9} = -3$$

X can be found using the values of Z and Y in any one of the three equations. Using Equation (1) gives

$$X = \frac{(17 - 3Y - 4Z)}{2}$$

$$= \frac{(17 - 3(-3) - (4)(4)}{2} = 5$$

The results can be checked by putting them into one of the original equations.

Two Equations with Phasors

The equations are

$$(1) \qquad (4 \angle 30°)X + (5 \angle 60°)Y = 27.81 \angle 109.3°$$

$$(2) \qquad (3 \angle 45°)X + (6 \angle 90°)Y = 30 \angle 135°$$

One of the unknowns must be written in terms of the other. From Equation (1),

$$X = \frac{27.81 \angle 109.3° - (5 \angle 60°)Y}{4 \angle 30°}$$

This expression must be substituted into the other equation. Doing so gives

$$\frac{(3\ \angle 45°)\ [27.81\ \angle 109.3° - (5\ \angle 60°)Y]}{4\ \angle 30°} + (6\ \angle 90°)Y = 30\ \angle 135°$$

or

$$Y(6\ \angle 90° - 3.75\ \angle 75°) = 30\ \angle 135° - 20.86\ \angle 124.3°$$

This reduces to

$$Y = \frac{10.245\ \angle 157.2°}{2.57\ \angle 112.17°} = 3.99\ \angle 45.03°$$

To solve for X, Y is written in terms of X. This gives

$$Y = \frac{27.81\ \angle 109.3° - (4\ \angle 30°)X}{5\ \angle 60°}$$

Putting this into Equation 2 gives

$$(3\ \angle 45°)X + \frac{(6\ \angle 90°)(27.81\ \angle 109.3° - (4\ \angle 30°)X)}{5\ \angle 60°} = 30\ \angle 135°$$

or

$$X(3\ \angle 45° - 4.8\ \angle 60°) = 30\ \angle 135° - 33.37\ \angle 139.3°$$

This reduces to

$$X = \frac{4.14\ \angle -7.78°}{2.06\ \angle -97.83°} = 2.0\ \angle 90.05°$$

Substitution can be used for three equations with phasors, but the solution can be lengthy and tedious. The procedure is the same as for the equations without phasors. However, the arithmetic must follow the rules for phasors.

A.3 Addition or Subtraction

Addition or subtraction has a format that is the same for all sets of equations. Also, the manipulation is less than that for determinants, so there is not as much chance for error.

In this method, the unknowns are eliminated by making their coefficients the same and adding or subtracting the equations to eliminate the unknown.

Two Equations with Two Unknowns

The equations are

$$(1) \qquad 2X + 8Y = 32$$

$$(2) \qquad -5X + 3Y = -11$$

Neither the X nor the Y terms have the same coefficients. Therefore, we arbitrarily select the Y terms and make their coefficients the same. To make them the same, Equation (1) must be multiplied by the coefficient of Y in Equation (2). Then Equation (2) is multiplied by the coefficient of Y in Equation (1). This gives

$$3(2X + 8Y = 32) \quad \text{or} \quad 6X + 24Y = 96$$
$$8(-5X + 3Y = -11) \quad \text{or} \quad -40X + 24Y = -88$$

The coefficients of Y have the same sign, so one equation must be subtracted from the other to eliminate Y. If they had opposite signs, the equations would be added. Doing so gives

$$6X + 24Y = 96$$

minus

$$(-40X + 24Y = -88)$$

gives

$$46X + 0Y = 184 \quad \text{or} \quad X = 4$$

Y can be found by eliminating X from the equations, but it is usually shorter to find it by substituting the value of X into one of the equations. Using the first equation, we get

$$(6)(4) + (24)Y = 96 \quad \text{or} \quad Y = 3$$

These values can be checked by putting them into the other equations. The result is

$$(-40)(4) + (24)(3) = -88$$

Three Equations with Three Unknowns

When there are more than two equations, the procedure is to eliminate the same unknown in pairs of equations. This is repeated until there is only one equation with one unknown.

The equations are

$$(1) \qquad 2X + 3Y - 4Z = 41$$
$$(2) \qquad 3X - 2Y + 2Z = 10$$
$$(3) \qquad -6X + 5Y - 3Z = -21$$

Equations (1) and (2) are taken as one pair, and Equations (2) and (3) are the other pair. The unknown X is selected for elimination in each pair.

In the first pair, Equation (1) is multiplied by 3, which is the coefficient of X in Equation (2). Equation (2) is multiplied by 2, which is the coefficient of X in Equation (1). This gives

$$6X + 9Y - 12Z = 123$$
$$6X - 4Y + 4Z = 20$$

Since the coefficients of X have the same sign, the second equation must be subtracted from the first. This gives

$$6X + 9Y - 12Z = 123$$

minus

$$(6X - 4Y + 4Z = 20)$$

gives

$$0X + 13Y - 16Z = 103$$

To eliminate X in the second pair of equations, Equation (2) must be multiplied by 6, and Equation (3) by 3. This gives

$$18X - 12Y + 12Z = 60$$
$$-18X + 15Y - 9Z = -63$$

Since the coefficients of X have opposite signs, the equations must be added to eliminate X. This gives

$$18X - 12Y + 12Z = 60$$

plus

$$(-18X + 15Y - 9Z = -63)$$

gives

$$0X + 3Y + 3Z = -3$$

We now have two equations

$$13Y - 16Z = 103$$
$$3Y + 3Z = -3$$

Multiplying the second equation by 13 and the first by 3 makes the coefficients of Y the same. Since they also have the same sign, subtraction must be used to eliminate Y.

$$39Y - 48Z = 309$$

minus

$$(39Y + 39Z = -39)$$

gives

$$0Y - 87Z = 348 \qquad \text{or} \qquad Z = -4$$

X and Y can be found by using addition or subtraction, but again, it is shorter to substitute the value of Z into one of the equations with two unknowns. Then Y and Z can be substituted into one of the original equations to find X.

From the equation

$$3Y + 3Z = -3$$

we get

$$3Y + (3)(-4) = -3 \qquad \text{or} \qquad Y = 3$$

Putting this value of Y and the value of Z into Equation (1) gives

$$(2)(X) + (3)(3) - (4)(-4) = 41 \qquad \text{or} \qquad X = 8$$

These values can be checked by putting them into one of the original equations. Using Equation (3) gives

$$(-6)(8) + (5)(3) - (3)(-4) = -21$$

Two Equations with Two Unknown Phasor Quantities

The difference between using the procedure for nonphasor quantities and phasor quantities is that the coefficient of the unknown has a magnitude and an angle. Therefore, both of these must be made the same, to eliminate the unknown.

The equations are

$$(1)\quad (2 \angle 30°)X - (3 \angle 60°)Y = 8.72 \angle 23.39°$$

$$(2)\quad (3 \angle 60°)X + (2 \angle -30°)Y = 17.35 \angle 78.48°$$

X will be eliminated from the equations by making the coefficients of the X terms the same. To do this, Equation (1) is multiplied by $3 \angle 60°$, which is the coefficient of X in Equation (2). Then Equation (2) will be multiplied by $2 \angle 30°$, which is the coefficient of X in Equation (1). This gives

$$(3 \angle 60°)(2 \angle 30°)X - (3 \angle 60°)(3 \angle 60°)Y = (3 \angle 60°)(8.72 \angle 23.39°)$$

which reduces to

$$(6 \angle 90°)X - (9 \angle 120°)Y = 26.16 \angle 83.39°$$

and

$$(2 \angle 30°)(3 \angle 60°)X + (2 \angle 30°)(2 \angle -30°)Y = (2 \angle 30°)(17.35 \angle 78.48°)$$

which reduces to

$$(6 \angle 90°)X + 4Y = 34.7 \angle 108.5°$$

Since the coefficients of X have the same sign, the equations must be subtracted to eliminate X. This gives

$$(6 \angle 90°)X - (9 \angle 120°)Y = 26.16 \angle 83.39°$$

minus

$$((6 \angle 90°)X + 4Y = 34.7 \angle 108.5°)$$

gives

$$0X - (9 \angle 120° + 4)Y = 26.16 \angle 83.39° - 34.7 \angle 108.5°$$

From which we get

$$Y = \frac{15.62 \angle 153.7°}{7.81 \angle 93.67°} \quad \text{or} \quad Y = 2 \angle 60°$$

Putting this value of Y into Equation (1) gives

$$X = \frac{8.72 \angle 23.41° + (3 \angle 60°)(2 \angle 60°)}{2 \angle 30°} = \frac{10 \angle 60°}{2 \angle 30°}$$

or $\quad X = 5 \angle 30°$

Three Equations with Three Unknown Phasor Quantities

The main difference between the solution of these and nonphasor equations is in the use of phasor arithmetic. The phasor arithmetic makes the solution longer than that for nonphasors.

The equations are grouped into pairs, as they were for the three nonphasor equations. The coefficients are made the same, as explained in the previous example. One unknown is eliminated in each pair of equations, leaving two equations with two unknowns. These are then solved, as shown in the previous example.

It should be apparent after seeing the solution to the equations by the three methods that once the equations exceed two, there is no quick and easy method for their solution. Therefore, one should try to use a method of analysis that requires a minimal number of equations. This is especially important in the analysis of ac circuits.

APPENDIX B

DERIVATION OF FORMULAS

B.1 Current Rise in the *R–L* Circuit

Kirchhoff's loop equation for the circuit is

$$E - v_R - v_L = 0$$

But

$$v_R = iR \qquad \text{and} \qquad v_L = L\frac{di}{dt}$$

so

$$E - iR - L\frac{di}{dt} = 0$$

This is a differential equation since it includes the differentials *di* and *dt*. These are infinitesimally small changes. The ratio *di*/*dt* is called the derivative of *i* with respect to *t*. It is the slope of the current versus time curve at any value of *t*.

Differential equations are solved by using integration to eliminate the differentials. Integration adds the area under the curve in each small section of *dt* to give the total area. Calculus texts and mathematics handbooks list standard forms, and the equations that result from their integration. Therefore, once an expression is put into one of the standard forms, its integral can be easily found.

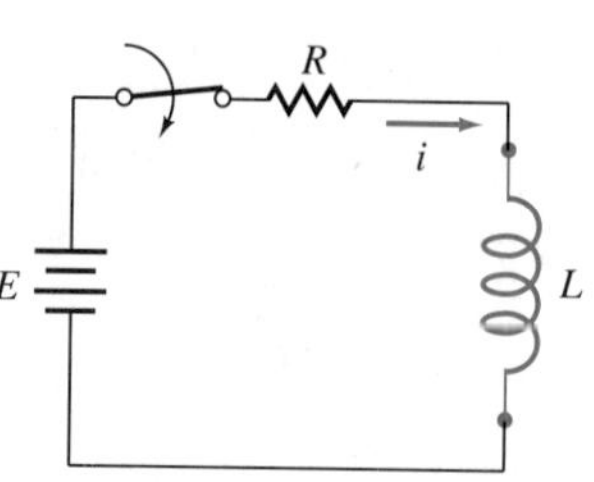

Figure B1 The *R–L* circuit for current rise.

Rearranging the differential equation for the R–L circuit gives

$$\frac{di}{\left(\frac{E}{R}\right) - i} = \frac{Rdt}{L}$$

The expressions are now in a standard form and can be integrated. Thus

$$\int \frac{di}{\left(\frac{E}{R}\right) - i} = \int \frac{Rdt}{L}$$

which integrates to $\ln(E/R - i) = -Rt/L + K$. K is a constant that results from the integration. Its value can be found by using the condition that $i = 0$ at $t = 0$.

This yields

$$\ln\left(\frac{E}{R} - 0\right) = -0 + K \quad \text{or} \quad K = \ln\left(\frac{E}{R}\right)$$

so

$$\ln\left(\frac{E}{R} - i\right) = -\frac{Rt}{L} + \ln\frac{E}{R}$$

Rearranging this expression gives

$$\ln\left(\frac{E}{R} - i\right) - \ln\left(\frac{E}{R}\right) = -\frac{Rt}{L}$$

But $\ln(A) - \ln(B) = \ln(A/B)$, so the expression becomes

$$\ln\left(\frac{\frac{E}{R} - i}{E/R}\right) = -\frac{Rt}{L} \quad \text{or} \quad e^{-Rt/L} = \frac{\frac{E}{R} - i}{E/R}$$

which reduces to

$$i = \frac{E}{R}(1 - e^{-t/L/R})$$

B.2 Current Decay in the *R–L* Circuit

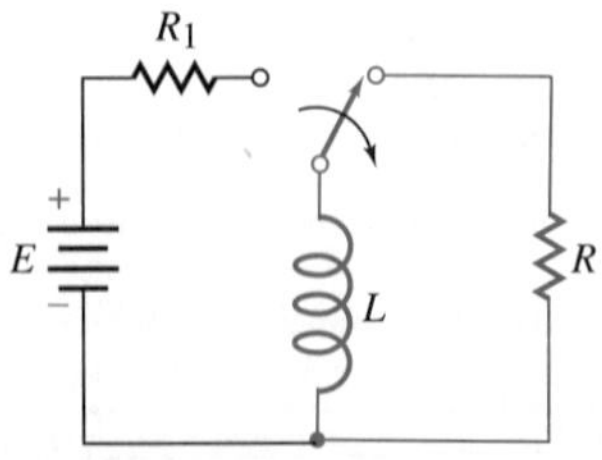

Figure B2 The R–L circuit for current decay.

During decay, v_L acts as a source. Kirchhoff's loop equation for the circuit is

$$-v_R + v_L = 0$$

But

$$v_L = L\frac{di}{dt} \quad \text{and} \quad v_R = iR$$

so

$$-iR + L\frac{di}{dt} = 0 \qquad \text{or} \qquad -Rdt = L\frac{di}{i}$$

Both sides of the differential equation are in a standard form, so

$$\int -Rdt = \int L\frac{di}{i}$$

This integrates to

$$-Rt = L\ln(i) + K$$

The value of K can be found by using the value of current at $t = 0$. This value will be called I_o.

So

$$0 = L\ln(I_o) + K \qquad \text{or} \qquad K = -L\ln(I_o)$$

Substituting this value for K gives

$$-Rt = L\ln(i) - L\ln(I_o)$$

or

$$\ln(i) - \ln(I_o) = \frac{-Rt}{L}$$

Then

$$\ln\left(\frac{i}{I_o}\right) = -\frac{Rt}{L}$$

or

$$e^{-Rt/L} = \frac{i}{I_o}$$

This can be arranged to give

$$i = I_o e^{-t/(L/R)}$$

If the circuit has reached steady state during the rise, $I_o = E/R$.

B.3 Energy Stored in the Inductor during Current Rise

The amount of energy in the narrow strip is

$$d\mathcal{W} = pdt$$

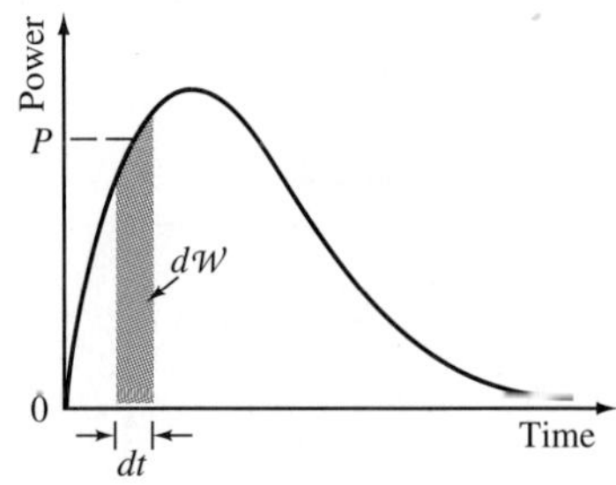

Figure B3

But

$$p = \upsilon i \qquad \text{and} \qquad \upsilon = L\frac{di}{dt}$$

so

$$d\mathcal{W} = \left(L\frac{di}{dt}\right)(i)\,dt = (Li)di$$

Both sides of the equation are in a standard form, so

$$\int d\mathcal{W} = \int_0^I (Li)di$$

This integrates to

$$\mathcal{W} = \left.\frac{Li^2}{2}\right|_0^I$$

or

$$\mathcal{W} = \frac{LI^2}{2} - \frac{L0^2}{2}$$

so

$$\mathcal{W} = \frac{LI^2}{2}$$

The values at the top and bottom of the integral sign mean that the integration is done from $i = 0$ to $i = I$.

B.4 Capacitor Voltage During Charging

Kirchhoff's Loop equation for the circuit is

$$E - \upsilon_R - \upsilon_C = 0$$

But

$$\upsilon_R = iR \quad \text{and} \quad i = C\frac{d\upsilon}{dt}$$

so

$$E - \left(C\frac{d\upsilon}{dt}\right)(R) - \upsilon_C = 0$$

or

$$\frac{d\upsilon}{E - \upsilon_C} = \frac{dt}{RC}$$

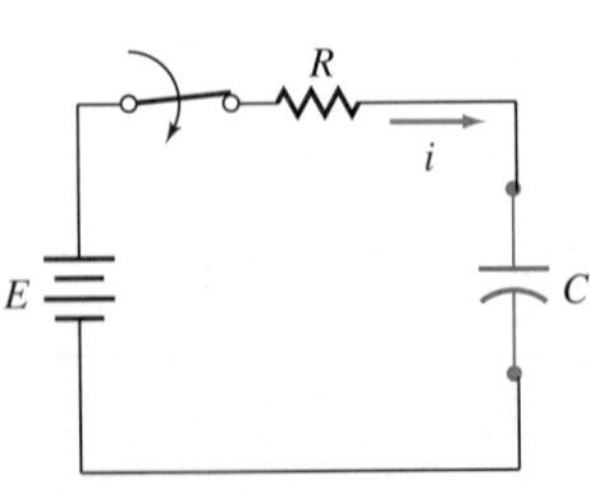

Figure B4 The R–L circuit for capacitor charge.

Both sides of the equation are in a standard form, so

$$\int\frac{d\upsilon}{E - \upsilon_C} = \int\frac{dt}{RC}$$

This integrates to

$$\ln(E - \upsilon_C) = -\frac{t}{RC} + K$$

At

$$t = 0, \upsilon_C = 0 \qquad \text{so} \qquad K = \ln(E)$$

Therefore,

$$\ln(E - \upsilon_C) = -\frac{t}{RC} + \ln(E)$$

or

$$\ln(E - \upsilon_C) - \ln(E) = -\frac{t}{RC}$$

Then

$$\ln\left(\frac{(E - \upsilon_C)}{E}\right) = -\frac{t}{RC}$$

so

$$e^{-t/RC} = \frac{E - \upsilon_C}{E}$$

which gives

$$\upsilon_C = E(1 - e^{-t/RC})$$

B.5 Capacitor Voltage during Discharge

Kirchhoff's loop equation for this circuit is

$$\upsilon_C - \upsilon_R = 0$$

But

$$\upsilon_R = iR \qquad \text{and} \qquad i = C\frac{d\upsilon_C}{dt}$$

so

$$\upsilon_C - \left(C\frac{d\upsilon_C}{dt}\right)(R) = 0$$

Rearranging this gives

$$\frac{d\upsilon_C}{V_C} = \frac{-1}{RC}\,\text{dt}$$

Both sides of the equation are in a standard form, so

$$\int\frac{d\upsilon_C}{V_C} = \int\frac{1}{RC}\,dt$$

This integrates to $\ln(\upsilon_C) = -\frac{1}{RC}t + K$

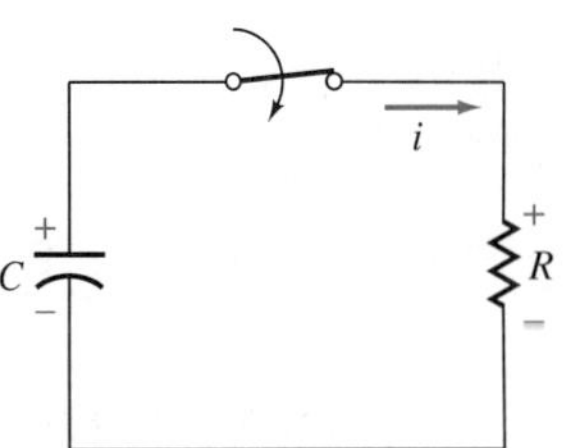

Figure B5 The *R–L* circuit for capacitor discharge.

At $t = 0$, υ_c equals some initial voltage, which will be called V_o. Therefore

$$K = \ln(V_o)$$

Substituting this value for K gives

$$\ln(\upsilon_C) = -\frac{t}{RC} + \ln(V_o)$$

or

$$\ln(\upsilon_C) - \ln(V_o) = -\frac{t}{RC}$$

so

$$e^{-t/RC} = \frac{\upsilon_C}{V_o}$$

This can be rearranged to give

$$\upsilon_C = V_o e^{-t/RC}$$

If the capacitor is fully charged to E volts, this becomes

$$-v_C = Ee^{-t/RC}$$

B.6 Energy Stored in the Capacitor During Charging

$$d\mathcal{W} = pdt$$

But

$$p = \upsilon i \quad \text{and} \quad i = C\frac{d\upsilon}{dt}$$

so

$$d\mathcal{W} = (\upsilon)\left(C\frac{d\upsilon}{dt}dt\right) = (C\upsilon)d\upsilon$$

Both sides of the equation are in a standard form, so

$$\int d\mathcal{W} = \int (C\upsilon)d\upsilon$$

This integrates to

$$\mathcal{W} = \frac{CV^2}{2} + K$$

at $V = 0$, $\mathcal{W} = 0$, so K must equal 0.

Therefore,

$$\mathcal{W} = \frac{CV^2}{2}$$

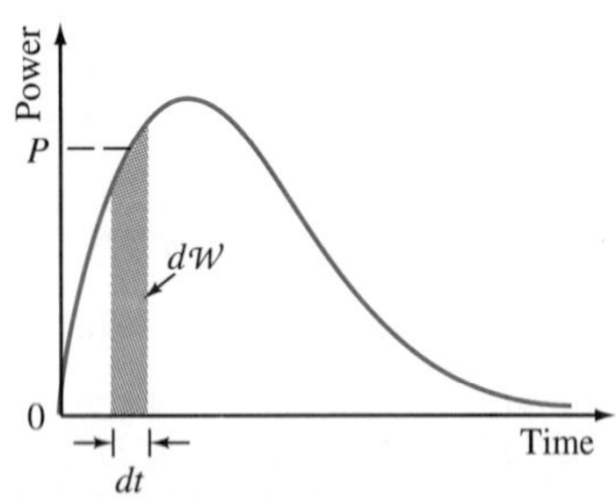

Figure B6 The energy stored in the capacitor is the area under the power curve.

B.7 (a) Average Value of the Half Wave for the Full Length of an Input Cycle

$$V_{ave} = \frac{\text{area}}{\text{length}} = \frac{\text{area}}{2\pi}$$
$$dA = (v)d(\omega t)$$

The total area is obtained by integrating $v\ d(wt)$.

$$A = \int_0^{\pi} (v)d(\omega t)$$

But

$$v = V_m \sin \omega t$$

so

$$A = \int_0^{\pi} (V_m \sin \omega t)[d(\omega t)]$$

which integrates to

$$A = -V_m \cos \omega t \big|_0^{\pi}$$
$$A = -[V_m \cos \pi - V_m \cos 0] = -[-V_m - V_m] = 2\ V_m$$

Then

$$V_{ave} = \frac{2V_m}{2\pi} = \frac{V_m}{\pi} = 0.318\ V_m$$

to three significant digits.

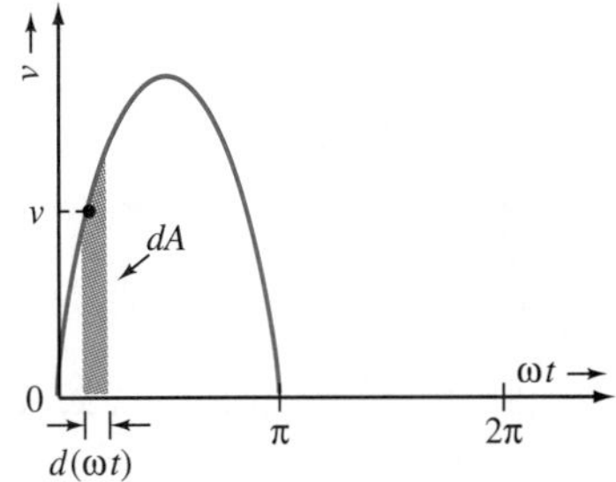

Figure B7 The waveform for the half-wave output.

B.7 (b) Average Value of the Half Wave for a Half Cycle of the Input

The area is the same as in (a), but the length is π instead of 2π. The average value then becomes

$$V_{ave} = \frac{2V_m}{\pi} \quad \text{or} \quad 0.637\ V_m$$

to three significant digits.

B.7 (c) Average Value of the Full-Wave Output

The area is now double that in (a), and the length is 2π, as in (a). So the area = $(2)(2V_m)$ or $4V_m$. Then

$$V_{ave} = \frac{4V_m}{2\pi} = \frac{2V_m}{\pi} \quad \text{or} \quad 0.637\ V_m$$

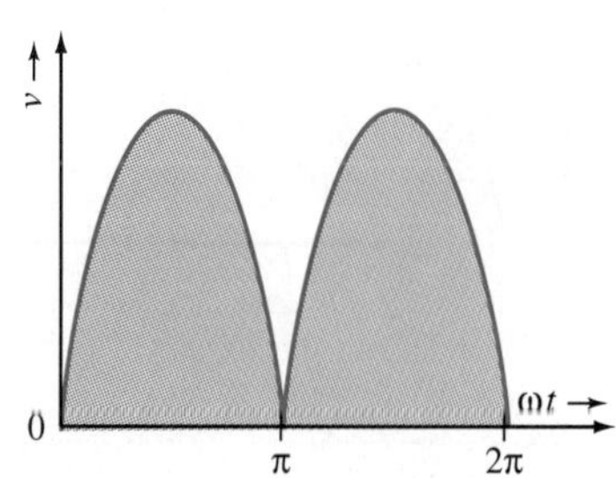

Figure B8 The waveform for the full-wave output.

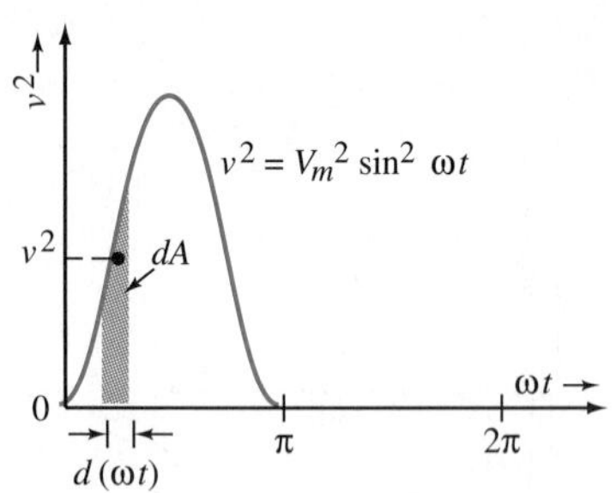

Figure B9 The voltage squared curve for the half-wave output.

B.8 (a) Effective Value of the Half Wave over a Full Cycle of the Input

$$V_{eff} = \sqrt{\frac{\text{area under the } v^2 \text{ curve curve}}{\text{length}}}$$

$$dA = (\upsilon^2)[d(\omega t)]$$

but

$$\upsilon = V_m \sin \omega t$$

so

$$dA = [V_m^2 \sin^2 \omega t)][d(\omega t)]$$

$$\int_0^\pi dA = \int (V_m^2 \sin^2 \omega t)[d(\omega t)]$$

This integrates into

$$A = V_m^2 \left[-\frac{1}{2} \cos \omega t \sin \omega t + \frac{\omega t}{2} \right]_0^\pi$$

or

$$A = \pi \frac{V_m^2}{2}$$

Then

$$V_{eff} = \sqrt{\frac{V_m^2(\pi/2)}{2\pi}} = 0.5 V_m$$

B.8 (b) Effective Value of the Half Wave over a Half Cycle of the Input

The area under the υ^2 curve is the same as in (a), but the length is π instead of 2π. So

$$V_{eff} = \sqrt{\frac{V_m^2 \pi/2}{\pi}} = \frac{V_m}{\sqrt{2}} \quad \text{or} \quad 0.707\ V_m$$

Figure B10 The voltage squared curve for the full-wave output.

B.8 (c) Effective Value of the Full Wave

The area of the υ^2 curve is now double that in (a), and the length is 2π as in (a). So

$$V_{eff} = \sqrt{\frac{V_m^2 \pi}{2\pi}} = \frac{V_m}{\sqrt{2}} \quad \text{or} \quad 0.707\ V_m$$

B.9 (a) Derivative of I_m sin ωt

When we find the derivative of a function, we are finding an expression for the instantaneous slope of the function.

If $i = I_m \sin \omega t$, the instantaneous slope is the derivative of i with respect to t, or di/dt. If $i = I_m \sin \omega t$, a small change of dt in t will cause a change of di in i. Then

$$i + di = I_m \sin(\omega t + \omega dt)$$

But

$$i = I_m \sin \omega t$$

so

$$I_m \sin \omega t + di = I_m \sin(\omega t + \omega dt)$$

or

$$di = I_m[\sin(\omega t + \omega dt) - \sin \omega t]$$

Using the identity $\sin A - \sin B = 2\left[\cos \frac{1}{2}(A + B)\right]$ $\left[\sin \frac{1}{2}(A - B)\right]$ gives

$$di = I_m\left[2 \cos\left(\omega t + \frac{\omega dt}{2}\right)\left(\sin \frac{\omega dt}{2}\right)\right]$$

then

$$\frac{di}{dt} = \frac{I_m\left[2 \cos\left(\omega t + \frac{\omega dt}{2}\right)\left(\sin \frac{\omega dt}{2}\right)\right]}{dt}$$

The sine of a small angle is equal to the angle, therefore $\sin \omega dt/2 = \omega dt/2$. Also, because dt is small, $\omega dt/2$ is small compared to ωt. Hence,

$$\frac{di}{dt} = I_m \frac{(2 \cos \omega t)(\omega dt)}{2dt} \quad \text{or} \quad \frac{di}{dt} = \omega I_m \cos \omega t$$

B.9 (b) Derivative of V_m cos ωt

Since this has the same form as the current equation, the derivative is found in the same way. This yields

$$\frac{dv}{dt} = -\omega V_m \sin \omega t$$

APPENDIX C

THE GREEK ALPHABET AND ITS USE FOR TEXT QUANTITIES

Name	Uppercase	Lowercase	Quantity it Represents
Alpha	Α	α	Angles, temperature coefficient of resistance
Beta	Β	β	Angles
Gamma	Γ	γ	——
Delta	Δ	δ	Small change in value
Epsilon	Ε	ϵ	Permittivity, relative permittivity, base of natural logarithm
Zeta	Ζ	ζ	——
Eta	Η	η	Efficiency
Theta	Θ	θ	Angles, phase angle
Iota	Ι	ι	——
Kappa	Κ	κ	——
Lambda	Λ	λ	Wavelength
Mu	Μ	μ	Permeability, relative permeability, prefix micro
Nu	Ν	ν	——
Xi	Ξ	ξ	——
Omicron	Ο	ο	——

Name	Uppercase	Lowercase	Quantity it Represents
Pi	Π	π	3.141593 (to seven digits)
Rho	P	ρ	Resistivity
Sigma	Σ	σ	Summation
Tau	T	τ	Time constant
Upsilon	Υ	υ	———
Phi	Φ	ϕ	Magnetic flux, phase quantities in three-phase system
Chi	X	χ	———
Psi	Ψ	ψ	Electric flux
Omega	Ω	ω	Angular velocity, ohm

APPENDIX D

READING METER SCALES

Figure D1 The digital display shows numbers.

The value of a quantity being measured with a meter is shown on either a digital display or an analog display.

Digital Display. This is a number display, as shown in Figure D1. The numbers are formed by exciting the proper sections of a seven-section, light-emitting diode or liquid crystal unit. Since obtaining the value involves only reading the numbers, little chance of error exists. For some measurements, the reading must be multiplied by a multiplier.

The display shown in Figure D1 is a $3\frac{1}{2}$ digit display. The $\frac{1}{2}$ results from the fact that the left digit shows only when it is a 1. This half digit almost doubles the range of the meter.

Analog Display. This display uses a scale and a pointer, as in Figure D2. The value is obtained by reading the number on the scale beneath the pointer.

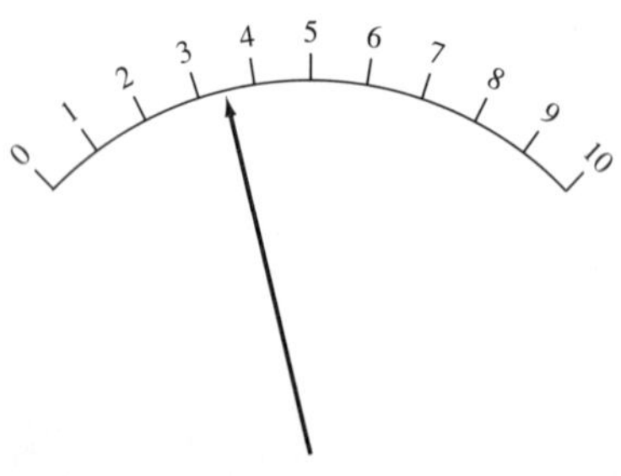

Figure D2 The analog display has a scale and a pointer.

The scale in Figure D2 is a linear scale. Linear scales have the same amount of space per unit quantity on all parts of the scale. On the other hand, the scales in Figure D3 are nonlinear scales. Nonlinear scales do not have the same space per unit quantity on all parts of the scale. The scales shown are the ohm scale and the decibel scale, two common nonlinear scales.

Two analog scales and the values indicated by the pointers are shown in Figure D4. The correct procedure for obtaining these values is as follows.

1. Set your line of vision at right angles to the scale surface. If the scale has a mirrored surface, view it so that the pointer image is

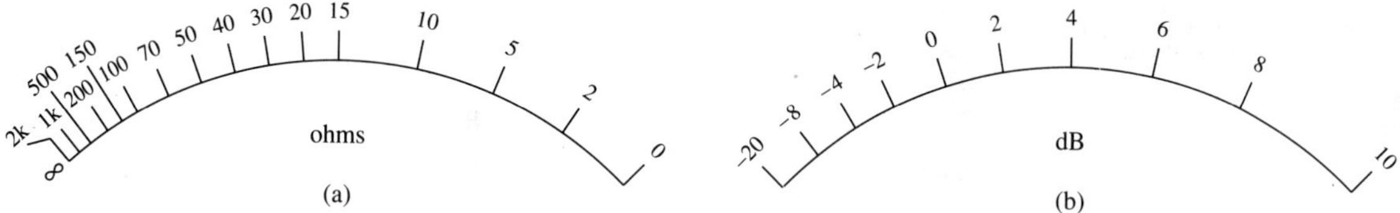

Figure D3 The amount of space per unit division is not constant for a nonlinear scale. (a) A nonlinear resistance scale. (b) A nonlinear decibel scale.

directly behind the pointer. Viewing at an angle results in an error called parallex.

2. Determine the number of units represented by the space between two adjacent lines on the scale. For Figure D4 (a), it is 0.1. For Figure D4 (b), it is 0.2.
3. Determine the value of the last line that the pointer passed. This is equal to: (last numbered line) + (number of divisions from the last numbered line to the last line passed) × (value of one division). For Figure D4 (a), it is 1 + (1)(0.1) = 1.1. For Figure D4 (b), it is 1 + (1)(0.2) = 1.2.
4. Estimate the value from the last line to the pointer. In general, only one nonzero digit should be used but two are acceptable for some scales. For Figure D4 (a), it is 0.02. For Figure D4 (b), it is 0.16.
5. Add the values of Steps 3 and 4. For Figure D4 (a), it is 1.1 + 0.02 or 1.12. For Figure D4 (b), it is 1.2 + 0.16 or 1.36.
6. If the meter is a multirange meter, multiply the scale reading in Step 5 by (range used/maximum value on scale). As an example, if the meter of Figure D4 (a) is used on the 20 range, the measured value is (1.12)(20/2) or 11.12.

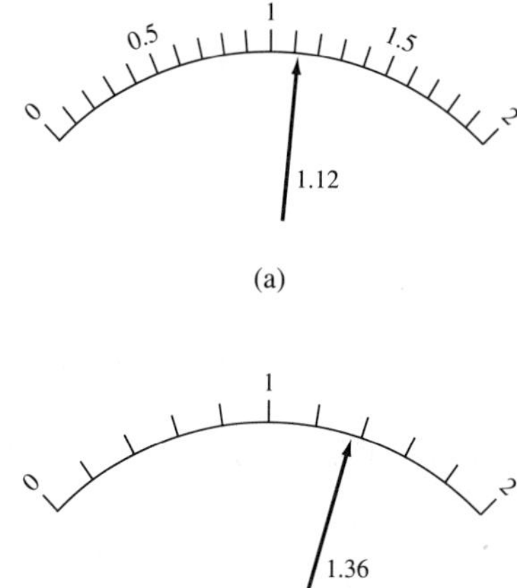

Figure D4 Two examples of analog display readings. (a) The reading can have only one estimated digit for this scale. (b) Each division represents 0.2, so two estimated digits can be used for this scale.

Some meters have several scales. With these meters, a scale should be used whose full-scale value is related to the range by a power of 10. This results in a more exact result in Step 6, since it involves only shifting the decimal. For instance, the scale selection for a meter with 0–10, 0–25, 0–50 scales, and 10, 25, 50, 100, 250, 500 ranges is:

Range	Scale	Multiplier
10	0–10	10/10 = 1
25	0–25	25/25 = 1
50	0–50	50/50 = 1
100	0–10	100/10 = 10
250	0–25	250/25 = 10
500	0–50	500/50 = 10

Although the procedure seems awkward and lengthy, it becomes second nature as one uses it.

APPENDIX E

PHASOR ARITHMETIC

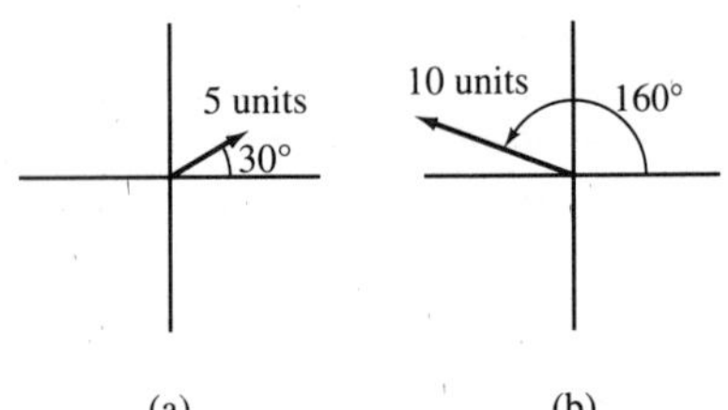

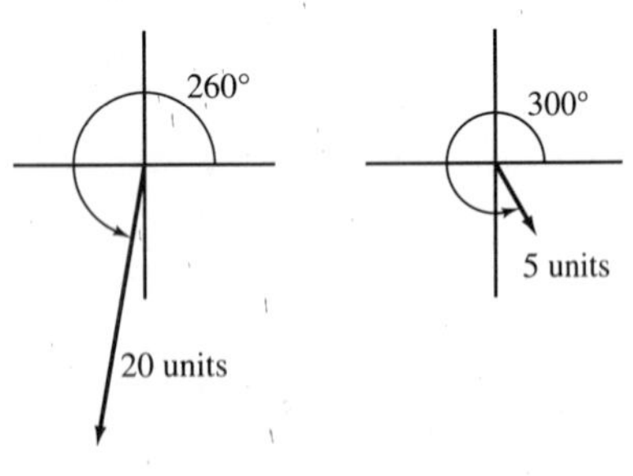

Figure E1 Graphic examples of the polar form (scale 1 cm = 2.5 units). (a) 5∠30°. (b) 10∠160°. (c) 20∠260°. (d) 5∠300°.

E.1 Polar and Rectangular Form

Phasors can be written in polar form or rectangular form. The two forms are as follows:

Polar Form

This form gives the magnitude and the angle that the phasor makes with the zero degree (x) axis. The general form of it is $C \angle \theta$. This is a phasor of C units in length at an angle of $\theta°$ with the (x) axis.

Examples

a. $5 \angle 30°$
b. $10 \angle 160°$
c. $20 \angle 260°$
d. $5 \angle 300°$

The graphical presentation of these phasors is shown in Figure E1.

Rectangular Form

This form gives the magnitudes of the real (x) and j (y) components. The general form is $A \pm jB$, where A is the component of the phasor on the real (x axis), and B is the component on the j (y axis). One or both of the components can be negative.

Examples

a. $4 + j6$
b. $-3 + j4$
c. $-3 - j6$
d. $3 - j6$

The graphical presentation of these phasors is shown in Figure E2.

(a) +j, j6, 4, −x, +x, −j

(b) +j, j4, −3, −x, +x, −j

E.2 Conversion Between Forms

In some arithmetic operations, the phasors must be converted from one form to the other. Although the conversion can readily be done with an electronic calculator, understanding the procedure is helpful in checking the results.

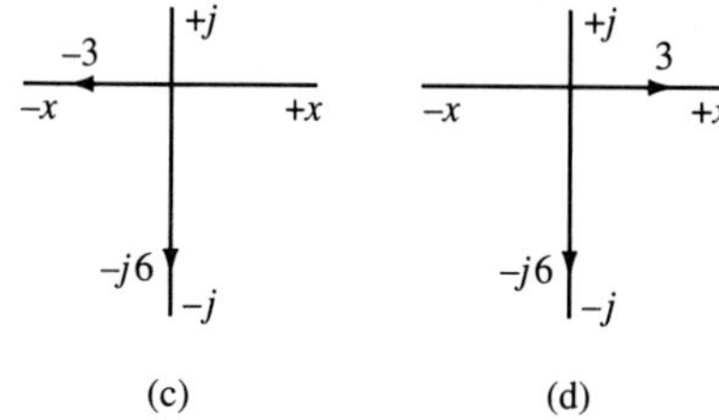

Figure E2 Graphic examples of the rectangular form (scale 1 cm = 1 unit). (a) $4 + j6$. (b) $-3 + j4$. (c) $-3 - j6$. 9d) $3 - j6$.

Polar to Rectangular

Given the polar form $C \angle \theta$, the rectangular components are $A = C\cos\theta$ and $B = C \sin \theta$.

Examples

a. $10\angle 45°$ $A = 10\cos(45°) = 10(0.707)$ or 7.07
$B = 10\sin(45°) = 10(0.707)$ or 7.07
$10\angle 45° = 7.07 + j7.07$

b. $8\angle 160°$ $A = 8\cos(160°) = 8(-0.940)$ or -7.52
$B = 8\sin(160°) = 8(0.342)$ or 2.74
$8\angle 160° = -7.52 + j2.74$

c. $6\angle -240°$ $A = 6\cos(-240°) = 6(-0.5)$ or -3
$B = 6\sin(-240°) = 6(0.866)$ or 5.2
$6\angle -240° = -3 + j5.2$

d. $10\angle 300°$ $A = 10\cos(300°) = 10(0.5)$ or 5
$B = 10\sin(300°) = 10(-0.866)$ or -8.66
$10\angle 300° = 5 - j8.66$

Rectangular to Polar

Given the rectangular forms $A + jB$ and $A - jB$, the angle and magnitude for the polar form equivalent are:

$$\theta = \tan^{-1}\left(\frac{B}{A}\right) \quad \text{if the sign of } A \text{ is } +$$

$$\theta = \pm 180° + \tan^{-1}\left(\frac{B}{A}\right) \quad \text{if the sign of } A \text{ is } -$$

where the signs of A and B must be included.

$$C = \sqrt{A^2 + B^2}$$

Also

$$C = A/\cos\theta \qquad \text{and} \qquad C = B/\sin\theta$$

where the magnitudes of the cosine and sine are used.

Examples

a. $3 + j4$

A is +, so $\theta = \tan^{-1}\left(\frac{4}{3}\right) = 53.13°$

$C = \sqrt{3^2 + 4^2} = 5$

$3 + j4 = 5\ \angle 53.13°$

b. $6 - j8$

A is +, so $\theta = \tan^{-1}\left(\frac{-8}{6}\right) = -53.13°$

$C = \sqrt{6^2 + 8^2} = 10$

$6 - j8 = 10\ \angle -53.13°$

c. $-5 - j5$

A is −, so $\theta = \pm 180° + \tan^{-1}\left(\frac{-5}{-5}\right) = \pm 180° + 45°$

$\theta = 225°$, also $-135°$

$C = \sqrt{5^2 + 5^2} = 7.07$

$-5 - j5 = 7.07\ \angle 225°$, also $7.07\ \angle -135°$

d. $-4 + j3$

A is −, so $\theta = \pm 180° + \tan^{-1}\left(\frac{3}{-4}\right) = \pm 180° - 36.87°$

$\theta = 143.13°$, also $-216.87°$

$C = \sqrt{4^2 + 3^2} = 5$

$-4 + j3 = 5\ \angle 143.13°$, also $5\ \angle -216.87°$

When using an electronic calculator to do the conversion, the angle is given directly, if the signs of A and B are entered.

E.3 Single Arithmetic Operations

Addition and Subtraction

The rectangular form is more convenient for these operations. To perform them, the real terms are added algebraically to give one real term. Then the j terms are added algebraically to give one j term. Real and j terms are never mixed.

Examples

a. $(3 + j4) + (6 - j2) = (3 + 6) + j(4 - 2) = 9 + j2$

b. $(6 + j8) - (2 + j4) = (6 - 2) + j(8 - 4) = 4 + j4$

c. $(-2 - j3) - (3 - j6) = (-2 - 3) + j(-3 + 6) = -5 + j3$

Multiplication

The polar form is most suitable for this operation. The magnitudes are multiplied to give one magnitude. The angles are added algebraically to give the angle.

Examples

a. $(5 \angle 30°)(6 \angle 20°) = (5)(6) \underline{/30° + 20°} = 30 \angle 50°$
b. $(5 \angle 60°)(6 \angle -40°) = (5)(6) \underline{/60° + (-40°)} = 30 \angle 20°$
c. $(4 \angle -100°)(8 \angle -30°) = (4)(8) \underline{/-100° + (-30°)} = 32 \angle -130°$

Division

The polar form is most suitable for division. The magnitude of the divisor (bottom) is divided into the magnitude of the dividend (top). The angle of the divisor is subtracted algebraically from the angle of the dividend.

Examples

a. $\dfrac{24 \angle 30°}{6 \angle 150°} = \left(\dfrac{24}{6}\right) \underline{/30° - 150°} = 4 \angle -120°$

b. $\dfrac{8 \angle 60°}{4 \angle -120°} = \left(\dfrac{8}{4}\right) \underline{/60° - (-120°)} = 2 \angle 180°$

c. $\dfrac{15 \angle -80°}{5 \angle 20°} = \left(\dfrac{15}{5}\right) \underline{/-80° - (20°)} = 3 \angle -100°$

d. $\dfrac{9 \angle -120°}{3 \angle -40°} = \left(\dfrac{9}{3}\right) \underline{/-120° - (-40°)} = 3 \angle -80°$

E.4 Combined Operations

Combined operations are done by dividing the problem into groups of single operations. The single operations are performed until there is no more than one real term and one *j* term. Doing the single operations might require converting from one form to the other.

Examples

a. $\dfrac{(5 \angle 30°)(8 \angle -20°)}{(3 + j4) + (4 - j2)} = \dfrac{40 \angle 10°}{7 + j2}$

$= \dfrac{40 \angle 10°}{7.28 \angle 15.94°} = 5.49 \angle -5.94°$

b. $\dfrac{(10 \angle 60°)(3 + j4)}{8 \angle 60° + (5 - j2)} = \dfrac{(10 \angle 60°)(5 \angle 53.13°)}{4 + j6.93 + 5 - j2}$

$= \dfrac{50 \angle 113.13°}{9 + j4.93} = \dfrac{50 \angle 113.13°}{10.26 \angle 28.7°} = 4.87 \angle 84.43°$

ANSWERS TO QUESTIONS AND PROBLEMS

The Field of Electricity and Electronics

Answers to Questions

1. There are many milestones that could be selected. For example:
   - The development of the battery by Alessandro Volta in 1800. This provided a useful power source that was transportable.
   - The development of the vacuum tube, which was pioneered by Fleming and DeForest in the early 1900's.
   - The development of the transistor by John Bardeen, Walter Brattain, and William Shockley at Bell Labs in 1947. This ushered in the age of semiconductors.
2. The invention of the microprocessor has had a tremendous impact on the field of electronics and our daily lives. We now have desktop, laptop, and palm-size computers. Microprocessors are embedded in many products, from toys to cars to household appliances.
3. The term "engineering technologist" refers to a graduate of a four-year curriculum in engineering technology. The term "engineering technician" refers to a graduate of a two-year program in engineering technology.
4. The skills needed by an engineering technologist and technician include hardware skills, needed to prototype and test electronic systems, software skills to create test programs and apply software-based engineering tools, and system level skills to install and implement electronics systems and networks.
5. Some of the tools-of-the-trade include electronics test instruments, such as digital multimeters, oscilloscopes, function generators, power supplies, logic and spectrum analyzers, and a host of other instruments. Increasingly important tools are the personal computer and engineering workstation, which provide simulation, analysis, design, and data acquisition capabilities.
6. Students can exercise their creativity in writing a news article. The article should answer the basic questions of who, when, what, etc.
7. This question is not answered directly in the brief history that is given in the chapter. A number of books on the history of technology contain information on the development of the vacuum tube. One book is *From Compass to Computer*, by W. A. Atherton, San Francisco, The San Francisco Press, Inc., 1984.
8. The engineering workstation has changed the way circuits are prototyped. Rather than soldering or wire-wrapping circuits to see if they are going to work, now circuits are designed and tested using engineering workstations. The advantage is the speed at which products can be developed and then manufactured.
9. This is an open-ended question that will allow students the opportunity to reflect on the impact of technology in their lives. There is no right or wrong answer, and each student will bring an individual perspective to the response to this question.

CHAPTER 1 Answer to Questions

1. A unit of measurement is a definite amount of a quantity that is recognized and accepted as a standard of measurement.
2. A system of units is a group of units that are related in some way.
3. International System of Units (SI)
4. Base, derived, and supplementary
5. Base: ampere, second, meter, mole, kilogram, candela, kelvin
   Derived: volt, ohm, joule, newton, farad
   Supplementary: radian, steradian
6. Prefix
7. 1,000,000, 0.000000001, 0.000001, 0.001, 0.01
8. meters per second, not meter per second; Celsius, not celsius; 0.15, not .15

9. Conversion factor
10. Incorrect; the units are ft/s.
11. Approximate
12. When it is between two nonzero digits.
13. A significant digit is one that has a reasonable amount of certainty.
14. Rounding off means dropping the unwanted digits so that the remaining number is as near to the original number as possible. Dropping digits simply results in a number with fewer digits, but the last remaining digit is unchanged.
15. The 5 should be left unchanged.
16. The last complete column.
17. The smallest number of significant digits in any of the numbers.
18. 5.5×10^3
19. Positive (+)
20. Negative (−)
21. It is equal to the number of places that the decimal is moved.
22. It is the algebraic sum of the exponents.
23. It is equal to the algebraic difference of the numerator exponent minus the denominator exponent.
24. Refer to your calculator or manual.
25. Refer to your calculator or manual.
26. They are used to simulate actual conditions and to test your ability to apply the theory.
27. 1. Read the problem statement.
    2. Draw a diagram.
    3. Form a path linking the knowns and the unknowns.
    4. Solve the equations.

CHAPTER 1 Answers to Problems

1. Base: kilogram, second, meter, ampere, candela
   Derived: joule, volt, coulomb, ohm
3. **(a)** m, meter **(b)** s, second **(c)** Hz, hertz **(d)** N, newton **(e)** J, joule
5. **(a)** 1 gigahertz, 1 GHz **(b)** 1 microampere, 1 μA **(c)** 1 picofarad, 1 pF **(d)** 1 centimeter, 1 cm
7. **(a)** Mega, M **(b)** kilo, k **(c)** giga, G **(d)** milli, m **(e)** micro, μ **(f)** nano, n
9. **(a)** joule per coulomb **(b)** volt per ampere **(c)** volt-second **(d)** ampere-second
11. **(a)** 1000 kHz **(b)** 5000 pF **(c)** 6.4 MW
13. 1006.5 m
15. **(a)** 0.0568 mile **(b)** 91.41 m
17. **(a)** 9.3 m^2 **(b)** 9.3×10^{-6} km^2
19. 29.4°C
21. **(a)** 3 **(b)** 2 **(c)** 2 **(d)** 4
23. **(a)** 0.027 **(b)** 5.2 **(c)** 3.1 **(d)** 37 **(e)** 2.6 **(f)** 3.7
25. In position A, 2.8; in position B, 6.4; only one estimated digit is acceptable.
27. **(a)** 285 **(b)** 0.34
29. **(a)** 6.7 **(b)** 245.35
31. **(a)** 29 **(b)** 0.15 **(c)** 0.00539 **(d)** 110
33. **(a)** 1.6×10^3 **(b)** 2.5×10^4 **(c)** 2.5×10^{-1}
35. **(a)** 3.5×10^{-3} **(b)** 1.55×10^5 **(c)** 2.58×10^1
37. **(a)** 5000 **(b)** 2,500,000 **(c)** 0. 04 **(d)** 0.00081
39. **(a)** 7.6×10^4 **(b)** 1.000×10^3 **(c)** 2.50000×10^5 **(d)** 1.2×10^2
41. **(a)** 2.7×10^6 **(b)** 3.96×10^7 **(c)** 1.225×10^{-11} **(d)** 5.04×10^4
43. **(a)** 7.68×10^2 **(b)** 2.016×10^{-15}
45. **(a)** 6×10^7 **(b)** 6×10^{-3} **(c)** 1.5×10^{-2} **(d)** 6×10^{-10}
47. **(a)** 3×10^8 **(b)** 4×10^{-8}
49. **(a)** 2.531×10^5 **(b)** 1.56×10^{-2}
51. **(a)** 3.4×10^6 **(b)** -2.4×10^{-3}
53. **(a)** Not enough; 3 unknowns; 2 equations **(b)** Enough; 3 of each
55. 200 of 25 Ω; 200 of 10 Ω; 1000 of 5 Ω
57. Box A is 5 N; box B is 2.5 N; box C is 7.5 N.
59. 2.61 minutes

CHAPTER 2 Answers to Questions

1. Matter is any substance that has weight and occupies space.
2. Elements are basic substances that cannot be chemically decomposed or formed by the chemical union of other substances. The atom is the smallest particle of an element that can enter into chemical reaction with other particles. A molecule is the smallest particle of a compound that can exist and retain its identity.
3. Electrons, protons, and neutrons.
4. Electrons whirl about the nucleus in set orbits called shells. Each shell represents an energy level. The shells can contain a set number of electrons. If the outer shell is completely filled, the element will not react with other elements.
5. Their whirling in orbit exerts an outward force on them to counteract the attraction of the nucleus.
6. The force increases as the separation is made smaller.
7. Coulomb's Law.
8. Valence electrons are electrons in the outer shell of an atom.
9. Free electrons are valence electrons that have been freed from the outer shell.

10. By rubbing a hard rubber rod with fur and by walking across a synthetic-material rug with leather-soled shoes.
11. **(a)** Preventing the buildup. **(b)** Draining off the charge. **(c)** Neutralizing the charge. **(d)** Reducing the effect of the charge.
12. The release of charge when a charged body comes in contact with another object.
13. The high voltage during discharge can damage semiconductor devices.
14. Electric current is the movement of charged particles, either positive or negative, in some general directions in a material.
15. Electric current was first considered a flow of fluid from the positive terminal, through the circuit, to the negative terminal. This was later shown to be incorrect, so it was given the name ''conventional current'' to distinguish it from electron current.
16. The 6-V battery has a greater potential difference than the 1.5-V battery.
17. The source of emf originated from the belief that the current was maintained by a force called electromotive force (emf).
18. Circular mil
19. CM is equal to the diameter in mils squared.
20. A mil is a thousandth of an inch.
21. $\pi/4$ circular mils
22. Resistivity is the resistance of a section of material one unit in length with a cross section of one unit of area.
23. Circular mil-ohms per foot and ohm-meter
24. The resistance will be doubled.
25. The resistance will be one half of what it was.
26. Increases
27. The temperature coefficient of resistance is the amount of resistance change per ohm of resistance per each degree change in temperature.
28. Superconductivity is the condition in which the resistance of a conductor approaches zero ohms.
29. A linear resistance has the same resistance within its current range. The resistance of a nonlinear resistance changes with current.
30. Thermistor and varistor
31. Carbon composition, wirewound, and thin film
32. A carbon composition resistor has a core of a carbon and ceramic mixture. A thin film resistor has a thin film of metal deposited on a ceramic core.
33. Fixed resistor, tapped resistor, variable resistor
34. The rheostat is a variable resistor used to control current. The potentiometer is a three-terminal variable resistor that is used to obtain different voltages from a fixed voltage.
35. A rheostat
36. Greater
37. First band–first digit; second band–second digit; third band–multiplier.
38. The fourth band
39. Black 0, brown 1, red 2, orange 3, yellow 4, green 5, blue 6, violet 7, gray 8, white 9
40. Red
41. No, because string is an insulating material.
42. There will be current.
43. The plastic is an insulator and keeps the conductors from making contact with each other, some person, or another object.
44. No
45. Because it is touching only one wire, so there is no potential difference across its body.
46. Refer to the end of Section 2.11.

CHAPTER 2 Answers to Problems

1. 6.24×10^{18}
3. 1.072×10^{-25} kg
5. **(a)** 4.5 N **(b)** repulsion
7. 2.63 N to the right
9. 8.223×10^{-8} N
11. 2 C
13. 4 s
15. 5760 C
17. 100 J
19. 144 J
21. **(a)** 22,500 **(b)** 40,000 **(c)** 8199.5 **(d)** 158.72
23. **(a)** 50 mils **(b)** 30 mils **(c)** 40 mils **(d)** 35 mils
25. 63,662 CM
27. 98,676.3 CM
29. 0.0998 ohms
31. 4861 m
33. Aluminum
35. 90.2 ohms
37. 1.141 ohms
39. 0.59 ohm
41. $\alpha = 0.003$; material is platinum.
43. Nonlinear
45. **(a)** 3.96 ohms **(b)** 3.93 ohms **(c)** 0.318 ohm
47. 0.159 ohm for #12 wire, 0.0999 ohm for #10 wire
49. 59.4 m
51. 10.375 CM-Ω/ft
53. **(a)** 18 MΩ ± 1.8 MΩ **(b)** 39 kΩ ± 1.95 kΩ **(c)** 2.7 kΩ ± 135 Ω
55. **(a)** Red, red, brown, no tolerance band **(b)** Brown, brown, orange, gold **(c)** Blue, gray, green, silver

57. **(a)** 4%, gold **(b)** 7%, silver
59. Minimum, 3510 Ω; maximum 4290 Ω

CHAPTER 3 Answers to Questions

1. A block diagram shows the function of sections of a circuit and the signal flow through the circuit. A circuit diagram shows the interconnection of the circuit components.
2. A graphic symbol is a simple line diagram that represents a part.
3. A letter symbol
4. To distinguish between letter symbols of the same kind.
5. Terminal B
6. A drop from A to B
7. By connecting the meter to obtain an upscale deflection or a positive readout. The positive terminal is the one that is connected to the (+) terminal of the meter.
8. Voltmeter
9. Ammeter
10. The 6-MΩ meter
11. ± 5% on the 10-V range
12. The 5.15-V meter
13. The 0.1-Ω meter
14. The meter will be damaged.
15. The lead from the + terminal of the meter is connected to the − terminal of the resistor.
16. $V = RI$
17. Brighter, and possibly burn out
18. Power is the rate at which work is done or energy is converted.
19. The watt
20. $P = VI$
21. In the long shunt connection, the voltmeter is connected across the series combination of the ammeter and the load. In the short shunt, it is connected across the load.
22. Long shunt
23. If the coils are connected so that the current enters the ± terminal of one and leaves the ± terminal of the other.
24. Energy is the product of power and time.
25. Joule and kilowatt-hour
26. The power output will be smaller.
27. For a given power input, an increase of losses will result in lesser efficiency.
28. No, because the system efficiency is the product of the part efficiency.
29. A short circuit
30. Open circuit
31. A fuse and a circuit breaker
32. Zero ohms
33. Infinite ohms

CHAPTER 3 Answers to Problems

1. **(a)**

(b)

(c) A

(d)

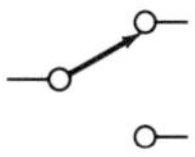

3. **(a)** 2 **(b)** A **(c)** 2 **(d)** 3
5. **(a)** 5(−), 3(+) **(b)** 2(−), 4(+) **(c)** 4(−), 5(+)
7. **(a)** 3(+), 4(−) **(b)** 1(+), 2(−) **(c)** 1(+), 2(−)
9. 1.16 A minimum; 1.24 A maximum
11. 116.8 V minimum; 123.2 V maximum
13. 0.2 A
15. 120 V
17. 200 Ω
19. (−) to tap 1, 36 Ω; tap 1 to tap 2, 24 Ω; tap 2 to (+), 8 Ω
21. 1150 W
23. 100 V
25. 288 Ω
27. 125 W
29. Power is 14.4 W, so it will be damaged.
31. 99.95 W
33. **(a)** 0.9999 **(b)** 0.3333
35. **(a)** 744 W **(b)** 1,339,200 J
37. 1.25 hours
39. $5.15
41. **(a)** 414.44 W **(b)** 3.45 A
43. 69.1%
45. 72 W
47. **(a)** 84.85% **(b)** 15.01 HP

CHAPTER 4 Answers to Questions

1. Secondary cell
2. Carbon-zinc, zinc-chloride, alkaline-manganese dioxide, mercuric oxide
3. Electrolyte, a negative electrode, a positive electrode, and a case
4. Nickel-cadmium
5. No, because its emf is 1.5 volts.

6. Ampere-hour expresses the number of hours that a cell or battery can supply a number of amperes.
7. Shelf time, cutoff voltage, current drain, duty cycle, and operating temperature
8. The 600 minutes at 1 ampere
9. The period is the reciprocal of the frequency.
10. The frequency will be halved.
11. The rms value is equal to 0.707 times the peak value.
12. The rise time is measured between the 10% and 90% points on the leading edge of the waveform.
13. Refer to the five steps in Section 4.6.
14. The oscilloscope shows a graph of voltage versus time.
15. DC coupling lets one view the dc and ac in a signal. AC coupling blocks out the dc and permits only the alternating components to be displayed.
16. Correct triggering prevents an overlay of images and determines at what point in the image the sweep starts.

CHAPTER 4 Answers to Problems

1. 80 hours
3. 8 hours
5. 75 W
7.

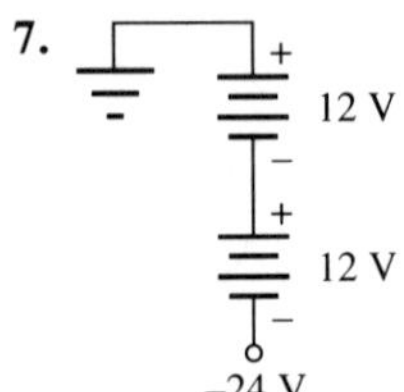

9. Peak-to-peak = 161.2 V
11. Volts

8
0
0
0.25
0.5
Time in ms

13. (a) Decrease (b) Remain the same (c) Decrease
15. Settings:

Frequency dial	2.5
Multiplier	10^3
Function	~
Amplitude	5 V
Offset	OFF or ON

17. Period = 20 ns
    Frequency = 50 MHz
    Rise time = 2 ns
    Fall time = 5 ns
    Duty cycle = 45%
19. V/div 2 V/div
    T/div 50 μs/div
    0 volt line center of graticule
    ac or dc coupling
21.

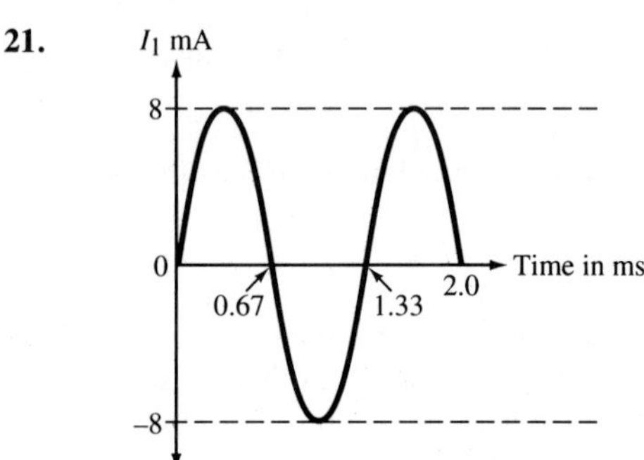

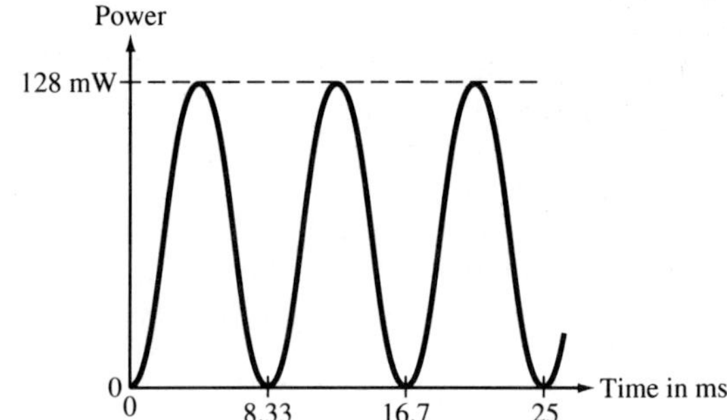

23.

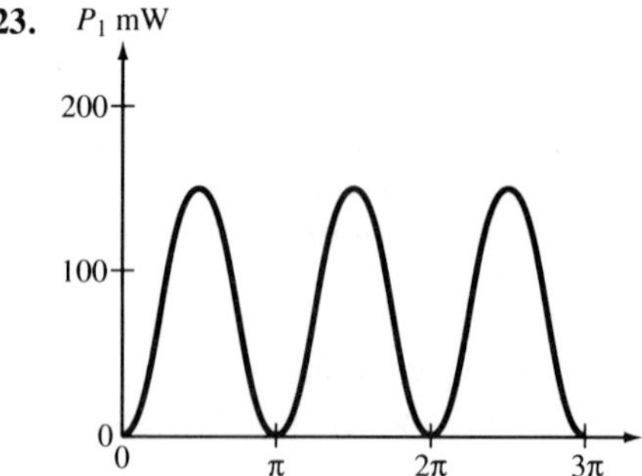

CHAPTER 5 Answers to Questions

1. A series circuit does not have any branch point between any two components.
2. No
3. Kirchhoff's Voltage Law

4. **(a)** Measure the resistance. **(b)** Measure voltage and current and use Ohm's Law. **(c)** Calculate the resistance.
5. A short circuit will make the equivalent resistance smaller.
6. A short circuit will cause an increase in the current. An open circuit will result in no current.
7. The current does not have to be known or calculated.
8. Total voltage, voltage across a part of the circuit, resistance of the part of the circuit, equivalent resistance of the circuit.
9. Total power is equal to the sum of the power in each part.
10. Since $P = VI$, a short circuit will result in an increase in power as long as V remains unchanged. If a resistor is open-circuited, the current will be zero, so the power will be zero.
11. The 75-W bulb
12. Energy is the product of power and time.
13. V_{BA} is -8 V.
14. Zero volts
15. Start at the second point and algebraically add the rises and drops in tracing a path through the circuit to the first point.
16. (1) Limit or interrupt current. (2) Protect devices from damage. (3) Reduce voltage.
17. A change in current in one part of a series circuit results in a change in all parts of the circuit.
18. Because the conductor will not produce an open circuit to protect the circuit from damage if the current is high.
19. The voltmeter will indicate a voltage equal to the circuit source voltage when connected across the open bulb. The voltage will be zero across the others.

CHAPTER 5 Answers to Problems

1. 40 V
3. **(a)** 30 V across R_1, 40 V across R_3 **(b)** $R_3 = 80\ \Omega$
5. **(a)** 24 V **(b)** 24 V for both values
7. **(a)** 20 V **(b)** 50 V **(c)** 40 V **(d)** 70 V
9. **(a)** 560 Ω **(b)** 68 Ω
11. 5
13. 30 V
15. 210 V
17. **(a)** 240 W **(b)** 0.18 kWh in R_1, 0.24 kWh in R_2, 0.3 kWh in R_3, 0.72 kWh total
19. **(a)** 4.8 W **(b)** 138,240 J
21. **(a)** 150 W in R_1, 200 W in R_2, 300 W in R_3 **(b)** 650 W
23. $V_{AB} = 10$ V, $V_{CA} = -18$ V, $V_B = 14$ V
25. **(a)** 0.2 A **(b)** $V_1 = 40$ V, $V_2 = 60$ V **(c)** 20 W **(d)** 20 W
27. **(a)** 117 V **(b)** 1170 W **(c)** 11.7 Ω
29. **(a)** 600 Ω **(b)** 5 V **(c)** 10 W **(d)** 5.56 V and 11.11 W
31. 481 Ω

CHAPTER 6 Answers to Questions

1. The ground or reference node is Node 0.
2. Names of components must begin with a letter. After that, they can contain either letters or numbers, up to 131 characters.
3. The suffixes have the following meanings:
   F = 10^{-15}
   P = 10^{-12}
   N = 10^{-9}
   U = 10^{-6}
   MIL = 25.4×10^{-6}
   M = 10^{-3}
   K = 10^{3}
   MEG = 10^{6}
   G = 10^{9}
   T = 10^{12}
4. The following parameters must be given in a resistor component line: name of the component, the (+) node identifier, the (−) node identifier, and the component value.
5. A dc voltage source component line includes the source name, the (+) node identifier, the (−) node identifier, and the voltage value.
6. The major parts of a PSpice source file include title line, component lines, command lines, and `.END` statement.
7. A dc analysis using PSpice yields the node voltages, the currents through each voltage source, and the total power dissipated in the circuit.
8. To describe an ac voltage or current source, the name of the source, the (+) node identifier, the (−) node identifier, and the magnitude and phase of the ac source must be given.
9. An ac command statement must include `.AC` followed by the type of sweep, the number of points to be calculated, the start frequency, and the end frequency.
10. The `.AC` command is used in conjunction with a `.PRINT`, `.PLOT`, or `.PROBE` command to produce the frequency response of a circuit over a range of frequencies.

11. The parameters required by the `.PRINT` command are the type (DC, AC, NOISE, or TRAN) and a list of output variables.

12. The parameters required by the `.TRAN` command include the print step value and the final time value.

CHAPTER 6 Answers to Problems

1. (a) 15MV (b) 75UA (c) 4.7KOHMS (d) 1NF (e) 25KHZ

3.
```
R3   2   3   100OHMS
R4   3   0   68OHMS
```

5.

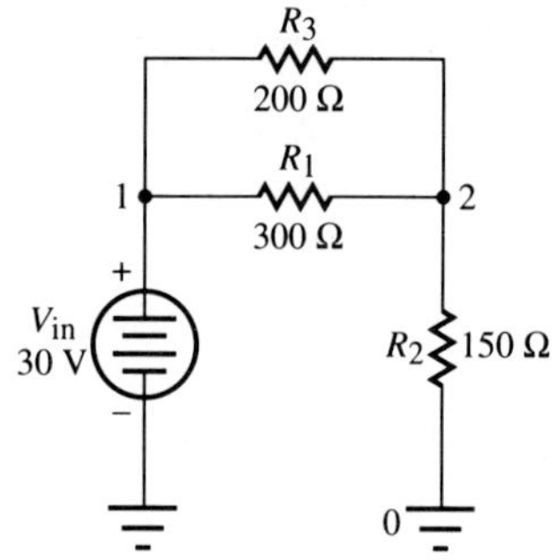

R3 200 Ω; R1 300 Ω; 1; 2; + V_{in} 30 V −; R_2 150 Ω; 0

7.
```
FIGURE 6.23
VS   1   0   52V
R1   1   2   500
R2   2   0   800
.END
```

9. Yes, the voltages at Node 2 and Node 3 can be almost the same, since there is a current of 1.428 mA flowing and a resistance of 5 ohms between the two nodes.

11. a. V_s; 2; 3; $5\text{ V}\angle 0°$

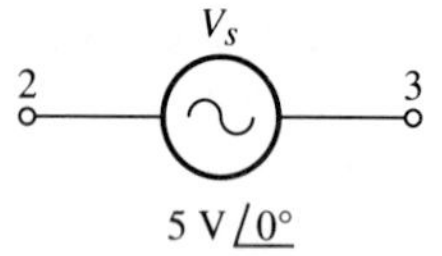

b. V_{in}; 1 +; − 0; 15 V

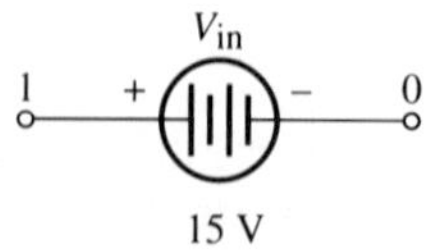

c. V_4; 4; 2; $5\angle 0°$ mV

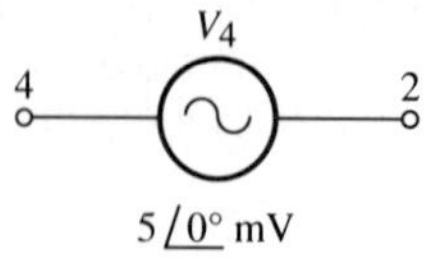

d. V_{AB}; 6; 3; $0.25\angle 45°$ V

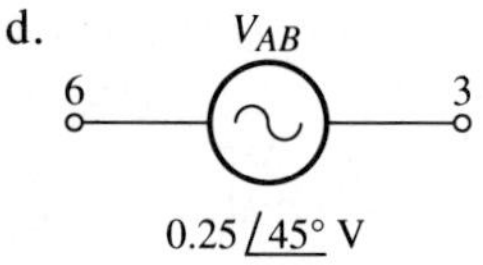

13. (a) 100, 110, 120, 130, 140, 150, 160, 170, 180, 190, 200
(b) 5000
(c) Six logarithmically spaced frequencies between 1 kHz and 10 kHz

15.
```
FIGURE 6.30
V1   1   0   2V
V2   3   0   AC   5V
R1   1   2   250
R2   2   3   750
.END
```

17.
```
FIGURE 6.31
VS   1   0   AC        300MV
R1   1   2    1KOHM
R2   2   3    2KOHM
R3   3   0   12KOHM
.AC    DEC    1KHZ    100KHZ
.PRINT    AC    VM(3)    VP(3)
.END
```

19.
```
FIGURE 6.33
VS   1   0   SIN(   0   60V   400HZ )
R1   1   2   350
R2   2   3   450
R3   3   0   1200
.TRAN   0.01MS   5.0MS
.PROBE
.END
```

CHAPTER 7 Answers to Questions

1. They will be connected across the same two points in a circuit.
2. Opening one path will not affect the voltage across the other path.
3. Opening one path will not affect the current in the other path.
4. A short circuit of one path will reduce the current to zero amperes in the other path.
5. A change in one path has little or no effect on the other parallel paths.
6. The smallest current is in the 15-ohm resistor. The largest current is in the 5-ohm resistor.
7. Kirchhoff's Current Law states that the current entering a parallel group must equal the sum of the currents in each branch.
8. Adding another parallel resistor decreases the equivalent resistance. Removing a resistor increases the equivalent resistance.

9. $I_x = \dfrac{I_T R_{eq}}{R_x}$
10. 30 W
11. The total energy is equal to the sum of the energy in each branch.
12. 1.999 A
13. 22.5 ohms
14. The increase could have been caused by a short circuit of some of the resistance in the branch.
15. The increase in current results in a larger voltage drop in the line. This reduces the voltage at the television set.
16. It should not affect the other lamps.
17. The lamps will be much dimmer when connected in series because the voltage across each lamp will be about 30 V.

CHAPTER 7 Answers to Problems

1. 6 A
3. (a) 0.5 A in the soldering iron, 2 A in the heater (b) 2.5 A (c) 0.5
5. 10 A
7. 240 Ω
9. 20 Ω
11. 27.43 Ω
13. 1200 Ω
15. (a) 4.5 A (b) 1.5 A
17. 0.75 A
19. 3 A
21. (a) 0.7 A (b) 0.4 A (c) 0.1 A
23. 300 Ω
25. 240 W
27. (a) 1300 W (b) 10.83 A
29. 150 W
31. (a) 4,536,000 J (b) 1.26 kWh
33. (a) 5.2 A (b) 124.8 W
    (c) $I_1 = 0.4$ A, $I_2 = 0.8$ A, $I_3 = 1.6$ A, $I_4 = 2.4$ A
    (d) $P_1 = 9.6$ W, $P_2 = 19.2$ W, $P_3 = 38.4$ W, $P_4 = 57.6$ W
35. (a) $I_1 = 2.5$ A, $I_2 = 0.5$ A, $I_3 = 1$ A (b) 125 V
    (c) 5.4×10^6 J
37.
```
FIGURE 7.41
VS    1   0   12V
R1    1   2   300
R2    1   3   900
VM1   2   0
VM2   3   0
.END
```
39. 5.555 Ω
41. 99.5 Ω

CHAPTER 8 Answers to Questions

1. Use the method explained in Section 8.1.
2. Because each circuit can have a different configuration.
3. In analysis, the component values are known and one determines the conditions. In synthesis, the conditions are known and one determines the component values.
4. Source A has the larger internal resistance and percent voltage regulation.
5. The top part
6. The load voltage would decrease if nothing else was changed.
7. The output voltage will decrease.
8. Volume control, tone control, light dimmer, and power supply output control
9. There would be a large power loss in the divider.
10. The output voltage will increase if the bleeder resistor is opened. The voltage will be zero volts for the short circuit.
11. Zero volts to 12 volts
12. The change in the circuit conditions caused by inserting the meter in the circuit.
13. Lower
14. High resistance
15. Low resistance
16. The loading can cause an increase in current in other parts of the circuit.
17. Lower
18. Voltmeter B because its loading effect must be smaller.
19. Using the relation $R_i = \dfrac{V_{NL} - V_L}{I_L}$
20. Decrease
21. The voltage can damage the ohmmeter.
22. It will cause a decrease in voltage.
23. Increases
24. It causes the terminal voltage to be lower when supplying current than at no load.
25. The armature winding resistance, brush resistance, contact resistance, and lead resistance
26. Electrolyte and electrode resistance
27. The internal resistance is constant.
28. The terminal voltage of each must be equal for all load conditions.
29. Because all of the current will go through the path of zero resistance.
30. A series connection
31. A parallel connection
32. The connection should be three parallel groups with three cells connected in series in each group.

CHAPTER 8 Answers to Problems

1. 100 Ω
3. 100 Ω
5. 30 Ω
7. 300 Ω
9. **(a)** 1.25 A **(b)** 112.5 V **(c)** 150 W
   (d) 87.89 W in the 144-Ω lamp, 52.73 W in the 240-Ω lamp
11. **(a)** 0.16 A **(b)** $V_1 = 8$ V, $V_2 = 16$ V, $V_3 = 12$ V, $V_4 = V_5 = 12$ V
    (c) $I = 0.08$ A, $I_2 = 0.08$ A, $I_3 = 0.08$ A, $I_4 = 0.04$ A, $I_5 = 0.04$ A
    (d) 3.84 W
13. **(a)** The lamp is not lit.
    (b) 192 W **(c)** 80 V
15. 1264.7 Ω
17. **(a)** 8 V **(b)** 0.107 A **(c)** 0.027 A
    (d) $P_A = 1.72$ W, $P_B = 0.219$ W
19. **(a)** 533.3 Ω **(b)** 8 V **(c)** 0.24 W
21. **(a)** $V_{L1} = 10.1$ V, $V_{L2} = 5$ V
    (b) $I_{L1} = 0.0168$ A, $I_{L2} = 0.0025$ A
    (c) $P_{R1} = 0.386$ W, $P_{R2} = 0.054$ W, $P_{R3} = 0.043$ W
    (d) 8.75 V
23. $R_A = 68.97$ Ω, $R_B = 500$ Ω
25. **(a)** 80 V **(b)** 79.67 V
27. **(a)** 0.17 A **(b)** 0.2 A
29. **(a)** 9.81 mA and 9.91 V **(b)** 10 mA and 10 V
31. 24 V
33. 4 Ω
35. 20% or less
37. 11.1%
39. $E = 24$ V top terminal (+), $R_i = 3$ Ω
41. $I = 3$ A up, $R_p = 4$ Ω
43. **(a)** $I = 10$ A, $R_p = 6$ Ω
    (b) $E = 60$ V, $R_i = 6$ Ω
45. **(a)** $E = 24$ V, $R_i = 1$ Ω
    (b) $I = 24$ A, $R_p = 1$ Ω
47. $I = 7.5$ A up, $R_p = 0.2$ Ω
49. **(a)** 35 V **(b)** 0.175 A **(c)** 0.8 A
51. **(a)** 10.8 V **(b)** 0.3 A **(c)** 266.7 hours
53. **(a)** $E_{eq} = 3$ V **(b)** 2.93 V **(c)** 2 A
55. $E_{eq} = 92.5$ V, top terminal (+); $R_{ieq} = 50$ Ω
57. 18.28 V
59.
```
FIGURE 8.41
VS    1    0    10V
R1    1    2    50
R2    2    0    100
R3    2    3    50
R4    3    0    100
R5    3    4    40
R6    4    0    60
.DC   VS   10   10   10
.PRINT DC I(R1), I(R2), I(R3)
.PRINT DC I(R4), I(R5), I(R6)
.END
```
61.
```
FIGURE 8.43
VS    1    0    10V
R1    1    2    68
R2    2    0    96
R3    2    3    48
R4    2    3    48
R5    3    0    60
R6    3    0    40
.DC   VS   10   10   10
.PRINT DC I(R1), I(R2), I(R3)
.PRINT DC I(R4), I(R5)
.END
```

CHAPTER 9 Answers to Questions

1. Loop current, branch current, and nodal analysis
2. **(a)** Three loops **(b)** Three loops
3. Yes
4. In general, the branch current is the algebraic sum of the loop currents in the branch.
5. The negative sign means that the current has a direction that is opposite to that which was assigned to it.
6. Calculate the source current and use Ohm's Law, $R_{eq} = E/I_s$.
7. Nodal analysis
8. Zero volts
9. From Node A to Node B
10. No
11. Four
12. No
13. Into the node in order to satisfy Kirchhoff's Current Law.

CHAPTER 9 Answers to Problems

1. For loop ABC: $70I_A - 10I_B - 40I_D - 120 = 0$
   For loop BDC: $60I_B - 10I_A - 30I_C = 0$
   For loop CDE: $100I_C - 30I_B - 50I_D + 60 = 0$
   For loop ACE: $150I_D - 40I_A - 50I_C = 0$
3. **(a)** $I_1 = 0.429$ A to right, $I_2 = 1.143$ A down, $I_3 = 0.714$ A up
   (b) $V_1 = 4.29$ V, $V_2 = 5.72$ V, $V_3 = 14.28$ V
   (c) $P = 18.57$ W
5. 0 V
7. **(a)** $I_1 = 2.25$ A right, $I_2 = 0.710$ A right, $I_3 = 0.065$ A right, $I_4 = 0.645$ A down

(b) $V_1 = 4.5$ V, $V_2 = 2.13$ V, $V_3 = 0.78$ V, $V_4 = 3.87$ V
(c) $I_{E1} = 2.96$ A, $I_{E2} = 2.25$ A, $I_{E3} = 2.315$ A

9. $I_1 = 0.588$ A right, $I_2 = 0.839$ A down, $I_3 = 1.427$ A down **(b)** 240 V

11. 405 W

13. **(a)** $I_1 = 0.2$ A down, $I_2 = 1.1$ A up, $I_3 = 0.1$ A down, $I_4 = 0.8$ A down
(b) $V_1 = 2$ V, $V_2 = 20$ V, $V_3 = 2$ V, $V_4 = 8$ V
(c) 31.2 W

15. 0 V

17. **(a)** $I_1 = 2.21$ A right, $I_2 = 0.710$ A right, $I_3 = 0.065$ A right, $I_4 = 0.645$ A down
(b) $V_1 = 4.5$ V, $V_2 = 2.13$ V, $V_3 = 0.78$ V, $V_4 = 3.87$ V
(c) $I_{E1} = 2.96$ A, $I_{E2} = 2.25$ A, $I_{E3} = 2.315$ A

19. Node B:
$$V_B(1/R_{s1} + 1/R_1 + 1/R_2) - V_C(1/R_1) - V_D(1/R_2) = E_1(1/R_{s1})$$
Node C:
$$V_C(1/R_1 + 1/R_4 + 1/R_3 + 1/R_5) - V_B(1/R_1) - V_D(1/R_3) - V_E(1/R_5) = 0$$
Node D:
$$V_D(1/R_2 + 1/R_3 + 1/R_{s2}) - V_B(1/R_2) - V_C(1/R_3) - V_E(1/R_{s2}) = E_2(1/R_{s2})$$
Node E:
$$V_E(1/R_6 + 1/R_5 + 1/R_{s2}) - V_C(1/R_5) + V_D(1/R_{s2}) = -E_2(1/R_{s2})$$

21. **(a)** $I_1 = 0.429$ A right, $I_2 = 1.142$ A down, $I_3 = 0.714$ A left
(b) $V_1 = 4.29$ V, $V_2 = 5.71$ V, $V_3 = 14.28$ V
(c) 18.56 W

23. $V_{L1} = 5.373$ V, $V_{L2} = 9.403$ V, $I_s = 0.26$ A

25. **(a)** $I_{E1} = 2.964$ A up, $I_{E3} = 2.32$ A up
(b) 3.86 V

27. 2 Ω, E_2 does not supply power

CHAPTER 10 Answers to Questions

1. Generally, because the resistance will not be the same value for the current from each source and for the total current in it from all sources.
2. Because the total current squared does not have the same value as the sum of the individual currents squared.
3. The other sources are replaced by their internal resistance.
4. The sources are replaced by their internal resistance.
5. It is useful for reducing calculations in which repetitive calculations are needed and for replacing electrical devices by an electrical circuit.
6. Yes
7. Yes
8. E_{TH} is equal to the open circuit voltage. I_N is equal to the short circuit current. R_{TH} and R_N are equal to the open circuit voltage divided by the short circuit current.
9. At maximum power transfer, R_L is equal to R_i.
10. 50%
11. It is lost in the internal resistance of the source.
12. Efficiency and power both decrease.
13. Because 50% of the generated energy will be lost in the lines.
14. 50%
15. 50%
16. The resistance across each pair of terminals must be the same for each network.
17. Pi
18. Tee

CHAPTER 10 Answers to Problems

1. **(a)** $I_1 = 0.428$ A to right, $I_2 = 1.142$ A down, $I_3 = 0.714$ A up
(b) $V_1 = 4.28$ V, $V_2 = 5.71$ V, $V_3 = 14.28$ V
(c) 18.56 W

3. **(a)** $I_1 = 0.685$ A down, $I_2 = 0.856$ A up, $I_3 = 0.171$ A to right, $I_4 = 0.257$ A down, $I_5 = 0.086$ A up
(b) 1.7 V **(c)** 21.03 W

5. **(a)** 24 V **(b)** 0.12 A up **(c)** 208.8 W

7. $E_{TH} = 16.65$ V top terminal (−), $R_{TH} = 3.33$ Ω

9. $E_{TH} = 16.63$ V, $R_{TH} = 10$ Ω, $V_5 = 11.09$ V, $P_5 = 6.15$ W

11. 2 Ω

13. $E_{TH} = 8$ V, $R_{TH} = 2$ Ω

15. $I_N = 5$ A down, $R_N = 3.33$ Ω

17. $V_5 = 11.14$ V, $P_5 = 6.2$ W

19. $I_N = 40$ A, $R_N = 3$ Ω

21. 30 Ω

23. **(a)** 4.2 μW **(b)** 3.7 μW

25. 4.73 W

27.
```
PROBLEM 10.27
V     1   0    12V
R1    1   2    100
R2    2   0     25
RL    2   0    RMOD 1
.MODEL    RMOD RES(R=10)
.DC    RES    RMOD(R)    10    100    2
.END
```

29. 10 Ω in the wye opposite R_3, 20 Ω in the wye opposite R_2, 20 Ω in the wye opposite R_1

31. 240 Ω

33. 12.08 Ω across R_1 and R_2 lines, 7.77 Ω across R_2 and R_3 lines, 12.66 Ω across R_1 and R_3 lines

CHAPTER 11 Answers to Questions

1. Capacitance is the property by which two conducting materials separated by an insulator can store charge.
2. The farad
3. Opposes a change in voltage, stores energy, and blocks dc current.
4. Two conducting materials and a dielectric.
5. Permittivity is a measure of how easily the lines of an electrostatic field can be set up in a material. Relative permittivity is the ratio of the permittivity of a material to the permittivity of a vacuum.
6. 1 mm of mica
7. If the breakdown voltage is smaller than the applied voltage, the dielectric will be damaged and the device will no longer be a capacitor.
8. The spacing between the plates and the type of dielectric material.
9. The plate area must be increased.
10. It will be a short circuit and there will no longer be any capacitance.
11. There will not be any effect.
12. Leakage resistance of the dielectric, and leakage because of dirt or other contaminants between the terminals.
13. A low leakage current.
14. No, because the peak voltage will be greater than 400 V.
15. Yes
16. More capacitance
17. The 2-microfarad capacitor
18. An increase of capacitance
19. Variable
20. Electrolytic
21. A varactor
22. Oil, mica, and ceramic
23. Oil
24. If the capacitor is not connected so that the terminal marked (+) is more positive than the one marked (−), the capacitor will break down.
25. They are smaller, the terminals are used both for contact and mounting, and holes for leads are not needed on the circuit board.

CHAPTER 11 Answers to Problems

1. 125 μC
3. 2.5 V
5. 72 μC
7. 16 μF
9. 10 V
11. 24,000 V/m
13. **(a)** 4×10^{-6} C **(b)** 0.2 μF
15. 1.2×10^{6} V/m
17. 3
19. 1.77×10^{-9} F
21. 9.04×10^{-2} m^2
23. 1.18×10^{-12} F
25. 2.223×10^{-9} C
27. 420 kV
29. 0.015 m^2
31. Glass
33. **(a)** 144 μC on each capacitor **(b)** 12 μF, 12 V: 16 μF, 9 V: 48 μF, 3 V **(c)** 6 μF
35. 18 μF
37. 6 μF, 40 V: 4 μF, 60 V: 12 μF, 20 V
39. 250 V
41. 114.29 V
43. **(a)** 36 μF **(b)** 12 μF, 288 μC: 16 μF, 384 μC: 8 μF, 192 μC **(c)** 864 μC
45. 8 μF
47. 3 μF, 250 μC: 6 μF, 500 μC: 9 μF, 750 μC: E = 83.33 V
49. **(a)** 2 μF **(b)** 12 μF and 6 μF, 240 μC: 3 μF, 180 μC: 1 μF, 60 μC **(c)** 12 μF, 20 V: 6 μF, 40 V: 3 μF, 60 V: 1 μF, 60 V

CHAPTER 12 Answers to Questions

1. In the transient state, the conditions are changing with time. In the steady state, they are not changing.
2. Yes
3. By making the resistance larger or the source voltage smaller.
4. The time constant is the time in which the capacitor voltage and current change 63.2% of the total. It is also the time in which the final voltage or current would be reached if the change continued at the initial rate.
5. The time can be increased by making R or C, or both, larger. Making them smaller will decrease the time.
6. The power is at a maximum at 0.693 time constants.
7. The capacitor was discharging through the resistance of the meter. It can be reduced by using a meter that has a larger resistance.

8. They show percent of maximum versus number of time constants instead of capacitor voltage or current versus time.
9. Because they can be used for all values of R and C.
10. The capacitor represents a short circuit at the start of charge and an open circuit in steady state.
11. Volt for E, ohm for R, farad for C, and second for t.
12. The polarity remains unchanged.
13. The current direction during discharge is opposite to that during charge.
14. Both circuits have the same time constant. The maximum current will not be the same. The maximum voltage will be the same.
15. The stored energy is the same since energy depends only on capacitance and voltage.
16. Reducing R after the capacitor has been charged will not have any effect on the voltage.
17. A change in R_2 will cause a voltage change across the capacitor, so a transient condition will exist until a new steady state is reached.
18. The short circuit can cause a high discharge current.
19. The times are not affected by the source voltage.
20. The stored energy can be increased by increasing the applied voltage or increasing the capacitance.
21. The frequency can be increased by making R or C smaller to obtain a smaller time constant.
22. The sawtooth wave would have a larger amplitude and a lower frequency.
23. Three applications are time delay circuits, camera flashes, and oscillators.

CHAPTER 12 Answers to Problems

1. **(a)** 48 mA, 0 V **(b)** 0.032 A, 16 V
3. **(a)** 54.5 μF **(b)** 120 mA
5. 0.25 mA
7. **(a)** 0.375 s **(b)** 10 kΩ
9. **(a)** 20 μF **(b)** 0.12 s **(c)** 11.68 V
11. **(a)** 0 A **(b)** 7.5×10^{-6} A **(c)** -12.5×10^{-6} A
13. **(a)** 60 A **(b)** 0.3 ms
15. **(a)** 2.4 mA **(b)** 0.866 s **(c)** 48 V/s
17. **(a)** 5 V **(b)** 250 V/s **(c)** 3.89 V
19. **(a)** 56.89 V, 77.85 V, 85.41 V **(b)** 7.47 V
21. 26.4 V, 0.66 mA
23. 0.1 A, 2 V
25. **(a)** 10 ms **(b)** 18.64 V, 10.7 mA, 0.8 V
27. **(a)** 15.08 mA **(b)** 750 V/s
29. 10.49 V, 0.602 mA
31. 28.39 V
33. **(a)** 2.25×10^{-3} J **(b)** 6.75×10^{-3} J
35. 0.047 J
37. **(a)** $V_{R1} = 120$ V, $V_{C1} = 120$ V, $V_{R2} = 0$ V, $V_{R3} = 0$ V, $V_{C2} = 0$ V **(b)** $W_{C1} = 0.144$ J, $W_{C2} = 0$ J
39. **(a)** $V_{R1} = 200$ V, $V_{R2} = 0$ V, $V_{R3} = 200$ V, $V_{R4} = 0$ V, $V_{C1} = 200$ V, $V_{C2} = 200$ V, $I_{R1} = 4$A, $I_{R2} = 0$ A, $I_{R3} = 4$ A, $I_{R4} = 0$ A, $I_{C1} = 0$ A, $I_{C2} = 0$ A **(b)** 400 V **(c)** $Q_{C1} = 0.004$ C, $Q_{C2} = 0.008$ C
41.
```
FIGURE 12.74
VS 1 0 PULSE(0 15V 0 0 0 12US 40US)
R1 1 2 10K
C1 2 0 200PF
.TRAN   0.5US   40US
.PROBE
.END
```
43.
```
FIGURE 12.75
VS 1 0 PULSE(0 10V 0 0 0 0.5MS 5MS)
R1 2 0 27K
C1 1 2 0.01UF
.TRAN   50US   5MS
.PROBE
.END
```
45.
```
FIGURE 12.75
VS 1 0 PULSE(10V -10 V 0 0 0 1MS 5MS)
R1 2 0 27K
C1 1 2 0.01UF
RL 2 0 100K
CL 2 0 470PF
.TRAN   50US   5MS
.PROBE
.END
```
47. **(a)** 200 V **(b)** 2.5 s **(c)** 0.2 A
49. 14,430 Ω

CHAPTER 13 Answers to Questions

1. The property by which it attracts iron and other magnetic materials.
2. The compass needle is a magnet whose north pole points to the North Pole of the earth.
3. No, because copper is not attracted by a magnet.
4. The magnetic field set up by the current will make the compass needle move.
5. In the temporary magnet, the domains remain aligned only while the coil around the material is energized. In the permanent magnet, they are always aligned.
6. The door latch should use a permanent magnet. The chimes should use a temporary one.
7. The doorbell will not ring because the aluminum core is nonmagnetic and will not become a magnet.

8. The permeability of iron is much greater than that of air.
9. To provide an easier path for the magnetic field and couple most of the primary field with the secondary winding.
10. A 1-meter length of iron.
11. The flux density and the area common to both surfaces.
12. Because the flux density is greater.
13. Electromagnetic induction is the generating of an emf in a conductor by exposing it to a changing magnetic field.
14. Faraday's Law
15. Lenz's Law
16. Zero volts because the movement is parallel to the lines.
17. In electromagnetic induction, the magnetic field is changed by a relative motion between the conductor and the field. In mutual induction, the conductors are stationary and the field is changed by a change in the current that establishes the field.
18. A generator uses electromagnetic induction. A transformer uses mutual induction.
19. Inductance is the property by which a device opposes a change in current.
20. The practical inductor has resistance, while the ideal one does not have resistance.
21. The inductance induces a large emf to try to maintain the current. This emf across the contacts causes arcing.
22. The diode is reverse-biased for the source voltage but is forward-biased for the induced emf. Thus there is a path consisting of the winding, the diode, and the resistor in which the energy can be dissipated.
23. The total inductance will increase.
24. Connecting a third inductor in parallel will result in a decrease in inductance.
25. If they are series aiding, the equivalent inductance will increase. If they are series opposing, the equivalent inductance will decrease.

CHAPTER 13 Answers to Problems

1. 1.27 T in section 1, 0.318 T in section 2
3. 6.28×10^{-5} Wb
5. 2387
7. (a) 15 A (b) 500 A (c) 100 A (d) 500 A
9. (a) 5×10^{-4} Wb (b) 2.5 T
11. 1000 A
13. (a) 720 A/m (b) 300 A/m
15. (a) 1.09 T (b) 0.92 T
17. (a) 0.625 T (b) 1004.1 A (c) 1255
19. 0.006 m
21. (a) 874.04 A (b) 437 turns
23. 8.67 A
25. 22.48 N
27. 0.008 N open, 0.796 N closed
29. 0.6 ms
31. (a) 24 mV (b) Front end is (+).
33. 1200
35. 25 V
37. (a) 5000 V (b) 0 V (c) −10,000 V
39. 4.17 ms
41. 36 mH for Figure 13.72, 20 H for Figure 13.73
43. 3 H
45. (a) 11 H (b) 3 H
47. 2.1×10^{-3} Wb
49. 1.25 H for Figure 13.77, 0.625 H for Figure 13.78
51. (a) 3.86 H (b) 1.42 H
53. 8 H

CHAPTER 14 Answers to Questions

1. The inductor current
2. (a) Stays the same (b) Increases (c) Stays the same (d) Increases (e) Increases
3. (a) Stays the same (b) Stays the same (c) Decreases (d) Increases (e) Increases
4. The time constant is a function of R and L, so changing E will not change the rate of current rise.
5. (a) R-C circuit (b) R-L circuit (c) R-C circuit (d) R-L circuit (e) R-L circuit (f) R-L circuit
6. (a) There will be no change. (b) Increases the energy (c) Increases the energy
7. Item (c) produces the largest voltage.
8. Doubling the current
9. The practical inductor has resistance, while the ideal inductor does not.
10. Zero ohms
11. Zero volts
12. It is dissipated in the arc that forms across the switch contacts.
13. $i_L = I_0(1 - e^{-t/(L/R)})$
14. $i_L = I_0 e^{-t/(L/R)}$
15. 10 ms
16. $v_L = L\,di/dt$, so the voltage is directly proportional to the slope.

CHAPTER 14 Answers to Problems

1. (a) 0 A, 12 V (b) 1.6 A, 9.58 V
3. (a) 0.139 s (b) 4 A
5. (a) 0.333 s (b) 8.45 V (c) 1.78 A

7. **(a)** Top terminal is (−). **(b)** 0.04 s
(c) 0.172 A, 8.6 V **(d)** 30 V
9. **(a)** 0.5 ms **(b)** 0.25 ms
(c)

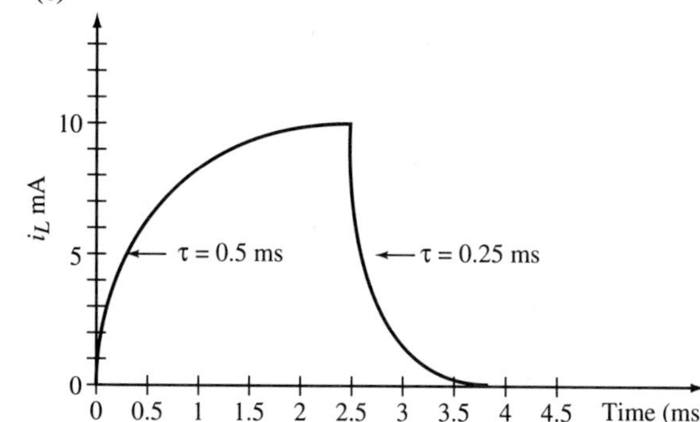

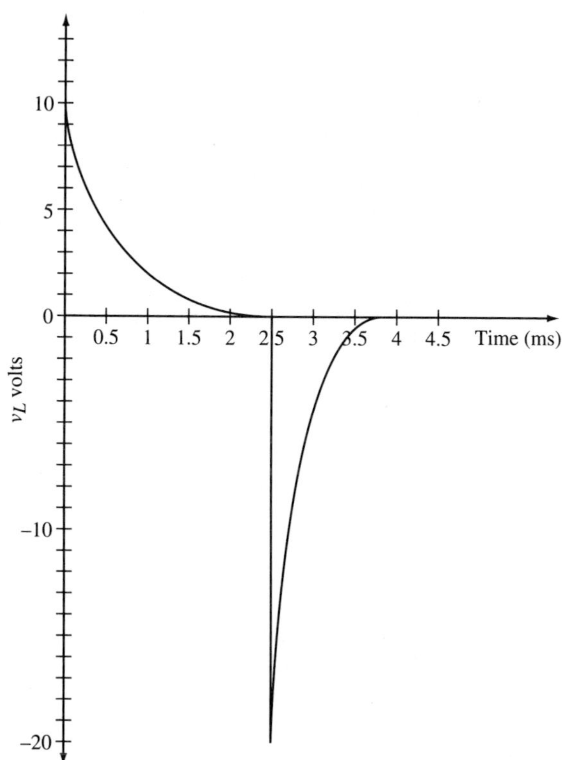

11. 40 Ω
13. 19.8 V, 0.1 A
15. **(a)** 2 Ω **(b)** 8.04 A **(c)** 0.81 s
17. **(a)** 1.4 s **(b)** 2 V
19. **(a)** 0.6 s **(b)** 0.398 V, 1.52 A, 2.36 A
21. 0.54 ms
23. **(a)** 0.0126 A, 2.02 V **(b)** 0.945 V
25. **(a)** 5.67 V **(b)** 3.17 A
27. 14.4 J
29. 6.92 J
31. $I_{R1} = 2$ A, $I_{L1} = I_{R2} = 0.8$ A, $I_{L2} = I_{R3} = 1.2$ A, $I_{R4} = 0$ A, $V_{R1} = 7.2$ V, $V_{R2} = 4.8$ V, $V_{R3} =$ 4.8 V, $V_{L1} = V_{L2} = V_{R4} = 0$ V
33. **(a)** $V_{R1} = 3$ V, $V_{R2} = 3$ V, $V_{R3} =$ 3V, $V_{L1} = V_{L2} = 0$ V, $I_{L1} = I_{R2} = 1$ A, $I_{L2} = I_{R3} = 0.5$ A, $I_{R1} =$ 1.5 A
(b) 6 V

CHAPTER 15 Answers to Questions

1. The direction and the magnitude of ac change with time.
2. The instantaneous value is the voltage at any instant of time.
3. Increasing the frequency decreases the period.
4. Frequency is the number of cycles per unit of time.
5. The positive amplitude will be +5 volts.
6. Time, degrees, and radians
7. The angular velocity is 377 radians per second. The amplitude is 10 volts.
8. A lagging wave is one in which a particular point of the wave occurs later in time, radians, or degrees than the point on the sine wave. A leading wave is one in which the point occurs earlier than that point on a sine wave.
9. If the wave is shifted to the right of the sine wave, it is lagging. If it is shifted to the left, it is leading.
10. The leading wave has a more positive phase angle, while the lagging wave has a more negative one than has the reference wave.
11. The phase angle is 20°.
12. The average value is the value of a constant voltage or current that would result in the same net movement of charge.
13. The effective value is the value of a constant voltage or current that would have the same energy content as the varying waveform.
14. The average value is 1.5 V. The effective value is 1.5 V.
15. RMS stands for root-mean-square. It is the square root of the average value of the v^2 or i^2 curve.
16. The average value of the sine wave is zero. Since the positive and negative halves are symmetrical, the net charge movement is zero and the area under the curve is zero.
17. The average value will be +2 volts.
18. A change in frequency does not affect the average and effective values.
19. Changing the amplitude will change the average and effective values.

20. A dc voltmeter will measure the average value. The effective value can be measured with an ac voltmeter that measures true RMS.
21. The ac component in the waveform accounts for the difference.
22. The effective value of the ac component can be measured by using an ac meter and either a blocking capacitor or a transformer coupling.
23. The effective value represents a constant voltage or current that will supply the same amount of energy as the ac.
24. The brightness should appear to be the same to the human eye.
25. When the shapes are not a complete approximation of the wave.
26. Using more strips will yield more accurate results.
27. The current and voltage are in phase for a resistor.
28. The voltage lags the current by 90° for the capacitor.
29. The voltage leads the current by 90° for the inductor.
30. Increasing the frequency will not change the phase angle of the capacitor or inductor. It still remains 90°.
31. Capacitive reactance is the opposition of a capacitor to ac current. Inductive reactance is the opposition of an inductor to ac current.
32. The current will increase for a doubling of frequency. The current will decrease if the frequency is halved.
33. The current will decrease if the frequency is doubled and increase if it is halved.
34. $X_C = 1/2\pi fC$ is correct.
35. $X_L = 2\pi fL$ is correct.

CHAPTER 15 Answers to Problems

1. Amplitude is 10 V, $T = 8$ ms, $f = 125$ Hz, $\upsilon = -7.1$ V
3. 0.25 ms
5. Two cycles, $T = 8$ s
7. 3π radians per second
9. **(a)** Amplitude is 20, $f = 200$ Hz, $\upsilon = 19.02$ V

(b)

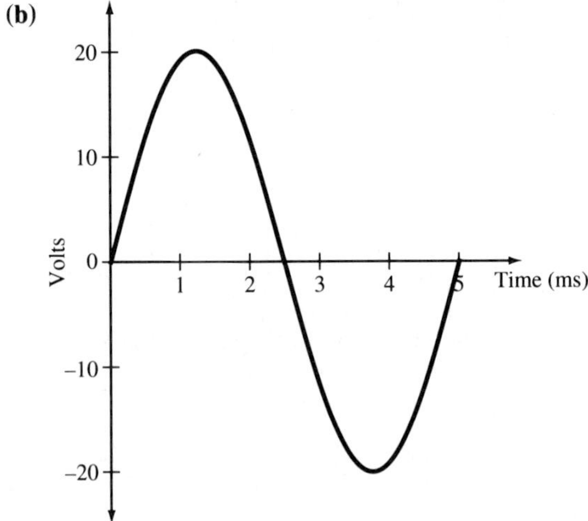

11. **(a)** $i = 20 \sin 392.5t$ **(b)** 62.5 Hz **(c)** 40 V
13. **(a)** $\upsilon = 50 \sin 62.83t$
(b)

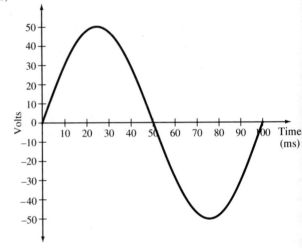

15. **(a)** $+15°$, lagging **(b)** $+\pi/4$ radians, leading
(c) $-\pi/6$ radians, lagging
(d) $+\pi/4$ radians, leading
17. 36°, leading

19.

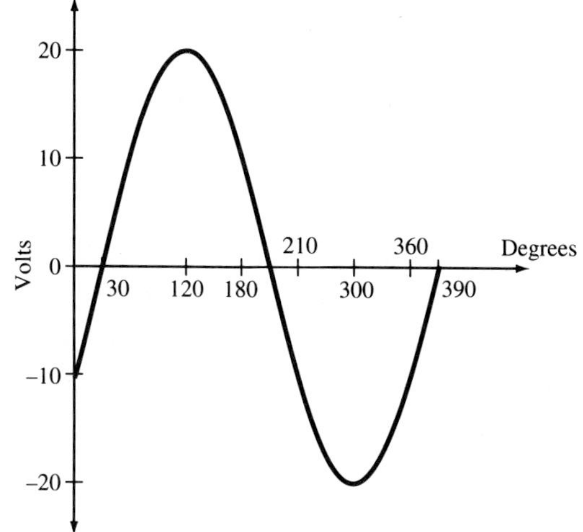

21. **(a)** υ leads by $3\pi/2$ radians **(b)** υ leads by 15°
(c) i leads by $\pi/2$ radians

23. B lags A by 120°

25. **(a)** Has a negative average value.
(b) Has a positive average value.
(c) Has a positive average value.

27. **(a)** Approximately 6.28 V
(b) Approximately 6.26 V

29. Average is 31.88 V, so the source is not satisfactory.

31. 2.08 V

33. 0.46 V

35. **(a)** dc and ac **(b)** dc and ac **(c)** dc only

37. $V_{ave} = +5$ V, $V_{eff} = 15.21$ V, V_{eff} of ac $= 14.36$ V

39. **(a)** Curve (a) 0 V; curve (b) +10 V; curve (c) +10 V
(b) Curve (a) − 10 V; curve (b) 14.14 V; curve (c) 10 V
(c) Curve (a) 10 V; curve (b) 14.14 V; curve (c) 10V

41.

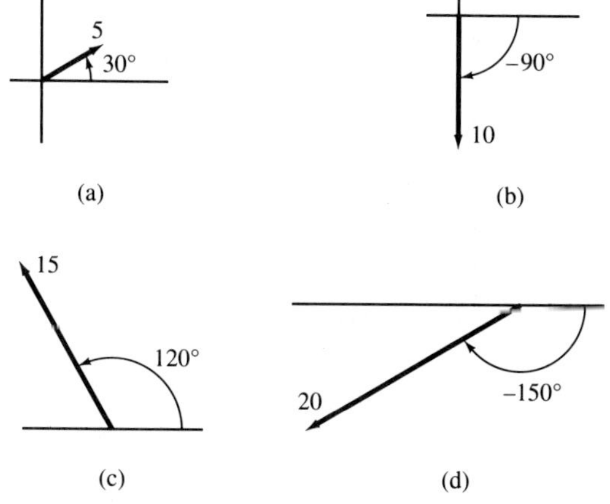

43. **(a)** $3.54 \angle 30^\circ$ **(b)** $7.07 \angle -90^\circ$
(c) $10.61 \angle 120^\circ$ **(d)** $14.14 \angle -150^\circ$

45. **(a)** $16.97 \sin(377t - 30^\circ)$
(b) $169.73 \sin(377t + 135^\circ)$ **(c)** $8.49 \sin 377t$
(d) $21.22 \sin(377t + 200^\circ)$

47.

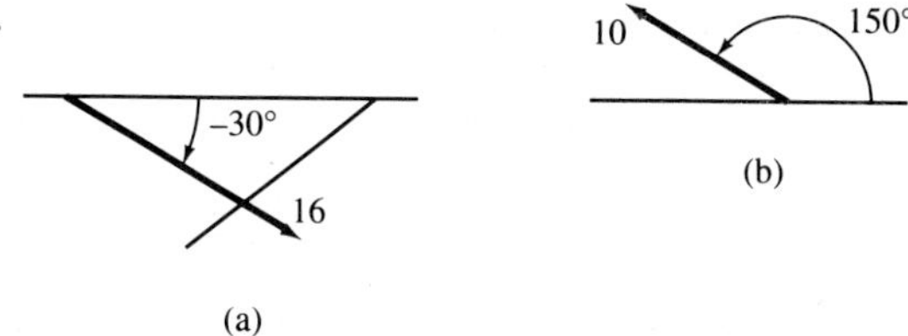

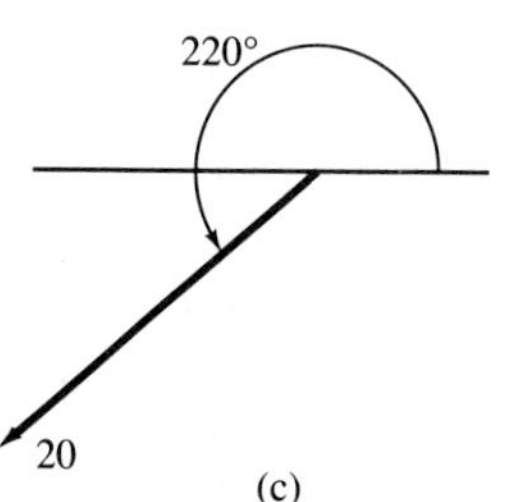

49. **(a)** $15 \angle 60^\circ$ **(b)** $10 \angle 180^\circ$ **(c)** $8 \angle -90^\circ$
(d) $15 \angle 240^\circ$

51. **(a)** $4 - j6$ **(b)** $-j6$ **(c)** $6 + j4$

53. $\upsilon = 25 \sin 754t$

55. 530.5 Ω at 60 Hz, 63.66 Ω at 500 Hz, 10.61 Ω at 3 kHz, 0.032 Ω at 1 MHz

57. $\upsilon = 25 \sin(754t - \pi/2)$

59. 450 Hz

61. $\upsilon = 18 \sin(377t + \pi/2)$

CHAPTER 16 Answers to Questions

1. The magnitude of the current will remain the same, but the current will lead the source voltage by 90°.
2. The two 5-ohm resistors will have the larger impedance.
3. The impedance is a 2.83-ohm resistor in series with a 2.83-ohm inductive reactance. Doubling the frequency doubles X_L, so the impedance is increased and the circuit becomes more inductive.
4. The impedance is a 2.83-ohm resistor in series with a 2.83-ohm capacitive reactance. Doubling the frequency halves X_c, so the impedance is decreased and the circuit becomes less capacitive.
5. The coil and capacitor voltages cancel because they are 180° out of phase. This leaves the sum as 12 V.
6. The combination of resistor and capacitor will have the larger impedance.

7. The combination is inductive, so doubling the frequency doubles the X_L branch. This makes the circuit less inductive and results in an increase in the impedance of the combination.
8. The combination is capacitive, so doubling the frequency halves the X_c branch. This makes the circuit more capacitive and results in a decrease in the impedance of the combination.
9. The phasor diagram provides the magnitudes and phase angles of the voltages and currents.
10. The angle is the angle between the 0° axis and the line of the phasor.
11. The magnitude is obtained by multiplying the scale used by the length of the phasor.
12. The phasor diagram shows voltages and currents; the impedance triangle shows resistance, reactance, and impedance.
13. The power factor is the cosine of the angle between the voltage and the current.
14. The larger power factor results in less current for the same power and less loss in the lines.
15. The power factor is 1 when V and I are in phase; 0.707 when V and I are out of phase by 45°; and 0 when V and I are out of phase by 90°.
16. $P = VI$ holds true for a resistor.
17. There will not be any change in power if the capacitor is replaced by an inductance. Replacing the capacitor by a 40-ohm resistor will result in an increase in power.
18. Connecting another capacitor in parallel will make the circuit more capacitive and result in a decrease in the power factor.
19. Connecting an inductance in the circuit can increase or decrease the power factor. If X_L is less than X_c, the power factor will be increased. If X_L is greater than X_c, the circuit will have a lagging power factor, which can be smaller or greater than before.
20. Analysis is the procedure whereby the conditions that exist in a circuit are determined. Synthesis is the procedure whereby the circuit is designed to provide specified conditions.

CHAPTER 16 Answers to Problems

1. **(a)** $22.36 \angle -63.43°\ \Omega$
   (b) $6.4 \angle 38.66°\ \Omega$
3. 18 Ω
5. **(a)** 6 Ω **(b)** capacitive
7. $20 \angle 90°$ V
9. $50 \angle 45°$ V
11. $50 \angle 37°$ V

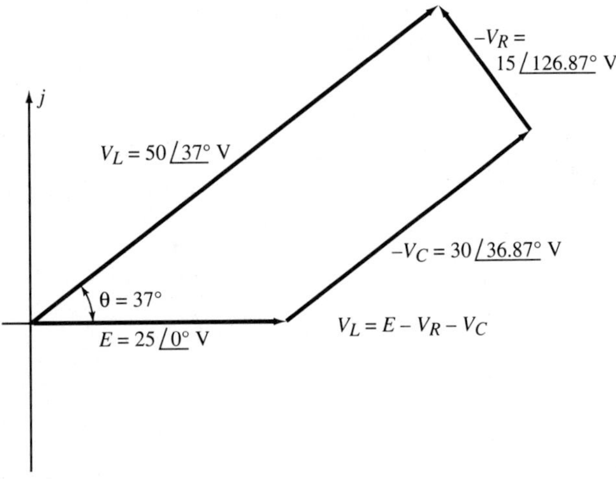

13. $42.43 \angle 45°$ V
15. $27.45 \angle 42.27°$ V
17. $2.4 \angle 36.87°\ \Omega$

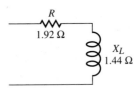

19. 40 Ω
21. **(a)** $0.2 \angle -30°$ S, 0.173 S, 0.10 S inductive
   (b) $0.1 \angle -45°$ S, 0.0707 S, 0.0707 S inductive
   (c) $0.1 \angle 53.13°$ S, 0.06 S, 0.08 S capacitive
   (d) $0.2 \angle 0°$ S, 0.2 S, 0 S
23. $Y_{eq} = 0.03 \angle 90°$ S, $I_s = 3.6 \angle 90°$ A, $I_1 = 6 \angle 90°$ A, $I_2 = 2.4 \angle -90°$ A
25. $0.022 \angle 63.43°$ S, 0.01 S, 0.02 S

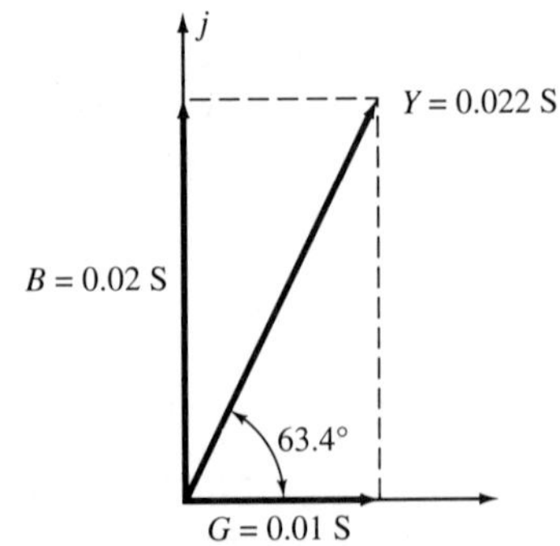

27. 2.47 $\angle 6.26°$ A
29. 2.83 $\angle -45°$ A

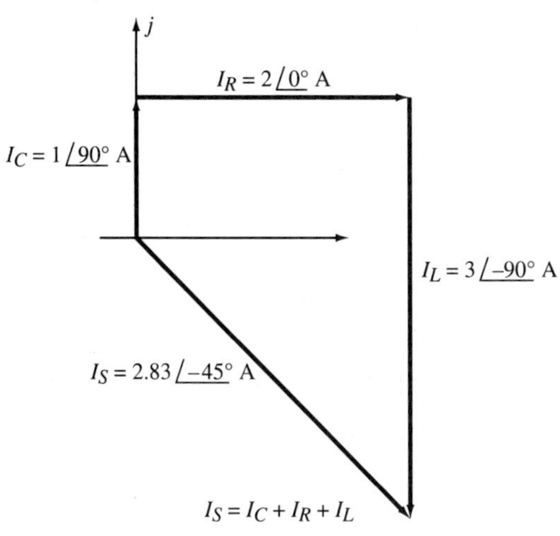

31. 1.2 $\angle -90°$ A
33. 2.87 $\angle 16.7°$

35.

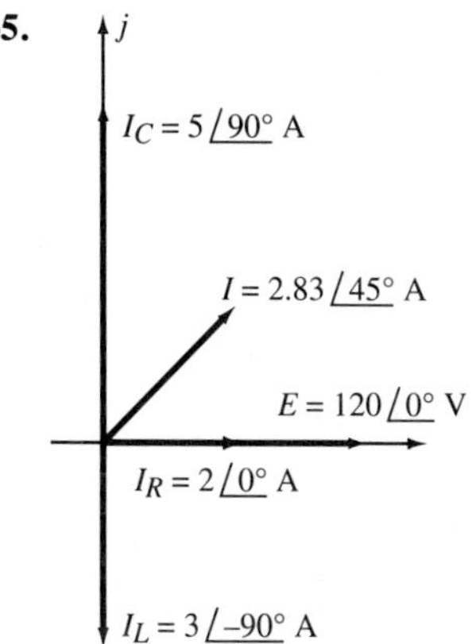

37. I = 1.2 $\angle 53.13°$ A, V_R = 72 $\angle 53.13°$ V, V_c = 96 $\angle -36.87°$

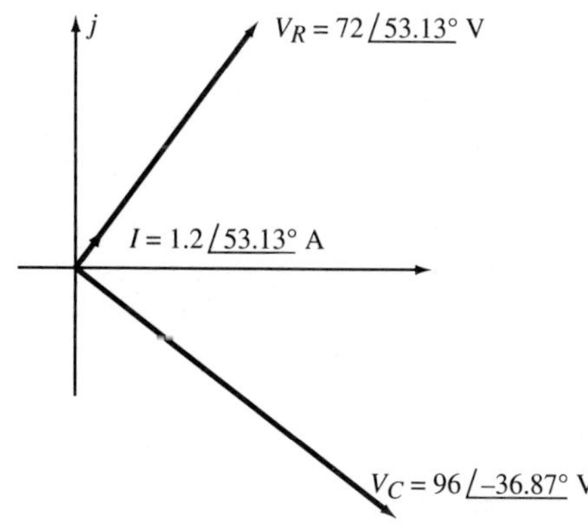

39. I = 5 $\angle 53.13°$ A, I_R = 3 $\angle 0°$ A, I_c = 4 $\angle 90°$ A. The circuit is capacitive.

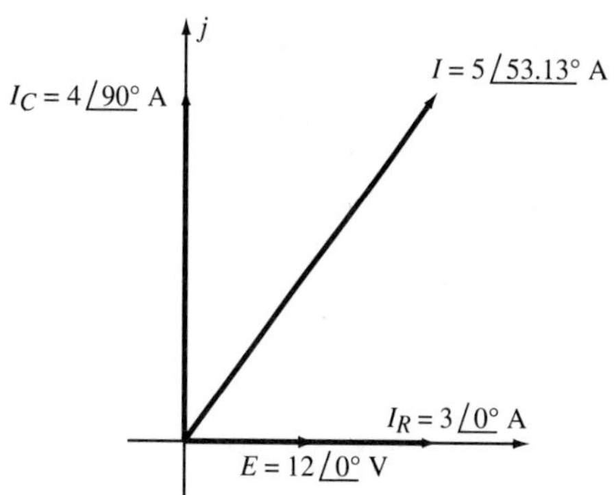

41. I = 3.16 $\angle -18.43°$ A, I_L = 4 $\angle -90°$ A, I_{R1} = 2 $\angle 0°$ A, I_c = 3 $\angle 90°$ A

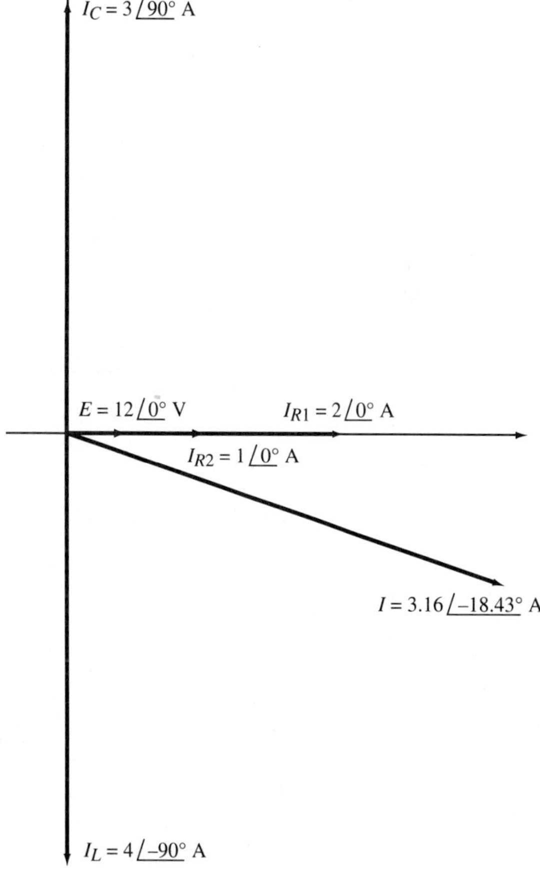

43. **(a)** 120 W **(b)** 1 A
45. 240 W, R_p = 0.707, lagging

47. **(a)** 10.42 A **(b)** 11.52 $\angle -36.87°$ Ω
49. **(a)** 0.894 **(b)** 288 W
51. **(a)** 2.5 A
(b) 60 Ω resistance and 80 Ω capacitive reactance
53. 960 W
55. **(a)** 2.15 $\angle 63.43°$ A **(b)** 115.16 W **(c)** 0.447

(d)

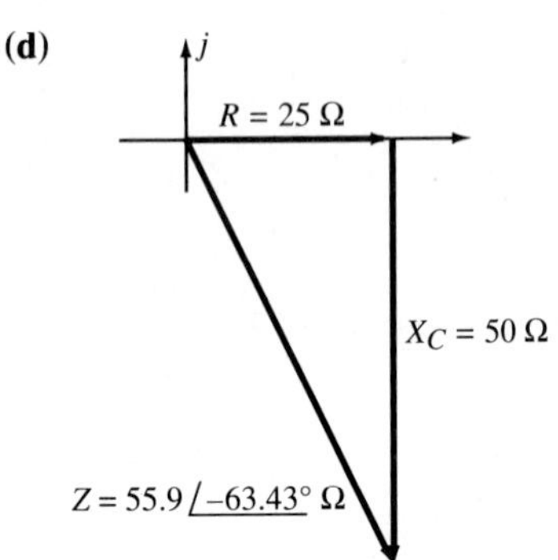

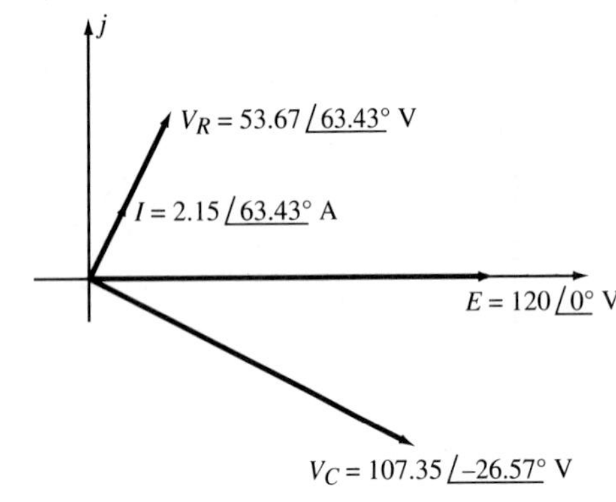

57. **(a)** 1.60 $\angle 53.13°$ **(b)** 23.04 W **(c)** 0.6

(d)

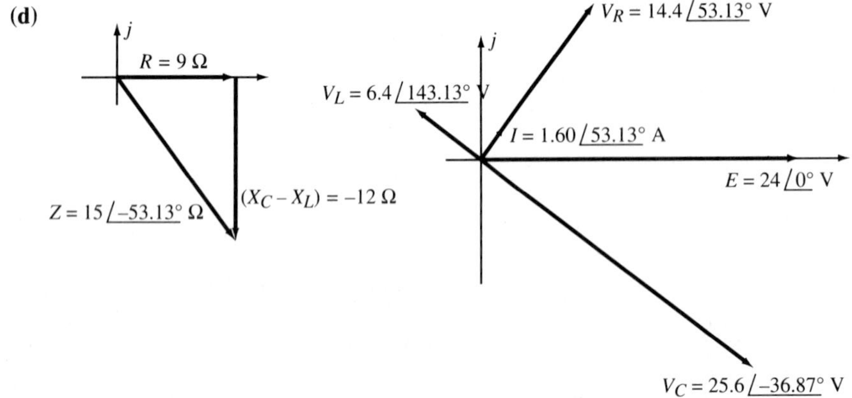

59. 56.6 Ω, 49.92 $\angle -23.13°$ V
61. **(a)** 1.05 $\angle -71.57°$ A **(b)** 0.32 **(c)** 40 W

(d)

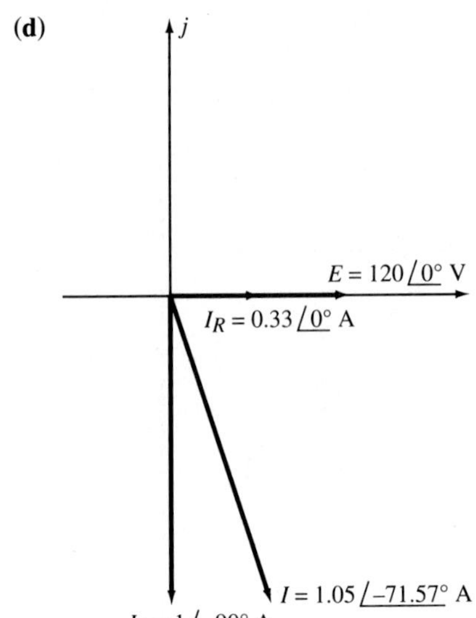

63. **(a)** 50 $\angle 0°$ V **(b)** 0.64 **(c)** 62.5 W

(d)

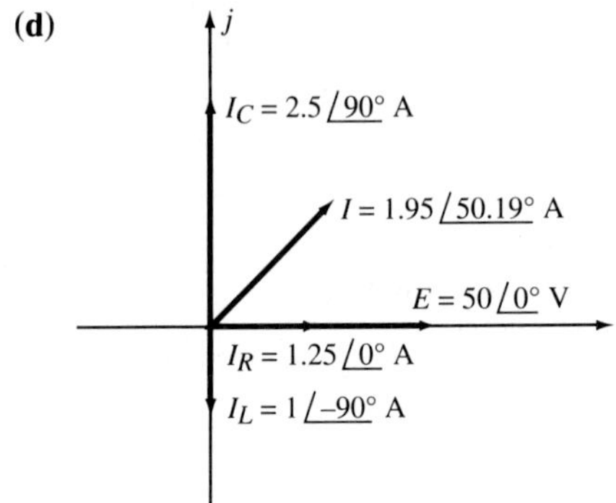

65. **(a)** $X_L = 250\ \Omega$, $R = 500\ \Omega$
(b) $I_c = 0.2\ \angle 90°$ A, $I_L = 0.4\ \angle -90°$ A, $I_{R2} = 0.2\ \angle 0°$ A

67. $38.14\ \angle -25.2°\ \Omega$

69. $I_{R2} = 2.83\ \angle -45°$ A, $V_L = 84.84\ \angle 45°$ V

71. $12\ \angle 80.4°$ V, $1.7\ \angle 35.4°$ A, $73.2\ \angle -90°$ V

73. 52.92 Ω, 0.96

75. 52.22 Ω and 7.66 Ω

77.
```
FIGURE 16.127A
VS    1    0    AC     20V    0DEG
C1    1    2    0.001UF
R1    2    0    10K
.AC    LIN    1    15KHZ    15KHZ
.PRINT    VM(R1)    VP(R1)
.END


FIGURE 16.127(B)
VS    1    0    SIN( 0 20V 15KHZ )
C1    1    2    0.001UF
R1    2    0    10K
.TRAN    1US    150US
.PROBE
.END
```

79.
```
FIGURE 16.95
VS    1    0    SIN( 0 12V 1KHZ )
R1    1    0    4
C1    1    0    53.1UF
.TRAN    10US    2MS
.PROBE
.END
```

CHAPTER 17 Answers to Questions

1. The apparent power can be calculated using the relation $S = \sqrt{P^2 + Q^2}$.
2. The power factor is equal to the average power divided by the apparent power.
3. The apparent power is equal to the reactive power for an inductance load and a capacitance load.
4. Capacitive vars must be added to increase the power factor of an inductive load.
5. The apparent power is equal to the average power for a resistance load.
6. The power rating at any power factor can be determined from the kVA rating.
7. Increasing the frequency decreases X_c and increases the capacitive vars. This will decrease the power factor.
8. Decreasing the frequency increases X_c and decreases the capacitive vars. This will increase the power factor.
9. If the power factor is leading, the load is capacitive, so inductive vars must be added.
10. Connecting the vars in series with the load will reduce the voltage at the load.
11. Residential loads are mostly resistance, while industrial loads are inductive.
12. The wattmeter measures average power.
13. The load must be all reactance, so the power factor and power are zero.
14. The solid core increases the eddy current losses.
15. The circuits can have the same apparent power since $S = VI$.
16. The voltage and current measurements will not give the correct resistance unless the magnitudes and frequency are exactly the same as those for which the coil is used.
17. Decreasing the frequency will increase the depth to which the rod is hardened.
18. The air core is used to reduce the losses and to prevent the characteristic changes that occur at various frequencies used in electronic applications.
19. One reason for the overheating of the transformer could be that the frequency of the source is higher than that for which the transformer is rated.
20. The conductors can be hollow because at higher frequencies the signal is concentrated near the surface of the conductor, so the center is not used.
21. The losses in the core of the transformer are eddy current losses and hysteresis losses.

CHAPTER 17 Answers to Problems

1. 480 W, 360 vars, 600 VA

3. **(a)** 300 W **(b)** 0.6

5. 24 W, 30.16 vars, 38.54 VA

7. 1.41 μF

9. **(a)**

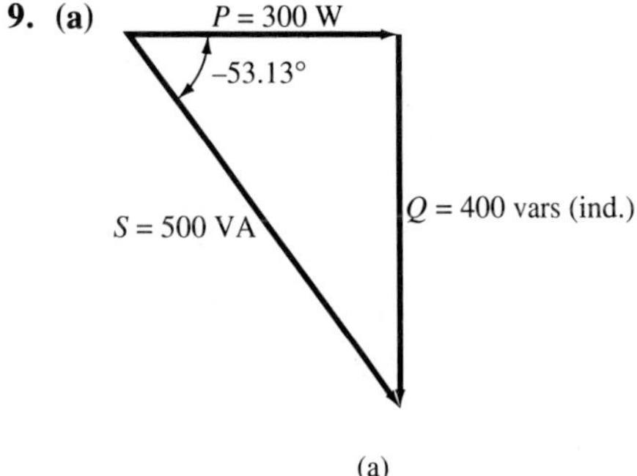

(a)

(b)

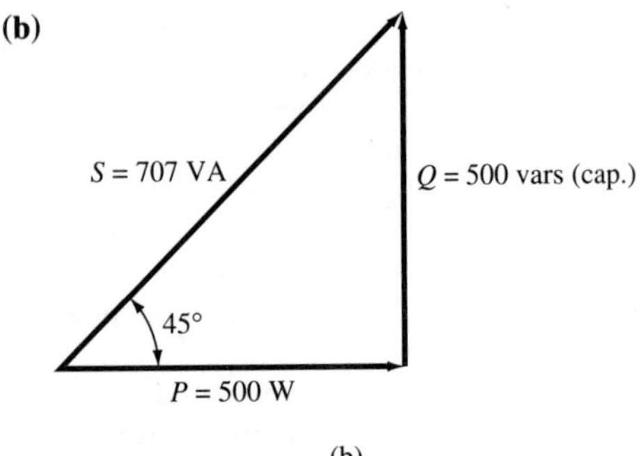

(b)

11.

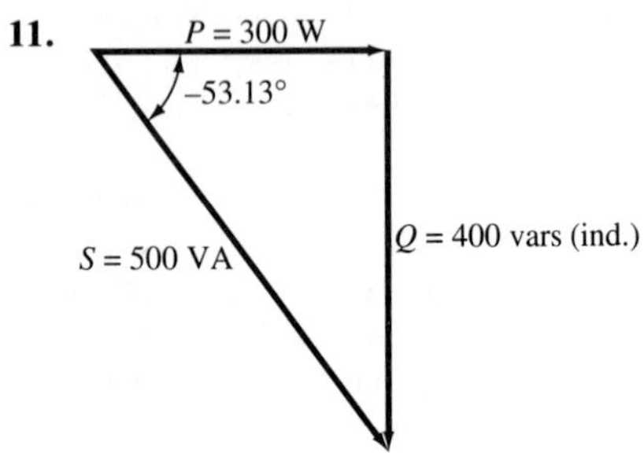

13. (a) R = 8.09 Ω, X_L = 12.95 Ω
(b) R = 28.8 Ω, X_L = 18 Ω

15. (a) 28.75 Ω (b) 14.38 Ω (c) 3.73 $\angle 26.56°$ A

17. (a) 800 W, 300 vars ind., 854.4 VA
(b) 0.936

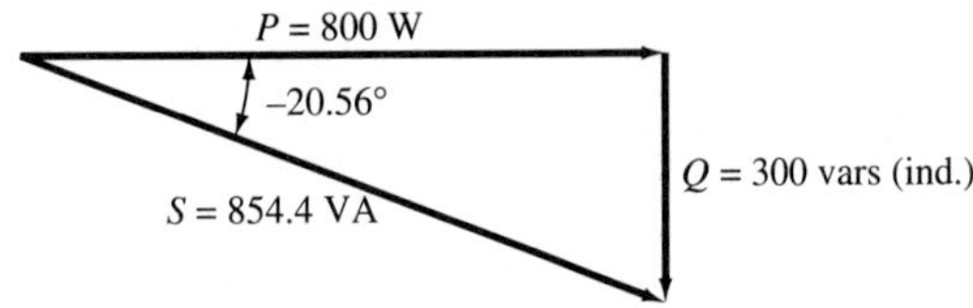

19. (a) For 200 W, 100 vars, I = 1.86 $\angle -26.56°$ A
For 500 W, 0 vars, I = 4.17 $\angle 0°$ A
For 0 W, 400 vars, I = 3.33 $\angle 90°$ A
(b) 6.35 $\angle 23.2°$ A
(c) 0.92 leading

21. (a) 2.57 $\angle -45°$ A, 0.707 lagging
(b) 400 W
(c) For 0 W, 100 vars, I = 2.57 $\angle -45°$ A,
V = 38.91 $\angle 45°$ V
For 200 W, 0 vars, I = 1.03 $\angle -8.13°$ A,
V = 194.55 $\angle -8.13°$ V
For 100 W, 100 vars, I = 1.85 $\angle -64.44°$ A,
V = 76.44 $\angle -109.44°$ V
For 100 W, 400 vars, I = 1.85 $\angle -64.44°$ A,
V = 222.87 $\angle 11.52°$ V

23. V_L goes from 107.37 V to 106.79 V.

25. 9.05 μF

27. 157.27 vars, 8.62 μF

29. 5 Ω

31. 3.6 W

CHAPTER 18 Answers to Questions

1. The equivalent impedance method cannot be used for circuits having more than one source because the impedance of the rest of the circuit cannot be calculated with sources in it.
2. Two loop equations are needed for a circuit that has three loops.
3. The branch current will be zero amperes if the magnitudes of the loop currents are equal and the currents are 180° out of phase.
4. No change is needed in the step if the loop currents are taken as counterclockwise.
5. Three loop equations are needed for a circuit that has four loops.
6. The node current equations are needed for N-1 nodes.
7. If the opposite direction were assigned to the current, the result would be 5 $\angle 240°$ or 5 $\angle -120°$.
8. Three nodal equations are needed for a circuit that has four nodes.
9. 5 $\angle 200°$ V for node C and 5.15 $\angle 39.7°$ V for node B.
10. The equivalent impedance can be obtained by calculating the source current first. Then $Z_{eq} = E/I_s$.
11. Loop, branch, and nodal analysis can be used for circuits tht have only one source.
12. Superposition will not provide correct results for circuits having nonlinear impedances. Since they are nonlinear, it is not possible to select a value to use for the impedance.
13. Superposition requires many calculations for circuits having several sources.
14. Yes, two different circuits can have the same Thevenin equivalent and the same Norton equivalent.
15. The ammeter resistance results in the measured current being smaller than the actual short circuit current, so Z_{TH} will be greater than the actual Z_{TH}.
16. If the circuit is all resistance that does not change with frequency, the equivalent will be the same for all frequencies. If the circuit includes reactance, then the equivalent can change.
17. The Norton equivalent uses a constant current source with a shunt impedance, while the Thevenin equivalent uses a constant voltage source with a series impedance.
18. Whether the power increases or decreases depends

on the reactance in the circuit and the circuit configuration.

19. The conjugate of an impedance is an impedance that has the opposite reactance of equal value. That is, the conjugate of $4 + j5$ is $4 - j5$.
20. The relation does not hold because I_1 does not equal I_2 and I_3 does not equal I_x when the bridge is not balanced.
21. The low resistance of the ammeter short circuits the two points, making the circuit no longer a bridge circuit.

CHAPTER 18 Answers to Problems

1. $I_R = 4.46 \angle 81.43°$ A right, $I_L = 12.69 \angle 102.03°$ A right, $I_C = 8.65 \angle -67.58°$ A down, $V_R = 13.38 \angle 81.43°$ V, $V_L = 63.45 \angle 192.03°$ V, $V_C = 34.60 \angle -157.58°$ V, $P = 59.67$ W
3. **(a)** $I_{E1} = 1.295 \angle -30.78°$ A up, $I_{E2} = 0.862 \angle 134.2°$ A down, **(b)** 183.18 W
5. $I_{Z1} = 4.88 \angle 91.11°$ A right, $I_{Z2} = 6.43 \angle 112.26°$ A down, $I_{Z3} = 2.58 \angle -24.88°$ A right, $I_{Z4} = 1.127 \angle 84.86°$ A down, $I_{Z5} = 3.14 \angle -44.35°$ A right, $V_{Z1} = 24.4 \angle 91.11°$ V, $V_{Z2} = 64.3 \angle 202.26°$ V, $V_{Z3} = 51.6 \angle -114.88°$ V, $V_{Z4} = 22.54 \angle 84.86°$ V, $V_{Z5} = 31.4 \angle 45.65°$ V, $P_{Z4} = 25.4$ W
7. $I_R = 4.45 \angle 81.47°$ A right, $I_L = 12.67 \angle 102.03°$ A right, $I_C = 8.65 \angle -67.57°$ A down, $V_R = 13.38 \angle 81.47°$ V, $V_L = 63.35 \angle 192.03°$ V, $V_C = 34.60 \angle -157.57°$ V, $P = 59.67$ W
9. **(a)** $I_{E1} = 1.3 \angle -30.78°$ A up, $I_{E2} = 0.864 \angle 134.22°$ A down **(b)** 184.5 W
11. $I_{Z1} = 4.89 \angle 90.7°$ A right, $I_{Z2} = 6.44 \angle 112°$ A down, $I_{Z3} = 2.58 \angle -24.63°$ A right, $I_{Z4} = 1.14 \angle 85°$ A down, $I_{Z5} = 3.16 \angle -44.62°$ right, $V_{Z1} = 24.45 \angle 90.7°$ V, $V_{Z2} = 64.4 \angle 202°$ V, $V_{Z3} = 51.6 \angle -114.63°$ V, $V_{Z4} = 22.8 \angle 85°$ V, $V_{Z5} = 31.6 \angle 45.38°$ V, $P_{Z4} = 25.99$ W
13. $I_R = 4.45 \angle 261.46°$ A left, $I_L = 12.67 \angle 102.02°$ right, $I_C = 8.65 \angle -67.57°$ A down, $V_R = 13.35 \angle 261.46°$ V, $V_L = 63.35 \angle 192.02°$ V, $V_C = 34.5 \angle -157.57°$ V, $P = 59.41$ W
15. **(a)** $I_{E1} = 1.28 \angle 150.0°$ A down, $I_{E2} = 0.854 \angle 135.0°$ A down, **(b)** 184.04 W
17. $I_{Z1} = 4.89 \angle 271.26°$ A left, $I_{Z2} = 6.43 \angle 112.33°$ A down, $I_{Z3} = 2.57 \angle 155.50°$ A left, $I_{Z4} = 1.12 \angle 85.45°$ A down, $I_{Z5} = 3.13 \angle -44.34°$ right, $V_{Z1} = 24.35 \angle 271.26°$ V, $V_{Z2} = 64.3 \angle 202.31°$ V, $V_{Z3} = 51.4 \angle 65.50°$ V, $V_{Z4} = 22.4 \angle 84.45°$ V, $V_{Z5} = 31.30 \angle 45.66°$ V, $P_{Z4} = 25.09$ W
19. $I_R = 4.46 \angle -98.51$ A left, $I_L = 12.69 \angle 102.1°$ A right, $I_C = 8.65 \angle -67.50°$ A down, $V_R = 13.38 \angle -98.51°$ V, $V_L = 63.45 \angle 192.04°$ V, $V_C = 34.60 \angle -157.46°$ V, $P = 59.67$ W
21. **(a)** $I_{E1} = 1.3 \angle -30.5°$ A up, $I_{E2} = 0.87 \angle 134.53°$ A down
    (b) 183.18 W
23. $I_{Z3} = 1.23 \angle 59.37°$ A down, $V_{Z3} = 12.3 \angle -30.63°$ V
25. $V_{Z4} = 9.99 \angle 19.43°$ V, $I_{Z4} = 0.333 \angle 19.43°$ A
27. $E_{TH} = 120 \angle 0°$ V, top terminal (+) $Z_{TH} = 24 \angle -30°$ ohms
29. $V_{Z1} = 7.09 \angle -81.13°$ V, left terminal (+), $P_{Z1} = 3.02$ W
31. $V_{Z4} = 9.97 \angle 19.43°$ V, $I_{Z4} = 0.332 \angle 19.43°$ A
33. $I_N = 5 \angle 30°$ A up, $Z_N = 24 \angle -30°$ ohms
35. $V_{Z1} = 7.07 \angle -81.86°$ V, $P_{Z1} = 3$ W
37. **(a)** $E_{TH} = 60 \angle 20°$ V top +, $Z_{TH} = 20 \angle 0°$ ohms
    (b) $E_{TH} = 20 \angle 58.13°$ V top +, $Z_{TH} = 10 \angle 53.13°$ ohms
39. **(a)** 20 ohms **(b)** 7.2 W **(c)** 12 V
41. $Z_L = 4 + j32$ or $32.25 \angle 82.88°$ ohms, $I_L = 0.159 \angle -161.57°$ A, $P = 0.10$ W
43. The rigid proof is to use calculus. It can also be shown by calculating power for Z_L at slightly smaller and slightly larger values of Z_{TH} with the angle at 30°.
45. 0.89 W per speaker
47. $L_x = 0.04$ H, $R_x = 200$ ohms
49. $R_x = \dfrac{R_2R_3}{R_1}$, $L_x = R_2R_3C_S$

CHAPTER 19 Answers to Questions

1. High-pass circuit
2. Bandpass circuit
3. Lower 3-dB point or cut-off frequency
4. The range of frequencies between the 3-dB or half-power points; the range of frequencies the circuit will pass
5. **(a)** Zero degrees at low frequencies to −90 degrees at high frequencies
   (b) +90 degrees at low frequencies to zero degrees at high frequencies
   (c) Zero degrees at low frequencies to −90 degrees at high frequencies
   (d) +90 degrees at low frequencies to zero degrees at high frequencies

6. Transfer function
7. A Bode plot is a graph of gain or attenuation of a circuit in dB vs. frequency. Frequency is plotted as the independent variable using a logarithmic scale.
8. −1 dB an octave above and below the corner frequency; −3 dB at the corner frequency
9. −20 dB per decade or −6 dB per octave
10. Decrease in frequency
11. Increase in frequency
12. Zero degrees
13.

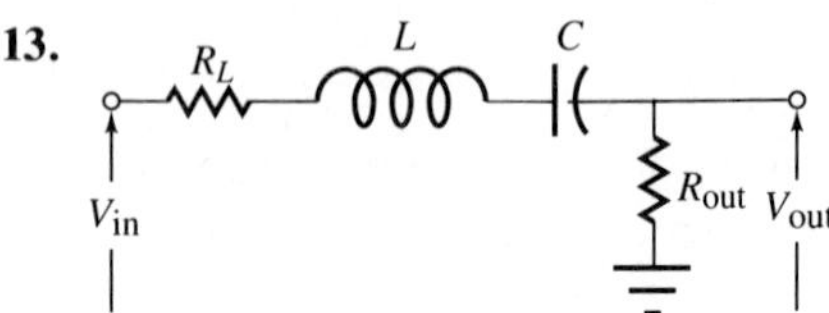

14. (a) Bandwidth will increase.
    (b) Bandwidth will increase.
    (c) Bandwidth will increase
15. (a) Peak output will remain unchanged.
    (b) Peak output will decrease.
16. `.AC DEC 20 1KHZ 10MEGHZ`
17. `.PROBE`
18. The calculation points may be too widely spaced, causing the resonant frequency to fall between calculation points.
19.

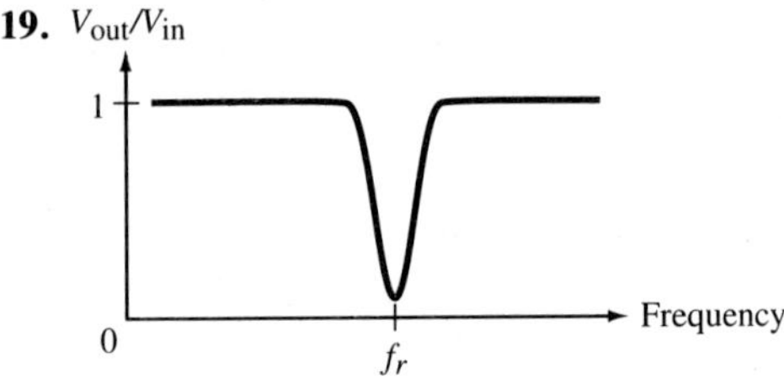

20. Because the low-pass first stage will filter out all of the high-frequency components.

CHAPTER 19 Answers to Problems

1. $\dfrac{V_{out}}{V_{in}} = \dfrac{jX_L}{R + jX_L}$

3. $\dfrac{V_{out}}{V_{in}} = \dfrac{(R_L + jX_L)\|(-jX_C)}{R_s + ((R_L + jX_L)\|(-jX_C))}$

5. $\text{Magnitude} = \dfrac{X_L}{\sqrt{R^2 + X_L^2}}$

   $\text{Phase} = 90° - \tan^{-1}\dfrac{X_L}{R}$

7. Lower 3-dB frequency = 18,512 Hz
   Upper 3-dB frequency = 20,672 Hz
9. BW = 159 kHz
11.

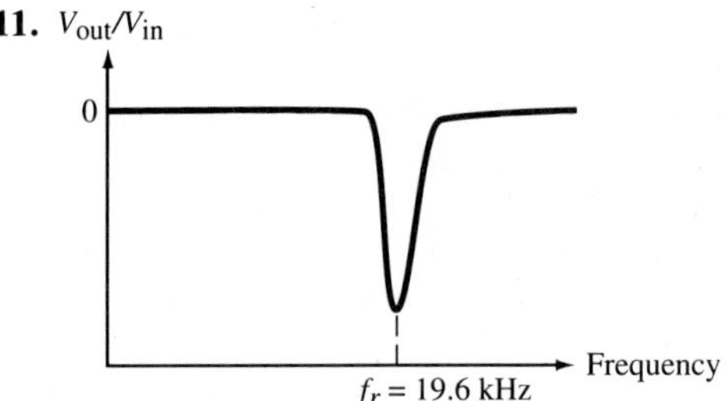

13.

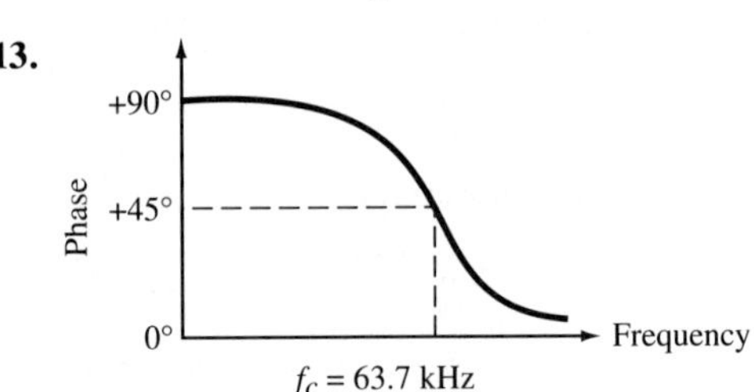

15.
```
PROBLEM 15 -- FIGURE 19.37
VIN   1   0   AC    20V    0.0DEG
R1    1   2   40K
L1    2   0   100MH
.AC   DEC   20   1KHZ   1000KHZ
.PROBE
.END
```
17.
```
PROBLEM 17 -- FIGURE 19.39
VIN   1   0   AC    30.0 V    0.0DEG
RS    1   2   2.2K
RL    2   3   50
L1    3   0   20MH
C1    2   0   330PF
.AC   DEC   20   1KHZ   1000KHZ
.PROBE
.END
```
19.
```
PROBLEM 19 -- FIGURE 19.41
VIN   1   0   AC    10.0 V    0.0DEG
C1    1   2   0.5UF
R1    2   0   12K
R2    2   3   33K
C2    3   0   500PF
.AC   DEC   20   10HZ   1000KHZ
.PROBE
.END
```
21. f_r = 503 kHz
    Q_o = 10.54

CHAPTER 20 Answers to Questions

1. Some examples of nonsinusoidal waveforms are square, triangular, exponential, and sawtooth waves.
2. The name of the series is the Fourier Series.

3. A harmonic is a sinusoidal component of a nonsinusoidal waveform whose frequency is a multiple of the fundamental frequency.
4. The fundamental term cannot be zero.
5. The sine terms are zero for the even function.
6. The cosine terms are zero for the odd function.
7. The A_0 term, which is the first term, represents the dc value of the waveform.
8. The A_0 term and all of the harmonic terms are zero for a 60-Hz sine wave.
9. If the higher-frequency terms are left out, the corners of the square wave will be rounded.
10. Two nonsinusoidal waves with different terms in the series can have the same effective value.
11. The dc value cannot be the same because the A_0 terms are different.
12. The capacitor in series will cause a greater attenuation of the low-frequency terms for the resistor load.
13. The capacitor in parallel will cause a greater attenuation of the high-frequency terms for the resistor load.
14. Increasing the value of the harmonics increases the effective value. Decreasing the value decreases the effective value.
15. The total instantaneous value is obtained by taking the algebraic sum of the individual instantaneous values.
16. The greater number of strips gives a more correct value.
17. Integration uses an infinite number of strips.
18. The superposition theorem is used to analyze circuits having a nonsinusoidal source.
19. The expression is the same with an A_0 term of +5 added to it.
20. The 5-ohm capacitor has more distortion than the 5-ohm resistor.
21. The spectrum analyzer is used to study the frequency content of a waveform.
22. The Fourier analyzer also measures the phase relation.

CHAPTER 20 Answers to Problems

1. **(a)** 1.937 V **(b)** 19 V
3. **(a)** 5 V, 60 Hz **(b)** −2 V, 40 Hz
5. **(a)** Has dc, even harmonics, odd harmonics, and cosine terms
   (b) Has dc cosine terms, sine terms, and odd harmonics
7. Has sine terms, even harmonics, and odd harmonics
9. **(a)** Section 1 start 0,1 end 270,5; section 2 start 270,2 end 360,2
   (b) Section 1 start 0,2 end 72,2; section 2 start 72,4 end 216,4; section 3 start 216,2 end 288,0; section 4 start 288, −1 end 360, −1
11. Section 1 start 0,0 end 30,3.7; section 2 start 30,3.7 end 60,5; section 3 start 60,5 end 90,5.8; section 4 start 90,5.8 end 120,6.4; section 5 start 120,6.4 end 150,6.8; section 6 start 150,6.8 end 180,6.9; section 7 start 180,6.9 end 210,7; section 8 start 210,0 end 240,0; section 9 start 240, −3 end 270, −1.8; section 10 start 270, −1.8 end 300, −1; section 11 start 300, −1 end 330, −0.5; section 12 start 330, −0.5 end 360,0
13. $\upsilon = V_m[0.637 - 0.423 \cos \omega t - 0.0835 \cos 2\omega t - 0.035 \cos 3\omega t - 0.0187 \cos 4\omega t - 0.0112 \cos 5\omega t]$ for 20 strips
15. $\upsilon = V_m[0.8072 \sin \omega t - 0.0865 \sin 3\omega t + 0.0283 \sin 5\omega t]$ for 20 strips
17. **(a)** 11.30 V, 0.565 A **(b)** 6.38 W
19. **(a)** 120 V, 4 A **(b)** 480 W
21. **(a)** 4681.3 W **(b)** 0.42
23. **(a)** 14.58 V **(b)** 0.729
25. **(a)** 11.54 V, 0.231 A **(b)** 2.66 W
27. **(a)** $i = 1.5 + 1.25 \sin(314t - 36.87°) + 0.52 \sin(628t - 56.3°) + 0.198 \sin(1256t - 71.56°)$
   (b) 32.42 V, 1.785 A **(c)** 50.97 W
29. **(a)** 32.42 V, 3.66 A **(b)** 87.58 W

CHAPTER 21 Answers to Questions

1. The iron core provides greater coupling but also has iron losses.
2. The current rating is obtained by dividing the kVA rating by the voltage rating.
3. The transformer can also supply 50 amperes at the 0.8 power factor.
4. The 5% transformer has a higher short circuit inrush current.
5. Short-circuiting some secondary turns decreases the output voltage. Short-circuiting some of the primary turns increases the output voltage.
6. The smaller diameter will not have any effect on the no-load voltage. It will reduce the voltage at load because of the higher resistance of the wire.
7. The insulation might be marginal. The ohmmeter will not cause it to break down, but the voltage of the insulation tester will.

8. There will not be any output with a 12-V dc input. There will be some primary current, but there will not be any secondary current.
9. Short-circuiting the secondary terminals will result in a high primary current.
10. The high-voltage terminals are marked with an H. The low-voltage terminals are marked with an X.
11. The windings are connected out of phase so that the voltages cancel, resulting in 0 volts.
12. The short-circuit test is used to provide data for copper losses.
13. The open-circuit test is used to provide data for the core loses.
14. The equation is

$$\frac{P_{\text{out}}}{P_{\text{out}} + P_{\text{lost}}} \times 100$$

15. The equation is

$$\frac{P_{\text{in}} - P_{\text{lost}}}{P_{\text{in}}} \times 100$$

16. Decreasing the current will decrease the copper losses. The copper losses at 3 amperes will be 2.5 watts.
17. Decreasing the voltage will decrease the core losses.
18. The core losses will increase if the lamination insulation is damaged.
19. There are no hysteresis and eddy current losses in the air-core transformer because there is no iron in it.
20. Turning the screw increased the coupling, resulting in a stronger signal.
21. The autotransformer is more efficient and has a smaller size than a regular transformer for the same power output.
22. The autotransformer transfers power inductively and conductively.
23. Increasing the frequency increases the core losses.
24. The potential transformer is used to step the voltage down, while the current transformer is used to step the current down.
25. The isolation transformer has a 1:1 turns ratio, while the turns ratio for the matching transformer is determined by the impedances being matched.

CHAPTER 21 Answers to Problems

1. 220 kVA
3. **(a)** 176 kW **(b)** 220 kW
5. 1136.4 A
7. **(a)** 1600 **(b)** 960 V **(c)** 4 A
9. **(a)** 2 **(b)** 10 A **(c)** 5.5 ohms **(d)** 22 ohms
11. 3
13. **(a)** 0.25 **(b)** 880 ohms, 55 W
15. 1000 turns total
17. **(a)** 99 V **(b)** 33 V
19. **(a)** X_2 to X_3, H_2 to H_3, input $X_1 - X_4$, output $H_1 - H_4$
    (b) X_1 to X_3, X_2 to X_4, H_1 to H_3, H_2 to H_4, input $X_1 - X_2$, output $H_1 - H_2$
21. **(a)** 3 A **(b)** 1.14 A, 366.5 V **(c)** 406.21 W
23. % eff = 95.2, % VR = −9.17
25. % eff = 95.47, % VR = 4.29
27. $23.74 \angle 19.72°$ ohms
29. **(a)** 3 A **(b)** 1 A, 365.73 V **(c)** 360.44 W
31. % eff = 95.27, % VR = −9.16
33. % eff = 95.83, % VR = 4.36
35. 90.9%, 1500 W, 7.1 A
37. % eff = 90, % VR = −9.2
39. **(a)** 95.8% **(b)** $R_p = 0.5\ \Omega$, $R_s = 0.2\ \Omega$, $X_p = 1.05\ \Omega$, $X_s = 0.26\ \Omega$, $R_c = 3996\ \Omega$, $X_m = 2000\ \Omega$
41. 225 turns
43. 4.17 A
45. **(a)** 400 V
    (b) $I_L = 20$ A, $I_{N1} = 13.33$ A, $I_{N2} = 20$ A
    (d) 8000 W
47. **(a)** 37.88 A in top section, 18.94 A in bottom section
    (b) 4.167 kVA
49. $3629.3 \angle 89.36°\ \Omega$
51. 2.78 V
53. 2.84 V

CHAPTER 22 Answers to Questions

1. The Wye connection should be used to get a greater line voltage.
2. The Delta connection should be used to get a greater line current.
3. The phase sequence gives the order in which the waves build up, with the leading wave being the first one in the sequence, followed by the others in the order of lag.
4. Reversing any two leads will reverse the direction of rotation of the motor shaft.
5. The proper polarity exists when a voltmeter connected across the two terminals indicates zero or near-zero volts.
6. There will be 208 volts across two pairs of lines and 120 volts across the third pair.
7. The frequency, voltage, or phase sequence is not the same for both alternators.
8. Since the loads are balanced, there is no difference in the phase voltages, so the neutral is not needed.

9. The neutral is needed to maintain the neutral of the load at the same potential as that of the source and thus to keep the voltage across the phases the same.
10. Measuring the neutral current will show if the loads are balanced.
11. The line impedance can affect the load voltage, current, and power.
12. The capacitance can change the phase angle of the current so that the load voltage, $E - IZ_{\text{line}}$, will be greater.
13. The voltage terminals should be connected to the top line.

9. **(a)** 220 V, **(b)** 3.46 A
11. **(a)** 120 V
    (b)

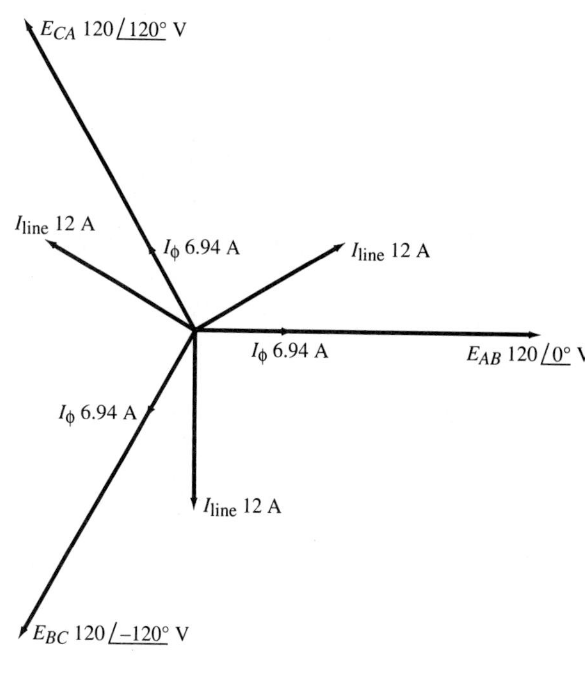

CHAPTER 22 Answers to Problems

1. $E_{ca} = 208 \angle 30°$ V, $E_{Ac} = 120 \angle -60°$ V
3. $E_{BC} = 208 \angle 90°$ V, $E_{Cb} = 120 \angle 180°$ V
5. **(a)** 254.03 V **(b)** 10 A
7. $E_{AC} = 360.27 \angle 30°$ V, $E_{BA} = 360.27 \angle 150°$ V, $E_{CB} = 360.27 \angle -90°$ V

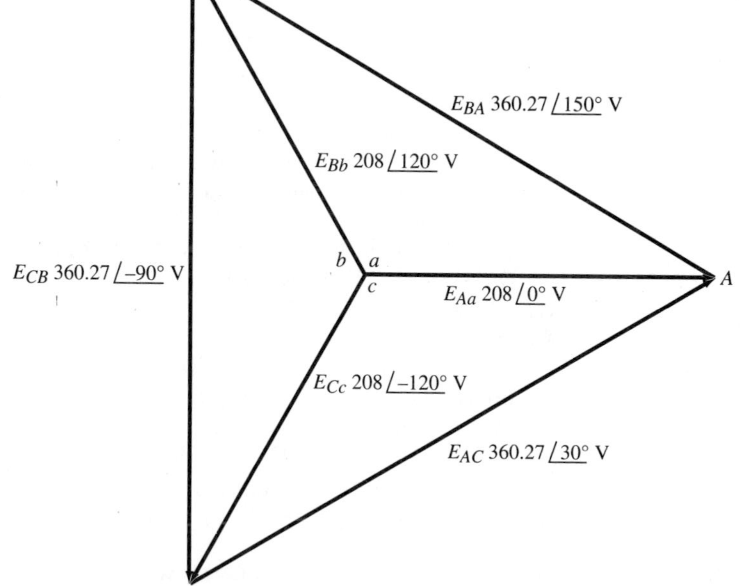

13. **(a)** *ACB* **(b)** *ABC*
15. **(a)** *ABC* **(b)** *ACB*
17. *ABC*
19. **(a)** 2.4 A **(b)** 120 V **(c)** 748 W
(d)

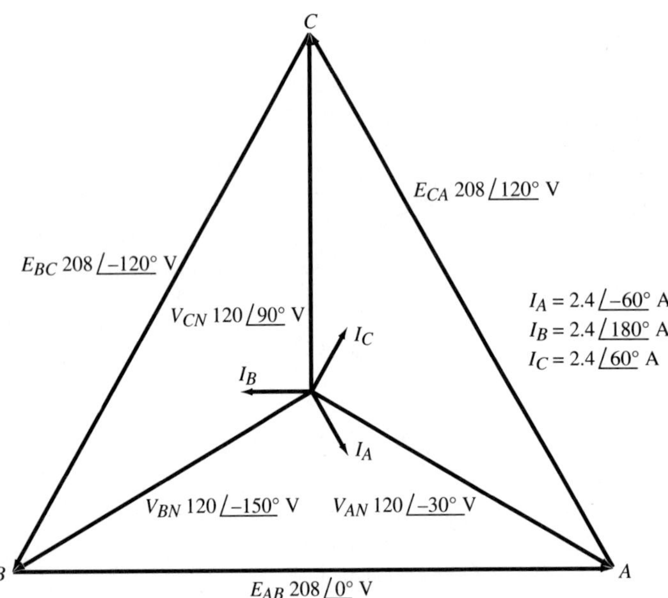

21. **(a)** $I_A = 2.4$ A, $I_B = 2.4$ A, $I_C = 2.4$ A, $I_N = 2.4$ A **(b)** 120 V **(c)** 786.8 W
23. $I_A = 2.4 \angle 100.53°$ A, $I_B = 0.8 \angle 183.56°$ A, $I_C = 1.2 \angle -144.31°$ A
25. **(a)** 138.6 V **(b)** 2.77 A **(c)** 814.1 W **(d)** 1151.5 VA
27. **(a)** $I_{BA} = 5.5 \angle 120°$ A, $I_{AC} = 5.5 \angle 240°$ A, $I_{CB} = 5.5 \angle 0°$A
(b) $I_A = 9.52 \angle -90°$ A, $I_B = 9.52 \angle 150°$ A, $I_C = 9.52 \angle 30°$ A **(c)** 1814 W
29. **(a)** 5.25 A, 1.6 kW **(b)** 3.03 A
31. $I_{BA} = 4.16 \angle 180°$ A, $I_{AC} = 10.4 \angle -30°$ A, $I_{CB} = 2.08 \angle 0°$ A
(b) $I_A = 14.16 \angle -21.55°$ A, $I_B = 6.24 \angle 180°$ A, $I_C = 8.66 \angle 143.1°$ A **(c)** 2955 W
33. **(a)** $I_A = 2.94 \angle -45°$ A, $I_B = 1.52 \angle -149.99°$ A, $I_C = 2.94 \angle 104.99°$ A
(b) $V_{AO} = 147.1 \angle -15°$ V, $V_{BO} = 76 \angle -149.99°$ V, $V_{CO} = 147 \angle 74.4°$ V
(c) 866.6 W
35. **(a)** $Z_B = 73.54 \angle -44.84°$ ohms, $Z_C = 137.91 \angle 5.99°$ ohms
(b) $V_{ZA} = 69 \angle 44°$ V, $V_{ZB} = 165.5 \angle 163.16°$ V, $V_{ZC} = 234.4 \angle 103.41°$ V **(c)** 739.69 W
37. 18.21 $\angle 33.33°$ ohms connected to point *B*, 13.66 $\angle 13.33°$ ohms connected to point *A*, 27.31 $\angle 43.33°$ ohms connected to point *C*
39. **(a)** $I_A = 11.79$ A, $I_B = 12.16$ A, $I_C = 12.62$ A
(b) 3381 W **(c)** 4.64 kVA
41. 2.91 A, 874.6 W, 1.11 kVA
43. **(a)** 112.24 A **(b)** 192.6 V **(c)** 0.987
(d) 2645.5 W
45. **(a)** At load, $V_{AC} = 190.74 \angle 117.84°$ V, $V_{CB} = 199.94 \angle 121.65°$ V, $V_{AB} = 194.01 \angle -179.9°$ V **(b)** 172.28 W
47. **(a)** $I_A = 9.29$ A, $I_B = 10.41$ A, $I_C = 9.17$ A
(b) At load, $V_{AB} = 192.35$ V, $V_{BC} = 194.53$ V, $V_{AC} = 195.13$ V
49. **(a)** $P_1 = 617.7$ W, $P_2 = 1097.1$ W
(b) $P_1 = -348.31$ W, $P_2 = 729.5$ W
51. **(a)** -150 W **(b)** 0.36

53.
```
THREE PHASE CIRCUIT, FIGURE 22.68
VAN   1   0   AC   120V      0DEG
VBN   2   0   AC   120V    120DEG
VCN   3   0   AC   120V   -120DEG
R2    4   0   16OHMS
R1    5   0   16OHMS
R3    6   0   16OHMS
L2    1   4   26.5MH
L1    2   5   26.5MH
L3    3   6   26.5MH
.AC   LIN   1   60HZ   60HZ
.PRINT   AC   IM(R2)   IP(R2)
.PRINT   AC   IM(R1)   IP(R1)
.PRINT   AC   IM(R3)   IP(R3)
.END
```

55.
```
FIGURE 22.69
VAN 1 0 SIN( 0 170V 60 0 0    0 )
VBN 2 0 SIN( 0 170V 60 0 0  120 )
VCN 3 0 SIN( 0 170V 60 0 0 -120 )
R1  4 0 16OHMS
R2  2 0 18OHMS
R3  5 0 6OHMS
C1  1 4 100UF
L1  3 5 8MH
.TRAN   50US   40MS
.PROBE
.END
```

GLOSSARY

absolute zero the temperature at which there is no molecular activity and the resistance is zero ohms.

ac a current whose value changes with time and whose direction reverses at regular intervals.

ac analysis an analysis in which the ac conditions in a circuit are determined when the components and their values are known.

ac resistance a value of resistance that accounts for power loss due to the opposition to the ac current along with that caused by the changing magnetic field.

admittance a measure of how well a circuit admits the ac current.

air core transformer a transformer that has the coils wound on a nonmagnetic core.

algebraic sum a summation in which the rises and drops are given opposite signs.

American Wire Gage a system that sets up the sizes of wire and assigns a number for each diameter size.

ammeter a meter that measures current.

ampere the SI unit for electric current. It is defined in terms of the force between two parallel current-carrying conductors.

ampere-hour capacity a measure of how long a battery can supply an amount of amperes.

amplitude the maximum positive or maximum negative value of a waveform.

analog display a meter display that uses a pointer and a scale to give the value of the quantity.

analysis a procedure in which the conditions in a circuit are determined when the components and their values are known.

angular velocity a measure of the radians displaced in one second.

apparent power the power that would be in a circuit if the power factor were one. It is given by $V \times I$.

approximate number a number that has some uncertainty in it.

atom the smallest particle of an element that can enter into chemical reaction with other particles.

audio transformer a transformer used to connect the output of an amplifier to a loudspeaker.

auto transformer a transformer that transfers power both by transformer action and directly.

auto transformer a transformer that transfers power both by transformer action and directly.

average current a value of current obtained by using $C\dfrac{\Delta v}{\Delta t}$ for a capacitor.

average emf the value of induced emf obtained by using the average change in current and time instead of the instantaneous changes.

average power that constant power that will result in the same area under the power-time curve as the area under the varying power-time curve.

average value the value of a constant voltage or current that causes an equal net amount of charge movement as the varying voltage or current causes.

balanced bridge the condition when the voltage across the two junctions that are not connected to the source is equal to zero volts.

balanced load a three-phase load in which the impedance values in the phases are equal.

band-pass filter a circuit that permits signals between two frequencies to pass with less than 3 dB attenuation.

band-reject filter a filter that attenuates signals between two frequencies at a level greater than 3 dB.

bandwidth the difference between the upper half-power frequency and the lower half-power frequency.

Bardeen, John an experimental and theoretical physicist at Bell Laboratories who shared in the development of the transistor.

base units well-defined units in the International System of Units.

battery a group of cells connected together to obtain a higher voltage or a higher current capacity.

bilateral resistance a resistance whose value is the same for both directions of current through it.

bleeder current the current in a bleeder resistor.

bleeder resistor the resistance part of the voltage divider that does not have any load current in it.

block diagram a diagram in which the sections of a circuit are divided into blocks and arrows are used to show the flow of power or signal through the circuit.

Bode plot a plot of dB gain versus frequency that uses straight line sections to approximate the exact response.

branch a path in a parallel group.

branch current method a method of circuit analysis in which a current is assigned to each branch.

branch point a junction of three or more parts of a circuit.

Brattain, Walter a physical chemist at Bell Laboratories who shared in the development of the transistor.

breakdown voltage the voltage that will cause electrical breakdown of a dielectric.

bridge circuit a circuit used to determine very accurately the value of an unknown impedance.

buck-boost transformer a transformer used to bring a voltage that is either too high or too low to the desired level.

candela the base unit for luminous intensity in SI.

capacitance the property by which two conducting materials separated by an insulator can store charge.

capacitive reactance the opposition of capacitance to an ac current.

capacitor a device designed to have a specific amount of capacitance.

capacitor charging the process whereby charge accumulates on the plates of a capacitor.

capacitor discharge the process during which charge is removed from the plates of a capacitor.

cell a source of emf in which chemical energy is changed to electrical energy.

center-tapped transformer a transformer that has a tap located at the midpoint of the secondary winding to provide two equal voltages.

ceramic capacitor a capacitor that uses ceramic for the dielectric.

charged capacitor a capacitor that has an accumulation of charge on the conducting plates.

charging a procedure in which the original condition of a cell is restored by passing a reverse current through the cell.

charging rate the rate at which energy is restored to a battery during charging.

circuit breaker a reusable device designed to open a circuit when the current exceeds a certain value.

circuit diagram a diagram in which graphic and letter symbols are used to show the parts and connections.

circular mil a unit for area that is equal to the area of a circle that has a diameter of one mil.

coefficient of coupling the fraction of the flux from one inductor that is linked with another inductor.

cold cranking performance the current that a lead acid battery can supply for 30 seconds at 0°F.

complex R–L circuits R–L circuits that include more than one resistor and/or source.

component line a line in the PSpice file that describes the component and gives its node connections.

compound matter formed by the chemical union of several elements.

conductance the real part of the admittance, which is the reciprocal of resistance for a resistor.

conductively transferred power the power transferred directly in an autotransformer.

conductivity a measure of how easily electrons can be made to move in a material.

conductor material a material that has little resistance and is used to connect parts in an electric circuit.

conjugate an impedance that has the same resistance but the opposite reactance of the impedance for which it is the conjugate.

constant voltage transformer a transformer that provides a constant voltage even through the input voltage or load condition changes.

control transformer a transformer used for operation of devices such as relays and solenoids.

conventional current a name given to electric current whose direction is taken as opposite to that of electron current.

conversion factor a number that specifies the amount of one unit that is equivalent to a second unit.

copper losses the transformer losses due to the resistance of the windings.

core losses the transformer losses in the core due to hysteresis and eddy current.

coulomb the unit for electric charge that represents the charge on 6.242×10^{18} electrons.

Coulomb's Law a relation developed by Charles Coulomb that relates the force between charged particles to the charge and the distance between the particles.

coupling the selection of ac, dc, or ac and dc input to the oscilloscope; also a measure of the flux linkage between coils.

current divider rule a relation that states that the current in a parallel branch is equal to the product of the current entering the parallel group and the equivalent resistance, divided by the resistance of the branch. Impedance is used in ac circuits.

current drain the current that a battery or cell is expected to supply.

current output the amount of current that a supply can provide under continuous use.
current source a source that maintains a constant current as the load changes.
current transformer a transformer used to step-down a high current to a value that can be measured with an ammeter.
cutoff voltage the voltage below which a battery is no longer useful.
cycle the part of a wave form between similar points on it.

dB stands for the term decibel and is equal to 10 $\log_{10}(P_{out}/P_{in})$, also 20 $\log_{10}(V_{out}/V_{in})$.
dc analysis an analysis in which the dc conditions in a circuit are determined when the components and their values are known.
dc offset the amount of dc that is included in a time-varying signal.
dc power supply a unit that converts an alternating voltage to provide dc with some ac in it.
dc resistance the resistance that a component offers to a dc current.
dc source a source of emf that provides a voltage that does not change in magnitude or direction with time.
dc voltage a voltage that does not change in magnitude or direction at regular time intervals.
Delta network a connection in which three resistors or impedances are connected in the form of a Delta.
demagnetizing the process whereby the magnetism is removed from a magnet.
derived units units that represent some combination of base units.
device a part in a circuit such as a resistor, a capacitor, and so forth.
dielectric the insulating material between the conducting materials of a capacitor.
dielectric constant another name for relative permittivity.
dielectric strength the breakdown voltage expressed in volts per unit thickness of a material.
digital display a display that uses numbers to give the value of the quantity being measured.
discharged capacitor a capacitor that does not have any accumulation of charge on its plates.
distortion the changing of the shape of an input wave, which results in the output wave being different from the input wave.
distribution transformer a large transformer used to step-down the line voltage to that needed by the customer.
division into known shapes a procedure for finding the area under a curve by using known shapes to approximate the area.
domains small bar magnets formed in a material by the arrangement of the electron fields in the material.
doping the addition of impurities to a semiconductor material so that the resulting current will be either positive charges or negative charges.
double subscript notation a notation that gives the voltage of the first subscript point as measured from the second subscript point.
duty cycle the sequence of ON periods and OFF periods during the operation of a battery; also the ratio of pulse width to pulse period for a pulse.

eddy current loss the loss caused by the circulating currents in a magnetic material.
effective value the value of a constant voltage or current that supplies the same amount of energy as the varying voltage or current.
efficiency the ratio of useful power or energy output to the power or energy input, usually expressed as a percent.
electric charge a quantity assigned to the protons and electrons to account for the force exerted between them.
electric circuit a group of electrical parts connected to form a path through which electric charge can move to perform some desired function.
electric current the movement of charged particles, either positive or negative, in some general direction.
electric shock the effect experienced by a person when electric current passes through the body.
electrodes conducting materials that react chemically with the electrolyte in a cell.
electrolyte a paste or liquid that permits ionic conduction between the electrodes.
electrolytic capacitor a capacitor that uses a thin oxide film for the dielectric and must be connected so that the (+) terminal is more positive than the (−) terminal.
electromagnet a magnet made by passing a current through a coil of wire that is wrapped around a magnetic material.
electromagnetic induction the generation of an emf when a conductor is exposed to a change in the lines of a magnetic field.
electromagnetic radiation the radiation of electrical energy into the atmosphere as the magnetic field alternates.
electromotive force a term used to represent the force that

early experimenters thought was used to maintain the current in a conductor.

electron a small, negatively charged particle that whirls in orbits around the nucleus of the atom.

electron current electric current in which the charged particles are electrons.

electrostatic discharge the discharge of the static electricity from an object.

electrostatic field the region around a charge in which a force is exerted on other charges.

element a substance that cannot be chemically decomposed or formed by a chemical union of other substances.

energy the capacity to do work.

engineering technician a graduate of a two-year associate of applied science degree program or a person who has received comparable education through other means.

engineering technologist a graduate of a four-year bachelor's degree program in engineering technology.

equipotential lines lines that join points in the electrostatic field that have the same voltage.

equivalent capacitance the value of capacitance that will store the same amount of charge as the group of capacitors for which it is equivalent.

equivalent impedance an impedance that presents the same overall characteristics to a source as the circuit to which it is equivalent.

equivalent inductance an inductance that induces the same emf as the circuit for which it is equivalent.

equivalent resistance the resistance that draws the same amount of current when it has the same voltage across it as the circuit that it replaces.

even function a wave for which the value at some positive angle, as measured from the dc value, is the same as that at an equal negative angle.

exact number a number that does not have any uncertainty in it.

exponential equation an equation that has the term e^{-x} in it, causing the curve to vary exponentially.

exponential wave a wave that varies exponentially with time.

falling edge the part of the pulse at which it drops to the less positive value.

farad the unit of capacitance.

Faraday's Law a relation that states that the generated emf is directly proportional to the rate at which the magnetic field changes with respect to time.

field intensity the force per unit charge in the electrostatic field.

fixed capacitor a capacitor that provides only one value of capacitance.

floating output a type of supply output that has ungrounded positive and negative terminals.

flux density the number of lines of electrostatic or magnetic field per unit area.

focus the degree of sharpness of the display on the oscilloscope screen.

Fourier analysis a theory that states that a nonsinusoidal wave can be expressed by a mathematical series of sine and cosine waves.

Fourier analyzer a spectrum analyzer that measures all the frequencies at one time as well as phase relation.

Fourier series the mathematical expression for the nonsinusoidal wave.

four-wire, three-phase Wye load a Wye load that has a neutral line.

free electron a valance electron that has been freed from the outer shell and that moves from atom to atom.

frequency the number of cycles that a waveform makes in one second. It is measured in hertz.

frequency response the manner in which the output signal of a circuit varies over a range of frequencies.

function generator an instrument that generates sine waves and repetitive pulse waveforms.

fundamental the term in the Fourier series that has the same frequency as the nonsinusoidal wave.

fuse a nonreusable device assigned to open a circuit when the current exceeds a certain value.

graphic symbol a simple line diagram that represents an electrical part.

hardware the physical components, such as computer, keyboard, and so forth, of a computer system.

harmonics the terms in the Fourier series that have frequencies that are multiples of the fundamental frequency.

Hay bridge a bridge for measuring inductance that has the capacitance in series with the resistance in the circuit and is suitable for use when the inductor resistance is small.

henry the unit of inductance.

hertz the unit for frequency. It represents one cycle per second.

high-pass filter a circuit that permits signals above a set frequency to pass with an attenuation of less than 3 dB.

horsepower the mechanical unit for power that is equal to 746 watts.

hysteresis loss the loss of energy in a magnetic material caused by reversing of the magnetic field.

impedance the opposition to the current in an ac circuit. It can be resistance, reactance, or both.

impedance matching the procedure in which the impedance of a load is made to look smaller or larger to obtain the desired operating characteristics.

impedance matching transformer a transformer that matches a load impedance to the source impedance.

impedance triangle a triangle in which resistance forms the base, reactance is the leg, and impedance is the hypotenuse.

inductance the property of a device whereby it opposes a change in current.

inductive reactance the opposition of inductance to an ac current.

inductively transferred power the power transferred by transformer action in an autotransformer.

inductor a device built to have a specific amount of inductance.

inductor stored energy the energy that is stored in the magnetic field of an inductor.

inert element an element that cannot react with other elements because it has eight electrons in its outer shell.

inferred absolute zero the temperature at which the resistance-temperature curve crosses the temperature axis when it is continued as a linear curve.

instantaneous value the value of the waveform at some instant of time.

insulator a material that for all practical purposes has no free electrons and has a high resistance.

integration a mathematical procedure that gives the exact area under a waveform.

intensity the brightness of the display on the oscilloscope screen.

internal resistance the resistance within a source of emf that accounts for the drop in voltage when a current is supplied.

International System of Units (SI) A metric system that is used by the majority of countries. It consists of base units, supplementary units, and derived units.

ion an electron-deficient atom that has a positive charge.

ionic current electric current made up of positive ions.

iron core transformer a transformer that has the coils wound on an iron core.

isolation the condition wherein there is no physical electrical connection between parts of a circuit.

isolation transformer a transformer that provides electrical isolation between circuits on the primary side and those on the secondary side.

joule the SI unit for energy.

kelvin the SI base unit for temperature.

kilogram The SI base unit for mass.

kilowatt-hour the unit for energy that is commonly used by electric utilities.

Kirchhoff's Current Law a relation developed by Gustav Kirchhoff, which states that the algebraic sum of the currents at a node is equal to zero. The phasor sum is used in ac circuits.

Kirchhoff's Voltage Law a relation developed by Gustav Kirchhoff, which states that the algebraic sum of the voltage rises and drops around a closed loop is equal to zero. The phasor sum is used in ac circuits.

lagging power factor the power factor for an impedance in which the current lags the voltage.

lagging wave a wave whose peaks, zeros, and other points occur at a larger number of radians or degrees than those on the wave that it lags.

lead-acid cell a secondary cell in which the electrodes are lead peroxide and metallic lead and the electrolyte is sulfuric acid.

leading power factor the power factor for an impedance in which the current leads the voltage.

leading wave a wave whose peaks, zeros, and other points occur at a smaller number of radians or degrees than those on the wave that it leads.

leakage current the current that exists in the dielectric of a capacitor because of impurities in it.

leakage resistance the ohmic resistance of the capacitor dielectric.

Lenz's Law a relation that states that the polarity of the induced emf is such that the emf opposes a change in the original field.

letter symbol a letter of the alphabet that represents an electrical quantity, such as current, resistance, and so forth.

line current the current in a line of a three-phase system.

line impedance the impedance of the line connecting the three-phase source to the three-phase load.

line of force the path along which an isolated north pole would move if placed in the magnetic field.

line regulation a measure of the output voltage change caused by a change in the supply input voltage.

line voltage the voltage across any pair of lines of a three-phase source.

linear resistance a resistance that has the same resistance within its current range.

linear scale a scale on which a given length represents the same amount on all parts of the scale.

loading effect the change in circuit conditions caused by the connection of an ammeter or a voltmeter in a circuit.

lodestone an oxide of iron that has magnetic properties.
long shunt the wattmeter connection in which the voltage coil is connected across the series combination of the load and the current coil.
loop a path formed by tracing a path from one point, around the circuit, and back to the initial point.
loop current method a procedure of circuit analysis that uses an assumed current in each loop.
lower half-power frequency the frequency below the resonant frequency at which the attenuation is 3 dB below the pass-band level.
low-pass filter a circuit that permits signals below a set frequency to pass with an attenuation of less than 3 dB.

magnet a piece of material that has the property of attracting iron.
magnetic field the space around a magnet in which a force is exerted on iron.
magnetic flux the magnetic lines of force.
magnetic materials materials, such as iron, that affect the magnetic field.
magnetization curves curves that plot flux density versus magnetizing force.
magnetizing force the emf needed to set up a flux density in one meter of a material.
matter any substance that has weight and occupies space.
maximum power transfer the condition when the most power is being delivered to a load by a source.
maximum power transfer theorem a theorem that states that maximum power is transferred from a dc source to a load when $R_L = R_i$, and for an ac source when Z_L is the conjugate of Z_i.
Maxwell-Wein bridge a bridge that measures inductance but uses capacitors instead of inductors in its circuit.
meter the SI base unit for length.
mica capacitor a capacitor that uses mica for the dielectric.
microprocessor the part of a computer unit that processes the data.
mil one-thousandth of an inch.
Millman's Theorem a theorem that states that a number of parallel current sources can be replaced by a single current source.
multiplier band the band of the color code that represents the number by which the first two digits are multiplied.
mole the SI base unit for an amount of substance.
molecular theory a theory that explains the magnetic field in terms of the field around the electron.
molecule the smallest particle of a compound that can exist and still retain the physical and chemical properties of the compound, such as density, weight, odor, taste, and hardness.
mutual induction the generation of an emf in one coil which is in a changing magnetic field that is set up by another coil.

neutral current the current in the neutral line of a four-wire, three-phase Wye load.
neutral line the line that is connected to the point at which the three windings of the Wye source are connected together.
neutron in the nucleus of the atom, a small particle that has no charge.
nickel-cadmium cell a secondary cell that has electrodes of cadmium and nickelic hydroxide, with an electrolyte of potassium hydroxide.
nodal method a procedure of circuit analysis in which the voltage at nodes, as measured from a reference node, is determined.
node for PSpice, a point where two or more circuit components are connected together; for Kirchhoff's Current Law, the junction of three or more components.
node voltage the voltage of a node as measured from a reference node.
nonlinear resistance a resistance whose resistance changes with current.
nonlinear scale a scale on which a given length does not represent the same amount on all parts of the scale.
nonmagnetic materials materials that do not affect the magnetic field.
nonsinusoidal waveforms waveforms that do not have a sinusoidal shape.
north pole the end of a magnet that is attracted to the North Pole of the earth.
Norton's Theorem a theorem that states that any two-terminal, linear circuit can be replaced by a constant current source and a parallel resistance. Impedance is used in ac circuits.
nucleus the central part of the atom, which is made up of neutrons, protons, and smaller particles.

odd function a wave for which the value at some positive angle, as measured from the dc value, is the negative of the value at an equal negative angle, as measured from the dc value.
ohm the unit of resistance.
Ohm's Law a relation developed by George Simon Ohm,

which states that the voltage across a resistance is equal to the product of the current and the resistance.

oil capacitor a capacitor that uses oil for the dielectric.

open circuit a condition whereby the circuit path is broken.

open-circuit test a test used to determine the core losses of a transformer by open-circuiting the transformer secondary.

operating point the dc conditions in a circuit being analyzed by PSpice.

operating temperature the temperature at which a device operates.

oscilloscope an instrument that presents a two-dimensional graph of the voltage applied to its input terminals.

paper capacitor a fixed capacitor that uses paper for the dielectric.

parallel ac circuit a group of components of an ac circuit that are connected across the same two points.

parallel capacitor connection a connection in which capacitors are connected across the same two points.

parallel circuit a circuit that has two or more components connected across the same two points.

parallel inductors inductors that are connected across the same two points of a circuit.

peak-to-peak the difference in value between a positive peak and a negative peak of an ac wave.

period the length of one cycle in degrees, radians, or seconds.

periodic wave a waveform that repeats itself at regular intervals.

permanent magnet a magnet that holds its magnetism for a long time.

permeability a measure of how easily a magnetic flux can be set up in a material.

permittivity a measure of how easily the lines of an electrostatic field can be set up in a dielectric.

phase angle the phase relation in degrees or radians.

phase current the current in one of the legs of a Wye load or source or in one of the sides of a Delta load or source.

phase relation the displacement, in seconds, degrees, or radians, between similar parts of two waves; also called phase shift.

phase sequence the order in which the phase voltages build up or reach their maximum point.

phase sequence indicator a device that shows the phase sequence of a three-phase source.

phase voltage the voltage across one coil of a three-phase source.

phasor the electrical circuit's equivalent of a vector that is used to represent voltage, current, and impedance.

phasor diagram a diagram that shows the phasors for voltages and/or currents.

piezolectric crystal a material that generates an emf from mechanical pressure.

plastic film capacitor a capacitor that uses plastic film for the dielectric.

polar form a form that expresses a phasor by its magnitude and angle.

polarity the designation of one point being more positive or more negative than another point.

potential difference a quantity used to represent the energy between two charged objects or parts of a circuit.

potential transformer a transformer connected across high potential lines to reduce the voltage to a level that is safe to measure with a meter.

potentiometer a three-terminal variable resistor that is used to obtain different voltages from a fixed voltage.

power the rate at which work is done.

power factor the ratio of power to apparent power, which is given by the cosine of the phase angle.

power factor correction a procedure in which vars are added to a circuit to change the power factor to the desired value.

power rating a measure of the heat-dissipating capacity (in watts) of a resistor.

power transformer a transformer used to supply the power in electronic equipment, such as radios, oscilloscopes, or signal generators.

power triangle a triangle in which real power forms the base, reactive power forms the leg, and apparent power forms the hypotenuse.

prefix a word placed before the name of unit to indicate that the unit is multiplied by ten raised to some power.

primary cell a chemical cell that cannot be recharged.

primary winding the transformer winding that is connected to the source.

probe a device used with an oscilloscope to attenuate the magnitude of the incoming signal.

probe compensation adjusting the capacitance of the probe to prevent rounding of the corners of the pulse.

proton a small, positively charged particle that is in the nucleus of the atom.

prototyping the process of building a model of a new design to test the characteristics and see if the design works.

PSpice a version of SPICE.

pulse a waveform that switches between two constant values at regular intervals.

pulse width the length of time at which a pulse remains at a constant value.

radian the SI supplementary unit that is used to measure a plane angle.

radio frequency transformer a transformer used at radio frequencies and higher, usually as part of a tuned circuit.

reactance the opposition of capacitance or inductance to ac current.

reactive power the rate at which energy is delivered to the magnetic or electrostatic field. It is given by $VI \sin \theta$.

rectangular form a form that expresses a phasor as real and imaginary components.

rectification a procedure in which an alternating waveform is changed to a unidirectional one.

reference node a node that is assigned a voltage of zero volts.

relative conductivity the ratio of a material's conductivity to that of a standard reference material.

relative permeability the ratio of a material's permeability to that of a vacuum.

relative permittivity the ratio of a material's permittivity to the permittivity of a vacuum.

reluctance the opposition of a material to the magnetic field.

reserve capacity the number of minutes that a lead acid battery takes to discharge to 1.75 volts per cell at a load of 25 amperes.

resistance the opposition that a material offers to the movement of charge that is current.

resistivity the resistance of a section one unit in length and having a cross-sectional area of one unit.

resistor a device designed to have a specific amount of resistance.

Resistor Color Code a system that uses colored bands to indicate the nominal resistance and tolerance of a resistor.

resonance the condition at which a circuit with resistance and reactance appears to be purely resistive.

resonant frequency the frequency at which a circuit of resistance and reactance is resonant.

rheostat a variable resistor that is used to control current.

right-hand rule a rule that states that the fingers of the right hand will point in the direction of the field around a conductor if they are wrapped around the conductor with the thumb pointing in the direction of the current.

ripple factor the ratio of the effective value of the ac component to the dc value.

rise time the time between the 10 and 90 percent points on the rising edge of a pulse.

rising edge the part of the pulse at which it rises to a more positive value.

rms value a value that is equal to the constant voltage or current that supplies the same amount of energy as the varying waveform.

rounding off dropping the unwanted digits in a number so that the remaining number is as near as possible in value to the original.

sawtooth wave a nonsinusoidal wave that has the shape of a sawtooth.

Schering bridge a bridge that is used for measuring capacitance.

scientific notation a form in which a number is expressed as a number between 1 and 10 multiplied by 10 raised to some power.

second the SI base unit for time.

secondary cell a chemical cell that can be recharged and used again.

secondary winding the transformer winding that is connected to the load.

selectivity a measure of how narrow a range of frequencies will be attenuated less than 3 dB.

semiconductor material a material that is neither a good conductor nor a good insulator but that can be made into a conductor under certain conditions, such as adding heat, light, and so forth.

series capacitor connection a connection in which there are no branch points between the capacitors.

series circuit in a circuit, a group of components that are connected so there are no branch points between them.

series inductors the connection of inductors so there are no branch points between them.

series-parallel circuit a circuit in which electrical components are connected in series and parallel groups.

series-parallel ac circuits ac circuits that include groups of series circuits and/or parallel circuits.

series *R-C* circuit a circuit of a resistor and a capacitor connected in series.

series *R-L* circuit a circuit of a resistance and an inductance connected in series.

shelf time how long a battery is in storage before it is used.

shell a group of orbits that are separated by a smaller distance than that between the shells.

shell diagram a two-dimensional diagram of the atom showing the relation of the orbits and the electrons in each orbit.

Shockley, William a circuit expert at Bell Laboratories who shared in the development of the transistor.
short circuit a condition wherein zero resistance is connected across two points of a circuit.
short-circuit test a test used to determine the copper losses of a transformer by short-circuiting the transformer secondary.
short shunt a wattmeter connection in which the voltage coil is connected directly across the load.
shunt resistor a resistor connected across the terminals of a current meter movement to increase its current range.
SI the abbreviation for International System of Units.
siemens the unit for admittance.
signal analyzers instruments that are used to measure and study the frequency content of a waveform.
significant figure in a number, a digit that has a reasonable amount of certainty.
simulation a procedure in which PSpice mathematically duplicates the operation of the circuit.
sinusoidal signal an emf that has a shape like a sine wave.
sinusoidal wave form a waveform that follows the relation given by the sine function.
skin effect an effect, caused by high frequencies, in which the current is concentrated near the surface of the conductor.
software the programs for a computer.
solar cell a semiconductor device that generates an emf from light.
source file a PSpice file that contains the description of the circuit and the PSpice commands.
south pole the end of a magnet that is repelled by the North Pole of the earth.
spectrum analyzer an instrument that displays the frequency content of a wave.
SPICE a circuit-simulating program that can be used to perform dc, ac, transient, and other forms of circuit analysis.
square wave a repetitive pulse waveform that has a duty cycle of 50 percent.
static electricity the accumulation of charged particles on an object.
steady state a condition in which there is no change of capacitor voltage or inductor current with time.
step-down transformer a transformer that has a smaller secondary than primary voltage.
step-up transformer a transformer that has a greater secondary than primary voltage.
steradian the SI supplementary unit that is used to measure a solid angle.
strip method a procedure for finding the area under a curve by using narrow strips to approximate the area.
subshells each orbit in a shell.
superconductivity the condition of a material having a resistance of zero ohm.
superposition a method of circuit analysis in which the currents and voltages from each source are combined to obtain the current and voltage in a part of a circuit with all sources in the circuit.
superposition theorem a theorem which states that the current and voltage at any point in a multisource circuit of linear bilateral resistances is the algebraic sum of the current and voltage from each source placed in a circuit one at a time, while the other sources are replaced by their internal resistance. Phasor sum and impedance are used in an ac circuit.
supplementary units two SI units that are used for plane and solid angles.
surge breakdown voltage the highest voltage that can be applied across the capacitor terminals for a short time without breakdown of the dielectric.
susceptance the "j" part of admittance, which is the reciprocal of reactance for a reactance.
switch a device designed to connect and disconnect parts in a circuit.
synthesis procedure in which the parts of a circuit are determined when the conditions are known.
system of units groups of units that are related in some way.

temperature coefficient the change in capacitance per degree of change in temperature.
temperature coefficient of resistance the resistance change per ohm of resistance for each degree of change in temperature.
temporary magnet a magnet that is a magnet only while acted upon by a magnetic field.
tesla the unit for magnetic flux density.
thermocouple a junction of two dissimilar metals that generates an emf from heat.
Thevenin's Theorem a theorem that states that any two-terminal, linear circuit can be replaced by a constant voltage source and a series resistance, or impedance in an ac circuit.
three-dB frequency the frequency at which the power in the output signal is one-half the power in the passband.
three-phase a system that has three emf's displaced by a fixed number of degrees, usually 120°.
three-phase Delta source a three-phase source that has the three coils connected in the shape of a Delta to generate three emf's that are 120° apart.
three-phase transformer a transformer designed to operate with three-phase systems.

three-phase Wye source a three-phase source that has the three coils connected in the shape of a Wye to generate three emf's that are 120° apart.

three-wire, three-phase Delta load a three-phase load made up of three impedances connected in the shape of a Delta.

three-wire, three-phase Wye load a Wye load that does not have a neutral line.

time constant the time required for capacitor voltage or inductor current to reach 63.21 percent of its maximum during rise and 36.79 percent during decay. Also, the time in which the capacitor voltage or inductor current will reach the final value if the change continues at the initial rate.

tolerance band the color band that represents the ± percent by which the resistance can vary from the nominal value.

trace rotation changing the angle of the display on the oscilloscope screen.

tractive force the amount of pull that a magnet exerts on a piece of iron or magnetic material.

transfer function the ratio of the output voltage of a circuit divided by the input voltage, expressed in terms of the parts of the circuit.

transformer a device that transfers electrical energy from one coil to another through the magnetic field that couples the coils.

transformer equivalent circuit a circuit of resistance and reactance that represents the transformer.

transformer impedance the inrush current-limiting characteristic of a transformer.

transient state the state of a capacitor or inductor during which the voltage and current change from one steady-state value to another.

triangular wave a wave that is shaped like a triangle.

trigger the electronic signal that tells the oscilloscope when to start drawing the graph of the input.

trimmer capacitor a variable capacitor that is used when the capacitance is changed infrequently.

tuned circuit a circuit in which a capacitor is connected across the transformer winding and either the capacitance or the transformer inductance is varied to obtain resonance.

turns ratio the ratio of the primary turns to the secondary turns in a transformer.

two-wattmeter method a connection in which the power in a three-phase load is measured using only two wattmeters.

unbalanced load a three-phase load in which the impedances are not the same in all phases.

unit of measurement a definite quantity that is recognized and accepted as a standard of measurement.

universal time constant curves curves that plot "percent of maximum" versus "number of time constants" and that are used to analyze *R–C* and *R–L* transient circuits.

upper half-power frequency the frequency above the resonant frequency at which the attenuation is 3 dB below the pass-band level.

valence electron an electron in the outer shell of the atom.

var the unit for reactive power.

varactor a semiconductor device whose capacitance changes with the voltage across it.

variable capacitor a capacitor that can be adjusted to give values of capacitance between some minimum and maximum values.

volt the unit in which potential difference is measured.

volt-ampere the unit for apparent power.

volt-ampere rating a transformer rating that specifies the maximum volt amperes for a rated voltage.

voltage divider a series circuit that is used to obtain one or more voltages from a fixed voltage.

voltage divider rule a rule that states that the voltage across a part of a series circuit is equal to the voltage across the circuit, multiplied by the resistance of the part, and divided by the resistance of the circuit. Impedance is used in ac circuits.

voltage drop the condition when one moves from one point in a circuit to a more negative point.

voltage output the maximum voltage that a source can provide at zero current.

voltage regulation a measure of the output voltage change caused by a change in load current.

voltage rise the condition when one moves from one point in a circuit to a more positive point.

voltmeter a meter that measures voltage or potential difference.

watt the unit for real power, which is the rate at which energy is converted.

wattmeter a meter that measures power.

wave analyzer an instrument that uses a voltmeter with a tuned filter to measure the strength of the frequency passed by the filter.

wave diagram a diagram that shows the variation of a quantity as a function of time or angular displacement.

weber the unit for magnetic flux.

working voltage the highest steady voltage that can be applied across the capacitor terminals without breakdown of the dielectric.

Wye network a network in which three resistances or impedances are connected in the form of a Wye.

Index

Note: Numbers prededed by A refer to the appendices.

D